Focus on essential genetic topics and explore the latest breakthroughs

Known for its focus on conceptual understanding, problem solving, and practical applications, the bestselling *Essentials of Genetics* strengthens problem-solving skills and explores the essential genetics topics that today's students need to understand. The 10th Edition has been extensively updated to provide comprehensive coverage of important, emerging topics such as CRISPR-Cas, epigenetics, and genetic testing and Mastering Genetics includes new tutorials on these topics, which prepare students for class and support the learning of key concepts.

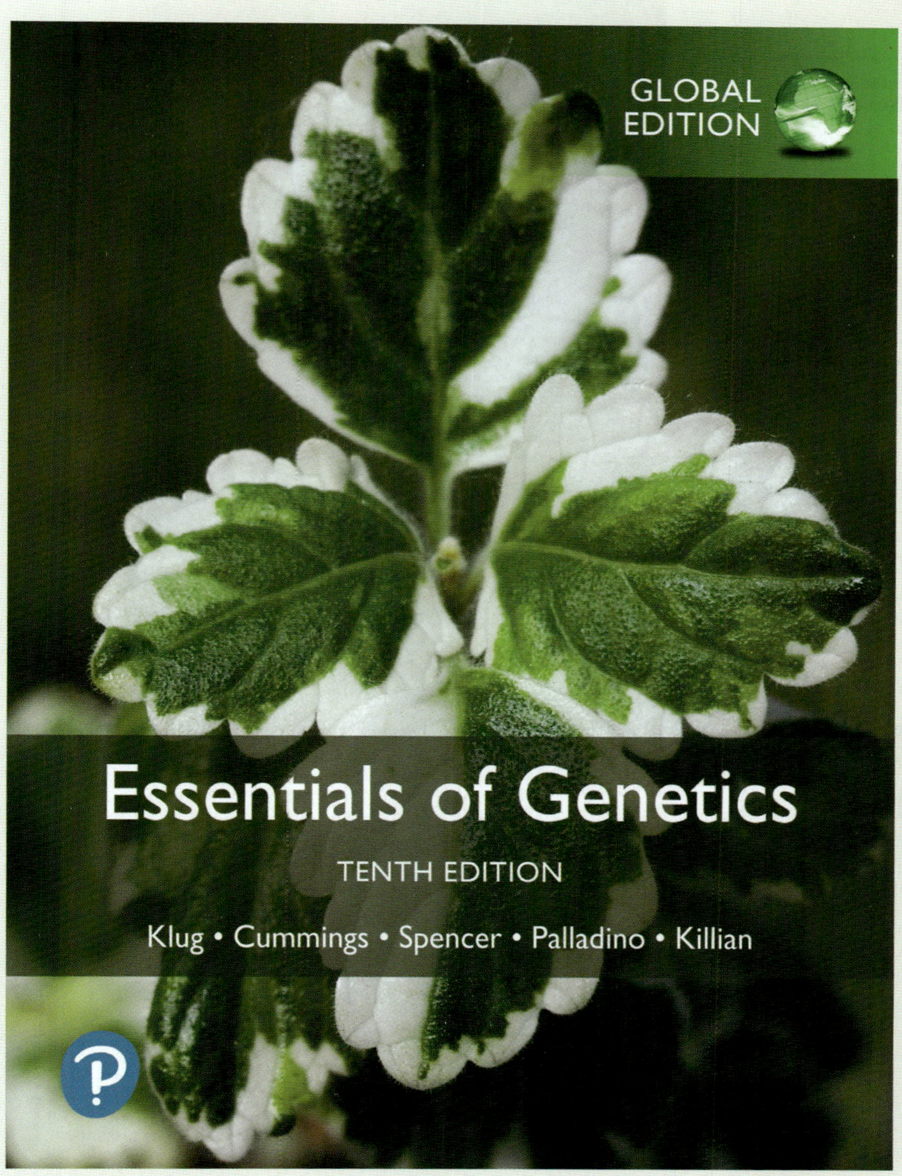

Make genetics relevant . . .

16

Regulation of Gene Expression in Eukaryotes

Chromosome territories in a human fibroblast cell nucleus. Each chromosome is stained with a different-colored probe.

CHAPTER CONCEPTS

- While transcription and translation are tightly coupled in bacteria, in eukaryotes, these processes are spatially and temporally separated, and thus independently regulated.
- Chromatin remodeling, as well as modifications to DNA and histones, play important roles in regulating gene expression in eukaryotes.
- Eukaryotic transcription initiation requires the assembly of transcription regulatory proteins on DNA sites known as promoters, enhancers, and silencers.
- Following transcription, there are several mechanisms that regulate gene expression, referred to as posttranscriptional regulation.
- Alternative splicing allows for a single gene to encode different protein isoforms with different functions.
- RNA-binding proteins regulate mRNA stability, degradation, localization, and translation.
- Noncoding RNAs may regulate gene

Virtually all cells in a multicellular eukaryotic organism contain a complete genome; however, such organisms often possess different cell types with diverse morphologies and functions. This simple observation highlights the importance of the regulation of gene expression in eukaryotes. For example, skin cells and muscle cells differ in appearance and function because they express different genes. Skin cells express keratins, fibrous structural proteins that bestow the skin with protective properties. Muscle cells express high levels of myosin II, a protein that mediates muscle contraction. Skin cells do not express myosin II, and muscle cells do not express keratins.

In addition to gene expression that is cell-type specific, some genes are only expressed under certain conditions or at certain times. For example, when oxygen levels in the blood are low, such as at high altitude or after rigorous exercise, expression of the hormone erythropoietin is upregulated, which leads to an increase in red blood cell production and thus oxygen-carrying capacity.

P. 326

NEW! Regulation of gene expression has been expanded and is now divided into coverage of bacteria in Chapter 15 and coverage of eukaryotes in Chapter 16.

Coverage of CRISPR-Cas is expanded and integrated in multiple chapters – Chapters 1, 15, 17, and Special Topics Chapters ST3 and ST6.

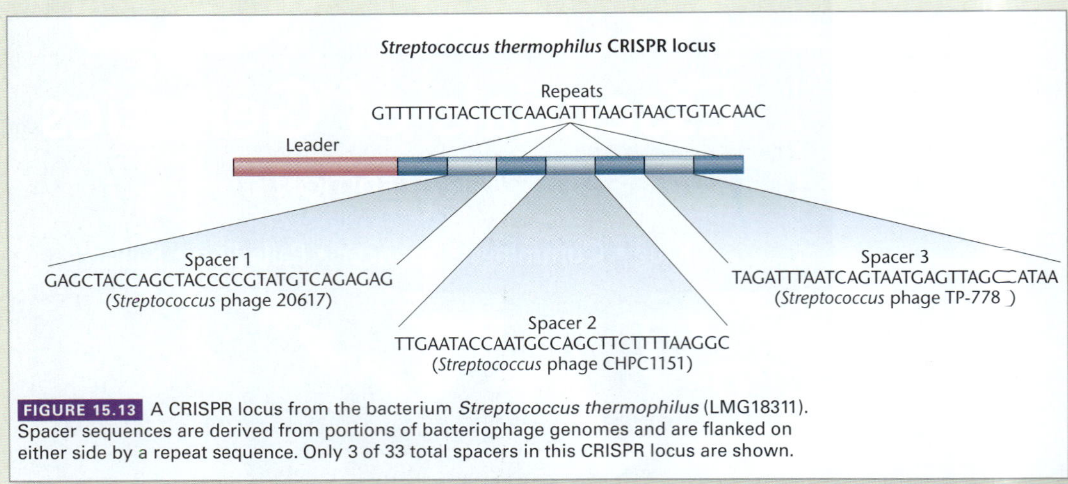

FIGURE 15.13 A CRISPR locus from the bacterium *Streptococcus thermophilus* (LMG18311). Spacer sequences are derived from portions of bacteriophage genomes and are flanked on either side by a repeat sequence. Only 3 of 33 total spacers in this CRISPR locus are shown.

P. 322

with current high interest topics

SPECIAL TOPICS IN MODERN GENETICS 2

Genetic Testing

Earlier in the text (see Chapters 17 and 18), we reviewed essential concepts of recombinant DNA technology and genomic analysis. Because of the Human Genome Project and related advances in genomics, researchers have been making rapid progress in identifying genes involved in both single-gene diseases and complex genetic traits. As a result, **genetic testing**—the ability to analyze DNA, and increasingly RNA, for the purposes of identifying specific genes or sequences associated with different genetic conditions—has advanced very rapidly.

Genetic testing, including genomic analysis by DNA sequencing, is transforming medical diagnostics. Technologies for genetic testing have had major impacts on the diagnosis of disease and are revolutionizing medical treatments based on the development of specific and effective pharmaceuticals. In this Special Topics chapter we provide an overview of applications that are effective for the genetic testing of children and adults and examine historical and modern methods. We consider the impact of different genetic technologies on the diagnosis of human diseases and dystrophy. Other tests have been developed for disorders that may involve multiple genes such as certain types of cancers.

Gene tests are used for prenatal, childhood, and adult prognosis and diagnosis of genetic diseases; to identify carriers; and to identify genetic diseases in embryos created by *in vitro* fertilization, among other applications. For genetic testing of adults, DNA from white blood cells is commonly used. Alternatively, many genetic tests can be carried out on cheek cells, collected by swabbing the inside of the mouth, or on hair cells. Some genetic testing can be carried out on gametes.

What does it mean when a genetic test is performed for *prognostic* purposes, and how does this differ from a *diagnostic* test? A prognostic test predicts a person's likelihood of developing a particular genetic disorder. A diagnostic test for a genetic condition

"Genetic testing, including genomic analysis by DNA sequencing, is transforming medical diagnostics. Technologies for genetic testing have had major

SPECIAL TOPIC X

NEW! Special Topics chapter on Genetic Testing guides students through the many contexts in which genetic testing is becoming prominent and explores many questions and ethical concerns related to its use.

P. 474

SPECIAL TOPICS IN MODERN GENETICS 4

Advances in Neurogenetics: The Study of Huntington Disease

As the result of groundbreaking advances in molecular genetics and genomics made since the 1970s, new fields in genetics and related disciplines have emerged. One new field is **neurogenetics**—the study of the genetic basis of normal and abnormal functioning of the nervous system, with emphasis on brain functions. Research in this field includes the genes associated with neurodegenerative disorders, with the ultimate goal of developing effective therapies to combat these devastating conditions. Of the many such diseases, including Alzheimer disease, Parkinson disease, and amyotrophic lateral sclerosis (ALS), **Huntington disease (HD)** stands out as a model for the genetic investigation of neurodegenerative disorders. Not only is it monogenic and 100 percent penetrant, but nearly all analytical approaches in molecular genetics have been successfully applied to the study of HD, validating its significance as a model for these diseases.

HD is an autosomal dominant disorder characterized by adult onset of defined and progressive behavioral changes, including uncontrolled movements (chorea), cognitive decline, and psychiatric disturbances, with death occurring within 10 to 15 years after symptoms appear. HD was one of the first examples of complete dominance in human inheritance, with no differences in phenotypes between homozygotes and heterozygotes. In the vast majority of cases, symptoms do not develop until about age 45. Overall, HD currently affects about 25,000 to 30,000 people in North America.

The disease is named after George Huntington, a nineteenth-century physician. He was not the first to describe the disorder,

"Driving with my father through a wooded road leading from Easthampton to Amagansett, we suddenly came upon two women, mother and daughter, both bowing, twisting, grimacing. I stared in wonderment, almost in fear. What could it mean?"

know about the molecular and cellular mechanisms associated with the disorder, particularly those discovered during the study of transgenic model systems. Finally, we will consider how this information is being used to develop a range of therapies.

ST 4.1 The Search for the Huntington Gene

Mapping the gene for Huntington disease was one of the first attempts to employ a method from a landmark 1980 paper by Botstein, White, and Davis in which the authors proposed that DNA sequence variations in humans could be detected as differences in the length of DNA fragments produced by cutting DNA with restriction enzymes. These differences, known as restriction fragment length polymorphisms (RFLPs), could be visualized using Southern blots (see Chapter 18 for a discussion of RFLPs, and Chapter 17 for a discussion of Southern blots). The authors estimated that a collection of about 150 RFLPs distributed across the genome could be used with pedigrees to detect linkage anywhere in the genome between an RFLP marker and a disease gene of interest. In practical terms, this meant that it would be possible to map a disease gene with no information about the gene, its gene product, or its function—an approach referred to as reverse genetics.

SPECIAL TOPIC 4

NEW! Special Topics chapter on Advances in Neurogenetics: The Study of Huntington Disease, explores how genetic analysis has informed scientists about the disease's causes, symptoms, and future treatment. All Special Topics chapters include a series of questions that help students review key ideas or facilitate personal contemplations and group discussions, and are assignable in Mastering Genetics.

P. 506

Explore the latest ethical considerations

GENETICS, ETHICS, AND SOCIETY

Down Syndrome and Prenatal Testing—The New Eugenics?

Down syndrome is the most common chromosomal abnormality seen in newborn babies. Prenatal diagnostic tests for Down syndrome have been available for decades, especially to older pregnant women who have an increased risk of bearing a child with Down syndrome. Scientists estimate that there is an abortion rate of about 30 percent for fetuses that test positive for Down syndrome in the United States, and rates of up to 85 percent in other parts of the world, such as Taiwan and France.

Some people agree that it is morally acceptable to prevent the birth of a genetically abnormal fetus. However, others argue that prenatal genetic testing, with the goal of eliminating congenital disorders, is unethical. In addition, some argue that prenatal genetic testing followed by selective abortion is eugenic. How does eugenics apply, if at all, to screening for Down syndrome and other human genetic disorders

The term *eugenics* was first defined by Francis Galton in 1883 as "the science which deals with all influences that improve the inborn qualities of a race; also with those that develop them to the utmost advantage." Galton believed that human traits such as intelligence and personality were hereditary and that humans could selectively mate with each other to create gifted groups of people—analogous to the creation of purebred dogs with specific traits. Galton did not propose coercion but thought that people would voluntarily select mates in order to enhance particular genetic outcomes for their offspring.

In the early to mid-twentieth century, countries throughout the world adopted eugenic policies with the aim of enhancing desirable human traits (positive eugenics) and eliminating undesirable ones (negative eugenics). Many countries, including Britain, Canada, and the United States, enacted compulsory sterilization programs for the "feebleminded," mentally ill, and criminals. The eugenic policies of Nazi Germany were particularly infamous, resulting in forced human genetic experimentation and the slaughter of tens of thousands of people with disabilities. The eugenics movement was discredited after World War II, and the evils perpetuated in its name have tainted the term *eugenics* ever since.

Given the history of the eugenics movement, is it fair to use the term *eugenics* when we speak about genetic testing for Down syndrome and other genetic disorders? Some people argue that it is not eugenic to select for healthy children because there is no coercion, the state is not involved, and the goal is the elimination of suffering. Others point out that such voluntary actions still constitute eugenics, since they involve a form of bioengineering for "better" human beings.

Now that we are entering an era of unprecedented knowledge about our genomes and our predisposition to genetic disorders, we must make decisions about whether our attempts to control or improve human genomes are ethical and what limits we should place on these efforts. The story of the eugenics movement provides us with a powerful cautionary tale about the potential misuses of genetic information.

Your Turn

Take time, individually or in groups, to consider the following questions. Investigate the references and links to help you discuss some of the ethical issues surrounding genetic testing and eugenics.

1. Do you think that modern prenatal and preimplantation genetic testing followed by selective abortion is eugenic? Why or why not?

   For background on these questions, see McCabe, L., and McCabe, E. (2011). Down syndrome: Coercion and eugenics. *Genet. Med.* 13:708–710. *Another useful discussion can be found in* Wilkinson, S., (2015). Prenatal screening, reproductive choice, and public health. *Bioethics* 29:26–35.

2. If genetic technologies were more advanced than today, and you could choose the traits of your children, would you take advantage of that option? Which traits would you choose—height, weight, intellectual abilities, athleticism, artistic talents? If so, would this be eugenic? Would it be ethical?

   To read about similar questions answered by groups of Swiss law and medical students, read Elger, B., and Harding, T., (2003). Huntington's disease: Do future physicians and lawyers think eugenically? *Clin. Genet.* 64:327–338.

P. 141

Genetics, Ethics, and Society essays provide synopses of ethical issues related to current findings in genetics that impact directly on society today. They include a section called *Your Turn*, which directs students to related resources of short readings and websites to support deeper investigation and discussion of the main topic of each essay.

Case Studies at the end of each chapter have been updated with new topics. Students can read and answer questions about a short scenario related to one of the chapter topics. Each Case Study links the coverage of formal genetic knowledge to everyday societal issues, and they include ethical considerations.

CASE STUDY To test or not to test

Thomas discovered a devastating piece of family history when he learned that his brother had been diagnosed with Huntington disease (HD) at age 49. This dominantly inherited autosomal condition usually begins around age 45 with progressive dementia, muscular rigidity, and seizures and ultimately leads to death when affected individuals are in their early 60s. There currently is no effective treatment or cure for this genetic disorder. Thomas, now 38, wonders what the chances are that he also has inherited the mutant allele for HD, leading him to discuss with his wife whether they should seek genetic counseling and whether he should undergo genetic testing. They have two teenage children, a boy and a girl.

1. If they seek genetic counseling, what issues would likely be discussed? Which of these pose grave ethical dilemmas?
2. If you were in Thomas's position, would you want to be tested and possibly learn that you were almost certain to develop the disorder sometime in the next 5–10 years?
3. If Thomas tests positive for the HD allele, should his children be told about the situation, and if so, at what age? Who should make the decision about having the son and daughter tested?

Fulda, K., and Lykens, K. (2006). Ethical issues in predictive genetic testing: A public health perspective. *J. Med. Ethics* 32:143–14

P. 74

Learn genetics concepts and problem solving in Mastering Genetics

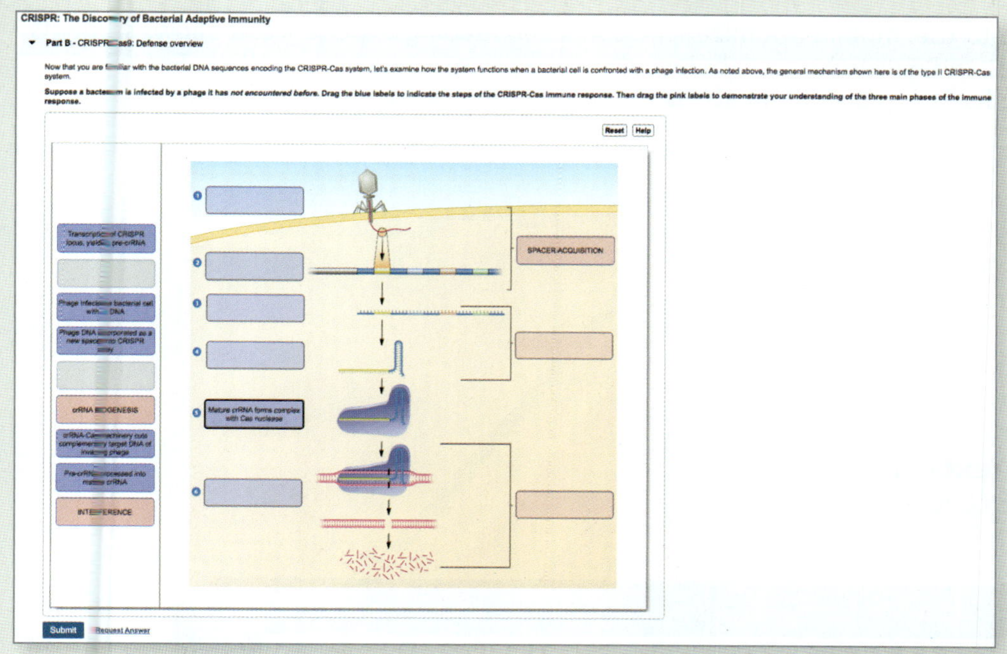

NEW! Tutorials have been added to the library on topics like CRISPR-Cas and epigenetics, to help students master important and challenging concepts.

A library of over 100 Practice Problems offers more opportunities to assign high quality problems for student homework or practice. These questions appear only in Mastering Genetics and include targeted wrong-answer feedback to help students learn from their mistakes. They are similar to end-of-chapter questions in terms of topic coverage and difficulty.

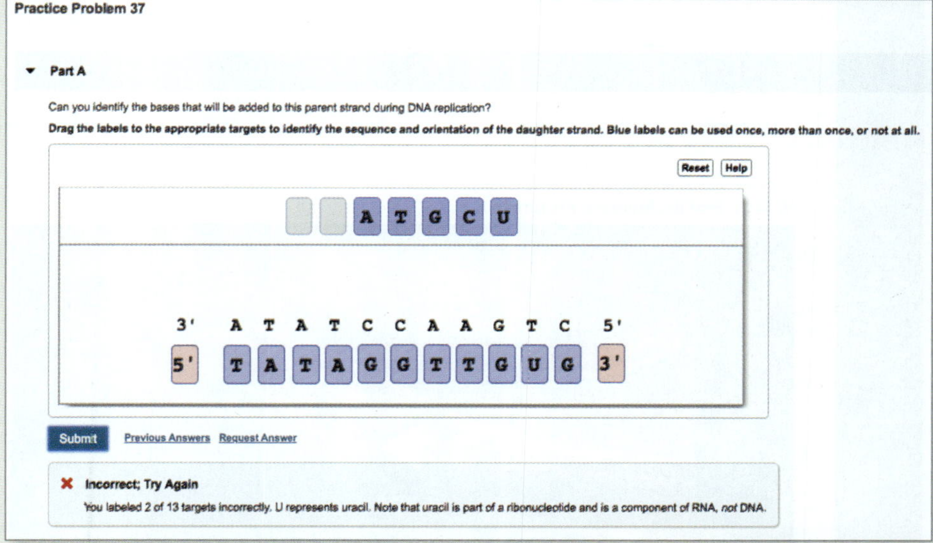

Give students anytime, anywhere access with Pearson eText

Pearson eText is a simple-to-use, mobile-optimized, personalized reading experience. It allows students to easily highlight, take notes, and review key vocabulary all in one place—even when offline. Seamlessly integrated videos engage students and give them access to the help they need, when they need it.

NEW! Pearson eText increases student engagement with embedded videos.

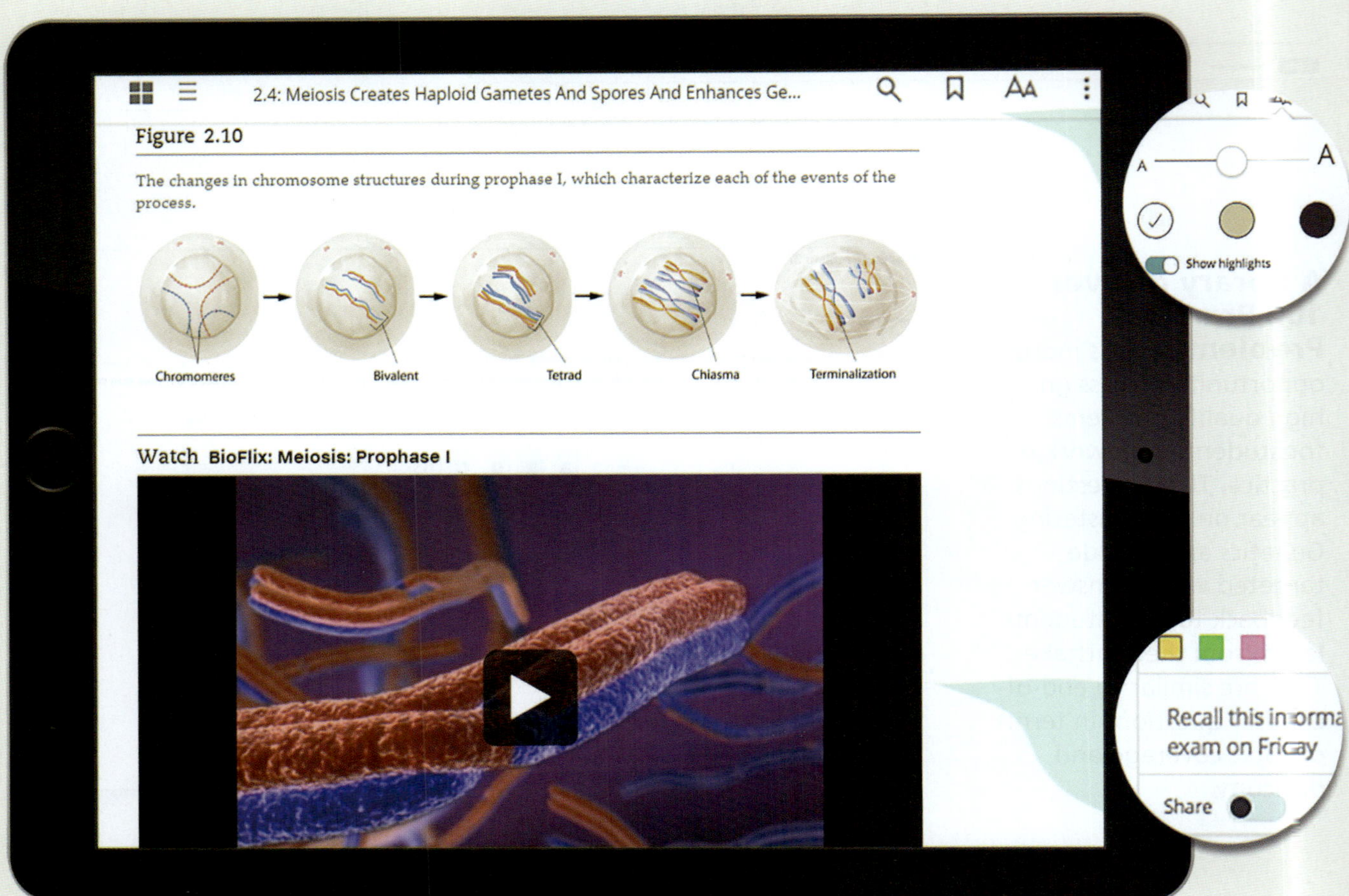

Instructor support you can rely on

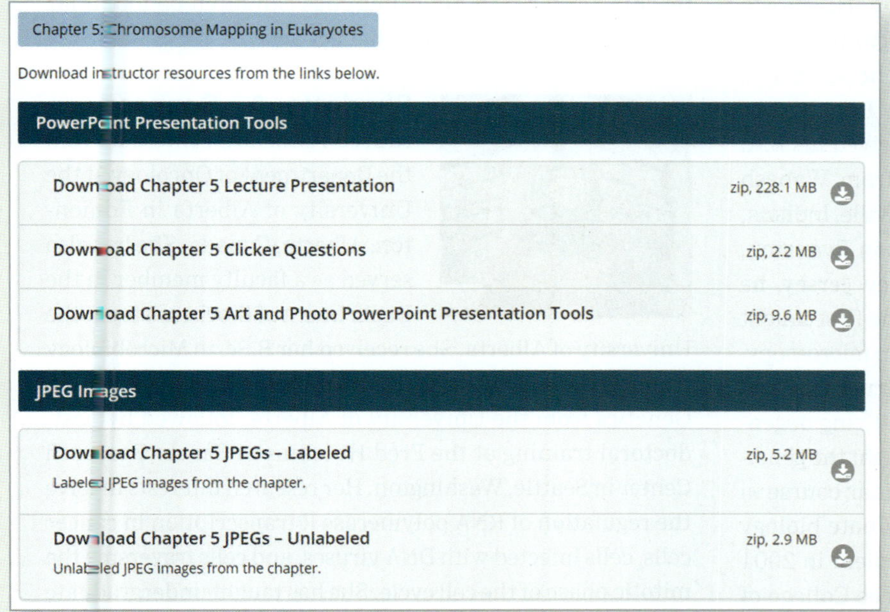

Essentials of Genetics includes a full suite of instructor support materials in the Instructor Resources area in Mastering Genetics. Resources include lecture presentations, clicker questions, and art and photos in PowerPoint®; labeled and unlabeled JPEGs of images from the text; and a test bank.

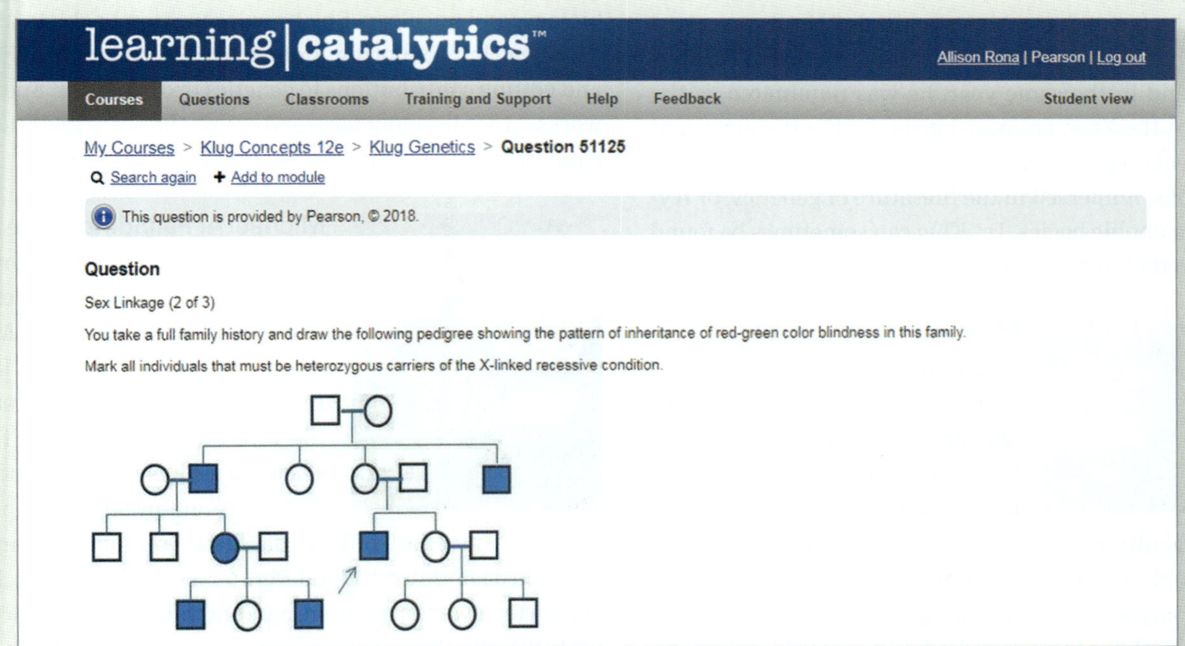

Instructors also have access to Learning Catalytics. With Learning Catalytics, you'll hear from every student when it matters most. You can pose a variety of questions in class that help students recall ideas, apply concepts, and develop critical-thinking skills. Your students respond using their own smartphones, tablets, or laptops. You can monitor responses with real-time analytics and find out what your students do—and don't—understand. Then, you can adjust your teaching accordingly and even facilitate peer-to-peer learning, helping students stay motivated and engaged. Write your own questions, pull from a shared library of community-generated questions, or use Pearson's content clusters, which pose 2-5 questions about a single data set or scenario.

About the Authors

William S. Klug is an Emeritus Professor of Biology at The College of New Jersey (formerly Trenton State College) in Ewing, New Jersey, where he served as Chair of the Biology Department for 17 years. He received his B.A. degree in Biology from Wabash College in Crawfordsville, Indiana, and his Ph.D. from Northwestern University in Evanston, Illinois. Prior to coming to The College of New Jersey, he was on the faculty of Wabash College, where he first taught genetics, as well as general biology and electron microscopy. His research interests have involved ultrastructural and molecular genetic studies of development, utilizing oogenesis in *Drosophila* as a model system. He has taught the genetics course as well as the senior capstone seminar course in Human and Molecular Genetics to undergraduate biology majors for over four decades. He was the recipient in 2001 of the first annual teaching award given at The College of New Jersey, granted to the faculty member who "most challenges students to achieve high standards." He also received the 2004 Outstanding Professor Award from Sigma Pi International, and in the same year, he was nominated as the Educator of the Year, an award given by the Research and Development Council of New Jersey. When not revising one of his textbooks, immersed in the literature of genetics, or trying to avoid double bogies, Dr. Klug can sometimes be found paddling in the Gulf of Mexico or in Maine's Penobscot Bay.

Michael R. Cummings is a Research Professor in the Department of Biological, Chemical, and Physical Sciences at Illinois Institute of Technology, Chicago, Illinois. For more than 25 years, he was a faculty member in the Department of Biological Sciences and in the Department of Molecular Genetics at the University of Illinois at Chicago. He has also served on the faculties of Northwestern University and Florida State University. He received his B.A. from St. Mary's College in Winona, Minnesota, and his M.S. and Ph.D. from Northwestern University in Evanston, Illinois. In addition to this text, he has written textbooks in human genetics and general biology. His research interests center on the molecular organization and physical mapping of the heterochromatic regions of human acrocentric chromosomes. At the undergraduate level, he teaches courses in molecular genetics, human genetics, and general biology, and has received numerous awards for teaching excellence given by university faculty, student organizations, and graduating seniors. When not teaching or writing, Dr. Cummings can often be found far offshore fishing for the one that got away.

Charlotte A. Spencer is a retired Associate Professor from the Department of Oncology at the University of Alberta in Edmonton, Alberta, Canada. She has also served as a faculty member in the Department of Biochemistry at the University of Alberta. She received her B.Sc. in Microbiology from the University of British Columbia and her Ph.D. in Genetics from the University of Alberta, followed by postdoctoral training at the Fred Hutchinson Cancer Research Center in Seattle, Washington. Her research interests involve the regulation of RNA polymerase II transcription in cancer cells, cells infected with DNA viruses, and cells traversing the mitotic phase of the cell cycle. She has taught undergraduate and graduate courses in biochemistry, genetics, molecular biology, and oncology. She has also written booklets in the Prentice Hall Exploring Biology series. When not writing and editing contributions to genetics textbooks, Dr. Spencer works on her hazelnut farm and enjoys the peace and quiet of a remote Island off the west coast of British Columbia.

Michael A. Palladino is Vice Provost for Graduate Studies, former Dean of the School of Science, and Professor of Biology at Monmouth University in West Long Branch, New Jersey. He received his B.S. degree in Biology from The College of New Jersey and his Ph.D. in Anatomy and Cell Biology from the University of Virginia. For more than 15 years he directed a laboratory of undergraduate student researchers supported by external funding from the National Institutes of Health, biopharma companies, and other agencies. He and his undergraduates studied molecular mechanisms involved in innate immunity of mammalian male reproductive organs and genes involved in oxygen homeostasis and ischemic injury of the testis. He has taught a wide range of courses including genetics, biotechnology, endocrinology, and cell and molecular biology. He has received several awards for research and teaching, including the 2009 Young Andrologist Award of the American Society of Andrology, the 2005 Distinguished Teacher Award from Monmouth University, and the 2005 Caring Heart Award from the New Jersey Association for Biomedical Research. He is co-author of the undergraduate textbook *Introduction to Biotechnology*. He was Series Editor for the Benjamin Cummings *Special Topics in Biology* booklet series, and author of the first booklet in the series, *Understanding the Human Genome Project*. When away from the university or authoring textbooks, Dr. Palladino can often be found watching or playing soccer or attempting to catch most any species of fish in freshwater or saltwater.

Darrell J. Killian is an Associate Professor in the Department of Molecular Biology at Colorado College in Colorado Springs, Colorado. He received his B.A. degree in Molecular Biology and Biochemistry from Wesleyan University in Middletown, Connecticut, prior to working as a Research Technician in Molecular Genetics at Rockefeller University in New York, New York. He earned his Ph.D. in Developmental Genetics from New York University in New York, New York, and received his postdoctoral training at the University of Colorado–Boulder in the Department of Molecular, Cellular, and Developmental Biology. Prior to joining Colorado College, he was an Assistant Professor of Biology at the College of New Jersey in Ewing, New Jersey. His research focuses on the genetic regulation of animal development, and he has received funding from the National Institutes of Health and the National Science Foundation. Currently, he and his undergraduate research assistants are investigating the molecular genetic regulation of nervous system development using *C. elegans* and *Drosophila* as model systems. He teaches undergraduate courses in genetics, molecular and cellular biology, stem cell biology, and developmental neurobiology. When away from the classroom and research lab, Dr. Killian can often be found on two wheels exploring trails in the Pike and San Isabel National Forests.

Dedication

We dedicate this edition to our long-time colleague and friend Harry Nickla, who sadly passed away in 2017. With decades of experience teaching Genetics to students at Creighton University, Harry's contribution to our texts included authorship of the Student Handbook and Solutions Manual and the test bank, as well as devising many of the data-based problems found near the end of each chapter. He was also a source of advice during the planning session for each new edition. We always appreciated his professional insights, friendship, and conviviality. We were lucky to have him as part of our team, and we miss him greatly.

Contents

SPECIAL TOPICS IN MODERN GENETICS 2

Genetic Testing 474

SPECIAL TOPICS IN MODERN GENETICS 3

Gene Therapy 492

SPECIAL TOPICS IN MODERN GENETICS 4

Advances in Neurogenetics: The Study of Huntington Disease 506

SPECIAL TOPICS IN MODERN GENETICS 5

DNA Forensics 515

SPECIAL TOPICS IN MODERN GENETICS 6

Genetically Modified Foods 524

SPECIAL TOPICS IN MODERN GENETICS 7

Genomics and Precision Medicine 534

Preface

Essentials of Genetics is written for courses requiring a text that is briefer and less detailed than its more comprehensive companion, *Concepts of Genetics*. While coverage is thorough and modern, *Essentials* is written to be more accessible to biology majors, as well as to students majoring in a number of other disciplines, including agriculture, animal husbandry, chemistry, nursing, engineering, forestry, psychology, and wildlife management. Because *Essentials of Genetics* is shorter than many other texts, it is also more manageable in one-quarter and trimester courses.

Goals

In this edition of *Essentials of Genetics*, the two most important goals have been to introduce pedagogic innovations that enhance learning and to provide carefully updated, highly accessible coverage of genetic topics of both historical and modern significance. As new tools and findings of genetics research continue to emerge rapidly and grow in importance in the study of all subdisciplines of biology, instructors face tough choices about what content is truly essential as they introduce the discipline to novice students. We have thoughtfully revised each chapter in light of this challenge, by selectively scaling back the detail or scope of coverage in the more traditional chapters in order to provide expanded coverage and broader context for the more modern, cutting-edge topics. Our aim is to continue to provide efficient coverage of the fundamental concepts in transmission and molecular genetics that lay the groundwork for more in-depth coverage of emerging topics of growing importance—in particular, the many aspects of the genomic revolution that is already relevant to our day-to-day lives.

While we have adjusted this edition to keep pace with changing content and teaching practices, we remain dedicated to the core principles that underlie this book. Specifically, we seek to

- Emphasize concepts rather than excessive detail.
- Write clearly and directly to students in order to provide understandable explanations of complex analytical topics.
- Emphasize problem solving, thereby guiding students to think analytically and to apply and extend their knowledge of genetics.
- Provide the most modern and up-to-date coverage of this exciting field.
- Propagate the rich history of genetics that so beautifully elucidates how information is acquired as the discipline develops and grows.
- Create inviting, engaging, and pedagogically useful figures enhanced by meaningful photographs to support student understanding.
- Provide outstanding interactive media support to guide students in understanding important concepts through animations, tutorial exercises, and assessment tools.

The above goals serve as the cornerstone of *Essentials of Genetics*. This pedagogic foundation allows the book to accommodate courses with many different approaches and lecture formats. While the book presents a coherent table of contents that represents one approach to offering a course in genetics, chapters are nevertheless written to be independent of one another, allowing instructors to utilize them in various sequences.

New to This Edition

In addition to updating information with new findings in all chapters throughout the text, four chapters are new to this edition.

- **Two new chapters expand the coverage of the regulation of gene expression** The topic of genetic regulation was previously covered in a single chapter, but has now been split into two new chapters. The first (Chapter 15) involves regulation in bacteria, while the second (Chapter 16) focuses on eukaryotes. The bacterial coverage represents the pioneering work in this field and then concludes with an introduction to CRISPR-Cas. The eukaryotic coverage focuses on the regulation of gene expression first at the level of transcription, and then post-transcriptionally, where the expanded coverage focuses on mechanisms that regulate RNA. Research into posttranscriptional regulation in the past 15 years has highlighted the importance of topics such as alternative splicing, mRNA stability and decay, and regulatory noncoding RNAs. Collectively, the addition of these two new chapters provides students and instructors with a thorough, up-to-date presentation of these important aspects of genetics.
- **Two new Special Topics in Modern Genetics chapters** Special Topics chapters are focused and flexible, providing abbreviated, cohesive coverage of important topics in genetics. There are seven Special Topics chapters in this edition, two of which are new. Special Topics Chapter 2—*Genetic Testing* explores how genetic testing is becoming prominent in many contexts and how its use raises many questions and ethical concerns. Special Topics Chapter 4—*Advances in Neurogenetics: The Study of Huntington Disease* illustrates the many advances that have been made in the study of Huntington disease, a

monogenic human disorder that has been subjected to analysis using multiple approaches involving molecular genetics. As such, the chapter exemplifies the growing body of information that has accrued regarding the causes, symptoms, and future treatment of this disorder.

- **Expanded coverage of CRISPR-Cas** Since the previous edition was published, techniques for genome editing have vastly improved due to CRISPR-Cas technology. Thus, we have integrated information about CRISPR-Cas in several different locations within the text. The impact of genome editing with CRISPR-Cas is briefly introduced in Chapter 1. Then, in Chapter 15, students learn how CRISPR-Cas was originally discovered as a bacterial system that regulates the gene expression of bacterial viruses (bacteriophages), providing an immunity against infection. The mechanism and applications to biotechnology are subsequently covered in Chapter 17. Finally, the use of CRISPR-Cas genome editing for gene therapy and the production of genetically modified foods is discussed in Special Topics Chapter 3—*Gene Therapy* and Special Topics Chapter 6—*Genetically Modified Foods.*
- **Increased emphasis on ethics** We recognize in this edition the importance of providing an increased emphasis on ethical considerations that genetics is bringing into everyday life. Regarding this point, we have converted the essay feature previously called *Genetics, Technology, and Society* to one with added emphasis on ethics and renamed it *Genetics, Ethics, and Society*. Approximately half the chapters have new or revised essays. In each case, a synopsis is presented of an ethical issue related to a current finding in genetics that impacts directly on society today. The feature then includes a section called *Your Turn*, which directs students to related resources of short readings and Web sites to support deeper investigation and discussion of the main topic of each essay. In addition, another feature called *Case Study*, which appears near the end of all chapters, has been recast with an increased focus on ethics. Both of these features increase the opportunities for active and cooperative learning as well.

New and Updated Coverage

Below is a chapter-by-chapter list of the most significant new and updated coverage present in this edition.

Ch. 1: Introduction to Genetics • New chapter introduction vignette emphasizing the significance of the discovery of CRISPR-Cas9, a powerful genome-editing system.

Ch. 2: Mitosis and Meiosis • New information on microtubules and microfilaments • Revised Figure 2.9 on Meiotic Prophase I • New Exploring Genomics (EG) entry: PubMed: Exploring and Retrieving Biomedical Literature • New Case Study (CS): Timing Is Everything

Ch. 3: Mendelian Genetics • New Table 3.2 on Dominant and Recessive Human Traits • New Now Solve This (NST) 3.5 on pedigree analysis

Ch. 4: Modification of Mendelian Ratios • New information in the "Mitochondria, Human Health, and Aging" section • New information on the *MERFF* mutation • New Genetics, Ethics, and Society (GES) entry: Mitochondrial Replacement and Three-Parent Babies

Ch. 5: Sex Determination and Sex Chromosomes • New information on Klinefelter syndrome • New GES: A Question of Gender: Sex Selection in Humans

Ch. 6: Chromosome Mutations: Variation in Number and Arrangement • Updated information on copy number variation • New GES: Down Syndrome and Prenatal Testing—The New Eugenics? • A new end of chapter problem involving mapping analysis in *Drosophila.*

Ch. 8: Genetic Analysis and Mapping in Bacteria and Bacteriophages • New GES: Multidrug-Resistant Bacteria: Fighting with Phage

Ch. 10: DNA Replication and Recombination • New details about DNA unwinding during replication • New section entitled "Telomeres in Disease, Aging, and Cancer" • Two new end of chapter problems involving telomeres and telomerase

Ch. 12: The Genetic Code and Transcription • Revised coverage of transcription and RNA processing in eukaryotes • New information on termination of transcription in bacteria • New section entitled "Why Do Introns Exist?" • New GES: Treating Duchene Muscular Dystrophy

Ch. 13: Translation and Proteins • Revised coverage of ribosome and tRNA structure • Revised coverage of translation in bacteria • Expanded coverage of translation in eukaryotes including new information on closed-loop translation, illustrated in a new figure (Fig. 13.10)

Ch. 14: Gene Mutation, DNA Repair, and Transposition • Reorganization of the section on mutation classification, including new table summaries • New and expanded coverage of human germ-line and somatic mutation rates • New, reorganized, and revised coverage of transposable elements, focusing on the major characteristics of retrotransposons and DNA transposons, as well as on how transposons create mutations • Three new figures and one new table

Ch. 15: Regulation of Gene Expression in Bacteria • New chapter that focuses specifically on gene regulation in bacteria • Expanded coverage on the roles of RNA in bacterial gene regulation • New coverage of CRISPR-Cas-mediated regulation of invading viral DNA sequences

Ch. 16: Regulation of Gene Expression in Eukaryotes • New chapter that focuses specifically on gene regulation in eukaryotes • Revised and expanded coverage of alternative splicing, including a new figure, and its relevance to human disease • Expanded coverage on RNA stability and RNA decay including a new figure (Fig. 16.11) • Updated information on noncoding RNAs that regulate gene expression • Enriched coverage of ubiquitin-mediated protein degradation, including a new figure (Fig. 16.14)

Ch. 17: Recombinant DNA Technology • Updated content on modern sequencing technologies including a new figure (Fig. 17.12) on third-generation sequencing (single-strand DNA sequencing) • New section, "Genome Editing with CRISPR-Cas," describes this system as a genome editing tool and includes a new figure (Fig. 17.16)

Ch. 18: Genomics, Bioinformatics, and Proteomics • A new section, "DNA Sequence Analysis Relies on Bioinformatics Applications and Genome Databases," integrating applications of bioinformatics, genome databases, and functional genomics for analyzing and understanding gene function by sequence analysis • Reorganized and revised content on the Human Genome Project, including a new end of chapter problem citing the PANTHER database as part of the Human Genome Project • Updated content on personal genome projects • New content on diploid genomes, mosaicism, and reference genomes and the pangenome to emphasize human genetic variations, including a new figure (Fig. 18.8) • Incorporated coverage of the Human Microbiome Project into a new section, "Metagenomics," and expanded content to include a new Figure (Fig. 18.9) displaying microbiome results of patients with different human disease conditions • A new section titled "RNA Sequencing" • A new section, "Synthetic Genomes and the Emergence of Synthetic Biology," including a new figure (Fig. 18.13) • New GES: Privacy and Anonymity in the Era of Genomic Big Data • Several new and revised end of chapter problems

Ch. 19: The Genetics of Cancer • Extended coverage of environmental agents that contribute to human cancers, including more information about both natural and human-made carcinogens • New subsection entitled "Tobacco Smoke and Cancer" explaining how a well-studied carcinogen induces a wide range of genetic effects that may lead to mutations and cancer

Ch. 20: Quantitative Genetics and Multifactorial Traits • Revised coverage of Expression QTLs (eQTLs) in the regulation of gene expression • New GES: Rice, Genes, and the Second Green Revolution • New CS: A Chance Discovery

Ch. 21: Population and Evolutionary Genetics • New figure (Fig. 21.7) on the relationship between genotype and allele frequency • Important modifications to Figures 21.8 and 21.9 illustrating allele selection • New figure (Fig. 21.13) on the impact of selection types on the phenotypic mean and variance • Revised text and figure (Fig. 21.24) on molecular clocks • Updated information about the origins of the human genome • New figure (Fig. 21.26) on hominin contributions to the genome of modern humans

Special Topic 1: Epigenetics • Revised, updated, and expanded coverage of epigenetic topics, including histone modifications, noncoding RNAs, assisted reproductive technologies, and the heritability of stress-induced behaviors • Updated coverage of epigenetics and cancer • New section on "Epigenetics and Monoallelic Gene Expression" • New figures on DNA methylation, chemical modification of histones, genomic imprinting, random autosomal monoallelic gene expression, imprinting in germ cells, and maternal behavior and stress responses in rat pups

Special Topic 2: Genetic Testing • New Special Topics chapter emphasizing modern approaches to genetic testing including prenatal genetic testing, noninvasive procedures for testing fetal DNA, testing using allele-specific oligonucleotides, microarrays, and genetic analysis by DNA and RNA sequencing • Includes coverage of the recommended uniform screening panel, undiagnosed diseases network, and genetic analysis for pathogen identification during infectious disease outbreaks • Section on genome-wide association studies incorporates approaches for genomic analysis of disease conditions at the population level • A range of ethical, social, and legal considerations are discussed

Special Topic 3: Gene Therapy • Updated information on gene therapy trials that are under way • An expanded section "Genome Editing" highlighting the application of the CRISPR-Cas system and describing some of the most promising trials under way in humans and animals • New ethical considerations of CRISPR-Cas and germ-line and embryo editing • New section, "RNA-Based Therapeutics," that includes coverage of antisense RNA; RNA interference; and updated trials for RNA-based therapeutics, including Spinraza as an antisense RNA modifying splicing for the treatment of spinal muscular atrophy • Updated content on roles of stem cells in gene therapy • New content on combining genome editing with immunotherapy

Special Topic 4: Advances in Neurogenetics: The Study of Huntington Disease • New Special Topics chapter that surveys the study of Huntington Disease (HD) from 1970 to the present • Coverage includes the genetic basis and progression of HD, the mapping and isolation of the gene responsible for the disorder, and information on the mutant gene product • Discussions include information on the molecular and cellular alterations caused by the mutant protein, the use of transgenic animal models of HD, and the molecular and cellular approaches to therapy

Special Topic 5: DNA Forensics • New section entitled "DNA Phenotyping," describing a controversial forensic method, including descriptions of how law-enforcement agencies currently use this new technology

Special Topic 6: Genetically Modified Foods • New section, entitled "Gene Editing and GM Foods," describing how scientists are using the new techniques of gene editing (including ZFN, TALENS, and CRISPR-Cas) to create GM food plants and animals, and how these methods are changing the way in which GM foods are being regulated • A new box, "The New CRISPR Mushroom," describing the development and regulatory approval of the first CRISPR-created GM food to be cleared for human consumption

Special Topic 7: Genomics and Precision Medicine • New section, entitled "Precision Oncology," describing two targeted cancer immunotherapies: adoptive cell transfer and engineered T-cell therapy • Updated section, "Pharmacogenomics," including a discussion of new trends in preemptive gene screening for pharmacogenomic variants • New box, "Preemptive Pharmacogenomic Screening: The pGEN4Kids Program," discussing preemptive gene screening that integrates DNA analysis into patient electronic health records

Emphasis on Concepts

Essentials of Genetics focuses on conceptual issues in genetics and uses problem solving to develop a deep understanding of them. We consider a concept to be a cognitive unit of meaning that encompasses a related set of scientifically derived findings and ideas. As such, a concept provides broad mental imagery, which we believe is a very effective way to teach science, in this case, genetics. Details that might be memorized, but soon forgotten, are instead subsumed within a conceptual framework that is more easily retained. Such a framework may be expanded in content as new information is acquired and may interface with other concepts, providing a useful mechanism to integrate and better understand related processes and ideas. An extensive set of concepts may be devised and conveyed to eventually encompass and represent an entire discipline—and this is our goal in this genetics textbook.

To aid students in identifying the conceptual aspects of a major topic, each chapter begins with a section called ***Chapter Concepts***, which identifies the most important ideas about to be presented. Then, throughout each chapter, ***Essential Points*** are provided that establish the key issues that have been discussed. And in the ***How Do We Know?*** question that starts each chapter's problem set, students are asked to identify the experimental basis of important genetic findings presented in the chapter. As an extension of the learning approach in biology called "Science as a Way of Knowing," this feature enhances students' understanding of many key concepts covered in each chapter. Finally, the second entry in each chapter's problem set is labeled as a **Concepts Question**, which asks the student to review and comment on specific aspects of the Chapter Concepts found at the beginning of each chapter.

Collectively, these features help to ensure that students easily become aware of and understand the major conceptual issues as they confront the extensive vocabulary and the many important details of genetics. Carefully designed figures also support this approach throughout the book.

Emphasis on Problem Solving

Helping students develop effective problem-solving skills is one of the greatest challenges of a genetics course. The feature called ***Now Solve This***, integrated throughout each chapter, asks students to link conceptual understanding in a more immediate way to problem solving. Each entry provides a problem for the student to solve that is closely related to the current text discussion. A pedagogic hint is then provided to aid in arriving at the correct solution. All chapters conclude with ***Insights and Solutions***, a popular and highly useful section that provides sample problems and solutions that demonstrate approaches useful in genetic analysis. These help students develop analytical thinking and experimental reasoning skills. Digesting the information in *Insights and Solutions* primes students as they move on to the lengthier ***Problems and Discussion Questions*** section that concludes each chapter. Here, we present questions that review topics in the chapter and problems that ask students to think in an analytical and applied way about genetic concepts. The addition of Mastering Genetics extends our focus on problem solving online, and it allows students to get help and guidance while practicing how to solve problems.

Continuing Features

The Tenth Edition has maintained several popular features that are pedagogically useful for students as they study genetics. Together, these create a platform that seeks to challenge students to think more deeply about, and thus understand more comprehensively, the information he or she has just finished studying.

- **Exploring Genomics** Appearing in numerous chapters, this feature illustrates the pervasiveness of genomics in the current study of genetics. Each entry asks students to access one or more genomics-related Web sites that collectively are among the best publicly available resources and databases. Students work through interactive exercises that ensure their familiarity with the type of genomic or proteomic information available. Exercises instruct students on how to explore specific topics and how to access significant data. Questions guide student exploration and challenge them to further explore the sites on their own. Importantly, *Exploring Genomics* integrates genomics information throughout the text, as this emerging field is linked to chapter content. This feature provides the basis for individual or group assignments in or out of the classroom.

- **Case Studies** This feature, with an increased emphasis on ethical considerations, appears at the end of each chapter and provides the basis for enhanced classroom interactions. In each entry, a short scenario related to one of the chapter topics is presented, followed by several questions. These ask students to apply their newly acquired knowledge to real-life issues that may be explored in small-group discussions or serve as individual assignments.

For the Instructor

Mastering Genetics http://www.masteringgenetics.com

Mastering Genetics engages and motivates students to learn and allows you to easily assign automatically graded activities. Tutorials provide students with personalized coaching and feedback. Using the gradebook, you can quickly monitor and display student results. Mastering Genetics easily captures data to demonstrate assessment outcomes. Resources include:

- In-depth tutorials that coach students with hints and feedback specific to their misconceptions.
- A new, robust library of **Practice Problems** offers more opportunities to assign challenging problems for student homework or practice. These questions include targeted wrong answer feedback to help students learn from their mistakes. They appear only in Mastering Genetics.
- An item library of assignable questions including end of chapter problems, test bank questions, and reading quizzes. You can use publisher-created prebuilt assignments to get started quickly. Each question can be easily edited to match the precise language you use.
- A gradebook that provides you with quick results and easy-to-interpret insights into student performance.

Instructor Resources

The Instructor Resources, available for download in the Instructor area of Mastering Genetics, offer adopters of the text convenient access to a comprehensive and innovative set of lecture presentation and teaching tools. Developed to meet the needs of veteran and newer instructors alike, these resources include:

- The JPEG files of all text line drawings with labels individually enhanced for optimal projection results (as well as unlabeled versions) and all text tables.
- Most of the text photos, including all photos with pedagogical significance, as JPEG files.
- The JPEG files of line drawings, photos, and tables preloaded into comprehensive PowerPoint presentations for each chapter.
- A second set of PowerPoint presentations consisting of a thorough lecture outline for each chapter augmented by key text illustrations.
- An impressive series of concise instructor animations adding depth and visual clarity to the most important topics and dynamic processes described in the text.
- The instructor animations preloaded into PowerPoint presentation files for each chapter.
- PowerPoint presentations containing a comprehensive set of in-class Classroom Response System (CRS) questions for each chapter.
- In Word files, a complete set of the assessment materials and study questions and answers from the test bank, the text's in-chapter text questions, and the student media practice questions.

TestGen EQ Computerized Testing Software

Test questions are available as part of the TestGen EQ Testing Software, a text-specific testing program that is networkable for administering tests. It also allows instructors to view and edit questions, export the questions as tests, and print them out in a variety of formats.

Mastering Genetics http://www.masteringgenetics.com

Used by over a million science students, the Mastering platform is the most effective and widely used online tutorial, homework, and assessment system for the sciences. Perform better on exams with Mastering Genetics. As an instructor-assigned homework system, Mastering Genetics is designed to provide students with a variety of assessments to help them understand key topics and concepts and to build problem-solving skills. Mastering Genetics tutorials guide students through the toughest topics in genetics with self-paced tutorials that provide individualized coaching with hints and feedback specific to a student's individual misconceptions. Students can also explore Mastering Genetics' Study Area, which includes animations, the eText, *Exploring Genomics* exercises, and other study aids. The interactive eText 2.0 allows students to access their text on mobile devices, highlight text, add study notes, review instructor's notes, and search throughout the text, 24/7.

Acknowledgments

Contributors

We begin with special acknowledgments to those who have made direct contributions to this text. We thank Christy Fillman of the University of Colorado–Boulder, Jutta Heller of the University of Washington–Tacoma, Christopher Halweg of North Carolina State University, Pamela Osenkowski of Loyola University–Chicago, Matthew Marcello of Pace University, Susan Wesmiller of University of Pittsburgh School of Nursing, Mandy Schmella of University of Pittsburgh School of Nursing, and Fiona Rawle of the University of Toronto–Mississauga for their work on the media program. Virginia McDonough of Hope College and Cindy Malone of California State University–Northridge contributed greatly to the instructor resources. We thank the following instructors for their work on the test bank: Mark Haefele of Community College of Denver, Scott Harrison of Georgia Southern University, David Kass of Eastern Michigan University, and Stephen Page of Community College of Baltimore County. We also express special thanks to eagle-eyed Michelle Gaudette, recently retired from Tufts University. In her role as author of the *Student Handbook and Solutions Manual* and the test bank, she has reviewed and edited the end of chapter problems and their corresponding solutions in the manual and in the Answers Appendix.

We are grateful to all of these contributors not only for sharing their genetic expertise, but for their dedication to this project as well as the pleasant interactions they provided.

Proofreaders and Accuracy Checking

Reading the detailed manuscript of a textbook deserves more thanks than words can offer. Our utmost appreciation is extended to Michelle Gaudette, Tufts University, and Ann Blakey, Ball State University, who provided accuracy checking of many chapters, and to Kerri Tomasso, who proofread the entire manuscript. They confronted this task with patience and diligence, contributing greatly to the quality of this text.

Reviewers

All comprehensive texts are dependent on the valuable input provided by many reviewers. While we take full responsibility for any errors in this book, we gratefully acknowledge the help provided by those individuals who reviewed the content and pedagogy of this edition:

Jessica Cottrell, Seton Hall University
Tamara Davis, Bryn Mawr College
Christy Fillman, University of Colorado–Boulder
Christy Fleet, Emory and Henry College
Donna-Marie Gardner, Middlesex County College
Christopher Harendza, Montgomery County Community College
Alfredo Leon, Miami Dade College
Tamara Mans, North Hennepin Community College
Holly Morris, Lehigh Carbon Community College
Isaiah G. Schauer, Brazosport College
Brenna Traver, Penn State Schuylkill
Susan Wesmiller, University of Pittsburgh School of Nursing
Michelle Wien, Bryn Mawr College

Special thanks go to Mike Guidry of LightCone Interactive and Karen Hughes of the University of Tennessee for their original contributions to the media program.

As these acknowledgments make clear, a text such as this is a collective enterprise. All of the above individuals deserve to share in any success this text enjoys. We want them to know that our gratitude is equaled only by the extreme dedication evident in their efforts. Many, many thanks to them all.

Editorial and Production Input

At Pearson, we express appreciation and high praise for the editorial guidance of Michael Gillespie, whose ideas and efforts have helped to shape and refine the features of this edition of the text. Brett Coker, our Content Producer, has worked tirelessly to keep the project on schedule and to maintain our standards of high quality. In addition, our editorial team—Ginnie Simione Jutson, Executive Director of Development, Robert Johnson, Rich Media Content Producer, and Sarah Jensen, Director of Editorial Content for Mastering Genetics—have provided valuable input into the current edition. They have worked creatively to ensure that the pedagogy and design of the book and media package are at the cutting edge of a rapidly changing discipline. Brett Coker and Heidi Aguiar supervised all of the production intricacies with great attention to detail and perseverance. Outstanding copyediting was performed by Lucy Mullins, for which we are most grateful. Allison Rona, Alysun Estes, and Kelly Galli have professionally and enthusiastically managed the marketing of the text. Finally, the beauty and consistent presentation of the art work are the product of Imagineering of Toronto. Without the work ethic and dedication of the above individuals, the text would never have come to fruition.

Acknowledgments for the Global Edition

Pearson would like to acknowledge and thank the following for their work on the Global Edition.

Contributor

Juan Pablo Labrador, Trinity College Dublin

Reviewers

Ayse Elif Erson Bensan, Middle East Technical University
Adriaan Engelbrecht, University of the Western Cape
Chris Finlay, University of Glasgow
Francisco Ramos Morales, Universidad de Sevilla
Preeti Srivastava, Indian Institute of Technology Delhi

1

Introduction to Genetics

Newer model organisms in genetics include the roundworm, *Caenorhabditis elegans*; the zebrafish, *Danio rerio*; and the mustard plant, *Arabidopsis thaliana*.

CHAPTER CONCEPTS

- Genetics in the twenty-first century is built on a rich tradition of discovery and experimentation stretching from the ancient world through the nineteenth century to the present day.
- Transmission genetics is the general process by which traits controlled by genes are transmitted through gametes from generation to generation.
- Mutant strains can be used in genetic crosses to map the location and distance between genes on chromosomes.
- The Watson–Crick model of DNA structure explains how genetic information is stored and expressed. This discovery is the foundation of molecular genetics.
- Recombinant DNA technology revolutionized genetics, was the foundation for the Human Genome Project, and has generated new fields that combine genetics with information technology.
- Biotechnology provides genetically modified organisms and their products that are used across a wide range of fields including agriculture, medicine, and industry.
- Model organisms used in genetics research are now utilized in combination with recombinant DNA technology and genomics to study human diseases.
- Genetic technology is developing faster than the policies, laws, and conventions that govern its use.

One of the small pleasures of writing a genetics textbook is being able to occasionally introduce in the very first paragraph of the initial chapter a truly significant breakthrough in the discipline that has started to have a major, diverse impact on human lives. In this edition, we are fortunate to be able to discuss the discovery of **CRISPR-Cas,** a molecular mechanism found in bacteria that has the potential to revolutionize our ability to rewrite the DNA sequence of genes from any organism. As such, it represents the ultimate tool in genetic technology, whereby the genome of organisms, including humans, may be precisely edited. Such gene modification represents the ultimate application of the many advances in biotechnology made in the last 35 years, including the sequencing of the human genome.

Although gene editing was first made possible with other methods, the CRISPR-Cas system is now the method of choice for gene modification because it is more accurate, more efficient, more versatile, and easier to use. CRISPR-Cas was initially discovered as a "seek and destroy" mechanism that bacteria use to fight off viral infection. CRISPR (clustered regularly interspersed short palindromic repeats) refers to part of the bacterial genome that produces RNA molecules, and Cas (CRISPR-associated) refers to a nuclease, or DNA-cutting enzyme. The CRISPR RNA binds to a matching sequence in the viral DNA (seek) and recruits the Cas nuclease to cut it (destroy). Researchers have harnessed this technology by synthesizing CRISPR RNAs that direct Cas nucleases to any chosen DNA sequence. In laboratory experiments, CRISPR-Cas has already been used to repair mutations in cells derived from individuals with genetic disorders, such as cystic fibrosis, Huntington disease, sickle-cell disease, and muscular dystrophy. In the United States a clinical trial using CRISPR-Cas for genome editing in cancer therapy is recruiting participants, while proposals for treating a genetic form of blindness and genetic blood disorders are in preparation. In China, at least 86 patients have already started receiving treatments in CRISPR-Cas clinical trials for cancer.

The application of this remarkable system goes far beyond developing treatments for human genetic disorders. In organisms of all kinds, wherever genetic modification may benefit human existence and our planet, the use of CRISPR-Cas will find many targets. For example, one research group edited a gene in mosquitoes, which prevents them from carrying the parasite that causes malaria in humans. Other researchers have edited the genome of algae to double their output for biofuel production. The method has also been used to create disease-resistant strains of wheat and rice.

The power of this system, like any major technological advance, has already raised ethical concerns. For example, genetic modification of human embryos would change the genetic information carried by future generations. These modifications may have unintended and significant negative consequences for our species. In 2017, an international panel of experts discussed the science, ethics, and governance of human genome editing. The panel recommended caution, but not a ban, stating that human embryo modification should "only be permitted for compelling reasons and under strict oversight."

CRISPR-Cas may turn out to be one of the most exciting genetic advances in decades. We will return later in the text to discuss its discovery in bacteria (Chapter 15), its development as a gene-editing tool (Chapter 17), its potential for gene therapy (Special Topic Chapter 3 Gene Therapy), and its uses in genetically edited foods (Special Topic Chapter 6 Genetically Modified Foods).

For now, we hope that this short introduction has stimulated your curiosity, interest, and enthusiasm for the study of genetics. The remainder of this chapter provides an overview of many important concepts of genetics and a survey of the major turning points in the history of the discipline.

1.1 Genetics Has an Interesting Early History

While as early as 350 B.C., Aristotle proposed that active "humors" served as bearers of hereditary traits, it was not until the 1600s that initial strides were made to understand the biological basis of life. In that century, the physician and anatomist William Harvey proposed the theory of **epigenesis,** which states that an organism develops from the fertilized egg by a succession of developmental events that eventually transform the egg into an adult. The theory of epigenesis directly conflicted with the theory of **preformationism,** which stated that the fertilized egg contains a complete miniature adult, called a **homunculus** (**Figure 1.1**). Around 1830, Matthias Schleiden and Theodor Schwann proposed the **cell theory,** stating that all organisms are composed of basic structural units called cells, which are derived from preexisting cells. The idea of **spontaneous generation,** the creation of living organisms from nonliving components, was disproved by Louis Pasteur later in the century, and living organisms were then considered to be derived from preexisting organisms and to consist of cells.

In the mid-1800s the work of Charles Darwin and Gregor Mendel set the stage for the rapid development of genetics in the twentieth and twenty-first centuries.

Darwin and Mendel

In 1859, Darwin published *On the Origin of Species,* describing his ideas about evolution. Darwin's geological, geographical, and biological observations convinced him that existing species arose by descent with modification from ancestral species. Greatly influenced by his voyage on the HMS *Beagle* (1831–1836), Darwin's thinking led him to formulate the theory of **natural selection,** which presented an explanation of the mechanism of evolutionary change. Formulated and proposed independently by Alfred Russel Wallace, natural selection is based on the observation

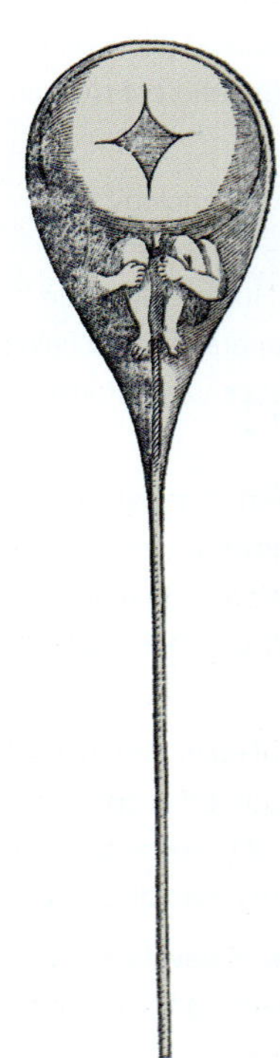

FIGURE 1.1 Depiction of the *homunculus,* a sperm containing a miniature adult, perfect in proportion and fully formed.

that populations tend to produce more offspring than the environment can support, leading to a struggle for survival among individuals. Those individuals with heritable traits that allow them to adapt to their environment are better able to survive and reproduce than those with less adaptive traits. Over time, advantageous variations, even very slight ones, will accumulate. If a population carrying these inherited variations becomes reproductively isolated, a new species may result.

Darwin, however, lacked an understanding of the genetic basis of variation and inheritance, a gap that left his theory open to reasonable criticism well into the twentieth century. Shortly after Darwin published his book, Gregor Johann Mendel published a paper in 1866 showing how traits were passed from generation to generation in pea plants and offered a general model of how traits are inherited. His research was little known until it was partially duplicated and brought to light by Carl Correns, Hugo de Vries, and Erich Tschermak around 1900.

By the early part of the twentieth century, it became clear that heredity and development were dependent on genetic information residing in genes contained in chromosomes, which were then contributed to each individual by gametes—the so-called *chromosome theory of inheritance*. The gap in Darwin's theory was closed, and Mendel's research now serves as the foundation of genetics.

1.2 Genetics Progressed from Mendel to DNA in Less Than a Century

Because genetic processes are fundamental to life itself, the science of genetics unifies biology and serves as its core. The starting point for this branch of science was a monastery garden in central Europe in the late 1850s.

Mendel's Work on Transmission of Traits

Gregor Mendel, an Augustinian monk, conducted a decade-long series of experiments using pea plants. He applied quantitative data analysis to his results and showed that traits are passed from parents to offspring in predictable ways. He further concluded that each trait in pea plants is controlled by a pair of factors (which we now call genes) and that members of a gene pair separate from each other during gamete formation (the formation of egg cells and sperm). Mendel's findings explained the transmission of traits in pea plants and all other higher organisms. His work forms the foundation for **genetics,** the branch of biology concerned with the study of heredity and variation. Mendelian genetics will be discussed later in the text (see Chapters 3 and 4).

The Chromosome Theory of Inheritance: Uniting Mendel and Meiosis

Mendel did his experiments before the structure and role of chromosomes were known. About 20 years after his work was published, advances in microscopy allowed researchers to identify chromosomes and establish that, in most eukaryotes, members of each species have a characteristic number of chromosomes called the **diploid number (2*n*)** in most of their cells. For example, humans have a diploid number of 46 (**Figure 1.2**). Chromosomes in diploid cells exist in pairs, called **homologous chromosomes.**

Researchers in the last decades of the nineteenth century also described chromosome behavior during two forms of cell division, **mitosis** and **meiosis.** In mitosis, chromosomes are copied and distributed so that each daughter cell receives a diploid set of chromosomes identical to those in the parental cell. Meiosis is associated with gamete formation. Cells produced by meiosis receive only one chromosome from each chromosome pair, and the resulting number of chromosomes is called the **haploid number (*n*).** This reduction in chromosome number is essential if the offspring arising from the fusion of egg and sperm are to maintain the constant number of chromosomes characteristic of their parents and other members of their species.

Early in the twentieth century, Walter Sutton and Theodor Boveri independently noted that the behavior of chromosomes during meiosis is identical to the behavior of genes

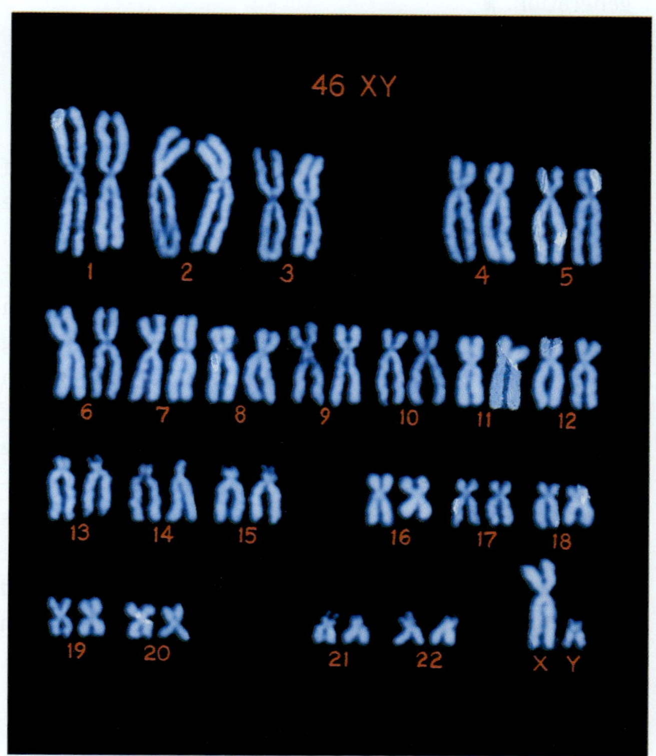

FIGURE 1.2 A colorized image of a replicated set of human male chromosomes. Arranged in this way, the set is called a karyotype.

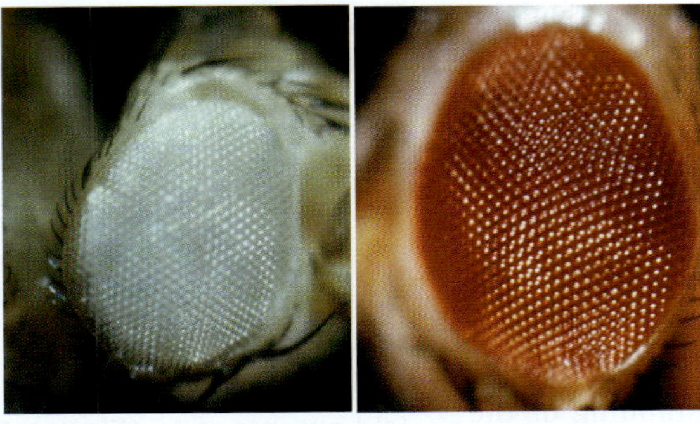

FIGURE 1.3 The white-eyed mutation in *D. melanogaster* (left) and the normal red eye color (right).

during gamete formation described by Mendel. For example, genes and chromosomes exist in pairs, and members of a gene pair and members of a chromosome pair separate from each other during gamete formation. Based on these and other parallels, Sutton and Boveri each proposed that genes are carried on chromosomes. They independently formulated the **chromosomal theory of inheritance,** which states that inherited traits are controlled by genes residing on chromosomes faithfully transmitted through gametes, maintaining genetic continuity from generation to generation.

ESSENTIAL POINT

The chromosome theory of inheritance explains how genetic information is transmitted from generation to generation. ■

Genetic Variation

About the same time that the chromosome theory of inheritance was proposed, scientists began studying the inheritance of traits in the fruit fly, *Drosophila melanogaster*. Early in this work, a white-eyed fly (**Figure 1.3**) was discovered among normal (wild-type) red-eyed flies. This variation was produced by a **mutation** in one of the genes controlling eye color. Mutations are defined as any heritable change in the DNA sequence and are the source of all genetic variation.

The white-eye variant discovered in *Drosophila* is an **allele** of a gene controlling eye color. Alleles are defined as alternative forms of a gene. Different alleles may produce differences in the observable features, or **phenotype,** of an organism. The set of alleles for a given trait carried by an organism is called the **genotype.** Using mutant genes as markers, geneticists can map the location of genes on chromosomes (Figure 1.5).

The Search for the Chemical Nature of Genes: DNA or Protein?

Work on white-eyed *Drosophila* showed that the mutant trait could be traced to a single chromosome, confirming the idea that genes are carried on chromosomes. Once this relationship was established, investigators turned their attention to identifying which chemical component of chromosomes carries genetic information. By the 1920s, scientists knew that proteins and DNA were the major chemical components of chromosomes. There are a large number of different proteins, present in both the nucleus and cytoplasm, and many researchers thought proteins carried genetic information.

In 1944, Oswald Avery, Colin MacLeod, and Maclyn McCarty, researchers at the Rockefeller Institute in New York, published experiments showing that DNA was the carrier of genetic information in bacteria. This evidence, though clear-cut, failed to convince many influential scientists. Additional evidence for the role of DNA as a carrier of genetic information came from Alfred Hershey and Martha Chase who worked with viruses. This evidence that DNA carries genetic information, along with other research over the next few years, provided solid proof that DNA, not protein, is the genetic material, setting the stage for work to establish the structure of DNA.

1.3 Discovery of the Double Helix Launched the Era of Molecular Genetics

Once it was accepted that DNA carries genetic information, efforts were focused on deciphering the structure of the DNA molecule and the mechanisms by which information stored in it produce a phenotype.

The Structure of DNA and RNA

One of the great discoveries of the twentieth century was made in 1953 by James Watson and Francis Crick, who described the structure of DNA. DNA is a long, ladder-like macromolecule that twists to form a double helix (**Figure 1.4**). Each linear strand of the helix is made up of subunits called **nucleotides.** In DNA, there are four different nucleotides, each of which contains a nitrogenous base, abbreviated A (adenine), G (guanine), T (thymine),

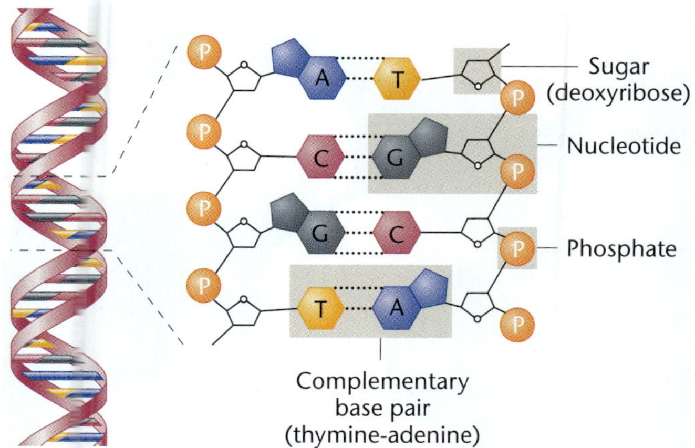

FIGURE 1.4 The structure of DNA showing the arrangement of the double helix (on the left) and the chemical components making up each strand (on the right). The dotted lines on the right represent weak chemical bonds, called hydrogen bonds, which hold together the two strands of the DNA helix.

or C (cytosine). These four bases, in various sequence combinations, ultimately encode genetic information. The two strands of DNA are exact complements of one another, so that the rungs of the ladder in the double helix always consist of A=T and G≡C base pairs. Along with Maurice Wilkins, Watson and Crick were awarded a Nobel Prize in 1962 for their work on the structure of DNA. We will discuss the structure of DNA later in the text (see Chapter 9).

Another nucleic acid, RNA, is chemically similar to DNA but contains a different sugar (ribose rather than deoxyribose) in its nucleotides and contains the nitrogenous base uracil in place of thymine. RNA, however, is generally a single-stranded molecule.

Gene Expression: From DNA to Phenotype

The genetic information encoded in the order of nucleotides in DNA is expressed in a series of steps that results in the formation of a functional gene product. In the majority of cases, this product is a protein. In eukaryotic cells, the process leading to protein production begins in the nucleus with **transcription,** in which the nucleotide sequence in one strand of DNA is used to construct a complementary RNA sequence (top part of **Figure 1.5**). Once an RNA molecule is produced, it moves to the cytoplasm, where the RNA—called **messenger RNA,** or **mRNA** for short—binds to a **ribosome.** The synthesis of proteins under the direction of mRNA is called **translation** (center part of Figure 1.5). The information encoded in mRNA (called the **genetic code**) consists of a linear series of nucleotide triplets. Each triplet, called a **codon,** is complementary to the information stored in DNA and specifies the insertion of a specific amino acid into a protein. Proteins (lower part of Figure 1.5) are polymers made up of amino acid monomers. There are 20 different amino acids commonly found in proteins.

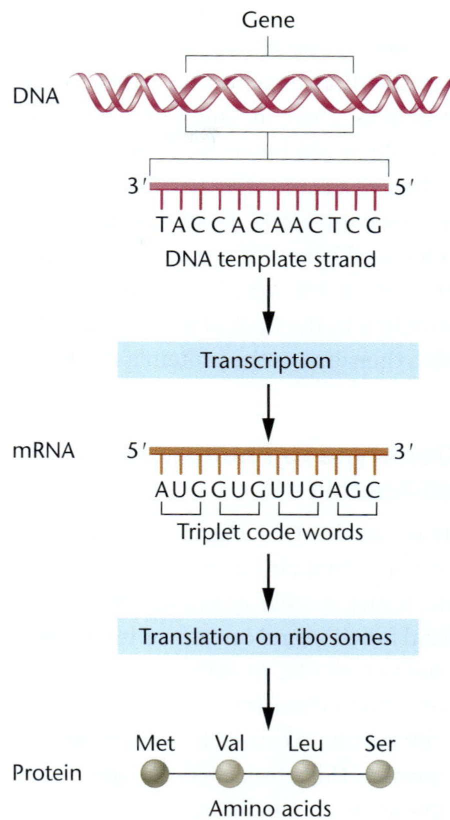

FIGURE 1.5 Gene expression consists of transcription of DNA into mRNA (top) and the translation (center) of mRNA (with the help of a ribosome) into a protein (bottom).

Protein assembly is accomplished with the aid of adapter molecules called **transfer RNA (tRNA).** Within the ribosome, tRNAs recognize the information encoded in the mRNA codons and carry the proper amino acids for construction of the protein during translation.

We now know that gene expression can be more complex than outlined here. Some of these complexities will be discussed later in the text (see Chapters 15 and 16).

Proteins and Biological Function

In most cases, proteins are the end products of gene expression. The diversity of proteins and the biological functions they perform—the diversity of life itself—arises from the fact that proteins are made from combinations of 20 different amino acids. Consider that a protein chain containing 100 amino acids can have at each position any one of 20 amino acids; the number of possible different 100-amino-acid proteins, each with a unique sequence, is therefore equal to

$$20^{100}$$

Obviously, proteins are molecules with the potential for enormous structural diversity and serve as a mainstay of biological systems.

Enzymes form the largest category of proteins. These molecules serve as biological catalysts, lowering the energy of activation in reactions and allowing cellular metabolism to proceed at body temperature.

Proteins other than enzymes are critical components of cells and organisms. These include hemoglobin, the oxygen-binding molecule in red blood cells; insulin, a pancreatic hormone; collagen, a connective tissue molecule; and actin and myosin, the contractile muscle proteins. A protein's shape and chemical behavior are determined by its linear sequence of amino acids, which in turn is dictated by the stored information in the DNA of a gene that is transferred to RNA, which then directs the protein's synthesis.

Linking Genotype to Phenotype: Sickle-Cell Anemia

Once a protein is made, its biochemical or structural properties play a role in producing a phenotype. When mutation alters a gene, it may modify or even eliminate the encoded protein's usual function and cause an altered phenotype. To trace this chain of events, we will examine sickle-cell anemia, a human genetic disorder.

Sickle-cell anemia is caused by a mutant form of hemoglobin, the protein that transports oxygen from the lungs to cells in the body. Hemoglobin is a composite molecule made up of two different proteins, α-globin and β-globin, each encoded by a different gene. In sickle-cell anemia, a mutation in the gene encoding β-globin causes an amino acid substitution in 1 of the 146 amino acids in the protein. **Figure 1.6** shows the DNA sequence, the corresponding mRNA codons, and the amino acids occupying positions 4–7 for the normal and mutant forms of β-globin. Notice that the mutation in sickle-cell anemia consists of a change in one DNA nucleotide, which leads to a change in codon 6 in mRNA from GAG to GUG, which in turn changes amino acid number 6 in β-globin from glutamic acid to valine. The other 145 amino acids in the protein are not changed by this mutation.

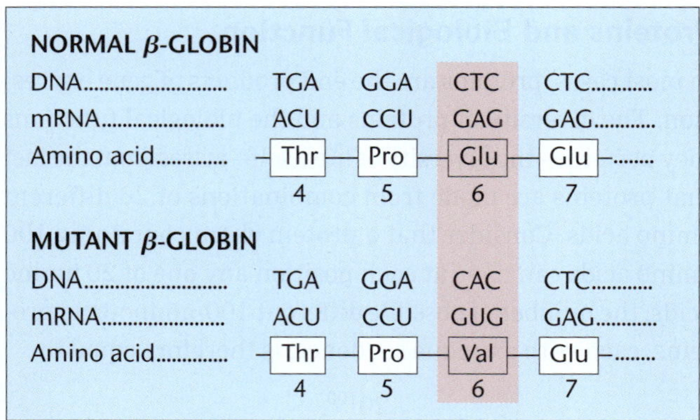

FIGURE 1.6 A single-nucleotide change in the DNA encoding β-globin (CTC → CAC) leads to an altered mRNA codon (GAG → GUG) and the insertion of a different amino acid (Glu → Val), producing the altered version of the β-globin protein that is responsible for sickle-cell anemia.

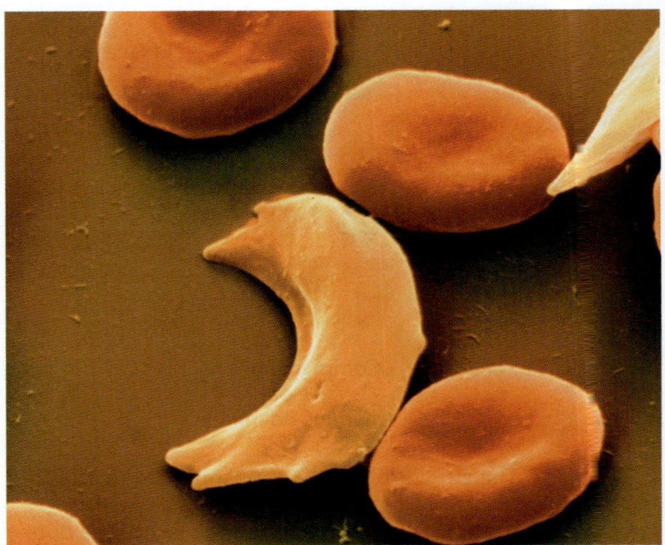

FIGURE 1.7 Normal red blood cells (round) and sickled red blood cells. The sickled cells block capillaries and small blood vessels.

ESSENTIAL POINT

The central dogma of molecular biology -- that DNA is a template for making RNA, which in turn directs the synthesis of proteins -- explains how genes control phenotype. ■

Individuals with two mutant copies of the β-globin gene have sickle-cell anemia. Their mutant β-globin proteins cause hemoglobin molecules in red blood cells to polymerize when the blood's oxygen concentration is low, forming long chains of hemoglobin that distort the shape of red blood cells (**Figure 1.7**). Deformed cells are fragile and break easily, reducing the number of circulating red blood cells (anemia is an insufficiency of red blood cells). Sickle-shaped cells block blood flow in capillaries and small blood vessels, causing severe pain and damage to the heart, brain, muscles, and kidneys. All the symptoms of this disorder are caused by a change in a single nucleotide in a gene that changes one amino acid out of 146 in the β-globin molecule, demonstrating the close relationship between genotype and phenotype.

1.4 Development of Recombinant DNA Technology Began the Era of DNA Cloning

The era of recombinant DNA began in the early 1970s when researchers discovered that **restriction enzymes**, used by bacteria to cut and inactivate the DNA of invading viruses, could be used to cut any organism's DNA at specific nucleotide sequences, producing a reproducible set of fragments.

Soon after, researchers discovered ways to insert the DNA fragments produced by the action of restriction enzymes into carrier DNA molecules called **vectors** to form recombinant DNA molecules. When transferred into bacterial cells, thousands of copies, or **clones,** of the combined vector and DNA fragments are produced during bacterial reproduction. Large amounts of cloned DNA fragments can be isolated from these bacterial host cells. These DNA fragments can be used to isolate genes, to study their organization and expression, and to study their nucleotide sequence and evolution.

Collections of clones that represent an organism's **genome,** defined as the complete haploid DNA content of a specific organism, are called genomic libraries. Genomic libraries are now available for hundreds of species.

Recombinant DNA technology has not only accelerated the pace of research but also given rise to the biotechnology industry, which has grown to become a major contributor to the U.S. economy.

FIGURE 1.8 Dolly, a Finn Dorset sheep cloned from the genetic material of an adult mammary cell, shown next to her first-born lamb, Bonnie.

1.5 The Impact of Biotechnology Is Continually Expanding

The use of recombinant DNA technology and other molecular techniques to make products is called **biotechnology.** In the United States, biotechnology has quietly revolutionized many aspects of everyday life; products made by biotechnology are now found in the supermarket, in health care, in agriculture, and in the court system. A later chapter (see Chapter 18) contains a detailed discussion of biotechnology, but for now, let's look at some everyday examples of biotechnology's impact.

Plants, Animals, and the Food Supply

The use of recombinant DNA technology to genetically modify crop plants has revolutionized agriculture. Genes for traits including resistance to herbicides, insects, and genes for nutritional enhancement have been introduced into crop plants. The transfer of heritable traits across species using recombinant DNA technology creates **transgenic organisms.** Herbicide-resistant corn and soybeans were first planted in the mid-1990s, and transgenic strains now represent about 88 percent of the U.S. corn crop and 93 percent of the U.S. soybean crop. It is estimated that more than 70 percent of the processed food in the United States contains ingredients from transgenic crops.

We will discuss the most recent findings involving genetically modified organisms later in the text. (Special Topics Chapter 6—Genetically Modified Foods).

New methods of cloning livestock such as sheep and cattle have changed the way we use these animals. In 1996, Dolly the sheep (**Figure 1.8**) was cloned by nuclear transfer, a method in which the nucleus of an adult cell is transferred into an egg that has had its nucleus removed. This makes it possible to produce dozens or hundreds of genetically identical offspring with desirable traits with many applications in agriculture and medicine.

Biotechnology has also changed the way human proteins for medical use are produced. Through use of gene transfer, transgenic animals now synthesize these therapeutic proteins. In 2009, an anticlotting protein derived from the milk of transgenic goats was approved by the U.S. Food and Drug Administration for use in the United States. Other human proteins from transgenic animals are now being used in clinical trials to treat several diseases. The biotechnology revolution will continue to expand as gene editing by CRISPR/Cas and other new methods are used to develop an increasing array of products.

Biotechnology in Genetics and Medicine

More than 10 million children or adults in the United States suffer from some form of genetic disorder, and every childbearing couple faces an approximately 3 percent risk of having a child with a genetic anomaly. The molecular basis for hundreds of genetic disorders is now known, and most of these genes have been mapped, isolated, and cloned. Biotechnology-derived genetic testing is now available to perform prenatal diagnosis of heritable disorders and to test parents for their status as heterozygous carriers of more than 100 inherited disorders. Newer methods now offer the possibility of scanning an entire genome to establish an individual's risk of developing a genetic disorder or having an affected child. The use of genetic testing and related technologies raises ethical concerns that have yet to be resolved.

ESSENTIAL POINT

Biotechnology has revolutionized agriculture and the pharmaceutical industry, while genetic testing has had a profound impact on the diagnosis of genetic diseases. ■

1.6 Genomics, Proteomics, and Bioinformatics Are New and Expanding Fields

The ability to create genomic libraries prompted scientists to consider sequencing all the clones in a library to derive the nucleotide sequence of an organism's genome. This sequence information would be used to identify each gene in the genome and establish its function.

One such project, the Human Genome Project (HGP), began in 1990 as an international effort to sequence the human genome. By 2003, the publicly funded HGP and a private, industry-funded genome project completed sequencing of the gene-containing portion of the genome.

As more genome sequences were acquired, several new biological disciplines arose. One, called **genomics** (the study of genomes), studies the structure, function, and evolution of genes and genomes. A second field, **proteomics,** identifies the set of proteins present in a cell under a given set of conditions, and studies their functions and interactions. To store, retrieve, and analyze the massive amount of data generated by genomics and proteomics, a specialized subfield of information technology called **bioinformatics** was created to develop hardware and software for processing nucleotide and protein data.

Geneticists and other biologists now use information in databases containing nucleic acid sequences, protein sequences, and gene-interaction networks to answer experimental questions in a matter of minutes instead of months and years. A feature called "Exploring Genomics," located at the end of many of the chapters in this textbook, gives you the opportunity to explore these databases for yourself while completing an interactive genetics exercise.

Modern Approaches to Understanding Gene Function

Historically, an approach referred to as **classical** or **forward genetics** was essential for studying and understanding gene function. In this approach geneticists relied on the use of naturally occurring mutations or intentionally induced mutations (using chemicals, X-rays, or UV light as examples) to cause altered phenotypes in model organisms, and then worked through the labor-intensive and time-consuming process of identifying the genes that caused these new phenotypes. Such characterization often led to the identification of the gene or genes of interest, and once the technology advanced, the gene sequence could be determined.

Classical genetics approaches are still used, but as whole genome sequencing has become routine, molecular approaches to understanding gene function have changed considerably in genetic research. These modern approaches are what we will highlight in this section.

For the past two decades or so, geneticists have relied on the use of molecular techniques incorporating an approach referred to as **reverse genetics.** In reverse genetics, the DNA sequence for a particular gene of interest is known, but the role and function of the gene are typically not well understood. For example, molecular biology techniques such as **gene knockout** render targeted genes nonfunctional in a model organism or in cultured cells, allowing scientists to investigate the fundamental question of "what happens if this gene is disrupted?" After making a knockout organism, scientists look for both apparent phenotype changes, as well as those at the cellular and molecular level. The ultimate goal is to determine the function of the gene being studied.

ESSENTIAL POINT

Recombinant DNA technology gave rise to several new fields, including genomics, proteomics, and bioinformatics, which allow scientists to explore the structure and evolution of genomes and the proteins they encode. ■

1.7 Genetic Studies Rely on the Use of Model Organisms

After the rediscovery of Mendel's work in 1900, research using a wide range of organisms confirmed that the principles of inheritance he described were of universal significance among plants and animals. Geneticists gradually came to focus attention on a small number of organisms, including the fruit fly (*Drosophila melanogaster*) and the mouse (*Mus musculus*) (**Figure 1.9**). This trend developed for two main reasons: First, it was clear that genetic mechanisms were the same in most organisms, and second, these organisms had characteristics that made them especially suitable for genetic research. They were easy to grow, had relatively short life cycles, produced many offspring, and their genetic analysis was fairly straightforward. Over time, researchers created a large catalog of mutant strains for these species,

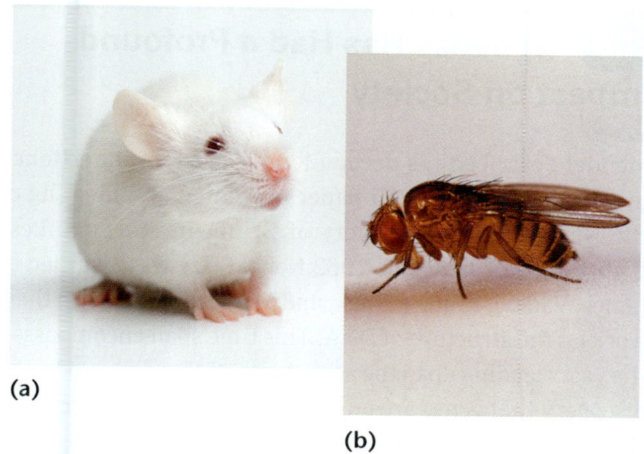

FIGURE 1.9 The first generation of model organisms in genetic analysis included (a) the mouse, *Mus musculus,* and (b) the fruit fly, *Drosophila melanogaster.*

and the mutations were carefully studied, characterized, and mapped. Because of their well-characterized genetics, these species became **model organisms,** defined as organisms used for the study of basic biological processes. In later chapters, we will see how discoveries in model organisms are shedding light on many aspects of biology, including aging, cancer, and behavior.

The Modern Set of Genetic Model Organisms

Gradually, geneticists added other species to their collection of model organisms: viruses (such as the T phages and lambda phage) and microorganisms (the bacterium *Escherichia coli* and the yeast *Saccharomyces cerevisiae*) (**Figure 1.10**).

More recently, additional species have been developed as model organisms, three of which are shown in the chapter opening photograph. Each species was chosen to allow study of some aspect of embryonic development. The nematode *Caenorhabditis elegans* was chosen as a model system to study the development and function of the nervous system because its nervous system contains only a few hundred cells and the developmental fate of these and all other cells in the body has been mapped out. *Arabidopsis thaliana*, a small plant with a short life cycle, has become a model organism for the study of many aspects of plant biology. The zebrafish, *Danio rerio*, is used to study vertebrate development: it is small, it reproduces rapidly, and its egg, embryo, and larvae are all transparent.

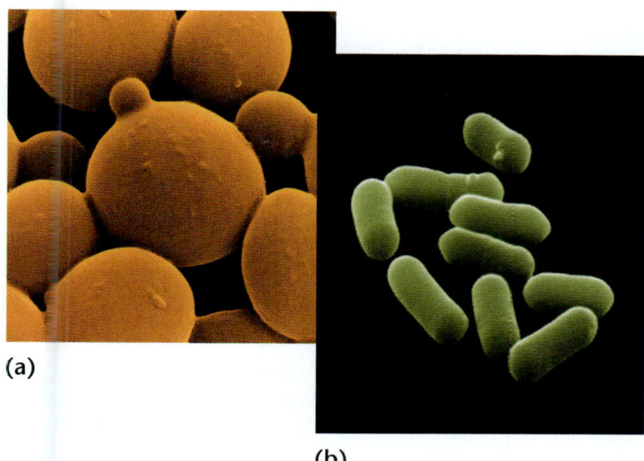

FIGURE 1.10 Microbes that have become model organisms for genetic studies include (a) the yeast *Saccharomyces cerevisiae* and (b) the bacterium *Escherichia coli.*

Model Organisms and Human Diseases

The development of recombinant DNA technology and the results of genome sequencing have confirmed that all life has a common origin. Because of this, genes with similar functions in different organisms tend to be similar or identical in structure and nucleotide sequence. Much of what scientists learn by studying the genetics of model organisms can therefore be applied to humans as the basis for understanding and treating human diseases. In addition, the ability to create transgenic organisms by transferring genes between species has enabled scientists to develop models of human diseases in organisms ranging from bacteria to fungi, plants, and animals (**Table 1.1**).

The idea of studying a human disease such as colon cancer by using *E. coli* may strike you as strange, but the basic steps of DNA repair (a process that is defective in some forms of colon cancer) are the same in both organisms, and a gene involved in DNA repair (*mutL* in *E. coli* and *MLH1* in humans) is found in both organisms. More importantly, *E. coli* has the advantage of being easier to grow (the cells divide every 20 minutes), and researchers can easily create and study new mutations in the bacterial *mutL* gene in order to figure out how it works. This knowledge may eventually lead to the development of drugs and other therapies to treat colon cancer in humans.

The fruit fly, *Drosophila melanogaster*, is also being used to study a number of human diseases. Mutant genes

TABLE 1.1 **Model Organisms Used to Study Some Human Diseases**

Organism	Human Diseases
E. coli	Colon cancer and other cancers
S. cerevisiae	Cancer, Werner syndrome
D. melanogaster	Disorders of the nervous system, cancer
C. elegans	Diabetes
D. rerio	Cardiovascular disease
M. musculus	Lesch–Nyhan syndrome, cystic fibrosis, fragile-X syndrome, and many other diseases

have been identified in *D. melanogaster* that produce phenotypes with structural abnormalities of the nervous system and adult-onset degeneration of the nervous system. The information from genome-sequencing projects indicates that almost all these genes have human counterparts. For example, genes involved in a complex human disease of the retina called retinitis pigmentosa are identical to *Drosophila* genes involved in retinal degeneration. Study of these mutations in *Drosophila* is helping to dissect this complex disease and identify the function of the genes involved.

Another approach to studying diseases of the human nervous system is to transfer mutant human disease genes into *Drosophila* using recombinant DNA technology. The transgenic flies are then used for studying the mutant human genes themselves, other genes that affect the expression of the human disease genes, and the effects of therapeutic drugs on the action of those genes—all studies that are difficult or impossible to perform in humans. This gene transfer approach is being used to study almost a dozen human neurodegenerative disorders, including Huntington disease, Machado–Joseph disease, myotonic dystrophy, and Alzheimer disease.

Throughout the following chapters, you will encounter these model organisms again and again. Remember each time you meet them that they not only have a rich history in basic genetics research but are also at the forefront in the study of human genetic disorders and infectious diseases.

ESSENTIAL POINT

The study of model organisms for understanding human health and disease is one of the many ways genetics and biotechnology are changing everyday life. ■

1.8 Genetics Has Had a Profound Impact on Society

Mendel described his decade-long project on inheritance in pea plants in an 1865 paper presented at a meeting of the Natural History Society of Brünn in Moravia. Less than 100 years later, the 1962 Nobel Prize was awarded to James Watson, Francis Crick, and Maurice Wilkins for their work on the structure of DNA. This time span encompassed the years leading up to the acceptance of Mendel's work, the discovery that genes are on chromosomes, the experiments that proved DNA encodes genetic information, and the elucidation of the molecular basis for DNA replication. The rapid development of genetics from Mendel's monastery garden to the Human Genome Project and beyond is summarized in a timeline in **Figure 1.11**.

The Nobel Prize and Genetics

No other scientific discipline has experienced the explosion of information and the level of excitement generated by the discoveries in genetics. This impact is especially apparent in the list of Nobel Prizes related to genetics, beginning with those awarded in the early and mid-twentieth century and continuing into the present (see inside back cover). Nobel Prizes in Medicine or Physiology and Chemistry have been consistently awarded for work in genetics and related fields. The first such prize awarded was given to Thomas H. Morgan in 1933 for his research on the chromosome theory of inheritance. That award was followed by many others, including prizes for the discovery of genetic recombination, the relationship between genes and proteins, the structure of DNA, and the genetic code. This trend has continued throughout

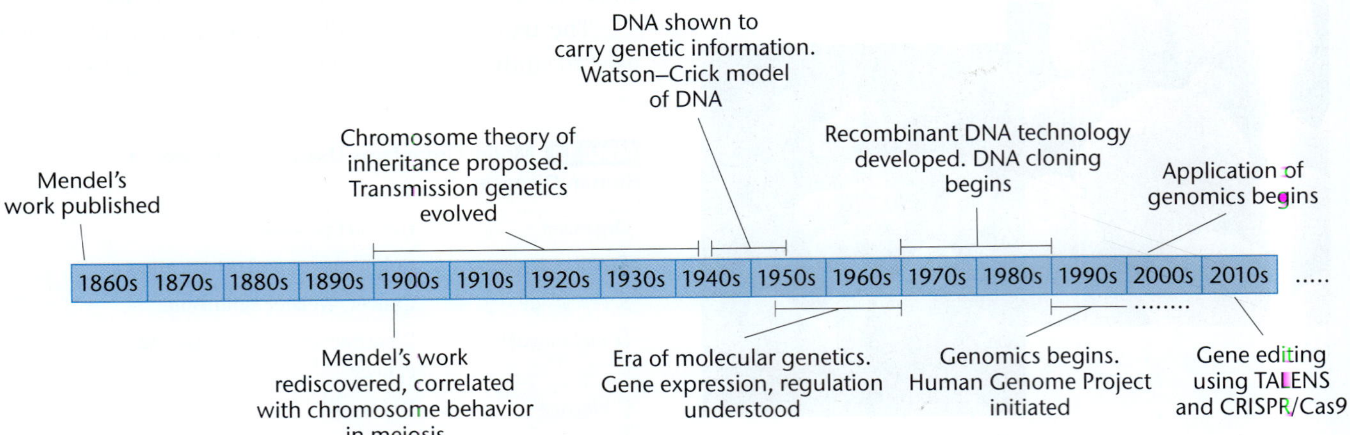

FIGURE 1.11 A timeline showing the development of genetics from Gregor Mendel's work on pea plants to the current era of genomics and its many applications in research, medicine, and society. Having a sense of the history of discovery in genetics should provide you with a useful framework as you proceed through this textbook.

the twentieth and twenty-first centuries. The advent of genomic studies and the applications of such findings will most certainly lead the way for future awards.

Genetics, Ethics, and Society

Just as there has never been a more exciting time to study genetics, the impact of this discipline on society has never been more profound. Genetics and its applications in biotechnology are developing much faster than the social conventions, public policies, and laws required to regulate their use. As a society, we are grappling with a host of sensitive genetics-related issues, including concerns about prenatal testing, genetic discrimination, ownership of genes, access to and safety of gene therapy, and genetic privacy. Two features appearing at the end of most chapters, "Case Study" and "Genetics, Ethics, and Society," consider ethical issues raised by the use of genetic technology. This emphasis on ethics reflects the growing concern and dilemmas that advances in genetics pose to our society and the future of our species. It is our hope that upon the completion of your study of genetics, you will become an informed, active participant in future debates that arise.

ESSENTIAL POINT

Genetic technology is having a profound effect on society, while raising many ethical dilemmas. ■

Problems and Discussion Questions

Mastering Genetics Visit for instructor-assigned tutorials and problems.

1. How does Mendel's work relate to our understanding of the transmission of traits?
2. **CONCEPT QUESTION** Review the Chapter Concepts list on p. 25. Most of these are related to the discovery of DNA as the genetic material and the subsequent development of recombinant DNA technology. Write a brief essay that discusses the impact of recombinant DNA technology on genetics as we perceive the discipline today.
3. What is the chromosome theory of inheritance, and how is it related to Mendel's findings?
4. Define genotype and phenotype. Describe how they are related and how alleles fit into your definitions.
5. Given the state of knowledge at the time of the Avery, MacLeod, and McCarty experiment, why was it difficult for some scientists to accept that DNA is the carrier of genetic information?
6. What is a gene?
7. What is the structure of DNA? How does it differ from that of RNA?
8. Describe the central dogma of molecular genetics and how it serves as the basis of modern genetics.
9. Until the mid-1940s, many scientists considered protein to be the likely candidate for genetic material. Why?
10. Outline the roles played by restriction enzymes and vectors in cloning DNA.
11. What are the impacts of biotechnology on human genetics?
12. Summarize the arguments for and against patenting genetically modified organisms.
13. We all carry about 20,000 genes in our genome. So far, patents have been issued for more than 6000 of these genes. Do you think that companies or individuals should be able to patent human genes? Why or why not?
14. How is it possible to study certain aspects of diseases such as cancer in simple model organisms like the *E. coli* bacteria?
15. If you knew that a devastating late-onset inherited disease runs in your family (in other words, a disease that does not appear until later in life) and you could be tested for it at the age of 20, would you want to know whether you are a carrier? Would your answer be likely to change when you reach age 40?
16. Why do you think discoveries in genetics have been recognized with so many Nobel Prizes?

2

Mitosis and Meiosis

Chromosomes in the prometaphase stage of mitosis, derived from a cell in the flower of *Haemanthus*.

CHAPTER CONCEPTS

- Genetic continuity between generations of cells and between generations of sexually reproducing organisms is maintained through the processes of mitosis and meiosis, respectively.
- Diploid eukaryotic cells contain their genetic information in pairs of homologous chromosomes, with one member of each pair being derived from the maternal parent and one from the paternal parent.
- Mitosis provides a mechanism by which chromosomes, having been duplicated, are distributed into progeny cells during cell reproduction.
- Mitosis converts a diploid cell into two diploid daughter cells.
- Meiosis provides a mechanism by which one member of each homologous pair of chromosomes is distributed into each gamete or spore, thus reducing the diploid chromosome number to the haploid chromosome number.
- Meiosis generates genetic variability by distributing various combinations of maternal and paternal members into gametes or spores.
- During the stages of mitosis and meiosis, the genetic material is condensed into discrete structures called chromosomes.

Every living thing contains a substance described as the genetic material. Except in certain viruses, this material is composed of the nucleic acid DNA. DNA has an underlying linear structure possessing segments called genes, the products of which direct the metabolic activities of cells. An organism's DNA, with its arrays of genes, is organized into structures called **chromosomes,** which serve as vehicles for transmitting genetic information. The manner in which chromosomes are transmitted from one generation of cells to the next and from organisms to their descendants must be exceedingly precise. In this chapter we consider exactly how genetic continuity is maintained between cells and organisms.

Two major processes are involved in the genetic continuity of nucleated cells: **mitosis** and **meiosis.** Although the mechanisms of the two processes are similar in many ways, the outcomes are quite different. Mitosis leads to the production of two cells, each with the same number of chromosomes as the parent cell. In contrast, meiosis reduces the genetic content and the number of chromosomes by precisely half. This reduction is essential if sexual reproduction is to occur without doubling the amount of genetic material in each new generation. Strictly speaking, mitosis is that portion of the cell cycle during which the hereditary components are equally partitioned into daughter cells. Meiosis is part of a special type of cell division that leads to the production of sex cells: **gametes** or **spores.** This process is an essential step in the transmission of genetic information from an organism to its offspring.

Normally, chromosomes are visible only during mitosis and meiosis. When cells are not undergoing division, the genetic material making up chromosomes unfolds and uncoils into a diffuse network within the nucleus, generally referred to as **chromatin.** Before describing mitosis and meiosis, we will briefly review the structure of cells, emphasizing components that are of particular significance to genetic function. We will also compare the structural differences between the nonnucleated cells of bacteria and the eukaryotic cells of higher organisms. We then devote the remainder of the chapter to the behavior of chromosomes during cell division.

2.1 Cell Structure Is Closely Tied to Genetic Function

Before 1940, our knowledge of cell structure was limited to what we could see with the light microscope. Around 1950, the transmission electron microscope was in its early stages of development, and by 1950, many details of cell ultrastructure had emerged. Under the electron microscope, cells were seen as highly varied, highly organized structures whose form and function are dependent on specific genetic expression by each cell type. A new world of whorled membranes, organelles, microtubules, granules, and filaments was revealed. These discoveries revolutionized thinking in the entire field of biology. Many cell components, such as the nucleolus, ribosome, and centriole, are involved directly or indirectly with genetic processes. Other components—the mitochondria and chloroplasts—contain their own unique genetic information. Here, we will focus primarily on those aspects of cell structure that relate to genetic study. The generalized animal cell shown in **Figure 2.1** illustrates most of the structures we will discuss.

All cells are surrounded by a *plasma membrane,* an outer covering that defines the cell boundary and delimits the cell from its immediate external environment. This membrane is not passive but instead actively controls the

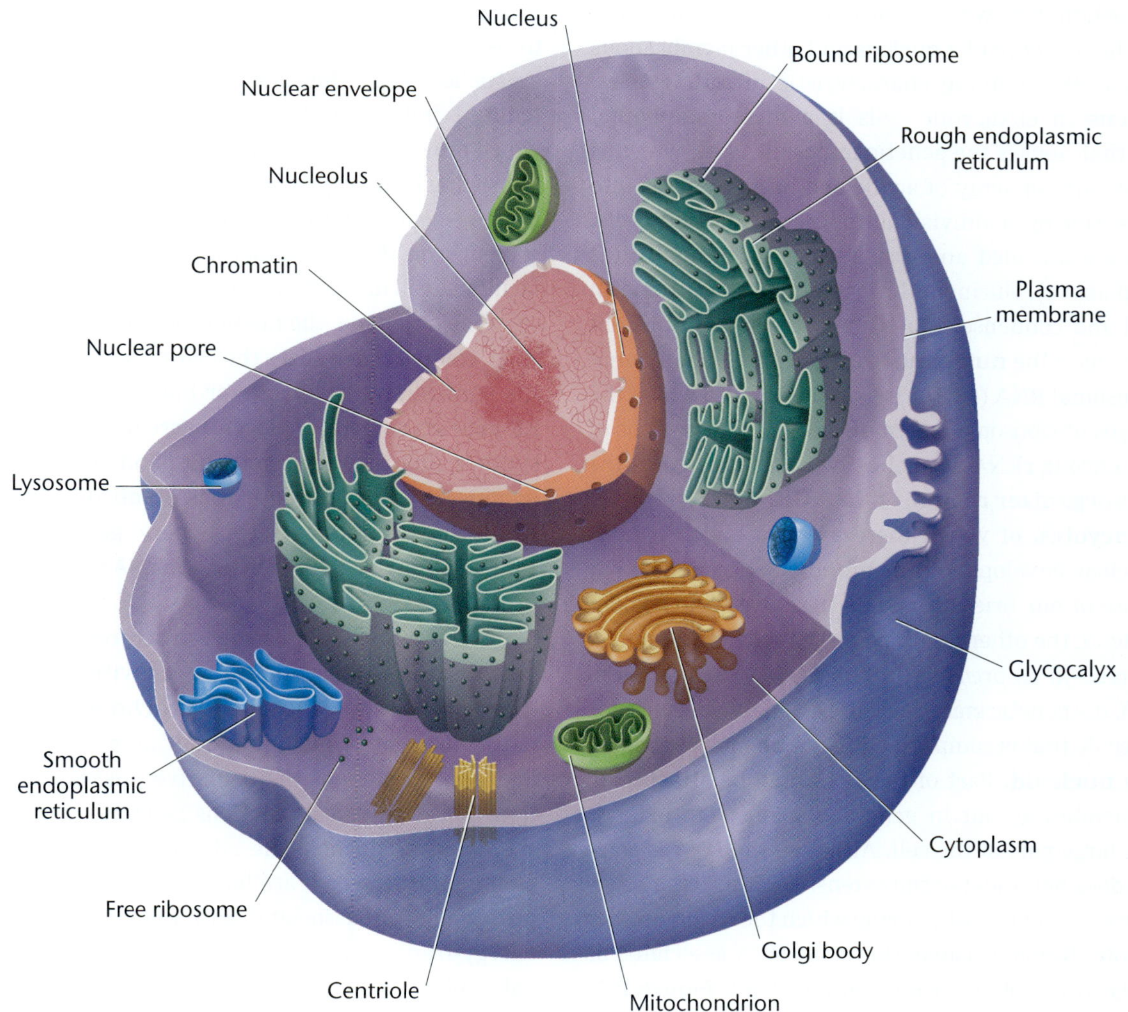

FIGURE 2.1 A generalized animal cell. The cellular components discussed in the text are emphasized here.

movement of materials into and out of the cell. In addition to this membrane, plant cells have an outer covering called the *cell wall* whose major component is a polysaccharide called *cellulose*.

Many, if not most, animal cells have a covering over the plasma membrane, referred to as the *glycocalyx*, or *cell coat*. Consisting of glycoproteins and polysaccharides, this covering has a chemical composition that differs from comparable structures in either plants or bacteria. The glycocalyx, among other functions, provides biochemical identity at the surface of cells, and the components of the coat that establish cellular identity are under genetic control. For example, various cell-identity markers that you may have heard of—the AB, Rh, and MN antigens—are found on the surface of red blood cells, among other cell types. On the surface of other cells, histocompatibility antigens, which elicit an immune response during tissue and organ transplants, are present. Various *receptor molecules* are also found on the surfaces of cells. These molecules act as recognition sites that transfer specific chemical signals across the cell membrane into the cell.

Living organisms are categorized into two major groups depending on whether or not their cells contain a nucleus. The presence of a nucleus and other membranous organelles is the defining characteristic of **eukaryotes.** The **nucleus** in eukaryotic cells is a membrane-bound structure that houses the genetic material, DNA, which is complexed with an array of acidic and basic proteins into thin fibers. During nondivisional phases of the cell cycle, the fibers are uncoiled and dispersed into chromatin (as mentioned above). During mitosis and meiosis, chromatin fibers coil and condense into chromosomes. Also present in the nucleus is the **nucleolus,** an amorphous component where ribosomal RNA (rRNA) is synthesized and where the initial stages of ribosomal assembly occur. The portions of DNA that encode rRNA are collectively referred to as the **nucleolus organizer region,** or the **NOR.**

Prokaryotes, of which there are two major groups, lack a nuclear envelope and membranous organelles. For the purpose of our brief discussion here, we will consider the *eubacteria,* the other group being the more ancient bacteria referred to as *archaea.* In eubacteria, such as *Escherichia coli,* the genetic material is present as a long, circular DNA molecule that is compacted into an unenclosed region called the **nucleoid.** Part of the DNA may be attached to the cell membrane, but in general the nucleoid extends through a large part of the cell. Although the DNA is compacted, it does not undergo the extensive coiling characteristic of the stages of mitosis, during which the chromosomes of eukaryotes become visible. Nor is the DNA associated as extensively with proteins as is eukaryotic DNA. **Figure 2.2**, which shows two bacteria forming by cell division, illustrates the nucleoid regions containing the bacterial chromosomes. Prokaryotic cells do not have a distinct nucleolus but do contain genes that specify rRNA molecules.

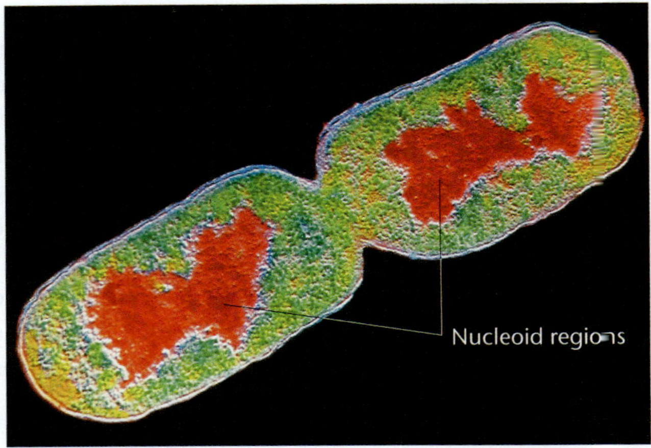

FIGURE 2.2 Color-enhanced electron micrograph of *E. coli* undergoing cell division. Particularly prominent are the two chromosomal areas (shown in red), called nucleoids, that have been partitioned into the daughter cells.

The remainder of the eukaryotic cell within the plasma membrane, excluding the nucleus, is referred to as *cytoplasm* and includes a variety of extranuclear cellular organelles. In the cytoplasm, a nonparticulate, colloidal material referred to as the *cytosol* surrounds and encompasses the cellular organelles. The cytoplasm also includes an extensive system of tubules and filaments, comprising the cytoskeleton, which provides a lattice of support structures within the cell. Consisting primarily of *microtubules*, which are made of the protein *tubulin*, and *microfilaments*, which derive from the protein *actin*, this structural framework maintains cell shape, facilitates cell mobility, and anchors the various organelles.

One organelle, the membranous *endoplasmic reticulum (ER)*, compartmentalizes the cytoplasm, greatly increasing the surface area available for biochemical synthesis. The ER appears smooth in places where it serves as the site for synthesizing fatty acids and phospholipids; in other places, it appears rough because it is studded with ribosomes. **Ribosomes** serve as sites where genetic information contained in messenger RNA (mRNA) is translated into proteins.

Three other cytoplasmic structures are very important in the eukaryotic cell's activities: mitochondria, chloroplasts, and centrioles. **Mitochondria** are found in most eukaryotes, including both animal and plant cells, and are the sites of the oxidative phases of cell respiration. These chemical reactions generate large amounts of the energy-rich molecule adenosine triphosphate (ATP). **Chloroplasts,** which are found in plants, algae, and some protozoans, are associated with photosynthesis, the major energy-trapping process on Earth. Both mitochondria and chloroplasts contain DNA in a form distinct from that found in the nucleus. They are able to duplicate themselves and transcribe and translate their own genetic information.

Animal cells and some plant cells also contain a pair of complex structures called **centrioles.** These cytoplasmic bodies, each located in a specialized region called the **centrosome,** are associated with the organization of spindle fibers that function in mitosis and meiosis. In some organisms, the centriole is derived from another structure, the basal body, which is associated with the formation of cilia and flagella (hair-like and whip-like structures for propelling cells or moving materials).

The organization of **spindle fibers** by the centrioles occurs during the early phases of mitosis and meiosis. These fibers play an important role in the movement of chromosomes as they separate during cell division. They are composed of arrays of microtubules consisting of polymers of the protein tubulin.

ESSENTIAL POINT

Most components of cells are involved directly or indirectly with genetic processes. ■

2.2 Chromosomes Exist in Homologous Pairs in Diploid Organisms

As we discuss the processes of mitosis and meiosis, it is important that you understand the concept of homologous chromosomes. Such an understanding will also be of critical importance in our future discussions of Mendelian genetics. Chromosomes are most easily visualized during mitosis. When they are examined carefully, distinctive lengths and shapes are apparent. Each chromosome contains a constricted region called the **centromere,** whose location establishes the general appearance of each chromosome. **Figure 2.3** shows chromosomes with centromere placements at different distances along their length. Extending from either side of the centromere are the arms of the chromosome. Depending on the position of the centromere, different arm ratios are produced. As Figure 2.3 illustrates, chromosomes are classified as **metacentric, submetacentric, acrocentric,** or **telocentric** on the basis of the centromere location. The shorter arm, by convention, is shown above the centromere and is called the **p arm** (p, for "petite"). The longer arm is shown below the centromere and is called the **q arm** (q because it is the next letter in the alphabet).

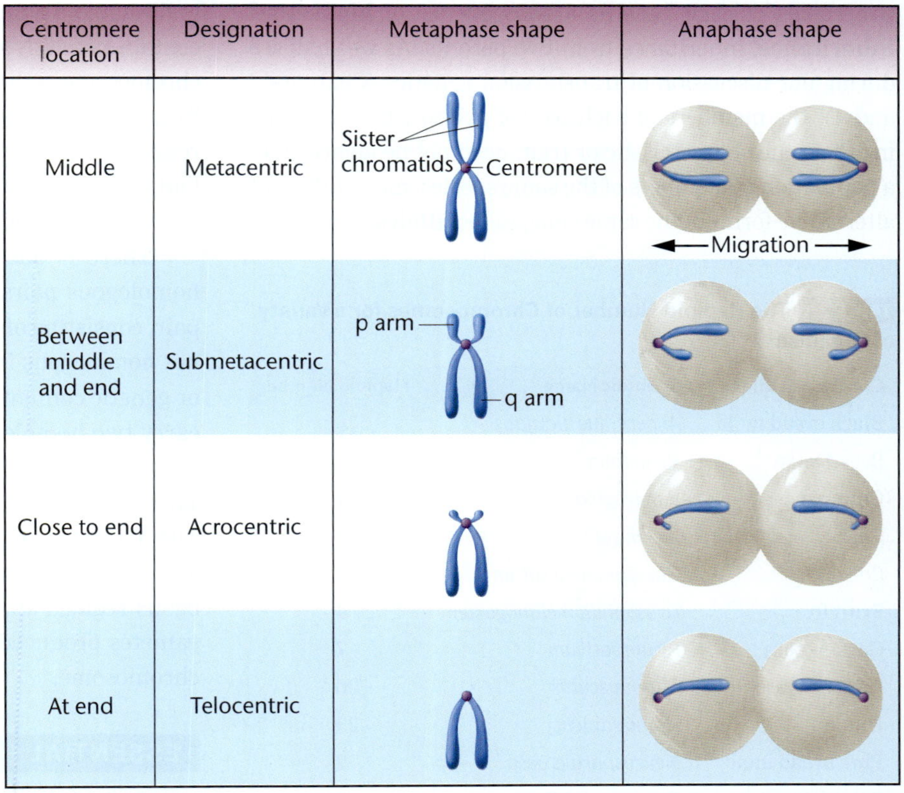

Centromere location	Designation	Metaphase shape	Anaphase shape
Middle	Metacentric	Sister chromatids; Centromere	Migration
Between middle and end	Submetacentric	p arm; q arm	
Close to end	Acrocentric		
At end	Telocentric		

FIGURE 2.3 Centromere locations and the chromosome designations that are based on them. Note that the shape of the chromosome during anaphase is determined by the position of the centromere during metaphase.

In the study of mitosis, several other observations are of particular relevance. First, all somatic cells derived from members of the same species contain an identical number of chromosomes. In most cases, this represents what is referred to as the **diploid number (2*n*).** When the lengths and centromere placements of all such chromosomes are examined, a second general feature is apparent. With the exception of sex chromosomes, they exist in pairs with regard to these two properties, and the members of each pair are called **homologous chromosomes.** So, for each chromosome exhibiting a specific length and centromere placement, another exists with identical features.

There are exceptions to this rule. Many bacteria and viruses have but one chromosome, and organisms such as yeasts and molds, and certain plants such as bryophytes (mosses), spend the predominant phase of their life cycle in the haploid stage. That is, they contain only one member of each homologous pair of chromosomes during most of their lives.

Figure 2.4 illustrates the physical appearance of different pairs of homologous chromosomes. There, the human mitotic chromosomes have been photographed, cut out of the print, and matched up, creating a display called a **karyotype.** As you can see, humans have a 2*n* number of 46 chromosomes, which on close examination exhibit a diversity of sizes and centromere placements. Note also that each of the 46 chromosomes in this karyotype is clearly a double structure consisting of two parallel *sister chromatids* connected by a common centromere. Had these chromosomes been allowed to continue dividing, the sister chromatids, which are replicas of one another, would have separated into the two new cells as division continued.

The **haploid number (*n*)** of chromosomes is equal to one-half the diploid number. Collectively, the genetic information contained in a haploid set of chromosomes constitutes the **genome** of the species. This, of course, includes copies of all genes as well as a large amount of noncoding DNA. The examples listed in **Table 2.1** demonstrate the wide range of *n* values found in plants and animals.

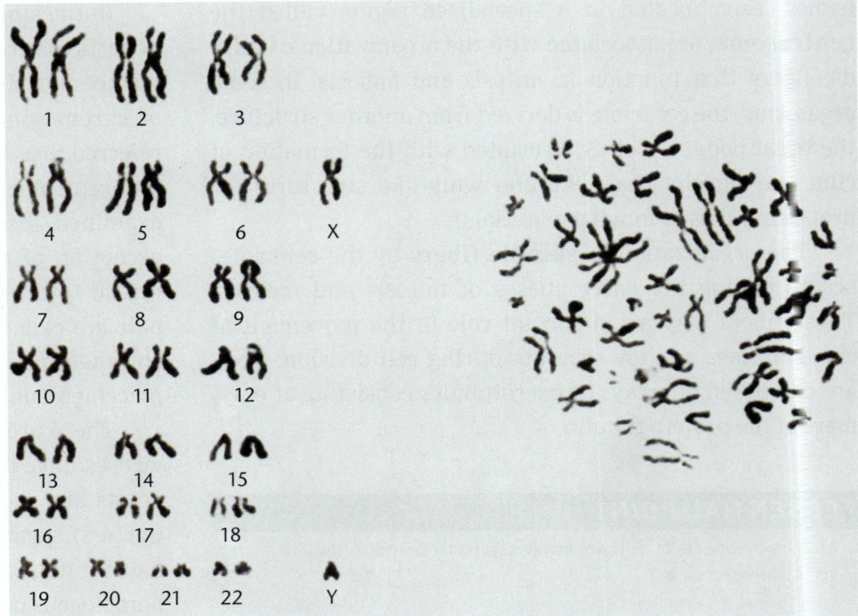

FIGURE 2.4 A metaphase preparation of chromosomes derived from a dividing cell of a human male (right), and the karyotype derived from the metaphase preparation (left). All but the X and Y chromosomes are present in homologous pairs. Each chromosome is clearly a double structure consisting of a pair of sister chromatids joined by a common centromere.

Homologous chromosomes have important genetic similarities. They contain identical gene sites along their lengths; each site is called a **locus** (pl. loci). Thus, they are identical in the traits that they influence and in their genetic potential. In sexually reproducing organisms, one member of each pair is derived from the maternal parent (through the ovum) and the other member is derived from the paternal parent (through the sperm). Therefore, each diploid organism contains two copies of each gene as a consequence of **biparental inheritance,** inheritance from two parents. As we shall see during our discussion of transmission genetics (Chapters 3 and 4), the members of each pair of genes, while influencing the same characteristic or trait, need not be identical. In a population of members of the same species, many different alternative forms of the same gene, called **alleles,** can exist.

TABLE 2.1 The Haploid Number of Chromosomes for a Variety of Organisms

Common Name	Scientific Name	Haploid Number
Black bread mold	*Aspergillus nidulans*	8
Broad bean	*Vicia faba*	6
Chimpanzee	*Pan troglodytes*	24
Corn	*Zea mays*	10
Cotton	*Gossypium hirsutum*	26
Fruit fly	*Drosophila melanogaster*	4
Garden pea	*Pisum sativum*	7
House mouse	*Mus musculus*	20
Human	*Homo sapiens*	23
Pink bread mold	*Neurospora crassa*	7
Roundworm	*Caenorhabditis elegans*	6
Yeast	*Saccharomyces cerevisiae*	16
Zebrafish	*Danio rerio*	25

The concepts of haploid number, diploid number, and homologous chromosomes are important for understanding the process of meiosis. During the formation of gametes or spores, meiosis converts the diploid number of chromosomes to the haploid number. As a result, haploid gametes or spores contain precisely one member of each homologous pair of chromosomes—that is, one complete haploid set. Following fusion of two gametes at fertilization, the diploid number is reestablished; that is, the zygote contains two complete haploid sets of chromosomes. The constancy of genetic material is thus maintained from generation to generation.

There is one important exception to the concept of homologous pairs of chromosomes. In many species, one pair, consisting of the *sex-determining chromosomes,* is often not homologous in size, centromere placement, arm ratio, or genetic content. For example, in humans, while females carry two homologous X chromosomes, males carry one Y chromosome in addition to one X chromosome (Figure 2.4). These X and Y chromosomes are not strictly homologous. The Y is considerably smaller and lacks most of the gene loci contained on the X. Nevertheless, they contain homologous regions and behave as homologs in meiosis so that gametes produced by males receive either one X or one Y chromosome.

ESSENTIAL POINT

In diploid organisms, chromosomes exist in homologous pairs, where each member is identical in size, centromere placement, and gene sites. One member of each pair is derived from the maternal parent, and one is derived from the paternal parent. ■

2.3 Mitosis Partitions Chromosomes into Dividing Cells

The process of mitosis is critical to all eukaryotic organisms. In some single-celled organisms, such as protozoans and some fungi and algae, mitosis (as a part of cell division) provides the basis for asexual reproduction. Multicellular diploid organisms begin life as single-celled fertilized eggs called **zygotes.** The mitotic activity of the zygote and the subsequent daughter cells is the foundation for the development and growth of the organism. In adult organisms, mitotic activity is the basis for wound healing and other forms of cell replacement in certain tissues. For example, the epidermal cells of the skin and the intestinal lining of humans are continuously sloughed off and replaced. Cell division also results in the continuous production of reticulocytes that eventually shed their nuclei and replenish the supply of red blood cells in vertebrates. In abnormal situations, somatic cells may lose control of cell division and form a tumor.

The genetic material is partitioned into daughter cells during nuclear division, or **karyokinesis.** This process is quite complex and requires great precision. The chromosomes must first be exactly replicated and then accurately partitioned. The end result is the production of two daughter nuclei, each with a chromosome composition identical to that of the parent cell.

Karyokinesis is followed by cytoplasmic division, or **cytokinesis.** This less complex process requires a mechanism that partitions the volume into two parts and then encloses each new cell in a distinct plasma membrane. As the cytoplasm is reconstituted, organelles replicate themselves, arise from existing membrane structures, or are synthesized *de novo* (anew) in each cell.

Following cell division, the initial size of each new daughter cell is approximately one-half the size of the parent cell. However, the nucleus of each new cell is not appreciably smaller than the nucleus of the original cell. Quantitative measurements of DNA confirm that there is an amount of genetic material in the daughter nuclei equivalent to that in the parent cell.

Interphase and the Cell Cycle

Many cells undergo a continuous alternation between division and nondivision. The events that occur from the completion of one division until the completion of the next division constitute the **cell cycle** (**Figure 2.5**). We will consider **interphase,** the initial stage of the cell cycle, as the interval between divisions. It was once thought that the biochemical activity during interphase was devoted solely to the cell's growth and its normal function. However, we now know that another biochemical step critical to the ensuing mitosis occurs during interphase: *the replication of the DNA of each chromosome.* This period, during which DNA is synthesized, occurs before the cell enters mitosis and is called the **S phase.** The initiation and completion of synthesis can be detected by monitoring the incorporation of radioactive precursors into DNA.

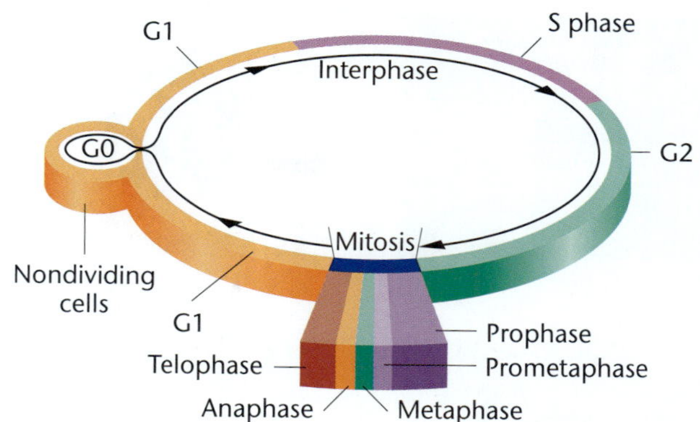

FIGURE 2.5 The stages comprising an arbitrary cell cycle. Following mitosis, cells enter the G1 stage of interphase, initiating a new cycle. Cells may become nondividing (G0) or continue through G1, where they become committed to begin DNA synthesis (S) and complete the cycle (G2 and mitosis). Following mitosis, two daughter cells are produced, and the cycle begins anew for both of them.

Investigations of this nature demonstrate two periods during interphase when no DNA synthesis occurs, one before and one after the S phase. These are designated **G1 (gap I)** and **G2 (gap II),** respectively. During both of these intervals, as well as during S, intensive metabolic activity, cell growth, and cell differentiation are evident. By the end of G2, the volume of the cell has roughly doubled, DNA has been replicated, and mitosis (M) is initiated. Following mitosis, continuously dividing cells then repeat this cycle (G1, S, G2, M) over and over, as shown in Figure 2.5.

Much is known about the cell cycle based on *in vitro* (literally, "in glass") studies. When grown in culture, many cell types in different organisms traverse the complete cycle in about 16 hours. The actual process of mitosis occupies only a small part of the overall cycle, often less than an hour. The lengths of the S and G2 phases of interphase are fairly consistent in different cell types. Most variation is seen in the length of time spent in the G1 stage. **Figure 2.6** shows the relative length of these intervals as well as the length of the stages of mitosis in a human cell in culture.

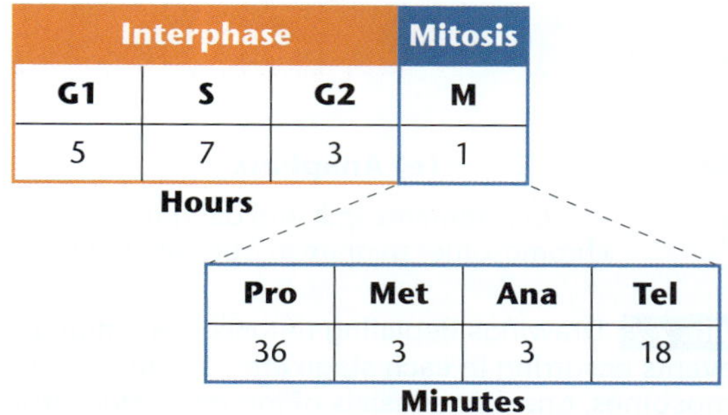

Interphase			Mitosis
G1	S	G2	M
5	7	3	1

Hours

Pro	Met	Ana	Tel
36	3	3	18

Minutes

FIGURE 2.6 The time spent in each interval of one complete cell cycle of a human cell in culture. Times vary according to cell types and conditions.

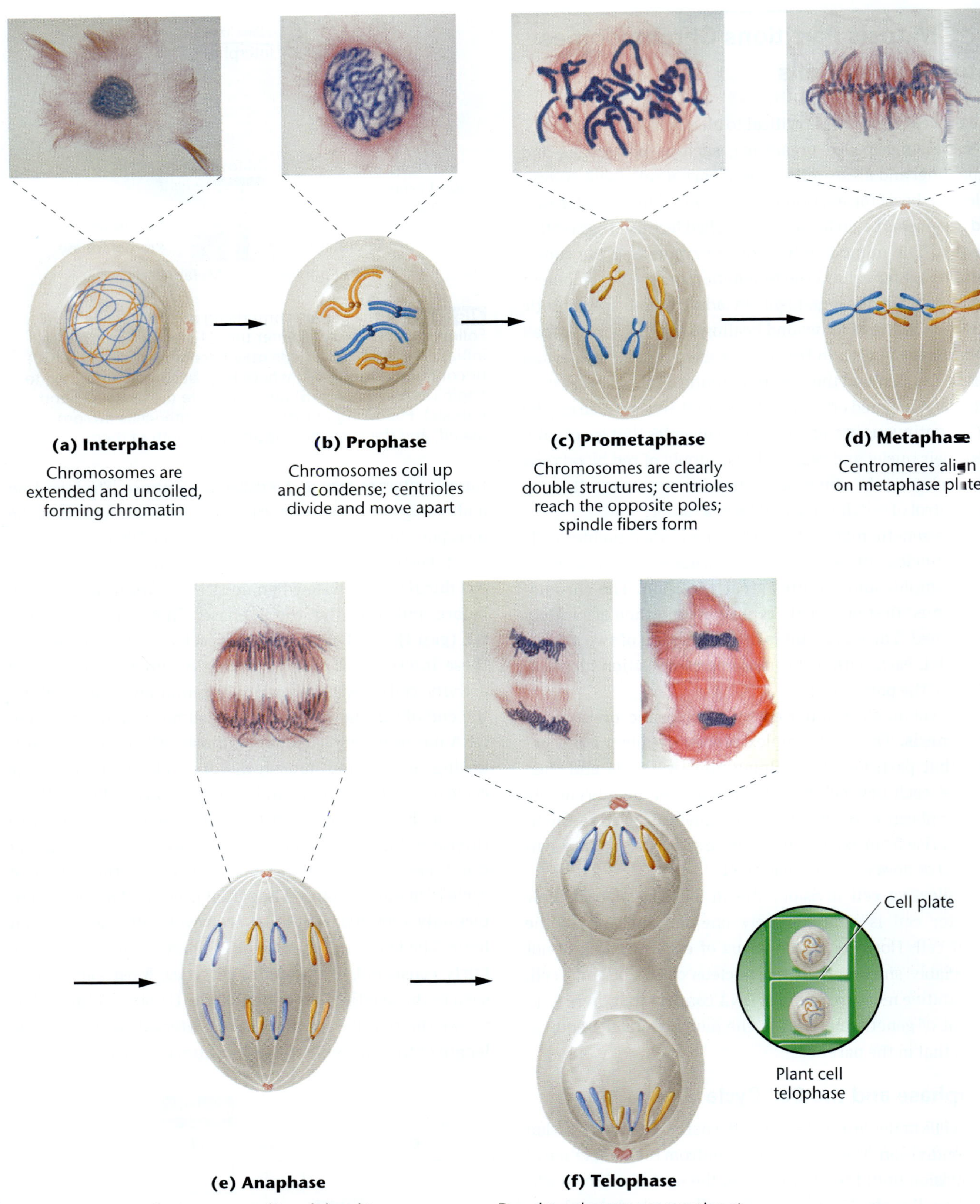

(a) Interphase
Chromosomes are extended and uncoiled, forming chromatin
(b) Prophase
Chromosomes coil up and condense; centrioles divide and move apart
(c) Prometaphase
Chromosomes are clearly double structures; centrioles reach the opposite poles; spindle fibers form
(d) Metaphase
Centromeres align on metaphase plate
(e) Anaphase
Centromeres split and daughter chromosomes migrate to opposite poles
(f) Telophase
Daughter chromosomes arrive at the poles; cytokinesis commences
Cell plate
Plant cell telophase

FIGURE 2.7 Drawings depicting mitosis in an animal cell with a diploid number of 4. The events occurring in each stage are described in the text. Of the two homologous pairs of chromosomes, one pair consists of longer, metacentric members and the other of shorter, submetacentric members. The maternal chromosome and the paternal chromosome of each pair are shown in different colors. To the right of (f), a drawing of late telophase in a plant cell shows the formation of the cell plate and lack of centrioles. The cells shown in the light micrographs came from the flower of *Haemanthus,* a plant that has a diploid number of 8.

G1 is of great interest in the study of cell proliferation and its control. At a point during G1, all cells follow one of two paths. They either withdraw from the cycle, become quiescent, and enter the **G0 stage** (see Figure 2.5), or they become committed to proceed through G1, initiating DNA synthesis, and completing the cycle. Cells that enter G0 remain viable and metabolically active but are not proliferative. Cancer cells apparently avoid entering G0 or pass through it very quickly. Other cells enter G0 and never reenter the cell cycle. Still other cells in G0 can be stimulated to return to G1 and thereby reenter the cell cycle.

Cytologically, interphase is characterized by the absence of visible chromosomes. Instead, the nucleus is filled with chromatin fibers that are formed as the chromosomes uncoil and disperse after the previous mitosis [**Figure 2.7(a)**]. Once G1, S, and G2 are completed, mitosis is initiated. Mitosis is a dynamic period of vigorous and continual activity. For discussion purposes, the entire process is subdivided into discrete stages, and specific events are assigned to each one. These stages, in order of occurrence, are prophase, prometaphase, metaphase, anaphase, and telophase.

Prophase

Often over half of mitosis is spent in **prophase** [**Figure 2.7(b)**], a stage characterized by several significant occurrences. One of the early events in prophase of all animal cells is the migration of two pairs of centrioles to opposite ends of the cell. These structures are found just outside the nuclear envelope in an area of differentiated cytoplasm called the centrosome (introduced in Section 2.1). It is believed that each pair of centrioles consists of one mature unit and a smaller, newly formed daughter centriole.

The centrioles migrate and establish poles at opposite ends of the cell. After migration, the centrosomes, in which the centrioles are localized, are responsible for organizing cytoplasmic microtubules into the spindle fibers that run between these poles, creating an axis along which chromosomal separation occurs. Interestingly, the cells of most plants (there are a few exceptions), fungi, and certain algae seem to lack centrioles. Spindle fibers are nevertheless apparent during mitosis.

As the centrioles migrate, the nuclear envelope begins to break down and gradually disappears. In a similar fashion, the nucleolus disintegrates within the nucleus. While these events are taking place, the diffuse chromatin fibers have begun to condense, until distinct thread-like structures, the chromosomes, become visible. It becomes apparent near the end of prophase that each chromosome is actually a double structure split longitudinally except at a single point of constriction, the centromere. The two parts of each chromosome are called **sister chromatids** because the DNA contained in each of them is genetically identical, having formed from a single replicative event. Sister chromatids are held together by a multi-subunit protein complex called **cohesin.** This molecular complex is originally formed between them during the S phase of the cell cycle when the DNA of each chromosome is replicated. Thus, even though we cannot see chromatids in interphase because the chromatin is uncoiled and dispersed in the nucleus, the chromosomes are already double structures, which becomes apparent in late prophase. In humans, with a diploid number of 46, a cytological preparation of late prophase reveals 46 chromosomes randomly distributed in the area formerly occupied by the nucleus.

Prometaphase and Metaphase

The distinguishing event of the two ensuing stages is the migration of every chromosome, led by its centromeric region, to the equatorial plane. The equatorial plane, also referred to as the *metaphase plate,* is the midline region of the cell, a plane that lies perpendicular to the axis established by the spindle fibers. In some descriptions, the term **prometaphase** refers to the period of chromosome movement [**Figure 2.7(c)**], and the term **metaphase** is applied strictly to the chromosome configuration following migration.

Migration is made possible by the binding of spindle fibers to the chromosome's **kinetochore,** an assembly of multilayered plates of proteins associated with the centromere. This structure forms on opposite sides of each paired centromere, in intimate association with the two sister chromatids. Once properly attached to the spindle fibers, cohesin is degraded by an enzyme, appropriately named *separase,* and the sister chromatid arms disjoin, except at the centromere region. A unique protein family called **shugoshin** (from the Japanese meaning "guardian spirit") protects cohesin from being degraded by separase at the centromeric regions. The involvement of the cohesin and shugoshin complexes with a pair of sister chromatids during mitosis is depicted in **Figure 2.8**.

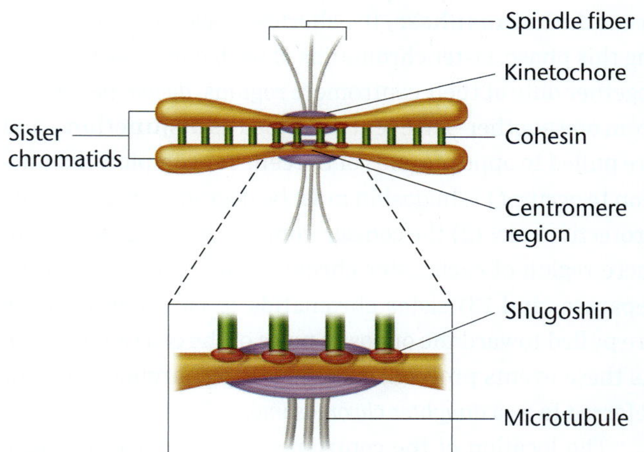

FIGURE 2.8 The depiction of the alignment, pairing, and disjunction of sister chromatids during mitosis, involving the molecular complexes cohesin and shugoshin and the enzyme separase.

We know a great deal about the molecular interactions involved in kinetechore assembly along the centromere. This is of great interest because of the consequences when mutations alter the proteins that make up the kinetechore complex. Altered kinetechore function potentially leads to errors during chromosome migration, altering the diploid content of daughter cells. A more detailed account will be presented later in the text, once we have provided more information about DNA and the proteins that make up chromatin (see Chapter 11).

We also know a great deal about spindle fibers and the mechanism responsible for their attachment to the kinetechore. Spindle fibers consist of microtubules, which themselves consist of molecular subunits of the protein tubulin. Microtubules seem to originate and "grow" out of the two centrosome regions at opposite poles of the cell. They are dynamic structures that lengthen and shorten as a result of the addition or loss of polarized tubulin subunits. The microtubules most directly responsible for chromosome migration make contact with, and adhere to, kinetochores as they grow from the centrosome region. They are referred to as *kinetochore microtubules* and have one end near the centrosome region (at one of the poles of the cell) and the other end anchored to the kinetochore. The number of microtubules that bind to the kinetochore varies greatly between organisms. Yeast (*Saccharomyces*) has only a single microtubule bound to each plate-like structure of the kinetochore. Mitotic cells of mammals, at the other extreme, reveal 30 to 40 microtubules bound to each portion of the kinetochore.

At the completion of metaphase, each centromere is aligned at the metaphase plate with the chromosome arms extending outward in a random array. This configuration is shown in **Figure 2.7(d)**.

Anaphase

Events critical to chromosome distribution during mitosis occur during **anaphase,** the shortest stage of mitosis. During this phase, sister chromatids of each chromosome, held together only at their centromere regions, *disjoin* (separate) from one another—an event described as **disjunction**—and are pulled to opposite ends of the cell. For complete disjunction to occur: (1) shugoshin must be degraded, reversing its protective role; (2) the cohesin complex holding the centromere region of each sister chromosome is then cleaved by separase; and (3) sister chromatids of each chromosome are pulled toward the opposite poles of the cell (Figure 2.8). As these events proceed, each migrating chromatid is now referred to as a *daughter chromosome*.

The location of the centromere determines the shape of the chromosome during separation, as you saw in Figure 2.3. The steps that occur during anaphase are critical in providing each subsequent daughter cell with an identical set of chromosomes. In human cells, there would now be 46 chromosomes at each pole, one from each original sister pair. **Figure 2.7(e)** shows anaphase prior to its completion.

Telophase

Telophase is the final stage of mitosis and is depicted in **Figure 2.7(f)**. At its beginning, two complete sets of chromosomes are present, one set at each pole. The most significant event of this stage is cytokinesis, the division or partitioning of the cytoplasm. Cytokinesis is essential if two new cells are to be produced from one cell. The mechanism of cytokinesis differs greatly in plant and animal cells, but the end result is the same: Two new cells are produced. In plant cells, a *cell plate* is synthesized and laid down across the region of the metaphase plate. Animal cells, however, undergo a constriction of the cytoplasm, much as a loop of string might be tightened around the middle of a balloon.

It is not surprising that the process of cytokinesis varies in different organisms. Plant cells, which are more regularly shaped and structurally rigid, require a mechanism for depositing new cell wall material around the plasma membrane. The cell plate laid down during telophase becomes a structure called the *middle lamella* Subsequently, the primary and secondary layers of the cell wall are deposited between the cell membrane and middle lamella in each of the resulting daughter cells. In animals, complete constriction of the cell membrane produces the *cell furrow* characteristic of newly divided cells.

Other events necessary for the transition from mitosis to interphase are initiated during late telophase. They generally constitute a reversal of events that occurred during prophase. In each new cell, the chromosomes begin to uncoil and become diffuse chromatin once again, while the nuclear envelope re-forms around them, the spindle fibers disappear, and the nucleolus gradually re-forms and becomes visible in the nucleus during early interphase. At the completion of telophase, the cell enters interphase.

Cell-Cycle Regulation and Checkpoints

The cell cycle, culminating in mitosis, is fundamentally the same in all eukaryotic organisms. This similarity in many diverse organisms suggests that the cell cycle is governed by a genetically regulated program that has been conserved throughout evolution. Because disruption of this regulation may underlie the uncontrolled cell division characterizing malignancy, interest in how genes regulate the cell cycle is particularly strong.

A mammoth research effort over the past 20 years has paid high dividends, and we now have knowledge of many genes involved in the control of the cell cycle. This work was recognized by the awarding of the 2001 Nobel Prize in Medicine or Physiology to Lee Hartwell, Paul Nurse, and Tim Hunt. As with other studies of genetic control over essential biological processes, investigation has focused on the discovery of mutations that interrupt the cell cycle and on the effects of those mutations. As we shall return to this subject in much greater detail later in the text during our consideration of the molecular basis of cancer (see Chapter 19), what follows is a very brief overview.

Many mutations are now known that exert an effect at one or another stage of the cell cycle. First discovered in yeast, but now evident in all organisms, including humans, such mutations were originally designated as *cell division cycle (cdc) mutations*. The normal products of many of the mutated genes are enzymes called **kinases** that can add phosphates to other proteins. They serve as "master control" molecules functioning in conjunction with proteins called **cyclins.** Cyclins bind to these kinases (creating *cyclin-dependent kinases*), activating them at appropriate times during the cell cycle. Activated kinases then phosphorylate other target proteins that regulate the progress of the cell cycle. The study of *cdc* mutations has established that the cell cycle contains at least three **cell-cycle checkpoints,** where the processes culminating in normal mitosis are monitored, or "checked," by these master control molecules before the next stage of the cycle is allowed to commence.

The importance of cell-cycle control and these checkpoints can be demonstrated by considering what happens when this regulatory system is impaired. Let's assume, for example, that the DNA of a cell has incurred damage leading to one or more mutations impairing cell-cycle control. If allowed to proceed through the cell cycle, this genetically altered cell would divide uncontrollably—a key step in the development of a cancer cell. If, instead, the cell cycle is arrested at one of the checkpoints, the cell can repair the DNA damage or permanently stop the cell from dividing, thereby preventing its potential malignancy.

ESSENTIAL POINT

Mitosis is subdivided into discrete stages that initially depict the condensation of chromatin into the diploid number of chromosomes, each of which is initially a double structure, each composed of a pair of sister chromatids. During mitosis, sister chromatids are pulled apart and directed toward opposite poles, after which cytoplasmic division creates two new cells with identical genetic information. ■

NOW SOLVE THIS

2.1 With the initial appearance of the feature we call "Now Solve This," a short introduction is in order. The feature occurs several times in this and all ensuing chapters, each time providing a problem related to the discussion just presented. A "Hint" is then offered that may help you solve the problem. Here is the first problem:

(a) If an organism has a diploid number of 16, how many chromatids are visible at the end of mitotic prophase?

(b) How many chromosomes are moving to each pole during anaphase of mitosis?

■ **HINT:** *This problem involves an understanding of what happens to each pair of homologous chromosomes during mitosis, asking you to apply your understanding of chromosome behavior to an organism with a diploid number of 16. The key to its solution is your awareness that throughout mitosis, the members of each homologous pair do not pair up, but instead behave independently.*

2.4 Meiosis Creates Haploid Gametes and Spores and Enhances Genetic Variation in Species

Whereas in diploid organisms, mitosis produces two daughter cells with full diploid complements, **meiosis** produces gametes or spores that are characterized by only one haploid set of chromosomes. During sexual reproduction, haploid gametes then combine at fertilization to reconstitute the diploid complement found in parental cells. Meiosis must be highly specific since haploid gametes or spores must contain precisely one member of each homologous pair of chromosomes. When successfully completed, meiosis provides the basis for maintaining genetic continuity from generation to generation.

Another major accomplishment of meiosis is to ensure that during sexual reproduction an enormous amount of genetic variation is produced among members of a species. Such variation occurs in two forms. First, meiosis produces haploid gametes with many unique combinations of maternally and paternally derived chromosomes. As we will see (Chapter 3), this process is the underlying basis of Mendel's principles of segregation and independent assortment. The second source of variation is created by the meiotic event referred to as **crossing over,** which results in genetic exchange between members of each homologous pair of chromosomes prior to one or the other finding its way into a haploid gamete or spore. This creates intact chromosomes that are mosaics of the maternal and paternal homologs. Sexual reproduction therefore significantly reshuffles the genetic material, producing highly diverse offspring.

Meiosis: Prophase I

As in mitosis, the process in meiosis begins with a diploid cell duplicating its genetic material in the interphase stage preceding chromosome division. To achieve haploidy, two divisions are thus required. The meiotic achievements are largely dependent on the behavior of chromosomes during the initial stage of the first division, called *prophase I*. Recall

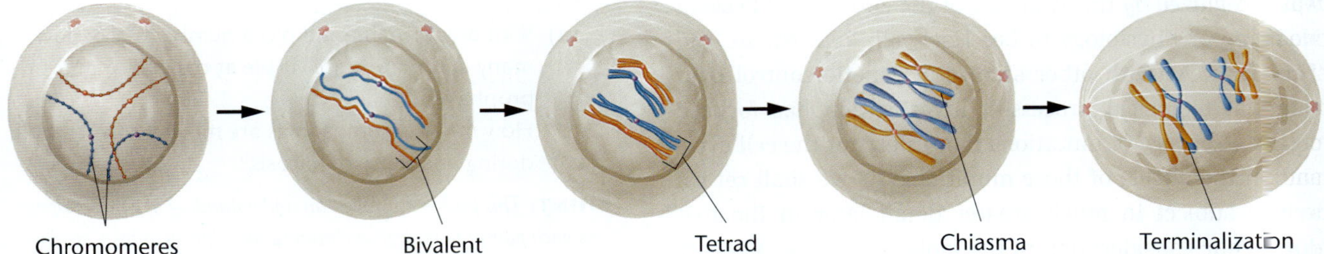

FIGURE 2.9 The events characterizing meiotic prophase I. In the first two frames, illustrating chromomeres and bivalents, each chromatid is actually a double structure, consisting of sister chromatids, which first becomes apparent in the ensuing tetrad stage.

that in mitosis the paternally and maternally derived members of each homologous pair of chromosomes behave autonomously during division. Each chromosome is duplicated, creating genetically identical **sister chromatids,** and subsequently, one chromatid of each pair is distributed to each new cell. The major difference in meiosis is that once the chromatin characterizing interphase has condensed into visible structures, the homologous chromosomes are not autonomous but are instead seen to be paired up, having undergone the process called **synapsis.** **Figure 2.9** illustrates this process as well as the ensuing events of prophase I. Each synapsed pair of homologs is initially called a **bivalent,** and the number of bivalents is equal to the haploid number. In Figure 2.9, we have

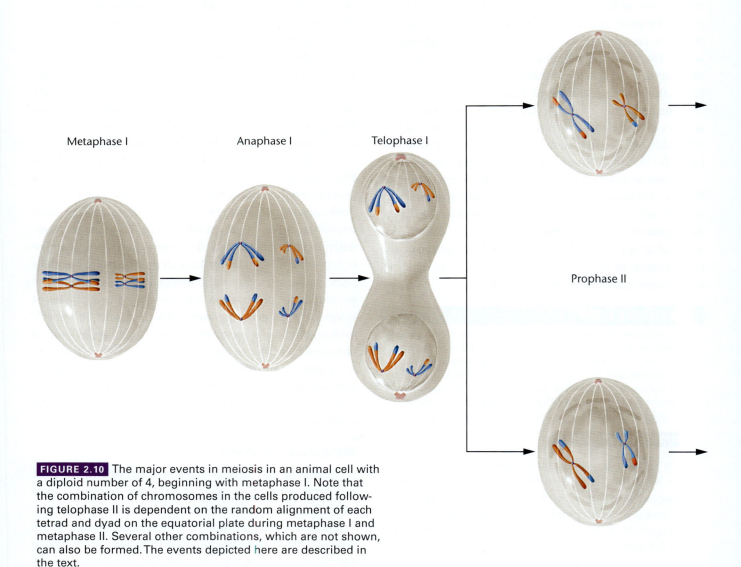

FIGURE 2.10 The major events in meiosis in an animal cell with a diploid number of 4, beginning with metaphase I. Note that the combination of chromosomes in the cells produced following telophase II is dependent on the random alignment of each tetrad and dyad on the equatorial plate during metaphase I and metaphase II. Several other combinations, which are not shown, can also be formed. The events depicted here are described in the text.

depicted two homologous pairs of chromosomes and thus two bivalents. As the homologs condense and shorten, each bivalent gives rise to a unit called a **tetrad,** consisting of two pairs of sister chromatids, each of which is joined at a common centromere. Remember that one pair of sister chromatids is maternally derived, and the other pair paternally derived. The presence of tetrads is visible evidence that *both* homologs have, in fact, duplicated. As prophase progresses within each tetrad, each pair of sister chromatids is seen to pull apart. However, one or more areas remain in contact where chromatids are intertwined. Each such area, called a **chiasma** (pl. chiasmata), is thought to represent a point where **nonsister chromatids** (one paternal and one maternal chromatid) have undergone genetic exchange through the process of crossing over. Since crossing over is thought to occur one or more times in each tetrad, mosaic chromosomes are routinely created during every meiotic event. During the final period of prophase I, the nucleolus and nuclear envelope break down, and the two centromeres of each tetrad attach to the recently formed spindle fibers.

Metaphase I, Anaphase I, and Telophase I

The remainder of the meiotic process is depicted in **Figure 2.10**. After meiotic prophase I, steps similar to those of mitosis occur. In the first division, *metaphase I,* the chromosomes have maximally shortened and thickened. The terminal chiasmata of each tetrad are visible and appear to be the only factor holding the nonsister chromatids together. Each tetrad interacts with spindle fibers, facilitating movement to the metaphase plate. The alignment of each tetrad prior to the first anaphase is random. Half of each tetrad is pulled randomly to one or the other pole, and the other half then moves to the opposite pole.

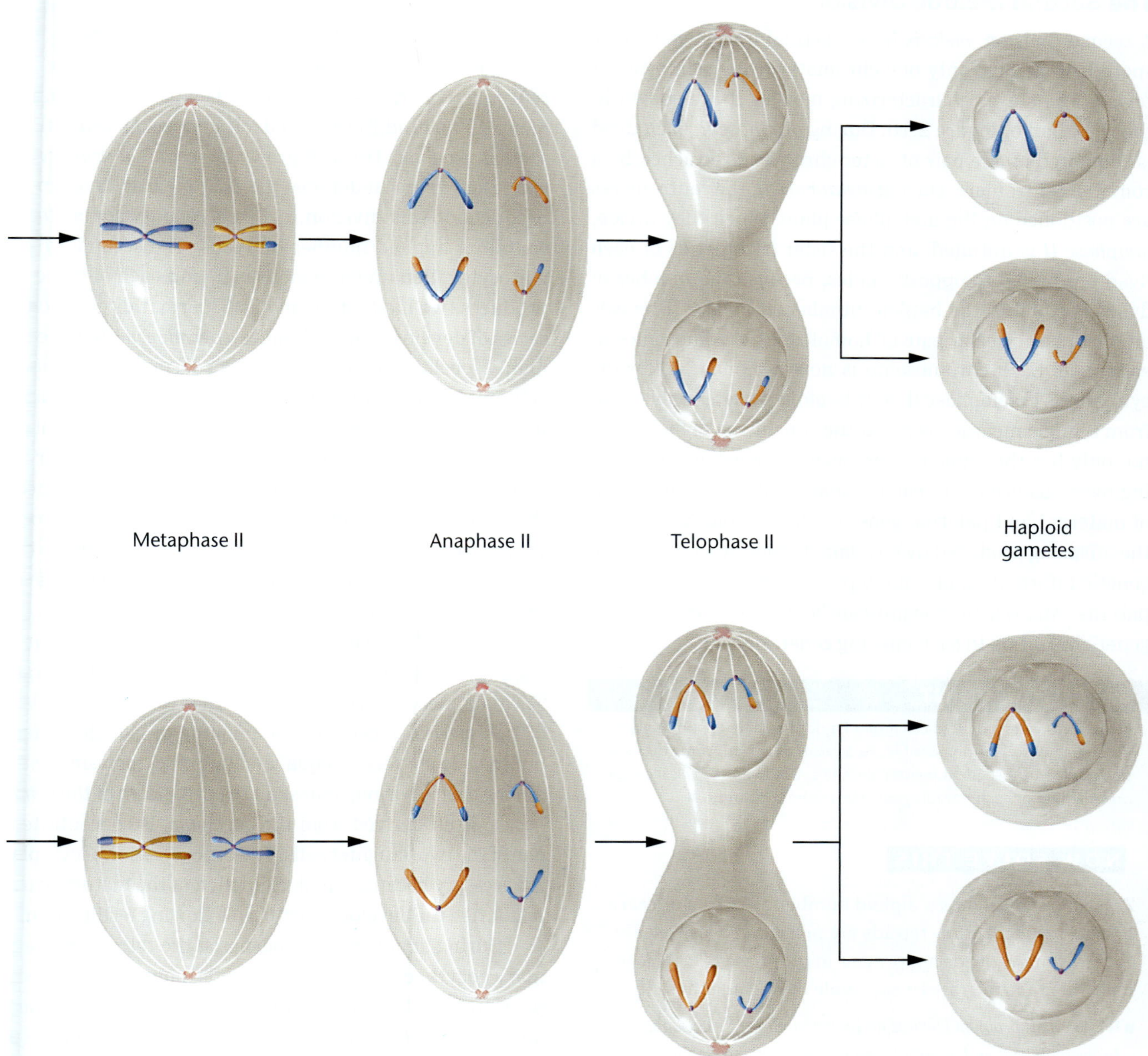

During the stages of meiosis I, a single centromere holds each pair of sister chromatids together. It does *not* divide. At *anaphase I,* one-half of each tetrad (the dyad) is pulled toward each pole of the dividing cell. This separation process is the physical basis of **disjunction,** the separation of chromosomes from one another. Occasionally, errors in meiosis occur and separation is not achieved. The term **nondisjunction** describes such an error. At the completion of a normal anaphase I, a series of dyads equal to the haploid number is present at each pole.

In many organisms, *telophase I* reveals a nuclear membrane forming around the dyads. Next, the nucleus enters into a short interphase period. If interphase occurs, the chromosomes do not replicate since they already consist of two chromatids. In other organisms, the cells go directly from anaphase I to meiosis II. In general, meiotic telophase is much shorter than the corresponding stage in mitosis.

The Second Meiotic Division

A second division, *meiosis II,* is essential if each gamete or spore is to receive only one chromatid from each original tetrad. The stages characterizing meiosis II are shown in the right half of Figure 2.10. During *prophase II,* each dyad is composed of one pair of sister chromatids attached by a common centromere. During *metaphase II,* the centromeres are positioned on the metaphase plate. When they divide, *anaphase II* is initiated, and the sister chromatids of each dyad are pulled to opposite poles. Because the number of dyads is equal to the haploid number, *telophase II* reveals one member of each pair of homologous chromosomes at each pole. Each chromosome is now a monad. Following cytokinesis in telophase II, four haploid gametes may result from a single meiotic event. At the conclusion of meiosis II, not only has the haploid state been achieved, but if crossing over has occurred, each monad is also a combination of maternal and paternal genetic information. As a result, the offspring produced by any gamete receives a mixture of genetic information originally present in his or her grandparents. Meiosis thus significantly increases the level of genetic variation in each ensuing generation.

ESSENTIAL POINT

Meiosis converts a diploid cell into a haploid gamete or spore, making sexual reproduction possible. As a result of chromosome duplication and two subsequent meiotic divisions, each haploid cell receives one member of each homologous pair of chromosomes. ■

NOW SOLVE THIS

2.2 An organism has a diploid number of 16 in a primary oocyte. (a) How many tetrads are present in prophase I? (b) How many dyads are present in prophase II? (c) How many monads migrate to each pole during anaphase II?

■ **HINT:** *This problem involves an understanding of what happens to the maternal and paternal members of each pair of homologous chromosomes during meiosis, asking you to extrapolate your understanding to chromosome behavior in an organism with a diploid number of 16. The major insight needed to solve this problem is to understand that maternal and paternal homologs synapse during meiosis. Once it is evident that each chromatid has duplicated, creating a tetrad in the early phases of meiosis, each original pair behaves as a unit and leads to two dyads during anaphase I.*

2.5 The Development of Gametes Varies in Spermatogenesis Compared to Oogenesis

Although events that occur during the meiotic divisions are similar in all cells participating in gametogenesis in most animal species, there are certain differences between the production of a male gamete (spermatogenesis) and a female gamete (oogenesis). **Figure 2.11** summarizes these processes.

Spermatogenesis takes place in the testes, the male reproductive organs. The process begins with the enlargement of an undifferentiated diploid germ cell called a *spermatogonium.* This cell grows to become a *primary spermatocyte,* which undergoes the first meiotic division. The products of this division, called *secondary spermatocytes,* contain a haploid number of dyads. The secondary spermatocytes then undergo meiosis II, and each of these cells produces two haploid *spermatids.* Spermatids go through a series of developmental changes, *spermiogenesis,* to become highly specialized, motile *spermatozoa,* or *sperm.* All sperm cells produced during spermatogenesis contain the haploid number of chromosomes and equal amounts of cytoplasm.

Spermatogenesis may be continuous or may occur periodically in mature male animals; its onset is determined by the species' reproductive cycles. Animals that reproduce year-round produce sperm continuously, whereas those whose breeding period is confined to a particular season produce sperm only during that time.

In animal *oogenesis,* the formation of *ova* (sing. *ovum*), or eggs, occurs in the ovaries, the female reproductive organs. The daughter cells resulting from the two meiotic divisions of this process receive equal amounts of genetic material, but they do *not* receive equal amounts of cytoplasm. Instead, during each division, almost all the cytoplasm of the *primary oocyte,* itself derived from the *oogonium,* is concentrated in one of the two daughter cells. The concentration of cytoplasm is necessary because a major function of the mature ovum is to nourish the developing embryo following fertilization.

During anaphase I in oogenesis, the tetrads of the primary oocyte separate, and the dyads move toward opposite poles. During telophase I, the dyads at one pole are pinched off with very little surrounding cytoplasm to form the *first polar body.* The first polar body may or may not divide again to produce two small haploid cells. The other daughter cell produced by this first meiotic division contains most of the cytoplasm and is called the *secondary oocyte*. The mature ovum will be produced from the secondary oocyte during the second meiotic division. During this division, the cytoplasm of the secondary oocyte again divides unequally, producing an *ootid* and a *second polar body*. The ootid then differentiates into the mature ovum.

Unlike the divisions of spermatogenesis, the two meiotic divisions of oogenesis may not be continuous. In some animal species, the second division may directly follow the first. In others, including humans, the first division of all oocytes begins in the embryonic ovary but arrests in prophase I. Many years later, meiosis resumes in each oocyte just prior to its ovulation. The second division is completed only after fertilization.

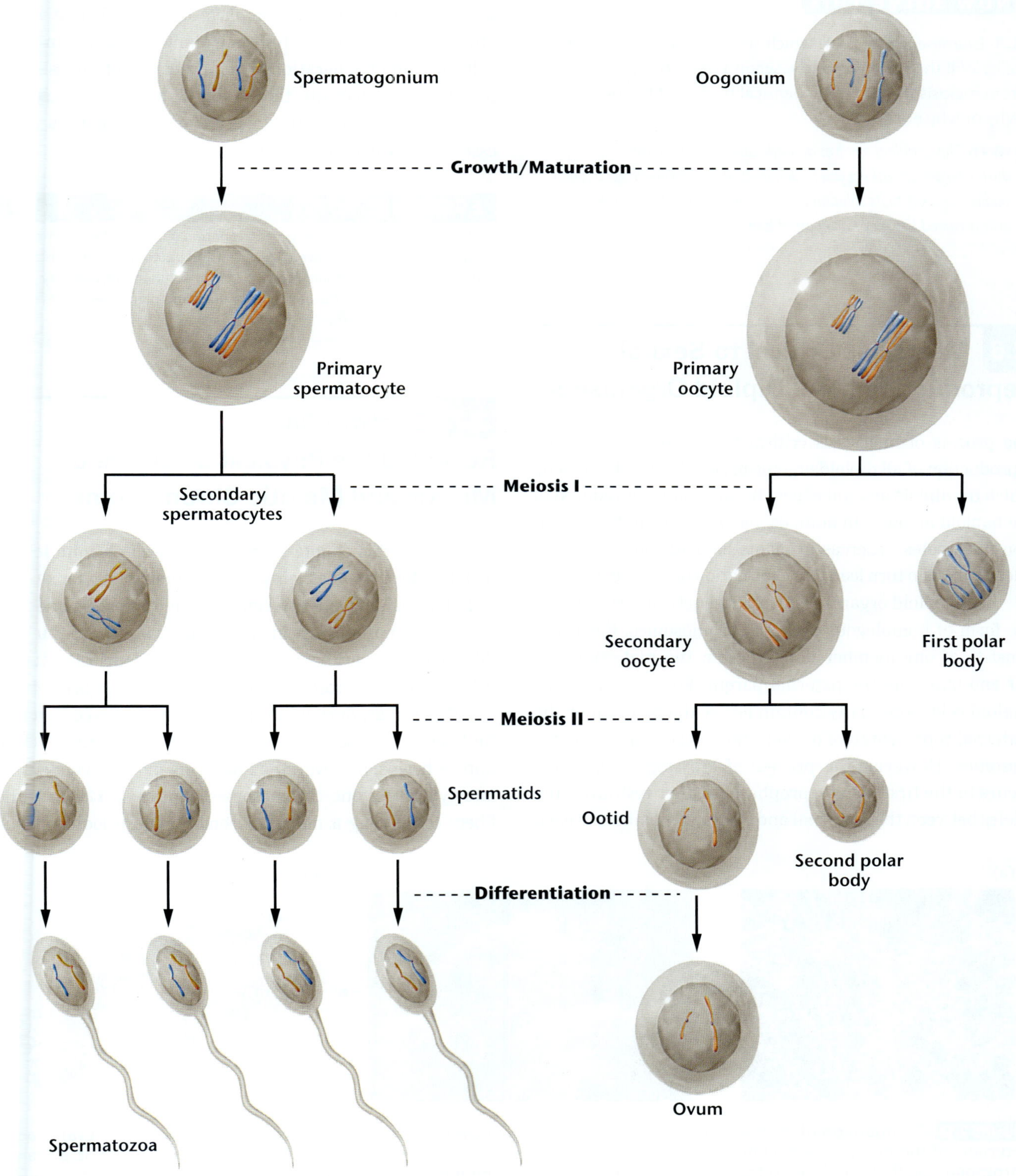

FIGURE 2.11 Spermatogenesis and oogenesis in animal cells.

ESSENTIAL POINT

There is a major difference between meiosis in males and in females. On the one hand, spermatogenesis partitions the cytoplasmic volume equally and produces four haploid sperm cells. Oogenesis, on the other hand, collects the bulk of cytoplasm in one egg cell and reduces the other haploid products to polar bodies. The extra cytoplasm in the egg contributes to zygote development following fertilization. ■

NOW SOLVE THIS

2.3 Examine Figure 2.11, which shows oogenesis in animal cells. Will the genotype of the second polar body (derived from meiosis II) always be identical to that of the ootid? Why or why not?

■ **HINT:** *This problem involves an understanding of meiosis during oogenesis, asking you to demonstrate your knowledge of polar bodies. The key to its solution is to take into account that crossing over occurred between each pair of homologs during meiosis I.*

2.6 Meiosis Is Critical to Sexual Reproduction in All Diploid Organisms

The process of meiosis is critical to the successful sexual reproduction of all diploid organisms. It is the mechanism by which the diploid amount of genetic information is reduced to the haploid amount. In animals, meiosis leads to the formation of gametes, whereas in plants haploid spores are produced, which in turn lead to the formation of haploid gametes.

Each diploid organism stores its genetic information in the form of homologous pairs of chromosomes. Each pair consists of one member derived from the maternal parent and one from the paternal parent. Following meiosis, haploid cells potentially contain either the paternal or the maternal representative of every homologous pair of chromosomes. However, the process of crossing over, which occurs in the first meiotic prophase, further reshuffles the alleles between the maternal and paternal members of each homologous pair, which then segregate and assort independently into gametes. These events result in the great amount of genetic variation present in gametes.

It is important to touch briefly on the significant role that meiosis plays in the life cycles of fungi and plants. In many fungi, the predominant stage of the life cycle consists of haploid vegetative cells. They arise through meiosis and proliferate by mitotic cell division. In multicellular plants, the life cycle alternates between the diploid *sporophyte stage* and the haploid *gametophyte stage*. While one or the other predominates in different plant groups during this "alternation of generations," the processes of meiosis and fertilization constitute the "bridges" between the sporophyte and gametophyte stages. Therefore, meiosis is an essential component of the life cycle of plants.

ESSENTIAL POINT

Meiosis results in extensive genetic variation by virtue of the exchange during crossing over between maternal and paternal chromatids and their random segregation into gametes. In addition, meiosis plays an important role in the life cycles of fungi and plants, serving as the bridge between alternating generations. ■

2.7 Electron Microscopy Has Revealed the Physical Structure of Mitotic and Meiotic Chromosomes

Thus far in this chapter, we have focused on mitotic and meiotic chromosomes, emphasizing their behavior during cell division and gamete formation. An interesting question is why chromosomes are invisible during interphase but visible during the various stages of mitosis and meiosis. Studies using electron microscopy clearly show why this is the case.

Recall that, during interphase, only dispersed chromatin fibers are present in the nucleus [**Figure 2.12(a)**]. Once mitosis begins, however, the fibers coil and fold, condensing into typical mitotic chromosomes [**Figure 2.12(b)**]. If the fibers comprising a mitotic chromosome are loosened, the

FIGURE 2.12 Comparison of (a) the chromatin fibers characteristic of the interphase nucleus with (b) metaphase chromosomes that are derived from chromatin during mitosis. Part (c) diagrams a mitotic chromosome, showing how chromatin is condensed to produce it. Part (a) is a transmission electron micrograph, and part (b) is a scanning electron micrograph.

areas of greatest spreading reveal individual fibers similar to those seen in interphase chromatin [**Figure 2.12(c)**]. Very few fiber ends seem to be present, and in some cases, none can be seen. Instead, individual fibers always seem to loop back into the interior. Such fibers are obviously twisted and coiled around one another, forming the regular pattern of folding in the mitotic chromosome. Starting in late telophase of mitosis and continuing during G1 of interphase, chromosomes unwind to form the long fibers characteristic of chromatin, which consist of DNA and associated proteins, particularly proteins called *histones*. It is in this physical arrangement that DNA can most efficiently function during transcription and replication.

Electron microscopic observations of metaphase chromosomes in varying degrees of coiling led Ernest DuPraw to postulate the **folded-fiber model,** shown in Figure 2.12(c). During metaphase, each chromosome consists of two sister chromatids joined at the centromeric region. Each arm of the chromatid appears to be a single fiber wound much like a skein of yarn. The fiber is composed of tightly coiled double-stranded DNA and protein. An orderly coiling–twisting–condensing process appears to facilitate the transition of the interphase chromatin into the more condensed mitotic chromosomes. Geneticists believe that during the transition from interphase to prophase, a 5000-fold compaction occurs in the length of DNA within the chromatin fiber! This process must be extremely precise given the highly ordered and consistent appearance of mitotic chromosomes in all eukaryotes. Note particularly in the micrographs the clear distinction between the sister chromatids constituting each chromosome. They are joined only by the common centromere that they share prior to anaphase. We will return to this general topic later in the text when we consider chromosome structure in further detail (see Chapter 11).

ESSENTIAL POINT

Mitotic chromosomes are produced as a result of the coiling and condensation of chromatin fibers characteristic of interphase and are thus visible only during cell division. ■

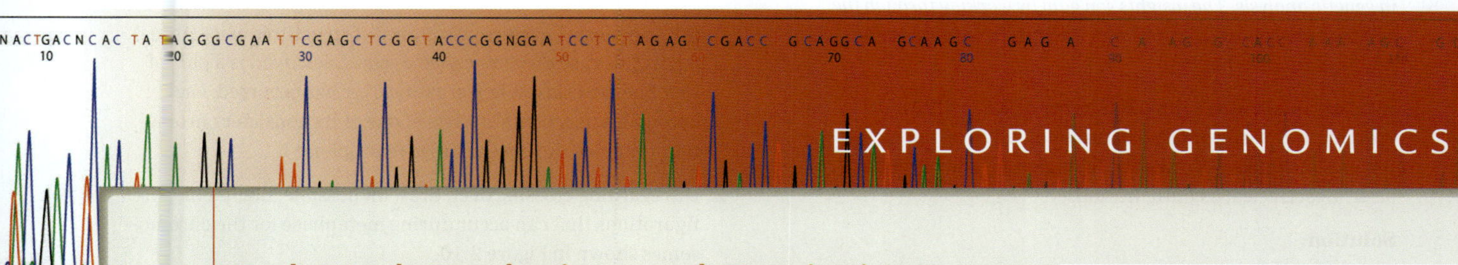

PubMed: Exploring and Retrieving Biomedical Literature

Mastering Genetics Visit the Study Area: Exploring Genomics

PubMed is an Internet-based search system developed by the National Center for Biotechnology Information (NCBI) at the National Library of Medicine. Using PubMed, one can access over 26 million citations for publications in over 5600 biomedical journals. The full text of many of the articles can be obtained electronically through college or university libraries, and some journals (such as *Proceedings of the National Academy of Sciences USA; Genome Biology;* and *Science*) provide free public access to articles within certain time frames.

In this exercise, we will explore PubMed to answer questions about relationships between tubulin, cancer, and cancer therapies.

■ **Exercise I – Tubulin, Cancer, and Cancer Therapies**

In this chapter we were introduced to tubulin and the dynamic behavior of microtubules during mitosis. Cancer cells are characterized by continuous and uncontrolled mitotic divisions.

Is it possible that tubulin and microtubules contribute to the development of cancer? Could these important structures be targets for cancer therapies?

1. To begin your search for the answers, access the PubMed site at http://www.ncbi.nlm.nih.gov/pubmed/.
2. In the search box, type "tubulin cancer" and then click the "Search" button to perform the search.
3. Select several research papers and read the abstracts.

To answer the question about tubulin's association with cancer, you may want to limit your search to fewer papers, perhaps those that are review articles. To do this, click the "Review" link under the Article Types category on the left side of the page.

Explore some of the articles, as abstracts or as full text, available in your library or by free public access. Prepare a brief report or verbally share your experiences with your class. Describe two of the most important things you learned during your exploration, and identify the information sources you encountered during the search.

CASE STUDY Timing is everything

Over a period of two years, a man in his early 20s received a series of intermittent chemotherapy and radiotherapy treatments for Hodgkin disease. During this therapy, he and his wife were unable to initiate a pregnancy. The man had a series of his semen samples examined at a fertility clinic. The findings revealed that shortly after each treatment very few mature sperm were present, and abnormal chromosome numbers were often observed in developing spermatocytes. However, such chromosome abnormalities disappeared about 40 days after treatment, and normal sperm reappeared about 74 days posttreatment.

1. How might a genetic counselor explain the time-related differences in sperm production and the appearance and subsequent disappearance of chromosomal abnormalities?
2. Do you think that exposure to chemotherapy and radiotherapy would cause more problems to spermatocytes than to mature sperm?
3. Prior to treatment, should the physician(s) involved have been ethically obligated to recommend genetic counseling? What advice regarding fertility might have been suggested?

For further reading, see: Harel, S., et al., (2011). Management of fertility in patients treated for Hodgkin's lymphoma. *Haematologica*. 96: 1692–1699.

INSIGHTS AND SOLUTIONS

This appearance of "Insights and Solutions" begins a feature that will have great value to you as a student. From this point on, "Insights and Solutions" precedes the "Problems and Discussion Questions" at each chapter's end to provide sample problems and solutions that demonstrate approaches you will find useful in genetic analysis. The insights you gain by working through the sample problems will improve your ability to solve the ensuing problems in each chapter.

1. In an organism with a diploid number of $2n = 6$, how many individual chromosomal structures will align on the metaphase plate during (a) mitosis, (b) meiosis I, and (c) meiosis II? Describe each configuration.

Solution:

(a) Remember that in mitosis, homologous chromosomes do not synapse, so there will be six double structures, each consisting of a pair of sister chromatids. In other words, the number of structures is equivalent to the diploid number.

(b) In meiosis I, the homologs have synapsed, reducing the number of structures to three. Each is called a tetrad and consists of two pairs of sister chromatids.

(c) In meiosis II, the same number of structures exist (three), but in this case they are called dyads. Each dyad is a pair of sister chromatids. When crossing over has occurred, each chromatid may contain parts of one of its nonsister chromatids, obtained during exchange in prophase I.

2. Disregarding crossing over, draw all possible alignment configurations that can occur during metaphase for the chromosomes shown in Figure 2.10.

Solution: As shown in the diagram below, four configurations are possible when $n = 2$.

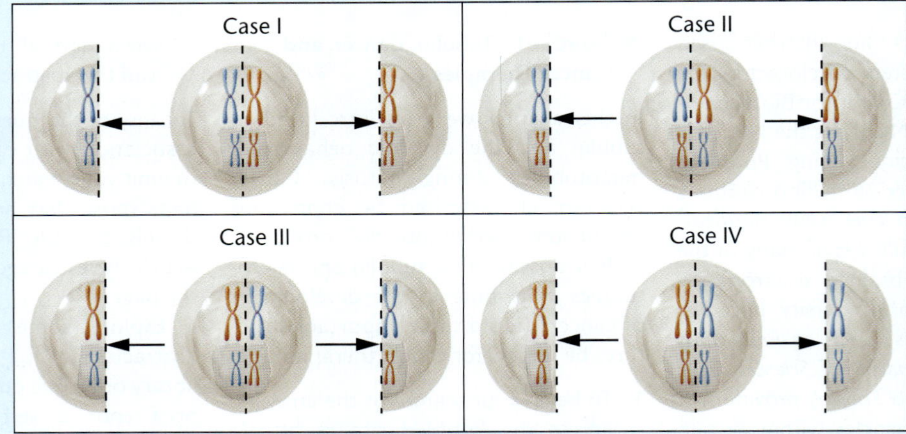

Solution for #2

3. For the chromosomes in Problem 2, assume that each of the larger chromosomes has a different allele for a given gene, *A* or *a*, as shown in the diagram below. Also assume that each of the smaller chromosomes has a different allele for a second gene, *B* or *b*. Calculate the probability of generating each possible combination of these alleles (*AB*, *Ab*, *aB*, *ab*) following meiosis I.

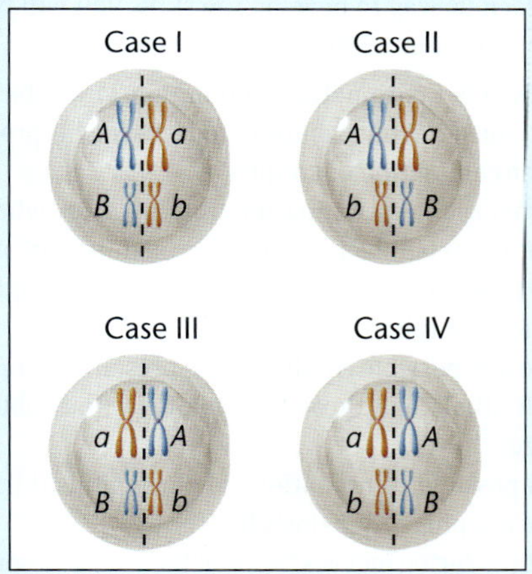

Solution:

Case I	*AB* and *ab*	**Total:**	$AB = 2\ (p = 1/4)$
Case II	*Ab* and *aB*		$Ab = 2\ (p = 1/4)$
Case III	*aB* and *Ab*		$aB = 2\ (p = 1/4)$
Case IV	*ab* and *AB*		$ab = 2\ (p = 1/4)$

Problems and Discussion Questions

Mastering Genetics Visit for instructor-assigned tutorials and problems.

1. **HOW DO WE KNOW?** In this chapter, we focused on how chromosomes are distributed during cell division, both in dividing somatic cells (mitosis) and in gamete- and spore-forming cells (meiosis). We found many opportunities to consider the methods and reasoning by which much of this information was acquired. From the explanations given in the chapter, answer the following questions.
   (a) How do we know that chromosomes exist in homologous pairs?
   (b) How do we know that DNA replication occurs during interphase, not early in mitosis?
   (c) How do we know that mitotic chromosomes are derived from chromatin?
2. **CONCEPT QUESTION** Review the Chapter Concepts list on page 36. All of these pertain to conceptual issues involving mitosis or meiosis. Based on these concepts, write a short essay that contrasts mitosis and meiosis, including their respective roles in organisms, the mechanisms by which they achieve their respective outcomes, and the consequences should either process fail to be executed with absolute fidelity. ■
3. What role do the following cellular components play in the storage, expression, or transmission of genetic information: (a) chromatin, (b) nucleolus, (c) ribosome, (d) mitochondrion, (e) centriole, (f) centromere?
4. Discuss the concepts of homologous chromosomes, diploidy, and haploidy. What characteristics do two homologous chromosomes share?
5. If two chromosomes of a species are the same length and have similar centromere placements and yet are not homologous, what is different about them?
6. Describe the events that characterize each stage of mitosis.
7. How are spindle fibers formed and how do chromosomes separate in animal cells?
8. Compare chromosomal separation in plant and animal cells.
9. Why might different cells of the same organism have cell cycles of different durations?
10. Define and discuss these terms: (a) synapsis, (b) bivalents, (c) chiasmata, (d) crossing over, (e) sister chromatids, (f) tetrads, (g) dyads, (h) monads.
11. A diploid organism has the alleles *T* and *t* for the same gene on a pair of homologous chromosomes. In what circumstances might both alleles segregate? At what stage of cell division would this occur?
12. Given the end results of the two types of division, why is it necessary for homologs to pair during meiosis and not desirable for them to pair during mitosis?
13. Contrast spermatogenesis and oogenesis. What is the significance of the formation of polar bodies?
14. How do the stages of mitosis and meiosis occur in a specific order and never alternate?
15. A diploid cell contains three pairs of homologous chromosomes designated C1 and C2, M1 and M2, and S1 and S2. No crossing over occurs. What combinations of chromosomes are possible in (a) daughter cells following mitosis, (b) cells undergoing the first meiotic metaphase, (c) haploid cells following both divisions of meiosis?
16. Predict the number of unique haploid gametes that could be produced through meiosis in an organism with a diploid number of $2n = 16$. Assume that crossing over does not occur.

17. To test the quality of eggs, polar bodies are routinely extracted from them. In a study, the first polar body dyads for all chromosomes were identified, but they were absent in Chromosome 13.
    (a) Provide a possible reason behind this oddity.
    (b) What chromosomal configuration would you expect in the secondary oocyte?
    (c) What would the consequences be in a zygote if it were fertilized?
18. Humans have a diploid number of 46. What is the probability that a sperm will be formed that contains all chromosomes whose centromeres are derived from maternal homologs?
19. Cattle (*Bos taurus*) have a diploid number of 60, and their haploid DNA content per cell is approximately 3.2 picogram. What would be the DNA content of a somatic cell (non-sex cell) at anaphase? What would be the nuclear DNA content of a secondary spermatocyte? What would be the nuclear DNA content of a spermatozoon?
20. Describe the role of meiosis in the life cycle of a vascular plant.
21. How many sister chromatids are seen in the metaphase for a single chromosome? How different are these structures from the interphase chromatin?
22. What is the significance of checkpoints in the cell cycle?
23. A metaphase chromosome preparation from an unknown organism clearly shows 40 chromosomes that can be easily paired. However, there are two unmatched chromosomes that differ from each other in size and centromere placement. What can you say about the chromosomes and the ploidy of this organism?
24. If one follows 50 primary oocytes in an animal through their various stages of oogenesis, how many secondary oocytes would be formed? How many first polar bodies would be formed? How many ootids would be formed? If one follows 50 primary spermatocytes in an animal through their various stages of spermatogenesis, how many secondary spermatocytes would be formed? How many spermatids would be formed?

For Problems **25–30,** consider a diploid cell that contains three pairs of chromosomes designated AA, BB, and CC. Each pair contains a maternal and a paternal member (e.g., A^m and A^p). Using these designations, demonstrate your understanding of mitosis and meiosis by drawing chromatid combinations as requested. Be sure to indicate when chromatids are paired as a result of replication and/or synapsis. You may wish to use a large piece of brown manila wrapping paper or a cut-up paper grocery bag for this project and to work in partnership with another student. We recommend cooperative learning as an efficacious way to develop the skills you will need for solving the problems presented throughout this text.

25. In mitosis, what chromatid combination(s) will be present during metaphase? What combination(s) will be present at each pole at the completion of anaphase?
26. During meiosis I, assuming no crossing over, what chromatid combination(s) will be present at the completion of prophase I? Draw all possible alignments of chromatids as migration begins during early anaphase.
27. Are there any possible combinations present during prophase of meiosis II other than those that you drew in Problem 26? If so, draw them.
28. Draw all possible combinations of chromatids during the early phases of anaphase in meiosis II.
29. Assume that during meiosis I none of the *C* chromosomes disjoin at metaphase, but they separate into dyads (instead of monads) during meiosis II. How would this change the alignments that you constructed during the anaphase stages in meiosis I and II? Draw them.
30. A normal gamete fuses with another that has undergone nondisjunction of B chromosome dyads into the secondary oocyte during meiosis I, but segregates into the daughter cells during meiosis II. What combinations could be obtained in the zygote?

3

Mendelian Genetics

Gregor Johann Mendel, who in 1866 put forward the major postulates of transmission genetics as a result of experiments with the garden pea.

CHAPTER CONCEPTS

- Inheritance is governed by information stored in discrete unit factors called genes.
- Genes are transmitted from generation to generation on vehicles called chromosomes.
- Chromosomes, which exist in pairs in diploid organisms, provide the basis of biparental inheritance.
- During gamete formation, chromosomes are distributed according to postulates first described by Gregor Mendel, based on his nineteenth-century research with the garden pea.
- Mendelian postulates prescribe that homologous chromosomes segregate from one another and assort independently with other segregating homologs during gamete formation.
- Genetic ratios, expressed as probabilities, are subject to chance deviation and may be evaluated statistically.
- The analysis of pedigrees allows predictions concerning the genetic nature of human traits.

Although inheritance of biological traits has been recognized for thousands of years, the first significant insights into how it takes place only occurred about 150 years ago. In 1866, Gregor Johann Mendel published the results of a series of experiments that would lay the foundation for the formal discipline of genetics. Mendel's work went largely unnoticed until the turn of the twentieth century, but eventually, the concept of the gene as a distinct hereditary unit was established. Since then, the ways in which genes, as segments of chromosomes, are transmitted to offspring and control traits have been clarified. Research continued unabated throughout the twentieth century and into the present—indeed, studies in genetics, most recently at the molecular level, have remained at the forefront of biological research since the early 1900s.

When Mendel began his studies of inheritance using *Pisum sativum,* the garden pea, chromosomes and the role and mechanism of meiosis were totally unknown. Nevertheless, he determined that discrete units of inheritance exist and predicted their behavior in the formation of gametes. Subsequent investigators, with access to cytological data, were able to relate their own observations of chromosome behavior during meiosis and Mendel's principles of inheritance. Once this correlation was recognized, Mendel's postulates were accepted as the basis for the study of what is known as **transmission genetics**—how genes are transmitted from parents to offspring. These principles were derived directly from Mendel's experimentation.

3.1 Mendel Used a Model Experimental Approach to Study Patterns of Inheritance

Johann Mendel was born in 1822 to a peasant family in the Central European village of Heinzendorf. An excellent student in high school, he studied philosophy for several years afterward and in 1843, taking the name Gregor, was admitted to the Augustinian Monastery of St. Thomas in Brno, now part of the Czech Republic. In 1849, he was relieved of pastoral duties, and from 1851 to 1853, he attended the University of Vienna, where he studied physics and botany. He returned to Brno in 1854, where he taught physics and natural science for the next 16 years. Mendel received support from the monastery for his studies and research throughout his life.

In 1856, Mendel performed his first set of hybridization experiments with the garden pea, launching the research phase of his career. His experiments continued until 1868, when he was elected abbot of the monastery. Although he retained his interest in genetics, his new responsibilities demanded most of his time. In 1884, Mendel died of a kidney disorder. The local newspaper paid him the following tribute:

> His death deprives the poor of a benefactor, and mankind at large of a man of the noblest character, one who was a warm friend, a promoter of the natural sciences, and an exemplary priest.

Mendel first reported the results of some simple genetic crosses between certain strains of the garden pea in 1865. Although his findings went unappreciated until the turn of the century, well after his death, his work was not the first attempt to provide experimental evidence pertaining to inheritance. Mendel's success where others had failed can be attributed, at least in part, to his elegant experimental design and analysis.

Mendel showed remarkable insight into the methodology necessary for good experimental biology. First, he chose an organism that was easy to grow and to hybridize artificially. The pea plant is self-fertilizing in nature, but it is easy to cross-breed experimentally. It reproduces well and grows to maturity in a single season. Mendel followed seven visible features (we refer to them as characters, or characteristics), each represented by two contrasting forms, or **traits** (**Figure 3.1**). For the character stem height, for example,

Character	Contrasting traits	F_1 results	F_2 results	F_2 ratio
Seed shape	round/wrinkled	 all round	5474 round 1850 wrinkled	2.96:1
Seed color	yellow/green	 all yellow	6022 yellow 2001 green	3.01:1
Pod shape	full/constricted	 all full	882 full 299 constricted	2.95:1
Pod color	green/yellow	 all green	428 green 152 yellow	2.82:1
Flower color	violet/white	 all violet	705 violet 224 white	3.15:1
Flower position	axial/terminal	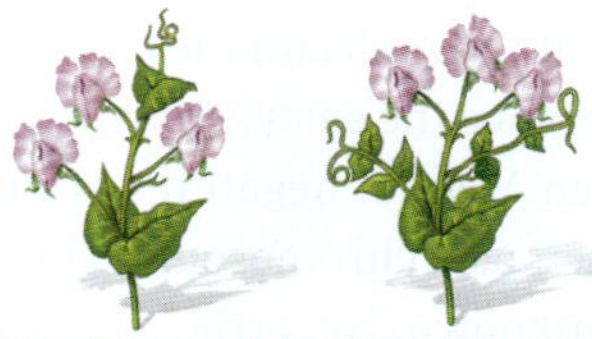 all axial	651 axial 207 terminal	3.14:1
Stem height	tall/dwarf	 all tall	787 tall 277 dwarf	2.84:1

FIGURE 3.1 Seven pairs of contrasting traits and the results of Mendel's seven monohybrid crosses of the garden pea (*Pisum sativum*). In each case, pollen derived from plants exhibiting one trait was used to fertilize the ova of plants exhibiting the other trait. In the F_1 generation, one of the two traits was exhibited by all plants. The contrasting trait reappeared in approximately 1/4 of the F_2 plants.

he experimented with the traits *tall* and *dwarf.* He selected six other contrasting pairs of traits involving seed shape and color, pod shape and color, and flower color and position. From local seed merchants, Mendel obtained true-breeding strains, those in which each trait appeared unchanged generation after generation in self-fertilizing plants.

There were several other reasons for Mendel's success. In addition to his choice of a suitable organism, he restricted his examination to one or very few pairs of contrasting traits in each experiment. He also kept accurate quantitative records, a necessity in genetic experiments. From the analysis of his data, Mendel derived certain postulates that have become the principles of transmission genetics.

3.2 The Monohybrid Cross Reveals How One Trait Is Transmitted from Generation to Generation

Mendel's simplest crosses involved only one pair of contrasting traits. Each such experiment is called a **monohybrid cross.** A monohybrid cross is made by mating true-breeding individuals from two parent strains, each exhibiting one of the two contrasting forms of the character under study. Initially, we examine the first generation of offspring of such a cross, and then we consider the offspring of **selfing,** that is, of self-fertilization of individuals from this first generation. The original parents constitute the $\mathbf{P_1}$, or **parental generation;** their offspring are the $\mathbf{F_1}$, or **first filial generation;** the individuals resulting from the selfed F_1 generation are the $\mathbf{F_2}$, or **second filial generation;** and so on.

The cross between true-breeding pea plants with tall stems and dwarf stems is representative of Mendel's monohybrid crosses. *Tall* and *dwarf* are contrasting traits of the character of stem height. Unless tall or dwarf plants are crossed together or with another strain, they will undergo self-fertilization and breed true, producing their respective traits generation after generation. However, when Mendel crossed tall plants with dwarf plants, the resulting F_1 generation consisted of only tall plants. When members of the F_1 generation were selfed, Mendel observed that 787 of 1064 F_2 plants were tall, while 277 of 1064 were dwarf. Note that in this cross (Figure 3.1), the dwarf trait disappeared in the F_1 generation, only to reappear in the F_2 generation.

Genetic data are usually expressed and analyzed as ratios. In this particular example, many identical P_1 crosses were made and many F_1 plants—all tall—were produced. As noted, of the 1064 F_2 offspring, 787 were tall and 277 were dwarf—a ratio of approximately 2.8:1.0, or about 3:1.

Mendel made similar crosses between pea plants exhibiting each of the other pairs of contrasting traits; the results of these crosses are shown in Figure 3.1. In every case, the outcome was similar to the tall/dwarf cross just described. For the character of interest, all F_1 offspring expressed the same trait exhibited by one of the parents, but in the F_2 offspring, an approximate ratio of 3:1 was obtained. That is, three-fourths looked like the F_1 plants, while one-fourth exhibited the contrasting trait, which had disappeared in the F_1 generation.

We note one further aspect of Mendel's monohybrid crosses. In each cross, the F_1 and F_2 patterns of inheritance were similar regardless of which P_1 plant served as the source of pollen (sperm) and which served as the source of the ovum (egg). The crosses could be made either way—pollination of dwarf plants by tall plants, or vice versa. Crosses made in both these ways are called **reciprocal crosses.** Therefore, the results of Mendel's monohybrid crosses were not sex dependent.

To explain these results, Mendel proposed the existence of particulate *unit factors* for each trait. He suggested that these factors serve as the basic units of heredity and are passed unchanged from generation to generation, determining various traits expressed by each individual plant. Using these general ideas, Mendel proceeded to hypothesize precisely how such factors could account for the results of the monohybrid crosses.

Mendel's First Three Postulates

Using the consistent pattern of results in the monohybrid crosses, Mendel derived the following three postulates, or principles, of inheritance.

1. UNIT FACTORS IN PAIRS
   Genetic characters are controlled by unit factors existing in pairs in individual organisms.
   In the monohybrid cross involving tall and dwarf stems, a specific **unit factor** exists for each trait. Each diploid individual receives one factor from each parent. Because the factors occur in pairs, three combinations are possible: two factors for tall stems, two factors for dwarf stems, or one of each factor. Every individual possesses one of these three combinations, which determines stem height.

2. DOMINANCE/RECESSIVENESS
   When two unlike unit factors responsible for a single character are present in a single individual, one unit factor is dominant to the other, which is said to be recessive.
   In each monohybrid cross, the trait expressed in the F_1 generation is controlled by the dominant unit factor. The trait not expressed is controlled by the recessive unit factor. The terms dominant and recessive are also used to designate traits. In this case, tall stems are said to be dominant over recessive dwarf stems.

3. SEGREGATION

During the formation of gametes, the paired unit factors separate, or segregate, randomly so that each gamete receives one or the other with equal likelihood.

If an individual contains a pair of like unit factors (e.g., both specific for tall), then all its gametes receive one of that same kind of unit factor (in this case, tall). If an individual contains unlike unit factors (e.g., one for tall and one for dwarf), then each gamete has a 50 percent probability of receiving either the tall or the dwarf unit factor.

These postulates provide a suitable explanation for the results of the monohybrid crosses. Let's use the tall/dwarf cross to illustrate. Mendel reasoned that P_1 tall plants contained identical paired unit factors, as did the P_1 dwarf plants. The gametes of tall plants all receive one tall unit factor as a result of segregation. Similarly, the gametes of dwarf plants all receive one dwarf unit factor. Following fertilization, all F_1 plants receive one unit factor from each parent—a tall factor from one and a dwarf factor from the other—reestablishing the paired relationship, but because tall is dominant to dwarf, all F_1 plants are tall.

When F_1 plants form gametes, the postulate of segregation demands that each gamete randomly receives either the tall *or* dwarf unit factor. Following random fertilization events during F_1 selfing, four F_2 combinations will result with equal frequency:

1. tall/tall
2. tall/dwarf
3. dwarf/tall
4. dwarf/dwarf

Combinations (1) and (4) will clearly result in tall and dwarf plants, respectively. According to the postulate of dominance/recessiveness, combinations (2) and (3) will both yield tall plants. Therefore, the F_2 is predicted to consist of 3/4 tall and 1/4 dwarf, or a ratio of 3:1. This is approximately what Mendel observed in his cross between tall and dwarf plants. A similar pattern was observed in each of the other monohybrid crosses (Figure 3.1).

ESSENTIAL POINT

Mendel's postulates help describe the basis for the inheritance of phenotypic traits. He hypothesized that unit factors exist in pairs and exhibit a dominant/recessive relationship in determining the expression of traits. He further postulated that unit factors segregate during gamete formation, such that each gamete receives one or the other factor, with equal probability. ■

Modern Genetic Terminology

To analyze the monohybrid cross and Mendel's first three postulates, we must first introduce several new terms as well as a symbol convention for the unit factors. Traits such as tall or dwarf are physical expressions of the information contained in unit factors. The physical expression of a trait is the **phenotype** of the individual. Mendel's unit factors represent units of inheritance called **genes** by modern geneticists. For any given character, such as plant height, the phenotype is determined by alternative forms of a single gene, called **alleles.** For example, the unit factors representing tall and dwarf are alleles determining the height of the pea plant.

Geneticists have several different systems for using symbols to represent genes. Later in the text (see Chapter 4), we will review a number of these conventions, but for now, we will adopt one to use consistently throughout this chapter. According to this convention, the first letter of the recessive trait symbolizes the character in question; in lowercase italic, it designates the allele for the recessive trait, and in uppercase italic, it designates the allele for the dominant trait. Thus for Mendel's pea plants, we use *d* for the *d*warf allele and *D* for the tall allele. When alleles are written in pairs to represent the two unit factors present in any individual (*DD*, *Dd*, or *dd*), the resulting symbol is called the **genotype.** The genotype designates the genetic makeup of an individual for the trait or traits it describes, whether the individual is haploid or diploid. By reading the genotype, we know the phenotype of the individual: *DD* and *Dd* are tall, and *dd* is dwarf. When both alleles are the same (*DD* or *dd*), the individual is **homozygous** for the trait, or a **homozygote;** when the alleles are different (*Dd*), we use the terms **heterozygous** and **heterozygote.** These symbols and terms are used in **Figure 3.2** to describe the monohybrid cross.

Punnett Squares

The genotypes and phenotypes resulting from combining gametes during fertilization can be easily visualized by constructing a diagram called a **Punnett square,** named after the person who first devised this approach, Reginald C. Punnett. **Figure 3.3** illustrates this method of analysis for our $F_1 \times F_1$ monohybrid cross. Each of the possible gametes is assigned a column or a row; the vertical columns represent those of the female parent, and the horizontal rows represent those of the male parent. After assigning the gametes to the rows and columns, we predict the new generation by entering the male and female gametic information into each box and thus producing every possible resulting genotype. By filling out the Punnett square, we are listing all possible random fertilization events. The genotypes and phenotypes of all potential offspring are ascertained by reading the combinations in the boxes.

The Punnett square method is particularly useful when you are first learning about genetics and how to solve genetics problems. Note the ease with which the 3:1 phenotypic ratio and the 1:2:1 genotypic ratio may be derived for the F_2 generation in Figure 3.3.

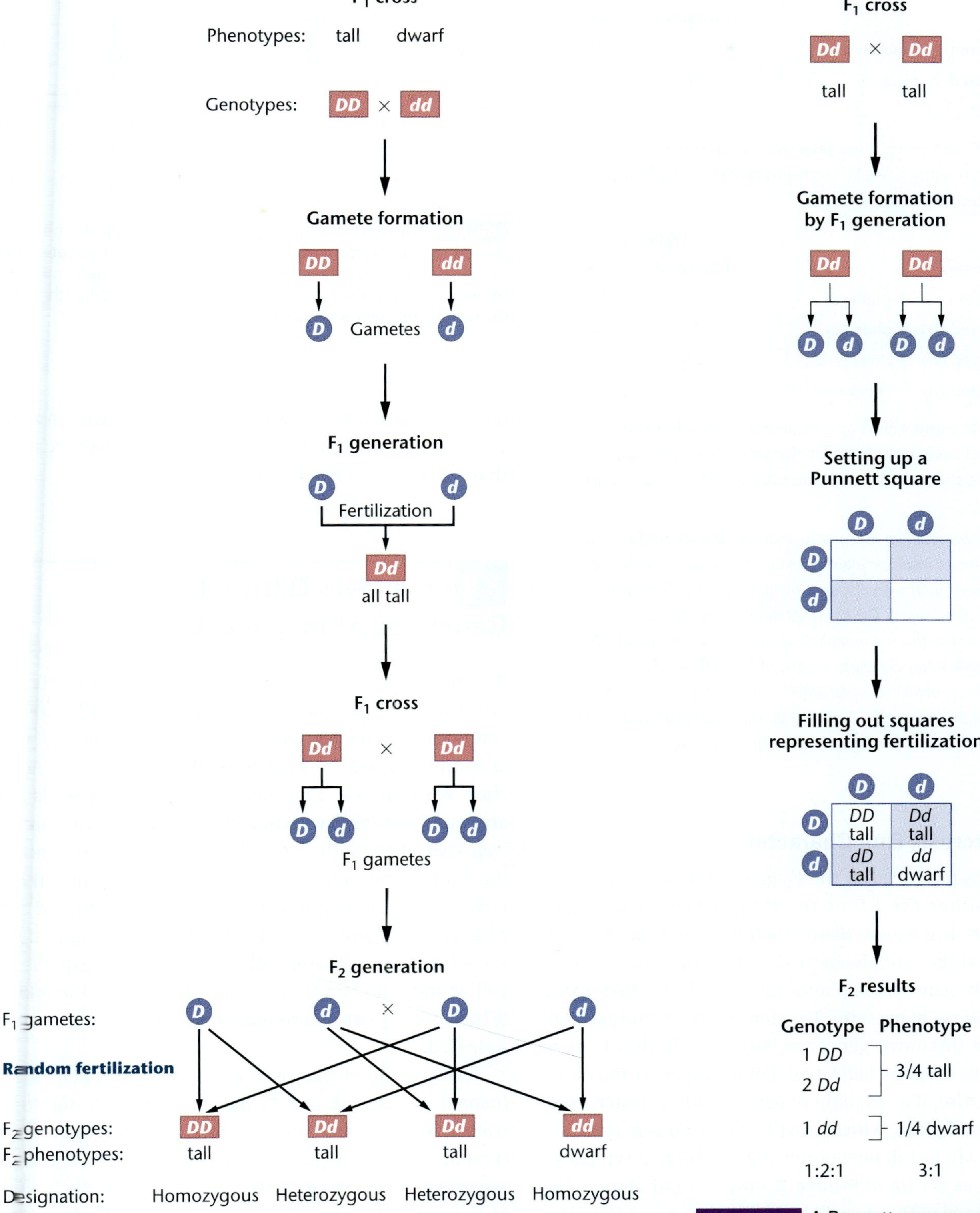

FIGURE 3.2 The monohybrid cross between tall (*D*) and dwarf (*d*) pea plants. Individuals are shown in rectangles, and gametes are shown in circles.

FIGURE 3.3 A Punnett square generating the F_2 ratio of the $F_1 \times F_1$ cross shown in Figure 3.2.

NOW SOLVE THIS

3.1 Pigeons may exhibit a checkered or plain color pattern. In a series of controlled matings, the following data were obtained.

P_1 Cross	F_1 Progeny Checkered	F_1 Progeny Plain
(a) checkered × checkered	36	0
(b) checkered × plain	38	0
(c) plain × plain	0	35

Then F_1 offspring were selectively mated with the following results. (The P_1 cross giving rise to each F_1 pigeon is indicated in parentheses.)

$F_1 \times F_1$ Crosses	F_2 Progeny Checkered	F_2 Progeny Plain
(d) checkered (a) × plain (c)	34	0
(e) checkered (b) × plain (c)	17	14
(f) checkered (b) × checkered (b)	28	9
(g) checkered (a) × checkered (b)	39	0

How are the checkered and plain patterns inherited? Select and assign symbols for the genes involved, and determine the genotypes of the parents and offspring in each cross.

■ **HINT:** *This problem asks you to analyze the data produced from several crosses involving pigeons and to determine the mode of inheritance and the genotypes of the parents and offspring in a number of instances. The key to its solution is to first determine whether or not this is a monohybrid cross. To do so, convert the data to ratios that are characteristic of Mendelian crosses. In the case of this problem, ask first whether any of the F_2 ratios match Mendel's 3:1 monohybrid ratio. If so, the second step is to determine which trait is dominant and which is recessive.*

The Testcross: One Character

Tall plants produced in the F_2 generation are predicted to have either the *DD* or the *Dd* genotype. You might ask if there is a way to distinguish the genotype. Mendel devised a rather simple method that is still used today to discover the genotype of plants and animals: the **testcross.** The organism expressing the dominant phenotype but having an unknown genotype is crossed with a known *homozygous recessive individual.* For example, as shown in **Figure 3.4(a)**, if a tall plant of genotype *DD* is testcrossed with a dwarf plant, which must have the *dd* genotype, all offspring will be tall phenotypically and *Dd* genotypically. However, as shown in **Figure 3.4(b)**, if a tall plant is *Dd* and is crossed with a dwarf plant (*dd*), then one-half of the offspring will be tall (*Dd*) and the other half will be dwarf (*dd*). Therefore, a 1:1 tall/dwarf ratio demonstrates the heterozygous nature of the tall plant of unknown genotype. The results of the testcross reinforced Mendel's conclusion that separate unit factors control traits.

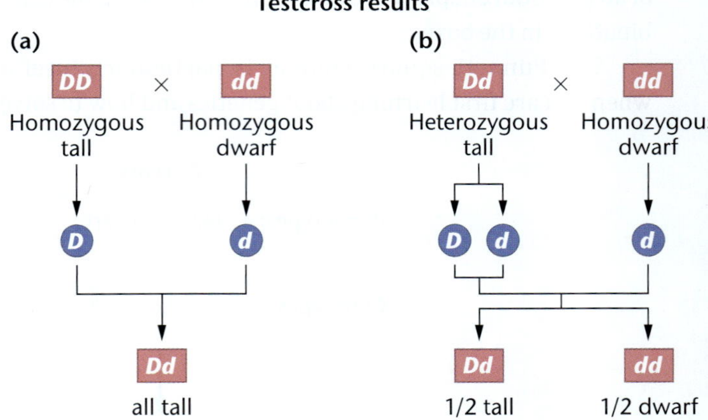

FIGURE 3.4 Testcross of a single character. In (a), the tall parent is homozygous, but in (b), the tall parent is heterozygous. The genotype of each tall P_1 plant can be determined by examining the offspring when each is crossed with the homozygous recessive dwarf plant.

3.3 Mendel's Dihybrid Cross Generated a Unique F_2 Ratio

As a natural extension of the monohybrid cross, Mendel also designed experiments in which he examined two characters simultaneously. Such a cross, involving two pairs of contrasting traits, is a **dihybrid cross,** or a *two-factor cross.* For example, if pea plants having yellow seeds that are round were bred with those having green seeds that are wrinkled, the results shown in **Figure 3.5** would occur: the F_1 offspring would all be yellow and round. It is therefore apparent that yellow is dominant to green and that round is dominant to wrinkled. When the F_1 individuals are selfed, approximately 9/16 of the F_2 plants express the yellow and round traits, 3/16 express yellow and wrinkled, 3/16 express green and round, and 1/16 express green and wrinkled.

A variation of this cross is also shown in Figure 3.5. Instead of crossing one P_1 parent with both dominant traits (yellow, round) to one with both recessive traits (green, wrinkled), plants with yellow, wrinkled seeds are crossed with those with green, round seeds. In spite of the change in the P_1 phenotypes, both the F_1 and F_2 results remain unchanged. Why this is so will become clear below.

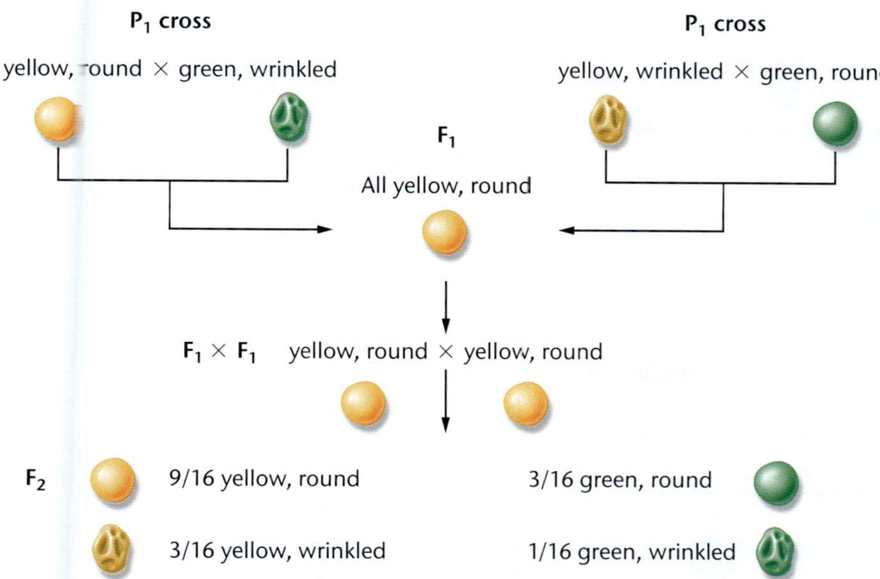

FIGURE 3.5 F_1 and F_2 results of Mendel's dihybrid crosses in which the plants on the top left with yellow, round seeds are crossed with plants having green, wrinkled seeds, and the plants on the top right with yellow, wrinkled seeds are crossed with plants having green, round seeds.

Mendel's Fourth Postulate: Independent Assortment

We can most easily understand the results of a dihybrid cross if we consider it theoretically as consisting of two monohybrid crosses conducted separately. Think of the two sets of traits as being inherited independently of each other; that is, the chance of any plant having yellow or green seeds is not at all influenced by the chance that this plant will have round or wrinkled seeds. Thus, because yellow is dominant to green, all F_1 plants in the first theoretical cross would have yellow seeds. In the second theoretical cross, all F_1 plants would have round seeds because round is dominant to wrinkled. When Mendel examined the F_1 plants of the dihybrid cross, all were yellow and round, as our theoretical crosses predict.

The predicted F_2 results of the first cross are 3/4 yellow and 1/4 green. Similarly, the second cross would yield 3/4 round and 1/4 wrinkled. Figure 3.5 shows that in the dihybric cross, 12/16 F_2 plants are yellow, while 4/16 are green, exhibiting the expected 3:1 (3/4:1/4) ratio. Similarly, 12/16 of all F_2 plants have round seeds, while 4/16 have wrinkled seeds, again revealing the 3:1 ratio.

These numbers demonstrate that the two pairs of contrasting traits are inherited independently, so we can predict the frequencies of all possible F_2 phenotypes by applying the **product law** of probabilities: *the probability of two or more independent events occurring simultaneously is equal to the product of their individual probabilities.* For example, the probability of an F_2 plant having yellow and round seeds is (3/4)(3/4), or 9/16, because 3/4 of all F_2 plants should be yellow and 3/4 of all F_2 plants should be round.

In a like manner, the probabilities of the other three F_2 phenotypes can be calculated: yellow (3/4) and wrinkled (1/4) are predicted to be present together 3/16 of the time; green (1/4) and round (3/4) are predicted 3/16 of the time; and green (1/4) and wrinkled (1/4) are predicted 1/16 of the time. These calculations are shown in **Figure 3.6**.

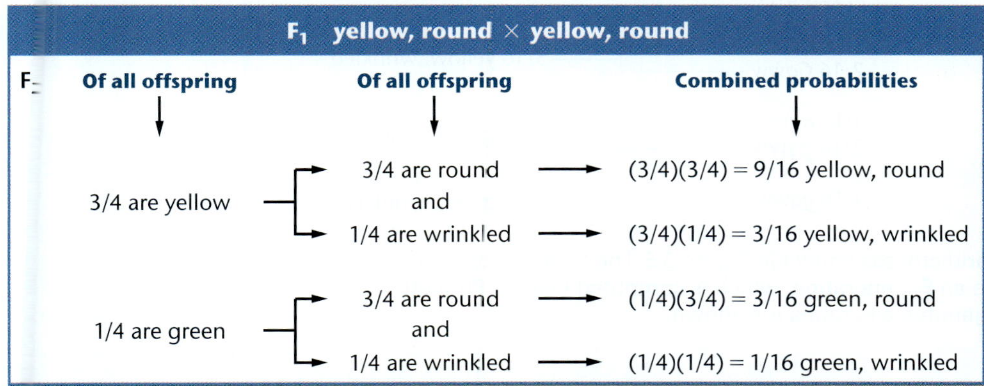

FIGURE 3.6 Computation of the combined probabilities of each F_2 phenotype for two independently inherited characters. The probability of each plant being yellow or green is independent of the probability of it bearing round or wrinkled seeds.

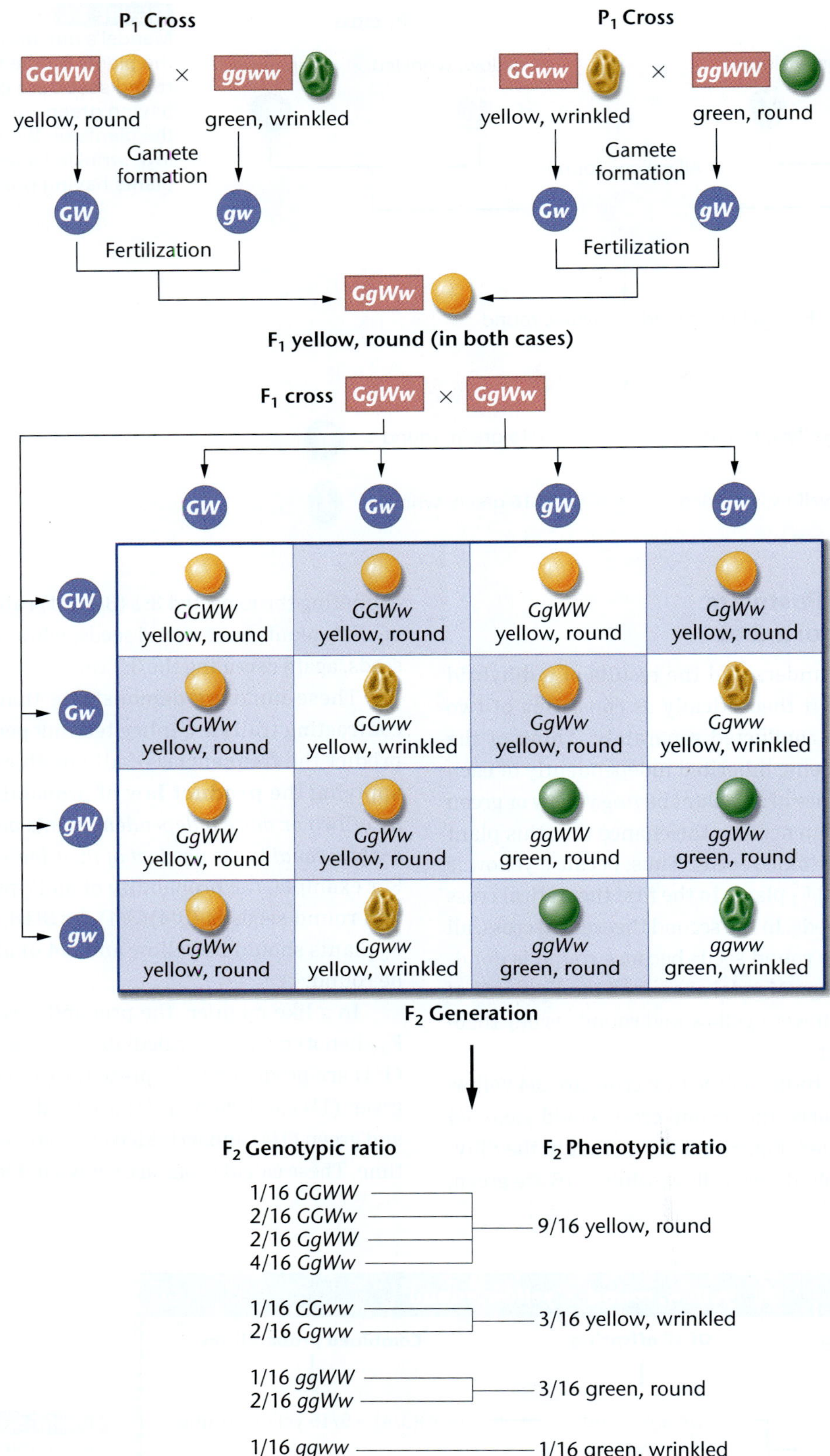

FIGURE 3.7 Analysis of the dihybrid crosses shown in Figure 3.5. The F_1 heterozygous plants are self-fertilized to produce an F_2 generation, which is computed using a Punnett square. Both the phenotypic and genotypic F_2 ratios are shown.

It is now apparent why the F_1 and F_2 results are identical whether the initial cross is yellow, round plants bred with green, wrinkled plants, or whether yellow, wrinkled plants are bred with green, round plants. In both crosses, the F_1 genotype of all offspring is identical. As a result, the F_2 generation is also identical in both crosses.

On the basis of similar results in numerous dihybrid crosses, Mendel proposed a fourth postulate:

4. INDEPENDENT ASSORTMENT

During gamete formation, segregating pairs of unit factors assort independently of each other.

This postulate stipulates that segregation of any pair of unit factors occurs independently of all others. As a result of random segregation, each gamete receives one member of every pair of unit factors. For one pair, whichever unit factor is received does not influence the outcome of segregation of any other pair. Thus, according to the postulate of independent assortment, all possible combinations of gametes should be formed in equal frequency.

The Punnett square in **Figure 3.7** shows how independent assortment works in the formation of the F_2 generation. Examine the formation of gametes by the F_1 plants; segregation prescribes that every gamete receives either a *G* or *g* allele and a *W* or *w* allele. Independent assortment stipulates that all four combinations (*GW*, *Gw*, *gW*, and *gw*) will be formed with equal probabilities.

In every $F_1 \times F_1$ fertilization event, each zygote has an equal probability of receiving one of the four combinations from each parent. If many offspring are produced, 9/16 have yellow, round seeds, 3/16 have yellow, wrinkled seeds, 3/16 have green, round seeds, and 1/16 have green, wrinkled seeds, yielding what is designated as **Mendel's 9:3:3:1 dihybrid ratio.** This is an ideal ratio based on probability events involving segregation, independent assortment, and random fertilization. Because of deviation due strictly to chance, particularly if small numbers of offspring are produced, actual results are highly unlikely to match the ideal ratio.

ESSENTIAL POINT

Mendel's postulate of independent assortment states that each pair of unit factors segregates independently of other such pairs. As a result, all possible combinations of gametes are formed with equal probability. ■

NOW SOLVE THIS

3.2 Considering the Mendelian traits round versus wrinkled and yellow versus green, consider the crosses below and determine the genotypes of the parental plants by analyzing the phenotypes of their offspring.

Parental Plants	Offspring
(a) round, yellow × round, yellow	3/4 round, yellow 1/4 wrinkled, yellow
(b) wrinkled, yellow × round, yellow	6/16 wrinkled, yellow 2/16 wrinkled, green 6/16 round, yellow 2/16 round, green
(c) round, yellow × round, yellow	9/16 round, yellow 3/16 round, green 3/16 wrinkled, yellow 1/16 wrinkled, green
(d) round, yellow × wrinkled, green	1/4 round, yellow 1/4 round, green 1/4 wrinkled, yellow 1/4 wrinkled, green

■ **HINT:** *This problem involves a series of Mendelian dihybrid crosses where you are asked to determine the genotypes of the parents in a number of instances. The key to its solution is to write down everything that you know for certain. This reduces the problem to its bare essentials, clarifying what you need to determine. For example, the wrinkled, yellow plant in case (b) must be homozygous for the recessive wrinkled alleles and bear at least one dominant allele for the yellow trait. Having established this, you need only determine the remaining allele for seed color.*

3.4 The Trihybrid Cross Demonstrates That Mendel's Principles Apply to Inheritance of Multiple Traits

Thus far, we have considered inheritance of up to two pairs of contrasting traits. Mendel demonstrated that the processes of segregation and independent assortment also apply to three pairs of contrasting traits, in what is called a **trihybrid cross,** or *three-factor cross.*

Although a trihybrid cross is somewhat more complex than a dihybrid cross, its results are easily calculated if the principles of segregation and independent assortment are followed. For example, consider the cross shown in **Figure 3.8** where the allele pairs of theoretical contrasting traits are represented by the symbols *A, a, B, b, C,* and *c.* In the cross between *AABBCC* and *aabbcc* individuals, all F_1 individuals are heterozygous for all three gene pairs. Their genotype, *AaBbCc*, results in the phenotypic expression of the dominant *A, B,* and *C* traits.

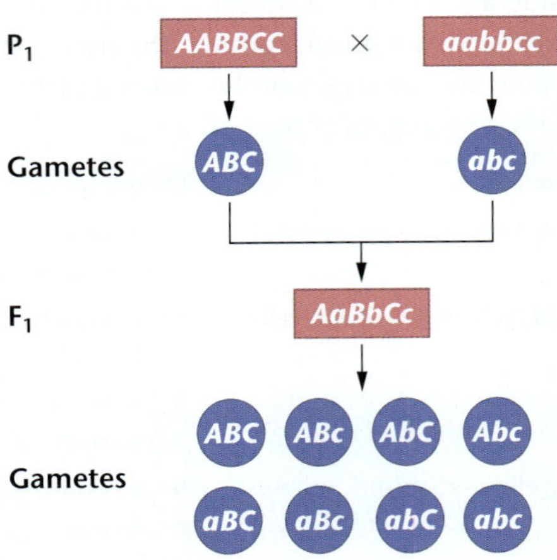

FIGURE 3.8 Formation of P_1 and F_1 gametes in a trihybrid cross.

Generation of F_2 trihybrid phenotypes

A or *a*	*B* or *b*	*C* or *c*	Combined proportion
3/4 *A*	3/4 *B*	3/4 *C* →	(3/4)(3/4)(3/4) *ABC* = 27/64 *ABC*
		1/4 *c* →	(3/4)(3/4)(1/4) *ABc* = 9/64 *ABc*
	1/4 *b*	3/4 *C* →	(3/4)(1/4)(3/4) *AbC* = 9/64 *AbC*
		1/4 *c* →	(3/4)(1/4)(1/4) *Abc* = 3/64 *Abc*
1/4 *a*	3/4 *B*	3/4 *C* →	(1/4)(3/4)(3/4) *aBC* = 9/64 *aBC*
		1/4 *c* →	(1/4)(3/4)(1/4) *aBc* = 3/64 *aBc*
	1/4 *b*	3/4 *C* →	(1/4)(1/4)(3/4) *abC* = 3/64 *abC*
		1/4 *c* →	(1/4)(1/4)(1/4) *abc* = 1/64 *abc*

FIGURE 3.9 Generation of the F_2 trihybrid phenotypic ratio using the forked-line method. This method is based on the expected probability of occurrence of each phenotype.

When F_1 individuals serve as parents, each produces eight different gametes in equal frequencies. At this point, we could construct a Punnett square with 64 separate boxes and read out the phenotypes—but such a method is cumbersome in a cross involving so many factors. Therefore, another method has been devised to calculate the predicted ratio.

The Forked-Line Method, or Branch Diagram

It is much less difficult to consider each contrasting pair of traits separately and then to combine these results by using the **forked-line method,** first shown in Figure 3.6. This method, also called a **branch diagram,** relies on the simple application of the laws of probability established for the dihybrid cross. Each gene pair is assumed to behave independently during gamete formation.

When the monohybrid cross *AA* × *aa* is made, we know that:

1. All F_1 individuals have the genotype *Aa* and express the phenotype represented by the *A* allele, which is called the *A* phenotype in the discussion that follows.
2. The F_2 generation consists of individuals with either the *A* phenotype or the *a* phenotype in the ratio of 3:1.

The same generalizations can be made for the *BB* × *bb* and *CC* × *cc* crosses. Thus, in the F_2 generation, 3/4 of all organisms will express phenotype *A*, 3/4 will express *B*, and 3/4 will express *C*. Similarly, 1/4 of all organisms will express *a*, 1/4 will express *b*, and 1/4 will express *c*. The proportions of organisms that express each phenotypic combination can be predicted by assuming that fertilization, following the independent assortment of these three gene pairs during gamete formation, is a random process. We apply the product law of probabilities once again. **Figure 3.9** uses the forked-line method to calculate the phenotypic proportions of the F_2 generation. They fall into the trihybrid ratio of 27:9:9:9:3:3:3:1. The same method can be used to solve crosses involving any number of gene pairs, *provided that all gene pairs assort independently from each other.* We shall see later that gene pairs do not always assort with complete independence. However, it appeared to be true for all of Mendel's characters.

ESSENTIAL POINT

The forked-line method is less complex than, but just as accurate as, the Punnett square in predicting the probabilities of phenotypes or genotypes from crosses involving two or more gene pairs. ■

NOW SOLVE THIS

3.3 Using the forked-line, or branch diagram, method, determine the genotypic and phenotypic ratios of these trihybrid crosses: (a) *AaBbCc* × *AaBBCC*, (b) *AaBBCc* × *aaBBCc*, and (c) *AaBbCc* × *AaBbCc*.

■ **HINT:** *This problem asks you to use the forked-line method to determine the outcome of a number of trihybrid crosses. The key to its solution is to realize that in using the forked-line method, you must consider each gene pair separately. For example, in this problem, first predict the outcome of each cross for the A/a genes, then for the B/b genes, and finally, for the C/c genes. Then you are prepared to pursue the outcome of each cross.*

3.5 Mendel's Work Was Rediscovered in the Early Twentieth Century

Mendel published his work in 1866. While his findings were often cited and discussed, their significance went unappreciated for about 35 years. Then, in the latter part of the nineteenth century, a remarkable observation set the scene for the recognition of Mendel's work: Walther Flemming's discovery of chromosomes in the nuclei of salamander cells. In 1879, Flemming described the behavior of these thread-like structures during cell division. As a result of his findings and the work of many other cytologists, the presence of discrete units within the nucleus soon became an integral part of scientists' ideas about inheritance.

In the early twentieth century, hybridization experiments similar to Mendel's were performed independently by three botanists, Hugo de Vries, Carl Correns, and Erich Tschermak. De Vries's work demonstrated the principle of segregation in several plant species. Apparently, he searched the existing literature and found that Mendel's work had anticipated his own conclusions! Correns and Tschermak also reached conclusions similar to those of Mendel.

About the same time, two cytologists, Walter Sutton and Theodor Boveri, independently published papers linking their discoveries of the behavior of chromosomes during meiosis to the Mendelian principles of segregation and independent assortment. They pointed out that the separation of chromosomes during meiosis could serve as the cytological basis of these two postulates. Although they thought that Mendel's unit factors were probably chromosomes rather than genes on chromosomes, their findings reestablished the importance of Mendel's work and led to many ensuing genetic investigations. Sutton and Boveri are credited with initiating the **chromosome theory of inheritance,** the idea that the genetic material in living organisms is contained in chromosomes, which was developed during the next two decades. As we will see in subsequent chapters, work by Thomas H. Morgan, Alfred H. Sturtevant, Calvin Bridges, and others established beyond a reasonable doubt that Sutton's and Boveri's hypothesis was correct.

ESSENTIAL POINT

The discovery of chromosomes in the late 1800s, along with subsequent studies of their behavior during meiosis, led to the rebirth of Mendel's work, linking the behavior of his unit factors to that of chromosomes during meiosis. ■

Unit Factors, Genes, and Homologous Chromosomes

Because the correlation between Sutton's and Boveri's observations and Mendelian postulates serves as the foundation for the modern description of transmission genetics, we will examine this correlation in some depth before moving on to other topics.

As we know, each species possesses a specific number of chromosomes in each somatic cell nucleus. For diploid organisms, this number is called the **diploid number (2*n*)** and is characteristic of that species. During the formation of gametes (meiosis), the number is precisely halved (*n*), and when two gametes combine during fertilization, the diploid number is reestablished. During meiosis, however, the chromosome number is not reduced in a random manner. It was apparent to early cytologists that the diploid number of chromosomes is composed of homologous pairs identifiable by their morphological appearance and behavior. The gametes contain one member of each pair—thus the chromosome complement of a gamete is quite specific, and the number of chromosomes in each gamete is equal to the haploid number.

With this basic information, we can correlate the behavior of unit factors and chromosomes and genes. Unit factors are really genes located on homologous pairs of chromosomes. Members of each pair of homologs separate, or segregate, during gamete formation.

To illustrate the principle of independent assortment, it is important to distinguish between members of any given homologous pair of chromosomes. One member of each pair is derived from the **maternal parent,** whereas the other comes from the **paternal parent.** Following independent segregation of each pair of homologs, each gamete receives one member from each pair of chromosomes. All possible combinations are formed with equal probability.

Observations of the phenotypic diversity of living organisms make it logical to assume that there are many more genes than chromosomes. Therefore, each homolog must carry genetic information for more than one trait. The currently accepted concept is that a chromosome is composed of a large number of linearly ordered, information-containing genes. Mendel's paired unit factors (which determine tall or dwarf stems, for example) actually constitute a pair of genes located on one pair of homologous chromosomes. The location on a given chromosome where any particular gene occurs is called its **locus** (pl. loci). The different alleles of a given gene (for example, *G* and *g*) contain slightly different genetic information (green or yellow) that determines the same character (seed color in this case). Although we have examined only genes with two alternative alleles, most genes have more than two allelic forms. We conclude this section by reviewing the criteria necessary to classify two chromosomes as a homologous pair:

1. During mitosis and meiosis, when chromosomes are visible in their characteristic shapes, both members of a homologous pair are the same size and exhibit identical centromere locations. The sex chromosomes (e.g., the X and the Y chromosomes in mammals) are an exception.
2. During early stages of meiosis, homologous chromosomes form pairs, or synapse.
3. Although it is not generally visible under the microscope, homologs contain the identical linear order of gene loci.

EVOLVING CONCEPT OF THE GENE

Based on the pioneering work of Gregor Mendel, the gene was viewed as a heritable unit factor that determines the expression of an observable trait, or phenotype. ■

3.6 Independent Assortment Leads to Extensive Genetic Variation

One consequence of independent assortment is the production by an individual of genetically dissimilar gametes. Genetic variation results because the two members of any homologous pair of chromosomes are rarely, if ever, genetically identical. As the maternal and paternal members of all pairs are distributed to gametes through independent assortment, all possible chromosome combinations are produced, leading to extensive genetic diversity.

We have seen that the number of possible gametes, each with different chromosome compositions, is 2^n, where n equals the haploid number. Thus, if a species has a haploid number of 4, then 2^4, or 16, different gamete combinations can be formed as a result of independent assortment. Although this number is not high, consider the human species, where $n = 23$. When 2^{23} is calculated, we find that in excess of 8×10^6, or over 8 million, different types of gametes are possible through independent assortment. Because fertilization represents an event involving only one of approximately 8×10^6 possible gametes from each of two parents, each offspring represents only one of $(8 \times 10^6)^2$ or one of only 64×10^{12} potential genetic combinations. Given that this probability is less than one in one trillion, it is no wonder that, except for identical twins, each member of the human species exhibits a distinctive set of traits—this number of combinations of chromosomes is far greater than the number of humans who have ever lived on Earth! Genetic variation resulting from independent assortment has been extremely important to the process of evolution in all sexually reproducing organisms.

3.7 Laws of Probability Help to Explain Genetic Events

Recall that genetic ratios—for example, 3/4 tall:1/4 dwarf—are most properly thought of as probabilities. These values predict the outcome of each fertilization event, such that the probability of each zygote having the genetic potential for becoming tall is 3/4, whereas the potential for its being a dwarf is 1/4. Probabilities range from 0.0, where an event *is certain not to occur,* to 1.0, where an event *is certain to occur.* In this section, we consider the relation of probability to genetics. When two or more events with known probabilities occur independently but at the same time, we can calculate the probability of their possible outcomes occurring together. This is accomplished by applying the **product law,** which states that *the probability of two or more independent events occurring simultaneously is equal to the product of their individual probabilities* (see Section 3.3). Two or more events are independent of one another if the outcome of each one does not affect the outcome of any of the others under consideration.

To illustrate the product law, consider the possible results if you toss a penny (P) and a nickel (N) at the same time and examine all combinations of heads (H) and tails (T) that can occur. There are four possible outcomes:

$$(P_H{:}N_H) = (1/2)(1/2) = 1/4$$
$$(P_T{:}N_H) = (1/2)(1/2) = 1/4$$
$$(P_H{:}N_T) = (1/2)(1/2) = 1/4$$
$$(P_T{:}N_T) = (1/2)(1/2) = 1/4$$

The probability of obtaining a head or a tail in the toss of either coin is 1/2 and is unrelated to the outcome for the other coin. Thus, all four possible combinations are predicted to occur with equal probability.

If we want to calculate the probability when the possible outcomes of two events are independent of one another but can be accomplished in more than one way, we can apply the **sum law.** For example, what is the probability of tossing our penny and nickel and obtaining one head and one tail? In such a case, we do not care whether it is the penny or the nickel that comes up heads, provided that the other coin has the alternative outcome. As we saw above, there are two ways in which the desired outcome can be accomplished, each with a probability of 1/4. The sum law states that *the probability of obtaining any single outcome, where that outcome* *can be achieved by two or more events, is equal to the sum of the individual probabilities of all such events.* Thus, according to the sum law, the overall probability in our example is equal to

$$(1/4) + (1/4) = 1/2$$

One-half of all two-coin tosses are predicted to yield the desired outcome.

These simple probability laws will be useful throughout our discussions of transmission genetics and for solving genetics problems. In fact, we already applied the product law when we used the forked-line method to calculate the phenotypic results of Mendel's dihybrid and trihybrid crosses. When we wish to know the results of a cross, we need only calculate the probability of each possible outcome. The results of this calculation then allow us to predict the proportion of offspring expressing each phenotype or each genotype.

An important point to remember when you deal with probability is that predictions of possible outcomes are based on large sample sizes. If we predict that 9/16 of the offspring of a dihybrid cross will express both dominant traits, it is very unlikely that, in a small sample, exactly 9 of every 16 will express this phenotype. Instead, our prediction is that, of a large number of offspring, approximately 9/16 will do so. The deviation from the predicted ratio in smaller sample sizes is attributed to chance, a subject we examine in our discussion of statistics in Section 3.8. As you shall see, the impact of deviation due strictly to chance diminishes as the sample size increases.

ESSENTIAL POINT

Since genetic ratios are expressed as probabilities, deriving outcomes of genetic crosses requires an understanding of the laws of probability. ■

3.8 Chi-Square Analysis Evaluates the Influence of Chance on Genetic Data

Mendel's 3:1 monohybrid and 9:3:3:1 dihybrid ratios are hypothetical predictions based on the following assumptions: (1) each allele is dominant or recessive, (2) segregation is unimpeded, (3) independent assortment occurs, and (4) fertilization is random. The final two assumptions are influenced by chance events and therefore are subject to random fluctuation. This concept of **chance deviation** is most easily illustrated by tossing a single coin numerous times and recording the number of heads and tails observed. In each toss, there is a probability of 1/2 that a head will occur and a probability of 1/2 that a tail will occur. Therefore, the expected ratio of many tosses is 1/2:1/2, or 1:1. If a coin is tossed 1000 times, usually *about* 500 heads and 500 tails will be observed. Any reasonable fluctuation from this hypothetical ratio (e.g., 486 heads and 514 tails) is attributed to chance.

As the total number of tosses is reduced, the impact of chance deviation increases. For example, if a coin is tossed only four times, you would not be too surprised if all four tosses resulted in only heads or only tails. For 1000 tosses, however, 1000 heads or 1000 tails would be most unexpected. In fact, you might believe that such a result would be impossible. Actually, all heads or all tails in 1000 tosses can be predicted to occur with a probability of $(1/2)^{1000}$. Since $(1/2)^{20}$ is less than one in a million times, an event occurring with a probability as small as $(1/2)^{1000}$ is virtually impossible. Two major points to keep in mind when predicting or analyzing genetic outcomes are:

1. The outcomes of independent assortment and fertilization, like coin tossing, are subject to random fluctuations from their predicted occurrences as a result of chance deviation.
2. As the sample size increases, the average deviation from the expected results decreases. Therefore, a larger sample size diminishes the impact of chance deviation on the final outcome.

Chi-Square Calculations and the Null Hypothesis

In genetics, being able to evaluate observed deviation is a crucial skill. When we assume that data will fit a given ratio such as 1:1, 3:1, or 9:3:3:1, we establish what is called the **null hypothesis (H_0).** It is so named because the hypothesis assumes that there is *no real difference* between the *measured values* (or ratio) and the *predicted values* (or ratio). Any apparent difference can be attributed purely to chance. The validity of the null hypothesis for a given set of data is measured using statistical analysis. Depending on the results of this analysis, the null hypothesis may either (1) *be rejected* or (2) *fail to be rejected*. If it is rejected, the observed deviation from the expected result is judged not to be attributable to chance alone. In this case, the null hypothesis and the underlying assumptions leading to it must be reexamined. If the null hypothesis fails to be rejected, any observed deviations are attributed to chance.

One of the simplest statistical tests for assessing the goodness of fit of the null hypothesis is **chi-square (χ^2) analysis.** This test takes into account the observed deviation in each component of a ratio (from what was expected) as well as the sample size and reduces them to a single numerical value. The value for χ^2 is then used to estimate how frequently the observed deviation can be expected to occur strictly as a result of chance. The formula used in chi-square analysis is

$$\chi^2 = \Sigma \frac{(o - e)^2}{e}$$

where o is the observed value for a given category, e is the expected value for that category, and Σ (the Greek letter sigma) represents the sum of the calculated values for each category in the ratio. Because $(o - e)$ is the deviation (d) in each case, the equation reduces to

$$\chi^2 = \Sigma \frac{d^2}{e}$$

Table 3.1(a) shows the steps in the χ^2 calculation for the F_2 results of a hypothetical monohybrid cross. To analyze the data obtained from this cross, work from left to right across the table, verifying the calculations as appropriate. Note that regardless of whether the deviation d is positive or negative, d^2 always becomes positive after the number is squared. In **Table 3.1(b)** F_2 results of a hypothetical dihybrid cross are analyzed. Make sure that you understand how each number was calculated in this example.

The final step in chi-square analysis is to interpret the χ^2 value. To do so, you must initially determine a value called the **degrees of freedom (*df*)**, which is equal to $n - 1$, where n is the number of different categories into which the data are divided, in other words, the number of possible outcomes. For the 3:1 ratio, $n = 2$, so $df = 1$. For the 9:3:3:1 ratio, $n = 4$ and $df = 3$. Degrees of freedom must be taken into account because the greater the number of categories, the more deviation is expected as a result of chance.

Once you have determined the degrees of freedom, you can interpret the χ^2 value in terms of a corresponding **probability value (*p*)**. Since this calculation is complex, we usually take the p value from a standard table or graph. **Figure 3.10** shows a wide range of χ^2 values and the corresponding p values for various degrees of freedom in both a graph and a table. Let's use the graph to explain how to determine the p value. The caption for Figure 3.10(b) explains how to use the table.

To determine p using the graph, execute the following steps:

1. Locate the χ^2 value on the abscissa (the horizontal axis, or x-axis).
2. Draw a vertical line from this point up to the line on the graph representing the appropriate df.
3. From there, extend a horizontal line to the left until it intersects the ordinate (the vertical axis, or y-axis).
4. Estimate, by interpolation, the corresponding p value.

We used these steps for the monohybrid cross in Table 3.1(a) to estimate the p value of 0.48, as shown in Figure 3.10(a). Now try this method to see if you can determine the p value for the dihybrid cross [Table 3.1(b)]. Since the χ^2 value is 4.16 and $df = 3$, an approximate p value is 0.26. Checking this result in the table confirms that p values for both the monohybrid and dihybrid crosses are between 0.20 and 0.50.

Interpreting Probability Values

So far, we have been concerned with calculating χ^2 values and determining the corresponding p values. These steps bring us to the most important aspect of chi-square analysis: understanding the meaning of the p value. It is simplest

TABLE 3.1 Chi-Square Analysis

(a) Monohybrid Cross Expected Ratio	Observed (o)	Expected (e)	Deviation ($o - e = d$)	Deviation^2	d^2/e
3/4	740	3/4(1000) = 750	740 − 750 = −10	$(-10)^2 = 100$	100/750 = 0.13
1/4	260	1/4(1000) = 250	260 − 250 = +10	$(+10)^2 = 100$	100/250 = 0.40
	Total = 1000				$\chi^2 = 0.53$
					$p = 0.48$

(b) Dihybrid Cross Expected Ratio	Observed (o)	Expected (e)	Deviation ($o - e = d$)	Deviation^2	d^2/e
9/16	587	567	+20	400	0.71
3/16	197	189	+8	64	0.34
3/16	168	189	−21	441	2.33
1/16	56	63	−7	49	0.78
	Total = 1008				$\chi^2 = 4.16$
					$p = 0.26$

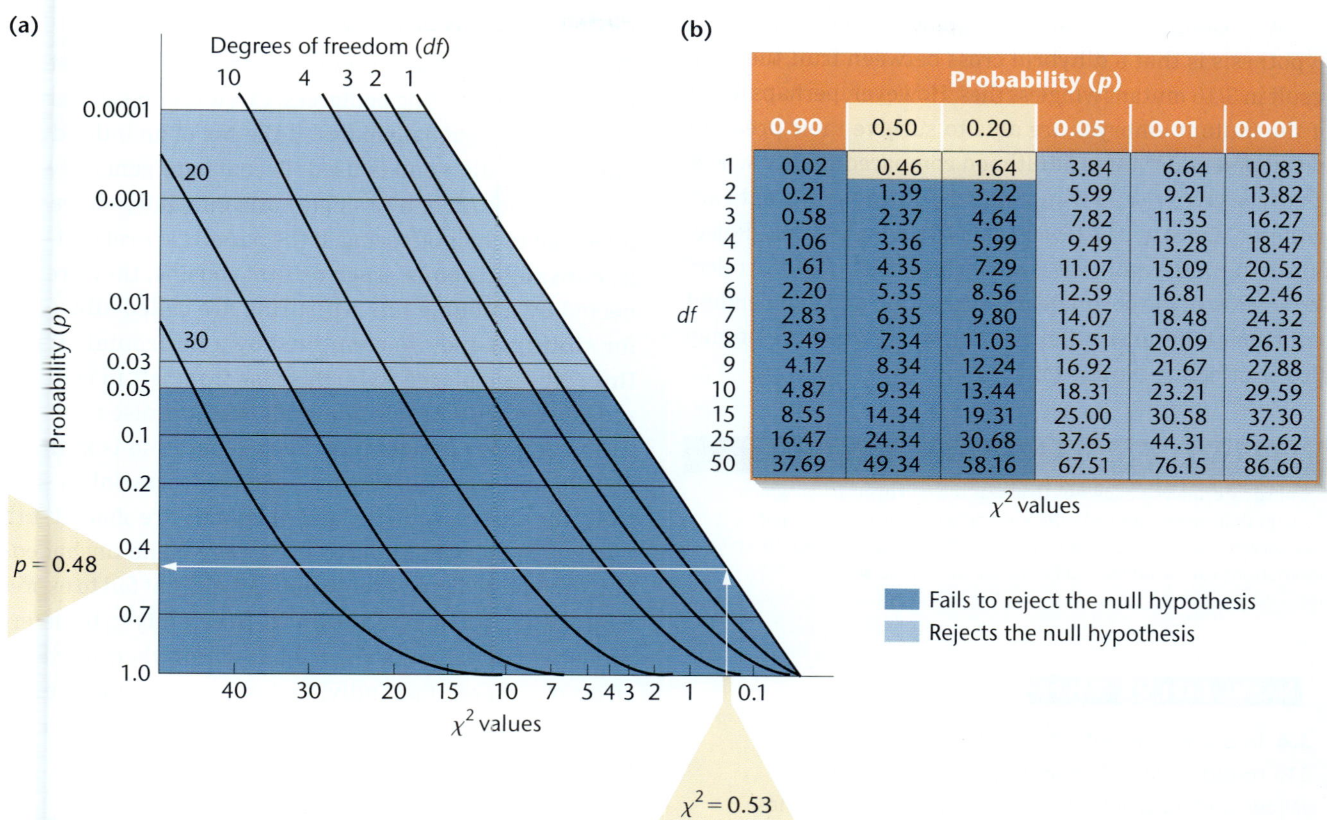

df	Probability (p) 0.90	0.50	0.20	0.05	0.01	0.001
1	0.02	0.46	1.64	3.84	6.64	10.83
2	0.21	1.39	3.22	5.99	9.21	13.82
3	0.58	2.37	4.64	7.82	11.35	16.27
4	1.06	3.36	5.99	9.49	13.28	18.47
5	1.61	4.35	7.29	11.07	15.09	20.52
6	2.20	5.35	8.56	12.59	16.81	22.46
7	2.83	6.35	9.80	14.07	18.48	24.32
8	3.49	7.34	11.03	15.51	20.09	26.13
9	4.17	8.34	12.24	16.92	21.67	27.88
10	4.87	9.34	13.44	18.31	23.21	29.59
15	8.55	14.34	19.31	25.00	30.58	37.30
25	16.47	24.34	30.68	37.65	44.31	52.62
50	37.69	49.34	58.16	67.51	76.15	86.60

χ^2 values

FIGURE 3.10 (a) Graph for converting χ^2 values to p values. (b) Table of χ^2 values for selected values of df and p. χ^2 values that lead to a p value of 0.05 or greater (darker blue areas) justify failure to reject the null hypothesis. Values leading to a p value of less than 0.05 (lighter blue areas) justify rejecting the null hypothesis. For example, the table in part (b) shows that for $\chi^2 = 0.53$ with 1 degree of freedom, the corresponding p value is between 0.20 and 0.50. The graph in (a) gives a more precise p value of 0.48 by interpolation. Thus, we fail to reject the null hypothesis.

to think of the *p* value as a percentage. Let's use the example of the dihybrid cross in Table 3.1(b) where $p = 0.26$, which can be thought of as 26 percent. In our example, the *p* value indicates that if we repeat the same experiment many times, 26 percent of the trials would be expected to exhibit chance deviation as great as or greater than that seen in the initial trial. Conversely, 74 percent of the repeats would show less deviation than initially observed as a result of chance. Thus, the *p* value reveals that a null hypothesis (concerning the 9:3:3:1 ratio, in this case) is never proved or disproved absolutely. Instead, a relative standard is set that we use to either *reject* or *fail to reject* the null hypothesis. This standard is most often a *p* value of 0.05. When applied to chi-square analysis, a *p* value less than 0.05 means that the observed deviation in the set of results will be obtained by chance alone less than 5 percent of the time. Such a *p* value indicates that the difference between the observed and predicted results is substantial and requires us to reject the null hypothesis.

On the other hand, *p* values of 0.05 or greater (0.05 to 1.0) indicate that the observed deviation will be obtained by chance alone 5 percent or more of the time. This conclusion allows us not to reject the null hypothesis (when we are using $p = 0.05$ as our standard). Thus, with its *p* value of 0.26, the null hypothesis that independent assortment accounts for the results fails to be rejected. Therefore, the observed deviation can be reasonably attributed to chance.

A final note is relevant here concerning the case where the null hypothesis is rejected, that is, where $p \leq 0.05$. Suppose we had tested a dataset to assess a possible 9:3:3:1 ratio, as in Table 3.1(b), but we rejected the null hypothesis based on our calculation. What are alternative interpretations of the data? Researchers will reassess the assumptions that underlie the null hypothesis. In our dyhibrid cross, we assumed that segregation operates faithfully for both gene pairs. We also assumed that fertilization is random and that the viability of all gametes is equal regardless of genotype—that is, all gametes are equally likely to participate in fertilization. Finally, we assumed that, following fertilization, all preadult stages and adult offspring are equally viable, regardless of their genotype. If any of these assumptions is incorrect, then the original hypothesis is not necessarily invalid.

An example will clarify this point. Suppose our null hypothesis is that a dihybrid cross between fruit flies will result in 3/16 mutant wingless flies. However, perhaps fewer of the mutant embryos are able to survive their preadult development or young adulthood compared to flies whose genotype gives rise to wings. As a result, when the data are gathered, there will be fewer than 3/16 wingless flies. Rejection of the null hypothesis is not in itself cause for us to reject the validity of the postulates of segregation and independent assortment, because other factors we are unaware of may also be affecting the outcome.

ESSENTIAL POINT

Chi-square analysis allows us to assess the null hypothesis, which states that there is no real difference between the expected and observed values. As such, it tests the probability of whether observed variations can be attributed to chance deviation. ■

NOW SOLVE THIS

3.4 In one of Mendel's dihybrid crosses, he observed 315 round, yellow; 108 round, green; 101 wrinkled, yellow; and 32 wrinkled, green F_2 plants. Analyze these data using the χ^2 test to see if

(a) they fit a 9:3:3:1 ratio.
(b) the round:wrinkled data fit a 3:1 ratio.
(c) the yellow:green data fit a 3:1 ratio.

■ **HINT:** *This problem asks you to apply χ^2 analysis to a set of data and to determine whether those data fit any of several ratios. The key to its solution is to first calculate χ^2 by initially determining the expected outcomes using the predicted ratios. Then follow a stepwise approach, determining the deviation in each case, and calculating d^2/e for each category. Once you have determined the χ^2 value, you must then determine and interpret the p value for each ratio.*

3.9 Pedigrees Reveal Patterns of Inheritance of Human Traits

We now explore how to determine the mode of inheritance of phenotypes in humans, where experimental matings are not made and where relatively few offspring are available for study. The traditional way to study inheritance has been to construct a family tree, indicating the presence or absence of the trait in question for each member of each generation. Such a family tree is called a **pedigree.** By analyzing a pedigree, we may be able to predict how the trait under study is inherited—for example, is it due to a dominant or recessive allele? When many pedigrees for the same trait are studied, we can often ascertain the mode of inheritance.

Pedigree Conventions

Figure 3.11 illustrates some of the conventions geneticists follow in constructing pedigrees. Circles represent females and squares designate males. If the sex of an individual is unknown, a diamond is used. Parents are generally connected to each other by a single horizontal line, and vertical lines lead to their offspring. If the parents are related—that is, **consanguineous**—such as first cousins, they are connected by a double line. Offspring are called **sibs** (short for **siblings**) and are connected by a horizontal **sibship line.** Sibs are placed in birth order from left to right and are labeled with Arabic numerals. Parents also receive an Arabic number designation. Each generation is indicated by a Roman numeral. When a pedigree traces only a single trait, the circles, squares, and diamonds are shaded if the phenotype being considered is expressed and unshaded if not. In some pedigrees, those individuals that fail to express a recessive trait but are known with certainty to be heterozygous carriers have a shaded dot within their unshaded circle or square. If an individual is deceased and the phenotype is unknown, a diagonal line is placed over the circle or square.

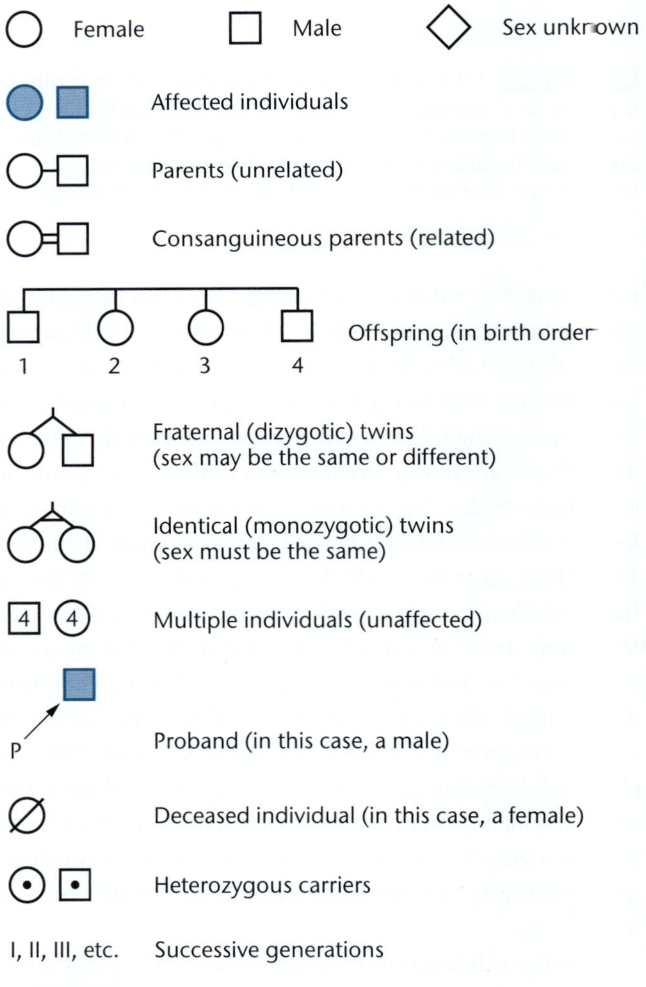

Female
Male
Sex unknown
Affected individuals
Parents (unrelated)
Consanguineous parents (related)
1 2 3 4 Offspring (in birth order)
Fraternal (dizygotic) twins
(sex may be the same or different)
Identical (monozygotic) twins
(sex must be the same)
4 4 Multiple individuals (unaffected)
P Proband (in this case, a male)
Deceased individual (in this case, a female)
Heterozygous carriers
I, II, III, etc. Successive generations

FIGURE 3.11 Conventions commonly encountered in human pedigrees.

Twins are indicated by diagonal lines stemming from a vertical line connected to the sibship line. For identical, or **monozygotic,** twins, the diagonal lines are linked by a horizontal line. Fraternal, or **dizygotic,** twins lack this connecting line. A number within one of the symbols represents that number of sibs of the same sex and of the same or unknown phenotypes. The individual whose phenotype first brought attention to the family is called the **proband** and is indicated by an arrow connected to the designation p. This term applies to either a male or a female.

Pedigree Analysis

In **Figure 3.12**, two pedigrees are shown. The first is a representative pedigree for a trait that demonstrates autosomal recessive inheritance, such as **albinism,** where synthesis of the pigment melanin in obstructed. The male parent of the first generation (I-1) is affected. Characteristic of a situation in which a parent has a rare recessive trait, the trait "disappears" in the offspring of the next generation. Assuming recessiveness, we might predict that the unaffected female parent (I-2) is a homozygous normal individual because none of the offspring show the disorder. Had she been heterozygous, one-half of the offspring would be expected to exhibit albinism, but none do. However, such a small sample (three offspring) prevents our knowing for certain.

Further evidence supports the prediction of a recessive trait. If albinism were inherited as a dominant trait, individual II-3 would have to express the disorder in order to pass it to his offspring (III-3 and III-4), but he does not. Inspection of the offspring constituting the third generation (row III) provides still further support for the hypothesis that albinism is a recessive trait. If it is, parents II-3 and II-4 are both heterozygous, and approximately one-fourth of their offspring should be affected. Two of the six offspring do show albinism. This deviation from the expected ratio is not unexpected in crosses with few offspring. Once we are confident that albinism is inherited as an autosomal recessive trait, we could portray the II-3 and II-4 individuals with a shaded dot within their larger square and circle. Finally, we can note that, characteristic of pedigrees for autosomal traits, both males and females are affected with equal probability. Later in the text (see Chapter 4), we will examine a pedigree representing a gene located on the sex-determining X chromosome. We will see certain patterns characteristic of the transmission of X-linked traits, such as that these traits are more prevalent in male offspring and are never passed from affected fathers to their sons.

The second pedigree illustrates the pattern of inheritance for a trait such as **Huntington disease,** which is caused by an autosomal dominant allele. The key to identifying a pedigree that reflects a dominant trait is that all affected offspring will have a parent that also expresses the trait. It is also possible, by chance, that none of the offspring will inherit the dominant allele. If so, the trait will cease to exist in future generations. Like recessive traits, provided

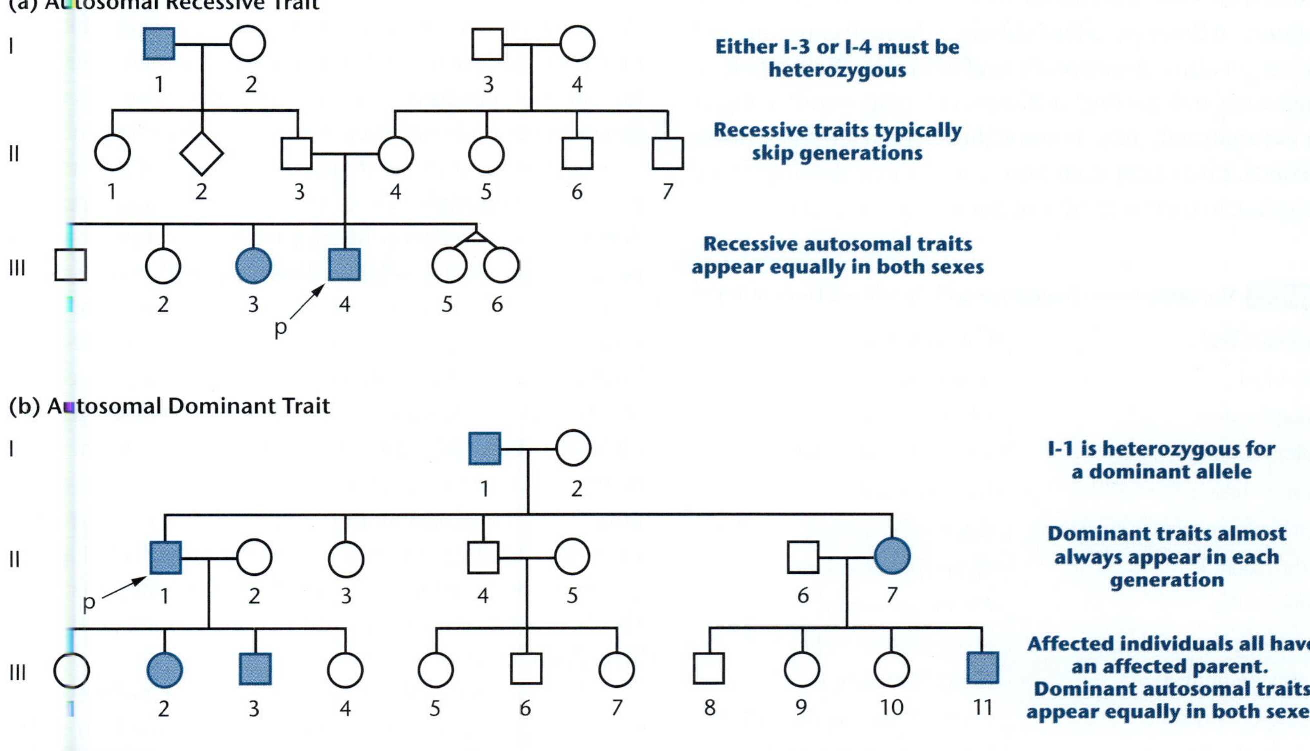

FIGURE 3.12 Representative pedigrees for two characteristics, one an autosomal recessive trait, and one an autosomal dominant trait, both followed through three generations.

that the gene is autosomal, both males and females are equally affected.

When a given autosomal dominant disease is rare within the population, and most are, then it is highly unlikely that affected individuals will inherit a copy of the mutant gene from both parents. Therefore, in most cases, affected individuals are heterozygous for the dominant allele. As a result, approximately one-half of the offspring inherit it. This is borne out in the second pedigree in Figure 3.12. Furthermore, when a mutation is dominant, and a single copy is sufficient to produce a mutant phenotype, homozygotes are likely to be even more severely affected, perhaps even failing to survive. An illustration of this is the dominant gene for **familial hypercholesterolemia.** Heterozygotes display a defect in their receptors for low-density lipoproteins, the so-called LDLs (known popularly as "bad cholesterol"). As a result, too little cholesterol is taken up by cells from the blood, and elevated plasma levels of LDLs result. Without intervention, such heterozygous individuals usually have heart attacks during the fourth decade of their life, or before. While heterozygotes have LDL levels about double that of a normal individual, rare homozygotes have been detected. They lack LDL receptors altogether, and their LDL levels are nearly ten times above the normal range. They are likely to have a heart attack very early in life, even before age 5, and almost inevitably before they reach the age of 20.

Pedigree analysis of many traits has historically been an extremely valuable research technique in human genetic studies. However, the approach does not usually provide the certainty of the conclusions obtained through experimental crosses yielding large numbers of offspring. Nevertheless, when many independent pedigrees of the same trait or disorder are analyzed, consistent conclusions can often be drawn. **Table 3.2** lists numerous human traits and classifies them according to their recessive or dominant expression.

TABLE 3.2 Representative Recessive and Dominant Human Traits

Recessive Traits	Dominant Traits
Albinism	Achondroplasia
Alkaptonuria	Brachydactyly
Color blindness	Ehler–Danlos syndrome
Cystic fibrosis	Hypotrichosis
Duchenne muscular dystrophy	Huntington disease
Galactosemia	Hypercholesterolemia
Hemophilia	Marfan syndrome
Lesch–Nyhan syndrome	Myotonic dystrophy
Phenylketonuria	Neurofibromatosis
Sickle-cell anemia	Phenylthiocarbamide tasting
Tay–Sachs disease	Porphyria (some forms)

NOW SOLVE THIS

3.5 The following pedigree is for myopia (nearsightedness) in humans.

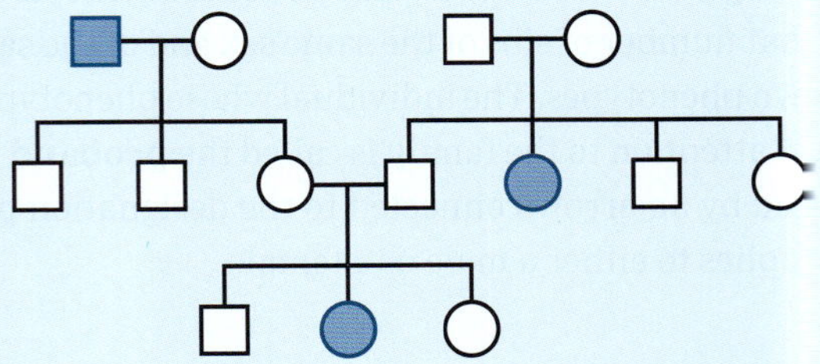

Predict whether the disorder is inherited as the result of a dominant or recessive trait. Determine the most probable genotype for each individual based on your prediction.

■ **HINT:** *This problem asks you to analyze a pedigree and determine the mode of inheritance of myopia. The key to its solution is to identify whether or not there are individuals who express the trait but neither of whose parents also express the trait. Such an observation is a powerful clue and allows you to rule out one mode of inheritance.*

3.10 Tay–Sachs Disease: The Molecular Basis of a Recessive Disorder in Humans

We conclude this chapter by examining a case where the molecular basis of normal and mutant genes and their resultant phenotypes have now been revealed. This discussion expands your understanding of how genes control phenotypes.

Of particular interest are cases where a single mutant gene causes multiple effects associated with a severe disorder in humans. Let's consider the modern explanation of the gene that causes **Tay–Sachs disease (TSD),** a devastating recessive disorder involving unalterable destruction of the central nervous system. Infants with TSD are unaffected at birth and appear to develop normally until they are about 6 months old. Then, a progressive loss of mental and physical abilities occurs. Afflicted infants eventually become blind, deaf, intellectually disabled, and paralyzed, often within only a year or two, seldom living beyond age 5. Typical of rare autosomal recessive disorders, two unaffected heterozygous parents, who most often have no family history of the disorder, have a probability of one in four of having a Tay–Sachs child.

We know that proteins are the end products of the expression of most all genes. The protein product involved in TSD has been identified, and we now have a clear understanding of the underlying molecular basis of the disorder. TSD results from the loss of activity of a single enzyme, **hexosaminidase A (Hex-A).** Hex-A, normally found in lysosomes within cells, is needed to break down the ganglioside GM2, a lipid component of nerve cell membranes. Without functional Hex-A, gangliosides accumulate within neurons in the brain and cause deterioration of the nervous system. Heterozygous carriers of TSD with one normal copy of the gene produce only about 50 percent of the normal amount of Hex-A, but they show no symptoms of the disorder. The observation that the activity of only one gene (one wild-type allele) is sufficient for the normal development and function of the nervous system explains and illustrates the molecular basis of recessive mutations. Only when both genes are disrupted by mutation is the mutant phenotype evident. The responsible gene is located on chromosome 15 and codes for the alpha subunit of the Hex-A enzyme. More than 50 different mutations within the gene have been identified that lead to TSD phenotypes.

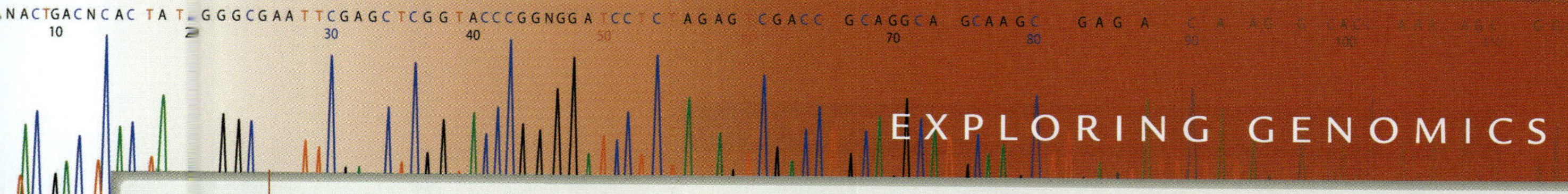

EXPLORING GENOMICS

Online Mendelian Inheritance in Man

Mastering Genetics Visit the Study Area: Exploring Genomics

The **Online Mendelian Inheritance in Man (OMIM) database** is a catalog of human genes and human disorders that are inherited in a Mendelian manner. Genetic disorders that arise from major chromosomal aberrations, such as monosomy or trisomy (the loss of a chromosome or the presence of a superfluous chromosome, respectively), are not included. The OMIM database, updated daily, is a version of the book *Mendelian Inheritance in Man*, conceived and edited by Dr. Victor McKusick of Johns Hopkins University, until he passed in 2008.

The OMIM entries provide links to a wealth of information, including DNA and protein sequences, chromosomal maps, disease descriptions, and relevant scientific publications. In this exercise, you will explore OMIM to answer questions about the recessive human disease sickle-cell anemia and other Mendelian inherited disorders.

■ Exercise I – Sickle-Cell Anemia

In this chapter, you were introduced to recessive and dominant human traits. You will now discover more about sickle-cell anemia as an autosomal recessive disease by exploring the OMIM database.

1. To begin the search, access the OMIM site at: www.omim.org.
2. In the "Search" box, type "sickle-cell anemia" and click on the "Search" button to perform the search.
3. Click on the link for the entry #603903.
4. Review the text that appears to learn about sickle-cell anemia. Examine the list of subject headings in the left-hand column and explore these links for more information about sickle-cell anemia.
5. Select one or two references at the bottom of the page and follow them to their abstracts in PubMed.
6. Using the information in this entry, answer the following questions:
   a. Which gene is mutated in individuals with sickle-cell anemia?
   b. What are the major symptoms of this disorder?
   c. What was the first published scientific description of sickle-cell anemia?
   d. Describe two other features of this disorder that you learned from the OMIM database, and state where in the database you found this information.

■ Exercise II – Other Recessive or Dominant Disorders

Select another human disorder that is inherited as either a dominant or recessive trait and investigate its features, following the general steps described in Exercise I. Follow links from OMIM to other databases if you choose.

Describe several interesting pieces of information you acquired during your exploration, and cite the information sources you encountered during the search.

CASE STUDY To test or not to test

Thomas discovered a devastating piece of family history when he learned that his brother had been diagnosed with Huntington disease (HD) at age 49. This dominantly inherited autosomal condition usually begins around age 45 with progressive dementia, muscular rigidity, and seizures and ultimately leads to death when affected individuals are in their early 60s. There currently is no effective treatment or cure for this genetic disorder. Thomas, now 38, wonders what the chances are that he also has inherited the mutant allele for HD, leading him to discuss with his wife whether they should seek genetic counseling and whether he should undergo genetic testing. They have two teenage children, a boy and a girl.

1. If they seek genetic counseling, what issues would likely be discussed? Which of these pose grave ethical dilemmas?
2. If you were in Thomas's position, would you want to be tested and possibly learn that you were almost certain to develop the disorder sometime in the next 5–10 years?
3. If Thomas tests positive for the HD allele, should his children be told about the situation, and if so, at what age? Who should make the decision about having the son and daughter tested?

Fulda, K., and Lykens, K. (2006). Ethical issues in predictive genetic testing: A public health perspective. *J. Med. Ethics* 32:143–147.

INSIGHTS AND SOLUTIONS

As a student, you will be asked to demonstrate your knowledge of transmission genetics by solving various problems. Success at this task requires not only comprehension of theory but also its application to more practical genetic situations. Most students find problem solving in genetics to be both challenging and rewarding. This section is designed to provide basic insights into the reasoning essential to this process.

1. Mendel found that full pea pods are dominant over constricted pods, while round seeds are dominant over wrinkled seeds. One of his crosses was between full, round plants and constricted, wrinkled plants. From this cross, he obtained an F_1 generation that was all full and round. In the F_2 generation, Mendel obtained his classic 9:3:3:1 ratio. Using this information, determine the expected F_1 and F_2 results of a cross between homozygous constricted, round and full, wrinkled plants.

Solution: First, assign gene symbols to each pair of contrasting traits. Use the lowercase first letter of each recessive trait to designate that trait, and use the same letter in uppercase to designate the dominant trait. Thus, *C* and *c* indicate full and constricted pods, respectively, and *W* and *w* indicate the round and wrinkled phenotypes, respectively.

Determine the genotypes of the P_1 generation, form the gametes, combine them in the F_1 generation, and read off the phenotype(s):

P_1:	*ccWW* constricted, round	×	*CCww* full, wrinkled
	↓		↓
Gametes:	*cW*		*Cw*
F_1 :		*CcWw* full, round	

You can immediately see that the F_1 generation expresses both dominant phenotypes and is heterozygous for both gene pairs. Thus, you expect that the F_2 generation will yield the classic Mendelian ratio of 9:3:3:1. Let's work it out anyway, just to confirm this expectation, using the forked-line method. Both gene pairs are heterozygous and can be expected to assort independently, so we can predict the F_2 outcomes from each gene pair separately and then proceed with the forked-line method.

The F_2 offspring should exhibit the individual traits in the following proportions:

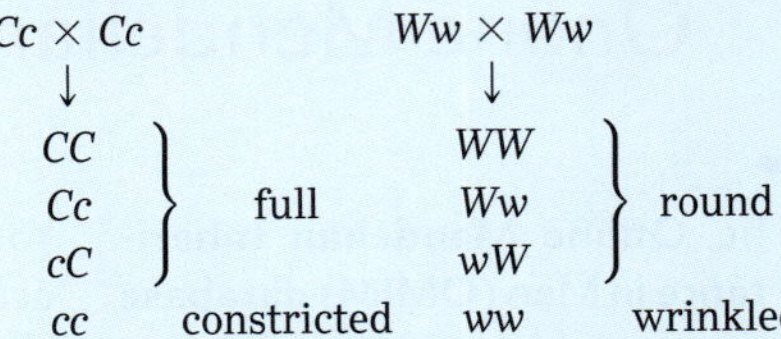

Using these proportions to complete a forked-line diagram confirms the 9:3:3:1 phenotypic ratio. (Remember that this ratio represents proportions of 9/16:3/16:3/16:1/16.) Note that we are applying the product law as we compute the final probabilities:

3/4 full
- 3/4 round $\xrightarrow{(3/4)(3/4)}$ 9/16 full, round
- 1/4 wrinkled $\xrightarrow{(3/4)(1/4)}$ 3/16 full, wrinkled

1/4 constricted
- 3/4 round $\xrightarrow{(1/4)(3/4)}$ 3/16 constricted, round
- 1/4 wrinkled $\xrightarrow{(1/4)(1/4)}$ 1/16 constricted, wrinkled

2. In the laboratory, a genetics student crossed flies with normal long wings with flies expressing the *dumpy* mutation (truncated wings), which she believed was a recessive trait. In the F_1 generation, all flies had long wings. The following results were obtained in the F_2 generation:

792 long-winged flies
208 dumpy-winged flies

The student tested the hypothesis that the dumpy wing is inherited as a recessive trait using χ^2 analysis of the F_2 data.

(a) What ratio was hypothesized?

(b) Did the analysis support the hypothesis?

(c) What do the data suggest about the *dumpy* mutation?

Solution:

(a) The student hypothesized that the F_2 data (792:208) fit Mendel's 3:1 monohybrid ratio for recessive genes.

(b) The initial step in χ^2 analysis is to calculate the expected results (*e*) for a ratio of 3:1. Then we can compute deviation $o - e$ (*d*) and the remaining numbers.

Ratio	*o*	*e*	*d*	d^2	d^2/e
3/4	792	750	42	1764	2.35
1/4	208	250	−42	1764	7.06
Total =	1000				

$$\chi^2 = \Sigma \frac{d^2}{e}$$
$$= 2.35 + 7.06$$
$$= 9.41$$

We consult Figure 3.10 to determine the probability (*p*) and to decide whether the deviations can be attributed to chance. There are two possible outcomes ($n = 2$), so the degrees of freedom (df) $= n - 1$, or 1. The table in Figure 3.10(b) shows that *p* is a value between 0.01 and 0.001; the graph in Figure 3.10(a) gives an estimate of about 0.001. Since $p < 0.05$, we reject the null hypothesis. The data do not fit a 3:1 ratio.

(c) When the student hypothesized that Mendel's 3:1 ratio was a valid expression of the monohybrid cross, she was tacitly making numerous assumptions. Examining these underlying assumptions may explain why the null hypothesis was rejected. For one thing, she assumed that all the genotypes resulting from the cross were equally viable—that genotypes yielding long wings are equally likely to survive from fertilization through adulthood as the genotype yielding dumpy wings. Further study would reveal that dumpy-winged flies are somewhat less viable than normal flies. As a result, we would expect *less* than 1/4 of the total offspring to express dumpy wings. This observation is borne out in the data, although we have not proven that this is true.

Problems and Discussion Questions

Mastering Genetics Visit for instructor-assigned tutorials and problems.

When working out genetics problems in this and succeeding chapters, always assume that members of the P_1 generation are homozygous, unless the information or data you are given require you to do otherwise.

1. **HOW DO WE KNOW?** In this chapter, we focused on the Mendelian postulates, probability, and pedigree analysis. We also considered some of the methods and reasoning by which these ideas, concepts, and techniques were developed. On the basis of these discussions, what answers would you propose to the following questions:
   (a) How was Mendel able to derive postulates concerning the behavior of "unit factors" during gamete formation, when he could not directly observe them?
   (b) How do we know whether an organism expressing a dominant trait is homozygous or heterozygous?
   (c) In analyzing genetic data, how do we know whether deviation from the expected ratio is due to chance rather than to another, independent factor?
   (d) Since experimental crosses are not performed in humans, how do we know how traits are inherited?
2. **CONCEPT QUESTION** Review the Chapter Concepts list on p. 55. The first five concepts provide a modern interpretation of Mendelian postulates. Based on these concepts, write a short essay that correlates Mendel's four postulates with what is now known about genes, alleles, and homologous chromosomes. ■
3. In a cross between a black and a white guinea pig, all members of the F_1 generation are black. The F_2 generation is made up of approximately 3/4 black and 1/4 white guinea pigs. Diagram this cross, and show the genotypes and phenotypes.
4. Albinism in humans is inherited as a simple recessive trait. Determine the genotypes of the parents and offspring for the following families. When two alternative genotypes are possible, list both.
   (a) Two parents without albinism have five children, four without albinism and one with albinism.
   (b) A male without albinism and a female with albinism have six children, all without albinism.
5. Explain how Mendel's first three postulates are demonstrated in the cross described in Problem 3.
6. What are the bases for Mendel's success in laying the principles of transmission genetics?
7. Mendel crossed peas having round seeds and yellow cotyledons with peas having wrinkled seeds and green cotyledons. All the F_1 plants had round seeds with yellow cotyledons. Diagram this cross through the F_2 generation, using both the Punnett square and forked-line methods.
8. Refer to the Now Solve This problem 3.2 on p. 63. Are any of the crosses in this problem testcrosses? If so, which one(s)?
9. What is the probability of obtaining an individual that is phenotypically recessive for all traits from a trihybrid cross? Which of Mendel's postulates supports your answer?
10. Correlate Mendel's four postulates with what is now known about homologous chromosomes, genes, alleles, and the process of meiosis.
11. Distinguish between segregation and independent assortment.
12. In *Drosophila*, gray body color is dominant over ebony body color, while long wings are dominant over vestigial wings. Work the following crosses through the F_2 generation, and determine the genotypic and phenotypic ratios for each generation. Assume that the P_1 individuals are homozygous: (a) gray, long × ebony, vestigial, and (b) gray, vestigial × ebony, long, and (c) gray, long × gray, vestigial.

13. How many different types of gametes can be formed by individuals of the following genotypes? What are they in each case? (a) *AaBb*, (b) *AaBB*, (c) *AaBbCc*, (d) *AaBBcc*, (e) *AaBbcc*, and (f) *AaBbCcDdEe*?
14. In crosses between true-breeding pea plants with inflated yellow seed pods and plants with pinched green pods, all the F_1 plants contain inflated green pods. Describe a testcross for these F_1 plants and determine the ratio of true-breeding plants with yellow pods in the progeny of this cross.
15. Shown are F_2 results of two of Mendel's monohybrid crosses. State a null hypothesis that you will test using chi-square analysis. Calculate the χ^2 value and determine the *p* value for both crosses; then interpret the *p* values. Which cross shows a greater amount of deviation?

(a)	Full pods	882
	Constricted pods	299
(b)	Violet flowers	705
	White flowers	224

16. A geneticist, in assessing data that fell into two phenotypic classes, observed values of 250:150. He decided to perform chi-square analysis using two different null hypotheses: (a) the data fit a 3:1 ratio; and (b) the data fit a 1:1 ratio. Calculate the χ^2 values for each hypothesis. What can you conclude about each hypothesis?
17. Define critical *p* value. Explain what significance this value has for predicting the reproducibility of an experiment involving crosses. Explain why the null hypothesis is generally rejected for *p* values lower than 0.05.
18. Consider three independently assorting gene pairs, *A*/*a*, *B*/*b*, and *C*/*c*, where each demonstrates typical dominance (*A*–, *B*–, *C*–) and recessiveness (*aa*, *bb*, *cc*). What is the probability of obtaining an offspring that is *AABbCc* from parents that are *AaBbCC* and *AABbCc*?
19. What is the probability of obtaining a triply recessive individual from the parents shown in Problem 18?
20. What proportion of the offspring would be dominant for *A* traits and recessive for *b* traits based on the parents mentioned in Problem 18?
21. The following pedigree follows the inheritance of myopia (nearsightedness) in humans. Predict whether the disorder is inherited as a dominant or a recessive trait. Based on your prediction, indicate the most probable genotype for each individual.

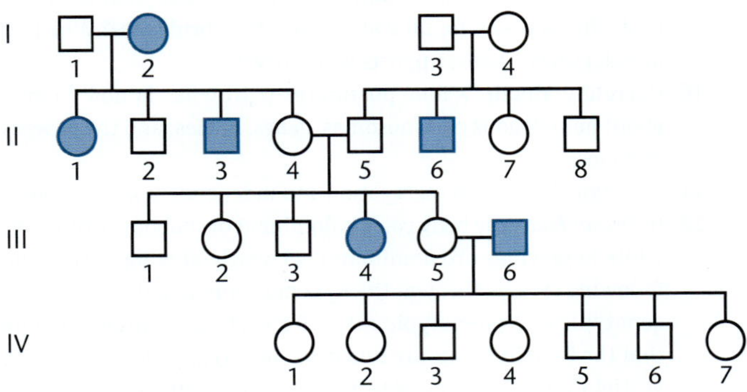

22. Draw all possible conclusions concerning the mode of inheritance of the trait expressed in each of the following limited pedigrees. (Each case is based on a different trait.)

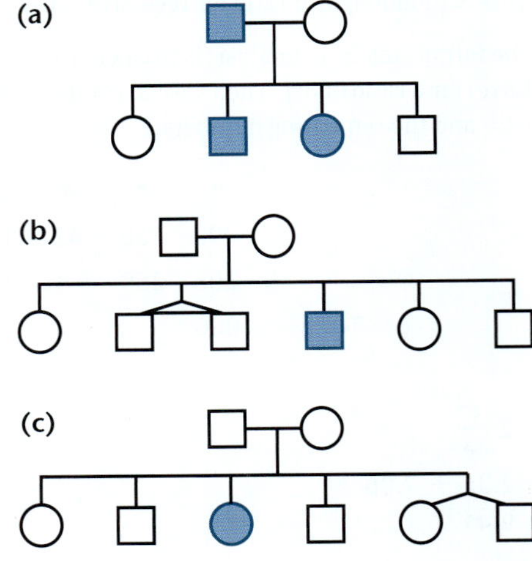

23. Two true-breeding pea plants are crossed. One parent is round, terminal, violet, constricted, while the other expresses the contrasting phenotypes of wrinkled, axial, white, full. The four pairs of contrasting traits are controlled by four genes, each located on a separate chromosome. In the F_1 generation, only round, axial, violet, and full are expressed. In the F_2 generation, all possible combinations of these traits are expressed in ratios consistent with Mendelian inheritance.
    (a) What conclusion can you draw about the inheritance of these traits based on the F_1 results?
    (b) Which phenotype appears most frequently in the F_2 results? Write a mathematical expression that predicts the frequency of occurrence of this phenotype.
    (c) Which F_2 phenotype is expected to occur least frequently? Write a mathematical expression that predicts this frequency.
    (d) How often is either P_1 phenotype likely to occur in the F_2 generation?
    (e) If the F_1 plant is testcrossed, how many different phenotypes will be produced?
24. To assess Mendel's law of segregation using tomatoes, a true-breeding tall variety (*SS*) is crossed with a true-breeding short variety (*ss*). The heterozygous tall plants (*Ss*) were crossed to produce the two sets of F_2 data as follows:

Set I	Set II
30 tall	300 tall
5 short	50 short

    (a) Using chi-square analysis, analyze the results for both data-sets. Calculate χ^2 values, and estimate the *p* values in both cases.
    (b) From the analysis in part (a), what can you conclude about the importance of generating large datasets in experimental settings?

4

Modification of Mendelian Ratios

Labrador retriever puppies, which may display brown (also called chocolate), golden (also called yellow), or black coats.

CHAPTER CONCEPTS

- While alleles are transmitted from parent to offspring according to Mendelian principles, they sometimes fail to display the clear-cut dominant/recessive relationship observed by Mendel.
- In many cases, in contrast to Mendelian genetics, two or more genes are known to influence the phenotype of a single characteristic.
- Still another exception to Mendelian inheritance is the presence of genes on sex chromosomes, whereby one of the sexes contains only a single member of that chromosome.
- Phenotypes are often the combined result of both genetics and the environment within which genes are expressed.
- The result of the various exceptions to Mendelian principles is the occurrence of phenotypic ratios that differ from those resulting from standard monohybrid, dihybrid, and trihybrid crosses.
- Extranuclear inheritance, resulting from the expression of genes present in the DNA found in mitochondria and chloroplasts, modifies Mendelian inheritance patterns. Such genes are most often transmitted through the female gamete.

In Chapter 3, we discussed the fundamental principles of transmission genetics. We saw that genes are present on homologous chromosomes and that these chromosomes segregate from each other and assort independently with other segregating chromosomes during gamete formation. These two postulates are the basic principles of gene transmission from parent to offspring. However, when gene expression does not adhere to a simple dominant/recessive mode or when more than one pair of genes influences the expression of a single character, the classic 3:1 and 9:3:3:1 ratios are usually modified. In this and the next several chapters, we consider more complex modes of inheritance. In spite of the greater complexity of these situations, the fundamental principles set down by Mendel still hold.

In this chapter, we restrict our initial discussion to the inheritance of traits controlled by only one set of genes. In diploid organisms, which have homologous pairs of chromosomes, two copies of each gene influence such traits. The copies need not be identical because alternative forms of genes (alleles) occur within populations. How alleles influence phenotypes is our primary focus. We will then consider gene interaction, a situation in which a single phenotype is affected by more than one set of genes. Numerous examples will be presented to illustrate a variety of heritable patterns observed in such situations.

Thus far, we have restricted our discussion to chromosomes other than the X and Y pair. By examining cases where genes are present on the X chromosome, illustrating X-linkage, we will see yet another modification of Mendelian ratios. Our discussion of modified ratios also includes the consideration of sex-limited and sex-influenced inheritance, cases where the sex of the individual, but not necessarily genes on the X chromosome, influences the phenotype. We will also consider how a given phenotype often varies depending on the overall environment in which a gene, a cell, or an organism finds itself. This discussion points out that phenotypic expression depends on more than just the genotype of an organism. Finally, we conclude with a discussion of extranuclear inheritance, cases where DNA within organelles influences an organism's phenotype.

4.1 Alleles Alter Phenotypes in Different Ways

After Mendel's work was rediscovered in the early 1900s, researchers focused on the many ways in which genes influence an individual's phenotype. Each type of inheritance was more thoroughly investigated when observations of genetic data did not conform precisely to the expected Mendelian ratios, and hypotheses that modified and extended the Mendelian principles were proposed and tested with specifically designed crosses. The explanations were in accord with the principle that a phenotype is under the control of one or more genes located at specific loci on one or more pairs of homologous chromosomes.

To understand the various modes of inheritance, we must first examine the potential function of alleles. Alleles are alternative forms of the same gene. The allele that occurs most frequently in a population, the one that we arbitrarily designate as normal, is called the **wild-type allele.** This is often, but not always, dominant. Wild-type alleles are responsible for the corresponding wild-type phenotype and are the standards against which all other mutations occurring at a particular locus are compared.

A mutant allele contains modified genetic information and often specifies an altered gene product. For example, in human populations, there are many known alleles of the gene that encodes the β chain of human hemoglobin. All such alleles store information necessary for the synthesis of the β-chain polypeptide, but each allele specifies a slightly different form of the same molecule. Once the allele's product has been manufactured, the function of the product may or may not be altered.

The process of mutation is the source of alleles. For a new allele to be recognized when observing an organism, it must cause a change in the phenotype. A new phenotype results from a change in functional activity of the cellular product specified by that gene. Often, the mutation causes the diminution or the loss of the specific wild-type function. For example, if a gene is responsible for the synthesis of a specific enzyme, a mutation in that gene may ultimately change the conformation of this enzyme and reduce or eliminate its affinity for the substrate. Such a case is designated as a **loss-of-function mutation.** If the loss is complete, the mutation has resulted in what is called a **null allele.**

Conversely, other mutations may enhance the function of the wild-type product. Most often when this occurs, it is the result of increasing the quantity of the gene product. In such cases, the mutation may be affecting the regulation of transcription of the gene under consideration. Such cases are designated **gain-of-function mutations,** which generally result in dominant alleles since one copy in a diploid organism is sufficient to alter the normal phenotype. Examples of gain-of-function mutations include the genetic conversion of proto-oncogenes, which regulate the cell cycle, to oncogenes, where regulation is overridden by excess gene product. The result is the creation of a cancerous cell.

Having introduced the concept of gain- or loss-of-function mutations, it is important to note the possibility that a mutation will create an allele where no change in function can be detected. In this case, the mutation would not be immediately apparent since no phenotypic variation would be evident. However, such a mutation could be detected if the DNA sequence of the gene was examined directly. These are sometimes referred to as **neutral mutations** because the gene product presents no change to either the phenotype or the evolutionary fitness of the organism.

Finally, we note here that while a phenotypic trait may be affected by a single mutation in one gene, traits are often influenced by more than one gene. For example, enzymatic reactions are most often part of complex metabolic pathways leading to the synthesis of an end product, such as an amino acid. Mutations in any of the various reactions have a common effect—the failure to synthesize the end product. Therefore, phenotypic traits related to the end product are often influenced by more than one gene.

4.2 Geneticists Use a Variety of Symbols for Alleles

In Chapter 3, we learned a standard convention that is used to symbolize alleles for very simple Mendelian traits. The initial letter of the name of a recessive trait, lowercased and italicized, denotes the recessive allele, and the same letter in uppercase refers to the dominant allele. Thus, in the case of *tall* and *dwarf,* where *dwarf* is recessive, *D* and *d* represent the alleles responsible for these respective traits. Mendel used upper- and lowercase letters such as these to symbolize his unit factors.

Another useful system was developed in genetic studies of the fruit fly *Drosophila melanogaster* to discriminate between wild-type and mutant traits. This system uses the initial letter, or a combination of two or three letters, of the name of the mutant trait. If the trait is recessive, lowercase is used; if it is dominant, uppercase is used. The contrasting wild-type trait is denoted by the same letter, but with a superscript +. For example, *ebony* is a recessive body color mutation in *Drosophila*. The normal wild-type body color is gray. Using this system, we denote *ebony* by the symbol *e*, and we denote gray by e^+. The responsible locus may be occupied by either the wild-type allele (e^+) or the mutant allele (e). A diploid fly may thus exhibit one of three possible genotypes:

e^+/e^+	gray homozygote (wild type)
e^+/e	gray heterozygote (wild type)
e/e	ebony homozygote (mutant)

The slash between the letters indicates that the two allele designations represent the same locus on two homologous chromosomes. If we instead consider a dominant wing mutation such as *Wrinkled* (*Wr*) wing in *Drosophila*, the three possible designations are Wr^+/Wr^+, Wr^+/Wr, and Wr/Wr. The latter two genotypes express the wrinkled-wing phenotype.

One advantage of this system is that further abbreviation can be used when convenient: the wild-type allele may simply be denoted by the + symbol. With *ebony* as an example, the designations of the three possible genotypes become

$+/+$	gray homozygote (wild type)
$+/e$	gray heterozygote (wild type)
e/e	ebony homozygote (mutant)

Another variation is utilized when no dominance exists between alleles. We simply use uppercase italic letters and superscripts to denote alternative alleles (e.g., R^1 and R^2, L^M and L^N, I^A and I^B). Their use will become apparent later in this chapter.

Many diverse systems of genetic nomenclature are used to identify genes in various organisms. Usually, the symbol selected reflects the function of the gene or even a disorder caused by a mutant gene. For example, the yeast *cdk* is the abbreviation for the *c*yclin *d*ependent *k*inase gene, whose product is involved in cell-cycle regulation. In bacteria, leu^- refers to a mutation that interrupts the biosynthesis of the amino acid leucine, where the wild-type gene is designated leu^+. The symbol *dnaA* represents a bacterial gene involved in DNA replication (and DnaA is the protein made by that gene). In humans, capital letters are used to name genes: *BRCA*1 represents the first gene associated with susceptibility to *br*east *ca*ncer. Although these different systems may seem complex, they are useful ways to symbolize genes.

4.3 Neither Allele Is Dominant in Incomplete, or Partial, Dominance

A cross between parents with contrasting traits may generate offspring with an intermediate phenotype. For example, if plants such as four-o'clocks or snapdragons with red flowers are crossed with white-flowered plants, the offspring have pink flowers. Some red pigment is produced in the F_1 intermediate pink-colored flowers. Therefore, neither red nor white flower color is dominant. This situation is known as **incomplete,** or **partial, dominance.**

If this phenotype is under the control of a single gene and two alleles where neither is dominant, the results of the F_1 (pink) × F_1 (pink) cross can be predicted. The resulting F_2 generation shown in **Figure 4.1** confirms the hypothesis that only one pair of alleles determines these phenotypes.

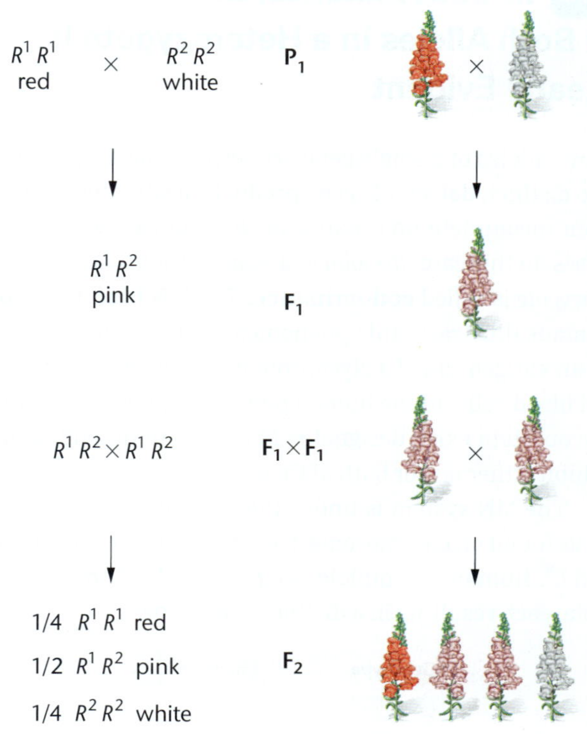

FIGURE 4.1 Incomplete dominance shown in the flower color of snapdragons.

The *genotypic ratio* (1:2:1) of the F_2 generation is identical to that of Mendel's monohybrid cross. However, because neither allele is dominant, the *phenotypic ratio* is identical to the *genotypic ratio*. Note that because neither allele is recessive, we have chosen not to use upper and lowercase letters as symbols. Instead, we denoted the red and white alleles as R^1 and R^2, respectively. We could have used W^1 and W^2 or still other designations such as C^W and C^R, where C indicates "color" and the W and R superscripts indicate white and red.

Clear-cut cases of incomplete dominance, which result in intermediate expression of the overt phenotype, are relatively rare. However, even when complete dominance seems apparent, careful examination of the gene product, rather than the phenotype, often reveals an intermediate level of gene expression. An example is the human biochemical disorder **Tay–Sachs disease,** in which homozygous recessive individuals are severely affected with a fatal lipid storage disorder (see Chapter 3, page 72). There is almost no activity of the enzyme **hexosaminidase A** in those with the disease. Heterozygotes, with only a single copy of the mutant gene, are phenotypically normal but express only about 50 percent of the enzyme activity found in homozygous normal individuals. Fortunately, this level of enzyme activity is adequate to achieve normal biochemical function—a situation not uncommon in enzyme disorders.

4.4 In Codominance, the Influence of Both Alleles in a Heterozygote Is Clearly Evident

If two alleles of a single gene are responsible for producing two distinct, detectable gene products, a situation different from incomplete dominance or dominance/recessiveness arises. In this case, *the joint expression of both alleles in a heterozygote* is called **codominance.** The **MN blood group** in humans illustrates this phenomenon and is characterized by an antigen called a glycoprotein, found on the surface of red blood cells. In the human population, two forms of this glycoprotein exist, designated M and N; an individual may exhibit either one or both of them.

The MN system is under the control of an autosomal locus found on chromosome 4 and two alleles designated L^M and L^N. Humans are diploid, so three combinations are possible, each resulting in a distinct blood type:

Genotype	Phenotype
$L^M L^M$	M
$L^M L^N$	MN
$L^N L^N$	N

As predicted, a mating between two heterozygous MN parents may produce children of all three blood types, as follows:

$$L^M L^N \times L^M L^N$$
$$\downarrow$$
$$1/4\ L^M L^M$$
$$1/2\ L^M L^N$$
$$1/4\ L^N L^N$$

Once again the genotypic ratio, 1:2:1, is upheld.

Codominant inheritance is characterized by *distinct expression of the gene products of both alleles*. This characteristic distinguishes it from incomplete dominance, where heterozygotes express an intermediate, blended phenotype. We shall see another example of codominance when we examine the ABO blood-type system in the following section.

4.5 Multiple Alleles of a Gene May Exist in a Population

The information stored in any gene is extensive, and mutations can modify this information in many ways. Each change produces a different allele. Therefore, for any specific gene, the number of alleles within members of a population need not be restricted to two. When three or more alleles of the same gene are found, **multiple alleles** are present that create a unique mode of inheritance. It is important to realize that *multiple alleles can be studied only in populations*. An individual diploid organism has, at most, two homologous gene loci that may be occupied by different alleles of the same gene. However, among many members of a species, numerous alternative forms of the same gene can exist.

The ABO Blood Group

The simplest case of multiple alleles is that in which three alternative alleles of one gene exist. This situation is illustrated by the **ABO blood group** in humans, discovered by Karl Landsteiner in the early 1900s. The ABO system, like the MN blood group, is characterized by the presence of antigens on the surface of red blood cells. The A and B antigens are distinct from MN antigens and are under the control of a different gene, located on chromosome 9. As in the MN system, one combination of alleles in the ABO system exhibits a codominant mode of inheritance.

When individuals are tested using antisera that contain antibodies against the A or B antigen, four phenotypes are revealed. Each individual has either the A antigen (A phenotype), the B antigen (B phenotype), the A and B antigens (AB phenotype), or neither antigen (O phenotype). In 1924, it was hypothesized that these phenotypes were inherited as the result of three alleles of a single gene. This hypothesis was based on studies of the blood types of many different families.

Although different designations can be used, we use the symbols I^A, I^B, and i to distinguish these three alleles; the i designation stands for *isoagglutinogen*, another term for antigen. If we assume that the I^A and I^B alleles are responsible for the production of their respective A and B antigens and that i is an allele that does not produce any detectable A or B antigens, we can list the various genotypic possibilities and assign the appropriate phenotype to each:

Genotype	Antigen	Phenotype
$I^A I^A$	A	A
$I^A i$	A	
$I^B I^B$	B	B
$I^B i$	B	
$I^A I^B$	A, B	AB
$i\,i$	Neither	O

In these assignments the I^A and I^B alleles are dominant to the i allele but are codominant to each other. Our knowledge of human blood types has several practical applications, the most important of which are compatible blood transfusions and organ transplantations.

The Bombay Phenotype

The biochemical basis of the ABO blood-type system has been carefully worked out. The A and B antigens are actually carbohydrate groups (sugars) that are bound to lipid molecules (fatty acids) protruding from the membrane of the red blood cell. The specificity of the A and B antigens is based on the terminal sugar of the carbohydrate group. Both the A and B antigens are derived from a precursor molecule called the **H substance,** to which one or two terminal sugars are added.

In extremely rare instances, first recognized in a woman in Bombay in 1952, the H substance is incompletely formed. As a result, it is an inadequate substrate for the enzyme that normally adds the terminal sugar. This condition results in the expression of blood type O and is called the **Bombay phenotype.** Research has revealed that this condition is due to a rare recessive mutation at a locus separate from that controlling the A and B antigens. The gene is now designated *FUT1* (encoding an enzyme, fucosyl transferase), and individuals that are homozygous for the mutation cannot synthesize the complete H substance. Thus, even though they may have the I^A and/or I^B alleles, neither the A nor B antigen can be added to the cell surface. This information explains why the woman in Bombay expressed blood type O, even though one of her parents was type AB (thus she should not have been type O), and why she was able to pass the I^B allele to her children (**Figure 4.2**).

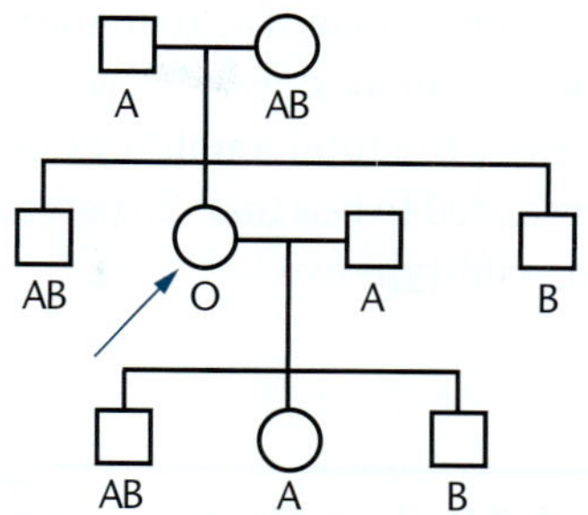

FIGURE 4.2 A partial pedigree of a woman with the Bombay phenotype. Functionally, her ABO blood group behaves as type O. Genetically, she is type B.

The *white* Locus in *Drosophila*

Many other phenotypes in plants and animals are known to be controlled by multiple allelic inheritance. In *Drosophila,* many alleles are known at practically every locus. The recessive mutation that causes white eyes, discovered by Thomas H. Morgan and Calvin Bridges in 1912, is one of over 100 alleles that can occupy this locus. In this allelic series, eye colors range from complete absence of pigment in the *white* allele to deep ruby in the *white-satsuma* allele, to orange in

NOW SOLVE THIS

4.1 In the guinea pig, one locus involved in the control of coat color may be occupied by any of four alleles: C (full color), c^k (sepia), c^d (cream), or c^a (albino), with an order of dominance of: $C > c^k > c^d > c^a$. (C is dominant to all others, c^k is dominant to c^d and c^a, but not C, etc.) In the following crosses, determine the parental genotypes and predict the phenotypic ratios that would result:

(a) sepia × cream, where both guinea pigs had an albino parent

(b) sepia × cream, where the sepia guinea pig had an albino parent and the cream guinea pig had two sepia parents

(c) sepia × cream, where the sepia guinea pig had two full-color parents and the cream guinea pig had two sepia parents

(d) sepia × cream, where the sepia guinea pig had a full-color parent and an albino parent and the cream guinea pig had two full-color parents

■ **HINT:** *This problem involves an understanding of multiple alleles. The key to its solution is to note particularly the hierarchy of dominance of the various alleles. Remember also that even though there can be more than two alleles in a population, an individual can have at most two of these. Thus, the allelic distribution into gametes adheres to the principle of segregation.*

the *white-apricot* allele, to a buff color in the *white-buff* allele. These alleles are designated w, w^{sat}, w^{a}, and w^{bf}, respectively. In each case, the total amount of pigment in these mutant eyes is reduced to less than 20 percent of that found in the brick-red, wild-type eye.

4.6 Lethal Alleles Represent Essential Genes

Many gene products are essential to an organism's survival. Mutations resulting in the synthesis of a gene product that is nonfunctional can often be tolerated in the heterozygous state; that is, one wild-type allele may be sufficient to produce enough of the essential product to allow survival. However, such a mutation behaves as a *recessive lethal allele,* and homozygous recessive individuals will not survive. The time of death will depend on when the product is essential. In mammals, for example, this might occur during development, early childhood, or even adulthood.

In some cases, the allele responsible for a lethal effect when homozygous may also result in a distinctive mutant phenotype when present heterozygously. It is behaving as a recessive lethal allele but is dominant with respect to the phenotype. For example, a mutation that causes a yellow coat in mice was discovered in the early part of this century. The yellow coat varies from the normal agouti (wild-type) coat phenotype, as shown in **Figure 4.3**. Crosses between the various combinations of the two strains yield unusual results:

		Crosses		
(A) agouti	×	agouti	⟶	all agouti
(B) yellow	×	yellow	⟶	2/3 yellow: 1/3 agouti
(C) agouti	×	yellow	⟶	1/2 yellow: 1/2 agouti

These results are explained on the basis of a single pair of alleles. With regard to coat color, the mutant *yellow* allele (A^{Y}) is dominant to the wild-type *agouti* allele (A), so heterozygous mice will have yellow coats. However, the *yellow* allele is also a homozygous recessive lethal. When present in two copies, the mice die before birth. Thus, there are no homozygous yellow mice. The genetic basis for these three crosses is shown in Figure 4.3.

In other cases, a mutation may behave as a *dominant lethal allele.* In such cases, the presence of just one copy of the allele results in the death of the individual. In humans, the disorder called **Huntington disease** is due to such an allele (H), where the onset of the disease and eventual lethality in heterozygotes (Hh) is delayed, usually well into

agouti mouse

yellow mouse

FIGURE 4.3 An agouti and a yellow mouse.

adulthood. Affected individuals then undergo gradual nervous and motor degeneration until they die. This lethal disorder is particularly tragic because it has such a late onset, typically at about age 40, often after the affected individual has produced a family, where each child has a 50 percent probability of inheriting the lethal allele, transmitting the allele to his or her offspring, and eventually developing the disorder. Note that Huntington disease is the subject of the Special Topics 4 chapter, where the disorder serves as the example of major advances in the study of neurogenetics in humans.

While Hungtington disease is an exception, most all dominant lethal alleles are rarely observed. For these alleles to exist in a population, the affected individuals must reproduce before the lethal allele is expressed. If all affected individuals die before reaching reproductive age, the mutant gene will not be passed to future generations, and the mutation will disappear from the population unless it arises again as a result of a new mutation.

EVOLVING CONCEPT OF THE GENE

Based on the work of many geneticists following the rediscovery of Mendel's work in the very early part of the twentieth century, the chromosome theory of inheritance was put forward, which hypothesized that chromosomes are the carriers of genes and that meiosis is the physical basis of Mendel's postulates. In the ensuing 40 years, the concept of a gene evolved to reflect the idea that this hereditary unit can exist in multiple forms, or alleles, each of which can have an impact on the phenotype in different ways, leading to incomplete dominance, codominance, and even lethality. It became clear that the process of mutation was the source of new alleles. ■

ESSENTIAL POINT

Since Mendel's work was rediscovered, transmission genetics has been expanded to include many alternative modes of inheritance, including the study of incomplete dominance, codominance, multiple alleles, and lethal alleles. ■

4.7 Combinations of Two Gene Pairs with Two Modes of Inheritance Modify the 9:3:3:1 Ratio

Each example discussed so far modifies Mendel's 3:1 F_2 monohybrid ratio. Therefore, combining any two of these modes of inheritance in a dihybrid cross will likewise modify the classical 9:3:3:1 ratio. Having established the foundation for the modes of inheritance of incomplete dominance, codominance, multiple alleles, and lethal alleles, we can now deal with the situation of two modes of inheritance occurring simultaneously. Mendel's principle of independent assortment applies to these situations, provided that the genes controlling each character are not linked on the same chromosome—in other words, that they do not demonstrate what is called *genetic linkage*.

Consider, for example, a mating that occurs between two humans who are both heterozygous for the autosomal recessive gene that causes albinism and who are both of blood type AB. What is the probability of a particular phenotypic combination occurring in each of their children? Albinism is inherited in the simple Mendelian fashion, and the blood types are determined by the series of three multiple alleles, I^A, I^B, and I. The solution to this problem is diagrammed in **Figure 4.4**, using the forked-line method. This

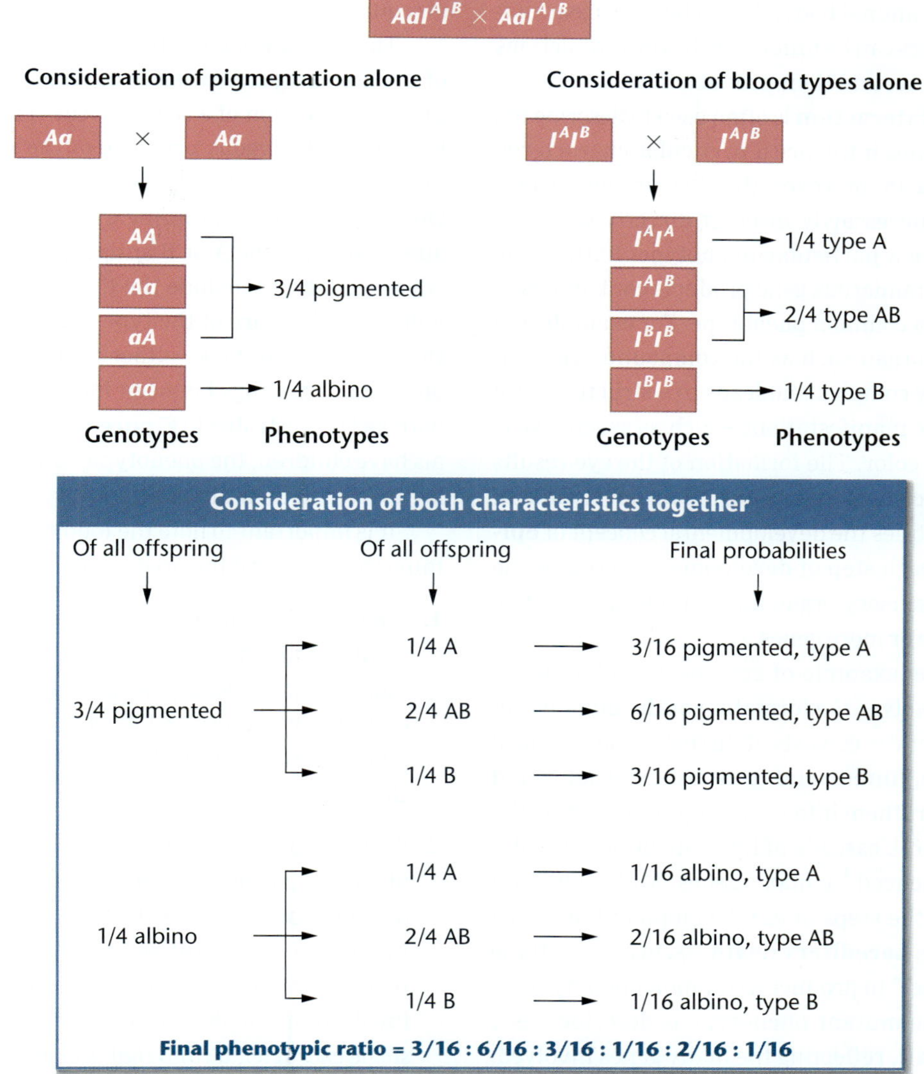

FIGURE 4.4 Calculation of the mating probabilities involving the ABO blood type and albinism in humans, using the forked-line method.

dihybrid cross does not yield the classical four phenotypes in a 9:3:3:1 ratio. Instead, six phenotypes occur in a 3:6:3:1:2:1 ratio, establishing the expected probability for each phenotype. This is just one of the many variants of modified ratios that are possible when different modes of inheritance are combined.

4.8 Phenotypes Are Often Affected by More Than One Gene

Soon after Mendel's work was rediscovered, experimentation revealed that individual characteristics displaying discrete phenotypes are often under the control of more than one gene. This was a significant discovery because it revealed that genetic influence on the phenotype is often much more complex than Mendel had envisioned. Instead of single genes controlling the development of individual parts of the plant or animal body, it soon became clear that phenotypic characters can be influenced by the interactions of many different genes and their products.

The term **gene interaction** is often used to describe the idea that several genes influence a particular characteristic. This does not mean, however, that two or more genes, or their products, necessarily interact directly with one another to influence a particular phenotype. Rather, the cellular function of numerous gene products contributes to the development of a common phenotype. For example, the development of an organ such as the compound eye of an insect is exceedingly complex and leads to a structure with multiple phenotypic manifestations—such as specific size, shape, texture, and color. The formation of the eye results from a complex cascade of events during its development. This process exemplifies the developmental concept of **epigenesis,** whereby each step of development increases the complexity of this sensory organ and is under the control and influence of one or more genes.

An enlightening example of epigenesis and multiple gene interaction involves the formation of the inner ear in mammals. The inner ear consists of distinctive anatomical features to capture, funnel, and transmit external sound waves and to convert them into nerve impulses. During the formation of the ear, a cascade of intricate developmental events occur, influenced by many genes. Mutations that interrupt many of the steps of ear development lead to a common phenotype: **hereditary deafness.** In a sense, these many genes "interact" to produce a common phenotype. In such situations, the mutant phenotype is described as a **heterogeneous trait,** reflecting the many genes involved. In humans, while a few common alleles are responsible for the vast majority of cases of hereditary deafness, over 50 genes are involved in development of the ability to discern sound.

Epistasis

Some of the best examples of gene interaction are those that reveal the phenomenon of **epistasis** where the expression of one gene or gene pair masks or modifies the expression of another gene or gene pair. Sometimes the genes involved control the expression of the same general phenotypic characteristic in an antagonistic manner, as when masking occurs. In other cases, however, the genes involved exert their influence on one another in a complementary, or cooperative, fashion.

For example, the homozygous presence of a recessive allele prevents or overrides the expression of other alleles at a second locus (or several other loci). In another example, a single dominant allele at the first locus influences the expression of the alleles at a second gene locus. In a third example, two gene pairs are said to complement one another such that at least one dominant allele at each locus is required to express a particular phenotype.

The Bombay phenotype discussed earlier is an example of the homozygous recessive condition at one locus masking the expression of a second locus. There, we established that the homozygous presence of the mutant form of the *FUT1* gene masks the expression of the I^A and I^B alleles. Only individuals containing at least one wild-type *FUT1* allele can form the A or B antigen. As a result, individuals whose genotypes include the I^A or I^B allele and who lack a wild-type allele are of the type O phenotype, regardless of their potential to make either antigen. An example of the outcome of matings between individuals heterozygous at both loci is illustrated in **Figure 4.5**. If many such individuals have children, the phenotypic ratio of 3 A: 6 AB: 3 B: 4 O is expected in their offspring.

It is important to note the following points when examining this cross and the predicted phenotypic ratio:

1. A key distinction exists in this cross compared to the modified dihybrid cross shown in Figure 4.4: *only one characteristic—blood type—is being followed.* In the modified dihybrid cross of Figure 4.4, blood type *and* skin pigmentation are followed as separate phenotypic characteristics.

2. Even though only a single character was followed, the phenotypic ratio is expressed in sixteenths. If we knew nothing about the H substance and the genes controlling it, we could still be confident that a second gene pair, other than that controlling the A and B antigens, is involved in the phenotypic expression. *When studying a single character, a ratio that is expressed in 16 parts (e.g., 3:6:3:4) suggests that two gene pairs are "interacting" during the expression of the phenotype under consideration.*

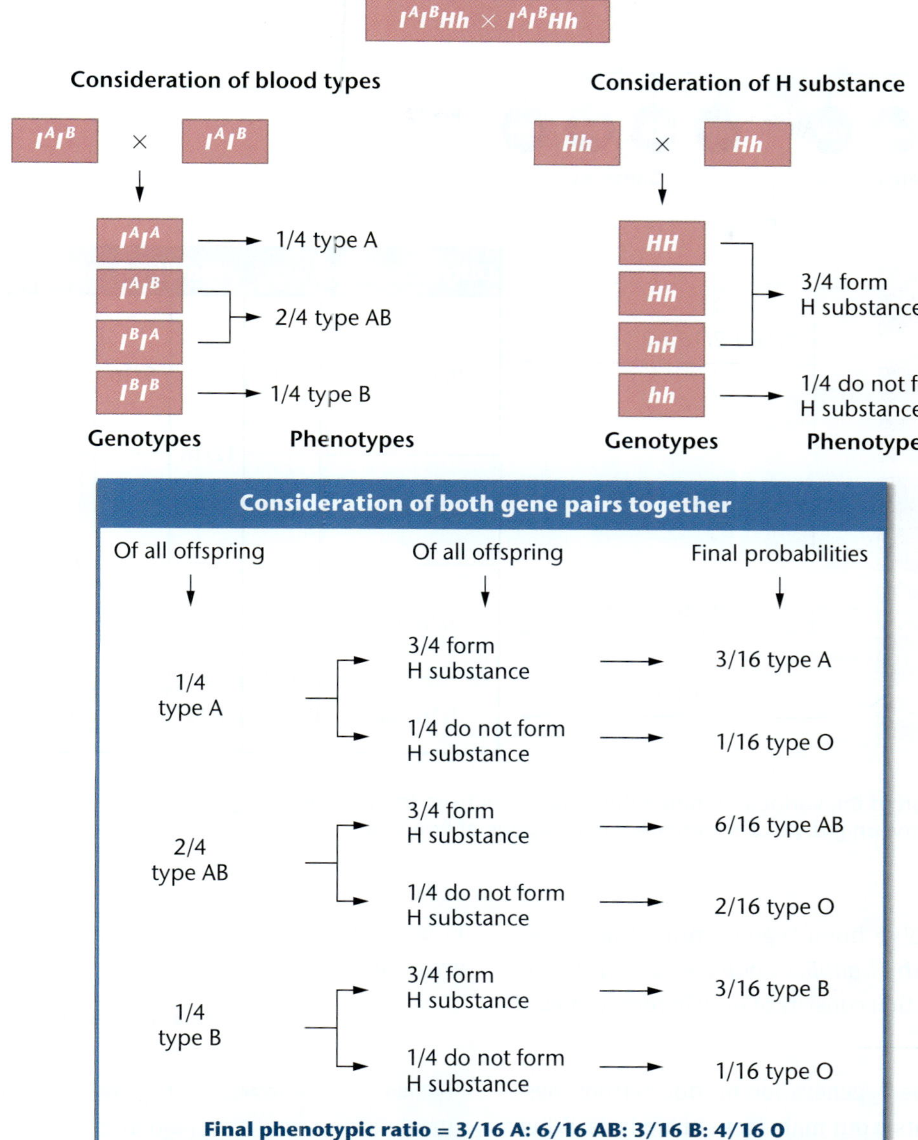

FIGURE 4.5 The outcome of a mating between individuals who are heterozygous at two genes determining their ABO blood type. Final phenotypes are calculated by considering both genes separately and then combining the results using the forked-line method.

The study of gene interaction reveals inheritance patterns that modify the classical Mendelian dihybrid F_2 ratio (9:3:3:1) in other ways as well. In these examples, epistasis combines one or more of the four phenotypic categories in various ways. The generation of these four groups is reviewed in **Figure 4.6**, along with several modified ratios.

As we discuss these and other examples, we will make several assumptions and adopt certain conventions:

1. In each case, distinct phenotypic classes are produced, each clearly discernible from all others. Such traits illustrate *discontinuous variation,* where phenotypic categories are discrete and qualitatively different from one another.
2. The genes considered in each cross are not linked and therefore assort independently of one another during gamete formation. To allow you to easily compare the results of different crosses, we designated alleles as *A*, *a* and *B*, *b* in each case.
3. When we assume that complete dominance exists between the alleles of any gene pair, such that *AA* and *Aa* or *BB* and *Bb* are equivalent in their genetic effects, we use the designations *A*– or *B*– for both combinations, where the dash (–) indicates that either allele may be present, without consequence to the phenotype.

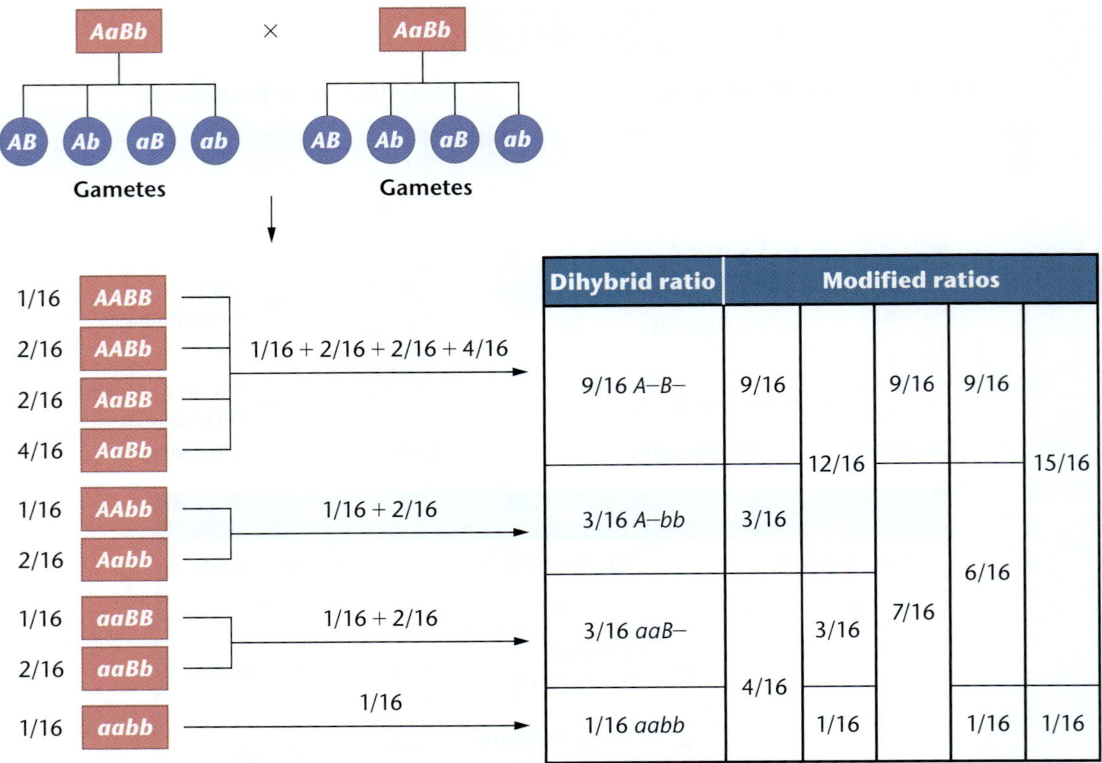

FIGURE 4.6 Generation of the various modified dihybrid ratios from the nine unique genotypes produced in a cross between individuals who are heterozygous at two genes.

4. All P_1 crosses involve homozygous individuals (e.g., *AABB* × *aabb*, *AAbb* × *aaBB*, or *aaBB* × *AAbb*). Therefore, each F_1 generation consists of only heterozygotes of genotype *AaBb*.

5. In each example, the F_2 generation produced from these heterozygous parents is our main focus of analysis. When two genes are involved (as in Figure 4.6), the F_2 genotypes fall into four categories: 9/16 *A–B–*, 3/16 *A–bb*, 3/16 *aaB–*, and 1/6 *aabb*. Because of dominance, all genotypes in each category have an equivalent effect on the phenotype.

Case 1 is the inheritance of coat color in mice (**Figure 4.7**). Normal wild-type coat color is agouti, a grayish pattern formed by alternating bands of pigment on each hair. Agouti is dominant to black (non-agouti) hair, which is caused by a recessive mutation, *a*. Thus, *A–* results in agouti, while *aa* yields black coat color. When it is homozygous, a recessive mutation, *b*, at a separate locus, eliminates pigmentation altogether, yielding albino mice (*bb*), regardless of the genotype at the other locus. The presence of at least one *B* allele allows pigmentation to occur in much the same way that the *H* allele in humans allows the expression of the ABO blood types. In a cross between agouti (*AABB*) and albino (*aabb*), members of the F_1 are all *AaBb* and have agouti coat color. In the F_2 progeny of a cross between two F_1 heterozygotes, the following genotypes and phenotypes are observed:

F_1: *AaBb* × *AaBb*
↓

F_2 Ratio	Genotype	Phenotype	Final Phenotypic Ratio
9/16	*A–B–*	agouti	9/16 agouti
3/16	*A–bb*	albino	3/16 black
3/16	*aaB–*	black	4/16 albino
1/16	*aabb*	albino	

We can envision gene interaction yielding the observed 9:3:4 F_2 ratio as a two-step process:

	Gene *B*		Gene *A*	
Precursor Molecule (colorless)	↓ ⟶ *B–*	Black Pigment	↓ ⟶ *A–*	Agouti Pattern

In the presence of a *B* allele, black pigment can be made from a colorless substance. In the presence of an *A* allele, the black pigment is deposited during the development of hair in a pattern that produces the agouti phenotype. If the *aa* genotype occurs, all of the hair remains black. If the *bb* genotype occurs, no black pigment is produced, regardless of the presence of the *A* or *a* alleles, and the mouse is albino. Therefore, the *bb* genotype masks or suppresses the

Case	Organism	Character	F_2 Phenotypes: 9/16	3/16	3/16	1/16	Modified ratio
1	Mouse	Coat color	agouti	albino	black	albino	9:3:4
2	Squash	Color	white		yellow	green	12:3:1
3	Pea	Flower color	purple	white			9:7
4	Squash	Fruit shape	disc	sphere		long	9:6:1
5	Chicken	Color	white		colored	white	13:3
6	Mouse	Color	white-spotted	white	colored	white-spotted	10:3:3
7	Shepherd's purse	Seed capsule	triangular			ovoid	15:1
8	Flour beetle	Color	6/16 sooty and 3/16 red	black	jet	black	6:3:3:4

FIGURE 4.7 The basis of modified dihybrid F_2 phenotypic ratios, resulting from crosses between doubly heterozygous F_1 individuals. The four groupings of the F_2 genotypes shown in Figure 4.6 and across the top of this figure are combined in various ways to produce these ratios.

expression of the *A* gene. As a result, this is referred to as *recessive epistasis*.

A second type of epistasis, called *dominant epistasis,* occurs when a dominant allele at one genetic locus masks the expression of the alleles at a second locus. For instance, Case 2 of Figure 4.7 deals with the inheritance of fruit color in summer squash. Here, the dominant allele *A* results in white fruit color regardless of the genotype at a second locus, *B*. In the absence of the dominant *A* allele (the *aa* genotype), *BB* or *Bb* results in yellow color, while *bb* results in green color. Therefore, if two white-colored double heterozygotes (*AaBb*) are crossed, this type of epistasis generates an interesting phenotypic ratio:

F_1: *AaBb* × *AaBb*

↓

F_2 Ratio	Genotype	Phenotype	Final Phenotypic Ratio
9/16	*A–B–*	white	12/16 white
3/16	*A–bb*	white	3/16 yellow
3/16	*aaB–*	yellow	1/16 green
1/16	*aabb*	green	

Of the offspring, 9/16 are *A–B–* and are thus white. The 3/16 bearing the genotypes *A–bb* are also white. Finally, 3/16 are yellow (*aaB–*) while 1/16 are green (*aabb*); and we obtain the modified ratio of 12:3:1.

Our third type of gene interaction (Case 3 of Figure 4.7) was first discovered by William Bateson and Reginald Punnett (of Punnett square fame). It is demonstrated in a cross between two true-breeding strains of white-flowered sweet peas. Unexpectedly, the results of this cross yield all purple F_1 plants, and the F_2 plants occur in a ratio of 9/16 purple to 7/16 white. The proposed explanation suggests that the presence of at least one dominant allele of each of two gene pairs is essential for flowers to be purple. Thus, this cross represents a case of *complementary gene interaction*. All other genotype combinations yield white flowers because the homozygous condition of *either* recessive allele masks the expression of the dominant allele at the other locus. The cross is shown as follows:

P_1: *AAbb* × *aaBB*

white white

↓

F_1: All *AaBb* purple

↓

F_2 Ratio	Genotype	Phenotype	Final Phenotypic Ratio
9/16	*A–B–*	purple	
3/16	*A–bb*	white	9/16 purple
3/16	*aa B–*	white	7/16 white
1/16	*aabb*	white	

We can now see how two gene pairs might yield such results:

	Gene *A*		Gene *B*	
Precursor Substance (colorless)	↓ ⟶ *A–*	Intermediate Product (colorless)	↓ ⟶ *B–*	Final Product (purple)

At least one dominant allele from each pair of genes is necessary to ensure both biochemical conversions to the final product, yielding purple flowers. In our cross, this will occur in 9/16 of the F_2 offspring. All other plants (7/16) have flowers that remain white.

The preceding examples illustrate how the products of two genes "interact" to influence the development of a common phenotype. In other instances, more than two genes and their products are involved in controlling phenotypic expression.

Novel Phenotypes

Other cases of gene interaction yield novel, or new, phenotypes in the F_2 generation, in addition to producing modified dihybrid ratios. Case 4 in Figure 4.7 depicts the inheritance of fruit shape in the summer squash *Cucurbita pepo*. When plants with disc-shaped fruit (*AABB*) are crossed to plants with long fruit (*aabb*), the F_1 generation all have disc fruit. However, in the F_2 progeny, fruit with a novel shape—sphere—appear, along with fruit exhibiting the parental phenotypes. A variety of fruit shapes are shown in **Figure 4.8**.

The F_2 generation, with a modified 9:6:1 ratio, is generated as follows:

F_1: *AaBb* × *AaBb*
disc disc
↓

F_2 Ratio	Genotype	Phenotype	Final Phenotypic Ratio
9/16	*A–B–*	disc	9/16 disc
3/16	*A–bb*	sphere	6/16 sphere
3/16	*aaB–*	sphere	1/16 long
1/16	*aabb*	long	

In this example of gene interaction, both gene pairs influence fruit shape equally. A dominant allele at either locus ensures a sphere-shaped fruit. In the absence of dominant alleles, the fruit is long. However, if both dominant alleles (*A* and *B*) are present, the fruit displays a flattened, disc shape.

FIGURE 4.8 Summer squash exhibiting the fruit-shape phenotypes disc, long, and sphere.

Other Modified Dihybrid Ratios

The remaining cases (5–8) in Figure 4.7 show additional modifications of the dihybrid ratio and provide still other examples of gene interactions. However, all eight cases have two things in common. First, we have not violated the principles of segregation and independent assortment to explain the inheritance pattern of each case. Therefore, the added complexity of inheritance in these examples does not detract from the validity of Mendel's conclusions. Second, the F_2 phenotypic ratio in each example has been expressed in sixteenths. When similar observations are made in crosses where the inheritance pattern is unknown, it suggests to geneticists that two gene pairs are controlling the observed phenotypes. You should make the same inference in your analysis of genetics problems.

NOW SOLVE THIS

4.2 In some plants a red flower pigment, cyanidin, is synthesized from a colorless precursor. The addition of a hydroxyl group (OH^-) to the cyanidin molecule causes it to become purple. In a cross between two randomly selected purple varieties, the following results were obtained:

94 purple
31 red
43 white

How many genes are involved in the determination of these flower colors? Which genotypic combinations produce which phenotypes? Diagram the purple × purple cross.

■ **HINT:** *This problem describes a plant in which flower color, a single characteristic, can take on one of three variations. The key to its solution is to first analyze the raw data and convert the numbers to a meaningful ratio. This will guide you in determining how many gene pairs are involved. Then you can group the genotypes in a way that corresponds to the phenotypic ratio.*

ESSENTIAL POINT

Mendel's classic F_2 ratio is often modified in instances when gene interaction controls phenotypic variation. Such instances can be identified when the final ratio is divided into eighths or sixteenths. ■

4.9 Complementation Analysis Can Determine if Two Mutations Causing a Similar Phenotype Are Alleles of the Same Gene

An interesting situation arises when two mutations, both of which produce a similar phenotype, are isolated independently. Suppose that two investigators independently

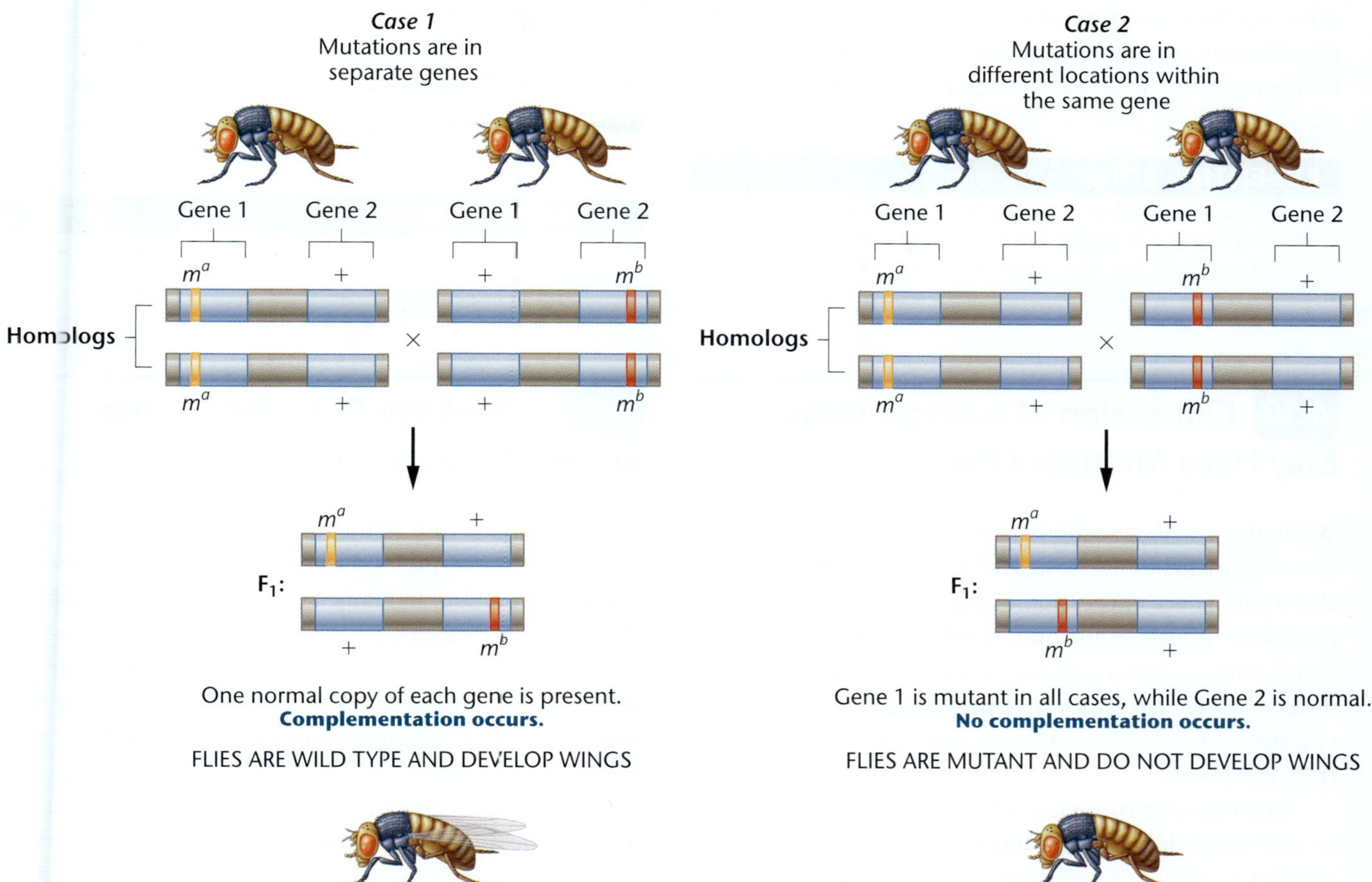

FIGURE 4.9 Complementation analysis of alternative outcomes of two wingless mutations in *Drosophila* (m^a and m^b). In Case 1, the mutations are not alleles of the same gene, whereas in Case 2, the mutations are alleles of the same gene.

isolate and establish a true-breeding strain of wingless *Drosophila* and demonstrate that each mutant phenotype is due to a recessive mutation. We might assume that both strains contain mutations in the same gene. However, since we know that many genes are involved in the formation of wings, mutations in any one of them might inhibit wing formation during development. The experimental approach called **complementation analysis** allows us to determine whether two such mutations are in the same gene—that is, whether they are alleles of the same gene or whether they represent mutations in separate genes.

To repeat, our analysis seeks to answer this simple question: *Are two mutations that yield similar phenotypes present in the same gene or in two different genes?* To find the answer, we cross the two mutant strains and analyze the F_1 generation. Two alternative outcomes and interpretations of this cross are shown in **Figure 4.9**. We discuss both cases, using the designations m^a for one of the mutations and m^b for the other one. Now we will determine experimentally whether or not m^a and m^b are alleles of the same gene.

Case 1. *All offspring develop normal wings.*
Interpretation: The two recessive mutations are in separate genes and are not alleles of one another. Following the cross, all F_1 flies are heterozygous for both genes. *Complementation* is said to occur. Since each mutation is in a separate gene and each F_1 fly is heterozygous at both loci, the normal products of both genes are produced (by the one normal copy of each gene), and wings develop.

Case 2. *All offspring fail to develop wings.*
Interpretation: The two mutations affect the same gene and are alleles of one another. Complementation does *not* occur. Since the two mutations affect the same gene, the F_1 flies are homozygous for the two mutant alleles (the m^a allele and the m^b allele). No normal product of the gene is produced, and in the absence of this essential product, wings do not form.

Complementation analysis, as originally devised by the Nobel Prize-winning *Drosophila* geneticist Edward B. Lewis, may be used to screen any number of individual mutations that result in the same phenotype. Such an analysis may reveal that only a single gene is involved or that two or more genes are involved. All mutations determined to be present in any single gene are said to fall into the same **complementation group,** and they will complement mutations in all other groups. When large numbers of mutations affecting the same trait are available and studied using complementation analysis, it is possible to predict the total number of genes involved in the determination of that trait.

ESSENTIAL POINT

Complementation analysis determines whether independently isolated mutations producing similar phenotypes are alleles of one another or whether they represent separate genes. ■

4.10 Expression of a Single Gene May Have Multiple Effects

While the previous sections have focused on the effects of two or more genes on a single characteristic, the converse situation, where expression of a single gene has multiple phenotypic effects, is also quite common. This phenomenon, which often becomes apparent when phenotypes are examined carefully, is referred to as **pleiotropy.** We will review two such cases involving human genetic disorders to illustrate this point.

Marfan syndrome is a human malady resulting from an autosomal dominant mutation in the gene encoding the connective tissue protein fibrillin. Because this protein is widespread in many tissues in the body, one would expect multiple effects of such a defect. In fact, fibrillin is important to the structural integrity of the lens of the eye, to the lining of vessels such as the aorta, and to bones, among other tissues. As a result, the phenotype associated with Marfan syndrome includes lens dislocation, increased risk of aortic aneurysm, and lengthened long bones in limbs. This disorder is of historical interest in that speculation abounds that Abraham Lincoln had Marfan syndrome.

Our second example involves another human autosomal dominant disorder, **porphyria variegata.** Individuals with this disorder cannot adequately metabolize the porphyrin component of hemoglobin when this respiratory pigment is broken down as red blood cells are replaced. The accumulation of excess porphyrins is immediately evident in the urine, which takes on a deep red color. The severe features of the disorder are due to the toxicity of the buildup of porphyrins in the body, particularly in the brain. Complete phenotypic characterization includes abdominal pain, muscular weakness, fever, a racing pulse, insomnia, headaches, vision problems (that can lead to blindness), delirium, and ultimately convulsions. As you can see, deciding which phenotypic trait best characterizes the disorder is impossible.

Like Marfan syndrome, porphyria variegata is also of historical significance. George III, king of England during the American Revolution, is believed to have suffered from episodes involving all of the above symptoms. He ultimately became blind and senile prior to his death. We could cite many other examples to illustrate pleiotropy, but suffice it to say that if one looks carefully, most mutations display more than a single manifestation when expressed.

ESSENTIAL POINT

Pleiotropy refers to multiple phenotypic effects caused by a single mutation. ■

4.11 X-Linkage Describes Genes on the X Chromosome

In many animal and some plant species, one of the sexes contains a pair of unlike chromosomes that are involved in sex determination. In many cases, these are designated as the X and Y. For example, in both *Drosophila* and humans, males contain an X and a Y chromosome, whereas females contain two X chromosomes. While the Y chromosome must contain a region of pairing homology with the X chromosome if the two are to synapse and segregate during meiosis, much of the remainder of the Y chromosome in humans and other species is considered to be relatively inert genetically. Thus, it lacks most genes that are present on the X chromosome. As a result, genes present on the X chromosome exhibit unique patterns of inheritance in comparison with autosomal genes. The term **X-linkage** is used to describe these situations.

In the following discussion, we will focus on inheritance patterns resulting from genes present on the X but absent from the Y chromosome. This situation results in a modification of Mendelian ratios, the central theme of this chapter.

X-Linkage in *Drosophila*

One of the first cases of X-linkage was documented by Thomas H. Morgan around 1920 during his studies of the *white* mutation in the eyes of *Drosophila*. The normal wild-type red eye color is dominant to white. We will use this case to illustrate X-linkage.

Morgan's work established that the inheritance pattern of the white-eye trait is clearly related to the sex of the parent carrying the mutant allele. Unlike the outcome of the typical monohybrid cross, reciprocal crosses between white- and red-eyed flies did not yield identical results. In contrast, in all of Mendel's monohybrid crosses, F_1 and F_2 data were similar regardless of which P_1 parent exhibited the recessive mutant trait. Morgan's analysis led to the conclusion that the *white* locus is present on the X chromosome rather than on one of the autosomes. As such, both the gene and the trait are said to be X-linked.

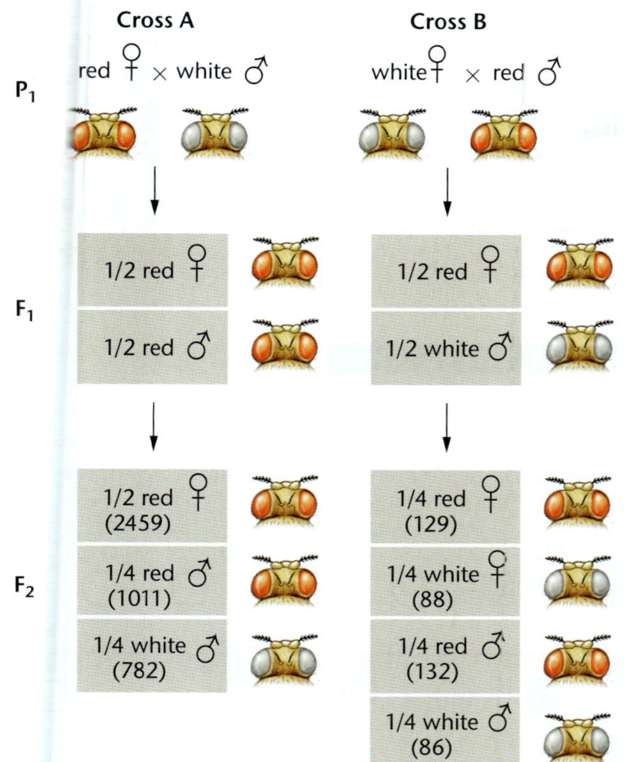

FIGURE 4.10 The F_1 and F_2 results of T. H. Morgan's reciprocal crosses involving the X-linked *white* mutation in *Drosophila melanogaster*. The actual F_2 data are shown in parentheses. Photographs of red and white eyes are shown in Chapter 1, Figure 1.3.

Results of reciprocal crosses between white-eyed and red-eyed flies are shown in **Figure 4.10**. The obvious differences in phenotypic ratios in both the F_1 and F_2 generations are dependent on whether or not the P_1 white-eyed parent was male or female.

Morgan was able to correlate these observations with the difference found in the sex-chromosome composition between male and female *Drosophila*. He hypothesized that the recessive allele for white eyes is found on the X chromosome, but its corresponding locus is absent from the Y chromosome. Females thus have two available gene sites, one on each X chromosome, whereas males have only one available gene site on their single X chromosome.

Morgan's interpretation of X-linked inheritance, shown in **Figure 4.11**, provides a suitable theoretical explanation for his results. Since the Y chromosome lacks homology with most genes on the X chromosome, whatever alleles are present on the X chromosome of the males will be expressed directly in their phenotype. Males cannot be homozygous or heterozygous for X-linked genes, and this condition is referred to as being **hemizygous.**

One result of X-linkage is the **crisscross pattern of inheritance,** whereby phenotypic traits controlled by recessive X-linked genes are passed from homozygous mothers to all sons. This pattern occurs because females exhibiting a recessive trait carry the mutant allele on both X chromosomes. Because male offspring receive one of their mother's

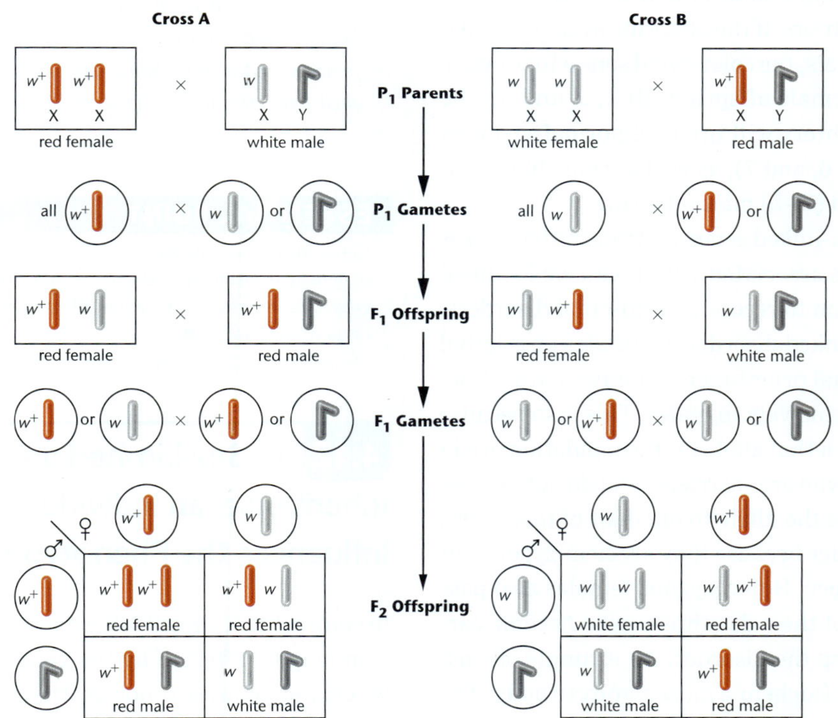

FIGURE 4.11 The chromosomal explanation of the results of the X-linked crosses shown in Figure 4.10.

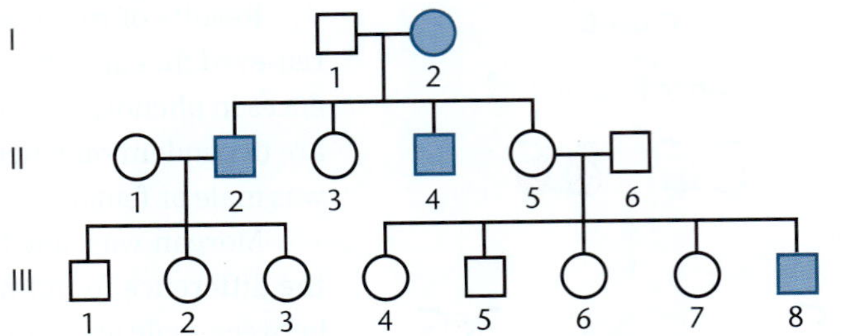

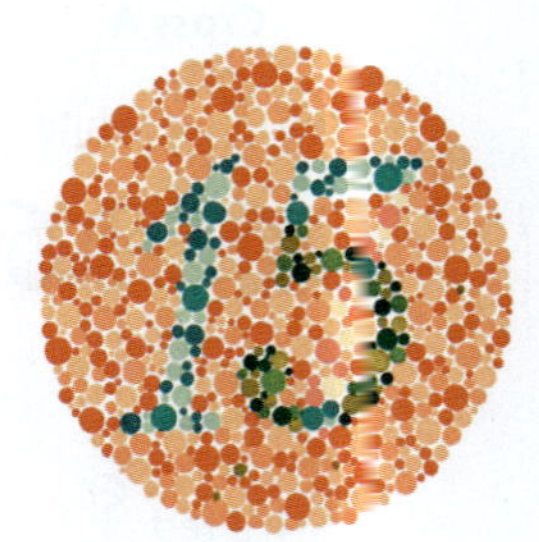

FIGURE 4.12 A human pedigree of the X-linked color-blindness trait. The photograph is of an Ishihara color-blindness chart. Those with normal vision will see the number 15, while those with red-green color blindness will see the number 17.

two X chromosomes and are hemizygous for all alleles present on that X, all sons will express the same recessive X-linked traits as their mother.

Morgan's work has taken on great historical significance. By 1910, the correlation between Mendel's work and the behavior of chromosomes during meiosis had provided the basis for the **chromosome theory of inheritance,** first introduced in Chapter 3. Work involving the X chromosome around 1920 is considered to be the first solid experimental evidence in support of this theory. In the ensuing two decades, these findings inspired further research, which provided indisputable evidence in support of this theory.

X-Linkage in Humans

In humans, many genes and the respective traits they control are recognized as being linked to the X chromosome. These X-linked traits can be easily identified in a pedigree because of the crisscross pattern of inheritance. A pedigree for one form of human **color blindness** is shown in **Figure 4.12**. The mother in generation I passes the trait to all her sons but to none of her daughters. If the offspring in generation II marry normal individuals, the color-blind sons will produce all normal male and female offspring (III-1, 2, and 3); the normal-visioned daughters will produce normal-visioned female offspring (III-4, 6, and 7), as well as color-blind (III-8) and normal-visioned (III-5) male offspring.

The way in which X-linked genes are transmitted causes unusual circumstances associated with recessive X-linked disorders, in comparison to recessive autosomal disorders. For example, if an X-linked disorder debilitates or is lethal to the affected individual prior to reproductive maturation, the disorder occurs exclusively in males. This is so because the only sources of the lethal allele in the population are in heterozygous females who are "carriers" and do not express the disorder. They pass the allele to one-half of their sons, who develop the disorder because they are hemizygous but rarely, if ever, reproduce. Heterozygous females also pass the allele to one-half of their daughters, who become carriers but do not develop the disorder. An example of such an X-linked disorder is Duchenne muscular dystrophy. The disease has an onset prior to age 6 and is often lethal around age 20. It normally occurs only in males.

NOW SOLVE THIS

4.3 Below are three pedigrees. For each trait, consider whether it is or is not consistent with X-linked recessive inheritance. In a sentence or two, indicate why or why not.

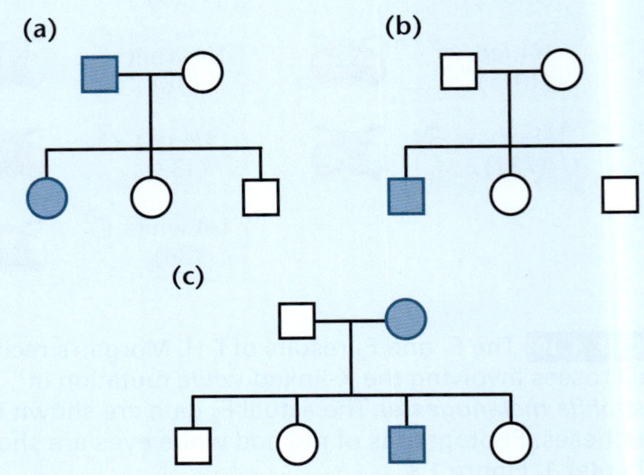

■ **HINT:** *This problem involves potential X-linked recessive traits as analyzed in pedigrees. The key to its solution is to focus on hemizygosity, where an X-linked recessive allele is always expressed in males, but never passed from a father to his sons. Homozygous females, on the other hand, pass the trait to all sons, but not to their daughters unless the father is also affected.*

ESSENTIAL POINT

Genes located on the X chromosome result in a characteristic mode of genetic transmission referred to as X-linkage, displaying so-called crisscross inheritance, whereby affected mothers pass X-linked traits to all of their sons. ■

4.12 In Sex-Limited and Sex-Influenced Inheritance, an Individual's Gender Influences the Phenotype

In contrast to X-linked inheritance, patterns of gene expression may be affected by the gender of an individual even when the genes are not on the X chromosome. In some cases, the expression of a specific phenotype is absolutely limited to one gender; in others, the gender of an individual influences the expression of a phenotype but the phenotype is expressed in both genders. This distinction differentiates sex-limited inheritance from sex-influenced inheritance. In both types of inheritance, autosomal genes are responsible for the existence of contrasting phenotypes, but the expression of these genes is dependent on the hormone constitution of the individual. Thus, a heterozygous genotype may exhibit one phenotype in males and the contrasting one in females. In domestic fowl, for example, tail and neck plumage is often distinctly different in males and females (**Figure 4.13**), demonstrating **sex-limited inheritance.** Cock feathering is longer, more curved, and pointed, whereas hen feathering is shorter and less curved. Inheritance of these feather phenotypes is controlled by a single pair of autosomal alleles whose expression is modified by the individual's sex hormones.

As shown in the following chart, hen feathering is due to a dominant allele, *H*, but regardless of the homozygous presence of the recessive *h* allele, all females remain hen-feathered. Only in males does the *hh* genotype result in cock feathering.

Genotype	Phenotype	
	Females	Males
HH	Hen-feathered	Hen-feathered
Hh	Hen-feathered	Hen-feathered
hh	Hen-feathered	Cock-feathered

In certain breeds of fowl, the hen-feathering or cock-feathering allele has become fixed in the population. In the Leghorn breed, all individuals are of the *hh* genotype; as a result, males always differ from females in their plumage. Sebright bantams are all *HH*, resulting in no sexual distinction in feathering phenotypes.

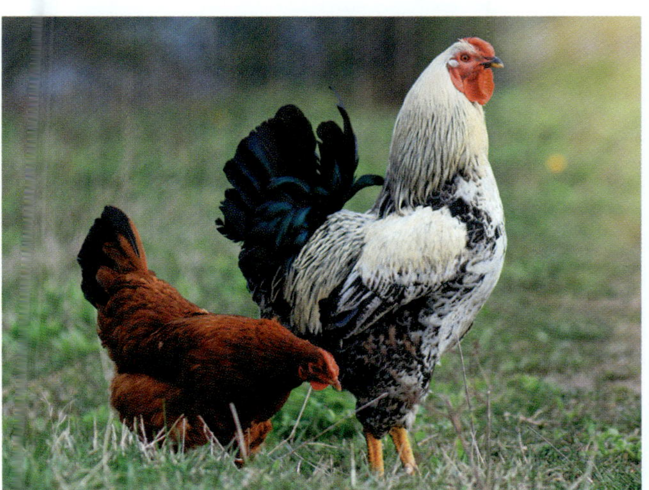

FIGURE 4.13 Hen feathering (left) and cock feathering (right) in domestic fowl. Note that the hen's feathers are shorter and less curved.

FIGURE 4.14 Pattern baldness, a sex-influenced autosomal trait in a young man.

Another example of sex-limited inheritance involves the autosomal genes responsible for milk yield in dairy cattle. Regardless of the overall genotype that influences the quantity of milk production, those genes are obviously expressed only in females.

Cases of **sex-influenced inheritance** include pattern baldness in humans, horn formation in certain breeds of sheep (e.g., Dorset Horn sheep), and certain coat-color patterns in cattle. In such cases, autosomal genes are again responsible for the contrasting phenotypes displayed by both males and females, but the expression of these genes is dependent on the hormonal constitution of the individual. Thus, the heterozygous genotype exhibits one phenotype in one sex and the contrasting one in the other. For example, **pattern baldness** in humans, where the hair is very thin on the top of the head (**Figure 4.14**), is inherited in this way:

Genotype	Phenotype	
	Females	Males
BB	Bald	Bald
Bb	Not bald	Bald
bb	Not bald	Not bald

Females can display pattern baldness, but this phenotype is much more prevalent in males. When females do inherit the *BB* genotype, the phenotype is less pronounced than in males and is expressed later in life.

ESSENTIAL POINT

Sex-limited and sex-influenced inheritance occurs when the sex of the organism affects the phenotype controlled by a gene located on an autosome. ■

4.13 Genetic Background and the Environment Affect Phenotypic Expression

We now focus on *phenotypic expression.* In previous discussions, we assumed that the genotype of an organism is always directly expressed in its phenotype. For example, pea plants homozygous for the recessive *d* allele (*dd*) will always be dwarf. We discussed gene expression as though the genes operate in a closed system in which the presence or absence of functional products directly determines the collective phenotype of an individual. The situation is actually much more complex. Most gene products function within the internal milieu of the cell, and cells interact with one another in various ways. Furthermore, the organism exists under diverse environmental influences. Thus, gene expression and the resultant phenotype are often modified through the interaction between an individual's particular genotype and the external environment. Here, we deal with several important variables that are known to modify gene expression.

Penetrance and Expressivity

Some mutant genotypes are always expressed as a distinct phenotype, whereas others produce a proportion of individuals whose phenotypes cannot be distinguished from normal (wild type). The degree of expression of a particular trait can be studied quantitatively by determining the *penetrance* and *expressivity* of the genotype under investigation. The percentage of individuals who show at least some degree of expression of a mutant genotype defines the **penetrance** of the mutation. For example, the phenotypic expression of many mutant alleles in *Drosophila* can overlap with wild type. If 15 percent of mutant flies show the wild-type appearance, the mutant gene is said to have a penetrance of 85 percent.

By contrast, **expressivity** reflects the *range of expression* of the mutant genotype. Flies homozygous for the recessive mutant *eyeless* gene yield phenotypes that range from the presence of normal eyes to a partial reduction in size to the complete absence of one or both eyes (**Figure 4.15**). Although the average reduction of eye size is one-fourth to one-half, expressivity ranges from complete loss of both eyes to completely normal eyes.

Examples such as the expression of the *eyeless* gene provide the basis for experiments to determine the causes of phenotypic variation. If, on one hand, a laboratory environment is held constant and extensive phenotypic variation is still observed, other genes may be influencing or modifying the *eyeless* phenotype. On the other hand, if the genetic background is not the cause of the phenotypic variation, environmental factors such as temperature, humidity, and nutrition may be involved. In the case of the *eyeless* phenotype, experiments have shown that both genetic background and environmental factors influence its expression.

Genetic Background: Position Effects

Although it is difficult to assess the specific effect of the **genetic background** and the expression of a gene responsible for determining a potential phenotype, one effect of genetic background has been well characterized, the **position effect.** In such instances, the physical location of a gene in relation to other genetic material may influence its expression. For example, if a region of a chromosome is relocated or rearranged (called a translocation or an inversion event), normal expression of genes in that chromosomal region may be modified. This is particularly true if the gene is relocated to or near certain areas of the chromosome that are prematurely condensed and genetically inert, referred to as **heterochromatin.** An example of a position effect involves female *Drosophila* heterozygous for the X-linked recessive eye color mutant *white* (w). The w^+/w genotype normally results in a wild-type brick-red eye color. However, if the region of the X chromosome containing the wild-type w^+ allele is translocated so that it is close to a heterochromatic region, expression of the w^+ allele is modified. Instead of having a red color, the eyes are variegated, or mottled with red and white patches. Apparently, heterochromatic regions inhibit the expression of adjacent genes.

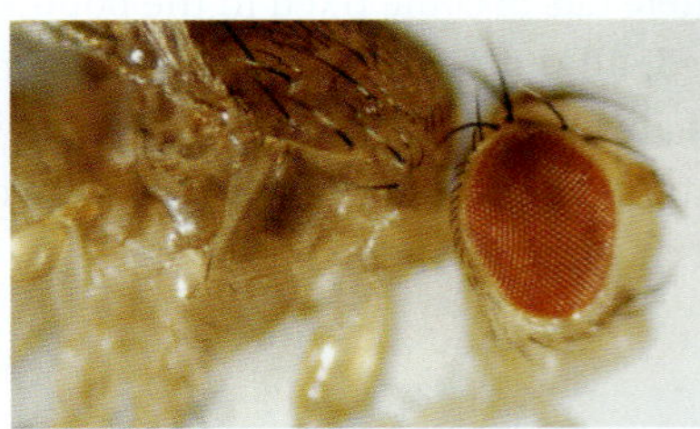

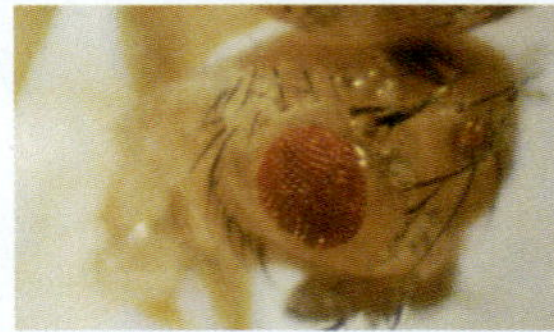

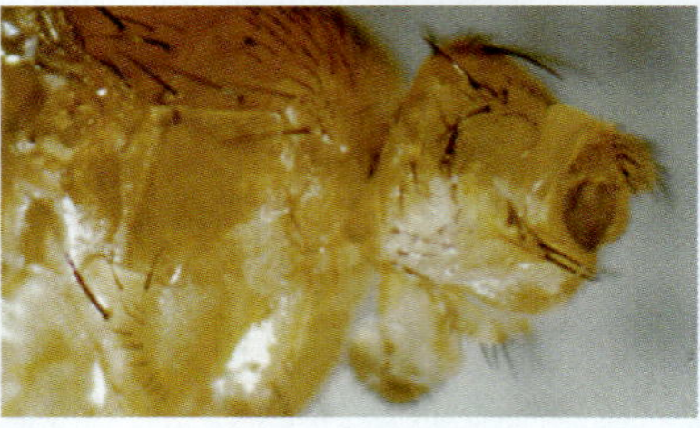

FIGURE 4.15 Variable expressivity, as shown in flies homozygous for the *eyeless* mutation in *Drosophila.* Gradations in phenotype range from wild type to partial reduction to eyeless.

Temperature Effects—An Introduction to Conditional Mutations

Chemical activity depends on the kinetic energy of the reacting substances, which in turn depends on the surrounding temperature. We can thus expect temperature to influence phenotypes. An example is seen in the evening primrose, which produces red flowers when grown at 23°C and white flowers when grown at 18°C. An even more striking example is seen in Siamese cats and Himalayan rabbits, which exhibit dark fur in certain regions where their body temperature is slightly cooler, particularly the nose, ears, and paws (**Figure 4.16**). In these cases, it appears that the enzyme normally responsible for pigment production is functional only at the lower temperatures present in the extremities, but it loses its catalytic function at the slightly higher temperatures found throughout the rest of the body.

Mutations whose expression is affected by temperature, called **temperature-sensitive mutations,** are examples of **conditional mutations,** whereby phenotypic expression is determined by environmental conditions. Examples of temperature-sensitive mutations are known in viruses and a variety of organisms, including bacteria, fungi, and *Drosophila*. In extreme cases, an organism carrying a mutant allele may express a mutant phenotype when grown at one temperature but express the wild-type phenotype when reared at another temperature. This type of temperature effect is useful in studying mutations that interrupt essential processes during development and are thus normally detrimental or lethal. For example, if bacterial viruses are cultured under *permissive conditions* of 25°C, the mutant gene product is functional, infection proceeds normally, and new viruses are produced and can be studied. However, if bacterial viruses carrying temperature-sensitive mutations infect bacteria cultured at 42°C—the *restrictive condition*—infection progresses up to the point where the essential gene product is required (e.g., for viral assembly) and then arrests. Temperature-sensitive mutations are easily induced and isolated in viruses, and have added immensely to the study of viral genetics.

Onset of Genetic Expression

Not all genetic traits become apparent at the same time during an organism's life span. In most cases, the age at which a mutant gene exerts a noticeable phenotype depends on events during the normal sequence of growth and development. In humans, the prenatal, infant, preadult, and adult phases require different genetic information. As a result, many severe inherited disorders are not manifested until after birth. For example, as we saw in Chapter 3, **Tay–Sachs disease,** inherited as an autosomal recessive, is a lethal lipid metabolism disease involving an abnormal enzyme, hexosaminidase A. Newborns appear to be phenotypically normal for the first few months. Then developmental disability, paralysis, and blindness ensue, and most affected children die around the age of 3.

Lesch–Nyhan syndrome, inherited as an X-linked recessive disease, is characterized by abnormal nucleic acid metabolism (biochemical salvage of nitrogenous purine bases), leading to the accumulation of uric acid in blood and tissues, intellectual disability, palsy, and self-mutilation of the lips and fingers. The disorder is due to a mutation in the gene encoding hypoxanthine-guanine phosphoribosyl transferase (HGPRT). Newborns are normal for six to eight months prior to the onset of the first symptoms.

Still another example involves **Duchenne muscular dystrophy (DMD),** an X-linked recessive disorder associated with progressive muscular wasting. It is not usually diagnosed until the child is 3 to 5 years old. Even with modern medical intervention, the disease is often fatal in the early 20s.

(a)

(b)

FIGURE 4.16 (a) A Himalayan rabbit. (b) A Siamese cat. Both species show dark fur color on the snout, ears, and paws. The patches are due to the temperature-sensitive allele responsible for pigment production.

Perhaps the most age-variable of all inherited human disorders is **Huntington disease.** Inherited as an autosomal dominant, Huntington disease affects the frontal lobes of the cerebral cortex, where progressive cell death occurs over a period of more than a decade. Brain deterioration is accompanied by spastic uncontrolled movements, intellectual and emotional deterioration, and ultimately death. Onset of this disease is varialble, but most frequently occurs around age 45.

These examples support the concept that the critical expression of genes varies throughout the life cycle of all organisms, including humans. Gene products may play more essential roles at certain life stages, and it is likely that the internal physiological environment of an organism changes with age.

Genetic Anticipation

Interest in studying the genetic onset of phenotypic expression has intensified with the discovery of heritable disorders that exhibit a progressively earlier age of onset and an increased severity of the disorder in each successive generation. This phenomenon is called **genetic anticipation.**

Myotonic dystrophy (DM), the most common type of adult muscular dystrophy, clearly illustrates genetic anticipation. Individuals with this autosomal dominant disorder exhibit extreme variation in the severity of symptoms. Mildly affected individuals develop cataracts as adults but have little or no muscular weakness. Severely affected individuals demonstrate more extensive myopathy and may be intellectually disabled. In its most extreme form, the disease is fatal just after birth. In 1989, C. J. Howeler and colleagues confirmed the correlation of increased severity and earlier onset with successive generations. They studied 61 parent–child pairs, and in 60 cases, age of onset was earlier and more severe in the child than in his or her affected parent.

In 1992, an explanation was put forward for the molecular cause of the mutation responsible for DM, as well as the basis of genetic anticipation. A particular region of the DM gene—a short trinucleotide DNA sequence—is repeated a variable number of times and is unstable. Normal individuals average about five copies of this region; minimally affected individuals have about 50 copies; and severely affected individuals have over 1000 copies. The most remarkable observation was that in successive generations, the size of the repeated segment increases. Although it is not yet clear how this expansion in size affects onset and phenotypic expression, the correlation is extremely strong. Several other inherited human disorders, including the fragile-X syndrome, Kennedy disease, and Huntington disease, also reveal an association between the size of specific regions of the responsible gene and disease severity.

ESSENTIAL POINT

Phenotypic expression is not always the direct reflection of the genotype. A percentage of organisms may not express the expected phenotype at all, the basis of the penetrance of a mutant gene. In addition, the phenotype can be modified by genetic background, temperature, and nutrition. The onset of expression of a gene may vary during the lifetime of an organism, and in future generations, the onset may occur earlier and the symptoms increased in severity (genetic anticipation). ■

4.14 Extranuclear Inheritance Modifies Mendelian Patterns

Throughout the history of genetics, occasional reports have challenged the basic tenet of Mendelian transmission genetics—that the phenotype is determined solely by nuclear genes located on the chromosomes of both parents. In this final section of the chapter, we consider several examples of inheritance patterns that vary from those predicted by the traditional biparental inheritance of nuclear genes, phenomena that are designated as **extranuclear inheritance.** In the following cases, we will focus on two broad categories. In the first, an organism's phenotype is affected by the expression of genes contained in the DNA of mitochondria or chloroplasts rather than the nucleus, generally referred to as organelle heredity. In the second category, referred to as a maternal effect, an organism's phenotype is determined by genetic information expressed in the gamete of the mother—such that, following fertilization, the developing zygote's phenotype is influenced not by the individual's genotype, but by gene products directed by the genotype of the mother.

Initially, such observations met with skepticism. However, with increasing knowledge of molecular genetics and the discovery of DNA in mitochondria and chloroplasts, the phenomenon of extranuclear inheritance came to be recognized as an important aspect of genetics.

Organelle Heredity: DNA in Chloroplasts and Mitochondria

We begin by examining examples of inheritance patterns related to chloroplast and mitochondrial function. Before DNA was discovered in these organelles, the exact mechanism of transmission of the traits was not clear, except that their inheritance appeared to be linked to something in the cytoplasm rather than to genes in the nucleus. Furthermore, transmission was most often from the maternal parent through the ooplasm, causing the results of reciprocal crosses to vary. Such an extranuclear pattern of inheritance is now appropriately called **organelle heredity.**

Source of Pollen	Location of Ovule		
	White branch	Green branch	Variegated branch
White branch	White	Green	White, green, or variegated
Green branch	White	Green	White, green, or variegated
Variegated branch	White	Green	White, green, or variegated

FIGURE 4.17 Offspring from crosses between flowers from various branches of four-o'clock plants. The photograph illustrates variegation in leaves of the madagascar spur.

Analysis of the inheritance patterns resulting from mutant alleles in chloroplasts and mitochondria has been difficult for two reasons. First, the function of these organelles is dependent on gene products from both nuclear and organelle DNA, making the discovery of the genetic origin of mutations affecting organelle function difficult. Second, many mitochondria and chloroplasts are contributed to each progeny. Thus, if only one or a few of the organelles contain a mutant gene in a cell among a population of mostly normal mitochondria, the corresponding mutant phenotype may not be revealed. This condition, referred to as **heteroplasmy,** may lead to normal cells since the organelles lacking the mutation provide the basis of wild-type function. Analysis is therefore much more complex than for Mendelian characters.

Chloroplasts: Variegation in Four-o'clock Plants

In 1908, Karl Correns (one of the rediscoverers of Mendel's work) provided the earliest example of inheritance linked to chloroplast transmission. Correns discovered a variant of the four-o'clock plant, *Mirabilis jalapa*, that had branches with either white, green, or variegated white-and-green leaves. The white areas in variegated leaves and in the completely white leaves lack chlorophyll that provides the green color to normal leaves. Chlorophyll is the light-absorbing pigment made within chloroplasts.

Correns was curious about how inheritance of this phenotypic trait occurred. As shown in **Figure 4.17**, inheritance in all possible combinations of crosses is strictly determined by the phenotype of the ovule source. For example, if the seeds (representing the progeny) were derived from ovules on branches with green leaves, all progeny plants bore only green leaves, regardless of the phenotype of the source of pollen. Correns concluded that inheritance was transmitted through the cytoplasm of the maternal parent because the pollen, which contributes little or no cytoplasm to the zygote, had no apparent influence on the progeny phenotypes.

Since leaf coloration is related to the chloroplast, genetic information contained either in that organelle or somehow present in the cytoplasm and influencing the chloroplast must be responsible for the inheritance pattern. It now seems certain that the genetic "defect" that eliminates the green chlorophyll in the white patches on leaves is a mutation in the DNA housed in the chloroplast.

Mitochondrial Mutations: *poky* in *Neurospora* and *petite* in *Saccharomyces*

Mutations affecting mitochondrial function have been discovered and studied, revealing that they too contain a distinctive genetic system. As with chloroplasts, mitochondrial mutations are transmitted through the cytoplasm. In our current discussion, we will emphasize the link between mitochondrial mutations and the resultant extranuclear inheritance patterns.

In 1952, Mary B. Mitchell and Hershel K. Mitchell studied the bread mold *Neurospora crassa*. They discovered a slow-growing mutant strain and named it *poky*. Slow growth is associated with impaired mitochondrial function, specifically in relation to certain cytochromes essential for electron transport. Results of genetic crosses between wild-type and *poky* strains suggest that *poky* is an extranuclear trait inherited through the cytoplasm. If one mating type is *poky* and the other is wild type, all progeny colonies are *poky*. The reciprocal cross, where *poky* is transmitted by the other mating type, produces normal wild-type colonies.

Another extensive study of mitochondrial mutations has been performed with the yeast *Saccharomyces cerevisiae*. The first such mutation, described by Boris Ephrussi and his

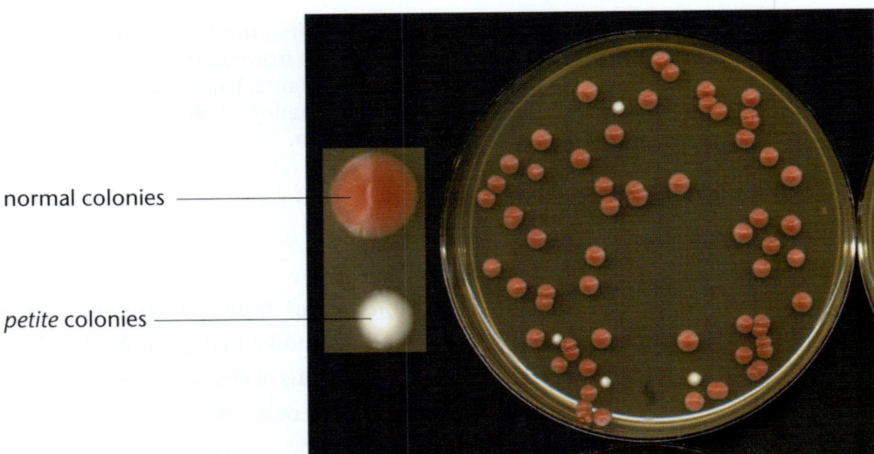

FIGURE 4.18 A photo comparing normal versus *petite* colonies of the yeast *Saccharomyces cerevisiae*. The larger normal colonies appear pink while the smaller mutant *petite* colonies appear white.

coworkers in 1956, was named *petite* because of the small size of the yeast colonies (**Figure 4.18**). Many independent *petite* mutations have since been discovered and studied, and all have a common characteristic—a deficiency in cellular respiration involving abnormal electron transport. The majority of them demonstrate cytoplasmic transmission, indicating mutations in the DNA of the mitochondria. This organism is a facultative anaerobe and can grow by fermenting glucose through glycolysis; thus, it may survive the loss of mitochondrial function by generating energy anaerobically.

The complex genetics of *petite* mutations has revealed that a small proportion are the result of nuclear DNA changes. They exhibit Mendelian inheritance and illustrate that mitochondria function depends on both nuclear and organellar gene products.

Mitochondrial Mutations: Human Genetic Disorders

Our knowledge of the genetics of mitochondria has now greatly expanded. The DNA found in human mitochondria has been completely sequenced and contains 16,569 base pairs. Mitochondrial gene products have been identified and include the following:

13 proteins, required for aerobic cellular respiration

22 transfer RNAs (tRNAs), required for translation

2 ribosomal RNAs (rRNAs), required for translation

Because a cell's energy supply is largely dependent on aerobic cellular respiration, disruption of any mitochondrial gene by mutation may potentially have a severe impact on that organism, such as we saw in our previous discussion of the *petite* mutation in yeast. In fact, mtDNA is particularly vulnerable to mutations for two possible reasons. First, the ability to repair mtDNA damage does not appear to be equivalent to that of nuclear DNA. Second, the concentration of highly mutagenic free radicals generated by cell respiration that accumulate in such a confined space very likely raises the mutation rate in mtDNA.

Fortunately, a zygote receives a large number of organelles through the egg, so if only one organelle or a few of them contain a mutation (an illustration of *heteroplasmy*), the impact is greatly diluted by the many mitochondria that lack the mutation and function normally. If a deleterious mutation arises or is present in the initial population of organelles, adults will have cells with a variable mixture of both normal and abnormal organelles. From a genetic standpoint, this condition of heteroplasmy makes analysis quite difficult.

Many disorders in humans are known to be due to mutations in mitochondrial genes. For example, **myoclonic epilepsy and ragged-red fiber disease (MERRF)** demonstrates a pattern of inheritance consistent with maternal transmission. Only the offspring of affected mothers inherit this disorder, while the offspring of affected fathers are normal. Individuals with this rare disorder express ataxia (lack of muscular coordination), deafness, dementia, and epileptic seizures. The disease is named for the presence of "ragged-red" skeletal-muscle fibers that exhibit blotchy red patches resulting from the proliferation of aberrant mitochondria. Brain function, which has a high energy demand, is also affected in this disorder, leading to the neurological symptoms described above.

Analysis of mtDNA from patients with MERRF has revealed a mutation in one of the 22 mitochondrial genes encoding a transfer RNA. Specifically, the gene encoding $tRNA^{Lys}$ (the tRNA that delivers lysine during translation) contains an altered DNA sequence. This genetic alteration interferes with the capacity for translation within the organelle, which in turn leads to the various manifestations of the disorder.

The cells of affected individuals exhibit the condition called heteroplasmy, containing a mixture of normal and abnormal mitochondria. Different patients contain different proportions of the two, and even different cells from the same patient exhibit various levels of abnormal mitochondria.

Were it not for heteroplasmy, the mutation would very likely be lethal, testifying to the essential nature of mitochondrial function and its reliance on the genes encoded by mtDNA within the organelle.

Mitochondria, Human Health, and Aging

The study of hereditary mitochondrial-based disorders provides insights into the critical importance of this organelle during normal development. In fact, mitochondrial dysfunction seems to be implicated in a large number of major human disease conditions, including anemia, blindness, type 2 (late-onset) diabetes, autism, atherosclerosis, infertility, neurodegenerative diseases such as Parkinson, Alzheimer, and Huntington disease, schizophrenia and bipolar disorders, and a variety of cancers. It is becoming evident, for example, that mutations in mtDNA are present in such human malignancies as skin, colorectal, liver, breast, pancreatic, lung, prostate, and bladder cancers.

Over 400 mtDNA mutations associated with more than 150 distinct mtDNA-based genetic syndromes have been identified. Genetic tests for detecting mutations in the mtDNA genome that may serve as early-stage disease markers have been developed. However, it is still unclear whether mtDNA mutations are causative effects contributing to development of malignant tumors or whether they are the consequences of tumor formation. Nonetheless, there is an interesting link between mtDNA mutations and cancer, including data suggesting that many chemical carcinogens have significant mutation effects on mtDNA.

The study of hereditary mitochondrial-based disorders has also suggested a link between the progressive decline of mitochondrial function and the aging process. It has been hypothesized that the accumulation of sporadic mutations in mtDNA leads to an increased prevalence of defective mitochondria (and the concomitant decrease in the supply of ATP) in cells over a lifetime. This condition in turn plays a significant role in aging.

Many studies have now documented that aging tissues contain mitochondria with increased levels of DNA damage. The major question is whether such changes are simply biomarkers of the aging process or whether they lead to a decline in physiological function, which in turn, contributes significantly to aging. In support of the latter hypothesis, one study links age-related muscle fiber atrophy in rats to deletions in mtDNA and electron transport abnormalities. Such deletions appear to be present in the mitochondria of atrophied muscle fibers, but are absent from fibers in regions of normal tissue. It is important to note that mutations in the nuclear genome also impact mitochondrial function and human disease and aging. For example, another study involving genetically altered mice is most revealing. Such mice have a nuclear gene altered that diminishes proofreading during the replication of mtDNA. These mice display reduced fertility and accumulate mutations over time at a much higher rate than is normal. These mice also show many characteristics of premature aging, as observed by loss and graying of hair, reduction in bone density and muscle mass, decline in fertility, anemia, and reduced life span.

These, and other studies, continue to speak to the importance of normal mitochondrial function. As cells undergo genetic damage, which appears to be a natural phenomenon, their function declines, which may be an underlying factor in aging as well as in the progression of age-related disorders.

Maternal Effect

In **maternal effect,** also referred to as maternal influence, an offspring's phenotype for a particular trait is under the control of the mother's *nuclear gene products* present in the egg. This is in contrast to biparental inheritance, where both parents transmit information on genes in the nucleus that determines the offspring's phenotype. In cases of maternal effect, the nuclear genes of the female gamete are transcribed, and the genetic products (either proteins or yet untranslated mRNAs) accumulate in the egg ooplasm. After fertilization, these products are distributed among newly formed cells and influence the patterns or traits established during early development. The following example will illustrate such an influence of the maternal genome on particular traits.

Embryonic Development in *Drosophila*

A recently documented example that illustrates maternal effect involves various genes that control embryonic development in *Drosophila melanogaster*. The genetic control of embryonic development in *Drosophila*, discussed in greater detail in Chapter 20, is a fascinating story. The protein products of the maternal-effect genes function to activate other genes, which may in turn activate still other genes. This cascade of gene activity leads to a normal embryo whose subsequent development yields a normal adult fly. The extensive work by Edward B. Lewis, Christiane Nüsslein-Volhard, and Eric Wieschaus (who shared the 1995 Nobel Prize for Physiology or Medicine for their findings) has clarified how these and other genes function. Genes that illustrate maternal effect have products that are synthesized by the developing egg and stored in the oocyte prior to fertilization. Following fertilization, these products specify molecular gradients that determine spatial organization as development proceeds.

For example, the gene *bicoid* (*bcd*) plays an important role in specifying the development of the anterior portion of the fly. Embryos derived from mothers who are homozygous for this mutation (bcd^-/bcd^-) fail to develop anterior areas that normally give rise to the head and thorax of the adult fly. Embryos whose mothers contain at least one wild-type allele (bcd^+) develop normally, even if the genotype of the embryo is homozygous for the mutation. Consistent with the concept of maternal effect, the *genotype of the female parent*, not the *genotype of the embryo*, determines the phenotype of the offspring. Nusslein-Volhard and Wieschaus, using large-scale mutant screens, discovered many other maternal-effect genes critical to early development in *Drosophila*.

ESSENTIAL POINT

When patterns of inheritance vary from that expected due to biparental transmission of nuclear genes, phenotypes are often found to be under the control of DNA present in mitochondria or chloroplasts, or are influenced during development by the expression of the maternal genotype in the egg. ■

GENETICS, ETHICS, AND SOCIETY

Mitochondrial Replacement and Three-Parent Babies

As many as 1 in 5000 people carry a potentially disease-causing mutation in most or all of their mitochondrial DNA (mtDNA). These mutations can lead to symptoms as varied as blindness, neurodegenerative defects, strokes, muscular dystrophies, and diabetes. In addition, mtDNA mutations can cause infertility, miscarriages, and the death of newborns and children. There are few treatments for mitochondrial diseases and, until recently, no potential cures.

In the 1990s, fertility clinics in the United States began using *cytoplasmic transfer* to treat some cases of infertility, including those caused by mitochondrial mutations. Cytoplasmic transfer involves the injection of cytoplasm containing normal mitochondria from a donor egg into a recipient egg prior to *in vitro* fertilization and implantation. Worldwide, more than 100 babies have been born following cytoplasmic transfer. In 2001, the U.S. Food and Drug Administration halted the procedure after two cases of X-chromosome abnormalities and one case of a developmental disorder appeared after the treatment. Since then, no research or treatments using cytoplasmic transfer have been permitted in the United States, although it is still used in other countries.

Since 2001, the United Kingdom and Japan have pioneered the use of other mitochondrial treatment methods. These methods are known generally as mitochondrial replacement therapies (MRTs). Presently, the two most commonly used MRTs are maternal spindle transfer (MST) and pronuclear transfer (PNT).

As described earlier in this chapter, MST is an *in vitro* fertilization method that involves the transfer of nuclear chromosomes at the metaphase II stage of meiosis from the patient's egg (containing defective mitochondria) into an enucleated donor egg (containing healthy mitochondria). The egg is then fertilized *in vitro*. PNT is a variation of MST. It involves *in vitro* fertilization of both the donor's egg and the patient's egg with sperm from the same donor. After fertilization, but before pronuclear fusion, the egg and sperm pronuclei from the patient's zygote are transferred to the donor's enucleated zygote. In both methods, the resulting hybrid zygotes are grown to the blastocyst stage, screened for genetic defects, and implanted into the patient. In both methods, the resulting offspring contain nuclear DNA from the mother and father, and mtDNA from the donor.

In 2015, the United Kingdom became the first country in the world to legalize the use of MRT methods for research in human embryos, and in 2016, it approved the limited use of MRT in humans. The worldwide media responded with headlines of "three-parent babies" and "children with three genetic parents," and the articles led to some confusion about the details of the techniques involved. Since then, the outburst of controversy has centered on the ethical and social implications of using MRT. Proponents of MRT welcome these mitochondrial treatments as significant contributions to the welfare of parents and children, who will no longer suffer from debilitating diseases. Opponents express reservations ranging from safety concerns to objections about human germ-line modification and damaging a child's identity.

Over the next decade, as more children are born with mitochondria from donor eggs, we will gain more perspective on the positive and negative aspects of MRT.

Your Turn

Take time, individually or in groups, to consider the following questions. Investigate the references dealing with the technical and ethical challenges surrounding mitochondrial replacement.

1. Summarize the ethical arguments used to support and oppose the use of MRT in humans. In your opinion, which arguments have validity, and why? How do the ethical arguments differ between PNT versus MST?

   These topics are discussed in Gómez-Tatay, L. et al. (2017). Mitochondrial Modification Techniques and Ethical Issues. *J. Clin. Med.* 6, 25; DOI:10.3390/jcm6030025. *Also, see* Baylis, F. (2013). The Ethics of Creating Children with Three Genetic Parents. *Reproductive BioMed Online*, 26:531–534.

2. Much of the controversy surrounding MRT methods has been triggered by the phrase "three-parent babies" in media headlines. Do you think that this phrase is an accurate description of children born following mitochondrial replacement?

   For a discussion of this question, see Dimond, R. (2015). Social and Ethical Issues in Mitochondrial Donation. *Br. Med. Bull.* 115:173–182.

CASE STUDY Is it all in the genes?

Marcia saw an ad on television for ancestry DNA testing and thought, "Why not?" She ordered a kit, swabbed her inner cheek, and returned the kit for analysis. Several weeks later, she was surprised to learn that she was 17 percent Native American. An elderly great aunt confirmed that her mother's family intermarried with members of Native American tribes in the Pacific Northwest in the early twentieth century. To investigate her maternal heritage, Marcia ordered a mitochondrial DNA (mtDNA) test, which confirmed her Native American ancestry. Based on these genetic results, she applied to several Native American tribes for enrollment as a tribal member. She was shocked when she was turned down. In discussions, tribal officials told her that DNA alone is not sufficient to define who is Native American. Tribal standards for enrollment vary, but usually cultural attributes such as knowledge of the language, customs, and history of the tribe are important considerations for enrollment decisions. Marcia was not satisfied and felt that she had a strong case based on biology alone. This series of events raises several questions:

1. Why did Marcia choose mitochondrial testing to determine her maternal heritage?
2. How many great-grandmothers does any individual (such as Marcia) have? How many of them contribute to the mitochondrial DNA that an individual (Marcia) carries?
3. How much importance should we place on the results of ancestral genetic testing especially when these results have social, political, and legal implications? Is it ethical to determine one's identity primarily or even partially on genetic considerations?

See Tallbear, K., and Blonick, D. A. (2004). "Native American" DNA Tests: What Are the Risks to Tribes? *The Native Voice*, Dec. 3–17.

INSIGHTS AND SOLUTIONS

Genetic problems take on added complexity if they involve two independent characters and multiple alleles, incomplete dominance, or epistasis. The most difficult types of problems are those that pioneering geneticists faced during laboratory or field studies. They had to determine the mode of inheritance by working backward from the observations of offspring to parents of unknown genotype.

1. Consider the problem of comb-shape inheritance in chickens, where walnut, pea, rose, and single are the observed distinct phenotypes (see the accompanying photographs). How is comb shape inherited, and what are the genotypes of the P_1 generation of each cross? Use the following data:

Cross 1:	single	×	single	⟶	all single
Cross 2:	walnut	×	walnut	⟶	all walnut
Cross 3:	rose	×	pea	⟶	all walnut
Cross 4:	$F_1 \times F_1$ of cross 3				
	walnut	×	walnut	⟶	93 walnut
					28 rose
					32 pea
					10 single

Walnut

Pea

Rose

Single

Solution: At first glance, this problem appears quite difficult. However, applying a systematic approach and breaking the analysis into steps usually simplifies it. Our approach involves two steps. First, analyze the data carefully for any useful information. Then, once you identify something that is clearly helpful, follow an empirical approach—that is, formulate a hypothesis and, in a sense, test it against the given data. Look for a pattern of inheritance that is consistent with all cases.

This problem gives two immediately useful facts. First, in cross 1, P_1 singles breed true. Second, while P_1 walnuts breed true in cross 2, a walnut phenotype is also produced in cross 3 between rose and pea. When these F_1 walnuts are crossed in cross 4, all four comb shapes are produced in a ratio that approximates 9:3:3:1. This observation immediately suggests a cross involving two gene pairs, because the resulting data display the same ratio as in Mendel's dihybrid crosses. Since only one trait is involved (comb shape), epistasis may be occurring. This could serve as your working hypothesis, and you must now propose how the two gene pairs "interact" to produce each phenotype.

If you call the allele pairs A, a and B, b, you can predict that because walnut represents 9/16 in cross 4, $A–B–$ will produce walnut. You might also hypothesize that in cross 2, the genotypes are $AABB \times AABB$, where walnut bred true. (Recall that $A-$ and $B-$ mean AA or Aa and BB or Bb, respectively.)

The phenotype representing 1/16 of the offspring of cross 4 is single; therefore, you could predict that this phenotype is the result of the *aabb* genotype. This is consistent with cross 1.

Now you have only to determine the genotypes for rose and pea. The most logical prediction is that at least one dominant *A* or *B* allele combined with the double recessive condition of the other allele pair accounts for these phenotypes. For example,

$$A-bb \longrightarrow \text{rose}$$
$$aaB- \longrightarrow \text{pea}$$

If *AAbb* (rose) is crossed with *aaBB* (pea) in cross 3, all offspring will be *AaBb* (walnut). This is consistent with the data, and you must now look at only cross 4. We predict these walnut genotypes to be *AaBb* (as above), and from the cross *AaBb* (walnut) × *AaBb* (walnut) we expect

9/16	$A-B-$	(walnut)
3/16	$A-bb$	(rose)
3/16	$aaB-$	(pea)
1/16	aabb	(single)

Our prediction is consistent with the information given. The initial hypothesis of the epistatic interaction of two gene pairs proves consistent throughout, and the problem is solved.

This problem demonstrates the need for a basic theoretical knowledge of transmission genetics. Then, you can search for appropriate clues that will enable you to proceed in a stepwise fashion toward a solution. Mastering problem solving requires practice but can give you a great deal of satisfaction. Apply this general approach to the following problems.

2. Consider the two very limited unrelated pedigrees shown at the right. Of the four combinations of X-linked recessive, X-linked dominant, autosomal recessive, and autosomal dominant, which modes of inheritance can be absolutely ruled out in each case?

Solution: For both pedigrees, X-linked recessive and autosomal recessive remain possible, provided that the maternal parent is heterozygous in pedigree (b). At first glance autosomal dominance seems unlikely in pedigree (a), since at least half of the offspring should express a dominant trait expressed by one of their parents. However, while it is true that if the affected parent carries an autosomal dominant gene heterozygously, each offspring has a 50 percent chance of inheriting and expressing the mutant gene, the sample size of four offspring is too small to rule out this possibility. In pedigree (b), autosomal dominance is clearly possible. In both cases, one can rule out X-linked dominance because the female offspring would inherit and express the dominant allele, and they do not express the trait in either pedigree.

(a)

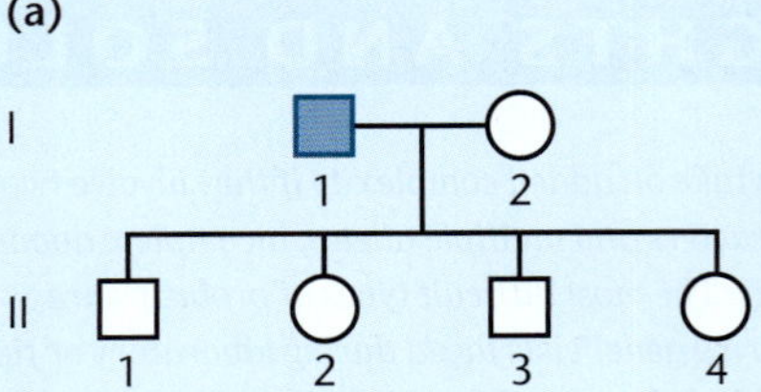

(b)

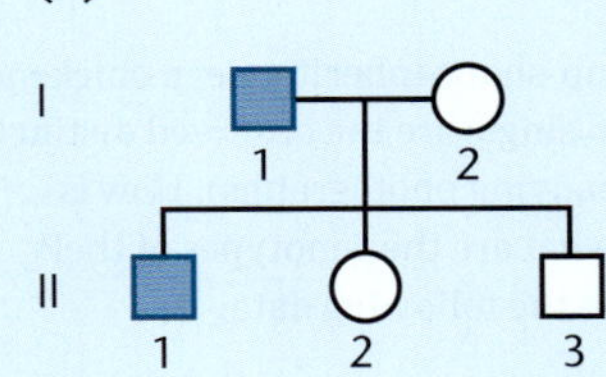

Problems and Discussion Questions

Mastering Genetics Visit for instructor-assigned tutorials and problems.

1. **HOW DO WE KNOW?** In this chapter, we focused on many extensions and modifications of Mendelian principles and ratios. In the process, we encountered many opportunities to consider how this information was acquired. Answer the following fundamental questions:
   (a) How were early geneticists able to ascertain inheritance patterns that did not fit typical Mendelian ratios?
   (b) How did geneticists determine that inheritance of some phenotypic characteristics involves the interactions of two or more gene pairs? How were they able to determine how many gene pairs were involved?
   (c) How do we know that specific genes are located on the sex-determining chromosomes rather than on autosomes?
   (d) For genes whose expression seems to be tied to the gender of individuals, how do we know whether a gene is X-linked in contrast to exhibiting sex-limited or sex-influenced inheritance?
   (e) How was extranuclear inheritance discovered?

2. **CONCEPT QUESTION** Review the Chapter Concepts list on page 77. These all relate to exceptions to the inheritance patterns encountered by Mendel. Write a short essay that explains why multiple and lethal alleles often result in a modification of the classic Mendelian monohybrid and dihybrid ratios. ■

3. In Shorthorn cattle, coat color may be red, white, or roan. Roan is an intermediate phenotype expressed as a mixture of red and white hairs. The following data are obtained from various crosses:

   red × red ⟶ all red
   white × white ⟶ all white
   red × white ⟶ all roan
   roan × roan ⟶ 1/4 red: 1/2 roan: 1/4 white

   (a) How is coat color inherited? What are the genotypes of parents and offspring for each cross?

(b) Does the roan phenotype illustrate a case of incomplete dominance or a case of codominance? Explain.

4. With regard to the ABO blood types in humans, determine the genotypes of the male parent and female parent:

> *Male parent:* blood type B whose mother was type O
> *Female parent:* blood type A whose father was type B

Predict the blood types of the offspring that this couple may have and the expected ratio of each.

5. In *Drosophila*, two alleles for the *Duox* gene may result in lethality (*Duox/Duox*), wings curled upwards (*Duox/duox*), or straight wings (*duox/duox*). What ratio is obtained when curly-winged flies are interbred? Is the *Duox* allele behaving dominantly or recessively in causing (a) lethality; (b) curled wings?

6. Three gene pairs located on separate autosomes determine flower color and shape as well as plant height. The first pair exhibits incomplete dominance, where color can be red, pink (the heterozygote), or white. The second pair leads to the dominant personate or recessive peloric flower shape, while the third gene pair produces either the dominant tall trait or the recessive dwarf trait. Homozygous plants that are red, personate, and tall are crossed with those that are white, peloric, and dwarf. Determine the F_1 genotype(s) and phenotype(s). If the F_1 plants are interbred, what proportion of the offspring will exhibit the same phenotype as the F_1 plants?

7. As in the plants of Problem 6, color may be red, white, or pink; and flower shape may be personate or peloric. Determine the P_1 and F_1 genotypes for the following crosses:

(a) red, peloric × white, personate → F_1: all pink, personate

(b) red, personate × white, peloric → F1: all pink, personate

(c) pink, personate × red, peloric → F_1:
- 1/4 red, personate
- 1/4 red, peloric
- 1/4 pink, personate
- 1/4 pink, peloric

(d) pink, personate × white, peloric → F_1:
- 1/ 4 white, personate
- 1/ 4 white, peloric
- 1/ 4 pink, personate
- 1/ 4 pink, peloric

(e) What phenotypic ratios woud result from crossing the F1 of (a) to the F_1 of (b)?

8. Two different genes, located on two different chromosomes, are responsible for color production in the aleurone layer of corn kernels. For color production (either purple or red), the dominant alleles of these two genes (*C* and *R*) must come together. Furthermore, a third gene, located on a third chromosome, interacts with the *C* and *R* alleles to determine whether the aleurone will be red or purple. While the dominant allele (*P*) ensures purple color, the homozygous recessive condition (*pp*) makes the aleurone red. Determine the F_1 phenotypic ratio of the following crosses: (a) *CCrrPP* × *ccRRpp*; (b) *CcRRpp* × *CCRrpp*; (c) *CcRrPp* × *CcRrpp*.

9. Given the inheritance pattern of aleurone color in corn kernels described in Problem 8, predict the genotype and phenotype of the parents that produced the following F_1 offspring: (a) 27/64 purple: 9/64 red: 28/64 colorless; (b) 9/32 purple: 0/32 red: 3/32 colorless.

10. In *Drosophila*, a locus responsible for eye color is X-linked, and the normal wild-type red eye color is dominant to white eye color. A cross is set up between *Drosophila* males and females who have both their eyes red and come from a previous cross of a white-eyed male and a red-eyed female. There is a progeny of 1000 individuals from this cross. Determine the number of red-eyed and white-eyed *Drosophila*, and note how many of these could be expected to be male or female.

11. In humans, the ABO blood type is under the control of autosomal multiple alleles. Red–green color blindness is a recessive X-linked trait. If two parents who are both type A and have normal vision produce a son who is color-blind and type O, what is the probability that their next child will be a female who has normal vision and is type O?

12. In goats, development of the beard is due to a recessive gene. The following cross involving true-breeding goats was made and carried to the F_2 generation:

> P_1: bearded female × beardless male
> ↓
> F_1: all bearded males and beardless females
>
> $F_1 \times F_1$ →
> - 1/8 beardless males
> - 3/8 bearded males
> - 3/8 beardless females
> - 1/8 bearded females

Offer an explanation for the inheritance and expression of this trait, diagramming the cross. Propose one or more crosses to test your hypothesis.

13. In cats, orange coat color is determined by the *b* allele, and black coat color is determined by the *B* allele. The heterozygous condition results in a coat pattern known as tortoiseshell. These genes are X-linked. What kinds of offspring would be expected from a cross of a black male and a tortoiseshell female? What are the chances of getting a tortoiseshell male?

14. In *Drosophila*, an X-linked recessive mutation, *scalloped* (*sd*), causes irregular wing margins. Diagram the F_1 and F_2 results if (a) a scalloped female is crossed with a normal male; (b) a scalloped male is crossed with a normal female. Compare these results to those that would be obtained if the *scalloped* gene were autosomal.

15. Another recessive mutation in *Drosophila*, *ebony* (*e*), is on an autosome (chromosome 3) and causes darkening of the body compared with wild-type flies. What phenotypic F_1 and F_2 male and female ratios will result if a scalloped-winged female with normal body color is crossed with a normal-winged ebony male? Work this problem by both the Punnett square method and the forked-line method.

16. While vermilion is X-linked in *Drosophila* and causes eye color to be bright red, *brown* is an autosomal recessive mutation that causes the eye to be brown. Flies carrying both mutations lose all pigmentation and are white-eyed. Predict the F_1 and F_2 results of the following crosses:
    (a) vermilion females × brown males
    (b) brown females × vermilion males
    (c) white females × wild males

17. In pigs, coat color may be sandy, red, or white. A geneticist spent several years mating true-breeding pigs of all different color combinations, even obtaining true-breeding lines from different parts of the country. For crosses 1 and 4 in the following table, she encountered a major problem: her computer crashed and she lost the F_2 data. She nevertheless persevered and, using the limited

data shown here, was able to predict the mode of inheritance and the number of genes involved, as well as to assign genotypes to each coat color. Attempt to duplicate her analysis, based on the available data generated from the crosses shown.

Cross	P_1	F_1	F_2
1	sandy × sandy	All red	Data lost
2	red × sandy	All red	3/4 red: 1/4 sandy
3	sandy × white	All sandy	3/4 sandy: 1/4 white
4	white × red	All red	Data lost

When you have formulated a hypothesis to explain the mode of inheritance and assigned genotypes to the respective coat colors, predict the outcomes of the F_2 generations where the data were lost.

18. A geneticist from an alien planet that prohibits genetic research brought with him two true-breeding lines of frogs. One frog line croaks by *uttering* "rib-it rib-it" and has purple eyes. The other frog line croaks by *muttering* "knee-deep knee-deep" and has green eyes. He mated the two frog lines, producing F_1 frogs that were all utterers with blue eyes. A large F_2 generation then yielded the following ratios:

27/64 blue, utterer
12/64 green, utterer
9/64 blue, mutterer
9/64 purple, utterer
4/64 green, mutterer
3/64 purple, mutterer

(a) How many total gene pairs are involved in the inheritance of both eye color and croaking?
(b) Of these, how many control eye color, and how many control croaking?
(c) Assign gene symbols for all phenotypes, and indicate the genotypes of the P_1, F_1, and F_2 frogs.
(d) After many years, the frog geneticist isolated true-breeding lines of all six F_2 phenotypes. Indicate the F_1 and F_2 phenotypic ratios of a cross between a blue, mutterer and a purple, utterer.

19. In another cross, the frog geneticist from Problem 18 mated two purple, utterers with the results shown here. What were the genotypes of the parents?

9/16 purple, utterers
3/16 purple, mutterers
3/16 green, utterers
1/16 green, mutterers

20. In cattle, coats may be solid white, solid black, or black-and-white spotted. When true-breeding solid whites are mated with true-breeding solid blacks, the F_1 generation consists of all solid white individuals. After many $F_1 \times F_1$ matings, the following ratio was observed in the F_2 generation:

12/16 solid white
3/16 black-and-white spotted
1/16 solid black

Explain the mode of inheritance governing coat color by determining how many gene pairs are involved and which genotypes yield which phenotypes. Is it possible to isolate a true-breeding strain of black-and-white spotted cattle? If so, what genotype would they have? If not, explain why not.

21. Consider the following three pedigrees, all involving the same human trait:

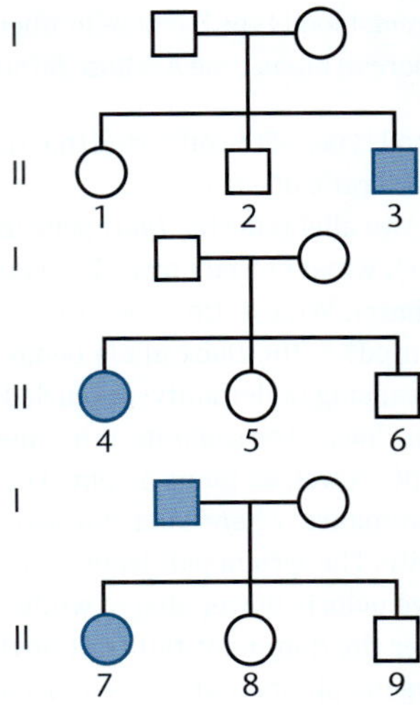

(a) Which sets of conditions, if any, can be excluded?

dominant and X-linked
dominant and autosomal
recessive and X-linked
recessive and autosomal

(b) For any set of conditions that you excluded, indicate the *single individual* in generation II (1–9) that was instrumental in your decision to exclude that condition. If none were excluded, answer "none apply."
(c) Given your conclusions in parts (a) and (b), indicate the *genotype* of individuals II-1, II-6, and II-9. If more than one possibility applies, list all possibilities. Use the symbols *A* and *a* for the genotypes.

22. Labrador retrievers may be black, brown, or golden in color (see the chapter opening photograph on p. 77). Although each color may breed true, many different outcomes occur if numerous litters are examined from a variety of matings, where the parents are not necessarily true-breeding. The following results show some of the possibilities. Propose a mode of inheritance that is consistent with these data, and indicate the corresponding genotypes of the parents in each mating. Indicate as well the genotypes of dogs that breed true for each color.

(a) black	× brown	⟶	all black
(b) black	× brown	⟶	1/2 black 1/2 brown
(c) black	× brown	⟶	3/4 black 1/4 golden
(d) black	× golden	⟶	all black
(e) black	× golden	⟶	4/8 golden 3/8 black 1/8 brown
(f) black	× golden	⟶	2/4 golden 1/4 black 1/4 brown
(g) brown	× brown	⟶	3/4 brown 1/4 golden
(h) black	× black	⟶	9/16 black 4/16 golden 3/16 brown

23. In a genetic screening in *Drosophila* for autosomal recessive mutations with bristle phenotypes, three independent mutants with forked bristles are identified ($f1$, $f2$, and $f3$). A partial complementation analysis is shown below. Which mutations are in the same complementation group? Would $f2$ and $f3$ complement each other?

Cross 1:	$f1 \times f2 \longrightarrow$	F_1: all forked bristles
Cross 2:	$f1 \times f3 \longrightarrow$	F_1: all wild-type bristles

24. Horses can be cremello (a light cream color), chestnut (a reddish brown color), or palomino (a golden color with white in the horse's tail and mane).

Chestnut

Palomino

Cremello

Of these phenotypes, only palominos never breed true. The following results have been observed:

cremello × palomino	⟶	1/2 cremello 1/2 palomino
chestnut × palomino	⟶	1/2 chestnut 1/2 palomino
palomino × palomino	⟶	1/4 chestnut 1/2 palomino 1/4 cremello

(a) From these results, determine the mode of inheritance by assigning gene symbols and indicating which genotypes yield which phenotypes.

(b) Predict the F_1 and F_2 results of many initial matings between cremello and chestnut horses.

25. Labrador Retrievers can have three coat colors: golden, brown, and black. The color of the coat is black if the *B* allele is present and brown when the *b* allele is homozygous. However, the colors black and brown are also dependent on the presence of the allele *A*. An *aa* animal would always have a golden coat.

(a) What phenotypic and genotypic ratios would you expect from a cross between *AaBb* × *AaBb* Labradors?

(b) Males with golden coats (*aabb*) are crossed with females of unknown genotype and the ratios obtained in the progenies are shown in the table. Determine the genotype of the females.

(1)	(2)	(3)
8 golden 8 black	9 black 10 brown	4 black 5 brown 10 golden

26. A genetic study was carried out on different families to determine their blood antigens. The antigens ABO and MN were identified in the parents and their children. Sibling information was recorded for the children, but it was erroneously not assigned to their parents. Parental information is in the following table:

Couple	Phenotype of the father	Phenotype of the mother
C1	AB, N	O, N
C2	O, N	AB, M
C3	AB, MN	AB, MN

Determine the expected genotypic and phenotypic ratios among the children for each couple in order to assign siblings to their parents.

27. Two boys are born at the same time in the same hospital. When they were about to start school, one of them is diagnosed with deuteranomaly (defective green color detection) despite both parents being normal. However, the other boy has normal vision although his father has deuteranomaly. As a genetic counselor, you know that the gene responsible for the condition lies in the X chromosome and has a 5 percent incidence among males. The family of the diagnosed child consults you worried that the children may have been swapped in the hospital. What information could you provide and what would you suggest?

28. Foxes can have platinum or silver coats. They can also have long or short tails. When short-tailed, platinum-coated foxes are mated with long-tailed, silver-coated ones, half of the offspring are long-tailed, platinum-coated and the other half, long-tailed, silver-coated. When long-tailed, platinum-coated foxes from this offspring are mated, the following results are obtained:

2/12 short-tailed, platinum-coated	6/12 long-tailed, platinum-coated
1/12 short-tailed, silver-coated	3/12 long-tailed, silver-coated

Fox breeders tell you that foxes with platinum coats never produce as many offspring as ones with silver coats. Knowing silver-coated foxes to be true-breeding, provide a genetic explanation for the phenotypes and ratios obtained crossing the F_1 platinum coated foxes.

29. What genetic criteria distinguish a case of extranuclear inheritance from (a) a case of Mendelian autosomal inheritance; (b) a case of X-linked inheritance?

30. The specification of the anterior–posterior axis in *Drosophila* embryos is initially controlled by various gene products that are synthesized and stored in the mature egg following oogenesis. Mutations in these genes result in abnormalities of the axis during embryogenesis, illustrating maternal effect. How do such mutations vary from those involved in organelle heredity that illustrate extranuclear inheritance? Devise a set of parallel crosses and expected outcomes involving mutant genes that contrast maternal effect and organelle heredity.

31. The gene *nanos* (*nos*) is recessive and its mutation can lead to maternal effect, since it is necessary to have mRNAs deposited maternally in the oocyte for embryonic development. Embryos derived from *nos* females lack abdominal segments and fail to complete embryogenesis.

(a) Describe the effect on the progeny for a male (nos^-/nos^-) and a female (nos^+/nos^-).

(b) What would happen in the reciprocal cross, a female (nos^-/nos^-) and a male (nos^+/nos^-)?

32. Students taking a genetics exam were expected to answer the following question by converting data to a "meaningful ratio" and then solving the problem. The instructor assumed that the final ratio would reflect two gene pairs, and most correct answers did. Here is the exam question:

"Flowers may be white, orange, or brown. When plants with white flowers are crossed with plants with brown flowers, all the F_1 flowers are white. For F_2 flowers, the following data were obtained:

48 white
12 orange
4 brown

Convert the F_2 data to a meaningful ratio that allows you to explain the inheritance of color. Determine the number of genes involved and the genotypes that yield each phenotype."

(a) Solve the problem for two gene pairs. What is the final F_2 ratio?

(b) A number of students failed to reduce the ratio for two gene pairs as described above and solved the problem using three gene pairs. When examined carefully, their solution was deemed a valid response by the instructor. Solve the problem using three gene pairs.

33. In four o'clock plants, many flower colors are observed. In a cross involving two true-breeding strains, one crimson and the other white, all of the F_1 generation were rose color. In the F_2, four new phenotypes appeared along with the P_1 and F_1 parental colors. The following ratio was obtained:

1/16 crimson
2/16 orange
1/16 yellow
2/16 magenta
4/16 rose
2/16 pale yellow
4/16 white

Propose an explanation for the inheritance of these flower colors.

34. Below is a partial pedigree of hemophilia in the British Royal Family descended from Queen Victoria, who is believed to be the original "carrier" in this pedigree. Analyze the pedigree and indicate which females are also certain to be carriers. What is the probability that Princess Irene is a carrier?

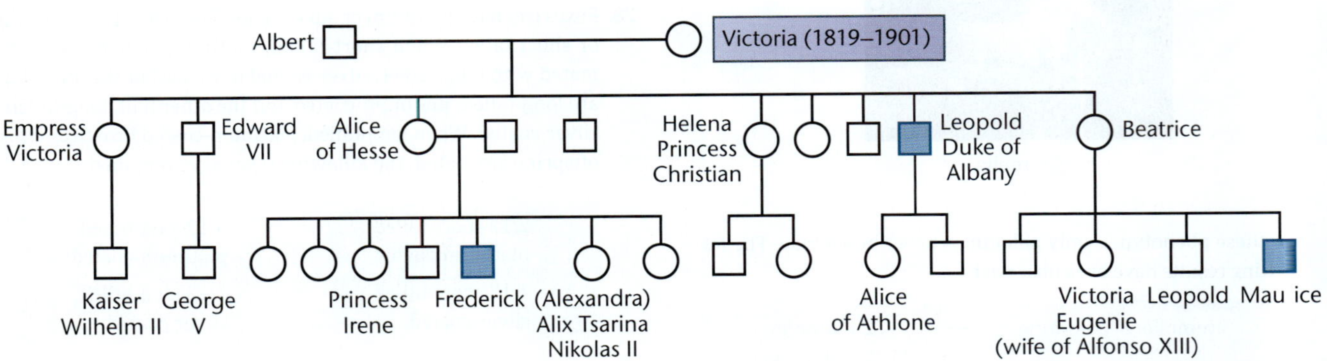

5

Sex Determination and Sex Chromosomes

A human X chromosome highlighted using fluorescence *in situ* hybridization (FISH), a method in which specific probes bind to specific sequences of DNA. The probe producing green fluorescence binds to DNA at the centromere of X chromosomes. The probe producing red fluorescence binds to the DNA sequence of the X-linked Duchenne muscular dystrophy (DMD) gene.

CHAPTER CONCEPTS

- A variety of mechanisms have evolved that result in sexual differentiation, leading to sexual dimorphism and greatly enhancing the production of genetic variation within species.
- Often, specific genes, usually on a single chromosome, cause maleness or femaleness during development.
- In humans, the presence of extra X or Y chromosomes beyond the diploid number may be tolerated but often leads to syndromes demonstrating distinctive phenotypes.
- While segregation of sex-determining chromosomes should theoretically lead to a one-to-one sex ratio of males to females, in humans the actual ratio favors males at conception.
- In mammals, females inherit two X chromosomes compared to one in males, but the extra genetic information in females is compensated for by random inactivation of one of the X chromosomes early in development.
- In some reptilian species, temperature during incubation of eggs determines the sex of offspring.

In the biological world, a wide range of reproductive modes and life cycles are observed. Some organisms are entirely asexual, displaying no evidence of sexual reproduction. Other organisms alternate between short periods of sexual reproduction and prolonged periods of asexual reproduction. In most diploid eukaryotes, however, sexual reproduction is the only natural mechanism for producing new members of the species. The perpetuation of all sexually reproducing organisms depends ultimately on an efficient union of gametes during fertilization. In turn, successful fertilization depends on some form of **sexual differentiation** in the reproductive organisms. Even though it is not overtly evident, this differentiation occurs in organisms as low on the evolutionary scale as bacteria and single-celled eukaryotic algae. In more complex forms of life, the differentiation of the sexes is more evident as phenotypic dimorphism of males and females. The ancient symbol for iron and for Mars, depicting a shield and spear (♂), and the ancient symbol for copper and for Venus, depicting a mirror (♀), have also come to symbolize maleness and femaleness, respectively.

Dissimilar, or **heteromorphic, chromosomes,** such as the XY pair in mammals, characterize one sex or the other in a wide range of species, resulting in their label as **sex chromosomes.** Nevertheless, it is genes, rather than chromosomes, that ultimately serve as the underlying basis of **sex determination.** As we will see, some of these genes are present on sex chromosomes, but others are autosomal. Extensive investigation has revealed a wide variation in sex-chromosome systems—even in closely related organisms—suggesting that mechanisms controlling sex determination have undergone rapid evolution many times in the history of life.

In this chapter, we delve more deeply into what is known about the genetic basis for the determination of sexual differences, with a particular emphasis on two organisms: our own species, representative of mammals; and *Drosophila*, on which pioneering sex-determining studies were performed.

5.1 X and Y Chromosomes Were First Linked to Sex Determination Early in the Twentieth Century

How sex is determined has long intrigued geneticists. In 1891, Hermann Henking identified a nuclear structure in the sperm of certain insects, which he labeled the X-body. Several years later, Clarence McClung showed that some of the sperm in grasshoppers contain an unusual genetic structure, called a *heterochromosome,* but the remainder of the sperm lack this structure. He mistakenly associated the presence of the heterochromosome with the production of male progeny. In 1906, Edmund B. Wilson clarified Henking and McClung's findings when he demonstrated that female somatic cells in the butterfly *Protenor* contain 14 chromosomes, including two X chromosomes. During oogenesis, an even reduction occurs, producing gametes with seven chromosomes, including one X chromosome. Male somatic cells, on the other hand, contain only 13 chromosomes, including one X chromosome. During spermatogenesis, gametes are produced containing either six chromosomes, without an X, or seven chromosomes, one of which is an X. Fertilization by X-bearing sperm results in female offspring, and fertilization by X-deficient sperm results in male offspring [**Figure 5.1(a)**].

The presence or absence of the X chromosome in male gametes provides an efficient mechanism for sex determination in this species and also produces a 1:1 sex ratio in the resulting offspring.

Wilson also experimented with the milkweed bug *Lygaeus turcicus*, in which both sexes have 14 chromosomes. Twelve of these are autosomes (A). In addition, the females have two X chromosomes, while the males have only a single X and a smaller heterochromosome labeled the **Y chromosome**. Females in this species produce only

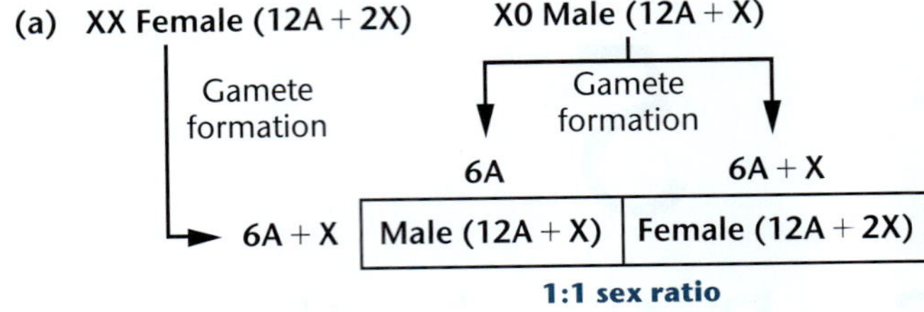

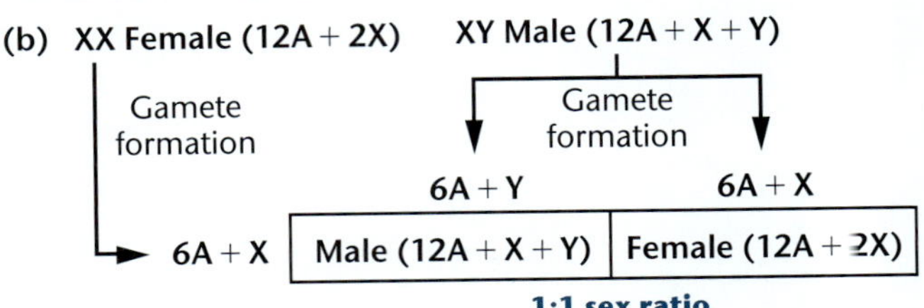

FIGURE 5.1 (a) Sex determination where the heterogametic sex (the male in this example) is XO and produces gametes with or without the X chromosome; (b) sex determination, where the heterogametic sex (again, the male in this example) is XY and produces gametes with either an X or a Y chromosome. In both cases, the chromosome composition of the offspring determines its sex.

gametes of the (6A + X) constitution, but males produce two types of gametes in equal proportions, (6A + X) and (6A + Y). Therefore, following random fertilization, equal numbers of male and female progeny will be produced with distinct chromosome complements [**Figure 5.1(b)**].

In *Protenor* and *Lygaeus* insects, males produce unlike gametes. As a result, they are described as the **heterogametic sex,** and in effect, their gametes ultimately determine the sex of the progeny in those species. In such cases, the female, which has like sex chromosomes, is the **homogametic sex,** producing uniform gametes with regard to chromosome numbers and types.

The male is not always the heterogametic sex. In some organisms, the female produces unlike gametes, exhibiting either the *Protenor* XX/XO or *Lygaeus* XX/XY mode of sex determination. Examples include certain moths and butterflies, some fish, reptiles, amphibians, at least one species of plants (*Fragaria orientalis*), and most birds. To immediately distinguish situations in which the female is the heterogametic sex, some geneticists use the notation **ZZ/ZW,** where ZZ is the homogametic male and ZW is the heterogametic female, instead of the XX/XY notation. For example, chickens are so denoted.

ESSENTIAL POINT

Specific sex chromosomes contain genetic information that controls sex determination and sexual differentiation. ■

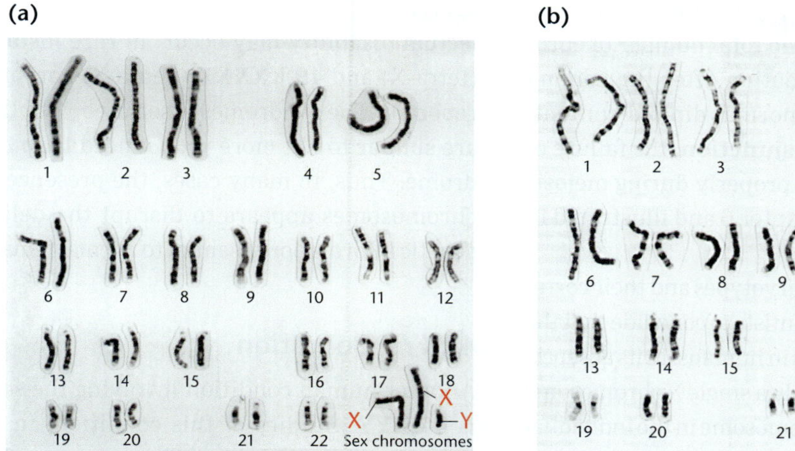

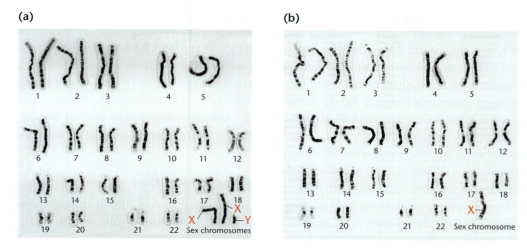

FIGURE 5.2 The karyotypes of individuals with (a) Klinefelter syndrome (47,XXY) and (b) Turner syndrome (45,X).

5.2 The Y Chromosome Determines Maleness in Humans

The first attempt to understand sex determination in our own species occurred almost 100 years ago and involved the visual examination of chromosomes in dividing cells. Efforts were made to accurately determine the diploid chromosome number of humans, but because of the relatively large number of chromosomes, this proved to be quite difficult. Then, in 1956, Joe Hin Tjio and Albert Levan discovered an effective way to prepare chromosomes for accurate viewing. This technique led to a strikingly clear demonstration of metaphase stages showing that 46 was indeed the human diploid number. Later that same year, C. E. Ford and John L. Hamerton, also working with testicular tissue, confirmed this finding. The familiar karyotypes of a human male (Figure 2.4) illustrate the difference in size between the human X and Y chromosomes.

Of the normal 23 pairs of human chromosomes, one pair was shown to vary in configuration in males and females. These two chromosomes were designated the X and Y sex chromosomes. The human female has two X chromosomes, and the human male has one X and one Y chromosome.

We might believe that this observation is sufficient to conclude that the Y chromosome determines maleness. However, several other interpretations are possible. The Y could play no role in sex determination; the presence of two X chromosomes could cause femaleness; or maleness could result from the lack of a second X chromosome. The evidence that clarified which explanation was correct came from study of the effects of human sex-chromosome variations, described in the following section. As such investigations revealed, the Y chromosome does indeed determine maleness in humans.

Klinefelter and Turner Syndromes

Around 1940, scientists identified two human abnormalities characterized by aberrant sexual development, **Klinefelter syndrome (47,XXY)** and **Turner syndrome (45,X).*** Individuals with Klinefelter syndrome are generally tall and often have long arms and legs. They usually have genitalia and internal ducts that are male, but their testes are reduced in size. Although 50 percent of affected individuals do produce sperm, a low sperm count renders most individuals sterile. At the same time, feminine sexual development is not entirely suppressed. Slight enlargement of the breasts (gynecomastia) is common, and the hips are often rounded. Individuals with Klinefelter syndrome most often show no cognitive reduction, and many individuals are unaware of having the disorder until they are treated for infertility.

In Turner syndrome, the affected individual has female external genitalia and internal ducts, but the ovaries are rudimentary. Other characteristic abnormalities include short stature (usually under 5 feet), skin flaps on the back of the neck, and underdeveloped breasts. A broad, shield-like chest is sometimes noted. Intelligence is usually normal.

In 1959, the karyotypes of individuals with these syndromes were determined to be abnormal with respect to the sex chromosomes. Individuals with Klinefelter syndrome have more than one X chromosome. Most often they have an XXY complement in addition to 44 autosomes [**Figure 5.2(a)**], which is why people with this karyotype are designated 47,XXY. Individuals with Turner syndrome most often have only 45 chromosomes, including just a single X chromosome; thus, they are designated 45,X [**Figure 5.2(b)**].

* Although the possessive form of the names of eponymous syndromes is sometimes used (e.g., Klinefelter's syndrome), the current preference is to use the nonpossessive form.

Note the convention used in designating these chromosome compositions. The number states the total number of chromosomes present, and the information after the comma indicates the deviation from the normal diploid content. Both conditions result from **nondisjunction,** the failure of the sex chromosomes to segregate properly during meiosis (nondisjunction is described in Chapter 6 and illustrated in Figure 6.1).

These Klinefelter and Turner karyotypes and their corresponding sexual phenotypes led scientists to conclude that the Y chromosome determines maleness in humans. In its absence, the person's sex is female, even if only a single X chromosome is present. The presence of the Y chromosome in the individual with Klinefelter syndrome is sufficient to determine maleness, even though male development is not complete. Similarly, in the absence of a Y chromosome, as in the case of individuals with Turner syndrome, no masculinization occurs. Note that we cannot conclude anything regarding sex determination under circumstances where a Y chromosome is present without an X because Y-containing human embryos lacking an X chromosome (designated 45,Y) do not survive.

Klinefelter syndrome occurs in about 1 of every 660 male births. The karyotypes **48,XXXY, 48,XXYY, 49,XXXXY,** and **49,XXXYY** are similar phenotypically to 47,XXY, but manifestations are often more severe in individuals with a greater number of X chromosomes.

Turner syndrome can also result from karyotypes other than 45,X, including individuals called **mosaics,** whose somatic cells display two different genetic cell lines, each exhibiting a different karyotype. Such cell lines result from a mitotic error during early development, the most common chromosome combinations being **45,X/46,XY** and **45,X/46,XX.** Thus, an embryo that began life with a normal karyotype can give rise to an individual whose cells show a mixture of karyotypes and who exhibits varying aspects of this syndrome.

Turner syndrome is observed in about 1 in 2000 female births, a frequency much lower than that for Klinefelter syndrome. One explanation for this difference is the observation that a substantial majority of 45,X fetuses die *in utero* and are aborted spontaneously. Thus, a similar frequency of the two syndromes may occur at conception.

47,XXX Syndrome

The abnormal presence of three X chromosomes along with a normal set of autosomes (**47,XXX**) results in female differentiation. The highly variable syndrome that accompanies this genotype, often called **triplo-X,** occurs in about 1 of 1000 female births. Frequently, 47,XXX women are perfectly normal and may remain unaware of their abnormality in chromosome number unless a karyotype is done. In other cases, underdeveloped secondary sex characteristics, sterility, delayed development of language and motor skills, and intellectual disability may occur. In rare instances, **48,XXXX** (tetra-X) and **49,XXXXX** (penta-X) karyotypes have been reported. The syndromes associated with these karyotypes are similar to but more pronounced than the 47,XXX syndrome. Thus, in many cases, the presence of additional X chromosomes appears to disrupt the delicate balance of genetic information essential to normal female development.

47,XYY Condition

Another human condition involving the sex chromosomes is **47,XYY.** Studies of this condition, in which the only deviation from diploidy is the presence of an additional Y chromosome in an otherwise normal male karyotype, were initiated in 1965 by Patricia Jacobs. She discovered that 9 of 315 males in a Scottish maximum security prison had the 47,XYY karyotype. These males were significantly above average in height and had been incarcerated as a result of dangerous, violent, or criminal propensities. Of the nine males studied, seven were of subnormal intelligence, and all suffered personality disorders. Several other studies produced similar findings.

The possible correlation between this chromosome composition and criminal behavior piqued considerable interest, and extensive investigation of the phenotype and frequency of the 47,XYY condition in both criminal and noncriminal populations ensued. Above-average height (usually over 6 feet) and subnormal intelligence were substantiated, and the frequency of males displaying this karyotype was indeed revealed to be higher in penal and mental institutions compared with unincarcerated populations (one study showed 29 XYY males when 28,366 were examined [0.10%]). A particularly relevant question involves the characteristics displayed by the XYY males who are not incarcerated. The only nearly constant association is that such individuals are over 6 feet tall.

A study to further address this issue was initiated in 1974 to identify 47,XYY individuals at birth and to follow their behavioral patterns during preadult and adult development. While the study was considered unethical and soon abandoned, it has became clear that there are many XYY males present in the population who do not exhibit antisocial behavior and who lead normal lives. Therefore, we must conclude that there is a high, but not constant, correlation between the extra Y chromosome and the predisposition of these males to exhibit behavioral problems.

Sexual Differentiation in Humans

Once researchers had established that, in humans, it is the Y chromosome that houses genetic information necessary for maleness, they attempted to pinpoint a specific gene or genes capable of providing the "signal" responsible for sex determination. Before we delve into this topic, it is useful to consider how sexual differentiation occurs in order to better comprehend how humans develop into sexually dimorphic males and females. During early development, every human embryo undergoes a period when it is potentially hermaphroditic. By the fifth week of gestation, gonadal primordia (the tissues that will form the gonad) arise as a pair of **gonadal (genital) ridges** associated with each embryonic kidney. The embryo is potentially hermaphroditic because at this stage its gonadal phenotype is sexually indifferent—male or female reproductive structures cannot be distinguished, and the gonadal ridge tissue can develop to form male or female gonads. As development progresses, primordial germ cells migrate to these ridges, where an outer cortex and inner medulla form (*cortex* and *medulla* are the outer and inner tissues of an organ, respectively). The cortex is capable of developing into an ovary, while the medulla may develop into a testis. In addition, two sets of undifferentiated ducts called the Wolffian and Müllerian ducts exist in each embryo. Wolffian ducts differentiate into other organs of the male reproductive tract, while Müllerian ducts differentiate into structures of the female reproductive tract.

Because gonadal ridges can form either ovaries or testes, they are commonly referred to as **bipotential gonads.** What switch triggers gonadal ridge development into testes or ovaries? The presence or absence of a Y chromosome is the key. If cells of the ridge have an XY constitution, development of the medulla into a testis is initiated around the seventh week. However, in the absence of the Y chromosome, no male development occurs, the cortex of the ridge subsequently forms ovarian tissue, and the Müllerian duct forms oviducts (Fallopian tubes), uterus, cervix, and portions of the vagina. Depending on which pathway is initiated, parallel development of the appropriate male or female duct system then occurs, and the other duct system degenerates. If testes differentiation is initiated, the embryonic testicular tissue secretes hormones that are essential for continued male sexual differentiation. As we will discuss in the next section, the presence of a Y chromosome and the development of the testes also inhibit formation of female reproductive organs.

In females, as the twelfth week of fetal development approaches, the oogonia within the ovaries begin meiosis, and primary oocytes can be detected. By the twenty-fifth week of gestation, all oocytes become arrested in meiosis and remain dormant until puberty is reached some 10 to 15 years later. In males, on the other hand, primary spermatocytes are not produced until puberty is reached (see Figure 2.11).

The Y Chromosome and Male Development

The human Y chromosome, unlike the X, was long thought to be mostly blank genetically. It is now known that this is not true, even though the Y chromosome contains far fewer genes than does the X. Data from the Human Genome Project indicate that the Y chromosome has at least 75 genes, compared to 900–1400 genes on the X. Current analysis of these genes and regions with potential genetic function reveals that some have homologous counterparts on the X chromosome and others do not. For example, present on both ends of the Y chromosome are so-called **pseudoautosomal regions (PARs)** that share homology with regions on the X chromosome and synapse and recombine with it during meiosis. The presence of such a pairing region is critical to segregation of the X and Y chromosomes during male gametogenesis. The remainder of the chromosome, about 95 percent of it, does not synapse or recombine with the X chromosome. As a result, it was originally referred to as the *nonrecombining region of the Y (NRY)*. More recently, researchers have designated this region as the **male-specific region of the Y (MSY).** Some portions of the MSY share homology with genes on the X chromosome, and others do not.

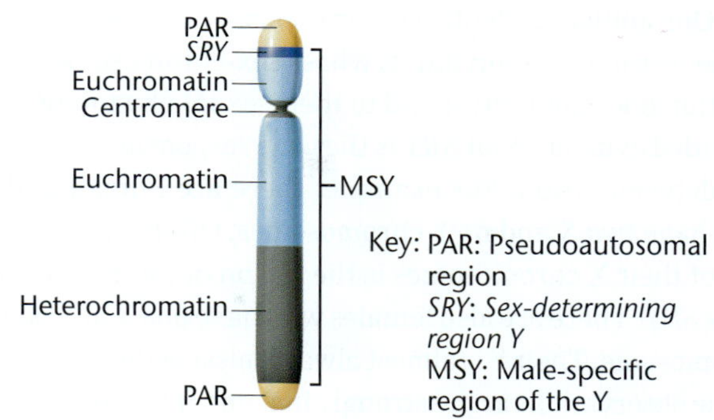

FIGURE 5.3 The regions of the human Y chromosome.

The human Y chromosome is diagrammed in **Figure 5.3**. The MSY is divided about equally between *euchromatic* regions, containing functional genes, and *heterochromatic* regions, lacking genes. Within euchromatin, adjacent to the PAR of the short arm of the Y chromosome, is a critical gene that controls male sexual development, called the ***sex-determining region Y (SRY).*** In humans, the absence of a Y chromosome almost always leads to female development; thus, this gene is absent from the X chromosome. At six to eight weeks of development, the *SRY* gene becomes active in XY embryos. *SRY* encodes a protein that causes the undifferentiated gonadal tissue of the embryo to form testes. This protein is called the **testis-determining factor (TDF).** *SRY* (or a closely related version) is present in all mammals thus far examined, indicative of its essential function throughout this diverse group of animals.*

* It is interesting to note that in chickens, a similar gene has recently been identified. Called *DMRT1*, it is located on the Z chromosome. This gene is the subject of Problem 26 in the Problems section at the end of the chapter.

Our ability to identify the presence or absence of DNA sequences in rare individuals whose sex-chromosome composition does not correspond to their sexual phenotype has provided evidence that *SRY* is the gene responsible for male sex determination. For example, there are human males who have two X and no Y chromosomes. Often, attached to one of their X chromosomes is the region of the Y that contains *SRY*. There are also females who have one X and one Y chromosome. Their Y is almost always missing the *SRY* gene. These observations argue strongly in favor of the role of *SRY* in providing the primary signal for male development.

Further support of this conclusion involves experiments using **transgenic mice.** These animals are produced from fertilized eggs injected with foreign DNA that is subsequently incorporated into the genetic composition of the developing embryo. In normal mice, a chromosome region designated *Sry* has been identified that is comparable to *SRY* in humans. When mouse DNA containing *Sry* is injected into normal XX mouse eggs, most of the offspring develop into males.

The question of how the product of this gene triggers development of embryonic gonadal tissue into testes rather than ovaries is the key question under investigation. TDF functions as a *transcription factor*, a DNA-binding protein that interacts directly with the regulatory sequences of other genes to stimulate their expression. Thus, TDF behaves as a master switch that controls other genes downstream in the process of sexual differentiation. Interestingly, many identified thus far reside on autosomes, including the human *SOX9* gene located on chromosome 17 and the subject of the preceding Now Solve This.

A more recent area of investigation has involved the Y chromosome and paternal age. For many years, it has been known that maternal age is correlated with an elevated rate of offspring with chromosomal defects, including Down syndrome (see Chapter 6). Advanced paternal age has now been associated with an increased risk in offspring of congenital disorders with a genetic basis, including certain cancers, schizophrenia, autism, and other conditions, collectively known as *paternal age effects (PAE)*. Studies in which the genomes of sperm have been sequenced have demonstrated the presence of specific PAE mutations, including numerous ones on the Y chromosome. Evidence suggests that PAE mutations are positively selected for and result in an enrichment of mutant sperm over time.

NOW SOLVE THIS

5.1 Campomelic dysplasia (CMD1) is a congenital human syndrome featuring malformation of bone and cartilage. It is caused by an autosomal dominant mutation of a gene located on chromosome 17. Consider the following observations in sequence, and in each case, draw whatever appropriate conclusions are warranted.

(a) Of those with the syndrome who are karyotypically 46,XY, approximately 75 percent are sex reversed, exhibiting a wide range of female characteristics.

(b) The nonmutant form of the gene, called *SOX9*, is expressed in the developing gonad of the XY male, but not the XX female.

(c) The *SOX9* gene shares 71 percent amino acid coding sequence homology with the Y-linked *SRY* gene.

(d) CMD1 patients who exhibit a 46,XX karyotype develop as females, with no gonadal abnormalities.

■ **HINT:** *This problem asks you to apply the information presented in this chapter to a real-life example. The key to its solution is knowing that some genes are activated and produce their normal product as a result of expression of products of other genes found on different chromosomes.*

ESSENTIAL POINT

The presence or absence of a Y chromosome that contains an intact *SRY* gene is responsible for causing maleness in humans. ■

5.3 The Ratio of Males to Females in Humans Is Not 1.0

The presence of heteromorphic sex chromosomes in one sex of a species but not the other provides a potential mechanism for producing equal proportions of male and female offspring. This potential depends on the segregation of the X and Y (or Z and W) chromosomes during meiosis, such that half of the gametes of the heterogametic sex receive one of the chromosomes and half receive the other one. As we learned in the previous section, small pseudoautosomal regions of pairing homology do exist at both ends of the human X and Y chromosomes, suggesting that the X and Y chromosomes do synapse and then segregate into different gametes. Provided that both types of gametes are equally successful in fertilization and that the two sexes are equally viable during development, a 1:1 ratio of male and female offspring should result.

The actual proportion of male to female offspring, referred to as the **sex ratio,** has been assessed in two ways. The **primary sex ratio (PSR)** reflects the proportion of males to females conceived in a population. The **secondary sex ratio** reflects the proportion of each sex that is born. The secondary sex ratio is much easier to determine but has the disadvantage of not accounting for any disproportionate embryonic or fetal mortality.

When the secondary sex ratio in the human population was determined in 1969 by using worldwide census data, it did not equal 1.0. For example, in the Caucasian population in the United States, the secondary ratio was a little less than 1.06, indicating that about 106 males were born for each 100 females. In 1995, this ratio dropped to slightly less than 1.05. In the African-American population in the United States, the ratio was 1.025. In other countries, the excess of male births is even greater than is reflected in these values. For example, in Korea, the secondary sex ratio was 1.15.

Despite these ratios, it is possible that the PSR is 1.0 and is altered between conception and birth. For the secondary ratio to exceed 1.0, then, prenatal female mortality would have to be greater than prenatal male mortality. However, when this hypothesis was first examined, it was deemed to be false. In a Carnegie Institute study, reported in 1948, the sex of approximately 6000 embryos and fetuses recovered from miscarriages and abortions was determined, and fetal mortality was actually higher in males. On the basis of the data derived from that study, the PSR in U.S. Caucasians was estimated to be 1.079, suggesting that more males than females are conceived in the human population.

To explain why, researchers examined the assumptions on which the theoretical ratio is based:

1. Because of segregation, males produce equal numbers of X- and Y-bearing sperm.
2. Each type of sperm has equivalent viability and motility in the female reproductive tract.
3. The egg surface is equally receptive to both X- and Y-bearing sperm.

No direct experimental evidence contradicts any of these assumptions.

A PSR favoring male conceptions remained dogma for many decades until, in 2015, a study using an extensive data set was published that concludes that the PSR is 1.0—suggesting that equal numbers of males and females are indeed conceived. Among other parameters, the examination of the sex of 3-day-old and 6-day-old embryos conceived using assisted reproductive technology provided the most direct assessment. Following conception, however, mortality was then shown to fluctuate between the sexes, until at birth, more males than females are born. Thus, female mortality during embryonic and fetal development exceeds that of males. Clearly, this is a difficult topic to investigate but one of continued interest. For now, the most recent findings are convincing and contradict the earlier studies.

ESSENTIAL POINT

In humans, the sex ratio at conception and birth remains an active area of research. The most current study shows that equal numbers of males and females are conceived, but that more males than females are born. ■

5.4 Dosage Compensation Prevents Excessive Expression of X-Linked Genes in Humans and Other Mammals

The presence of two X chromosomes in normal human females and only one X in normal human males is unique compared with the equal numbers of autosomes present in the cells of both sexes. On theoretical grounds alone, it is possible to speculate that this disparity should create a "genetic dosage" difference between males and females, with attendant problems, for all X-linked genes. There is the potential for females to produce twice as much of each product of all X-linked genes. The additional X chromosomes in both males and females exhibiting the various syndromes discussed earlier in this chapter are thought to compound this dosage problem. Embryonic development depends on proper timing and precisely regulated levels of gene expression. Otherwise, disease phenotypes or embryonic lethality can occur. In this section, we will describe research findings regarding X-linked gene expression that demonstrate a genetic mechanism of **dosage compensation** that balances the dose of X chromosome gene expression in females and males.

Barr Bodies

Murray L. Barr and Ewart G. Bertram's experiments with cats, as well as Keith Moore and Barr's subsequent study in humans, demonstrate a genetic mechanism in mammals that compensates for X chromosome dosage disparities. Barr and Bertram observed a darkly staining body in the interphase nerve cells of female cats that was absent in similar cells of males. In humans, this body can be easily demonstrated in female cells derived from the buccal mucosa (cheek cells) or in fibroblasts (undifferentiated connective tissue cells), but not in similar male cells (**Figure 5.4**). This highly condensed structure, about 1 μm in diameter, lies against the nuclear

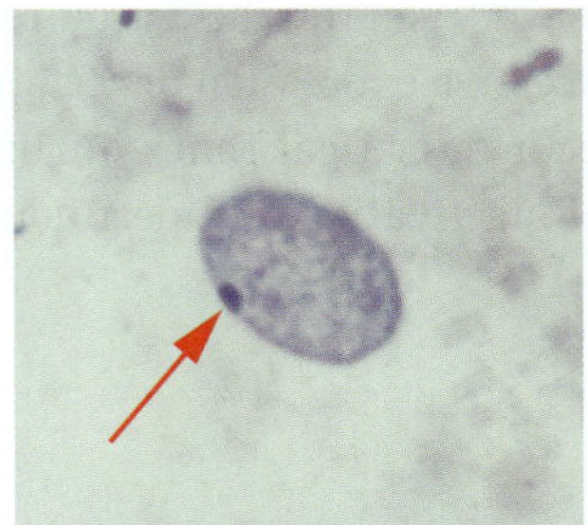

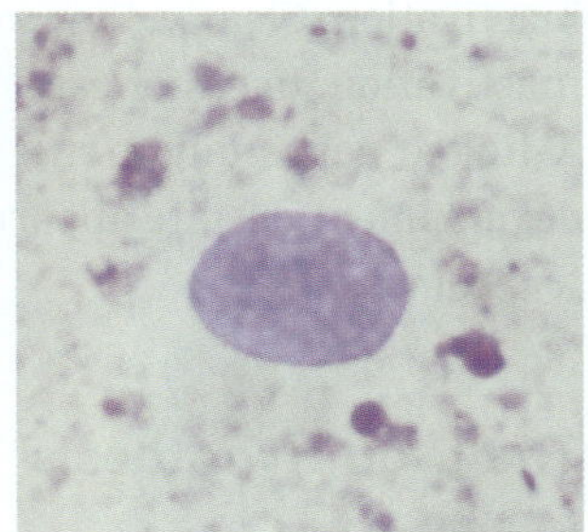

FIGURE 5.4 Photomicrographs comparing cheek epithelial cell nuclei from a male that fails to reveal Barr bodies (right) with a nucleus from a female that demonstrates a Barr body (indicated by the arrow in the left image). This structure, also called a sex chromatin body, represents an inactivated X chromosome.

envelope of interphase cells, and it stains positively for a number of different DNA-binding dyes.

This chromosome structure, called a **sex chromatin body,** or simply a **Barr body,** is an inactivated X chromosome. Susumo Ohno was the first to suggest that the Barr body arises from one of the two X chromosomes. This hypothesis is attractive because it provides a possible mechanism for dosage compensation. If one of the two X chromosomes is inactive in the cells of females, the dosage of genetic information that can be expressed in males and females will be equivalent. Convincing, though indirect, evidence for this hypothesis comes from study of the sex-chromosome syndromes described earlier in this chapter. Regardless of how many X chromosomes a somatic cell possesses, all but one of them appear to be inactivated and can be seen as Barr bodies. For example, no Barr body is seen in the somatic cells of Turner 45,X females; one is seen in Klinefelter 47,XXY males; two in 47,XXX females; three in 48,XXXX females; and so on (**Figure 5.5**). Therefore, the number of Barr bodies follows an $N - 1$ rule, where N is the total number of X chromosomes present.

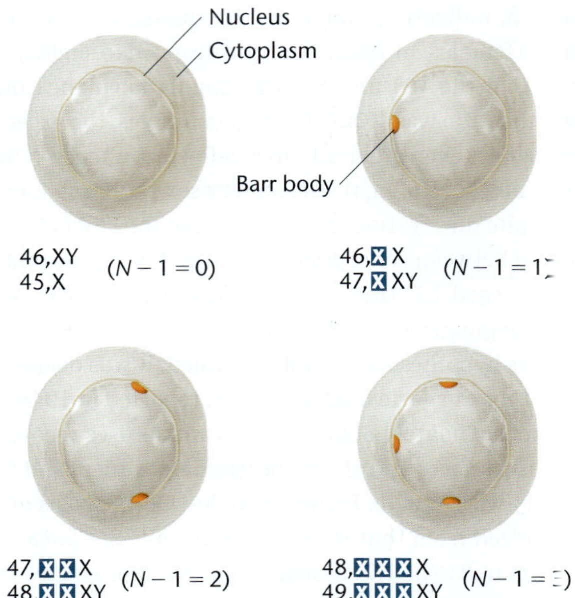

FIGURE 5.5 Occurrence of Barr bodies in various human karyotypes, where all X chromosomes except one ($N - 1$) are inactivated.

Although this apparent inactivation of all but one X chromosome increases our understanding of dosage compensation, it further complicates our perception of other matters. For example, because one of the two X chromosomes is inactivated in normal human females, why then is the Turner 45,X individual not entirely normal? Why aren't females with the triplo-X and tetra-X karyotypes (47,XXX and 48,XXXX) completely unaffected by the additional X chromosome? Furthermore, in Klinefelter syndrome (47,XXY), X chromosome inactivation effectively renders the person 46,XY. Why aren't these males unaffected by the extra X chromosome in their nuclei?

One possible explanation is that chromosome inactivation does not normally occur in the very early stages of development of those cells destined to form gonadal tissues. Another possible explanation is that not all genes on each X chromosome forming a Barr body are inactivated. Recent studies have indeed demonstrated that as many as 15 percent of the human X chromosomal genes actually escape inactivation. Clearly, then, not every gene on the X requires inactivation. In either case, excessive expression of certain X-linked genes might still occur at critical times during development despite apparent inactivation of superfluous X chromosomes.

The Lyon Hypothesis

In mammalian females, one X chromosome is of maternal origin, and the other is of paternal origin. Which one is inactivated? Is the inactivation random? Is the same chromosome inactive in all somatic cells? In the early 1960s, Mary Lyon, Liane Russell, and Ernest Beutler independently proposed a hypothesis that answers these questions. They postulated that the inactivation of X chromosomes occurs randomly in somatic cells at a point early in embryonic development, most likely sometime during the blastocyst stage of development. Once inactivation has occurred, all descendant cells have the same X chromosome inactivated as their initial progenitor cell.

This explanation, which has come to be called the **Lyon hypothesis,** was initially based on observations of female mice heterozygous for X-linked coat-color genes. The pigmentation of these heterozygous females was mottled, with large patches expressing the color allele on one X and other patches expressing the allele on the other X. This is the phenotypic pattern that would be expected if different X chromosomes were inactive in adjacent patches of cells. Similar mosaic patterns occur in the black and yellow-orange patches of female tortoiseshell and calico cats (**Figure 5.6**). Such X-linked coat-color patterns do not occur in male cats because all their cells contain the single maternal X chromosome and are therefore hemizygous for only one X-linked coat-color allele.

The most direct evidence in support of the Lyon hypothesis comes from studies of gene expression in clones of human fibroblast cells. Individual cells are isolated following biopsy and cultured *in vitro*. A culture of cells derived from a single cell is called a **clone.** The synthesis of the enzyme glucose-6-phosphate dehydrogenase (G6PD) is controlled by an X-linked gene. Numerous mutant alleles of this gene have been detected, and their gene products can be differentiated from the wild-type enzyme by their migration pattern in an electrophoretic field.

Fibroblasts have been taken from females heterozygous for different allelic forms of *G6PD* and studied. The Lyon hypothesis predicts that if inactivation of an X chromosome

(a)

(b)

FIGURE 5.6 (a) The random distribution of orange and black patches in a calico cat illustrates the Lyon hypothesis. The white patches are due to another gene, distinguishing calico cats from tortoiseshell cats (b), which lack the white patches.

occurs randomly early in development, and thereafter all progeny cells have the same X chromosome inactivated as their progenitor, such a female should show two types of clones, each containing only one electrophoretic form of *G6PD* product, in approximately equal proportions. This prediction has been confirmed experimentally, and studies involving modern techniques in molecular biology have clearly established that X chromosome inactivation occurs.

One ramification of X-inactivation is that mammalian females are mosaics for all heterozygous X-linked alleles—some areas of the body express only the maternally derived alleles, and others express only the paternally derived alleles. An especially interesting example involves **red–green color blindness,** an X-linked recessive disorder. In humans, hemizygous males are fully color-blind in all retinal cells. However, heterozygous females display mosaic retinas, with patches of defective color perception and surrounding areas with normal color perception. In this example, random inactivation of one or the other X chromosome early in the development of heterozygous females has led to these phenotypes.

The Mechanism of Inactivation

The least understood aspect of the Lyon hypothesis is the mechanism of X chromosome inactivation. Somehow, either DNA, the attached histone proteins, or both DNA and histone proteins are chemically modified, silencing most genes that are part of that chromosome. Once silenced, a memory is created that keeps the same homolog inactivated following chromosome replications and cell divisions. Such a process, whereby expression of genes on one homolog, but not the other, is affected, is referred to as **imprinting.** This term also applies to a number of other examples in which genetic information is modified and gene expression is repressed. Collectively, such events are part of the growing field of **epigenetics** (see Special Topics Chapter 1-Epigenetics).

NOW SOLVE THIS

5.2 Carbon Copy (CC), the first cat produced from a clone, was created from an ovarian cell taken from her genetic donor, Rainbow, a calico cat. The diploid nucleus from the cell was extracted and then injected into an enucleated egg. The resulting zygote was then allowed to develop in a petri dish, and the cloned embryo was implanted in the uterus of a surrogate mother cat, who gave birth to CC. CC's surrogate mother was a tabby (see the photo below). Geneticists were very interested in the outcome of cloning a calico cat because they were not certain if the cloned cat would have patches of orange and black, just orange, or just black. Taking into account the Lyon hypothesis, explain the basis of the uncertainty. Would you expect CC to appear identical to Rainbow? Explain why or why not.

Carbon Copy with her surrogate mother.

■ **HINT:** *This problem involves an understanding of the Lyon hypothesis. The key to its solution is to realize that the donor nucleus was from a differentiated ovarian cell of an adult female cat, which itself had inactivated one of its X chromosomes.*

Ongoing investigations are beginning to clarify the mechanism of inactivation. A region of the mammalian X chromosome is the major control unit. This region, located on the proximal end of the p arm in humans, is called the **X-inactivation center (*Xic*),** and its genetic expression *occurs only on the X chromosome that is inactivated*. The *Xic* is about 1 Mb (10^6 base pairs) in length and is known to contain several putative regulatory units and four genes. One of these, ***X-inactive specific transcript (XIST),*** is now known to be a critical gene for X-inactivation.

Interesting observations have been made regarding the RNA that is transcribed from the *XIST* gene, many coming from experiments that focused on the equivalent gene in the mouse (*Xist*). First, the RNA product is quite large and does not encode a protein, and thus is not translated. The RNA products of *Xist* spread over and coat the X chromosome *bearing the gene that produced them*. Two other noncoding genes at the *Xic* locus, *Tsix* (an antisense partner of *Xist*) and *Xite,* are also believed to play important roles in X-inactivation.

A second observation is that transcription of *Xist* initially occurs at low levels on all X chromosomes. As the inactivation process begins, however, transcription continues, and is enhanced, only on the X chromosome that becomes inactivated. In 1996, a research group led by Neil Brockdorff and Graeme Penny provided convincing evidence that transcription of *Xist* is the critical event in chromosome inactivation. These researchers introduced a targeted deletion (7 kb) into this gene, disrupting its sequence. As a result, the chromosome bearing the deletion lost its ability to become inactivated.

ESSENTIAL POINT

In mammals, female somatic cells randomly inactivate one of two X chromosomes during early embryonic development, a process important for balancing the expression of X chromosome-linked genes in males and females. ■

5.5 The Ratio of X Chromosomes to Sets of Autosomes Can Determine Sex

We now discuss two interesting cases where the Y chromosome does not play a role in sex determination. First, in the fruit fly, *Drosophila melanogaster*, even though most males contain a Y chromosome, the Y plays no role. Second, in the roundworm, *Caenorhabditis elegans,* the organism lacks a Y chromosome altogether. In both cases, we shall see that the critical factor is the ratio of X chromosomes to the number of sets of autosomes.

Drosophila melanogaster

Because males and females in *Drosophila melanogaster* (and other *Drosophila* species) have the same general sex-chromosome composition as humans (males are XY and females are XX), we might assume that the Y chromosome also causes maleness in these flies. However, the elegant work of Calvin Bridges in 1921 showed this not to be true. His studies of flies with quite varied chromosome compositions led him to the conclusion that the Y chromosome is not involved in sex determination in this organism. Instead, Bridges proposed that the X chromosomes and autosomes together play a critical role in sex determination.

Bridges' work can be divided into two phases: (1) A study of offspring resulting from nondisjunction of the X chromosomes during meiosis in females and (2) subsequent work with progeny of females containing three copies of each chromosome, called triploid (3*n*) females. As we have seen previously in this chapter (and as you will see in Figure 6.1), nondisjunction is the failure of paired chromosomes to segregate or separate during the anaphase stage of the first or second meiotic divisions. The result is the production of two types of abnormal gametes, one of which contains an extra chromosome ($n + 1$) and the other of which lacks a chromosome ($n - 1$). Fertilization of such gametes with a haploid gamete produces ($2n + 1$) or ($2n - 1$) zygotes. As in humans, if nondisjunction involves the X chromosome, in addition to the normal complement of autosomes, both an XXY and an XO sex-chromosome composition may result. (The "O" signifies that neither a second X nor a Y chromosome is present, as occurs in XO genotypes of individuals with Turner syndrome.)

Contrary to what was later discovered in humans, Bridges found that the XXY flies were normal females and the XO flies were sterile males. The presence of the Y chromosome in the XXY flies did not cause maleness, and its absence in the XO flies did not produce femaleness. From these data, Bridges concluded that the Y chromosome in *Drosophila* lacks male-determining factors, but since the XO males were sterile, it does contain genetic information essential to male fertility.

Bridges was able to clarify the mode of sex determination in *Drosophila* by studying the progeny of triploid females (3*n*), which have three copies each of the haploid complement of chromosomes. *Drosophila* has a haploid number of 4, thereby possessing three pairs of autosomes in addition to its pair of sex chromosomes. Triploid females apparently originate from rare diploid eggs fertilized by normal haploid sperm. Triploid females have heavy-set bodies, coarse bristles, and coarse eyes, and they may be fertile. Because of the odd number of each chromosome (3), during meiosis, a variety of different chromosome complements are distributed into gametes that give rise to offspring with a variety of

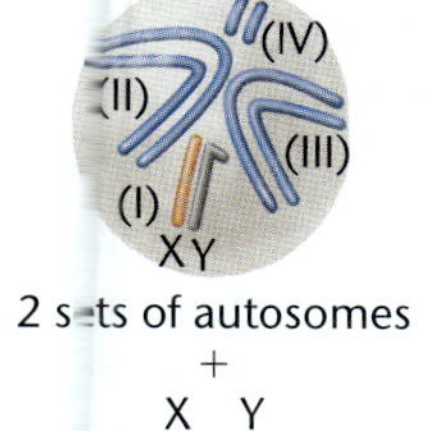

Chromosome formulation	Ratio of X chromosomes to autosome sets	Sexual morphology
3X:2A	1.5	Metafemale
3X:3A	1.0	Female
2X:2A	1.0	Female
3X:4A	0.75	Intersex
2X:3A	0.67	Intersex
X:2A	0.50	Male
XY:2A	0.50	Male
XY:3A	0.33	Metamale

FIGURE 5.7 The ratios of X chromosomes to sets of autosomes and the resultant sexual morphology seen in *Drosophila melanogaster.*

abnormal chromosome constitutions. Correlations between the sexual morphology and chromosome composition, along with Bridges' interpretation, are shown in **Figure 5.7**.

Bridges realized that the critical factor in determining sex is the ratio of X chromosomes to the number of haploid sets of autosomes (A) present. Normal (2X:2A) and triploid (3X:3A) females each have a ratio equal to 1.0, and both are fertile. As the ratio exceeds unity (3X:2A, or 1.5, for example), what was once called a *superfemale* is produced. Because such females are most often inviable, they are now more appropriately called **metafemales.**

Normal (XY:2A) and sterile (XO:2A) males each have a ratio of 1:2, or 0.5. When the ratio decreases to 1:3, or 0.33, as in the case of an XY:3A male, infertile **metamales** result. Other flies recovered by Bridges in these studies had an (X:A) ratio intermediate between 0.5 and 1.0. These flies were generally larger, and they exhibited a variety of morphological abnormalities and rudimentary bisexual gonads and genitalia. They were invariably sterile and expressed both male and female morphology, thus being designated as **intersexes.**

Bridges' results indicate that in *Drosophila*, factors that cause a fly to develop into a male are not located on the sex chromosomes but are instead found on the autosomes. Some female-determining factors, however, are located on the X chromosomes. Thus, with respect to primary sex determination, male gametes containing one of each autosome plus a Y chromosome result in male offspring not because of the presence of the Y but because they fail to contribute an X chromosome. This mode of sex determination is explained by the **genic balance theory.** Bridges proposed that a threshold for maleness is reached when the X:A ratio is 1:2 (X:2A), but that the presence of an additional X (XX:2A) alters the balance and results in female differentiation.

Numerous genes involved in sex determination in *Drosophila* have been identified. The recessive autosomal gene *transformer* (*tra*), discovered over 50 years ago by Alfred H. Sturtevant, clearly demonstrated that a single autosomal gene could have a profound impact on sex determination. Females homozygous for *tra* are transformed into sterile males, but homozygous males are unaffected. More recently, another gene, *Sex-lethal* (*Sxl*), has been shown to play a critical role, serving as a "master switch" in sex determination. Activation of the X-linked *Sxl* gene, which relies on a ratio of X chromosomes to sets of autosomes that equals 1.0, is essential to female development. In the absence of activation—as when, for example, the X:A ratio is 0.5—male development occurs.

Although it is not yet exactly clear how this ratio influences the *Sxl* locus, we do have some insights into the question. The *Sxl* locus is part of a hierarchy of gene expression and exerts control over other genes, including *tra* (discussed in the previous paragraph) and *dsx* (*doublesex*). The wild-type allele of *tra* is activated by the product of *Sxl* only in females and in turn influences the expression of *dsx*. Depending on how the initial RNA transcript of *dsx* is processed (spliced, as explained below), the resultant dsx protein activates either male- or female-specific genes required for sexual differentiation. Each step in this regulatory cascade requires a form of processing called **RNA splicing,** in which portions of the RNA are removed and the remaining fragments are "spliced" back together prior to translation into a protein. In the case of the *Sxl* gene, the RNA transcript may be spliced in different ways, a phenomenon called **alternative splicing.** A different RNA transcript is produced in females than in males. In potential females, the transcript is active and initiates a cascade of regulatory gene expression, ultimately leading to female differentiation. In potential males, the transcript is inactive, leading

(a)

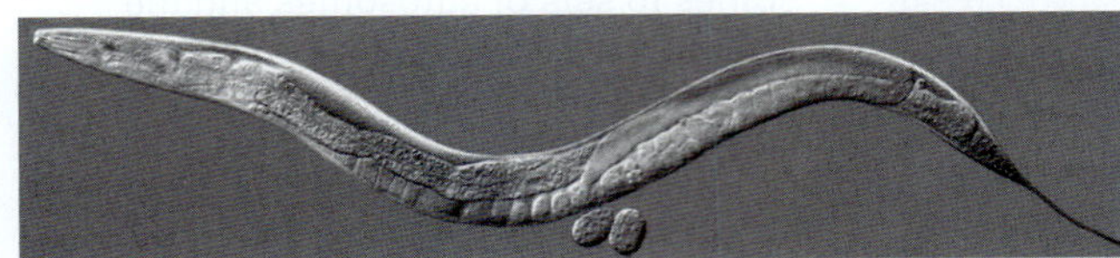

(b)

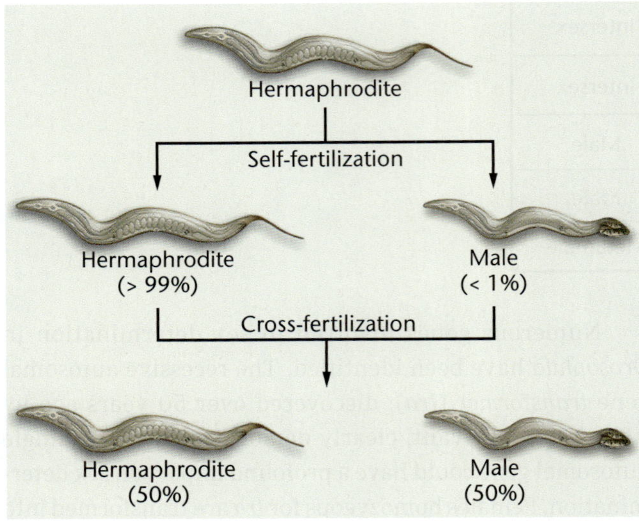

FIGURE 5.8 (a) Photomicrograph of a hermaphroditic nematode, *C. elegans;* (b) the outcomes of self-fertilization in a hermaphrodite, and a mating of a hermaphrodite and a male worm.

to a different pattern of gene activity, whereby male differentiation occurs. We will return to this topic in Chapter 16, where alternative splicing is again addressed as one of the mechanisms involved in the regulation of genetic expression in eukaryotes.

Caenorhabditis elegans

The nematode worm *C. elegans* [**Figure 5.8(a)**] has become a popular organism in genetic studies, particularly for investigating the genetic control of development. Its usefulness is based on the fact that adults consist of approximately 1000 cells, the precise lineage of which can be traced back to specific embryonic origins. There are two sexual phenotypes in these worms: males, which have only testes, and hermaphrodites, which contain both testes and ovaries. During larval development of hermaphrodites, testes form that produce sperm, which is stored. Ovaries are also produced, but oogenesis does not occur until the adult stage is reached several days later. The eggs that are produced are fertilized by the stored sperm in a process of self-fertilization.

The outcome of this process is quite interesting [**Figure 5.8(b)**]. The vast majority of organisms that result are hermaphrodites, like the parental worm; less than 1 percent of the offspring are males. As adults, males can mate with hermaphrodites, producing about half male and half hermaphrodite offspring.

The genetic signal that determines maleness in contrast to hermaphroditic development is provided by genes located on both the X chromosome and autosomes. *C. elegans* lacks a Y chromosome altogether—hermaphrodites have two X chromosomes, while males have only one X chromosome. It is believed that, as in *Drosophila*, it is the ratio of X chromosomes to the number of sets of autosomes that ultimately determines the sex of these worms. A ratio of 1.0 (two X chromosomes and two copies of each autosome) results in hermaphrodites, and a ratio of 0.5 results in males. The absence of a heteromorphic Y chromosome is not uncommon in organisms.

5.6 Temperature Variation Controls Sex Determination in Reptiles

We conclude this chapter by discussing several cases where the environment—specifically temperature—is the major factor in sex determination. In contrast to situations where sex is determined genetically (as is true of all examples thus far presented in the chapter), the cases that we will now discuss are categorized as **temperature-dependent sex determination (TSD).**

In many species of reptiles, sex is predetermined at conception by sex-chromosome composition. For example, in many snakes, including vipers, a ZZ/ZW mode is in effect, in which the female is the heterogamous sex (ZW). However, in boas and pythons, it is impossible to distinguish one sex chromosome from the other in either sex. In many lizards, both the XX/XY and ZZ/ZW systems are found, depending on the species.

In still other reptilian species, however, TSD is the norm, including all crocodiles, most turtles, and some lizards, where sex determination is achieved according to the incubation temperature of eggs during a critical period of embryonic development. Three distinct patterns of TSD emerge (cases I–III in **Figure 5.9**). In case I, low temperatures yield 100 percent females, and high temperatures yield 100 percent males. Just the opposite occurs in case II. In case III, low *and* high temperatures yield 100 percent females, while intermediate temperatures yield various proportions of males. The third pattern is seen in various species of crocodiles, turtles, and lizards, although other members of these groups are known to exhibit the other patterns.

Two observations are noteworthy. First, in all three patterns, certain temperatures result in both male and female offspring; second, this pivotal temperature T_P range is fairly narrow, usually spanning less than 5°C, and sometimes only 1°C. The central question raised by these observations is:

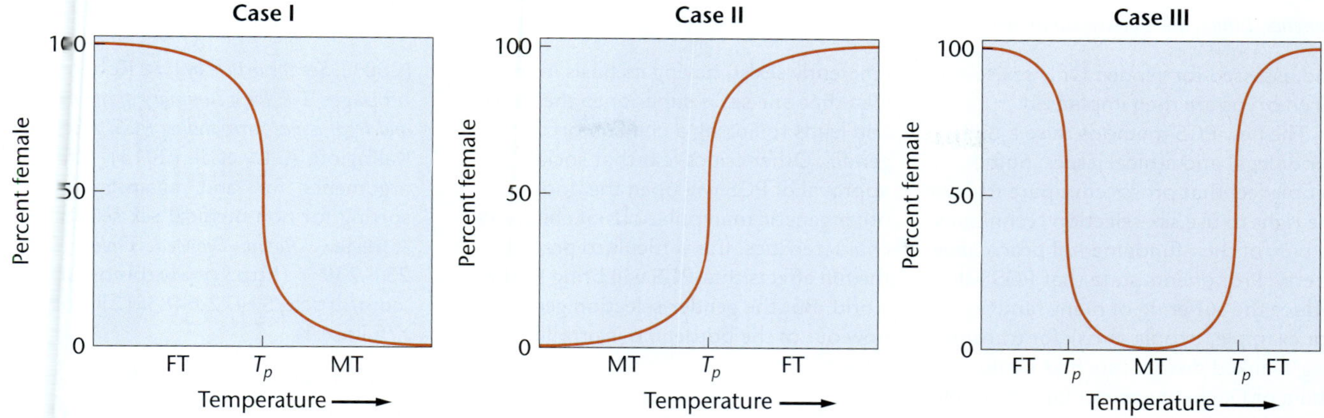

FIGURE 5.9 Three different patterns of temperature-dependent sex determination (TSD) in reptiles, as described in the text. The relative pivotal temperature T_p is crucial to sex determination during a critical point during embryonic development. FT = Female-determining temperature; MT = male-determining temperature

What are the metabolic or physiological parameters affected by temperature that lead to the differentiation of one sex or the other?

The answer is thought to involve steroids (mainly estrogens) and the enzymes involved in their synthesis. Studies clearly demonstrate that the effects of temperature on estrogens, androgens, and inhibitors of the enzymes controlling their synthesis are involved in the sexual differentiation of ovaries and testes. One enzyme in particular, **aromatase,** converts androgens (male hormones such as testosterone) to estrogens (female hormones such as estradiol). The activity of this enzyme is correlated with the pathway of reactions that occurs during gonadal differentiation activity and is high in developing ovaries and low in developing testes. Researchers in this field, including Claude Pieau and colleagues, have proposed that a thermosensitive factor mediates the transcription of the reptilian aromatase gene, leading to temperature-dependent sex determination. Several other genes are likely to be involved in this mediation.

The involvement of sex steroids in gonadal differentiation has also been documented in birds, fishes, and amphibians. Thus, sex-determining mechanisms involving estrogens seem to be characteristic of numerous nonmammalian vertebrates.

ESSENTIAL POINT

Although chromosome composition determines the sex of some reptiles, many others show that temperature-dependent effects during egg incubation are critical for sex determination. ■

GENETICS, ETHICS, AND SOCIETY

A Question of Gender: Sex Selection in Humans

Throughout history, people have attempted to influence the gender of their unborn offspring by following varied and sometimes bizarre procedures. In medieval Europe, prospective parents would place a hammer under the bed to help them conceive a boy, or a pair of scissors, to conceive a girl. Other practices were based on the ancient belief that semen from the right testicle created male offspring and that from the left testicle created females. In some cultures, efforts to control the sex of offspring has had a darker side—female infanticide. In ancient Greece, the murder of female infants was so common that the male:female ratio in some areas approached 4:1. In some parts of rural India, female infanticide continued up to the 1990s. In 1997, the World Health Organization reported that about 50 million women were "missing" in China, likely because of selective abortion of female fetuses and institutionalized neglect of female children. In recent times, amniocentesis and ultrasound testing, followed by sex-specific abortion, have replaced much of the traditional female infanticide.

New genetic and reproductive technologies offer parents ways to select their children's gender prior to implantation of the embryo in the uterus—methods called *preimplantation gender selection* (*PGS*). Following *in vitro* fertilization, embryos can be biopsied

(continued)

Genetics, Ethics, and Society, continued

and assessed for gender. Only sex-selected embryos are then implanted.

The new PGS methods raise a number of legal and ethical issues. Some people feel that prospective parents have the right to use sex-selection techniques as part of their fundamental procreative liberty. Proponents state that PGS will reduce the suffering of many families. For example, people at risk for transmitting X-linked diseases such as hemophilia or Duchenne muscular dystrophy would be able to enhance their chance of conceiving a female child, who would not express the disease.

The majority of people who undertake PGS, however, do so for nonmedical reasons—to "balance" their families. One argument in favor of this use is that the intentional selection of the sex of an offspring may reduce overpopulation and economic burdens for families who would repeatedly reproduce to get the desired gender. Also, PGS may increase the happiness of both parents and children, as the children would be more "wanted."

On the other hand, some argue that PGS serves neither the individual nor the common good. They argue that PGS is inherently sexist, having its basis in the idea that one sex is superior to the other, and leads to linking a child's worth to gender. Other critics fear that social approval of PGS will open the door to other genetic manipulations of children's characteristics. It is difficult to predict the full effects that PGS will bring to the world. But the gender-selection genie is now out of the bottle and is unwilling to return.

Your Turn

Take time, individually or in groups, to answer the following questions. Investigate the references and links to help you understand some of the ethical issues that surround the topic of gender selection.

1. A generally accepted moral and legal concept is that of reproductive autonomy—the freedom to make individual reproductive decisions without external interference. Are there circumstances in which reproductive autonomy should be restricted?

   This question is explored in a series of articles in the American Journal of Bioethics, Vol. 1 (2001). *See the article by* J. A. Robertson on pages 2–9 *for a summary of the moral and legal issues surrounding PGS. Also see* Kalfoglou, A. L., et al. (2013). Ethical arguments for and against sperm sorting for non-medical sex selection: A review. *Repro. BioMed. Online* 26: 231–239 (http://www.rbmojournal.com/article/S1472-6483(12)00602-X/fulltext#).

2. If safe and efficient methods of PGS were available, would you use them to help you with family planning? Under what circumstances might you use them?

   A discussion of PGS ethics and methods is presented in Use of reproductive technology for sex selection for nonmedical reasons, Ethics Committee of the American Society for Reproductive Medicine (2015) (http://www.reproductivefacts.org/globalassets/asrm/asrm-content/news-and-publications/ethics-committee-opinions_use_of_reproductive_technology_for_sex_selection_for_nonmedical_reasons-pdfmembers.pdf).

CASE STUDY Is the baby a boy or a girl?

Gender is someone's conscious and unconscious feelings of belonging to one sex or another. Each year, about 1 in 4500 children are born with a disorder involving sexual development, where the chromosomal, gonadal, or anatomical sex is atypical. Here we will consider two similar cases with different outcomes. In case 1, a 2-year-old child displayed a mosaic chromosome composition of 45,X/46,XY, with one ovary, one testis, a uterus, and ambiguous genitalia. In case 2, a fetus was diagnosed with a mosaic chromosome composition of 46,XX/47,XXY, and after birth, also displayed one testis, one ovary, a uterus, and ambiguous genitalia. The child in case 1 was adopted from an orphanage and raised as a girl. After consultation with the medical team, the parents decided to continue raising the child as a girl and requested surgery that would assign the child female sex characteristics. In case 2, the parents decided to forego treatment and let the child make the choice about gender later in life and to remain neutral about the child's present condition. These cases raise questions about sex determination and the ethics of sex and gender assignment.

1. In humans, what is the role of the MSY region of the Y chromosome in sex determination and gender development?
2. Compare and contrast the ethical decisions faced by the parents in both cases 1 and 2. Should parents be allowed to make the decision about the gender of their child? If not, at what age should the child be allowed to make this decision?

See Kipnis, K., and Diamond, M. (1998). Pediatric ethics and the surgical assignment of sex. *J. Clin. Ethics* 9(4):398–410.

INSIGHTS AND SOLUTIONS

1. In *Drosophila*, the X chromosomes may become attached to one another ($\widehat{XX}$) such that they always segregate together. Some flies thus contain a set of attached X chromosomes plus a Y chromosome.

   (a) What sex would such a fly be? Explain why this is so.

   (b) Given the answer to part (a), predict the sex of the offspring that would occur in a cross between this fly and a normal one of the opposite sex.

   (c) If the offspring described in part (b) are allowed to interbreed, what will be the outcome?

Solution:

(a) The fly will be a female. The ratio of X chromosomes to sets of autosomes—which determines sex in *Drosophila*—will be 1.0, leading to normal female development. The Y chromosome has no influence on sex determination in *Drosophila*.

(b) All progeny flies will have two sets of autosomes along with one of the following sex-chromosome compositions:

(1) $\widehat{XX}X$ → a metafemale with 3 X's (called a trisomic)

(2) $\widehat{XX}Y$ → a female like her mother

(3) XY → a normal male

(4) YY → no development occurs

(c) A stock will be created that maintains attached-X females generation after generation.

2. The Xg cell-surface antigen is coded for by a gene located on the X chromosome. No equivalent gene exists on the Y chromosome. Two codominant alleles of this gene have been identified: *Xg1* and *Xg2*. A woman of genotype *Xg2/Xg2* bears children with a man of genotype *Xg1/Y*, and they produce a son with Klinefelter syndrome of genotype *Xg1/Xg2Y*. Using proper genetic terminology, briefly explain how this individual was generated. In which parent and in which meiotic division did the mistake occur?

Solution: Because the son with Klinefelter syndrome is *Xg1/Xg2Y*, he must have received both the *Xg1* allele and the Y chromosome from his father. Therefore, nondisjunction must have occurred during meiosis I in the father.

Problems and Discussion Questions

Mastering Genetics Visit for instructor-assigned tutorials and problems.

1. **HOW DO WE KNOW?** In this chapter, we have focused on sex differentiation, sex chromosomes, and genetic mechanisms involved in sex determination. At the same time, we found many opportunities to consider the methods and reasoning by which much of this information was acquired. From the explanations given in the chapter, you should answer the following fundamental questions:
   (a) How do we know that in humans the X chromosomes play no role in sex determination, while the Y chromosome causes maleness and its absence causes femaleness?
   (b) How did we originally (in the late 1940s) analyze the sex ratio at conception in humans, and how has our approach to studying this issue changed in 2015?
   (c) How do we know that X chromosomal inactivation of either the paternal or maternal homolog is a random event during early development in mammalian females?
   (d) How do we know that *Drosophila* utilizes a different sex-determination mechanism than mammals, even though it has the same sex-chromosome compositions in males and females?
2. **CONCEPT QUESTION** Review the Chapter Concepts list on p. 107. These all center on sex determination or the expression of genes encoded on sex chromosomes. Write a short essay that discusses sex chromosomes as they contrast with autosomes. ■
3. As related to sex determination, what is meant by
   (a) homomorphic and heteromorphic chromosomes; and
   (b) homogametic sex and heterogametic sex?
4. Distinguish between the concepts of sex determination and sexual differentiation.
5. Describe how temperature variation controls sex determination in crocodiles.
6. Describe the major differences between XO individuals in *Drosophila* and those in humans.
7. How do mammals, including humans, solve the "dosage problem" caused by the presence of an X and Y chromosome in one sex and two X chromosomes in the other sex?
8. What is the first evidence that the Y chromosome determines maleness in humans?
9. Describe how nondisjunction in human female gametes can give rise to Klinefelter and Turner syndrome offspring following fertilization by a normal male gamete.
10. An insect species is discovered in which the heterogametic sex is unknown. An X-linked recessive mutation for *reduced wing* (*rw*) is discovered. Contrast the F_1 and F_2 generations from a cross between a female with reduced wings and a male with normal-sized wings when
    (a) the female is the heterogametic sex.
    (b) the male is the heterogametic sex.
11. Given your answers to Problem 10, is it possible to distinguish between the *Protenor* and *Lygaeus* mode of sex determination based on the outcome of these crosses?
12. A group of scientists developing an XX zygote in vitro are curious to see the impact of certain chemicals on the development of the organism. They incubate the zygote with the help of testosterone and some transcription factors, which are usually produced by the activity of the Y chromosome. They discover that the zygote develops into a sterile female with masculinized reproductive organs. Explain why this happens.
13. An attached-X female fly, $\widehat{XX}Y$ (see the Insights and Solutions box), expresses the recessive X-linked white-eye phenotype. It is

crossed to a male fly that expresses the X-linked recessive miniature wing phenotype. Determine the outcome of this cross in terms of sex, eye color, and wing size of the offspring.

14. Assume that on rare occasions the attached X chromosomes in female gametes become unattached. Based on the parental phenotypes in Problem 13, what outcomes in the F_1 generation would indicate that this has occurred during female meiosis?
15. It is believed that any male-determining genes contained on the Y chromosome in humans are not located in the limited region that synapses with the X chromosome during meiosis. What might be the outcome if such genes were located in this region?
16. What is a Barr body, and where is it found in a cell?
17. Indicate the expected number of Barr bodies in interphase cells of individuals with Klinefelter syndrome; Turner syndrome; and karyotypes 47,XYY, 47,XXX, and 48,XXXX.
18. Define the Lyon hypothesis.
19. Could the Lyon hypothesis be tested in a Klinefelter male that is heterozygous for an X-linked gene X?
20. An undiagnosed triplo-X female is a carrier of a recessive X-linked mutation that leads to red–green color blindness. Predict the effects of the mutation based on the Lyon hypothesis.
21. Under what circumstances can a male cat exhibit a tortoiseshell coat pattern?
22. What does the apparent need for dosage compensation mechanisms suggest about the expression of genetic information in normal diploid individuals?
23. A color-blind, chromatin-positive male child (one Barr body) has a maternal grandfather who was color blind. The boy's mother and father are phenotypically normal. Construct and support a rationale whereby the chromosomal and genetic attributes of the chromatin-positive male are fully explained.
24. In Drosophila, an individual female fly was observed to be of the XXY chromosome complement (normal autosomal complement) and to have white eyes as contrasted with the normal red eye color of wild type. The female's father had red eyes, and the mother had white eyes. Knowing that white eyes are X-linked and recessive, present an explanation for the genetic and chromosomal constitution of the XXY, white-eyed individual. It is important that you state in which parent and at what stage the chromosomal event occurred that caused the genetic and cytogenetic abnormality.
25. In mice, the X-linked dominant mutation *Testicular feminization* (*Tfm*) eliminates the normal response to the testicular hormone testosterone during sexual differentiation. An XY mouse bearing the *Tfm* allele on the X chromosome develops testes, but no further male differentiation occurs—the external genitalia of such an animal are female. From this information, what might you conclude about the role of the *Tfm* gene product and the X and Y chromosomes in sex determination and sexual differentiation in mammals? Can you devise an experiment, assuming you can "genetically engineer" the chromosomes of mice, to test and confirm your explanation?
26. In chickens, a key gene involved in sex determination has recently been identified. Called *DMRT1*, it is located on the Z chromosome and is absent on the W chromosome. Like *SRY* in humans, it is male determining. Unlike *SRY* in humans, however, female chickens (ZW) have a single copy while males (ZZ) have two copies of the gene. Nevertheless, it is transcribed only in the developing testis. Working in the laboratory of Andrew Sinclair (a co-discoverer of the human *SRY* gene), Craig Smith and colleagues were able to "knock down" expression of *DMRT1* in ZZ embryos using RNA interference techniques (see Chapter 16). In such cases, the developing gonads look more like ovaries than testes [*Nature* 461: 267 (2009)]. What conclusions can you draw about the role that the *DMRT1* gene plays in chickens in contrast to the role the *SRY* gene plays in humans?
27. Shown here are graphs that plot the percentage of fertilized eggs containing males against the atmospheric temperature during early development in (a) snapping turtles and (b) most lizards. Interpret these data as they relate to the effect of temperature on sex determination.

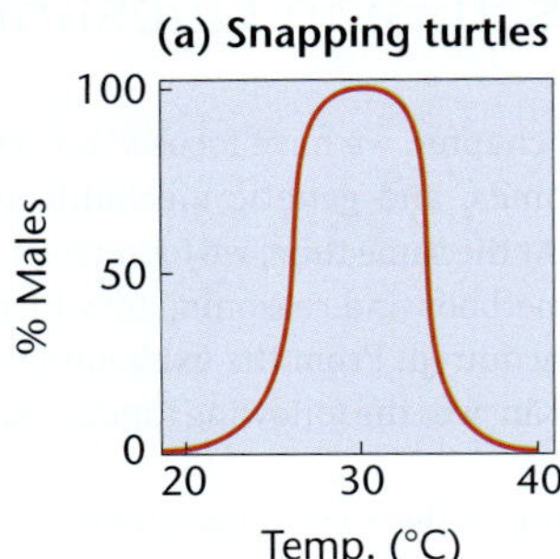

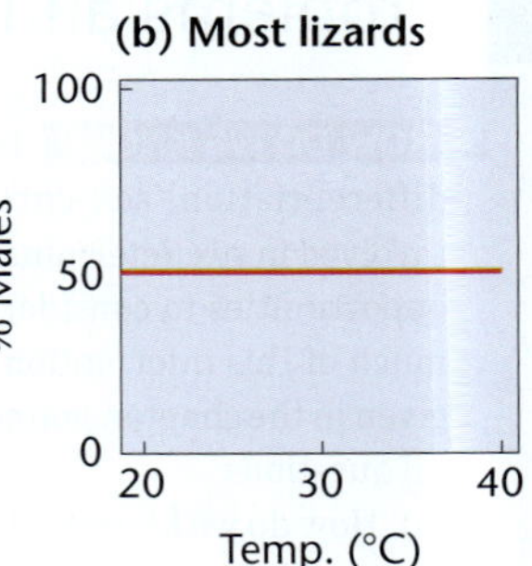

6

Chromosome Mutations: Variation in Number and Arrangement

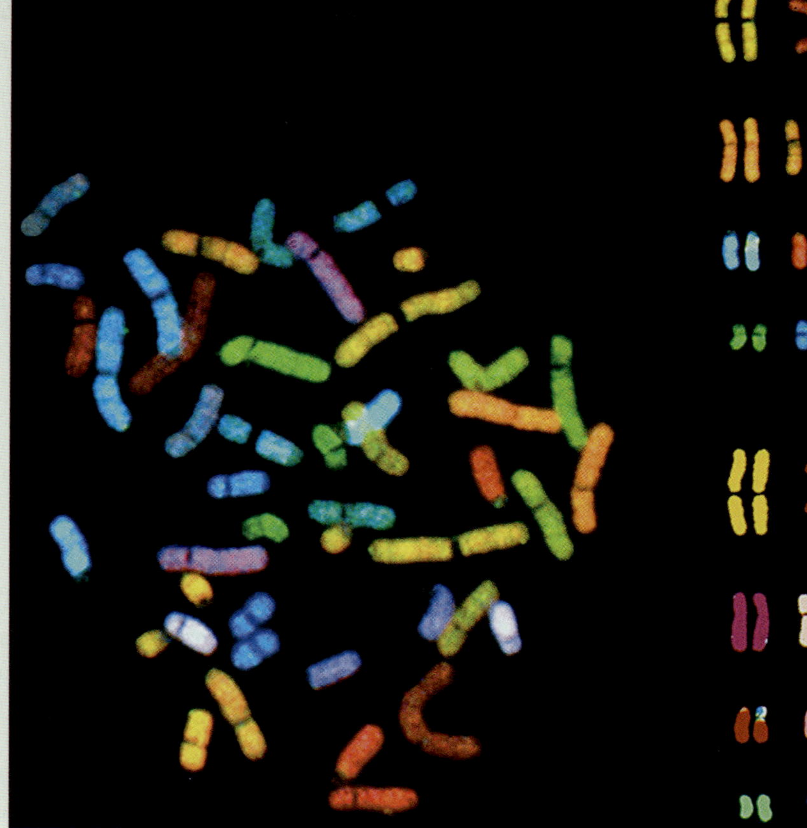

Spectral karyotyping of human chromosomes utilizing differentially labeled "painting" probes.

CHAPTER CONCEPTS

- The failure of chromosomes to properly separate during meiosis results in variation in the chromosome content of gametes and subsequently in offspring arising from such gametes.
- Plants often tolerate an abnormal genetic content, but, as a result, they often manifest unique phenotypes. Such genetic variation has been an important factor in the evolution of plants.
- In animals, genetic information is in a delicate equilibrium whereby the gain or loss of a chromosome, or part of a chromosome, in an otherwise diploid organism often leads to lethality or to an abnormal phenotype.
- The rearrangement of genetic information within the genome of a diploid organism may be tolerated by that organism but may affect the viability of gametes and the phenotypes of organisms arising from those gametes.
- Chromosomes in humans contain fragile sites—regions susceptible to breakage, which lead to abnormal phenotypes.

In previous chapters, we have emphasized how mutations and the resulting alleles affect an organism's phenotype and how traits are passed from parents to offspring according to Mendelian principles. In this chapter, we look at phenotypic variation that results from more substantial changes than alterations of individual genes—modifications at the level of the chromosome.

Although most members of diploid species normally contain precisely two haploid chromosome sets, many known cases vary from this pattern. Modifications include a change in the total number of chromosomes, the deletion or duplication of genes or segments of a chromosome, and rearrangements of the genetic material either within or among chromosomes. Taken together, such changes are called **chromosome mutations** or **chromosome aberrations** in order to distinguish them from gene mutations. Because the chromosome is the unit of genetic transmission, according to Mendelian laws, chromosome aberrations are passed to offspring in a predictable manner, resulting in many unique genetic outcomes.

Because the genetic component of an organism is delicately balanced, even minor alterations of either content or location of genetic information within the genome can result in some form of phenotypic variation. More substantial changes may be lethal, particularly in animals. Throughout this chapter, we consider many types of chromosomal aberrations, the phenotypic consequences for the organism that harbors an aberration, and the impact of the aberration on the offspring of an affected individual. We will also discuss the role of chromosome aberrations in the evolutionary process.

6.1 Variation in Chromosome Number: Terminology and Origin

Variation in chromosome number ranges from the addition or loss of one or more chromosomes to the addition of one or more haploid sets of chromosomes. Before we embark on our discussion, it is useful to clarify the terminology that describes such changes. In the general condition known as **aneuploidy,** an organism gains or loses one or more chromosomes but not a complete set. The loss of a single chromosome from an otherwise diploid genome is called *monosomy.* The gain of one chromosome results in *trisomy.* These changes are contrasted with the condition of **euploidy,** where complete haploid sets of chromosomes are present. If more than two sets are present, the term **polyploidy** applies. Organisms with three sets are specifically *triploid,* those with four sets are *tetraploid,* and so on. **Table 6.1** provides an organizational framework for you to follow as we discuss each of these categories of aneuploid and euploid variation and the subsets within them.

As we consider cases that include the gain or loss of chromosomes, it is useful to examine how such aberrations originate. For instance, how do the syndromes arise where the number of sex-determining chromosomes in humans is altered, as described in Chapter 5? As you may recall, the gain (47,XXY) or loss (45,X) of an X chromosome from an otherwise diploid genome affects the phenotype, resulting in **Klinefelter syndrome** or **Turner syndrome,** respectively (see Figure 5.2). Human females may contain extra X chromosomes (e.g., 47,XXX, 48,XXXX), and some males contain an extra Y chromosome (47,XYY).

Such chromosomal variation originates as a random error during the production of gametes, a phenomenon referred to as **nondisjunction,** whereby paired homologs fail to disjoin during segregation. This process disrupts the normal distribution of chromosomes into gametes. The results of nondisjunction during meiosis I and meiosis II for a single chromosome of a diploid organism are shown in **Figure 6.1**.

TABLE 6.1 Terminology for Variation in Chromosome Numbers

Term	Explanation
Aneuploidy	$2n \pm x$ chromosomes
Monosomy	$2n - 1$
Disomy	$2n$
Trisomy	$2n + 1$
Tetrasomy, pentasomy, etc.	$2n + 2$, $2n + 3$, etc.
Euploidy	Multiples of n
Diploidy	$2n$
Polyploidy	$3n, 4n, 5n, \ldots$
Triploidy	$3n$
Tetraploidy, pentaploidy, etc.	$4n$, $5n$, etc.
Autopolyploidy	Multiples of the same genome
Allopolyploidy (amphidiploidy)	Multiples of closely related genomes

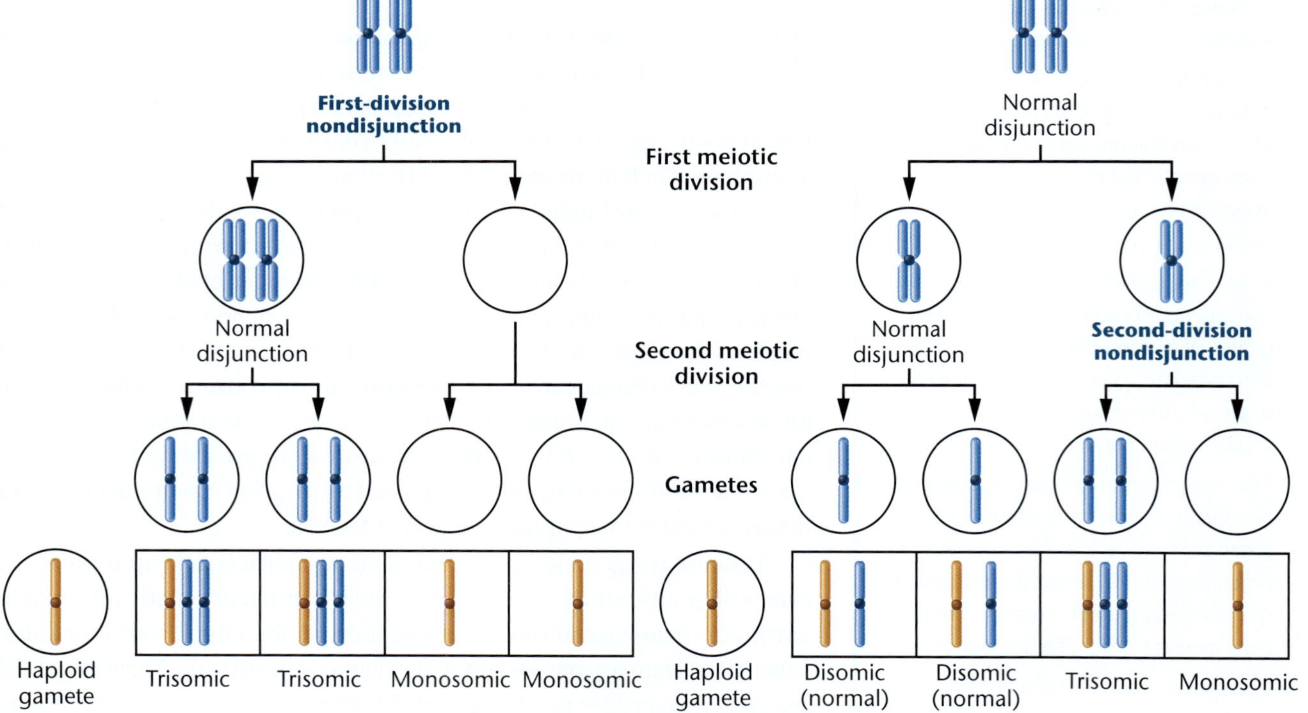

FIGURE 6.1 Nondisjunction during the first and second meiotic divisions. In both cases, some of the gametes that are formed either contain two members of a specific chromosome or lack that chromosome. After fertilization by a gamete with normal haploid content, monosomic, disomic (normal), or trisomic zygotes are produced.

As you can see, abnormal gametes can form that contain either two members of the affected chromosome or none at all. Fertilizing these with a normal haploid gamete produces a zygote with either three members (trisomy) or only one member (monosomy) of this chromosome. Nondisjunction leads to a variety of aneuploid conditions in humans and other organisms.

NOW SOLVE THIS

6.1 A human female with Turner syndrome (45,X) also expresses the X-linked trait hemophilia, as did her father. Which of her parents underwent nondisjunction during meiosis, giving rise to the gamete responsible for the syndrome?

■ **HINT:** *This problem involves an understanding of how nondisjunction leads to aneuploidy. The key to its solution is first to review Turner syndrome, discussed above and in more detail in Chapter 5, then to factor in that she expresses hemophilia, and finally, to consider which parent contributed a gamete with an X chromosome that underwent normal meiosis.*

ESSENTIAL POINT

Alterations of the precise diploid content of chromosomes are referred to as chromosomal aberrations or chromosomal mutations. ■

6.2 Monosomy and Trisomy Result in a Variety of Phenotypic Effects

We turn now to a consideration of variations in the number of autosomes and the genetic consequence of such changes. The most common examples of aneuploidy, where an organism has a chromosome number other than an exact multiple of the haploid set, are cases in which a single chromosome is either added to, or lost from, a normal diploid set.

Monosomy

The loss of one chromosome produces a $2n - 1$ complement called **monosomy.** Although monosomy for the X chromosome occurs in humans, as we have seen in 45,X Turner syndrome, monosomy for any of the autosomes is not usually tolerated in humans or other animals. In *Drosophila,* flies that are monosomic for the very small chromosome IV (containing less than 5 percent of the organism's genes) develop more slowly, exhibit reduced body size, and have impaired viability. Monosomy for the larger chromosomes II and III is apparently lethal because such flies have never been recovered.

The failure of monosomic individuals to survive is at first quite puzzling, since at least a single copy of every gene is present in the remaining homolog. However, one explanation is that if just one of those genes is represented by a lethal allele, monosomy unmasks the recessive lethal allele that is tolerated in heterozygotes carrying the corresponding wild-type allele, leading to the death of the organism. In other cases, a single copy of a recessive gene due to monosomy may be insufficient to provide life-sustaining function for the organism, a phenomenon called **haploinsufficiency.**

Aneuploidy is better tolerated in the plant kingdom. Monosomy for autosomal chromosomes has been observed in maize, tobacco, the evening primrose (*Oenothera*), and the jimson weed (*Datura*), among many other plants. Nevertheless, such monosomic plants are usually less viable than their diploid derivatives. Haploid pollen grains, which undergo extensive development before participating in fertilization, are particularly sensitive to the lack of one chromosome and are seldom viable.

Trisomy

In general, the effects of **trisomy** ($2n + 1$) parallel those of monosomy. However, the addition of an extra chromosome produces somewhat more viable individuals in both animal and plant species than does the loss of a chromosome. In animals, this is often true, provided that the chromosome involved is relatively small. However, the addition of a large autosome to the diploid complement in both *Drosophila* and humans has severe effects and is usually lethal during development.

In plants, trisomic individuals are viable, but their phenotype may be altered. A classical example involves the jimson weed, *Datura,* whose diploid number is 24. Twelve primary trisomic conditions are possible, and examples of each one have been recovered. Each trisomy alters the phenotype of the plant's capsule sufficiently to produce a unique phenotype. These capsule phenotypes were first thought to be caused by mutations in one or more genes.

Still another example is seen in the rice plant (*Oryza sativa*), which has a haploid number of 12. Trisomic strains for each chromosome have been isolated and studied—the plants of 11 strains can be distinguished from one another and from wild-type plants. Trisomics for the longer chromosomes are the most distinctive, and the plants grow more slowly. This is in keeping with the belief that larger chromosomes cause greater genetic imbalance than smaller ones. Leaf structure, foliage, stems, grain morphology, and plant height also vary among the various trisomies.

Down Syndrome: Trisomy 21

The only human autosomal trisomy in which a significant number of individuals survive longer than a year past birth was discovered in 1866 by John Langdon Down. The

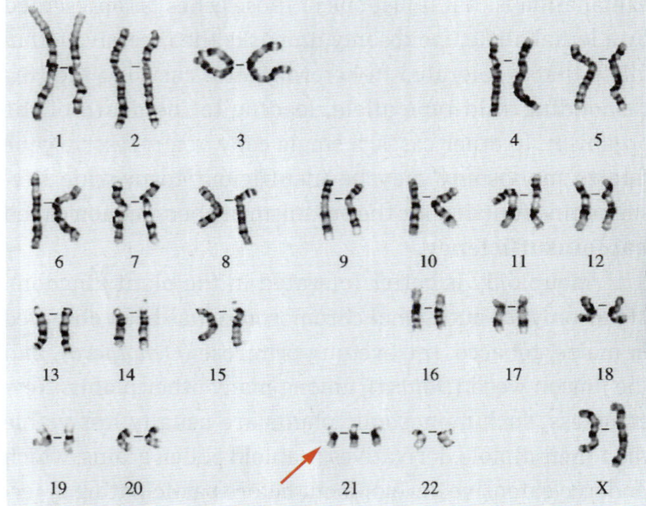

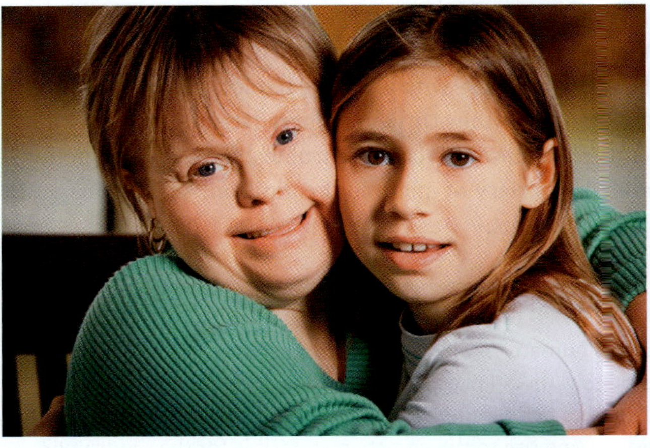

FIGURE 6.2 The karyotype and a photograph of a child with Down syndrome (hugging her unaffected sister on the right). In the karyotype, three members of the G-group chromosome 21 are present, creating the 47, 21+ condition.

condition is now known to result from trisomy of chromosome 21, one of the G group* (**Figure 6.2**), and is called **Down syndrome** or simply **trisomy 21** (designated 47, 21+). This trisomy is found in approximately 1 infant in every 800 live births. While this might seem to be a rare, improbable event, there are approximately 4000–5000 such births annually in the United States, and there are currently over 250,000 individuals with Down syndrome.

Typical of other conditions classified as syndromes, many phenotypic characteristics *may* be present in trisomy 21, but any single affected individual usually exhibits only a subset of these. In the case of Down syndrome, there are 12 to 14 such characteristics, with each individual, on average, expressing 6 to 8 of them. Nevertheless, the outward appearance of these individuals is very similar, and they bear a striking resemblance to one another. This is, for the most part, due to a prominent epicanthic fold in each eye** and the typically flat face and round head. People with Down syndrome are also characteristically short and may have a protruding, furrowed tongue (which causes the mouth to remain partially open) and short, broad hands with characteristic palm and fingerprint patterns. Physical, psychomotor, and cognitive disabilities are evident, and poor muscle tone is characteristic. While life expectancy is shortened to an average of about 50 years, individuals are known to survive into their 60s.

Children with Down syndrome are prone to respiratory disease and heart malformations, and they show an incidence of leukemia approximately 20 times higher than that of the normal population. However, careful medical scrutiny and treatment throughout their lives can extend their survival significantly. A striking observation is that death in older adults with Down syndrome is frequently due to Alzheimer disease. The onset of this disease occurs at a much earlier age than in the normal population.

Because Down syndrome is common in our population, a comprehensive understanding of the underlying genetic basis has long been a research goal. Investigations have given rise to the idea that a critical region of chromosome 21 contains the genes that are dosage sensitive in this trisomy and responsible for the many phenotypes associated with the syndrome. This hypothetical portion of the chromosome has been called the **Down syndrome critical region (DSCR).** A mouse model was created in 2004 that is trisomic for the DSCR, although some mice do not exhibit the characteristics of the syndrome. Nevertheless, this remains an important investigative approach.

Current studies of the DSCR region in both humans and mice have led to several interesting findings. We now believe that the three copies of the genes present in this region are necessary, but themselves not sufficient, for the cognitive deficiencies characteristic of the syndrome. Another finding involves the important observation that Down syndrome individuals have a decreased risk of developing a number of cancers involving solid tumors, including lung cancer and melanoma. This health benefit has been correlated with the presence of an extra copy of the *DSCR1* gene,

*On the basis of size and centromere placement, human autosomal chromosomes are divided into seven groups: A (1–3), B (4–5), C (6–12), D (13–15), E (16–18), F (19–20), and G (21–22).

**The epicanthic fold, or epicanthus, is a skin fold of the upper eyelid, extending from the nose to the inner side of the eyebrow. It covers and appears to lower the inner corner of the eye, giving the eye an almond-shaped appearance.

which encodes a protein that suppresses *vascular endothelial growth factor* (*VEGF*). This suppression, in turn, blocks the process of angiogenesis. As a result, the overexpression of this gene inhibits tumors from forming proper vascularization, diminishing their growth. A 14-year study published in 2002 involving 17,800 Down syndrome individuals revealed an approximate 10 percent reduction in cancer mortality in contrast to a control population.

The Origin of the Extra 21st Chromosome in Down Syndrome

Most frequently, this trisomic condition occurs through nondisjunction of chromosome 21 during meiosis. Failure of paired homologs to disjoin during either anaphase I or II may lead to gametes with the $n + 1$ chromosome composition. About 75 percent of these errors leading to Down syndrome are attributed to nondisjunction during the first meiotic division. Subsequent fertilization with a normal gamete creates the trisomic condition.

Chromosome analysis has shown that, while the additional chromosome may be derived from either the mother or father, the ovum is the source in about 95 percent of 47, 21+ trisomy cases. Before the development of techniques using polymorphic markers to distinguish paternal from maternal homologs, this conclusion was supported by the more indirect evidence derived from studies of the age of mothers giving birth to infants with Down syndrome. **Figure 6.3** shows the relationship between the incidence of children born with Down syndrome and maternal age, illustrating the dramatic increase as the age of the mother increases. While the frequency is about 1 in 1000 at maternal age 30, a tenfold increase to a frequency of 1 in 100 is noted at age 40. The frequency increases still further to about 1 in 30 at age 45. A very alarming statistic is that as the age of childbearing women exceeds 45, the probability of a child born with Down syndrome continues to increase substantially. In spite of this high probability, substantially more than half of such births occur to women younger than 35 years, because the overwhelming proportion of pregnancies in the general population involve women under that age.

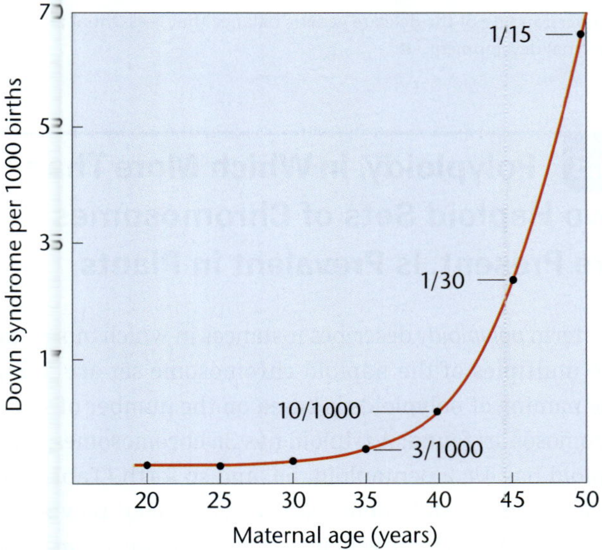

FIGURE 6.3 Incidence of children born with Down syndrome related to maternal age.

Although the nondisjunctional event that produces Down syndrome seems more likely to occur during oogenesis in women over the age of 35, we do not know with certainty why this is so. However, one observation may be relevant. Meiosis is initiated in all the eggs of a human female when she is still a fetus, until the point where the homologs synapse and recombination has begun. Then oocyte development is arrested in meiosis I. Thus, all primary oocytes have been formed by birth. When ovulation begins at puberty, meiosis is reinitiated in one egg during each ovulatory cycle and continues into meiosis II. The process is once again arrested after ovulation and is not completed unless fertilization occurs.

The end result of this progression is that each ovum that is released has been arrested in meiosis I for about a month longer than the one released during the preceding cycle. As a consequence, women 30 or 40 years old produce ova that are significantly older and that have been arrested longer than those they ovulated 10 or 20 years previously. In spite of the logic underlying this hypothesis explaining the cause of the increased incidence of Down syndrome as women age, it remains difficult to prove directly.

These statistics obviously pose a serious problem for the woman who becomes pregnant late in her reproductive years. Genetic counseling early in such pregnancies is highly recommended. Counseling informs prospective parents about the probability that their child will be affected and educates them about Down syndrome. Although some individuals with Down syndrome experience moderate to severe cognitive delays, most experience only mild to moderate delays. These individuals are increasingly integrated into society, including school, the work force, and social and recreational activities. A genetic counselor may also recommend a prenatal diagnostic technique in which fetal cells are isolated and cultured.

In **amniocentesis** and **chorionic villus sampling (CVS),** the two most familiar approaches, fetal cells are obtained from the amniotic fluid or the chorion of the placenta, respectively. In a newer approach, fetal cells and DNA are derived directly from the maternal circulation, a technique referred to as **noninvasive prenatal genetic diagnosis (NIPGD).** Requiring only a 10-mL maternal blood sample, this procedure will become increasingly more common because it poses no risk to the fetus. After fetal cells are obtained and cultured, the karyotype can be determined by cytogenetic analysis. If the fetus is diagnosed as being affected, further counseling for parents will be offered regarding the options open to them, one of which is abortion of the fetus. Obviously, this is a difficult decision involving both religious and ethical issues.

Since Down syndrome is caused by a random error—nondisjunction of chromosome 21 during maternal or paternal meiosis—the occurrence of the disorder is *not* expected to be inherited. Nevertheless, Down syndrome occasionally runs in families. These instances, referred to as familial Down syndrome, involve a translocation of chromosome 21, another type of chromosomal aberration, which we will discuss later in the chapter.

Human Aneuploidy

Besides Down syndrome, only two human trisomies, and no autosomal monosomies, survive to term: **Patau** and **Edwards syndromes** (47, 13+ and 47, 18+, respectively). Even so, these individuals manifest severe malformations and early lethality. **Figure 6.4** illustrates the abnormal karyotype and the many defects characterizing infants with Patau syndrome.

The preceding observation leads us to ask whether many other aneuploid conditions arise but that the affected fetuses do not survive to term. That this is the case has been confirmed by karyotypic analysis of spontaneously aborted fetuses. These studies reveal two striking statistics: (1) Approximately 20 percent of all conceptions terminate in spontaneous abortion (some estimates are considerably higher); and (2) about 30 percent of all spontaneously aborted fetuses demonstrate some form of chromosomal imbalance. This suggests that at least 6 percent (0.20×0.30) of conceptions contain an abnormal chromosome complement. A large percentage of fetuses demonstrating chromosomal abnormalities are aneuploids.

An extensive review of this subject by David H. Carr has revealed that a significant percentage of aborted fetuses are trisomic for one of the chromosome groups. Trisomies for every human chromosome have been recovered. Interestingly, the monosomy with the highest incidence among abortuses is the 45,X condition, which produces an infant with Turner syndrome if the fetus survives to term. Autosomal monosomies are seldom found, however, even though nondisjunction should produce $n - 1$ gametes with a frequency equal to $n + 1$ gametes. This finding suggests that gametes lacking a single chromosome are functionally impaired to a serious degree or that the embryo dies so early in its development that recovery occurs infrequently. We discussed the potential causes of monosomic lethality earlier in this chapter. Carr's study also found various forms of polyploidy and other miscellaneous chromosomal anomalies.

These observations support the hypothesis that normal embryonic development requires a precise diploid complement of chromosomes to maintain the delicate equilibrium in the expression of genetic information. The prenatal mortality of most aneuploids provides a barrier against the introduction of these genetic anomalies into the human population.

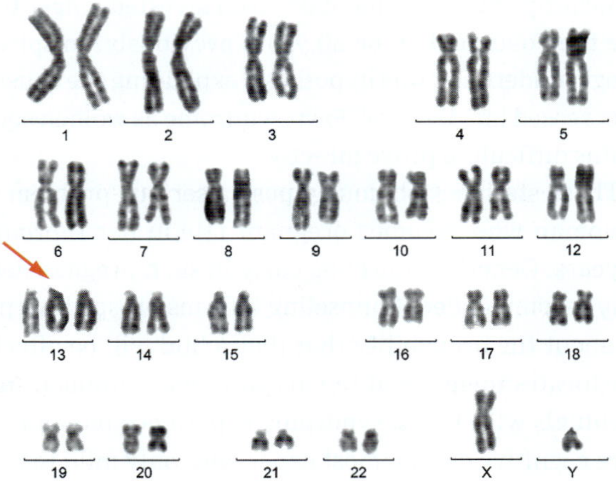

Intellectual disability	Microcephaly
Growth failure	Cleft lip and palate
Low-set, deformed ears	Polydactyly
Deafness	Deformed finger nails
Atrial septal defect	Kidney cysts
Ventricular septal defect	Double ureter
Abnormal polymorphonuclear granulocytes	Umbilical hernia
	Developmental uterine abnormalities
	Cryptorchidism

FIGURE 6.4 The karyotype and phenotypic description of an infant with Patau syndrome, where three members of the D-group chromosome 13 are present, creating the 47, 13+ condition.

ESSENTIAL POINT

Studies of monosomic and trisomic disorders are increasing our understanding of the delicate genetic balance that is essential for normal development. ■

6.3 Polyploidy, in Which More Than Two Haploid Sets of Chromosomes Are Present, Is Prevalent in Plants

The term *polyploidy* describes instances in which more than two multiples of the haploid chromosome set are found. The naming of polyploids is based on the number of sets of chromosomes found: A triploid has $3n$ chromosomes; a tetraploid has $4n$; a pentaploid, $5n$; and so forth (Table 6.1). Several general statements can be made about polyploidy. This condition is relatively infrequent in many animal species but is well known in lizards, amphibians, and fish, and is much more common in plant species. Usually, odd numbers of chromosome sets are not reliably maintained from generation to generation because a polyploid organism with an uneven number of homologs often does not produce genetically balanced gametes. For this reason, triploids, pentaploids, and so on, are not usually found in plant species that depend solely on sexual reproduction for propagation.

Polyploidy originates in two ways: (1) The addition of one or more extra sets of chromosomes, identical to the normal haploid complement of the same species, resulting in **autopolyploidy**; or (2) the combination of chromosome sets from different species occurring as a consequence of hybridization, resulting in **allopolyploidy** (from the Greek word *allo*, meaning "other" or "different"). The distinction between auto- and allopolyploidy is based on the genetic origin of the extra chromosome sets, as shown in **Figure 6.5**.

In our discussion of polyploidy, we use the following symbols to clarify the origin of additional chromosome sets. For example, if A represents the haploid set of chromosomes of any organism, then

$$A = a_1 + a_2 + a_3 + a_4 + \cdots + a_n$$

where a_1, a_2, and so on, are individual chromosomes and n is the haploid number. A normal diploid organism is represented simply as AA.

Autopolyploidy

In autopolyploidy, each additional set of chromosomes is identical to the parent species. Therefore, triploids are represented as AAA, tetraploids are $AAAA$, and so forth.

Autotriploids arise in several ways. A failure of all chromosomes to segregate during meiotic divisions can produce a diploid gamete. If such a gamete is fertilized by a haploid gamete, a zygote with three sets of chromosomes is produced. Or, rarely, two sperm may fertilize an ovum, resulting in a triploid zygote. Triploids are also produced under experimental conditions by crossing diploids with tetraploids. Diploid organisms produce gametes with n chromosomes, while tetraploids produce $2n$ gametes. Upon fertilization, the desired triploid is produced.

Because they have an even number of chromosomes, **autotetraploids** ($4n$) are theoretically more likely to be found in nature than are autotriploids. Unlike triploids, which often produce genetically unbalanced gametes with odd numbers of chromosomes, tetraploids are more likely to produce balanced gametes when involved in sexual reproduction.

How polyploidy arises naturally is of great interest to geneticists. In theory, if chromosomes have replicated, but the parent cell never divides and instead reenters interphase, the chromosome number will be doubled. That this very likely occurs is supported by the observation that tetraploid cells can be produced experimentally from diploid cells. This is accomplished by applying cold or heat shock to meiotic cells or by applying colchicine to somatic cells undergoing mitosis. Colchicine, an alkaloid derived from the autumn crocus, interferes with spindle formation, and thus replicated chromosomes cannot separate at anaphase and do not migrate to the poles. When colchicine is removed, the cell can reenter interphase. When the paired sister chromatids separate and uncoil, the nucleus contains twice the diploid number of chromosomes and is therefore $4n$. This process is shown in **Figure 6.6**.

In general, autopolyploids are larger than their diploid relatives. This increase seems to be due to larger cell size rather than greater cell number. Although autopolyploids do not contain new or unique information compared with their diploid relatives, the flower and fruit of plants are often increased in size, making such varieties of greater horticultural or commercial value. Economically important triploid plants include several potato species of the genus *Solanum*, Winesap apples, commercial bananas, seedless watermelons, and the cultivated tiger lily *Lilium tigrinum*. These plants are propagated asexually. Diploid bananas contain hard seeds, but the commercial, triploid, "seedless" variety has edible seeds. Tetraploid alfalfa, coffee, peanuts,

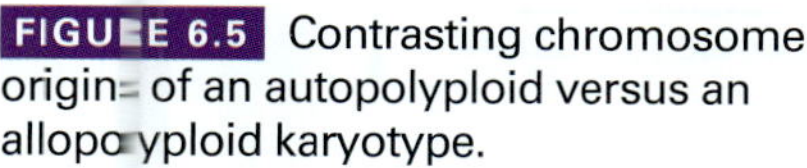

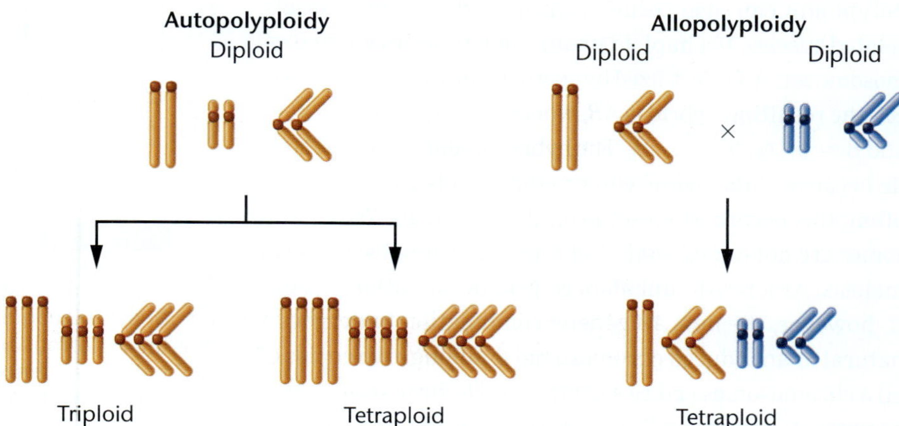

FIGURE 6.5 Contrasting chromosome origins of an autopolyploid versus an allopolyploid karyotype.

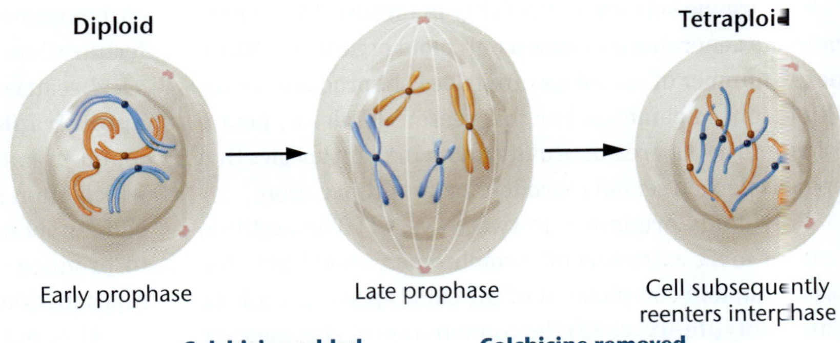

FIGURE 6.6 The potential involvement of colchicine in doubling the chromosome number. Two pairs of homologous chromosomes are shown. While each chromosome had replicated its DNA earlier during interphase, the chromosomes do not appear as double structures until late prophase. When anaphase fails to occur normally, the chromosome number doubles if the cell reenters interphase.

and McIntosh apples are also of economic value because they are either larger or grow more vigorously than do their diploid or triploid counterparts. Many of the most popular varieties of hosta plant are tetraploid. In each case, leaves are thicker and larger, the foliage is more vivid, and the plant grows more vigorously. The commercial strawberry is an octoploid.

How cells with increased ploidy values express different phenotypes from their diploid counterparts has been investigated. Gerald Fink and his colleagues created strains of the yeast *Saccharomyces cerevisiae* with one, two, three, or four copies of the genome and then examined the expression levels of all genes during the cell cycle. Using the stringent standards of at least a tenfold increase or decrease of gene expression, Fink and coworkers identified numerous cases where, as ploidy increased, gene expression either increased or decreased at least tenfold. Among these cases are two genes that encode **G1 cyclins,** which are repressed when ploidy increases. G1 cyclins facilitate the cell's movement through G1 of the cell cycle, which is thus delayed when expression of these genes is repressed. The polyploid cell stays in the G1 phase longer and, on average, grows to a larger size before it moves beyond the G1 stage of the cell cycle, providing a clue as to how other polyploids demonstrate increased cell size.

Allopolyploidy

Polyploidy can also result from hybridizing two closely related species. If a haploid ovum from a species with chromosome sets *AA* is fertilized by sperm from a species with sets *BB*, the resulting hybrid is *AB*, where $A = a_1, a_2, a_3, \ldots, a_n$ and $B = b_1, b_2, b_3, \ldots, b_n$. The hybrid organism may be sterile because of its inability to produce viable gametes. Most often, this occurs when some or all of the *a* and *b* chromosomes are not homologous and therefore cannot synapse in meiosis. As a result, unbalanced genetic conditions result. If, however, the new *AB* genetic combination undergoes a natural or an induced chromosomal doubling, two copies of all *a* chromosomes and two copies of all *b* chromosomes will be present, and they will pair during meiosis. As a result, a fertile *AABB* tetraploid is produced. These events are shown in **Figure 6.7**. Since this polyploid contains the equivalent of four haploid genomes derived from separate species, such an organism is called an **allotetraploid.** When both original species are known, an equivalent term, **amphidiploid,** is preferred in describing the allotetraploid.

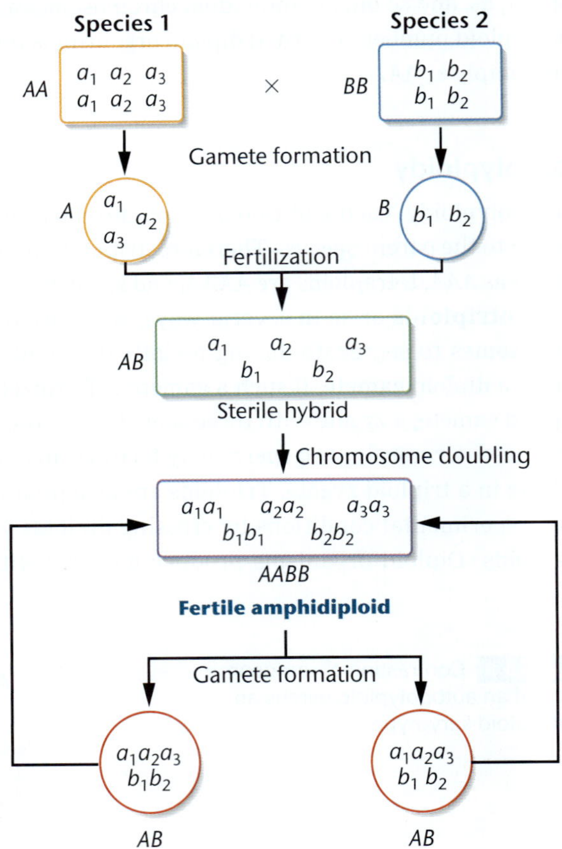

FIGURE 6.7 The origin and propagation of an amphidiploid. Species 1 contains genome *A* consisting of three distinct chromosomes, a_1, a_2, and a_3. Species 2 contains genome *B* consisting of two distinct chromosomes, b_1 and b_2. Following fertilization between members of the two species and chromosome doubling, a fertile amphidiploid containing two complete diploid genomes (*AABB*) is formed.

Amphidiploid plants are often found in nature. Their reproductive success is based on their potential for forming balanced gametes. Since two homologs of each specific chromosome are present, meiosis occurs normally (Figure 6.7) and fertilization successfully propagates the plant sexually. This discussion assumes the simplest situation, where none of the chromosomes in set *A* are homologous to those in set *B*. In amphidiploids formed from closely related species, some homology between *a* and *b* chromosomes is likely. Allopolyploids are rare in most animals because mating behavior is most often species-specific, and thus the initial step in hybridization is unlikely to occur.

A classical example of amphidiploidy in plants is the cultivated species of American cotton, *Gossypium* (**Figure 6.8**). This species has 26 pairs of chromosomes: 13 are large and 13 are much smaller. When it was discovered that Old World cotton had only 13 pairs of large chromosomes, allopolyploidy was suspected. After an examination of wild American cotton revealed 13 pairs of small chromosomes, this speculation was strengthened. J. O. Beasley reconstructed the origin of cultivated cotton experimentally by crossing the Old World strain with the wild American strain and then treating the hybrid with colchicine to double the chromosome number. The result of these treatments was a fertile amphidiploid variety of cotton. It contained 26 pairs of chromosomes as well as characteristics similar to the cultivated variety.

Amphidiploids often exhibit traits of both parental species. An interesting example involves the grasses wheat and rye. Wheat (genus *Triticum*) has a basic haploid genome of seven chromosomes. In addition to normal diploids ($2n = 14$), cultivated autopolyploids exist, including tetraploid ($4n = 28$) and hexaploid ($6n = 42$) species. Rye (genus *Secale*) also has a genome consisting of seven chromosomes. The only cultivated species is the diploid plant ($2n = 14$).

FIGURE 6.8 The pods of the amphidiploid form of *Gossypium*, the cultivated American cotton plant.

Using the technique outlined in Figure 6.7, geneticists have produced various hybrids. When tetraploid wheat is crossed with diploid rye and the F_1 plants are treated with colchicine, a hexaploid variety ($6n = 42$) is obtained; the hybrid, designated *Triticale,* represents a new genus. Other *Triticale* varieties have been created. These hybrid plants demonstrate characteristics of both wheat and rye. For example, they combine the high-protein content of wheat with rye's high content of the amino acid lysine, which is low in wheat and thus is a limiting nutritional factor. Wheat is considered to be a high-yielding grain, whereas rye is noted for its versatility of growth in unfavorable environments. *Triticale* species that combine both traits have the potential of significantly increasing grain production. This and similar programs designed to improve crops through hybridization have long been under way in several developing countries.

NOW SOLVE THIS

6.2 When two plants belonging to the same genus but different species are crossed, the F_1 hybrid is viable and has more ornate flowers. Unfortunately, this hybrid is sterile and can only be propagated by vegetative cuttings. Explain the sterility of the hybrid and what would have to occur for the sterility of this hybrid to be reversed.

■ **HINT:** *This problem involves an understanding of allopolyploid plants. The key to its solution is to focus on the origin and composition of the chromosomes in the F_1 and how they might be manipulated.*

ESSENTIAL POINT

When complete sets of chromosomes are added to the diploid genome, these sets can have an identical or a diverse genetic origin, creating either autopolyploidy or allopolyploidy, respectively. ■

6.4 Variation Occurs in the Composition and Arrangement of Chromosomes

The second general class of chromosome aberrations includes changes that delete, add, or rearrange substantial portions of one or more chromosomes. Included in this broad category are deletions and duplications of genes or part of a chromosome and rearrangements of genetic material in which a chromosome segment is inverted, exchanged with a segment of a nonhomologous chromosome, or merely transferred to another chromosome.

Exchanges and transfers are called translocations, in which the locations of genes are altered within the genome. These types of chromosome alterations are illustrated in **Figure 6.9**.

In most instances, these structural changes are due to one or more breaks along the axis of a chromosome, followed by either the loss or rearrangement of genetic material. Chromosomes can break spontaneously, but the rate of breakage may increase in cells exposed to chemicals or radiation. The ends produced at points of breakage are "sticky" and can rejoin other broken ends. If breakage and rejoining do not reestablish the original relationship and if the alteration occurs in germ plasm, the gametes will contain the structural rearrangement, which is heritable.

If the aberration is found in one homolog but not the other, the individual is said to be *heterozygous for the aberration.* In such cases, unusual but characteristic pairing configurations are formed during meiotic synapsis. These patterns are useful in identifying the type of change that has occurred. If no loss or gain of genetic material occurs, individuals bearing the aberration "heterozygously" are likely to be unaffected phenotypically. However, the unusual pairing arrangements often lead to gametes that are duplicated or deficient for some chromosomal regions. When this occurs, the offspring of "carriers" of certain aberrations have an increased probability of demonstrating phenotypic changes.

6.5 A Deletion Is a Missing Region of a Chromosome

When a chromosome breaks in one or more places and a portion of it is lost, the missing piece is called a **deletion** (or a **deficiency**). The deletion can occur either near one end or within the interior of the chromosome. These are **terminal** and **intercalary deletions,** respectively [**Figure 6.10(a) and (b)**]. The portion of the chromosome that retains the centromere region is usually maintained when the cell divides, whereas the segment without the centromere is eventually lost in progeny cells following mitosis or meiosis. For synapsis to occur between a chromosome with a large intercalary deletion and a normal homolog, the unpaired region of the normal homolog must

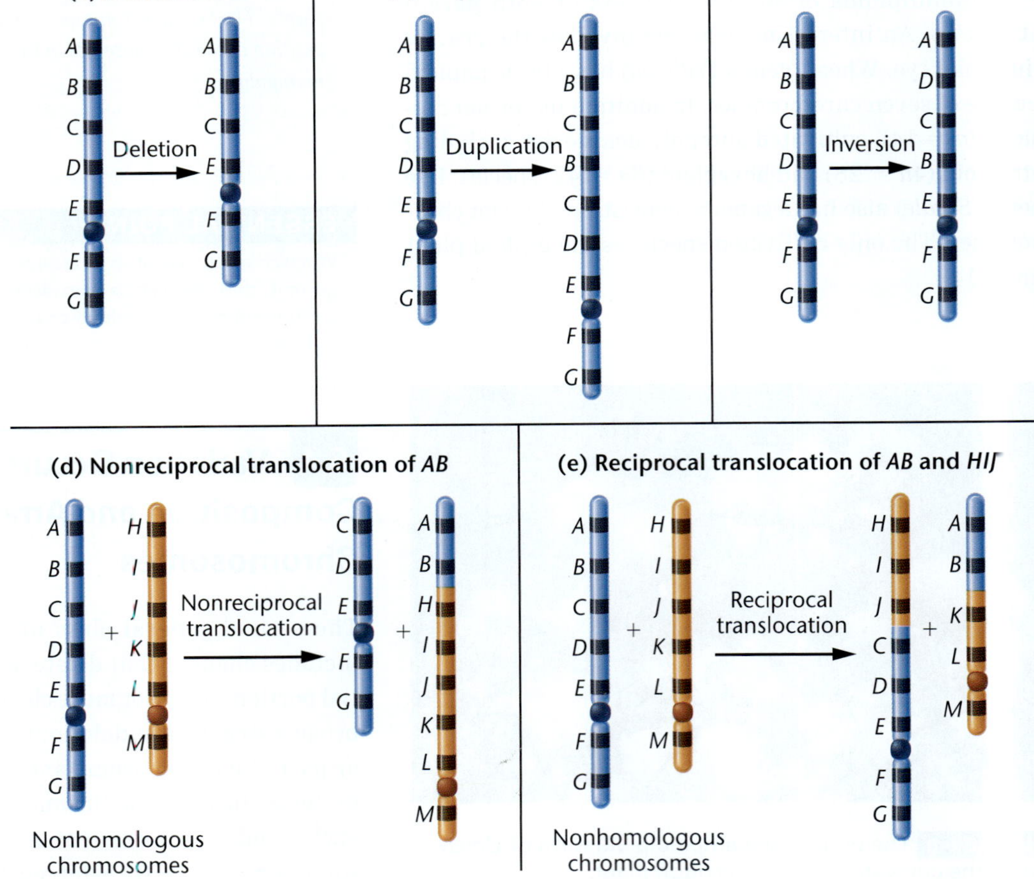

FIGURE 6.9 An overview of the five different types of gain, loss, or rearrangement of chromosome segments.

(a) Origin of terminal deletion

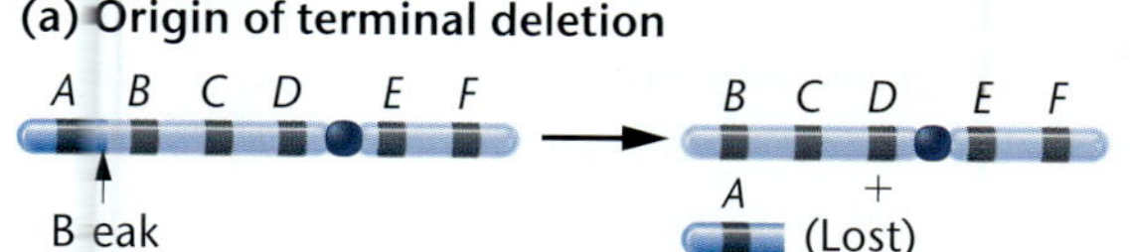

(b) Origin of intercalary deletion

D C
Break Break
A B E F
A B E F
D C
A B E F +
Deleted chromosome
D C
(Lost)

(c) Formation of deletion loop

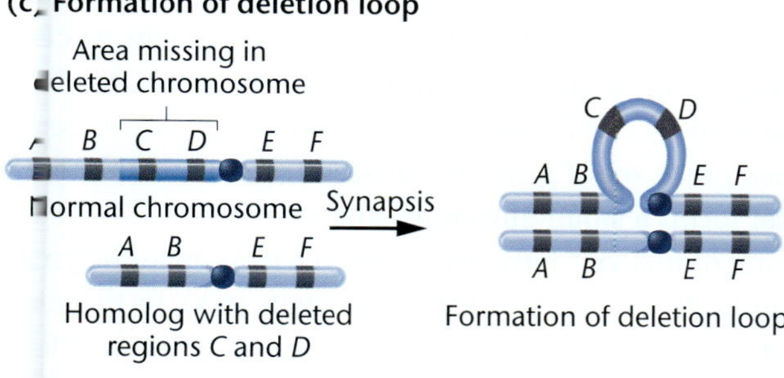

FIGURE 6.10 Origins of (a) a terminal and (b) an intercalary deletion. In (c), pairing occurs between a normal chromosome and one with an intercalary deletion by looping out the undeleted portion to form a deletion (or compensation) loop.

"buckle out" into a **deletion,** or **compensation, loop** [**Figure 6.10(c)**].

If only a small part of a chromosome is deleted, the organism might survive. However, a deletion of a portion of a chromosome need not be very great before the effects become severe. We see an example of this in the following discussion of the cri du chat syndrome in humans. If even more genetic information is lost as a result of a deletion, the aberration is often lethal, in which case the chromosome mutation never becomes available for study.

Cri du Chat Syndrome in Humans

In humans, the **cri du chat syndrome** results from the deletion of a small terminal portion of chromosome 5. It might be considered a case of *partial monosomy,* but since the region that is missing is so small, it is better referred to as a *segmental deletion.* This syndrome was first reported by Jérôme Lejeune in 1963, when he described the clinical symptoms, including an eerie cry similar to the meowing of a cat, after which the syndrome is named. This syndrome is associated with the loss of a small, variable part of the short arm of chromosome 5 (**Figure 6.11**). Thus, the genetic constitution may be designated as 46,5p–, meaning that the individual has all 46 chromosomes but that some or all of the p arm (the petite, or short, arm) of one member of the chromosome 5 pair is missing.

Infants with this syndrome exhibit intellectual disability, delayed development, small head size, and distinctive facial features in addition to abnormalities in the glottis and larynx, leading to the characteristic crying sound.

Since 1963, hundreds of cases of cri du chat syndrome have been reported worldwide. An incidence of 1 in 20,000–50,000 live births has been estimated. Most often, the condition is not inherited but instead results from the sporadic loss of chromosomal material in gametes. The length of the short arm that is deleted varies somewhat; longer deletions tend to result in more severe intellectual disability and developmental delay. Although the effects of the syndrome are severe, most individuals achieve motor and language skills. The deletion of several genes, including the *telomerase reverse transcriptase (TERT)* gene, has been implicated in various phenotypic changes in cri du chat syndrome.

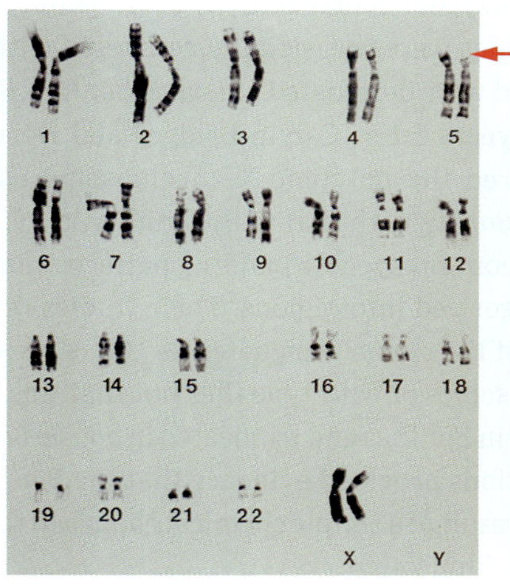

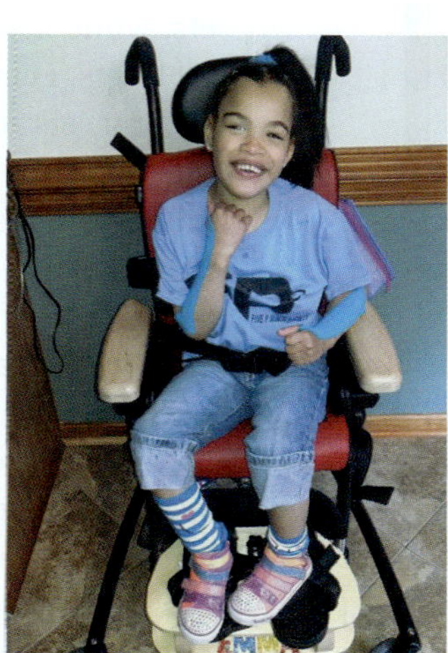

FIGURE 6.11 A representative karyotype and a photograph of a child with cri du chat syndrome (46,5p–). In the karyotype, the arrow identifies the absence of a small piece of the short arm of one member of the chromosome 5 homologs.

6.6 A Duplication Is a Repeated Segment of a Chromosome

When any part of the genetic material—a single locus or a large piece of a chromosome—is present more than once in the genome, it is called a **duplication.** As in deletions, pairing in heterozygotes can produce a compensation loop. Duplications may arise as the result of unequal crossing over between synapsed chromosomes during meiosis (**Figure 6.12**) or through a replication error prior to meiosis. In the former case, both a duplication and a deletion are produced.

We consider three interesting aspects of duplications. First, they may result in gene redundancy. Second, as with deletions, duplications may produce phenotypic variation. Third, according to one convincing theory, duplications have also been an important source of genetic variability during evolution.

Gene Redundancy—Ribosomal RNA Genes

Although many gene products are not needed in every cell of an organism, other gene products are known to be essential components of all cells. For example, ribosomal RNA must be present in abundance to support protein synthesis. The more metabolically active a cell is, the higher the demand for this molecule. We might hypothesize that a single copy of the gene encoding rRNA is inadequate in many cells. Studies using the technique of molecular hybridization, which enables us to determine the percentage of the genome that codes for specific RNA sequences, show that our hypothesis is correct. Indeed, multiple copies of genes code for rRNA.

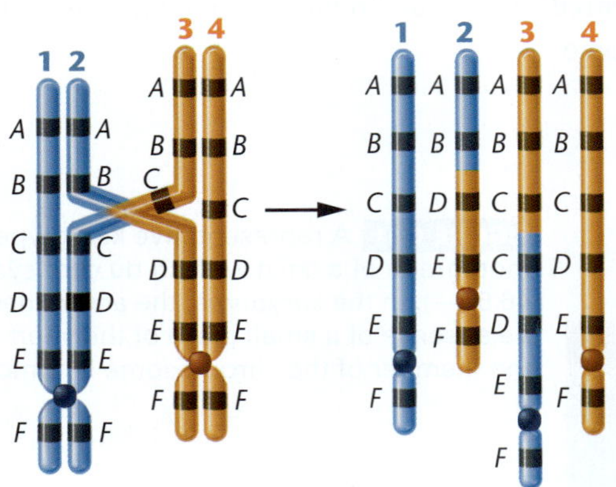

FIGURE 6.12 The origin of duplicated and deficient regions of chromosomes as a result of unequal crossing over. The tetrad on the left is mispaired during synapsis. A single crossover between chromatids 2 and 3 results in the deficient (chromosome 2) and duplicated (chromosome 3) chromosomal regions shown on the right. The two chromosomes uninvolved in the crossover event remain normal in gene sequence and content.

Such DNA is called **rDNA,** and the general phenomenon is referred to as **gene redundancy.** For example, in the common intestinal bacterium *Escherichia coli* (*E. coli*), about 0.7 percent of the haploid genome consists of rDNA—the equivalent of seven copies of the gene. In *Drosophila melanogaster,* 0.3 percent of the haploid genome, equivalent to 130 gene copies, consists of rDNA. Although the presence of multiple copies of the same gene is not restricted to those coding for rRNA, we will focus on them in this section.

In some cells, particularly oocytes, even the normal amplification of rDNA is insufficient to provide adequate amounts of rRNA needed to construct ribosomes. For example, in the amphibian *Xenopus laevis,* 400 copies of rDNA are present per haploid genome. These genes are all found in a single area of the chromosome known as the **nucleolar organizer region (NOR).** In *Xenopus* oocytes, the NOR is selectively replicated to further increase rDNA copies, and each new set of genes is released from its template. Each set forms a small nucleolus, and as many as 1500 of these "micronucleoli" have been observed in a single oocyte. If we multiply the number of micronucleoli (1500) by the number of gene copies in each NOR (400), we see that amplification in *Xenopus* oocytes can result in over half a million gene copies! If each copy is transcribed only 20 times during the maturation of the oocyte, in theory, sufficient copies of rRNA are produced to result in well over 12 million ribosomes.

The *Bar* Mutation in *Drosophila*

Duplications can cause phenotypic variation that might at first appear to be caused by a simple gene mutation. The *Bar*-eye phenotype in *Drosophila* (**Figure 6.13**) is a classic example. Instead of the normal oval-eye shape, *Bar*-eyed flies have narrow, slit-like eyes. This phenotype is inherited in the same way as a dominant X-linked mutation.

In the early 1920s, Alfred H. Sturtevant and Thomas H. Morgan discovered and investigated this "mutation." Normal wild-type females (B^+/B^+) have about 800 facets in each eye. Heterozygous females (B/B^+) have about 350 facets, while homozygous females (B/B) average only about 70 facets. Females were occasionally recovered with even fewer facets and were designated as *double Bar* (B^D/B^+).

About 10 years later, Calvin Bridges and Herman J. Muller compared the polytene X chromosome banding pattern of the *Bar* fly with that of the wild-type fly. These chromosomes contain specific banding patterns that have been well categorized into regions. Their studies revealed that one copy of the region designated as 16A is present on both X chromosomes of wild-type flies but that this region was duplicated in *Bar* flies and triplicated in *double Bar* flies. These observations provided evidence that the *Bar* phenotype is not the result of a simple chemical change in the gene but is instead a duplication.

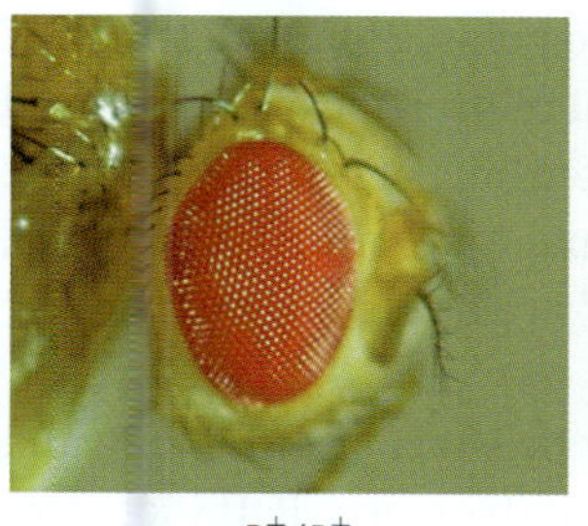
B^+/B^+

B/B^+

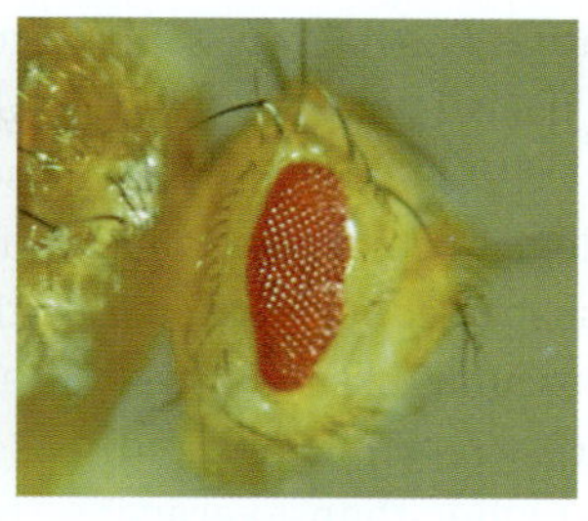
B/B

FIGURE 6.13 *Bar*-eye phenotypes in contrast to the wild-type eye in *Drosophila* (shown in the left panel).

The Role of Gene Duplication in Evolution

During the study of evolution, it is intriguing to speculate on the possible mechanisms of genetic variation. The origin of unique gene products present in more recently evolved organisms but absent in ancestral forms is a topic of particular interest. In other words, how do "new" genes arise?

In 1970, Susumo Ohno published a provocative monograph, *Evolution by Gene Duplication,* in which he suggested that gene duplication is essential to the origin of new genes during evolution. Ohno's thesis is based on the supposition that the gene products of many genes, present as only a single copy in the genome, are indispensable to the survival of members of any species during evolution. Therefore, unique genes are not free to accumulate mutations sufficient to alter their primary function and give rise to new genes.

However, if an essential gene is duplicated in the germ line, major mutational changes in this extra copy will be tolerated in future generations because the original gene provides the genetic information for its essential function. The duplicated copy will be free to acquire many mutational changes over extended periods of time. Over short intervals, the new genetic information may be of no practical advantage. However, over long evolutionary periods, the duplicated gene may change sufficiently so that its product assumes a divergent role in the cell. The new function may impart an "adaptive" advantage to organisms, enhancing their fitness. Ohno has outlined a mechanism through which sustained genetic variability may have originated.

Ohno's thesis is supported by the discovery of genes that have a substantial amount of their organization and DNA sequence in common, but whose gene products are distinct. For example, trypsin and chymotrypsin fit this description, as do myoglobin and the various forms of hemoglobin. The DNA sequence is so similar (homologous) in each case that we may conclude that members of each pair of genes arose from a common ancestral gene through duplication. During evolution, the related genes diverged sufficiently that their products became unique.

Other support includes the presence of **multigene families**—groups of contiguous genes whose products perform the same, or very similar functions. Again, members of a family show DNA sequence homology sufficient to conclude that they share a common origin and arose through the process of gene duplication. One of the most interesting supporting examples is the case of the *SRGAP2* gene in primates. This gene is known to be involved in the development of the brain. Humans have at least four similar copies of the gene, while all nonhuman primates have only a single copy. Several duplication events can be traced back to 3.4 million years ago, to 2.4 million years ago, and finally to 1 million years ago, resulting in distinct forms of *SRGAP2* labeled A–D. These evolutionary periods coincide with the emergence of the human lineage in primates. The function of these genes has now been related to the regulation and formation of dendritic spines in the brain, which is believed to contribute to the evolution of expanded brain function in humans, including the development of language and social cognition.

Other examples of gene families arising from duplication during evolution include the various types of human hemoglobin polypeptide chains, as well as the immunologically important T-cell receptors and antigens encoded by the major histocompatibility complex.

Duplications at the Molecular Level: Copy Number Variations (CNVs)

As we entered the era of genomics and became capable of sequencing entire genomes (see Chapter 17), we quickly realized that duplications of *portions of genes,* most often involving thousands of base pairs, occur on a regular basis. When individuals in the same species are compared, the number of copies of any given duplicated sequence within a given gene is found to differ—sometimes there are larger and sometimes smaller numbers of copies, a condition described as **copy number variation (CNV)**. Such duplications are found in both coding and noncoding regions of the genome.

CNVs are of major interest in genetics because they are now believed to play crucial roles in the expression of many of our individual traits, in both normal and diseased individuals. Currently, when CNVs of sizes ranging from 50 bp to 3 Mb are considered, it is estimated that they occupy between 5–10 percent of the human genome. Current studies have focused on finding associations with human diseases. CNVs appear to have both positive and negative associations with many diseases in which the genetic basis is not yet fully understood. For example, pathogenic CNVs have been associated with autism and other neurological disorders, and with cancer. Additionally, CNVs are suspected to be associated with Type I diabetes and cardiovascular disease.

In some cases, entire gene sequences are duplicated and impact individuals. For example, a higher-than-average copy number of the gene *CCL3L1* imparts an HIV-suppressive effect during viral infection, diminishing the progression to AIDS. Another finding has associated specific mutant CNV sites with certain subset populations of individuals with lung cancer—the greater number of copies of the *EGFR* (*Epidermal Growth Factor Receptor*) gene, the more responsive are patients with non-small-cell lung cancer to treatment. Finally, the greater the reduction in the copy number of the gene designated *DEFB,* the greater the risk of developing Crohn's disease, a condition affecting the colon. Relevant to this chapter, these findings reveal that duplications and deletions are no longer restricted to textbook examples of these chromosomal mutations. We will return to this interesting topic later in the text (see Chapter 18), when genomics is discussed in detail.

ESSENTIAL POINT

Deletions or duplications of segments of a gene or a chromosome may be the source of mutant phenotypes such as cri du chat syndrome in humans and *Bar* eyes in *Drosophila,* while duplications can be particularly important as a source of redundant or new genes. ■

6.7 Inversions Rearrange the Linear Gene Sequence

The **inversion,** another class of structural variation, is a type of chromosomal aberration in which a segment of a chromosome is turned around 180 degrees within a chromosome. An inversion does not involve a loss of genetic information but simply rearranges the linear gene sequence. An inversion requires breaks at two points along the length of the chromosome and subsequent reinsertion of the inverted segment. **Figure 6.14** illustrates how an inversion might arise. By forming a chromosomal loop prior to breakage, the newly created "sticky ends" are brought close together and rejoined.

The inverted segment may be short or quite long and may or may not include the centromere. If the centromere is not part of the rearranged chromosome segment, it is a **paracentric inversion,** which is the type shown in Figure 6.14. If the centromere is part of the inverted segment, it is described as a **pericentric inversion.**

Consequences of Inversions during Gamete Formation

If only one member of a homologous pair of chromosomes has an inverted segment, normal *linear synapsis* during meiosis is not possible. Organisms with one inverted chromosome and one noninverted homolog are called **inversion heterozygotes.** Pairing between two such chromosomes in meiosis is accomplished only if they form an **inversion loop** (**Figure 6.15**).

If crossing over does not occur within the inverted segment of the inversion loop, the homologs will segregate, which results in two normal and two inverted chromatids that are distributed into gametes. However, if crossing over does occur within the inversion loop, abnormal chromatids are produced. The effect of a single crossover (SCO) event within a paracentric inversion is diagrammed in Figure 6.15.

In any meiotic tetrad, a single crossover between nonsister chromatids produces two parental chromatids and two recombinant chromatids. When the crossover occurs within a paracentric inversion, however, one recombinant **dicentric chromatid** (two centromeres) and one recombinant **acentric chromatid** (lacking a centromere) are produced. Both contain duplications and deletions of chromosome segments as well. During anaphase, an acentric chromatid moves randomly to one pole or the other or may be lost, while a dicentric chromatid is pulled in two directions.

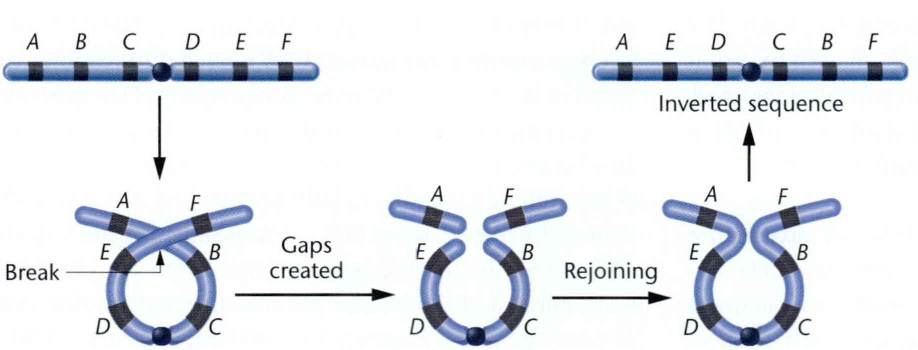

FIGURE 6.14 One possible origin of a paracentric inversion.

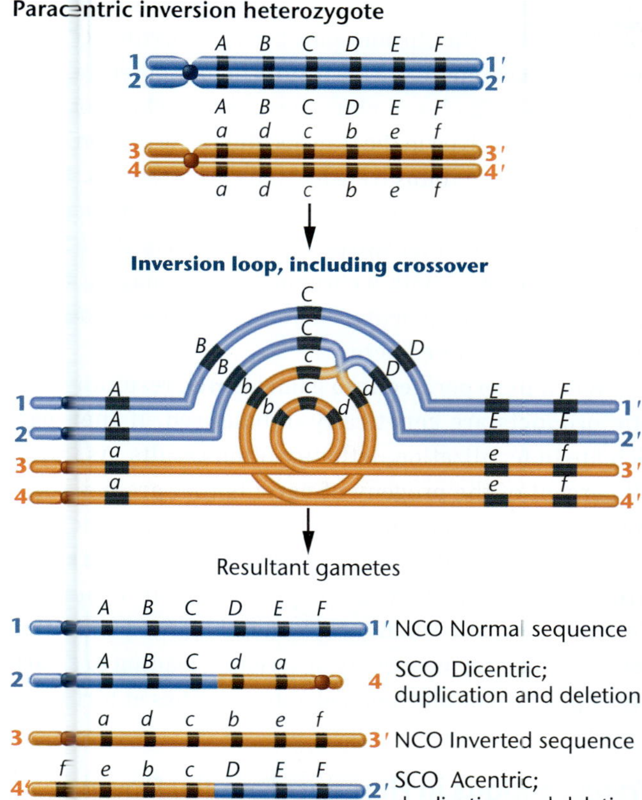

FIGURE 6.15 The effects of a single crossover (SCO) within an inversion loop in a paracentric inversion heterozygote, where two altered chromosomes are produced, one acentric and one dicentric. Both chromosomes also contain duplicated and deficient regions.

This polarized movement produces *dicentric bridges* that are cytologically recognizable. A dicentric chromatid usually breaks at some point so that part of the chromatid goes into one gamete and part into another gamete during the meiotic divisions. Therefore, gametes containing either recombinant chromatid are deficient in genetic material. In animals, when such a gamete participates in fertilization, the zygote most often develops abnormally, if at all.

Because offspring bearing crossover gametes are inviable and not recovered, it *appears* as if the inversion suppresses crossing over. Actually, in inversion heterozygotes, the inversion has the effect of *suppressing the recovery of crossover products* when chromosome exchange occurs within the inverted region. Moreover, up to one-half of the viable gametes have the inverted chromosome, and the inversion will be perpetuated within the species. The cycle will be repeated continuously during meiosis in future generations.

Evolutionary Advantages of Inversions

Because recovery of crossover products is suppressed in inversion heterozygotes, groups of specific alleles at adjacent loci within inversions may be preserved from generation to generation. If the alleles of the involved genes confer a survival advantage on the organisms maintaining them, the inversion is beneficial to the evolutionary survival of the species. For example, if a set of alleles *ABcDef* is more adaptive than sets *AbCdeF* or *abcdEF*, effective gametes will contain this favorable set of genes, undisrupted by crossing over.

In laboratory studies, the same principle is applied using **balancer chromosomes,** which contain inversions. When an organism is heterozygous for a balancer chromosome, desired sequences of alleles are preserved during experimental work.

NOW SOLVE THIS

6.3 What is the effect of a rare double crossover within a chromosome segment that is heterozygous for a paracentric inversion?

■ **HINT:** *This problem involves an understanding of how homologs synapse in the presence of a heterozygous paracentric inversion. The key to its solution is to draw out the tetrad and follow the chromatids undergoing a double crossover.*

6.8 Translocations Alter the Location of Chromosomal Segments in the Genome

Translocation, as the name implies, is the movement of a chromosomal segment to a new location in the genome. Reciprocal translocation, for example, involves the exchange of segments between two nonhomologous chromosomes. The least complex way for this event to occur is for two nonhomologous chromosome arms to come close to each other so that an exchange is facilitated. **Figure 6.16(a)** shows a simple reciprocal translocation in which only two breaks are required. If the exchange includes internal chromosome segments, four breaks are required, two on each chromosome.

The genetic consequences of reciprocal translocations are, in several instances, similar to those of inversions. For example, genetic information is not lost or gained. Rather, there is only a rearrangement of genetic material. The presence of a translocation does not, therefore, directly alter the viability of individuals bearing it.

Homologs that are heterozygous for a reciprocal translocation undergo unorthodox synapsis during meiosis. As shown in **Figure 6.16(b)**, pairing results in a cross-like configuration. As with inversions, genetically unbalanced gametes are also produced as a result of this unusual alignment during meiosis. In the case of translocations, however, aberrant gametes are not necessarily the result of crossing over. To see how unbalanced gametes are produced, focus on the homologous centromeres in Figure 6.16(b) and **Figure 6.16(c)**. According to the principle of independent assortment, the chromosome containing centromere 1 migrates randomly toward one pole of the spindle during the first meiotic anaphase; it travels along with *either* the chromosome having centromere 3 *or* the chromosome having centromere 4.

(a) Possible origin of a reciprocal translocation between two nonhomologous chromosomes

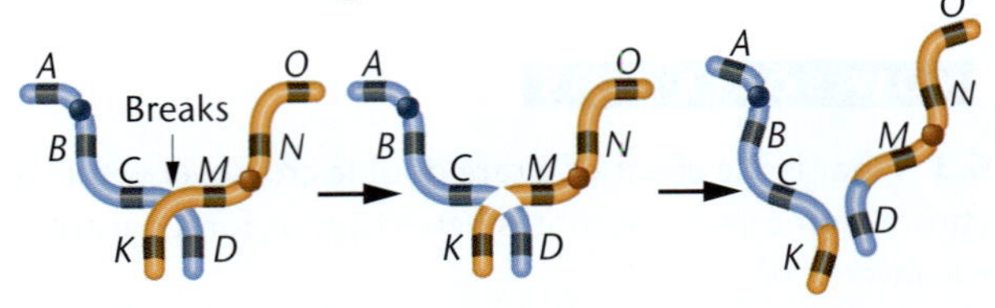

(b) Synapsis of translocation heterozygote

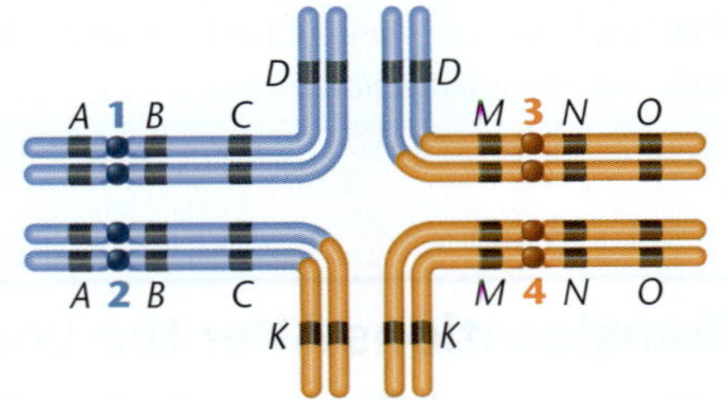

(c) Two possible segregation patterns leading to gamete formation

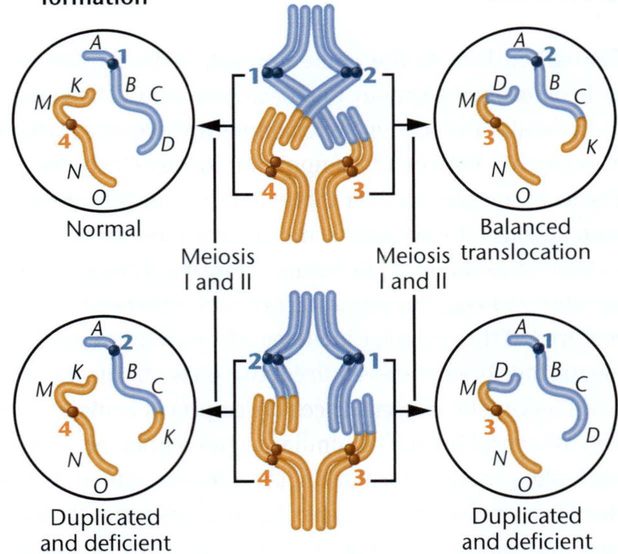

FIGURE 6.16 (a) Possible origin of a reciprocal translocation. (b) Synaptic configuration formed during meiosis in an individual that is heterozygous for the translocation. (c) Two possible segregation patterns, one of which leads to a normal and a balanced gamete (called alternate segregation) and one that leads to gametes containing duplications and deficiencies (called adjacent segregation).

The chromosome with centromere 2 moves to the other pole along with the chromosome containing *either* centromere 3 *or* centromere 4. This results in four potential meiotic products. The 1,4 combination contains chromosomes that are not involved in the translocation. The 2,3 combination, however, contains translocated chromosomes. These contain a complete complement of genetic information and are balanced. The other two potential products, the 1,3 and 2,4 combinations, contain chromosomes displaying duplicated and deleted segments. To simplify matters, crossover exchanges are ignored here.

When incorporated into gametes, the resultant meiotic products are genetically unbalanced. If they participate in fertilization, lethality often results. As few as 50 percent of the progeny of parents that are heterozygous for a reciprocal translocation survive. This condition, called **semisterility,** has an impact on the reproductive fitness of organisms, thus playing a role in evolution. Furthermore, in humans, such an unbalanced condition results in partial monosomy or trisomy, leading to a variety of physical or biochemical abnormalities at birth.

Translocations in Humans: Familial Down Syndrome

Research conducted since 1959 has revealed numerous translocations in members of the human population. One common type of translocation involves breaks at the extreme ends of the short arms of two nonhomologous acrocentric chromosomes. These small segments are lost, and the larger segments fuse at their centromeric region. This type of translocation produces a new, large submetacentric or metacentric chromosome, often called a **Robertsonian translocation.**

One such translocation accounts for cases in which Down syndrome is familial (inherited). Earlier in this chapter, we pointed out that most instances of Down syndrome are due to trisomy 21. This chromosome composition results from nondisjunction during meiosis in one parent. Trisomy accounts for over 95 percent of all cases of Down syndrome. In such instances, the chance of the same parents producing a second affected child is extremely low. However, in the remaining families with a child with Down syndrome, it occurs with a much higher frequency over several generations; that is, it "runs in families.".

Cytogenetic studies of the parents and their offspring from these unusual cases explain the cause of **familial Down syndrome.** Analysis reveals that one of the parents contains a 14/21, D/G translocation (**Figure 6.17**). That is, one parent has the majority of the G-group chromosome 21 translocated to one end of the D-group chromosome 14. This individual is phenotypically normal even

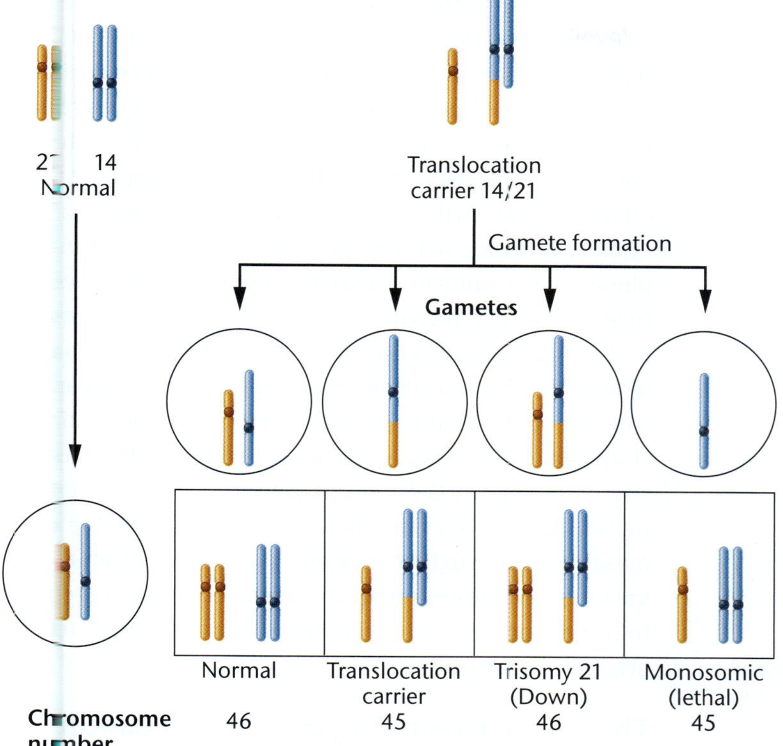

FIGURE 6.17 Chromosomal involvement and translocation in familial Down syndrome.

though he or she has only 45 chromosomes. During meiosis, one-fourth of the individual's gametes have two copies of chromosome 21: a normal chromosome and a second copy translocated to chromosome 14. When such a gamete is fertilized by a standard haploid gamete, the resulting zygote has 46 chromosomes but three copies of chromosome 21. These individuals exhibit Down syndrome. Other potential surviving offspring contain either the standard diploid genome (without a translocation) or the balanced translocation like the parent. Both cases result in normal individuals. Although not illustrated in Figure 6.17, two other gametes may be formed, though rarely. Such gametes are unbalanced, and upon fertilization, lethality occurs.

The above findings have allowed geneticists to resolve the seeming paradox of an inherited trisomic phenotype in an individual with an apparent diploid number of chromosomes. It is also unique that the "carrier," who has 45 chromosomes and exhibits a normal phenotype, does not contain the *complete* diploid amount of genetic material. A small region is lost from both chromosomes 14 and 21 during the translocation event. This occurs because the ends of both chromosomes have broken off prior to their fusion. These specific regions are known to be two of many chromosomal locations housing multiple copies of the genes encoding rRNA, the major component of ribosomes. Despite the loss of up to 20 percent of these genes, the carrier is unaffected.

ESSENTIAL POINT

Inversions and translocations may initially cause little or no loss of genetic information or deleterious effects. However, heterozygous combinations of the involved chromosome segments may result in genetically abnormal gametes following meiosis, with lethality or inviability often ensuing. ■

6.9 Fragile Sites in Human Chromosomes Are Susceptible to Breakage

We conclude this chapter with a brief discussion of the results of an intriguing discovery made around 1970 during observations of metaphase chromosomes prepared following human cell culture. In cells derived from certain individuals, a specific area along one of the chromosomes failed to stain, giving the appearance of a gap. In other individuals whose chromosomes displayed such morphology, the gaps appeared at other positions within the set of chromosomes. Such areas eventually became known as **fragile sites,** since they appeared to be susceptible to chromosome breakage when cultured in the absence of certain chemicals such as folic acid, which is normally present in the culture medium.

Because they represent points along the chromosome that are susceptible to breakage, these sites may indicate regions where the chromatin is not tightly coiled. Note that even though almost all studies of fragile sites have been carried out *in vitro* using mitotically dividing cells, clear associations have been established between several of these sites and the corresponding altered phenotype, including intellectual disability and cancer.

Fragile-X Syndrome

Most fragile sites do not appear to be associated with any clinical syndrome. However, individuals bearing a folate-sensitive site on the X chromosome (**Figure 6.18**) may exhibit the **fragile-X syndrome (FXS)**, the most common form of inherited intellectual disability. This syndrome affects about 1 in 4000 males and 1 in 8000 females. All males bearing this X chromosome exhibit the syndrome, while about 60 percent of females bearing one affected chromosome exhibit the syndrome. In addition to intellectual disability,

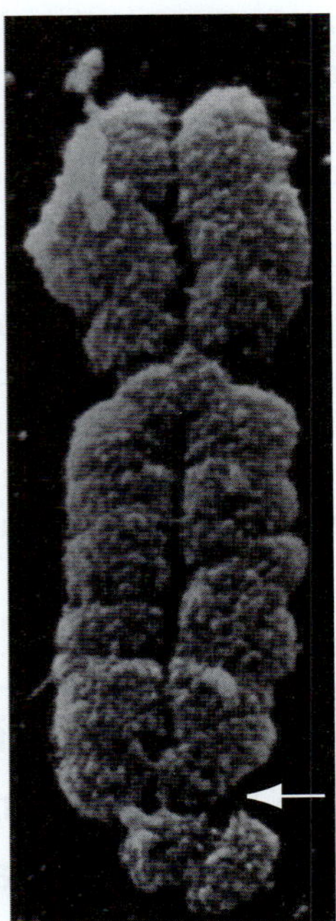

FIGURE 6.18 A fragile human X chromosome. The "gap" region, identified by the arrow, is associated with the fragile-X syndrome.

affected males and females have characteristic long, narrow faces with protruding chins and enlarged ears.

A gene that spans the fragile site is now known to be responsible for this syndrome. Named *FMR1,* it is one of a growing number of genes that have been discovered in which a sequence of three nucleotides is repeated many times, expanding the size of the gene. Such **trinucleotide repeats** are also recognized in other human disorders, including Huntington disease and myotonic dystrophy. In *FMR1,* the trinucleotide sequence CGG is repeated in an untranslated area adjacent to the coding sequence of the gene (called the "upstream" region). The number of repeats varies immensely within the human population, and a high number correlates directly with expression of fragile-X syndrome. Normal individuals have between 6 and 54 repeats, whereas those with 55 to 230 repeats are considered "carriers" of the disorder. More than 230 repeats lead to expression of the syndrome.

It is thought that, once the gene contains this increased number of repeats, it becomes chemically modified so that the bases within and around the repeats are methylated, an epigenetic process that inactivates the gene. The normal product of the *FMR1* gene is an RNA-binding protein, FMRP, known to be produced in the brain. Evidence is now accumulating that directly links the absence of the protein in the brain with the cognitive defects associated with the syndrome.

From a genetic standpoint, an interesting aspect of fragile-X syndrome is the instability of the number of CGG repeats. An individual with 6 to 54 repeats transmits a gene containing the same number of copies to his or her offspring. However, carrier individuals with 55 to 230 repeats, though not at risk to develop the syndrome, may transmit to their offspring a gene with an increased number of repeats. This number increases in future generations, demonstrating the phenomenon known as **genetic anticipation.** Once the threshold of 230 repeats is exceeded, retardation becomes more severe in each successive generation as the number of trinucleotide repeats increases. Interestingly, expansion from the carrier status (55 to 230 repeats) to the syndrome status (over 230 repeats) occurs only during the transmission of the gene by the maternal parent, not by the paternal parent. Thus, a "carrier" male may transmit a stable chromosome to his daughter, who may subsequently transmit an unstable chromosome with an increased number of repeats to her offspring. Their grandfather was the source of the original chromosome.

The Link between Fragile Sites and Cancer

While the study of the fragile-X syndrome first brought unstable chromosome regions to the attention of geneticists, a link between an autosomal fragile site and lung cancer was reported in 1996 by Carlo Croce, Kay Huebner, and their colleagues. They have subsequently postulated that the defect is associated with the formation of a variety of different tumor types. Croce and Huebner first showed that the *FHIT* gene (standing for *f*ragile *hi*stidine *t*riad), located within the well-defined fragile site designated as *FRA3B* on the p arm of chromosome 3, is often altered or missing in cells taken from tumors of individuals with lung cancer. More extensive studies have now revealed that the normal protein product of this gene is absent in cells of many other cancers, including those of the esophagus, breast, cervix, liver, kidney, pancreas, colon, and stomach. Genes such as *FHIT* that are located within fragile regions undoubtedly have an increased susceptibility to mutations and deletions.

The study of this and still other fragile sites is but one example of how chromosomal abnormalities of many sorts are linked to cancer. We will return to this discussion in Chapter 19.

ESSENTIAL POINT

Fragile sites in human mitotic chromosomes have sparked research interest because one such site on the X chromosome is associated with the most common form of inherited mental retardation, while other autosomal sites have been linked to various forms of cancer. ■

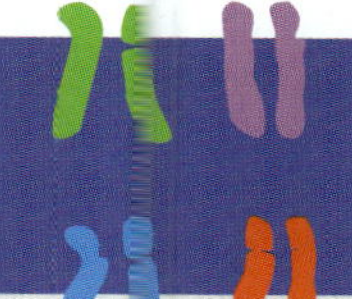

GENETICS, ETHICS, AND SOCIETY

Down Syndrome and Prenatal Testing—The New Eugenics?

Down syndrome is the most common chromosomal abnormality seen in newborn babies. Prenatal diagnostic tests for Down syndrome have been available for decades, especially to older pregnant women who have an increased risk of bearing a child with Down syndrome. Scientists estimate that there is an abortion rate of about 30 percent for fetuses that test positive for Down syndrome in the United States, and rates of up to 85 percent in other parts of the world, such as Taiwan and France.

Some people agree that it is morally acceptable to prevent the birth of a genetically abnormal fetus. However, others argue that prenatal genetic testing, with the goal of eliminating congenital disorders, is unethical. In addition, some argue that prenatal genetic testing followed by selective abortion is eugenic. How does eugenics apply, if at all, to screening for Down syndrome and other human genetic disorders

The term *eugenics* was first defined by Francis Galton in 1883 as "the science which deals with all influences that improve the inborn qualities of a race; also with those that develop them to the utmost advantage." Galton believed that human traits such as intelligence and personality were hereditary and that humans could selectively mate with each other to create gifted groups of people—analogous to the creation of purebred dogs with specific traits. Galton did not propose coercion but thought that people would voluntarily select mates in order to enhance particular genetic outcomes for their offspring.

In the early to mid-twentieth century, countries throughout the world adopted eugenic policies with the aim of enhancing desirable human traits (positive eugenics) and eliminating undesirable ones (negative eugenics). Many countries, including Britain, Canada, and the United States, enacted compulsory sterilization programs for the "feebleminded," mentally ill, and criminals. The eugenic policies of Nazi Germany were particularly infamous, resulting in forced human genetic experimentation and the slaughter of tens of thousands of people with disabilities. The eugenics movement was discredited after World War II, and the evils perpetuated in its name have tainted the term *eugenics* ever since.

Given the history of the eugenics movement, is it fair to use the term *eugenics* when we speak about genetic testing for Down syndrome and other genetic disorders? Some people argue that it is not eugenic to select for healthy children because there is no coercion, the state is not involved, and the goal is the elimination of suffering. Others point out that such voluntary actions still constitute eugenics, since they involve a form of bioengineering for "better" human beings.

Now that we are entering an era of unprecedented knowledge about our genomes and our predisposition to genetic disorders, we must make decisions about whether our attempts to control or improve human genomes are ethical and what limits we should place on these efforts. The story of the eugenics movement provides us with a powerful cautionary tale about the potential misuses of genetic information.

Your Turn

Take time, individually or in groups, to consider the following questions. Investigate the references and links to help you discuss some of the ethical issues surrounding genetic testing and eugenics.

1. Do you think that modern prenatal and preimplantation genetic testing followed by selective abortion is eugenic? Why or why not?

   For background on these questions, see McCabe, L., and McCabe, E. (2011). Down syndrome: Coercion and eugenics. *Genet. Med.* 13:708–710. *Another useful discussion can be found in* Wilkinson, S., (2015). Prenatal screening, reproductive choice, and public health. *Bioethics* 29:26–35.

2. If genetic technologies were more advanced than today, and you could choose the traits of your children, would you take advantage of that option? Which traits would you choose—height, weight, intellectual abilities, athleticism, artistic talents? If so, would this be eugenic? Would it be ethical?

   To read about similar questions answered by groups of Swiss law and medical students, read Elger, B., and Harding, T., (2003). Huntington's disease: Do future physicians and lawyers think eugenically? *Clin. Genet.* 64:327–338.

CASE STUDY Fish tales

Controlling the overgrowth of invasive aquatic vegetation is a significant problem in the waterways of most U.S. states. Originally, herbicides and dredging were used for control, but in 1963, diploid Asian carp were introduced in Alabama and Arkansas. Unfortunately, through escapes and illegal introductions, the carp spread rapidly and became serious threats to aquatic ecosystems in 45 states. Beginning in 1983, many states began using triploid, sterile grass carp as an alternative, because of their inability to reproduce, their longevity, and their voracious appetite. On the other hand, this genetically modified exotic species, if not used properly, can reduce or eliminate desirable plants and outcompete native fish, causing more damage than good. The use of one exotic species to control other exotic species has had a problematic history across the globe, generating controversy and criticism. Newer methods for genetic modification of organisms to achieve specific outcomes will certainly become more common in the future and raise several interesting questions.

1. Why would the creation and use of a tetraploid carp species be unacceptable in the above situation?
2. If you were a state official in charge of a particular waterway, what questions would you ask before approving the use of a laoraory-produced, triploid species in this waterway?
3. What ethical responsibilities accompany the ecological and economic risks and benefits of releasing exotic species into the environment? Who pays the costs if ecosystems and food supplies are damaged?

See Seastedt, T. R. (2015). Biological control of invasive plant species: A reassessment for the Anthropocene. *New Phytologist* 205:490–502.

INSIGHTS AND SOLUTIONS

1. In a cross using maize that involves three genes, *a*, *b*, and *c*, a heterozygote (*abc*/+++) is testcrossed to *abc*/*abc*. Even though the three genes are separated along the chromosome, thus predicting that crossover gametes and the resultant phenotypes should be observed, only two phenotypes are recovered: *abc* and +++. In addition, the cross produced significantly fewer viable plants than expected. Can you propose why no other phenotypes were recovered and why the viability was reduced?

Solution: One of the two chromosomes may contain an inversion that overlaps all three genes, effectively precluding the recovery of any "crossover" offspring. If this is a paracentric inversion and the genes are clearly separated (ensuring that a significant number of crossovers occurs between them), then numerous acentric and dicentric chromosomes will form, resulting in the observed reduction in viability.

2. A male *Drosophila* from a wild-type stock is discovered to have only seven chromosomes, whereas normally $2n = 8$. Close examination reveals that one member of chromosome IV (the smallest chromosome) is attached to (translocated to) the distal end of chromosome II and is missing its centromere, thus accounting for the reduction in chromosome number.

(a) Diagram all members of chromosomes II and IV during synapsis in meiosis I.

Solution:

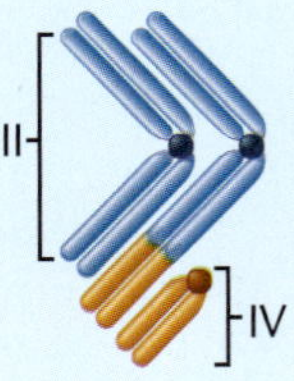

(b) If this male mates with a female with a normal chromosome composition who is homozygous for the recessive chromosome IV mutation *eyeless* (*ey*), what chromosome compositions will occur in the offspring regarding chromosomes II and IV?

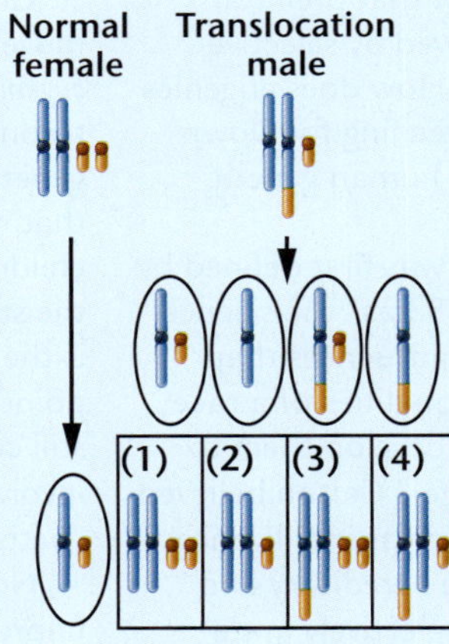

Solution:

(c) Referring to the diagram in the solution to part (b), what phenotypic ratio will result regarding the presence of eyes, assuming all abnormal chromosome compositions survive?

Solution:

1. normal (heterozygous)
2. eyeless (monosomic, contains chromosome IV from mother)
3. normal (heterozygous; trisomic and may die)
4. normal (heterozygous; balanced translocation)

The final ratio is 3/4 normal: 1/4 eyeless.

Problems and Discussion Questions

Mastering Genetics Visit for instructor-assigned tutorials and problems.

1. **HOW DO WE KNOW?** In this chapter, we focused on chromosomal mutations resulting from a change in number or arrangement of chromosomes. In our discussions, we found many opportunities to consider the methods and reasoning by which much of this information was acquired. From the explanations given in the chapter, what answers would you propose to the following fundamental questions?
   (a) How do we know that the extra chromosome causing Down syndrome is usually maternal in origin?
   (b) How do we know that human aneuploidy for each of the 22 autosomes occurs at conception, even though most often human aneuploids do not survive embryonic or fetal development and thus are never observed at birth?
   (c) How do we know that specific mutant phenotypes are due to changes in chromosome number or structure?
   (d) How do we know that the mutant *Bar*-eye phenotype in *Drosophila* is due to a duplicated gene region rather than to a change in the nucleotide sequence of a gene?
2. **CONCEPT QUESTION** Review the Chapter Concepts list on page 123. These all center on chromosome aberrations that create variations from the "normal" diploid genome. Write a short essay that discusses five altered phenotypes that result from specific chromosomal aberrations. ■
3. Define these pairs of terms, and distinguish between them.
   aneuploidy/euploidy
   monosomy/trisomy
   Patau syndrome/Edwards syndrome
   autopolyploidy/allopolyploidy
   autotetraploid/amphidiploid
   paracentric inversion/pericentric inversion
4. Identify the chromosomal abnormality, the associated syndrome, and the sex of an individual carrying the following karyotypes:
   (a) 45,X; (b) 47,XXY; (c) 47,XX,21+; (d) 47,XY,13+.
5. Compare partial monosomy with haploinsufficiency.
6. What are the possible reasons behind translocations?
7. Why do human monosomics most often fail to survive prenatal development?
8. What advantages and disadvantages do polyploid plants have?
9. A couple goes through multiple miscarriages, and fetal karyotyping indicates the same trisomy every time. This trisomy is not compatible with life. Obviously, neither of the parents has this trisomy. Can you explain this situation?
10. What are inversion heterozygotes? How can meiotic pairing occur in these organisms? What will be the consequence?
11. An organism produces a fraction of gametes with acentric and dicentric chromosomes because of a single crossover event that occurred in the inversion loop of an inversion heterozygote. Explain whether the nature of the inversion is pericentric or paracentric.
12. What are the possible consequences of genome and gene duplication in the evolutionary process?
13. *Spartina anglica*, a species of cordgrass that has 122 chromosomes, appeared near the English coastline around 1870 from an autochthonous grass, *Spartina maritima* ($2n = 60$), and a non-native grass, *Spartina alterniflora* ($2n = 62$). Describe *Spartina anglica* in genetic terms and explain how it arose from the two species.
14. Certain varieties of chrysanthemums contain 18, 36, 54, 72, and 90 chromosomes; all are multiples of a basic set of nine chromosomes. How would you describe these varieties genetically? What feature do the karyotypes of each variety share? A variety with 27 chromosomes has been discovered, but it is sterile. Why?
15. *Drosophila* may be monosomic for chromosome 4, yet remain fertile. Contrast the F_1 and F_2 results of the following crosses involving the recessive chromosome 4 trait, *bent* bristles:
   (a) monosomic IV, bent bristles × normal bristles;
   (b) monosomic IV, normal bristles × bent bristles.
16. Mendelian ratios are modified in crosses involving autotetraploids. Assume that one plant expresses the dominant trait green seeds and is homozygous (*WWWW*). This plant is crossed to one with white seeds that is also homozygous (*wwww*). If only one dominant allele is sufficient to produce green seeds, predict the F_1 and F_2 results of such a cross. Assume that synapsis between chromosome pairs is random during meiosis.
17. Having correctly established the F_2 ratio in Problem 16, predict the F_2 ratio of a "dihybrid" cross involving two independently assorting characteristics (e.g., $P_1 = WWWWAAAA \times wwwwaaaa$).
18. In an autotetraploid plant, flower color depends on a single gene with two alleles (B_1 and B_2). Since only B_1 produces color, its intensity depends on the number of B_1 alleles: from purple ($4B_1$ alleles), dark blue, blue, light blue, to white ($4B_2$ alleles). Determine the progeny of a plant with blue and a plant with white flowers.
19. A couple planning their family are aware that through the past three generations on the husband's side a substantial number of stillbirths have occurred and several malformed babies were born who died early in childhood. The wife has studied genetics and urges her husband to visit a genetic counseling clinic, where a complete karyotype-banding analysis is performed. Although the tests show that he has a normal complement of 46 chromosomes, banding analysis reveals that one member of the chromosome 1 pair (in group A) contains an inversion covering 70 percent of its length. The homolog of chromosome 1 and all other chromosomes show the normal banding sequence.
   (a) How would you explain the high incidence of past stillbirths?
   (b) What can you predict about the probability of abnormality/normality of their future children?
   (c) Would you advise the woman that she will have to bring each pregnancy to term to determine whether the fetus is normal? If not, what else can you suggest?
20. A woman who sought genetic counseling is found to be heterozygous for a chromosomal rearrangement between the second and third chromosomes. Her chromosomes, compared to those in a normal karyotype, are diagrammed on the next page:
   (a) What kind of chromosomal aberration is shown?

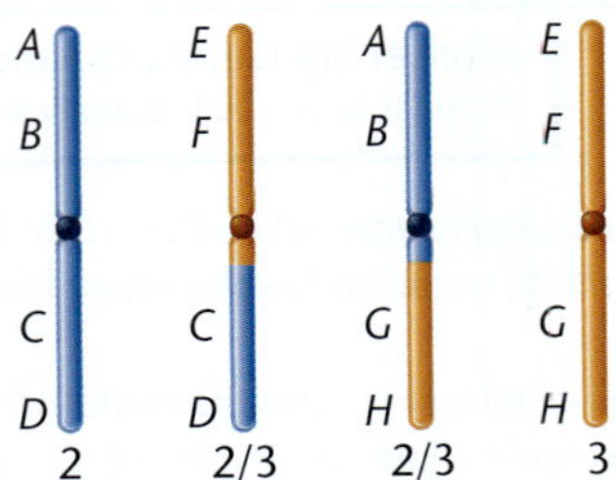

(b) Using a drawing, demonstrate how these chromosomes would pair during meiosis. Be sure to label the different segments of the chromosomes.

(c) This woman is phenotypically normal. Does this surprise you? Why or why not? Under what circumstances might you expect a phenotypic effect of such a rearrangement?

21. The woman in Problem 20 has had two miscarriages. She has come to you, an established genetic counselor, with these questions:
    (a) Is there a genetic explanation of her frequent miscarriages?
    (b) Should she abandon her attempts to have a child of her own?
    (c) If not, what is the chance that she could have a normal child? Provide an informed response to her concerns.

22. Chromosomes 1 and 4 of a species have gene orders *ABCDE* and *VWXYZ*, respectively. A reciprocal translocation between these chromosomes produces heterozygotes with translocated gene orders *ABCYZ* and *VWXDE*, respectively. After gametogenesis in a particular meiotic division, none of the gametes has a normal genotype for chromosomes 1 and 4. Describe the segregation pattern that might occur during this meiosis.

23. A boy with Klinefelter syndrome (47,XXY) is born to a mother who is phenotypically normal and a father who has the X-linked skin condition called anhidrotic ectodermal dysplasia. The mother's skin is completely normal with no signs of the skin abnormality. In contrast, her son has patches of normal skin and patches of abnormal skin.
    (a) Which parent contributed the abnormal gamete?
    (b) Using the appropriate genetic terminology, describe the meiotic mistake that occurred. Be sure to indicate in which division the mistake occurred.
    (c) Using the appropriate genetic terminology, explain the son's skin phenotype.

24. In a human genetic study, a family with five phenotypically normal children was investigated. Two children were "homozygous" for a Robertsonian translocation between chromosomes 19 and 20 (they contained two identical copies of the fused chromosome). They have only 44 chromosomes but a complete genetic complement. Three of the children were "heterozygous" for the translocation and contained 45 chromosomes, with one translocated chromosome plus a normal copy of both chromosomes 19 and 20. Two other pregnancies resulted in stillbirths. It was later discovered that the parents were first cousins. Based on this information, determine the chromosome compositions of the parents. What led to the stillbirths? Why was the discovery that the parents were first cousins a key piece of information in understanding the genetics of this family?

25. A species "X" always produces diploid (disomic) progenies upon selfing. When "X" is crossed with a related species "Y", only monosomic and trisomic progenies are produced. Describe the pattern of nondisjunction in species "Y".

26. A normal female is discovered with 45 chromosomes, one of which exhibits a Robertsonian translocation containing most of chromosomes 18 and 21. Discuss the possible outcomes in her offspring when her husband contains a normal karyotype

27. In a cross in *Drosophila*, a female heterozygous for the autosomally linked genes *a*, *b*, *c*, *d*, and *e* (*abcde*/+++++) was testcrossed with a male homozygous for all recessive alleles (abcde/abcde). Even though the distance between each of the loci was at least 3 map units, only four phenotypes were recovered, yielding the following data:

Phenotype	No. of Flies
+++++	440
a b c d e	460
++++ e	48
a b c d+	52
Total =	1000

Why are many expected crossover phenotypes missing? Can any of these loci be mapped from the data given here? If so, determine map distances.

7

Linkage and Chromosome Mapping in Eukaryotes

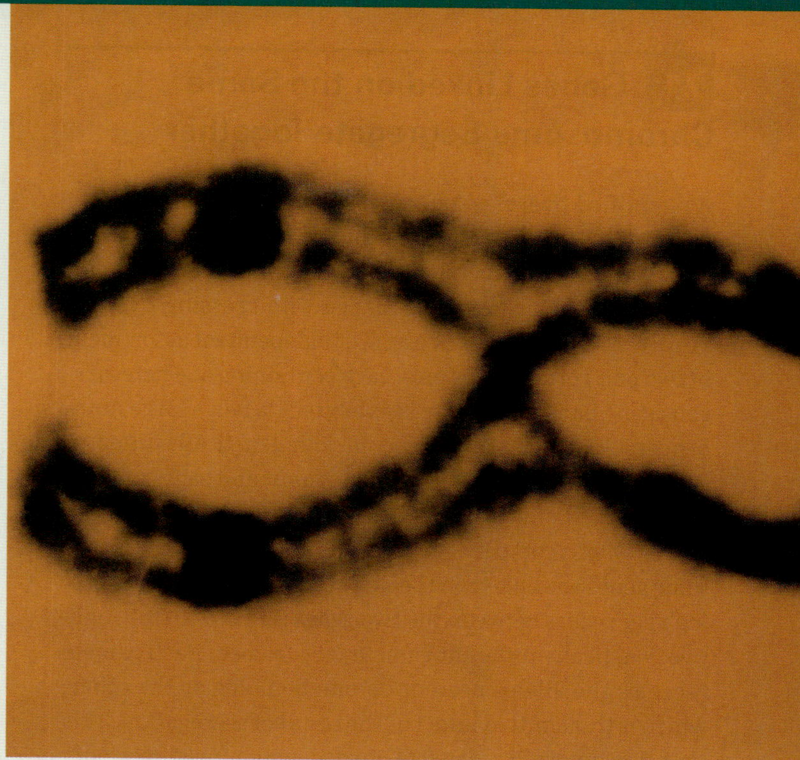

A single chiasma between synapsed homologs during the first meiotic prophase.

CHAPTER CONCEPTS

- Chromosomes in eukaryotes contain many genes whose locations are fixed along the length of the chromosomes.
- Unless separated by crossing over, alleles present on a chromosome segregate as a unit during gamete formation.
- Crossing over between homologs is a process of genetic recombination during meiosis that creates gametes with new combinations of alleles that enhance genetic variation within species.
- Crossing over between homologs serves as the basis for the construction of chromosome maps.
- While exchange occurs between sister chromatids during mitosis, no new recombinant chromatids are created.

Walter Sutton, along with Theodor Boveri, was instrumental in uniting the fields of cytology and genetics. As early as 1903, Sutton pointed out the likelihood that there must be many more "unit factors" than chromosomes in most organisms. Soon thereafter, genetics investigations revealed that certain genes segregate as if they were somehow joined or linked together. Further investigations showed that such genes are part of the same chromosome and may indeed be transmitted as a single unit. We now know that most chromosomes contain a very large number of genes. Those that are part of the same chromosome are said to be *linked* and to demonstrate **linkage** in genetic crosses.

Because the chromosome, not the gene, is the unit of transmission during meiosis, linked genes are not free to undergo independent assortment. Instead, the alleles at all loci of one chromosome should, in theory, be transmitted as a unit during gamete formation. However, in many instances this does not occur. During the first meiotic prophase, when homologs are paired or synapsed, a reciprocal exchange of chromosome segments can take place. This **crossing over** event results in the reshuffling, or **recombination,** of the alleles between homologs, and it always occurs during the tetrad stage.

The frequency of crossing over between any two loci on a single chromosome is proportional to the distance between them. Therefore, depending on which loci are being studied, the percentage of recombinant gametes varies. This correlation allows us to construct chromosome maps, which give the relative locations of genes on chromosomes.

In this chapter, we will discuss linkage, crossing over, and chromosome mapping in more detail.

7.1 Genes Linked on the Same Chromosome Segregate Together

A simplified overview of the major theme of this chapter is given in **Figure 7.1**, which contrasts the meiotic consequences of (a) independent assortment, (b) linkage *without* crossing over, and (c) linkage *with* crossing over. In **Figure 7.1(a)**, we see the results of independent assortment of two pairs of chromosomes, each containing one heterozygous gene pair. No linkage is exhibited. When a large number of meiotic events are observed, four genetically different gametes are formed in equal proportions, and each contains a different combination of alleles of the two genes.

Now let's compare these results with what occurs if the same genes are linked on the same chromosome. If no crossing over occurs between the two genes [**Figure 7.1(b)**], only two genetically different gametes are formed. Each gamete receives the alleles present on one homolog or the other, which is transmitted intact as the result of segregation. This case demonstrates *complete linkage,* which produces only **parental** or **noncrossover gametes.** The two parental gametes are formed in equal proportions. Though complete linkage between two genes seldom occurs, it is useful to consider the theoretical consequences of this concept.

Figure 7.1(c) shows the results of crossing over between two linked genes. As you can see, this crossover involves only two nonsister chromatids of the four chromatids present in the tetrad. This exchange generates two new allele combinations, called **recombinant** or **crossover gametes.** The two chromatids not involved in the exchange result in noncrossover gametes, like those in Figure 7.1(b). The frequency with which crossing over occurs between any two linked genes is generally proportional to the distance separating the respective loci along the chromosome. In theory, two randomly selected genes can be so close to each other that crossover events are too infrequent to be detected easily. As shown in Figure 7.1(b), this complete linkage produces only parental gametes. On the other hand, if a small but distinct distance separates two genes, few recombinant and many parental gametes will be formed. As the distance between the two genes increases, the proportion of recombinant gametes increases and that of the parental gametes decreases.

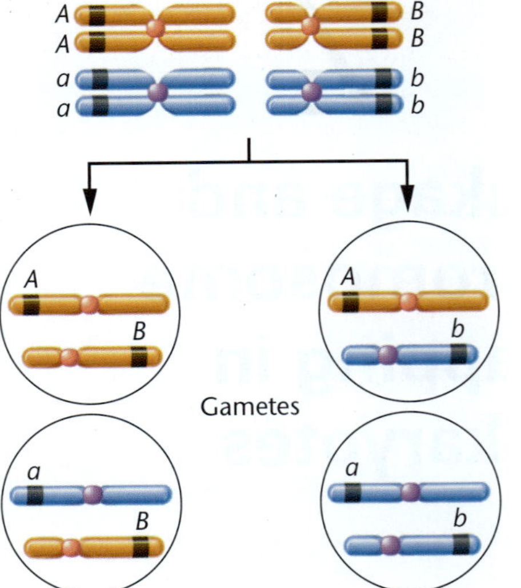

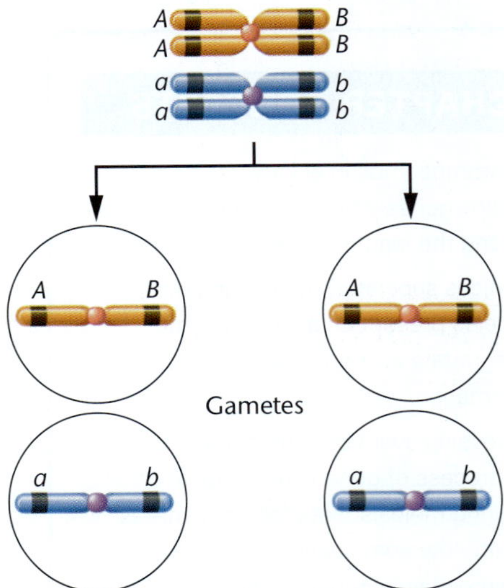

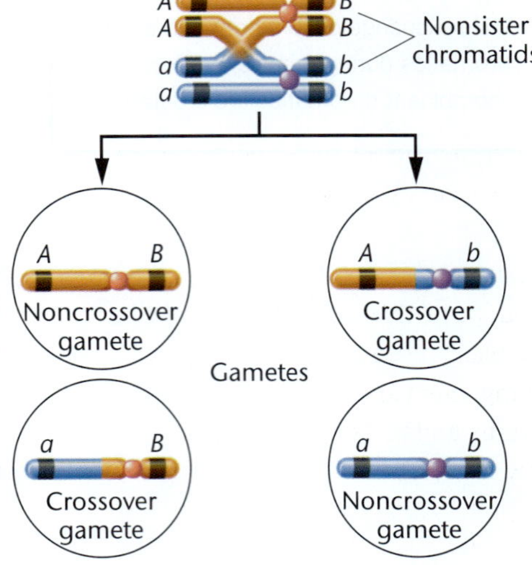

FIGURE 7.1 Results of gamete formation when two heterozygous genes are (a) on two different pairs of chromosomes; (b) on the same pair of homologs, but with no exchange occurring between them; and (c) on the same pair of homologs, but with an exchange occurring between two nonsister chromatids. Note in this and the following figures that members of homologous pairs of chromosomes are shown in two different colors. This convention was established in Chapter 2 (see, for example, Figure 2.7 and Figure 2.11).

As we will discuss later in this chapter, when the loci of two linked genes are far apart, the number of recombinant gametes approaches, but does not exceed, 50 percent. If 50 percent recombinants occur, the result is a 1:1:1:1 ratio of the four types (two parental and two recombinant gametes). In this case, transmission of two linked genes is indistinguishable from that of two unlinked, independently assorting genes. That is, the proportion of the four possible genotypes is identical, as shown in Figure 7.1(a) and (c).

The Linkage Ratio

If complete linkage exists between two genes because of their close proximity, and organisms heterozygous at both loci are mated, a unique F_2 phenotypic ratio results, which we designate the **linkage ratio.** To illustrate this ratio, let's consider a cross involving the closely linked, recessive, mutant genes *heavy* wing vein (*hv*) and *brown* eye (*bw*) in *Drosophila melanogaster* (**Figure 7.2**). The normal, wild-type alleles hv^+ and bw^+ are both dominant and result in thin wing veins and red eyes, respectively.

In this cross, flies with normal thin wing veins and mutant brown eyes are mated to flies with mutant heavy wing veins and normal red eyes. In more concise terms, brown-eyed flies are crossed with heavy-veined flies. If we extend the system of genetic symbols established in Chapter 4, linked genes are represented by placing their allele designations above and below a single or double horizontal line. Those above the line are located at loci on one homolog, and those below are located at the homologous loci on the other homolog. Thus, we represent the P_1 generation as follows:

$$P_1\!: \frac{hv^+\ bw}{hv^+\ bw} \times \frac{hv\ bw^+}{hv\ bw^+}$$

thin, brown   heavy, red

These genes are located on an autosome, so no distinction between males and females is necessary.

In the F_1 generation, each fly receives one chromosome of each pair from each parent. All flies are heterozygous for both gene pairs and exhibit the dominant traits of thin wing veins and red eyes:

$$F_1\!: \frac{hv^+\ bw}{hv\ \ bw^+}$$

thin, red

As shown in **Figure 7.2a**, when the F_1 generation is interbred, each F_1 individual forms only parental gametes because of complete linkage. After fertilization, the F_2 generation is produced in a 1:2:1 phenotypic and genotypic ratio. One-fourth of this generation shows thin wing veins and brown eyes; one-half shows both wild-type traits, namely, thin wing veins and red eyes; and one-fourth shows heavy wing veins and red eyes. In more concise terms, the ratio is 1 heavy:2 wild:1 brown. Such a 1:2:1 ratio is characteristic of complete linkage. Complete linkage is usually observed only when genes are very close together and the number of progeny is relatively small.

Figure 7.2(b) demonstrates the results of a testcross with the F_1 flies. Such a cross produces a 1:1 ratio of thin, brown and heavy, red flies. Had the genes controlling these traits been incompletely linked or located on separate autosomes, the testcross would have produced four phenotypes rather than two.

When large numbers of mutant genes present in any given species are investigated, genes located on the same chromosome show evidence of linkage to one another. As a result, **linkage groups** can be established, one for each chromosome. In theory, the number of linkage groups should correspond to the haploid number of chromosomes. In diploid organisms in which large numbers of mutant genes are available for genetic study, this correlation has been confirmed.

NOW SOLVE THIS

7.1 Consider two hypothetical recessive autosomal genes *a* and *b*, where a heterozygote is testcrossed to a double-homozygous mutant. Predict the phenotypic ratios under the following conditions:

(a) *a* and *b* are located on separate autosomes.

(b) *a* and *b* are linked on the same autosome but are so far apart that a crossover always occurs between them.

(c) *a* and *b* are linked on the same autosome but are so close together that a crossover almost never occurs.

■ **HINT:** *This problem involves an understanding of linkage, crossing over, and independent assortment. The key to its solution is to be aware that results are indistinguishable when two genes are unlinked compared to the case where they are linked but so far apart that crossing over always intervenes between them during meiosis.*

ESSENTIAL POINT

Genes located on the same chromosome are said to be linked. Alleles of linked genes located close together on the same homolog are usually transmitted together during gamete formation. ■

7.2 Crossing Over Serves as the Basis of Determining the Distance between Genes during Mapping

It is highly improbable that two randomly selected genes linked on the same chromosome will be so close to one another along the chromosome that they demonstrate complete linkage. Instead, crosses involving two such genes

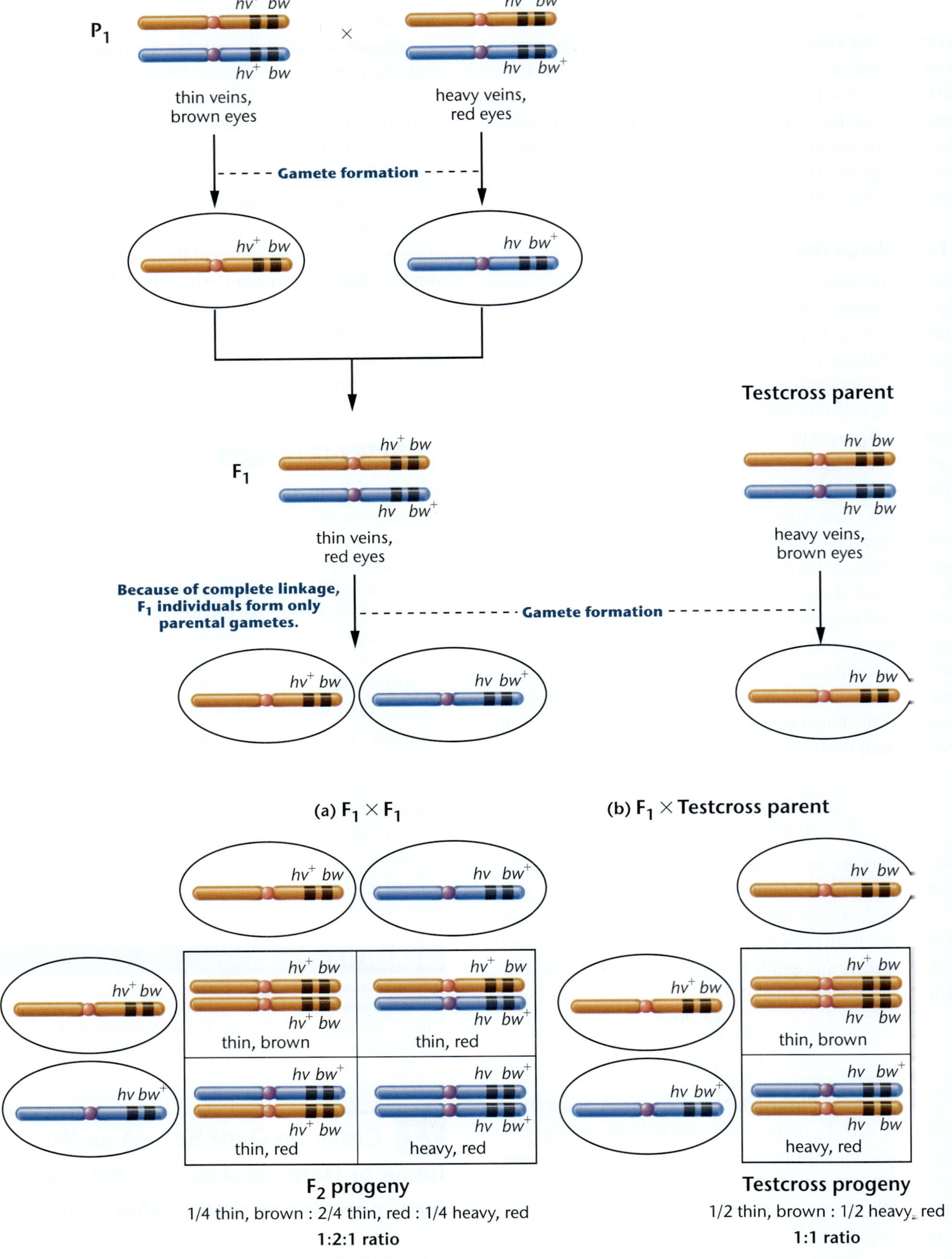

FIGURE 7.2 Results of a cross involving two genes located on the same chromosome and demonstrating complete linkage. (a) The F_2 results of the cross. (b) The results of a testcross involving the F_1 progeny.

almost always produce a percentage of offspring resulting from recombinant gametes. This percentage is variable and depends on the distance between the two genes along the chromosome. This phenomenon was first explained around 1910 by two *Drosophila* geneticists, Thomas H. Morgan and his undergraduate student, Alfred H. Sturtevant.

Morgan, Sturtevant, and Crossing Over

In his studies, Morgan investigated numerous *Drosophila* mutations located on the X chromosome. When he analyzed crosses involving only one trait, he deduced the mode of X-linked inheritance. However, when he made crosses involving two X-linked genes, his results were initially puzzling. For example, female flies expressing the mutant *yellow* body (*y*) and *white* eyes (*w*) alleles were crossed with wild-type males (gray bodies and red eyes). The F_1 females were wild type, while the F_1 males expressed both mutant traits. In the F_2 generation, the vast majority of the offspring showed the expected parental phenotypes—either yellow-bodied, white-eyed flies or wild-type flies (gray-bodied, red-eyed). However, the remaining flies, less than 1.0 percent, were either yellow-bodied with red eyes or gray-bodied with white eyes. It was as if the two mutant alleles had somehow separated from each other on the homolog during gamete formation in the F_1 female flies. This cross is illustrated in cross A of **Figure 7.3**, using data later compiled by Sturtevant.

When Morgan studied other X-linked genes, the same basic pattern was observed, but the proportion of the unexpected F_2 phenotypes differed. For example, in a cross involving the mutant *white* eye (*w*), *miniature* wing (*m*) alleles, the majority of the F_2 again showed the parental phenotypes, but a much higher proportion of the offspring appeared as if the mutant genes had separated during gamete formation. This is illustrated in cross B of Figure 7.3, again using data subsequently compiled by Sturtevant.

Morgan was faced with two questions: (1) What was the source of gene separation, and (2) why did the frequency of the apparent separation vary depending on the genes being studied? The answer he proposed for the first question was based on his knowledge of earlier cytological observations made by F. A. Janssens and others. Janssens had observed that synapsed homologous chromosomes in meiosis wrapped around each other, creating **chiasmata** (sing., *chiasma*) where points of overlap are evident (see the photo on p. 145). Morgan proposed that chiasmata could represent points of genetic exchange.

In the crosses shown in Figure 7.3, Morgan postulated that if an exchange occurs during gamete formation between the mutant genes on the two X chromosomes of the F_1 females, the unique phenotypes will occur. He suggested that such exchanges led to *recombinant gametes* in both the *yellow–white* cross and the *white–miniature* cross, in contrast to the *parental gametes* that have undergone no exchange. On the basis of this and other experiments, Morgan concluded that linked genes exist in a linear order along the chromosome and that a variable amount of exchange occurs between any two genes during gamete formation.

In answer to the second question, Morgan proposed that two genes located relatively close to each other along a chromosome are less likely to have a chiasma form between them than if the two genes are farther apart on the chromosome. Therefore, the closer two genes are, the less likely a genetic exchange will occur between them. Morgan was the first to propose the term **crossing over** to describe the physical exchange leading to recombination.

Sturtevant and Mapping

Morgan's student, Alfred H. Sturtevant, was the first to realize that his mentor's proposal could be used to map the sequence of linked genes. According to Sturtevant,

> *In a conversation with Morgan . . . I suddenly realized that the variations in strength of linkage, already attributed by Morgan to differences in the spatial separation of the genes, offered the possibility of determining sequences in the linear dimension of a chromosome. I went home and spent most of the night (to the neglect of my undergraduate homework) in producing the first chromosomal map.*

Sturtevant compiled data from numerous crosses made by Morgan and other geneticists involving recombination between the genes represented by the *yellow*, *white*, and *miniature* mutants. These data are shown in Figure 7.3. The following recombination between each pair of these three genes, published in Sturtevant's paper in 1913, is as follows:

(1)	*yellow–white*	0.5%
(2)	*white–miniature*	34.5%
(3)	*yellow–miniature*	35.4%

Because the sum of (1) and (2) approximately equals (3), Sturtevant suggested that the recombination frequencies between linked genes are additive. On this basis, he predicted that the order of the genes on the X chromosome is *yellow–white–miniature*. In arriving at this conclusion, he reasoned as follows: the *yellow* and *white* genes are apparently close to each other because the recombination frequency is low. However, both of these genes are much farther apart from *miniature* genes because the *white–miniature* and *yellow–miniature* combinations show larger recombination frequencies. Because *miniature* shows more recombination with *yellow* than with *white* (35.4 percent versus 34.5 percent), it follows that *white* is located between the other two genes, not outside of them.

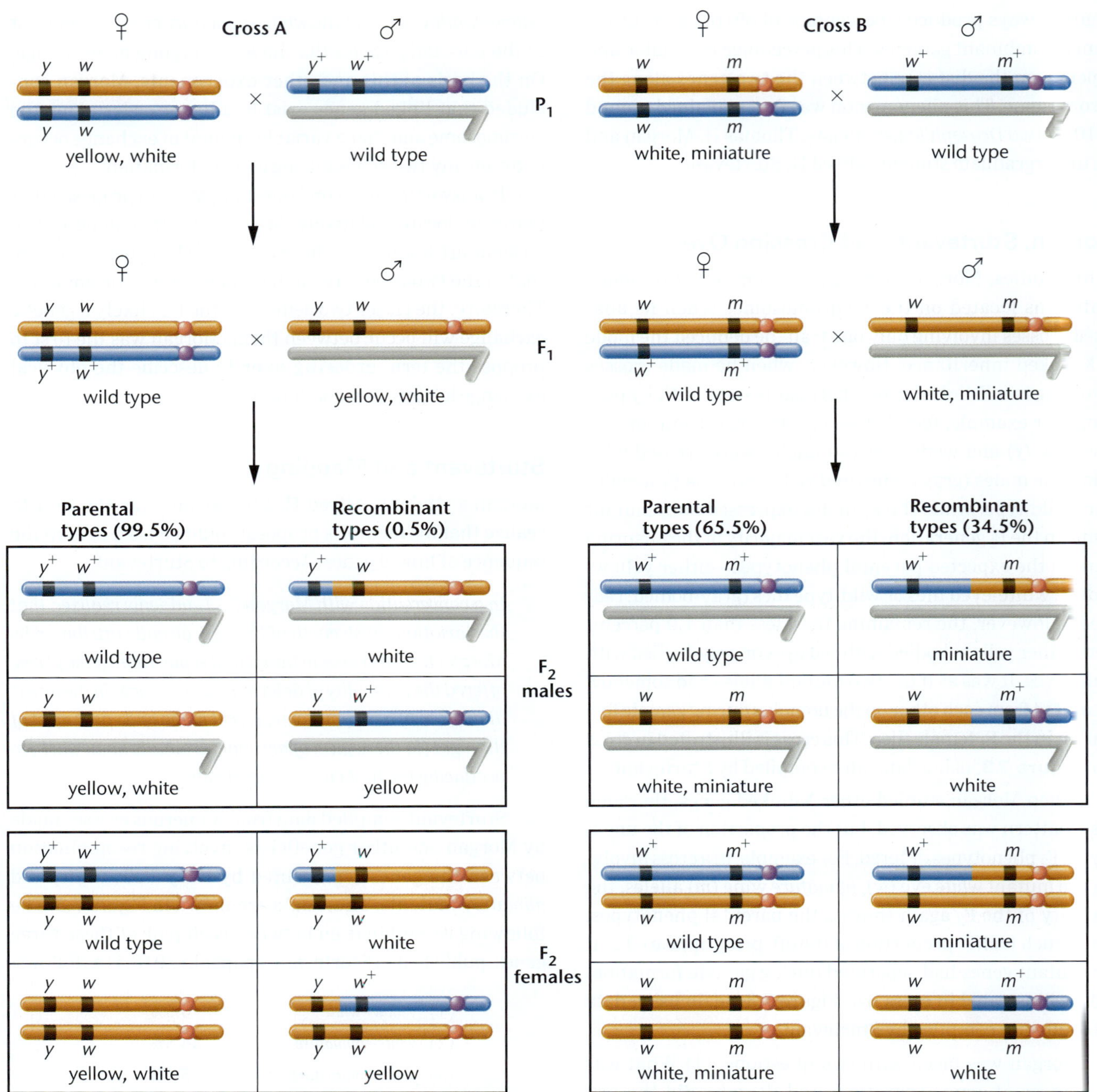

FIGURE 7.3 The F$_1$ and F$_2$ results of crosses involving the *yellow* (*y*), *white* (*w*) mutations (cross A), and the *white, miniature* (*m*) mutations (cross B), as compiled by Sturtevant. In cross A, 0.5 percent of the F$_2$ flies (males and females) demonstrate recombinant phenotypes, which express either *white* or *yellow*. In cross B, 34.5 percent of the F$_2$ flies (males and females) demonstrate recombinant phenotypes, which are either *miniature* or *white* mutants.

Sturtevant knew from Morgan's work that the frequency of exchange could be used as an estimate of the distance between two genes or loci along the chromosome. He constructed a **chromosome map** of the three genes on the X chromosome, setting 1 map unit (mu) equal to 1 percent recombination between two genes.* In the preceding example, the distance between *yellow* and *white* is thus 0.5 mu, and the distance between *yellow* and *miniature* is 35.4 mu. It follows that the distance between *white* and *miniature* should be 35.4 − 0.5 = 34.9 mu. This estimate is close to the actual frequency of recombination between *white* and *miniature* (34.5 percent). The map for these three genes is shown in **Figure 7.4**. The fact that these do not add up perfectly is due to the imprecision of mapping experiments, particularly as the distance between genes increases.

In addition to these three genes, Sturtevant considered two other genes on the X chromosome and produced a more

* In honor of Morgan's work, map units are often referred to as centiMorgans (cM).

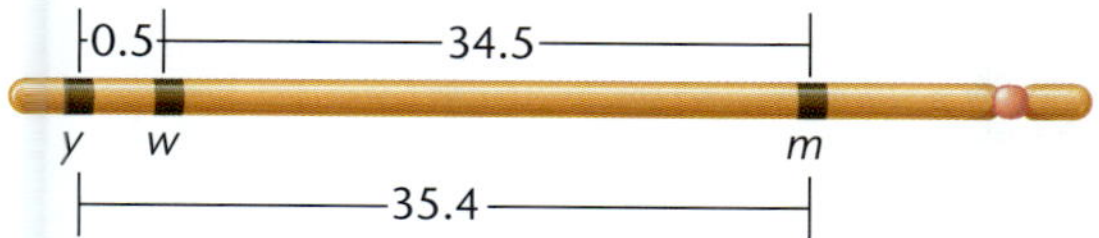

FIGURE 7.4 A map of the *yellow (y)*, *white (w)*, and *miniature (m)* genes on the X chromosome of *Drosophila melanogaster*. Each number represents the percentage of recombinant offspring produced in one of three crosses, each involving two different genes.

extensive map that included all five genes. He and a colleague, Calvin Bridges, soon began a search for autosomal linkage in *Drosophila*. By 1923, they had clearly shown that linkage and crossing over are not restricted to X-linked genes but can also be demonstrated with autosomes. During this work they made another interesting observation. Crossing over in *Drosophila* was shown to occur only in females. The fact that no crossing over occurs in males made genetic mapping much less complex to analyze in *Drosophila*. However, crossing over does occur in both sexes in most other organisms.

Although many refinements in chromosome mapping have been developed since Sturtevant's initial work, his basic principles are considered to be correct. These principles are used to produce detailed chromosome maps of organisms for which large numbers of linked mutant genes are known. Sturtevant's findings are also historically significant to the broader field of genetics. In 1910, the **chromosomal theory of inheritance** was still widely disputed—even Morgan was skeptical of this theory before he conducted his experiments. Research has now firmly established that chromosomes contain genes in a linear order and that these genes are the equivalent of Mendel's unit factors.

Single Crossovers

Why should the relative distance between two loci influence the amount of recombination and crossing over observed between them? During meiosis, a limited number of crossover events occur in each tetrad. These recombinant events occur randomly along the length of the tetrad. Therefore, the closer two loci reside along the axis of the chromosome, the less likely any single-crossover event will occur between them. The same reasoning suggests that the farther apart two linked loci are, the more likely a random crossover event will occur between them.

In **Figure 7.5(a)**, a **single crossover** occurs between two nonsister chromatids but not between the two loci; therefore the crossover is not detected because no recombinant gametes are produced. In **Figure 7.5(b)**, where two loci are quite far apart, a crossover does occur between them, yielding gametes in which the traits of interest are recombined.

When a single crossover occurs between two nonsister chromatids, the other two chromatids of the tetrad are not

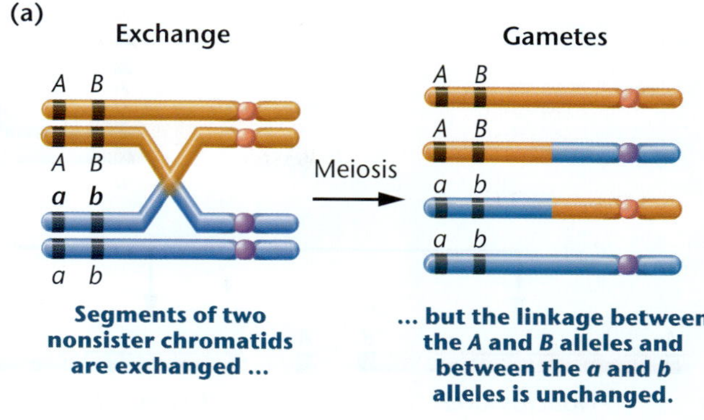

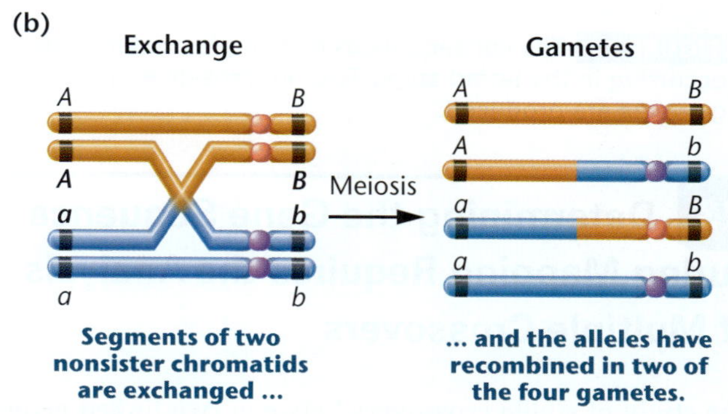

FIGURE 7.5 Two examples of a single crossover between two nonsister chromatids and the gametes subsequently produced. In (a) the exchange does not alter the linkage arrangement between the alleles of the two genes, only parental gametes are formed, and the exchange goes undetected. In (b) the exchange separates the alleles, resulting in recombinant gametes, which are detectable.

involved in this exchange and enter the gamete unchanged. Even if a single crossover occurs 100 percent of the time between two linked genes, recombination is subsequently observed in only 50 percent of the potential gametes formed. This concept is diagrammed in **Figure 7.6**. Theoretically, if we consider only single exchanges and observe 20 percent recombinant gametes, crossing over actually occurred in 40 percent of the tetrads. Under these conditions, the general rule is that the percentage of tetrads involved in an exchange between two genes is twice the percentage of recombinant gametes produced. Therefore, the theoretical limit of observed recombination due to crossing over is 50 percent.

When two linked genes are more than 50 mu apart, a crossover can theoretically be expected to occur between them in 100 percent of the tetrads. If this prediction were achieved, each tetrad would yield equal proportions of the four gametes shown in Figure 7.6, just as if the genes were on different chromosomes and assorting independently. However, this theoretical limit is seldom achieved.

ESSENTIAL POINT

Crossover frequency between linked genes during gamete formation is proportional to the distance between genes, providing the experimental basis for mapping the location of genes relative to one another along the chromosome. ■

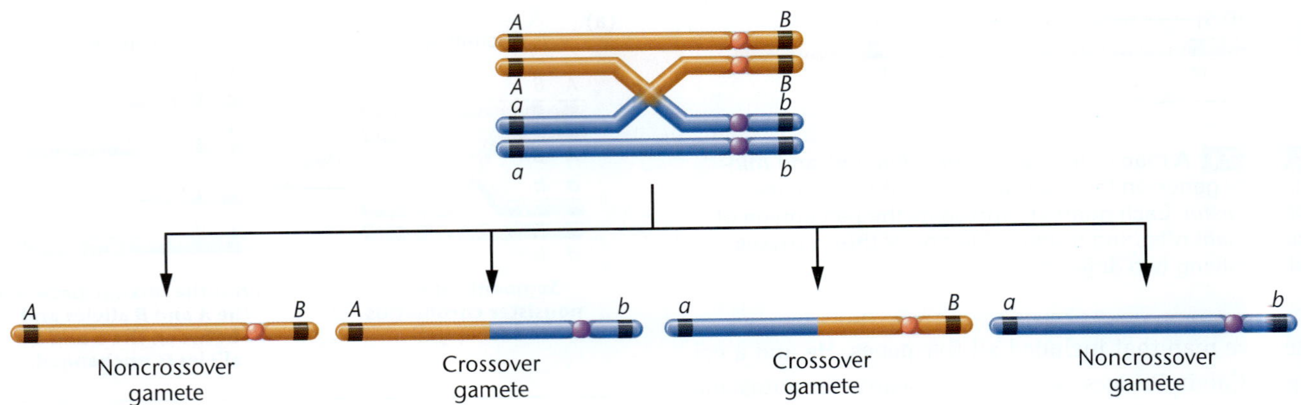

FIGURE 7.6 The consequences of a single exchange between two nonsister chromatids occurring in the tetrad stage. Two noncrossover (parental) and two crossover (recombinant) gametes are produced.

7.3 Determining the Gene Sequence during Mapping Requires the Analysis of Multiple Crossovers

The study of single crossovers between two linked genes provides the basis of determining the *distance* between them. However, when many linked genes are studied, their *sequence* along the chromosome is more difficult to determine. Fortunately, the discovery that multiple exchanges occur between the chromatids of a tetrad has facilitated the process of producing more extensive chromosome maps. As we shall see next, when three or more linked genes are investigated simultaneously, it is possible to determine first the sequence of the genes and then the distances between them.

Multiple Crossovers

It is possible that in a single tetrad, two, three, or more exchanges will occur between nonsister chromatids as a result of several crossover events. Double exchanges of genetic material result from **double crossovers (DCOs),** as shown in **Figure 7.7**. For a double exchange to be studied, three gene pairs must be investigated, each heterozygous for two alleles. Before we determine the frequency of recombination among all three loci, let's review some simple probability calculations.

As we have seen, the probability of a single exchange occurring between the *A* and *B* or the *B* and *C* genes relates directly to the distance between the respective loci. The closer *A* is to *B* and *B* is to *C*, the less likely a single exchange will occur between either of the two sets of loci. In the case of a double crossover, two separate and independent events or exchanges must occur simultaneously. The mathematical probability of two independent events occurring simultaneously is equal to the product of the individual probabilities (the **product law**).

Suppose that crossover gametes resulting from single exchanges are recovered 20 percent of the time ($p = 0.20$) between *A* and *B*, and 30 percent of the time ($p = 0.30$) between *B* and *C*. The probability of recovering a double-crossover gamete arising from two exchanges (between *A* and *B*, and between *B* and *C*) is predicted to be $(0.20)(0.30) = 0.06$, or 6 percent. It is apparent from this calculation that the frequency of double-crossover gametes is always expected to be much lower than that of either single-crossover class of gametes.

If three genes are relatively close together along one chromosome, the expected frequency of double-crossover gametes is extremely low. For example, suppose the *A–B* distance in Figure 7.7 is 3 mu and the *B–C* distance is 2 mu. The expected double-crossover frequency is $(0.03)(0.02) = 0.0006$, or 0.06 percent. This translates to only 6 events in 10,000. Thus, in a mapping experiment where closely linked genes are involved, very large

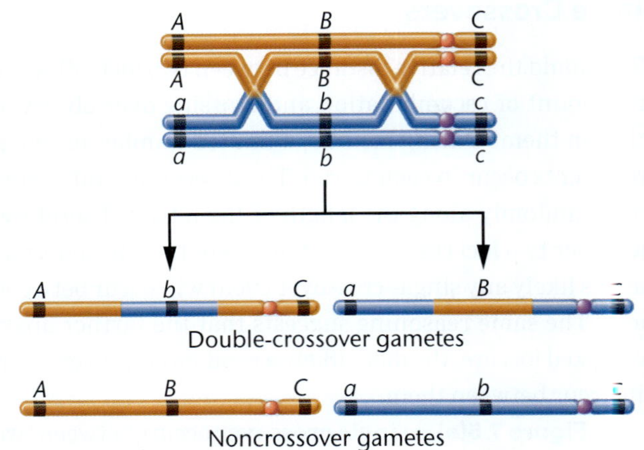

FIGURE 7.7 Consequences of a double exchange occurring between two nonsister chromatids. Because the exchanges involve only two chromatids, two noncrossover gametes and two double-crossover gametes are produced. The chapter opening photograph on p. 145 illustrates a single chiasma present in a tetrad isolated during the first meiotic prophase stage.

numbers of offspring are required to detect double-crossover events. In this example, it is unlikely that a double crossover will be observed even if 1000 offspring are examined. Thus, it is evident that if four or five genes are being mapped, even fewer triple and quadruple crossovers can be expected to occur.

NOW SOLVE THIS

7.2 With two pairs of genes involved (*P/p* and *Z/z*), a testcross (*ppzz*) with an organism of unknown genotype indicated that the gametes produced were in the following proportions

PZ, 42.4%; *Pz*, 6.9%; *pZ*, 7.1%; *pz*, 43.6%

Draw all possible conclusions from these data.

■ **HINT:** *This problem involves an understanding of the proportionality between crossover frequency and distance between genes. The key to its solution is to be aware that noncrossover and crossover gametes occur in reciprocal pairs of approximately equal proportions.*

ESSENTIAL POINT

Determining the sequence of genes in a three-point mapping experiment requires analysis of the double-crossover gametes, as reflected in the phenotype of the offspring receiving those gametes. ■

Three-Point Mapping in *Drosophila*

The information in the preceding section enables us to map three or more linked genes in a single cross. To illustrate the mapping process in its entirety, we examine two situations involving three linked genes in two quite different organisms.

To execute a successful mapping cross, three criteria must be met:

1. The genotype of the organism producing the crossover gametes must be heterozygous at all loci under consideration.
2. The cross must be constructed so that genotypes of all gametes can be determined accurately by observing the phenotypes of the resulting offspring. This is necessary because the gametes and their genotypes can never be observed directly. To overcome this problem, each phenotypic class must reflect the genotype of the gametes of the parents producing it.
3. A sufficient number of offspring must be produced in the mapping experiment to recover a representative sample of all crossover classes.

These criteria are met in the three-point mapping cross from *Drosophila* shown in **Figure 7.8**. In this cross, three X-linked recessive mutant genes—*yellow* body color (*y*), *white* eye color (*w*), and *echinus* eye shape (*ec*)—are considered. To diagram the cross, *we must assume some theoretical sequence, even though we do not yet know if it is correct.* In Figure 7.8, we initially assume the sequence of the three genes to be *y–w–ec*. If this assumption is incorrect, our analysis will demonstrate this and reveal the correct sequence.

In the P_1 generation, males hemizygous for all three wild-type alleles are crossed to females that are homozygous for all three recessive mutant alleles. Therefore, the P_1 males are wild type with respect to body color, eye color, and eye shape. They are said to have a *wild-type phenotype.* The females, on the other hand, exhibit the three mutant traits—yellow body color, white eyes, and echinus eye shape.

This cross produces an F_1 generation consisting of females that are heterozygous at all three loci and males that, because of the Y chromosome, are hemizygous for the three mutant alleles. Phenotypically, all F_1 females are wild type, while all F_1 males are yellow, white, and echinus. The genotype of the F_1 females fulfills the first criterion for mapping; that is, it is heterozygous at the three loci and can serve as the source of recombinant gametes generated by crossing over. Note that because of the genotypes of the P_1 parents, all three mutant alleles in the F_1 female are on one homolog and all three wild-type alleles are on the other homolog. With other females, *other arrangements are possible that could produce a heterozygous genotype.* For example, a heterozygous female could have the *y* and *ec* mutant alleles on one homolog and the *w* allele on the other. This would occur if, in the P_1 cross, one parent was yellow, echinus and the other parent was white.

In our cross, the second criterion is met by virtue of the gametes formed by the F_1 males. Every gamete contains either an X chromosome bearing the three mutant alleles or a Y chromosome, which is genetically inert for the three loci being considered. Whichever type participates in fertilization, the genotype of the gamete produced by the F_1 female will be expressed phenotypically in the F_2 male and female offspring derived from it. Thus, all F_1 noncrossover and crossover gametes can be detected by observing the F_2 phenotypes.

With these two criteria met, we can now construct a chromosome map from the crosses shown in Figure 7.8. First, we determine which F_2 phenotypes correspond to the various noncrossover and crossover categories. To determine the noncrossover F_2 phenotypes, we must identify individuals derived from the parental gametes formed by the F_1 female. Each such gamete contains an X chromosome *unaffected by crossing over.* As a result of segregation, approximately equal proportions of the two types of gametes and, subsequently, the F_2 phenotypes, are produced. Because they derive from a heterozygote, the genotypes of the two parental gametes and

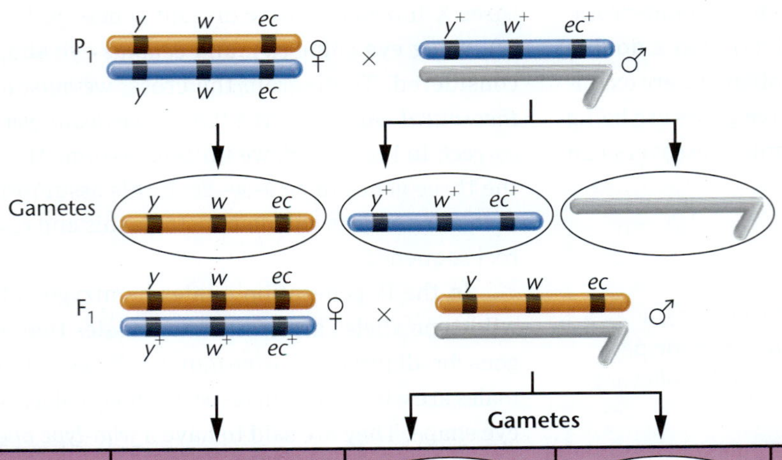

Origin of female gametes	Gametes	Male gamete: y w ec	Male gamete: Y	F_2 phenotype	Observed number	Category, total, and percentage
NCO	1) y w ec	y w ec / y w ec	y w ec	y w ec	4685	Non-crossover
	2) y^+ w^+ ec^+	y^+ w^+ ec^+ / y w ec	y^+ w^+ ec^+	y^+ w^+ ec^+	4759	9444 94.44%
SCO	3) y w^+ ec^+	y w^+ ec^+ / y w ec	y w^+ ec^+	y w^+ ec^+	80	Single crossover between *y* and *w*
	4) y^+ w ec	y^+ w ec / y w ec	y^+ w ec	y^+ w ec	70	150 1.50%
SCO	5) y w ec^+	y w ec^+ / y w ec	y w ec^+	y w ec^+	193	Single crossover between *w* and *ec*
	6) y^+ w^+ ec	y^+ w^+ ec / y w ec	y^+ w^+ ec	y^+ w^+ ec	207	400 4.00%
DCO	7) y w^+ ec	y w^+ ec / y w ec	y w^+ ec	y w^+ ec	3	Double crossover between *y* and *w* and between *w* and *ec*
	8) y^+ w ec^+	y^+ w ec^+ / y w ec	y^+ w ec^+	y^+ w ec^+	3	6 0.06%

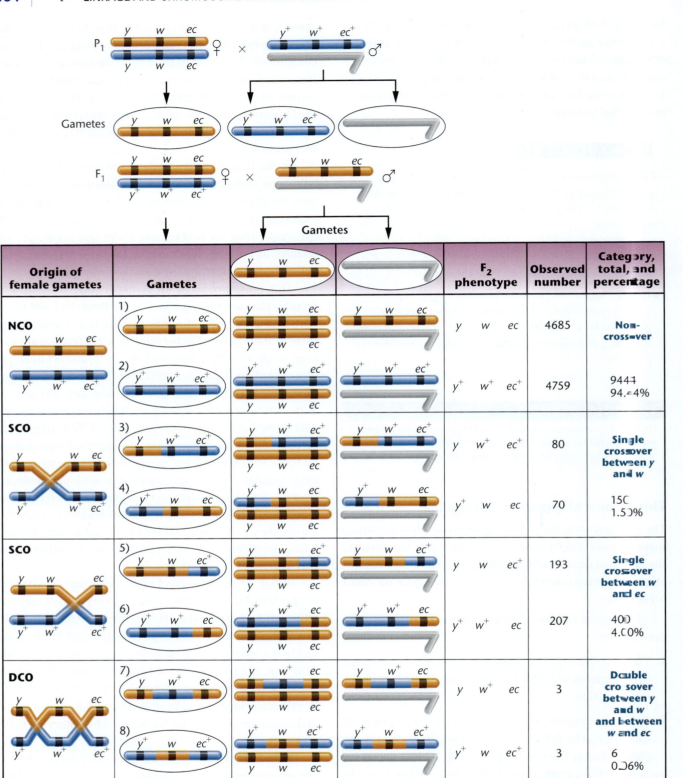

FIGURE 7.8 A three-point mapping cross involving the *yellow* (y or y^+), white (w or w^+), and *echinus* (ec or ec^+) genes in *Drosophila melanogaster*. NCO, SCO, and DCO refer to noncrossover, single-crossover, and double-crossover groups, respectively. Centromeres are not drawn on the chromosomes, and only two nonsister chromatids are initially shown in the left-hand column.

the resultant F_2 phenotypes complement one another. For example, if one is wild type, the other is completely mutant. This is the case in the cross being considered. In other situations, if one chromosome shows one mutant allele, the second chromosome shows the other two mutant alleles, and so on. They are therefore called *reciprocal classes* of gametes and phenotypes.

The two noncrossover phenotypes are most easily recognized because *they exist in the greatest proportion.* Figure 7.8 shows that gametes 1 and 2 are present in the greatest numbers. Therefore, flies that express yellow, white, and echinus phenotypes and flies that are normal (or wild type) for all three characters constitute the noncrossover category and represent 94.44 percent of the F_2 offspring.

The second category that can be easily detected is represented by the double-crossover phenotypes. Because of their low probability of occurrence, *they must be present in the least numbers.* Remember that this group represents two independent but simultaneous single-crossover events. Two reciprocal phenotypes can be identified: gamete 7, which shows the mutant traits yellow, echinus but normal eye color; and gamete 8, which shows the mutant trait white but normal body color and eye shape. Together these double-crossover phenotypes constitute only 0.06 percent of the F_2 offspring.

The remaining four phenotypic classes represent two categories resulting from single crossovers. Gametes 3 and 4, reciprocal phenotypes produced by single-crossover events occurring between the *yellow* and *white* loci, are equal to 1.50 percent of the F_2 offspring; gametes 5 and 6, constituting 4.00 percent of the F_2 offspring, represent the reciprocal phenotypes resulting from single-crossover events occurring between the *white* and *echinus* loci.

The map distances separating the three loci can now be calculated. The distance between *y* and *w* or between *w* and *ec* is equal to the percentage of all detectable exchanges occurring between them. For any two genes under consideration, this includes all appropriate single crossovers as well as all double crossovers. *The latter are included because they represent two simultaneous single crossovers.* For the *y* and *w* genes, this includes gametes 3, 4, 7, and 8, totaling 1.50% + 0.06%, or 1.56 mu. Similarly, the distance between *w* and *ec* is equal to the percentage of offspring resulting from an exchange between these two loci: gametes 5, 6, 7, and 8, totaling 4.00% + 0.06%, or 4.06 mu. The map of these three loci on the X chromosome is shown at the bottom of Figure 7.8.

Determining the Gene Sequence

In the preceding example, the sequence (or order) of the three genes along the chromosome was assumed to be *y—w—ec*. Our analysis shows this sequence to be consistent with the data. However, in most mapping experiments the gene sequence is not known, and this constitutes another variable in the analysis. In our example, had the gene sequence been unknown, it could have been determined using a straightforward method.

This method is based on the fact that there are only three possible arrangements, each containing one of the three genes between the other two:

(I)	*w—y—ec*	(*y* in the middle)
(II)	*y—ec—w*	(*ec* in the middle)
(III)	*y—w—ec*	(*w* in the middle)

Use the following steps during your analysis to determine the gene order:

1. Assuming any one of the three orders, first determine the *arrangement of alleles* along each homolog of the heterozygous parent giving rise to noncrossover and crossover gametes (the F_1 female in our example).
2. Determine whether a double-crossover event occurring within that arrangement will produce the *observed double-crossover phenotypes.* Remember that these phenotypes occur least frequently and are easily identified.
3. If this order does not produce the predicted phenotypes, try each of the other two orders. One must work!

In **Figure 7.9**, the above steps are applied to each of the three possible arrangements (I, II, and III above). A full analysis can proceed as follows:

1. Assuming that *y* is between *w* and *ec*, arrangement I of alleles along the homologs of the F_1 heterozygote is

$$\frac{w \quad y \quad ec}{w^+ \quad y^+ \quad ec^+}$$

We know this because of the way in which the P_1 generation was crossed: The P_1 female contributes an X chromosome bearing the *w*, *y*, and *ec* alleles, while the P_1 male contributes an X chromosome bearing the w^+, y^+, and ec^+ alleles.

2. A double crossover within that arrangement yields the following gametes

$$\underline{w \quad y^+ \quad ec} \quad \text{and} \quad \underline{w^+ \quad y \quad ec^+}$$

Following fertilization, if *y* is in the middle, the F_2 double-crossover phenotypes will correspond to these gametic genotypes, yielding offspring that express the white, echinus phenotype and offspring that express the yellow phenotype. Instead, determination of the actual double-crossover phenotypes reveals them to be yellow, echinus flies and white flies. *Therefore, our assumed order is incorrect.*

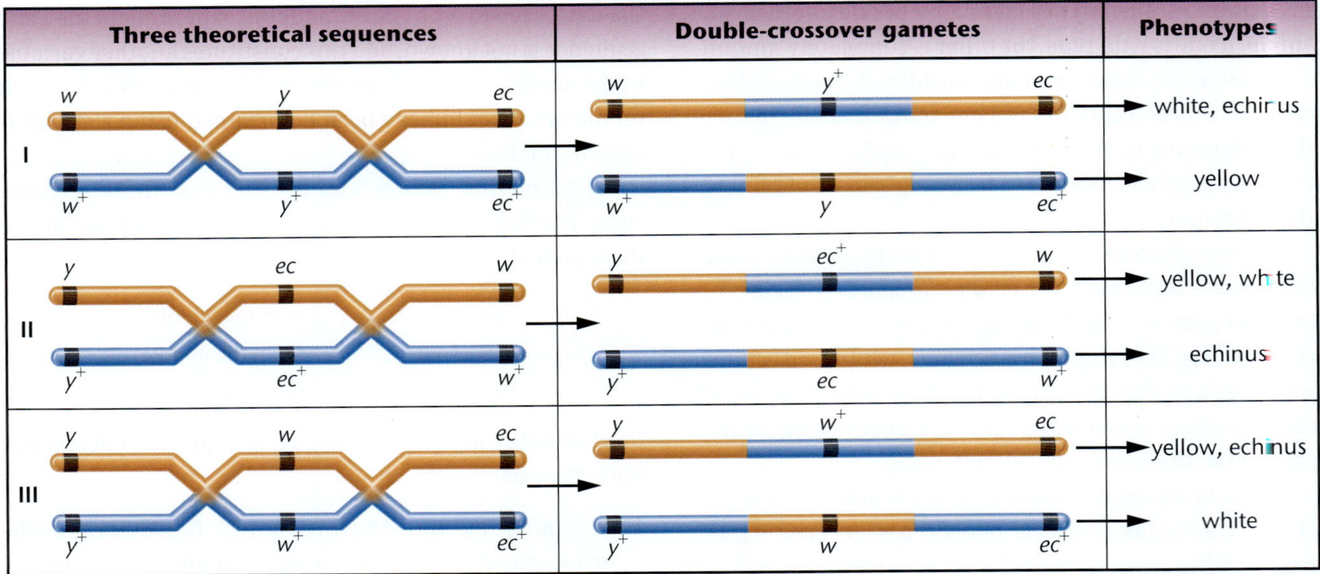

FIGURE 7.9 The three possible sequences of the *white, yellow,* and *echinus genes,* the results of a double crossover in each case, and the resulting phenotypes produced in a testcross. For simplicity, the two noncrossover chromatids of each tetrad are omitted.

3. If we consider arrangement II with the ec/ec^+ alleles in the middle or arrangement III with the w/w^+ alleles in the middle

$$\text{(II)}\ \frac{w \quad ec \quad w}{y^+ \quad ec^+ \quad w^+} \quad \text{or} \quad \text{(III)}\ \frac{y \quad w \quad ec}{y^+ \quad w^+ \quad ec^+}$$

we see that arrangement II again provides *predicted* double-crossover phenotypes that *do not* correspond to the *actual* (observed) double-crossover phenotypes. The predicted phenotypes are yellow, white flies and echinus flies in the F_2 generation. *Therefore, this order is also incorrect.* However, arrangement III produces the observed phenotypes—yellow, echinus flies and white flies. *Therefore, this arrangement, with the w gene in the middle, is correct.*

To summarize, first determine the arrangement of alleles on the homologs of the heterozygote yielding the crossover gametes by locating the reciprocal noncrossover phenotypes. Then, test each of three possible orders to determine which yields the observed double-crossover phenotypes—*the one that does so represents the correct order.*

Solving an Autosomal Mapping Problem

Having established the basic principles of chromosome mapping, we will now consider a related problem in maize (corn). This analysis differs from the preceding example in two ways. First, the previous mapping cross involved X-linked genes. Here, we consider autosomal genes. Second, in the discussion of this cross we have changed our use of symbols, as first suggested in Chapter 4. Instead of using the gene symbols and superscripts (e.g., bm^+, v^+, and pr^+), we simply use + to denote each wild-type allele. This system is easier to manipulate but requires a better understanding of mapping procedures.

When we look at three autosomally linked genes in maize, the experimental cross must still meet the same three criteria we established for the X-linked genes in *Drosophila:* (1) One parent must be heterozygous for all traits under consideration; (2) the gametic genotypes produced by the heterozygote must be apparent from observing the phenotypes of the offspring; and (3) a sufficient sample size must be available for complete analysis.

In maize, the recessive mutant genes *brown* midrib (*bm*), *virescent* seedling (*v*), and *purple* aleurone (*pr*) are linked on chromosome 5. Assume that a female plant is known to be heterozygous for all three traits, but we do not know (1) the arrangement of the mutant alleles on the maternal and paternal homologs of this heterozygote, (2) the sequence of genes, or (3) the map distances between the genes. What genotype must the male plant have to allow successful mapping? To meet the second criterion, the male must be homozygous for all three recessive mutant alleles. Otherwise, offspring of this cross showing a given phenotype might represent more than one genotype, making accurate mapping impossible.

Figure 7.10 diagrams this cross. As shown, we know neither the arrangement of alleles nor the sequence of loci in the heterozygous female. Several possibilities are shown, but we have yet to determine which is correct. We don't know the sequence in the testcross male parent either, and so we must designate it randomly. Note that we have initially placed *v* in the middle. *This may or may not be correct.*

(a) Some possible allele arrangements and gene sequences in a heterozygous female

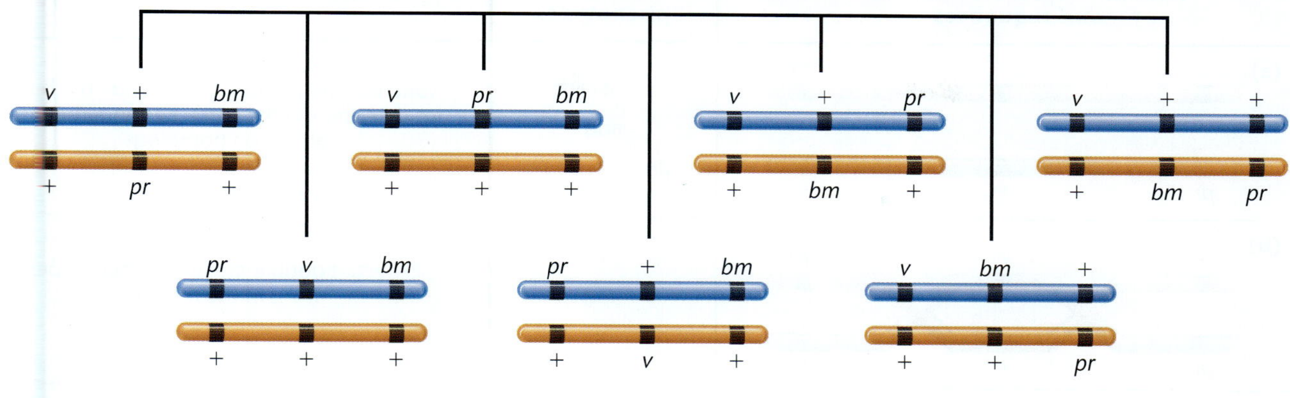

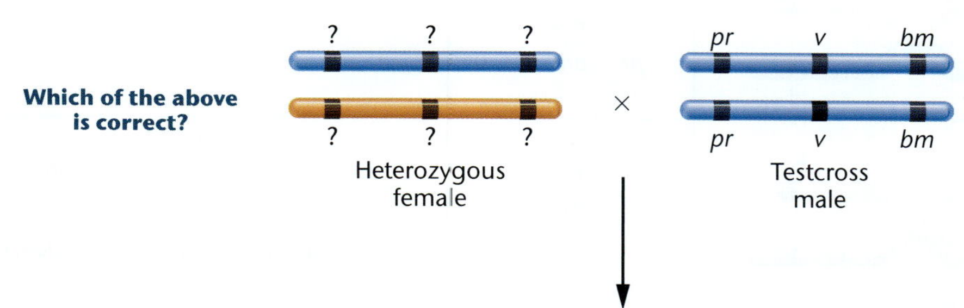

(b) Actual results of mapping cross*

Phenotypes of offspring	Number	Total and percentage	Exchange classification
+ *v* *bm*	230	467 42.1%	Noncrossover (NCO)
pr + +	237		
+ + *bm*	82	161 14.5%	Single crossover (SCO)
pr *v* +	79		
+ *v* +	200	395 35.6%	Single crossover (SCO)
pr + *bm*	195		
pr *v* *bm*	44	86 7.8%	Double crossover (DCO)
+ + +	42		

* The sequence *pr* – *v* – *bm* may or may not be correct.

FIGURE 7.10 (a) Some possible allele arrangements and gene sequences in a heterozygous female. The data from a three-point mapping cross, depicted in (b), where the female is testcrossed, provide the basis for determining which combination of arrangement and sequence is correct. [See Figure 7.11(d).]

The offspring are arranged in groups of two for each pair of reciprocal phenotypic classes. The two members of each reciprocal class are derived from no crossing over (NCO), one of two possible single-crossover events (SCO), or a double crossover (DCO).

To solve this problem, refer to Figures 7.10 and 7.11 as you consider the following questions.

1. *What is the correct heterozygous arrangement of alleles in the female parent?*
   Determine the two noncrossover classes, those that occur with the highest frequency. In this case, they are + *v* *bm* and *pr* + +. Therefore, the alleles on the homologs of the female parent must be arranged as shown in **Figure 7.11(a)**. These homologs segregate into

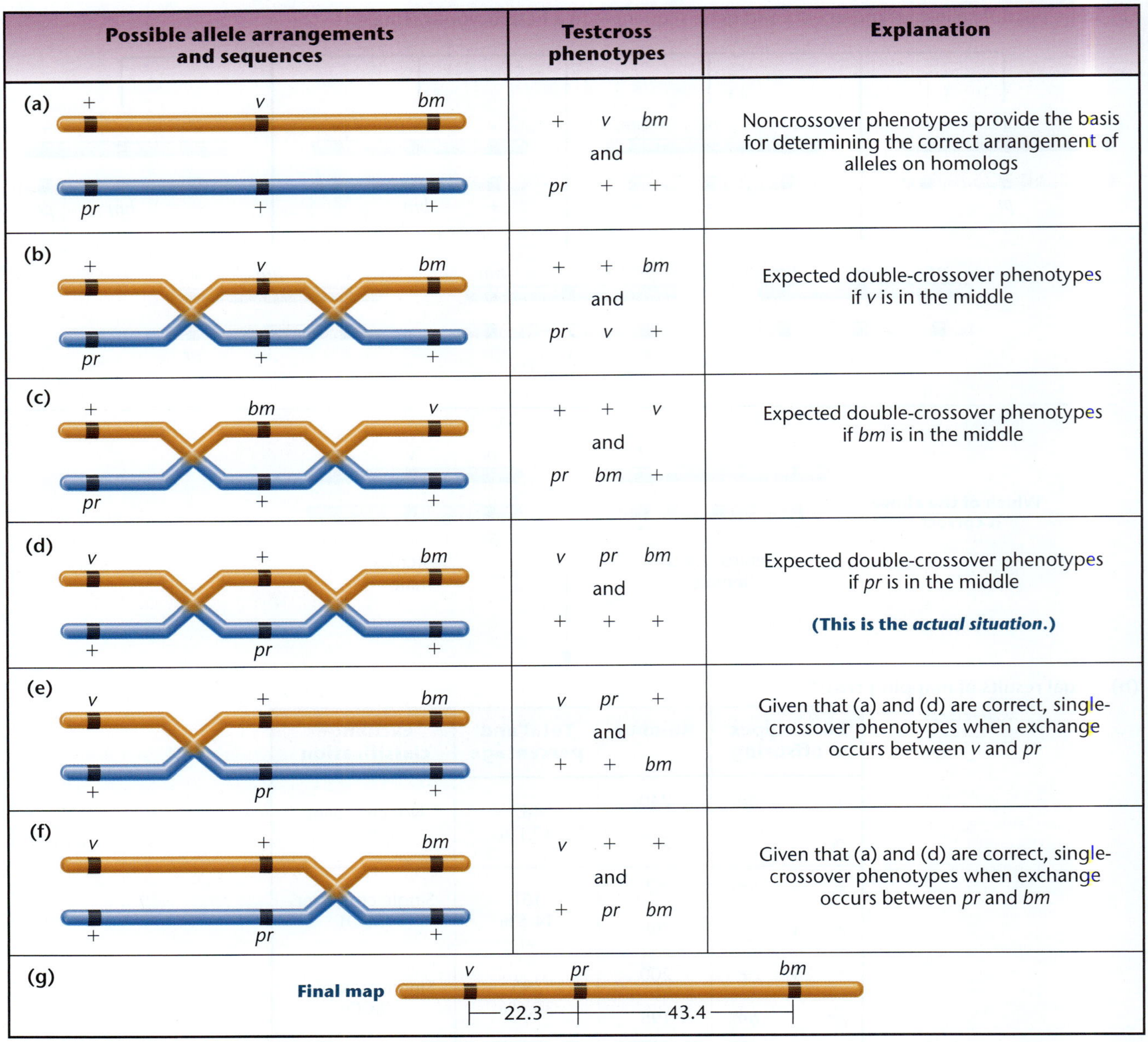

Possible allele arrangements and sequences	Testcross phenotypes	Explanation
(a) + v bm / pr + +	+ v bm and pr + +	Noncrossover phenotypes provide the basis for determining the correct arrangement of alleles on homologs
(b) + v bm / pr + +	+ + bm and pr v +	Expected double-crossover phenotypes if *v* is in the middle
(c) + bm v / pr + +	+ + v and pr bm +	Expected double-crossover phenotypes if *bm* is in the middle
(d) v + bm / + pr +	v pr bm and + + +	Expected double-crossover phenotypes if *pr* is in the middle **(This is the *actual situation.*)**
(e) v + bm / + pr +	v pr + and + + bm	Given that (a) and (d) are correct, single-crossover phenotypes when exchange occurs between *v* and *pr*
(f) v + bm / + pr +	v + + and + pr bm	Given that (a) and (d) are correct, single-crossover phenotypes when exchange occurs between *pr* and *bm*
(g) **Final map** v — 22.3 — pr — 43.4 — bm		

FIGURE 7.11 Steps utilized in producing a map of the three genes in the cross in Figure 7.10, where neither the arrangement of alleles nor the sequence of genes in the heterozygous female parent is known.

gametes, unaffected by any recombination event. Any other arrangement of alleles will not yield the observed noncrossover classes. (Remember that $\underline{+ \quad v \quad bm}$ is equivalent to $\underline{pr^{+} \quad v \quad bm}$ and that $\underline{pr \quad + \quad +}$ is equivalent to $\underline{pr \quad v^{+} \quad bm^{+}}$.)

2. *What is the correct sequence of genes?*

We know that the arrangement of alleles is

$$\frac{+ \quad v \quad bm}{pr \quad + \quad +}$$

But is the gene sequence correct? That is, will a double-crossover event yield the observed double-crossover phenotypes after fertilization? *Observation shows that it will not* [**Figure 7.11(b)**]. Now try the other two orders [**Figure 7.11(c)** and **(d)**] *maintaining the same arrangement of alleles:*

$$\frac{+ \quad bm \quad v}{pr \quad + \quad +} \quad \text{or} \quad \frac{v \quad + \quad bm}{+ \quad pr \quad +}$$

Only the order on the right yields the observed double-crossover gametes [Figure 7.11(d)]. Therefore, the *pr* gene is in the middle. From this point on, work the problem using this arrangement and sequence, with the *pr* locus in the middle.

3. *What is the distance between each pair of genes?*

Having established the sequence of loci as *v–pr–bm*, we can determine the distance between *v* and *pr* and between *pr* and *bm*. Remember that the map distance between two genes is calculated on the basis of all detectable recombination events occurring between them. This includes both single- and double-crossover events.

Figure 7.11(e) shows that the phenotypes *v pr +* and *+ + bm* result from single crossovers between the *v* and *pr* loci, accounting for 14.5 percent of the offspring [according to data in Figure 7.10(b)]. By adding the percentage of double crossovers (7.8 percent) to the number obtained for single crossovers, the total distance between the *v* and *pr* loci is calculated to be 22.3 mu.

Figure 7.11(f) shows that the phenotypes *v + +* and *+ pr bm* result from single crossovers between the *pr* and *bm* loci, totaling 35.6 percent. Added to the double crossovers (7.8 percent), the distance between *pr* and *bm* is calculated to be 43.4 mu. The final map for all three genes in this example is shown in **Figure 7.11(g)**.

NOW SOLVE THIS

7.3 In *Drosophila*, a heterozygous female for the X-linked recessive traits *a*, *b*, and *c* was crossed to a male that phenotypically expressed *a*, *b*, and *c*. The offspring occurred in the following phenotypic ratios.

+	*b*	*c*	460
a	+	+	450
a	*b*	*c*	32
+	+	+	38
a	+	*c*	11
+	*b*	+	9

No other phenotypes were observed.

(a) What is the genotypic arrangement of the alleles of these genes on the X chromosome of the female?

(b) Determine the correct sequence and construct a map of these genes on the X chromosome.

(c) What progeny phenotypes are missing? Why?

HINT: *This problem involves a three-point mapping experiment where only six phenotypic categories are observed, even though eight categories are typical of such a cross. The key to its solution is to be aware that if the distances between the loci are relatively small, the sample size may be too small for the predicted number of double crossovers to be recovered, even though reciprocal pairs of single crossovers are seen. You should write the missing gametes down as double crossovers and record zeros for their frequency of appearance.*

7.4 As the Distance between Two Genes Increases, Mapping Estimates Become More Inaccurate

So far, we have assumed that crossover frequencies are directly proportional to the distance between any two loci along the chromosome. However, it is not always possible to detect all crossover events. A case in point is a double exchange that occurs between the two loci in question. As shown in **Figure 7.12(a)**, if a double exchange occurs, the original arrangement of alleles on each nonsister homolog is recovered. Therefore, even though crossing over has occurred, it is impossible to detect. This phenomenon is true for all even-numbered exchanges between two loci.

Furthermore, as a result of complications posed by *multiple-strand exchanges*, mapping determinations usually underestimate the actual distance between two genes. The farther apart two genes are, the greater the probability that undetected crossovers will occur. While the discrepancy is minimal for two genes relatively close together, the degree of inaccuracy increases as the distance increases, as shown in the graph of map distance versus recombination frequency in **Figure 7.12(b)**. There, the theoretical frequency where a direct correlation between recombination and map distance exists is contrasted with the actual frequency observed as the distance between two genes increases. The most accurate maps are constructed from experiments where genes are relatively close together.

Interference and the Coefficient of Coincidence

As shown in our maize example, we can predict the expected frequency of multiple exchanges, such as double crossovers, once the distance between genes is established. For example, in the maize cross, the distance between *v* and *pr* is 22.3 mu, and the distance between *pr* and *bm* is 43.4 mu. If the two single crossovers that make up a double crossover occur independently of one another, we can calculate the expected frequency of double crossovers (DCO_{exp}):

$$DCO_{exp} = (0.223) \times (0.434) = 0.097 = 9.7\%$$

Often in mapping experiments, the observed DCO frequency is less than the expected number of DCOs. In the maize cross, for example, only 7.8 percent DCOs are observed when 9.7 percent are expected. **Interference (*I*)**, the phenomenon through which a crossover event in one region of the chromosome inhibits a second event in nearby regions, causes this reduction.

To quantify the disparities that result from interference, we calculate the **coefficient of coincidence (*C*)**:

(a) Two-strand double exchange

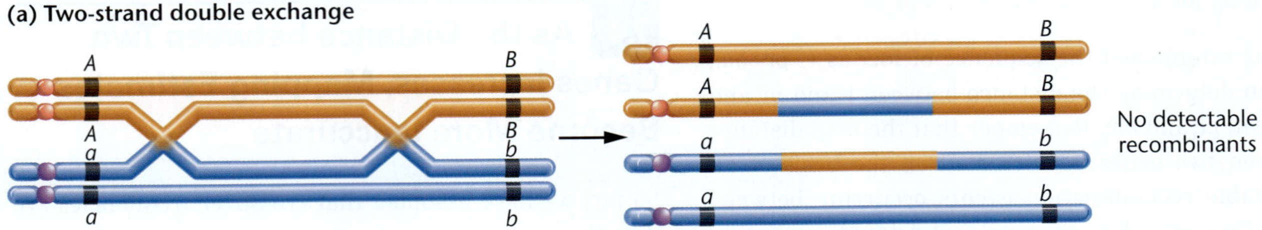

(b)

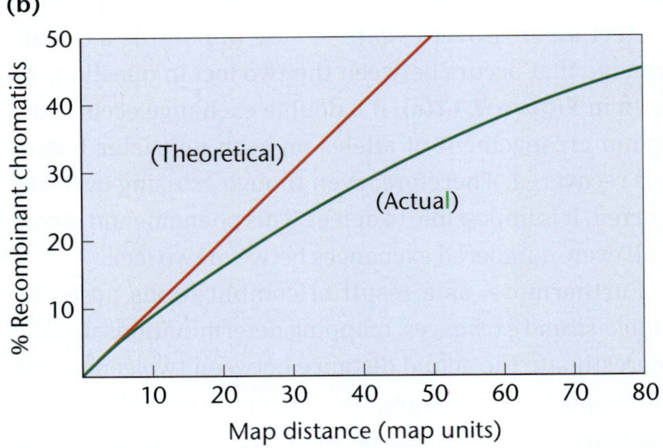

FIGURE 7.12 (a) A double crossover is undetected because no rearrangement of alleles occurs. (b) The theoretical and actual percentage of recombinant chromatids versus map distance. The straight line shows the theoretical relationship if a direct correlation between recombination and map distance exists. The curved line is the actual relationship derived from studies of *Drosophila*, *Neurospora*, and *Zea mays*.

$$C = \frac{\text{Observed DCO}}{\text{Expected DCO}}$$

In the maize cross, we have

$$C = \frac{0.078}{0.097} = 0.804$$

Once we have found C, we can quantify interference using the simple equation

$$I = 1 - C$$

In the maize cross, we have

$$I = 1.000 - 0.804 = 0.196$$

If interference is complete and no double crossovers occur, then $I = 1.0$. If fewer DCOs than expected occur, I is a positive number and positive interference has occurred. If more DCOs than expected occur, I is a negative number and negative interference has occurred. In the maize example, I is a positive number (0.196), indicating that 19.6 percent fewer double crossovers occurred than expected.

Positive interference is most often the rule in eukaryotic systems. In general, the closer genes are to one another along the chromosome, the more positive interference occurs. In fact, interference in *Drosophila* is often complete within a distance of 10 mu, and no multiple crossovers are recovered. This observation suggests that physical constraints preventing the formation of closely aligned chiasmata contribute to interference. This interpretation is consistent with the finding that interference decreases as the genes in question are located farther apart. In the maize cross in Figures 7.10 and 7.11, the three genes are relatively far apart, and 80 percent of the expected double crossovers are observed.

ESSENTIAL POINT

Interference describes the extent to which a crossover in one region of a chromosome influences the occurrence of a crossover in an adjacent region of the chromosome and is quantified by calculating the coefficient of coincidence (C). ■

EVOLVING CONCEPT OF THE GENE

Based on the gene-mapping studies in *Drosophila* and many other organisms from the 1920s through the mid-1950s, geneticists regarded genes as hereditary units organized in a specific sequence along chromosomes, between which recombination could occur. Genes were thus viewed as indivisible "beads on a string." ■

7.5 Chromosome Mapping Is Now Possible Using DNA Markers and Annotated Computer Databases

Although traditional methods based on recombination analysis have produced detailed chromosomal maps in several organisms, such maps in other organisms (including humans) that do not lend themselves to such studies are greatly limited. Fortunately, the development of technology allowing direct analysis of DNA has greatly enhanced mapping in those organisms. We will address this topic using humans as an example.

Progress has initially relied on the discovery of **DNA markers** that have been identified during recombinant DNA and genomic studies. These markers are short segments of DNA whose sequence and location are known, making them useful *landmarks* for mapping purposes. The analysis of human genes in relation to these markers has extended our knowledge of the location within the genome of countless genes, which is the ultimate goal of mapping.

The earliest examples are the DNA markers referred to as **restriction fragment length polymorphisms (RFLPs)** (see Special Topics Chapter 2—Genetic Testing) and **microsatellites** (see Chapter 11). RFLPS are polymorphic sites generated when specific DNA sequences are recognized and cut by restriction enzymes. Microsatellites are short repetitive sequences that are found throughout the genome, and they vary in the number of repeats at any given site. For example, the two-nucleotide sequence CA is repeated 5–50 times per site [$(CA)_n$] and appears throughout the genome approximately every 10,000 bases, on average. Microsatellites may be identified not only by the number of repeats but by the DNA sequences that flank them. More recently, variation in single nucleotides (called **single-nucleotide polymorphisms** or **SNPs**) has been utilized. Found throughout the genome, up to several million of these variations may be screened for an association with a disease or trait of interest, thus providing geneticists with a means to identify and locate related genes.

Cystic fibrosis offers an early example of a gene located by using DNA markers. It is a life-shortening autosomal recessive exocrine disorder resulting in excessive, thick mucus that impedes the function of organs such as the lung and pancreas. After scientists established that the gene causing this disorder is located on chromosome 7, they were then able to pinpoint its exact location on the long arm (the q arm) of that chromosome.

In 2007, using SNPs as DNA markers, associations between 24 genomic locations were established with seven common human diseases: *Type 1* (insulin dependent) and *Type 2 diabetes, Crohn disease* (inflammatory bowel disease), *hypertension, coronary artery disease, bipolar disorder*, and *rheumatoid arthritis*. In each case, an inherited susceptibility effect was mapped to a specific location on a specific chromosome within the genome. In some cases, this either confirmed or led to the identification of a specific gene involved in the cause of the disease.

During the past 15 years or so, dramatic improvements in DNA sequencing technology have resulted in a proliferation of **sequence maps** for humans and many other species. Sequence maps provide the finest level of mapping detail because they pinpoint the nucleotide sequence of genes (and noncoding sequences) on a chromosome. The Human Genome Project resulted in sequence maps for all human chromosomes, providing an incredible level of detail about human gene sequences, the specific location of genes on a chromosome, and the proximity of genes and noncoding sequences to each other, among other details. For instance, when human chromosome sequences were analyzed by software programs, an approach called **bioinformatics,** to be discussed later in the text (see Chapter 18), geneticists could utilize such data to map possible protein-coding sequences in the genome. This led to the identification of thousands of potential genes that were previously unknown.

The many Human Genome Project databases that are now available make it possible to map genes along a human chromosome in base-pair distances rather than recombination frequency. This distinguishes what is referred to as a *physical map* of the genome from the *genetic maps* described above. When the genome sequence of a species is available, mapping by linkage or other genetic mapping approaches becomes obsolete.

ESSENTIAL POINT

Human linkage studies have been enhanced by the use of newly discovered molecular DNA markers. ■

7.6 Other Aspects of Genetic Exchange

Careful analysis of crossing over during gamete formation allows us to construct chromosome maps in many organisms. However, we should not lose sight of the real biological significance of crossing over, which is to generate genetic variation in gametes and, subsequently, in the offspring derived from the resultant eggs and sperm. Many unanswered questions remain, which we consider next.

Crossing Over—A Physical Exchange between Chromatids

Once genetic mapping was understood, it was of great interest to investigate the relationship between chiasmata observed in meiotic prophase I and crossing over. Are chiasmata visible manifestations of crossover events? If so, then crossing over in higher organisms appears to result from an actual physical exchange between homologous chromosomes. That this is the case was demonstrated independently in the 1930s by Harriet Creighton and Barbara McClintock in *Zea mays* and by Curt Stern in *Drosophila*.

Since the experiments are similar, we will consider only the work with maize. Creighton and McClintock studied two linked genes on chromosome 9. At one locus, the alleles *colorless* (*c*) and *colored* (*C*) control endosperm coloration. At the other locus, the alleles *starchy* (*Wx*) and *waxy* (*wx*) control the carbohydrate characteristics of the endosperm. The maize plant studied is heterozygous at both loci. The key to this experiment is that one of the homologs contains two unique cytological markers. The markers consist of a densely stained knob at one end of the chromosome and a translocated piece of another chromosome (8) at the other end. The arrangements of alleles and cytological markers can be detected cytologically and are shown in **Figure 7.13**.

Creighton and McClintock crossed this plant to one homozygous for the *colored* allele (*c*) and heterozygous for the endosperm alleles. They obtained a variety of different phenotypes in the offspring, but they were most interested in a crossover result involving the chromosome with the unique cytological markers. They examined the chromosomes of this plant with the colorless, waxy phenotype (Case I in Figure 7.13) for the presence of the cytological markers. If physical exchange between homologs accompanies genetic crossing over, the translocated chromosome will still be present, but the knob will not—this is exactly what happened. In a second plant (Case II), the phenotype colored, starchy should result from either nonrecombinant gametes or crossing over. Some of the plants then ought to contain chromosomes with the dense knob but not the translocated chromosome. This condition was also found, and the conclusion that a physical exchange takes place was again supported. Along with Stern's findings with *Drosophila*, this work clearly established that crossing over has a cytological basis.

ESSENTIAL POINT

Cytological investigations of both maize and *Drosophila* reveal that crossing over involves a physical exchange of segments between nonsister chromatids. ■

Sister Chromatid Exchanges between Mitotic Chromosomes

Considering that crossing over occurs between synapsed homologs in meiosis, we might ask whether a similar physical exchange occurs between homologs during mitosis. While homologous chromosomes do not usually pair up or synapse in somatic cells (*Drosophila* is an exception), each individual chromosome in prophase and metaphase of mitosis consists of two identical sister chromatids, joined at a common centromere. Surprisingly, several experimental approaches have demonstrated that reciprocal exchanges similar to crossing over occur between sister chromatids. These **sister chromatid exchanges (SCEs)** do not produce new allelic combinations, but evidence is accumulating that attaches significance to these events.

Identification and study of SCEs are facilitated by several modern staining techniques. In one technique, cells replicate for two generations in the presence of the thymidine analog **bromodeoxyuridine (BrdU).** Following two rounds of replication, each pair of sister chromatids has one member with one strand of DNA "labeled" with BrdU

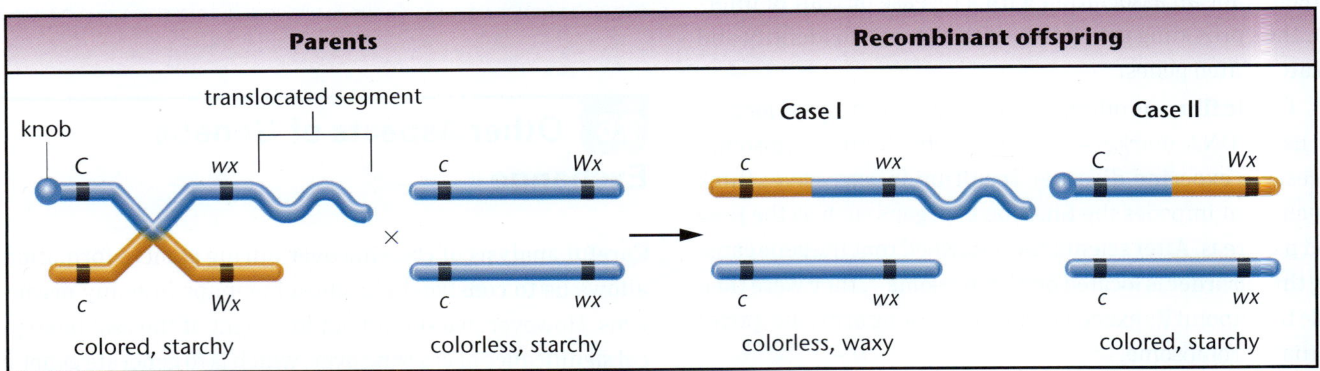

FIGURE 7.13 The phenotypes and chromosome compositions of parents and recombinant offspring in Creighton and McClintock's experiment in maize. The knob and translocated segment served as cytological markers, which established that crossing over involves an actual exchange of chromosome arms.

and the other member with both strands labeled with BrdU. Using a differential stain, chromatids with the analog in both strands stain differently than chromatids with BrdU in only one strand. As a result, SCEs are readily detectable if they occur. In **Figure 7.14**, numerous instances of SCE events are clearly evident. These sister chromatids are sometimes referred to as **harlequin chromosomes** because of their patchlike appearance.

The significance of SCEs is still uncertain, but several observations have generated great interest in this phenomenon. We know, for example, that agents that induce chromosome damage (viruses, X rays, ultraviolet light, and certain chemical mutagens) increase the frequency of SCEs. The frequency of SCEs is also elevated in **Bloom syndrome,** a human disorder caused by a mutation in the *BLM* gene on chromosome 15. This rare, recessively inherited disease is characterized by prenatal and postnatal delays in growth, a great sensitivity of the facial skin to the sun, immune deficiency, a predisposition to malignant and benign tumors, and abnormal behavior patterns. The chromosomes from cultured leukocytes, bone marrow cells, and fibroblasts derived from homozygotes are very fragile and unstable compared to those of homozygous and heterozygous normal individuals. Increased breaks and rearrangements between nonhomologous chromosomes are observed in addition to excessive amounts of sister chromatid exchanges. Work by James German and colleagues suggests that the *BLM* gene encodes an enzyme called **DNA helicase,** which is best known for its role in DNA replication (see Chapter 10).

ESSENTIAL POINT

Recombination events between sister chromatids in mitosis, referred to as sister chromatid exchanges (SCEs), occur at an elevated frequency in the human disorder, Bloom syndrome. ■

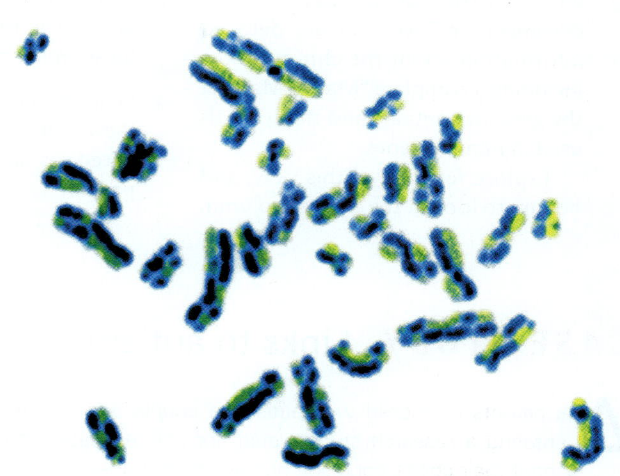

FIGURE 7.14 Demonstration of sister chromatid exchanges (SCEs) in mitotic chromosomes from a Bloom syndrome patient. Using a differential staining technique that involves growing cells for two rounds of DNA replication in the presence of a base analog, regions of sister chromatids stained blue have one strand of the DNA labeled with the base analog while regions of sister chromatids stained green/yellow have both strands of the DNA labeled. The SCE events cause the alternating patterns along each sister chromatid.

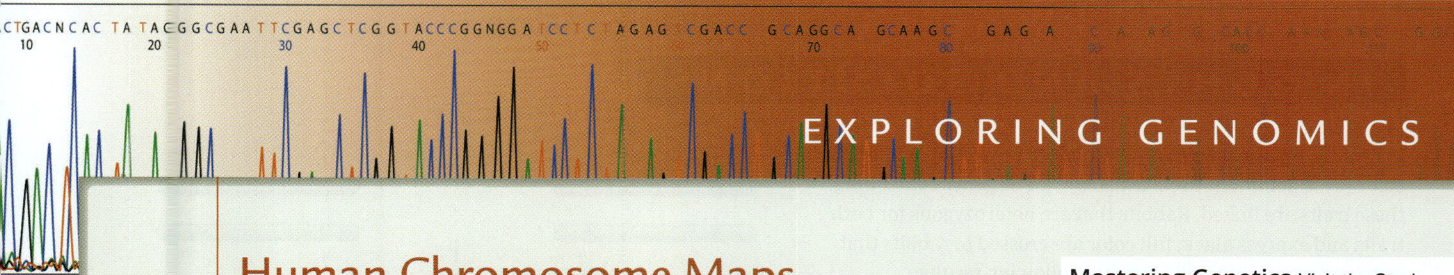

EXPLORING GENOMICS

Human Chromosome Maps on the Internet

Mastering Genetics Visit the Study Area: Exploring Genomics

In this chapter we discussed how recombination data can be analyzed to develop chromosome maps based on linkage. Increasingly, chromosome maps are being developed using genomics techniques. As a result of the Human Genome Project, maps of human chromosomes are now freely available on the Internet. In this exercise we will explore the **National Center for Biotechnology Information (NCBI) Genes and Disease** Web site to learn more about human chromosome maps.

■ NCBI Genes and Disease

Here we explore the Genes and Disease site, which presents human chromosome maps that show the locations of specific disease genes.

1. Access the Genes and Disease site at https://www.ncbi.nlm.nih.gov/books/NBK22185/
2. Under contents, click on "chromosome map" to see a page with an image of a karyotype of human chromosomes. Click on a chromosome in the chromosome map image, scroll down the page to view a chromosome, or click on a chromosome listed on the right side of the page. For example, click on chromosome 7. Notice that the number of genes on the chromosomes and the number of base pairs the chromosome contains are displayed above the image.

(continued)

Exploring Genomics—continued

3. Look again at chromosome 7. At first you might think there are only five disease genes on this chromosome because the initial view shows only selected disease genes. However, if you click the "MapViewer" link for chromosome 7, you will see detailed information about the chromosome, including a complete "Master Map" of the genes it contains and the symbols used in naming genes.

   Explore features of this view, and be sure to look at the "Links" column, which provides access to OMIM (Online Mendelian Inheritance in Man, discussed in the Exploring Genomics feature for Chapter 3) data for a particular gene, as well as to protein information (*pr*) and lists of homologous genes (*hm*; these are other genes that have similar sequences).

4. Scan the chromosome maps in Map Viewer until you see one of the genes listed as a "hypothetical gene or protein."

   a. What does it mean if a gene or protein is referred to as hypothetical?

   b. What information do you think genome scientists use to assign a gene locus for a gene encoding a hypothetical protein?

Visit the **NCBI Genome Data Viewer** homepage (https://www.ncbi.nlm.nih.gov/genome/gdv/) for an excellent database containing chromosome maps for a wide variety of different organisms.

CASE STUDY Links to autism

As parents of a child with autism, a couple decided that entering a research study would not only educate them about their child's condition but also help further research into this complex, behaviorally defined disorder. Researchers explained to the parents that autism results from the action of hundreds of genes as well as nongenetic factors. Modern DNA analysis techniques, including mapping studies, have identified a set of 9–18 genes with a high likelihood of involvement, referred to as candidate genes. Probing the genome at a deeper level has revealed as many as 2500 genes that might be risk factors. Generally unaware of the principles of basic genetics, the couple wondered if any future children they might have would be at risk of having autism and if prenatal diagnosis for autism is possible. The interviewer explained that if one child has autism, there is an approximate 25 percent risk of a future child having this condition. A prenatal test for autism is possible, but because the condition involves a potentially large number of genes and environmental conditions, such tests can only estimate, without any certainty, the likelihood of a second child with autism.

1. In a family with a child with autism, the risk for another affected child is approximately 25 percent. This is the same level of risk that a couple who are each heterozygous for a recessive allele will have an affected child. What are the similarities and differences in these two situations?

2. Given that the prenatal test can provide only a probability estimate that the fetus will develop autism, what ethical issues should be discussed with the parents?

See Hens, K, et al., 2018. The ethics of patenting autism genes. *Nat. Rev. Genet.* 19:247–248.

INSIGHTS AND SOLUTIONS

1. In rabbits, black color (B) is dominant to brown (b), while full color (C) is dominant to *chinchilla* (c^{ch}). The genes controlling these traits are linked. Rabbits that are heterozygous for both traits and express black, full color are crossed to rabbits that express brown, chinchilla with the following results:

31	brown, chinchilla	34	black, full
16	brown, full	19	black, chinchilla

Determine the arrangement of alleles in the heterozygous parents and the map distance between the two genes.

Solution: This is a two-point map problem, where the two most prevalent reciprocal phenotypes are the noncrossovers. The less frequent reciprocal phenotypes arise from a single crossover. The arrangement of alleles is derived from the non-crossover phenotypes because they enter gametes intact.

The single crossovers give rise to 35 of 100 offspring (35 percent). Therefore, the distance between the two genes is 35 mu.

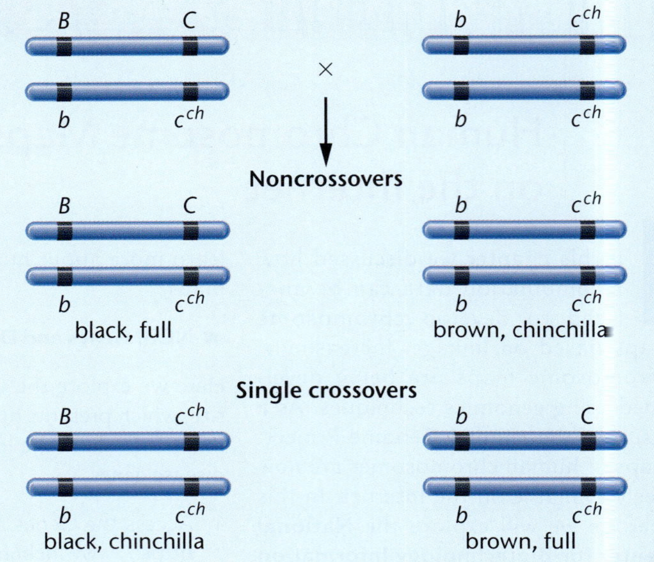

2. Examine the set of three-point mapping data from *Drosophila* involving two dominant mutations [*Stubble* (*Sb*) and *Lyra* (*Ly*)] and one recessive mutation [*bright* (*br*)]. Identify the categories of data below (1–8) that are essential in establishing: (a) the arrangement of alleles in the crossover parent; (b) the sequence of the three genes along the chromosome; and (c) the interlocus distance between the genes.

Phenotype				Number
(1)	*Ly*	*Sb*	*br*	404
(2)	+	+	+	422
(3)	*Ly*	+	+	18
(4)	+	*Sb*	*br*	16
(5)	*Ly*	+	*br*	75
(6)	+	*Sb*	+	59
(7)	*Ly*	*Sb*	+	4
(8)	+	+	*br*	2
				Total = 1000

Solution:

(a) Categories 1 and 2, which represent the noncrossover gametes.

(b) Categories 7 and 8, which represent the double crossover gametes.

(c) Categories 3, 4, 5, 6, 7, and 8, all of which arise from crossover events.

Problems and Discussion Questions

Mastering Genetics™ Visit for instructor-assigned tutorials and problems.

1. **HOW DO WE KNOW?** In this chapter, we focused on linkage, chromosomal mapping, and many associated phenomena. In the process, we found many opportunities to consider the methods and reasoning by which much of this information was acquired. From the explanations given in the chapter, what answers would you propose to the following fundamental questions?
   (a) How was it established experimentally that the frequency of recombination (crossing over) between two genes is related to the distance between them along the chromosome?
   (b) How do we know that specific genes are linked on a single chromosome, in contrast to being located on separate chromosomes?
   (c) How do we know that crossing over results from a physical exchange between chromatids?
   (d) How do we know that sister chromatids undergo recombination during mitosis?
2. **CONCEPT QUESTION** Review the Chapter Concepts list on p. 145. Most of these center on the process of crossing over between linked genes. Write a short essay that discusses how crossing over can be detected and how the resultant data provide the basis of chromosome mapping. ■
3. How does the process of crossing over take place and when does it occur?
4. Why does more crossing over occur between two distantly linked genes than between two genes that are very close together on the same chromosome?
5. Why is a 50 percent recovery of single-crossover products the upper limit, even when crossing over *always* occurs between two linked genes?
6. Why is crossover frequency not always directly proportional to the distance between two loci?
7. Explain the meaning of the term *interference*.
8. What three essential criteria must be met in order to execute a successful mapping cross?
9. The *Drosophila* genes *ebony body* (*e*), *hairless body* (*H*), and *groucho bristles* (*gro*) are linked on chromosome III. The genetic distances obtained in a series of two-point mapping experiments are shown below. Construct a map of these genes.

H–e	2 *mu*
H–gro	22 *mu*
gro–e	20 *mu*

10. Colored aleurone in the kernels of corn is due to the dominant allele *R*. The recessive allele *r*, when homozygous, produces colorless aleurone. The plant color (not kernel color) is controlled by another gene with two alleles, *Y* and *y*. The dominant *Y* allele results in green color, whereas the homozygous presence of the recessive *y* allele causes the plant to appear yellow. In a testcross between a plant of unknown genotype and phenotype and a plant that is homozygous recessive for both traits, the following progeny were obtained:

colored, green	88
colored, yellow	12
colorless, green	8
colorless, yellow	92

Explain how these results were obtained by determining the exact genotype and phenotype of the unknown plant, including the precise association of the two genes on the homologs (i.e., the arrangement).

11. In the cross shown here, involving two linked genes, *ebony* (*e*) and *claret* (*ca*), in *Drosophila,* where crossing over does not occur in males, offspring were produced in a 2+:1ca:1e phenotypic ratio:

$$\overset{\female}{\frac{e\ \ ca^{+}}{e^{+}\ ca}} \times \overset{\male}{\frac{e\ \ ca^{+}}{e^{+}\ ca}}$$

These genes are 30 mu apart on chromosome III. What did crossing over in the female contribute to these phenotypes?

12. In a series of two-point map crosses involving five genes located on chromosome II in *Drosophila,* the following recombinant (single-crossover) frequencies were observed:

pr–adp	29
pr–vg	13
pr–c	21
pr–b	6
adp–b	35
adp–c	8
adp–vg	16
vg–b	19
vg–c	8
c–b	27

(a) If the *adp* gene is present near the end of chromosome II (locus 83), construct a map of these genes.
(b) In another set of experiments, a sixth gene (*d*) was tested against *b* and *pr*, and the results were $d - b = 17\%$ and $d - pr = 23\%$ Predict the results of two-point maps between *d* and *c*, *d* and *vg*, and *d* and *adp*.

13. Two different female *Drosophila* were isolated, each heterozygous for the autosomally linked genes *black* body (*b*), *dachs* tarsus (*d*), and *curved* wings (*c*). These genes are in the order *d–b–c*, with *b* closer to *d* than to *c*. Shown in the following table is the genotypic arrangement for each female, along with the various gametes formed by both. Identify which categories are noncrossovers (NCO), single crossovers (SCO), and double crossovers (DCO) in each case. Then, indicate the relative frequency with which each will be produced.

Female A		Female B	
$\frac{d\ b\ +}{+\ +\ c}$		$\frac{d\ +\ +}{+\ b\ c}$	
↓	*Gamete formation*	↓	
(1) *d b c*	(5) *d* + +	(1) *d b* +	(5) *d b c*
(2) + + +	(6) + *b c*	(2) + + *c*	(6) + + +
(3) + + *c*	(7) *d* + *c*	(3) *d* + *c*	(7) *d* + +
(4) *d b* +	(8) + *b* +	(4) + *b* +	(8) + *b* C

14. In *Drosophila*, a cross was made between females—all expressing autosomal recessive traits *purple eyes* (*pr*), *corrugated* wings (*corr*), and *black body* (*b*)—and wild-type males. A testcross was performed with a female from the F_1, and 1652 offspring were counted, with the results shown in the following table.

Phenotype	Offspring
pr corr b	669
+ + +	683
pr + *b*	96
+ *corr* +	100
+ *corr* +	49
pr + +	54
pr corr +	1
+ + *b*	0

No determination of sex was made in the data.
(a) Using proper nomenclature, determine the genotypes of the P_1 females, the F_1 females, and the F_1 tester male.
(b) Map the three genes and determine the distance between them.
(c) Calculate the coefficient of coincidence.
(d) Is there interference? Explain.

15. Suppose we are analyzing three different genes in an imaginary insect with a diploid number of 8. Trait 1 pertains to wing shape and has two different alleles: *stout* and *dented*. Trait 2 pertains to tarsus shape and has two different alleles: *trimmed* and normal (wild-type). Trait 3 pertains to wing size and has two different alleles: *shrunken* and normal (wild-type). A cross is first set up with a dented, wild-type, *shrunken* male and a *stout*, trimmed, wild-type female. From the F_1, we obtain 132 male insects with *stout, trimmed,* and wild-type phenotypes and 127 female insects with *stout*, wild-type, and wild-type phenotypes.
(a) Based on the previous results, we can unequivocally assign a gene to a specific chromosome. Determine which alleles are dominant, and note the genotype of the F_1 female. Explain your reasoning.
(b) A second cross is set up with a female insect from the F_1 and a male insect homozygous for *dented, trimmed,* and *shrunken* alleles. The progeny of the cross is as follows.

Wing Shape	Tarsus Shape	Wing Size	Male	Female	Total
stout	trimmed	shrunken	6	3	9
stout	trimmed	wild type	30	26	56
stout	wild type	shrunken	9	4	13
stout	wild type	wild type	27	22	49
dented	trimmed	shrunken	28	30	58
dented	trimmed	wild type	6	4	10
dented	wild type	shrunken	26	25	51
dented	wild type	wild type	2	4	6

How can these results be explained and what conclusion can be derived from the data?

16. *Drosophila* females homozygous for the third chromosomal genes *pink* eye (*p*) and *ebony* body (*e*) were crossed with males homozygous for the second chromosomal gene *dumpy* wings (*dp*). Because these genes are recessive, all offspring were wild type (normal). F_1 females were testcrossed to triply recessive males. If we assume that the two linked genes (*p* and *e*) are 20 mu apart, predict the results of this cross. If the reciprocal cross were made (F_1 males—where no crossing over occurs—with triply recessive females), how would the results vary, if at all?

17. In *Drosophila,* the two mutations *Stubble* bristles (*Sb*) and *curled* wings (*cu*) are linked on chromosome III. *Sb* is a dominant gene

that is lethal in a homozygous state, and *cu* is a recessive gene. If a female of the genotype

$$\frac{Sb \quad cu}{+ \quad +}$$

is to be mated to detect recombinants among her offspring, what male genotype would you choose as her mate?

18. A female of genotype

$$\frac{a \quad b \quad c}{+ \quad + \quad +}$$

produces 100 meiotic tetrads. Of these, 68 show no crossover events. Of the remaining 32, 20 show a crossover between *a* and *b*, 10 show a crossover between *b* and *c*, and 2 show a double crossover between *a* and *b* and between *b* and *c*. Of the 400 gametes produced, how many of each of the eight different genotypes will be produced? Assuming the order *a*–*b*–*c* and the allele arrangement shown above, what is the map distance between these loci?

19. In a plant, fruit color is either red or yellow, and fruit shape is either oval or long. Red and oval are the dominant traits. Two plants, both heterozygous for these traits, were testcrossed, with the results shown in the following table. Determine the location of the genes relative to one another and the genotypes of the two parental plants.

	Progeny	
Phenotype	**Plant A**	**Plant B**
red, long	46	4
yellow, oval	44	6
red, oval	5	43
yellow, long	5	47
Total	100	100

20. In *Drosophila*, *Dichaete* (*D*) is a mutation on chromosome III with a dominant effect on wing shape. It is lethal when homozygous. The genes *ebony* body (*e*) and *pink* eye (*p*) are recessive mutations on chromosome III. Flies from a Dichaete stock were crossed to homozygous ebony, pink flies, and the F_1 progeny with a Dichaete phenotype were backcrossed to the ebony, pink homozygotes.
    (a) Using the results of this backcross shown in the following table, diagram the cross, showing the genotypes of the parents and offspring of both crosses.
    (b) What is the sequence and interlocus distance between these three genes?

Phenotype	Number
Dichaete	401
ebony, pink	389
Dichaete, ebony	84
pink	96
Dichaete, pink	2
ebony	3
Dichaete, ebony, pink	12
wild type	13

21. An organism of the genotype *AaBbCc* was testcrossed to a triply recessive organism (*aabbcc*). The genotypes of the progeny are in the following table.

AaBbCc	20	*AaBbcc*	20
aabbCc	20	*aabbcc*	20
AabbCc	5	*Aabbcc*	5
aaBbCc	5	*aaBbcc*	5

(a) Assuming simple dominance and recessiveness in each gene pair, if these three genes were all assorting independently, how many genotypic and phenotypic classes would result in the offspring, and in what proportion?
(b) Answer part (a) again, assuming the three genes are so tightly linked on a single chromosome that no crossover gametes were recovered in the sample of offspring.
(c) What can you conclude from the actual data about the location of the three genes in relation to one another?

22. Based on our discussion of the potential inaccuracy of mapping [see Figure 7.12, part (b)], would you revise your answer to Problem 21(c)? If so, how?
23. In Creighton and McClintock's experiment demonstrating that crossing over involves physical exchange between chromosomes (see Section 7.6), explain the importance of the cytological markers (the translocated segment and the chromosome knob) in the experimental rationale.
24. Explain why restriction fragment length polymorphisms and microsatellites are important landmarks for mapping purposes.
25. Consider a nonsister chromatid exchange between homologs in mitosis involving a heterozygous locus. Describe the result.

8

Genetic Analysis and Mapping in Bacteria and Bacteriophages

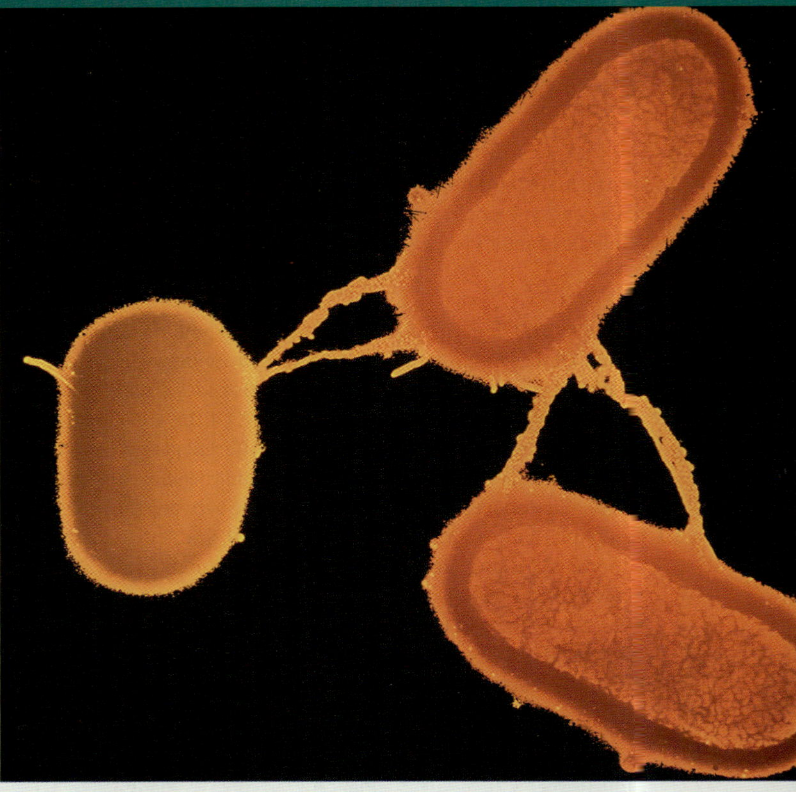

An electron micrograph showing the sex pilus between two conjugating *Escherichia coli* cells.

CHAPTER CONCEPTS

- Bacterial genomes are most often contained in a single circular chromosome.
- Bacteria have developed numerous ways in which they can exchange and recombine genetic information between individual cells, including conjugation, transformation, and transduction.
- The ability to undergo conjugation and to transfer a portion or all of the bacterial chromosome from one cell to another is governed by the presence of genetic information contained in the DNA of a "fertility," or F factor.
- The F factor can exist autonomously in the bacterial cytoplasm as a plasmid, or it can integrate into the bacterial chromosome, where it facilitates the transfer of the host chromosome to the recipient cell, leading to genetic recombination.
- Genetic recombination during conjugation provides a means of mapping bacterial genes.
- Bacteriophages are viruses that have bacteria as their hosts. During infection of the bacterial host, bacteriophage DNA is injected into the host cell, where it is replicated and directs the reproduction of the bacteriophage.
- Rarely, following infection, bacteriophage DNA integrates into the host chromosome, becoming a prophage, where it is replicated along with the bacterial DNA.

In this chapter, we shift from consideration of mapping genetic information in eukaryotes to discussion of the analysis and mapping of genes in **bacteria** (prokaryotes) and **bacteriophages,** viruses that use bacteria as their hosts. The study of bacteria and bacteriophages has been essential to the accumulation of knowledge in many areas of genetic study. For example, much of what we know about molecular genetics, recombinational phenomena, and gene structure was initially derived from experimental work with them. Furthermore, our extensive knowledge of bacteria and their resident plasmids has led to their widespread use in DNA cloning and other recombinant DNA studies.

Bacteria and their viruses are especially useful research organisms in genetics for several reasons. They have extremely short reproductive cycles—literally hundreds of generations, giving rise to billions of genetically identical bacteria or phages, can be produced in short periods of time. Furthermore, they can be studied in pure cultures. That is, a single species or mutant strain of bacteria or one type of virus can be isolated and investigated independently of other similar organisms.

In this chapter, we focus on genetic recombination and chromosome mapping. Complex processes have evolved in bacteria and bacteriophages that facilitate the transfer of genetic information between individual cells within populations. As we shall see, these processes are the basis for the chromosome mapping analysis that forms the cornerstone of molecular genetic investigations of bacteria and the viruses that invade them.

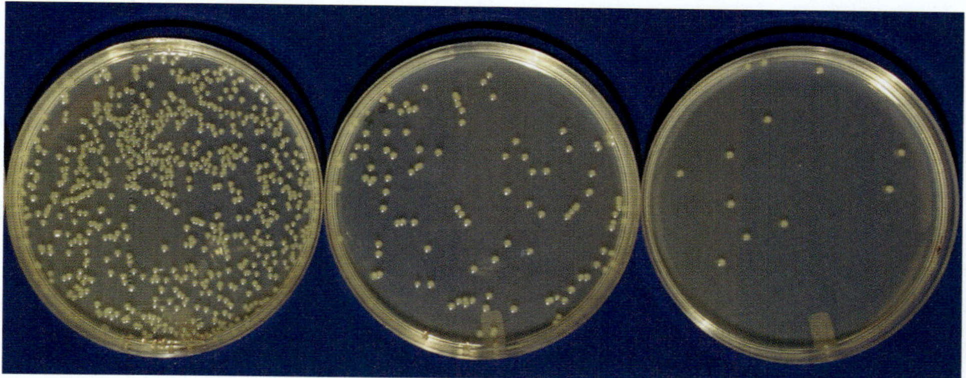

FIGURE 8.1 Results of the serial dilution technique and subsequent culture of bacteria. Each dilution varies by a factor of 10. Each colony is derived from a single bacterial cell.

8.1 Bacteria Mutate Spontaneously and Are Easily Cultured

It has long been known that pure cultures of bacteria give rise to cells that exhibit heritable variation, particularly with respect to growth under unique environmental conditions. Mutant cells that arise spontaneously in otherwise pure cultures can be isolated and established independently from the parent strain by using established selection techniques. As a result, mutations for almost any desired characteristic can now be isolated. Because bacteria and viruses usually contain only a single chromosome and are therefore haploid, all mutations are expressed directly in the descendants of mutant cells, adding to the ease with which these microorganisms can be studied.

Bacteria are grown in a liquid culture medium or in a petri dish on a semisolid agar surface. If the nutrient components of the growth medium are simple and consist only of an organic carbon source (such as glucose or lactose) and a variety of ions, including Na^+, K^+, Mg^{2+}, Ca^{2+}, and NH_4^+, present as inorganic salts, it is called **minimal medium.** To grow on such a medium, a bacterium must be able to synthesize all essential organic compounds (e.g., amino acids, purines, pyrimidines, vitamins, and fatty acids). A bacterium that can accomplish this remarkable biosynthetic feat—one that we ourselves cannot duplicate—is a **prototroph.** It is said to be wild-type for all growth requirements. On the other hand, if a bacterium loses the ability to synthesize one or more organic components through mutation, it is an **auxotroph.** For example, if a bacterium loses the ability to make histidine, then this amino acid must be added as a supplement to the minimal medium for growth to occur. The resulting bacterium is designated as an *his*$^-$ auxotroph, in contrast to its prototrophic *his*$^+$ counterpart.

To study bacterial growth quantitatively, an inoculum of bacteria is placed in liquid culture medium. Cells grown in liquid medium can be quantified by transferring them to a semisolid medium in a petri dish. Following incubation and many divisions, each cell gives rise to a visible colony on the surface of the medium. If the number of colonies is too great to count, then a series of successive dilutions (a technique called **serial dilution**) of the original liquid culture is made and plated, until the colony number is reduced to the point where it can be counted (**Figure 8.1**). This technique allows the number of bacteria present in the original culture to be calculated.

As an example, let's assume that the three dishes in Figure 8.1 represent serial dilutions of 10^{-3}, 10^{-4}, and 10^{-5} (from left to right). We need only select the dish in which the number of colonies can be counted accurately. Assuming that a 1-mL sample was used, and because each colony arose from a single bacterium, the number of colonies multiplied by the dilution factor represents the number of bacteria in each milliliter of the initial inoculum used to start the serial dilutions. In Figure 8.1, the rightmost dish has 12 colonies. The dilution factor for a 10^{-5} dilution is 10^5. Therefore, the initial number of bacteria was 12×10^5 per mL.

8.2 Genetic Recombination Occurs in Bacteria

Development of techniques that allowed the identification and study of bacterial mutations led to detailed investigations of the transfer of genetic information between individual organisms. As we shall see, as with meiotic crossing over in eukaryotes, the process of genetic recombination in bacteria provided the basis for the development of chromosome mapping methodology. It is important to note at the outset of our discussion that the term *genetic recombination,* as applied to bacteria, refers to the *replacement* of one or more genes present in the chromosome of one cell with those from the chromosome of a genetically distinct cell. While this is somewhat different from our use of the term in eukaryotes—where it describes *crossing over resulting in a reciprocal exchange*—the overall effect is the same: Genetic information is transferred, and it results in an altered genotype.

We will discuss three processes that result in the transfer of genetic information from one bacterium to another: *conjugation, transformation,* and *transduction*. Collectively, knowledge of these processes has helped us understand the origin of genetic variation between members of the same bacterial species, and in some cases, between members of different species. When transfer of genetic information occurs between generations of the same species, the term **vertical gene transfer** applies. When transfer occurs between unrelated cells, the term **horizontal gene transfer** is used. The horizontal gene transfer process has played a significant role in the evolution of bacteria. Often, the genes discovered to be involved in horizontal transfer are those that also confer survival advantages to the recipient species. For example, one species may transfer antibiotic resistance genes to another species. Or genes conferring enhanced pathogenicity may be transferred. Thus, the potential for such transfer is a major concern in the medical community. In addition, horizontal gene transfer has been a major factor in the process of speciation in bacteria. Many, if not most, bacterial species have been the recipient of genes from other species.

Conjugation in Bacteria: The Discovery of F^+ and F^- Strains

Studies of bacterial recombination began in 1946, when Joshua Lederberg and Edward Tatum showed that bacteria undergo **conjugation,** a process by which genetic information from one bacterium is transferred to and recombined with that of another bacterium. Their initial experiments were performed with two multiple auxotrophs (nutritional mutants) of *E. coli* strain K12. As shown in **Figure 8.2**, strain A required methionine (met) and biotin (bio) in order to grow, whereas strain B required threonine (thr), leucine (leu), and thiamine (thi). Neither strain would grow on minimal medium. The two strains were first grown separately in supplemented media, and then cells from both were mixed and grown together for several more generations. They were then plated on minimal medium. Any cells that grew on minimal medium were prototrophs. It is highly improbable that any of the cells containing two or three mutant genes would undergo spontaneous mutation simultaneously at two or three independent locations to become wild-type cells. Therefore, the researchers assumed that any prototrophs recovered must have arisen as a result of some form of genetic exchange and recombination between the two mutant strains.

In this experiment, prototrophs were recovered at a rate of $1/10^7$ (or 10^{-7}) cells plated. The controls for this experiment involved separate plating of cells from strains A and B on minimal medium. No prototrophs were recovered.

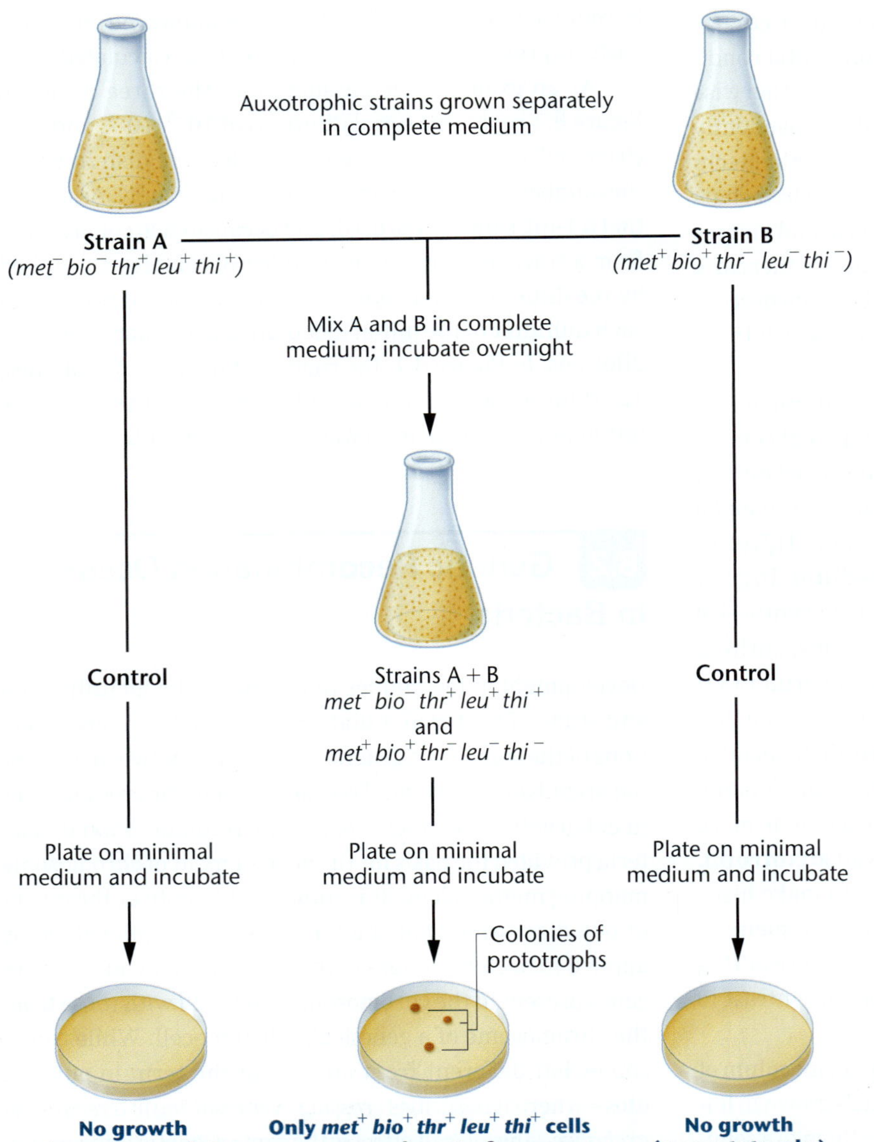

FIGURE 8.2 Genetic recombination of two auxotrophic strains producing prototrophs. Neither auxotroph grows on minimal medium, but prototrophs do, suggesting that genetic recombination has occurred.

Based on these observations, Lederberg and Tatum proposed that genetic exchange had occurred. Lederberg and Tatum's findings were soon followed by numerous experiments that elucidated the genetic basis of conjugation. It quickly became evident that different strains of bacteria are involved in a unidirectional transfer of genetic material. When cells serve as donors of parts of their chromosomes, they are designated as **F^- cells** (F for "fertility"). Recipient bacteria receive the donor chromosome material (now known to be DNA), and recombine it with part of their own chromosome. They are designated as **F^- cells.**

Experimentation subsequently established that cell contact is essential for chromosome transfer to occur. Support for this concept was provided by Bernard Davis, who designed the Davis U-tube for growing F^+ and F^- cells shown in **Figure 8.3**. At the base of the tube is a sintered glass filter with a pore size that allows passage of the liquid medium but that is too small to allow the passage of bacteria. The F^+ cells are placed on one side of the filter, and F^- cells on the other side. The medium is moved back and forth across the filter so that the cells share a common medium during bacterial incubation. When Davis plated samples from both sides of the tube on minimal medium, no prototrophs were found, and he logically concluded that *physical contact between cells of the two strains is essential to genetic recombination.* We now know that this physical interaction is the initial step in the process of conjugation established by a structure called the **F pilus** (or **sex pilus;** pl. pili). Bacteria often have many pili, which are tubular extensions of the cell. After contact is initiated between mating pairs (see the chapter opening photograph on p. 168), chromosome transfer is then possible.

Later evidence established that F^+ cells contain a **fertility factor (F factor)** that confers the ability to donate part of their chromosome during conjugation. Experiments by Joshua and Esther Lederberg and by William Hayes and Luca Cavalli-Sforza showed that certain conditions eliminate the F factor in otherwise fertile cells. However, if these "infertile" cells are then grown with fertile donor cells, the F factor is regained.

The conclusion that the F factor is a mobile element is further supported by the observation that, following conjugation and genetic recombination, recipient cells always become F^+. Thus, in addition to the *rare* cases of gene transfer from the bacterial chromosome (genetic recombination), the F factor itself is passed to *all* recipient cells. On this basis, the initial cross of Lederberg and Tatum (see Figure 8.2) can be interpreted as follows:

Strain A		Strain B
F^+	×	F^-
Donor		Recipient

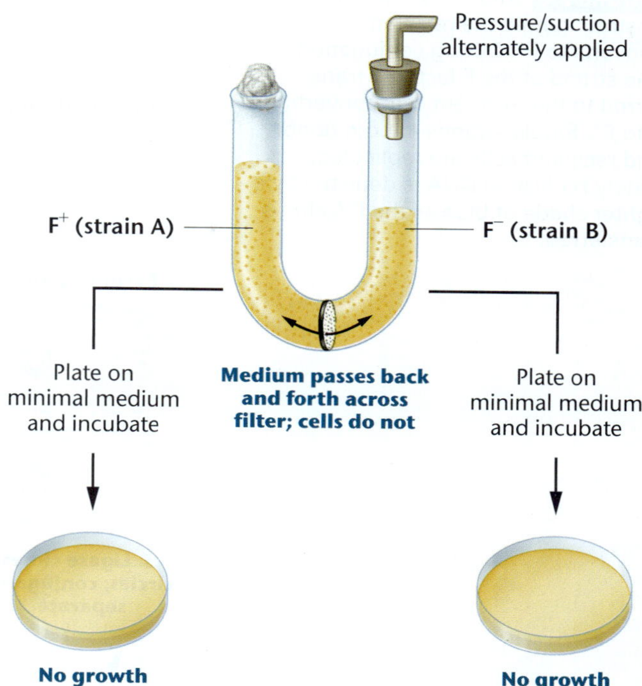

FIGURE 8.3 When strain A and B auxotrophs are grown in a common medium but separated by a filter, as in this Davis U-tube apparatus, no genetic recombination occurs and no prototrophs are produced.

Characterization of the F factor confirmed these conclusions. Like the bacterial chromosome, though distinct from it, the F factor has been shown to consist of a circular, double-stranded DNA molecule, equivalent to about 2 percent of the bacterial chromosome (about 100,000 nucleotide pairs). There are 19 genes contained within the F factor whose products are involved in the transfer of genetic information, excluding those involved in the formation of the sex pilus.

Geneticists believe that transfer of the F factor during conjugation involves separation of the two strands of its double helix and movement of one of the two strands into the recipient cell. Both strands, one moving across the conjugation tube and one remaining in the donor cell, are replicated. The result is that both the donor *and* the recipient cells become F^+. This process is diagrammed in **Figure 8.4**.

To summarize, an *E. coli* cell may or may not contain the F factor. When this factor is present, the cell is able to form a sex pilus and potentially serve as a donor of genetic information. During conjugation, a copy of the F factor is almost always transferred from the F^+ cell to the F^- recipient, converting the recipient to the F^+ state. The question remained as to exactly why such a low proportion of cells involved in these matings (10^{-7}) also results in genetic recombination. The answer awaited further experimentation.

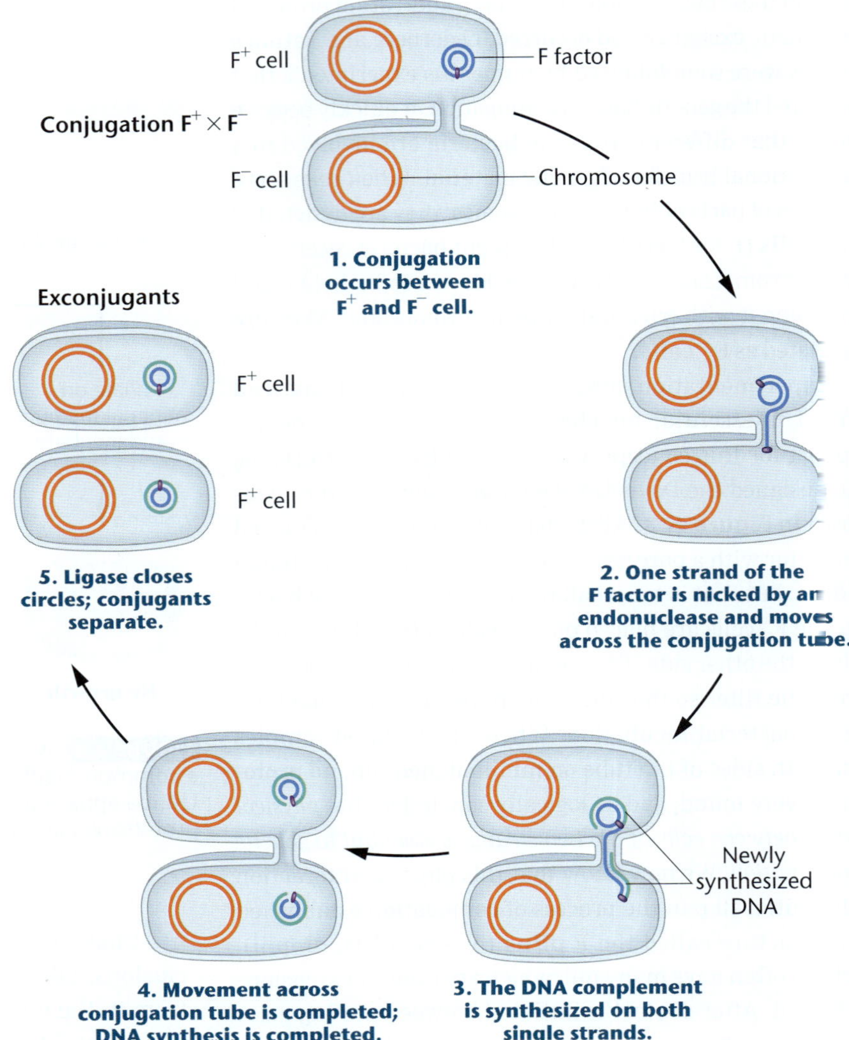

F⁺ cell
F factor
Conjugation F⁺ × F⁻
F⁻ cell
Chromosome
1. Conjugation occurs between F⁺ and F⁻ cell.
2. One strand of the F factor is nicked by an endonuclease and moves across the conjugation tube.
Newly synthesized DNA
3. The DNA complement is synthesized on both single strands.
4. Movement across conjugation tube is completed; DNA synthesis is completed.
Exconjugants
F⁺ cell
F⁺ cell
5. Ligase closes circles; conjugants separate.

FIGURE 8.4 An $F^+ \times F^-$ mating demonstrating how the recipient F^- cell converts to F^+. During conjugation, one strand of the F factor is transferred to the recipient cell, converting it to F^+. Single strands in both donor and recipient cells are replicated. Newly replicated DNA is depicted by a lighter shade of blue as the F factor is transferred.

As you soon shall see, the F factor is in reality an autonomous genetic unit called a *plasmid*. However, in our historical coverage of its discovery, we will continue to refer to it as a factor.

ESSENTIAL POINT

Conjugation may be initiated by a bacterium housing a plasmid called the F factor in its cytoplasm, making it a donor (F^+) cell. Following conjugation, the recipient (F^-) cell receives a copy of the F factor and is converted to the F^+ status. ■

Hfr Bacteria and Chromosome Mapping

Subsequent discoveries not only clarified how genetic recombination occurs but also defined a mechanism by which the *E. coli* chromosome could be mapped. Let's address chromosome mapping first.

In 1950, Cavalli-Sforza treated an F^+ strain of *E. coli* K12 with nitrogen mustard, a chemical known to induce mutations. From these treated cells, he recovered a genetically altered strain of donor bacteria that underwent recombination at a rate of $1/10^4$ (or 10^{-4}), 1000 times more frequently than the original F^+ strains. In 1953, Hayes isolated another strain that demonstrated an elevated frequency. Both strains were designated **Hfr,** for **high-frequency recombination.** Because Hfr cells behave as donors, they are a special class of F^+ cells.

Another important difference was noted between Hfr strains and the original F^+ strains. If the donor is from an Hfr strain, recipient cells, though sometimes displaying genetic recombination, almost never become Hfr; that is, they remain F^-. In comparison, then,

$$F^+ \times F^- \longrightarrow F^+ \text{ (low rate of recombination)}$$

$$\text{Hfr} \times F^- \longrightarrow F^- \text{ (higher rate of recombination)}$$

Perhaps the most significant characteristic of Hfr strains is the *nature of recombination.* In any given strain, certain genes are more frequently recombined than others, and some not at all. This *nonrandom* pattern was shown to vary between Hfr strains. Although these results were puzzling, Hayes interpreted them to mean that some physiological alteration of the F factor had occurred, resulting in the production of Hfr strains of *E. coli*.

In the mid-1950s, experimentation by Elie Wollman and François Jacob elucidated the difference between Hfr and F^+ strains and showed how Hfr strains allow genetic mapping of the *E. coli* chromosome. In their experiments, Hfr and F^- strains with suitable marker genes were mixed, and recombination of specific genes was assayed at different times. To accomplish this, a culture containing a mixture of an Hfr and an F^- strain was first incubated, and samples were removed at various intervals and placed in a blender. The shear forces in the blender separated conjugating bacteria so that the transfer of the chromosome was terminated. The cells were then assayed for genetic recombination.

This process, called the **interrupted mating technique,** demonstrated that specific genes of a given Hfr strain were transferred and recombined sooner than others. The graph in **Figure 8.5** illustrates this point. During the first 8 minutes after the two strains were mixed, no genetic recombination was detected. At about 10 minutes, recombination of the azi^R gene was detected, but no transfer of the ton^s, lac^+, or gal^+ genes was noted. By 15 minutes, 50 percent

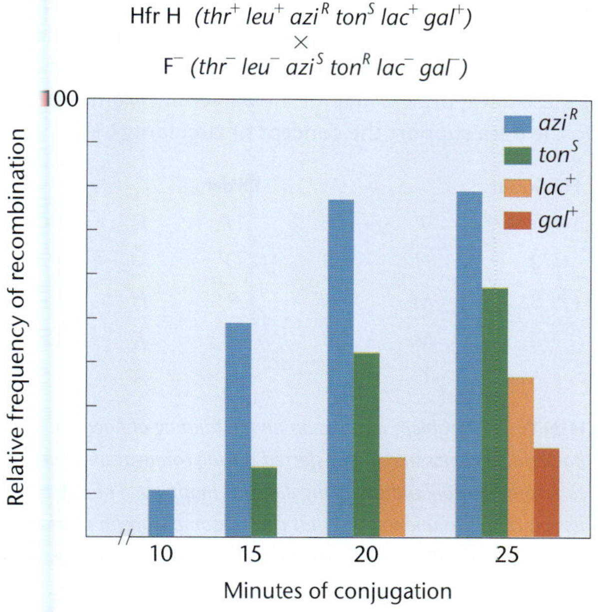

FIGURE 8.5 The progressive transfer during conjugation of various genes from a specific Hfr strain of *E. coli* to an F^- strain. Certain genes (*azi* and *ton*) transfer more quickly than others and recombine more frequently. Others (*lac* and *gal*) take longer to transfer, and recombinants are found at a lower frequency.

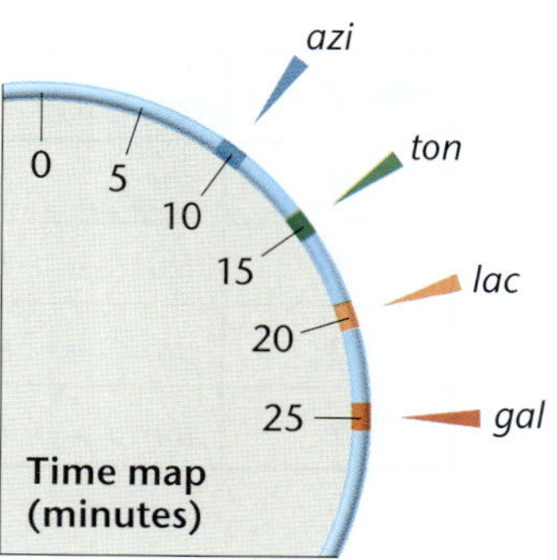

FIGURE 8.6 A time map of the genes studied in the experiment depicted in Figure 8.5.

of the recombinants were azi^R, and 15 percent were ton^s; but none were lac^+ or gal^+. Within 20 minutes, the lac^+ was found among the recombinants; and within 25 minutes, gal^+ was also being transferred. Wollman and Jacob had demonstrated *an ordered transfer of genes* that correlated with the length of time conjugation proceeded.

It appeared that the chromosome of the Hfr bacterium was transferred linearly and that the gene order and distance between genes, as measured in minutes, could be predicted from such experiments (**Figure 8.6**). This process, sometimes referred to as **time mapping,** served as the basis for the first genetic map of the *E. coli* chromosome. Minutes in bacterial mapping are similar to map units in eukaryotes.

Wollman and Jacob then repeated the same type of experiment with other Hfr strains, obtaining similar results with one important difference. Although genes were always transferred linearly with time, as in their original experiment, the order in which genes entered seemed to vary from Hfr strain to Hfr strain [**Figure 8.7(a)**]. When they reexamined the entry rate of genes, and thus the genetic maps for each strain, a definite pattern emerged. The major difference between each strain was simply the point of origin (*O*) and the direction in which entry proceeded from that point [**Figure 8.7(b)**].

To explain these results, Wollman and Jacob postulated that the *E. coli* chromosome is circular (a closed circle, with no free ends). If the point of origin (*O*) varies from strain to strain, a different sequence of genes will be transferred in each case. But what determines *O*? They proposed that *in various Hfr strains, the F factor integrates into the chromosome at different points and that its position determines the site of O.* A case of integration is shown in step 1 of **Figure 8.8.** During conjugation between an Hfr and an F^- cell, the position of the F factor determines the initial point of transfer (steps 2 and 3). Those genes adjacent to *O* are transferred first, and the F factor becomes the last part that can be transferred (step 4).

(a)

Hfr strain	(earliest)			Order of transfer				(latest)
H	*thr* –	*leu* –	*azi* –	*ton* –	*pro* –	*lac* –	*gal* –	*thi*
1	*leu* –	*thr* –	*thi* –	*gal* –	*lac* –	*pro* –	*ton* –	*azi*
2	*pro* –	*ton* –	*azi* –	*leu* –	*thr* –	*thi* –	*gal* –	*lac*
7	*ton* –	*azi* –	*leu* –	*thr* –	*thi* –	*gal* –	*lac* –	*pro*

(b)

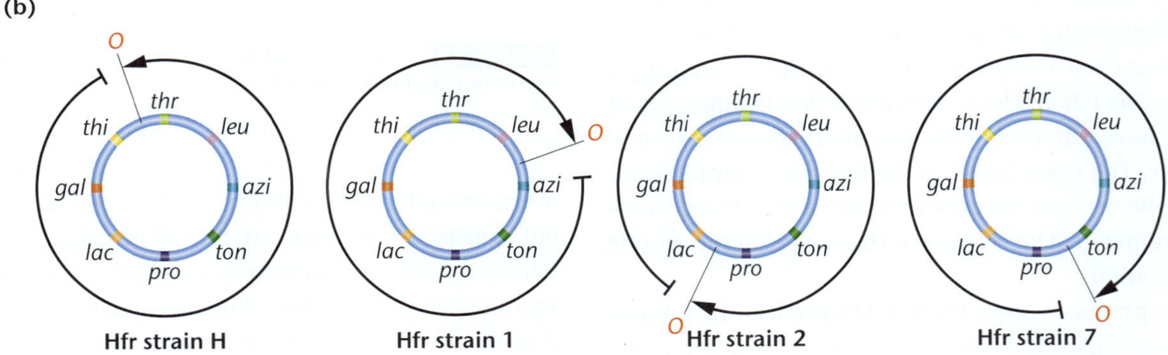

FIGURE 8.7 (a) The order of gene transfer in four Hfr strains, suggesting that the *E. coli* chromosome is circular. (b) The point where transfer originates (*O*) is identified in each strain. The origin is determined by the point of integration into the chromosome of the F factor, and the direction of transfer is determined by the orientation of the F factor as it integrates. The arrowheads indicate the points of initial transfer.

However, conjugation rarely, if ever, lasts long enough to allow the entire chromosome to pass across the conjugation tube (step 5). *This proposal explains why most recipient cells, when mated with Hfr cells, remain* F^-.

Figure 8.8 also depicts the way in which the two strands making up a DNA molecule behave during transfer, allowing for the entry of one strand of DNA into the recipient (see step 3). Following replication, the entering DNA now has the potential to recombine with its homologous region of the host chromosome. The DNA strand that remains in the donor also undergoes replication.

Use of the interrupted mating technique with different Hfr strains allowed researchers to map the entire *E. coli* chromosome. Mapped in time units, strain K12 (or *E. coli* K12) was shown to be 100 minutes long. While modern genome analysis of the *E. coli* chromosome has now established the presence of just over 4000 protein-coding sequences, this original mapping procedure established the location of approximately 1000 genes.

ESSENTIAL POINT

When the F factor is integrated into the donor cell chromosome (making it Hfr), the donor chromosome moves unidirectionally into the recipient, initiating recombination and providing the basis for time mapping of the bacterial chromosome. ■

NOW SOLVE THIS

8.1 When the interrupted mating technique was used with five different strains of Hfr bacteria, the following orders of gene entry and recombination were observed. On the basis of these data, draw a map of the bacterial chromosome. Do the data support the concept of circularity?

Hfr Strain			Order		
1	*T*	*C*	*H*	*R*	*O*
2	*H*	*R*	*O*	*M*	*B*
3	*M*	*O*	*R*	*H*	*C*
4	*M*	*B*	*A*	*K*	*T*
5	*C*	*T*	*K*	*A*	*B*

■ **HINT:** *This problem involves an understanding of how the bacterial chromosome is transferred during conjugation, leading to recombination and providing data for mapping. The key to its solution is to understand that chromosome transfer is strain-specific and depends on where in the chromosome, and in which orientation, the F factor has integrated.*

Recombination in $F^+ \times F^-$ Matings: A Reexamination

The preceding experiment helped geneticists better understand how genetic recombination occurs during $F^+ \times F^-$

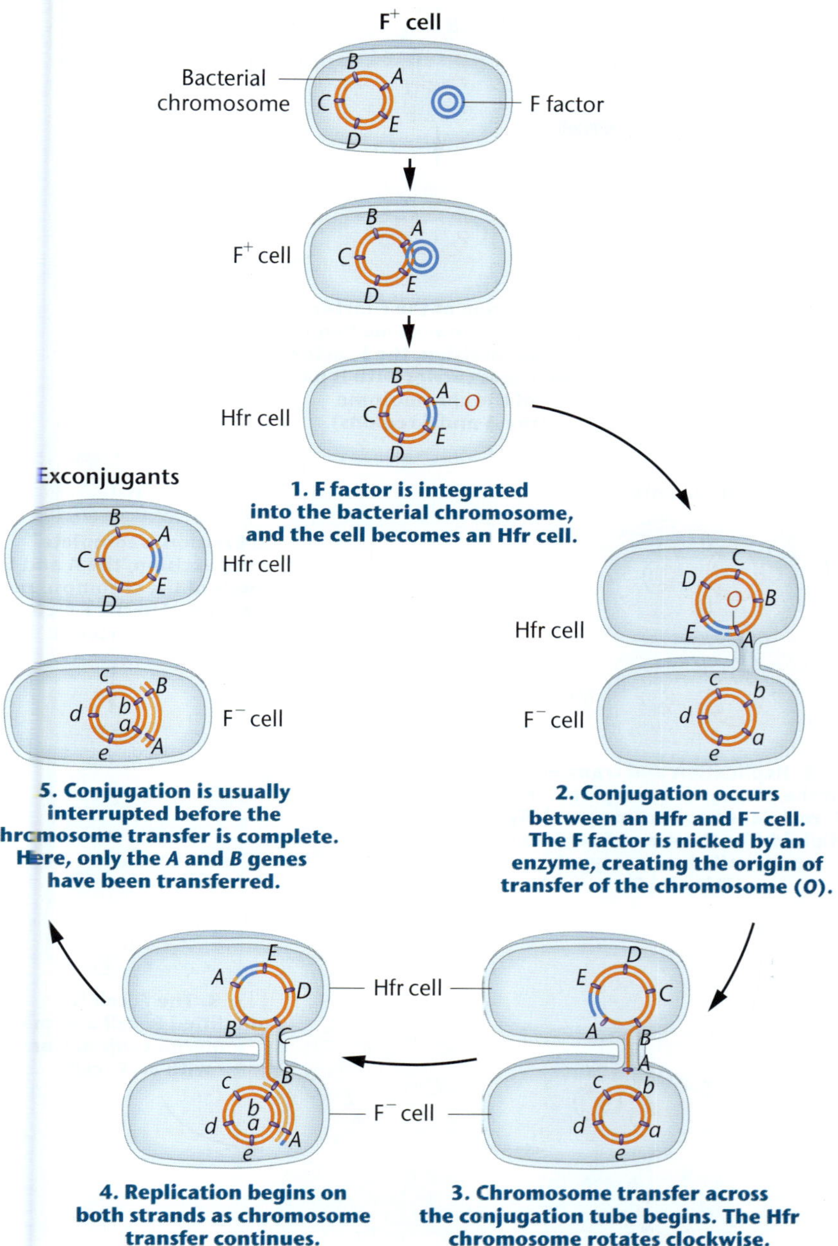

F⁺ cell
Bacterial chromosome
F factor
F⁺ cell
Hfr cell
1. F factor is integrated into the bacterial chromosome, and the cell becomes an Hfr cell.
Hfr cell
F⁻ cell
2. Conjugation occurs between an Hfr and F⁻ cell. The F factor is nicked by an enzyme, creating the origin of transfer of the chromosome (O).
Hfr cell
F⁻ cell
3. Chromosome transfer across the conjugation tube begins. The Hfr chromosome rotates clockwise.
Hfr cell
F⁻ cell
4. Replication begins on both strands as chromosome transfer continues.
Exconjugants
Hfr cell
F⁻ cell
5. Conjugation is usually interrupted before the chromosome transfer is complete. Here, only the A and B genes have been transferred.

FIGURE 8.8 Conversion of F$^+$ to an Hfr state occurs by integrating the F factor into the bacterial chromosome. The point of integration determines the origin (O) of transfer. During conjugation, an enzyme nicks the F factor, now integrated into the host chromosome, initiating transfer of the chromosome at that point. Conjugation is usually interrupted prior to complete transfer. Only the A and B genes are transferred to the F$^-$ cell, which may recombine with the host chromosome. Newly replicated DNA of the chromosome is depicted by a lighter shade of orange.

matings. Recall that recombination occurs much less frequently than in Hfr $\times$ F$^-$ matings and that random gene transfer is involved. The current belief is that when F$^+$ and F$^-$ cells are mixed, conjugation occurs readily and each F$^-$ cell involved in conjugation with an F$^+$ cell receives a copy of the F factor, *but no genetic recombination occurs.* However, at an extremely low frequency in a population of F$^+$ cells, the F factor integrates spontaneously from the cytoplasm to a random point in the bacterial chromosome, converting the F$^+$ cell to the Hfr state, as we saw in Figure 8.8. Therefore, in F$^+$ $\times$ F$^-$ matings, the extremely low frequency of genetic recombination (10^{-7}) is attributed to the rare, newly formed Hfr cells, which then undergo conjugation with F$^-$ cells. Because the point of integration of the F factor is random, the gene or genes that are transferred by any newly formed Hfr donor *will also appear to be random within the larger F$^+$/F$^-$ population.* The recipient bacterium will appear as a recombinant but will remain F$^-$. If it subsequently undergoes conjugation with an F$^+$ cell, it will then be converted to F$^+$.

The F′ State and Merozygotes

In 1959, during experiments with Hfr strains of *E. coli,* Edward Adelberg discovered that the F factor could lose its integrated status, causing the cell to revert to the F$^+$ state (**Figure 8.9**, Step 1). When this occurs, the F factor frequently carries several adjacent bacterial genes along with

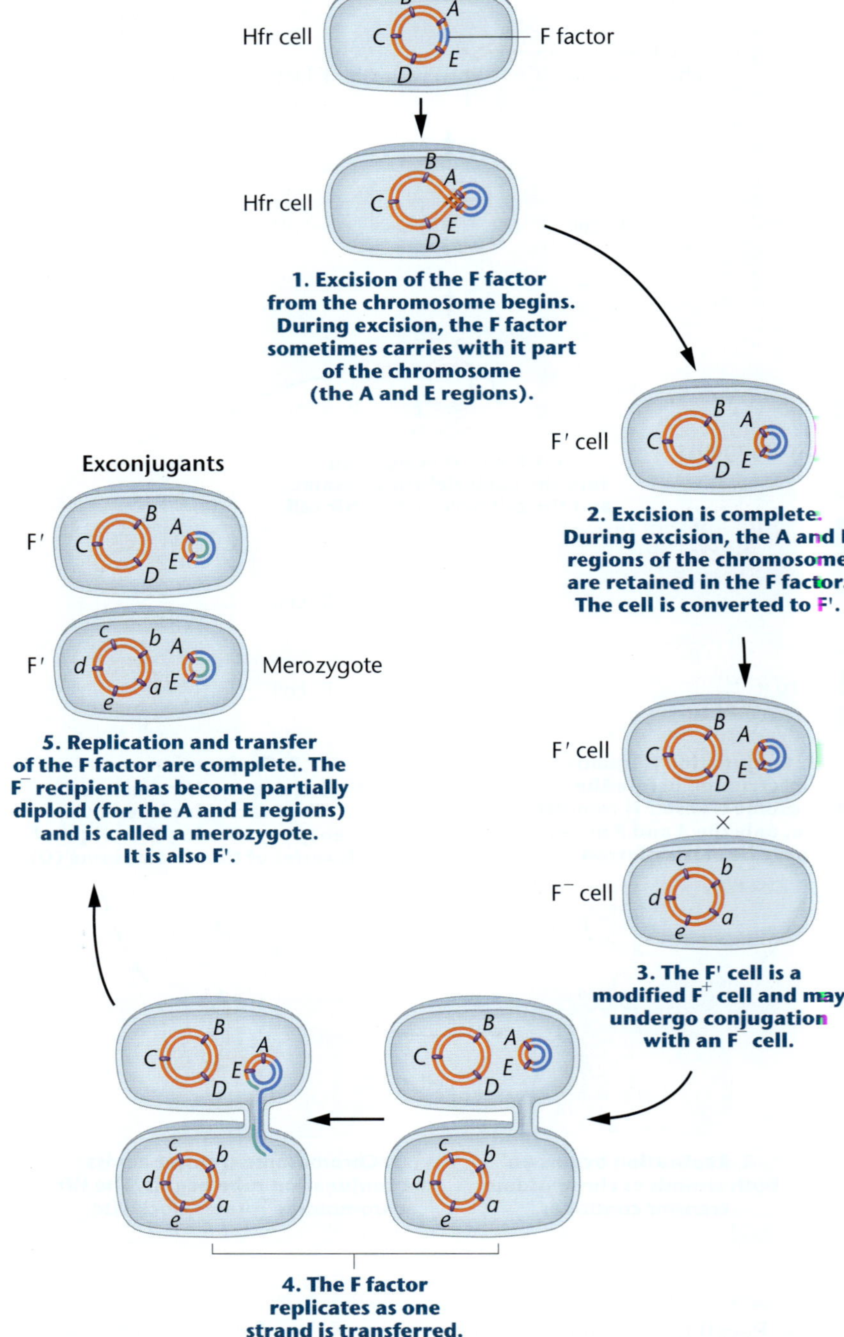

FIGURE 8.9 Conversion of an Hfr bacterium to F′ and its subsequent mating with an F⁻ cell. The conversion occurs when the F factor loses its integrated status. During excision from the chromosome, it carries with it one or more chromosomal genes (*A* and *E*). Following conjugation with an F⁻ cell, the recipient cell becomes partially diploid and is called a merozygote; it also behaves as an F⁺ donor cell.

it (Step 2). Adelberg labeled this condition **F′** to distinguish it from F⁺ and Hfr. F′, like Hfr, is thus another special case of F⁺, but this conversion is from Hfr to F′.

The presence of bacterial genes within a cytoplasmic F factor creates an interesting situation. An F′ bacterium behaves like an F⁺ cell by initiating conjugation with F⁻ cells (Figure 8.9, Step 3). When this occurs, the F factor, containing chromosomal genes, is transferred to the F⁻ cell (Step 4). As a result, whatever chromosomal genes are part of the F factor are now present as duplicates in the recipient cell (Step 5) because the recipient still has a complete chromosome. This creates a partially diploid cell called a **merozygote.** Pure cultures of F′ merozygotes can be established. They have been extremely useful in the study of bacterial genetics, particularly in genetic regulation.

8.3 The F Factor Is an Example of a Plasmid

The preceding sections introduced the extrachromosomal heredity unit required for conjugation called the F factor. When it exists autonomously in the bacterial cytoplasm,

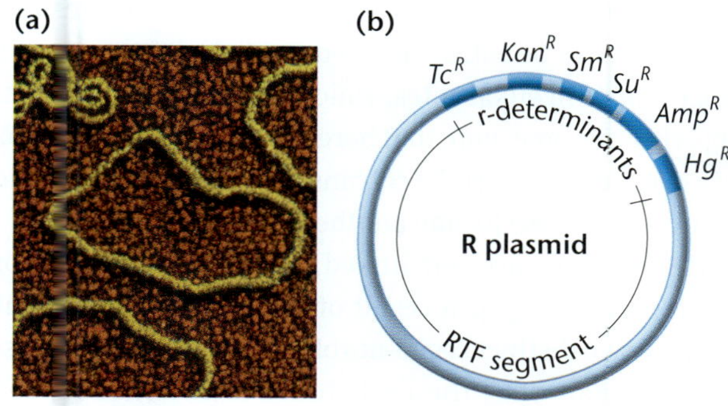

FIGURE 8.10 (a) Electron micrograph of a plasmid isolated from *E. coli.* (b) An R plasmid containing resistance transfer factors (RTFs) and multiple r-determinants (Tc, tetracycline; Kan, kanamycin; Sm, streptomycin; Su, sulfonamide; Amp, ampicillin; and Hg, mercury).

the F factor is composed of a double-stranded closed circle of DNA. These characteristics place the F factor in the more general category of genetic structures called **plasmids** [**Figure 8.10(a)**]. These structures contain one or more genes and often, quite a few. Their replication depends on the same enzymes that replicate the chromosome of the host cell, and they are distributed to daughter cells along with the host chromosome during cell division.

Plasmids are generally classified according to the genetic information specified by their DNA. The F factor plasmid confers fertility and contains the genes essential for sex pilus formation, on which genetic recombination depends. Other examples of plasmids include the R and Col plasmids.

Most **R plasmids** consist of two components: the **resistance transfer factor (RTF)** and one or more **r-determinants** [**Figure 8.10(b)**]. The RTF encodes genetic information essential to transferring the plasmid between bacteria, and the r-determinants are genes that confer resistance to antibiotics or mercury. While RTFs are similar in a variety of plasmids from different bacterial species, r-determinants are specific for resistance to one class of antibiotic and vary widely. Resistance to tetracycline, streptomycin, ampicillin, sulfonamide, kanamycin, or chloramphenicol is most frequently encountered. Sometimes several r-determinants occur in a single plasmid, conferring multiple resistance to several antibiotics [**Figure 8.10(b)**]. Bacteria bearing these plasmids are of great medical significance not only because of their multiple resistance but because of the ease with which the plasmids can be transferred to other bacteria.

The first known case of such a plasmid occurred in Japan in the 1950s in the bacterium *Shigella*, which causes dysentery. In hospitals, bacteria were isolated that were resistant to as many as five of the above antibiotics. Obviously, this phenomenon represents a major health threat. Fortunately, a bacterial cell sometimes contains r-determinant plasmids but no RTF. Although such a cell is resistant, it cannot transfer the genetic information for resistance to recipient cells. The most commonly studied plasmids, however, contain the RTF as well as one or more r-determinants.

The **Col plasmid,** ColE1 (derived from *E. coli*), is clearly distinct from the R plasmid. It encodes one or more proteins that are highly toxic to bacterial strains that do not harbor the same plasmid. These proteins, called **colicins,** can kill neighboring bacteria, and bacteria that carry the plasmid are said to be *colicinogenic*. Present in 10 to 20 copies per cell, a gene in the Col plasmid encodes an immunity protein that protects the host cell from the toxin. Unlike an R plasmid, the Col plasmid is not usually transmissible to other cells.

Interest in plasmids has increased dramatically because of their role in recombinant DNA research. As we will see in Chapter 17, specific genes from any source can be inserted into a plasmid, which may then be inserted into a bacterial cell. As the altered cell replicates its DNA and undergoes division, the foreign gene is also replicated, thus cloning the foreign genes.

ESSENTIAL POINT

Plasmids, such as the F factor, are autonomously replicating DNA molecules found in the bacterial cytoplasm, sometimes containing unique genes conferring antibiotic resistance as well as the genes necessary for plasmid transfer during conjugation. ■

8.4 Transformation Is Another Process Leading to Genetic Recombination in Bacteria

Transformation provides another mechanism for recombining genetic information in some bacteria. Small pieces of extracellular DNA are taken up by a living bacterium, potentially leading to a stable genetic change in the recipient cell. We discuss transformation in this chapter because in those bacterial species where it occurs, the process can be used to map bacterial genes, though in a more limited way than conjugation. Transformation has also played a central role in experiments proving that DNA is the genetic material.

The process of transformation consists of numerous steps divided into two categories: (1) entry of DNA into a recipient cell and (2) recombination of the donor DNA with its homologous region in the recipient chromosome. In a population of bacterial cells, only those in the particular physiological state of **competence** take up DNA. Entry is thought to occur at a limited number of receptor sites on the surface of the bacterial cell. Passage into the cell is an active process that requires energy and specific transport molecules. This model is supported by the fact that substances that inhibit energy production or protein synthesis in the recipient cell also inhibit the transformation process.

During entry, one of the two strands of the double helix is digested by nucleases, leaving only a single strand to participate in transformation. The surviving strand of DNA then aligns with its complementary region of the bacterial chromosome. In a process involving several enzymes, the segment replaces its counterpart in the chromosome, which is excised and degraded. For recombination to be detected, the transforming DNA must be derived from a different strain of bacteria that bears some genetic variation, such as a mutation. Once it is integrated into the chromosome, the recombinant region contains one host strand (present originally) and one mutant strand. Because these strands are from different sources, this helical region is referred to as a **heteroduplex.** Following one round of DNA replication, one chromosome is restored to its original configuration, and the other contains the mutant gene. Following cell division, one untransformed cell (nonmutant) and one transformed cell (mutant) are produced.

Transformation and Linked Genes

In early transformation studies, the most effective exogenous DNA contained 10,000 to 20,000 nucleotide pairs, a length sufficient to encode several genes.* Genes adjacent to or very close to one another on the bacterial chromosome can be carried on a single segment of this size. Consequently, a single transfer event can result in the **cotransformation** of several genes simultaneously. Genes that are close enough to each other to be cotransformed are *linked*. In contrast to *linkage groups* in eukaryotes, which consist of all genes on a single chromosome, note that here *linkage* refers to the proximity of genes that permits cotransformation (i.e., the genes are next to, or close to, one another).

If two genes are not linked, simultaneous transformation occurs only as a result of two independent events involving two distinct segments of DNA. As in double crossovers in eukaryotes, the probability of two independent events occurring simultaneously is equal to the product of the individual probabilities. Thus, the frequency of two unlinked genes being transformed simultaneously is much lower than if they are linked.

NOW SOLVE THIS

8.2 In a transformation experiment involving a recipient bacterial strain of genotype a^-b^-, the following results were obtained. What can you conclude about the location of the *a* and *b* genes relative to each other?

	Transformants (%)		
Transforming DNA	a^+b^-	a^-b^+	a^+b^+
a^+b^+	3.1	1.2	0.04
a^+b^- and a^-b^+	2.4	1.4	0.03

■ **HINT:** *This problem involves an understanding of how transformation can be used to determine if bacterial genes are closely "linked." You are asked to predict the location of two genes relative to one another. The key to its solution is to understand that cotransformation (of two genes) occurs according to the laws of probability. Two "unlinked" genes are transformed only as a result of two independent events. In such a case, the probability of that occurrence is equal to the product of the individual probabilities.*

* Today, we know that a 2000 nucleotide pair length of DNA is highly effective in gene cloning experiments.

ESSENTIAL POINT

Transformation in bacteria, which does not require cell-to-cell contact, involves exogenous DNA that enters a recipient bacterium and recombines with the host's chromosome. Linkage mapping of closely aligned genes is possible during the analysis of transformation. ■

8.5 Bacteriophages Are Bacterial Viruses

Bacteriophages, or **phages** as they are commonly known, are viruses that have bacteria as their hosts. During their reproduction, phages can be involved in still another mode of bacterial genetic recombination called *transduction*. To understand this process, we must consider the genetics of bacteriophages, which themselves undergo recombination.

A great deal of genetic research has been done using bacteriophages as a model system, making them a worthy subject of discussion. In this section, we will first examine the structure and life cycle of one type of bacteriophage. We then discuss how these phages are studied during their infection of bacteria. Finally, we contrast two possible modes of behavior once the initial phage infection occurs. This information is background for our discussion of *transduction*.

Phage T4: Structure and Life Cycle

Bacteriophage T4, which has *E. coli* as its host, is one of a group of related bacterial viruses referred to as T-even phages. It exhibits the intricate structure shown in **Figure 8.11**. Its genetic material, DNA, is contained within an icosahedral (referring to a polyhedron with 20 faces) protein coat, making up the head of the virus. The DNA is sufficient in quantity to encode more than 150 average-sized genes. The head is connected to a complex tail structure consisting of a collar, an outer contractile sheath that surrounds

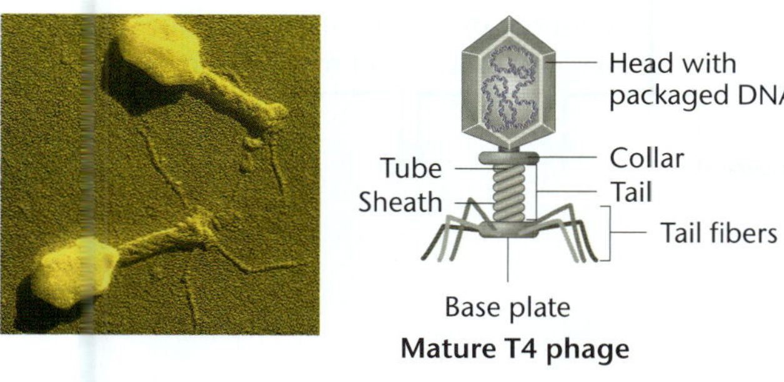
Head with packaged DNA
Collar
Tail
Tube
Sheath
Tail fibers
Base plate
Mature T4 phage

FIGURE 8.11 The structure of bacteriophage T4 includes an icosahedral head filled with DNA, a tail consisting of a collar, tube, sheath, base plate, and tail fibers. During assembly, the tail components are added to the head, and then tail fibers are added.

an inner spike-like tube, which sits atop a base plate from which tail fibers protrude. The base plate is an extremely complex structure, consisting of 15 different proteins, most present in multiple copies. The base plate coordinates the host cell recognition and is involved in providing the signal whereby the outer sheath contracts, propelling the inner tube across the cell membrane of the host cell.

The life cycle of phage T4 (**Figure 8.12**) is initiated when the virus binds to the bacterial host cell. Then, during contraction of the outer sheath, the DNA in the head is extruded, and it moves across the cell membrane into the bacterial cytoplasm. Within minutes, all bacterial DNA, RNA, and protein synthesis in the host cell is inhibited, and synthesis of viral molecules begins. At the same time, degradation of the host DNA is initiated.

A period of intensive viral gene activity characterizes infection. Initially, phage DNA replication occurs, leading to a pool of viral DNA molecules. Then, the components of the head, tail, and tail fibers are synthesized. The assembly of mature viruses is a complex process that has been well studied by William Wood, Robert Edgar, and others. Three sequential pathways occur: (1) DNA packaging as the viral heads are assembled, (2) tail assembly, and (3) tail fiber assembly. Once DNA is packaged into the head, it combines with the tail components, to which tail fibers are added. Total construction is a combination of self-assembly and enzyme-directed processes.

When approximately 200 new viruses have been constructed, the bacterial cell is ruptured by the action of the enzyme lysozyme (a phage gene product), and the mature phages are released from the host cell. The new phages infect other available bacterial cells, and the process repeats itself over and over again.

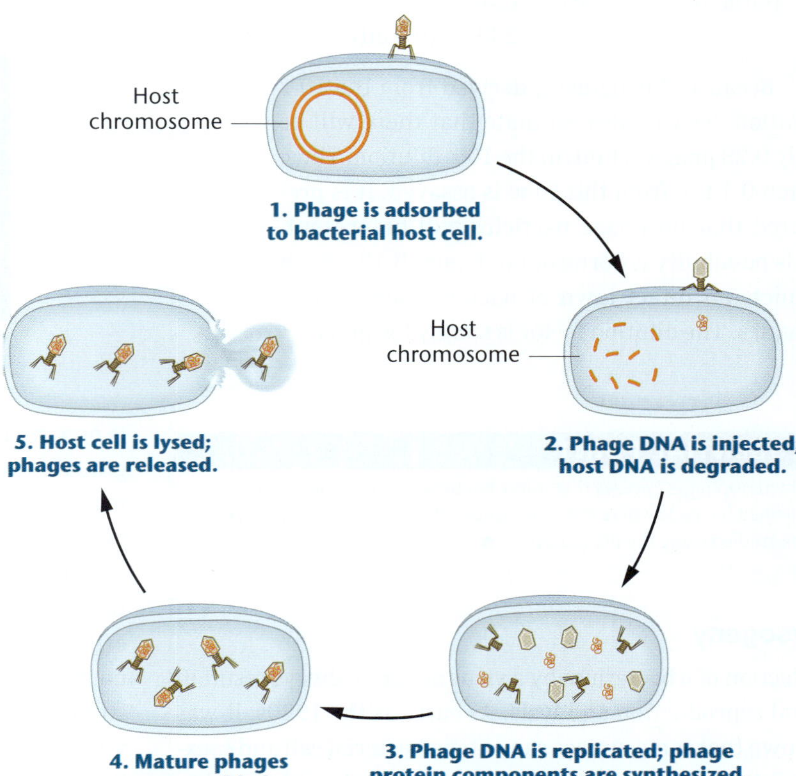
Host chromosome
1. Phage is adsorbed to bacterial host cell.
Host chromosome
2. Phage DNA is injected; host DNA is degraded.
3. Phage DNA is replicated; phage protein components are synthesized.
4. Mature phages are assembled.
5. Host cell is lysed; phages are released.

FIGURE 8.12 Life cycle of bacteriophage T4.

The Plaque Assay

Bacteriophages and other viruses have played a critical role in our understanding of molecular genetics. During infection of bacteria, enormous quantities of bacteriophages can be obtained for investigation. Often, over 10^{10} viruses are produced per milliliter of culture medium. Many genetic studies rely on the ability to quantify the number of phages produced following infection under specific culture conditions. The **plaque assay** is a routinely used technique, which is invaluable in mutational and recombinational studies of bacteriophages.

This assay is shown in **Figure 8.13**, where actual plaque morphology is also illustrated. A serial dilution of the original virally infected bacterial culture is performed first. Then, a 0.1-mL sample (an *aliquot*) from a dilution is added to melted nutrient agar (about 3 mL) into which a few drops of a healthy bacterial culture have been added. The solution is then poured evenly over a base of solid nutrient agar in a petri dish and allowed to solidify before incubation. A clear area called a **plaque** occurs wherever a single virus initially infected one bacterium in the culture (the lawn) that has grown up during incubation. The plaque represents clones of the single infecting bacteriophage, created as reproduction cycles are

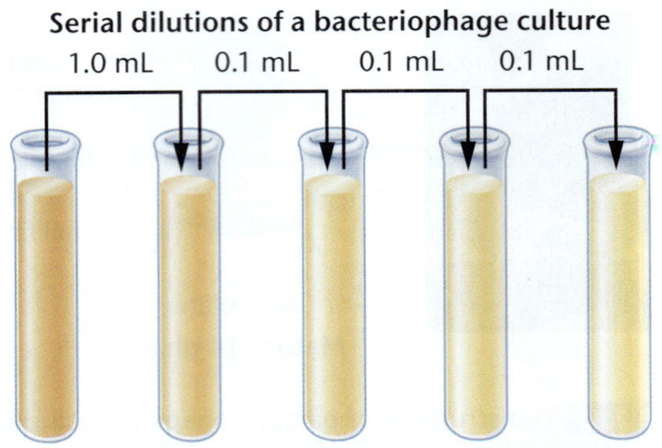

Total volume	10 mL	10 mL	10 mL	10 mL	10 mL
Dilution	0	10^{-1}	10^{-3}	10^{-5}	10^{-7}
Dilution factor	0	10	10^3	10^5	10^7

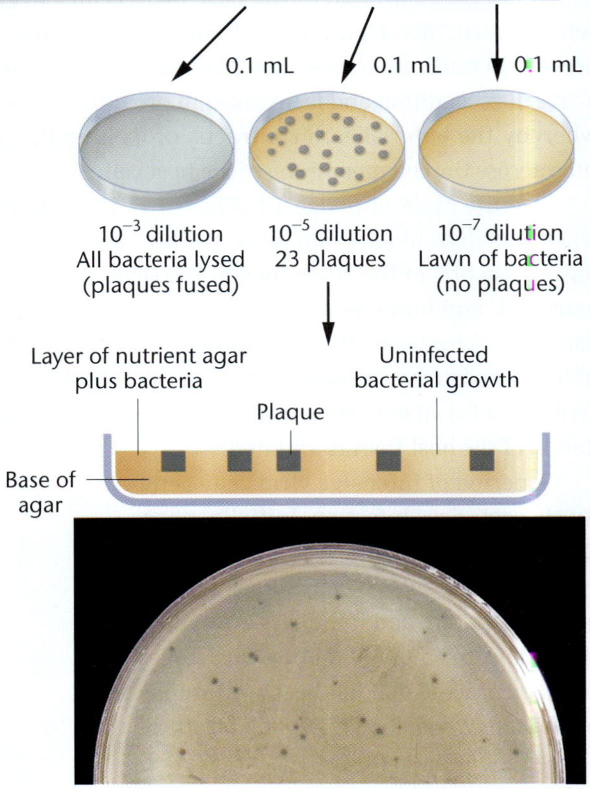

FIGURE 8.13 A plaque assay for bacteriophage analysis. Serial dilutions of a bacterial culture infected with bacteriophages are first made. Then three of the dilutions (10^{-3}, 10^{-5}, and 10^{-7}) are analyzed using the plaque assay technique. Each plaque represents the initial infection of one bacterial cell by one bacteriophage. In the 10^{-3} dilution, so many phages are present that all bacteria are lysed. In the 10^{-5} dilution, 23 plaques are produced. In the 10^{-7} dilution, the dilution factor is so great that no phages are present in the 0.1-mL sample, and thus no plaques form. From the 0.1-mL sample of the 10^{-5} dilution, the original bacteriophage density is calculated to be (230/mL) × (10^5) phages/mL (23 × 10^6, or 2.3 × 10^7). The photograph shows phage plaques on a lawn of *E. coli*.

repeated. If the dilution factor is too low, the plaques are plentiful, and they will fuse, lysing the entire lawn—which has occurred in the 10^{-3} dilution of Figure 8.13. On the other hand, if the dilution factor is increased, plaques can be counted and the density of viruses in the initial culture can be estimated as

$$\text{initial phage density} = (\text{plaque number/mL}) \times (\text{dilution factor})$$

Using the results shown in Figure 8.13, 23 phage plaques are derived from the 0.1-mL aliquot of the 10^{-5} dilution. Therefore, we estimate that there are 230 phages/mL *at this dilution* (since the initial aliquot was 0.1 mL). The initial phage density in the undiluted sample, factoring in the 10^{-5} dilution, is then calculated as

$$\begin{aligned}\text{Initial phage density} &= (230/\text{mL}) \times (10^5)\\ &= 230 \times 10^5/\text{mL}\end{aligned}$$

Because this figure is derived from the 10^{-5} dilution, we can also estimate that there will be only 0.23 phage/0.1 mL in the 10^{-7} dilution. Thus, when 0.1 mL from this tube is assayed, it is predicted that no phage particles will be present. This possibility is borne out in Figure 8.13, which depicts an intact lawn of bacteria lacking any plaques. The dilution factor is simply too great.

ESSENTIAL POINT

Bacteriophages (viruses that infect bacteria) demonstrate a well-defined life cycle where they reproduce within the host cell and can be studied using the plaque assay. ■

Lysogeny

Infection of a bacterium by a virus does not always result in viral reproduction and lysis. As early as the 1920s, it was known that some viruses can enter a bacterial cell and coexist with it. The precise molecular basis of this relationship is now well understood. Upon entry, the viral DNA is integrated into the bacterial chromosome instead of replicating in the bacterial cytoplasm, a step that characterizes the developmental stage referred to as **lysogeny.** Subsequently, each time the bacterial chromosome is replicated, the viral DNA is also replicated and passed to daughter bacterial cells following division. No new viruses are produced, and no lysis of the bacterial cell occurs. However, under certain stimuli, such as chemical or ultraviolet light treatment, the viral DNA loses its integrated status and initiates replication, phage reproduction, and lysis of the bacterium.

Several terms are used to describe this relationship. The viral DNA that integrates into the bacterial chromosome is called a **prophage.** Viruses that either lyse the cell or behave as a prophage are **temperate phages.** Those that only lyse the cell are referred to as **virulent phages.** A bacterium harboring a prophage is said to be **lysogenic;** that is, it is capable of being lysed as a result of induced viral reproduction.

ESSENTIAL POINT

Bacteriophages can be lytic, meaning they infect the host cell, reproduce, and then lyse it, or in contrast, they can lysogenize the host cell, where they infect it and integrate their DNA into the host chromosome, but do not reproduce. ■

8.6 Transduction Is Virus-Mediated Bacterial DNA Transfer

In 1952, Norton Zinder and Joshua Lederberg were investigating possible recombination in the bacterium *Salmonella typhimurium.* Although they recovered prototrophs from mixed cultures of two different auxotrophic strains, investigation revealed that recombination was occurring in a manner different from that attributable to the presence of an F factor, as in *E. coli.* What they had discovered was a process of bacterial recombination mediated by bacteriophages and now called **transduction.**

The Lederberg–Zinder Experiment

Lederberg and Zinder mixed the *Salmonella* auxotrophic strains LA-22 and LA-2 together, and when the mixture was plated on minimal medium, they recovered prototrophic cells. The LA-22 strain was unable to synthesize the amino acids phenylalanine and tryptophan (*phe*$^-$, *trp*$^-$), and LA-2 could not synthesize the amino acids methionine and histidine (*met*$^-$, *his*$^-$). Prototrophs (*phe*$^+$, *trp*$^+$, *met*$^+$, *his*$^+$) were recovered at a rate of about $1/10^5$ (10^{-5}) cells.

Although these observations at first suggested that the recombination involved was the type observed earlier in conjugative strains of *E. coli,* experiments using the Davis U-tube soon showed otherwise (**Figure 8.14**). The two auxotrophic strains were separated by a sintered glass filter, thus preventing cell contact but allowing growth to occur in a common medium. Surprisingly, when samples were removed from both sides of the filter and plated independently on minimal medium, prototrophs *were* recovered, but only from the side of the tube containing LA-22 bacteria. Recall that if conjugation were responsible, the conditions in the Davis U-tube would be expected to *prevent* recombination altogether (see Figure 8.3).

Since LA-2 cells appeared to be the source of the new genetic information (*phe*$^+$ and *trp*$^+$), how that information crossed the filter from the LA-2 cells to the LA-22 cells, allowing recombination to occur, was a mystery. The unknown source was designated simply as a *filterable agent (FA).*

Three observations were used to identify the FA:

1. The FA was produced by the LA-2 cells only when they were grown in association with LA-22 cells. If LA-2 cells were grown independently and that culture medium was then added to LA-22 cells, recombination did not occur. Therefore, LA-22 cells play some role in the production of FA by LA-2 cells and do so only when they share a common growth medium.
2. The addition of DNase, which enzymatically digests DNA, did not render the FA ineffective. Therefore, the FA is not naked DNA, ruling out transformation.
3. The FA could not pass across the filter of the Davis U-tube when the pore size was reduced below the size of bacteriophages.

Aided by these observations and aware that temperate phages can lysogenize *Salmonella,* researchers proposed that the genetic recombination event is mediated by bacteriophage P22, present initially as a prophage in the chromosome of the LA-22 *Salmonella* cells. They hypothesized that

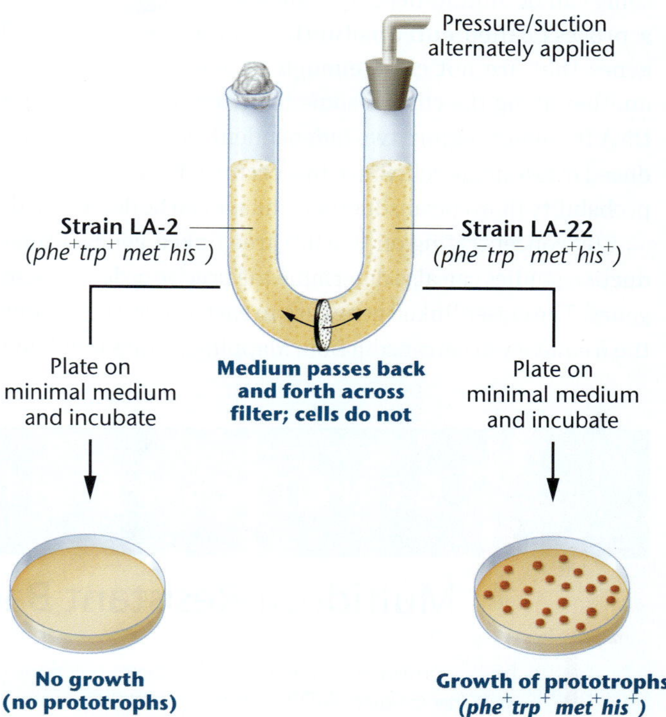

FIGURE 8.14 The Lederberg–Zinder experiment using *Salmonella.* After placing two auxotrophic strains on opposite sides of a Davis U-tube, Lederberg and Zinder recovered prototrophs from the side with the LA-22 strain, but not from the side containing the LA-2 strain.

P22 prophages rarely enter the vegetative or lytic phase, reproduce, and are released by the LA-22 cells. Such P22 phages, being much smaller than a bacterium, then cross the filter of the U-tube and subsequently infect and lyse some of the LA-2 cells. In the process of lysis of LA-2, these P22 phages occasionally package a region of the LA-2 chromosome in their heads. If this region contains the phe^+ and trp^+ genes and the phages subsequently pass back across the filter and infect LA-22 cells, these newly lysogenized cells will behave as prototrophs. This process of transduction, whereby bacterial recombination is mediated by bacteriophage P22, is diagrammed in **Figure 8.15**.

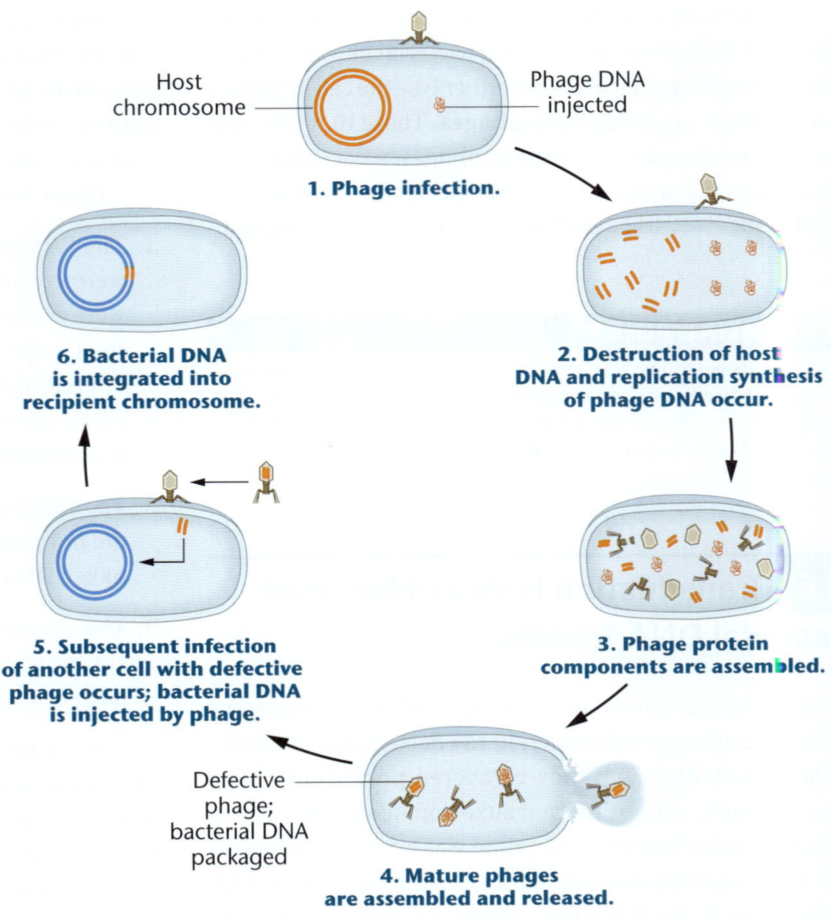
Host chromosome
Phage DNA injected
1. Phage infection.
2. Destruction of host DNA and replication synthesis of phage DNA occur.
3. Phage protein components are assembled.
4. Mature phages are assembled and released.
Defective phage; bacterial DNA packaged
5. Subsequent infection of another cell with defective phage occurs; bacterial DNA is injected by phage.
6. Bacterial DNA is integrated into recipient chromosome.

FIGURE 8.15 The process of transduction, where bacteriophages mediate bacterial recombination.

Transduction and Mapping

Like transformation, transduction was used in linkage and mapping studies of the bacterial chromosome. The fragment of bacterial DNA involved in a transduction event is large enough to include numerous genes. As a result, two genes that closely align (are linked) on the bacterial chromosome can be simultaneously transduced, a process called **cotransduction.** Two genes that are not close enough to one another along the chromosome to be included on a single DNA fragment require two independent events to be transduced into a single cell. Since this occurs with a much lower probability than cotransduction, linkage can be determined.

By concentrating on two or three linked genes, transduction studies can also determine the precise order of these genes. The closer linked genes are to each other, the greater the frequency of cotransduction. Mapping studies involving three closely aligned genes can thus be executed, and the analysis of such an experiment is predicated on the same rationale underlying other mapping techniques.

ESSENTIAL POINT

Transduction is virus-mediated bacterial DNA transfer and can be used to map phage genes. ■

GENETICS, ETHICS, AND SOCIETY

Multidrug-Resistant Bacteria: Fighting with Phage

The worldwide spread of *multidrug-resistant (MDR)* pathogenic bacteria has become an urgent threat to human and animal health. More than two million people in the United States become infected with antibiotic-resistant bacteria each year, and more than 23,000 of them will die from their infections. In 2015, approximately 480,000 cases of MDR tuberculosis occurred worldwide and another 100,000 cases were resistant to at least one antibiotic. In the United States, cases of drug-resistant enterobacteriaceae infections increased three-fold between 2001 and 2012. In 2016, a woman in Nevada died of a *Klebsiella pneumoniae* infection caused by a strain that was resistant to 26 different antibiotics, including colistin, which is considered the "last resort" antibiotic.

One factor leading to the spread of MDR bacteria is the selective pressure brought about by repeated exposure to antibiotics. Worldwide, livestock consume as much as 80 percent of all antibiotics, used as feed supplements. The routine use of antibiotics in livestock feed and the overuse of human antibiotic prescriptions are thought to be the most significant contributors to the spread of MDR bacteria.

A second factor leading to the new "post-antibiotic era" is the reduction in antibiotic drug development by pharmaceutical companies. Economic issues are significant. Drug companies spend hundreds of millions of dollars to develop and test a new drug. However, they receive less profit from antibiotics than from more expensive drugs such as chemotherapies or diabetes drugs.

Several alternative approaches are in development and early clinical trials. One very unique approach is the use of therapeutic bacteriophages (phages). Phages have been used to treat bacterial infections since the early 1900s, especially in Europe, but were abandoned in the mid-twentieth century after the introduction of antibiotics. Researchers are returning to phage, using modern molecular tools to modify phage and phage-derived products for use as antibacterial drugs. No phage or phage products are yet approved for human therapies in the United States or Europe; however, several phage preparations, targeted at pathogens such as *Listeria*, are approved for topical use on fresh and prepared foods, and at least one phage therapy is in clinical trials.

Although scientific and regulatory challenges must still be overcome, we may be on the verge of the Age of the Phage.

Your Turn

Take time, individually or in groups, to consider the following questions. Investigate the references dealing with the technical and ethical challenges of combating drug-resistant bacteria.

1. How do phage therapies work, and what are the main advantages and disadvantages of using phage to treat bacterial infections?

These topics are discussed in Potera, C. (2013). Phage renaissance: New hope against antibiotic resistance (https://ehp.niehs.nih.gov/121-a48) *and* Cooper, C. J., et al. (2016). Adapting drug approval pathways for bacteriophage-based therapeutics. https://doi.org/10.3389/fmicb.2016.01209

2. Two significant reasons for the spread of MDR bacteria are the overuse of agricultural antibiotics and the reluctance of pharmaceutical companies to develop new antibiotics. Discuss the ethical concerns surrounded these two situations. For example, how do we balance our need for both abundant food and infection control? Also, how can we resolve the ethical disconnect between private-sector profits and the public good?

These topics are discussed in Littmann, J., and Viens, A. M. (2015). The ethical significance of antimicrobial resistance. *Public Health Ethics* 8:209–224.

CASE STUDY To test or not to test

A 4-month-old infant had been running a moderate fever for 36 hours, and a nervous mother made a call to her pediatrician. Examination and tests revealed no outward signs of infection or cause of the fever. The anxious mother wanted a prescription for antibiotics, but the pediatrician recommended watching the infant for two days before making a decision. He explained that decades of rampant use of antibiotics in medicine and agriculture has caused a global surge in antibiotic-resistant bacteria, drastically reducing the effectiveness of antibiotic therapy for infections. He pointed out that bacteria can exchange antibiotic resistance traits and that many pathogenic strains are now resistant to several antibiotics. The mother was not placated by these explanations and insisted that her baby receive antibiotics immediately. This situation raises several issues.

1. Was the pediatrician correct in stating that bacteria can exchange antibiotic resistance genes? If so, how is this possible?
2. If the infant was given antibiotics, how might this have contributed to the production of resistant bacteria?
3. If you were an anxious parent of the patient, would it change your mind if you learned that a woman died in 2016 from a bacterial infection that was resistant to all 26 antibiotics available in the United States?
4. How should the pediatrician balance his ethical responsibility to provide effective treatment to the present patient with his ethical responsibility to future patients who may need antibiotics for effective treatment?

See Garau, J. (2006). Impact of antibiotic restrictions: The ethical perspective. *Clin. Microbiol. Infect.* 12 (Supplement 5):16–24. See also the Genetics, Ethics, and Society essay above.

INSIGHTS AND SOLUTIONS

1. Time mapping is performed in a cross involving the genes *his*, *leu*, *mal*, and *xyl*. The recipient cells are auxotrophic for all four genes. After 25 minutes, mating is interrupted, with the results in recipient cells shown below. Diagram the positions of these genes relative to the origin (*O*) of the F factor and to one another.

   (a) 90% are xyl^+

   (b) 80% are mal^+

   (c) 20% are his^+

   (d) None are leu^+

Solution: The *xyl* gene is transferred most frequently, so it is closest to *O* (very close). The *mal* gene is next and reasonably close to *xyl*, followed by the more distant *his* gene. The *leu* gene is far beyond these three, since no recovered recombinants include it. The diagram shows these relative locations along a piece of the circular chromosome.

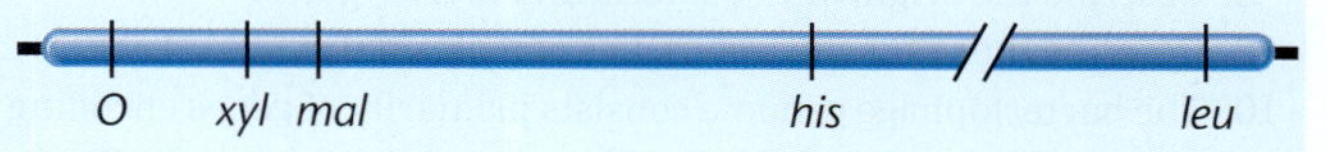

(continued)

Insights and Solutions continued

2. In four Hfr strains of bacteria, all derived from an original F^+ culture grown over several months, a group of hypothetical genes is studied and shown to transfer in the orders shown in the following table.

   (a) Assuming *b* is the first gene along the chromosome, determine the sequence of all genes shown. (b) One strain creates an apparent dilemma. Which one is it? Explain why the dilemma is only apparent, not real.

Hfr Strain	Order of Transfer					
1	*e*	*r*	*i*	*u*	*m*	*b*
2	*u*	*m*	*b*	*a*	*c*	*t*
3	*c*	*t*	*e*	*r*	*i*	*u*
4	*r*	*e*	*t*	*c*	*a*	*b*

Solution:

(a) The sequence is found by overlapping the genes in each strain.

Strain 2	*u*	*m*	***b***	***a***	***c***	*t*					
Strain 3					*c*	***t***	***e***	***r***	*i*	*u*		
Strain 1							*e*	*r*	***i***	***u***	***m***	*b*

Starting with *b* in strain 2, the gene sequence is *bacterium*.

(b) Strain 4 creates a dilemma, which is resolved when we realize that the F factor is integrated in the opposite orientation. Thus, the genes enter in the opposite sequence, starting with gene *r*.

retcab →

Problems and Discussion Questions

Mastering Genetics Visit for instructor-assigned tutorials and problems.

1. **HOW DO WE KNOW?** In this chapter, we have focused on genetic systems present in bacteria and the viruses that use bacteria as hosts (bacteriophages). In particular, we discussed mechanisms by which bacteria and their phages undergo genetic recombination, the basis of chromosome mapping. Based on your knowledge of these topics, answer several fundamental questions:
   (a) How do we know that bacteria undergo genetic recombination, allowing the transfer of genes from one organism to another?
   (b) How do we know that conjugation leading to genetic recombination between bacteria involves cell contact, which precedes the transfer of genes from one bacterium to another?
   (c) How do we know that during transduction bacterial cell-to-cell contact is not essential?

2. **CONCEPT QUESTION** Review the Chapter Concepts list on p. 168. A number of these center around the findings that genetic recombination occurs in bacteria. Write a short summary that describes ways in which recombination may occur in bacteria. ■
3. Distinguish between vertical and horizontal gene transfer.
4. With respect to F^+ and F^- bacterial matings,
   (a) How was it established that physical contact was necessary?
   (b) How was it established that chromosome transfer was unidirectional?
   (c) What is the genetic basis of a bacterium being F^+?
5. What is the F factor and why does its behavior differ in F^+ and Hfr bacterial strains?
6. Describe the basis for chromosome mapping in the Hfr $\times$ F^- crosses.
7. What is the consequence of the excision of the F factor from the chromosome in an Hfr strain?
8. Describe the origin of F' bacteria and merozygotes.
9. Describe the main characteristics of a plasmid.
10. The bacteriophage genome consists primarily of genes encoding proteins that make up the head, collar and tail, and tail fibers. When these genes are transcribed following phage infection, how are these proteins synthesized, since the phage genome lacks genes essential to ribosome structure?
11. Describe the structure of the T4 phage.
12. In the plaque assay, what is the precise origin of a single plaque?
13. Explain the basis of the plaque assay. How is it used?
14. A plaque assay is performed beginning with 1.0 mL of a solution containing bacteriophages. This solution is serially diluted three times by taking 0.1 mL and adding it to 9.9 mL of liquid medium. 0.1mL of the final dilution is plated and yields 17 plaques. What is the initial density of bacteriophages in the original 1.0 mL?
15. Why are the lytic and the lysogenic cycles both replicative phage strategies?
16. Define the term *prophage*.
17. Explain the observations that led Zinder and Lederberg to conclude that the prototrophs recovered in their transduction experiments were not the result of Hfr-mediated conjugation.
18. Describe the execution of and rationale behind linkage and mapping studies of bacterial genes during transduction experiments.
19. If a single bacteriophage infects one *E. coli* cell present in a culture of bacteria and, upon lysis, yields 200 viable viruses, how many phages will exist in a single plaque if three more lytic cycles occur?
20. A phage-infected bacterial culture was subjected to a series of dilutions, and a plaque assay was performed in each case, with the following results. What conclusion can be drawn in the case of each dilution?

	Dilution Factor	Assay Results
(a)	10^4	All bacteria lysed
(b)	10^6	14 plaques
(c)	10^8	0 plaques

9

DNA Structure and Analysis

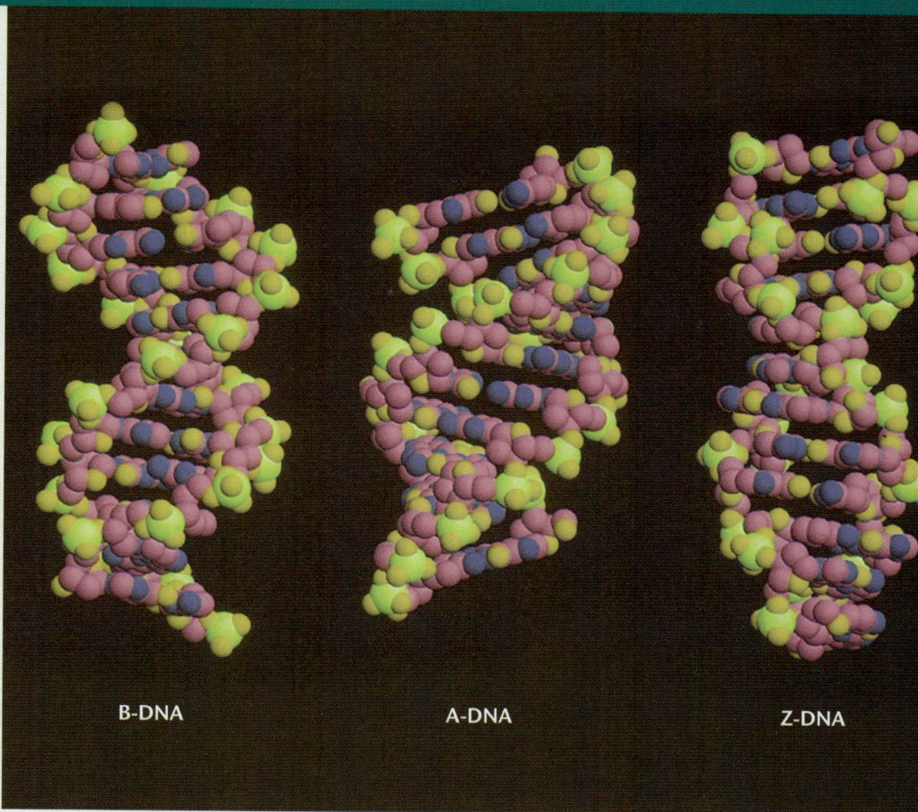

Computer-generated space-filling models of alternative forms of DNA.

CHAPTER CONCEPTS

- With the exception of some viruses, DNA serves as the genetic material in all living organisms on Earth.
- According to the Watson–Crick model, DNA exists in the form of the right-handed double helix.
- The strands of the double helix are antiparallel and held together by hydrogen bonding between complementary nitrogenous bases.
- The structure of DNA provides the basis for storing and expressing genetic information.
- RNA has many similarities to DNA but exists mostly as a single-stranded molecule.
- In some viruses, RNA serves as the genetic material.
- Many techniques have been developed that facilitate the analysis of nucleic acids, most based on detection of the complementarity of nitrogenous bases.

Up to this point in the text, we have described chromosomes as structures containing genes that control phenotypic traits that are transmitted through gametes to future offspring. Logically, genes must contain some sort of information that, when passed to a new generation, influences the form and characteristics of each individual. We refer to that information as the **genetic material.** Logic also suggests that this same information in some way directs the many complex processes that lead to an organism's adult form.

Until 1944, it was not clear what chemical component of the chromosome makes up genes and constitutes the genetic material. Because chromosomes were known to have both a nucleic acid and a protein component, both were candidates. In 1944, however, direct experimental evidence emerged showing that the nucleic acid DNA serves as the informational basis for heredity.

Once the importance of DNA in genetic processes was realized, work intensified with the hope of discerning not only the structural basis of this molecule but also the relationship of its structure to its function. Between 1944 and 1953, many scientists sought information that might answer the most significant and intriguing question in the history of biology: How does DNA serve as the genetic basis for the living process? Researchers believed the answer depended strongly on the chemical structure of the DNA molecule, given the complex but orderly functions ascribed to it.

These efforts were rewarded in 1953 when James Watson and Francis Crick set forth their hypothesis for the double-helical nature of DNA. The assumption that the molecule's functions would be clarified more easily once its general structure was determined proved to be correct. In this chapter, we initially review the evidence that DNA is the genetic material and then discuss the elucidation of its structure. We conclude the chapter with a discussion of several analytical techniques useful during the study of nucleic acids, DNA and RNA.

9.1 The Genetic Material Must Exhibit Four Characteristics

For a molecule to serve as the genetic material, it must possess four major characteristics: **replication, storage of information, expression of information,** and **variation by mutation.** Replication of the genetic material is one facet of the cell cycle, a fundamental property of all living organisms. Once the genetic material of cells replicates and is doubled in amount, it must then be partitioned equally into daughter cells. During the formation of gametes, the genetic material is also replicated but is partitioned so that each cell gets only one-half of the original amount of genetic material—the process of meiosis. Although the products of mitosis and meiosis differ, both of these processes are part of the more general phenomenon of cellular reproduction.

Storage of information requires the molecule to act as a repository of genetic information that may or may not be expressed by the cell in which it resides. It is clear that while most cells contain a complete copy of the organism's genome, at any point in time they express only a part of this genetic potential. For example, in bacteria many genes "turn on" in response to specific environmental conditions and "turn off" when conditions change. In vertebrates, skin cells may display active melanin genes but never activate their hemoglobin genes; in contrast, digestive cells activate many genes specific to their function but do not activate their melanin genes.

Expression of the stored genetic information is the basis of the process of **information flow** within the cell (**Figure 9.1**). The initial event is the **transcription** of DNA, in which three main types of RNA molecules are synthesized: messenger RNA (mRNA), ribosomal RNA (rRNA), and transfer RNA (tRNA). Of these, mRNAs are translated into proteins. Each mRNA is the product of a specific gene and directs the synthesis of a different protein. In **translation,** the chemical information in mRNA directs the construction of a chain of amino acids, called a polypeptide, which then folds into a protein. Collectively, these processes form the **central dogma of molecular genetics:** "DNA makes RNA, which makes proteins."

The genetic material is also the source of variation among organisms through the process of mutation. If a mutation—a change in the chemical composition of DNA—occurs, the alteration may be reflected during transcription and translation, affecting the specific protein. If such a mutation is present in gametes, it may be passed to future generations and, with time, become distributed throughout the population. Genetic variation, which also includes alterations of chromosome number and rearrangements within and between chromosomes, provides the raw material for the process of evolution.

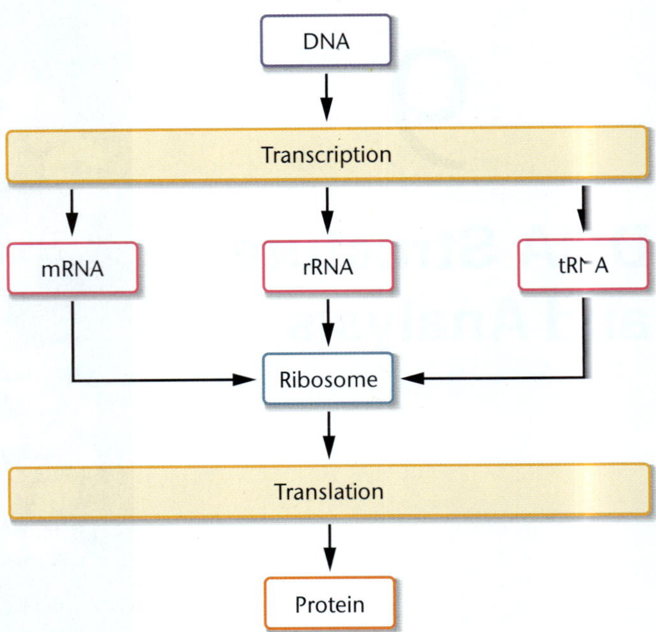

FIGURE 9.1 Simplified view of information flow (the central dogma) involving DNA, RNA, and proteins within cells.

9.2 Until 1944, Observations Favored Protein as the Genetic Material

The idea that genetic material is physically transmitted from parent to offspring has been accepted for as long as the concept of inheritance has existed. Beginning in the late nineteenth century, research into the structure of biomolecules progressed considerably, setting the stage for describing the genetic material in chemical terms. Although both proteins and nucleic acid were major candidates for the role of the genetic material, until the 1940s many geneticists favored proteins. This is not surprising because a diversity of proteins was known to be abundant in cells, and much more was known about protein chemistry.

DNA was first studied in 1868 by a Swiss chemist, Friedrich Miescher. He isolated cell nuclei and derived an acid substance containing DNA that he called **nuclein.** As investigations progressed, however, DNA, which was shown to be present in chromosomes, seemed to lack the chemical diversity necessary to store extensive genetic information. This conclusion was based largely on Phoebus A. Levene's observations in 1910 that DNA contained approximately equal amounts of four similar molecules called *nucleotides*. Levene postulated incorrectly that identical groups of these four components were repeated over and over, which was the basis of his **tetranucleotide hypothesis** for DNA structure. Attention was thus directed away from DNA, favoring proteins. However, in the 1940s, Erwin Chargaff showed that Levene's proposal was incorrect when he demonstrated that most organisms do not contain precisely equal proportions of the four nucleotides. We shall see later that the structure of DNA accounts for Chargaff's observations.

ESSENTIAL POINT

Although both proteins and nucleic acids were initially considered as possible candidates, proteins were initially favored to serve as the genetic material. ■

9.3 Evidence Favoring DNA as the Genetic Material Was First Obtained during the Study of Bacteria and Bacteriophages

Oswald Avery, Colin MacLeod, and Maclyn McCarty's 1944 publication on the chemical nature of a "transforming principle" in bacteria was the initial event that led to the acceptance of DNA as the genetic material. Their work, along with subsequent findings of other research teams, constituted the first direct experimental proof that DNA, and not protein, is the biomolecule responsible for heredity. It marked the beginning of the era of molecular genetics, a period of discovery in biology that made biotechnology feasible and has moved us closer to understanding the basis of life. The impact of the initial findings on future research and thinking paralleled that of the publication of Darwin's theory of evolution and the subsequent rediscovery of Mendel's postulates of transmission genetics. Together, these events constitute the three great revolutions in biology.

Transformation Studies

The research that provided the foundation for Avery, MacLeod, and McCarty's work was initiated in 1927 by Frederick Griffith, a medical officer in the British Ministry of Health. He experimented with several different strains of the bacterium *Diplococcus pneumoniae*.* Some were *virulent strains*, which cause pneumonia in certain vertebrates (notably humans and mice), while others were *avirulent strains*, which do not cause illness.

The difference in virulence depends on the existence of a polysaccharide capsule; virulent strains have this capsule, whereas avirulent strains do not. The nonencapsulated bacteria are readily engulfed and destroyed by phagocytic cells in the animal's circulatory system. Virulent bacteria, which possess the polysaccharide coat, are not easily engulfed; they multiply and cause pneumonia.

The presence or absence of the capsule causes a visible difference between colonies of virulent and avirulent strains. Encapsulated bacteria form *smooth colonies* (*S*) with a shiny surface when grown on an agar culture plate; nonencapsulated strains produce *rough colonies* (*R*). Thus, virulent and avirulent strains are easily distinguished by standard microbiological culture techniques.

Each strain of *Diplococcus* may be one of dozens of different types called *serotypes*. The specificity of the serotype is due to the detailed chemical structure of the polysaccharide constituent of the thick, slimy capsule. Serotypes are identified by immunological techniques and are usually designated by Roman numerals. Griffith used the avirulent type II*R* and the virulent type III*S* in his critical experiments. **Table 9.1** summarizes the characteristics of these strains.

Griffith knew from the work of others that only living virulent cells produced pneumonia in mice. If heat-killed virulent bacteria were injected into mice, no pneumonia resulted, just as living avirulent bacteria failed to produce the disease. Griffith's critical experiment involved injecting mice with living II*R* (avirulent) cells combined with heat-killed III*S* (virulent) cells. Since neither cell type caused death in mice when injected alone, Griffith expected that the double injection would not kill the mice. But, after five days, all of the mice that had received both types of cells were dead. Paradoxically, analysis of their blood revealed large numbers of living type III*S* bacteria.

As far as could be determined, these III*S* bacteria were identical to the III*S* strain from which the heat-killed cell preparation had been made. Control mice, injected only with living avirulent II*R* bacteria, did not develop pneumonia and remained healthy. This ruled out the possibility that the avirulent II*R* cells simply changed (or mutated) to virulent III*S* cells in the absence of the heat-killed III*S* bacteria. Instead, some type of interaction had taken place between living II*R* and heat-killed III*S* cells.

TABLE 9.1 Strains of *Diplococcus pneumoniae* Used by Frederick Griffith in His Original Transformation Experiments

Serotype	Colony Morphology	Capsule	Virulence
II*R*	Rough	Absent	Avirulent
III*S*	Smooth	Present	Virulent

*This organism is now named *Streptococcus pneumoniae*.

Griffith concluded that the heat-killed III*S* bacteria somehow converted live avirulent II*R* cells into virulent III*S* cells. Calling the phenomenon **transformation,** he suggested that the **transforming principle** might be some part of the polysaccharide capsule or a compound required for capsule synthesis, although the capsule alone did not cause pneumonia. To use Griffith's term, the transforming principle from the dead III*S* cells served as a "pabulum" for the II*R* cells.

Griffith's work led others to explore the phenomenon of transformation. By 1931, Henry Dawson and his coworkers showed that transformation could occur *in vitro* (in a test tube containing only bacterial cells). That is, injection into mice was not necessary for transformation to occur. By 1933, Lionel J. Alloway had refined the *in vitro* experiments using extracts from *S* cells added to living *R* cells. The soluble filtrate from the heat-killed III*S* cells was as effective in inducing transformation as were the intact cells. Alloway and others did not view transformation as a genetic event, but rather as a physiological modification of some sort. Nevertheless, the experimental evidence that a chemical substance was responsible for transformation was quite convincing.

Then, in 1944, after 10 years of work, Avery, MacLeod, and McCarty published their results in what is now regarded as a classic paper in the field of molecular genetics. They reported that they had obtained the transforming principle in a highly purified state and that they strongly believed that it was DNA.

The details of their work are illustrated in **Figure 9.2**. The researchers began their isolation procedure with large quantities (50–75 L) of liquid cultures of type III*S* virulent cells. The cells were centrifuged, collected, and heat-killed. Following various chemical treatments, a soluble filtrate was derived from these cells, which retained the ability to induce transformation of type II*R* avirulent cells. The soluble filtrate was treated with a protein-digesting enzyme, called a protease, and an RNA-digesting enzyme called **ribonuclease.** Such treatment destroyed the activity of any remaining protein and RNA. Nevertheless, transforming activity still remained. They concluded that neither protein

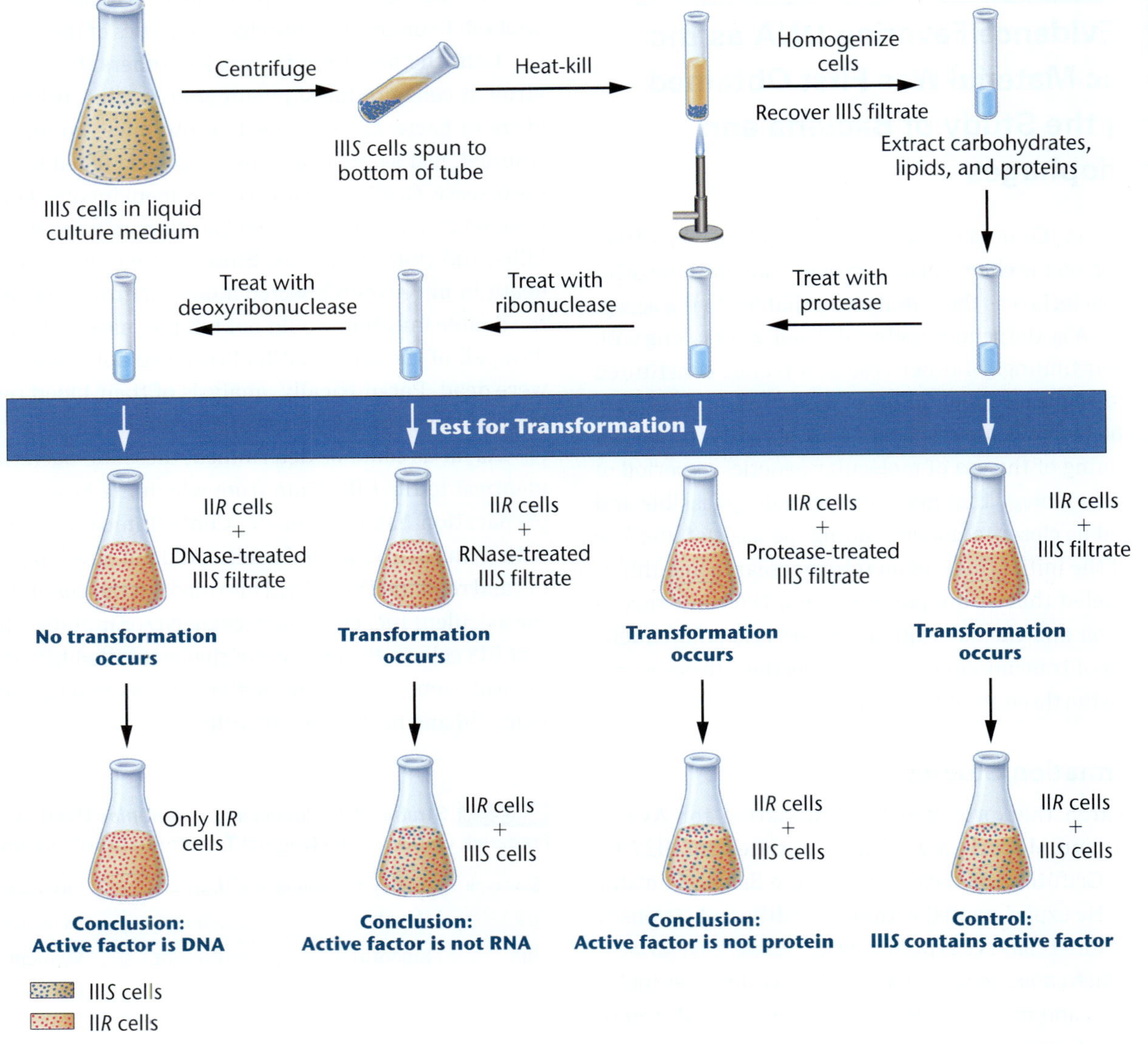

FIGURE 9.2 Summary of Avery, MacLeod, and McCarty's experiment demonstrating that DNA is the transforming principle.

nor RNA was responsible for transformation. The final confirmation came with experiments using crude samples of the DNA-digesting enzyme **deoxyribonuclease,** isolated from dog and rabbit sera. Digestion with this enzyme destroyed transforming activity present in the filtrate; thus, Avery and his coworkers were certain that the active transforming principle in these experiments was DNA.

The great amount of work, the confirmation and reconfirmation of the conclusions, and the logic of the experimental design involved in the research of these three scientists are truly impressive. Their conclusion in the 1944 publication, however, was stated very simply: "The evidence presented supports the belief that a nucleic acid of the desoxyribose* type is the fundamental unit of the transforming principle of *Pneumococcus* type III."

The researchers also immediately recognized the genetic and biochemical implications of their work. They suggested that the transforming principle interacts with the II*R* cell and gives rise to a coordinated series of enzymatic reactions that culminates in the synthesis of the type III*S* capsular polysaccharide. They emphasized that, once transformation occurs, the capsular polysaccharide is produced in successive generations. Transformation is therefore heritable, and the process affects the genetic material.

Transformation, originally introduced in Chapter 8, has now been shown to occur in *Haemophilus influenzae, Bacillus subtilis, Shigella paradysenteriae,* and *Escherichia coli,* among many other microorganisms. Transformation of numerous genetic traits other than colony morphology has been demonstrated, including those that resist antibiotics. These observations further strengthened the belief that transformation by DNA is primarily a genetic event rather than simply a physiological change.

The Hershey–Chase Experiment

The second major piece of evidence supporting DNA as the genetic material was provided during the study of the bacterium *E. coli* and one of its infecting viruses, **bacteriophage T2.** Often referred to simply as a **phage,** the virus consists of a protein coat surrounding a core of DNA. Electron micrographs reveal that the phage's external structure is composed of a hexagonal head plus a tail. **Figure 9.3** shows the life cycle of a T-even bacteriophage such as T2, as

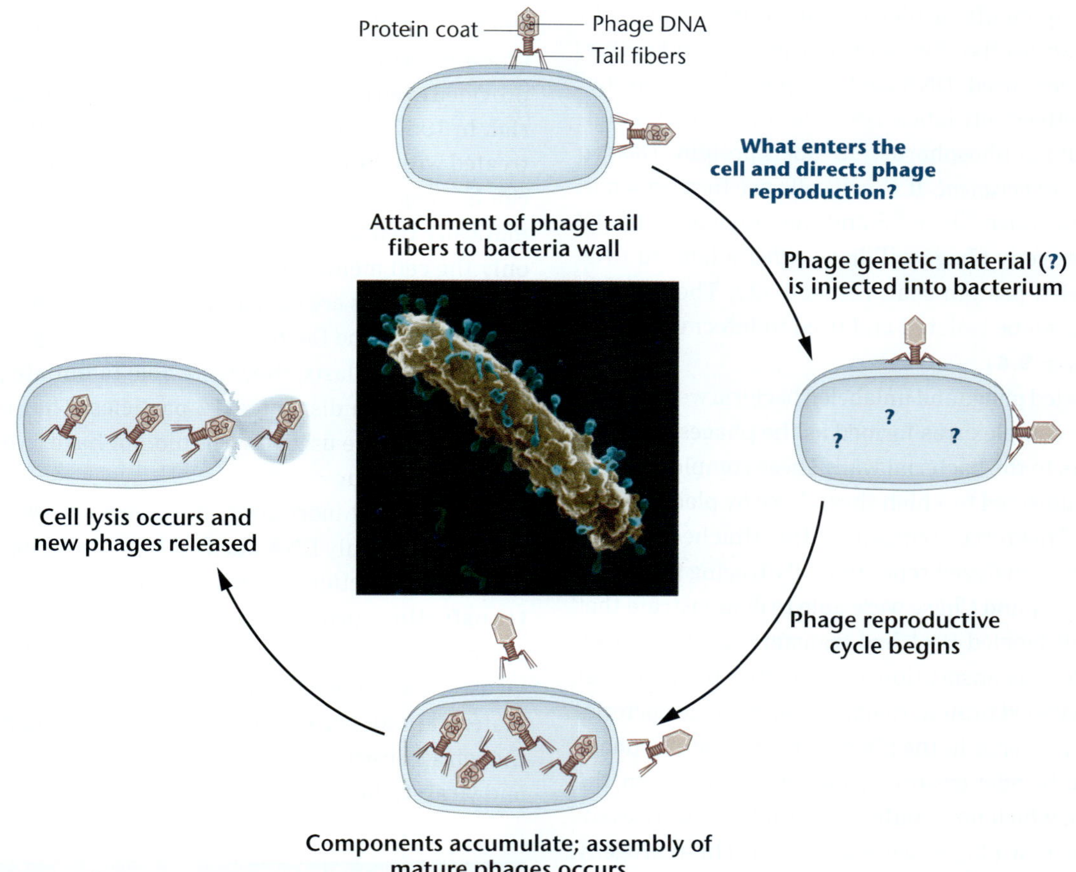

FIGURE 9.3 Life cycle of a T-even bacteriophage. The electron micrograph shows an *E. coli* cell during infection by numerous T2 phages (shown in blue).

*Desoxyribose is now spelled deoxyribose.

it was known in 1952. Recall that the phage adsorbs to the bacterial cell and that some component of the phage enters the bacterial cell. Following infection, the viral information "commandeers" the cellular machinery of the host and directs viral reproduction. In a reasonably short time, many new phages are constructed and the bacterial cell is lysed, releasing the progeny viruses.

In 1952, Alfred Hershey and Martha Chase published the results of experiments designed to clarify the events leading to phage reproduction. Several of the experiments clearly established the independent functions of phage protein and nucleic acid in the reproduction process of the bacterial cell. Hershey and Chase knew from this existing data that:

1. T2 phages consist of approximately 50 percent protein and 50 percent DNA.
2. Infection is initiated by adsorption of the phage by its tail fibers to the bacterial cell.
3. The production of new viruses occurs within the bacterial cell.

It appeared that some molecular component of the phage, DNA and/or protein, entered the bacterial cell and directed viral reproduction. Which was it?

Hershey and Chase used radioisotopes to follow the molecular components of phages during infection. Both ^{32}P and ^{35}S, radioactive forms of phosphorus and sulfur, respectively, were used. DNA contains phosphorus but not sulfur, so ^{32}P effectively labels DNA. Because proteins contain sulfur, but not phosphorus, ^{35}S labels protein. *This is a key point in the experiment.* If *E. coli* cells are first grown in the presence of *either* ^{32}P *or* ^{35}S and then infected with T2 viruses, the progeny phage will have *either* a labeled DNA core *or* a labeled protein coat, respectively. These radioactive phages can be isolated and used to infect unlabeled bacteria (**Figure 9.4**).

When labeled phage and unlabeled bacteria were mixed, an adsorption complex was formed as the phages attached their tail fibers to the bacterial wall. These complexes were isolated and subjected to a high shear force by placing them in a blender. This force stripped off the attached phages, which were then analyzed separately. By tracing the radioisotopes, Hershey and Chase were able to demonstrate that most of the ^{32}P-labeled DNA had transferred into the bacterial cell following adsorption; on the other hand, almost all of the ^{35}S-labeled protein remained outside the bacterial cell and was recovered in the phage "ghosts" (empty phage coats) after the blender treatment. Following separation, the bacterial cells, which now contained viral DNA, were eventually lysed as new phages were produced. These progeny contained ^{32}P but not ^{35}S.

Hershey and Chase interpreted these results as indicating that the protein of the phage coat remains outside the host cell and is not involved in the production of new phages. On the other hand, and most important, phage DNA enters the host cell and directs phage reproduction. Hershey and Chase had demonstrated that the genetic material in phage T2 is DNA, not protein.

These experiments, along with those of Avery and his colleagues, provided convincing evidence that DNA is the molecule responsible for heredity. This conclusion has since served as the cornerstone of the field of molecular genetics.

NOW SOLVE THIS

9.1 Would an experiment similar to that performed by Hershey and Chase work if the basic design were applied to the phenomenon of transformation? Explain why or why not.

■ **HINT:** *This problem involves an understanding of the protocol of the Hershey–Chase experiment as applied to the investigation of transformation. The key to its solution is to remember that in transformation, exogenous DNA enters the soon-to-be transformed cell and that no cell-to-cell contact is involved in the process.*

Transfection Experiments

During the eight years following publication of the Hershey–Chase experiment, additional research with bacterial viruses provided even more solid proof that DNA is the genetic material. In 1957, several reports demonstrated that if *E. coli* is treated with the enzyme lysozyme, the outer wall of the cell can be removed without destroying the bacterium. Enzymatically treated cells are naked, so to speak, and contain only the cell membrane as the outer boundary of the cell; these structures are called **protoplasts** (or **spheroplasts**). John Spizizen and Dean Fraser independently reported that by using protoplasts, they were able to initiate phage multiplication with disrupted T2 particles. That is, provided protoplasts were used, a virus did not have to be intact for infection to occur.

Similar but more refined experiments were reported in 1960 using only DNA purified from bacteriophages. This process of infection by only the viral nucleic acid, called **transfection,** proves conclusively that phage DNA alone contains all the necessary information for producing mature viruses. Thus, the evidence that DNA serves as the genetic material in all organisms was further strengthened, even though all direct evidence had been obtained from bacterial and viral studies.

ESSENTIAL POINT

By 1952, transformation studies and experiments using bacteria infected with bacteriophages strongly suggested that DNA is the genetic material in bacteria and most viruses. ■

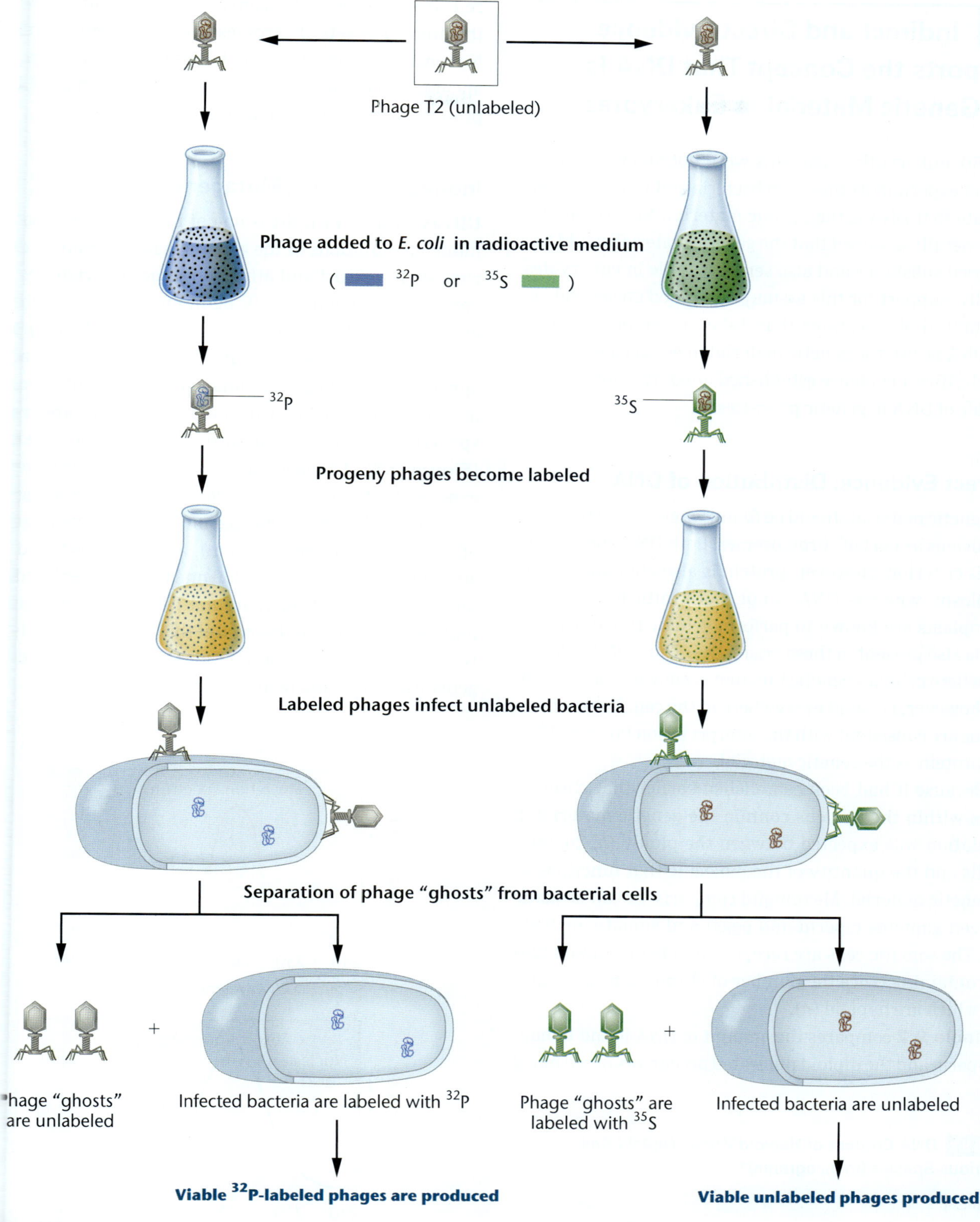

FIGURE 9.4 Summary of the Hershey–Chase experiment demonstrating that DNA, not protein, is responsible for directing the reproduction of phage T2 during the infection of *E. coli*.

9.4 Indirect and Direct Evidence Supports the Concept That DNA Is the Genetic Material in Eukaryotes

In 1950, eukaryotic organisms were not amenable to the types of experiments that used bacteria and viruses to demonstrate that DNA is the genetic material. Nevertheless, it was generally assumed that the genetic material would be a universal substance and also serve this role in eukaryotes. Initially, support for this assumption relied on several circumstantial observations that, taken together, indicated that DNA is also the genetic material in eukaryotes. Subsequently, direct evidence established unequivocally the central role of DNA in genetic processes.

Indirect Evidence: Distribution of DNA

The genetic material should be found where it functions—in the nucleus as part of chromosomes. Both DNA and protein fit this criterion. However, protein is also abundant in the cytoplasm, whereas DNA is not. Both mitochondria and chloroplasts are known to perform genetic functions, and DNA is also present in these organelles. Thus, DNA is found only where primary genetic function is known to occur. Protein, however, is found everywhere in the cell. These observations are consistent with the interpretation favoring DNA over protein as the genetic material.

Because it had been established earlier that chromosomes within the nucleus contain the genetic material, a correlation was expected between the ploidy (n, $2n$, etc.) of cells and the quantity of the molecule that functions as the genetic material. Meaningful comparisons can be made between gametes (sperm and eggs) and somatic or body cells. The somatic cells are recognized as being diploid ($2n$) and containing twice the number of chromosomes as gametes, which are haploid (n).

Table 9.2 compares the amount of DNA found in haploid sperm and the diploid nucleated precursors of red blood cells from a variety of organisms. The amount of DNA and the number of sets of chromosomes are closely correlated. No consistent correlation can be observed between gametes and diploid cells for proteins, thus again favoring DNA over proteins as the genetic material of eukaryotes.

TABLE 9.2 DNA Content of Haploid Versus Diploid Cells of Various Species (in picograms)*

Organism	n	$2n$
Human	3.25	7.30
Chicken	1.26	2.49
Trout	2.67	5.79
Carp	1.65	3.49
Shad	0.91	1.97

*Sperm (n) and nucleated precursors to red blood cells ($2n$) were used to contrast ploidy levels.

Indirect Evidence: Mutagenesis

Ultraviolet (UV) light is one of many agents capable of inducing mutations in the genetic material. Simple organisms such as yeast and other fungi can be irradiated with various wavelengths of UV light, and the effectiveness of each wavelength can then be measured by the number of mutations it induces. When the data are plotted, an **action spectrum** of UV light as a mutagenic agent is obtained. This action spectrum can then be compared with the **absorption spectrum** of any molecule suspected to be genetic material (**Figure 9.5**). *The molecule serving as the genetic material is expected to absorb at the wavelengths shown to be mutagenic.*

UV light is most mutagenic at the wavelength (λ) of about 260 nanometers (nm), and both DNA and RNA strongly absorb UV light at 260 nm. On the other hand, protein absorbs most strongly around 280 nm, yet no significant mutagenic effects are observed at this wavelength. This indirect evidence also supports the idea that a nucleic acid is the genetic material and tends to exclude protein.

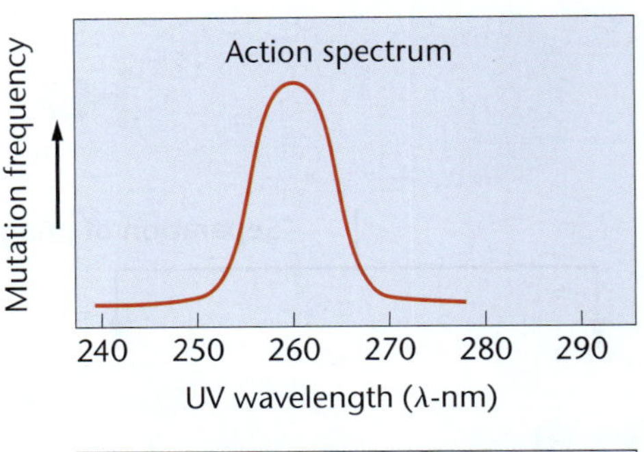

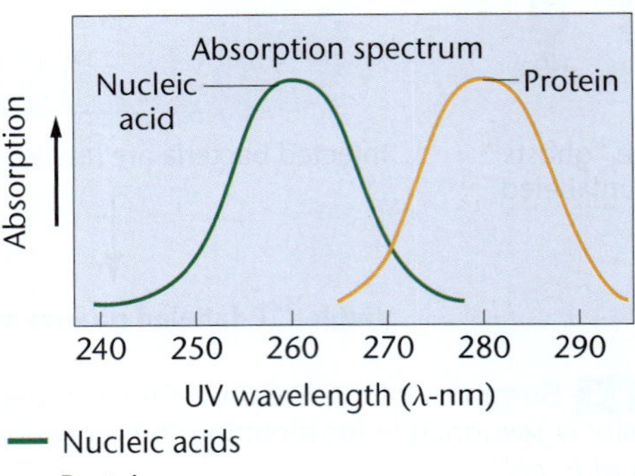

FIGURE 9.5 Comparison of the action spectrum, which determines the most effective mutagenic UV wavelength, and the absorption spectrum, which shows the range of wavelengths where nucleic acids and proteins absorb UV light.

Direct Evidence: Recombinant DNA Studies

Although the circumstantial evidence just described does not constitute direct proof that DNA is the genetic material in eukaryotes, over a half century of research has provided irrefutable evidence that DNA serves this role. In fact, this simple concept of genetics is at the foundation of modern genetic research and its applications.

For example, **recombinant DNA technology,** which involves splicing together DNA sequences from different organisms (see Chapter 17), has combined the DNA sequence encoding the human hormone insulin with bacterial DNA sequences. When this recombinant DNA molecule is introduced into bacteria, the bacteria replicate the DNA and pass it to daughter cells at each cell division. In addition, the bacteria express the recombinant DNA molecule and thus produce human insulin. This example of biotechnology clearly establishes how a specific DNA sequence confers heritable information responsible for a product of a gene.

Genomics (see Chapter 18), which can provide the full set of DNA sequences of organisms, is the basis of still another example in support of this concept. We have known the full sequence of the human genome since 2001. However, geneticists are still uncovering new clues as to how the 3.2 billion base pairs of human DNA serve as the basis for human life. For example, the sequencing of genomes from individuals with specific heritable disorders and their comparison to genomes of healthy individuals has provided many insights about which DNA sequences (or genes) harbor mutations responsible for these genetic disorders. The underlying premise of such studies is that DNA is the genetic material.

By the mid-1970s the concept that DNA is the genetic material in eukaryotes was accepted, and since then, no information has been forthcoming to dispute that conclusion. In the upcoming chapters, we will see exactly how DNA is stored, replicated, mutated, repaired, and expressed.

ESSENTIAL POINT

Although at first only indirect observations supported the hypothesis that DNA controls inheritance in eukaryotes, subsequent studies involving recombinant DNA techniques and transgenic mice provided direct experimental evidence that the eukaryotic genetic material is DNA. ■

9.5 RNA Serves as the Genetic Material in Some Viruses

Some viruses contain an RNA core rather than a DNA core. In these viruses, it appears that RNA serves as the genetic material—an exception to the general rule that DNA performs this function. In 1956, it was demonstrated that when purified RNA from **tobacco mosaic virus (TMV)** was spread on tobacco leaves, the characteristic lesions caused by viral infection subsequently appeared on the leaves. Thus, it was concluded that RNA is the genetic material of this virus.

In 1965 and 1966, Norman Pace and Sol Spiegelman demonstrated further that RNA from the phage $Q\beta$ can be isolated and replicated *in vitro*. Replication depends on an enzyme, **RNA replicase,** which is isolated from host *E. coli* cells following normal infection. When the RNA replicated *in vitro* is added to *E. coli* protoplasts, infection and viral multiplication (transfection) occur. Thus, RNA synthesized in a test tube serves as the genetic material in these phages by directing the production of all the components necessary for viral reproduction.

One other group of RNA-containing viruses bears mention. These are the **retroviruses,** which replicate in an unusual way. Their RNA serves as a template for the synthesis of the complementary DNA molecule. The process, **reverse transcription,** occurs under the direction of an RNA-dependent DNA polymerase enzyme called **reverse transcriptase.** This DNA intermediate can be incorporated into the genome of the host cell, and when the host DNA is transcribed, copies of the original retroviral RNA chromosomes are produced. Retroviruses include the human immunodeficiency virus (HIV), which causes AIDS, as well as RNA tumor viruses.

ESSENTIAL POINT

RNA serves as the genetic material in some bacteriophages as well as some plant and animal viruses. ■

9.6 The Structure of DNA Holds the Key to Understanding Its Function

Having established that DNA is the genetic material in all living organisms (except certain viruses), we turn now to the structure of this nucleic acid. In 1953, James Watson and Francis Crick proposed that the structure of DNA is in the form of a double helix. Their proposal was published in a short paper in the journal *Nature*. In a sense, this publication was the finish of a highly competitive scientific race to obtain what some consider to be the most significant finding in the history of biology. This race, as recounted in Watson's book *The Double Helix,* demonstrates the human interaction, genius, frailty, and intensity involved in the scientific effort that eventually led to elucidation of the DNA structure.

The data available to Watson and Crick, crucial to the development of their proposal, came primarily from two sources: (1) base composition analysis of hydrolyzed samples of DNA and (2) X-ray diffraction studies of DNA. Watson and Crick's analytical success can be attributed to model building that conformed to the existing data. If the correct solution to the structure of DNA is viewed as a puzzle, Watson and Crick, working in the Cavendish Laboratory in Cambridge, England, were the first to fit the pieces together successfully.

Nucleic Acid Chemistry

Before turning to this work, a brief introduction to nucleic acid chemistry is in order. This chemical information was well known to Watson and Crick during their investigation and served as the basis of their model building.

DNA is a nucleic acid, and nucleotides are the building blocks of all nucleic acid molecules. Sometimes called mononucleotides, these structural units have three essential components: a **nitrogenous base,** a **pentose sugar** (a five-carbon sugar), and a **phosphate group.** There are two kinds of nitrogenous bases: the nine-member double-ring **purines** and the six-member single-ring **pyrimidines.**

Two types of purines and three types of pyrimidines are found in nucleic acids. The two purines are **adenine** and **guanine,** abbreviated A and G, respectively. The three pyrimidines are **cytosine, thymine,** and **uracil** (respectively, C, T, and U). The chemical structures of the five bases are shown in **Figure 9.6(a)**. Both DNA and RNA contain A, G, and C, but only DNA contains the base T and only RNA contains the base U. Each nitrogen or carbon atom of the ring structures of purines and pyrimidines is designated by a number. Note that corresponding atoms in the purine and pyrimidine rings are numbered differently.

The pentose sugars found in nucleic acids give them their names. **Ribonucleic acids (RNA)** contain **ribose,** while **deoxyribonucleic acids (DNA)** contain **deoxyribose. Figure 9.6(b)** shows the chemical structures for these two pentose sugars. Each carbon atom is

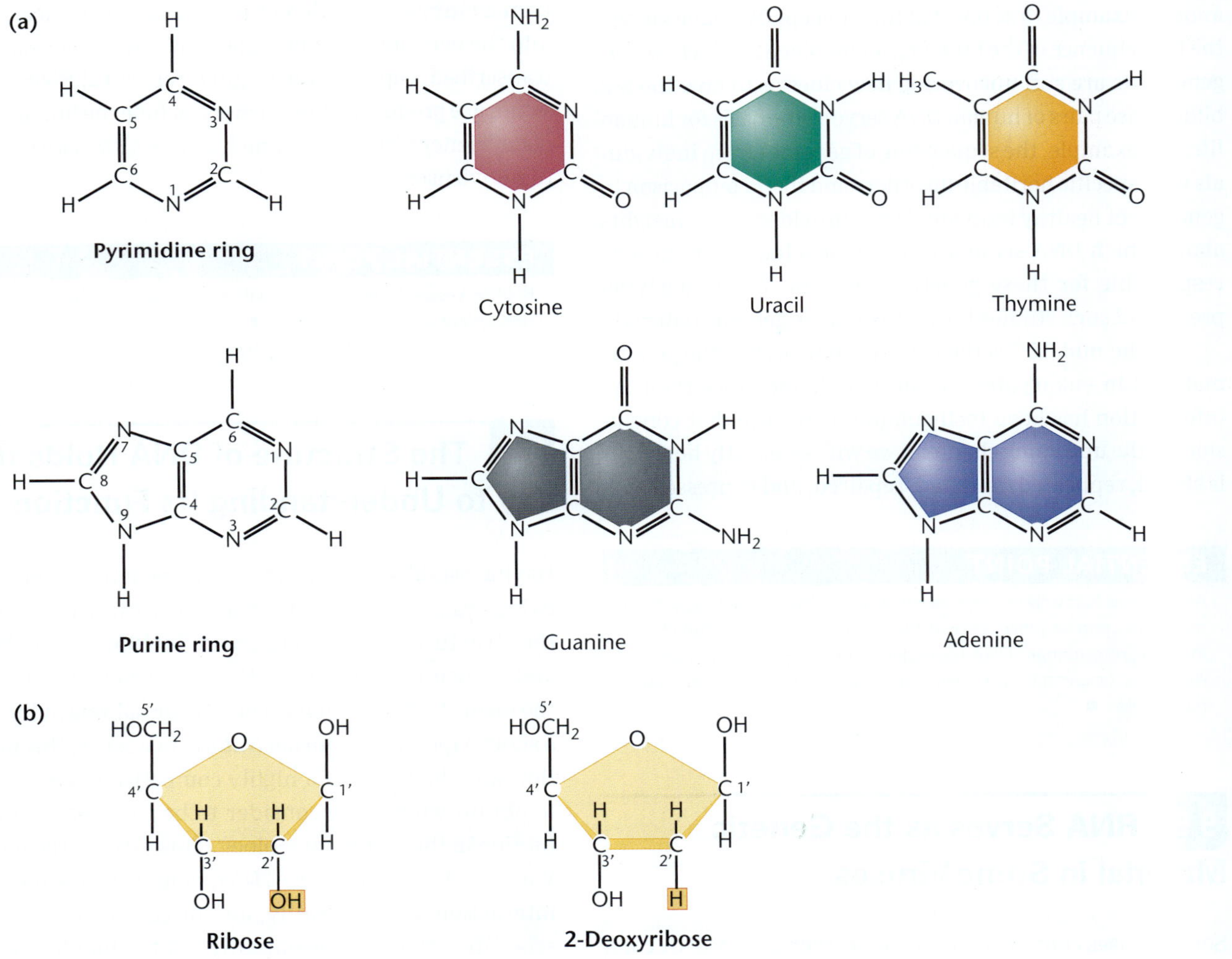

FIGURE 9.6 (a) Chemical structures of the pyrimidines and purines that serve as the nitrogenous bases in RNA and DNA. (b) Chemical ring structures of ribose and 2-deoxyribose, which serve as the pentose sugars in RNA and DNA, respectively.

distinguished by a number with a prime sign ('). As you can see in Figure 9.6(b), compared with ribose, deoxyribose has a hydrogen atom rather than a hydroxyl group at the (C-2′) position. The absence of a hydroxyl group at the (C-2′) position thus distinguishes DNA from RNA. In the absence of the (C-2′) hydroxyl group, the sugar is more specifically named **2-deoxyribose.**

If a molecule is composed of a purine or pyrimidine base and a ribose or deoxyribose sugar, the chemical unit is called a **nucleoside.** If a phosphate group is added to the nucleoside, the molecule is now called a **nucleotide.** Nucleosides and nucleotides are named according to the specific nitrogenous base (A, G, C, T, or U) that is part of the molecule. The structure of a nucleotide and the nomenclature used in naming DNA nucleotides and nucleosides are shown in **Figure 9.7**.

The bonding between components of a nucleotide is highly specific. The (C-1′) atom of the sugar is involved in the chemical linkage to the nitrogenous base. If the base is a purine, the N-9 atom is covalently bonded to the sugar; if the base is a pyrimidine, the N-1 atom bonds to the sugar. In deoxyribonucleotides, the phosphate group may be bonded to the (C-2′), (C-3′), or (C-5′) atom of the sugar. The (C-5′)-phosphate configuration is shown in Figure 9.7. It is by far the most prevalent one in biological systems and the one found in DNA and RNA.

Nucleotides are also described by the term **nucleoside monophosphate (NMP).** The addition of one or two phosphate groups results in **nucleoside diphosphates (NDPs)** and **triphosphates (NTP),** respectively, as shown in **Figure 9.8**. The triphosphate form is significant because it is the precursor molecule during nucleic acid synthesis within the cell. In addition, **adenosine triphosphate (ATP)** and **guanosine triphosphate (GTP)** are important in cell bioenergetics because of the large amount of energy involved in adding or removing the terminal phosphate group. The hydrolysis of ATP or GTP to ADP or GDP and inorganic phosphate (P_i) is accompanied by the release of a large amount of energy in the cell. When these chemical conversions are coupled to other reactions, the energy produced is used to drive them. As a result, ATP and GTP are involved in many cellular activities.

The linkage between two mononucleotides involves a phosphate group linked to two sugars. A **phosphodiester bond** is formed as phosphoric acid is joined to two alcohols (the hydroxyl groups on the two sugars) by an

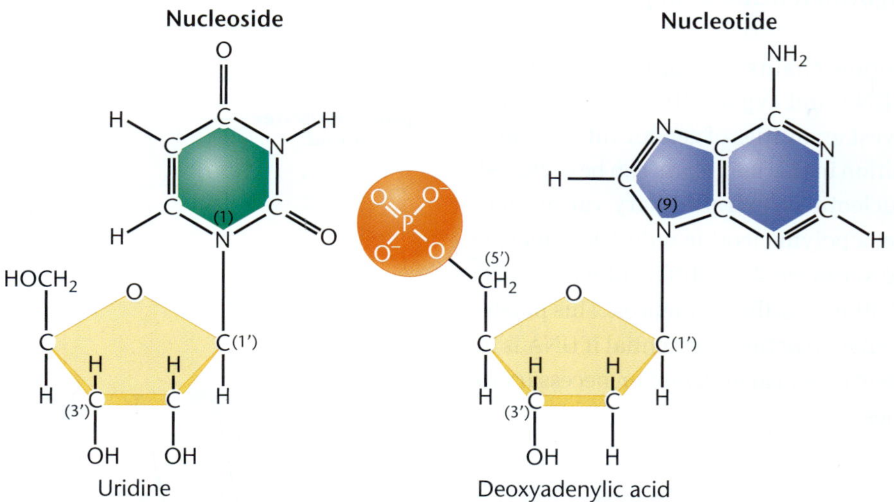

Ribonucleosides	Ribonucleotides
Adenosine Cytidine Guanosine Uridine	Adenylic acid Cytidylic acid Guanylic acid Uridylic acid
Deoxyribonucleosides	**Deoxyribonucleotides**
Deoxyadenosine Deoxycytidine Deoxyguanosine Deoxythymidine	Deoxyadenylic acid Deoxycytidylic acid Deoxyguanylic acid Deoxythymidylic acid

FIGURE 9.7 Structures and names of the nucleosides and nucleotides of RNA and DNA.

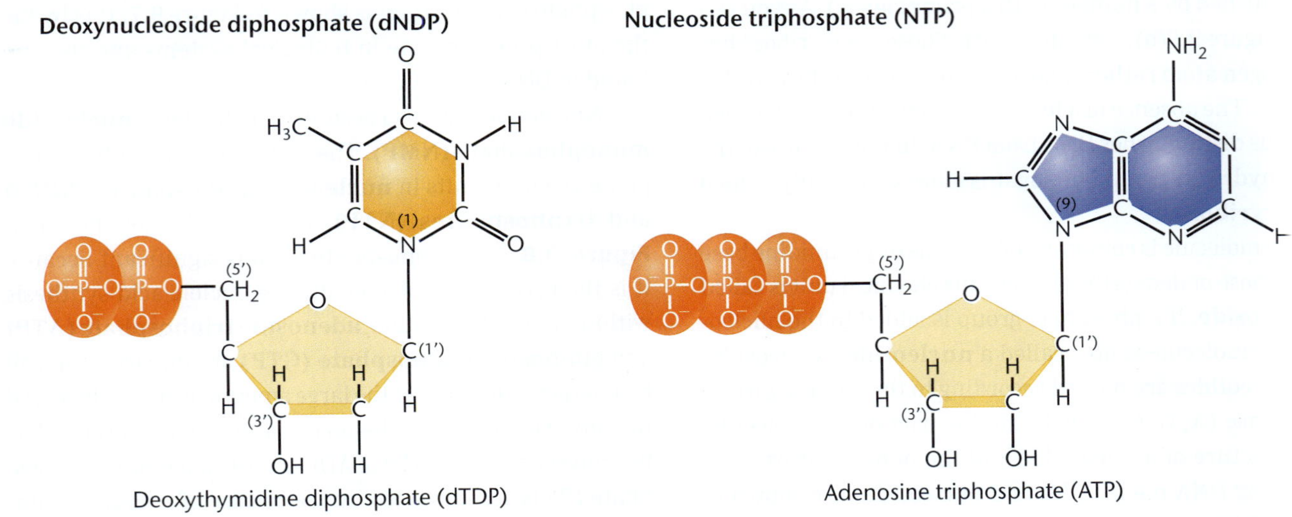

Deoxynucleoside diphosphate (dNDP)

Deoxythymidine diphosphate (dTDP)

Nucleoside triphosphate (NTP)

Adenosine triphosphate (ATP)

FIGURE 9.8 Basic structures of nucleoside diphosphates and triphosphates. Deoxythymidine diphosphate and adenosine triphosphate are diagrammed here.

ester linkage on both sides. **Figure 9.9** shows the resultant phosphodiester bond in DNA. Each structure has a (C-3′) end and a (C-5′) end. Two joined nucleotides form a dinucleotide; three nucleotides, a trinucleotide; and so forth. Short chains consisting of up to 20 nucleotides or so are called **oligonucleotides;** longer chains are **polynucleotides.**

Long polynucleotide chains account for the large molecular weight of DNA and explain its most important property—storage of vast quantities of genetic information. If each nucleotide position in this long chain can be occupied by any one of four nucleotides, extraordinary variation is possible. For example, a polynucleotide only 1000 nucleotides in length can be arranged 4^{1000} different ways, each one different from all other possible sequences. This potential variation in molecular structure is essential if DNA is to store the vast amounts of chemical information necessary to direct cellular activities.

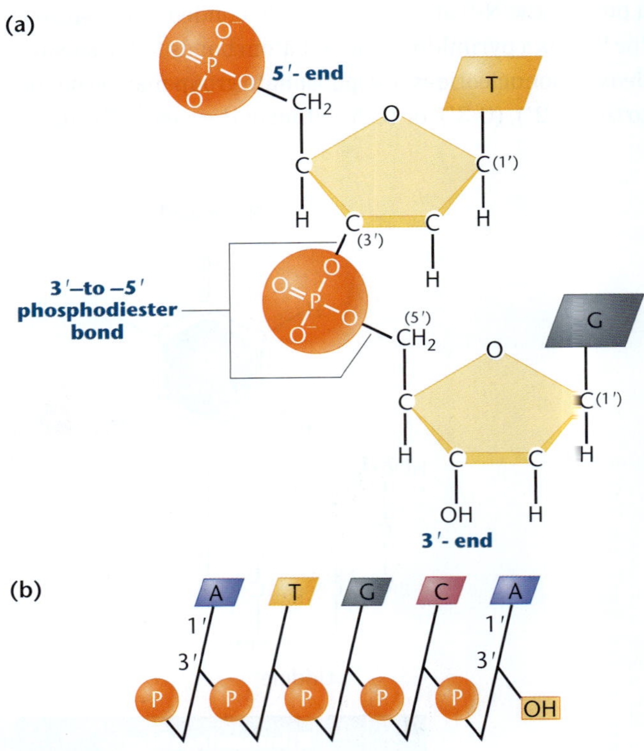

FIGURE 9.9 (a) Linkage of two nucleotides by the formation of a C-3′ to C-5′ (3′–5′) phosphodiester bond, producing a dinucleotide. (b) Shorthand notation for a polynucleotide chain.

Base-Composition Studies

Between 1949 and 1953, Erwin Chargaff and his colleagues used chromatographic methods to separate the four bases in DNA samples from various organisms. Quantitative methods were then used to determine the amounts of the four nitrogenous bases from each source. On the basis of these data, the following conclusions may be drawn:

1. The amount of adenine residues is proportional to the amount of thymine residues in DNA. Also, the amount of guanine residues is proportional to the amount of cytosine residues.
2. Based on this proportionality, the sum of the purines (A + G) equals the sum of the pyrimidines (C + T).
3. The percentage of (G + C) does not necessarily equal the percentage of (A + T). Instead, this ratio varies greatly between different organisms.

These conclusions indicate a definite pattern of base composition of DNA molecules. The data were critical to Watson and Crick's successful model of DNA. They also directly refuted Levene's tetranucleotide hypothesis, which stated that all four bases are present in equal amounts.

X-Ray Diffraction Analysis

When fibers of a DNA molecule are subjected to X-ray bombardment, the X rays scatter according to the molecule's atomic structure. The pattern of scatter can be captured as spots on photographic film and analyzed, particularly for the overall shape of and regularities within the molecule. This process, **X-ray diffraction analysis,** was applied successfully to the study of protein structure by Linus Pauling and other chemists. The technique had been attempted on DNA as early as 1938 by William Astbury. By 1947, he had detected a periodicity of 3.4 angstroms (Å)* within the structure of the molecule, which suggested to him that the bases were stacked like coins on top of one another.

Between 1950 and 1953, Rosalind Franklin, working in the laboratory of Maurice Wilkins, obtained improved X-ray data from more purified samples of DNA (**Figure 9.10**). Her work confirmed the 3.4-Å periodicity seen by Astbury and suggested that the structure of DNA was some sort of helix. However, she did not propose a definitive model. Pauling had analyzed the work of Astbury and others and proposed incorrectly that DNA is a triple helix.

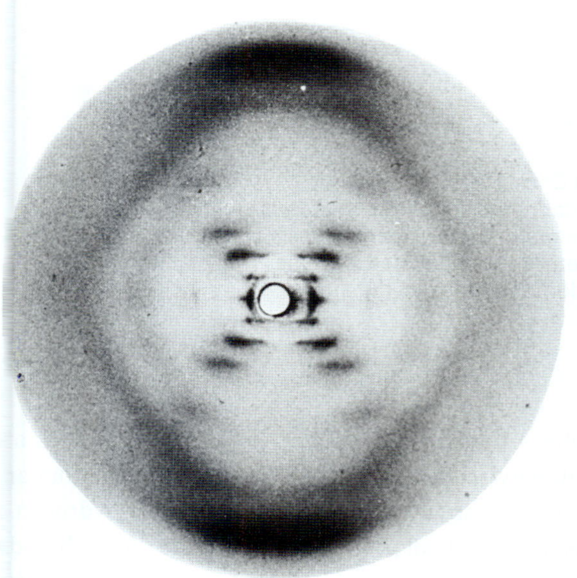

FIGURE 9.10 X-ray diffraction photograph of purified DNA fibers. The strong arcs on the periphery show closely spaced aspects of the molecule, providing an estimate of the periodicity of nitrogenous bases, which are 3.4 Å apart. The inner cross pattern of spots shows the grosser aspect of the molecule, indicating its helical nature.

*Today, measurement in nanometers (nm) is favored (1 nm = 10 Å).

The Watson–Crick Model

Watson and Crick published their analysis of DNA structure in 1953. By building models under the constraints of the information just discussed, they proposed the double-helical form of DNA shown in **Figure 9.11(a)**. This model has the following major features:

1. Two long polynucleotide chains are coiled around a central axis, forming a right-handed double helix.
2. The two chains are antiparallel; that is, their (C-5′) to (C-3′) orientations run in opposite directions.
3. The bases of both chains are flat structures, lying perpendicular to the axis; they are "stacked" on one another, 3.4 Å (0.34 nm) apart, and located on the inside of the structure.
4. The nitrogenous bases of opposite chains are *paired* as the result of hydrogen bonds; in DNA, only A-T and C-G pairs occur.
5. Each complete turn of the helix is 34 Å (3.4 nm) long; thus, 10 bases exist per turn in each chain.
6. In any segment of the molecule, alternating larger **major grooves** and smaller **minor grooves** are apparent along the axis.
7. The double helix measures 20 Å (2.0 nm) in diameter.

The nature of *base pairing* (point 4 above) is the most genetically significant feature of the model. Before we discuss it in detail, several other important features warrant emphasis. First, the antiparallel nature of the two chains is a key part of the double helix model. While one chain runs in the 5′-to-3′ orientation (what seems right side up to us), the other chain goes in the 3′-to-5′ orientation (and thus appears upside down). This is illustrated in **Figure 9.11(b)** and **(c)**. Given the constraints of the bond angles of the various nucleotide components, the double helix could not be constructed easily if both chains ran parallel to one another.

The key to the model proposed by Watson and Crick is the specificity of base pairing. Chargaff's data suggested that the amounts of A equaled T and that the amounts of G equaled C. Watson and Crick realized that if A pairs with T and C pairs with G, this would account for these proportions and that such pairing could occur as a result of hydrogen bonding between base pairs [Figure 9.11(c)], providing the

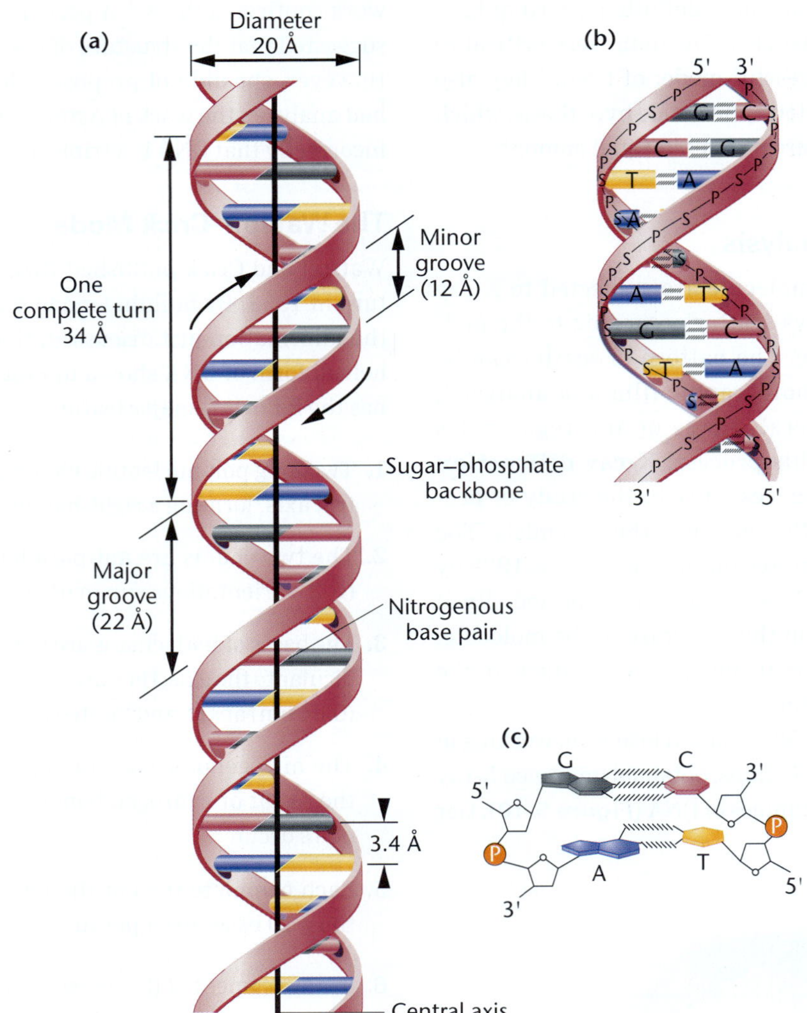

FIGURE 9.11 (a) The DNA double helix as proposed by Watson and Crick. The ribbon-like strands constitute the sugar–phosphate backbones, and the horizontal rungs constitute the nitrogenous base pairs, of which there are 10 per complete turn. The major and minor grooves are shown. The solid vertical bar represents the central axis. (b) A detailed view labeled with the bases, sugars, phosphates, and hydrogen bonds of the helix. (c) A demonstration of the antiparallel nature of the helix and the horizontal stacking of the bases.

chemical stability necessary to hold the two chains together. Arranged in this way, both major and minor grooves become apparent along the axis. Further, a purine (A or G) opposite a pyrimidine (T or C) on each "rung of the spiral staircase" of the proposed double helix accounts for the 20-Å (2-nm) diameter suggested by X-ray diffraction studies.

The specific A-T and C-G base pairing is the basis for **complementarity.** This term describes the chemical affinity provided by hydrogen bonding between the bases. As we shall see, complementarity is very important in DNA replication and gene expression.

It is appropriate to inquire into the nature of a hydrogen bond and to ask whether it is strong enough to stabilize the helix. A **hydrogen bond** is a very weak electrostatic attraction between a covalently bonded hydrogen atom and an atom with an unshared electron pair. The hydrogen atom assumes a partial positive charge, while the unshared electron pair—characteristic of covalently bonded oxygen and nitrogen atoms—assumes a partial negative charge. These opposite charges are responsible for the weak chemical attractions. As oriented in the double helix, adenine forms two hydrogen bonds with thymine, and guanine forms three hydrogen bonds with cytosine. Although two or three individual hydrogen bonds are energetically very weak, 2000 to 3000 bonds in tandem (typical of two long polynucleotide chains) provide great stability to the helix.

Another stabilizing factor is the arrangement of sugars and bases along the axis. In the Watson–Crick model, the *hydrophobic* ("water-fearing") nitrogenous bases are stacked almost horizontally on the interior of the axis and are thus shielded from water. The *hydrophilic* ("water-loving") sugar–phosphate backbone is on the outside of the axis, where both components can interact with water. These molecular arrangements provide significant chemical stabilization to the helix.

A more recent and accurate analysis of the form of DNA that served as the basis for the Watson–Crick model has revealed a minor structural difference between the substance and the model. A precise measurement of the number of base pairs per turn has demonstrated a value of 10.4, rather than the 10.0 predicted by Watson and Crick. In the classic model, each base pair is rotated 36° around the helical axis relative to the adjacent base pair, but the new finding requires a rotation of 34.6°. This results in slightly more than 10 base pairs per 360° turn.

The Watson–Crick model had an immediate effect on the emerging discipline of molecular biology. Even in their initial 1953 article, the authors noted, "It has not escaped our notice that the specific pairing we have postulated immediately suggests a possible copying mechanism for the genetic material." Two months later, Watson and Crick pursued this idea in a second article in *Nature,* suggesting a specific mechanism of replication of DNA—the **semiconservative mode of replication.** The second article alluded to two new concepts: (1) the storage of genetic information in the sequence of the bases, and (2) the mutations or genetic changes that would result from an alteration of the bases. These ideas have received vast amounts of experimental support since 1953 and are now universally accepted.

Watson and Crick's synthesis of ideas was highly significant with regard to subsequent studies of genetics and biology. The nature of the gene and its role in genetic mechanisms could now be viewed and studied in biochemical terms. Recognition of their work, along with that of Wilkins, led to their receipt of the Nobel Prize in Physiology or Medicine in 1962. Unfortunately, Rosalind Franklin had died in 1958 at the age of 37, making her contributions ineligible for consideration since the award is not given posthumously. The Nobel Prize was to be one of many such awards bestowed for work in the field of molecular genetics.

ESSENTIAL POINT

As proposed by Watson and Crick, DNA exists in the form of a right-handed double helix composed of two long antiparallel polynucleotide chains held together by hydrogen bonds formed between complementary, nitrogenous base pairs. ■

NOW SOLVE THIS

9.2 In sea urchin DNA, which is double stranded, 17.5 percent of the bases were shown to be cytosine (C). What percentages of the other three bases are expected to be present in this DNA?

■ **HINT:** *This problem asks you to extrapolate from one measurement involving a unique DNA molecule to three other values characterizing the molecule. The key to its solution is to understand the base-pairing rules in the Watson–Crick model of DNA.*

EVOLVING CONCEPT OF THE GENE

Based on the model of DNA put forward by Watson and Crick in 1953, the gene was viewed for the first time in molecular terms as a sequence of nucleotides in a DNA helix that encodes genetic information. ■

9.7 Alternative Forms of DNA Exist

Under different conditions of isolation, several conformational forms of DNA have been recognized. At the time Watson and Crick performed their analysis, two forms—**A-DNA** and **B-DNA**—were known. Watson and Crick's analysis was based on X-ray studies of B-DNA performed by Franklin, which is present under aqueous, low-salt conditions and is believed to be the biologically significant conformation.

While DNA studies around 1950 relied on the use of X-ray diffraction, more recent investigations have been performed using **single-crystal X-ray analysis.** The earlier studies achieved limited resolution of about 5 Å, but single crystals diffract X rays at about 1 Å, near atomic resolution. As a result, every atom is "visible" and much greater structural detail is available during analysis.

With this modern technique, A-DNA, which is prevalent under high-salt or dehydration conditions, has now been scrutinized. In comparison to B-DNA, A-DNA is slightly more compact, with 11 bp in each complete turn of the helix, which is 23 Å (2.3 nm) in diameter. It is also a right-handed helix, but the orientation of the bases is somewhat different—they are tilted and displaced laterally in relation to the axis of the helix. As a result, the appearance of the major and minor grooves is modified. It seems doubtful that A-DNA occurs *in vivo* (under physiological conditions).

Other forms of DNA (e.g., C-, D-, E-, and most recently, P-DNA) are now known, but it is **Z-DNA** that has drawn the most attention. Discovered by Andrew Wang, Alexander

Rich, and their colleagues in 1979 when they examined a small synthetic DNA fragment containing only C-G base pairs, Z-DNA takes on the rather remarkable configuration of a *left-handed double helix.* Like A- and B-DNA, Z-DNA consists of two antiparallel chains held together by Watson–Crick base pairs. Beyond these characteristics, Z-DNA is quite different. The left-handed helix is 18 Å (1.8 nm) in diameter, contains 12 bp per turn, and assumes a zigzag conformation (hence its name). The major groove present in B-DNA is nearly eliminated. Z-DNA is compared with the A and B forms in the chapter opening photograph on p. 185.

Speculation abounds over the possibility that regions of Z-DNA exist in the chromosomes of living organisms. The unique helical arrangement could provide an important recognition point for the interaction with other molecules. However, it is still not clear whether Z-DNA occurs *in vivo.*

Still other forms of DNA have been studied, including P-DNA, named after Linus Pauling. It is produced by artificial "stretching" of DNA, creating a longer, narrower version with the phosphate groups on the interior.

9.8 The Structure of RNA Is Chemically Similar to DNA, but Single Stranded

The structure of RNA molecules resembles DNA, with several important exceptions. Although RNA also has nucleotides linked with polynucleotide chains, the sugar ribose replaces deoxyribose, and the nitrogenous base uracil replaces thymine. Another important difference is that most RNA is single stranded, but there are important exceptions. RNA molecules often fold back on themselves to form double-stranded regions of complementary base pairs. In addition, some animal viruses that have RNA as their genetic material contain double-stranded helices. More recently, we have learned that double-stranded RNA molecules can regulate gene expression in eukaryotes (see Chapter 16).

As established earlier (see Figure 9.1), three major classes of cellular RNA molecules function during the expression of genetic information: **ribosomal RNA (rRNA), messenger RNA (mRNA),** and **transfer RNA (tRNA).** These molecules all originate as complementary copies of deoxyribonucleotide sequences of DNA. Because uracil replaces thymine in RNA, uracil is complementary to adenine during transcription and RNA base pairing.

Different RNAs are distinguished by their sedimentation behavior in a centrifugal field. Sedimentation behavior depends on a molecule's density, mass, and shape, and its measure is called the **Svedberg coefficient** (S). Although higher S values almost always designate molecules of greater molecular weight, the correlation is not direct; that is, a two-fold increase in molecular weight does not lead to a two-fold increase in S.

Ribosomal RNAs are generally the largest of these molecules and usually constitute about 80 percent of all RNA in the cell. Ribosomal RNAs are important structural components of **ribosomes,** which function as nonspecific workbenches where proteins are synthesized during translation. The various forms of rRNA found in bacteria and eukaryotes differ distinctly in size.

Messenger RNA molecules carry genetic information from the DNA of the gene to the ribosome. The mRNA molecules vary considerably in size, which reflects the variation in the size of the protein encoded by the mRNA as well as the gene serving as the template for transcription of mRNA.

Transfer RNA, the smallest class of these RNA molecules, carries amino acids to the ribosome during translation. Since more than one tRNA molecule interacts simultaneously with the ribosome, the molecule's smaller size facilitates these interactions.

Other unique RNAs exist that perform various genetic roles, especially in eukaryotes. For example, *telomerase RNA* is involved in DNA replication at the ends of chromosomes (the telomeres). *Small nuclear RNA (snRNA)* participates in processing mRNAs, and *antisense RNA, microRNA (miRNA), short interfering RNA (siRNA),* and *long noncoding RNA (lncRNA)* are involved in gene regulation (see Chapter 16).

ESSENTIAL POINT

The second category of nucleic acids important in genetic function is RNA, which is similar to DNA with the exceptions that it is usually single stranded, the sugar ribose replaces the deoxyribose, and the pyrimidine uracil replaces thymine. ■

NOW SOLVE THIS

9.3 German measles results from an infection of the rubella virus, which can cause a multitude of health problems in newborns. What conclusions can you reach from a nucleic acid analysis of the virus that reveals an A + G/U + C ratio of 1.13?

■ **HINT:** *This problem asks you to analyze information about the chemical composition of a nucleic acid serving as the genetic material of a virus. The key to its solution is to apply your knowledge of nucleic acid chemistry, in particular your understanding of base pairing.*

9.9 Many Analytical Techniques Have Been Useful during the Investigation of DNA and RNA

Since 1953, the role of DNA as the genetic material and the role of RNA in transcription and translation have been clarified through detailed analysis of nucleic acids. Several important methods of analysis are based on the unique nature of the hydrogen bond that is so integral to the structure of nucleic acids. For example, if DNA is subjected to heat, the double helix is denatured and unwinds. During unwinding, the viscosity of DNA decreases and UV absorption increases (called the **hyperchromic shift**). A melting profile, in which OD_{260} is plotted against temperature, is shown for two DNA molecules in **Figure 9.12**. The midpoint of each curve is called the **melting temperature** T_m where 50 percent of the strands have unwound. The molecule with a higher T_m has a higher percentage of C-G base pairs than A-T base pairs since C-G pairs share three hydrogen bonds compared to the two bonds between A-T pairs.

The denaturation/renaturation of nucleic acids is the basis for one of the most useful techniques in molecular genetics—**molecular hybridization.** Provided that a reasonable degree of base complementarity exists between any two nucleic acid strands, denaturation can be reversed whereby molecular hybridization is possible. Duplexes can be re-formed between DNA strands, even from different organisms, and between DNA and RNA strands. For example, an RNA molecule will hybridize with the segment of DNA from which it was transcribed. As a result, nucleic acid **probes** are often used to identify complementary sequences.

The technique can even be performed using the DNA present in chromosomal preparations as the "target" for hybrid formation. This process is called ***in situ* molecular hybridization.** Mitotic cells are first fixed to slides and then subjected to hybridization conditions. Single-stranded DNA or RNA is added (a probe), and hybridization is monitored. The nucleic acid that is added may be either radioactive or contain a fluorescent label to allow its detection. In the former case, autoradiography is used.

Figure 9.13 illustrates the use of a fluorescent label. A short fragment of DNA that is complementary to DNA in the chromosomes' centromere regions has been hybridized. Fluorescence occurs only in the centromere regions and thus identifies each one along its chromosome. Because fluorescence is used, the technique is known by the acronym **FISH (fluorescence *in situ* hybridization).** The use of this technique to identify chromosomal locations housing specific genetic information has been a valuable addition to geneticists' repertoire of experimental techniques.

Electrophoresis

Another technique essential to the analysis of nucleic acids is **electrophoresis.** This technique may be adapted to separate different-sized fragments of DNA and RNA chains and is invaluable in current research investigations in molecular genetics.

Electrophoresis separates the molecules in a mixture by causing them to migrate under the influence of an electric

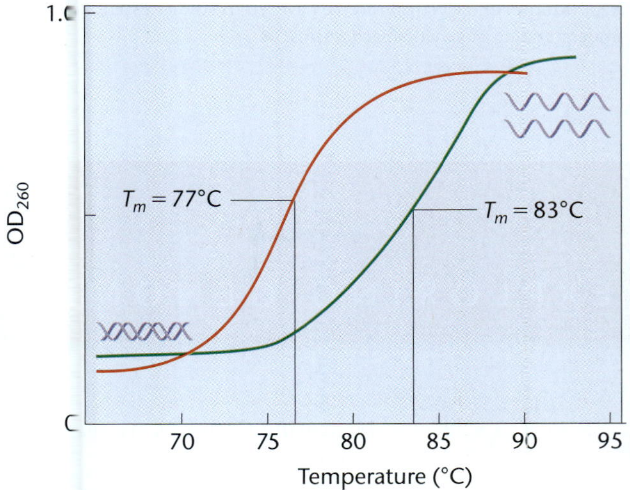

FIGURE 9.12 A melting profile shows the increase in UV absorption versus temperature (the hyperchromic effect) for two DNA molecules with different C-G contents. The molecule with a melting point (T_m) of 83°C has a greater C-G content than the molecule with a T_m of 77°C.

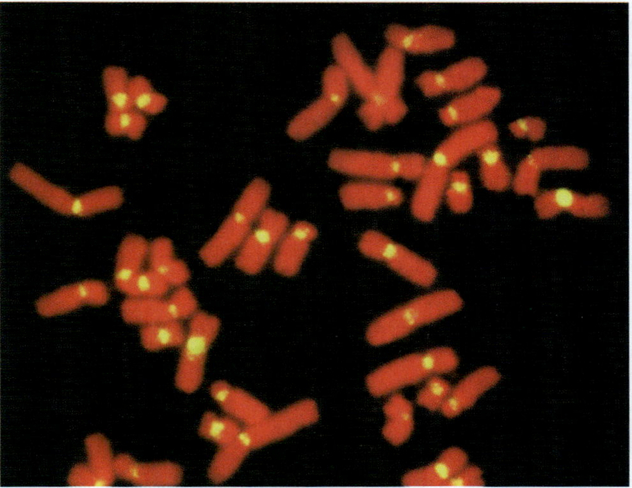

FIGURE 9.13 Fluorescence *in situ* hybridization (FISH) of human metaphase chromosomes. The probe, specific to centromeric DNA, produces a yellow fluorescence signal indicating hybridization. The red fluorescence is produced by propidium iodide counterstaining of chromosomal DNA.

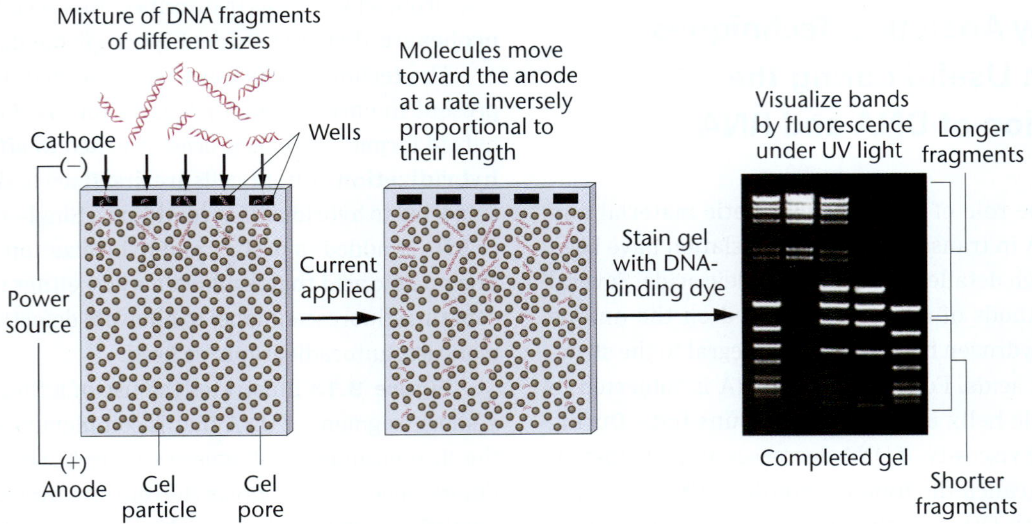

FIGURE 9.14 Electrophoretic separation of a mixture of DNA fragments that vary in length. The photograph at the right shows the results derived from an agarose gel that reveals DNA bands stained with ethidium bromide.

field. A sample is placed on a porous substance, such as a semisolid gel, which is then placed in a solution that conducts electricity. Mixtures of molecules with a similar charge–mass ratio but of different sizes will migrate at different rates through the gel based on their size. For example, two polynucleotide chains of different lengths, such as 10 versus 20 nucleotides, are both negatively charged (based on the phosphate groups of the nucleotides) and will both move to the positively charged pole (the anode), but at different rates. Using a medium such as an **agarose gel,** which can be prepared with various pore sizes, the *shorter chains migrate at a faster rate through the gel than larger chains* (**Figure 9.14**). Once electrophoresis is complete, bands representing the variously sized molecules are identified either by autoradiography (if a component of the molecule is radioactive) or by use of a fluorescent dye that binds to nucleic acids. The resolving power is so great that polynucleotides that vary by just one nucleotide in length may be separated.

Electrophoretic separation of nucleic acids is at the heart of a variety of other commonly used research techniques. Of particular note are the various "blotting" techniques (e.g., Southern blots and Northern blots), as well as DNA sequencing methods, which we will discuss in detail later in Chapter 17.

ESSENTIAL POINT

Various methods of analysis of nucleic acids, particularly molecular hybridization and electrophoresis, have led to studies essential to our understanding of genetic mechanisms. ■

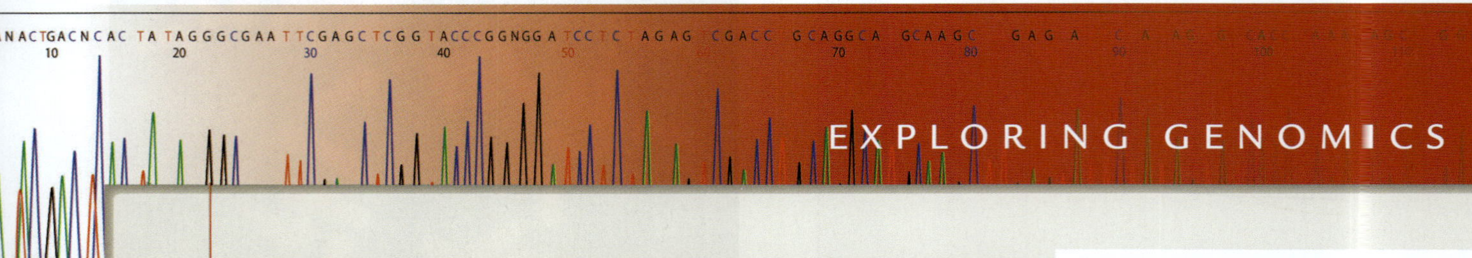

Introduction to Bioinformatics: BLAST

Mastering Genetics Visit the Study Area: Exploring Genomics

In this chapter, we focused on the structural details of DNA. Later, you will learn how scientists can clone and sequence DNA (see Chapter 17). The explosion of DNA and protein sequence data that has occurred in the last 20 years has launched the field of *bioinformatics,* an interdisciplinary science that applies mathematics and computing technology to develop hardware and software for storing, sharing, comparing, and analyzing nucleic acid and protein sequence data.

A large number of sequence databases that make use of bioinformatics have been developed. An example is **GenBank** (http://www.ncbi.nlm.nih.gov/genbank/), which is the National Institutes of Health database of all publicly available sequence data. This global resource, with access to databases in Europe and Japan, currently contains more than 220 billion base pairs of sequence data!

Earlier (in the Exploring Genomics exercises for Chapter 7), you were introduced to the National Center for Biotechnology Information (NCBI) Genes and Disease site. Now we will use an NCBI application called **BLAST, Basic Local Alignment Search Tool**. BLAST is an invaluable program for searching through GenBank and other databases to find DNA and protein-sequence similarities between cloned substances. It has many additional functions that we will explore in other exercises.

■ **Exercise I – Introduction to BLAST**

1. Access BLAST from the NCBI Web site at **http://blast.ncbi.nlm.nih.gov/Blast.cgi**.
2. Click on "nucleotide blast." This feature allows you to search DNA databases to look for a similarity between a sequence you enter and other sequences in the database. Do a nucleotide search with the following sequence:

   CCAGAGTCCAGCTGCTGCTCATA
   CTACTGATACTGCTGGG
3. Imagine that this sequence is a short part of a gene you cloned in your laboratory. You want to know if this gene or others with similar sequences have been discovered. Enter this sequence into the "Enter Query Sequence" text box at the top of the page. Near the bottom of the page, under the "Program Selection" category, choose "blastn"; then click on the "BLAST" button at the bottom of the page to run the search. It may take several minutes for results to be available because BLAST is using powerful algorithms to scroll through billions of bases of sequence data! A new page will appear with the results of your search.
4. On the search results page, below the Graphic Summary you will see a category called Descriptions and a table showing significant matches to the sequence you searched with (called the query sequence). BLAST determines significant matches based on statistical measures that consider the length of the query sequence, the number of matches with sequences in the database, and other factors. Significant *alignments,* regions of significant similarity in the query and subject sequences, typically have E values less than 1.0.
5. The top part of the table lists matches to transcripts (mRNA sequences), and the lower part lists matches to genomic DNA sequences, in order of highest to lowest number of matches.
6. Alignments are indicated by horizontal lines. BLAST adjusts for gaps in the sequences, that is, for areas that may not align precisely because of missing bases in otherwise similar sequences. Scroll below the table to see the aligned sequences from this search, and then answer the following questions:
   a. What were the top three matches to your query sequence?
   b. For each alignment, BLAST also indicates the percent *identity* and the number of gaps in the match between the query and subject sequences. What was the percent identity for the top three matches? What percentage of each aligned sequence showed gaps indicating sequence differences?
   c. Click on the links for the first matched sequence (far-right column). These will take you to a wealth of information, including the size of the sequence; the species it was derived from; a PubMed-linked chronology of research publications pertaining to this sequence; the complete sequence; and if the sequence encodes a polypeptide, the predicted amino acid sequence coded by the gene. Skim through the information presented for this gene. What is the gene's function?
7. A BLAST search can also be done by entering the *accession number* for a sequence, which is a unique identifying number assigned to a sequence before it can be put into a database. For example, search with the accession number NM_007305. What did you find?

CASE STUDY Credit where credit is due

In the early 1950s, it became clear to many researchers that DNA was the cellular molecule that carries genetic information. However, an understanding of the genetic properties of DNA could only be achieved through a detailed knowledge of its structure. To this end, several laboratories began a highly competitive race to discover the three-dimensional structure of DNA, which ended when Watson and Crick published their now classic paper in 1953. Their model was based, in part, on an X-ray diffraction photograph of DNA taken by Rosalind Franklin (Figure 9.10). Two ethical issues surround this photo. First, the photo was given to Watson and Crick by Franklin's co-worker, Maurice Wilkins, without her knowledge or consent. Second, in their paper, Watson and Crick did not credit Franklin's contribution. The fallout from these lapses lasted for decades and raises some basic questions about ethics in science.

1. What vital clues were provided by Franklin's work to Watson and Crick about the molecular structure of DNA?
2. Was it ethical for Wilkins to show Franklin's unpublished photo to Watson and Crick without her consent? Would it have been more ethical for Watson and Crick to have offered Franklin co-authorship on this paper?
3. Given that these studies were conducted in the 1950s, how might gender have played a role in the fact that Rosalind Franklin did not receive due credit for her X-ray diffraction work?

See the Understanding Science: How Science Really Works Web site: "Credit and debt"(http://undsci.berkeley.edu/article/0_0_0/dna_13).

INSIGHTS AND SOLUTIONS

This chapter recounts some of the initial experimental analyses that launched the era of molecular genetics. Quite fittingly, then, our "Insights and Solutions" section shifts its emphasis from problem solving to experimental rationale and analytical thinking.

1. Based strictly on the transformation analysis of Avery, MacLeod, and McCarty, what objection might be made to the conclusion that DNA is the genetic material? What other conclusion might be considered?

Solution: Based solely on their results, we could conclude that DNA is essential for transformation. However, DNA might have been a substance that caused capsular formation by converting nonencapsulated cells *directly* to cells with a capsule. That is, DNA may simply have played a catalytic role in capsular synthesis, leading to cells that display smooth, type III colonies.

2. What observations argue against this objection?

Solution: First, transformed cells pass the trait on to their progeny cells, thus supporting the conclusion that DNA is responsible for heredity, not for the direct production of polysaccharide coats. Second, subsequent transformation studies over the next five years showed that other traits, such as antibiotic resistance, could be transformed. Therefore, the transforming factor has a broad general effect, not one specific to polysaccharide synthesis.

3. If RNA were the universal genetic material, how would this have affected the Avery experiment and the Hershey–Chase experiment?

Solution: In the Avery experiment, ribonuclease (RNase) rather than deoxyribonuclease (DNase), would have eliminated transformation. Had this occurred, Avery and his colleagues would have concluded that RNA was the transforming factor. Hershey and Chase would have obtained identical results, since ^{32}P would also label RNA but not protein.

Problems and Discussion Questions

Mastering Genetics Visit for instructor-assigned tutorials and problems.

1. HOW DO WE KNOW? In this chapter, we have focused on DNA, the molecule that stores genetic information in all living things. In particular, we discussed its structure and delved into how we analyze this molecule. Based on your knowledge of these topics, answer several fundamental questions:
   (a) How were we able to determine that DNA, and not some other molecule, serves as the genetic material in bacteria, bacteriophages, and eukaryotes?
   (b) How do we know that the structure of DNA is in the form of a right-handed double-helical molecule?
   (c) How do we know that in DNA G pairs with C and that A pairs with T as complementary strands are formed?
2. CONCEPTS QUESTION Review the Chapter Concepts list on p. 185. Most center on DNA and RNA and their role of serving as the genetic material. Write a short essay that contrasts these molecules, including a comparison of advantages conferred by their structure that each of them has over the other in serving in this role. ■
3. What was Erwin Chargaff's contribution to the understanding of the DNA structure? What was its impact in the 1940s?
4. Contrast the contributions made to an understanding of transformation by Griffith and by Avery and his colleagues.
5. When Avery and his colleagues had obtained what was concluded to be the transforming factor from the IIIS virulent cells, they treated the fraction with proteases, ribonuclease, and deoxyribonuclease, followed by the assay for retention or loss of transforming ability. What were the purpose and results of these experiments? What conclusions were drawn?
6. Describe the differences between the original transformation and transfection experiments and discuss the implications of their results.
7. Does the design of the Hershey–Chase experiment distinguish between DNA and RNA as the molecule serving as the genetic material? Why or why not?
8. What observations are consistent with the conclusion that DNA serves as the genetic material in eukaryotes? List and discuss them.
9. What are the exceptions to the general rule that DNA is the genetic material in all organisms? What evidence supports these exceptions?
10. Draw the chemical structure of the three components of a nucleotide, and then link them together. What atoms are removed from the structures when the linkages are formed?
11. What are the structural differences between (a) purines and pyrimidines, and (b) ribose and deoxyribose sugars?
12. Cytosine may also be named 2-oxy-4-amino pyrimidine. How would you name the other four nitrogenous bases, using this alternative system? (CH_3 is methyl.)
13. Draw the chemical structure of a dinucleotide composed of A and G. Opposite this structure, draw the dinucleotide composed of T and C in an antiparallel (or upside-down) fashion. Form the possible hydrogen bonds.
14. Describe the various characteristics of the Watson–Crick double helix model for DNA.
15. What was known about the structure of the DNA before Watson and Crick proposed the double helix model in 1953?
16. What might Watson and Crick have concluded, had Chargaff's data from a single source indicated the following base composition?

	A	T	G	C
%	29	19	21	31

Why would this conclusion be contradictory to Wilkins and Franklin's data?

17. If the GC content of a DNA molecule is 60%, what are the molar percentages of the four bases (G, C, T, A)?
18. If an RNA strand generates its complementary strand, thus producing a double helix, will a molecule of this be structurally identical to that of DNA? Explain.
19. What are the three major types of RNA molecules? How is each related to the concept of information flow?
20. Describe Z-DNA, its structure, and its possible functions in a living organism.
21. What is the physical state of DNA after being denatured by heat?
22. What is the hyperchromic effect? How is it measured? What does T_m imply?
23. Why is T_m related to base composition?
24. What is the basis of electrophoretic separation of nucleic acids?
25. What did the Watson–Crick model suggest about the replication of DNA?
26. A genetics student was asked to draw the chemical structure of an adenine- and thymine-containing dinucleotide derived from DNA. His answer is shown below. The student made more than six major errors. One of them is circled, numbered 1, and explained. Find five others. Circle them, number them 2 to 6, and briefly explain each by following the example given.

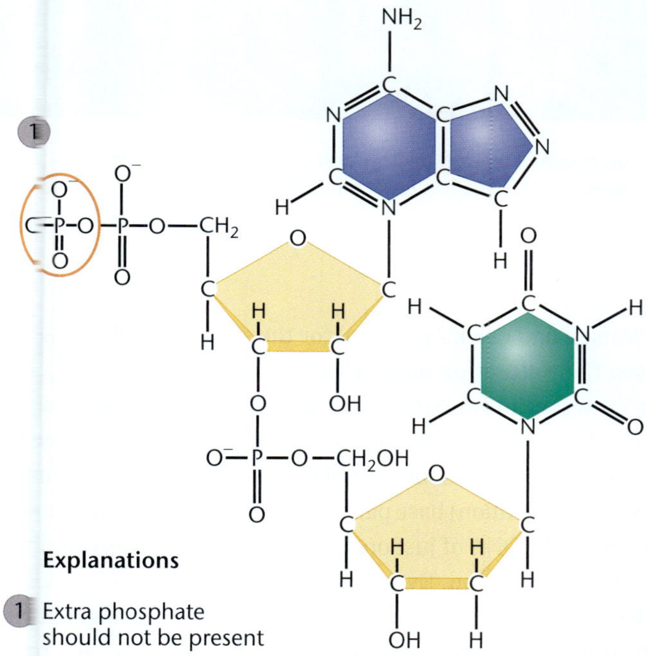

Explanations

1 Extra phosphate should not be present

27. A primitive eukaryote was discovered that displayed a unique nucleic acid as its genetic material. Analysis revealed the following observations:
    (a) X-ray diffraction studies display a general pattern similar to DNA, but with somewhat different dimensions and more irregularity.
    (b) A major hyperchromic shift is evident upon heating and monitoring UV absorption at 260 nm.
    (c) Base-composition analysis reveals four bases in the following proportions:

Adenine	= 8%	Hypoxanthine	= 18%
Guanine	= 37%	Xanthine	= 37%

    (d) About 75 percent of the sugars are deoxyribose, whereas 25 percent are ribose.

    Attempt to solve the structure of this molecule by postulating a model that is consistent with the foregoing observations.
28. While demethylation can convert thymine to uracil, deamination can convert cytosine to uracil. Suppose these two mutations occur in a cell. What would be the impact on the DNA structure?
29. Consider the structure of double-stranded DNA. When DNA is placed into distilled water, it denatures; however, by adding NaCl, the DNA renatures. Why?
30. *Newsdate: March 1, 2030.* A unique creature has been discovered during exploration of outer space. Recently, its genetic material has been isolated and analyzed, and has been found to be similar in some ways to DNA in chemical makeup. It contains in abundance the 4-carbon sugar erythrose and a molar equivalent of phosphate groups. In addition, it contains six nitrogenous bases: adenine (A), guanine (G), thymine (T), cytosine (C), hypoxanthine (H), and xanthine (X). These bases exist in the following relative proportion:

$$A = T = H \quad \text{and} \quad C = G = X$$

    X-ray diffraction studies have established a regularity in the molecule and a constant diameter of about 30 Å. Together, these data have suggested a model for the structure of this molecule.
    (a) Propose a general model of this molecule, and briefly describe it.
    (b) What base-pairing properties must exist for H and for X in the model?
    (c) Given the constant diameter of 30 Å, do you think *either* (i) both H and X are purines or both pyrimidines, *or* (ii) one is a purine and one is a pyrimidine?
31. You are provided with DNA samples from two newly discovered bacterial viruses. Based on the various analytical techniques discussed in this chapter, construct a research protocol that would be useful in characterizing and contrasting the DNA of both viruses. Indicate the type of information you hope to obtain for each technique included in the protocol.
32. During electrophoresis, DNA molecules can easily be separated according to size because all DNA molecules have the same charge–mass ratio and the same shape (long rod). Would you expect RNA molecules to behave in the same manner as DNA during electrophoresis? Why or why not?

10

DNA Replication

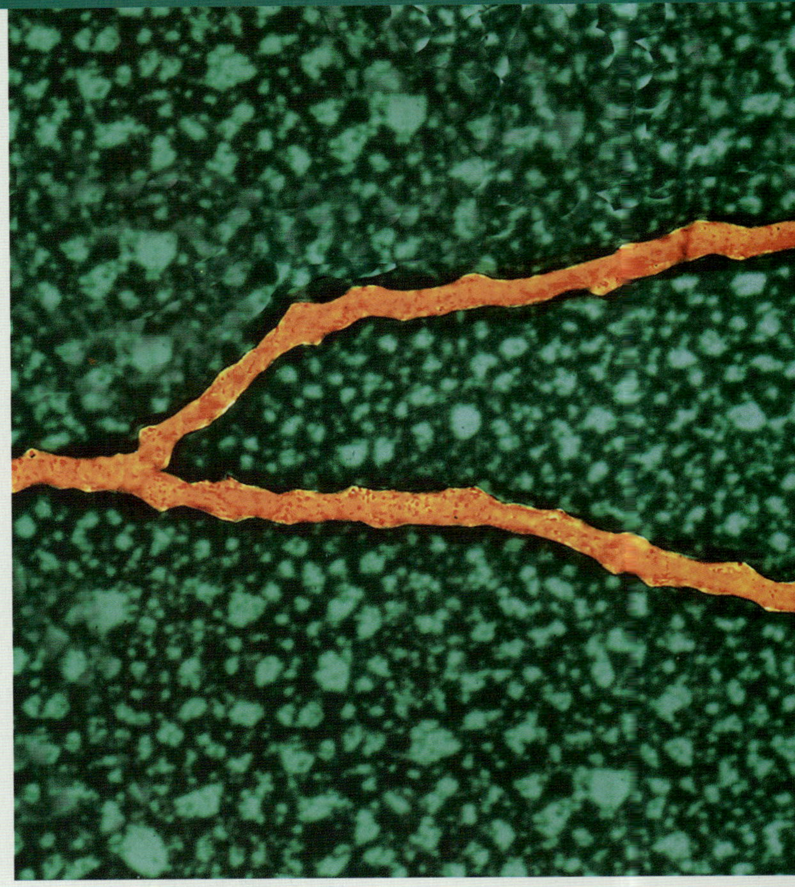

Transmission electron micrograph of human DNA from a HeLa cell, illustrating a replication fork characteristic of active DNA synthesis.

CHAPTER CONCEPTS

- Genetic continuity between parental and progeny cells is maintained by semiconservative replication of DNA, as predicted by the Watson–Crick model.
- Semiconservative replication uses each strand of the parent double helix as a template, and each newly replicated double helix includes one "old" and one "new" strand of DNA.
- DNA synthesis is a complex but orderly process, occurring under the direction of a myriad of enzymes and other proteins.
- DNA synthesis involves the polymerization of nucleotides into polynucleotide chains.
- DNA synthesis is similar in bacteria and eukaryotes, but more complex in eukaryotes.
- In eukaryotes, DNA synthesis at the ends of chromosomes (telomeres) poses a special problem, overcome by a unique RNA-containing enzyme, telomerase.

Following Watson and Crick's proposal for the structure of DNA, scientists focused their attention on how this molecule is replicated. Replication is an essential function of the genetic material and must be executed precisely if genetic continuity between cells is to be maintained following cell division. It is an enormous, complex task. Consider for a moment that more than 3×10^9 (3 billion) base pairs exist within the human genome. To duplicate faithfully the DNA of just one of these chromosomes requires a mechanism of extreme precision. Even an error rate of only 10^{-6} (one in a million) will still create 3000 errors (obviously an excessive number) during each replication cycle of the genome. Although it is not error free, and much of evolution would not have occurred if it were, an extremely accurate system of DNA replication has evolved in all organisms.

As Watson and Crick wrote at the end of their classic 1953 paper that announced the double-helical model of DNA, "It has not escaped our notice that the specific pairing (A-T and C-G) we have postulated immediately suggests a copying mechanism for the genetic material." Called semiconservative replication, this mode of DNA duplication was soon to receive strong support from numerous studies of viruses, bacteria, and eukaryotes. Once the general *mode* of replication was clarified, research to determine the precise details of the *synthesis* of DNA intensified. What has since been discovered is that numerous enzymes and other proteins are needed to copy a DNA helix. Because of the complexity of the chemical events during synthesis, this subject remains an extremely active area of research.

In this chapter, we will discuss the general mode of replication, as well as the specific details of DNA synthesis. The research leading to such knowledge is another link in our understanding of life processes at the molecular level.

10.1 DNA Is Reproduced by Semiconservative Replication

Watson and Crick recognized that, because of the arrangement and nature of the nitrogenous bases, each strand of a DNA double helix could serve as a template for the synthesis of its complement (**Figure 10.1**). They proposed that, if the helix were unwound, each nucleotide along the two parent strands would have an affinity for its complementary nucleotide. As we learned in Chapter 9, the complementarity is due to the potential hydrogen bonds that can be formed. If

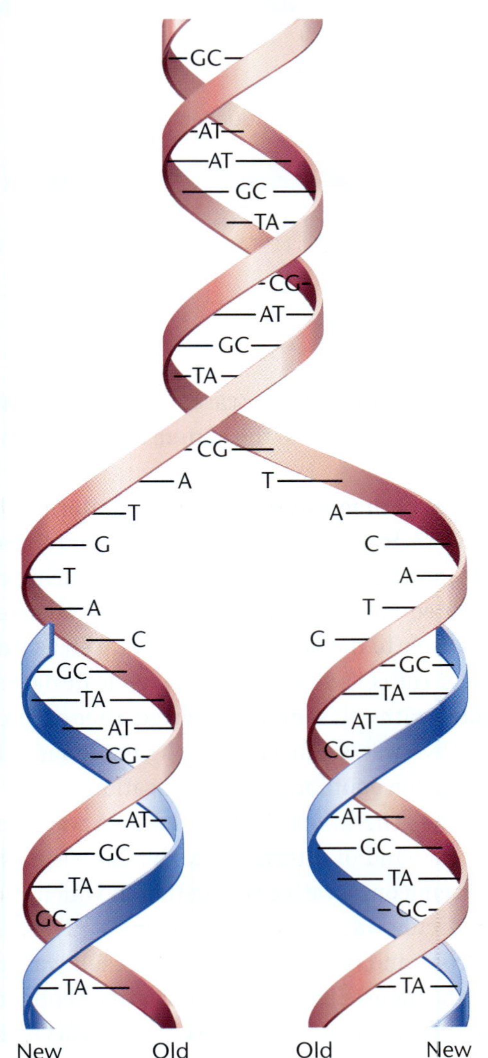

FIGURE 10.1 Generalized model of semiconservative replication of DNA. New synthesis is shown in blue.

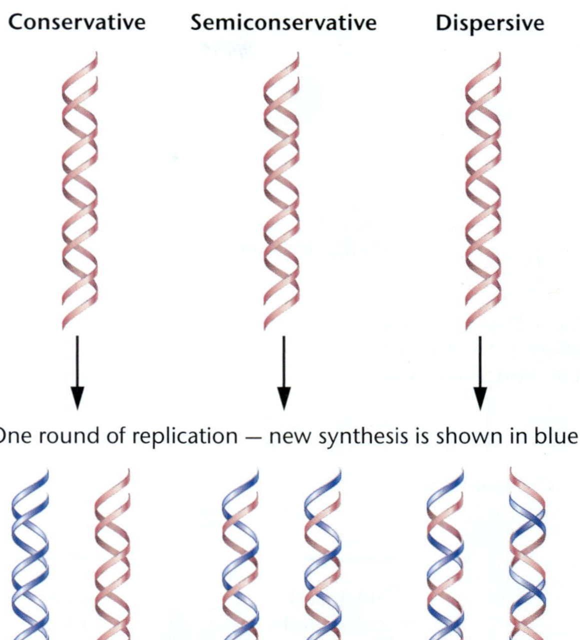

FIGURE 10.2 Results of one round of replication of DNA for each of the three possible modes by which replication could be accomplished.

thymidylic acid (T) were present, it would "attract" adenylic acid (A); if guanylic acid (G) were present, it would attract cytidylic acid (C); likewise, A would attract T, and C would attract G. If these nucleotides were then covalently linked into polynucleotide chains along both templates, the result would be the production of two identical double strands of DNA. Each replicated DNA molecule would consist of one "old" and one "new" strand, hence the reason for the name **semiconservative replication.**

Two other theoretical modes of replication are possible that also rely on the parental strands as a template (**Figure 10.2**). In **conservative replication,** complementary polynucleotide chains are synthesized as described earlier. Following synthesis, however, the two newly created strands then come together and the parental strands reassociate. The original helix is thus "conserved."

In the second alternative mode, called **dispersive replication,** the parental strands are dispersed into two new double helices following replication. Hence, each strand consists of both old and new DNA. This mode would involve cleavage of the parental strands during replication. It is the most complex of the three possibilities and is therefore considered to be least likely to occur. It could not, however, be ruled out as an experimental model. Figure 10.2 shows the theoretical results of a single round of replication by each of the three different modes.

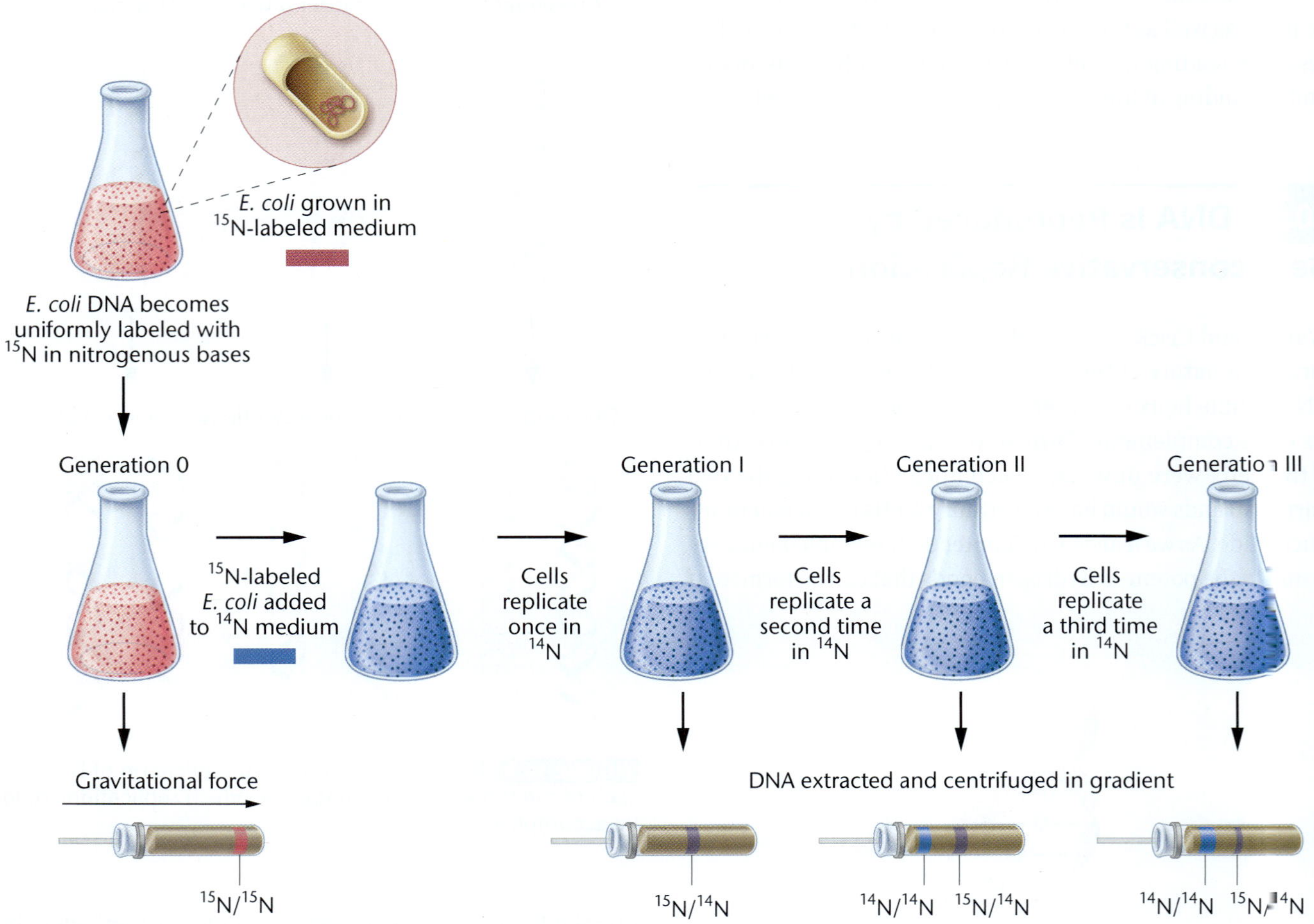

FIGURE 10.3 The Meselson–Stahl experiment.

The Meselson–Stahl Experiment

In 1958, Matthew Meselson and Franklin Stahl published the results of an experiment providing strong evidence that semiconservative replication is the mode used by bacterial cells to produce new DNA molecules. They grew *Escherichia coli* cells for many generations in a medium that had $^{15}NH_4Cl$ (ammonium chloride) as the only nitrogen source. A "heavy" isotope of nitrogen, ^{15}N contains one more neutron than the naturally occurring ^{14}N isotope; thus, molecules containing ^{15}N are more dense than those containing ^{14}N. Unlike radioactive isotopes, ^{15}N is stable. After many generations in this medium, almost all nitrogen-containing molecules in the *E. coli* cells, including the nitrogenous bases of DNA, contained the heavier isotope.

Critical to the success of this experiment, DNA containing ^{15}N can be distinguished from DNA containing ^{14}N. The experimental procedure involves the use of a technique referred to as *sedimentation equilibrium centrifugation* (also called buoyant density gradient centrifugation). Samples are forced by centrifugation through a density gradient of a heavy metal salt, such as cesium chloride. Molecules of DNA will reach equilibrium when their density equals the density of the gradient medium. In this case, ^{15}N-DNA will reach this point at a position closer to the bottom of the tube than will ^{14}N-DNA.

In this experiment (**Figure 10.3**), uniformly labeled ^{15}N cells were transferred to a medium containing only $^{14}NH_4Cl$. Thus, all "new" synthesis of DNA during replication contained only the "lighter" isotope of nitrogen. The time of transfer to the new medium was taken as time zero $t = 0$. The *E. coli* cells were allowed to replicate over several generations, with cell samples removed after each replication cycle. DNA was isolated from each sample and subjected to sedimentation equilibrium centrifugation.

After one generation, the isolated DNA was present in only a single band of intermediate density—the expected result for semiconservative replication in which each replicated molecule was composed of one new ^{14}N-strand and one old ^{15}N-strand (**Figure 10.4**). This result was not consistent with the prediction of conservative replication, in which two distinct bands would occur; thus this mode may be ruled out.

After two cell divisions, DNA samples showed two density bands—one intermediate band and one lighter

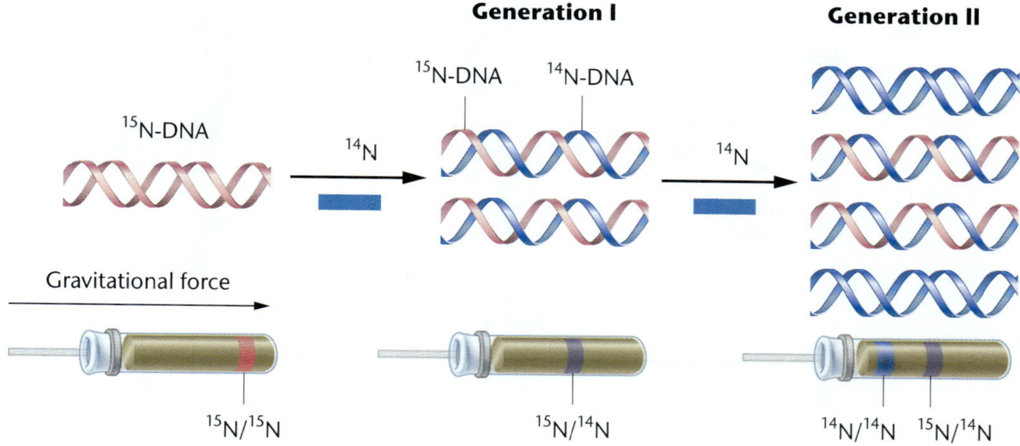

FIGURE 10.4 The expected results of two generations of semiconservative replication in the Meselson–Stahl experiment.

band corresponding to the ^{14}N position in the gradient. Similar results occurred after a third generation, except that the proportion of the lighter band increased. This was again consistent with the interpretation that replication is semiconservative.

You may have realized that a molecule exhibiting intermediate density is also consistent with dispersive replication. However, Meselson and Stahl also ruled out this mode of replication on the basis of two observations. First, after the first generation of replication in an ^{14}N-containing medium, they isolated the hybrid molecule and heat denatured it.

Recall from Chapter 9 that heating will separate a duplex into single strands. When the densities of the single strands of the hybrid were determined, they exhibited *either* an ^{15}N profile *or* an ^{14}N profile, but *not* an intermediate density. This observation is consistent with the semiconservative mode but inconsistent with the dispersive mode.

Furthermore, if replication were dispersive, *all* generations after $t = 0$ would demonstrate DNA of an intermediate density. In each generation after the first, the ratio of $^{15}N/^{14}N$ would decrease, and the hybrid band would become lighter and lighter, eventually approaching the ^{14}N band. This result was not observed. The Meselson–Stahl experiment provided conclusive support for semiconservative replication in bacteria and tended to rule out both the conservative and dispersive modes.

ESSENTIAL POINT

In 1958 Meselson and Stahl resolved the question of which of three potential modes of replication is utilized by *E. coli* during the duplication of DNA in favor of semiconservative replication, showing that newly synthesized DNA consists of one old strand and one new strand. ■

NOW SOLVE THIS

10.1 In the Meselson–Stahl experiment, which of the three modes of replication could be ruled out after one round of replication? After two rounds?

■ **HINT:** *This problem involves an understanding of the nature of the experiment as well as the difference between the three possible modes of replication. The key to its solution is to determine which mode will not create "hybrid" helices after one round of replication.*

Semiconservative Replication in Eukaryotes

In 1957, the year before the work of Meselson and Stahl was published, J. Herbert Taylor, Philip Woods, and Walter Hughes presented evidence that semiconservative replication also occurs in eukaryotic organisms. They experimented with root tips of the broad bean *Vicia faba*, which are an excellent source of dividing cells. These researchers were able to monitor the process of replication by labeling DNA with ^{3}H-thymidine, a radioactive precursor of DNA, and performing autoradiography.

Autoradiography is a common technique that, when applied cytologically, pinpoints the location of a radioisotope in a cell. In this procedure, a photographic emulsion is placed over a histological preparation containing cellular material (root tips, in this experiment), and the preparation is stored in the dark. The slide is then developed, much as photographic film is processed. Because the radioisotope emits energy, upon development the emulsion turns black at the approximate point of emission. The end result is the presence of dark spots or "grains" on the surface of the section, identifying the location of newly synthesized DNA within the cell.

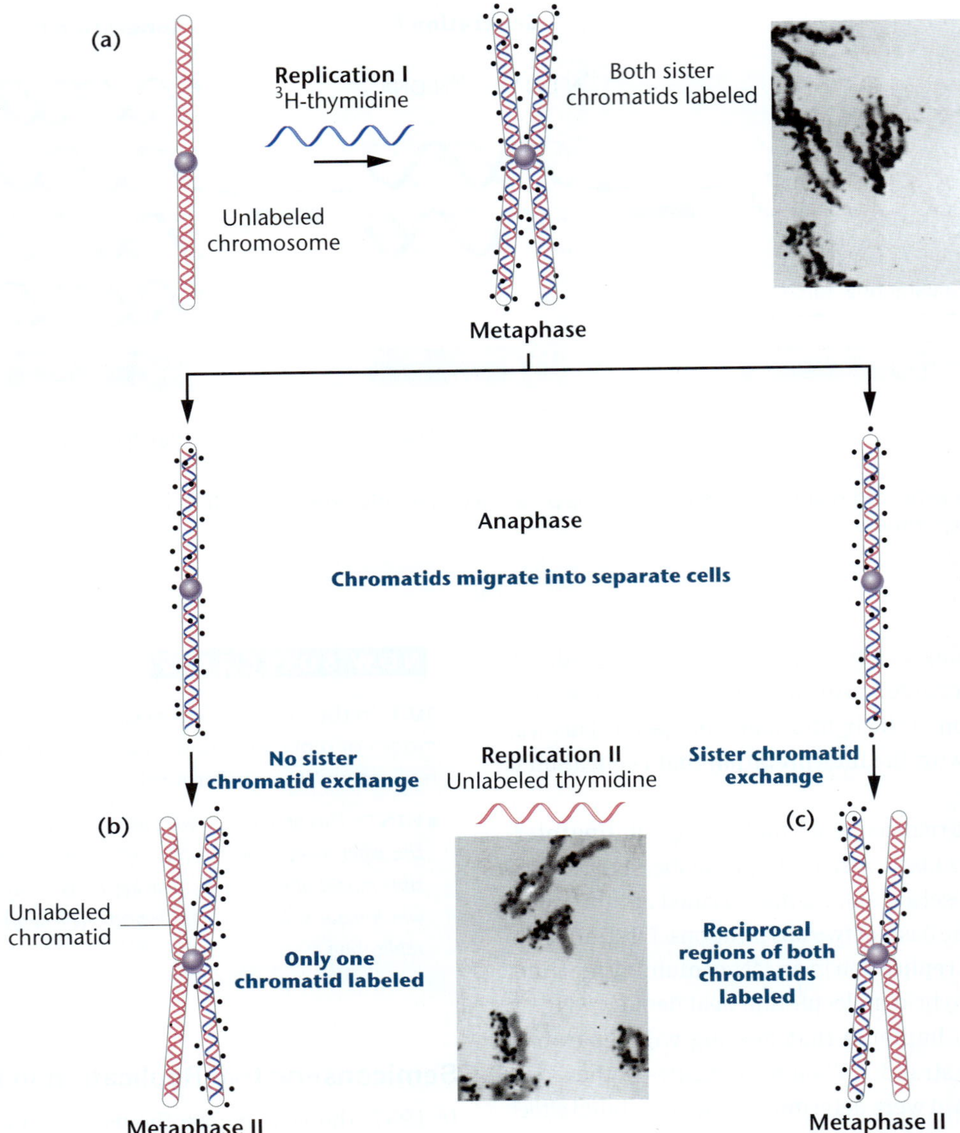

FIGURE 10.5 The Taylor–Woods–Hughes experiment, demonstrating the semiconservative mode of replication of DNA in the root tips of *Vicia faba*. (a) An unlabeled chromosome proceeds through the cell cycle in the presence of ^{3}H-thymidine. As it enters mitosis, both sister chromatids of the chromosome are labeled, as shown by autoradiography. After a second round of replication (b), this time in the absence of ^{3}H-thymidine, only one chromatid of each chromosome is expected to be surrounded by grains. Except where a reciprocal exchange has occurred between sister chromatids (c), the expectation was upheld.

Taylor and his colleagues grew root tips for approximately one generation in the presence of the radioisotope and then placed them in unlabeled medium in which cell division continued. At the conclusion of each generation, they arrested the cultures at metaphase by adding colchicine (a chemical derived from the crocus plant that poisons the spindle fibers) and then examined the chromosomes by autoradiography. They found radioactive thymidine only in association with chromatids that contained newly synthesized DNA. **Figure 10.5** illustrates the replication of a single chromosome over two division cycles, including the distribution of grains.

These results are compatible with the semiconservative mode of replication. After the first replication cycle in the presence of the isotope, both sister chromatids show radioactivity, indicating that each chromatid contains one new radioactive DNA strand and one old unlabeled strand. After the second replication cycle, *which takes place in unlabeled medium*, only one of the two sister chromatids of each chromosome should be radioactive because half of the parent strands are unlabeled. With only the minor exceptions of *sister chromatid exchanges* (discussed in Chapter 7), this result was observed.

Together, the Meselson–Stahl experiment and the experiment by Taylor, Woods, and Hughes soon led to the general acceptance of the semiconservative mode of replication. Later studies with other organisms reached the same conclusion and also strongly supported Watson and Crick's proposal for the double helix model of DNA.

ESSENTIAL POINT

Taylor, Woods, and Hughes demonstrated semiconservative replication in eukaryotes using the root tips of the broad bean as the source of dividing cells. ■

Origins, Forks, and Units of Replication

To enhance our understanding of semiconservative replication, let's briefly consider a number of relevant issues. The first concerns the *origin of replication*. Where along the chromosome is DNA replication initiated? Is there only a single origin, or does DNA synthesis begin at more than one point? Is any given point of origin random, or is it located at a specific region along the chromosome? Second, once replication begins, does it proceed in a single direction or in both directions away from the origin? In other words, is replication *unidirectional* or *bidirectional*?

To address these issues, we need to introduce two terms. First, at each point along the chromosome where replication is occurring, the strands of the helix are unwound, creating what is called a **replication fork** (see the chapter opening photograph on p. 206). Such a fork will initially appear at the point of origin of synthesis and then move along the DNA duplex as replication proceeds. If replication is bidirectional, two such forks will be present, migrating in opposite directions away from the origin. The second term refers to the length of DNA that is replicated following one initiation event at a single origin. This is a unit referred to as the **replicon.**

The evidence is clear regarding the origin and direction of replication. John Cairns tracked replication in *E. coli*, using radioactive precursors of DNA synthesis and autoradiography. He was able to demonstrate that in *E. coli* there is only a single region, called *oriC*, where replication is initiated. The presence of only a single origin is characteristic of bacteria, which have only one circular chromosome. Since DNA synthesis in bacteriophages and bacteria originates at a single point, the entire chromosome constitutes one replicon. In *E. coli*, the replicon consists of the entire genome of 4.6 Mb (4.6 million base pairs).

Figure 10.6 illustrates Cairns's interpretation of DNA replication in *E. coli*. This interpretation and the accompanying micrograph do not answer the question of unidirectional versus bidirectional synthesis. However, other results, derived from studies of bacteriophage lambda, have demonstrated that replication is bidirectional, moving away

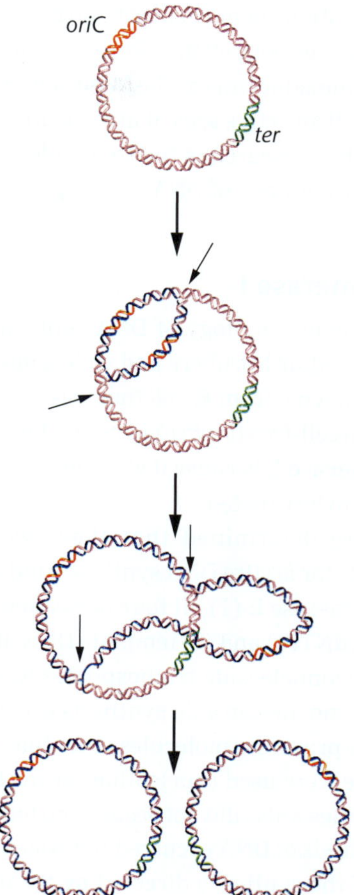

FIGURE 10.6 Bidirectional replication of the *E. coli* chromosome. The thin black arrows identify the advancing replication forks.

from *oriC* in both directions. Figure 10.6 therefore interprets Cairns's work with that understanding. Bidirectional replication creates two replication forks that migrate farther and farther apart as replication proceeds. These forks eventually merge, as semiconservative replication of the entire chromosome is completed, at a termination region, called *ter*.

Later in this chapter, we will see that in eukaryotes, each chromosome contains multiple points of origin.

10.2 DNA Synthesis in Bacteria Involves Five Polymerases, as Well as Other Enzymes

To say that replication is semiconservative and bidirectional describes the overall *pattern* of DNA duplication and the association of finished strands with one another once synthesis is completed. However, it says little about the more complex issue of how the actual *synthesis* of long complementary polynucleotide chains occurs on a DNA template.

Like most questions in molecular biology, this one was first studied using microorganisms. Research on DNA synthesis began about the same time as the Meselson–Stahl work, and the topic is still an active area of investigation. What is most apparent in this research is the tremendous complexity of the biological synthesis of DNA.

DNA Polymerase I

Studies of the enzymology of DNA replication were first reported by Arthur Kornberg and colleagues in 1957. They isolated an enzyme from *E. coli* that was able to direct DNA synthesis in a cell-free (*in vitro*) system. The enzyme is called **DNA polymerase I,** because it was the first of several similar enzymes to be isolated.

Kornberg determined that there were two major requirements for *in vitro* DNA synthesis under the direction of DNA polymerase I: (1) all four deoxyribonucleoside triphosphates (dNTPs) and (2) template DNA. If any one of the four deoxyribonucleoside triphosphates was omitted from the reaction, no measurable synthesis occurred. If derivatives of these precursor molecules other than the nucleoside triphosphate were used (nucleotides or nucleoside diphosphates), synthesis also did not occur. If no template DNA was added, synthesis of DNA occurred but was reduced greatly.

Most of the synthesis directed by Kornberg's enzyme appeared to be exactly the type required for semiconservative replication. The reaction is summarized in **Figure 10.7**, which depicts the addition of a single nucleotide. The enzyme has since been shown to consist of a single polypeptide containing 928 amino acids.

The way in which each nucleotide is added to the growing chain is a function of the specificity of DNA polymerase I. As shown in **Figure 10.8**, the precursor dNTP contains the three phosphate groups attached to the 5′-carbon of deoxyribose. As the two terminal phosphates are cleaved during synthesis, the remaining phosphate attached to the 5′-carbon is covalently linked to the 3′-OH group of the deoxyribose to which it is added. Thus, **chain elongation** occurs in the **5′ to 3′ direction** by the addition of one nucleotide at a time to the growing 3′-end. Each step provides a newly exposed 3′-OH group that can participate in the next addition of a nucleotide as DNA synthesis proceeds.

Having isolated DNA polymerase I and demonstrated its catalytic activity, Kornberg next sought to demonstrate the accuracy, or fidelity, with which the enzyme replicated the DNA template. Because technology for ascertaining the nucleotide sequences of the template and newly synthesized strand was not yet available in 1957, he initially had to rely on several indirect methods.

One of Kornberg's approaches was to compare the nitrogenous base compositions of the DNA template with those of the recovered DNA product. Using several sources of DNA (phage T2, *E. coli*, and calf thymus), he discovered that, within experimental error, the base composition of each product agreed with the template DNA used. This suggested that the templates were replicated faithfully.

ESSENTIAL POINT

Arthur Kornberg isolated the enzyme DNA polymerase I from *E. coli* and showed that it is capable of directing *in vitro* DNA synthesis, provided that a template and precursor nucleoside triphosphates were supplied. ■

DNA Polymerase II, III, IV, and V

While DNA polymerase I clearly directs the synthesis of DNA, a serious reservation about the enzyme's true biological role was raised in 1969. Paula DeLucia and John Cairns discovered a mutant strain of *E. coli* that was deficient in polymerase I activity. The mutation was designated *polA1*. In the absence of the functional enzyme, this mutant strain of *E. coli* still duplicated its DNA and successfully reproduced. However, the cells were deficient in their ability to repair DNA. For example, the mutant strain is highly sensitive to ultraviolet light (UV) and radiation, both of which damage DNA and are mutagenic. Nonmutant bacteria are able to repair a great deal of UV-induced damage.

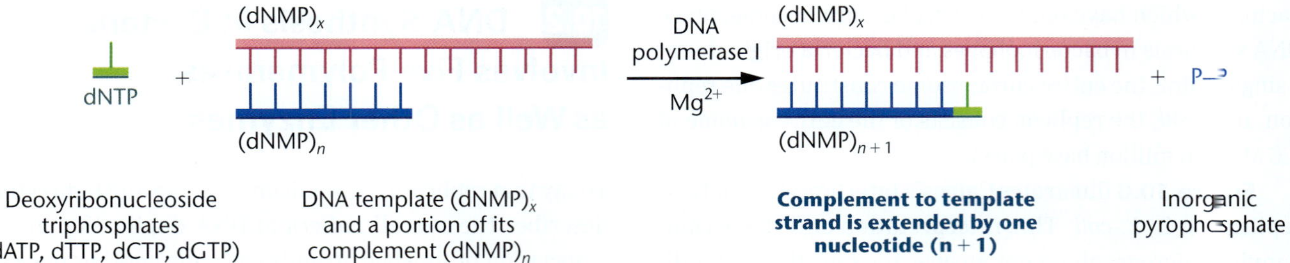

FIGURE 10.7 The chemical reaction catalyzed by DNA polymerase I. During each step, a single nucleotide is added to the growing complement of the DNA template using a nucleoside triphosphate as the substrate. The release of inorganic pyrophosphate drives the reaction energetically.

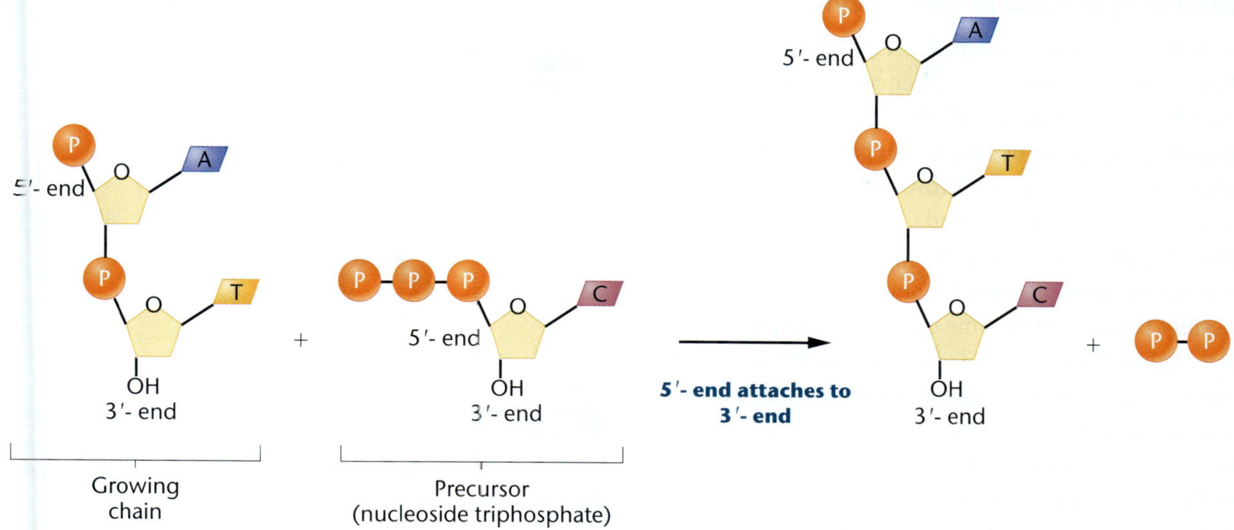

FIGURE 10.8 Demonstration of 5′ to 3′ synthesis of DNA.

These observations led to two conclusions:

1. At least one other enzyme that is responsible for replicating DNA *in vivo* is present in *E. coli* cells.
2. DNA polymerase I serves a secondary function *in vivo*, now believed to be critical to the maintenance of fidelity of DNA synthesis.

To date, four other unique DNA polymerases have been isolated from cells lacking polymerase I activity and from normal cells that contain polymerase I. **Table 10.1** contrasts several characteristics of DNA polymerase I with **DNA polymerase II** and **III.** Although none of the three can *initiate* DNA synthesis on a template, all three can *elongate* an existing DNA strand, called a **primer,** and all three possess 3′ to 5′ exonuclease activity, which means that they have the potential to polymerize in one direction and then pause, reverse their direction, and excise nucleotides just added. As we will later discuss, this activity provides a capacity to proofread newly synthesized DNA and to remove and replace incorrect nucleotides.

DNA polymerase I also demonstrates 5′ to 3′ exonuclease activity. This activity allows the enzyme to excise nucleotides, starting at the end at which synthesis begins and proceeding in the same direction of synthesis. Two final observations probably explain why Kornberg isolated polymerase I and not polymerase III: polymerase I is present in greater amounts than is polymerase III, and it is also much more stable.

What then are the roles of the polymerases *in vivo*? Polymerase III is the enzyme responsible for the 5′ to 3′ polymerization essential to *in vivo* replication. Its 3′ to 5′ exonuclease activity also provides a proofreading function that is activated when it inserts an incorrect nucleotide. When this occurs, synthesis stalls and the polymerase "reverses course," excising the incorrect nucleotide. Then, it proceeds back in the 5′ to 3′ direction, synthesizing the complement of the template strand. Polymerase I is believed to be responsible for removing the primer, as well as for the synthesis that fills gaps produced after this removal. Its exonuclease activities also allow for its participation in DNA repair. Polymerase II, as well as **polymerase IV and V,** are involved in various aspects of repair of DNA that has been damaged by external forces, such as ultraviolet light. Polymerase II is encoded by a gene whose transcription is activated by disruption of DNA synthesis at the replication fork.

TABLE 10.1 Properties of Bacterial DNA Polymerases I, II, and III

Properties	I	II	III
Initiation of chain synthesis	−	−	−
5′–3′ polymerization	+	+	+
3′–5′ exonuclease activity	+	+	+
5′–3′ exonuclease activity	+	−	−
Molecules of polymerase/cell	400	?	15

ESSENTIAL POINT

The discovery of the *polA1* mutant strain of *E. coli*, capable of DNA replication despite its lack of polymerase I activity, cast doubt on the enzyme's hypothesized *in vivo* replicative function. Polymerase III has been identified as the enzyme responsible for DNA replication *in vivo*. ■

The DNA Polymerase III Holoenzyme

We conclude this section by emphasizing the complexity of the DNA polymerase III enzyme, henceforth referred to as DNA Pol III. The active form of DNA Pol III, referred to as the **holoenzyme,** is made up of unique polypeptide subunits, ten of which have been identified (**Table 10.2**). The largest subunit, α, along with subunits ε and θ, form a complex called the **core enzyme,** which imparts the catalytic function to the holoenzyme. In *E. coli*, each holoenzyme contains two, and possibly three, core enzyme complexes. As part of each core, the α subunit is responsible for DNA synthesis along the template strands, whereas the ε subunit possesses 3′ to 5′ exonuclease capability, essential to proofreading. The need for more than one core enzyme will soon become apparent. A second group of five subunits (γ, δ, δ', χ, and υ) are complexed to form what is called the **sliding clamp loader,** which pairs with the core enzyme and facilitates the function of a critical component of the holoenzyme, called the **sliding DNA clamp.** The enzymatic function of the sliding clamp loader is dependent on energy generated by the hydrolysis of ATP. The sliding DNA clamp links to the core enzyme and is made up of multiple copies of the β subunit, taking on the shape of a donut, whereby it can open and shut, to encircle the unreplicated DNA helix. By doing so, and being linked to the core enzyme, the clamp leads the way during synthesis, maintaining the binding of the core enzyme to the template during polymerization of nucleotides. Thus, the length of DNA that is replicated by the core enzyme before it detaches from the template, a property referred to as **processivity,** is vastly increased. There is one sliding clamp per core enzyme. Finally, one τ subunit interacts with each core enzyme, linking it to the sliding clamp loader.

The DNA Pol III holoenzyme is diagrammatically illustrated in **Figure 10.9**. You should compare the diagram to the description of each component above. Note that we have shown the holoenzyme to contain two core enzyme complexes, although as stated above, a third one may be present. The components of the DNA Pol III holoenzyme will be referred to in the discussion that follows.

TABLE 10.2 Subunits of the DNA Polymerase III Holoenzyme

Subunit	Function	Groupings
α	5′–3′ polymerization	Core enzyme: Elongates polynucleotide chain and proofreads
ε	3′–5′ exonuclease	
θ	Core assembly	
γ	Loads enzyme on template (serves as clamp loader)	γ complex
δ		
δ'		
χ		
υ		
β	Sliding clamp structure (processivity factor)	
τ	Dimerizes core complex	

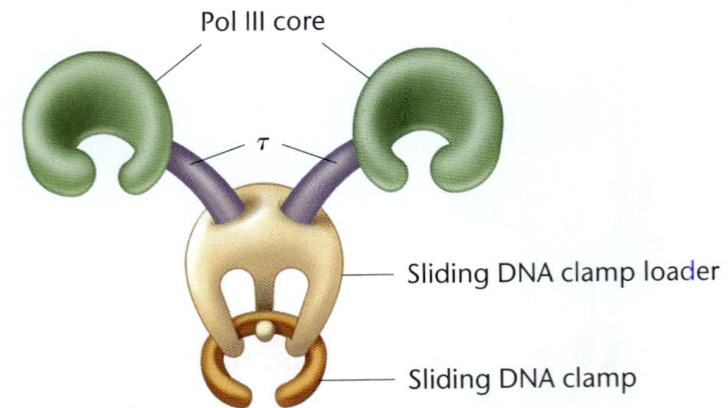

FIGURE 10.9 The components making up the DNA Pol III holoenzyme, as described in the text. While there may be three core enzyme complexes present in the holoenzyme, for simplicity, we illustrate only two.

10.3 Many Complex Issues Must Be Resolved during DNA Replication

We have thus far established that in bacteria and viruses replication is semiconservative and bidirectional along a single replicon. We also know that synthesis is catalyzed by DNA polymerase III and occurs in the 5′ to 3′ direction. Bidirectional synthesis creates two replication forks that move in opposite directions away from the origin of synthesis. As we can see from the following list, many issues remain to be resolved in order to provide a comprehensive understanding of DNA replication:

1. The helix must undergo localized unwinding, and the resulting "open" configuration must be stabilized so that synthesis may proceed along both strands.
2. As unwinding and subsequent DNA synthesis proceed, increased coiling creates tension further down the helix, which must be reduced.
3. A primer of some sort must be synthesized so that polymerization can commence under the direction of DNA polymerase III. Surprisingly, RNA, not DNA, serves as the primer.
4. Once the RNA primers have been synthesized, DNA polymerase III begins to synthesize the DNA complement of both strands of the parent molecule. Because the two strands are antiparallel to one another, continuous synthesis in the direction that the replication fork moves is possible along only one of the two strands. On the other

strand, synthesis must be discontinuous and thus involves a somewhat different process.

5. The RNA primers must be removed prior to completion of replication. The gaps that are temporarily created must be filled with DNA complementary to the template at each location.

6. The newly synthesized DNA strand that fills each temporary gap must be joined to the adjacent strand of DNA.

7. While DNA polymerases accurately insert complementary bases during replication, they are not perfect, and, occasionally, incorrect nucleotides are added to the growing strand. A proofreading mechanism that also corrects errors is an integral process during DNA synthesis.

As we consider these points, examine Figures 10.10, 10.11, and 10.12 to see how each issue is resolved. Figure 10.13 summarizes the model of DNA synthesis.

Unwinding the DNA Helix

As discussed earlier, there is a single point of origin along the circular chromosome of most bacteria and viruses at which DNA replication is initiated. This region in *E. coli* has been particularly well studied and is called ***oriC***. It consists of 245 DNA base pairs and is characterized by five repeating sequences of 9 base pairs, and three repeating sequences of 13 base pairs, called **9mers** and **13mers,** respectively. Both 9mers and 13mers are AT-rich, which renders them relatively less stable than an average double-helical sequence of DNA, which no doubt enhances helical unwinding. A specific initiator protein, called **DnaA** (because it is encoded by the *dnaA* gene), is responsible for initiating replication by binding to a region of 9mers. This newly formed complex then undergoes a slight conformational change and associates with the region of 13mers, which causes the helix to destabilize and open up, exposing single-stranded regions of DNA (ssDNA). This step facilitates the subsequent binding of another key player in the process—a protein called **DNA helicase** (made up of multiple copies of the DnaB polypeptide). DNA helicase is assembled as a hexamer of subunits around one of the exposed single-stranded DNA molecules. The helicase subsequently recruits the holoenzyme to bind to the newly formed replication fork to formally initiate replication, and it then proceeds to move along the ssDNA, opening up the helix as it progresses. Helicases require energy supplied by the hydrolysis of ATP, which aids in denaturing the hydrogen bonds that stabilize double helix.

Once the helicase has opened up the helix and ssDNA is available, base pairing must be inhibited until it can serve as a template for synthesis. This is accomplished by proteins that bind specifically to single strands of DNA, appropriately called **single-stranded binding proteins (SSBs).**

As unwinding proceeds, a coiling tension is created ahead of the replication fork, often producing **supercoiling.** In circular molecules, supercoiling may take the form of added twists and turns of the DNA, much like the coiling you can create in a rubber band by stretching it out and then twisting one end. Such supercoiling can be relaxed by **DNA gyrase,** a member of a larger group of enzymes referred to as **DNA topoisomerases.** The gyrase makes either single- or double-stranded "cuts" and also catalyzes localized movements that have the effect of "undoing" the twists and knots created during supercoiling. The strands are then resealed. These various reactions are driven by the energy released during ATP hydrolysis.

Together, the DNA, the polymerase complex, and associated enzymes make up an array of molecules that initiate DNA synthesis and are part of what we have previously called the *replisome*.

ESSENTIAL POINT

During the initiation of DNA synthesis, the double helix unwinds, forming a replication fork at which synthesis begins. Proteins stabilize the unwound helix and assist in relaxing the coiling tension created ahead of the replication fork. ■

Initiation of DNA Synthesis Using an RNA Primer

Once a small portion of the helix is unwound, what else is needed to initiate synthesis? As we have seen, DNA polymerase III requires a primer with a free 3′-hydroxyl group in order to elongate a polynucleotide chain. Since none is available in a circular chromosome, this absence prompted researchers to investigate how the first nucleotide could be added. It is now clear that RNA serves as the primer that initiates DNA synthesis.

A short segment of RNA (about 10 to 12 ribonucleotides long), complementary to DNA, is first synthesized on the DNA template. Synthesis of the RNA is directed by a form of RNA polymerase called **primase,** which does not require a free 3′-end to initiate synthesis. It is to this short segment of RNA that DNA polymerase III begins to add deoxyribonucleotides, initiating DNA synthesis. A conceptual diagram of initiation on a DNA template is shown in **Figure 10.10**. Later, the RNA primer is clipped out and replaced with DNA. This is thought to occur under the direction of DNA polymerase I. Recognized in viruses, bacteria, and several eukaryotic organisms, RNA priming is a universal phenomenon during the initiation of DNA synthesis.

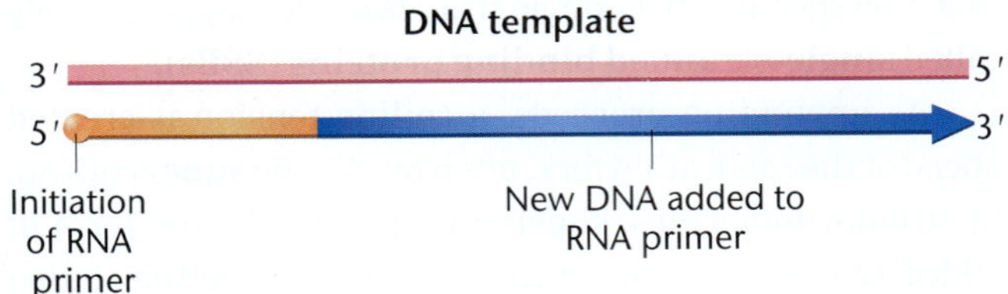

FIGURE 10.10 The initiation of DNA synthesis. A complementary RNA primer is first synthesized, to which DNA is added. All synthesis is in the 5′ to 3′ direction. Eventually, the RNA primer is replaced with DNA under the direction of DNA polymerase I.

ESSENTIAL POINT

DNA synthesis is initiated at specific sites along each template strand by the enzyme primase, resulting in short segments of RNA that provide suitable 3′-ends upon which DNA polymerase III can begin polymerization. ■

Continuous and Discontinuous DNA Synthesis

We must now revisit the fact that the two strands of a double helix are **antiparallel** to each other—that is, one runs in the 5′ to 3′ direction, while the other has the opposite 3′ to 5′ polarity. Because DNA polymerase III synthesizes DNA in only the 5′ to 3′ direction, synthesis along an advancing replication fork occurs in one direction on one strand and in the opposite direction on the other.

As a result, as the strands unwind and the replication fork progresses down the helix (**Figure 10.11**), only one strand can serve as a template for **continuous DNA synthesis.** This newly synthesized DNA is called the **leading strand.** As the fork progresses, many points of initiation are necessary on the opposite DNA template, resulting in **discontinuous DNA synthesis*** of the **lagging strand.**

Evidence supporting the occurrence of discontinuous DNA synthesis was first provided by Tsuneko and Reiji Okazaki award. They discovered that when bacteriophage DNA is replicated in *E. coli*, some of the newly formed DNA that is hydrogen bonded to the template strand is present as small fragments containing 1000 to 2000 nucleotides. RNA primers are part of each such fragment. These pieces, now called **Okazaki fragments,** are converted into longer and longer DNA strands of higher molecular weight as synthesis proceeds.

Discontinuous synthesis of DNA requires enzymes that both remove the RNA primers and unite the Okazaki fragments into the lagging strand. As we have noted, DNA polymerase I removes the primers and replaces the missing nucleotides. Joining the fragments is the work of **DNA ligase,**

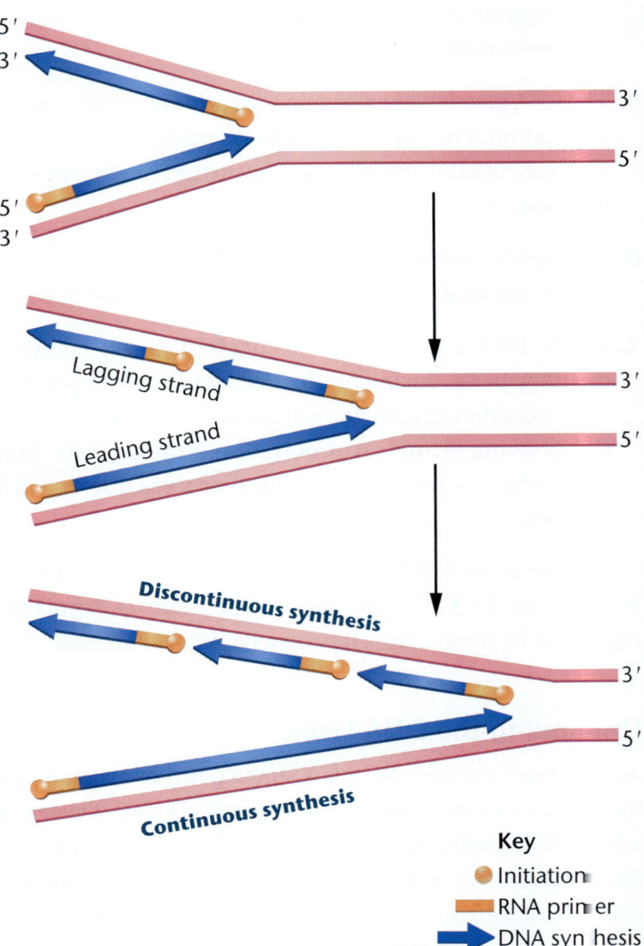

FIGURE 10.11 Opposite polarity of synthesis along the two strands of DNA is necessary because they run antiparallel to one another, and because DNA polymerase III synthesizes in only one direction (5′ to 3′). On the lagging strand, synthesis must be discontinuous, resulting in the production of Okazaki fragments. On the leading strand, synthesis is continuous. RNA primers are used to initiate synthesis on both strands.

which is capable of catalyzing the formation of the phosphodiester bond that seals the nick between the discontinuously synthesized strands. The evidence that DNA ligase performs this function during DNA synthesis is strengthened by the observation of a ligase-deficient mutant strain (*lig*) of *E. coli*, in which a large number of unjoined Okazaki fragments accumulate.

Concurrent Synthesis Occurs on the Leading and Lagging Strands

Given the model just discussed, we might ask how the holoenzyme of DNA Pol III synthesizes DNA on both the leading and lagging strands. Can both strands be replicated simultaneously at the same replication fork, or are the events distinct, involving two separate copies of the enzyme? Evidence suggests that both strands are replicated simultaneously, with each strand acted upon by one of the two core enzymes

*Because DNA synthesis is continuous on one strand and discontinuous on the other, the term **semidiscontinuous synthesis** is sometimes used to describe the overall process.

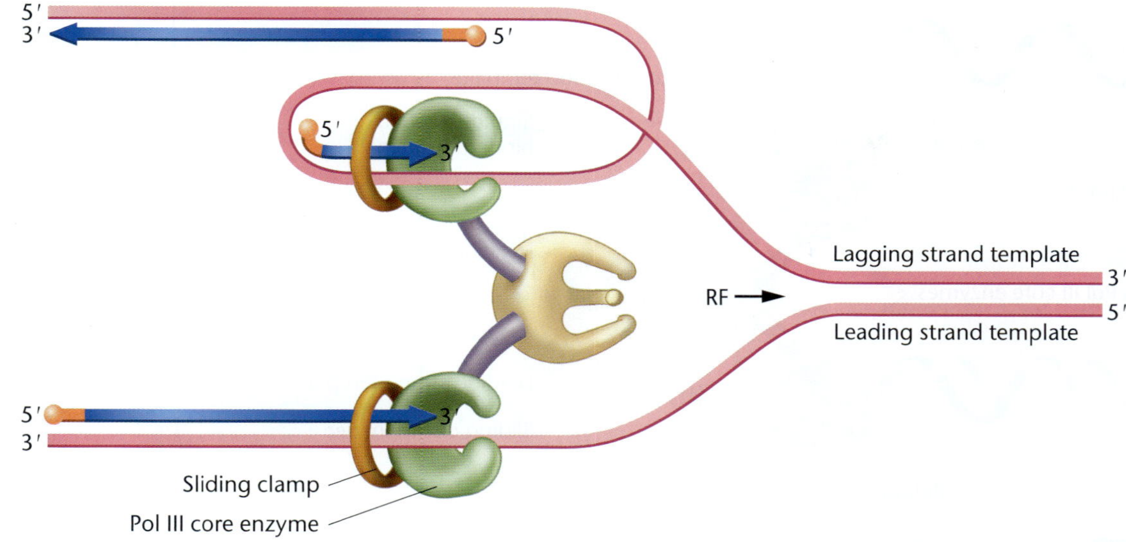

FIGURE 10.12 Illustration of how concurrent DNA synthesis may be achieved on both the leading and lagging strands at a single replication fork (RF). The lagging template strand is "looped" in order to invert the physical direction of synthesis, but not the biochemical direction. The enzyme functions as a dimer, with each core enzyme achieving synthesis on one or the other strand.

that are part of the DNA Pol III holoenzyme. As **Figure 10.12** illustrates, if the lagging strand template is spooled out, forming a loop, nucleotide polymerization can occur simultaneously on both template strands under the direction of the holoenzyme. After the synthesis of 1000 to 2000 nucleotides, the monomer of the enzyme on the lagging strand will encounter a completed Okazaki fragment, at which point it releases the lagging strand. A new loop of the lagging strand is spooled out, and the process is repeated. Looping inverts the orientation of the template but not the direction of actual synthesis on the lagging strand, which is always in the 5′ to 3′ direction. As mentioned above, it is believed that there is a third core enzyme associated with the DNA Pol III holoenzyme, and that it functions in the synthesis of Okazaki fragments. For simplicity, we will include only two core enzymes in this and subsequent figures.

Another important feature of the holoenzyme that facilitates synthesis at the replication fork is the donut-shaped sliding DNA clamp that surrounds the unreplicated double helix and is linked to the advancing core enzyme. This clamp prevents the core enzyme from dissociating from the template as polymerization proceeds. By doing so, the clamp is responsible for vastly increasing the processivity of the core enzyme—that is, the number of nucleotides that may be continually added prior to dissociation from the template. This function is critical to the rapid *in vivo* rate of DNA synthesis during replication.

ESSENTIAL POINT

Concurrent DNA synthesis occurs continuously on the leading strand and discontinuously on the opposite lagging strand, resulting in short Okazaki fragments that are later joined by DNA ligase. ■

Proofreading and Error Correction Occur during DNA Replication

The immediate purpose of DNA replication is the synthesis of a new strand that is precisely complementary to the template strand at each nucleotide position. Although the action of DNA polymerases is very accurate, synthesis is not perfect and a noncomplementary nucleotide is occasionally inserted erroneously. To compensate for such inaccuracies, the DNA polymerases all possess 3′ to 5′ exonuclease activity. This property imparts the potential for them to detect and excise a mismatched nucleotide (in the 3′ to 5′ direction). Once the mismatched nucleotide is removed, 5′ to 3′ synthesis can again proceed. This process, called **proofreading,** increases the fidelity of synthesis by a factor of about 100. In the case of the holoenzyme form of DNA polymerase III, the epsilon (ε) subunit is directly involved in the proofreading step. In strains of *E. coli* with a mutation that has rendered the ε subunit nonfunctional, the error rate (the mutation rate) during DNA synthesis is increased substantially.

10.4 A Coherent Model Summarizes DNA Replication

We can now combine the various aspects of DNA replication occurring at a single replication fork into a coherent model, as shown in **Figure 10.13**. At the advancing fork, a helicase is unwinding the double helix. Once unwound, single-stranded binding proteins associate with the strands, preventing the re-formation of the helix. In advance of the replication fork,

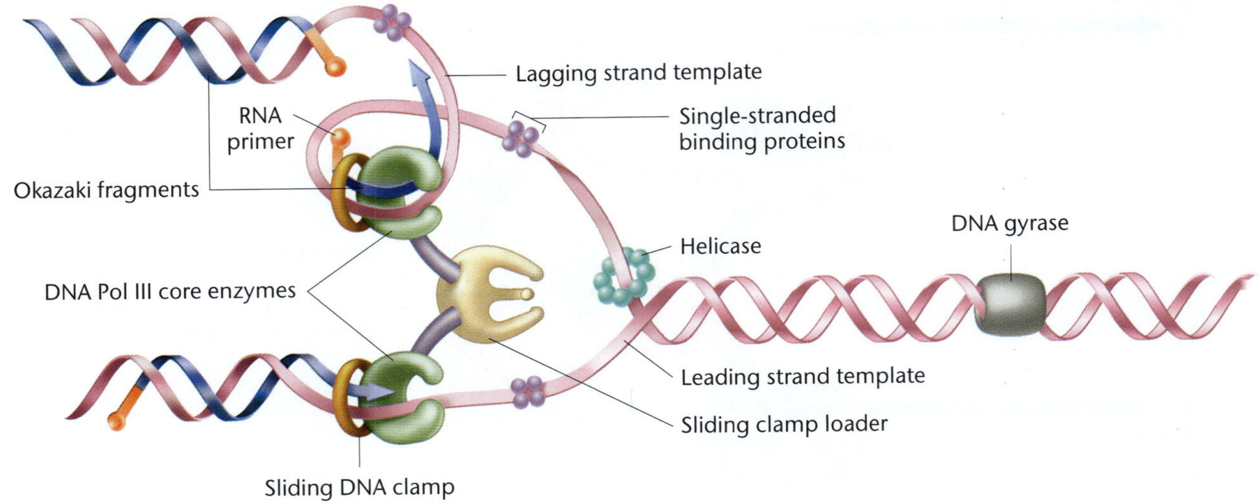

FIGURE 10.13 Summary of DNA synthesis at a single replication fork. Various enzymes and proteins essential to the process are shown.

DNA gyrase functions to diminish the tension created as the helix supercoils. Each half of the dimeric polymerase is a core enzyme bound to one of the template strands by a β-subunit sliding clamp. Continuous synthesis occurs on the leading strand, while the lagging strand must loop out and around the polymerase in order for simultaneous (concurrent) synthesis to occur on both strands. Not shown in the figure, but essential to replication on the lagging strand, is the action of DNA polymerase I and DNA ligase, which together replace the RNA primers with DNA and join the Okazaki fragments, respectively.

The preceding model provides a summary of DNA synthesis against which genetic phenomena can be interpreted.

NOW SOLVE THIS

10.2 An alien organism was investigated. When DNA replication was studied, a unique feature was apparent: No Okazaki fragments were observed. Create a model of DNA that is consistent with this observation.

■ **HINT:** *This problem involves an understanding of the process of DNA synthesis in bacteria, as depicted in Figure 10.12. The key to its solution is to consider why Okazaki fragments are observed during DNA synthesis and how their formation relates to DNA structure, as described in the Watson–Crick model.*

10.5 Replication Is Controlled by a Variety of Genes

Much of what we know about DNA replication in viruses and bacteria is based on genetic analysis of the process. For example, we have already discussed studies involving the *polA1* mutation, which revealed that DNA polymerase I is not the major enzyme responsible for replication. Many other mutations interrupt or seriously impair some aspect of replication, such as the ligase-deficient and the proofreading-deficient mutations mentioned previously. Because such mutations are lethal, genetic analysis frequently uses **conditional mutations,** which are expressed under one condition but not under a different condition. For example, a **temperature-sensitive mutation** may not be expressed at a particular *permissive* temperature. When mutant cells are grown at a *restrictive* temperature, the mutant phenotype is expressed and can be studied. By examining the effect of the loss of function associated with the mutation, the investigation of such temperature-sensitive mutants can provide insight into the product and the associated function of the normal, nonmutated gene.

As shown in **Table 10.3**, a variety of genes in *E. coli* specify the subunits of the DNA polymerases and encode products involved in specification of the origin of synthesis, helix

TABLE 10.3 Some of the Various *E. coli* Genes and Their Products or Role in Replication

Gene	Product or Role
polA	DNA polymerase I
polB	DNA polymerase II
dnaE, N, Q, X, Z	DNA polymerase III subunits
dnaG	Primase
dnaA, I, P	Initiation
dnaB, C	Helicase at *oriC*
gyrA, B	Gyrase subunits
lig	DNA ligase
rep	DNA helicase
ssb	Single-stranded binding proteins
rpoB	RNA polymerase subunit

unwinding and stabilization, initiation and priming, relaxation of supercoiling, repair, and ligation. The discovery of such a large group of genes attests to the complexity of the process of replication, even in this relatively simple organism. Given the enormous quantity of DNA that must be unerringly replicated in a very brief time, this level of complexity is not unexpected. As we will see in Section 10.6, the process is even more involved and therefore more difficult to investigate in eukaryotes.

10.6 Eukaryotic DNA Replication Is Similar to Replication in Bacteria, but Is More Complex

Eukaryotic DNA replication shares many features with replication in bacteria. In both systems, double-stranded DNA is unwound at replication origins, replication forks are formed, and bidirectional DNA synthesis creates leading and lagging strands from single-stranded DNA templates under the direction of DNA polymerase. Eukaryotic polymerases have the same fundamental requirements for DNA synthesis as do bacterial polymerases: four deoxyribonucleoside triphosphates, a template, and a primer. However, eukaryotic DNA replication is more complex, due to several features of eukaryotic DNA. For example, eukaryotic cells contain much more DNA, this DNA is complexed with nucleosomes, and eukaryotic chromosomes are linear rather than circular. In this section, we will describe some of the ways in which eukaryotes deal with this added complexity.

Initiation at Multiple Replication Origins

The most obvious difference between eukaryotic and bacterial DNA replication is that eukaryotic replication must deal with greater amounts of DNA. For example, yeast cells contain three times as much DNA, and *Drosophila* cells contain 40 times as much as *E. coli* cells. In addition, eukaryotic DNA polymerases synthesize DNA at a rate 25 times slower (about 2000 nucleotides per minute) than that in bacteria. Under these conditions, replication from a single origin on a typical eukaryotic chromosome would take days to complete. However, replication of entire eukaryotic genomes is usually accomplished in a matter of minutes to hours.

To facilitate the rapid synthesis of large quantities of DNA, eukaryotic chromosomes contain multiple replication origins. Yeast genomes contain between 250 and 400 origins and mammalian genomes have as many as 25,000. Multiple origins are visible under the electron microscope as "replication bubbles" that form as the DNA helix opens up, each bubble providing two potential replication forks (**Figure 10.14**). Origins in yeast, called **autonomously replicating sequences (ARSs),** consist of approximately 120 base pairs containing a **consensus sequence** (meaning a sequence that is the same, or nearly the same, in all yeast ARSs) of 11 base pairs. Origins in mammalian cells appear to be unrelated to specific sequence motifs and may be defined more by chromatin structure over a 6- to 55-kb region.

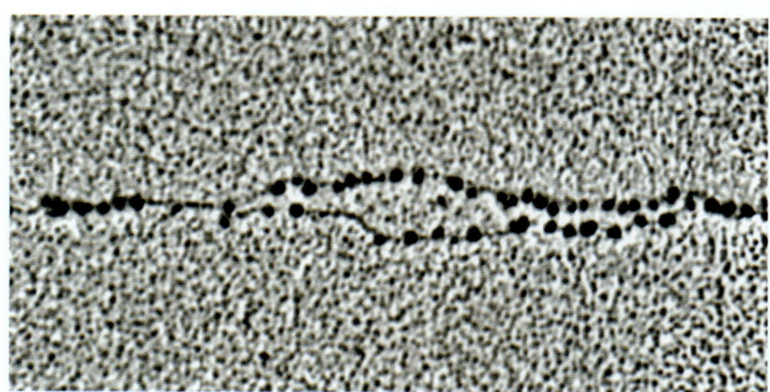

FIGURE 10.14 An electron micrograph of a eukaryotic replicating fork demonstrating the presence of histone-protein-containing nucleosomes on both branches.

Multiple Eukaryotic DNA Polymerases

To accommodate the large number of replicons, eukaryotic cells contain many more DNA polymerase molecules than do bacterial cells. For example, a single *E. coli* cell contains about 15 molecules of DNA polymerase III, but a mammalian cell contains tens of thousands of DNA polymerase molecules.

Eukaryotes also utilize a larger number of different DNA polymerase types than do bacteria. The human genome contains genes that encode at least 14 different DNA polymerases, only three of which are involved in the majority of nuclear genome DNA replication.

Pol α, δ, and ε are the major forms of the enzyme involved in initiation and elongation during eukaryotic nuclear DNA synthesis, so we will concentrate our discussion on these. Two of the four subunits of the **Pol α enzyme** synthesize RNA primers on both the leading and lagging strands. After the RNA primer reaches a length of about 10 ribonucleotides, another subunit adds 10 to 20 complementary deoxyribonucleotides. Pol α is said to possess low **processivity,** a term that refers to the strength of the association between the enzyme and its substrate, and thus the length of DNA that is synthesized before the enzyme dissociates from the template. Once the primer is in place, an event known as **polymerase switching** occurs, whereby Pol α dissociates from the template and is replaced by Pol δ or ε. These enzymes extend the primers on opposite strands of DNA, possess much greater processivity, and exhibit 3′ to 5′ exonuclease activity, thus having the potential to proofread. Pol ε synthesizes DNA on the leading strand, and Pol δ synthesizes the lagging strand. Both Pol δ and ε participate in other DNA synthesizing events in the cell, including several types of DNA repair and recombination. All three DNA polymerases are essential for viability.

As in bacterial DNA replication, the final stages in eukaryotic DNA replication involve replacing the RNA primers with DNA and ligating the Okazaki fragments on the lagging strand. In eukaryotes, the Okazaki fragments are about ten times smaller (100 to 150 nucleotides) than in bacteria.

Included in the remainder of DNA-replicating enzymes is Pol γ, which is found exclusively in mitochondria, synthesizing the DNA present in that organelle. Other DNA polymerases are involved in DNA repair and replication through regions of the DNA template that contain damage or distortions.

Replication through Chromatin

One of the major differences between bacterial and eukaryotic DNA is that eukaryotic DNA is complexed with DNA-binding proteins, existing in the cell as *chromatin*. As we will discuss later in the text (see Chapter 11), chromatin consists of regularly repeating units called nucleosomes, each of which consists of about 200 base pairs of DNA wrapped around eight histone protein molecules. Before DNA polymerases can begin synthesis, nucleosomes and other DNA-binding proteins must be stripped away or otherwise modified to allow the passage of replication proteins. As DNA synthesis proceeds, the histones and non-histone proteins must rapidly reassociate with the newly formed duplexes, reestablishing the characteristic nucleosome pattern. Electron microscopy studies, such as the one shown in Figure 10.14, show that nucleosomes form immediately after new DNA is synthesized at replication forks.

In order to re-create nucleosomal chromatin on replicated DNA, the synthesis of new histone proteins is tightly coupled to DNA synthesis during the S phase of the cell cycle. Research data suggest that nucleosomes are disrupted just ahead of the replication fork and that the preexisting histone proteins can assemble with newly synthesized histone proteins into new nucleosomes. The new nucleosomes are assembled behind the replication fork, onto the two daughter strands of DNA. The assembly of new nucleosomes is carried out by **chromatin assembly factors (CAFs)** that move along with the replication fork.

ESSENTIAL POINT

DNA replication in eukaryotes is more complex than replication in bacteria, using multiple replication origins, multiple forms of DNA polymerases, and factors that disrupt and assemble nucleosomal chromatin. ■

10.7 Telomeres Solve Stability and Replication Problems at Eukaryotic Chromosome Ends

A final difference between bacterial and eukaryotic DNA synthesis stems from the structural differences in their chromosomes. Unlike the closed, circular DNA of bacteria and most bacteriophages, eukaryotic chromosomes are linear. The presence of linear DNA "ends" on eukaryotic chromosomes creates two potential problems.

The first problem is that the double-stranded "ends" of DNA molecules at the termini of linear chromosomes potentially resemble the **double-stranded breaks (DSBs)** that can occur when a chromosome becomes fragmented internally as a result of DNA damage. Such double-stranded DNA ends are recognized by the cell's DNA repair mechanisms that join the "loose ends" together, leading to chromosome fusions and translocations. If the ends do not fuse, they are vulnerable to degradation by nucleases. The second problem occurs during DNA replication, because DNA polymerases cannot synthesize new DNA at the tips of single-stranded 5′-ends.

To deal with these two problems, linear eukaryotic chromosomes end in distinctive sequences called telomeres, as we will describe next.

Telomere Structure and Chromosome Stability

In 1978, Elizabeth Blackburn and Joe Gall reported the presence of unexpected structures at the ends of chromosomes of the ciliated protozoan *Tetrahymena*. They showed that the protozoan's chromosome ends consisted of the short sequence 5′-TTGGGG-3′, tandemly repeated from 30 to 60 times. This strand is referred to as the G-rich strand, in contrast to its complementary strand, the so-called C-rich strand, which displays the repeated sequence 5′-AACCCC-3′. Since then, researchers have discovered similar tandemly repeated DNA sequences at the ends of linear chromosomes in most eukaryotes. These repeat regions make up the chromosome's **telomeres.** In humans, the telomeric sequence 5′-TTAGGG-3′ is repeated several thousand times and telomeres can vary in length from 5 to 15 kb. In contrast, yeast telomeres are several 100 base pairs long and mouse telomeres are between 20 and 50 kb long. Since each linear chromosome ends with two DNA strands running antiparallel to one another, one strand has a 3′-ending and the other has a 5′-ending. It is the 3′-strand that is the G-rich one. This has special significance during telomere replication.

Two features of telomeric DNA help to explain how telomeres protect the ends of linear chromosomes. First, at the end of telomeres, a stretch of single-stranded DNA extends out from the 3′ G-rich strand. This single-stranded tail varies in length between organisms. In *Tetrahymena*, the tail is between 12 and 16 nucleotides long, whereas in mammals, it varies between 30 and 400 nucleotides long. The 3′-ends of G-rich single-stranded tails are capable of interacting with upstream sequences within the tail, creating loop structures. The loops, called **t-loops,** resemble those created when you tie your shoelaces into a bow. Second, a complex of six proteins binds and stabilizes telomere t-loops, forming the **shelterin complex.** It is believed that t-loop structures, in combination with the shelterin proteins, close off the ends of chromosomes and make them resistant to nuclease digestion and DNA fusions. Shelterin proteins also help to recruit telomerase enzymes to telomeres during telomere replication, which we discuss next.

Telomeres and Chromosome End Replication

Now let's consider the problem that semiconservative replication poses the ends of double-stranded DNA molecules. As we have learned previously in this chapter, DNA replication initiates from short RNA primers, synthesized on both leading and lagging strands (**Figure 10.15**). Primers are necessary because DNA polymerase requires a free 3′-OH on which to initiate synthesis. After replication is completed, these RNA primers are removed. The resulting gaps within the new daughter strands are filled by DNA polymerase and sealed by ligase. These internal gaps have free 3′-OH groups available at the ends of the Okazaki fragments for DNA polymerase to initiate synthesis. The problem arises at the gaps left at the 5′-ends of the newly synthesized DNA [gaps (b) and (c) in Figure 10.15]. These gaps cannot be filled by DNA polymerase because no free 3′-OH groups are available for the initiation of synthesis.

Thus, in the situation depicted in Figure 10.15, gaps remain on newly synthesized DNA strands at each successive round of synthesis, shortening the double-stranded ends of the chromosome by the length of the RNA primer. With each round of replication, the shortening becomes more severe in each daughter cell, eventually extending beyond the telomere and potentially deleting gene-coding regions.

The solution to this so-called **end-replication problem** is provided by a unique eukaryotic enzyme called **telomerase.** Telomerase was first discovered by Elizabeth Blackburn and her graduate student, Carol Greider, in studies of *Tetrahymena*. As noted earlier, telomeric DNA in eukaryotes consists of many short, repeated nucleotide sequences, with the G-rich strand overhanging in the form of a single-stranded tail. In *Tetrahymena* the tail contains several repeats of the sequence 5′-TTGGGG-3′. As we will see, telomerase is capable of adding several more repeats of this six-nucleotide sequence to the 3′-end of the G-rich strand. Detailed investigation by Blackburn and Greider of

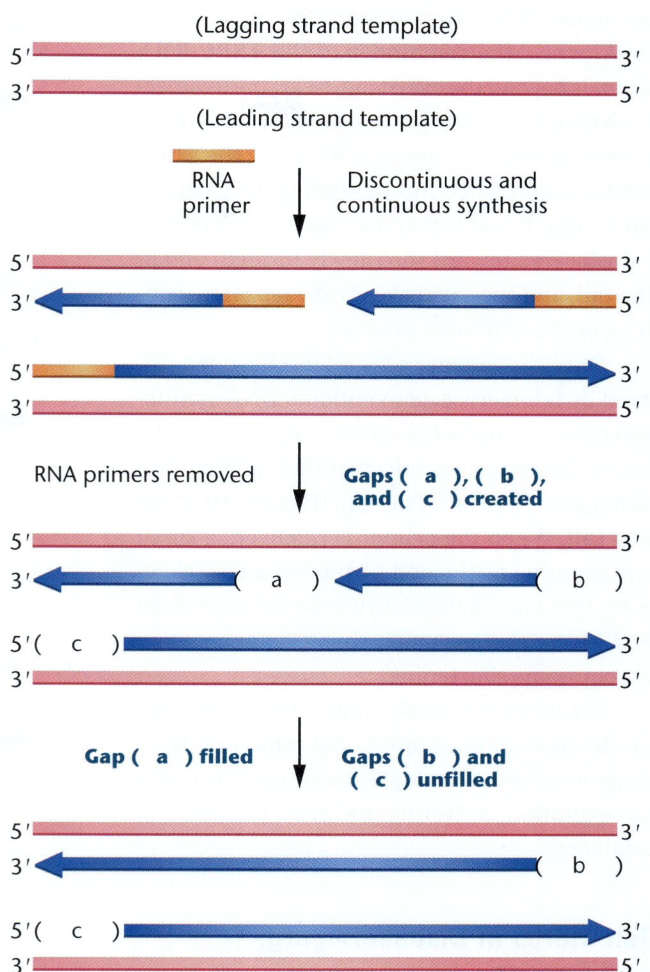

FIGURE 10.15 Diagram illustrating the difficulty encountered during the replication of the ends of linear chromosomes. Of the three gaps, (b) and (c) are left following synthesis on both the leading and lagging strands.

how the *Tetrahymena* telomerase enzyme accomplishes this synthesis yielded an extraordinary finding. The enzyme is a *ribonucleoprotein*, containing within its molecular structure a short piece of RNA that is essential to its catalytic activity. The **telomerase RNA component (TERC)** serves as both a "guide" to proper attachment of the enzyme to the telomere and a "template" for synthesis of its DNA complement. Synthesis of DNA using RNA as a template is called **reverse transcription**. The **telomerase reverse transcriptase (TERT)** is the catalytic subunit of the telomerase enzyme. In addition to TERC and TERT, telomerase contains a number of accessory proteins. In *Tetrahymena*, TERC contains the sequence CAACCCCAA, within which is found the complement of the repeating telomeric DNA sequence that must be synthesized (TTGGGG).

Figure 10.16 shows a model of how researchers envision the enzyme working. Part of the TERC RNA sequence of the enzyme (shown in green) base pairs with the ending sequence of the single-stranded overhanging DNA, while the remainder of the TERC RNA extends beyond the overhang. Next, the telomerase's reverse transcription activity synthesizes a stretch of single-stranded DNA using the TERC RNA as a template, thus increasing the length of the 3′-tail. It is believed that the enzyme is then translocated toward the (newly formed) end of the tail, and the same events are repeated, continuing the extension process.

Once the telomere 3′-tail has been lengthened by telomerase, conventional DNA synthesis ensues. Primase lays down a primer near the end of the telomere tail; then DNA polymerase and ligase fill most of the gap [**Figure 10.16(d)** and **(e)**]. When the primer is removed, a small gap remains at the end of the telomere. However, this gap is located well beyond the original end of the chromosome, thus preventing any chromosome shortening.

Telomerase function has now been found in all eukaryotes studied, and telomeric DNA sequences have been highly conserved throughout evolution, reflecting the critical function of telomeres.

Telomeres in Disease, Aging, and Cancer

Despite the importance of maintaining telomere length for chromosome integrity, only some cell types express telomerase. In humans, these include embryonic stem cells, some types of adult stem cells, and other cell types that need to divide repeatedly such as epidermal cells and cells of the immune system. In contrast, most normal somatic cells do not express telomerase. As a result, after many cell divisions, somatic cell telomeres become seriously eroded, leading to chromosome damage that can either kill the cell or cause it to cease dividing and enter a state called senescence.

Several rare human diseases have been associated with loss of telomerase activity and abnormally short telomeres. For example, patients with the inherited form of dyskeratosis congenita have mutations in genes encoding telomerase or shelterin subunits. These mutations bring about many different symptoms that are also seen in premature aging, and patients suffer early deaths due to stem cell failure. Many studies show a correlation between telomere length or telomerase activity and common diseases such as diabetes or heart disease.

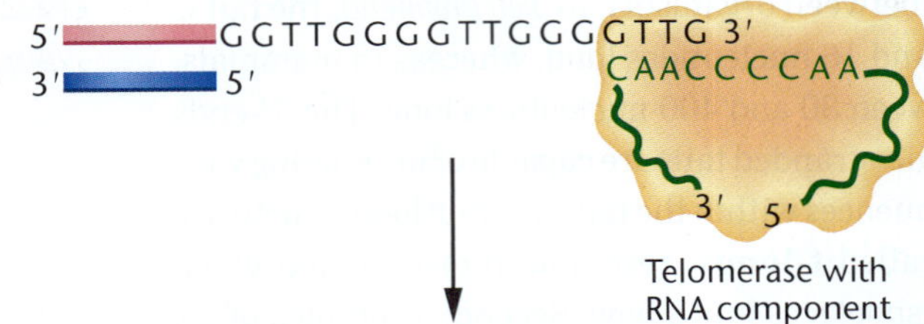

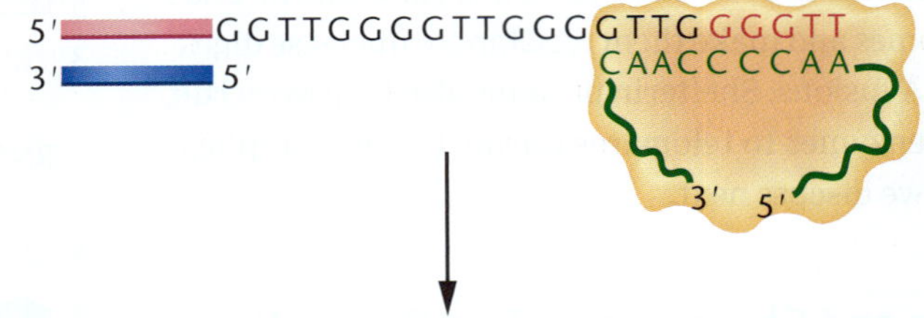

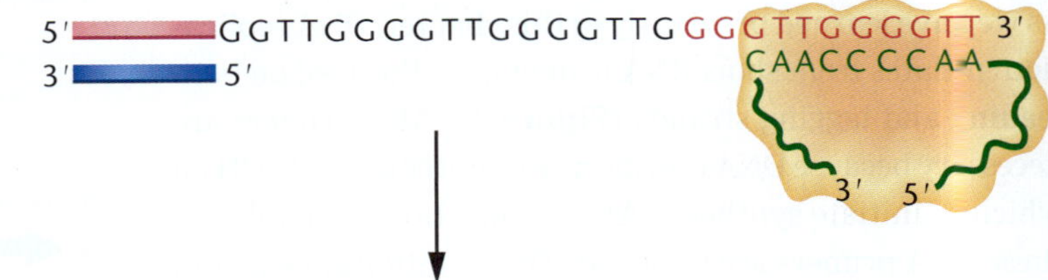

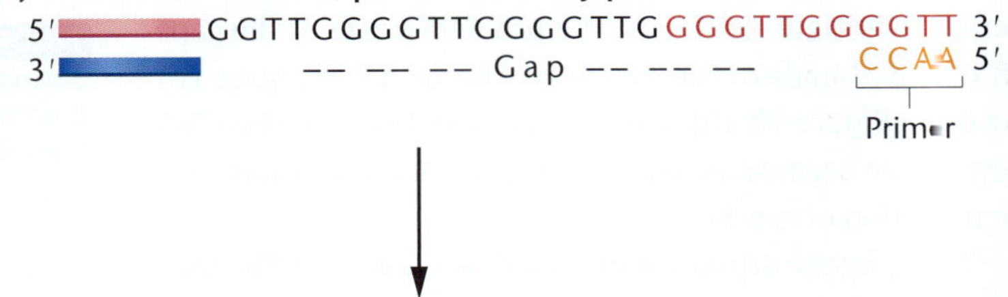

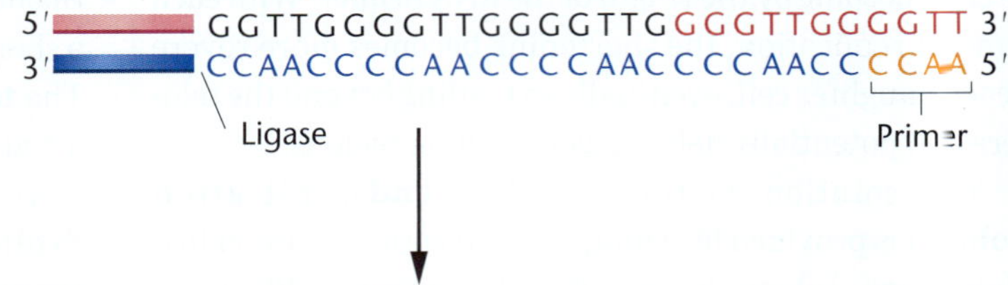

FIGURE 10.16 The predicted solution to the problem posed in Figure 10.15. The enzyme telomerase (with its TERC RNA component shown in green) directs synthesis of repeated TTGGGG sequences, resulting in the formation of an extended 3′-overhang. This facilitates DNA synthesis on the opposite strand, filling in the gap that would otherwise have been increased in length with each replication cycle.

The connection between telomere length and aging has also been the subject of much research and speculation. When telomeres become critically short, a cell may suffer chromosome damage and enter **senescence,** a state in which cell division ceases and the cell undergoes metabolic changes that cause it to function less efficiently. Some scientists propose that the presence of shortened telomeres in such senescent cells is directly related to changes associated with aging. This topic is discussed in the "Genetics, Ethics, and Society" essay below.

While most somatic cells contain little if any telomerase, shorten their telomeres, and undergo senescence after multiple cell divisions, most all human cancer cells retain telomerase activity and thus maintain telomere length through multiple cell divisions, suggesting that this may be the key to their "immortality." Those cancer cells that do not contain telomerase use a different telomere-lengthening method called *alternative lengthening of telomeres (ALT)*. And, in tissue cultures, cells can be transformed into cancer cells by introducing telomerase activity, as long as at least two other types of genes (proto-oncogenes and tumor-suppressor genes) are mutated or abnormally expressed. These findings do not suggest that telomerase activity is in itself sufficient to create a cancer cell. We return to the multistep nature of the genesis of cancer later (see Chapter 19).

Nevertheless, the requirement for telomerase activity in cancer cells suggests that researchers may be able to develop cancer drugs that repress tumor growth by inhibiting telomerase activity. Because most human somatic cells do not express telomerase, such a therapy might be relatively tumor specific and less toxic than many current anticancer drugs. A number of such anti-telomerase drugs are currently being developed, and some have entered phase III clinical trials.

ESSENTIAL POINT

Replication at the ends of linear chromosomes in eukaryotes poses a special problem that can be solved by the presence of telomeres and by a unique RNA-containing enzyme called telomerase. ■

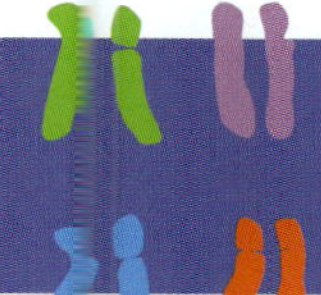

GENETICS, ETHICS, AND SOCIETY

Telomeres: The Key to a Long Life?

We humans, like all multicellular organisms, grow old and die. As we age, our immune systems become less efficient, wound healing is impaired, and tissues lose resilience. Why do we go through these age-related declines, and can we reverse the march to mortality? Some researchers suggest that the answers to these questions may lie at the ends of our chromosomes.

Human cells, both those in our bodies and those growing in culture dishes, have a finite life span. When placed into tissue culture dishes, normal human fibroblasts become senescent, losing their ability to grow and divide after about 50 cell divisions. Eventually, they die. Although we don't know whether cellular senescence directly causes aging in a multicellular organism, the evidence is suggestive. For example, cultured cells derived from young people undergo more divisions than those from older people; cells from short-lived species stop growing after fewer divisions than those from longer-lived species; and cells from patients with premature aging syndromes undergo fewer divisions than those from normal patients.

One of the many characteristics of aging cells involves telomeres. As described in this chapter, most mammalian somatic cells do not contain telomerase activity and telomeres shorten with each DNA replication. Some epidemiological studies show a correlation between telomere length in humans and their life spans. In addition, some common diseases such as cardiovascular disease and some lifestyle factors such as smoking, poor diets, and stress, correlate with shorter than average telomere lengths. Despite these correlations, the data linking telomere length and longevity in humans are not consistent and remain the subject of scientific debate.

Telomerase activity has also been correlated with aspects of aging in multicellular organisms. In one study, investigators introduced cloned telomerase genes into normal human cells in culture. The increase in telomerase activity caused the telomeres' lengths to increase, and the cells continued to grow past their typical senescence point. In another study, researchers created a strain of mice that was defective in the TERT subunit of telomerase. These mice developed extremely short telomeres and showed the classic symptoms of aging, including tissue atrophy, neurodegeneration, and a shortened life span. When the researchers reactivated telomerase function in these prematurely aging adult mice, tissue atrophies and neurodegeneration were reversed and their life spans increased. Similar studies led to the conclusion that overexpression of telomerase in normal mice also increased their life spans, although it was not clear that telomere lengths were altered. These studies suggest that some of the symptoms that accompany old age in humans might be reversed by activating telomerase genes. However, some scientists still debate whether telomerase activation or telomere lengthening directly cause these effects or may simply accompany other, unknown, causative mechanisms.

YOUR TURN

Take time, individually or in groups, to answer the following questions. Investigate the references and links to help you understand some of the research and ethical questions that surround the links between telomeres and aging.

1. The connection between telomeres and aging has been of great interest to both

(continued)

Genetics, Technology, and Society, continued

scientists and the public. Studies have used model organisms, cultured cells, and data from epidemiological surveys to try to determine a correlation. Although these studies suggest links between telomeres and aging, the conclusions from these studies have also been the subject of debate. How would you assess the current status of research on the link between telomeres and human aging?

Begin your exploration of current telomere research with two reviews that draw differing conclusions from the same data: Bär, C. and Blasco, M. A. (2016). Telomeres and telomerase as therapeutic targets to prevent and treat age-related diseases. *F1000Res* 5 2016 Jan 20; 5:89, and Simons, M. J. P. (2015). Questioning causal involvement of telomeres in aging. *Ageing Res. Rev.* 24:191–196.

2. A large number of private companies now offer telomere length diagnostic tests and telomere lengthening supplements to the public. Discuss the ethics of offering such tests and supplements when some scientists argue that it is too early to sufficiently understand the mechanisms that may (or may not) link telomere length or telomerase activity with aging. What are the potential benefits and harms of these tests and treatments? Would you purchase these tests or supplements? Why or why not?

This topic is discussed in Leslie, M. (2011). Are telomere tests ready for prime time? *Science* 332:414–415.

CASE STUDY At loose ends

Dyskeratosis congenita (DKC) is a rare human genetic disorder affecting telomere replication. Mutations in the genes encoding the protein or RNA subunits of telomerase result in very short telomeres. DKC symptoms include bone marrow failure (reduced production of blood cells) and anemia. If symptoms are severe, a bone marrow transplant may be the only form of effective treatment. In one case, clinicians recommended that a 27-year-old woman with a dominant form of DKC undergo a bone marrow transplant to treat the disorder. Her four siblings were tested, and her 13-year-old brother was identified as the best immunologically matched donor. However, before being tested, he was emphatic that he did not want to know if he had DKC. During testing, it was discovered that he had unusually short telomeres and would most likely develop symptoms of DKC.

1. Why might mutations in genes encoding telomerase subunits lead to bone marrow failure?
2. Although the brother is an immunologically matched donor for his sister, it would be unethical for the clinicians to transplant bone marrow from the brother to the sister. Why?
3. The clinicians are faced with another ethical dilemma. How can they respect the brother's desire to not know if he has DKC while also revealing that he is not a suitable donor for his sister? In addition, what should the clinicians tell the sister about her brother?

See Denny, C., et al. (2008). All in the family: Disclosure of "unwanted" information to an adolescent to benefit a relative. *Am. J. Med. Genet.* 146A(21):2719–2724.

INSIGHTS AND SOLUTIONS

1. Predict the theoretical results of conservative and dispersive replication of DNA under the conditions of the Meselson–Stahl experiment. Follow the results through two generations of replication after cells have been shifted to an ^{14}N-containing medium, using the following sedimentation pattern.

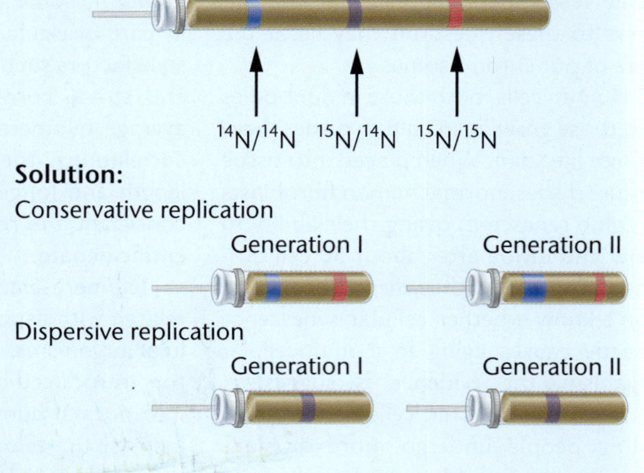

2. Mutations in the *dnaA* gene of *E. coli* are lethal and can only be studied following the isolation of conditional, temperature-sensitive mutations. Such mutant strains grow nicely and replicate their DNA at the permissive temperature of 18°C, but they do not grow or replicate their DNA at the restrictive temperature of 37°C. Two observations were useful in determining the function of the DnaA protein product. First, *in vitro* studies using DNA templates that have unwound do not require the DnaA protein. Second, if intact cells are grown at 18°C and are then shifted to 37°C, DNA synthesis continues at this temperature until one round of replication is completed and then stops. What do these observations suggest about the role of the *dnaA* gene product?

Solution: At 18°C (the permissive temperature), the mutation is not expressed and DNA synthesis begins. Following the shift to the restrictive temperature, the already initiated DNA synthesis continues, but no new synthesis can begin. Because the DnaA protein is not required for synthesis of unwound DNA, these observations suggest that, *in vivo*, the DnaA protein plays an essential role in DNA synthesis by interacting with the intact helix and somehow facilitating the localized denaturation necessary for synthesis to proceed.

Problems and Discussion Questions

Mastering Genetics Visit for instructor-assigned tutorials and problems.

1. **HOW DO WE KNOW?** In this chapter, we focused on how DNA is replicated and synthesized. In particular, we elucidated the general mechanism of replication and described how DNA is synthesized when it is copied. Based on your study of these topics, answer the following fundamental questions:
   (a) What is the experimental basis for concluding that DNA replicates semiconservatively in both bacteria and eukaryotes?
   (b) How was it demonstrated that DNA synthesis occurs under the direction of DNA polymerase III and not polymerase I?
   (c) How do we know that *in vivo* DNA synthesis occurs in the 5′ to 3′ direction?
   (d) How do we know that DNA synthesis is discontinuous on one of the two template strands?
   (e) What observations reveal that a "telomere problem" exists during eukaryotic DNA replication, and how did we learn of the solution to this problem?
2. **CONCEPT QUESTION** Review the Chapter Concepts list on p. 206. These are concerned with the replication and synthesis of DNA. Write a short essay that distinguishes between the terms *replication* and *synthesis*, as applied to DNA. Which of the two is most closely allied with the field of biochemistry? ■
3. Compare conservative, semiconservative, and dispersive modes of DNA replication.
4. Describe sedimentation equilibrium centrifugation and its possible applications.
5. Predict the results of the experiment by Taylor, Woods, and Hughes if replication were (a) conservative and (b) dispersive.
6. Reconsider Problem 30 in Chapter 9. In the model you proposed, could the molecule be replicated semiconservatively? Why? Would other modes of replication work?
7. What are the requirements for *in vitro* synthesis of DNA under the direction of DNA polymerase I?
8. How did Kornberg assess the fidelity of DNA polymerase I in copying a DNA template?
9. How does the exonuclease activity of certain polymerases relate to proofreading and primer removal?
10. You have two strains of bacteria, one in which the process of replication is sensitive to RNase and the other in which the presence of RNase does not affect replication. Speculate why the replications in these two strains are differentially sensitive to RNase.
11. During DNA replication, which enzyme can be disposed of in an organism with a mutant DNA polymerase that does not require a free 3′-OH
12. Summarize and compare the properties of DNA polymerase I, II, and III.
13. List and describe the function of the ten subunits constituting DNA polymerase III. Distinguish between the holoenzyme and the core enzyme.
14. Distinguish between (a) unidirectional and bidirectional synthesis, and (b) continuous and discontinuous synthesis of DNA.
15. List the proteins that unwind DNA during *in vivo* DNA synthesis. How do they function?
16. Define and indicate the significance of (a) Okazaki fragments, (b) DNA ligase, and (c) primer RNA during DNA replication.
17. What would be the impact of the loss of processivity on DNA Pol III?
18. What are the replication origins in bacteria, yeast, and mammalian cells?
19. What is the advantage of conditional mutations in genetics analysis?
20. Several temperature-sensitive mutant strains of *E. coli* display the following characteristics. Predict what enzyme or function is being affected by each mutation.
   (a) Newly synthesized DNA contains many mismatched base pairs.
   (b) Okazaki fragments accumulate, and DNA synthesis is never completed.
   (c) No initiation occurs.
   (d) Synthesis is very slow.
   (e) Supercoiled strands remain after replication, which is never completed.
21. Many of the gene products involved in DNA synthesis were initially defined by studying mutant *E. coli* strains that could not synthesize DNA.
   (a) The *dnaE* gene encodes the α subunit of DNA polymerase III. What effect is expected from a mutation in this gene? How could the mutant strain be maintained?
   (b) The *dnaQ* gene encodes the ε subunit of DNA polymerase. What effect is expected from a mutation in this gene?
22. Assume a hypothetical organism in which DNA replication is conservative. Design an experiment similar to that of Taylor, Woods, and Hughes that will unequivocally establish this fact. Using the format established in Figure 10.5, draw sister chromatids and illustrate the expected results depicting this mode of replication.
23. Describe the major difference between bacterial and eukaryotic replication with respect to the DNA structure.
24. In 1994, telomerase activity was discovered in human cancer cell lines. Although telomerase is not active in human somatic tissue, these cells do contain the genes for telomerase proteins and telomerase RNA. Since inappropriate activation of telomerase may contribute to cancer, why do you think the genes coding for this enzyme have been maintained in the human genome throughout evolution? Are there any types of human body cells where telomerase activation would be advantageous or even necessary? Explain.

11

Chromosome Structure and DNA Sequence Organization

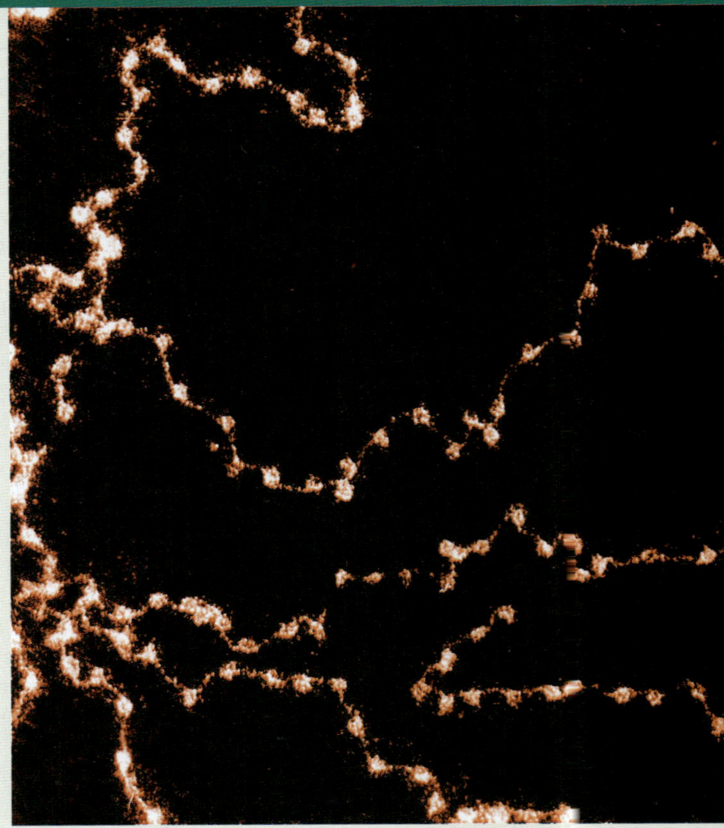

A chromatin fiber viewed using a scanning transmission electron microscope (STEM)

CHAPTER CONCEPTS

- Genetic information in viruses, bacteria, mitochondria, and chloroplasts is most often contained in a short, circular DNA molecule, relatively free of associated proteins.
- Eukaryotic cells, in contrast to viruses and bacteria, contain relatively large amounts of DNA organized into nucleosomes and present during most of the cell cycle as chromatin fibers.
- Uncoiled chromatin fibers characteristic of interphase coil up and condense into chromosomes during eukaryotic cell division.
- Eukaryotic genomes are characterized by both unique and repetitive DNA sequences.
- Eukaryotic genomes consist mostly of noncoding DNA sequences.

Once geneticists understood that DNA houses genetic information, it became very important to determine how DNA is organized into genes and how these basic units of genetic function are organized into chromosomes. In short, the major question had to do with how the genetic material was organized as it makes up the genome of organisms. There has been much interest in this question because knowledge of the organization of the genetic material and associated molecules is important to understanding many other areas of genetics. For example, the way in which the genetic information is stored, expressed, and regulated must be related to the molecular organization of the genetic molecule DNA.

In this chapter, we focus on the various ways DNA is organized into chromosomes. These structures have been studied using numerous techniques, instruments, and approaches, including analysis by light microscopy and electron microscopy. More recently, molecular analysis has provided significant insights into chromosome organization. In the first half of the chapter, after surveying what we know about chromosomes in viruses and bacteria, we examine the large specialized eukaryotic structures called polytene and lampbrush chromosomes. Then, in the second half, we discuss how eukaryotic chromosomes are organized at the molecular level—for example, how DNA is complexed with proteins to form chromatin and how the chromatin fibers characteristic of interphase are condensed into chromosome structures visible during mitosis and meiosis. We conclude the chapter by examining certain aspects of DNA sequence organization in eukaryotic genomes.

11.1 Viral and Bacterial Chromosomes Are Relatively Simple DNA Molecules

The chromosomes of viruses and bacteria are much less complicated than those of eukaryotes. They usually consist of a single nucleic acid molecule, unlike the multiple chromosomes comprising the genome of higher forms. Compared to eukaryotes, the chromosomes contain much less genetic information and the DNA is not as extensively bound to proteins. These characteristics have greatly simplified analysis, and we now have a fairly comprehensive view of the structure of viral and bacterial chromosomes.

The chromosomes of viruses consist of a nucleic acid molecule—either DNA or RNA—that can be either single- or double-stranded. They can exist as circular structures (closed loops), or they can take the form of linear molecules. For example, the single-stranded DNA of the **ϕX174 bacteriophage** and the double-stranded DNA of the **polyoma virus** are closed loops housed within the protein coat of the mature viruses. The **bacteriophage lambda** (λ), on the other hand, possesses a linear double-stranded DNA molecule prior to infection, which closes to form a ring upon its infection of the host cell. Still other viruses, such as the T-even series of bacteriophages, have linear double-stranded chromosomes of DNA that do not form circles inside the bacterial host. Thus, circularity is not an absolute requirement for replication in viruses.

Viral nucleic acid molecules have been seen with the electron microscope. **Figure 11.1** shows a mature bacteriophage λ and its double-stranded DNA molecule in the circular configuration. One constant feature shared by viruses, bacteria, and eukaryotic cells is the ability to package an exceedingly long DNA molecule into a relatively small volume. In λ, the DNA is 17 μm long and must fit into the phage head, which is less than 0.1 μm on any side. **Table 11.1** compares the length of the chromosomes of several viruses to the size of their head structure. In each case, a similar packaging feat must be accomplished. Compare the dimensions given for phage T2 with the micrograph of both the DNA and the viral particle shown in **Figure 11.2.** Seldom does the space available in the head of a virus exceed the chromosome volume by more than a factor of two. In many cases, almost all of the space is filled, indicating nearly perfect packing. Once packed within the head, the genetic material is functionally inert until it is released into a host cell.

Bacterial chromosomes are also relatively simple in form. They generally consist of a double-stranded DNA molecule, compacted into a structure sometimes referred to as the **nucleoid.** *Escherichia coli*, the most extensively studied bacterium, has a large circular chromosome measuring approximately 1200 μm (1.2 mm) in length that may occupy up to one-third of the volume of the cell. When the cell is gently lysed and the chromosome is released, it can be visualized under the electron microscope (**Figure 11.3**).

DNA in bacterial chromosomes is found to be associated with several types of DNA-binding proteins. Two, called **HU** and **H-NS (Histone-like Nucleoid Structuring) proteins,** are small but abundant in the cell and contain a high percentage of positively charged amino acids that can bond ionically to the negative charges of the phosphate groups in DNA. These proteins function to fold and bend DNA. As such, coils are created that have the effect of compacting the DNA constituting the nucleoid. Additionally, H-NS proteins, like histones in eukaryotes, have been implicated in regulating gene activity in a nonspecific way.

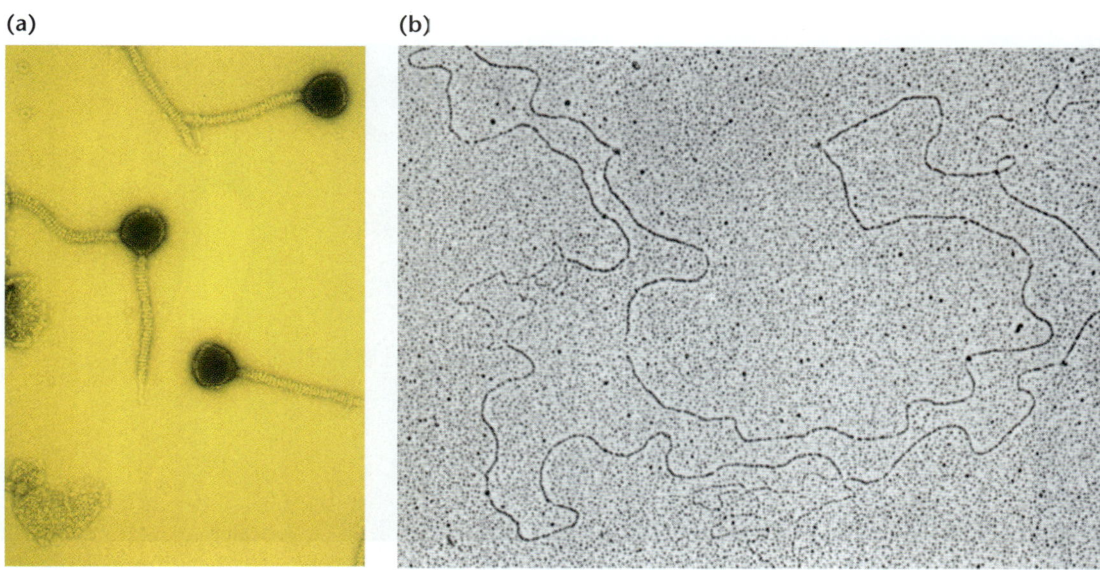

FIGURE 11.1 Electron micrographs of phage λ (a) and the DNA that was isolated from it (b). The chromosome is 17 μm long. Note that the phages are magnified about five times more than the DNA.

TABLE 11.1 The Genetic Material of Representative Viruses and Bacteria

Organism		Nucleic Acid Type	SS or DS*	Length (μm)	Overall Size of Viral Head or Bacteria (μm)
Viruses	ϕX174	DNA	SS	2.0	0.025×0.025
	Tobacco mosaic virus	RNA	SS	3.3	0.30×0.02
	Phage λ	DNA	DS	17.0	0.07×0.07
	T2 phage	DNA	DS	52.0	0.07×0.10
Bacteria	*Haemophilus influenzae*	DNA	DS	832.0	1.00×0.30
	Escherichia coli	DNA	DS	1200.0	2.00×0.50

*SS = single-stranded, DS = double-stranded.

ESSENTIAL POINT

In contrast to eukaryotes, bacteriophage and bacterial chromosomes are largely devoid of associated proteins, are of much smaller size, and most often consist of circular DNA. ■

NOW SOLVE THIS

11.1 In bacteriophages and bacteria, the DNA is almost always organized into circular (closed loops) chromosomes. Phage λ is an exception, maintaining its DNA in a linear chromosome within the viral particle. However, as soon as this DNA is injected into a host cell, it circularizes before replication begins. What advantage exists in replicating circular DNA molecules compared to linear molecules, characteristic of eukaryotic chromosomes?

■ **HINT:** *This problem involves an understanding of eukaryotic DNA replication, as discussed in Chapter 10. The key to its solution is to consider why the enzyme telomerase is essential in eukaryotic DNA replication, and why bacterial and viral chromosomes can be replicated without encountering the "telomere problem."*

11.2 Mitochondria and Chloroplasts Contain DNA Similar to Bacteria and Viruses

That both **mitochondria** and **chloroplasts** contain their own DNA and a system for expressing genetic information was first suggested by the discovery of mutations and the resultant inheritance patterns in plants, yeast, and other fungi. Because both mitochondria and chloroplasts are inherited through the maternal cytoplasm in most organisms, and because each of the above-mentioned examples of mutations could be linked hypothetically to the altered function of either chloroplasts or mitochondria, geneticists set out to look for more direct evidence of DNA in these organelles. Not only was unique DNA found to be a normal component of both mitochondria and chloroplasts, but careful examination of the nature of this genetic information revealed a remarkable similarity to that found in viruses and bacteria.

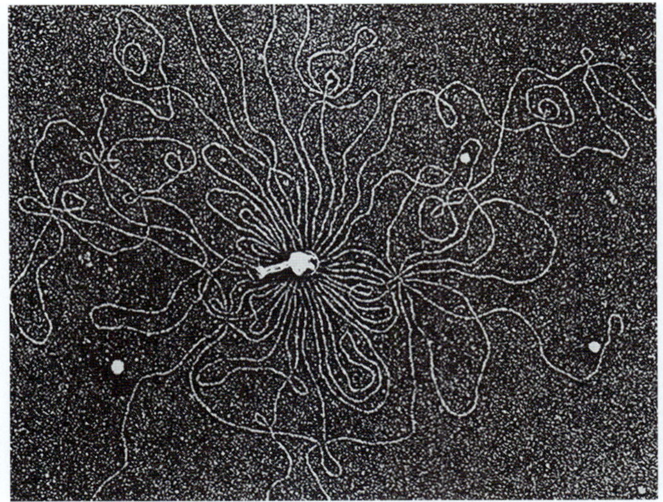

FIGURE 11.2 Electron micrograph of bacteriophage T2, which has had its DNA released by osmotic shock. The chromosome is 52 μm long.

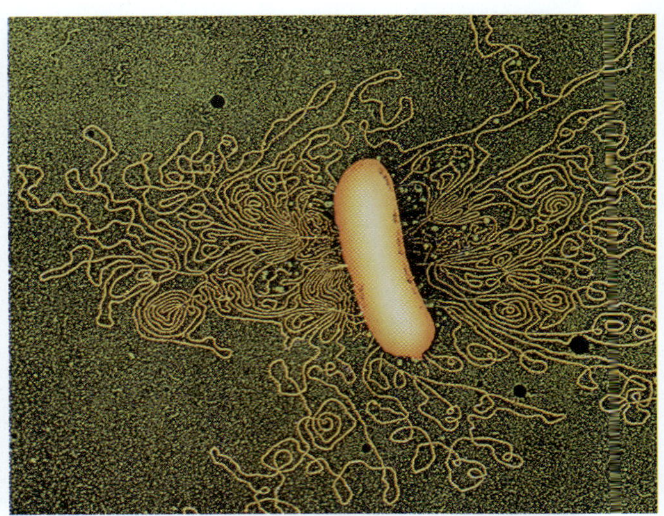

FIGURE 11.3 Electron micrograph of the bacterium *E. coli*, which has had its DNA released by osmotic shock. The chromosome is 1200 μm long.

Molecular Organization and Gene Products of Mitochondrial DNA

Extensive information is also available concerning the structure and gene products of **mitochondrial DNA (mtDNA).** In most eukaryotes, mtDNA exists as a double-stranded, closed circle (**Figure 11.4**) that is free of the chromosomal proteins characteristic of eukaryotic chromosomal DNA. An exception is found in some ciliated protozoans, in which the DNA is linear.

In size, mtDNA varies greatly among organisms. In a variety of animals, including humans, mtDNA consists of about 16,000 to 18,000 bp (16 to 18 kb). However, yeast (*Saccharomyces*) mtDNA consists of 75 kb. Plants typically exceed this amount—367 kb is present in mitochondria in the mustard plant, *Arabidopsis*. Vertebrates have 5 to 10 such DNA molecules per organelle, whereas plants have 20 to 40 copies per organelle.

There are several other noteworthy aspects of mtDNA. With only rare exceptions, *introns* (noncoding regions within genes) are absent from mitochondrial genes, and gene repetitions are seldom present. Nor is there usually much in the way of intergenic spacer DNA. This is particularly true in species whose mtDNA is fairly small in size, such as humans. In *Saccharomyces*, with a much larger mtDNA molecule, introns and intergenic spacer DNA account for much of the excess DNA. As will be discussed in Chapter 12, the expression of mitochondrial genes uses several modifications of the otherwise standard genetic code. Also of interest is the fact that replication in mitochondria is dependent on enzymes encoded by nuclear DNA.

Another interesting observation is that in vertebrate mtDNA, the two strands vary in density, as revealed by centrifugation. This provides researchers with a way to isolate the strands for study, designating one heavy (H) and the other light (L). While most of the mitochondrial genes are encoded by the H strand, several are encoded by the complementary L strand.

Molecular Organization and Gene Products of Chloroplast DNA

Chloroplasts provide the photosynthetic function specific to plants. Like mitochondria, they contain an autonomous genetic system distinct from that found in the nucleus and cytoplasm, which has as its foundation a unique DNA molecule **(cpDNA). Chloroplast DNA,** shown in **Figure 11.5**, is fairly uniform in size among different organisms, ranging between 100 and 225 kb in length. It shares many similarities to DNA found in bacterial cells. It is circular and double stranded, and it is free of the associated proteins characteristic of eukaryotic DNA.

The size of cpDNA is much larger than that of mtDNA. To some extent, this can be accounted for by a larger number of genes. However, most of the difference appears to be due to the presence in cpDNA of many long noncoding nucleotide sequences both between and within genes, the latter being called *introns*. Duplications of many DNA sequences are also present.

In the green alga *Chlamydomonas*, there are about 75 copies of the chloroplast DNA molecule per organelle. In higher plants, such as the sweet pea, multiple copies of the DNA molecule are also present in each organelle, but the molecule is considerably smaller (134 kb) than that in *Chlamydomonas* (195 kb).

ESSENTIAL POINT

Mitochondria and chloroplasts contain DNA that is remarkably similar in form and appearance to some bacterial and bacteriophage DNA. ■

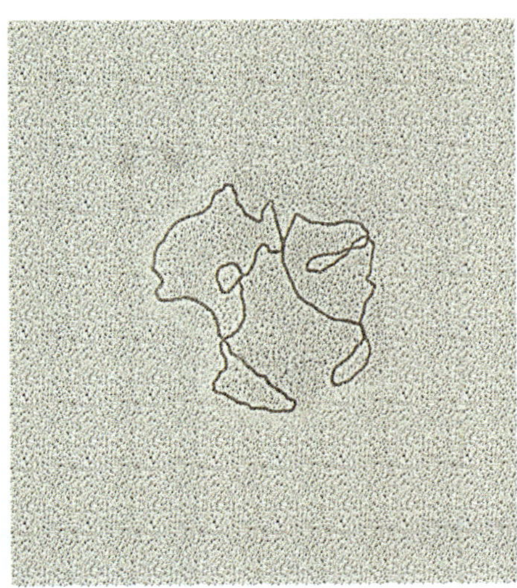

FIGURE 11.4 Electron micrograph of mitochondrial DNA (mtDNA) derived from *Xenopus laevis*.

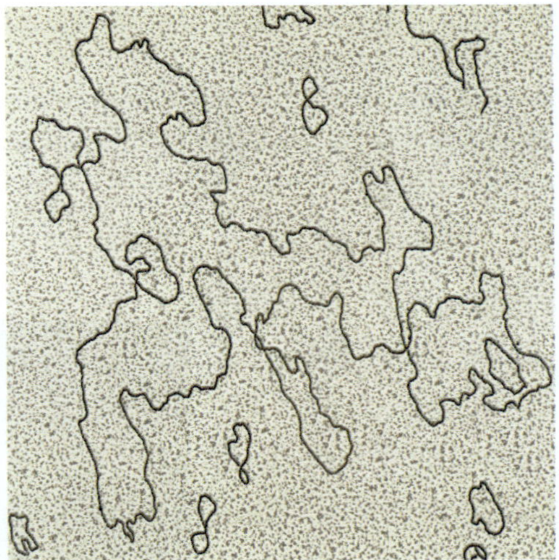

FIGURE 11.5 Electron micrograph of chloroplast DNA (cpDNA) derived from lettuce.

11.3 Specialized Chromosomes Reveal Variations in the Organization of DNA

We now consider two cases of genetic organization that demonstrate the specialized forms that eukaryotic chromosomes can take. Both types—*polytene chromosomes* and *lampbrush chromosomes*—are so large that their organization was discerned using light microscopy long before we understood how mitotic chromosomes form from interphase chromatin. The study of these chromosomes provided many of our initial insights into the arrangement and function of the genetic information. It is important to note that polytene and lampbrush chromosomes are unusual and not typically found in most eukaryotic cells, but the study of their structure has revealed many common themes of chromosome organization.

Polytene Chromosomes

Giant **polytene chromosomes** are found in various tissues (salivary, midgut, rectal, and malpighian excretory tubules) in the larvae of some flies and in several species of protozoans and plants. Such structures, first observed by E. G. Balbiani in 1881, provided a model system for subsequent investigations of chromosomes. What is particularly intriguing about polytene chromosomes is that they can be seen in the nuclei of interphase cells.

Each polytene chromosome is 200 to 600 μm long, and when they are observed under the light microscope, they reveal a linear series of alternating bands and interbands (**Figure 11.6**). The banding pattern is distinctive for each chromosome in any given species. Individual bands are sometimes called **chromomeres,** a generalized term describing lateral condensations of material along the axis of a chromosome.

Extensive study using electron microscopy and radioactive tracers led to an explanation for the unusual appearance of these chromosomes. First, polytene chromosomes represent paired homologs. This is highly unusual because they are present in somatic cells, where in most organisms,

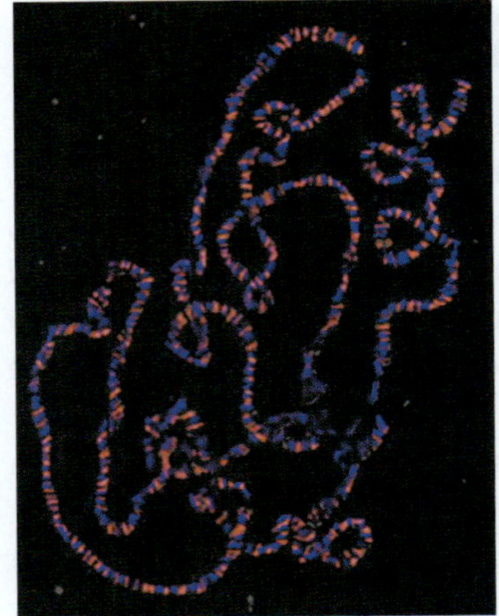
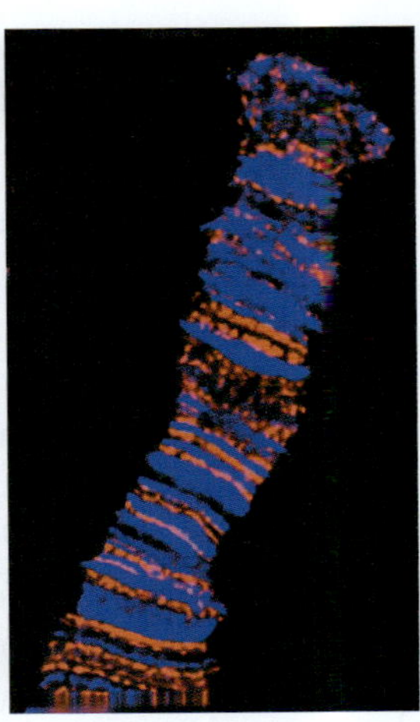

FIGURE 11.6 Polytene chromosomes derived from larval salivary gland cells of *Drosophila*.

chromosomal material is normally dispersed as chromatin and homologs are not paired. Second, their large size and distinctiveness result from the many DNA strands that compose them. The DNA of these paired homologs undergoes many rounds of replication, *but without strand separation or cytoplasmic division*. As replication proceeds, chromosomes contain 1000 to 5000 DNA strands that remain in precise parallel alignment with one another, giving rise to the distinctive band pattern along the axis of the chromosome.

The presence of bands on polytene chromosomes was initially interpreted as the visible manifestation of individual genes. The discovery that the strands present in bands undergo localized uncoiling during genetic activity further strengthened this view. Each such uncoiling event results in what is called a **puff** because of its appearance (**Figure 11.7**). That puffs are visible manifestations of gene activity (transcription that produces RNA) is evidenced by their high rate of incorporation of radioactively labeled RNA precursors, as assayed by autoradiography. Bands that are not extended into puffs incorporate fewer radioactive precursors or none at all.

The study of bands during development in insects such as *Drosophila* and the midge fly *Chironomus* reveals *differential gene activity*. A characteristic pattern of band formation that is equated with gene activation is observed

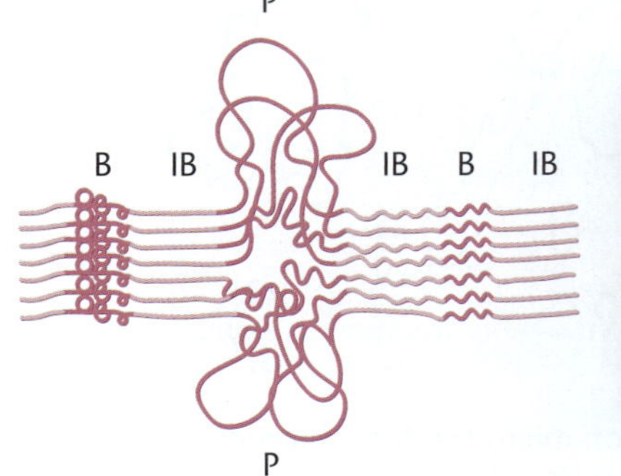

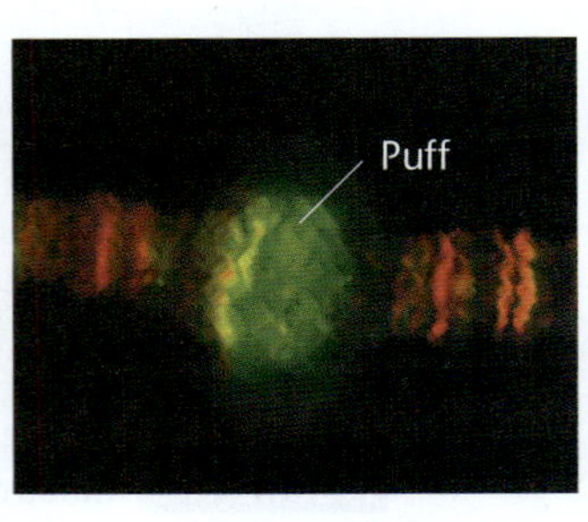

FIGURE 11.7 Photograph of a puff within a polytene chromosome. The diagram depicts the uncoiling of strands within a band (B) region to produce a puff (P) in polytene chromosomes. Interband regions (IB) are also labeled.

as development proceeds. Despite attempts to resolve the issue it is not yet clear how many genes are contained in each band. However, we do know that in *Drosophila*, which contains about 15,000 genes, there are approximately 5000 bands. Interestingly, a band may contain up to 10^7 base pairs of DNA, enough to encode 50 to 100 average-size genes.

NOW SOLVE THIS

11.2 After salivary gland cells from *Drosophila* are isolated and cultured in the presence of radioactive thymidylic acid, autoradiography is performed, revealing the presence of thymidine within polytene chromosomes. Predict the distribution of the grains along the chromosomes.

■ **HINT:** *This problem involves an understanding of the organization of DNA in polytene chromosomes. The key to its solution is to be aware that* ^{3}*H-thymidine, as a molecular tracer, will only be incorporated into DNA during its replication.*

Lampbrush Chromosomes

Another specialized chromosome that has given us insight into chromosomal structure is the **lampbrush chromosome,** so named because it resembles the brushes used to clean kerosene-lamp chimneys in the nineteenth century. Lampbrush chromosomes were first studied in detail in 1892 in the oocytes of sharks and are now known to be characteristic of most vertebrate oocytes as well as the spermatocytes of some insects. Therefore, they are meiotic chromosomes. Most experimental work has been done with material taken from amphibian oocytes.

These unique chromosomes are easily isolated from oocytes in the first prophase stage of meiosis, where they are active in directing the metabolic activities of the developing cell. The homologs are seen as synapsed pairs held together by chiasmata. However, instead of condensing, as most meiotic chromosomes do, lampbrush chromosomes often extend to lengths of 500 to 800 μm. Later in meiosis, they revert to their normal length of 15 to 20 μm. Based on these observations, lampbrush chromosomes are interpreted as extended, uncoiled versions of the normal meiotic chromosomes.

The two views of lampbrush chromosomes in **Figure 11.8** provide significant insights into their morphology. Part (a) shows the meiotic configuration under the light microscope. The linear axis of each structure contains a large number of condensed areas, and as with polytene chromosomes, these are referred to as *chromomeres*. Emanating from each chromomere is a pair of lateral loops, which give the chromosome its distinctive appearance. In part (b), the scanning electron micrograph (SEM) reveals adjacent loops present along one of the two axes of the chromosome. As with bands in polytene chromosomes, much more DNA is present in each loop than is needed to encode a single gene. Such an SEM provides a clear view of the chromomeres and the chromosomal fibers emanating from them. Each chromosomal loop is thought to be composed of one DNA double helix, while the central axis is made up of two DNA helices. This hypothesis is consistent with the belief that each meiotic chromosome is composed of a pair of sister chromatids. Studies using radioactive RNA precursors reveal that the loops are active in the synthesis of RNA. The lampbrush loops, in a manner similar to puffs in polytene chromosomes, represent DNA that has been reeled out from the central chromomere axis during transcription.

ESSENTIAL POINT

Polytene and lampbrush chromosomes are examples of specialized structures that extended our knowledge of genetic organization and function well in advance of the technology available to the modern-day molecular biologist. ■

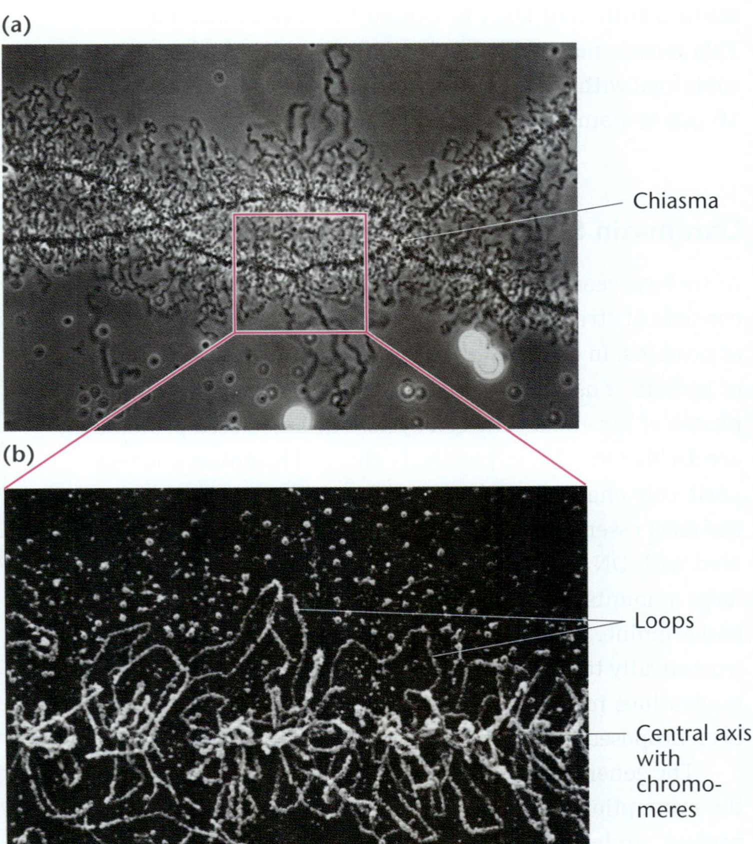

FIGURE 11.8 Lampbrush chromosomes derived from amphibian oocytes. Part (a) is a photomicrograph; part (b) is a scanning electron micrograph.

11.4 DNA Is Organized into Chromatin in Eukaryotes

We now turn our attention to the way DNA is organized in eukaryotic chromosomes. Our focus will be on eukaryotic cells, in which chromosomes are visible only during mitosis. After chromosome separation and cell division, cells enter the interphase stage of the cell cycle, during which time the components of the chromosome uncoil and are present in the form referred to as **chromatin.** While in interphase, the chromatin is dispersed in the nucleus, and the DNA of each chromosome is replicated. As the cell cycle progresses, most cells reenter mitosis, whereupon chromatin coils into visible chromosomes once again. This condensation represents a length contraction of some 10,000 times for each chromatin fiber.

The organization of DNA during the transitions just described is much more intricate and complex than in viruses or bacteria, which never exhibit a process similar to mitosis. This is due to the greater amount of DNA per chromosome, as well as the presence of a large number of proteins associated with eukaryotic DNA. For example, while DNA in the *E. coli* chromosome is 1200 μm long, the DNA in each human chromosome ranges from 19,000 to 73,000 μm in length. In a single human nucleus, all 46 chromosomes contain sufficient DNA to extend to more than 2 meters. This genetic material, along with its associated proteins, is contained within a nucleus that usually measures about 5 to 10 μm in diameter.

Chromatin Structure and Nucleosomes

As we have seen, the genetic material of viruses and bacteria consists of strands of DNA or RNA that are nearly devoid of proteins. In eukaryotic chromatin, a substantial amount of protein is associated with the chromosomal DNA in all phases of the eukaryotic cell cycle. The associated proteins are divided into basic, positively charged **histones** and less positively charged nonhistones. The histones clearly play the most essential structural role of all the proteins associated with DNA. There are five types, and they all contain large amounts of the positively charged amino acids lysine and arginine. This makes it possible for them to bond electrostatically to the negatively charged phosphate groups of nucleotides in DNA. Recall that a similar interaction has been proposed for several bacterial proteins.

The general model for chromatin structure is based on the assumption that chromatin fibers, composed of DNA and protein, undergo extensive coiling and folding as they are condensed within the cell nucleus. X-ray diffraction studies confirm that histones play an important role in chromatin structure. Chromatin produces regularly spaced diffraction rings, suggesting that repeating structural units occur along the chromatin axis. If the histone molecules are chemically removed from chromatin, the regularity of this diffraction pattern is disrupted.

A basic model for chromatin structure was worked out in the mid-1970s. Several observations were particularly relevant to the development of this model:

1. Digestion of chromatin by certain endonucleases such as micrococcal nuclease, yields DNA fragments that are approximately 200 bp in length or multiples thereof. This demonstrates that enzymatic digestion is not random, for if it were, we would expect a wide range of fragment sizes. Thus, chromatin consists of some type of repeating unit, each of which is protected from enzymatic cleavage, except where any two units are joined. It is the area between units that is attacked and cleaved by the endonuclease.

2. Electron microscopic observations of chromatin reveal that chromatin fibers are composed of linear arrays of spherical particles (**Figure 11.9**). Discovered by Ada and Donald Olins, the particles occur regularly along the axis of a chromatin strand and resemble beads on a string. This conforms nicely to the earlier observation, which suggests the existence of repeating units. These particles, initially referred to as ν-bodies (ν is the Greek letter nu), are now called **nucleosomes.**

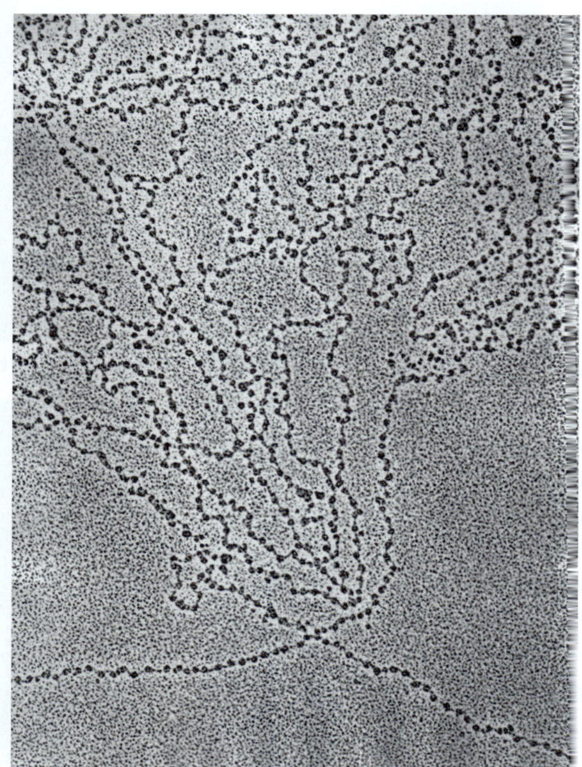

FIGURE 11.9 An electron micrograph revealing nucleosomes appearing as "beads on a string" along chromatin strands derived from *Drosophila melanogaster*.

3. Studies of precise interactions of histone molecules and DNA in the nucleosomes constituting chromatin show that histones H2A, H2B, H3, and H4 occur as two types of tetramers, $(H2A)_2 \cdot (H2B)_2$ and $(H3)_2 \cdot (H4)_2$. Roger Kornberg predicted that each repeating nucleosome unit consists of one of each tetramer (creating an octamer) in association with about 200 bp of DNA. Such a structure is consistent with previous observations and provides the basis for a model that explains the interaction of histones and DNA in chromatin.

4. When nuclease digestion time is extended, some of the 200 bp of DNA are removed from the nucleosome, creating a **nucleosome core particle** consisting of 147 bp. The DNA lost in this prolonged digestion is responsible for linking nucleosomes together. This linker DNA is associated with the fifth histone, H1.

On the basis of this information, as well as on X-ray and neutron-scattering analyses of crystallized core particles by John T. Finch, Aaron Klug, and others, a detailed model of the nucleosome was put forward in 1984, providing a basis for predicting chromatin structure and its condensation into chromosomes. In this model, illustrated in **Figure 11.10,** a 147-bp length of the 2-nm-diameter DNA molecule coils around an octamer of histones in a left-handed superhelix that completes about 1.7 turns per nucleosome. Each nucleosome, ellipsoidal in shape, measures about 11 nm at its longest point [**Figure 11.10(a)**]. Significantly, the formation of the nucleosome represents the first level of packing, whereby the DNA helix is reduced to about one-third of its original length by winding around the histones.

In the nucleus, the chromatin fiber seldom, if ever, exists in the extended form described in the previous paragraph (that is, as an extended chain of nucleosomes). Instead,

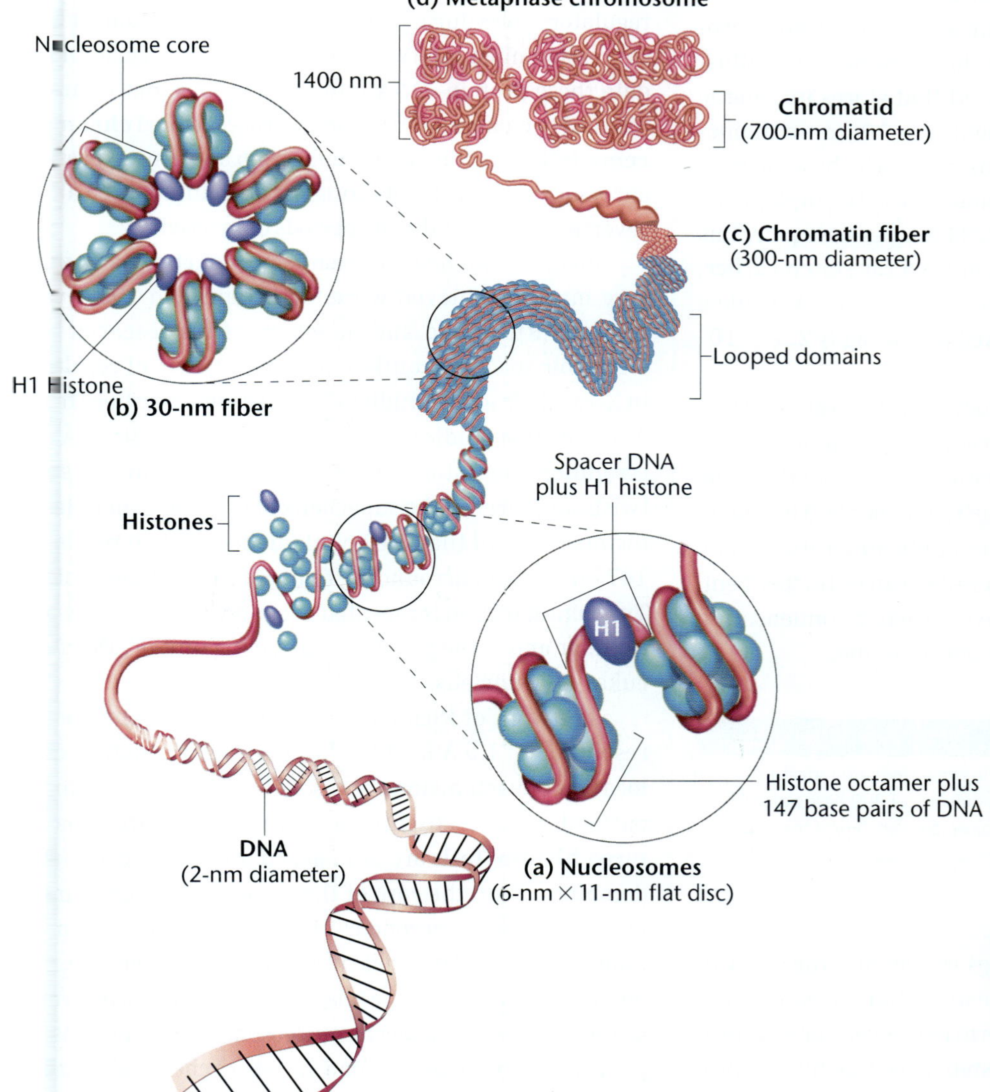

FIGURE 11.10 General model of the association of histones and DNA to form nucleosomes, illustrating the way in which each thickness of fiber may be coiled into a more condensed structure, ultimately producing a metaphase chromosome.

the 11-nm-diameter fiber is further packed into a thicker, 30-nm-diameter structure that was initially called a *solenoid* [**Figure 11.10(b)**]. This thicker structure, which is dependent on the presence of histone H1, consists of numerous nucleosomes coiled around and stacked upon one another, creating a second level of packing. This provides a six-fold increase in compaction of the DNA. It is this structure that is characteristic of an uncoiled chromatin fiber in interphase of the cell cycle. In the transition to the mitotic chromosome, still further compaction must occur. The 30-nm structures are folded into a series of *looped domains*, which further condense the chromatin fiber into a structure that is 300 nm in diameter [**Figure 11.10(c)**]. These *coiled chromatin fibers* are then compacted into the chromosome arms that constitute a chromatid, one of the longitudinal subunits of the metaphase chromosome [**Figure 11.10(d)**]. While Figure 11.10 shows the chromatid arms to be 700 nm in diameter, this value undoubtedly varies among different organisms. At a value of 700 nm, a pair of sister chromatids comprising a chromosome measures about 1400 nm.

The importance of the organization of DNA into chromatin and chromatin into mitotic chromosomes can be illustrated by considering a human cell that stores its genetic material in a nucleus that is about 5 to 10 μm in diameter. The haploid genome contains 3.2×10^9 base pairs of DNA distributed among 23 chromosomes. The diploid cell contains twice that amount. At 0.34 nm per base pair, this amounts to an enormous length of DNA (as stated earlier, to more than 2 m). One estimate is that the DNA inside a typical human nucleus is complexed with roughly 2.5×10^7 nucleosomes.

In the overall transition from a fully extended DNA helix to the extremely condensed status of the mitotic chromosome, a packing ratio (the ratio of DNA length to the length of the structure containing it) of about 500 to 1 must be achieved. In fact, our model accounts for a ratio of only about 50 to 1. Obviously, the larger fiber can be further bent, coiled, and packed to achieve even greater condensation during the formation of a mitotic chromosome.

ESSENTIAL POINT

Eukaryotic chromatin is a nucleoprotein organized into repeating units called nucleosomes, which are composed of 200 base pairs of DNA, an octamer of four types of histones, plus one linker histone. ■

Chromatin Remodeling

As with many significant findings in genetics, the study of nucleosomes has answered some important questions, but at the same time it has also led us to new ones. For example, in the preceding discussion, we established that histone proteins play an important structural role in packaging DNA into the nucleosomes that make up chromatin. While solving the structural problem of how to organize a huge amount of DNA within the eukaryotic nucleus, a new problem was apparent: *the chromatin fiber, when complexed with histones and folded into various levels of compaction, makes the DNA inaccessible to interaction with important nonhistone proteins.* For example, the proteins that function in enzymatic and regulatory roles during the processes of replication and gene expression must interact directly with DNA. To accommodate these protein–DNA interactions, chromatin must be induced to change its structure, a process called **chromatin remodeling.** In the case of replication and gene expression, chromatin must relax its compact structure but be able to reverse the process during periods of inactivity.

NOW SOLVE THIS

11.3 If a human nucleus is 10 μm in diameter, and it must hold as much as 2 m of DNA, which is complexed into nucleosomes that during full extension are 11 nm in diameter, what percentage of the volume of the nucleus does the genetic material occupy?

■ **HINT:** *This problem asks you to make some numerical calculations in order to see just how "filled" the eukaryotic nucleus is with a diploid amount of DNA. The key to its solution is the use of the formula* $V = (4/3)\pi r^3$, *which calculates the volume of a sphere.*

Insights into how different states of chromatin structure may be achieved were forthcoming in 1997, when Timothy Richmond and members of his research team were able to significantly improve the level of resolution in X-ray diffraction studies of nucleosome crystals (from 7 Å in the 1984 studies to 2.8 Å in the 1997 studies). At this resolution, most atoms are visible, thus revealing the subtle twists and turns of the superhelix of DNA that encircles the histones. Recall that the double-helical ribbon represents 147 bp of DNA surrounding four pairs of histone proteins. This configuration is repeated over and over in the chromatin fiber and is the principal packaging unit of DNA in the eukaryotic nucleus.

The work of Richmond and colleagues, extended to a resolution of 1.9 Å in 2003, has revealed the details of the location of each histone entity within the nucleosome. Of particular interest to chromatin remodeling is that unstructured **histone tails** are not packed into the folded histone domains within the core of the nucleosome. For example, tails devoid of any secondary structure extending from histones H3 and H2B protrude through the minor groove channels of the DNA helix. The tails of histone H4 appear to make a connection with adjacent nucleosomes. Histone tails also provide potential targets for a variety of chemical modifications that may be linked to genetic functions along the chromatin fiber, including the regulation of gene expression.

Several of these potential chemical modifications are now recognized as important to genetic function. One of the most well-studied histone modifications involves **acetylation** by the action of the enzyme *histone acetyltransferase (HAT)*. The addition of an acetyl group to the positively charged amino group present on the side chain of the amino acid lysine effectively changes the net charge of the protein by neutralizing the positive charge. Lysine is in abundance in histones, and geneticists have known for some time that acetylation is linked to gene activation. It appears that high levels of acetylation open up, or remodel, the chromatin fiber, an effect that increases in regions of active genes and decreases in inactive regions. In a well-studied example, the inactivation of the X chromosome in mammals, forming a Barr body (Chapter 5), histone H4 is known to be greatly underacetylated.

Two other important chemical modifications are the **methylation** and **phosphorylation** of amino acids that are part of histones. These chemical processes result from the action of enzymes called *methyltransferases* and *kinases*, respectively. Methyl groups may be added to both arginine and lysine residues of histones, and these changes can either increase or decrease transcription depending on which amino acids are methylated. Phosphate groups can be added to the hydroxyl groups of the amino acids serine and histidine introducing a negative charge on the protein. During the cell cycle, increased phosphorylation, particularly of histone H3, is known to occur at characteristic times. Such chemical modification is believed to be related to the cycle of chromatin unfolding and condensation that occurs during and after DNA replication. It is important to note that the above chemical modifications (acetylation, methylation, and phosphorylation) are all reversible, under the direction of specific enzymes.

Not to be confused with histone methylation, the nitrogenous base cytosine within the DNA itself can also be methylated, forming *5-methyl cytosine*. Cytosine methylation is usually negatively correlated with gene activity and occurs most often when the nucleotide cytidylic acid is next to the nucleotide guanylic acid, forming what is called a **CpG island.**

The research described above has extended our knowledge of nucleosomes and chromatin organization and serves here as a general introduction to the concept of chromatin remodeling. A great deal more work must be done, however, to elucidate the specific involvement of chromatin remodeling during genetic processes. In particular, the way in which the modifications are influenced by regulatory molecules within cells will provide important insights into the mechanisms of gene expression. What is clear is that the dynamic forms in which chromatin exists are vitally important to the way that all genetic processes directly involving DNA are executed. We will return to a more detailed discussion of the role of chromatin remodeling when we consider the regulation of eukaryotic gene expression later in the text (see Chapter 16). In addition, chromatin remodeling is an important topic in the discussion of **epigenetics,** the study of modifications of an organism's genetic and phenotypic expression that are *not* attributable to alteration of the DNA sequence making up a gene. This topic is discussed in depth in a future chapter (Special Topics Chapter 1—Epigenetics).

Heterochromatin

Although we know that the DNA of the eukaryotic chromosome consists of one continuous double-helical fiber along its entire length, we also know that the whole chromosome is not structurally uniform from end to end. In the early part of the twentieth century, it was observed that some parts of the chromosome remain condensed and stain deeply during interphase, while most parts are uncoiled and do not stain. In 1928, the terms **euchromatin** and **heterochromatin** were coined to describe the parts of chromosomes that are uncoiled and those that remain condensed, respectively.

Subsequent investigation revealed a number of characteristics that distinguish heterochromatin from euchromatin. Heterochromatic areas are genetically inactive because they either lack genes or contain genes that are repressed. Also, heterochromatin replicates later during the S phase of the cell cycle than euchromatin does. The discovery of heterochromatin provided the first clues that parts of eukaryotic chromosomes do not always encode proteins. For example, one heterochromatic region of the chromosome, the *telomere*, maintains the chromosome's structural integrity during DNA replication, while another, the *centromere*, facilitates chromosome movement during cell division.

The presence of heterochromatin is unique to and characteristic of the genetic material of eukaryotes. In some cases, whole chromosomes are heterochromatic. A case in point is the mammalian Y chromosome, much of which is genetically inert. And, as we discussed in Chapter 5, the inactivated X chromosome in mammalian females is condensed into an inert heterochromatic Barr body. In some species, such as mealy bugs, all chromosomes of one entire haploid set are heterochromatic.

When certain heterochromatic areas from one chromosome are translocated to a new site on the same or another nonhomologous chromosome, genetically active areas sometimes become genetically inert if they lie adjacent to the translocated heterochromatin. This influence on existing euchromatin is one example of what is more generally referred to as a **position effect.** That is, the position of a gene or group of genes relative to all other genetic material may affect their expression.

Chromosome Banding

Until about 1970, mitotic chromosomes viewed under the light microscope could be distinguished only by their relative sizes and the positions of their centromeres. In karyotypes, two or more chromosomes are often visually indistinguishable from one another. Numerous cytological techniques, referred to as **chromosome banding,** have now made it possible to distinguish such chromosomes from one another as a result of differential staining along the longitudinal axis of mitotic chromosomes.

The most useful of these techniques, called **G-banding,** involves the digestion of the mitotic chromosomes with the proteolytic enzyme trypsin, followed by Giemsa staining. This procedure stains regions of DNA that are rich in A-T base pairs. Another technique, called **C-banding,** uses chromosome preparations that are heat denatured. Subsequent Giemsa staining reveals only the heterochromatic regions of the centromeres.

These, and other chromosome-banding techniques, reflect the heterogeneity and complexity of the chromosome along its length. So precise is the banding pattern that when a segment of one chromosome has been translocated to another chromosome, its origin can be determined with great precision.

ESSENTIAL POINT

Heterochromatin, prematurely condensed in interphase and for the most part genetically inert, is illustrated by the centromeric and telomeric regions of eukaryotic chromosomes, the Y chromosome, and the Barr body. ■

11.5 Eukaryotic Genomes Demonstrate Complex Sequence Organization Characterized by Repetitive DNA

Thus far, we have looked at how DNA is organized into chromosomes in bacteriophages, bacteria, and eukaryotes. We now begin an examination of what we know about the organization of DNA sequences within the chromosomes making up an organism's genome, placing our emphasis on eukaryotes.

In addition to single copies of unique DNA sequences that make up genes, many DNA sequences within eukaryotic chromosomes are repetitive in nature. Various levels of repetition occur within the genomes of organisms. Many studies have now provided insights into **repetitive DNA,** demonstrating various classes of sequences and organization. **Figure 11.11** schematizes these categories. Some functional genes are present in more than one copy (they are referred to as **multiple-copy genes**) and so are repetitive in nature. However, the majority of repetitive sequences do not encode proteins. Nevertheless, many are transcribed, and the resultant RNAs play multiple roles in eukaryotes, including chromatin remodeling and regulation of gene expression, as discussed in greater detail in Chapter 16. We will explore three main categories of repetitive sequences: (1) heterochromatin found to be associated with centromeres and making up telomeres, (2) tandem repeats of both short and long DNA sequences, and (3) transposable sequences that are interspersed throughout the genome of eukaryotes.

Satellite DNA

The nucleotide composition of the DNA (e.g., the percentage of C-G versus A-T pairs) of a particular species is reflected in its density, which can be measured with sedimentation equilibrium centrifugation. When eukaryotic DNA is analyzed in this way, the majority is present as a single main band, or peak, of fairly uniform density. However, one or more additional peaks represent DNA that differs slightly in density. This component, called **satellite DNA,** represents a variable proportion of the total DNA, depending on the species. A profile of main-band and satellite DNA from the mouse is shown in **Figure 11.12.** By contrast, bacteria contain only main-band DNA.

The significance of satellite DNA remained an enigma until the mid-1960s, when Roy Britten and David Kohne developed the technique for measuring the reassociation kinetics of DNA that had previously been dissociated into single strands. They demonstrated that certain portions of DNA reassociated more rapidly than others. They concluded that

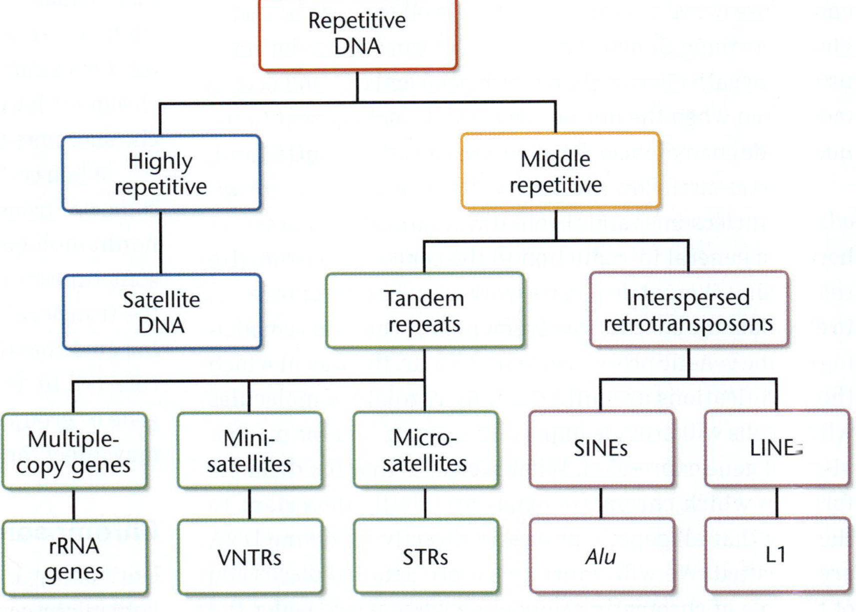

FIGURE 11.11 An overview of the categories of repetitive DNA.

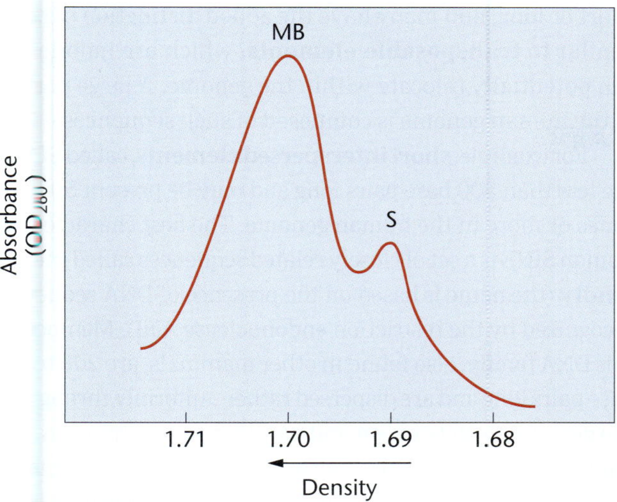

FIGURE 11.12 Separation of main-band (MB) and satellite (S) DNA from the mouse, using ultracentrifugation in a CsCl gradient.

rapid reassociation was characteristic of multiple DNA fragments composed of identical or nearly identical nucleotide sequences—the basis for the descriptive term repetitive DNA.

When satellite DNA is subjected to analysis by reassociation kinetics, it falls into the category of **highly repetitive DNA,** which is known to consist of relatively short sequences repeated a large number of times. Further evidence suggests that these sequences are present as tandem repeats clustered in very specific chromosomal areas known to be heterochromatic—the regions flanking centromeres. This was discovered in 1969 when several researchers, including Mary Lou Pardue and Joe Gall, applied ***in situ* hybridization** to the study of satellite DNA. This technique involves the molecular hybridization between an isolated fraction of radioactively labeled DNA or RNA probes and the DNA contained in the chromosomes of a cytological preparation. Following the hybridization procedure, autoradiography is performed to locate the chromosome areas complementary to the fraction of DNA or RNA.

Pardue and Gall demonstrated that radioactive probes made from mouse satellite DNA hybridize with the DNA of centromeric regions of mouse mitotic chromosomes, which are all telocentric (**Figure 11.13**). Several conclusions were drawn: Satellite DNA differs from main-band DNA in its molecular composition, as established by buoyant density studies. It is composed of repetitive sequences. Finally, satellite DNA is found in the heterochromatic centromeric regions of chromosomes.

Centromeric DNA Sequences

The separation of homologs during mitosis and meiosis depends on **centromeres,** described cytologically as the *primary constrictions* along eukaryotic chromosomes. In this role, it is believed that the DNA sequence contained within the centromere is critical. Careful analysis has confirmed this prediction. The minimal region of the centromere that supports the function of chromosomal segregation is designated the **CEN region.** Within this heterochromatic region of the chromosome, the DNA binds a platform of proteins, which in multicellular organisms includes the **kinetochore** that binds to the spindle fiber during division (see Figure 2.8).

The CEN regions of the yeast *Saccharomyces cerevisiae* were the first to be studied. Each centromere serves an identical function, so it is not surprising that CENs from different chromosomes were found to be remarkably similar in their organization. The CEN region of yeast chromosomes consists of about 120 bp. Mutational analysis suggests that portions near the 3′-end of this DNA region are most critical to centromere function since mutations in them, but not those nearer the 5′-end, disrupt centromere function. Thus, the DNA of this region appears to be essential to the eventual binding to the spindle fiber.

Centromere sequences of multicellular eukaryotes are much more extensive than in yeast and vary considerably in size. For example, in *Drosophila* the CEN region is found within some 200 to 600 kb of DNA, much of which is highly repetitive. Recall from our prior discussion that highly repetitive satellite DNA is localized in the centromere regions of mice. In humans, one of the most recognized satellite DNA sequences is the **alphoid family,** found mainly in the centromere regions. Alphoid sequences, each about 170 bp in length, are present in tandem arrays of up to 1 million base pairs. Embedded within this repetitive DNA are more specific sequences that are critical to centromere function.

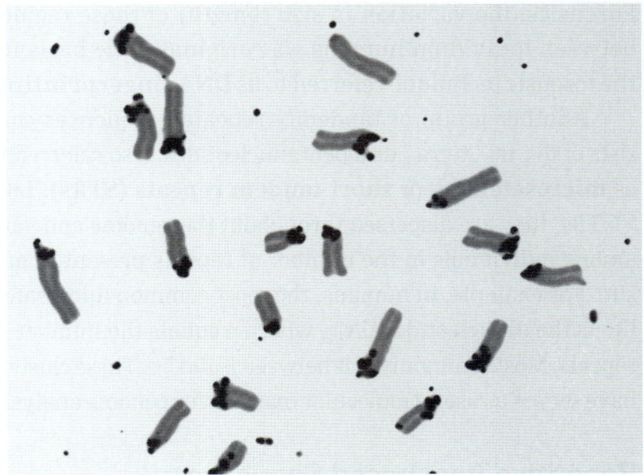

FIGURE 11.13 *In situ* molecular hybridization between RNA transcribed from mouse satellite DNA and mitotic chromosomes. The grains in the autoradiograph localize the chromosome regions (the centromeres) containing satellite DNA sequences.

One final observation of interest is that the H3 histone, a normal part of most all eukaryotic nucleosomes, is substituted by a variant histone designated CENP-A in centromeric heterochromatin. It is believed that the N-terminal protein tails that make CENP-A unique are involved in the binding of kinetechore proteins that are essential to the microtubules of spindle fibers. This finding supports the supposition that the DNA sequence found only in centromeres is related to the function of this unique chromosomal structure.

Middle Repetitive Sequences: VNTRs and STRs

A brief look at still another prominent category of repetitive DNA sheds additional light on the organization of the eukaryotic genome. In addition to highly repetitive DNA, which constitutes about 5 percent of the human genome (and 10 percent of the mouse genome), a second category, **middle** (or **moderately**) **repetitive DNA,** is fairly well characterized. Because we now know a great deal about the human genome, we will use our own species to illustrate this category of DNA in genome organization.

Although middle repetitive DNA does include some duplicated genes (such as those encoding ribosomal RNA), most prominent in this category are either noncoding tandemly repeated sequences or noncoding interspersed sequences. No function has been ascribed to these components of the genome. An example is DNA described as **variable number tandem repeats (VNTRs).** These repeating DNA sequences may be 15 to 100 bp long and are found within and between genes. Many such clusters are dispersed throughout the genome and are often referred to as **minisatellites.**

The number of tandem copies of each specific sequence at each location varies from one individual to the next, creating localized regions of 1000 to 20,000 bp (1 to 20 kb) in length. As we will see in Special Topics Chapter 6—DNA Forensics, the variation in size (length) of these regions between individual humans was originally the basis for the forensic technique referred to as **DNA fingerprinting.**

Another group of tandemly repeated sequences consists of di-, tri-, tetra-, and pentanucleotides, also referred to as **microsatellites** or **short tandem repeats (STRs).** Like VNTRs, they are dispersed throughout the genome and vary among individuals in the number of repeats present at any site. For example, in humans, the most common microsatellite is the dinucleotide $(CA)_n$, where n equals the number of repeats. Most commonly, n is between 5 and 50. These clusters have served as useful molecular markers for genome analysis.

Repetitive Transposed Sequences: SINEs and LINEs

Still another category of repetitive DNA consists of sequences that are interspersed individually throughout the genome, rather than being tandemly repeated. They can be either short or long, and many have the added distinction of being similar to **transposable elements,** which are mobile and can potentially relocate within the genome. A large portion of the human genome is composed of such sequences.

For example, **short interspersed elements,** called **SINEs,** are less than 500 base pairs long and may be present 500,000 times or more in the human genome. The best characterized human SINE is a set of closely related sequences called the ***Alu* family** (the name is based on the presence of DNA sequences recognized by the restriction endonuclease *AluI*). Members of this DNA family, also found in other mammals, are 200 to 300 base pairs long and are dispersed rather uniformly throughout the genome, both between and within genes. In humans this family encompasses more than 5 percent of the entire genome.

Alu sequences are of particular interest because some members of the *Alu* family are transcribed into RNA, although the specific role of this RNA is not certain. Even so, the consequence of *Alu* sequences is their potential for transposition within the genome, which is related to chromosome rearrangements during evolution. *Alu* sequences are thought to have arisen from an RNA element whose DNA complement was dispersed throughout the genome as a result of the activity of reverse transcriptase (an enzyme that synthesizes DNA on an RNA template).

The group of **long interspersed elements (LINEs)** represents yet another category of repetitive transposable DNA sequences. LINEs are usually about 6 kb in length and in the human genome are present approximately 850,000 times. The most prominent example in humans is the **L1 family.** Members of this sequence family are about 6400 base pairs long and are present up to 500,000 times. Their 5′-end is highly variable, and their role within the genome has yet to be defined.

The general mechanism for transposition of L1 elements is now clear. The L1 DNA sequence is first transcribed into an RNA molecule. The RNA then serves as the template for the synthesis of the DNA complement using the enzyme *reverse transcriptase*. This enzyme is encoded by a portion of the L1 sequence. The new L1 copy then integrates into the DNA of the chromosome at a new site. Because of the similarity of this transposition mechanism to that used by retroviruses, LINEs are referred to as **retrotransposons.**

SINEs and LINEs represent a significant portion of human DNA. SINEs constitute about 13 percent of the human genome, whereas LINEs constitute up to 21 percent. Within both types of elements, repeating sequences of DNA are present in combination with unique sequences.

Middle Repetitive Multiple-Copy Genes

In some cases, middle repetitive DNA includes functional genes present tandemly in multiple copies. For example, many copies exist of the genes encoding ribosomal RNA. *Drosophila* has 120 copies per haploid genome. Single genetic units encode a large precursor molecule that is processed into the 5.8*S*, 18*S*, and 28*S* rRNA components. In humans, multiple copies of this gene are clustered on the p arm of the acrocentric chromosomes 13, 14, 15, 21, and 22. Multiple copies of the genes encoding 5*S* rRNA are transcribed separately from multiple clusters found together on the terminal portion of the p arm of chromosome 1.

ESSENTIAL POINT

Eukaryotic genomes demonstrate complex sequence organization characterized by numerous categories of repetitive DNA, consisting of either tandem repeats clustered in various regions of the genome or single sequences repeatedly interspersed at random throughout the genome. ■

11.6 The Vast Majority of a Eukaryotic Genome Does Not Encode Functional Genes

Given the preceding information concerning various forms of repetitive DNA in eukaryotes, it is of interest to pose an important question: *What proportion of the eukaryotic genome actually encodes functional genes?*

We have seen that, taken together, the various forms of highly repetitive and moderately repetitive DNA comprise a substantial portion of the human genome. In addition to repetitive DNA, a large amount of the DNA consists of single-copy sequences as defined by reassociation kinetic analysis (Chapter 9) that appear to be noncoding. Included are many instances of what we call **pseudogenes.** These are DNA sequences representing evolutionary vestiges of duplicated copies of genes that have undergone significant mutational alteration. As a result, although they show some homology to their parent gene, they are usually not transcribed because of insertions and deletions throughout their structure.

Although the proportion of the genome consisting of repetitive DNA varies among organisms, one feature seems to be shared: *Only a very small part of the genome actually codes for proteins.* For example, the 20,000 to 30,000 genes encoding proteins in the sea urchin occupy less than 10 percent of the genome. In *Drosophila*, only 5 to 10 percent of the genome is occupied by genes coding for proteins. In humans, it appears that the estimated 20,000 protein-coding genes occupy only about 2 percent of the total DNA sequence making up the genome.

Related to the above observation, we are currently discovering many cases where DNA sequences are transcribed into RNA molecules that are not translated into proteins, and that play important cellular roles, such as regulation of genetic activity. This topic is explored in more depth later in the text; see Chapter 16.

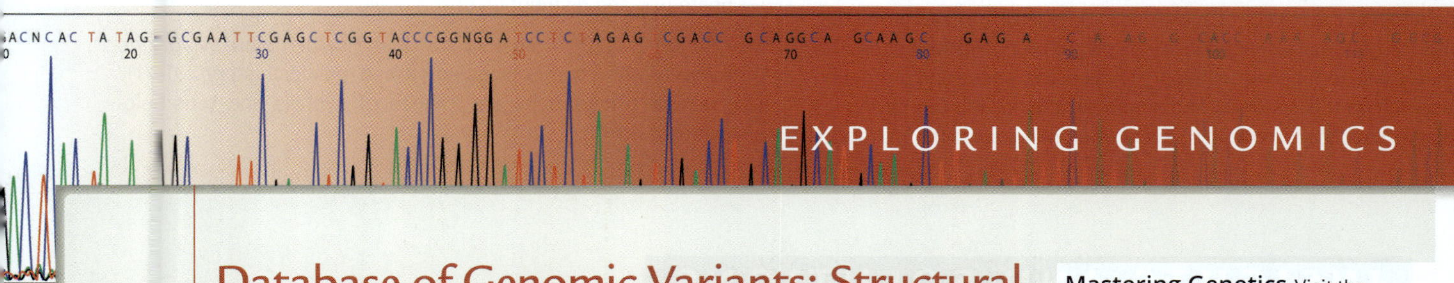

Database of Genomic Variants: Structural Variations in the Human Genome

Mastering Genetics Visit the Study Area: Exploring Genomics

In this chapter, we focused on structural details of chromosomes and DNA sequence organization in chromosomes. A related finding is that large segments of DNA and a number of genes can vary greatly in copy number due to duplications, creating **copy number variations (CNVs)**. Many studies are under way to identify and map CNVs and to find possible disease conditions associated with them.

Several thousand CNVs have been identified in the human genome, and estimates suggest there may be thousands more within human populations. In this Exploring Genomics exercise we will visit the **Database of Genomic Variants (DGV)**, which provides a quickly expanding summary of structural variations in the human genome including CNVs.

■ **Exercise I - Database of Genomic Variants**

1. Access the DGV at http://dgv.tcag.ca/dgv/app/home. Click the "About the Project" tab to learn more about the purpose of the DGV.
2. Information in the DGV is easily viewed by clicking on a chromosome of interest using the "Find DGV Variants by Chromosome" feature. Using this feature, click on a chromosome of interest to you. A table will appear under the "Variants" tab showing several columns of data including:

■ Start and Stop: Shows the locus for the CNV, including the base pairs that span the variation.

■ Variant Accession: Provides a unique identifying number for each variation. Click on the variant accession number to reveal a separate page of specific details about each CNV, including the chromosomal banding location for the variation and known genes that are located in the CNV.

(continued)

Exploring Genomics—continued

- Variant Type: Most variations in this database are CNVs. Variant subtypes, such as deletions or insertions, and duplications, are shown in an adjacent column.

3. Let's analyze a particular group of CNVs. Many CNVs are unlikely to affect phenotype because they involve large areas of non-protein-coding or nonregulatory sequences. But gene-containing CNVs have been identified, including variants containing genes associated with Alzheimer disease, Parkinson disease, and other conditions.

Defensin (*DEF*) genes are part of a large family of highly duplicated genes. To learn more about *DEF* genes and CNVs, use the Keyword Search box and search for *DEF*. A results page for the search will appear with a listing of relevant CNVs. Click on the name for any of the different *DEF* genes listed, which will take you to a wealth of information [including links to Online Mendelian Inheritance in Man (OMIM)] about these genes so that you can answer the following questions. Do this for several *DEF*-containing CNVs on different chromosomes to find the information you will need for your answers.

a. On what chromosome(s) did you find CNVs containing *DEF* genes?

b. What did you learn about the function of *DEF* gene products? What do DEF proteins do?

c. Variations in *DEF* genotypes and *DEF* gene expression in humans have been implicated in a number of different human disease conditions. Give examples of the kinds of disorders affected by variations in *DEF* genotypes.

CASE STUDY Helping or hurting?

Roberts syndrome is a rare inherited disorder characterized by facial defects as well as severe limb shortening, extra digits, and deformities of the knees and ankles. A cytogenetic analysis of patients with Roberts syndrome, using Giemsa staining or C-banding, reveals that there is premature separation of centromeres and other heterochromatic regions during mitotic metaphase instead of anaphase. A couple who has an infant with Roberts syndrome is contacted by a local organization dedicated to promoting research on rare genetic diseases, asking if they can photograph the infant as part of a campaign to obtain funding for these conditions. The couple learned that the privacy of such medical images is not well protected, and they often are subsequently displayed on public Web sites. The couple was torn between helping to raise awareness and promoting research on this condition and sheltering their child from having his images used inappropriately. Several interesting questions are raised.

1. In Roberts syndrome, how could premature separation of centromeres during mitosis cause the wide range of phenotypic deficiencies?
2. What ethical obligations do the parents owe to their child in this situation and to helping others with Roberts syndrome by allowing images of their child to be used in raising awareness of this disorder?
3. If the parents decide to allow their infant to be photographed, what steps should the local organization take to ensure appropriate use and distribution of the photos?

See Onion R. (2014). History, or Just Horror? *Slate Magazine* (http://www.slate.com/articles/news_and_politics/history/2014/11/old_medical_photographs_are_images_of_syphilis_and_tuberculosis_patients.html).

INSIGHTS AND SOLUTIONS

A previously undiscovered single-cell organism was found living at a great depth on the ocean floor. Its nucleus contained only a single linear chromosome with 7×10^6 nucleotide pairs of DNA coalesced with three types of histone-like proteins. Consider the following questions:

1. A short micrococcal nuclease digestion yielded DNA fractions of 700, 1400, and 2100 bp. Predict what these fractions represent. What conclusions can be drawn?

   Solution: The chromatin fiber may consist of a variation of nucleosomes containing 700 bp of DNA. The 1400- and 2100-bp fractions, respectively, represent two and three linked nucleosomes. Enzymatic digestion may have been incomplete, leading to the latter two fractions.

2. The analysis of individual nucleosomes reveals that each unit contained one copy of each protein and that the short linker DNA contained no protein bound to it. If the entire chromosome consists of nucleosomes (discounting any linker DNA), how many are there, and how many total proteins are needed to form them?

   Solution: Since the chromosome contains 7×10^6 bp of DNA, the number of nucleosomes, each containing 7×10^2 bp, is equal to

   $$7 \times 10^6 / 7 \times 10^2 = 10^4 \text{ nucleosomes}$$

   The chromosome contains 10^4 copies of each of the three proteins, for a total of 3×10^4 proteins.

3. Further analysis revealed the organism's DNA to be a double helix similar to the Watson–Crick model but containing 20 bp per complete turn of the right-handed helix. The physical size of the nucleosome was exactly double the volume occupied by that found in all other known eukaryotes, by virtue of increasing the distance along the fiber axis by a factor of two

Compare the degree of compaction of this organism's nucleosome to that found in other eukaryotes.

Solution: The unique organism compacts a length of DNA consisting of 35 complete turns of the helix (700 bp per nucleosome/20 bp per turn) into each nucleosome. The normal eukaryote compacts a length of DNA consisting of 20 complete turns of the helix (200 bp per nucleosome/10 bp per turn) into a nucleosome one-half the volume of that in the unique organism. The degree of compaction is therefore less in the unique organism.

Problems and Discussion Questions

Mastering Genetics Visit for instructor-assigned tutorials and problems.

1. **HOW DO WE KNOW?** In this chapter, we focused on how DNA is organized at the chromosomal level. Along the way, we found many opportunities to consider the methods and reasoning by which much of this information was acquired. From the explanations given in the chapter, propose answers to the following fundamental questions:
   (a) How do we know that viral and bacterial chromosomes most often consist of circular DNA molecules devoid of protein?
   (b) What is the experimental basis for concluding that puffs in polytene chromosomes and loops in lampbrush chromosomes are areas of intense transcription of RNA?
   (c) How did we learn that eukaryotic chromatin exists in the form of repeating nucleosomes, each consisting of about 200 base pairs and an octamer of histones?
   (d) How do we know that satellite DNA consists of repetitive sequences and has been derived from regions of the centromere?
2. **CONCEPT QUESTION** Review the Chapter Concepts list on p. 226. These all relate to how DNA is organized in viral, bacterial, and eukaryote chromosomes. Write a short essay that contrasts the major differences between the organization of DNA in viruses and bacteria versus eukaryotes. ■
3. How does the 1200-μm-long chromosome of *E. coli* fit into a bacterial cell 2.0×0.5 μm long?
4. Describe how giant polytene chromosomes are formed.
5. What are the commonalities between polytene chromosomes and lampbrush chromosomes?
6. Describe the structure of lampbrush chromosomes. Where are they located?
7. What chemical and structural properties of histones enable them to successfully package eukaryotic DNA? What is chromatin remodeling, and how is it controlled within eukaryotic cells?
8. Describe the sequence of research findings that led to the development of the model of chromatin structure.
9. What are the molecular composition and arrangement of the components in the nucleosome?
10. Describe the transitions that occur as nucleosomes are coiled and folded, ultimately forming a chromatid.
11. Provide a comprehensive definition of heterochromatin, and list as many examples as you can.
12. Contrast the various categories of repetitive DNA.
13. Compare satellite and centromeric DNAs and determine their locations and roles in eukaryotic chromosomes.
14. What do SINE and LINE mean in terms of chromosome structure? Why are they called "repetitive"?
15. Mammals contain a diploid genome consisting of at least 10^9 bp. If this amount of DNA is present as chromatin fibers, where each group of 200 bp of DNA is combined with nine histones into a nucleosome and each group of six nucleosomes is combined into a solenoid, achieving a final packing ratio of 50, determine:
    (a) the total number of nucleosomes in all fibers.
    (b) the total number of histone molecules combined with DNA in the diploid genome.
    (c) the combined length of all fibers.
16. Assume that a viral DNA molecule is a 50-μm-long circular strand of a uniform 2-nm diameter. If this molecule is contained in a viral head that is a 0.08-μm-diameter sphere, will the DNA molecule fit into the viral head, assuming complete flexibility of the molecule? Justify your answer mathematically.
17. A particular variant of the lambda bacteriophage has a DNA double-stranded genome of 51,365 base pairs. How long would this DNA be?
18. The human genome contains approximately 10^6 copies of an *Alu* sequence, one of the best-studied classes of short interspersed elements (SINEs), per haploid genome. Individual *Alu*s share a 282-nucleotide consensus sequence followed by a 3′-adenine-rich tail region. Given that there are approximately 3×10^9 bp per human haploid genome, about how many base pairs are spaced between each *Alu* sequence?
19. Below is a diagram of the general structure of the bacteriophage λ chromosome. Speculate on the mechanism by which it forms a closed ring upon infection of the host cell.

5′GGGCGGCGACCT—double-stranded region—3′
3′—double-stranded region—CCCGCCGCTGGA5′

12

The Genetic Code and Transcription

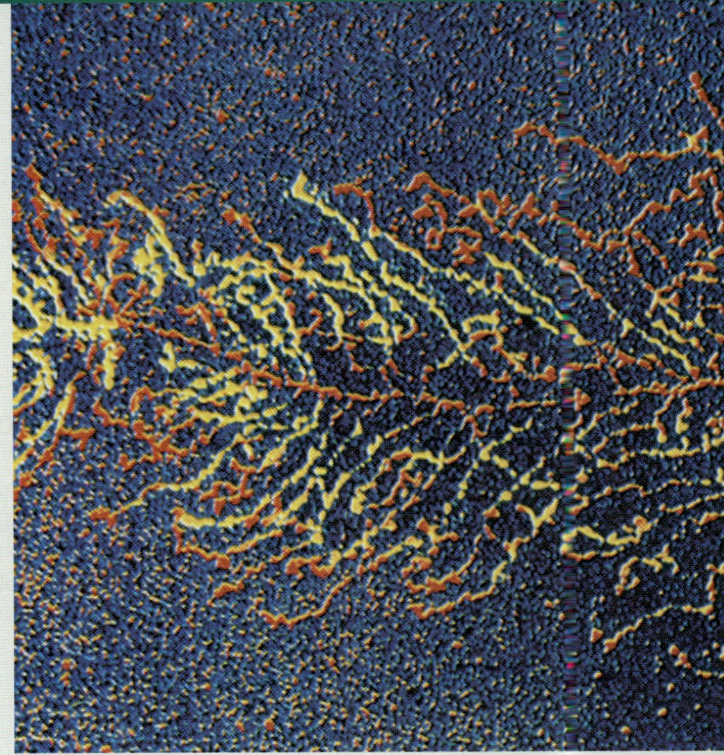

Electron micrograph visualizing the process of transcription.

CHAPTER CONCEPTS

- Genetic information is stored in DNA using a triplet code that is nearly universal to all living things on Earth.
- The encoded genetic information stored in DNA is initially copied into an RNA transcript.
- Once transferred from DNA to RNA, the genetic code exists as triplet codons, using the four ribonucleotides in RNA as the letters composing it.
- By using four different letters taken three at a time, 64 triplet sequences are possible. Most encode one of the 20 amino acids present in proteins, which serve as the end products of most genes.
- Several codons provide signals that initiate or terminate protein synthesis.
- The process of transcription is similar but more complex in eukaryotes compared to bacteria and bacteriophages that infect them.

The linear sequence of deoxyribonucleotides making up DNA ultimately dictates the components constituting proteins, the end product of most genes. The central question is how such information stored as a nucleic acid is decoded into a protein. **Figure 12.1** gives a simplified overview of how this transfer of information occurs. In the first step in gene expression, information on one of the two strands of DNA (the template strand) is copied into an RNA complement through transcription. Once synthesized, this RNA acts as a "messenger" molecule bearing the coded information—hence its name, *messenger RNA (mRNA)*. The mRNAs then associate with ribosomes, where decoding into proteins takes place.

In this chapter, we focus on the initial phases of gene expression by addressing two major questions. First, how is genetic information encoded? Second, how does the transfer from DNA to RNA occur, thus defining the process of transcription? As you shall see, ingenious analytical research established that the genetic code is written in units of three letters—ribonucleotides present in mRNA that reflect the stored information in genes. Most all triplet code words direct the incorporation of a specific amino acid into a protein as it is synthesized. As we can predict based on our prior discussion of the replication of DNA, transcription is also a complex process dependent on a major polymerase enzyme and a cast of supporting proteins. We will explore what is known about transcription in bacteria and then contrast this model with the differences found in eukaryotes. Together, the information in this and Chapter 13 provides a comprehensive picture of

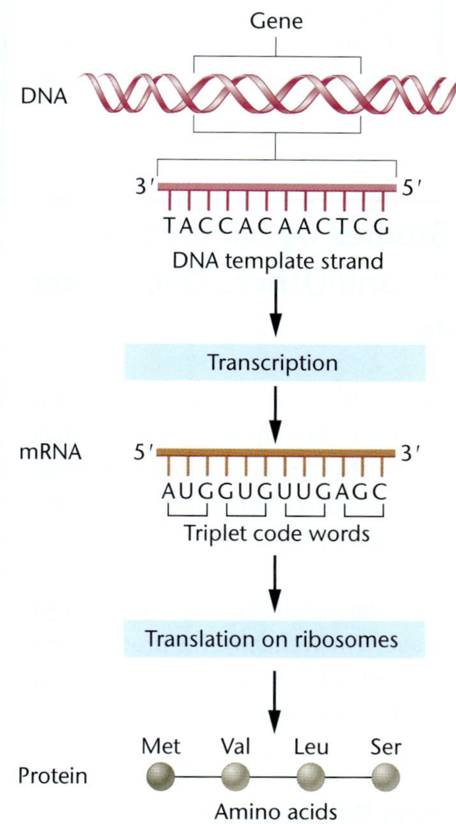

FIGURE 12.1 Flowchart illustrating how genetic information encoded in DNA produces protein.

molecular genetics, which serves as the most basic foundation for understanding living organisms. In Chapter 13, we will address how translation occurs and discuss the structure and function of proteins.

12.1 The Genetic Code Exhibits a Number of Characteristics

Before we consider the various analytical approaches that led to our current understanding of the genetic code, let's summarize the general features that characterize it.

1. The genetic code is written in linear form, using the ribonucleotide bases that compose mRNA molecules as "letters." The ribonucleotide sequence is derived from the complementary nucleotide bases in DNA.
2. Each "word" within the mRNA consists of three ribonucleotide letters, thus representing a triplet code. With several exceptions, each group of *three* ribonucleotides, called a codon, specifies *one* amino acid.
3. The code is **unambiguous**—each triplet specifies only a single amino acid.
4. The code is **degenerate;** that is, a given amino acid can be specified by more than one triplet codon. This is the case for 18 of the 20 amino acids.
5. The code contains one "start" and three "stop" signals, triplets that **initiate** and **terminate** translation.
6. No internal punctuation (such as a comma) is used in the code. Thus, the code is said to be **commaless.** Once translation of mRNA begins, the codons are read one after the other, with no breaks between them.
7. The code is **nonoverlapping.** Once translation commences, any single ribonucleotide at a specific location within the mRNA is part of only one triplet.
8. The code is nearly **universal.** With only minor exceptions, almost all viruses, bacteria, archaea, and eukaryotes use a single coding dictionary.

12.2 Early Studies Established the Basic Operational Patterns of the Code

In the late 1950s, before it became clear that mRNA is the intermediate that transfers genetic information from DNA to proteins, researchers thought that DNA itself might directly encode proteins during their synthesis. Because ribosomes had already been identified, the initial thinking was that information in DNA was transferred in the nucleus to the RNA of the ribosome, which served as the template for protein synthesis in the cytoplasm. This concept soon became untenable as accumulating evidence indicated the existence of an unstable intermediate template. The RNA of ribosomes, on the other hand, was extremely stable. As a result, in 1961 François Jacob and Jacques Monod postulated the existence of **messenger RNA (mRNA).** Once mRNA was discovered, it was clear that even though genetic information is stored in DNA, the code that is translated into proteins resides in RNA. The central question then was how only four letters—the four ribonucleotides—could specify 20 words (the amino acids).

The Triplet Nature of the Code

In the early 1960s, Sydney Brenner argued on theoretical grounds that the code had to be a triplet since three-letter words represent the minimal use of four letters to specify

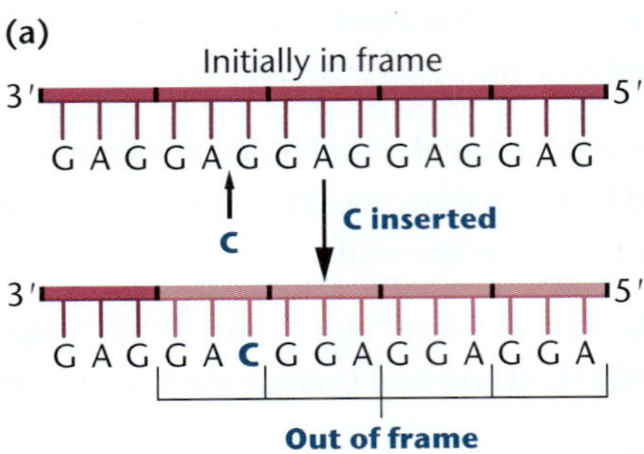

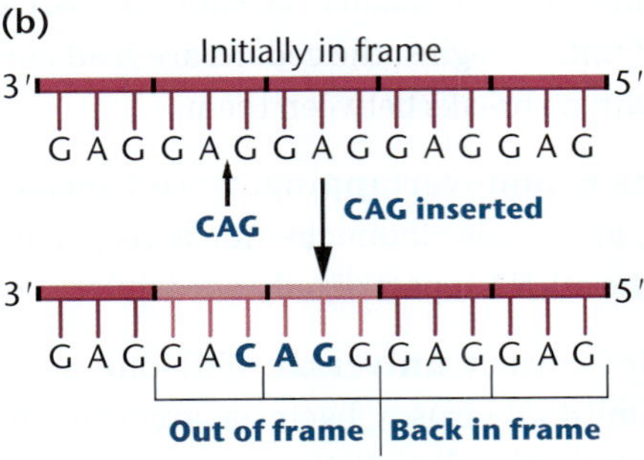

FIGURE 12.2 The effect of frameshift mutations on a DNA sequence with the repeating triplet sequence GAG. (a) The insertion of a single nucleotide shifts all subsequent triplet reading frames. (b) The insertion of three nucleotides changes only two triplets, but the frame of reading is then reestablished to the original sequence.

20 amino acids. A code of four nucleotides, taken two at a time, for example, provides only 16 unique code words (4^2). A triplet code yields 64 words (4^3)—clearly more than the 20 needed—and is much simpler than a four-letter code (4^4), which specifies 256 words.

Experimental evidence supporting the triplet nature of the code was subsequently derived from research by Francis Crick and his colleagues. Using phage T4, they studied **frameshift mutations,** which result from the addition or deletion of one or more nucleotides within a gene and subsequently the mRNA transcribed by it. The gain or loss of letters shifts the *frame of reading* during translation. Crick and his colleagues found that the gain or loss of one or two nucleotides caused a frameshift mutation, but when three nucleotides were involved, the frame of reading was reestablished (**Figure 12.2**). This would not occur if the code was anything other than a triplet. This work also suggested that most triplet codes are not blank, but rather encode amino acids, supporting the concept of a degenerate code.

12.3 Studies by Nirenberg, Matthaei, and Others Deciphered the Code

In 1961, Marshall Nirenberg and J. Heinrich Matthaei deciphered the first specific coding sequences, which served as a cornerstone for the complete analysis of the genetic code. Their success, as well as that of others who made important contributions to breaking the code, was dependent on the use of two experimental tools—an *in vitro (cell-free) protein-synthesizing system* and an enzyme, **polynucleotide phosphorylase,** which enabled the production of synthetic mRNAs. These mRNAs are templates for polypeptide synthesis in the cell-free system.

Synthesizing Polypeptides in a Cell-Free System

In the cell-free system, amino acids are incorporated into polypeptide chains. This *in vitro* mixture must contain the essential factors for protein synthesis in the cell: ribosomes, tRNAs, amino acids, and other molecules essential to translation (see Chapter 13). In order to follow (or trace) protein synthesis, one or more of the amino acids must be radioactive. Finally, an mRNA must be added, which serves as the template that will be translated.

In 1961, mRNA had yet to be isolated. However, use of the enzyme polynucleotide phosphorylase allowed the artificial synthesis of RNA templates, which could be added to the cell-free system. This enzyme, isolated from bacteria, catalyzes the reaction shown in **Figure 12.3.** Discovered in 1955 by Marianne Grunberg-Manago and Severo Ochoa, the enzyme functions metabolically in bacterial cells to degrade RNA. However, *in vitro*, with high concentrations

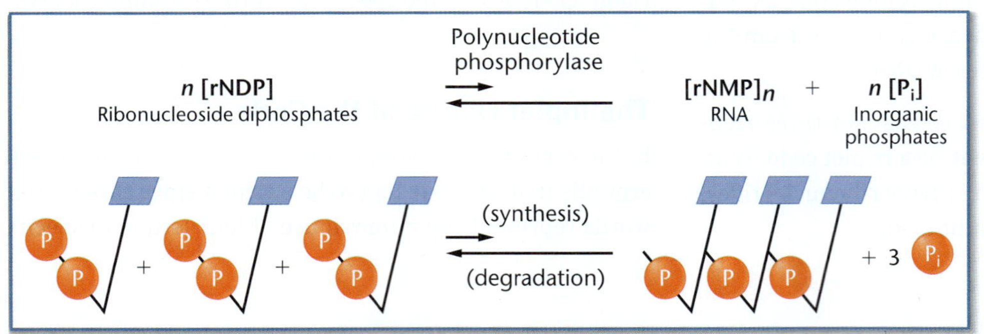

FIGURE 12.3 The reaction catalyzed by the enzyme polynucleotide phosphorylase. Note that the equilibrium of the reaction favors the degradation of RNA but can be "forced" in the direction favoring synthesis.

of ribonucleoside diphosphates, the reaction can be "forced" in the opposite direction to synthesize RNA, as shown.

In contrast to RNA polymerase, polynucleotide phosphorylase does not require a DNA template. As a result, each addition of a ribonucleotide is random, based on the relative concentration of the four ribonucleoside diphosphates added to the reaction mixtures. The probability of the insertion of a specific ribonucleotide is proportional to the availability of that molecule, relative to other available ribonucleotides. *This point is absolutely critical to understanding the work of Nirenberg and others in the ensuing discussion.*

Together, the cell-free system and the availability of synthetic mRNAs provided a means of deciphering the ribonucleotide composition of various triplets encoding specific amino acids.

The Use of Homopolymers

In their initial experiments, Nirenberg and Matthaei synthesized **RNA homopolymers**, each with only one type of ribonucleotide. Therefore, the mRNA added to the *in vitro* system was UUUUUU . . . , AAAAAA . . . , CCCCCC . . . , or GGGGGG . . . . They tested each mRNA and were able to determine which, if any, amino acids were incorporated into newly synthesized proteins. To do this, the researchers labeled 1 of the 20 amino acids added to the *in vitro* system and conducted a series of experiments, each with a different radioactively labeled amino acid.

For example, in experiments using ^{14}C-phenylalanine (**Table 12.1**), Nirenberg and Matthaei concluded that the message poly U (polyuridylic acid) directed the incorporation of only phenylalanine into the homopolymer polyphenylalanine. Assuming the validity of a triplet code, they determined the first specific codon assignment—UUU codes for phenylalanine. Using similar experiments, they quickly found that AAA codes for lysine and CCC codes for proline. Poly G was not an adequate template, probably because the molecule folds back upon itself. Thus, the assignment for GGG had to await other approaches.

Note that the *specific triplet codon assignments* were possible only because homopolymers were used. This method yields only the *composition of triplets*, but since three identical letters can have only one possible sequence (e.g., UUU), the actual codons were identified.

TABLE 12.1 Incorporation of ^{14}C-phenylalanine **into Protein**

Artificial mRNA	Radioactivity (counts/min)
None	44
Poly U	39,800
Poly A	50
Poly C	38

Source: After Nirenberg and Matthaei (1961).

Mixed Heteropolymers

With these techniques in hand, Nirenberg and Matthaei, and Ochoa and coworkers turned to the use of **RNA heteropolymers.** In this type of experiment, two or more different ribonucleoside diphosphates are added in combination to form the synthetic mRNA. The researchers reasoned that if they knew the relative proportion of each type of ribonucleoside diphosphate, they could predict the frequency of any particular triplet codon occurring in the synthetic mRNA. If they then added the mRNA to the cell-free system and ascertained the percentage of any particular amino acid present in the new protein, they could analyze the results and predict the composition (not the specific sequence) of triplets specifying particular amino acids.

This approach is shown in **Figure 12.4**. Suppose that A and C are added in a ratio of 1A:5C. The insertion of a ribonucleotide at any position along the RNA molecule during its synthesis is determined by the ratio of A:C. Therefore, there is a 1/6 chance for an A and a 5/6 chance for a C to occupy each position. On this basis, we can calculate the frequency of any given triplet appearing in the message.

For AAA, the frequency is $(1/6)^3$, or about 0.4 percent. For AAC, ACA, and CAA, the frequencies are identical—that is, $(1/6)^2(5/6)$, or about 2.3 percent for each triplet. Together, all three 2A:1C triplets account for 6.9 percent of the total three-letter sequences. In the same way, each of three 1A:2C triplets accounts for $(1/6)(5/6)^2$, or 11.6 percent (or a total of 34.8 percent); CCC is represented by $(5/6)^3$, or 57.9 percent of the triplets.

By examining the percentages of any given amino acid incorporated into the protein synthesized under the direction of this message, we can propose probable base compositions for each amino acid (Figure 12.4). Since proline appears 69 percent of the time, we could propose that proline is encoded by CCC (57.9 percent) and one triplet of 2C:1A (11.6 percent). Histidine, at 14 percent, is probably coded by one 2C:1A (11.6 percent) and one 1C:2A (2.3 percent). Threonine, at 12 percent, is likely coded by only one 2C:1A. Asparagine and glutamine each appear to be coded by one of the 1C:2A triplets, and lysine appears to be coded by AAA.

Using as many as all four ribonucleotides to construct the mRNA, the researchers conducted many similar experiments. Although determining the *composition* of the triplet code words for all 20 amino acids represented a significant breakthrough, the *specific sequences* of triplets were still unknown—other approaches were needed.

ESSENTIAL POINT

The use of RNA homopolymers and mixed heteroploymers in a cell-free system allowed the determination of the composition, but not the sequence, of triplet codons designating specific amino acids. ■

RNA Heteropolymer with Ratio of 1A:5C			
Possible compositions	Possible triplets	Probability of occurrence of any triplet	Final %
3A	AAA	$(1/6)^3 = 1/216 = 0.5\%$	0.5
2A:1C	AAC ACA CAA	$(1/6)^2(5/6) = 5/216 = 2.3\%$	$3 \times 2.3 = 6.9$
1A:2C	ACC CAC CCA	$(1/6)(5/6)^2 = 25/216 = 11.6\%$	$3 \times 11.6 = 34.8$
3C	CCC	$(5/6)^3 = 125/216 = 57.9\%$	57.9
			~100

Chemical synthesis of message

RNA

C C C C C C C C A C C C C C C A A C C A C C C C C A C C C C C A C C C A A

Translation of message

Percentage of amino acids in protein		Probable base-composition assignments
Lysine	<1	AAA
Glutamine	2	2A:1C
Asparagine	2	2A:1C
Threonine	12	1A:2C
Histidine	14	1A:2C, 2A:1C
Proline	69	CCC, 1A:2C

FIGURE 12.4 Results and interpretation of a mixed heteropolymer experiment where a ratio of 1A:5C is used (1/6A:5/6C).

The Triplet Binding Assay

It was not long before more advanced techniques were developed. In 1964, Nirenberg and Philip Leder developed the **triplet binding assay,** which led to specific assignments of triplets. The technique took advantage of the observation that ribosomes, when presented *in vitro* with an RNA sequence as short as three ribonucleotides, will bind to it and form a complex similar to that found *in vivo*. The triplet sequence acts like a codon in mRNA, attracting a **transfer RNA** molecule containing the complementary sequence and carrying a specific amino acid (**Figure 12.5**). The triplet sequence in tRNA that is complementary to a codon of mRNA is called an **anticodon.**

Although it was not yet feasible to chemically synthesize long stretches of RNA, triplets of known sequence could be synthesized in the laboratory to serve as templates. All that was needed was a method to determine which tRNA–amino acid was bound to the triplet RNA–ribosome complex. The test system Nirenberg and Leder devised was quite simple. The amino acid to be tested was made radioactive, and a charged tRNA was produced. Because codon compositions

NOW SOLVE THIS

12.1 In a mixed heteropolymer experiment using polynucleotide phosphorylase, 3/4G:1/4C was used to form the synthetic message. The amino acid composition of the resulting protein was determined to be:

Glycine	36/64	(56 percent)
Alanine	12/64	(19 percent)
Arginine	12/64	(19 percent)
Proline	4/64	(6 percent)

From this information,

(a) Indicate the percentage (or fraction) of the time each possible codon will occur in the message.

(b) Determine one consistent base-composition assignment for the amino acids present.

■ **HINT:** *This problem asks you to analyze a mixed heteropolymer experiment and to predict codon composition assignments for the amino acids encoded by the synthetic message. The key to its solution is to first calculate the proportion of each triplet codon in the synthetic RNA and then match these to the proportions of amino acids that are synthesized.*

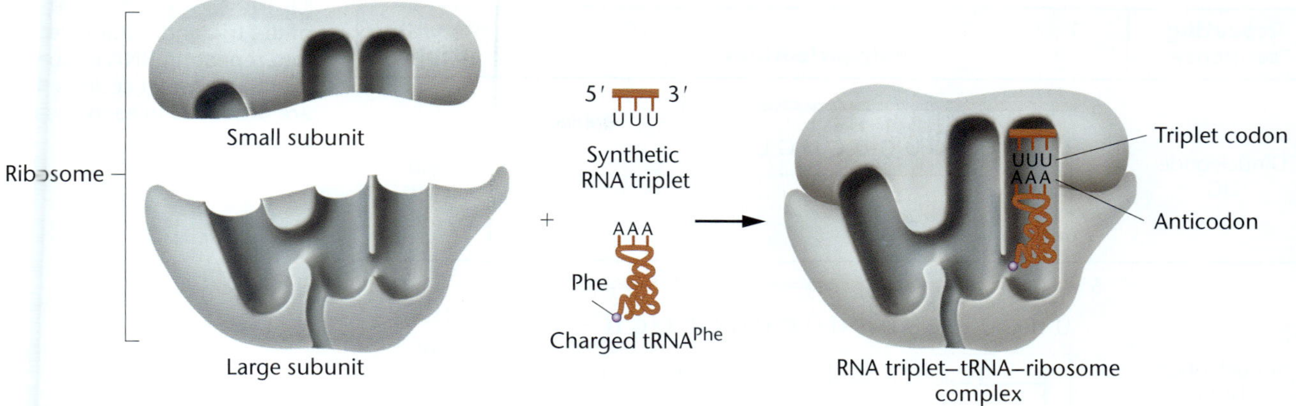

FIGURE 12.5 Illustration of the behavior of the components during the triplet binding assay. The synthetic UUU triplet RNA sequence acts as a codon, attracting the complementary AAA anticodon of the charged $tRNA^{Phe}$, which together are bound by the subunits of the ribosome.

were known, researchers could narrow the range of amino acids that should be tested for each specific triplet.

The radioactively charged tRNA, the RNA triplet, and ribosomes were incubated together and then passed through a nitrocellulose filter, which retains the larger ribosomes but not the other smaller components, such as unbound charged tRNA. If radioactivity is not retained on the filter, an incorrect amino acid has been tested. But if radioactivity remains on the filter, it is retained because the charged tRNA has bound to the triplet associated with the ribosome. When this occurs, a specific codon assignment can be made.

Work proceeded in several laboratories, and in many cases clear-cut, unambiguous results were obtained. **Table 12.2**, for example, shows 26 triplets assigned to 9 amino acids. However, in some cases, the degree of triplet binding was inefficient and assignments were not possible. Eventually, about 50 of the 64 triplets were assigned. These specific assignments of triplets to amino acids led to two major conclusions. First, the genetic code is *degenerate*; that is, one amino acid can be specified by more than one triplet. Second, the code is *unambiguous*. That is, a single triplet specifies only one amino acid. As you shall see later in this chapter, these conclusions have been upheld with only minor exceptions. The triplet binding technique was a major innovation in deciphering the genetic code.

TABLE 12.2 Amino Acid Assignments to Specific Trinucleotides Derived from the Triplet Binding Assay

Trinucleotides	Amino Acid
AAA AAG	Lysine
AUG	Methionine
AUU AUC AUA	Isoleucine
CCG CCA CCU CCC	Proline
CUC CUA CUG CUU	Leucine
GAA GAG	Glutamic acid
UCA UCG UCU UCC	Serine
UGU UGC	Cysteine
UUA UUG	Leucine
UUU UUC	Phenylalanine

Repeating Copolymers

Yet another innovative technique used to decipher the genetic code was developed in the early 1960s by Har Gobind Khorana, who chemically synthesized long RNA molecules consisting of short sequences repeated many times. First, he created shorter sequences (e.g., di-, tri-, and tetranucleotides), which were then replicated many times and finally joined enzymatically to form the long polynucleotides, referred to as copolymers. As shown in **Figure 12.6**, a dinucleotide made in this way is converted to an mRNA with two repeating triplets. A trinucleotide is converted to an mRNA with three potential triplets, depending on the point at which initiation occurs, and a tetranucleotide creates four repeating triplets.

When synthetic mRNAs were added to a cell-free system, the predicted number of amino acids incorporated into polypeptides was upheld. Several examples are shown in **Table 12.3**. When the data were combined with those on composition assignment and triplet binding, specific assignments were possible.

One example of specific assignments made in this way will illustrate the value of Khorana's approach. Consider the following experiments in concert with one another:

(1) The *repeating trinucleotide sequence* UUCUUCUUC . . . can be read as three possible repeating triplets—UUC, UCU, and CUU—depending on the initiation point. When placed in a cell-free translation system, three different polypeptide homopolymers—containing either phenylalanine, serine, or leucine—are produced. Thus, we know that each of the three triplets encodes one of the

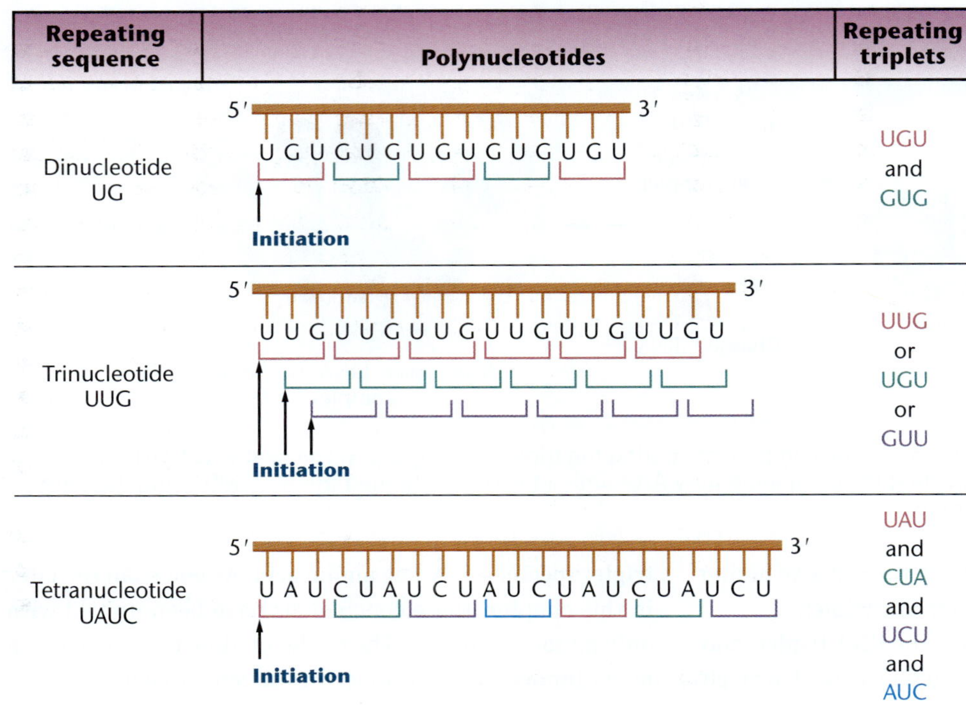

FIGURE 12.6 The conversion of di-, tri-, and tetranucleotides into repeating RNA copolymers. The triplet codons that are produced in each case are shown.

three amino acids, but we do not know which codes which;

(2) On the other hand, the *repeating dinucleotide sequence* UCUCUCUC . . . produces the triplets UCU and CUC and, when used in an experiment, leads to the incorporation of leucine and serine into a polypeptide. Thus, the triplets UCU and CUC specify leucine and serine, but we still do not know which triplet specifies which amino acid. However, when considering both sets of results in concert, we can conclude that UCU, which is common to both experiments, must encode either leucine or serine but not phenylalanine. Thus, either CUU *or* UUC encodes leucine *or* serine, while the other encodes phenylalanine;

TABLE 12.3 Amino Acids Incorporated Using Repeated Synthetic Copolymers of RNA

Repeating Copolymer	Codons Produced	Amino Acids in Polypeptides
UG	UGU	Cysteine
	GUG	Valine
AC	ACA	Threonine
	CAC	Histidine
UUC	UUC	Phenylalanine
	UCU	Serine
	CUU	Leucine
AUC	AUC	Isoleucine
	UCA	Serine
	CAU	Histidine
UAUC	UAU	Tyrosine
	CUA	Leucine
	UCU	Serine
	AUC	Isoleucine
GAUA	GAU	None
	AGA	None
	UAG	None
	AUA	None

(3) To derive more specific information, we can examine the results of using the repeating tetranucleotide sequence UUAC, which produces the triplets UUA, UAC, ACU, and CUU. The CUU triplet is one of the two in which we are interested. Three amino acids are incorporated by this experiment: leucine, threonine, and tyrosine. Because CUU must specify only serine or leucine, and because, of these two, only leucine appears in the resulting polypeptide, we may conclude that CUU specifies leucine. Once this assignment is established, we can logically determine all others. Of the two triplet pairs remaining (UUC and UCU from the first experiment *and* UCU and CUC from the second experiment), whichever triplet is common to both must encode serine. This is UCU. By elimination, UUC is determined to encode phenylalanine and CUC is determined to encode leucine. Thus, through painstaking logical analysis, four specific triplets encoding three different amino acids have been assigned from these experiments.

From these and similar interpretations, Khorana reaffirmed the identity of triplets that had already been deciphered and filled in gaps left from other approaches. A number of examples are shown in Table 12.3. Of great interest, the use of two tetranucleotide sequences, GAUA and

GUAA, suggested that at least two triplets were *termination codons*. Khorana reached this conclusion because neither of these repeating sequences directed the incorporation of more than a few amino acids into a polypeptide, too few for him to detect. There are no triplets common to both messages, and both seemed to contain at least one triplet that terminates protein synthesis. Of the possible triplets in the poly-(GAUA) sequence shown in Table 12.3, UAG was later shown to be a termination codon.

ESSENTIAL POINT

Use of the triplet binding assay and of repeating copolymers allowed the determination of the specific sequences of triplet codons designating specific amino acids. ■

NOW SOLVE THIS

12.2 When repeating copolymers are used to form synthetic mRNAs, dinucleotides produce a single type of polypeptide that contains only two different amino acids. On the other hand, using a trinucleotide sequence produces three different polypeptides, each consisting of only a single amino acid. Why? What will be produced when a repeating tetranucleotide is used?

■ **HINT:** *This problem asks you to consider different outcomes of repeating copolymer experiments. The key to its solution is to be aware that when using a repeating copolymer of RNA, translation can be initiated at different ribonucleotides. You must simply determine the number of triplet codons produced by initiation at each of the different ribonucleotides.*

12.4 The Coding Dictionary Reveals the Function of the 64 Triplets

The various techniques used to decipher the genetic code have yielded a dictionary of 61 triplet codons assigned to amino acids. The remaining three triplets are termination signals and do not specify any amino acid.

Degeneracy and the Wobble Hypothesis

A general pattern of triplet codon assignments becomes apparent when we look at the genetic coding dictionary. **Figure 12.7** designates the assignments in a particularly illustrative form first suggested by Francis Crick.

Most evident is that the code is degenerate, as the early researchers predicted. That is, almost all amino acids are specified by two, three, or four different codons. Three amino acids (serine, arginine, and leucine) are each encoded by six different codons. Only tryptophan and methionine are encoded by single codons.

Second position of codon

First position of codon (5′- end)	U	C	A	G	Third position of codon (3′- end)
U	UUU Phe	UCU Ser	UAU Tyr	UGU Cys	U
U	UUC Phe	UCC Ser	UAC Tyr	UGC Cys	C
U	UUA Leu	UCA Ser	UAA *Stop*	UGA *Stop*	A
U	UUG Leu	UCG Ser	UAG *Stop*	UGG Trp	G
C	CUU Leu	CCU Pro	CAU His	CGU Arg	U
C	CUC Leu	CCC Pro	CAC His	CGC Arg	C
C	CUA Leu	CCA Pro	CAA Gln	CGA Arg	A
C	CUG Leu	CCG Pro	CAG Gln	CGG Arg	G
A	AUU Ile	ACU Thr	AAU Asn	AGU Ser	U
A	AUC Ile	ACC Thr	AAC Asn	AGC Ser	C
A	AUA Ile	ACA Thr	AAA Lys	AGA Arg	A
A	AUG Met	ACG Thr	AAG Lys	AGG Arg	G
G	GUU Val	GCU Ala	GAU Asp	GGU Gly	U
G	GUC Val	GCC Ala	GAC Asp	GGC Gly	C
G	GUA Val	GCA Ala	GAA Glu	GGA Gly	A
G	GUG Val	GCG Ala	GAG Glu	GGG Gly	G

Initiation (AUG) Termination (UAA, UAG, UGA)

FIGURE 12.7 The coding dictionary. AUG encodes methionine, which initiates most polypeptide chains. All other amino acids except tryptophan, which is encoded only by UGG, are encoded by two to six triplets. The triplets UAA, UAG, and UGA are termination signals and do not encode any amino acids.

Also evident is the *pattern of degeneracy*. Most often, in a set of codons specifying the same amino acid, the first two letters are the same, with only the third differing. Crick discerned a pattern in the degeneracy at the third position, and in 1966, he postulated the **wobble hypothesis.**

Crick's hypothesis first predicted that the initial two ribonucleotides of triplet codes are more critical than the third in attracting the correct tRNA during translation. He postulated that hydrogen bonding at the third position of the codon–anticodon interaction is less constrained and need not adhere as specifically to the established base-pairing rules. The wobble hypothesis thus proposes a more flexible set of base-pairing rules at the third position of the codon (**Table 12.4**).

This relaxed base-pairing requirement, or "wobble," allows the anticodon of a single form of tRNA to pair with more than one triplet in mRNA. Consistent with the wobble hypothesis and degeneracy, U at the first position (the 5′-end) of the tRNA anticodon may pair with A or G at the third position (the 3′-end) of the mRNA codon, and G may likewise pair with U or C. Inosine (I), one of the modified bases found in tRNA, may pair with C, U, or A. Applying these wobble rules, a minimum of about 30 different tRNA species is necessary to accommodate the 61 triplets specifying an amino acid. If nothing more, wobble can be considered a potential economy measure, provided that the

TABLE 12.4 Codon–Anticodon Base-Pairing Rules

Base at First Position (5′-end) of tRNA	Base at Third Position (3′-end) of mRNA
A	U
C	G
G	C or U
U	A or G
I	A, U, or C

fidelity of translation is not compromised. Current estimates are that 30 to 40 tRNA species are present in bacteria and up to 50 tRNA species exist in animal and plant cells.

The Ordered Nature of the Code

Still another observation has become apparent in the pattern of codon sequences and their corresponding amino acids, leading to the description referred to as an **ordered genetic code.** Chemically similar amino acids often share one or two "middle" bases in the different triplets encoding them. For example, either U or C is often present in the second position of triplets that specify hydrophobic amino acids, including valine and alanine, among others. Two codons (AAA and AAG) specify the positively charged amino acid lysine. If only the middle letter of these codons is changed from A to G (AGA and AGG), the positively charged amino acid arginine is specified.

The chemical properties of amino acids will be discussed in more detail in Chapter 13. The end result of an "ordered" code is that it buffers the potential effect of mutation on protein function. While many mutations of the second base of triplet codons result in a change of one amino acid to another, the change is often to an amino acid with similar chemical properties. In such cases, protein function may not be noticeably altered.

Initiation and Termination

In contrast to the *in vitro* experiments discussed earlier, initiation of protein synthesis *in vivo* is a highly specific process. In bacteria, the initial amino acid inserted into all polypeptide chains is a modified form of methionine—**N-formylmethionine (fMet).** Only one codon, AUG, codes for methionine, and it is sometimes called the **initiator codon.** However, when AUG appears internally in mRNA, rather than at an initiating position, unformylated methionine is inserted into the polypeptide chain. Rarely, another codon, GUG, specifies methionine during initiation, though it is not clear why this happens, since GUG normally encodes valine.

In bacteria, either the formyl group is removed from the initial methionine upon the completion of protein synthesis or the entire formylmethionine residue is removed. In eukaryotes, methionine is also the initial amino acid during polypeptide synthesis. However, it is not formylated.

As mentioned in the preceding section, three other triplets (UAG, UAA, and UGA) serve as **termination codons,** punctuation signals that do not code for any amino acid. They are not recognized by a tRNA molecule, and translation terminates when they are encountered. Mutations that produce any of the three triplets internally in a gene will also result in termination. Consequently, only a partial polypeptide has been synthesized when it is prematurely released from the ribosome. When such a change occurs in the DNA, it is called a **nonsense mutation.**

ESSENTIAL POINT

The complete coding dictionary reveals that of the 64 possible triplet codons, 61 encode the 20 amino acids found in proteins, while three triplets terminate translation. ■

12.5 The Genetic Code Has Been Confirmed in Studies of Bacteriophage MS2

The various aspects of the genetic code discussed thus far yield a fairly complete picture, suggesting that it is triplet in nature, degenerate, unambiguous, and commaless, and that it contains punctuation start and stop signals. That these features are correct was confirmed by analysis of the RNA-containing **bacteriophage MS2** by Walter Fiers and his coworkers.

MS2 is a bacteriophage that infects *E. coli*. Its nucleic acid (RNA) contains only about 3500 ribonucleotides, making up only four genes, specifying a coat protein, an RNA replicase, a lysis protein, and a maturation protein. The small genome and a few gene products enabled Fiers and his colleagues to sequence the genes and their products. When the chemical constitution of these genes and their encoded proteins were compared, they were found to exhibit **colinearity.** That is, based on the coding dictionary, *the linear sequence of triplet codons corresponds precisely with the linear sequence of amino acids in each protein.* Punctuation was also confirmed. For example, in the coat protein gene, the codon for the first amino acids is AUG, the common initiator codon. The codon for the last amino acid is followed by two consecutive termination codons, UAA and UAG. The analysis clearly showed that the genetic code in this virus was identical to that established experimentally in bacterial systems.

12.6 The Genetic Code Is Nearly Universal

Between 1960 and 1978, it was generally assumed that the genetic code would be found to be universal, applying equally to viruses, bacteria, archaea, and eukaryotes. Certainly, the nature of mRNA and the translation machinery seemed to be very similar in these organisms. For example, cell-free systems derived from bacteria can translate eukaryotic mRNAs. Poly U stimulates synthesis of polyphenylalanine in cell-free systems when the components are derived from eukaryotes. Many recent studies involving recombinant DNA technology (see Chapter 17) reveal that eukaryotic genes can be inserted into bacterial cells, which are then transcribed and translated. Within eukaryotes, mRNAs from mice and rabbits have been injected into amphibian eggs and efficiently translated. For the many eukaryotic genes that have been sequenced, notably those for hemoglobin molecules, the amino acid sequence of the encoded proteins adheres to the coding dictionary established from bacterial studies.

However, several 1979 reports on the coding properties of DNA derived from mitochondria (**mtDNA**) of yeast and humans undermined the principle of the universality of the genetic language. Since then, mtDNA has been examined in many other organisms.

Cloned mtDNA fragments have been sequenced and compared with the amino acid sequences of various mitochondrial proteins, revealing several exceptions to the coding dictionary (**Table 12.5**). Most surprising is that the codon UGA, normally specifying termination, encodes tryptophan during translation in yeast and human mitochondria. In yeast mitochondria, threonine is inserted instead of leucine when CUA is encountered in mRNA. In human mitochondria, AUA, which normally specifies isoleucine, directs the internal insertion of methionine.

In 1985, several other exceptions to the standard coding dictionary were discovered in the bacterium *Mycoplasma capricolum* and in the nuclear genes of the protozoan ciliates *Paramecium, Tetrahymena*, and *Stylonychia*. For example, as shown in **Table 12.5**, one alteration converts the termination codon UGA to tryptophan, yet several others convert the normal termination codons UAA and UAG to glutamine. These changes are significant because bacteria and several eukaryotes are involved, representing distinct species that have evolved separately over a long period of time.

Note the apparent pattern in several of the altered codon assignments. The change in coding capacity involves only a shift in recognition of the third, or wobble, position. For example, AUA specifies isoleucine in the cytoplasm and methionine in the mitochondrion, but in the cytoplasm, methionine is specified by AUG. Similarly, UGA calls for termination in the cytoplasm, but it specifies tryptophan in the mitochondrion; in the cytoplasm, tryptophan is specified by UGG. It has been suggested that such changes in codon recognition may represent an evolutionary trend toward reducing the number of tRNAs needed in mitochondria; only 22 tRNA species are encoded in human mitochondria, for example. However, until more examples are found, the differences must be considered to be exceptions to the previously established general coding rules.

TABLE 12.5 Exceptions to the Universal Code

Codon	Normal Code Word	Altered Code Word	Source
UGA	Termination	Trp	Human and yeast mitochondria; *Mycoplasma*
CUA	Leu	Thr	Yeast mitochondria
AUA	Ile	Met	Human mitochondria
AGA	Arg	Termination	Human mitochondria
AGG	Arg	Termination	Human mitochondria
UAA	Termination	Gln	*Paramecium, Tetrahymena*, and *Stylonychia*
UAG	Termination	Gln	*Paramecium*

12.7 Different Initiation Points Create Overlapping Genes

Earlier we stated that the genetic code is nonoverlapping—each ribonucleotide in an mRNA is part of only one codon. However, this characteristic of the code does not rule out the possibility that a single mRNA may have multiple initiation points for translation. If so, these points could theoretically create several different reading frames within the same mRNA, thus specifying more than one polypeptide and leading to the concept of **overlapping genes.**

That this might actually occur in some viruses was suspected when phage ϕX174 was carefully investigated. The circular DNA chromosome consists of 5386 nucleotides, which should encode a maximum of 1795 amino acids, sufficient for five or six proteins. However, this small virus in fact synthesizes 11 proteins consisting of more than 2300 amino acids. A comparison of the nucleotide sequence of the DNA and the amino acid sequences of the polypeptides synthesized has clarified the apparent paradox. At least four cases of multiple initiation have been discovered, creating overlapping genes.

For example, in one case, the coding sequences for the initiation of two polypeptides are found at separate positions within the reading frame that specifies the sequence of a third polypeptide. In one case, seven different polypeptides may be created from a DNA sequence that might otherwise have specified only three polypeptides.

A similar situation has been observed in other viruses, including phage G4 and the animal virus SV40. Like ϕX174, phage G4 contains a circular single-stranded DNA molecule. The use of overlapping reading frames optimizes the use of a limited amount of DNA present in these small viruses. However, such an approach to storing information has a distinct disadvantage in that a single mutation may affect more than one protein and thus increase the chances that the change will be deleterious or lethal.

12.8 Transcription Synthesizes RNA on a DNA Template

Even while the genetic code was being studied, it was quite clear that proteins were the end products of many genes. Thus, while some geneticists attempted to elucidate the code, other research efforts focused on the nature of genetic expression. The central question was how DNA, a nucleic acid, could specify a protein composed of amino acids.

The complex multistep process begins with the transfer of genetic information stored in DNA to RNA. The process by which RNA molecules are synthesized on a DNA template is called **transcription.** It results in an mRNA molecule complementary to the gene sequence of one of the double helix's two strands. Each triplet codon in the mRNA is, in turn, complementary to the anticodon region of its corresponding tRNA as the amino acid is correctly inserted into the polypeptide chain during translation. The significance of transcription is enormous, for it is the initial step in the process of *information flow* within the cell. The idea that RNA is involved as an intermediate molecule in the process of information flow between DNA and protein was suggested by the following observations:

1. DNA is, for the most part, associated with chromosomes in the nucleus of the eukaryotic cell. However, protein synthesis occurs in association with ribosomes located outside the nucleus in the cytoplasm. Therefore, DNA does not appear to participate directly in protein synthesis.

2. RNA is synthesized in the nucleus of eukaryotic cells, where DNA is found, and is chemically similar to DNA.

3. Following its synthesis, most messenger RNA migrates to the cytoplasm, where protein synthesis (translation) occurs.

Collectively, these observations suggested that genetic information, stored in DNA, is transferred to an RNA intermediate, which directs the synthesis of proteins. As with most new ideas in molecular genetics, the initial supporting evidence was based on experimental studies of bacteria and their phages. It was clearly established that during initial infection, RNA synthesis preceded phage protein synthesis and that the RNA is complementary to phage DNA.

The results of these experiments agree with the concept of a messenger RNA (mRNA) being made on a DNA template and then directing the synthesis of specific proteins in association with ribosomes. This concept was formally proposed by François Jacob and Jacques Monod in 1961 as part of a model for gene regulation in bacteria. Since then, mRNA has been isolated and studied thoroughly. There is no longer any question about its role in genetic processes.

12.9 RNA Polymerase Directs RNA Synthesis

To establish that RNA can be synthesized on a DNA template, it was necessary to demonstrate that there is an enzyme capable of directing this synthesis. By 1959, several investigators, including Samuel Weiss, had independently isolated such a molecule from rat liver. Called **RNA polymerase,** it has the same general substrate requirements as does DNA polymerase, the major exception being that the substrate nucleotides contain the ribose rather than the deoxyribose form of the sugar. Unlike DNA polymerase, no primer is required to initiate synthesis. The overall reaction summarizing the synthesis of RNA on a DNA template can be expressed as

$$n(\text{NTP}) \xrightarrow[\text{polymerase}]{\text{RNA}} (\text{NMP})_n + n(\text{PP}_i)$$

As this equation reveals, nucleoside triphosphates (NTPs) are substrates for the enzyme, which catalyzes the polymerization of nucleoside monophosphates (NMPs), or nucleotides, into a polynucleotide chain $(\text{NMP})_n$. Nucleotides are linked during synthesis by 3′ to 5′ phosphodiester bonds (see Figure 9.9). The energy created by cleaving the triphosphate precursor into the monophosphate form drives the reaction, and inorganic pyrophosphates (PP_i) are produced.

A second equation summarizes the sequential addition of each ribonucleotide as the process of transcription progresses

$$(\text{NMP})_n + \text{NTP} \xrightarrow[\text{polymerase}]{\text{RNA}} (\text{NMP})_{n+1} + \text{PP}_i$$

As this equation shows, each step of transcription involves the addition of one ribonucleotide (NMP) to the growing polyribonucleotide chain $(NMP)_{n+1}$, using a nucleoside triphosphate (NTP) as the precursor.

RNA polymerase from *E. coli* has been extensively characterized and shown to consist of subunits designated α, β, β', and ω. The **core enzyme** contains the subunits $\alpha_{(\text{two copies})}, \beta, \beta'$, and ω. A slightly more complex form of the enzyme, the **holoenzyme,** contains an additional subunit called the **sigma (σ) factor** and has a molecular weight of almost 500,000 Da. The β and β' polypeptides provide the catalytic basis and active site for transcription, while the σ factor [**Figure 12.8(a)**] plays a regulatory function in the initiation of RNA transcription [Figure 12.8(a)].

Although there is but a single form of the core enzyme in *E. coli*, there are several different σ factors, creating variations of the polymerase holoenzyme. On the other hand, eukaryotes display three distinct forms of RNA polymerase, each consisting of a greater number of polypeptide subunits than in bacteria.

ESSENTIAL POINT

Transcription—the initial step in gene expression—is the synthesis, under the direction of RNA polymerase, of a strand of RNA complementary to a DNA template. ■

Promoters, Template Binding, and the σ Factor

Transcription results in the synthesis of a single-stranded RNA molecule complementary to a region along only one strand of the DNA double helix. When discussing transcription, the DNA strand that serves as a template for RNA polymerase is denoted as the **template strand** and the complementary DNA strand is called the **coding strand.** Note that the complementary strand is called the coding strand because it and the RNA molecule transcribed from the template strand have the same 5′ to 3′ nucleotide sequence, but with uridine (U) substituted for thymidine (T) in the RNA.

The initial step is **template binding** [**Figure 12.8(b)**]. In bacteria, the site of this initial binding is established when the RNA polymerase σ factor recognizes specific DNA sequences called **promoters.** These regions are located in the region upstream (5′) from the point of initial transcription of a gene. It is believed that the enzyme "explores" a length of DNA until it recognizes the promoter region and binds to about 60 nucleotide pairs of the helix, 40 of which are upstream from the point of initial transcription. Once this occurs, the helix is denatured or unwound locally, making the DNA template accessible to the action of the enzyme. The point at which transcription actually begins is called the **transcription start site,** often indicated as position +1.

The importance of promoter sequences cannot be overemphasized. They govern the efficiency of the initiation of transcription. In bacteria, both strong promoters and weak promoters have been discovered. Because the interaction of promoters with RNA polymerase governs transcription, the nature of the binding between them is at the heart of discussions concerning genetic regulation, the subject of Chapters 15 and 16. While we will later pursue more

(a) Transcription components

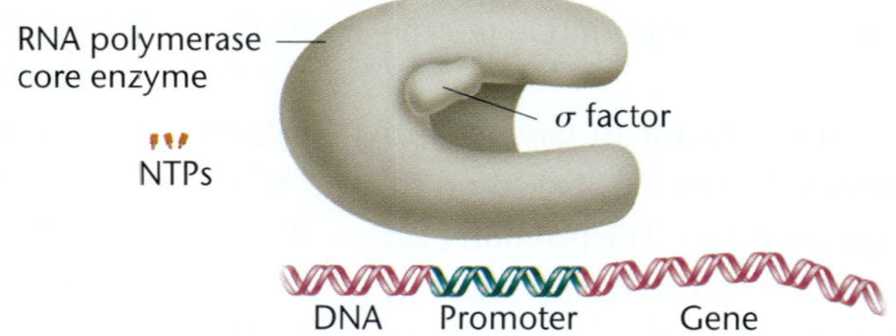

(b) Template binding and initiation of transcription

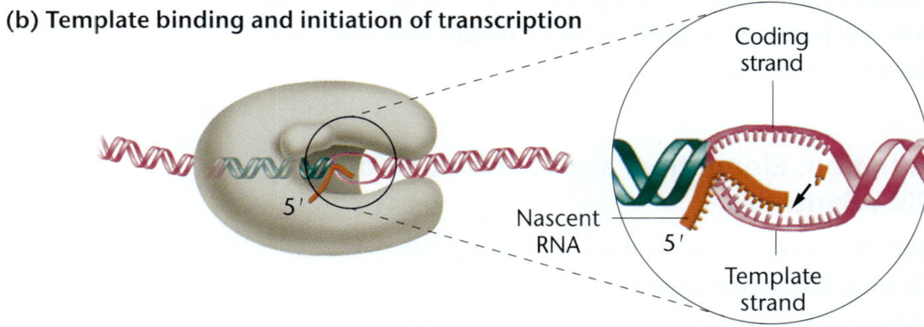

(c) Chain elongation

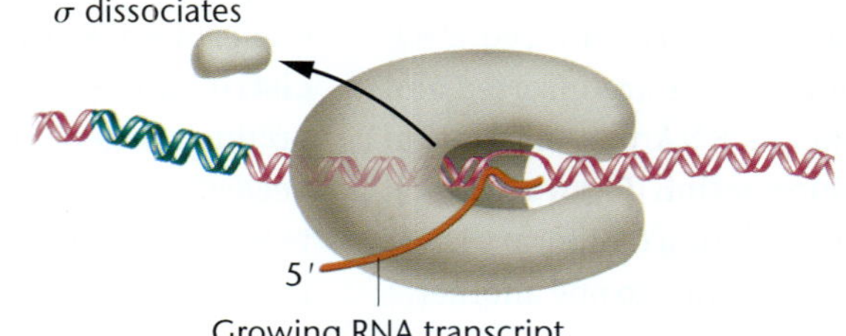

FIGURE 12.8 The early stages of transcription in bacteria, showing (a) the components of the process; (b) template binding at the −10 site involving the σ factor of RNA polymerase and subsequent initiation of RNA synthesis; and (c) chain elongation, after the σ factor has dissociated from the transcription complex and the enzyme moves along the DNA template.

detailed information involving promoter–enzyme interactions, we must address three points here.

The first point is the concept of **consensus sequences** of DNA. These sequences are similar (homologous) in different genes of the same organism or in one or more genes of related organisms. Their conservation throughout evolution attests to the critical nature of their role in biological processes. Two such sequences have been found in bacterial promoters. One, TATAAT, is located 10 nucleotides upstream from the site of initial transcription (the −10 region, or **Pribnow box**). The other, TTGACA, is located 35 nucleotides upstream (the −35 region). Mutations in either region diminish transcription, often severely.

Sequences such as these are said to be ***cis*-acting DNA elements.** Use of the term *cis* is drawn from organic chemistry nomenclature, meaning "next to" or on the same side as, in contrast to being "across from," or *trans*, to other functional groups. In molecular genetics, then, *cis*-elements are adjacent parts of the same DNA molecule. This is in contrast to ***trans*-acting factors,** molecules that bind to these DNA elements to influence gene expression.

The second point is that the degree of RNA polymerase binding to different promoters varies greatly, causing variable gene expression. Currently, this is attributed to sequence variation in the promoters. In bacteria, both strong promoters and weak promoters have been discovered, causing a variation in time of initiation from once every 1 to 2 seconds to as little as once every 10 to 20 minutes. Mutations in promoter sequences may severely reduce the initiation of gene expression.

The third point involves the σ factor in bacteria. In *E. coli*, the major form is designated as σ^{70}, based on its molecular weight of 70 kilodaltons (kDa). The promoters of most bacterial genes are recognized by this form; however, several alternative σ factors (e.g., σ^{32}, σ^{54}, σ^{S}, and σ^{E}) are called upon to regulate other genes. Each σ factor recognizes different promoter sequences, which in turn provides specificity to the initiation of transcription.

Initiation, Elongation, and Termination of RNA Synthesis

Once it has recognized and bound to the promoter [Figure 12.8(b)], RNA polymerase catalyzes **initiation,** the insertion of the first 5′-ribonucleoside triphosphate, which is complementary to the first nucleotide at the start site of the DNA template strand. As we noted earlier, no primer is required. Subsequent ribonucleotide complements are inserted and linked by phosphodiester bonds as RNA polymerization proceeds. This process of **chain elongation** [**Figure 12.8(c)**] continues in a 5′ to 3′ extension, creating a temporary DNA/RNA duplex whose chains run antiparallel to one another.

After a few ribonucleotides have been added to the growing RNA chain, the σ factor dissociates from the holoenzyme and elongation proceeds under the direction of the core enzyme. In *E. coli*, this process proceeds at the rate of about 50 nucleotides/second at 37°C.

Eventually, the enzyme traverses the entire gene until it encounters a specific nucleotide sequence that acts as a termination signal. Such termination sequences are extremely important in bacteria because of the close proximity of the end of one gene and the upstream sequences of the adjacent gene. An interesting aspect of termination in bacteria is that the termination sequence alluded to above is actually transcribed into RNA. The unique sequence of nucleotides in this termination region causes the newly formed transcript to fold back on itself, forming what is called a **hairpin secondary structure,** held together by hydrogen bonds. There are two different types of transcription termination mechanisms in bacteria, both of which are dependent on the formation of a hairpin structure in the RNA being transcribed.

Most transcripts in *E. coli* are terminated by **intrinsic termination**. In intrinsic termination, a hairpin structure encoded by the termination sequence causes RNA polymerase to stall. Immediately after the hairpin is a string of uracil residues. The U bases of the transcript have a relatively weak interaction with the A bases on the template strand of the DNA because there are only two hydrogen bonds per base pair. This leads to dissociation of RNA polymerase and the transcript is released.

Other bacterial transcripts are terminated by **rho-dependent termination,** which involves a termination factor called rho (ρ). Rho is a large hexameric protein with RNA helicase activity—it can dissociate RNA hairpins and DNA–RNA interactions. Rho binds to a specific sequence on the transcript and moves in the 3′ direction chasing after RNA polymerase. When RNA polymerase reaches the hairpin structure encoded by the termination sequence, it pauses and rho catches up. Rho moves through the hairpin and then causes dissociation of RNA polymerase by breaking the hydrogen bonds between the DNA template and the transcript.

In bacteria, groups of genes whose products function together are often clustered along the chromosome. In many such cases, they are contiguous, and all but the last gene lack the termination sequence. The result is that during transcription, a large mRNA is produced that encodes more than one protein. Since genes in bacteria are sometimes called **cistrons,** the RNA is called a **polycistronic mRNA.** The products of genes transcribed in this fashion are usually all needed by the cell at the same time, so this is an efficient way to transcribe and subsequently translate the needed genetic information. Polycistronic mRNAs are rare in eukaryotes.

NOW SOLVE THIS

12.3 The following represent deoxyribonucleotide sequences in the template strand of DNA:

Sequence 1:	5′-CTTTTTTGCCAT-3′
Sequence 2:	5′-ACATCAATAACT-3′
Sequence 3:	5′-TACAAGGGTTCT-3′

(a) For each strand, determine the mRNA sequence that would be derived from transcription.

(b) Using Figure 12.7, determine the amino acid sequence that is encoded by these mRNAs.

(c) For Sequence 1, what is the sequence of the coding DNA strand?

■ **HINT:** *This problem asks you to consider the outcome of the transfer of complementary information from DNA to RNA and to determine the amino acids encoded by this information. The key to its solution is to remember that in RNA, uracil is complementary to adenine, and that while DNA stores genetic information in the cell, the code that is translated is contained in the RNA complementary to the template strand of DNA.*

12.10 Transcription in Eukaryotes Differs from Bacterial Transcription in Several Ways

Much of our knowledge of transcription has been derived from studies of bacteria. The general aspects of the mechanics of these processes are mostly similar in eukaryotes, but there are several notable differences:

1. Transcription in eukaryotes occurs within the nucleus. Thus, unlike the bacterial process, in eukaryotes the RNA transcript is not free to associate with ribosomes prior to the completion of transcription. For the mRNA to be translated, it must move out of the nucleus into the cytoplasm.

2. Transcription in eukaryotes occurs under the direction of *three separate forms* of RNA polymerase, rather than the single form seen in bacteria.

3. Initiation of transcription of eukaryotic genes requires that compact chromatin fiber, characterized by nucleosome coiling, be uncoiled to make the DNA helix accessible to RNA polymerase and other regulatory proteins. This transition, referred to as *chromatin remodeling* (Chapter 11), reflects the dynamics involved in the conformational change that occurs as the DNA helix is opened.

4. Initiation and regulation of transcription entail a more extensive interaction between *cis*-acting DNA sequences and *trans*-acting protein factors. For example, while bacterial RNA polymerase requires only a σ factor to bind the promoter and initiate transcription, in eukaryotes, several *general transcription factors* (*GTFs*) are required to bind the promoter, recruit RNA polymerase, and initiate transcription. Furthermore, in addition to promoters, eukaryotic genes often have other *cis*-acting control units called *enhancers* and *silencers* (discussed below, and in more detail in Chapter 16), which greatly influence transcriptional activity.

5. In bacteria, transcription termination is often dependent upon the formation of a hairpin secondary structure in the transcript. However, eukaryotic transcription termination is more complex. Transcriptional termination for protein-coding genes involves sequence-specific cleavage of the transcript, which then leads to eventual dissociation of RNA polymerase from the DNA template.

6. In eukaryotes, the initial (or primary) transcripts of protein-coding mRNAs, called **pre-mRNAs,** undergo complex alterations, generally referred to as "processing," to produce a mature mRNA. Processing often involves the addition of a 5′-cap and a 3′-tail, and the removal of intervening sequences that are not a part of the mature mRNA. In the remainder of this chapter we will look at the basic details of transcription and mRNA processing in eukaryotic cells. The process of transcription is highly regulated, determining which DNA sequences are copied into RNA and when and how frequently they are transcribed. We will return to topics directly related to the regulation of eukaryotic gene transcription later in the text (see Chapter 16).

Initiation, Elongation, and Termination of Transcription in Eukaryotes

As noted earlier, eukaryotic RNA polymerase exists in three distinct forms. Each eukaryotic RNA polymerase is larger and more complex than the single form of RNA polymerase found in bacteria. For example, yeast and human RNA polymerase II enzymes consist of 12 subunits. While the three forms of the enzyme share certain protein subunits, each nevertheless transcribes different types of genes, as indicated in **Table 12.6**.

TABLE 12.6 RNA Polymerases in Eukaryotes

Form	Product	Location
I	rRNA	Nucleolus
II*	mRNA, snRNA	Nucleoplasm
III	5SrRNA, tRNA	Nucleoplasm

* RNAP II also synthesizes a variety of other RNAs, including miRNAs and lncRNAs (see Chapter 16).

RNA polymerases I and III (RNAP I and RNAP III) transcribe transfer RNAs (tRNAs) and ribosomal RNAs (rRNAs), which are needed in essentially all cells at all times for the basic process of protein synthesis. In contrast, **RNA polymerase II (RNAP II),** which transcribes protein-coding genes, is highly regulated. Protein-coding genes are often expressed at different times, in response to different signals, and in different cell types. Thus, RNAP II activity is tightly regulated on a gene-by-gene basis. For this reason, most studies of transcription in eukaryotes have focused on RNAP II.

The activity of RNAP II is dependent on both the *cis*-acting regulatory elements of the gene and a number of *trans*-acting transcription factors that bind to these DNA elements. (We will consider *cis* elements first and then turn to *trans* factors.)

At least four different types of *cis*-acting DNA elements regulate the initiation of transcription by RNAP II. The first of these, the **core promoter,** includes the transcription start site. It determines where RNAP II binds to the DNA and where it begins transcribing the DNA into RNA. Another promoter element, called a **proximal-promoter element,** is located upstream of the start site and helps modulate the level of transcription. The last two types of *cis*-acting elements, called **enhancers** and **silencers,** influence the efficiency or the rate of transcription initiation by RNAP II from the core-promoter element.

In some eukaryotic genes, a *cis*-acting element within the core promoter is the **Goldberg–Hogness box,** or **TATA box.** Located about 30 nucleotide pairs upstream (−30) from the start point of transcription, TATA boxes share a consensus sequence TATA$^{A}/_{T}$AAR, where R indicates any purine nucleotide. The sequence and function of TATA boxes are analogous to those found in the −10 promoter region of bacterial genes. However, recall that in bacteria, RNA polymerase binds directly to the −10 promoter region. As we will see below, the same is not the case in eukaryotes.

Although eukaryotic promoter elements can determine the site and general efficiency of initiation, other elements—known as *enhancers* and *silencers*—have more dramatic effects on eukaryotic gene transcription. As their names suggest, enhancers increase transcription levels and silencers decrease them. The locations of these elements can vary from immediately upstream of a promoter to downstream, within, or kilobases away, from a gene. In other words, they can modulate transcription from a distance. Each eukaryotic gene has its own unique arrangement of promoter, enhancer, and silencer elements.

Complementing the *cis*-acting regulatory sequences are various *trans*-acting factors that facilitate RNAP II binding and, therefore, the initiation of transcription. These proteins are referred to as **transcription factors.** There are two broad categories of transcription factors: the **general transcription factors (GTFs)** that are absolutely required for all RNAP II–mediated transcription, and the **transcriptional activators** and **transcriptional repressors** that influence the efficiency or the rate of RNAP II transcription initiation.

The general transcription factors are essential because RNAP II cannot bind directly to eukaryotic core-promoter sites and initiate transcription without their presence. The general transcription factors involved with human RNAP II binding are well characterized and are designated **TFIIA, TFIIB,** and so on. One of these, **TFIID,** binds directly to the TATA-box sequence. Once initial binding of TFIID to DNA occurs, the other general transcription factors, along with RNAP II, bind sequentially to TFIID, forming an extensive **pre-initiation complex.**

Transcriptional activators and repressors bind to enhancer and silencer elements and regulate transcription initiation by aiding or preventing the assembly of pre-initiation complexes and the release of RNAP II from pre-initiation into full transcription elongation. They appear to supplant the role of the σ factor seen in the bacterial enzyme and are important in eukaryotic gene regulation. We will consider the roles of general and specific transcription factors in eukaryotic gene regulation, as well as the various DNA elements to which they bind, in more detail later in the text (Chapter 16).

Unlike in bacteria, there is no specific sequence that signals for the termination of transcription. In fact, RNAP II often continues transcription well beyond what will be the eventual 3′-end of the mature mRNA. Once transcription has incorporated a specific sequence AAUAAA, known as the **polyadenylation signal sequence** (discussed below), the transcript is enzymatically cleaved roughly 10 to 35 bases further downstream in the 3′ direction. Cleavage of the transcript destabilizes RNAP II, and both DNA and RNA are released from the enzyme as transcription is terminated. This completes the cycle that constitutes transcription.

Processing Eukaryotic RNA: Caps and Tails

Although the base sequence of DNA in bacteria is transcribed into an mRNA that is immediately and directly translated into the amino acid sequence as dictated by the genetic code, eukaryotic mRNAs require significant alteration before they are transported to the cytoplasm and translated. By 1970, evidence showed that eukaryotic mRNA is transcribed initially as a precursor molecule much larger than that which is translated into protein. It was proposed that this *primary transcript* of a gene (a *pre-mRNA*) must be processed in the nucleus before it appears in the cytoplasm as a *mature mRNA* molecule. The various processing steps, discussed in the sections that follow, are summarized in **Figure 12.9**.

An important **posttranscriptional modification** of eukaryotic RNA transcripts destined to become

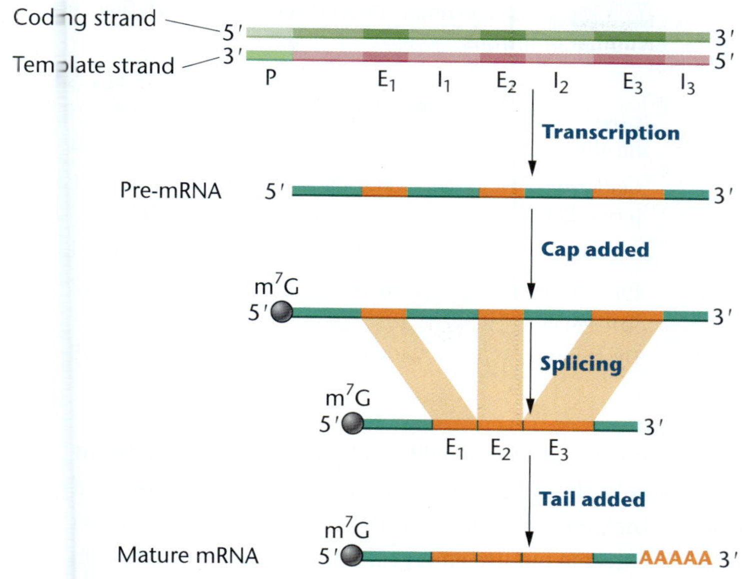

FIGURE 12.9 Posttranscriptional RNA processing in eukaryotes. Beginning at the promoter (P) of a gene, transcription produces a pre-mRNA containing several introns (I) and exons (E), as identified under the DNA template strand. Shortly after transcription begins, a m^7G cap is added to the 5′-end. Next, and during transcription elongation, the introns are spliced out and the exons joined. Finally, a poly-A tail is added to the 3′-end. While this figure depicts these steps sequentially, in some eukaryotic transcripts, the poly-A tail is added before splicing of all introns has been completed.

mRNAs occurs at the 5′-end of these molecules, where a **7-methylguanosine (m^7G) cap** is added. This cap is added shortly after synthesis of the initial RNA transcript has begun. The cap stabilizes the mRNA by protecting the 5′-end of the molecule from nuclease attack. Subsequently, the cap facilitates the transport of mature mRNAs from the nucleus into the cytoplasm and is required for the initiation of translation of the mRNA into protein. Chemically, the cap is a guanosine residue with a methyl group (CH_3) at position 7 of the base. The cap is also distinguished by a unique 5′ to 5′ triphosphate bridge that connects it to the initial ribonucleotide of the RNA.

Further insights into the processing of RNA transcripts during the maturation of mRNA came from the discovery that mRNAs contain, at their 3′-end, a stretch of as many as 250 adenylic acid residues. As discussed earlier in the context of eukaryotic transcription termination, the transcript is cleaved roughly 10 to 35 ribonucleotides after the highly conserved AAUAAA polyadenylation signal sequence. An enzyme known as **poly-A polymerase** then catalyzes the addition of a poly-A tail to the free 3′-OH group at the end of the transcript. Poly-A tails are found at the 3′-end of almost all mRNAs studied in a variety of eukaryotic organisms. The exceptions in eukaryotes seem to be mRNAs that encode histone proteins.

While the AAUAAA signal sequence is not found on all eukaryotic mRNAs, it appears to be essential to those that have it. If the sequence is changed as a result of a mutation, those transcripts that would normally have it cannot add the poly-A tail. In the absence of this tail, these RNA transcripts are rapidly degraded by nucleases. The **poly-A binding protein,** as the name suggests, binds to poly-A tails and prevents nucleases from degrading the 3′-end of the mRNA. In addition, the poly-A tail is important for export of the mRNA from the nucleus to the cytoplasm and for translation of the mRNA.

Poly-A tails are also found on mRNAs in bacteria and archaea. However, these bacterial poly-A tails are generally much shorter and found on only a small fraction of mRNA molecules. In addition, whereas poly-A tails are protective in eukaryotes, poly-A tails are generally associated with mRNA degradation in bacteria.

ESSENTIAL POINT

The process of creating the initial transcript during transcription is more complex in eukaryotes than in bacteria, including the addition of a 5′m^7G cap and a 3′ poly-A tail to the pre-mRNA. ■

12.11 The Coding Regions of Eukaryotic Genes Are Interrupted by Intervening Sequences Called Introns

As mentioned above, the primary mRNA transcript, or pre-mRNA, is often longer than the mature mRNA in eukaryotes. An explanation for this phenomenon emerged in 1977 when research groups led by Phillip Sharp and Richard Roberts independently published direct evidence that the genes of animal viruses contain *internal* (also referred to as *intervening* or *intragenic*) nucleotide sequences that do not encode for amino acids in the final protein product. These noncoding internal sequences are also present in pre-mRNAs, but they are removed during RNA processing to produce the mature mRNA (Figure 12.9), which is then translated. Such nucleotide sequences—ones that intervene between sequences that code for amino acids—are called **introns** (derived from *intragenic region*). Sequences that are retained in the mature mRNA and expressed are called **exons** (for *ex*pressed region). The process of removing introns from a pre-mRNA and joining together exons is called **RNA splicing.**

One of the first intron-containing genes identified was the **β-globin gene** in mice and rabbits, studied independently by Philip Leder and Richard Flavell. The mouse gene contains an intron 550 nucleotides long, beginning

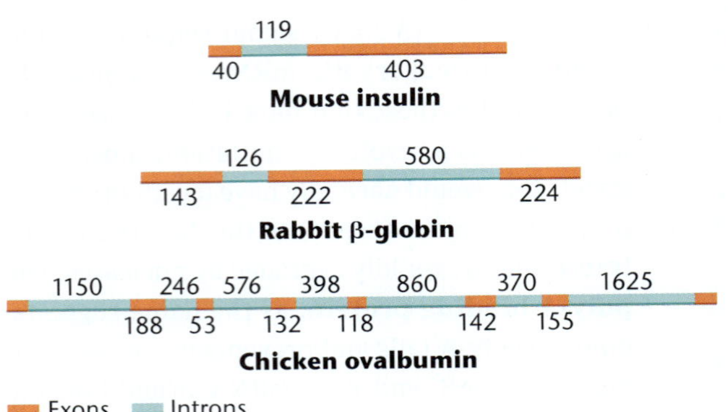

FIGURE 12.10 Intron and exon sequences in various eukaryotic genes. The numbers indicate the number of nucleotides present in various intron and exon regions.

immediately after the sequence specifying the 104th amino acid. In rabbits, there is an intron of 580 base pairs near the sequence for the 110th amino acid—a strikingly similar pattern to that seen in mice. In addition, another intron of about 120 nucleotides exists earlier in both genes. Similar introns have been found in the β-globin gene in all mammals examined thus far.

The **ovalbumin gene** of chickens, as shown in **Figure 12.10**, contains seven introns. In fact, the majority of the gene's DNA sequence is composed of introns and is thus noncoding. The pre-mRNA is nearly three times the length of the mature spliced mRNA.

The identification of introns in eukaryotic genes involves a direct comparison of nucleotide sequences of DNA with those of mRNA and their correlation with amino acid sequences. Such an approach allows the precise identification of all intervening sequences. By identifying common sequences that appear at intron/exon boundaries, scientists are now able to identify introns with excellent accuracy using only the genomic DNA sequence and computational tools. We will return to this topic when we consider genomic analysis (Chapter 18).

We now have a fairly comprehensive view of intron-containing eukaryotic genes from many species. In the budding yeast *Saccharomyces cervisiae*, 283 out of the roughly 6000 protein-coding genes have introns. However, introns are far more common in humans; roughly 94 percent of human protein-coding genes contain introns with an average of nine exons and eight introns per gene. An extreme example of the number of introns present in a single gene is provided by the gene coding for one of the subunits of collagen, the major connective tissue protein in vertebrates. The *pro-α-2(1) collagen* gene contains 51 introns. The precision of RNA splicing must be extraordinary if errors are not to be introduced into the mature mRNA.

Equally noteworthy is the difference between the length of a typical gene and the length of the final mRNA after

TABLE 12.7 Contrasting Human Gene Size, mRNA Size, and Number of Introns

Gene	Gene Size (kb)	mRNA Size (kb)	Number of Introns
Insulin	1.7	0.4	2
Collagen [*pro-α-2(1)*]	38.0	5.0	51
Albumin	25.0	2.1	14
Phenylalanine hydroxylase	90.0	2.4	12
Dystrophin	2400.0	17.0	79

introns are removed by splicing. As shown in **Table 12.7**, only about 13 percent of the collagen gene consists of exons that appear in mature mRNA. For other genes, an even more extreme picture emerges. Only about 8 percent of the albumin gene codes for the amino acids in the albumin protein, and in the largest human gene known, dystrophin (which is the protein product absent in Duchenne muscular dystrophy), less than 1 percent of the gene sequence is retained in the mRNA.

Although the vast majority of mammalian genes examined thus far contain introns, there are several exceptions. Notably, the genes coding for histones and for interferon, a signaling protein of the immune system, appear to contain no introns.

Why Do Introns Exist?

A curious genetics student who first learns about the concept of introns and RNA splicing often wonders why introns exist. If intron sequences are destined for removal, then why are they there in the first place? Wouldn't it be more efficient if introns were absent and hence never transcribed? Indeed, scientists asked these same questions shortly after introns were discovered in 1977. However, we know now that introns indeed serve several functions:

1. Some genes can encode for more than one protein product through the alternative use of exons. This process, known as **alternative splicing** (described in more detail in Chapter 16), produces different mature mRNAs from the same pre-mRNA by splicing out introns and ligating together different combinations of exons. This means that a eukaryotic genome can encode a greater number of proteins than it has protein-coding genes.

2. Introns may also be important to the evolution of genes. On evolutionary time scales, DNA sequences may be moved around within the genome. The modular exon/intron gene structure allows for new genes to evolve when, for example, an exon is introduced into an existing gene.

3. Once an intron is excised from a pre-mRNA, it is generally degraded. However, there are many documented cases where an intron actually contains a functional noncoding RNA (see Chapter 16). In such cases, the excised intron is processed to liberate the noncoding RNA, which then functions within the cell.

4. Introns can also regulate transcription. For example, intronic sequences in the DNA frequently harbor *cis*-regulatory elements, such as enhancers and silencers that upregulate or downregulate transcription, respectively.

ESSENTIAL POINT

Many eukaryotic genes contain intervening sequences, or introns, which are transcribed into the pre-mRNA and must be spliced out to create the mature mRNA. ■

Splicing Mechanisms: Self-Splicing RNAs

The discovery of introns led to intensive attempts to elucidate the mechanism by which they are excised and exons are spliced back together. A great deal of progress has been made, relying heavily on *in vitro* studies. Interestingly, it appears that somewhat different mechanisms exist for different classes of transcripts, as well as for RNAs produced in mitochondria and chloroplasts.

We might envision the simplest possible mechanism for removing an intron to involve two steps: (1) the intron is cut at both ends by an endonuclease and (2) the adjacent exons are joined, or ligated, by a ligase. This is, apparently, what happens to the introns present in transfer RNAs (tRNAs) in bacteria. However, in studies of other RNAs—tRNAs in higher eukaryotes and rRNAs and pre-mRNAs in all eukaryotes—precise excision of introns is much more complex and a much more interesting story.

Introns in eukaryotes can be categorized into several groups based on their splicing mechanisms. *Group I introns*, such as those in the primary transcript of rRNAs, require no outside help for intron excision; the intron itself is the source of the enzymatic activity necessary for splicing. This amazing discovery was made in 1982 by Thomas Cech and colleagues during a study of the ciliate protozoan *Tetrahymena*. RNAs that are capable of such catalytic activity are referred to as **ribozymes.**

The self-excision process for group I introns is illustrated in **Figure 12.11**. Chemically, two reactions take place. The first is an interaction between a free guanosine (symbolized as "G"), which acts as a cofactor in the reaction, and the primary transcript [**Figure 12.11(a)**]. After guanosine is positioned in the active site of the intron, its 3′-OH group attacks

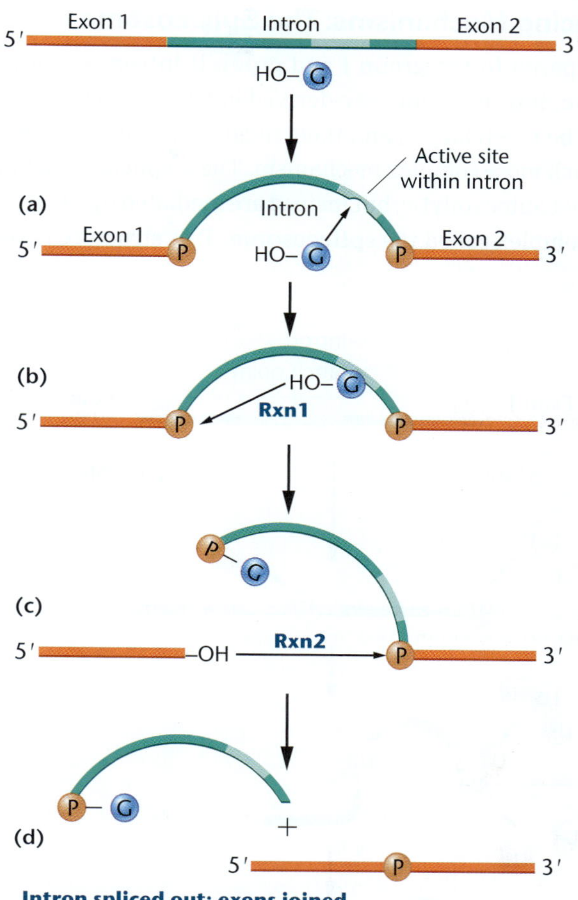

FIGURE 12.11 Splicing mechanism for removal of a group I intron. The process is one of self-excision involving two transesterification reactions.

and breaks the phosphodiester bond ("P") between the nucleotides at the 5′-end of the intron and the 3′-end of the left-hand exon [**Figure 12.11(b)**]. The second reaction involves the interaction of the newly formed 3′-OH group on the left-hand exon and the phosphodiester bond at the right intron/exon boundary [**Figure 12.11(c)**]. The intron is spliced out and the two exons are ligated, leading to the mature RNA [**Figure 12.11(d)**].

Self-excision of group I introns, as described above, is known to occur in preliminary transcripts for mRNAs, tRNAs, and rRNAs in bacteria, lower eukaryotes, and higher plants.

Self-excision also governs the removal of introns from the primary mRNA and tRNA transcripts produced in mitochondria and chloroplasts; these are examples of *group II introns*. Splicing of group II introns is somewhat different than for group I, but also involves two autocatalytic reactions leading to the excision of introns. Group II introns are found in fungi, plants, protists, and bacteria.

Splicing Mechanisms: The Spliceosome

Compared to the group I and group II introns discussed above, introns in nuclear-derived eukaryotic pre-mRNAs may be much larger, and their removal appears to require a much more complex mechanism. These splicing reactions are not autocatalytic, but instead are mediated by a molecular complex called the **spliceosome.** This structure is very large, 40*S* in yeast and 60*S* in mammals, being the same size as ribosomal subunits! Introns removed by the spliceosome are known as *spliceosomal introns.*

One set of essential components of spliceosomes are the **small nuclear RNAs (snRNAs).** These RNAs are usually 80 to 400 nucleotides long and, because they are rich in uridine residues, have been arbitrarily named U1, U2, . . . , U6. The snRNAs are complexed with proteins to form **small nuclear ribonucleoproteins (snRNPs),** pronounced "snurps," which are named after the specific snRNAs contained within them (the U2 snRNA is contained within the U2 snRNP).

Figure 12.12 depicts a model of the steps involved in the removal of one spliceosomal intron. Keep in mind that while this figure shows separate components, the process involves the huge spliceosome that envelopes the RNA being spliced. The nucleotide sequences near the ends of the intron begin at the 5′-end with a GU dinucleotide sequence, called the *donor sequence*, and terminate at the 3′-end with an AG dinucleotide, called the *acceptor sequence*. These, as well as other consensus sequences shared by introns, attract specific snRNAs of the spliceosome. For example, the snRNA U1 bears a nucleotide sequence that is complementary to the 5′-donor sequence end of the intron. Base pairing resulting from this homology promotes the binding that represents the initial step in the formation of the spliceosome. After the other snRNPs (U2, U4, U5, and U6) are added, splicing commences. As with group I splicing, two reactions occur. The first involves the interaction of the 3′-OH group from an adenine (A) residue present within the branch point region of the intron. The A residue attacks the 5′-splice site, cutting the RNA chain. In a subsequent step involving several other snRNPs, an intermediate structure is formed and the second reaction ensues, linking the cut 5′-end of the intron to the A. This results in the formation of a characteristic loop structure called a *lariat,* which contains the excised intron. The exons are then ligated and the snRNPs are released.

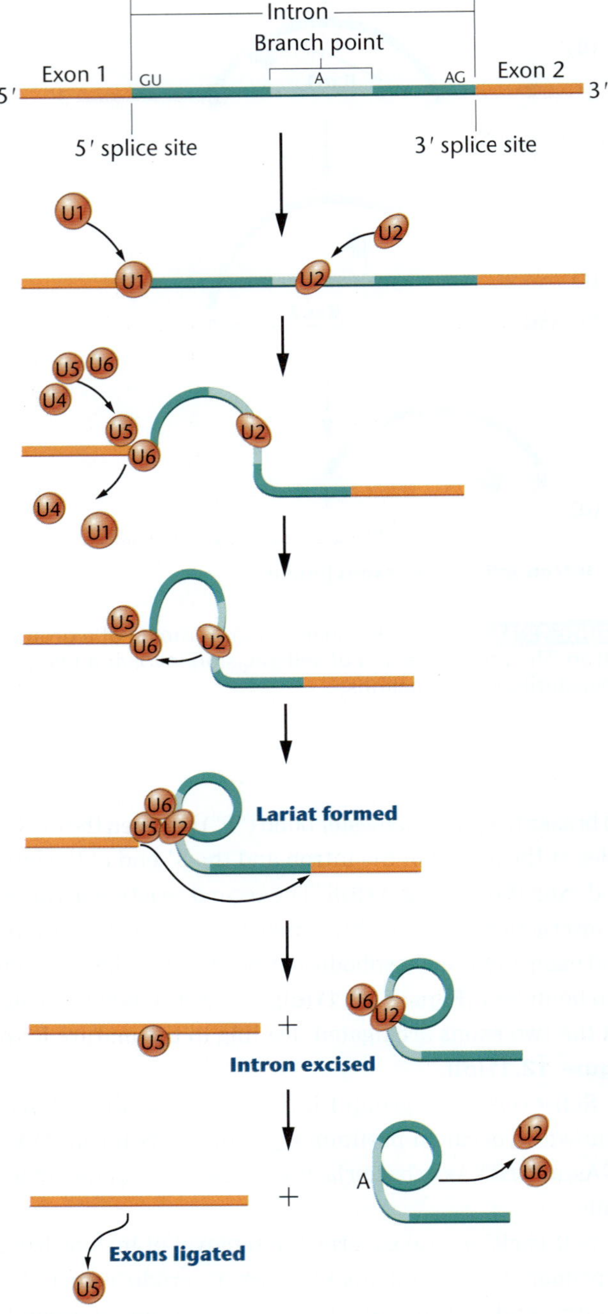

FIGURE 12.12 A model of the splicing mechanism for removal of a spliceosomal intron. Excision is dependent on snRNPs (U1, U2, etc.). The lariat structure is characteristic of this mechanism.

EVOLVING CONCEPT OF THE GENE

The elucidation of the genetic code in the 1960s supported the concept that the gene is composed of a linear series of triplet nucleotides encoding the amino acid sequence of a protein. While this is indeed the case in bacteria and viruses, in 1977, it became apparent that in eukaryotes, the gene is divided into coding sequences, called exons, which are interrupted by noncoding sequences, called introns (intervening sequences), which must be spliced out during production of the mature mRNA. ■

12.12 RNA Editing May Modify the Final Transcript

In the late 1980s, still another unexpected form of posttranscriptional RNA processing was discovered in several organisms. In this form, referred to as **RNA editing,** the nucleotide sequence of a pre-mRNA is actually changed prior to translation. As a result, the ribonucleotide sequence of the mature RNA differs from the sequence encoded in the exons of the DNA from which the RNA was transcribed.

Although other variations exist, there are two main types of RNA editing: **insertion/deletion editing,** in which nucleotides are added to or subtracted from the total number of bases; and **substitution editing,** in which the identities of individual nucleotide bases are altered. Substitution editing is used in some nuclear-derived eukaryotic RNAs and is prevalent in mitochondrial and chloroplast RNAs transcribed in plants.

Trypanosoma, a parasite that causes African sleeping sickness, uses extensive insertion/deletion editing in mitochondrial RNAs. The uridines added to an individual transcript can make up more than 60 percent of the coding sequence, usually forming the initiation codon and bringing the rest of the sequence into the proper reading frame. Insertion/deletion editing in trypanosomes is directed by **gRNA (guide RNA)** templates, which are also transcribed from the mitochondrial genome. These small RNAs share a high degree of complementarity to the edited region of the final mRNAs. They base-pair with the pre-edited mRNAs to direct the editing machinery to make the correct changes.

An excellent example of substitutional editing involves the subunits constituting the *glutamate receptor channels* (*GluR*) in mammalian brain tissue. In this case, adenosine (A) to inosine (I) editing occurs in pre-mRNAs prior to their translation, during which I is read as guanosine (G). A family of three ADAR (*a*denosine *d*eaminase *a*cting on *R*NA) enzymes is responsible for this editing. The double-stranded RNAs required for editing by the ADAR enzymes are provided by intron/exon pairing of the GluR mRNA transcripts. The editing alters the physiological parameters (solute permeability and desensitization response time) of the receptors containing the subunits.

Findings such as these have established that RNA editing provides still another important mechanism of *posttranscriptional modification*. These discoveries have important implications for the regulation of gene expression.

12.13 Transcription Has Been Visualized by Electron Microscopy

We conclude our coverage of transcription by referring you back to the chapter opening photograph (p. 242), which is a striking visualization of transcription occurring in the oocyte nucleus of *Xenopus laevis*, the clawed frog. Note the central axis that runs horizontally from left to right and from which threads appear to be emanating vertically. This axis, appearing as a thin thread, is the DNA of most of one gene encoding ribosomal RNA (rDNA). Each of the emanating threads, which grows longer the farther to the right it is found, is an rRNA molecule being transcribed. What is apparent is that multiple copies of RNA polymerase have initiated transcription at a point near the left end and that transcription by each of them has proceeded to the right. Simultaneous transcription by many of these polymerases results in the electron micrograph that has captured an image of the entire process.

It is fascinating to visualize the process and to confirm our expectations based on the biochemical analysis of this process.

CASE STUDY Treatment dilemmas

A 30-year-old woman was undergoing therapy for **β-thalassemia,** a recessive trait caused by absence of or reduced synthesis of the hemoglobin β chain, a subunit of the oxygen-carrying molecule in red blood cells. In this condition, red blood cells are rapidly destroyed, freeing a large amount of iron, which is deposited in tissues and organs. The blood transfusions the patient had received every 2 or 3 weeks since the age of 7 to stave off anemia were further aggravating iron buildup. Her major organs were showing damage, and she was in danger of death from cardiac disease. Her physician suggested that she consider undergoing a hematopoietic (bone marrow) stem cell transplant (HSCT). Since these stem cells give rise to red blood cells, such a transplant could potentially restore her health. While this might seem like an easy decision, it is not. Advanced cases have a high risk (almost 30 percent) for transplantation-related death. At this point, the woman is faced with a difficult and important decision.

1. Consider different ways in which a mutation, a single base-pair change or small deletion, in the gene encoding hemoglobin β chain could lead to **β-thalassemia.** For example, how might mutations in promoter, enhancer, or coding regions yield this outcome?
2. Why is it important that the physician emphasize to the patient that she must bear the responsibility for the final decision (i.e., that once she has considered all aspects of the decision, she act autonomously)?
3. If you were faced with this decision, what further input might you seek?

For related reading, see Caocci, G., et al. (2011). Ethical issues of unrelated hematopoietic stem cell transplantation in adult thalassemia patients. *BMC Medical Ethics* 12(4):1–7.

GENETICS, ETHICS, AND SOCIETY

Treating Duchenne Muscular Dystrophy with Exon-Skipping Drugs

One in every 3500 newborn males is afflicted by a serious X-linked recessive disease known as Duchenne muscular dystrophy (DMD). This disease is progressive, resulting in muscle degeneration, heart disease, and premature death.

DMD is caused by mutations in the *dystrophin* gene, a 2400-kb gene containing 79 exons. The DMD primary transcript is 2100 kb long and takes about 16 hours to transcribe. Many DMD mutations are frameshift mutations that shift the reading frame in the mRNA so that at least one codon downstream of the frameshift mutation becomes a stop codon. This stop codon causes premature termination of translation of the mRNA. Most of the resulting truncated dystrophin proteins are not functional. There are no cures for DMD and few effective treatments.

Recently, a new DMD drug called eteplirsen completed clinical trials and received accelerated approval by the U.S. Food and Drug Administration (FDA). This drug uses a technique called *exon skipping* to target mutations in exon 51 of *dystrophin*.

In exon skipping, the exon containing the mutation is removed during pre-mRNA splicing. If the exons that precede and follow the skipped exon are spliced together in such a way that the reading frame is restored, the translated protein may retain some activity, even though it lacks the amino acids encoded by the skipped exon.

Eteplirsen is a molecule known as an *antisense oligonucleotide* (*ASO*). ASOs are short synthetic single-stranded DNA molecules that have specific sequences complementary to a portion of a targeted mRNA. When ASOs enter cells, they bind to their complementary sequence in the target mRNA. This results in the mRNA's degradation, or interferes with its splicing or translation.

Once inside a cell, eteplirsen binds to the *dystrophin* pre-mRNA exon 51 splice junctions, interfering with normal pre-mRNA splicing. As a result, exon 50 is spliced to exon 52, thereby eliminating exon 51 and its mutation from the mature mRNA. This exon deletion restores the correct reading frame to the *dystrophin* mRNA. Upon translation, the new dystrophin protein, although lacking some amino acids, has enough activity to restore partial function to the patient's muscles.

Your Turn

Take time, individually or in groups, to consider the ethical and technical issues that surround DMD exon-skipping drugs.

1. In 2016, the FDA gave accelerated approval to eteplirsen. Although the FDA's advisory panel initially voted against approval, intense lobbying by DMD patients and their families may have successfuly pressured the FDA to approve this new drug. Discuss this case of accelerated approval, considering ethical arguments on both sides of the controversy.

   Read about the FDA's approval of eteplirsen at: Tavernise, S., F.D.A. approves muscular dystrophy drug that patients lobbied for. *New York Times*, September 19, 2016 (http://www.nytimes.com/2016/09/20/business/fda-approves-muscular-dystrophy-drug-that-patients-lobbied-for.html).

2. Only about 13 percent of DMD patients have exon 51 mutations leading to premature translation termination. Describe several other exon-skipping drugs that are under development for the treatment of DMD. Are any of these new drugs in clinical trials?

   Investigate the links on the Muscular Dystrophy Association Web site (https://www.mda.org/quest/article/exon-skipping-dmd-what-it-and-whom-can-it-help).

INSIGHTS AND SOLUTIONS

1. Calculate how many triplet codons would be possible had evolution seized on six bases (three complementary base pairs) rather than four bases within the structure of DNA. Would six bases accommodate a two-letter code, assuming 20 amino acids and start and stop codons?

   Solution: Six things taken three at a time will produce $(6)^3$ or 216 triplet codes. If the code was a doublet, there would be $(6)^2$ or 36 two-letter codes, more than enough to accommodate 20 amino acids and start and stop signals.

2. In a heteropolymer experiment using 1/2C:1/4A:1/4G, how many different triplets will occur in the synthetic RNA molecule? How frequently will the most frequent triplet occur?

   Solution: There will be $(3)^3$ or 27 triplets produced. The most frequent will be CCC, present $(1/2)^3$ or 1/8 of the time.

3. In a regular copolymer experiment, where UUAC is repeated over and over, how many different triplets will occur in the synthetic RNA, and how many amino acids will occur in the polypeptide when this RNA is translated? (Consult Figure 12.7.)

   Solution: The synthetic RNA will repeat four triplets—UUA, CUU, ACU, UAC—over and over. Because both UUA and CUU encode leucine, while ACU and UAC encode threonine and tyrosine, respectively, the polypeptides synthesized under the directions of this RNA would contain three amino acids in the repeating sequence Leu-Leu-Thr-Tyr.

4. Actinomycin D inhibits DNA-dependent RNA synthesis. This antibiotic is added to a bacterial culture where a specific protein is being monitored. Compared to a control culture, where no antibiotic is added, translation of the protein declines over a period of 20 minutes, until no further protein is made. Explain these results.

Solution: The mRNA, which is the basis for translation of the protein, has a lifetime of about 20 minutes. When actinomycin D is added, transcription is inhibited and no new mRNAs are made. Those already present support the translation of the protein for up to 20 minutes.

Problems and Discussion Questions

Mastering Genetics™ Visit for instructor-assigned tutorials and problems.

1. **HOW DO WE KNOW?** In this chapter, we focused on the genetic code and the transcription of genetic information stored in DNA into complementary RNA molecules. Along the way, we found many opportunities to consider the methods and reasoning by which much of this information was acquired. From the explanations given in the chapter, what answers would you propose to the following fundamental questions:
   (a) How did we determine the *compositions* of codons encoding specific amino acids?
   (b) How were the specific sequences of triplet codes determined experimentally?
   (c) How were the experimentally derived triplet codon assignments verified in studies using bacteriophage MS2?
   (d) How do we know that mRNA exists and serves as an intermediate between information encoded in DNA and its concomitant gene product?
   (e) How do we know that the initial transcript of a eukaryotic gene contains noncoding sequences that must be removed before accurate translation into proteins can occur?
2. **CONCEPT QUESTION** Review the Chapter Concepts list on p. 242. These all center on how genetic information is stored in DNA and transferred to RNA prior to translation into proteins. Write a short essay that summarizes the key properties of the genetic code and the process by which RNA is transcribed on a DNA template. ■
3. In studies of frameshift mutations, Crick, Barnett, Brenner, and Watts-Tobin found that either three nucleotide insertions or deletions restored the correct reading frame.
   (a) Assuming the code is a triplet, what effect would the addition or loss of six nucleotides have on the reading frame?
   (b) If the code were a sextuplet (consisting of six nucleotides), would the reading frame be restored by the addition or loss of three, six, or nine nucleotides?
4. When the repeating trinucleotide sequence UUCUUCUUC is added to a cell-free translation system, three different polypeptide homopolymers are produced. Why?
5. From the late 1950s to the mid-1960s, numerous experiments using *in vitro* cell-free systems provided information on the nature of the genetic code. Briefly outline significant experiments in the determination of the genetic code.
6. In a coding experiment using repeating copolymers (as shown in Table 12.3), the following data were obtained.

Copolymer	Codons Produced	Amino Acids in Polypeptide
AG	AGA, GAG	Arg, Glu
AAG	AGA, AAG, GAA	Lys, Arg, Glu

   AGG is known to code for arginine. Taking into account the wobble hypothesis, assign each of the four remaining different triplet codes to its correct amino acid.
7. "Breaking the genetic code" has been referred to as one of the most significant scientific achievements in modern times. Describe (in outline or brief statement form) the procedures used to break the code.
8. The HIV gp120 protein binds to CD4 on its host. Sequencing different HIV-1 strains lead to changes within the CD4 binding loop. Leu could be substituted by Ile, Ser and Lys by Arg, and Glu by Gln in the same positions. List the single-base changes that could occur in triplets to produce these amino acid changes.
9. In studies of the amino acid sequence of wild-type and mutant forms of tryptophan synthetase in *E. coli*, the following changes have been observed:

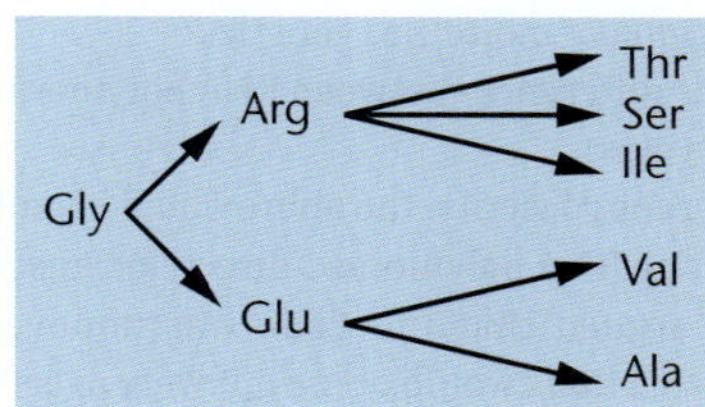

   Determine a set of triplet codes in which only a single-nucleotide change produces each amino acid change.
10. Why doesn't polynucleotide phosphorylase (Ochoa's enzyme) synthesize RNA *in vivo*?
11. Refer to Table 12.1. Can you hypothesize why a mixture of (Poly U)+(Poly A) would not stimulate incorporation of ^{14}C-phenylalanine into protein?
12. A candidate gene, suspected of being responsible for a disease, is sequenced. While sequencing the mRNA, a short stretch within the sequences seems different between unaffected individuals (sequence 1) and affected individuals (sequence 2). Predict the protein sequences for both (note that the reading frame starts in the first nucleotide):

   Sequence 1: 5′-AUGGUUGUUCGAAGAAGAAGAAGAUGA-3′
   Sequence 2: 5′-AUGGUUGUUGAAGAAGAAGAAGAUGAA-3′

   Explain the difference between the two sequences.
13. A short RNA molecule was isolated that demonstrated a hyperchromic shift indicating secondary structure (see p. 201 in Chapter 9). Its sequence was determined to be

   5′-AGGCGCCGACUCUACU-3′

(a) Propose a two-dimensional model for this molecule.
(b) What DNA sequence would give rise to this RNA molecule through transcription?
(c) If the molecule were a tRNA fragment containing a CGA anticodon, what would the corresponding codon be?
(d) If the molecule were an internal part of a message, what amino acid sequence would result from it following translation? (Refer to the code chart in Figure 12.7.)

14. A serine residue exists at position 33 of the human insulin protein precursor. If the codon specifying serine is UCA, how many single-base substitutions results in an amino acid substitution at position 33, and what are they? How many if the wild-type codon is AGU?

15. Shown here is a theoretical viral mRNA sequence

5′-AUGCAUACCUAUGAGACCCUUGGA-3′

(a) Assuming that it could arise from overlapping genes, how many different polypeptide sequences can be produced? Using the chart in Figure 12.7, what are the sequences?
(b) A base-substitution mutation that altered the sequence in part (a) eliminated the synthesis of all but one polypeptide. The altered sequence is shown below. Use Figure 12.7 to determine why it was altered.

5′-AUGCAUACCUAUGUGACCCUUGGA-3′

16. A novel protein discovered in a certain plant has many leucine-rich regions, fewer alanine-rich regions, and even fewer tyrosine residues. Correlate the number of codons for these three amino acids with this information.

17. Define the process of transcription. Where does this process fit into the central dogma of molecular genetics?

18. Discuss some of the differences between RNA polymerase in bacteria and in eukaryotes.

19. In a written paragraph, describe the abbreviated chemical reactions that summarize RNA polymerase-directed transcription.

20. Sydney Brenner argued that the Code was nonoverlapping because coding restrictions would occur if it were overlapping. A second major argument against an overlapping code involved the effect of a single nucleotide change. In an overlapping code, how many adjacent amino acids would be affected by a point mutation? In a nonoverlapping code, how many amino acid(s) would be affected?

21. One form of posttranscriptional modification of most eukaryotic RNA transcripts is the addition of a poly-A tail at the 3′-end. The absence of a poly-A tail leads to rapid degradation of the transcript. Poly-A tails of various lengths are also added to many bacterial RNA transcripts where, instead of promoting stability, they enhance degradation. In both cases, RNA secondary structures, stabilizing proteins, or degrading enzymes interact with poly-A tails. Considering the activities of RNAs, what might be the general functions of 3′-polyadenylation??

22. In a mixed heteropolymer experiment, messages were created with either 4/5C:1/5A or 4/5A:1/5C. These messages yielded proteins with the amino acid compositions shown in the following table. Using these data, predict the most specific *coding composition* for each amino acid.

4/5C:1/5A		4/5A:1/5C	
Pro	63.0%	Pro	3.5%
His	13.0%	His	3.0%
Thr	16.0%	Thr	16.6%
Glu	3.0%	Glu	13.0%
Asp	3.0%	Asp	13.0%
Lys	0.5%	Lys	50.0%
	98.5%		99.1%

23. Shown in this problem are the amino acid sequences of the wild type and three mutant forms of a short protein.
(a) Using Figure 12.7, predict the type of mutation that created each altered protein.
(b) Determine the specific ribonucleotide change that led to the synthesis of each mutant protein.
(c) The wild-type RNA consists of nine triplets. What is the role of the ninth triplet?
(d) For the first eight wild-type triplets, which, if any, can you determine specifically from an analysis of the mutant proteins? In each case, explain why or why not.
(e) Another mutation (mutant 4) is isolated. Its amino acid sequence is unchanged, but mutant cells produce abnormally low amounts of the wild-type proteins. As specifically as you can, predict where this mutation exists in the gene.

Wild type:	Met-Trp-Tyr-Arg-Gly-Ser-Pro-Thr
Mutant 1:	Met-Trp
Mutant 2:	Met-Trp-His-Arg-Gly-Ser-Pro-Thr
Mutant 3:	Met-Cys-Ile-Val-Val-Val-Gln-His

24. Alternative splicing is a common mechanism for eukaryotes to expand their repertoire of gene functions. At least one estimate indicates that approximately 50 percent of human genes use alternative splicing, and approximately 15 percent of disease-causing mutations involve aberrant alternative splicing. Different tissues show remarkably different frequencies of alternative splicing, with the brain accounting for approximately 18 percent of such events.
(a) Define alternative splicing and speculate on the evolutionary strategy alternative splicing offers to organisms.
(b) Why might some tissues engage in more alternative splicing than others?

13

Translation and Proteins

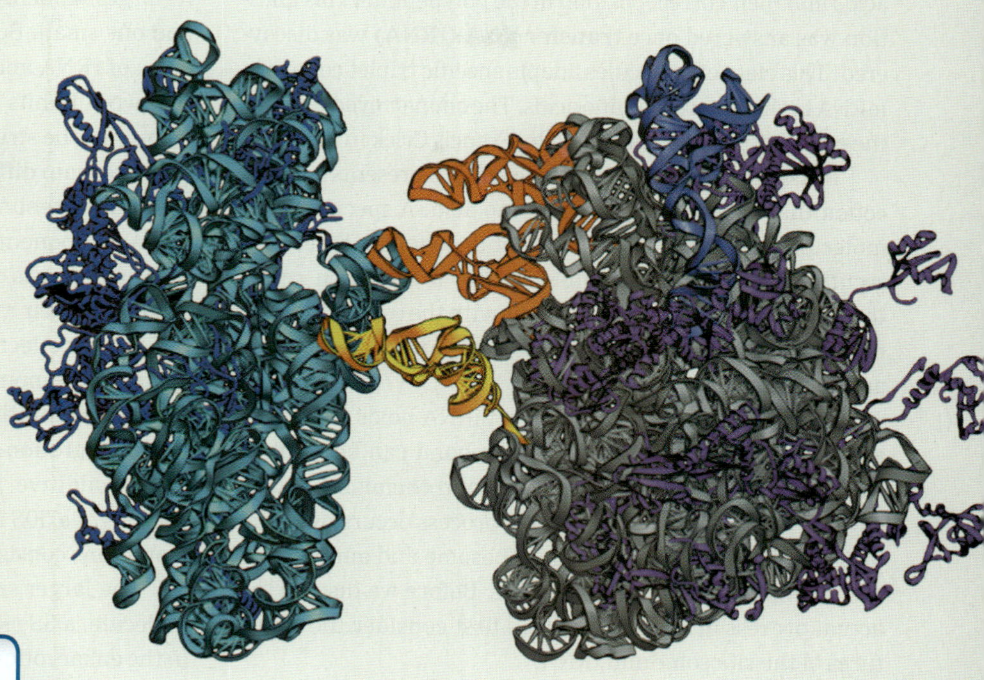

Crystal structure of a *Thermus thermophilus* 70*S* ribosome containing three bound transfer RNAs.

CHAPTER CONCEPTS

- The ribonucleotide sequence of messenger RNA (mRNA) reflects genetic information stored in DNA that makes up genes and corresponds to the amino acid sequences in proteins encoded by those genes.
- The process of translation decodes the information in mRNA, leading to the synthesis of polypeptide chains.
- Translation involves the interactions of mRNA, tRNA, ribosomes, and a variety of translation factors essential to the initiation, elongation, and termination of the polypeptide chain.
- Proteins, the final product of many genes, achieve a three-dimensional conformation that is based on the primary amino acid sequences of the polypeptide chains making up each protein.
- The function of any protein is closely tied to its three-dimensional structure, which can be disrupted by mutation.

In Chapter 12, we established that a genetic code stores information in the form of triplet nucleotides in DNA and that this information is initially expressed through the process of transcription into a messenger RNA that is complementary to one strand of the DNA helix. However, the final product of gene expression, in the case of protein-coding genes, is a polypeptide chain consisting of a linear series of amino acids whose sequence has been prescribed by the genetic code. In this chapter, we will examine how the information present in mRNA is utilized to create polypeptides, which then fold into protein molecules. We will also review the evidence confirming that proteins are the end products of many genes, and we will briefly discuss the various levels of protein structure, diversity, and function. This information extends our understanding of gene expression and provides an important foundation for interpreting how the mutations that arise in DNA can result in the diverse phenotypic effects observed in organisms.

13.1 Translation of mRNA Depends on Ribosomes and Transfer RNAs

Translation of mRNA is the biological polymerization of amino acids into polypeptide chains. This process, alluded to in our discussion of the genetic code in Chapter 12, occurs only in association with **ribosomes,** which serve as nonspecific workbenches. The central question in translation is how triplet ribonucleotides of mRNA direct specific amino acids into their correct position in the polypeptide. This question was answered once **transfer RNA (tRNA)** was discovered. This class of molecules adapts specific triplet codons in mRNA to their correct amino acids. The *adaptor hypothesis* for the role of tRNA was postulated by Francis Crick in 1957.

In association with a ribosome, mRNA presents a triplet codon that calls for a specific amino acid. A specific tRNA molecule contains within its nucleotide sequence three consecutive ribonucleotides complementary to the codon, called the **anticodon,** which can base-pair with the codon. Another region of this tRNA is covalently bonded to its corresponding amino acid.

Hydrogen bonding of tRNAs to mRNA holds the amino acids in proximity so that a peptide bond can be formed. (Amino acid structure and peptide bond chemistry will be covered later in this chapter.) This process occurs over and over as mRNA runs through the ribosome and amino acids are polymerized into a polypeptide. Before we discuss the actual process of translation, let's first consider the structures of the ribosome and tRNA.

Ribosomal Structure

Because of its essential role in the expression of genetic information, the ribosome has been extensively analyzed. One bacterial cell contains about 60,000 ribosomes, and a eukaryotic cell contains many times more. Electron microscopy reveals that the bacterial ribosome is about 40 nm at its largest dimension and consists of two subunits, one large and one small. Both subunits consist of one or more molecules of rRNA and an array of **ribosomal proteins.** When the two subunits are associated with each other in a single ribosome, the structure is sometimes called a **monosome.**

The main differences between bacterial and eukaryotic ribosomes are summarized in **Figure 13.1**. The subunit and rRNA components are most easily isolated and characterized on the basis of their sedimentation behavior in sucrose gradients [their rate of migration, or *Svedberg coefficient (S)*, which is a reflection of their density, mass, and shape]. In bacteria, the monosome is a 70*S* particle; in eukaryotes, it is 80*S*. Sedimentation coefficients, which reflect the variable rate of migration of different-sized particles and molecules, are not additive. For example, the bacterial 70*S* monosome consists of a 50*S* and a 30*S* subunit, and the eukaryotic 80*S* monosome consists of a 60*S* and a 40*S* subunit.

The larger subunit in bacteria consists of a 23*S* rRNA molecule, a 5*S* rRNA molecule, and 33 ribosomal proteins. In the eukaryotic equivalent, a 28*S* rRNA molecule is accompanied by a 5.8*S* and 5*S* rRNA molecule and 47 proteins. The smaller bacterial subunits consist of a 16*S* rRNA component and 21 proteins. In the eukaryotic equivalent, an 18*S* rRNA component and 33 proteins are found. The approximate molecular weights (MWs) and the number of nucleotides of these components are also shown in **Figure 13.1**.

Bacteria
Monosome 70*S* (2.3×10^6 Da)

Eukaryotes
Monosome 80*S* (4.3×10^6 Da)

Large subunit	**Small subunit**	**Large subunit**	**Small subunit**
50*S* 1.5×10^6 Da	30*S* 0.8×10^6 Da	60*S* 2.9×10^6 Da	40*S* 1.4×10^6 Da
23*S* rRNA (2904 nucleotides) + 33 proteins + 5*S* rRNA (121 nucleotides)	16*S* rRNA (1542 nucleotides) + 21 proteins	28*S* rRNA (5034 nucleotides) + 47 proteins 5*S* rRNA (121 nucleotides) + 5.8*S* rRNA (156 nucleotides)	18*S* rRNA (1870 nucleotides) + 33 proteins

FIGURE 13.1 A comparison of the components of bacterial and eukaryotic ribosomes. Specific values are given for *E. coli* and human ribosomes.

It is now clear that the rRNA molecules perform the all-important catalytic functions associated with translation. The many proteins, whose functions were long a mystery, are thought to promote the binding of the various molecules involved in translation and, in general, to fine-tune the process. This conclusion is based on the observation that some of the catalytic functions in ribosomes still occur in experiments involving "ribosomal protein-depleted" ribosomes.

Genes coding for the rRNA components are redundant in the genome. For example, the *E. coli* genome contains seven copies of a single sequence that encodes all three rRNAs—23*S*, 16*S*, and 5*S*. The initial transcript of each set of these genes produces a 30*S* RNA molecule that is enzymatically cleaved into these smaller components. Coupling of the genetic information encoding these three rRNA components ensures that after multiple transcription events, equal quantities of all three will be present as ribosomes are assembled.

In eukaryotes, many more copies of a precursor sequence are present. Each copy is initially transcribed into an RNA molecule of about 45*S* that is subsequently processed into 28*S*, 18*S*, and 5.8*S* rRNA components. These species are homologous to the three rRNA components of *E. coli*. In *Drosophila*, approximately 120 copies per haploid genome are present, while in *Xenopus laevis*, more than 500 copies of the larger precursor sequence are present per haploid genome. In mammalian cells, the initial transcript is also 45*S*. The unique 5*S* rRNA component of eukaryotes is not part of this larger transcript. Instead, copies of the gene coding for the 5*S* rRNA are distinct and located separately.

The rRNA genes, called **rDNA,** are present in clusters at various chromosomal sites. Each cluster in eukaryotes consists of **tandem repeats,** with each repeat unit separated by a noncoding **spacer DNA** sequence. In humans, these gene clusters have been localized near the ends of chromosomes 13, 14, 15, 21, and 22. A separate gene cluster encoding 5*S* rRNA has been located on human chromosome 1.

Despite a detailed knowledge of the structure and genetic origin of the ribosomal components, a complete understanding of the function of these components has, to date, eluded geneticists. This is not surprising; the ribosome is the largest and perhaps the most intricate of all cellular structures. For example, the human monosome has a combined molecular weight of 4.3 million Da!

ESSENTIAL POINT

Translation is the synthesis of polypeptide chains under the direction of mRNA in association with ribosomes. ■

tRNA Structure

Because of their small size and stability in the cell, transfer RNAs (tRNAs) have been investigated extensively and are the best characterized RNA molecules. They are composed of only 75 to 90 nucleotides, displaying a nearly identical structure in bacteria and eukaryotes. In both types of organisms, tRNAs are transcribed from DNA as larger precursors, which are cleaved into mature 4*S* tRNA molecules. Take, for example, $tRNA^{Tyr}$ (the superscript identifies the specific tRNA by the amino acid that binds to it, called its *cognate amino acid*). In *E. coli*, mature $tRNA^{Tyr}$ is composed of 77 nucleotides, yet its precursor contains 126 nucleotides.

In 1965, Robert Holley and his colleagues reported the complete sequence of $tRNA^{Ala}$ isolated from yeast. Of great interest was their finding that a number of nucleotides in the tRNA contain a *modified base* not typically found in mRNA. Two of these nucleotides, inosinic acid and pseudouridylic acid, are illustrated in **Figure 13.2**. These modified bases serve several functions. For example, they confer structural stability and are important for hydrogen bonding between the tRNA and the mRNA being translated.

Holley's sequence analysis led him to propose the two-dimensional **cloverleaf model of tRNA.** It was known that tRNA demonstrates a secondary structure due to base pairing. Holley discovered that he could arrange the linear model in such a way that several stretches of base pairing would result. This arrangement created a series of paired stems and unpaired loops resembling the shape of

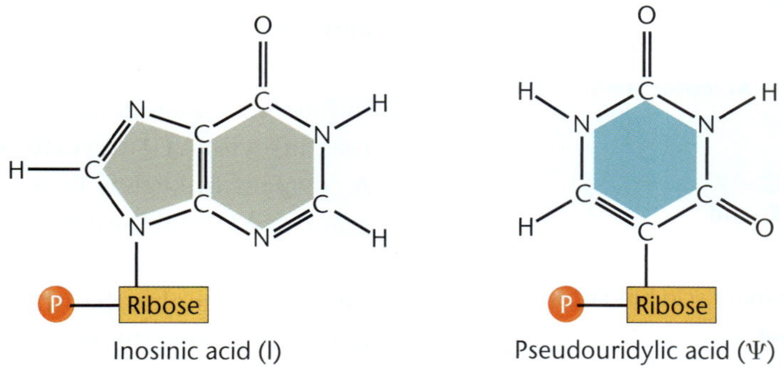

FIGURE 13.2 Ribonucleotides containing two unusual nitrogenous bases found in transfer RNA.

a cloverleaf. Loops consistently contained modified bases that did not generally form base pairs. Holley's model is shown in **Figure 13.3**.

The triplets GCU, GCC, and GCA specify alanine; therefore, Holley looked for an anticodon sequence complementary to one of these codons in his $tRNA^{Ala}$ molecule. He found it in the form of CGI (the 3′ to 5′ direction) in one loop of the cloverleaf. The nitrogenous base I (inosinic acid) can form hydrogen bonds with U, C, or A, the third members of the alanine triplets. Thus, the **anticodon loop** was established.

Studies of other tRNA species reveal many constant features. At the 3′-end, all tRNAs contain the sequence . . . pCpCpA-3′. This is the end of the molecule where the amino acid is covalently joined to the terminal adenosine residue. All tRNAs contain the nucleotide 5′-Gp . . . at the other end of the molecule. In addition, the lengths of various stems and loops are very similar. Each tRNA that has been examined also contains an anticodon complementary to the known amino acid codon for which it is specific, and all anticodon loops are present in the same position of the cloverleaf.

Because the cloverleaf model was predicted strictly on the basis of nucleotide sequence, there was great interest in the X-ray crystallographic examination of tRNA, which reveals a three-dimensional structure. By 1974, Alexander Rich, Jon Roberts, Brian Clark, Aaron Klug, and their colleagues had succeeded in crystallizing tRNA and performing X-ray crystallography at a resolution of 3 Å. At this resolution, the pattern formed by individual nucleotides is discernible.

As a result of these studies, a complete three-dimensional model of tRNA was proposed, as shown in **Figure 13.4**. At one end of the molecule is the anticodon loop, and at the other end is the 3′-acceptor region where the amino acid is bound. Geneticists speculate that the shapes of the intervening loops may be recognized by the specific enzymes responsible for adding amino acids to tRNAs—a subject to which we now turn our attention.

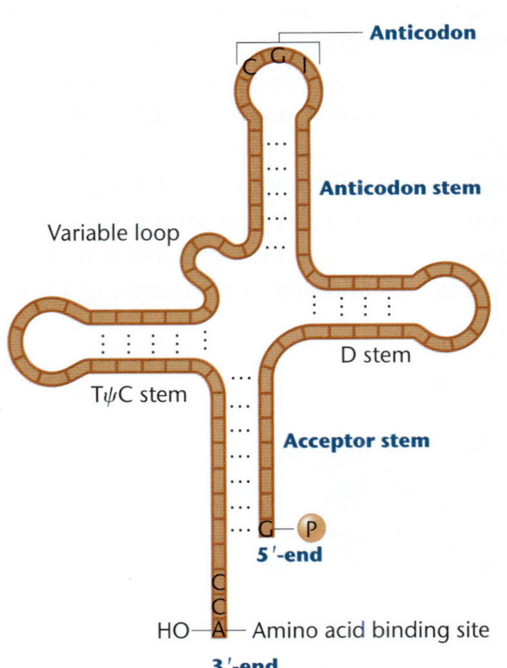

FIGURE 13.3 Holley's two-dimensional cloverleaf model of transfer RNA. Blocks represent nitrogenous bases.

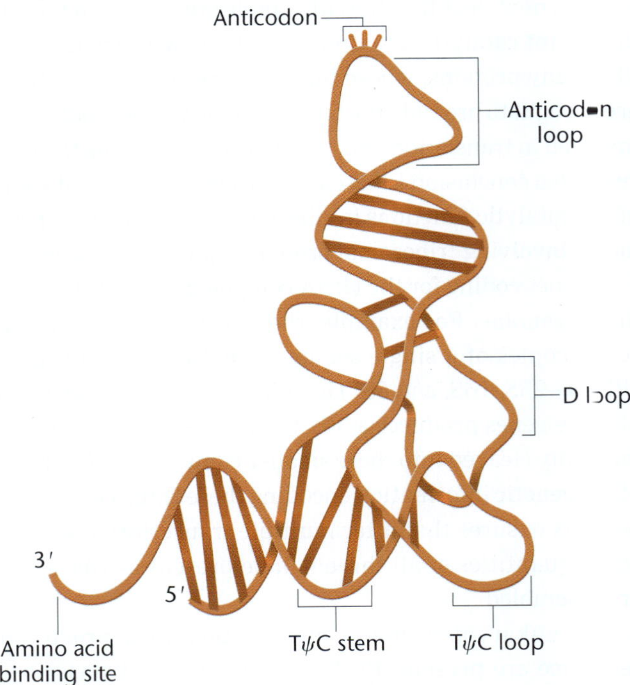

FIGURE 13.4 A three-dimensional model of transfer RNA.

Charging tRNA

Before translation can proceed, the tRNA molecules must be chemically linked to their respective amino acids. This activation process, called **charging,** occurs under the direction of enzymes called **aminoacyl tRNA synthetases.** There are 20 different amino acids, thus there are 20 different aminoacyl tRNA synthetases. Recall from Chapter 12 that the third position of the triplet code has a relaxed base-pairing requirement, or 'wobble.' Based on this, only 30 to 40 different tRNAs (depending on the type of organism) are necessary to accommodate the 61 amino acid specifying codons of the genetic code.

The charging process is outlined in **Figure 13.5**. In the initial step, the amino acid is converted to an activated form, reacting with ATP to create an **aminoacyladenylic acid.** A covalent linkage is formed between the 5′-phosphate group of ATP and the carboxyl end of the amino acid. This molecule remains associated with the synthetase enzyme, forming a complex that then reacts with a specific tRNA molecule. During this next step, the amino acid is attached to the appropriate tRNA through a high-energy ester bond between the 3′-end of the tRNA and the carboxyl group of

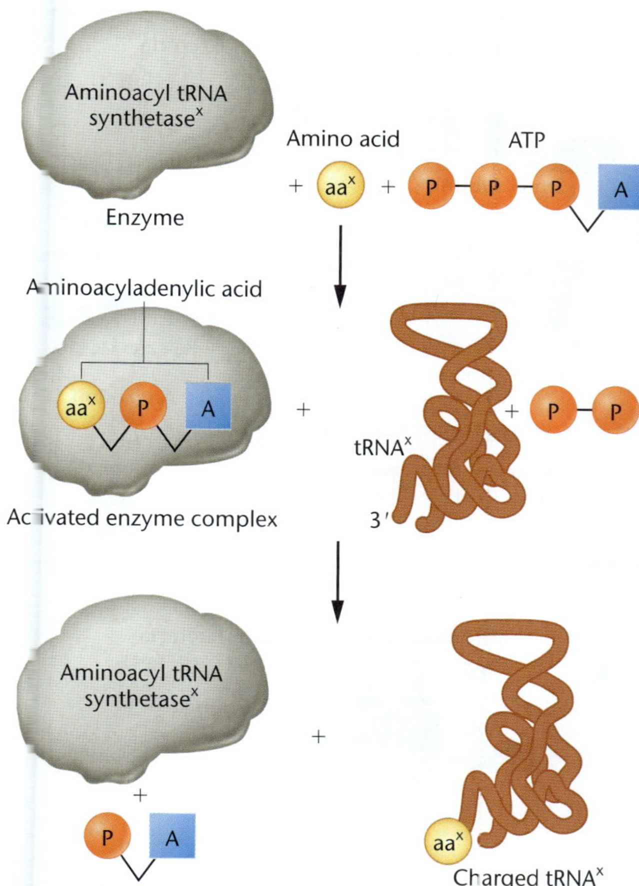

FIGURE 13.5 Steps involved in charging tRNA. The superscript x denotes that only the corresponding specific tRNA and specific aminoacyl tRNA synthetase enzyme are involved in the charging process for each amino acid.

the amino acid. The charged tRNA may now participate directly in protein synthesis. Aminoacyl tRNA synthetases are highly specific enzymes because they recognize only one amino acid and the subset of corresponding tRNAs called **isoaccepting tRNAs.** Accurate charging is crucial if fidelity of translation is to be maintained.

ESSENTIAL POINT

Translation depends on tRNA molecules that serve as adaptors between triplet codons in mRNA and the corresponding amino acids. ■

NOW SOLVE THIS

13.1 In 1962, F. Chapeville and others reported an experiment in which they isolated radioactive ^{14}C-cysteinyl-tRNA^{Cys} (charged tRNA^{Cys} + cysteine). They then removed the sulfur group from the cysteine, creating alanyl-tRNA^{Cys} (charged tRNA^{Cys} + alanine). When alanyl-tRNA^{Cys} was added to a synthetic mRNA calling for cysteine, but not alanine, a polypeptide chain was synthesized containing alanine. What can you conclude from this experiment?

■ **HINT:** *This problem is concerned with establishing whether tRNA or the amino acid added to the tRNA during charging is responsible for attracting the charged tRNA to mRNA during translation. The key to its solution is the observation that in this experiment, when the triplet codon in mRNA calls for cysteine, alanine is inserted during translation, even though it is the "incorrect" amino acid.*

13.2 Translation of mRNA Can Be Divided into Three Steps

Much like transcription, the process of translation can best be described by breaking it into discrete phases. We will consider three such phases, each with its own set of illustrations (Figures 13.6, 13.7, and 13.8), but keep in mind that translation is a dynamic, continuous process. As you read the following discussion, keep track of the step-by-step events depicted in the figures. While the core concepts of translation are common for bacterial and eukaryotic cells, the process is simpler in bacteria and is discussed in this section. Many of the protein factors involved in bacterial translation, and their roles, are summarized in **Table 13.1**.

TABLE 13.1 Various Protein Factors Involved during Translation in *E. coli*

Process	Factor	Role
Initiation of translation	IF1	Binds to 30S subunit and prevents aminoacyl tRNA from binding to the A site prematurely
	IF2	Binds to the initiator fMet-tRNA and transfers it to the P site of the 30S-mRNA complex; releases from complex upon GTP hydrolysis, which is required for 50S subunit binding
	IF3	Binds to 30S subunit, preventing it from associating with the 50S subunit prematurely
Elongation of polypeptide	EF-Tu	Binds GTP; brings aminoacyl tRNA to the A site of the ribosome
	EF-Ts	Regulates EF-Tu activity
	EF-G	Stimulates translocation; GTP-dependent
Termination of translation and release of polypeptide	RF1	Catalyzes release of the polypeptide chain from tRNA and dissociation of the translocation complex; specific for UAA and UAG termination codons
	RF2	Behaves like RF1; specific for UGA and UAA codons
	RF3	Stimulates RF1 and RF2 release

Initiation

Initiation of bacterial translation is depicted in **Figure 13.6**, which at the top, in a box, shows all the individual components involved in the process. Recall that the ribosome serves as a workbench for the translation process. Ribosomes, when they are not involved in translation, are dissociated into their large and small subunits. Note that ribosomes contain three sites, the **aminoacyl (A) site,** the **peptidyl (P) site,** and the **exit (E) site,** the roles of which will soon become apparent. The initiation phase of translation in bacteria requires the association of a small ribosomal subunit, an mRNA molecule, a specific charged initiator tRNA, GTP, Mg^{2+}, and three proteinaceous **initiation factors (IFs).** In bacteria, the initiation codon of mRNA—AUG—calls for the modified amino acid **N-formylmethionine (fMet).**

The three initiation factors first bind to the small ribosomal subunit, and this complex in turn binds to mRNA (Step 1). In bacteria, a short sequence on the mRNA, called the **Shine–Dalgarno sequence,** base-pairs with a region of the 16*S* rRNA of the small ribosomal subunit, facilitating initiation.

While IF1 primarily blocks the A site from being bound to a tRNA and IF3 serves to inhibit the small subunit from associating with the large subunit prematurely, IF2 plays a more direct role in initiation. Essentially a GTPase, IF2 interacts with the mRNA and $tRNA^{fMet}$, stabilizing them in the P site (Step 2). This step "sets" the reading frame so that all subsequent groups of three ribonucleotides are translated accurately. Upon release of IF3, the small subunit (and its associated mRNA and $tRNA^{fMet}$) then combines with the large ribosomal subunit to create the 70*S* initiation complex. In this process, a molecule of GTP covalently linked to IF2 is hydrolyzed to GDP, causing a conformational change in IF2, and IF1 and IF2 are subsequently released (Step 3).

Elongation

The second phase of translation, elongation, is depicted in **Figure 13.7**. As per our prior discussion, the initiation complex is now poised for the insertion into the A site of the second aminoacyl tRNA bearing the amino acid corresponding to the second triplet sequence on the mRNA. Charged tRNAs are transported into the complex by one of the **elongation factors (EFs),** EF-Tu (Step 1). Like IF2 during initiation, EF-Tu is a GTPase and is bound by a GTP. Hydrolysis of GTP causes a conformational change of EF-Tu such that it releases the bound aminoacyl tRNA.

The next step is for the terminal amino acid in the P site (methionine in this case) to be linked to the amino acid now present on the tRNA in the A site by the formation of a peptide bond. Such lengthening of a growing polypeptide chain by one amino acid is called **elongation.** The newly formed

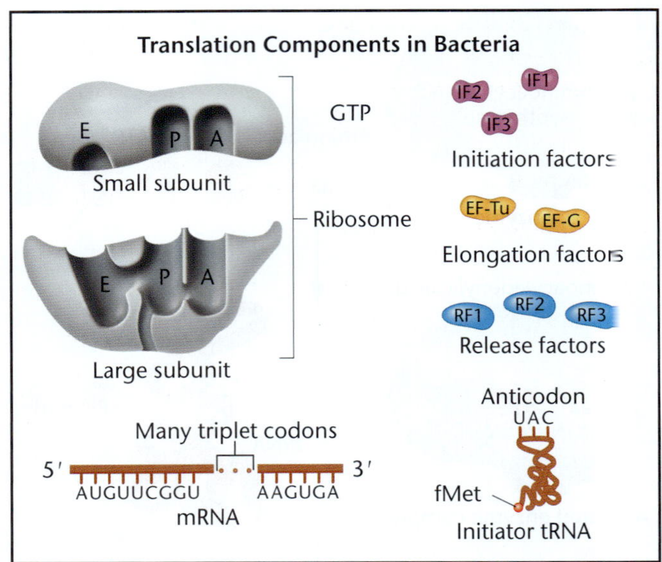

Initiation of Translation in Bacteria

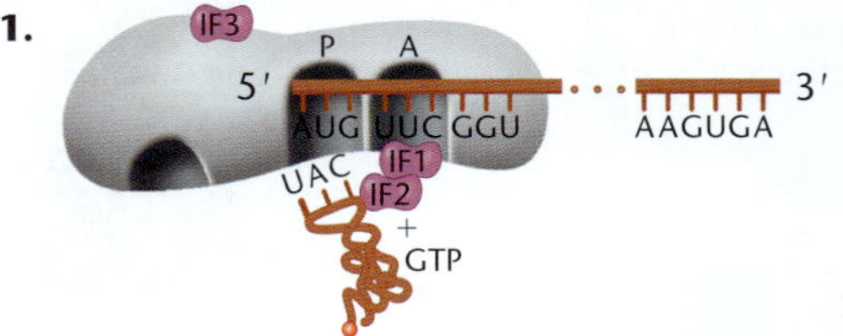

Initiation factors bind to small subunit and attract mRNA.

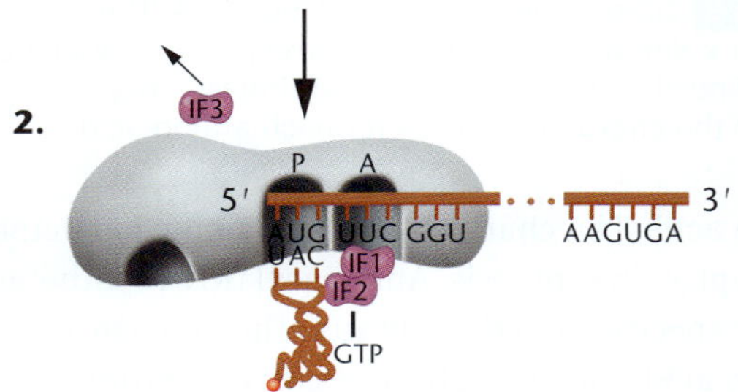

$tRNA^{fMet}$ binds to AUG codon of mRNA in P site, forming initiation complex; IF3 is released.

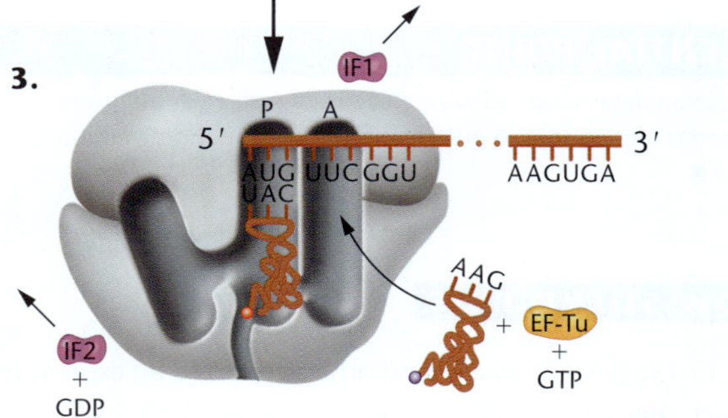

Large subunit binds to complex; IF1 and IF2 are released. Subsequent aminoacyl tRNA is poised to enter the A site.

FIGURE 13.6 Initiation of translation in bacteria. (The separate components required for all three phases of translation are depicted in the box at the top of the figure.)

dipeptide remains attached to the end of the tRNA still residing in the A site. The energy to form a new peptide bond is supplied by breaking the high-energy ester bond between a tRNA and its cognate amino acid. These reactions were initially believed to be catalyzed by an enzyme called **peptidyl transferase,** embedded in the large subunit of the ribosome. However, it is now clear that the catalytic activity is actually a function of the 23*S* rRNA of the large subunit. In such a case, as we saw with splicing of pre-mRNAs (see Chapter 12), we refer to the complex as a **ribozyme,** recognizing the catalytic role that RNA plays in the process.

Before elongation can be repeated, the tRNA attached to the P site, which is now uncharged, must be released from the large subunit. The uncharged tRNA moves briefly into a third site on the ribosome, the E (exit) site. The entire mRNA–tRNA–aa2–aa1 complex then shifts in the direction of the P site by a distance of three nucleotides (Step 2). This event, called *translocation*, requires elongation factor G (EF-G), a GTPase (Step 3). After peptide bond formation, EF-G hydrolyzes GTP, which causes a conformational change of EF-G such that it elongates. This change causes a ratchet-like movement of the small subunit relative to the large subunit. The end result is that the third codon of mRNA is now positioned in the A site and is ready to accept its specific charged tRNA (Step 4). One simple way to distinguish the A and P sites in your mind is to remember that, *following translocation*, the P site (*P* for peptide) contains a tRNA attached to a peptide chain (a peptidyl tRNA), whereas the A site (*A* for amino acid) contains a charged tRNA with its amino acid attached (an aminoacyl tRNA).

The sequence of elongation and translocation is repeated over and over (Steps 4 and 5). An additional amino acid is added to the growing polypeptide chain each time the mRNA advances by three nucleotides through the ribosome. Once a polypeptide chain of sufficient size is assembled (about 30 amino acids), it begins to emerge from the base of the large subunit through an exit tunnel, as illustrated in Step 6.

As we have seen, the role of the small subunit during elongation is to "decode" the codons in the mRNA, while the role of the large subunit is peptide-bond synthesis. The efficiency of the process is remarkably high: The observed error rate is only about 10^{-4}. At this rate, an incorrect amino acid will occur only once in every 20 polypeptides of an average length of 500 amino acids! In a species such as *E. coli*, elongation proceeds at a rate of about 15 amino acids per second at 37°C.

Elongation during Translation in Bacteria

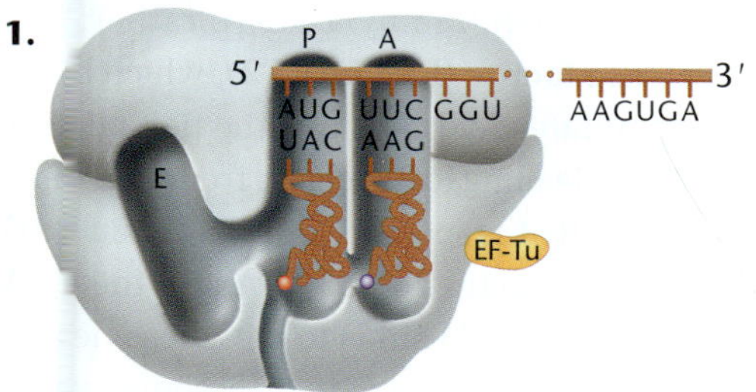

Second charged tRNA has entered the A site, facilitated by EF-Tu; first elongation step commences.

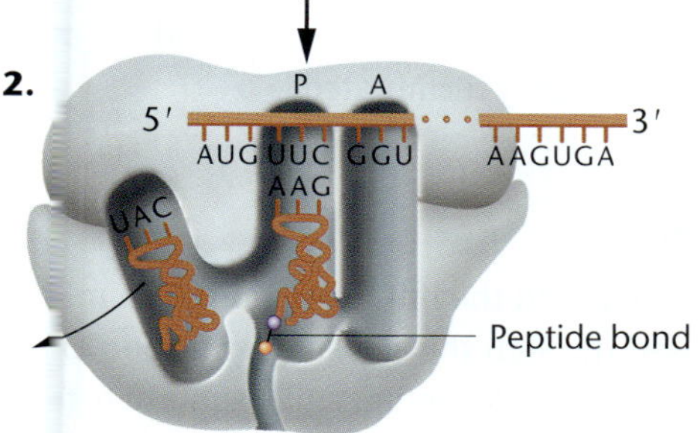

Peptide bond forms; uncharged tRNA moves to the E site and subsequently out of the ribosome; the mRNA has been translocated three bases to the left, causing the tRNA bearing the dipeptide to shift into the P site.

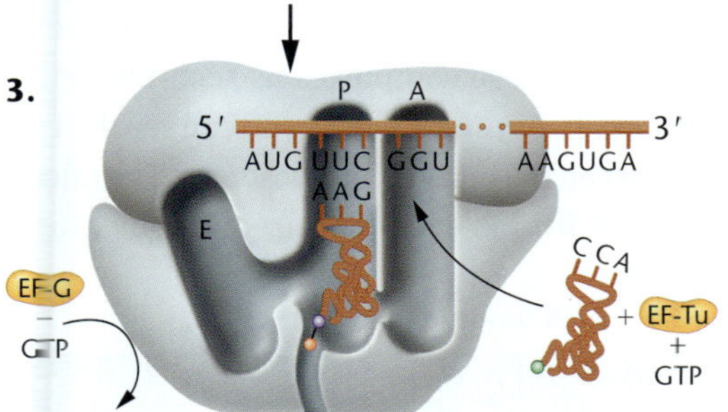

The first elongation step is complete, facilitated by EF-G. The third charged tRNA is ready to enter the A site.

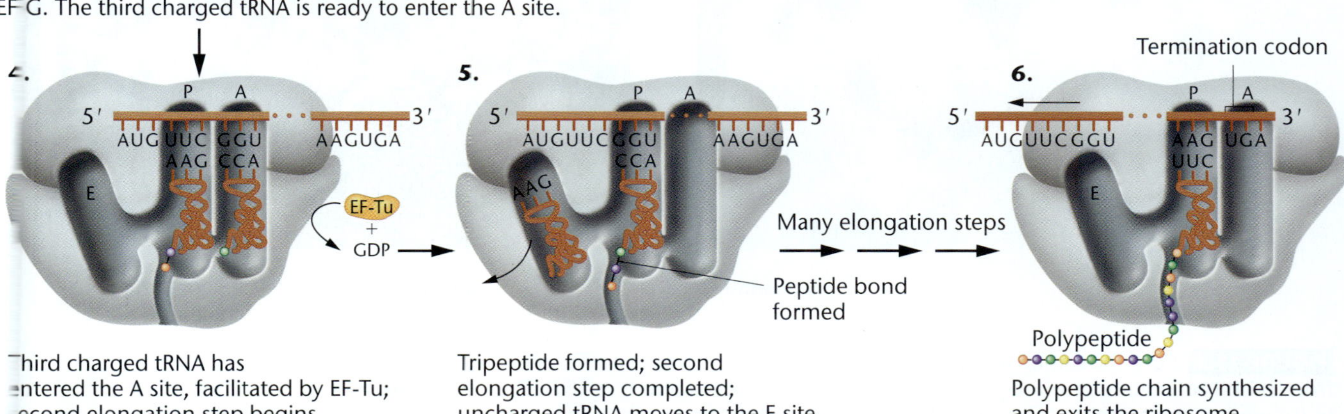

FIGURE 13.7 Elongation of the growing polypeptide chain during translation in bacteria.

Termination of Translation in Bacteria

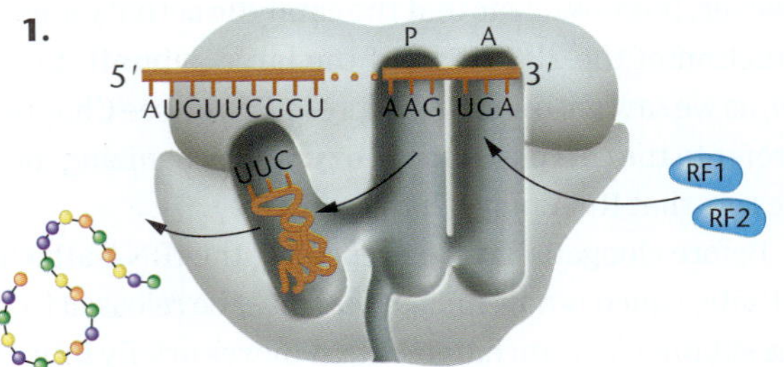

Termination codon enters the A site; RF1 or RF2 stimulates hydrolysis of the polypeptide from peptidyl tRNA.

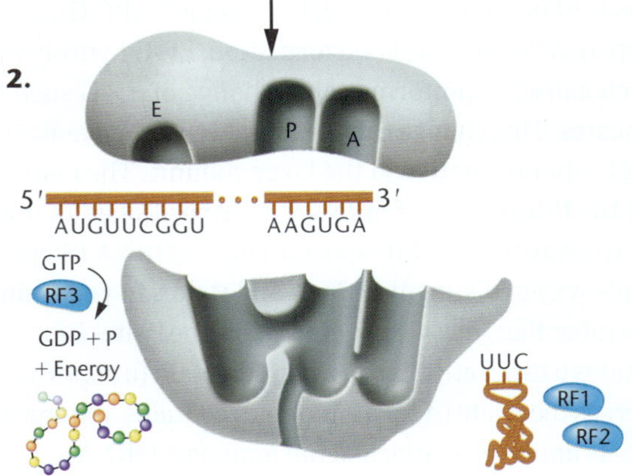

Ribosomal subunits dissociate and mRNA is released; polypeptide folds into native 3D conformation of protein; tRNA is released.

FIGURE 13.8 Termination of the process of translation in bacteria.

Termination

Termination, the third phase of translation, is depicted in **Figure 13.8**. The process is signaled by the presence of any one of three possible triplet codons appearing in the A site: UAG, UAA, or UGA. These codons do not specify an amino acid, nor do they call for a tRNA in the A site. They are called **stop codons, termination codons,** or **nonsense codons.** Often, several consecutive stop codons are part of an mRNA. When one such stop codon is encountered, the polypeptide, now completed, is still connected to the peptidyl tRNA in the P site, and the A site is empty. The termination codon is not recognized by a tRNA; rather, it is recognized by a **release factor (RF1 or RF2),** which binds to the A site and stimulates hydrolysis of the polypeptide from the peptidyl tRNA, leading to its release from the translation complex (Step 1). RF1 is specific to the UAA and UAG stop codons, and RF2 is specific to the UAA and the UGA codons. Then, release factor RF3 binds to the ribosome and the tRNA is released from the P site of the ribosome, which then dissociates into its subunits (Step 2). If a termination codon should appear in the middle of an mRNA molecule as a result of mutation, the same process occurs, and the polypeptide chain is prematurely terminated.

Polyribosomes

As elongation proceeds and the initial portion of an mRNA molecule has passed through the ribosome, this portion of mRNA is free to associate with another small subunit to form a second initiation complex. The process can be repeated several times with a single mRNA and results in what are called **polyribosomes,** or just **polysomes.**

After cells are gently lysed in the laboratory, polyribosomes can be isolated from them and analyzed. The photos in **Figure 13.9** show these complexes as seen under the electron microscope. In **Figure 13.9(a)**, you can see the thin lines of mRNA between the individual ribosomes. The micrograph in **Figure 13.9(b)** is even more remarkable, for it

(a)

mRNA

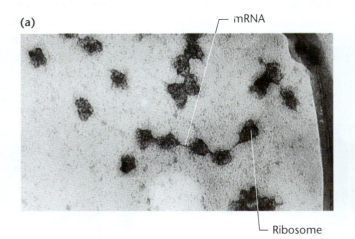

Ribosome

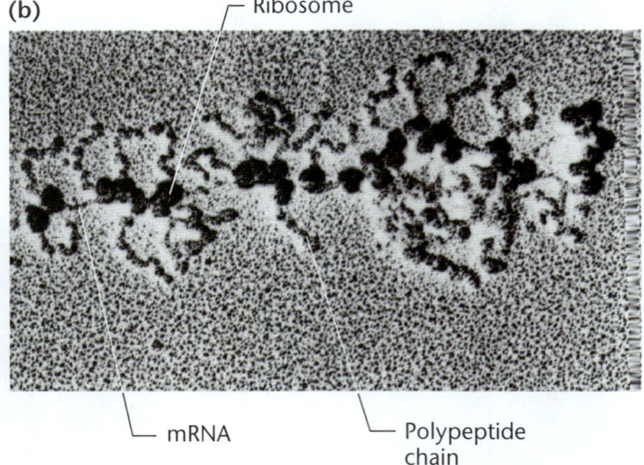

FIGURE 13.9 Polyribosomes as seen under the electron microscope. Those in (a) were derived from rabbit reticulocytes engaged in the translation of hemoglobin mRNA. The polyribosomes in (b) were taken from the giant salivary gland cells of the midgefly, *Chironomus thummi.* Note that the nascent polypeptide chains are apparent as they emerge from each ribosome. Their length increases as translation proceeds from left (5′) to right (3′) along the mRNA.

shows the polypeptide chains emerging from the ribosomes during translation. The formation of polysome complexes represents an efficient use of the components available for protein synthesis during a unit of time. It is as if the mRNA is threaded through numerous ribosomes that are side-by-side such that translation is occurring simultaneously in each one, but each subsequent ribosome is a bit behind its neighbor in the amount of mRNA that has been translated.

ESSENTIAL POINT

Translation, like transcription, is subdivided into the stages of initiation, elongation, and termination and relies on base-pairing affinities between complementary nucleotides. ■

13.3 High-Resolution Studies Have Revealed Many Details about the Functional Bacterial Ribosome

Our knowledge of the process of translation and the structure of the ribosome is based primarily on biochemical and genetic observations, in addition to the visualization of ribosomes under the electron microscope. To confirm and refine this information, the next step is to examine the ribosome at even higher levels of resolution. For example, X-ray diffraction analysis of ribosome crystals is one way to achieve this. However, because of its tremendous size and the complexity of molecular interactions occurring in the functional ribosome, it was extremely difficult to obtain the crystals necessary to perform X-ray diffraction studies. Nevertheless, great strides have been made over the past decade. First, the individual ribosomal subunits were crystallized and examined in several laboratories, most prominently that of Venkatraman Ramakrishnan. Then, the crystal structure of the intact 70*S* ribosome, complete with associated mRNA and tRNAs, was examined by Harry Noller and colleagues. In essence, the entire translational complex was seen at the atomic level. Both Ramakrishnan and Noller derived the ribosomes from the bacterium *Thermus thermophilus.*

Many noteworthy observations have come from these investigations. For example, the shape of the ribosome changes during different functional states, attesting to the dynamic nature of the process of translation. A great deal has also been learned about the location of the RNA components of the subunits. About one-third of the 16*S* RNA is responsible for producing a flat projection, referred to as the *platform,* within the smaller 30*S* subunit, and it modulates movement of the mRNA–tRNA complex during translocation. One of the models based on Noller's findings is shown in the opening photograph of this chapter (p. 265).

Crystallographic analysis also supports the concept that rRNA is the real "player" in the ribosome during translation. The interface between the two subunits, considered to be the location in the ribosome where polymerization of amino acids occurs, is composed almost exclusively of RNA. In contrast, the numerous ribosomal proteins are found mostly on the periphery of the ribosome. These observations confirm what has been predicted on genetic grounds—the catalytic steps that join amino acids during translation occur under the direction of RNA, not proteins.

Another interesting finding involves the actual location of the various sites predicted to house tRNAs during translation. All three sites (A, P, and E) have been identified in X-ray diffraction studies, and in each case, the RNA of the ribosome makes direct contact with the various loops and domains of the tRNA molecule. This observation helps us understand why the distinctive three-dimensional conformation that is characteristic of all tRNA molecules has been preserved throughout evolution.

Still another noteworthy observation takes us back almost 50 years, to when Francis Crick proposed the *wobble hypothesis*, as introduced in Chapter 12. The Ramakrishnan group has identified the precise location along the 16*S* rRNA of the 30*S* subunit involved in the decoding step that connects mRNA to the proper tRNA. At this location, two particular nucleotides of the 16*S* rRNA actually flip out and probe the codon:anticodon region, and are believed to check for accuracy of base pairing during this interaction. According to the wobble hypothesis, the stringency of this step is high for the first two base pairs but less so for the third (or wobble) base pair.

As our knowledge of the translation process in bacteria has continued to grow, a remarkable study was reported in 2010 by Niels Fischer and colleagues. Using a unique high-resolution approach—the technique of *time-resolved single particle cryo-electron microscopy (cryo-EM)*—the 70*S E. coli* ribosome was captured while in the process of translation and examined at a resolution of 5.5 Å. This research team examined how tRNA is translocated during elongation of the polypeptide chain. They demonstrated that the trajectories are coupled with dynamic conformational changes in the components of the ribosome. Surprisingly, the work has revealed that during translation, the ribosome behaves as a complex molecular machine *powered by Brownian movement driven by thermal energy*. That is, the energetic requirements for achieving the various conformational changes essential to translocation are inherent to the ribosome itself.

Numerous questions about ribosome structure and function still remain. In particular, the precise role of the many ribosomal proteins is yet to be clarified. Nevertheless, the models that are emerging from the above research provide us with a much better understanding of the mechanism of translation.

13.4 Translation Is More Complex in Eukaryotes

The general features of the model of translation we just discussed were initially derived from investigations of the process in bacteria. Conceptually, the most significant difference between translation in bacteria and eukaryotes is that in bacteria transcription and translation both take place in the cytoplasm and therefore are coupled, whereas in eukaryotes these two processes are separated both spatially and temporally. In eukaryotic cells transcription occurs in the nucleus and translation in the cytoplasm. This separation provides multiple opportunities for the regulation of gene expression in eukaryotic cells (a topic we will turn to in Chapter 16).

Another central difference between bacterial and eukaryotic translation, as we have already seen (Figure 13.1), is that eukaryotes have larger ribosomes composed of a greater number of proteins and RNAs. Interestingly, bacterial and eukaryotic rRNAs share what is called a *core sequence*, but in eukaryotes, rRNAs are lengthened by the addition of *expansion segments (ESs)*, which are important for ribosome assembly and may also contribute to the regulation and specificity of translation.

The initiation phase is particularly rich in differences between eukaryotes and bacteria. For example, recall that bacterial translation initiation is dependent upon the small subunit pairing with a short sequence upstream of the start codon—the *Shine–Dalgarno sequence*. Eukaryotes lack this sequence. Instead, in a process termed *cap-dependent translation*, initiation in eukaryotes begins with the small subunit associating with the 7-methylguanosine (m^7G) cap located at the 5′-end of eukaryotic mRNAs (Chapter 12). In this context, a specific sequence on bacterial mRNAs and the physical cap structure on eukaryotic mRNAs serve analogous functions.

Recall, too, that bacterial translation initiation requires initiation factors (IF1, IF2, and IF3; Figure 13.6) to attract mRNA to the small subunit and prevent premature association of the large subunit. In eukaryotes, a suite of *eukaryotic* initiation factors (eIFs) carry out these processes. A complex consisting of several eIFs, the initiator tRNA, and the small subunit of the ribosome assemble adjacent to the m^7G cap. The assembly then slides along the mRNA searching for the start codon in a process known as *scanning*. While the bacterial start codon is recognized by an initiator tRNA carrying an N-formylmethionine (fMet), the eukaryotic start codon encodes unformylated methionine. It turns out that a unique transfer RNA ($tRNA_i^{Met}$) is used during eukaryotic initiation, one different from the $tRNA^{Met}$ used for AUG codons after the start. Another eukaryote-specific feature of translation initiation is that many mRNAs contain a purine (A or G) three bases upstream from the AUG initiator codon, which is often followed by a G ($^{A}/_{G}$NN**AUG**G). Named after its discoverer, Marilyn Kozak, this **Kozak sequence** is considered to increase the efficiency of translation initiation in eukaryotes.

Interestingly, the poly-A tail at the 3′-end of eukaryotic mRNAs also plays an important role in translation initiation. Recall from Chapter 12 that **poly-A binding proteins** bind to the poly-A tail to protect the mRNA from degradation. The poly-A-binding proteins also bind to one of the eukaryotic initiation factors, eIF4G, which in turn binds to eIF4E, also known as the *cap-binding protein*. As this name suggests, eIF4G binds the m^7G cap. The resulting complex is required for translation initiation for many eukaryotic mRNAs. Because the mRNA forms a loop that is closed where the cap and tail are brought together, the process is called **closed-loop translation** (**Figure 13.10**). One possible advantage of closed-loop translation is that the cell will not waste energy translating a partially degraded mRNA lacking either a cap or poly-A tail, features necessary for the closed loop. Another advantage that has been proposed is that the closed-loop structure allows for efficient *ribosome recycling* whereby ribosomes that complete synthesis of one polypeptide can dissociate from the mRNA and then reinitiate translation adjacent to the cap, which is a short distance away in the loop.

Still other differences between translation in bacteria and eukaryotes are noteworthy. Eukaryotic mRNAs are much longer lived than are their bacterial counterparts. Bacterial mRNAs, which lack a cap and poly-A tail, are often translated immediately after transcription and degraded within five minutes. On the other hand eukaryotic mRNAs,

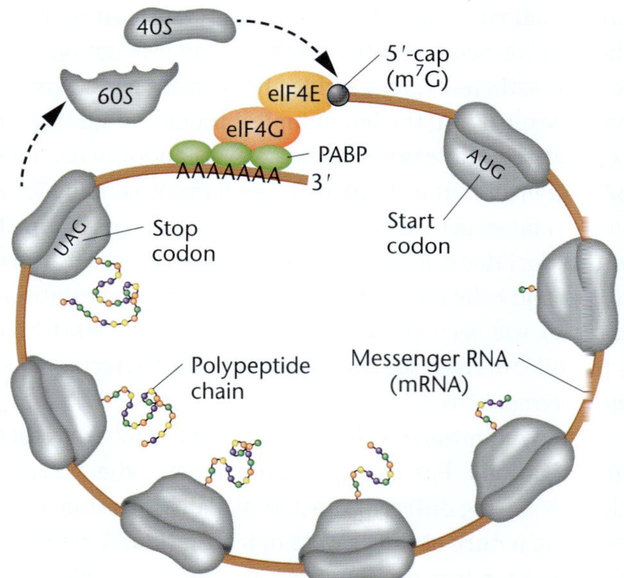

FIGURE 13.10 Eukaryotic closed-loop translation. eIF4E binds to the cap on the mRNA and to a scaffold protein, eIF4G, which binds to poly-A binding proteins (PABPs) on the poly-A tail of the mRNA. Ribosomes assemble at the cap, scan for the start codon, translate around the loop terminating at a stop codon, and may then reinitiate translation in a process called ribosome recycling.

with the protection of a cap and poly-A tail, can persist far longer with an average of a 10-hour half-life for mRNAs in cultured human cells. Thus, eukaryotic mRNAs are often available for translation for much longer periods of time.

After translation initiation, proteins similar to those in bacteria guide the elongation and termination of translation in eukaryotes. Many of these **eukaryotic elongation factors (eEFs)** and **eukaryotic release factors (eRFs)** are clearly homologous to their counterparts in bacteria. For example, the role of bacterial EF-Tu, which guides aminoacyl tRNAs into the A site of the ribosome, is fulfilled by eEF-1α. Unlike bacteria, there is a single release factor, eRF1, which recognizes each of the three stop codons in eukaryotes.

We conclude this section by noting that in 2015, the crystal structure of the highly complex 80*S* human ribosome was visualized by Bruno Klaholz and colleagues at the remarkable resolution of 3.6 Å. This research reveals the interactions between the rRNAs and proteins of the ribosome at an atomic level of detail. Their images reveal that the interface of the large and small subunits remodels during translation, reflecting a rotational movement of the subunits as the ribosome translocates, or moves along the mRNA. Many antibiotics target the bacterial ribosome to block its activity, but have some negative side effects when used as drugs to fight bacterial infections in humans due to partially inhibiting the activity of the human ribosome. This study provides an important model that may assist in reducing the side effects of antibiotics by increasing their specificity for bacterial ribosomes. In addition, this study may enable the design of drugs to slow down the rate of translation of the highly active ribosomes in human cancer cells, thus starving these cells of the protein synthesis on which they are dependent.

13.5 The Initial Insight That Proteins Are Important in Heredity Was Provided by the Study of Inborn Errors of Metabolism

Let's consider how we know that proteins are the end products of genetic expression. The first insight into the role of proteins in genetic processes was provided by observations made by Sir Archibald Garrod and William Bateson early in the twentieth century. Garrod was born into an English family of medical scientists. His father was a physician with a strong interest in the chemical basis of rheumatoid arthritis, and his eldest brother was a leading zoologist in London. It is not surprising, then, that as a practicing physician, Garrod became interested in several human disorders that seemed to be inherited. Although he also studied albinism and cystinuria, we shall describe his investigation of the disorder **alkaptonuria.** Individuals with this disorder have an important metabolic pathway blocked. As a result, they cannot metabolize the alkapton 2,5-dihydroxyphenylacetic acid, also known as homogentisic acid. Homogentisic acid accumulates in cells and tissues and is excreted in the urine. The molecule's oxidation products are black and easily detectable in the diapers of newborns. The products tend to accumulate in cartilaginous areas, causing the ears and nose to darken. The deposition of homogentisic acid in joints leads to a benign arthritic condition. This rare disease is not serious, but it persists throughout an individual's life.

Garrod studied alkaptonuria by looking for patterns of inheritance of this benign trait. Eventually he concluded that it was genetic in nature. Of 32 known cases, he ascertained that 19 were confined to seven families, with one family having four affected siblings. In several instances, the parents were unaffected but known to be related as first cousins. Parents who are so related have a higher probability than unrelated parents of producing offspring that express recessive traits because such parents are both more likely to be heterozygous for some of the same recessive traits. Garrod concluded that this inherited condition was the result of an alternative mode of metabolism, thus implying that hereditary information controls chemical reactions in the body. While *genes* and *enzymes* were not familiar terms during Garrod's time (1902), he used the corresponding concepts of *unit factors* and *ferments.*

Only a few geneticists, including Bateson, were familiar with or referred to Garrod's work. Garrod's ideas fit nicely with Bateson's belief that inherited conditions are caused by the lack of some critical substance. In 1909, Bateson published *Mendel's Principles of Heredity,* in which he linked Garrod's ferments with heredity. However, for almost 30 years, most geneticists failed to see the relationship between genes and enzymes. Garrod and Bateson, like Mendel, were ahead of their time.

13.6 Studies of *Neurospora* Led to the One-Gene: One-Enzyme Hypothesis

In two separate investigations beginning in 1933, George Beadle provided the first convincing experimental evidence that genes are directly responsible for the synthesis of enzymes. The first investigation, conducted in collaboration with Boris Ephrussi, involved *Drosophila* eye pigments. Together, they confirmed that mutant genes that alter the eye color of fruit flies could be linked to biochemical errors that, in all likelihood, involved the loss of enzyme function. Encouraged by these findings, Beadle then joined with Edward Tatum to investigate nutritional mutations in the pink bread mold *Neurospora crassa.* This investigation led to the **one-gene:one-enzyme hypothesis.**

Analysis of *Neurospora* Mutants by Beadle and Tatum

In the early 1940s, Beadle and Tatum chose to work with *Neurospora* because much was known about its biochemistry and because mutations could be induced and isolated with relative ease. By inducing mutations, they produced strains that had genetic blocks of reactions essential to the growth of the organism.

Beadle and Tatum knew that this mold could manufacture nearly everything necessary for normal development. For example, using rudimentary carbon and nitrogen sources, this organism can synthesize nine water-soluble vitamins, 20 amino acids, numerous carotenoid pigments, and all essential purines and pyrimidines. Beadle and Tatum irradiated asexual conidia (spores) with X rays to increase the frequency of mutations and allowed them to be grown on "complete" medium containing all the necessary growth factors (e.g., vitamins and amino acids). Under such growth conditions, a mutant strain unable to grow on minimal medium was able to grow by virtue of supplements present in the enriched complete medium. All the cultures were then transferred to minimal medium. If growth occurred on the minimal medium, the organisms were able to synthesize all the necessary growth factors themselves, and the researchers concluded that the culture did not contain a nutritional mutation. If no growth occurred on minimal medium, they concluded that the culture contained a nutritional mutation, and the only task remaining was to determine its type. These results are shown in **Figure 13–11(a)**.

Many thousands of individual spores from this procedure were isolated and grown on complete medium. In subsequent tests on minimal medium, many cultures failed

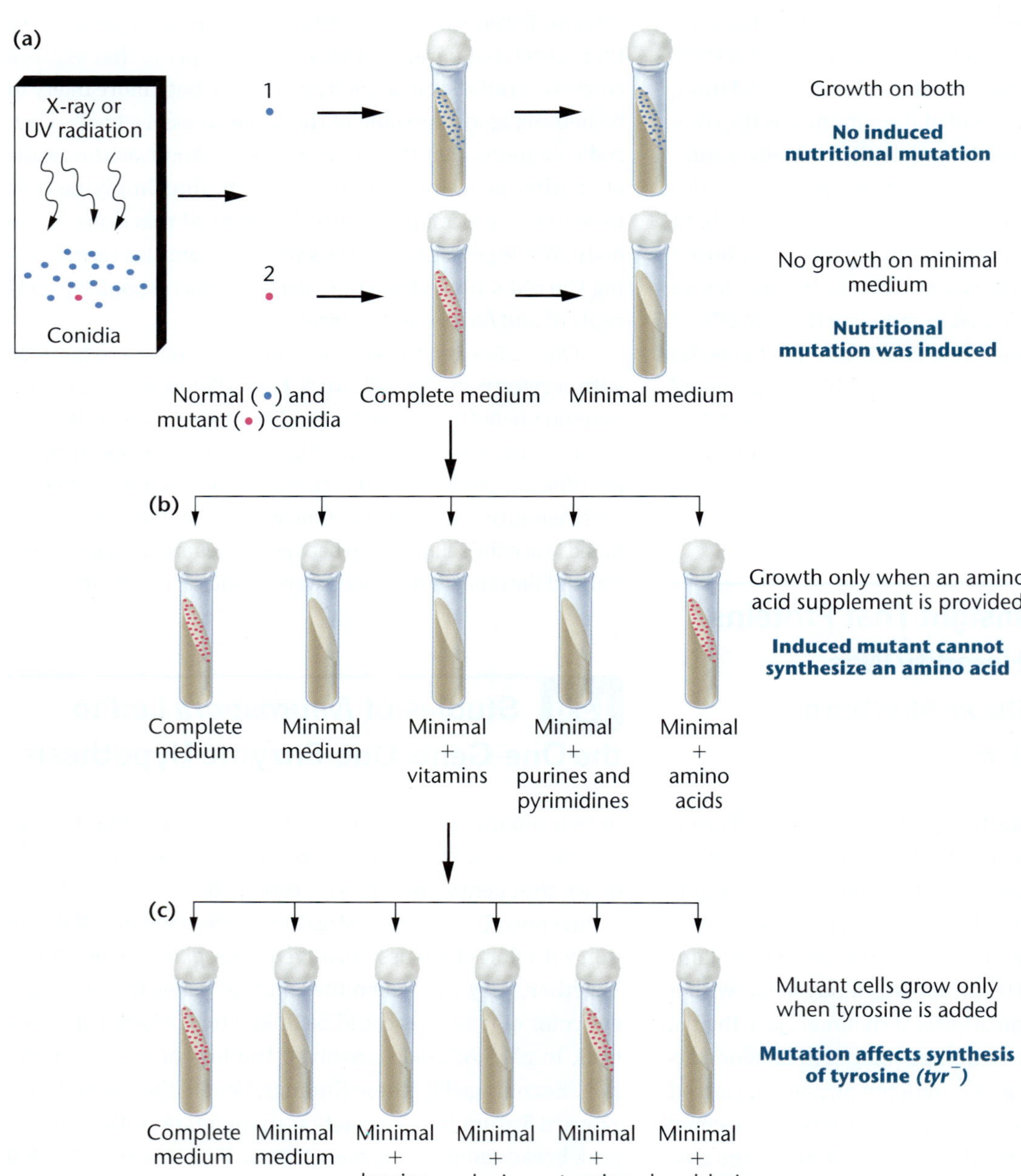

FIGURE 13.11 Induction, isolation, and characterization of a nutritional auxotrophic mutation in *Neurospora*. (a) Most conidia are not affected, but one conidium (shown in red) contains a mutation. In (b) and (c), the precise nature of the mutation is established and found to involve the biosynthesis of tyrosine.

to grow, indicating that a nutritional mutation had been induced. To identify the mutant type, the mutant strains were then tested on a series of different minimal media [Figure 13–11(b)], each containing groups of supplements, and subsequently on media containing single vitamins, purines, pyrimidines, or amino acids [Figure 13–11(c)] until one specific supplement that permitted growth was found. Beadle and Tatum reasoned that *the supplement that restored growth would be the molecule that the mutant strain could not synthesize.*

The first mutant strain they isolated required vitamin B_6 (pyridoxine) in the medium, and the second required vitamin B_1 (thiamine). Using the same procedure, Beadle and Tatum eventually isolated and studied hundreds of mutants deficient in the ability to synthesize other vitamins, amino acids, or other substances.

The findings derived from testing over 80,000 spores convinced Beadle and Tatum that genetics and biochemistry have much in common. It seemed likely that each nutritional mutation caused the loss of the enzymatic activity that facilitated an essential reaction in wild-type organisms. It also appeared that a mutation could be found for nearly any enzymatically controlled reaction. Beadle and Tatum had thus provided sound experimental evidence for the hypothesis that *one gene specifies one enzyme,* an idea alluded to over 30 years earlier by Garrod and Bateson. With modifications, this concept was to become another major principle of genetics.

ESSENTIAL POINT

Beadle and Tatum's work with nutritional mutations in *Neurospora* led them to propose that one gene encodes one enzyme. ■

13.7 Studies of Human Hemoglobin Established That One Gene Encodes One Polypeptide

The one-gene:one-enzyme hypothesis that was developed in the early 1940s was not immediately accepted by all geneticists. This is not surprising because it was not yet clear how mutant enzymes could cause variation in many phenotypic traits. For example, *Drosophila* mutants demonstrate altered eye size, wing shape, wing-vein pattern, and so on. Plants exhibit mutant varieties of seed texture, height, and fruit size. How an inactive mutant enzyme could result in such phenotypes puzzled many geneticists.

Two factors soon modified the one-gene:one-enzyme hypothesis. First, although *nearly all enzymes are proteins, not all proteins are enzymes.* As the study of biochemical genetics progressed, it became clear that all proteins are specified by the information stored in genes, leading to the more accurate phraseology, **one-gene:one-protein hypothesis.** Second, proteins often show a substructure consisting of two or more polypeptide chains. This is the basis of the quaternary protein structure, which we will discuss later in this chapter.

Because each distinct polypeptide chain is encoded by a separate gene, a more accurate statement of Beadle and Tatum's basic tenet is **one-gene:one-polypeptide chain hypothesis.** These modifications of the original hypothesis became apparent during the analysis of hemoglobin structure in individuals with sickle-cell anemia.

Sickle-Cell Anemia

The first direct evidence that genes specify proteins other than enzymes came from work on mutant hemoglobin molecules found in humans with the disorder **sickle-cell anemia.** Affected individuals have erythrocytes that, under low oxygen tension, become elongated and curved because of the polymerization of hemoglobin. The sickle shape of these erythrocytes is in contrast to the biconcave disc shape characteristic in unaffected individuals (Figure 13.12). Those with the disease suffer attacks when red blood cells aggregate in the venous side of capillary systems, where oxygen tension is very low. As a result, a variety of tissues are deprived of oxygen and suffer severe damage. When this occurs, an individual is said to experience a *sickle-cell crisis.* If left untreated, a crisis can be fatal. The kidneys, muscles, joints, brain, gastrointestinal tract, and lungs can be affected.

In addition to undergoing crises, these individuals are anemic because their erythrocytes are destroyed more rapidly than are normal red blood cells. Compensatory physiological mechanisms include increased red blood cell production by bone marrow, along with accentuated heart

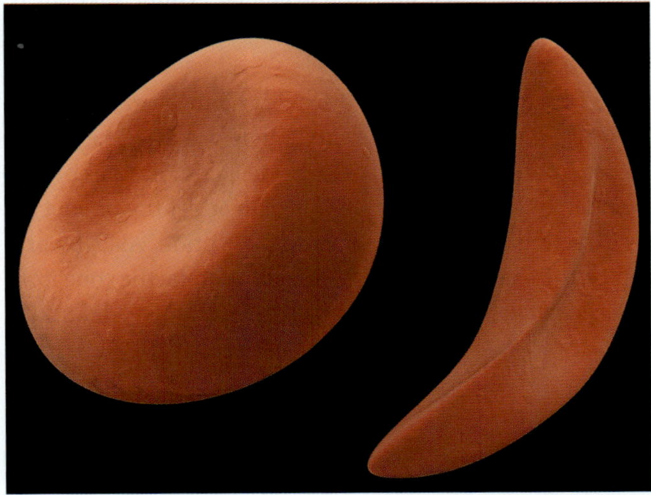

FIGURE 13.12 A comparison of an erythrocyte from a healthy individual (left) and from an individual afflicted with sickle-cell anemia (right).

action. These mechanisms lead to abnormal bone size and shape, as well as dilation of the heart.

In 1949, James Neel and E. A. Beet demonstrated that the disease is inherited as a Mendelian trait. Pedigree analysis revealed three genotypes and phenotypes controlled by a single pair of alleles, Hb^A and Hb^S. Unaffected and affected individuals result from the homozygous genotypes Hb^AHb^A and Hb^SHb^S, respectively. The red blood cells of heterozygotes, who exhibit the **sickle-cell trait** but not the disease, undergo much less sickling because over half of their hemoglobin is normal. Although they are largely unaffected, heterozygotes are "carriers" of the defective gene, which is transmitted on average to 50 percent of their offspring.

In the same year, Linus Pauling and his coworkers provided the first insight into the molecular basis of the disease. They showed that hemoglobins isolated from people with and without sickle-cell anemia differ in their rates of electrophoretic migration. In this technique, charged molecules migrate in an electric field. If the net charge of two molecules is different, their rates of migration will be different. On this basis, Pauling and his colleagues concluded that a chemical difference exists between normal **(HbA)** and sickle-cell **(HbS)** hemoglobin.

Pauling's findings suggested two possibilities. It was known that hemoglobin consists of four nonproteinaceous, iron-containing *heme groups* and a *globin portion* that contains four polypeptide chains. The alteration in net charge in HbS had to be due, theoretically, to a chemical change in one of these components. Work carried out between 1954 and 1957 by Vernon Ingram demonstrated that the chemical change occurs in the primary structure of the globin portion of the hemoglobin molecule. Ingram showed that HbS differs in amino acid composition compared to HbA. Human adult hemoglobin contains two identical α chains of 141 amino acids and two identical β chains of 146 amino acids. Analysis revealed just a single amino acid change: Valine was substituted for glutamic acid at the sixth position of the β chain (**Figure 13.13**).

The significance of this discovery has been multifaceted. It clearly establishes that a single gene provides the genetic information for a single polypeptide chain. Studies of HbS also demonstrate that a mutation can affect the phenotype by directing a single amino acid substitution. Also, by providing the explanation for sickle-cell anemia, the concept of *inherited molecular disease* was firmly established. Finally, this work has led to a thorough study of human hemoglobins, which has provided valuable genetic insights.

ESSENTIAL POINT

Pauling and Ingram's investigations of hemoglobin from patient with sickle-cell anemia led to the modification of the one-gene:one-enzyme hypothesis to indicate that one gene encodes one polypeptide chain. ■

EVOLVING CONCEPT OF THE GENE

In the 1940s, a time when the molecular nature of the gene had yet to be defined, groundbreaking work of Beadle and Tatum provided the first experimental evidence concerning the product of genes, their "one-gene:one-enzyme" hypothesis. This idea received further support and was later modified to indicate that one gene specifies one polypeptide chain. ■

13.8 Variation in Protein Structure Is the Basis of Biological Diversity

Having established that the genetic information is stored in DNA and influences cellular activities through the proteins it encodes, we turn now to a brief discussion of protein structure. How can these molecules play such a critical role in determining the complexity of cellular activities? As we shall see, the fundamental aspects of the structure of proteins provide the basis for incredible complexity and diversity. At the outset, we should differentiate between **polypeptides** and **proteins.** Both are molecules composed of amino acids. They differ, however, in their state of assembly and functional capacity. Polypeptides are the precursors of proteins. As it is assembled on the ribosome during translation, the molecule is called a *polypeptide.* When released from the ribosome following translation, a polypeptide folds and assumes a higher order of structure. When this occurs, a three-dimensional conformation emerges. In many cases, several polypeptides interact to produce this conformation. When the final conformation is achieved, the molecule is now fully functional and is appropriately called a *protein.* Its three-dimensional conformation is essential to the function of the molecule.

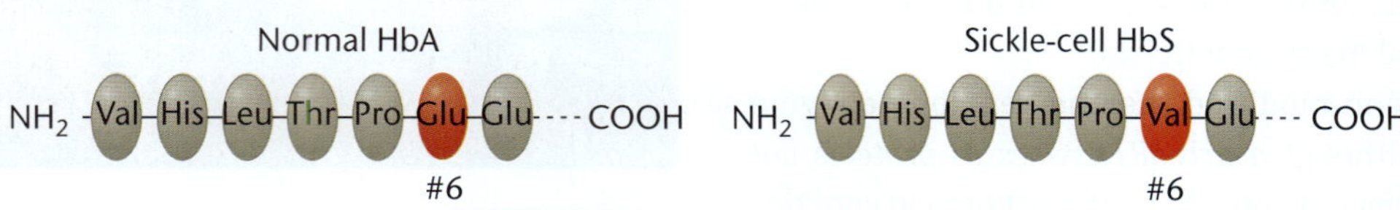

FIGURE 13.13 A comparison of the amino acid sequence of the β chain found in HbA and HbS.

The polypeptide chains of proteins, like nucleic acids, are linear nonbranched polymers. There are 20 commonly occurring amino acids that serve as the subunits (the building blocks) of proteins. Each amino acid has a **carboxyl group,** an **amino group,** and an **R (radical) group** (a side chain) bound covalently to a **central carbon (C) atom.** The R group gives each amino acid its chemical identity exhibiting a variety of configurations that can be divided into four main classes: *nonpolar* (hydrophobic), *polar* (hydrophilic), *positively charged,* and *negatively charged.* **Figure 13.14** shows the chemical structure of all 20 amino acids and the class to which each belongs. Because polypeptides are often long polymers and because each position may be occupied by any 1 of the 20 amino acids with their unique chemical properties, enormous variation in chemical conformation and activity is possible. For example, if an average polypeptide is composed of 200 amino acids, 20^{200} different molecules, each with a unique sequence, can be created using the 20 different building blocks.

Around 1900, German chemist Emil Fischer determined the manner in which the amino acids are bonded together. He showed that the amino group of one amino acid reacts with the carboxyl group of another amino acid during a dehydration reaction, releasing a molecule of H_2O. The resulting covalent bond is a **peptide bond** [**Figure 13.15(a)**]. Two amino acids linked together constitute a **dipeptide,** three a **tripeptide,** and so on. Once 10 or more amino acids are linked by peptide bonds, the chain is referred to as a polypeptide. Generally, no matter how long a polypeptide is, it will contain a free amino group at one end (the N-terminus) and a free carboxyl group at the other end (the C-terminus).

Four levels of protein structure are recognized: primary, secondary, tertiary, and quaternary. The sequence of amino acids in the linear backbone of the polypeptide constitutes its **primary structure.** It is specified by the sequence in DNA via an mRNA intermediate. The primary structure of a polypeptide helps determine the specific characteristics of the higher orders of organization as a protein is formed.

Secondary structures are certain regular or repeating configurations in space assumed by amino acids lying close to one another in the polypeptide chain. In 1951, Linus Pauling, Herman Branson, and Robert Corey predicted, on theoretical grounds, an **α-helix** as one type of secondary structure. The α-helix model [**Figure 13.15(b)**] has since been confirmed by X-ray crystallographic studies. The helix is composed of a right-handed spiral chain of amino acids

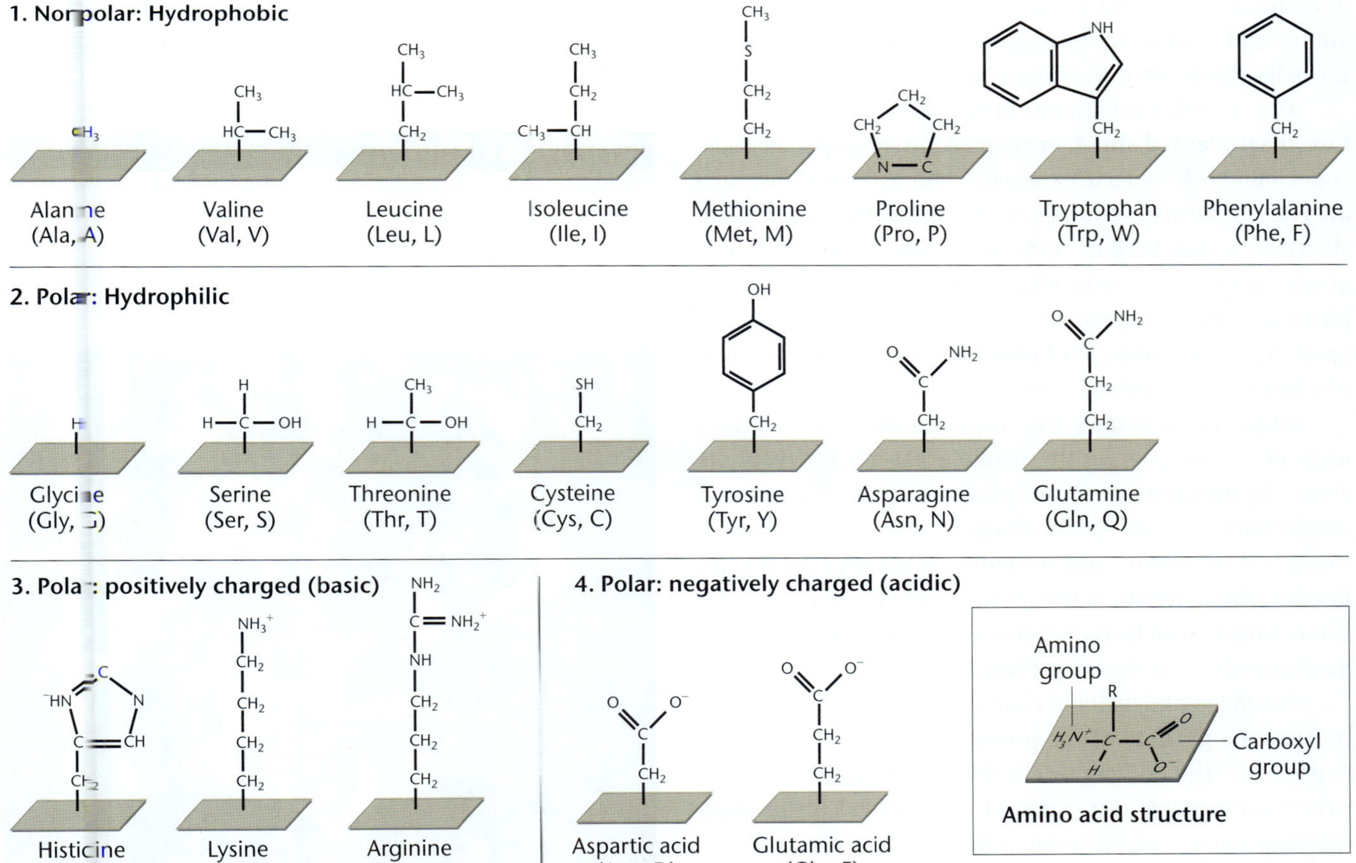

FIGURE 13.14 Chemical structures of the 20 amino acids encoded by organisms, divided into four major classes. Each amino acid has two abbreviations in universal use; for example alanine is designated by Ala or A.

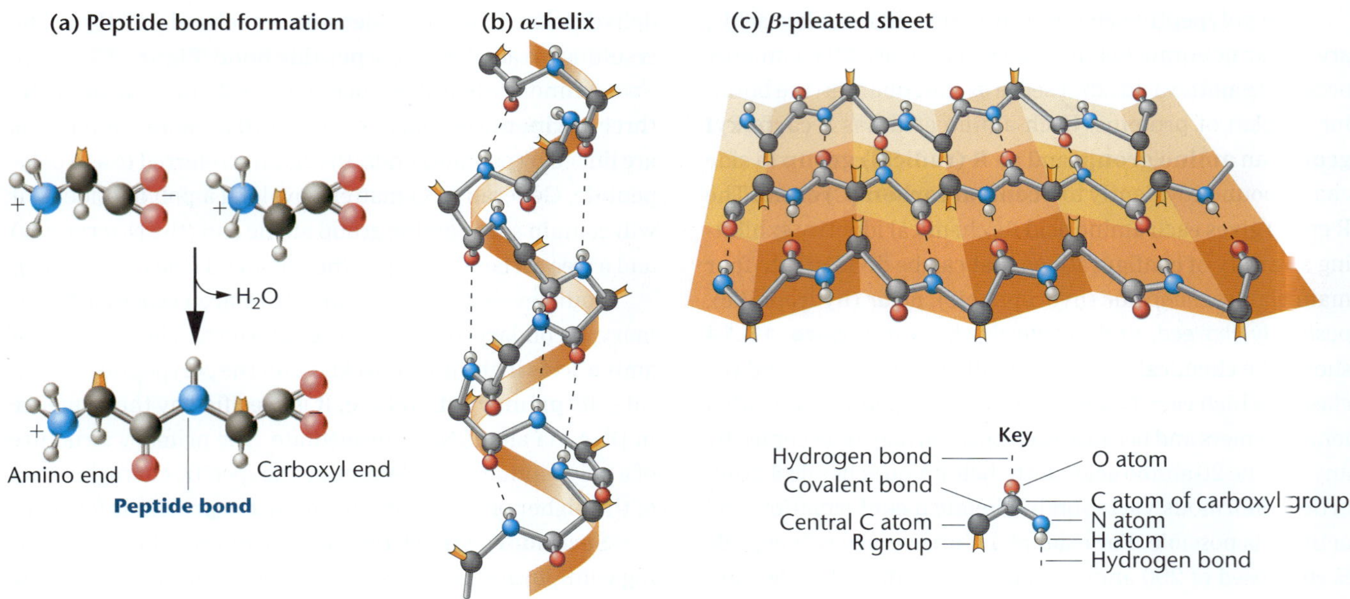

FIGURE 13.15 (a) Peptide bond formation between two amino acids, resulting from a dehydration reaction. (b) The righthanded α-helix, which represents one form of secondary structure of a polypeptide chain. (c) The β-pleated sheet, an alternative form of secondary structure of polypeptide chains. To maintain clarity, not all atoms are shown.

stabilized by hydrogen bonds. The side chains (the R groups) of amino acids extend outward from the helix, and each amino acid residue occupies a vertical distance of 1.5 Å in the helix. There are 3.6 residues per turn.

Also in 1951, Pauling and Corey proposed a second structure, the **β-pleated sheet.** In this model, a single-polypeptide chain folds back on itself, or several chains run in either parallel or antiparallel fashion next to one another. Each such structure is stabilized by hydrogen bonds formed between atoms on adjacent chains [**Figure 13.15(c)**]. A zigzagging plane is formed in space with adjacent amino acids 3.5 Å apart. As a general rule, most proteins demonstrate a mixture of α-helix and β-pleated-sheet structures.

While the secondary structure describes the arrangement of amino acids within certain areas of a polypeptide chain, the **tertiary structure** defines the three-dimensional conformation of the entire chain in space. Each protein twists and turns and loops around itself in a very particular fashion, characteristic of the specific protein. A model of the three-dimensional tertiary structure of the respiratory pigment myoglobin is shown in **Figure 13.16**.

The three-dimensional conformation achieved by any protein is a product of the *primary structure* of the polypeptide. As the polypeptide is folded, the most thermodynamically stable conformation is created. This level of organization is essential because the specific function of any protein is directly related to its tertiary structure.

The concept of **quaternary structure** applies to those proteins composed of more than one polypeptide chain and indicates the position of the various chains in relation to one another. Hemoglobin, a protein consisting of four polypeptide chains, has been studied in great detail. Many enzymes, including DNA and RNA polymerase, demonstrate quaternary structure.

ESSENTIAL POINT

Proteins, the end products of genes, demonstrate four levels of structural organization that together describe their three-dimensional conformation, which is the basis of each molecule's function. ■

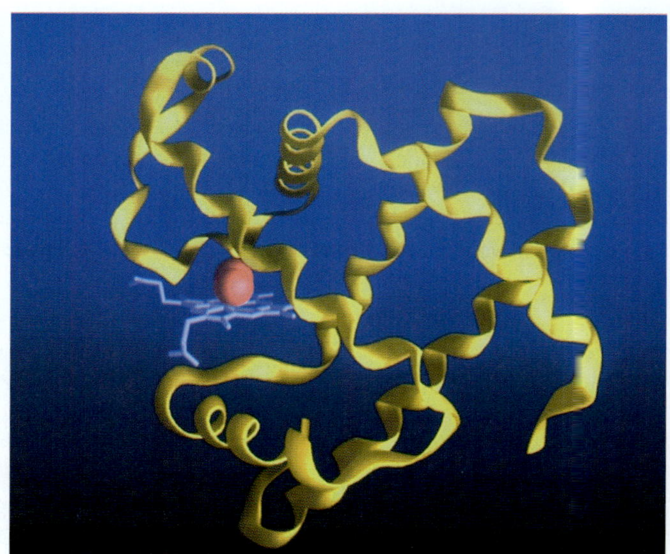

FIGURE 13.16 The tertiary level of protein structure in a respiratory pigment, myoglobin. The bound oxygen atom is shown in red.

NOW SOLVE THIS

13.2 HbS results from the substitution of valine for glutamic acid at the number 6 position in the β chain of human hemoglobin. HbC is the result of a change at the same position in the β chain, but in this case lysine replaces glutamic acid. Return to the genetic code table (Figure 12.7) and determine whether single-nucleotide changes can account for these mutations. Then view Figure 13.14 and examine the R groups in the amino acids glutamic acid, valine, and lysine. Describe the chemical differences between the three amino acids. Predict how the changes might alter the structure of the molecule and lead to altered hemoglobin function.

■ **HINT:** *This problem asks you to consider the potential impact of several amino acid substitutions that result from mutations in one of the genes encoding one of the chains making up human hemoglobin. The key to its solution is to consider and compare the structure of the three amino acids (glutamic acid, lysine, and valine) and their net charge (see Figure 13.14).*

Protein Folding and Misfolding

It was long thought that **protein folding** was a spontaneous process whereby a linear molecule exiting the ribosome achieved a three-dimensional, thermodynamically stable conformation based solely on the combined chemical properties inherent in the amino acid sequence. This indeed is the case for many proteins. However, numerous studies have shown that for other proteins, correct folding is dependent on members of a family of molecules called **chaperones.** Chaperones are themselves proteins (sometimes called *molecular chaperones* or *chaperonins*) that function by mediating the folding process by excluding the formation of alternative, incorrect patterns. While they may initially interact with the protein in question, like enzymes, they do not become part of the final product. Chaperones have been identified in all organisms and are even present in mitochondria and chloroplasts.

In eukaryotic cells, chaperones are particularly important when translation occurs on membrane-bound ribosomes, where the newly translated polypeptide is extruded into the lumen of the endoplasmic reticulum. Even in their presence, misfolding may still occur, and one more system of "quality control" exists. As misfolded proteins are transported out of the endoplasmic reticulum to the cytoplasm, they are "tagged" by another class of small proteins called **ubiquitins.** The protein–ubiquitin complex moves to a cellular structure called the **proteasome,** which cleaves off ubiquitin and degrades the protein (see Chapter 16 for additional information on this process).

Protein folding is a critically important process, not only because misfolded proteins may be nonfunctional, but also because improperly folded proteins can accumulate and be detrimental to cells and the organisms that contain them. For example, a group of transmissible brain disorders in mammals—**scrapie** in sheep, **bovine spongiform encephalopathy** (**mad cow disease**) in cattle, and **Creutzfeldt–Jakob disease** in humans—are caused by the presence in the brain of **prions,** which are aggregates of a misfolded protein. The misfolded protein (called PrP^{Sc}) is an altered version of a normal cellular protein (called PrP^{C}) synthesized in neurons and found in the brains of all adult mammals. The difference between PrP^{C} and PrP^{Sc} lies in their secondary protein structures. Normal, noninfectious PrP^{c} folds into an α-helix, whereas infectious PrP^{Sc} folds into a β-pleated sheet. When an abnormal PrP^{Sc} molecule contacts a PrP^{C} molecule, the normal protein refolds into the abnormal conformation. The process continues as a chain reaction, with potentially devastating results—the formation of prion particles that eventually destroy the brain. Hence, this group of disorders can be considered diseases of secondary protein structure.

Currently, many laboratories are studying protein folding and misfolding, particularly as related to genetics. Numerous inherited human disorders are caused by misfolded proteins that form aggregates. Sickle-cell anemia, discussed earlier in this chapter, is a case in point, where the β chains of hemoglobin are altered as the result of a single amino acid change, causing the molecules to aggregate within erythrocytes, with devastating results. Various progressive neurodegenerative diseases such as **Huntington disease, Alzheimer disease,** and **Parkinson disease** are linked to the formation of abnormal protein aggregates in the brain. Huntington disease is inherited as an autosomal dominant trait, whereas less clearly defined genetic components are associated with Alzheimer and Parkinson diseases.

13.9 Proteins Function in Many Diverse Roles

The essence of life on Earth rests at the level of diverse cellular function. While DNA and mRNA serve as vehicles to store and express genetic information, proteins are at the heart of cellular function. And it is the capability of cells to assume diverse structures and functions that distinguishes most eukaryotes from simpler organisms such as bacteria. Therefore, an introductory understanding of protein function is critical to a complete view of genetic processes.

Proteins are the most diverse macromolecules found in cells and serve many different functions. For example, the respiratory pigments **hemoglobin** and **myoglobin,** transport oxygen, which is essential for cellular metabolism.

Collagen and **keratin** are structural proteins associated with skin, connective tissue, and hair. **Actin** and **myosin** are contractile proteins, found in abundance in muscle tissue, while **tubulin** is the basis of the function of microtubules in mitotic and meiotic spindles. Still other examples are the **immunoglobulins,** which function in the immune system of vertebrates; **transport proteins,** involved in the movement of molecules across membranes; some of the hormones and their receptors, which regulate various types of chemical activity; **histones,** which bind to DNA in eukaryotic organisms; and **transcription factors** that regulate gene expression.

Nevertheless, the most diverse and extensive group of proteins (in terms of function) are the enzymes, to which we have referred throughout this chapter. Enzymes specialize in catalyzing chemical reactions within living cells. Like all catalysts, they increase the rate at which a chemical reaction reaches equilibrium, but they do not alter the end-point of the chemical equilibrium. Their remarkable, highly specific catalytic properties largely determine the metabolic capacity of any cell type and provide the underlying basis of what we refer to as biochemistry. The specific functions of many enzymes involved in the genetic and cellular processes are described throughout the text.

ESSENTIAL POINT

Of the myriad functions performed by proteins, the most influential role belongs to enzymes, which serve as highly specific biological catalysts that play a central role in the production of all classes of molecules in living systems. ■

Protein Domains Impart Function

We conclude this chapter by briefly discussing the important finding that regions made up of specific amino acid sequences are associated with specific functions in protein molecules. Such sequences, usually between 50 and 300 amino acids, constitute **protein domains** and represent modular portions of the protein that fold into stable, unique conformations independently of the rest of the molecule. Different domains impart different functional capabilities. Some proteins contain only a single domain, while others contain two or more.

The significance of domains resides in the tertiary structures of proteins. Each domain can contain a mixture of secondary structures, including α-helices and β-pleated sheets. The unique conformation of a given domain imparts a specific function to the protein. For example, a domain may serve as the catalytic site of an enzyme, or it may impart an ability to bind to a specific ligand. Thus, discussions of proteins may mention *catalytic domains, DNA-binding domains,* and so on. In short, a protein must be seen as being composed of a series of structural and functional modules. Obviously, the presence of multiple domains in a single protein increases the versatility of each molecule and adds to its functional complexity.

CASE STUDY Crippled ribosomes

Diamond–Blackfan anemia (DBA) is a rare, dominant genetic disorder characterized by bone marrow malfunction, birth defects, and a predisposition to certain cancers. Infants with DBA usually develop anemia in the first year of life, have lower than normal production of red blood cells in their bone marrow, and have a high risk of developing leukemia and bone cancer. At the molecular level, DBA is caused by mutations in any one of 10 genes that encode ribosomal proteins. The first-line therapy for DBA is steroid treatment, but more than half of affected children develop resistance to the drugs and in these cases, treatment is halted. DBA can be treated successfully with bone marrow or stem cell transplants from donors with closely matching immune system markers. Transplants from unrelated donors have significant levels of complications and mortality.

1. Given that a faulty ribosomal protein is the culprit and causes DBA, discuss the possible role of normal ribosomal proteins. Why might bone marrow cells be more susceptible to such a mutation than other cells?
2. A couple with a child affected with DBA undergoes *in vitro* fertilization (IVF) and genetic testing of the resulting embryos to ensure that the embryos will not have DBA. However, they also want the embryos screened to ensure that the one implanted can serve as a suitable donor for their existing child. Their plan is to have stem cells from the umbilical cord of the new baby transplanted to their existing child with DBA, thereby curing the condition. What are the ethical pros and cons of this situation?
3. While a stem cell transplant from an unaffected donor is currently the only cure for DBA, genome-editing technologies may one day enable the correction of a mutation in a patient's own bone marrow stem cells. However, what specific information would be needed, beyond a symptom-based diagnosis of DBA, in order to accomplish this?

For related reading, see Penning, G., et al. (2002). Ethical considerations on preimplantation genetic diagnosis for HLA typing to match a future child as a donor of haematopoietic stem cells to a sibling. *Hum. Reprod.* 17(3):534–538.

INSIGHTS AND SOLUTIONS

1. As an extension of Beadle and Tatum's work with *Neurospora,* it is possible to study multiple mutations whose impact is on the same biochemical pathway. The growth responses in the following chart were obtained using four mutant strains of *Neurospora* and the chemically related compounds A, B, C, and D. None of the mutants grow on minimal medium. Draw all possible conclusions from these data.

	Growth Supplement			
Mutation	A	B	C	D
1	–	–	–	–
2	+	+	–	+
3	+	+	–	–
4	–	+	–	–

Solution: Nothing can be concluded about mutation *1* except that it lacks some essential growth factor, perhaps even unrelated to the biochemical pathway represented by mutations *2, 3,* and *4*. Nor can anything be concluded about compound C. If it is involved in the pathway, it is a product that was synthesized prior to compounds A, B, and D.

We now analyze these three compounds and the control of their synthesis by the enzymes encoded by mutations *2, 3,* and *4*. Because product B allows growth in all three cases, it may be considered the "end product"—it bypasses the block in all three instances. Using similar reasoning, product A precedes B in the pathway because it bypasses the block in two of the three steps, and product D precedes B yielding a partial solution

$$C(?) \longrightarrow D \longrightarrow A \longrightarrow B$$

Now let's determine which mutations control which steps. Since mutation *2* can be alleviated by products D, B, and A, it must control a step prior to all three products, perhaps the direct conversion to D (although we cannot be certain). Mutation *3* is alleviated by B and A, so its effect must precede them in the pathway. Thus, we assign it as controlling the conversion of D to A. Likewise, we can assign mutation *4* to the conversion of A to B, leading to a more complete solution

$$C(?) \xrightarrow{2(?)} D \xrightarrow{3} A \xrightarrow{4} B$$

Problems and Discussion Questions

Mastering Genetics Visit for instructor-assigned tutorials and problems.

1. **HOW DO WE KNOW?** In this chapter, we focused on the translation of mRNA into proteins as well as on protein structure and function. Along the way, we found many opportunities to consider the methods and reasoning by which much of this information was acquired. From the explanations given in the chapter, what answers would you propose to the following fundamental questions:
   (a) What experimentally derived information led to Holley's proposal of the two-dimensional cloverleaf model of tRNA?
   (b) What experimental information verifies that certain codons in mRNA specify chain termination during translation?
   (c) How do we know, based on studies of *Neurospora* nutritional mutations, that one gene specifies one enzyme?
   (d) On what basis have we concluded that proteins are the end products of genetic expression?
2. **CHAPTER CONCEPTS** Review the Chapter Concepts list on p. 26[illegible]. These all relate to the translation of genetic information stored in mRNA into proteins and how chemical information in proteins imparts function to those molecules. Write a brief essay that discusses the role of ribosomes in the process of translation as it relates to these concepts. ■
3. List the structural similarities and differences between a eukaryotic and a prokaryotic ribosome.
4. Contrast the roles of tRNA and mRNA during translation, and list all enzymes that participate in the translation processes.
5. tRNA adapts specific triplet codons in mRNA to their correct amino acids. Do you agree with this statement? Justify your answer.
6. During translation, what molecule bears the anticodon? The codon?
7. Summarize the steps involved in elongation during translation in bacteria.
8. Based on the cloverleaf model and the three-dimensional structure of tRNA, mention the different regions present in a tRNA molecule.
9. What are the major differences between translation in bacteria and translation in eukaryotes?
10. Consider the structure of a ribosome and discuss the statement "one-gene:one-RNA".
11. Hemoglobin is a tetramer consisting of two α and two β chains. What level of protein structure is described in this statement?
12. Using sickle-cell anemia as a basis, describe what is meant by a genetic or inherited molecular disease. What are the similarities and dissimilarities between this type of a disorder and a disease caused by an invading microorganism?
13. Describe Pauling's experiment that set the molecular basis of sickle-cell anemia.
14. Assuming that each nucleotide is 0.34 nm long in mRNA, how many triplet codes can simultaneously occupy space in a ribosome that is 20 nm in diameter?
15. Review the concept of colinearity in Section 12.5 (p. 250) and consider the following question: Certain mutations called *amber* in bacteria and viruses result in premature termination of polypeptide chains during translation. Many *amber* mutations have been detected at different points along the gene that codes for a head protein in phage T4. How might this system be

further investigated to demonstrate and support the concept of colinearity?

16. In your opinion, which of the four levels of protein organization is the most critical to a protein's function? Defend your choice.
17. List and describe the function of as many nonenzymatic proteins as you can that are unique to eukaryotes.
18. Define protein domains and their relevance to protein structure and function.
19. Shown in the following table are several amino acid substitutions in the α and β chains of human hemoglobin. Use the genetic code table in Figure 12.7 to determine how many of them can occur as a result of a single nucleotide change.

Hb Type	Normal Amino Acid	Substituted Amino Acid
HbJ Toronto	Ala	Asp (α-5)
HbJ Oxford	Gly	Asp (α-15)
Hb Mexico	Gln	Glu (α-54)
Hb Bethesda	Tyr	His (β-145)
Hb Sydney	Val	Ala (β-67)
HbM Saskatoon	His	Tyr (β-63)

20. Three independently assorting genes are known to control the biochemical pathway below that provides the basis for flower color in a hypothetical plant

$$\text{colorless} \xrightarrow{A-} \text{yellow} \xrightarrow{B-} \text{green} \xrightarrow{C-} \text{speckled}$$

Homozygous recessive mutations, which disrupt enzyme function controlling each step, are known. Determine the phenotypic results in the F_1 and F_2 generations resulting from the P_1 crosses involving true-breeding plants given here.
(a) speckled (*AABBCC*) × yellow (*AAbbCC*)
(b) yellow (*AAbbCC*) × green (*AABBcc*)
(c) colorless (*aaBBCC*) × green (*AABBcc*)

21. How would the results in cross (a) of Problem 20 vary if genes *A* and *B* were linked with no crossing over between them? How would the results of cross (a) vary if genes *A* and *B* were linked and 20 map units apart?
22. A series of mutations in the bacterium *Salmonella typhimurium* results in the requirement of either tryptophan or some related molecule in order for growth to occur. From the data shown here, suggest a biosynthetic pathway for tryptophan.

Mutation	Growth Supplement: Minimal Medium	Anthranilic Acid	Indole Glycerol Phosphate	Indole	Tryptophan
trp-8	−	+	+	+	+
trp-2	−	−	+	+	+
trp-3	−	−	−	+	+
trp-1	−	−	−	−	+

23. Name the elements within eukaryotic mRNAs that are important for efficient translation. Describe their function.

14

Gene Mutation, DNA Repair, and Transposition

Pigment mutations within an ear of corn, caused by transposition of the *Ds* element.

CHAPTER CONCEPTS

- Mutations comprise any change in the nucleotide sequence of an organism's genome.
- Mutations are a source of genetic variation and provide the raw material for natural selection. They are also the source of genetic damage that contributes to cell death, genetic diseases, and cancer.
- Mutations have a wide range of effects on organisms depending on the type of base-pair alteration, the location of the mutation within the chromosome, and the function of the affected gene product.
- Mutations can occur spontaneously as a result of natural biological and chemical processes, or they can be induced by external factors, such as chemicals or radiation.
- Single-gene mutations cause a wide variety of human diseases.
- Organisms rely on a number of DNA repair mechanisms to counteract mutations. These mechanisms range from proofreading and correction of replication errors to base excision and homologous recombination repair.
- Mutations in genes whose products control DNA repair lead to genome hypermutability, human DNA repair diseases, and cancers.
- Transposable elements may move into and out of chromosomes, causing chromosome breaks and inducing mutations both within coding regions and in gene-regulatory regions.

The ability of DNA molecules to store, replicate, transmit, and decode information is the basis of genetic function. But equally important are the changes that occur to DNA sequences. Without the variation that arises from changes in DNA sequences, there would be no phenotypic variability, no adaptation to environmental changes, and no evolution. Gene mutations are the source of new alleles and are the origin of genetic variation within populations. On the downside, they are also the source of genetic changes that can lead to cell death, genetic diseases, and cancer.

Mutations also provide the basis for genetic analysis. The phenotypic variations resulting from mutations allow geneticists to identify and study the genes responsible for the modified trait. In genetic investigations, mutations act as identifying "markers" for genes so that they can be followed during their transmission from parents to offspring. Without phenotypic variability, classical genetic analysis would be impossible. For example, if all pea plants displayed a uniform phenotype, Mendel would have had no foundation for his research.

We have examined mutations in large regions of chromosomes—chromosomal mutations (see Chapter 6). In contrast, the mutations we will now explore are those occurring primarily in the base-pair sequence of DNA within and surrounding individual genes—**gene mutations.** We will also describe how the cell defends itself from mutations using various mechanisms of DNA repair.

14.1 Gene Mutations Are Classified in Various Ways

A mutation can be defined as an alteration in the nucleotide sequence of an organism's genome. Any base-pair change in any part of a DNA molecule can be considered a mutation. A mutation may comprise a single base-pair substitution, a deletion or insertion of one or more base pairs, or a major alteration in the structure of a chromosome.

Mutations may occur within regions of a gene that code for protein or within noncoding regions of a gene such as introns and regulatory sequences, including promoters, enhancers, and splicing signals. Mutations may or may not bring about a detectable change in phenotype. The extent to which a mutation changes the characteristics of an organism depends on which type of cell suffers the mutation and the degree to which the mutation alters the function of a gene product or a gene-regulatory region.

Because of the wide range of types and effects of mutations, geneticists classify mutations according to several different schemes. These organizational schemes are not mutually exclusive. In this section, we outline some of the ways in which gene mutations are classified.

Classification Based on Type of Molecular Change

Geneticists often classify gene mutations in terms of the nucleotide changes that constitute the mutation. A change of one base pair to another in a DNA molecule is known as a **point mutation,** or **base substitution** (see **Figure 14.1**). A change of one nucleotide of a triplet within a protein-coding portion of a gene may result in the creation of a new triplet that codes for a different amino acid in the protein product. If this occurs, the mutation is known as a **missense mutation.** A second possible outcome is that the triplet will be changed into a stop codon, resulting in the termination of translation of the protein. This is known as a **nonsense mutation.** If the point mutation alters a codon but does not result in a change in the amino acid at that position in the protein (due to degeneracy of the genetic code), it can be considered a **silent mutation.**

Because eukaryotic genomes consist of so much more noncoding DNA than coding DNA (see Chapter 11), the vast majority of mutations are likely to occur in noncoding regions. These mutations may be considered **neutral mutations** if they do not affect gene products or gene expression. Most silent mutations, which do not change the amino acid sequence of the encoded protein, can also be considered neutral mutations.

You will often see two other terms used to describe base substitutions. If a pyrimidine replaces a pyrimidine or a purine replaces a purine, a **transition** has occurred. If a purine replaces a pyrimidine, or vice versa, a **transversion** has occurred.

Another type of change is the insertion or deletion of one or more nucleotides at any point within the gene. As illustrated in Figure 14.1, the loss or addition of a single nucleotide causes all of the subsequent three-letter codons to be changed. These are called **frameshift mutations** because the frame of triplet reading during translation is altered. A frameshift mutation will occur when any number of bases are added or deleted, except multiples of three, which would reestablish the initial frame of reading (see Figure 12.2). It is possible that one of the many altered triplets will be UAA, UAG, or UGA, the translation termination codons. When one of these triplets is encountered during translation, polypeptide synthesis is terminated at that point. Obviously, the results of frameshift mutations can be very severe, such as producing a truncated protein or defective enzymes, especially if they occur early in the coding sequence.

Classification Based on Effect on Function

As discussed earlier (see Chapter 4), a **loss-of-function mutation** is one that reduces or eliminates the function of the gene product. Mutations that result in complete loss of

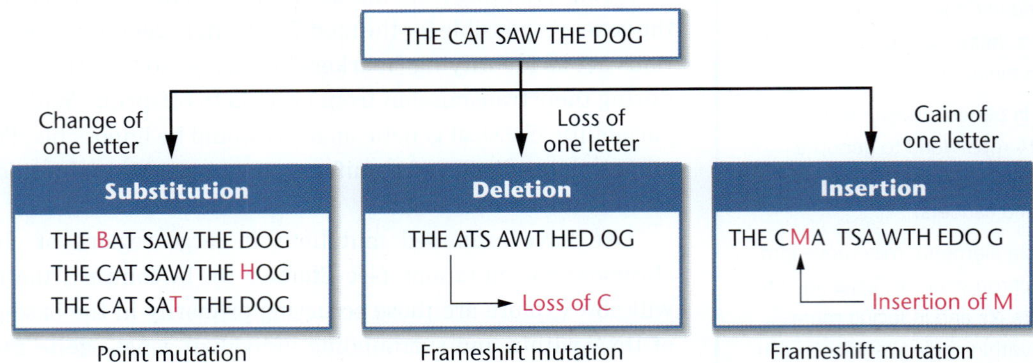

FIGURE 14.1 Analogy showing the effects of substitution, deletion, and insertion of one letter in a sentence composed of three-letter words to demonstrate point and frameshift mutations.

function are known as **null mutations.** Any type of mutation, from a point mutation to deletion of the entire gene, may lead to a loss of function.

Most loss-of-function mutations are recessive. A **recessive mutation** results in a wild-type phenotype when present in a diploid organism and the other allele is wild type. In this case, the presence of less than 100 percent of the gene product is sufficient to bring about the wild-type phenotype.

Some loss-of-function mutations can be dominant. A **dominant mutation** results in a mutant phenotype in a diploid organism, even when the wild-type allele is also present. Dominant mutations in diploid organisms can have several different types of effects. A **dominant negative mutation** in one allele may encode a gene product that is inactive and directly interferes with the function of the product of the wild-type allele. For example, this can occur when the nonfunctional gene product binds to the wild-type gene product in a homodimer, inactivating or reducing the activity of the homodimer.

A dominant negative mutation can also result from **haploinsufficiency,** which occurs when one allele is inactivated by mutation, leaving the individual with only one functional copy of a gene. The active allele may be a wild-type copy of the gene but does not produce enough wild-type gene product to bring about a wild-type phenotype. In humans, Marfan syndrome is an example of a disorder caused by haploinsufficiency—in this case as a result of a loss-of-function mutation in one copy of the *fibrillin-1* (*FBN1*) gene.

In contrast, a **gain-of-function mutation** codes for a gene product with enhanced, negative, or new functions. This may be due to a change in the amino acid sequence of the protein that confers a new activity, or it may result from a mutation in a regulatory region of the gene, leading to expression of the gene at higher levels or at abnormal times or places. Typically, gain-of-function mutations are dominant.

A **suppressor mutation** is a second mutation that either reverts or relieves the effects of a previous mutation. A suppressor mutation can occur within the same gene that suffered the first mutation (**intragenic mutation**) or elsewhere in the genome (**intergenic mutation**).

Depending on their type and location, mutations can have a wide range of phenotypic effects, from none to severe. Some examples of mutation types based on their phenotypic outcomes are listed in **Table 14.1**.

Classification Based on Location of Mutation

Mutations may be classified according to the cell type or chromosomal locations in which they occur. **Somatic mutations** are those occurring in any cell in the body except germ cells, whereas germ-line mutations occur only in germ cells. **Autosomal mutations** are mutations within genes located on the autosomes, whereas **X-linked** and **Y-linked mutations** are those within genes located on the X or Y chromosome, respectively.

Mutations arising in somatic cells are not transmitted to future generations. When a recessive autosomal mutation occurs in a somatic cell of an adult multicellular diploid organism, it is unlikely to result in a detectable phenotype. The expression of most such mutations is likely to be masked by expression of the wild-type allele within that cell and the presence of nonmutant cells in the remainder of the organism. Somatic mutations will have a greater impact if they are dominant or, in males, if they are X-linked, since such mutations are most likely to be immediately expressed. In addition, the impact of dominant or X-linked somatic mutations will be more noticeable if they occur early in development, when a small number of undifferentiated cells replicate to give rise to several differentiated tissues or organs.

Mutations in germ cells have the potential of being expressed in all cells of an offspring. Inherited dominant autosomal mutations will be expressed phenotypically in the first generation. X-linked recessive mutations arising in the gametes of a female (the **homogametic sex;** having

TABLE 14.1 Classifications of Mutations by Phenotypic Effects

Classification	Phenotype	Example
Visible	Visible morphological trait	Mendel's pea characteristics
Nutritional	Altered nutritional characteristics	Loss of ability to synthesize an essential amino acid in bacteria
Biochemical	Changes in protein function	Defective hemoglobin leading to sickle-cell anemia in humans
Behavioral	Behavior pattern changes	Brain mutations affecting *Drosophila* mating behaviors
Regulatory	Altered gene expression	Regulatory gene mutations affecting expression of the *lac* operon in *E. coli*
Lethal	Altered organism survival	Tay-Sachs and Huntington disease in humans
Conditional	Phenotype expressed only under certain environmental conditions	Temperature-sensitive mutations affecting coat color in Siamese cats

two X chromosomes) may be expressed in male offspring, who are by definition **hemizygous** for the gene mutation because they have one X and one Y chromosome. This will occur provided that the male offspring receives the affected X chromosome. Because of heterozygosity, the occurrence of an autosomal recessive mutation in the gametes of either males or females (even one resulting in a lethal allele) may go unnoticed for many generations, until the resultant allele has become widespread in the population. Usually, the new allele will become evident only when a chance mating brings two copies of it together into the homozygous condition.

ESSENTIAL POINT

Mutations can have many different effects on gene function, depending on the type of nucleotide changes that comprise the mutation and their locations. Phenotypic effects can range from neutral or silent to loss of function, gain of function, or lethality. ■

NOW SOLVE THIS

14.1 If a point mutation occurs within a human egg cell genome that changes an A to a T, what is the most likely effect of this mutation on the phenotype of an offspring that develops from this mutated egg?

■ **HINT:** *This problem asks you to predict the effects of a single base-pair mutation on phenotype. The key to its solution involves an understanding of the organization of the human genome as well as the effects of mutations on coding and noncoding regions of genes and the effects of mutations on development.*

For more practice, see Problems 4–7.

14.2 Mutations Can Be Spontaneous or Induced

Mutations can be classified as either spontaneous or induced, although these two categories overlap to some degree. **Spontaneous mutations** are changes in the nucleotide sequence of genes that appear to occur naturally. No specific agents are associated with their occurrence. Many of these mutations arise as a result of normal biological or chemical processes in the organism that alter the structure of nitrogenous bases. Often, spontaneous mutations occur during the enzymatic process of DNA replication, as we discuss later in this chapter.

In contrast to spontaneous mutations, the mutations that result from the influence of extraneous factors are considered to be **induced mutations.** Induced mutations may be the result of either natural or artificial agents. For example, radiation from cosmic and mineral sources and ultraviolet radiation from the sun are energy sources to which most organisms are exposed and, as such, may be considered natural agents that cause induced mutations.

We will next describe several aspects of spontaneous mutations, including mutation rates.

Spontaneous Mutation Rates in Nonhuman Organisms

Several generalizations can be made regarding spontaneous mutation rates. The **mutation rate** is defined as the likelihood that a gene will undergo a mutation in a single generation or in forming a single gamete. First, the rate of spontaneous mutation is exceedingly low for all organisms. Second, the rate varies between different organisms. Third, even within the same species, the spontaneous mutation rate varies from gene to gene.

Viral and bacterial genes undergo spontaneous mutation at an average of about 1 in 100 million (10^{-8}) replications or cell divisions. Maize and *Drosophila* demonstrate rates several orders of magnitude higher. The genes studied in these groups average between 1 in 1,000,000 (10^{-6}) and 1 in 100,000 (10^{-5}) mutations per gamete formed. Some mouse genes are another order of magnitude higher in their spontaneous mutation rate:1 in 100,000 to 1 in 10,000 (10^{-5} to 10^{-4}). It is not clear why such large variations occur in mutation rates.

The variation in rates between organisms may, in part, reflect the relative efficiencies of their DNA proofreading and repair systems. We will discuss these systems later in the chapter. Variation between genes in a given organism may be due to inherent differences in mutability in different regions of the genome. Some DNA sequences appear to be highly susceptible to mutation and are known as **mutation hot spots.**

Spontaneous Mutation Rates in Humans

Now that whole-genome sequencing is becoming both rapid and economical, it is possible to examine entire genomes, both coding and noncoding regions, and to compare genomes from parents and offspring and estimate spontaneous germ-line mutation rates.

In 2012, a research group in Iceland sequenced the genomes of 78 parent/offspring sets, comprising 219 individuals, and compared the **single-nucleotide polymorphisms (SNPs)** (see Chapter 7) throughout their genomes.[1] Their data revealed that a newborn baby's genome contains an average of 60 new mutations, compared with those of his or her parents. Their research also revealed that the number of new mutations depends significantly on the age of the father at the time of conception.

[1] Kong, A., et al. (2012). Rate of de novo mutations and the importance of father's age to disease risk. *Nature* 488:471–475.

For example, when the father is 20 years old, he contributes approximately 25 new mutations to the child. When he is 40 years old, he contributes approximately 65 new mutations. In contrast, the mother contributes about 15 new mutations, at any age. The researchers estimated that the father contributes approximately 2 mutations per year of his age, with the mutation rate doubling every 16.5 years. The large proportion of mutations contributed by fathers is likely due to the fact that male germ cells go through more cell divisions during a lifetime than do female germ cells.

Of the 4933 new SNP mutations that were identified in this study, only 73 occurred within gene exons. Other studies have suggested that about 10 percent of single-nucleotide mutations lead to negative phenotypic changes. If so, then an average spontaneous mutation rate of 60 new mutations might yield about six deleterious phenotypic effects per generation.

It is estimated that somatic cell mutation rates are between 4 and 25 times higher than those in germ-line cells. It is well accepted that somatic mutations are responsible for the development of most cancers. We will discuss the effects of somatic mutations on the development of cancer in more detail later (see Chapter 19).

ESSENTIAL POINT

Spontaneous mutations can occur naturally without the action of extraneous agents. Induced mutations occur as a result of extraneous agents which can be either natural or human-made. Spontaneous mutation rates vary between organisms and between different regions of genome. ■

14.3 Spontaneous Mutations Arise from Replication Errors and Base Modifications

In this section, we will outline some of the processes that lead to spontaneous mutations. Many of the DNA changes that occur during spontaneous mutagenesis also occur, at a higher rate, during induced mutagenesis.

DNA Replication Errors and Slippage

As we learned earlier (see Chapter 10), the process of DNA replication is imperfect. Occasionally, DNA polymerases insert incorrect nucleotides during replication of a strand of DNA. Although DNA polymerases can correct most of these replication errors using their inherent 3′ to 5′ exonuclease proofreading capacity, misincorporated nucleotides may persist after replication. If these errors are not detected and corrected by DNA repair mechanisms, they may lead to mutations. Replication errors due to mispairing predominantly lead to point mutations. The fact that bases can take several forms, known as **tautomers**, increases the chance of mispairing during DNA replication, as we will explain shortly.

In addition to mispairing and point mutations, DNA replication can lead to the introduction of small insertions or deletions. These mutations can occur when one strand of the DNA template loops out and becomes displaced during replication, or when DNA polymerase slips or stutters during replication—events termed **replication slippage.** If a loop occurs in the template strand during replication, DNA polymerase may miss the looped-out nucleotides, and a small deletion in the new strand will be introduced. If DNA polymerase repeatedly introduces nucleotides that are not present in the template strand, an insertion of one or more nucleotides will occur. Insertions and deletions may lead to frameshift mutations or amino acid insertions or deletions in the gene product.

Replication slippage can occur anywhere in the DNA but seems more common in regions containing tandemly repeated sequences. Repeat sequences are hot spots for DNA mutation and in some cases contribute to hereditary diseases, such as fragile-X syndrome and Huntington disease. The hypermutability of repeat sequences in noncoding regions of the genome is the basis for several current methods of forensic DNA analysis.

Tautomeric Shifts

Purines and pyrimidines can exist in tautomeric forms—that is, in alternate chemical forms that differ by the shift of a single proton in the molecule. The biologically important tautomers are the keto–enol forms of thymine and guanine and the amino–imino forms of cytosine and adenine. **Tautomeric shifts** change the covalent structure of the molecule, allowing hydrogen bonding with noncomplementary bases, and hence, may lead to permanent base-pair changes and mutations. **Figure 14.2** compares normal base-pairing arrangements with rare unorthodox pairings. Anomalous T-G and C-A pairs, among others, may be formed.

A mutation occurs during DNA replication when a transiently formed tautomer in the template strand pairs with a noncomplementary base. In the next round of replication, the "mismatched" members of the base pair are separated, and each becomes the template for its normal complementary base. The end result is a point mutation (**Figure 14.3**).

Depurination and Deamination

Some of the most common causes of spontaneous mutations are two forms of DNA base damage: depurination and deamination. **Depurination** is the loss of one of the nitrogenous bases in an intact double-helical DNA molecule. Most frequently, the base is either guanine or adenine—in

(a) Standard base-pairing arrangements

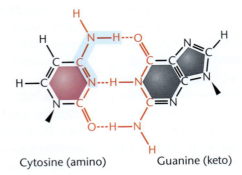

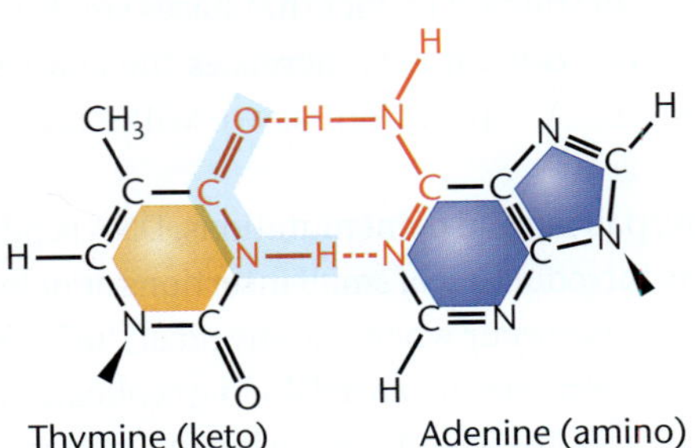

Cytosine (amino) Guanine (keto)

(b) Anomalous base-pairing arrangements

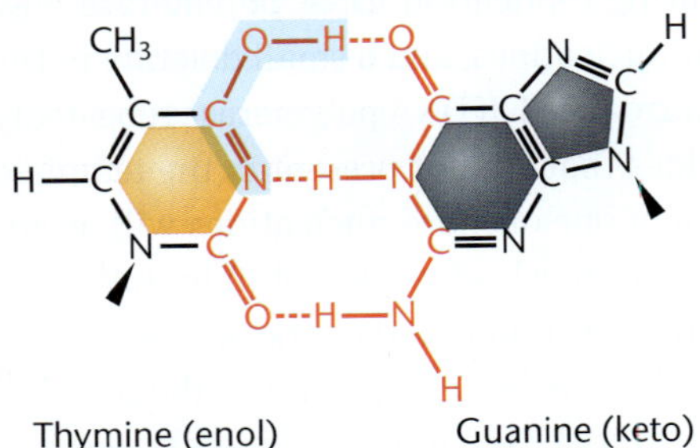

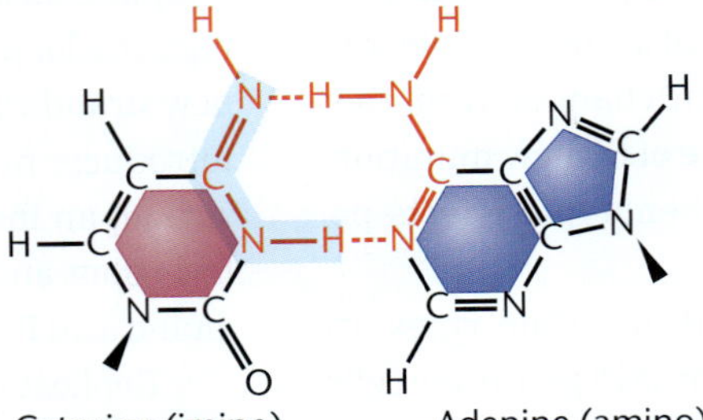

FIGURE 14.2 Examples of standard base-pairing arrangements (a) compared with examples of the anomalous base pairing that occurs as a result of tautomeric shifts (b). The long triangles indicate the point at which each base bonds to a backbone sugar.

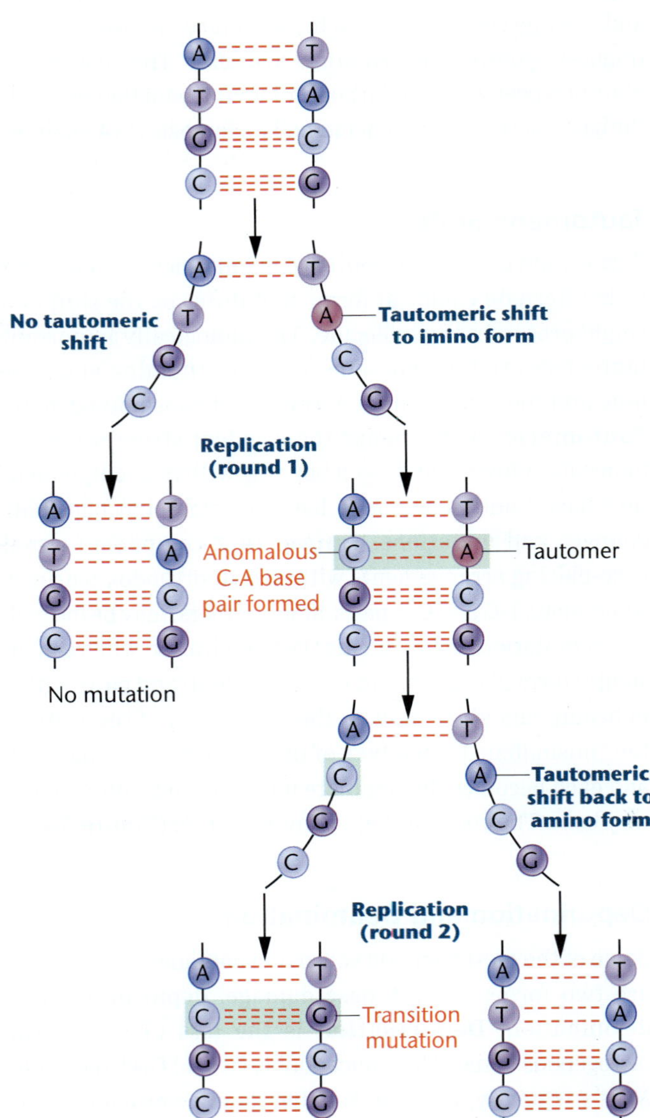

FIGURE 14.3 Formation of an A-T to G-C transition mutation as a result of a transient tautomeric shift in adenine.

other words, a purine. These bases may be lost if the glycosidic bond linking the 1′-C of the deoxyribose and the number 9 position of the purine ring is broken, leaving an **apurinic site** on one strand of the DNA. Geneticists estimate that thousands of such spontaneous lesions are formed daily in the DNA of mammalian cells in culture. If apurinic sites are not repaired, there will be no base at that position to act as a template during DNA replication. As a result, DNA polymerase may introduce a nucleotide at random at that site.

In **deamination,** an amino group in cytosine or adenine is converted to a keto group. In these cases, cytosine is converted to uracil, and adenine is changed to the guanine-resembling compound hypoxanthine (**Figure 14.4**). The major effect of these changes is an alteration in the base-pairing specificities of these two bases during DNA replication. For example, cytosine normally pairs with guanine. Following its conversion to uracil, which pairs with adenine, the original G-C pair is converted to an A-U pair and then, in the next replication, is converted to an A-T pair. When adenine is deaminated, the original A-T pair is ultimately converted to a G-C pair because hypoxanthine pairs naturally with cytosine, which then pairs with guanine in the next replication.

Oxidative Damage

DNA may also suffer damage from the by-products of normal cellular processes. These by-products include reactive oxygen species (electrophilic oxidants) that are generated during normal aerobic respiration. For example,

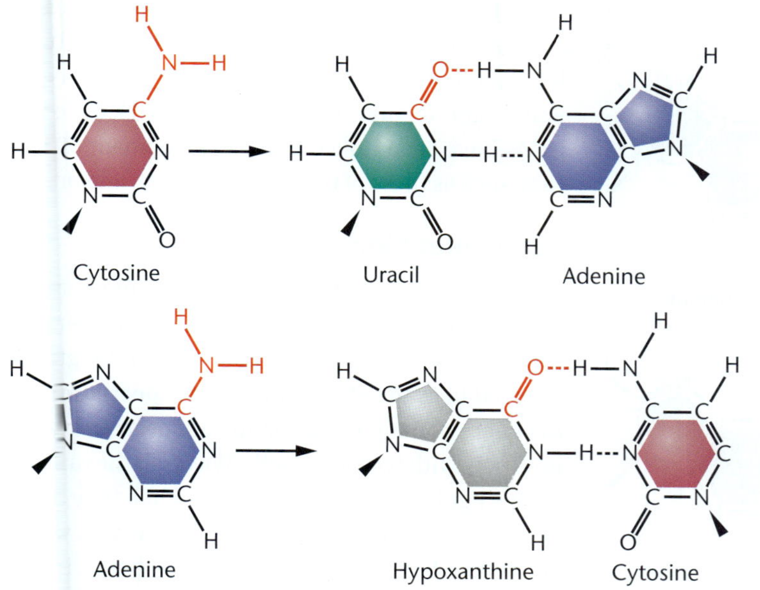

FIGURE 14.4 Deamination of cytosine and adenine, leading to new base pairing and mutation. Cytosine is converted to uracil, which base-pairs with adenine. Adenine is converted to hypoxanthine, which base-pairs with cytosine.

superoxides (O_2^-), hydroxyl radicals ($\cdot OH$), and hydrogen peroxide (H_2O_2) are created during cellular metabolism and are constant threats to the integrity of DNA. Such **reactive oxidants,** also generated by exposure to high-energy radiation, can produce more than 100 different types of chemical modifications in DNA, including modifications to bases, loss of bases, and single-stranded breaks.

ESSENTIAL POINT

Spontaneous mutations result from many different causes including errors during DNA replication and changes in DNA base pairing related to tautomeric shifts, depurinations, deaminations, and reactive oxidant damage. ■

NOW SOLVE THIS

14.2 One of the most famous cases of an X-linked recessive mutation in humans is that of hemophilia found in the descendants of Britain's Queen Victoria. The pedigree of the royal family indicates that Victoria was heterozygous for the trait; however, her father was not affected, and no other member of her maternal line appeared to carry the mutation. What are some possible explanations of how the mutation arose? What types of mutations could lead to the disease?

■ **HINT:** *This problem asks you to determine the sources of new mutations. The key to its solution is to consider the ways in which mutations occur, the types of cells in which they can occur, and how they are inherited.*

14.4 Induced Mutations Arise from DNA Damage Caused by Chemicals and Radiation

All cells on Earth are exposed to a plethora of agents called **mutagens,** which have the potential to damage DNA and cause induced mutations. Some of these agents, such as some fungal toxins, cosmic rays, and UV light, are natural components of our environment. Others, including some industrial pollutants, medical X rays, and chemicals within tobacco smoke, can be considered as unnatural or human-made additions to our modern world. On the positive side, geneticists have harnessed some mutagens for use in analyzing genes and gene functions. The mechanisms by which some of these natural and unnatural agents lead to mutations are outlined in this section.

Base Analogs

One category of mutagenic chemicals is **base analogs,** compounds that can substitute for purines or pyrimidines during nucleic acid biosynthesis. For example, the synthetic chemical **5-bromouracil (5-BU),** a derivative of uracil, behaves as a thymine analog but with a bromine atom substituted at the number 5 position of the pyrimidine ring. If 5-BU is chemically linked to deoxyribose, the nucleoside analog **bromodeoxyuridine (BrdU)** is formed. **Figure 14.5** compares the structure of 5-BU with that of thymine. The presence of the bromine atom in place of the methyl group increases the probability that a tautomeric shift will occur. If BrdU is incorporated into DNA in place of thymidine and a tautomeric shift to the enol form of 5-BU occurs, 5-BU base-pairs with guanine. After one round of replication, an A-T to G-C transition results. Furthermore, the presence of 5-BU within DNA increases the sensitivity of the molecule to UV light, which itself is mutagenic.

Alkylating, Intercalating, and Adduct-Forming Agents

A number of naturally occurring and human-made chemicals alter the structure of DNA and cause mutations. The sulfur-containing mustard gases, used during World War I, were some of the first chemical mutagens identified in chemical warfare studies. Mustard gases are **alkylating agents**—that is, they donate an alkyl group, such as CH_3 or CH_2CH_3, to amino or keto groups in nucleotides. Ethylmethane sulfonate (EMS), for example, alkylates the keto groups in the number 6 position of guanine and in the number 4 position of thymine. As with base analogs, base-pairing affinities are altered, and transition mutations result.

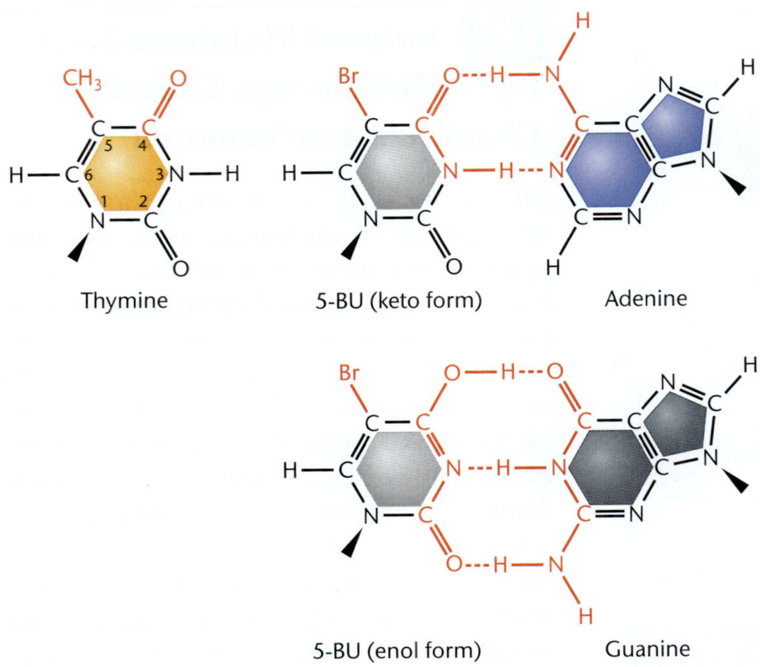

FIGURE 14.5 Similarity of the chemical structure of 5-bromouracil (5-BU) and thymine. In the common keto form, 5-BU base-pairs normally with adenine, behaving as a thymine analog. In the rare enol form, it pairs anomalously with guanine.

For example, 6-ethylguanine acts as an analog of adenine and pairs with thymine (**Figure 14.6**).

Intercalating agents are chemicals that have dimensions and shapes that allow them to wedge between the base pairs of DNA. Wedged intercalating agents cause base pairs to distort and DNA strands to unwind. These changes in DNA structure affect many functions including transcription, replication, and repair. Deletions and insertions occur during DNA replication and repair, leading to frame-shift mutations.

Another group of chemicals that cause mutations are known as **adduct-forming agents.** A DNA adduct is a substance that covalently binds to DNA, altering its conformation and interfering with replication and repair. Two examples of adduct-forming substances are acetaldehyde (a component of cigarette smoke) and heterocyclic amines (HCAs). HCAs are cancer-causing chemicals that are created during the cooking of meats such as beef, chicken, and fish. HCAs are formed at high temperatures from amino acids and creatine. Many HCAs covalently bind to guanine bases. At least 17 different HCAs have been linked to the development of cancers, such as those of the stomach, colon, and breast.

Ultraviolet Light

All electromagnetic radiation consists of energetic waves that we define by their different wavelengths (**Figure 14.7**). The full range of wavelengths is referred to as the **electromagnetic spectrum**, and the energy of any radiation in the spectrum varies inversely with its wavelength. Waves in the range of visible light and longer are benign when they interact with most organic molecules. However, waves of shorter length than visible light, being inherently more energetic, have the potential to disrupt organic molecules.

Purines and pyrimidines absorb **ultraviolet (UV) radiation** most intensely at a wavelength of about 260 nanometers (nm). Although Earth's ozone layer absorbs the most dangerous types of UV radiation, sufficient UV radiation can induce thousands of DNA lesions per hour in any cell exposed to this radiation. One major effect of UV radiation on DNA is the creation of **pyrimidine dimers**—chemical species consisting of two identical pyrimidines—particularly ones consisting of two thymidine residues (**Figure 14.8**). The dimers distort the DNA conformation and inhibit normal replication. As a result, errors can be introduced in the base sequence of DNA during replication through the actions of error-prone DNA polymerases. When UV-induced dimerization is extensive, it is responsible (at least in part) for the killing effects of UV radiation on cells.

Ionizing Radiation

As noted above, the energy of radiation varies inversely with wavelength. Therefore, **X rays, gamma rays,** and **cosmic rays** are more energetic than UV radiation (Figure 14.7).

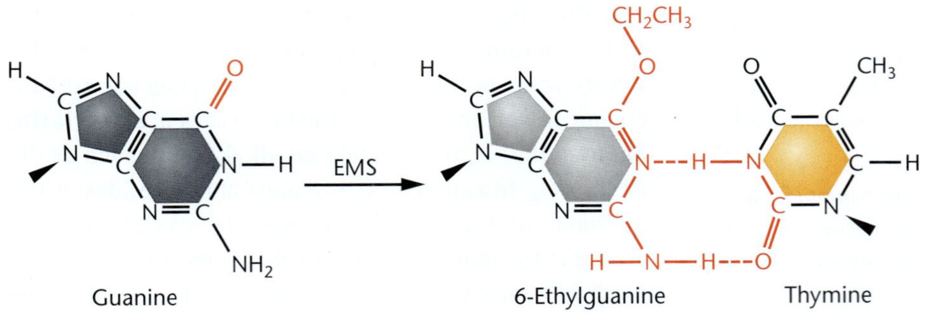

FIGURE 14.6 Conversion of guanine to 6-ethylguanine by the alkylating agent ethylmethane sulfonate (EMS). The 6-ethylguanine base-pairs with thymine.

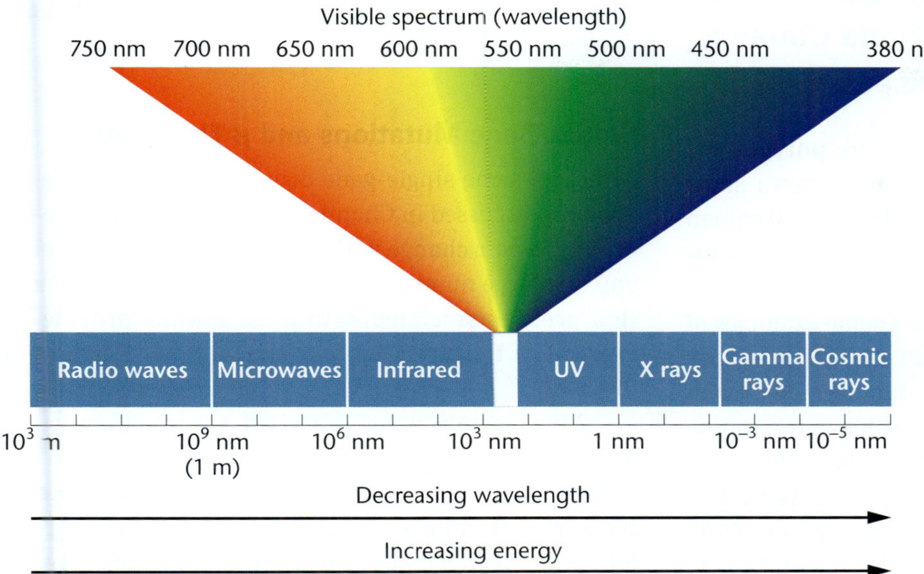

FIGURE 14.7 The regions of the electromagnetic spectrum and their associated wavelengths.

As a result, they penetrate deeply into tissues, causing ionization of the molecules encountered along the way. Hence, this type of radiation is called **ionizing radiation.**

As ionizing radiation penetrates cells, stable molecules and atoms are transformed into **free radicals**—chemical species containing one or more unpaired electrons. Free radicals can directly or indirectly affect the genetic material, altering purines and pyrimidines in DNA, breaking phosphodiester bonds, disrupting the integrity of chromosomes, and producing a variety of chromosomal aberrations, such as deletions, translocations, and chromosomal fragmentation.

Although it is often assumed that radiation from artificial sources such as nuclear power plant waste and medical X rays are the most significant sources of radiation exposure for humans, scientific data indicate otherwise. Scientists estimate that less than 20 percent of human radiation exposure arises from human-made sources. The greatest radiation exposure comes from radon gas, cosmic rays, and natural soil radioactivity. More than half of human-made radiation exposure comes from medical X rays and radioactive pharmaceuticals.

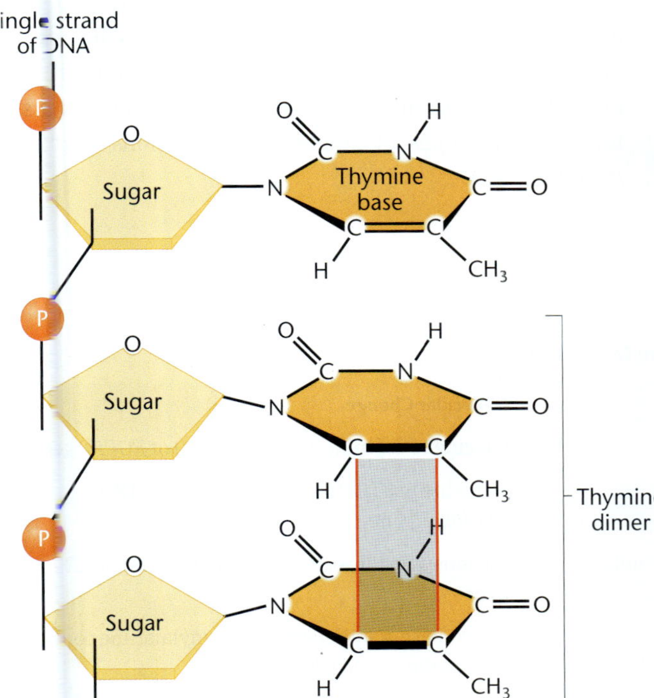

FIGURE 14.8 Depiction of a thymine dimer induced by UV radiation. The covalent crosslinks (shown in red) occur between carbon atoms of the pyrimidine rings.

ESSENTIAL POINT

Mutations can be induced by many types of chemicals and radiations. These agents can damage both DNA bases and the sugar-phosphate backbones of DNA molecules. ■

NOW SOLVE THIS

14.3 The cancer drug melphalan is an alkylating agent of the mustard gas family. It acts in two ways: by causing alkylation of guanine bases and by cross linking DNA strands together. Describe two ways in which melphalan might kill cancer cells. What are two ways in which cancer cells could repair the DNA-damaging effects of melphalan?

■ **HINT:** *This problem asks you to consider the effect of the alkylation of guanine on base pairing during DNA replication. The key to its solution is to consider the effects of mutations on cellular processes that allow cells to grow and divide. In Section 14.6, you will learn about the ways in which cells repair the types of mutations introduced by alkylating agents.*

14.5 Single-Gene Mutations Cause a Wide Range of Human Diseases

Although most human genetic diseases are **polygenic**—that is, caused by variations in several genes—even a single base-pair change in one of the approximately 20,000 human genes can lead to a serious inherited disorder. These **monogenic** diseases can be caused by many different types of single-gene mutations. **Table 14.2** lists some examples of the types of single-gene mutations that can lead to serious genetic diseases. A comprehensive database of human genes, mutations, and disorders is available in the Online Mendelian Inheritance in Man (OMIM) database (described in the Exploring Genomics feature in Chapter 3). As of 2018, the OMIM database has cataloged approximately 5000 human phenotypes for which the molecular basis is known.

Geneticists estimate that approximately 30 percent of mutations that cause human diseases are single base-pair changes that create nonsense mutations. These mutations not only code for a prematurely terminated protein product, but also trigger rapid decay of the mRNA. Many more mutations are missense mutations that alter the amino acid sequence of a protein and frameshift mutations that alter the protein sequence and create internal nonsense codons. Other common disease-associated mutations affect the sequences of gene promoters, mRNA splicing signals, and other noncoding sequences that affect transcription, processing, and stability of mRNA or protein. One recent study showed that about 15 percent of all point mutations that cause human genetic diseases result in abnormal mRNA splicing. Approximately 85 percent of these splicing mutations alter the sequence of 5′ and 3′ splice signals. The remainder create new splice sites within the gene. Splicing defects often result in degradation of the abnormal mRNA or creation of abnormal protein products.

Single-Gene Mutations and β Thalassemia

Although some single-gene diseases, such as sickle cell anemia (discussed in Chapter 13), are caused by one specific base-pair change within a gene, most are caused by any of a large number of different mutations. The mutation profile associated with β-thalassemia provides an example of the latter, more common, type of monogenic disease.

β-thalassemia is an inherited autosomal recessive blood disorder resulting from a reduction or absence of hemoglobin. It is the most common single-gene disease in the world, affecting people worldwide, but especially populations in Mediterranean, North African, Middle Eastern, Central Asian, and Southeast Asian countries.

People with β-thalassemia have varying degrees of anemia—from severe to mild—with symptoms including weakness, delayed development, jaundice, enlarged organs, and often a need for frequent blood transfusions.

Mutations in the *β-globin* gene (*HBB* gene) cause β-thalassemia. The *HBB* gene encodes the 146-amino-acid β-globin polypeptide. Two β-globin polypeptides associate with two α-globin polypeptides to form the adult hemoglobin tetramer. The *HBB* gene spans 1.6 kilobases of DNA on the short arm of chromosome 11. It is made up of three exons and two introns.

Scientists have discovered approximately 400 different mutations in the *HBB* gene that cause β-thalassemia, although most cases worldwide are associated with only about 20 of these mutations. **Table 14.3** provides a summary of the types of single-gene mutations that cause β-thalassemia.

TABLE 14.2 Examples of Human Disorders Caused by Single-Gene Mutations

Type of Mutation	Disorder	Molecular Change
Missense	Achondroplasia	Glycine to arginine at position 380 of *FGFR3* gene
Nonsense	Marfan syndrome	Tyrosine to STOP codon at position 2113 of *fibrillin-1* gene
Insertion	Familial hypercholesterolemia	Various short insertions throughout the *LDLR* gene
Deletion	Cystic fibrosis	Three-base-pair deletion of phenylalanine codon at position 508 of *CFTR* gene
Trinucleotide repeat expansions	Huntington disease	>40 repeats of (CAG) sequence in coding region of *Huntingtin* gene

TABLE 14.3 Types of Mutations in the *HBB* Gene That Cause β-Thalassemia

Gene Region Affected	Number of Mutations Known	Description
5′ upstream region	22	Single base-pair mutations occur between −101 and −25 upstream from transcription start site. For example, a T → A transition in the TATA sequence at −30 results in decreased gene transcription and severe disease.
mRNA CAP site	1	Single base-pair mutation (A → C transversion) at +1 position leads to decreased levels of mRNA.
5′ untranslated region	3	Single base-pair mutations at +20, +22, and +33 cause decreases in transcription and translation and mild disease.
ATG translation initiation codon	7	Single base-pair mutations alter the mRNA AUG sequence, resulting in no translation and severe disease.
Exons 1, 2, and 3 coding regions	36	Single base-pair missense and nonsense mutations, and mutations that create abnormal mRNA splice sites. Disease severity varies from mild to extreme.
Introns 1 and 2	38	Single base-pair transitions and transversions that reduce or abolish mRNA splicing and create abnormal splice sites that affect mRNA stability. Most cause severe disease.
Polyadenylation site	6	Single base-pair changes in the AATAAA sequence reduce the efficiency of mRNA cleavage and polyadenylation, yielding long mRNAs or unstable mRNAs. Disease is mild.
Throughout and surrounding the *HBB* gene	>100	Short insertions, deletions, and duplications that alter coding sequences, create frameshift stop codons, and alter mRNA splicing.

14.6 Organisms Use DNA Repair Systems to Counteract Mutations

Living systems have evolved a variety of elaborate repair systems that counteract both spontaneous and induced DNA damage. These **DNA repair** systems are absolutely essential to the maintenance of the genetic integrity of organisms and, as such, to the survival of organisms on Earth. The balance between mutation and repair results in the observed mutation rates of individual genes and organisms. Of foremost interest in humans is the ability of these systems to counteract genetic damage that would otherwise result in genetic diseases and cancer. The link between defective DNA repair and cancer susceptibility is described later (see Chapter 19).

We now embark on a review of these and other DNA repair mechanisms, with the emphasis on the major approaches that organisms use to counteract genetic damage.

Proofreading and Mismatch Repair

Some of the most common types of mutations arise during DNA replication when an incorrect nucleotide is inserted by DNA polymerase. The major DNA synthesizing enzyme in bacteria (**DNA polymerase III**) makes an error approximately once every 100,000 insertions, leading to an error rate of 10^{-5}. Fortunately, DNA polymerase proofreads each step, catching 99 percent of those errors. If an incorrect nucleotide is inserted during polymerization, the enzyme can recognize the error and "reverse" its direction. It then behaves as a 3′ to 5′ exonuclease, cutting out the incorrect nucleotide and replacing it with the correct one. This improves the efficiency of replication 100-fold, creating only 1 mismatch in every 10^7 insertions, for a final error rate of 10^{-7}.

To cope with errors such as base–base mismatches, small insertions, and deletions that remain after proofreading, another mechanism, called **mismatch repair (MMR),** may be activated. During MMR, the mismatches are detected, the incorrect nucleotide is removed, and the correct nucleotide is inserted in its place.

Following replication, the repair enzymes mentioned below are able to recognize any mismatch that is introduced on the newly synthesized DNA strand and bind to the strand. An **endonuclease** enzyme creates a nick in the backbone of the newly synthesized DNA strand, either 5′ or 3′ to the mismatch. An **exonuclease** unwinds and degrades the nicked DNA strand, until the region of the mismatch is reached. Finally, DNA polymerase fills in the gap created by the exonuclease, using the correct DNA strand as a template. DNA ligase then seals the gap.

A series of *E. coli* gene products, MutH, MutL, and MutS, as well as exonucleases, DNA polymerase III, and DNA ligase, are involved in MMR. Mutations in the *mutH*, *mutL*, and *mutS* genes result in bacterial strains deficient in MMR.

In humans, mutations in genes that code for DNA MMR proteins (such as *hMSH2* and *hMLH1*, which are the human equivalents of the *mutS* and *mutL* genes of *E. coli*) are associated with the hereditary nonpolyposis colon cancer. MMR defects are commonly found in other cancers, such as leukemias, lymphomas, and tumors of the ovary, prostate, and endometrium. Cells from these cancers show genome-wide increases in the rate of spontaneous mutation. The link between defective MMR and cancer is supported by experiments with mice. Mice that are engineered to have deficiencies in MMR genes accumulate large numbers of mutations and are cancer-prone.

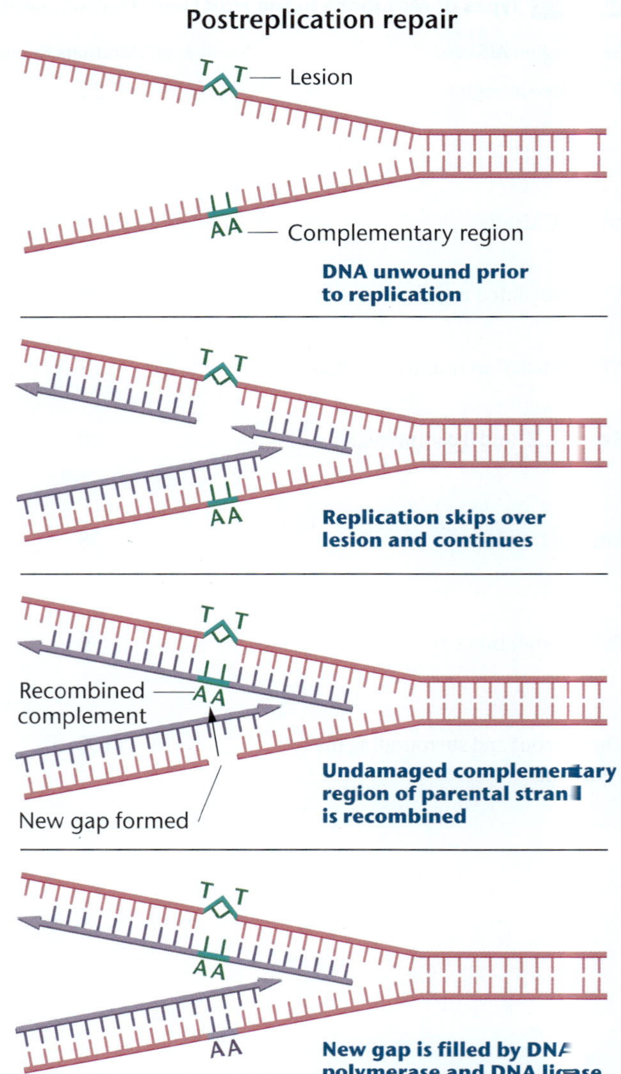
Postreplication repair
T T — Lesion
AA — Complementary region
DNA unwound prior to replication
T T
AA
Replication skips over lesion and continues
T T
Recombined complement — AA
New gap formed
Undamaged complementary region of parental strand is recombined
T T
AA
AA
New gap is filled by DNA polymerase and DNA ligase

FIGURE 14.9 Postreplication repair occurs if DNA replication has skipped over a lesion such as a thymine dimer. Through the process of recombination, the correct complementary sequence is recruited from the parental strand and inserted into the gap opposite the lesion. The new gap is filled by DNA polymerase and DNA ligase.

Postreplication Repair and the SOS Repair System

Another type of DNA repair system, called **postreplication repair,** responds *after* damaged DNA has escaped repair and has failed to be completely replicated. As illustrated in **Figure 14.9**, when DNA bearing a lesion of some sort (such as a pyrimidine dimer) is being replicated, DNA polymerase may stall at the lesion and then skip over it, leaving an unreplicated gap on the newly synthesized strand. To correct the gap, RecA protein directs a recombinational exchange with the corresponding region on the undamaged parental strand of the same polarity (the "donor" strand). When the undamaged segment of the donor strand DNA replaces the gapped segment, a gap is created on the donor strand. The gap can be filled by repair synthesis as replication proceeds. Because a recombinational event is involved in this type of DNA repair, it is considered to be a form of **homologous recombination repair.**

Another postreplication repair pathway, the *E. coli* **SOS repair system,** also responds to damaged DNA, but in a different way. In the presence of a large number of unrepaired DNA mismatches and gaps, the bacteria can induce expression of about 20 genes (including *lexA, recA,* and *uvr*) whose products allow DNA replication to occur even in the presence of DNA lesions. This type of repair is a last resort to minimize DNA damage, hence its name. During SOS repair, DNA synthesis becomes error-prone, inserting random and possibly incorrect nucleotides in places that would normally stall DNA replication. As a result, SOS repair itself becomes mutagenic—although it may allow the cell to survive DNA damage that would otherwise kill it.

Photoreactivation Repair: Reversal of UV Damage

As was illustrated in Figure 14.8, UV light introduces mutations by the creation of pyrimidine dimers. UV-induced damage to *E. coli* DNA can be partially reversed if, following irradiation, the cells are exposed briefly to visible light, especially in the blue range of the visible spectrum. The process is dependent on the activity of a protein called **photoreactivation enzyme (PRE)** or **photolyase.** The enzyme's mode of action is to cleave the cross-linking bonds between thymine dimers. Although the enzyme will associate with a thymine dimer in the dark, it must absorb a photon of blue light to cleave the dimer. The enzyme is also detectable in many organisms, including other bacteria, fungi, plants, and some vertebrates—though not in humans. Humans and other organisms that lack photoreactivation repair must rely on other repair mechanisms to reverse the effects of UV radiation.

Base and Nucleotide Excision Repair

A number of light-independent DNA repair systems exist in all bacteria and eukaryotes. The basic mechanisms involved in these types of repair—collectively referred to as **excision repair** or cut-and-paste mechanisms—consist of the following three steps.

1. The damage, distortion, or error present on one of the two strands of the DNA helix is recognized and enzymatically clipped out by an endonuclease. Excisions in the phosphodiester backbone usually include a number of nucleotides adjacent to the error as well, leaving a gap on one strand of the helix.

2. A DNA polymerase fills in the gap by inserting nucleotides complementary to those on the intact strand, which it uses as a replicative template. The enzyme adds these nucleotides to the free 3′-OH end of the clipped DNA. In *E. coli*, this step is usually performed by DNA polymerase I.

3. DNA ligase seals the final "nick" that remains at the 3′-OH end of the last nucleotide inserted, closing the gap.

There are two types of excision repair: base excision repair and nucleotide excision repair. **Base excision repair (BER)** corrects DNA that contains incorrect base pairings due to the presence of chemically modified bases or uridine nucleosides that are inappropriately incorporated into DNA or created by deamination of cytosine. The first step in the BER pathway involves the recognition of an inappropriately paired base by enzymes called **DNA glycosylases.** There are a number of DNA glycosylases, each of which recognizes a specific base. For example, the enzyme uracil DNA glycosylase recognizes the presence of uracil in DNA (**Figure 14.10**). DNA glycosylases first cut the glycosidic bond between the target base and its sugar, creating an **apyrimidinic** (or apurinic) **site.** The sugar with the missing base is then recognized by an enzyme called **AP endonuclease.** The AP endonuclease makes cuts in the phosphodiester backbone at the apyrimidinic or apurinic site. The gap is filled by DNA polymerase and DNA ligase.

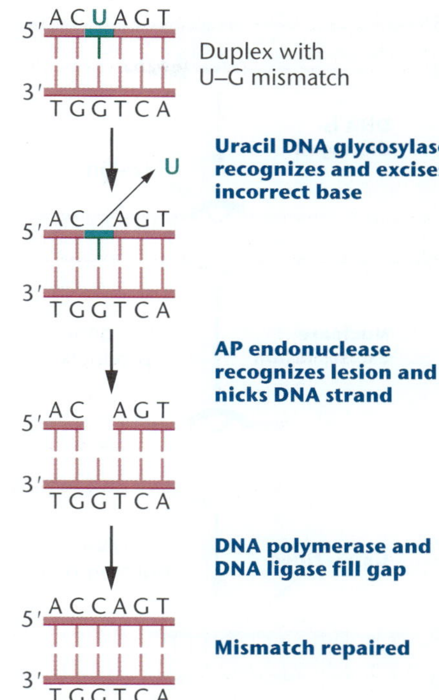

FIGURE 14.10 Base excision repair (BER) accomplished by uracil DNA glycosylase, AP endonuclease, DNA polymerase, and DNA ligase. Uracil is recognized as a noncomplementary base, excised, and replaced with the complementary base (C).

Although much has been learned about the mechanisms of BER in *E. coli*, BER systems have also been detected in eukaryotes from yeast to humans. Experimental evidence shows that both mouse and human cells that are defective in BER activity are hypersensitive to the killing effects of gamma rays and oxidizing agents.

Nucleotide excision repair (NER) pathways repair "bulky" lesions in DNA that alter or distort the double helix. These lesions include the UV-induced pyrimidine dimers and DNA adducts discussed previously.

The NER pathway (**Figure 14.11**) was first discovered in 1964 by Paul Howard-Flanders and coworkers, who isolated several independent *E. coli* mutants that are sensitive to UV radiation. One group of genes was designated *uvr* (ultraviolet repair) and included the *uvrA, uvrB,* and *uvrC* mutations. In the NER pathway, the *uvr* gene products are involved in recognizing and clipping out lesions in the DNA. Usually, a specific number of nucleotides are clipped out around both sides of the lesion. In *E. coli,* usually a total of 13 nucleotides are removed, including the lesion. The repair is then completed by DNA polymerase I and DNA ligase, in a manner similar to that occurring in BER. The undamaged strand opposite the lesion is used as a template for the replication, resulting in repair.

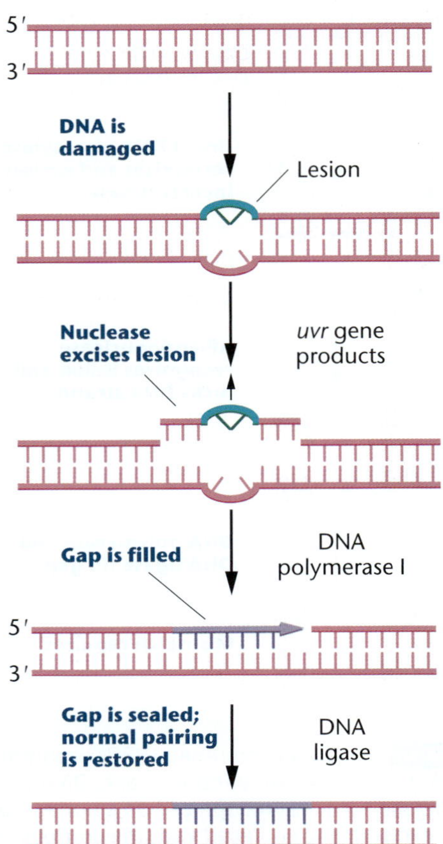

FIGURE 14.11 Nucleotide excision repair (NER) of a UV-induced thymine dimer. During repair, 13 nucleotides are excised in bacteria, and 28 nucleotides are excised in eukaryotes.

Nucleotide Excision Repair and Xeroderma Pigmentosum in Humans

The mechanism of NER in eukaryotes is much more complicated than that in bacteria and involves many more proteins, encoded by about 30 genes. Much of what is known about the system in humans has come from detailed studies of individuals with **xeroderma pigmentosum (XP),** a rare recessive genetic disorder that predisposes individuals to severe skin abnormalities, skin cancers, and a wide range of other symptoms including developmental and neurological defects. Patients with XP are extremely sensitive to UV radiation in sunlight. In addition, they have a 2000-fold higher rate of cancer, particularly skin cancer, than the general population. The condition is severe and may be lethal, although early detection and protection from sunlight can arrest it (**Figure 14.12**).

The repair of UV-induced lesions in XP has been investigated *in vitro*, using human fibroblast cell cultures derived from normal individuals and those with XP. (Fibroblasts are undifferentiated connective tissue cells.)

The involvement of multiple genes in NER and XP has been investigated using **somatic cell hybridization.** Fibroblast cells from any two unrelated XP patients, when grown together in tissue culture, can fuse together, forming heterokaryons. A **heterokaryon** is a single cell with two nuclei from different organisms but a common cytoplasm. If the mutation in each of the two XP cells occurs in the same gene, the heterokaryon, like the cells that fused to form it, will still be unable to undergo NER. This is because there is no normal copy of the relevant gene present in the heterokaryon.

However, if NER does occur in the heterokaryon, the mutations in the two XP cells must have been present in two different genes. Hence, the two mutants are said to demonstrate **complementation,** a concept discussed earlier (see Chapter 4). Complementation occurs because the heterokaryon has at least one normal copy of each gene in the fused cell. By fusing XP cells from a large number of XP patients, researchers were able to determine how many genes contribute to the XP phenotype. Based on these and other studies, XP patients were divided into seven complementation groups, indicating that at least seven different genes code for proteins that are involved in nucleotide excision repair in humans. A gene representing each of these complementation groups, *XPA* to *XPG* (*X*eroderma *P*igmentosum gene *A* to *G*), has now been identified, and a homologous gene for each has been identified in yeast.

Approximately 20 percent of XP patients do not fall into any of the seven complementation groups. Cells from most

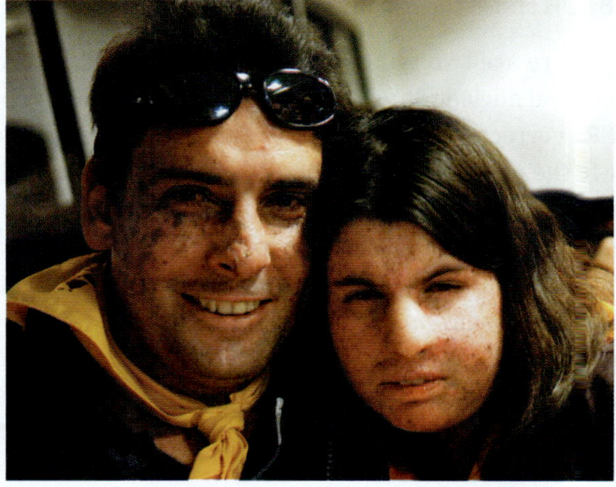

FIGURE 14.12 Two individuals with xeroderma pigmentosum. These XP patients show characteristic XP skin lesions induced by sunlight, as well as mottled redness (erythema) and irregular pigment changes to the skin, in response to cellular injury.

of these patients have mutations in the gene coding for DNA polymerase eta (Pol η), which is a lower-fidelity DNA polymerase that allows DNA replication to proceed past damaged DNA. Approximately another 6 percent of XP patients do not have mutations in either the seven complementation group genes or the *DNA polymerase eta* (*POLH*) gene, suggesting that other genes or mutations outside of coding regions may be involved in XP.

Double-Strand Break Repair in Eukaryotes

Thus far, we have discussed repair pathways that deal with damage or errors within one strand of DNA. We conclude our discussion of DNA repair by considering what happens in eukaryotic cells when both strands of the DNA helix are cleaved—as a result of exposure to ionizing radiation, for example. These types of damage are extremely dangerous to cells, leading to chromosome rearrangements, cancer, or cell death.

Specialized forms of DNA repair, the DNA **double-strand break (DSB) repair** pathways, are activated and are responsible for reattaching two broken DNA strands. Defects in these pathways are associated with X-ray hypersensitivity and immune deficiencies, as well as familial dispositions to breast and ovarian cancer. Several human disease syndromes, such as Fanconi anemia and ataxia telangiectasia, result from defects in DSB repair.

One pathway involved in double-strand break repair is **homologous recombination repair.** The first step in this process involves the activity of an enzyme that recognizes the double-strand break and then digests back the 5′-ends of the broken DNA helix, leaving overhanging 3′-ends (**Figure 14.13**). One overhanging end searches for a region of sequence complementarity on the sister chromatid and then invades the homologous DNA duplex, aligning the complementary sequences.

Once aligned, DNA synthesis proceeds from the 3′ overhanging ends, using the undamaged homologous DNA strands as templates. The interaction of two sister chromatids is necessary because, when both strands of one helix are broken, there is no undamaged parental DNA strand available to use as a template DNA sequence during repair. After DNA repair synthesis, the resulting heteroduplex molecule is resolved and the two chromatids separate.

DSB repair usually occurs during the late S or early G2 phase of the cell cycle, after DNA replication, a time when sister chromatids are available to be used as repair templates. Because an undamaged template is used during repair synthesis, homologous recombination repair is an accurate process.

A second pathway, called **nonhomologous end joining,** also repairs double-strand breaks. However, as the name implies, the mechanism does not recruit a homologous region of DNA during repair. This system is activated in G1, prior to DNA replication. End joining involves a complex of many proteins and may include the DNA-dependent protein kinase and the breast cancer susceptibility gene product,

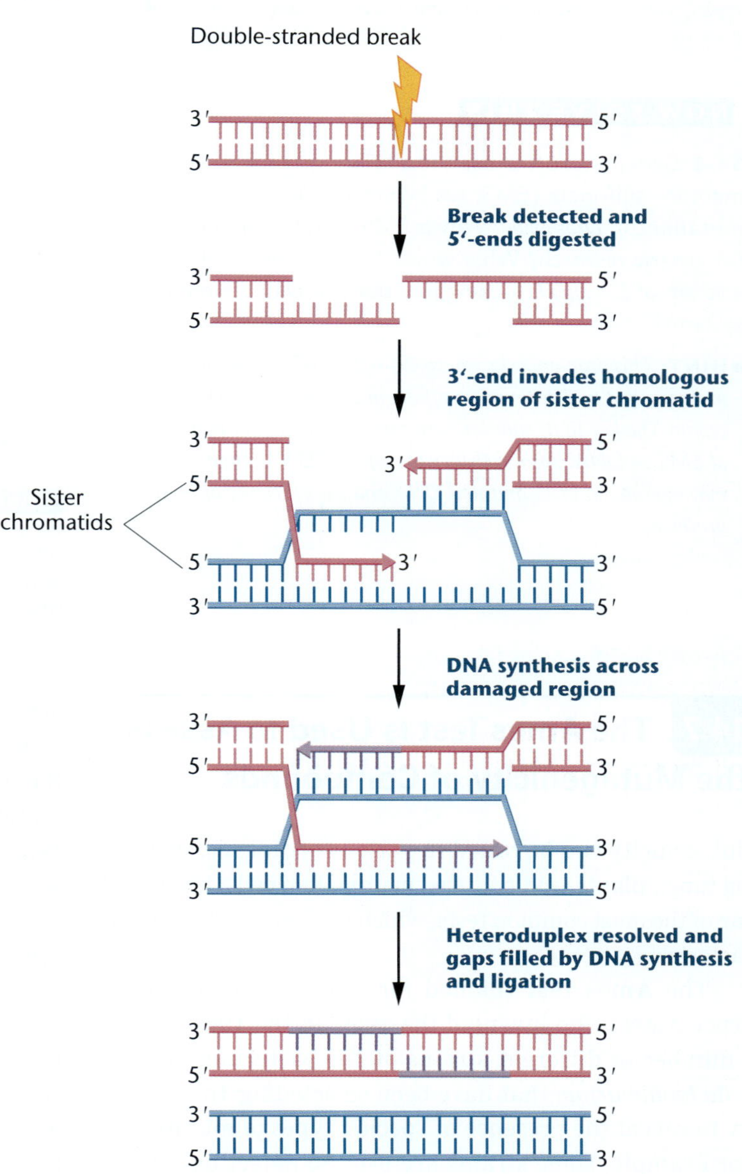

FIGURE 14.13 Steps in homologous recombination repair of double-stranded breaks.

BRCA1. These and other proteins bind to the free ends of the broken DNA, trim the ends, and ligate them back together. Because some nucleotide sequences are lost in the process of end joining, it is an error-prone repair system. In addition, if more than one chromosome suffers a double-strand break, the wrong ends could be joined together, leading to abnormal chromosome structures, such as those discussed earlier (see Chapter 6).

ESSENTIAL POINT

Organisms counteract mutations by using a range of DNA repair mechanisms. Errors in DNA synthesis can be repaired by proofreading, mismatch repair, and postreplication repair. DNA damage can be repaired by photoreactivation repair, SOS repair, base excision repair, nucleotide excision repair, and double-strand break repair. ■

NOW SOLVE THIS

14.4 Geneticists often use the alkylating agent ethylmethane sulfonate (EMS; see Figure 14.6) to induce mutations in *Drosophila*. Why is EMS a mutagen of choice for genetic research? What would be the effects of EMS in a strain of *Drosophila* lacking functional mismatch repair systems?

■ **HINT:** *This problem asks you to evaluate EMS as a useful mutagen and to determine its effects in the absence of DNA repair. The key to its solution is to consider the chemical effects of EMS on DNA. Also, consider the types of DNA repair that may operate on EMS-mutated DNA and the efficiency of these processes.*

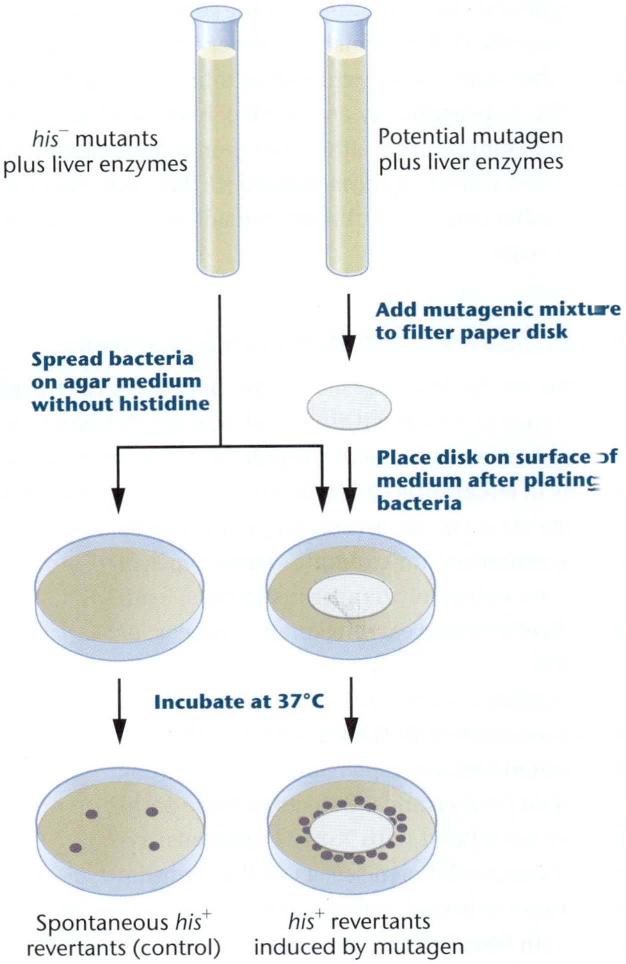

FIGURE 14.14 The Ames test, which screens compounds for potential mutagenicity. The high number of *his*$^+$ revertant colonies on the right side of the figure confirms that the substance being tested was indeed mutagenic.

14.7 The Ames Test Is Used to Assess the Mutagenicity of Compounds

Mutagenicity can be tested in various organisms, including fungi, plants, and cultured mammalian cells; however, one of the most common tests, which we describe here, uses bacteria.

The **Ames test** (named for American biochemist Bruce Ames, who invented the assay in the 1960s) uses a number of different strains of the bacterium *Salmonella typhimurium* that have been selected for their ability to reveal the presence of specific types of mutations. For example, some strains are used to detect base-pair substitutions, and other strains detect various frameshift mutations. Each strain contains a mutation in one of the genes of the histidine operon. The mutant strains are unable to synthesize histidine (*his*$^-$ strains) and therefore require histidine for growth. The assay measures the frequency of reverse mutations that occur within the mutant gene, yielding wild-type bacteria (*his*$^+$ revertants) (**Figure 14.14**). The *his*$^-$ strains also have an increased sensitivity to mutagens due to the presence of mutations in genes involved in both DNA damage repair and the synthesis of the lipopolysaccharide barrier that coats these bacteria and protects them from external substances.

Many substances entering the human body are relatively innocuous until activated metabolically, usually in the liver, to more chemically reactive products. Thus, the Ames test includes a step in which the test compound is incubated *in vitro* in the presence of a mammalian liver extract. Alternatively, test compounds may be injected into a mouse where they are modified by liver enzymes and then recovered for use in the Ames test.

In the initial use of Ames testing in the 1970s, a large number of known **carcinogens,** or cancer-causing agents, were examined, and more than 80 percent of these were shown to be strong mutagens. This is not surprising, as the transformation of cells to the malignant state occurs as a result of mutations. For example, more than 60 compounds found in cigarette smoke test positive in the Ames test and cause cancer in animal tests. Although a positive response in the Ames test does not prove that a compound is carcinogenic, the Ames test is useful as a preliminary screening device, as it is a rapid, convenient way to assess mutagenicity. Other tests of potential mutagens and carcinogens use laboratory animals such as rats and mice; however, these tests can take several years to complete and are more expensive. The Ames test is used extensively during the development of industrial and pharmaceutical chemical compounds.

14.8 Transposable Elements Move Within the Genome and May Create Mutations

Transposable elements (TEs), informally known as "jumping genes," are DNA sequences that can move or transpose within and between chromosomes, inserting themselves into various locations within the genome. They can range from 50 to 10,000 base pairs in length.

TEs are present in the genomes of all organisms from bacteria to humans. Not only are they ubiquitous, but they also make up large portions of some eukaryotic genomes. For example, almost 50 percent of the human genome is derived from TEs. Some organisms with unusually large genomes, such as salamanders and barley, contain hundreds of thousands of copies of various types of TEs constituting as much as 85 percent of these genomes.

The movement of TEs from one place in the genome to another has the capacity to disrupt genes and cause mutations, as well as to create chromosomal damage such as double-strand breaks. TEs also act as sites of genome rearrangement events, when homologous recombination occurs between DNA sequences with sequence similarities.

Since their discovery, TEs have also become valuable tools in genetic research. Geneticists harness these DNA elements as mutagens, as cloning tags, and as vehicles for introducing foreign DNA into model organisms.

TEs can be classified into two groups, based on their methods of transposition. *Retrotransposons* move using an RNA intermediate, and *DNA transposons* move in and out of the genome as DNA elements. We will look at both groups in the sections that follow.

DNA Transposons

DNA transposons move from one location to another without going through an RNA intermediate stage. They are abundant in many organisms from bacteria to humans.

DNA transposons share several structural features that are important for their function (**Figure 14.15**). Inverted terminal repeats (ITRs) are located on each end of the TE, and an open reading frame (ORF) codes for the enzyme transposase; both are required for movement of the TE in and out of the genome. ITRs are DNA sequences of between 9 and 40 bp long that are identical in sequence, but inverted relative to each other. ITRs are essential for transposition and are recognized and bound by the transposase enzyme. Short direct repeats (DRs) are present in the host DNA, flanking each TE insertion. These flanking DRs are created as a consequence of the TE insertion process.

DNA transposons vary considerably in length and are classified as either autonomous or nonautonomous. *Autonomous transposons* are able to transpose by themselves, as they encode a functional transposase enzyme and have intact ITRs. *Nonautonomous transposons* cannot move on their own because they do not encode their own functional transposase enzyme. They require the presence of an autonomous transposon elsewhere in the genome, so that the transposase synthesized by the autonomous element can be used by the nonautonomous element for transposition.

Most DNA transposons move through the genome using "cut-and-paste" mechanisms, in which the transposon is physically cut out of the genome and then inserted into a new position in the same or a different chromosome. Usually, but not always, the site from which the DNA transposon was cut is repaired accurately, leaving no trace of the original DNA transposon.

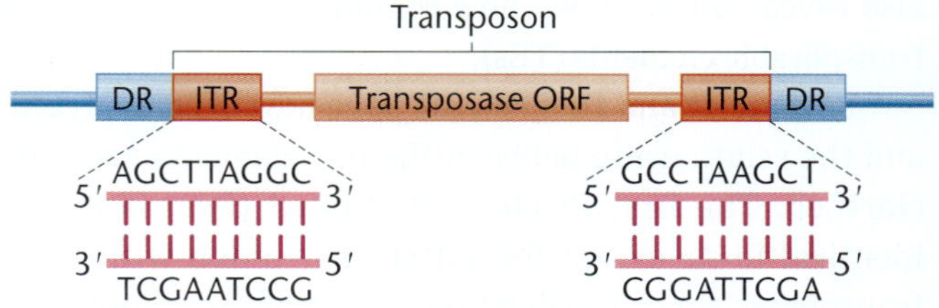

FIGURE 14.15 Structural features of DNA transposons. DNA transposons, shown in red, contain an open reading frame (ORF) that encodes the enzyme transposase. Some DNA transposons also contain ORFs encoding other proteins in addition to transposase. Inverted terminal repeats (ITRs), shown in detail below the main diagram, are short DNA sequences that are inverted relative to each other. Direct repeats (DRs, shown in blue) flank the DNA transposon in the chromosomal DNA.

Examples of two DNA transposons and the ways in which their movements can affect gene expression are described next.

DNA Transposons—the *Ac–Ds* System in Maize

DNA transposons were first discovered by Barbara McClintock in the late 1940s as a result of her research on the genetics of maize. Her work involved analysis of the genetic behavior of two mutations, ***Dissociation (Ds)*** and ***Activator (Ac),*** expressed in either the endosperm or aleurone layers of maize seeds. She then correlated her genetic observations with cytological examinations of maize chromosomes. Initially, McClintock determined that *Ds* was located on chromosome 9. If *Ac* was also present in the genome, *Ds* induced breakage at a point on the chromosome adjacent to its own location. If chromosome breakage occurred in somatic cells during their development, progeny cells often lost part of the broken chromosome, causing a variety of phenotypic effects. The chapter opening photo illustrates the types of phenotypic effects caused by *Ds* mutations in kernels of corn.

Subsequent analysis suggested to McClintock that both *Ds* and *Ac* elements sometimes moved to new chromosomal locations. While *Ds* moved only if *Ac* was also present, *Ac* was capable of autonomous movement. Where *Ds* came to reside determined its genetic effects—that is, it might cause chromosome breakage, or it might inhibit expression of a certain gene. In cells in which *Ds* caused a gene mutation, *Ds* might move again, restoring the gene mutation to wild type. **Figure 14.16** illustrates the types of movements and effects brought about by *Ds* and *Ac* elements.

In McClintock's original observation, pigment synthesis was restored in cells in which the *Ds* element jumped out of chromosome 9. McClintock concluded that *Ds* was a **mobile controlling element.** Similar mobility was later also revealed for *Ac*. We now commonly refer to these as transposable elements (TEs).

Several *Ac* and *Ds* elements have now been analyzed, and the relationship between the two elements has been clarified. The first *Ds* element studied (*Ds*9) is nearly identical to *Ac* except for a 194-bp deletion within the transposase gene. The deletion of part of the transposase gene in the *Ds*9 element explains its dependence on the *Ac* element for transposition. Several other *Ds* elements have also been sequenced, and each contains an even larger deletion within the transposase gene. In each case, however, the ITRs are retained.

Although the significance of Barbara McClintock's mobile controlling elements was not fully appreciated following her initial observations, molecular analysis has since verified her conclusions. She was awarded the Nobel Prize in Physiology or Medicine in 1983.

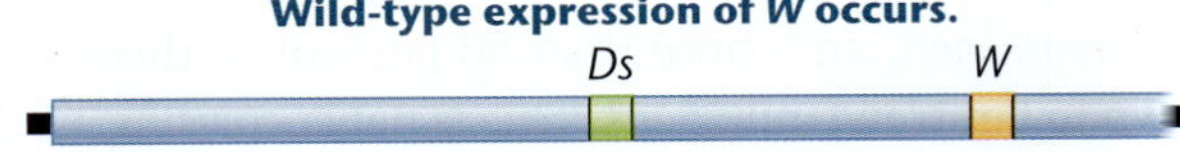
(a) In absence of *Ac*, *Ds* is not transposable.
Wild-type expression of *W* occurs.
Ds
W

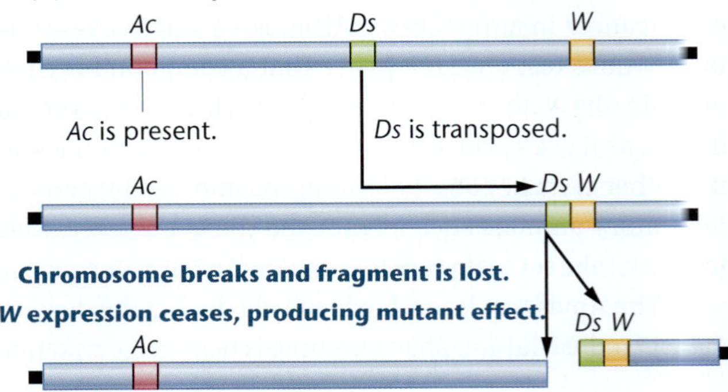
(b) When *Ac* is present, *Ds* may be transposed.
Ac
Ds
W
Ac is present.
Ds is transposed.
Ac
Ds W
Chromosome breaks and fragment is lost.
W expression ceases, producing mutant effect.
Ac
Ds W

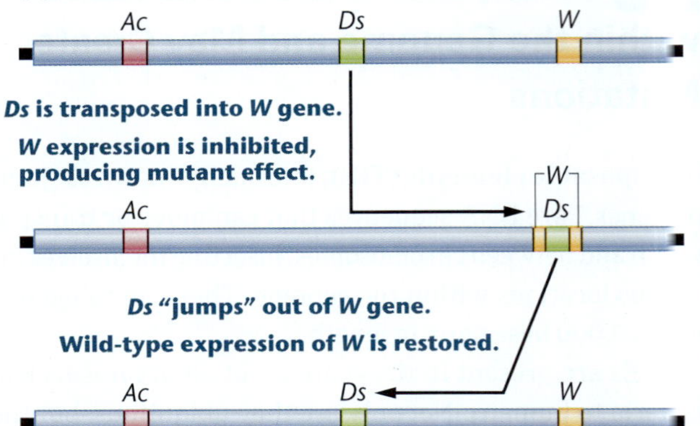
(c) *Ds* can move into and out of another gene.
Ac
Ds
W
Ds is transposed into *W* gene.
W expression is inhibited, producing mutant effect.
Ac
W
Ds
Ds "jumps" out of *W* gene.
Wild-type expression of *W* is restored.
Ac
Ds
W

FIGURE 14.16 Effects of *Ac* and *Ds* elements on gene expression. (a) If *Ds* is present in the absence of *Ac*, there is normal expression of a distantly located hypothetical gene *W*. (b) In the presence of *Ac*, *Ds* may transpose to a region adjacent to *W*. *Ds* can induce chromosome breakage, which may lead to loss of a chromosome fragment bearing the *W* gene. (c) In the presence of *Ac*, *Ds* may transpose into the *W* gene, disrupting *W*-gene expression. If *Ds* subsequently transposes out of the *W* gene, *W*-gene expression may return to normal.

Retrotransposons

Retrotransposons are TEs that amplify and move within the genome using RNA as an intermediate. Their methods of transposition are sometimes described as "copy-and-paste" mechanisms. In many ways, retrotransposons resemble retroviruses, which replicate using similar mechanisms. However, retrotransposons do not encode all of the proteins that are required to form mature virus particles and therefore are not infective.

Retrotransposons can be very abundant in some organisms. For example, maize genomes are made up of as much as 78 percent retrotransposon DNA. Approximately 42 percent of the human genome consists of retrotransposons or their remnants, whereas only approximately 3 percent of the human genome consists of DNA transposons.

There are two types of retrotransposons—the long-terminal-repeat (LTR) retrotransposons and the non-LTR retrotransposons. In addition, retrotransposons, like DNA transposons, can be either autonomous or nonautonomous.

Like DNA transposons, retrotransposons encode proteins that are required for their transposition and are flanked by direct repeats at their insertion sites. The structure of an LTR retrotransposon and its gene products is shown in **Figure 14.17**.

The steps in retrotransposon transposition involve the actions of both retrotransposon-encoded proteins and those that are part of the cell's normal transcriptional and translational machinery. In the first step, the cell's RNA polymerases transcribe the retrotransposon DNA into one or more RNA copies. In the second step, the RNA copies are translated into the two enzymes required for transposition—reverse transcriptase and integrase. The retrotransposon RNAs are then converted to double-stranded DNA copies through the actions of reverse transcriptase. The ends of the double-stranded DNAs are recognized by integrase, which then inserts the retrotransposons into the genome. Because many RNA copies can be converted to DNA and transposed in this way, retrotransposons are able to accumulate rapidly and may create mutations at many sites in the genome. In addition, the original retrotransposon is not excised during transposition.

Next, we will look at one well-studied example of a retrotransposon—*copia*—and describe its effects on the *white* locus in *Drosophila*.

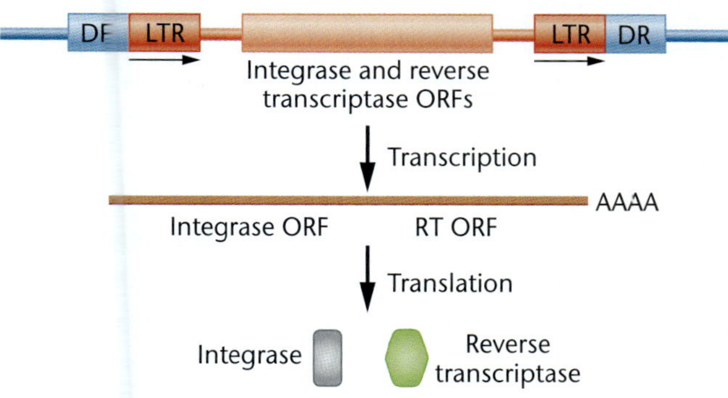

FIGURE 14.17 Structure of an LTR retrotransposon. Open reading frames (ORFs) encode the enzymes integrase and reverse transcriptase (RT). Transcription promoters and polyadenylation sites are located within 5′ and 3′ long terminal repeats (LTRs). The bottom part of the diagram shows transcription of the LTR retrotransposon and translation into integrase and reverse transcriptase. Non-LTR transposons lack LTRs, and their promoter and polyadenylation sites are located adjacent to the retrotransposon ORFs.

Retrotransposons—the *Copia*-White-Apricot System in *Drosophila*

Copia elements are transcribed into "copious" amounts of RNA (hence their name). They are present in 10 to 100 copies in the genomes of *Drosophila* cells. Mapping studies show that they are transposable to different chromosomal locations and are dispersed throughout the genome.

Each *copia* element consists of approximately 5000 to 8000 bp of DNA, including a **long terminal repeat (LTR)** sequence of 267 bp at each end. Like other LTR retrotransposons, transcription of the *copia* element begins in the 5′ LTR, which contains a promoter and transcription start site. The transcript is cleaved and polyadenylated within the 3′ LTR, which contains a polyadenylation site. These features allow the retrotransposon to be transcribed by the cell's RNA polymerase.

One of the earliest descriptions of *copia* effects came from research into the *white-apricot* mutation in *Drosophila*. This mutation changes the *Drosophila* eye color from a wild-type red to an orange-yellow color [**Figure 14.18(a)**]. DNA sequencing studies demonstrate that the mutation is caused by an insertion of *copia* into the second intron of the *white* gene. As a result of this insertion, most of the transcripts that originate from the *white* gene promoter terminate prematurely within the 3′ LTR of the *copia* retrotransposon. These prematurely terminated transcripts do not encode functional *white* gene product, resulting in a loss of red pigment in the eye [**Figure 14.18(b)**]. Because some *white* gene transcripts read through the *copia* element, enough white gene product is produced to yield a light-orange colored eye.

Transposable Elements in Humans

Recent genomic sequencing data reveal that almost half of the human genome is composed of TE DNA. As discussed earlier (see Chapter 11), the major families of human TEs are the **long interspersed elements (LINEs)** and **short interspersed elements (SINEs),** both of which are non-LTR retrotransposons. Together, they make up 34 percent of the human genome. Other families of TEs account for a further 11 percent (**Table 14.4**). As coding sequences comprise only about 1 percent of the human genome, there is about 40 to 50 times more TE DNA in the human genome than DNA in functional genes.

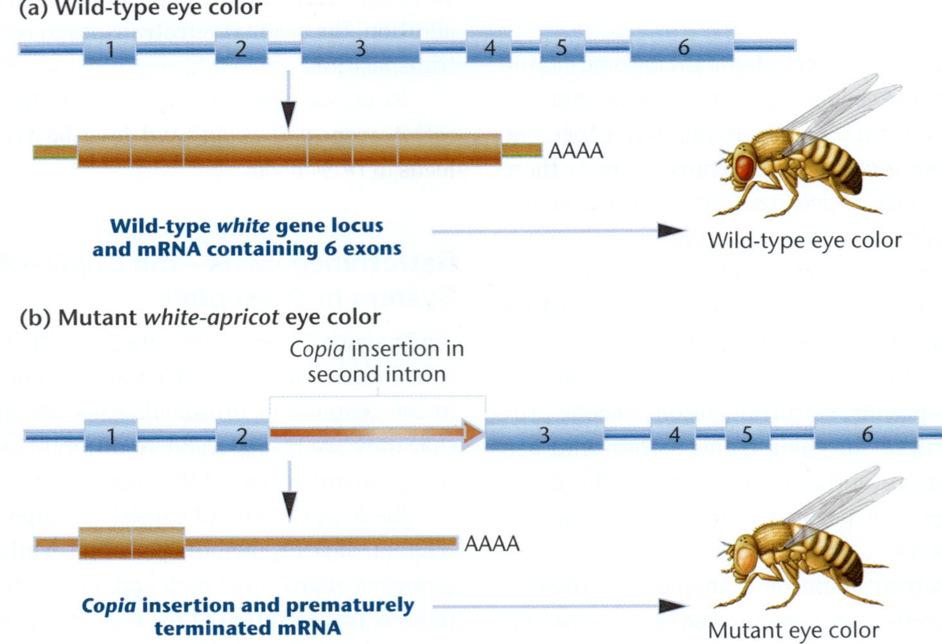

FIGURE 14.18 Effects of *copia* insertion into the *white* gene of *Drosophila*. (a) The *white* gene (top, blue) in wild-type *Drosophila* contains six exons, all of which are present in the mRNA (bottom, orange). (b) The *white* gene in mutant *white-apricot Drosophila* contains an insertion of *copia* (red) in the second intron and a prematurely terminated mRNA containing only two exons.

Although most human TEs appear to be inactive, the potential mobility and mutagenic effects of these elements have far-reaching implications for human genetics, as can be seen in an example of a TE "caught in the act." The case involves a male child with hemophilia. One cause of hemophilia is a defect in blood-clotting factor VIII, the product of an X-linked gene. Haig Kazazian and colleagues found LINEs inserted at two points within the gene. Researchers were interested in determining if one of the mother's X chromosomes also contained this specific LINE. If so, the unaffected mother would be heterozygous and could pass the LINE-containing chromosome to her son. The surprising finding was that the LINE sequence was *not* present on either of her X chromosomes but *was* detected on chromosome 22 of both parents. This suggests that this mobile element may have transposed from one chromosome to another in the gamete-forming cells of the mother, prior to being transmitted to the son.

LINE insertions into the human *dystrophin* gene (another X-linked gene) have resulted in at least two separate cases of Duchenne muscular dystrophy. In one case, a LINE inserted into exon 48, and in another case, it inserted into exon 44, both leading to frameshift mutations and premature termination of translation of the dystrophin protein. There are also reports that LINEs have inserted into the *APC* and *c-myc* genes, leading to mutations that may have contributed to the development of some colon and breast cancers.

SINE insertions are also responsible for more than 30 cases of human disease. In one case, an ***Alu* element** integrated into the *BRCA2* gene, inactivating this tumor-suppressor gene and leading to a familial case of breast cancer.

TABLE 14.4 Transposable Elements in the Human Genome

Element Type	Length	Copies in Genome	% of Genome
LINEs	1–6 kb	850,000	2[illegible]
SINEs	100–500 bp	1,500,000	1[illegible]
LTR elements	<5 kb	443,000	[illegible]
DNA transposons	80–300 bp	294,000	3
Unclassified	—	3,000	0.1

Other genes that have been mutated by *Alu* integrations are the *factor IX* gene (leading to hemophilia B), the *ChE* gene (leading to acholinesterasemia), and the *NF1* gene (leading to neurofibromatosis).

Transposable Elements, Mutations, and Evolution

TEs can have a wide range of effects on genes, based on where they are inserted and their composition. Here are a few examples:

- The insertion of a TE into one of a gene's coding regions may disrupt the gene's normal translation reading frame or may induce premature termination of translation of the gene's mRNA.
- The insertion of a TE containing polyadenylation or transcription termination signals into a gene's intron may bring about termination of the gene's transcription within the element. In addition, it can cause aberrant splicing of an RNA transcribed from the gene.
- Insertions of a TE into a gene's transcription regulatory region may disrupt the gene's normal regulation or may cause the gene to be expressed differently as a result of the presence of the TE's own promoter or enhancer sequences.
- The presence of two or more identical TEs in a genome creates the potential for recombination between the transposons, leading to duplications, deletions, inversions, or chromosome translocations. Any of these rearrangements may bring about phenotypic changes or disease.

It is thought that about 0.2 percent of detectable human mutations may be due to TE insertions. Other organisms appear to suffer more damage due to transposition. For example, about 10 percent of new mouse mutations and 50 percent of *Drosophila* mutations are caused by insertions of TEs in or near genes.

Because of their ability to alter genes and chromosomes, TEs contribute to evolution. Some mutations caused by TE insertions or deletions may be beneficial to the organism under certain circumstances. These mutations may be selected for and maintained through evolution.

In some cases, TEs themselves may be modified to perform functions that become beneficial to the organism. One example of TEs that contributed to evolution is provided by *Drosophila* telomeres. LINE-like elements are present at the ends of *Drosophila* chromosomes, and these elements have evolved to act as telomeres, maintaining the length of *Drosophila* chromosomes over successive cell divisions.

ESSENTIAL POINT

Transposable elements can move within a genome, creating mutations and altering gene expression. They may also contribute to evolution. ■

CASE STUDY An unexpected diagnosis

Six months pregnant, an expectant mother had a routine ultrasound that showed that the limbs of the fetus were unusually short. Her physician suspected that the baby might have a genetic form of dwarfism called achondroplasia, an autosomal dominant trait occurring with a frequency of about 1 in 27,000 births. The parents were directed to a genetic counselor to discuss this diagnosis. In the conference, they learned that achondroplasia is caused by a mutant allele. Sometimes it is passed from one generation to another, but in 80 percent of all cases it is the result of a spontaneous mutation that arises in a gamete of one of the parents. They also learned that most children with achondroplasia have normal intelligence and a normal life span.

1. What information would be most relevant to concluding which of the two mutation origins, inherited or new, most likely pertains in this case? How does this conclusion impact on this couple's decision to have more children?
2. It has been suggested that prenatal genetic testing for achondroplasia be made available and offered to all women. Would you agree with this initiative? What ethical considerations would you consider when evaluating the medical and societal consequences of offering such testing?

For related reading see Radoi, V., et al. (2016). How to provide a genetic counseling in a simple case of antenatal diagnosis of achondroplasia. *Gineco.eu.*12:56–58. DOI:10.18643/gieu.2016.56.

INSIGHTS AND SOLUTIONS

1. A rare dominant mutation expressed at birth was studied in humans. Records showed that six cases were discovered in 40,000 live births. Family histories revealed that in two cases, the mutation was already present in one of the parents. Calculate the spontaneous mutation rate for this mutation. What are some underlying assumptions that may affect our conclusions?

   Solution: Only four cases represent a new mutation. Because each live birth represents two gametes, the sample size is from 80,000 meiotic events. The rate is equal to

   $$4/80{,}000 = 1/20{,}000 = 5 \times 10^{-5}$$

   We have assumed that the mutant gene is fully penetrant and is expressed in each individual bearing it. If it is not fully penetrant, our calculation may be an underestimate because one or more mutations may have gone undetected. We have also assumed that the screening was 100 percent accurate. One or more mutant individuals may have been "missed," again leading to an underestimate. Finally, we assumed that the viabilities of the mutant and nonmutant individuals are equivalent and that they survive equally *in utero*. Therefore, our assumption is that the number of mutant individuals at birth is equal to the number at conception. If this were not true, our calculation would again be an underestimate.

2. Consider the following estimates:

   (a) There are 7×10^9 humans living on this planet.

   (b) Each individual has about 20,000 (0.2×10^5) genes.

   (c) The average mutation rate at each locus is 10^{-5}.

   How many spontaneous mutations are currently present in the human population? Assuming that these mutations are equally distributed among all genes, how many new mutations have arisen in each gene in the human population?

   Solution: First, since each individual is diploid, there are two copies of each gene per person, each arising from a separate gamete. Therefore, the total number of spontaneous mutations is

   $$(2 \times 0.2 \times 10^5 \text{ genes/individual}) \times (7 \times 10^9 \text{ individuals}) \times (10^{-5} \text{ mutations/gene})$$
   $$= (0.4 \times 10^5) \times (7 \times 10^9) \times (10^{-5}) \text{ mutations}$$
   $$= 2.8 \times 10^9 \text{ mutations in the population}$$

   $$2.8 \times 10^9 \text{ mutations} / 0.2 \times 10^5 \text{ genes}$$
   $$= 14 \times 10^4 \text{ mutations per gene in the population}$$

3. The base analog 2-amino purine (2-AP) substitutes for adenine during DNA replication, but it may base-pair with cytosine. The base analog 5-bromouracil (5-BU) substitutes for thymine, but it may base-pair with guanine. Follow the double-stranded trinucleotide sequence shown at the top of the figure through three rounds of replication, assuming that, in the first round, both analogs are present and become incorporated wherever possible. Before the second and third round of replication, any unincorporated base analogs are removed. What final sequences occur?

   Solution:

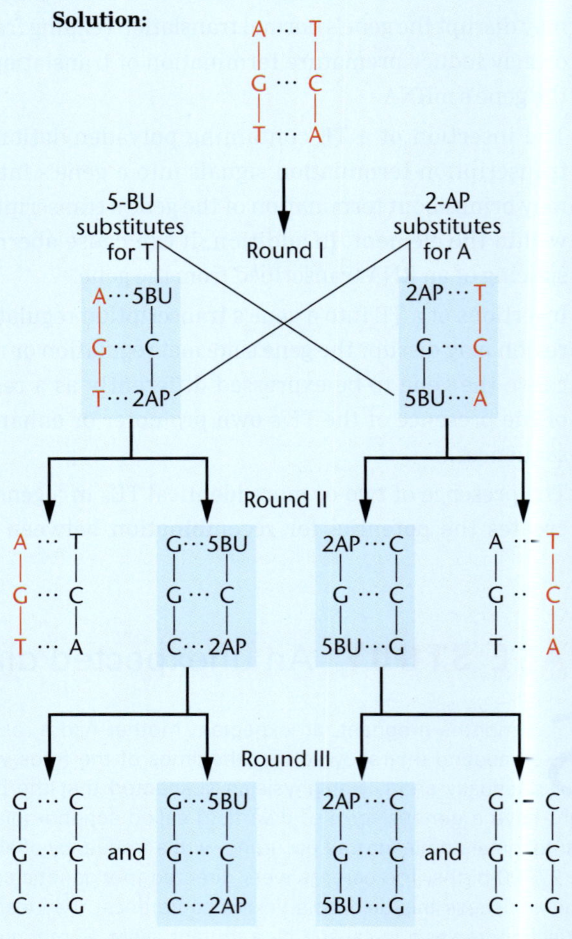

Problems and Discussion Questions

Mastering Genetics Visit for instructor-assigned tutorials and problems.

1. **HOW DO WE KNOW?** In this chapter, we focused on how gene mutations arise and how cells repair DNA damage. At the same time, we found opportunities to consider the methods and reasoning by which much of this information was acquired. From the explanations given in the chapter,
   (a) How do we know that many cancer-causing agents (carcinogens) are also mutagenic?
   (b) How do we know that certain chemicals and wavelengths of radiation induce mutations in DNA?
   (c) How do we know that DNA repair mechanisms detect and correct the majority of spontaneous and induced mutations?
2. **CONCEPT QUESTION** Review the Chapter Concepts list on p. 285. These concepts relate to how gene mutations occur, their phenotypic effects, and how mutations can be repaired. Write a short essay contrasting how these concepts may differ between bacteria and eukaryotes.
3. Distinguish between spontaneous and induced mutations. Give some examples of mutagens that cause induced mutations.
4. Why would a mutation in a germ-line cell not necessarily result in a hereditary phenotype?
5. Discuss why germ-line mutations provide the raw material for natural selection while somatic mutations do not.
6. Why is a random mutation more likely to be deleterious than beneficial?
7. Most mutations in a diploid organism are recessive. Why?
8. What is the difference between a silent mutation and a neutral mutation?
9. Compare tautomeric shifts with deamination mutations.
10. Contrast and compare the mutagenic effects of deaminating agents, alkylating agents, and base analogs.
11. Why are frameshift mutations likely to be more detrimental than point mutations, in which a single pyrimidine or purine has been substituted?
12. In which phases of the cell cycle would you expect double-strand break repair and nonhomologous end joining to occur and why?
13. Postreplication repair and the nucleotide excision repair can both correct thymine dimers. Describe both systems in detail and contrast their mode of action.
14. Mammography is an accurate screening technique for the early detection of breast cancer in humans. Because this technique uses X rays diagnostically, it has been highly controversial. Can you explain why? What reasons justify the use of X rays for such a medical screening technique?
15. A significant number of mutations in the *HBB* gene that cause human β-thalassemia occur within introns or in upstream noncoding sequences. Explain why mutations in these regions often lead to severe disease, although they may not directly alter the coding regions of the gene.
16. A chemist has synthesized a novel chemical, which he suspects to be a potential mutagen. Name and explain a popular test that can be used to test the mutagenicity of this product in bacteria.
17. What genetic defects result in the disorder xeroderma pigmentosum (XP) in humans? How do these defects create the phenotypes associated with the disorder?
18. Describe the structures and modes of action of the different types of retrotransposons.
19. In maize, a *Ds* or *Ac* transposon can alter the function of genes at or near the site of transposon insertion. It is possible for these elements to transpose away from their original insertion site, causing a reversion of the mutant phenotype. In some cases, however, even more severe phenotypes appear, due to events at or near the mutant allele. What might be happening to the transposon or the nearby gene to create more severe mutations?
20. Suppose you are studying a DNA repair system, such as the nucleotide excision repair *in vitro*. By mistake, you add DNA ligase from a tube that has already expired. What would be the result?
21. In a bacterial culture in which all cells are unable to synthesize leucine (leu^-), a potent mutagen is added, and the cells are allowed to undergo one round of replication. At that point, samples are taken, a series of dilutions are made, and the cells are plated on either minimal medium or minimal medium containing leucine. The first culture condition (minimal medium) allows the growth of only leu^+ cells, while the second culture condition (minimal medium with leucine added) allows growth of all cells. The results of the experiment are as follows:

Culture Condition	Dilution	Colonies
Minimal medium	10^{-1}	18
Minimal medium + leucine	10^{-7}	6

What is the rate of mutation at the locus associated with leucine biosynthesis?

22. Presented here are hypothetical findings from studies of heterokaryons formed from seven human xeroderma pigmentosum cell strains:

	XP1	XP2	XP3	XP4	XP5	XP6	XP7
XP1	−						
XP2	−	−					
XP3	−	−	−				
XP4	+	+	+	−			
XP5	+	+	+	+	−		
XP6	+	+	+	+	−	−	
XP7	+	+	+	+	−	−	−

Note: + = complementation; − = no complementation

These data are measurements of the occurrence or nonoccurrence of unscheduled DNA synthesis in the fused heterokaryon. None of the strains alone shows any unscheduled DNA synthesis. Which strains fall into the same complementation groups? How many different groups are revealed based on these data? What can we conclude about the genetic basis of XP from these data?

23. Skin cancer carries a lifetime risk nearly equal to that of all other cancers combined. Following is a graph [modified from K. H. Kraemer (1997). *Proc. Natl. Acad. Sci. (USA)* 94:11–14] depicting the age of onset of skin cancers in patients with or without XP, where the cumulative percentage of skin cancer is

plotted against age. The non-XP curve is based on 29,757 cancers surveyed by the National Cancer Institute, and the curve representing those with XP is based on 63 skin cancers from the Xeroderma Pigmentosum Registry.

(a) Provide an overview of the information contained in the graph.

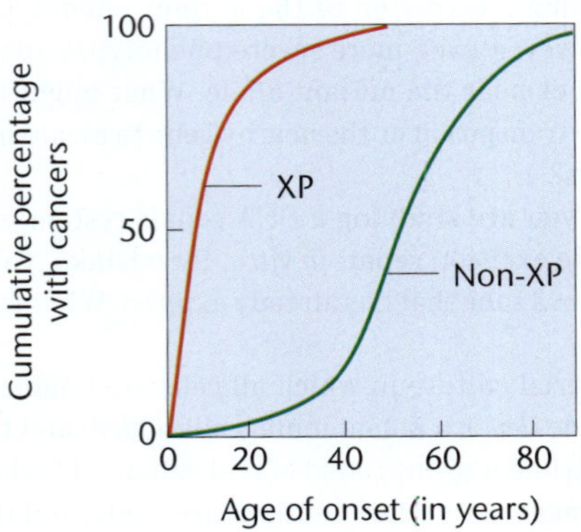

(b) Explain why individuals with XP show such an early age of onset.

24. It has been noted that most transposons in humans and other organisms are located in noncoding regions of the genome—regions such as introns, pseudogenes, and stretches of particular types of repetitive DNA. There are several ways to interpret this observation. Describe two possible interpretations. Which interpretation do you favor? Why?

25. Mutations in the *IL2RG* gene cause approximately 30 percent of severe combined immunodeficiency disorder (SCID) cases in humans. These mutations result in alterations to a protein component of cytokine receptors that are essential for proper development of the immune system. The *IL2RG* gene is composed of eight exons and contains upstream and downstream sequences that are necessary for proper transcription and translation. Below are some of the mutations observed. For each, explain its likely influence on the *IL2RG* gene product (assume its length to be 375 amino acids).

(a) Nonsense mutation in a coding region
(b) Insertion in exon 1, causing frameshift
(c) Insertion in exon 7, causing frameshift
(d) Missense mutation
(e) Deletion in exon 2, causing frameshift
(f) Deletion in exon 2, in frame
(g) Large deletion covering exons 2 and 3

15

Regulation of Gene Expression in Bacteria

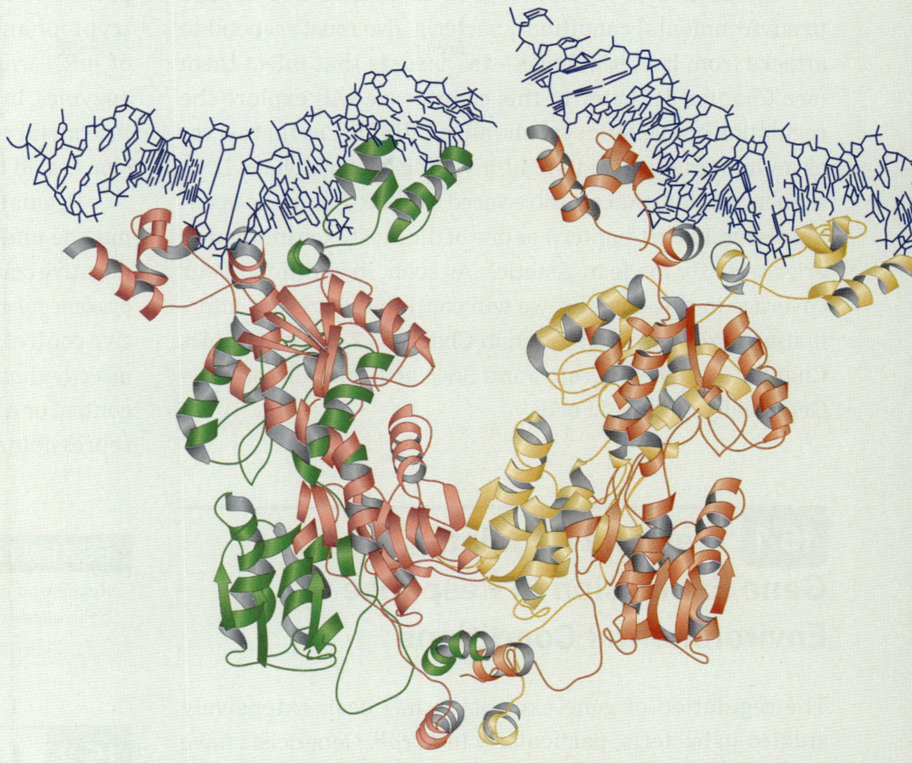

A model generated using crystal structure analysis depicting the *lac* repressor tetramer bound to two operator DNA sequences (shown in blue).

CHAPTER CONCEPTS

- In bacteria, regulation of gene expression is often linked to the metabolic needs of the cell.
- Efficient expression of genetic information in bacteria is dependent on intricate regulatory mechanisms that exert control over transcription.
- Mechanisms that regulate transcription are categorized as exerting either positive or negative control of gene expression.
- Bacterial genes that encode proteins with related functions tend to be organized in clusters and are often under coordinated control. Such clusters, including their adjacent regulatory sequences, are called operons.
- Transcription of genes within operons is either inducible or repressible.
- Often, the end product of a metabolic pathway induces or represses gene expression in that pathway.

Previous chapters have discussed how DNA is organized into genes, how genes store genetic information, and how this information is expressed through the processes of transcription and translation. We now consider one of the most fundamental questions in molecular genetics: *How is genetic expression regulated?* It is now clear that gene expression varies widely in bacteria under different environmental conditions. For example, detailed analysis of proteins in *Escherichia coli* shows that concentrations of the 4000 or so polypeptide chains encoded by the genome vary widely. Some proteins may be present in as few as 5 to 10 molecules per cell, whereas others, such as ribosomal proteins and the many proteins involved in the glycolytic pathway, are present in as many as 100,000 copies per cell. Although most bacterial gene products are present continuously at a basal level (a few copies), the concentration of these products can increase dramatically when required. Clearly, fundamental regulatory mechanisms must exist to control the expression of the genetic information.

In this chapter, we will explore regulation of gene expression in bacteria. As we have seen in several previous chapters, bacteria have been especially useful research organisms in genetics for a number of reasons. For one thing, they have extremely short reproductive cycles, and literally hundreds of generations, giving rise to billions of genetically identical bacteria, can be produced in overnight cultures. In addition, they can be studied in "pure culture," allowing mutant strains of genetically unique bacteria to be isolated and investigated separately.

In addition to regulating gene expression in response to environmental conditions, bacteria also must respond to attacks from bacteriophages—the viruses that infect them (see Chapter 8). Later in this chapter we will explore the regulation of a genetic system called CRISPR-Cas that serves as an immune system to fight invading bacteriophage DNA sequences. Note that we introduced CRISPR-Cas in the opening discussion in Chapter 1 as one of the most exciting discoveries recently made in genetics. As such, in addition to our coverage in this chapter, we will continue to present information involving this system in Chapter 17, Special Topics Chapter 3—Gene Therapy, and Special Topics Chapter 6—Genetically Modified Foods.

15.1 Bacteria Regulate Gene Expression in Response to Environmental Conditions

The regulation of gene expression has been extensively studied in bacteria, particularly in *E. coli*. Geneticists have learned that highly efficient genetic mechanisms have evolved in these organisms to turn transcription of specific genes on and off, depending on the cell's metabolic need for the respective gene products. Not only do bacteria respond to changes in their environment, but they also regulate gene activity associated with a variety of cellular activities, including the replication, recombination, and repair of their DNA, and with cell division.

The idea that microorganisms regulate the synthesis of their gene products is not a new one. As early as 1900, it was shown that when lactose (a galactose and glucose-containing disaccharide) is present in the growth medium of yeast, the organisms synthesize enzymes required for lactose metabolism. When lactose is absent, synthesis diminishes to a basal level. Soon thereafter, investigators generalized that bacteria adapt to their environment, producing certain enzymes only when specific chemical substrates are present. These are now referred to as **inducible enzymes.** In contrast, enzymes that are produced continuously, regardless of the chemical makeup of the environment, are called **constitutive enzymes.**

More recent investigation has revealed a contrasting system, whereby the presence of a specific molecule *inhibits* gene expression. Such molecules are usually end products of anabolic biosynthetic pathways. For example, utilizing a multistep metabolic pathway, the amino acid tryptophan can be synthesized by bacterial cells. If a sufficient supply of tryptophan is present in the environment or culture medium, then it is inefficient for the organism to expend energy to synthesize the enzymes necessary for tryptophan production. A mechanism has therefore evolved whereby tryptophan plays a role in repressing the transcription of mRNA needed for producing tryptophan-synthesizing enzymes. In contrast to the inducible system controlling lactose metabolism, the system governing tryptophan expression is said to be **repressible.**

Regulation, whether of the inducible or repressible type, may be under either **negative** or **positive control.** Under negative control, genetic expression occurs *unless it is shut off by some form of a regulator molecule*. In contrast, under positive control, transcription occurs *only if a regulator molecule directly stimulates RNA production*. In theory, either type of control or a combination of the two can govern inducible or repressible systems.

ESSENTIAL POINT

Regulation of gene expression in bacteria is often linked to the metabolic needs of the organism. ■

15.2 Lactose Metabolism in *E. coli* Is Regulated by an Inducible System

Beginning in 1946 with the studies of Jacques Monod and continuing through the next decade with significant contributions by Joshua Lederberg, François Jacob, and André Lwoff, genetic and biochemical evidence concerning lactose metabolism was amassed. Research provided insights into the way in which gene activity is repressed when lactose is absent but induced when it is available. In the presence of lactose, the concentration of the enzymes responsible for its metabolism increases rapidly from a few molecules to thousands per cell. The enzymes responsible for lactose metabolism are thus inducible, and lactose serves as the inducer.

In bacteria, genes that code for enzymes with related functions (for example, the set of genes involved with lactose metabolism) tend to be organized in clusters on the bacterial chromosome, and transcription of these genes is often under the coordinated control of a single regulatory region. Such clusters, including their adjacent regulatory sequences, are called **operons.** The location of the regulatory region is almost always upstream (5′) of the gene cluster it controls. Because the regulatory region is on the same strand as those genes, we refer to it as a ***cis*-acting site.** *Cis*-acting regulatory regions are bound by molecules that control transcription of the gene cluster. Such molecules are called ***trans*-acting factors.** Events

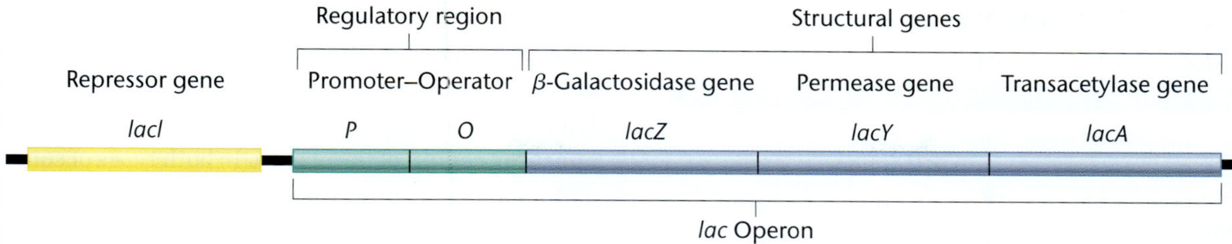

FIGURE 15.1 A simplified overview of the genes and regulatory units involved in the control of lactose metabolism. (The regions within this stretch of DNA are not drawn to scale.)

at the regulatory site determine whether the genes are transcribed into mRNA and thus whether the corresponding enzymes or other protein products may be synthesized from the genetic information in the mRNA. Binding of a *trans*-acting element at a *cis*-acting site can regulate the gene cluster either negatively (by turning off transcription) or positively (by turning on transcription of genes in the cluster). In this section, we discuss how transcription of such bacterial gene clusters is coordinately regulated.

The discovery of a regulatory gene and a regulatory site that are part of the gene cluster was paramount to the understanding of how gene expression is controlled in the system. Neither of these regulatory elements encodes enzymes necessary for lactose metabolism—the function of the three genes in the cluster. As illustrated in **Figure 15.1**, the three structural genes and the adjacent regulatory site constitute the **lactose (*lac*) operon.** Together, the entire gene cluster functions in an integrated fashion to provide a rapid response to the presence or absence of lactose.

Structural Genes

Genes coding for the primary structure of an enzyme are called **structural genes.** There are three structural genes in the *lac* operon. The *lacZ* gene encodes **β-galactosidase,** an enzyme whose primary role is to convert the disaccharide lactose to the monosaccharides glucose and galactose (**Figure 15.2**). This conversion is essential if lactose is to serve as the primary energy source in glycolysis. The second gene, *lacY*, specifies the primary structure of **permease,** an enzyme that facilitates the entry of lactose into the bacterial cell. The third gene, *lacA*, codes for the enzyme **transacetylase.** While its physiological role is still not completely clear, it may be involved in the removal of toxic by-products of lactose digestion from the cell.

To study the genes coding for these three enzymes, researchers isolated numerous mutations that lacked the function of one or the other enzyme. Such *lac*$^-$ mutants were first isolated and studied by Joshua Lederberg. Mutant cells that fail to produce active β-galactosidase (*lacZ*$^-$) or permease (*lacY*$^-$) are unable to use lactose as an energy source. Mutations were also found in the transacetylase gene. Mapping studies by Lederberg established that all three genes are closely linked or contiguous to one another on the bacterial chromosome, in the order *Z–Y–A* (see Figure 15.1).

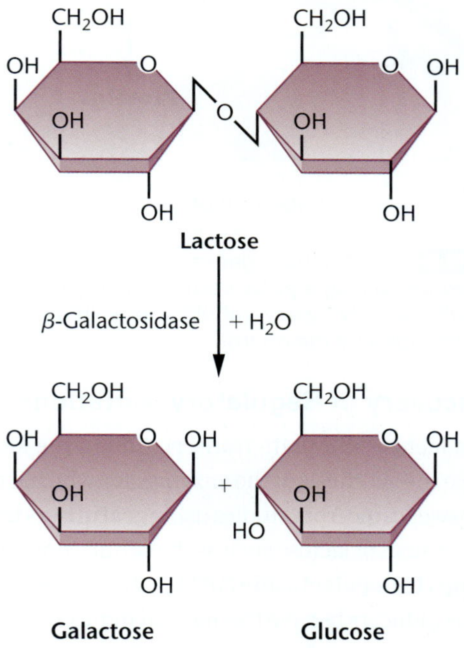

FIGURE 15.2 The catabolic conversion of the disaccharide lactose into its monosaccharide units, galactose and glucose.

Knowledge of their close linkage led to another discovery relevant to what later became known about the regulation of structural genes: All three genes are transcribed as a single unit, resulting in a so-called *polycistronic mRNA* (**Figure 15.3**) (recall that *cistron* refers to the part of a nucleotide sequence coding for a single gene). This results in the coordinate regulation of all three genes, since a single-message RNA is simultaneously translated into all three gene products.

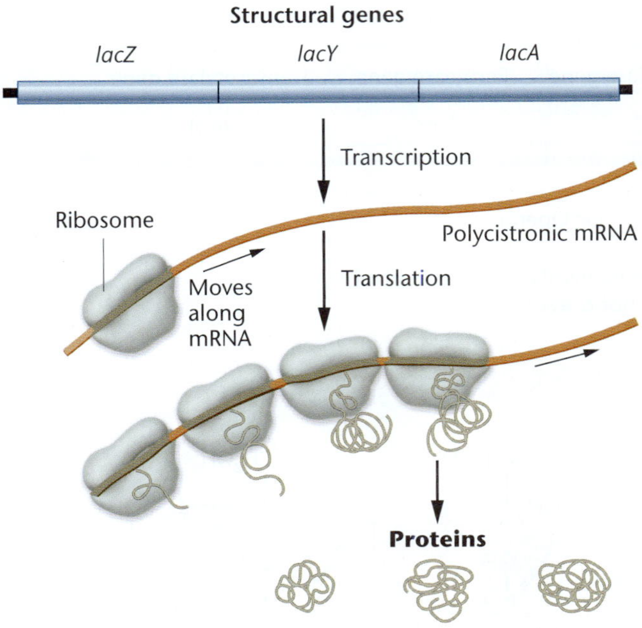

FIGURE 15.3 The structural genes of the *lac* operon are transcribed into a single polycistronic mRNA, which is translated simultaneously by several ribosomes into the three enzymes encoded by the operon.

The Discovery of Regulatory Mutations

How does lactose stimulate transcription of the *lac* operon and induce the synthesis of the enzymes for which it codes? A partial answer came from studies using **gratuitous inducers,** chemical analogs of lactose such as the sulfur-containing analog **isopropylthiogalactoside (IPTG),** shown in **Figure 15.4**. Gratuitous inducers behave like natural inducers, but they do not serve as substrates for the enzymes that are subsequently synthesized. Their discovery provides strong evidence that the primary induction event does *not* depend on the interaction between the inducer and the enzyme.

What, then, is the role of lactose in induction? The answer to this question requires the study of another class of mutations described as **constitutive mutations.** In cells bearing these types of mutations, enzymes are produced regardless of the presence or absence of lactose. Studies of the constitutive mutation *lacI*$^-$ mapped the mutation to a site on the bacterial chromosome close to, but not part of, the structural genes *lacZ, lacY,* and *lacA*. This mutation led researchers to discover the *lacI* gene, which is appropriately called a **repressor gene.** A second set of constitutive mutations producing effects identical to those of *lacI*$^-$ is present in a region immediately adjacent to the structural genes. This class of mutations, designated *lacO*^{C}, is located in the **operator region** of the operon. In both types of constitutive mutants, the enzymes are produced continually, inducibility is eliminated, and gene regulation has been lost.

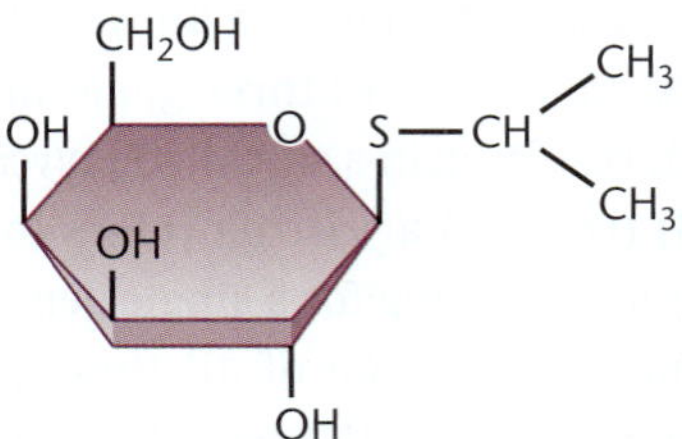

FIGURE 15.4 The gratuitous inducer isopropylthiogalactoside (IPTG).

The Operon Model: Negative Control

Around 1960, Jacob and Monod proposed a hypothetical mechanism involving negative control that they called the **operon model,** in which a group of genes is regulated and expressed together as a unit. As we saw in Figure 15.1, the *lac* operon they proposed consists of the *Z, Y,* and *A* structural genes, as well as the adjacent sequences of DNA referred to as the *operator region*. They argued that the *lacI* gene regulates the transcription of the structural genes by producing a **repressor molecule** and that the repressor is **allosteric,** meaning that the molecule reversibly interacts with another molecule, undergoing both a conformational change in three-dimensional shape and a change in chemical activity. **Figure 15.5** illustrates the components of the *lac* operon as well as the action of the *lac* repressor in the presence and absence of lactose.

Jacob and Monod suggested that the repressor normally binds to the DNA sequence of the operator region. When it does so, it inhibits the action of RNA polymerase, effectively repressing the transcription of the structural genes [**Figure 15.5(b)**]. However, when lactose is present, this sugar binds to the repressor and causes an allosteric (conformational) change. The change alters the binding site of the repressor, rendering it incapable of interacting with operator DNA [**Figure 15.5(c)**]. In the absence of the repressor–operator interaction, RNA polymerase transcribes the structural genes, and the enzymes necessary for lactose metabolism are produced. Because transcription occurs only when the repressor *fails* to bind to the operator region, regulation is said to be under *negative control*.

To summarize, the operon model invokes a series of molecular interactions between proteins, inducers, and DNA to explain the efficient regulation of structural gene expression. In the absence of lactose, the enzymes encoded by the genes are not needed and the expression of genes encoding these enzymes is repressed. When lactose is present, it indirectly induces the activation of the genes by binding with the repressor.* If all lactose is metabolized, none is available to bind to the repressor, which is again free to bind to operator DNA and to repress transcription.

Both the I^- and O^C constitutive mutations interfere with these molecular interactions, allowing continuous

*Technically, the inducer is allolactose, an isomer of lactose. When lactose enters the bacterial cell, some of it is converted to allolactose by the β-galactosidase enzyme.

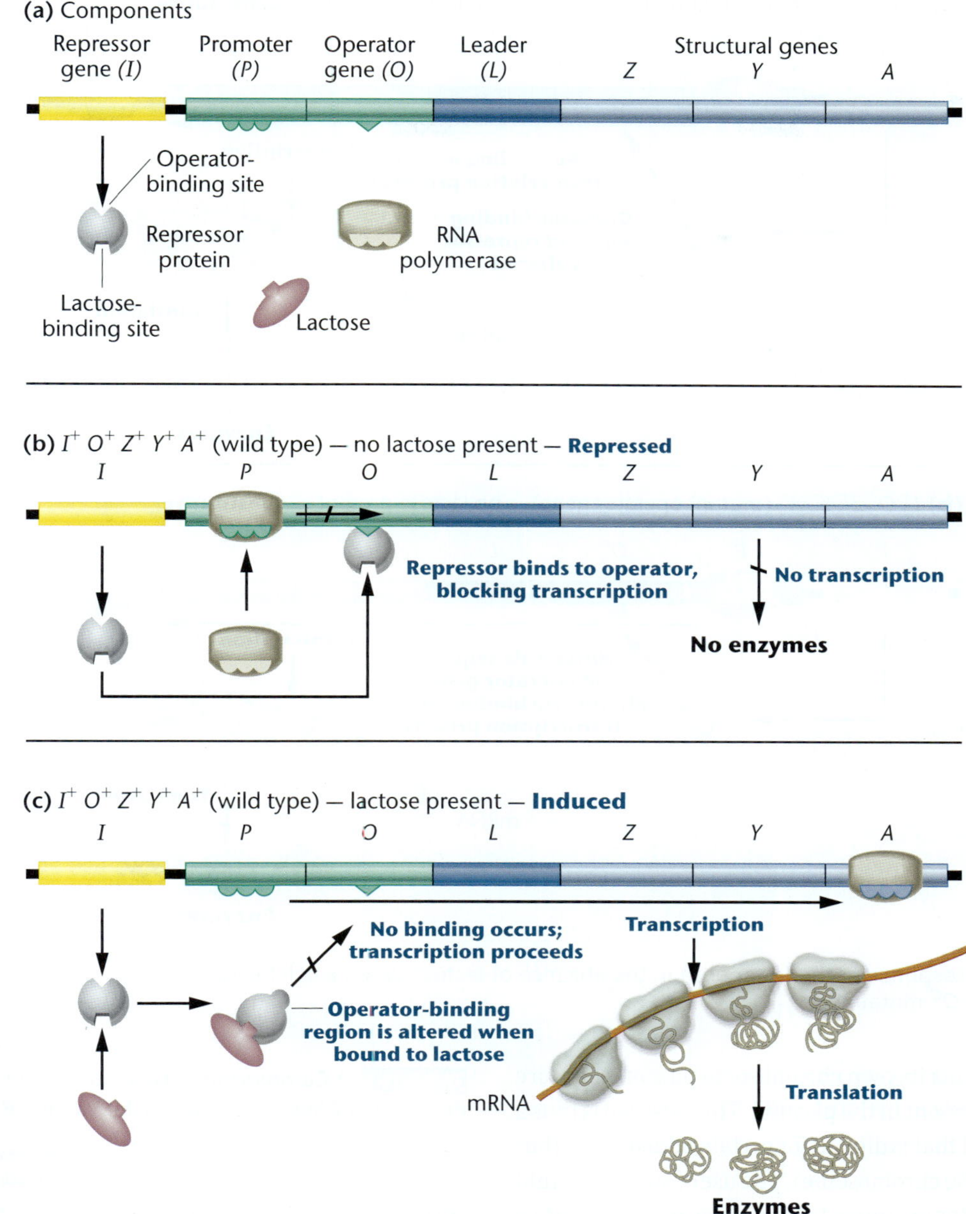

FIGURE 15.5 The components of the wild-type *lac* operon and the response in the absence and presence of lactose.

transcription of the structural genes. In the case of the I^- mutant, seen in **Figure 15.6(a)**, the repressor protein is altered or absent and cannot bind to the operator region, so the structural genes are always turned on. In the case of the O^C mutant [**Figure 15.6(b)**], the nucleotide sequence of the operator DNA is altered and will not bind with a normal repressor molecule. The result is the same: The structural genes are always transcribed.

ESSENTIAL POINT

Genes involved in the metabolism of lactose are coordinately regulated by a negative control system that responds to the presence or absence of lactose. ■

Genetic Proof of the Operon Model

The operon model is a good one because it leads to three major predictions that can be tested to determine its validity. The major predictions to be tested are that (1) the *I* gene produces a diffusible product (that is, a *trans*-acting product), (2) the *O* region is involved in regulation but does not produce a product (it is *cis*-acting), and (3) the *O* region must be adjacent to the structural genes to regulate transcription.

The creation of partially diploid bacteria allows us to assess these assumptions, particularly those that predict the presence of *trans*-acting regulatory elements. For example, the F plasmid may contain chromosomal genes (Chapter 8), in which case it is designated F′. When an F^- cell acquires such

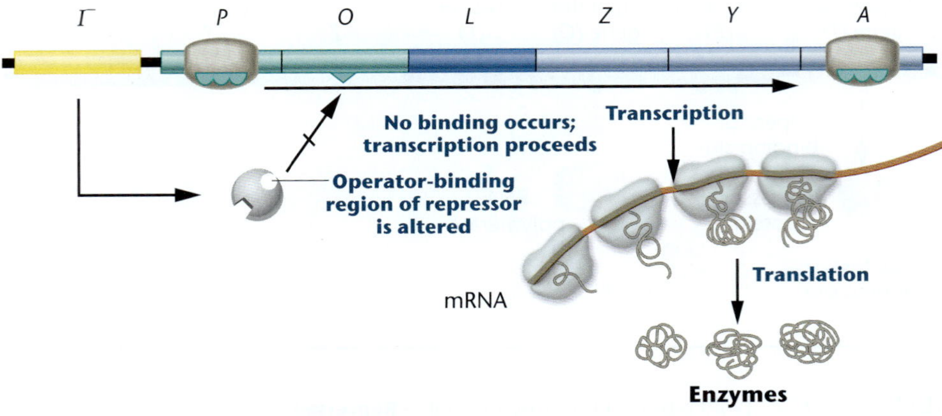

(b) $I^+ O^c Z^+ Y^+ A^+$ (mutant operator gene) — no lactose present — **Constitutive**

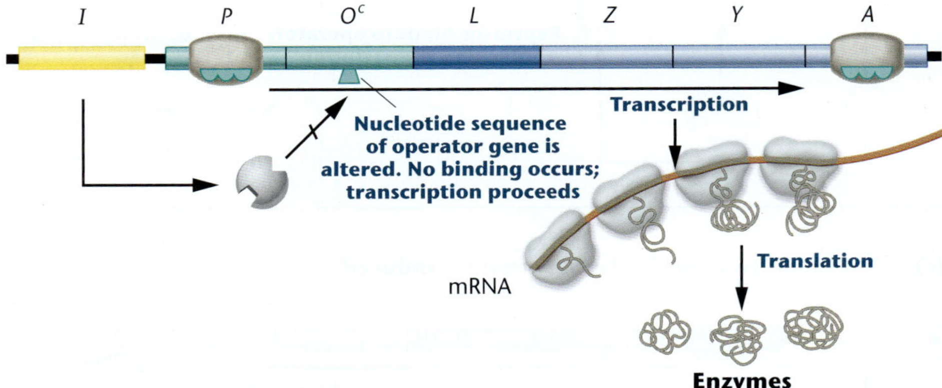

FIGURE 15.6 The response of the *lac* operon in the absence of lactose when a cell bears either the I^- or the O^C mutation.

a plasmid, it contains its own chromosome plus one or more additional genes present in the plasmid. This host cell is thus a **merozygote,** a cell that is diploid for certain added genes (but not for the rest of the chromosome). The use of such a plasmid makes it possible, for example, to introduce an I^+ gene into a host cell whose genotype is I^-, or to introduce an O^+ region into a host cell of genotype O^C. The Jacob–Monod operon model predicts how regulation should be affected in such cells. Adding an I^+ gene to an I^- cell should restore inducibility, because the normal wild-type repressor, which is a *trans*-acting factor, would be produced by the inserted I^+ gene. In contrast, adding an O^+ region to an O^C cell should have no effect on constitutive enzyme production, since regulation depends on an O^+ region being located immediately adjacent to the structural genes—that is, O^+ is a *cis*-acting element.

Results of these experiments are shown in **Table 15.1**, where Z represents the structural genes (and the inserted genes are listed after the designation F′). In both cases described above, the Jacob–Monod model is upheld (part B of Table 15.1). Part C of the table shows the reverse experiments, where either an I^- gene or an O^C region is added to cells of normal inducible genotypes. As the model predicts, inducibility is maintained in these partial diploids.

TABLE 15.1 A Comparison of Gene Activity (+ or −) in the Presence or Absence of Lactose for Various *E. coli* Genotypes

	Presence of β-Galactosidase Activity	
Genotype	Lactose Present	Lactose Absent
A. $I^+O^+Z^+$	+	−
$I^+O^+Z^-$	−	−
$I^-O^+Z^+$	+	+
$I^+O^CZ^+$	+	+
B. $I^-O^+Z^+/F'I^+$	+	−
$I^+O^CZ^+/F'O^+$	+	+
C. $I^+O^+Z^+/F'I^-$	+	−
$I^+O^+Z^+/F'O^C$	+	−
D. $I^SO^+Z^+$	−	−
$I^SO^+Z^+/F'I^+$	−	−

*Note: In parts B to D, most genotypes are partially diploid, containing an F factor plus attached genes (F′).

Another prediction of the operon model is that certain mutations in the I gene should have the opposite effect of I^-. That is, instead of being constitutive because the repressor cannot bind the operator, mutant repressor molecules should

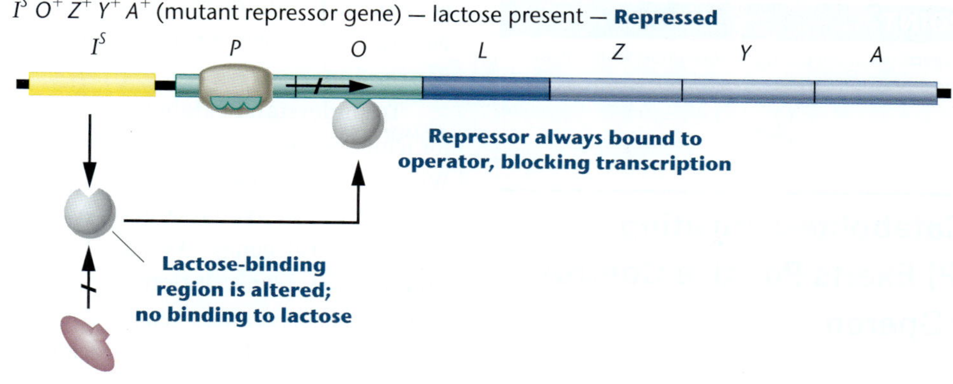

FIGURE 15.7 The response of the *lac* operon in the presence of lactose in a cell bearing the I^S mutation.

be produced that cannot interact with the inducer, lactose. Thus, these repressors would always bind to the operator sequence, and the structural genes would be permanently repressed. In cases like this, the presence of an additional I^+ gene would have little or no effect on repression.

In fact, such a mutation, I^S, was discovered wherein the operon, as predicted, is "superrepressed," as shown in part D of Table 15.1 (and depicted in **Figure 15.7**). An additional I^+ gene does not effectively relieve repression of gene activity. These observations are consistent with the idea that the repressor contains separate DNA-binding domains and inducer-binding domains.

Isolation of the Repressor

Although Jacob and Monod's operon theory succeeded in explaining many aspects of genetic regulation in bacteria, the nature of the repressor molecule was not known when their landmark paper was published in 1961. Although they had assumed that the allosteric repressor was a protein, RNA was also a viable candidate because the activity of the molecule required the ability to bind to DNA. Despite many attempts to isolate and characterize the hypothetical repressor molecule, no direct chemical evidence was forthcoming. A single *E. coli* cell contains no more than ten or so copies of the *lac* repressor, and direct chemical identification of ten molecules in a population of millions of proteins and RNAs in a single cell presented a tremendous challenge.

In 1966, Walter Gilbert and Benno Müller-Hill reported the isolation of the *lac* repressor in partially purified form. To facilitate the isolation, they used a *regulator quantity* (I^q) mutant strain that contains about ten times as much repressor as do wild-type *E. coli* cells. Also instrumental in their success was the use of the gratuitous inducer IPTG, which binds to the repressor, and the technique of **equilibrium dialysis.** In this technique, extracts of I^q cells were placed in a dialysis bag and allowed to attain equilibrium with an external solution of radioactive IPTG, a molecule small enough to diffuse freely in and out of the bag. At equilibrium, the concentration of radioactive IPTG was higher inside the bag than in the external solution, indicating that an IPTG-binding material was present in the cell extract and was too large to diffuse across the wall of the bag.

NOW SOLVE THIS

15.1 Even though the *lac Z, Y,* and *A* structural genes are transcribed as a single polycistronic mRNA, each gene contains the initiation and termination signals essential for translation. Predict what will happen when a cell growing in the presence of lactose contains a deletion of one nucleotide (a) early in the *Z* gene and (b) early in the *A* gene.

■ **HINT:** *This problem requires you to combine your understanding of the genetic expression of the lac operon with that of the genetic code, frameshift mutations, and termination of transcription. The key to its solution is to consider the effect of the loss of one nucleotide within a polycistronic mRNA.*

Ultimately, the IPTG-binding material was purified and shown to have various characteristics of a protein. In contrast, extracts of I^- constitutive cells having no *lac* repressor *activity* did not exhibit IPTG binding, strongly suggesting that the isolated protein was the repressor molecule.

To confirm this thinking, Gilbert and Müller-Hill grew *E. coli* cells in a medium containing radioactive sulfur and then isolated the IPTG-binding protein, which was labeled in its sulfur-containing amino acids. Next, this protein was mixed with DNA from a strain of phage lambda (λ) carrying the *lacO*$^+$ gene. When the two substances are separate, the DNA sediments at 40*S* and the IPTG-binding protein sediments at 7*S*. However, when the DNA and protein were mixed and sedimented in a gradient, using ultracentrifugation, the radioactive protein sedimented at the same rate as did the DNA, indicating that the protein binds to the DNA. Further experiments showed that the IPTG-binding, or repressor, protein binds only to DNA containing the *lac* region and does not bind to *lac* DNA containing an operator-constitutive O^C mutation.

ESSENTIAL POINT

Research on the *lac* operon in *E. coli* pioneered our understanding of the regulation of gene expression. ■

15.3 The Catabolite-Activating Protein (CAP) Exerts Positive Control over the *lac* Operon

As described in the preceding discussion of the *lac* operon, the role of β-galactosidase is to cleave lactose into its components, glucose and galactose. Then, to be used by the cell, the galactose, too, must be converted to glucose. What if the cell found itself in an environment that contained ample amounts of both lactose *and* glucose? Given that glucose is the preferred carbon source for *E. coli,* it would not be energetically efficient for a cell to induce transcription of the *lac* operon, since what it really needs—glucose—is already present. As we will see next, still another molecular component, called the **catabolite-activating protein (CAP),** is involved in diminishing the expression of the *lac* operon when glucose is present. This inhibition is called **catabolite repression.**

To understand catabolite repression, let's backtrack for a moment to review the system depicted in Figure 15.5. When the *lac* repressor is bound to the inducer (lactose), the *lac* operon is activated, and RNA polymerase transcribes the structural genes. As stated earlier in the text (see Chapter 12), transcription is initiated as a result of the binding that occurs between RNA polymerase and the nucleotide sequence of the promoter region, found upstream (5′) from the initial coding sequences. Within the *lac* operon, the promoter is found between the *I* gene and the operator region (*O*) (see Figure 15.1). *Careful examination has revealed that polymerase binding is never very efficient unless CAP is also present to facilitate the process.*

The mechanism is summarized in **Figure 15.8**. In the absence of glucose and under inducible conditions, CAP exerts positive control by binding to the CAP site, facilitating RNA-polymerase binding at the promoter, and thus transcription. Therefore, for maximal transcription of the structural genes to occur, the repressor must be bound by

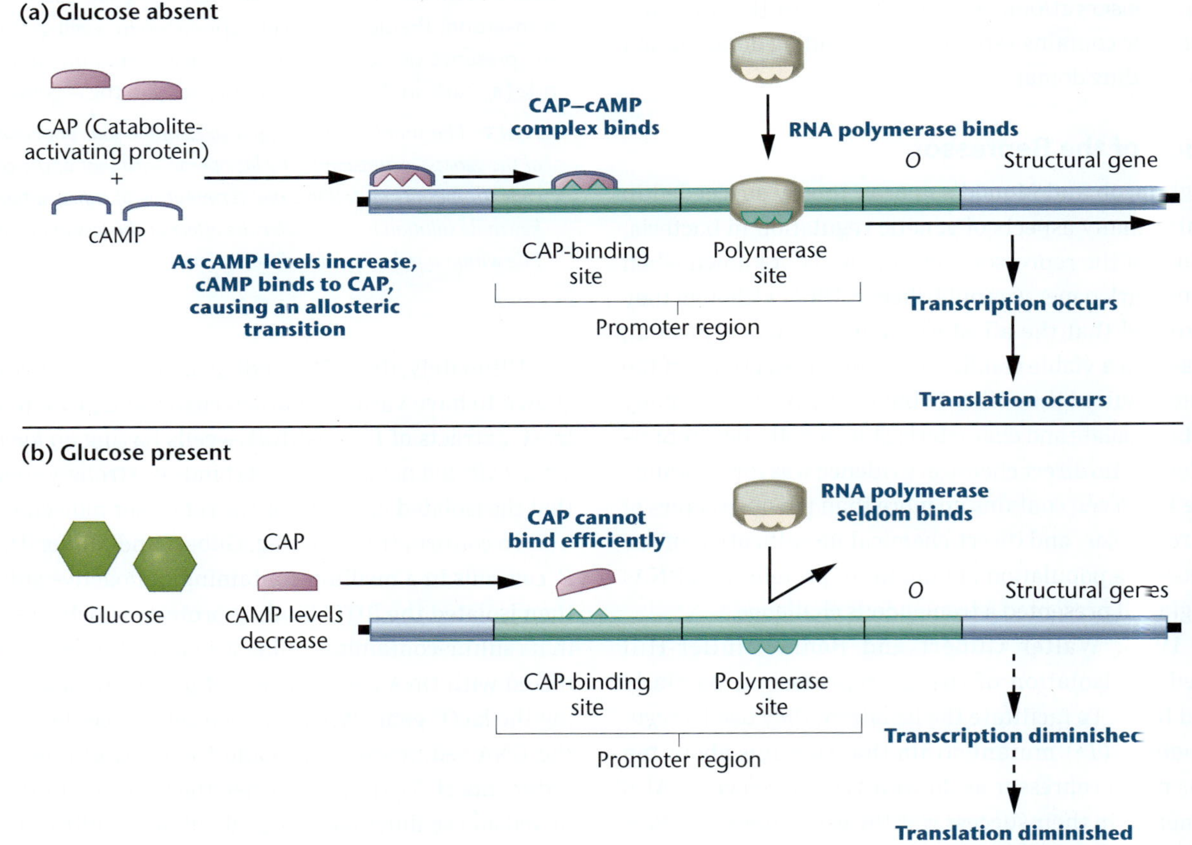

(a) Glucose absent
CAP (Catabolite-activating protein)
+
cAMP
As cAMP levels increase, cAMP binds to CAP, causing an allosteric transition
CAP–cAMP complex binds
RNA polymerase binds
O
Structural gene
CAP-binding site
Polymerase site
Promoter region
Transcription occurs
Translation occurs
(b) Glucose present
Glucose
CAP
cAMP levels decrease
CAP cannot bind efficiently
RNA polymerase seldom binds
O
Structural genes
CAP-binding site
Polymerase site
Promoter region
Transcription diminished
Translation diminished

FIGURE 15.8 Catabolite repression. (a) In the absence of glucose, cAMP levels increase, resulting in the formation of a cAMP–CAP complex, which binds to the CAP site of the promoter, stimulating transcription. (b) In the presence of glucose, cAMP levels decrease, cAMP–CAP complexes are not formed, and transcription is not stimulated.

lactose (so as not to repress operon expression) *and* CAP must be bound to the CAP-binding site.

This leads to the central question about CAP: How does the presence of glucose inhibit CAP binding? The answer involves still another molecule, **cyclic adenosine monophosphate (cAMP),** which is a nucleotide with an adenine base, a ribose sugar, and a single phosphate bound to the sugar at both the 5′ and 3′ positions. *In order to bind to the promoter, CAP must first be bound to cAMP.* The level of cAMP is itself dependent on an enzyme, **adenyl cyclase,** which catalyzes the conversion of ATP to cAMP. The role of glucose in catabolite repression is now clear. It inhibits the activity of adenyl cyclase, causing a decline in the level of cAMP in the cell. Under this condition, CAP cannot form the cAMP—CAP complex essential to the positive stimulation of transcription of the *lac* operon.

Regulation of the *lac* operon by catabolite repression results in efficient energy use, because the presence of glucose will override the need for the metabolism of lactose, if lactose is also available to the cell. In contrast to the negative regulation conferred by the *lac* repressor, the action of cAMP–CAP constitutes positive regulation. Thus, a combination of positive and negative regulatory mechanisms determines transcription levels of the *lac* operon. Catabolite repression involving CAP has also been observed for other inducible operons, including those controlling the metabolism of galactose and arabinose.

ESSENTIAL POINT

The catabolite-activating protein (CAP) exerts positive control over *lac* gene expression by interacting with RNA polymerase at the *lac* promoter and by responding to the levels of cyclic AMP in the bacterial cell. ■

NOW SOLVE THIS

15.2 Predict the level of genetic activity of the *lac* operon as well as the status of the *lac* repressor and the CAP protein under the cellular conditions listed in the accompanying table.

	Lactose	Glucose
(a)	−	−
(b)	+	−
(c)	−	+
(d)	+	+

■ **HINT:** *This problem asks you to combine your knowledge of the Jacob–Monod model of the regulation of the lac operon with your understanding of how catabolite repression impacts on this model. The key to its solution is to keep in mind that regulation involving lactose is a negative control system, while regulation involving glucose and catabolite repression is a positive control system.*

15.4 The Tryptophan (*trp*) Operon in *E. coli* Is a Repressible Gene System

Although the process of induction had been known for some time, it was not until 1953 that Monod and colleagues discovered a repressible operon. When grown in minimal medium (see Chapter 8), wild-type *E. coli* produce the enzymes necessary for the biosynthesis of amino acids as well as many other essential macromolecules. Focusing his studies on the amino acid tryptophan and the enzyme **tryptophan synthetase,** Monod discovered that if tryptophan is present in sufficient quantity in the growth medium, the enzymes necessary for its synthesis are not produced. It is energetically advantageous for bacteria to repress expression of genes involved in tryptophan synthesis when ample tryptophan is present in the growth medium.

Further investigation showed that a series of enzymes encoded by five contiguous genes on the *E. coli* chromosome are involved in tryptophan synthesis. These genes are part of an operon, and in the presence of tryptophan, all are coordinately repressed and none of the enzymes are produced. Because of the great similarity between this repression and the induction of enzymes for lactose metabolism, Jacob and Monod proposed a model of gene regulation analogous to the *lac* system.

To account for repression, Jacob and Monod suggested the presence of a *normally inactive repressor* that alone cannot interact with the operator region of the operon. However, the repressor is an allosteric molecule that can bind to tryptophan. When tryptophan is present, the resultant complex of repressor and tryptophan attains a new conformation that binds to the operator, repressing transcription. Thus, when tryptophan, the end product of this anabolic pathway, is present, the system is repressed and enzymes are not made. Since the regulatory complex inhibits transcription of the operon, this repressible system is under negative control. And as tryptophan participates in repression, it is referred to as a **corepressor** in this regulatory scheme.

Evidence for the *trp* Operon

Support for the concept of a repressible operon was soon forthcoming, based primarily on the isolation of two distinct categories of constitutive mutations. The first class, *trpR*$^-$, maps at a considerable distance from the structural genes. This locus represents the gene coding for the repressor. Presumably, the mutation inhibits either the repressor's interaction with tryptophan or repressor formation entirely. Whichever the case, repression never occurs in cells with the *trpR*$^-$ mutation. As expected, if the *trpR*$^+$ gene encodes a functional repressor molecule, the presence of a copy of this gene will restore repressibility.

The second constitutive mutation is analogous to that of the operator of the lactose operon, because it maps immediately adjacent to the structural genes. Furthermore, the insertion of a plasmid bearing a wild-type operator gene into mutant cells does not restore repression. This is what would be predicted if the mutant operator no longer interacts with the repressor–tryptophan complex.

A detailed model of the *trp* operon and its regulation is shown in **Figure 15.9**. The five contiguous structural genes (*trpE*, *D*, *C*, *B*, and *A*) are transcribed as a polycistronic message directing translation of the enzymes that catalyze the biosynthesis of tryptophan. As in the *lac* operon, a promoter region (*trpP*) represents the binding site for RNA polymerase, and an operator region (*trpO*) is bound by the repressor. In the absence of binding, transcription is initiated within the *trpP–trpO* region and proceeds along a **leader sequence** 162 nucleotides prior to the first structural gene (*trpE*). Within that leader sequence, still another regulatory site has been demonstrated, called an *attenuator*—the subject of Section 15.5. As we will see, this regulatory unit is an integral part of this operon's control mechanism.

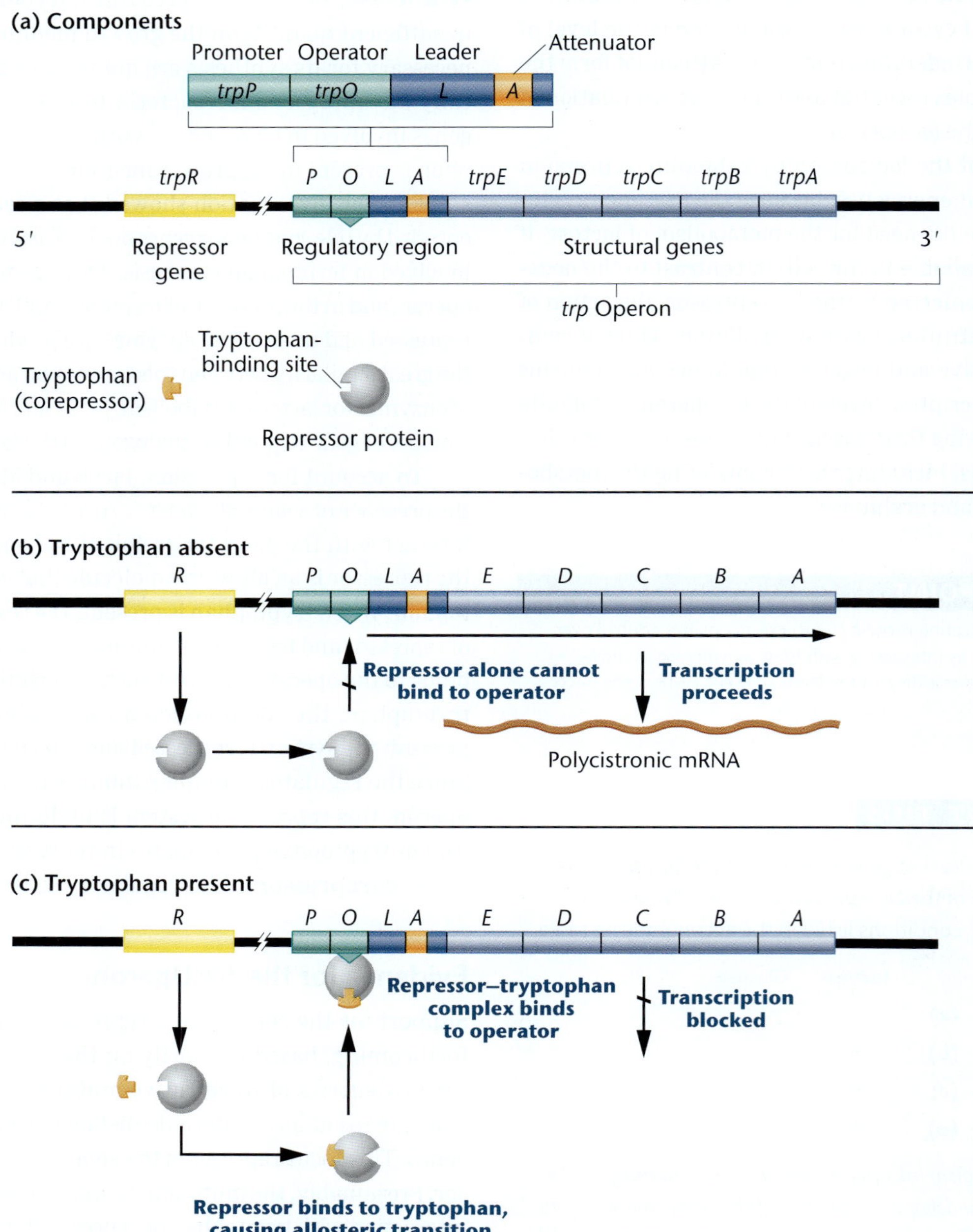

FIGURE 15.9 A repressible operon. (a) The components involved in regulation of the tryptophan operon. (b) In the absence of tryptophan, an inactive repressor is made that cannot bind to the operator (O), thus allowing transcription to proceed. (c) When tryptophan is present, it binds to the repressor, causing an allosteric transition to occur. This complex binds to the operator region, leading to repression of the operon.

ESSENTIAL POINT

Unlike the inducible *lac* operon, the *trp* operon is repressible. In the presence of tryptophan, the repressor binds to the regulatory region of the *trp* operon and represses transcription initiation. ■

EVOLVING CONCEPT OF THE GENE

The groundbreaking work of Jacob, Monod, and Lwoff in the early 1960s, which established the operon model for the regulation of gene expression in bacteria, expanded the concept of the gene to include noncoding regulatory sequences that are present upstream (5′) from the coding region. In bacterial operons, the transcription of several contiguous structural genes whose products are involved in the same biochemical pathway is regulated in a coordinated fashion. ■

15.5 RNA Plays Diverse Roles in Regulating Gene Expression in Bacteria

In the preceding sections of this chapter we focused on gene regulation brought about by DNA-binding regulatory proteins that interact with promoter and operator regions of the genes to be regulated. These regulatory proteins, such as the *lac* repressor and the CAP protein, act to decrease or increase transcription initiation from their target promoters by affecting the binding of RNA polymerase to the promoter.

Gene regulation in bacteria can also occur through the interactions of regulatory molecules with specific regions of a nascent mRNA, after transcription has been initiated. The binding of these regulatory molecules alters the secondary structure of the mRNA, leading to premature transcription termination or repression of translation. We will discuss three types of regulation involving RNA—*attenuation*, *riboswitches*, and *small noncoding RNAs*, abbreviated in bacteria as *sRNAs*. These types of regulation fine-tune levels of gene expression in bacteria.

Attenuation

Charles Yanofsky, Kevin Bertrand, and their colleagues observed that, when tryptophan is present and the *trp* operon is repressed, initiation of transcription still occurs at a low level but is subsequently terminated at a point about 140 nucleotides along the transcript. They called this process **attenuation,** as it "weakens or impairs" expression of the operon. In contrast, when tryptophan is absent or present in very low concentrations, transcription is initiated but is *not* subsequently terminated, instead continuing beyond the leader sequence into the structural genes.

Based on these observations, Yanofsky and colleagues presented a model to explain how attenuation occurs (**Figure 15.10**). They proposed that the initial DNA sequence that is transcribed gives rise to an mRNA sequence that has

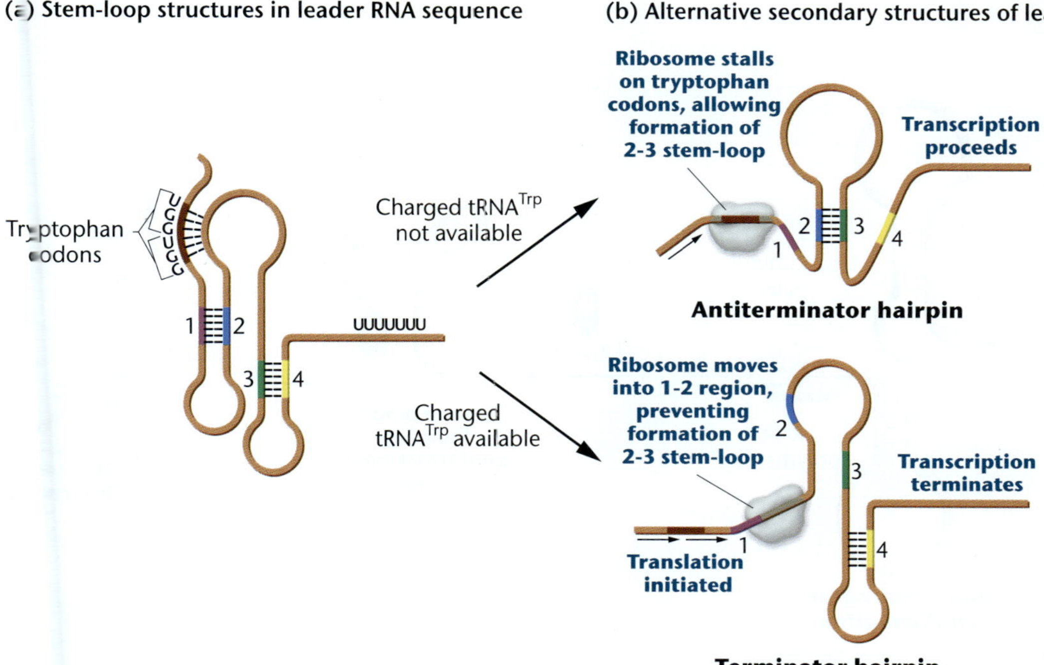

FIGURE 15.10 The attenuation model regulating the tryptophan operon.

the potential to fold into two mutually exclusive stem-loop structures referred to as *hairpins*. If tryptophan is scarce, an mRNA secondary structure referred to as the **antiterminator hairpin** is formed. Transcription proceeds past the antiterminator hairpin region, and the entire mRNA is subsequently produced. Alternatively, in the presence of excess tryptophan, the mRNA structure that is formed is referred to as a **terminator hairpin,** and transcription is almost always terminated prematurely, just beyond a region called the **attenuator.**

A key point in Yanofsky's model is that the transcript of the leader sequence (see Figure 15.9) must be translated for the antiterminator hairpin to form. This leader transcript contains two triplets (UGG) that encode tryptophan, and these are present just downstream of the AUG sequence that signals the initiation of translation by ribosomes. When adequate tryptophan is present, charged $tRNA^{Trp}$ is present in the cell, whereby ribosomes translate these UGG triplets, proceed through the attenuator, and allow the *terminator hairpin* to form, and the operon is not transcribed. If cells are starved of tryptophan, charged $tRNA^{Trp}$ will be unavailable and ribosomes "stall" during translation of the UGG triplets. Thus, the antiterminator hairpin forms within the leader transcript, and as a result, transcription proceeds, leading to expression of the entire set of structural genes.

Many other bacterial operons use attenuation to control gene expression. These include operons that encode enzymes involved in the biosynthesis of amino acids such as threonine, histidine, leucine, and phenylalanine. As with the *trp* operon, attenuation occurs in a leader sequence that contains an attenuator region.

Riboswitches

Since the elucidation of attenuation in the *trp* operon, numerous cases of gene regulation that also depend on alternative forms of mRNA secondary structure have been documented. These are examples of what are more generally referred to as **riboswitches.** As with attenuation of the *trp* operon discussed earlier, the mechanism of riboswitch regulation involves short ribonucleotide sequences (or elements) present in the 5′-untranslated regions of mRNAs (UTRs). These RNA elements are capable of binding with small molecule ligands, such as metabolites, whose synthesis or activity is controlled by the genes encoded by the mRNA. Such binding causes a conformational change in one domain of the riboswitch element, which induces another change at a second RNA domain, most often creating a transcription *terminator structure*. This terminator structure interfaces directly with the transcriptional machinery and shuts it down.

Riboswitches can recognize a broad range of ligands, including amino acids, purines, vitamin cofactors, amino sugars, and metal ions. They are widespread in bacteria. In *Bacillus subtilis,* for example, approximately 5 percent of this bacterium's genes are regulated by riboswitches. They are also found in archaea, fungi, and plants and may prove to be present in animals as well.

The two important domains within a riboswitch are the ligand-binding site, called the **aptamer,** and the **expression platform,** which is capable of forming the terminator structure. **Figure 15.11** illustrates the principles involved in riboswitch control. The 5′-UTR of an mRNA is shown on the left side of the figure in the absence of the ligand (metabolite). RNA polymerase has transcribed the unbound ligand-binding site,

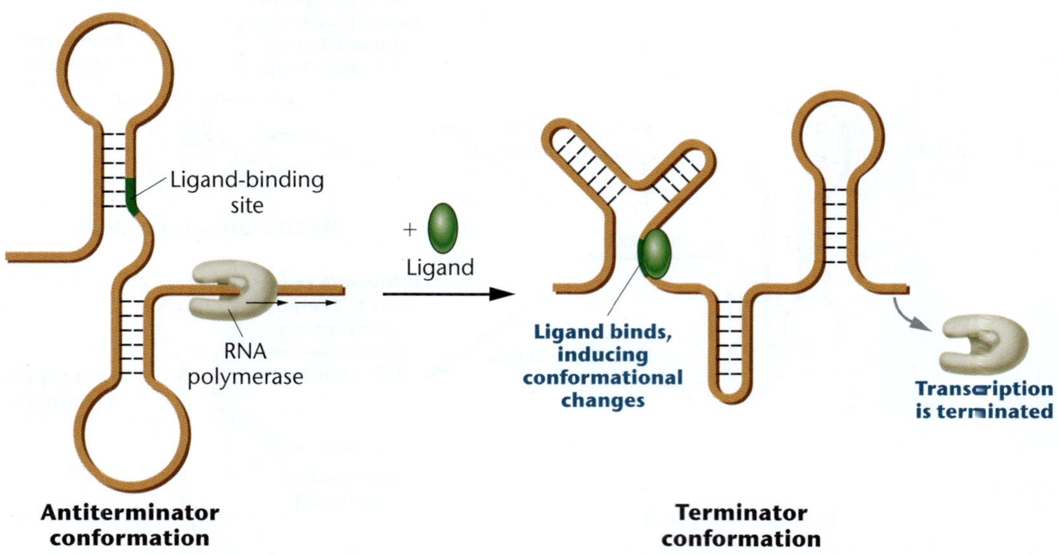

FIGURE 15.11 An illustration of the mechanism of riboswitch regulation of gene expression, where the default position (left) is in the antiterminator conformation. Upon binding by the ligand, the mRNA adopts the terminator conformation (right).

and in the *default conformation,* the expression domain adopts an *antiterminator conformation.* Thus, transcription continues through the expression platform and into the coding region. On the right side of the figure, the presence of the ligand on the ligand-binding site induces an alternative conformation in the expression platform, creating the *terminator conformation*. RNA polymerase is effectively blocked, and transcription ceases.

Small Noncoding RNAs Play Regulatory Roles in Bacteria

Bacterial **small noncoding RNAs (sRNAs)** were discovered decades ago, and it is thought that *E. coli* contain roughly 80 to 100 different sRNAs, while other species are reported to have three times that number. sRNAs are generally between 50 and 500 nucleotides long and are involved in gene regulation and the modification of protein function. sRNAs involved in gene regulation are often transcribed from loci that partially overlap the coding genes that they regulate. However, they are transcribed from the opposite strand of DNA and in the opposite direction, making them complementary to mRNAs transcribed from that locus. In other cases, sRNAs are complementary to target mRNAs but are transcribed from loci that do not overlap target genes. sRNAs regulate gene expression by binding to mRNAs (usually at the 5′-end) that are being transcribed. In some cases, the binding of sRNAs to mRNAs blocks translation of the mRNA by masking the *ribosome-binding site* (*RBS*). In other cases, binding enhances translation by preventing secondary structures from forming in the mRNA that would block translation, often by masking the RBS (see **Figure 15.12**). Thus, sRNAs can be both negative and positive regulators of gene expression.

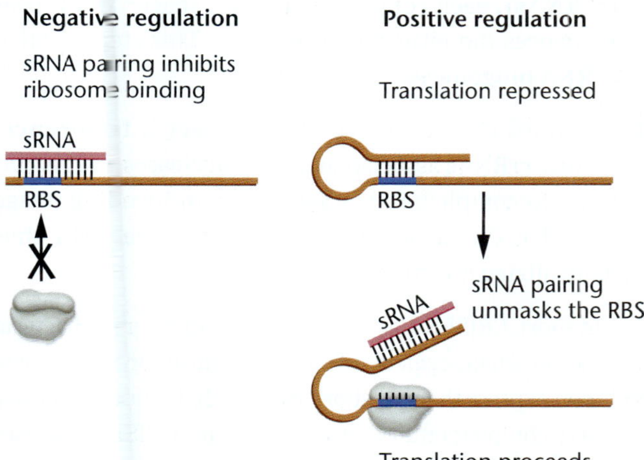

FIGURE 15.12 Bacterial small noncoding RNAs regulate gene expression. Bacterial sRNAs can be negative regulators of gene expression by binding to mRNAs and preventing translation by masking the ribosome-binding site (RBS), or they can be positive regulators of gene expression by binding to mRNAs and preventing secondary structures (that would otherwise mask an RBS) and enable translation.

sRNAs have been shown to play important roles in gene regulation in response to changing environmental conditions or stress. For example, the sRNA *DsrA* of *E. coli* is upregulated in response to low temperature and promotes the expression of genes that enable the long-term survival of the cell under stressful conditions. In contrast, *RyhB* sRNA from *E. coli* is a negative regulator of gene expression. In response to low iron levels, *RyhB* is transcribed to inhibit the translation of several nonessential iron-containing enzymes so that the more critical iron-containing enzymes can utilize what little iron is present in the cytoplasm.

ESSENTIAL POINT

RNAs are sometimes involved in the regulation of gene expression in bacteria, including the process of attenuation, the involvement of riboswitches, and interactions involving small noncoding RNAs (sRNAs). ■

15.6 CRISPR-Cas Is an Adaptive Immune System in Bacteria

In the final section of our discussion of the regulation of gene expression in bacteria, we introduce a molecular mechanism by which bacteria respond to a specific bacteriophage (or simply phage) attack by expressing genetic information that leads to the destruction of that phage's invading DNA.

While bacteria have evolved mechanisms to ward off phage infection, most of these mechanisms provide **innate immunity** because they are not tailored to a specific phage. For example, some bacteria possess mechanisms to prevent phage binding to the cell surface, block phage DNA from entering the cell, or induce suicide in infected cells to prevent the spread of infection. In contrast, **adaptive immunity** refers to a defense mechanism whereby past exposure to a pathogen stimulates a robust defense against future exposure to the same pathogen. For example, adaptive immunity in humans enables vaccines to provide protection against specific disease-causing viruses. While it was once thought that bacteria were not advanced enough to possess such immunity, we now know that bacteria possess an immune system called CRISPR-Cas that can be adapted to fight specific types of phage by preventing phage gene expression and subsequent phage reproduction.

The Discovery of CRISPR

A **CRISPR** is a genomic locus in bacteria that contains clustered regularly interspaced short palindromic repeats. Prior to the coining of this term, CRISPR loci were first identified in 1987 in the *Escherichia coli* genome based on a simple

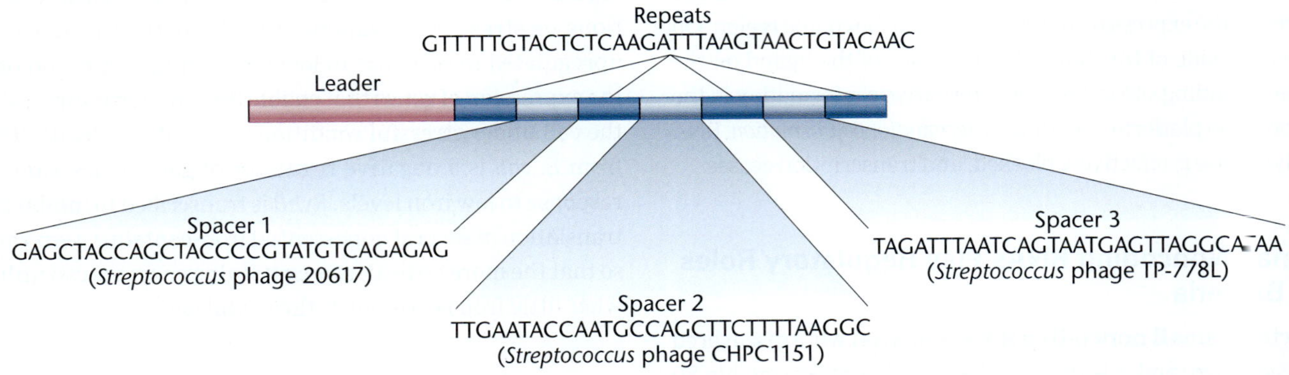

FIGURE 15.13 A CRISPR locus from the bacterium *Streptococcus thermophilus* (LMG18311). Spacer sequences are derived from portions of bacteriophage genomes and are flanked on either side by a repeat sequence. Only 3 of 33 total spacers in this CRISPR locus are shown.

description of repeated DNA sequences with nonrepetitive *spacer* sequences between them. Since then, CRISPR loci have been identified in ~50 percent of bacteria species and in ~90 percent of archaea, another type of prokaryote (**Figure 15.13**). The spacers remained a mystery until 2005 when three independent studies demonstrated that CRISPR spacer sequences were identical to fragments of phage genomes. This insight led to speculation that viral sequences within CRISPR loci serve as a "molecular memory" of previous viral attacks.

The first experimental evidence that CRISPRs are important for adaptive immunity came from an unexpected place. Danisco, a Danish food science company, sought to create a strain of *Streptococcus thermophilus* that was more resistant to phage, thus making it more efficient for use in the production of yogurt and cheese. Philippe Horvath's lab at Danisco, in collaboration with others, found that when they exposed *S. thermophilus* to a specific phage, bacterial cells that survived became resistant to the same phage strain, but not to other phage strains. Furthermore, the resistant bacteria possessed new spacers within their CRISPR loci with an exact sequence match to portions of the genome of the phages by which they had been challenged.

Next, the Horvath lab showed that deletion of new spacers in the resistant strains abolished their phage resistance. Remarkably, the converse was also true; experimental insertion of new viral sequence-derived spacers into the CRISPR loci of sensitive bacteria rendered them resistant!

The CRISPR-Cas Mechanism for RNA-Guided Destruction of Invading DNA

Studies from several labs have elucidated the mechanism underlying the bacterial adaptive immune system. In addition to the CRISPR loci, adaptive immunity is dependent on a set of adjacent **CRISPR-associated (*cas*) genes.** The *cas* genes encode a wide variety of Cas proteins such as DNases, RNases, and proteins of unknown function. The **CRISPR-Cas** mechanism includes three steps outlined in **Figure 15.14**.

1. The first step is known as **spacer acquisition.** Invading phage DNA is cleaved into small fragments, which are directly inserted into the CRISPR locus to become new spacers. The Cas1 nuclease and an associated Cas2 protein are required for spacer acquisition. New spacers are inserted proximal to the *leader sequence* of the CRISPR locus, with older spacers being located progressively more distal. When new spacers are added, repeat sequences are duplicated such that each spacer is flanked by repeats.

2. In the second step, CRISPR loci are transcribed, starting at the promoter within the leader. These long transcripts are then processed into short **CRISPR-derived RNAs (crRNAs),** each containing a single spacer and repeat sequences (on either or both sides). This step is called **crRNA biogenesis.**

3. The third step is referred to as **target interference.** Mature crRNAs associate with Cas nucleases and recruit them to complementary sequences in invading phage DNA. The Cas nucleases then cleave the viral DNA, thus neutralizing infection.

In short, CRISPR-Cas is a simple system whereby bacteria incorporate small segments of viral DNA into their own genome and then express them as short crRNAs that guide a nuclease to cleave complementary viral DNA. Thus, CRISPR-Cas uses viral DNA sequences to specifically fight that same virus. While the three steps of the CRISPR-Cas mechanism are conserved across bacterial species, the molecular details and Cas proteins involved are varied. For example, the well-studied CRISPR-Cas system of *Streptococcus pyogenes* uses a single nuclease called **Cas9** for target interference while many other species require a large complex of Cas proteins.

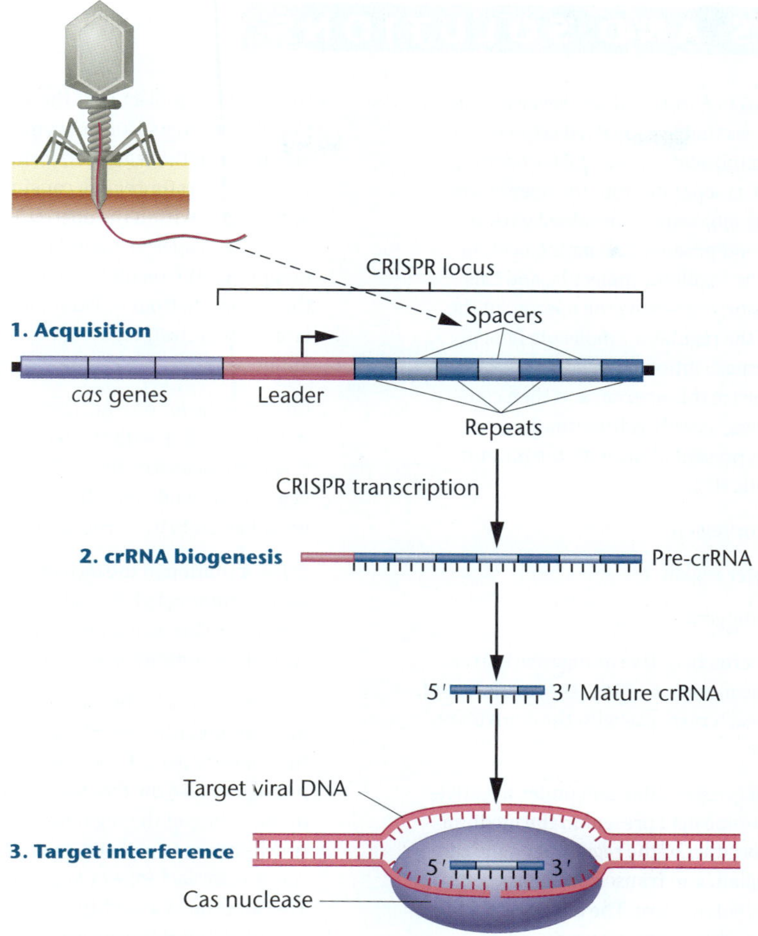

FIGURE 15.14 The mechanism of CRISPR-Cas adaptive immunity in bacteria.

Later in the text (see Chapter 17), we will see how CRISPR-Cas has been adapted as a powerful tool for genome editing with applications in gene therapy (see Special Topics Chapter 3—Gene Therapy) and gene-edited foods (see Special Topics Chapter 6—Genetically Modified Foods).

ESSENTIAL POINT

Bacteria have an adaptive immune system that uses an RNA-guided nuclease to cleave invading viral DNA in a sequence-specific manner. ■

CASE STUDY MRSA in the National Football League (NFL)

In 2013, there was an outbreak of methicillin-resistant *Staphylococcus aureus* (MRSA) at an NFL training facility. One player suffered a career-ending infection to his foot and sued the team owners for $20 million for unsanitary conditions that contributed to the bacterial infection. A settlement with undisclosed terms was reached in 2017. MRSA is highly contagious and is spread by direct skin contact or by airborne transmission and can result in amputation or death. In addition, MRSA is very difficult to treat because it is resistant to many antibiotics. For example, β-lactam antibiotics, such as penicillin, function by binding to and inactivating bacterial penicillin-binding proteins (PBPs), which synthesize the bacterial cell wall. However, MRSA expresses an alternative type of PBP, called PBP2a encoded by the *mecA* gene. β-lactam antibiotics only weakly bind PBP2a, and thus cell wall synthesis can continue in their presence. Moreover, in a system somewhat analogous to the regulation of the *lac* operon, *mecA* is induced by the presence of β-lactam antibiotics and repressed in their absence. This "on-demand" expression of *mecA* means that when the infection is treated with antibiotics, the cells ramp up their resistance.

1. Speculate on how *mecA* expression is inducible and repressible based on what you know about the *lac* operon.
2. Chen et al. [*Antimicrob. Agents Chemother.* (2014) 58(2): 1047–1054] studied several strains of MRSA isolated in Taiwan and found that some single point mutations in the *mecA* promoter were linked to increased antibiotic resistance while other point mutations were linked to decreased antibiotic resistance. Why might *mecA* promoter mutations have these opposing effects?
3. What ethical responsibility do team owners have with respect to preventing the spread of pathogenic bacteria? What responsibilities must players assume to prevent infecting other athletes?

See CDC page: Methicillin-resistant *Staphyloccocus aureus* (MRSA) (https://www.cdc.gov/mrsa/community/team-hc-providers/advice-for-athletes.html).

INSIGHTS AND SOLUTIONS

A hypothetical operon (*theo*) in *E. coli* contains several structural genes encoding enzymes that are involved sequentially in the biosynthesis of an amino acid. Unlike the *lac* operon, in which the repressor gene is separate from the operon, the gene encoding the regulator molecule is contained within the *theo* operon. When the end product (the amino acid) is present, it combines with the regulator molecule, and this complex binds to the operator, repressing the operon. In the absence of the amino acid, the regulatory molecule fails to bind to the operator, and transcription proceeds.

Categorize and characterize this operon, and then consider the following mutations, as well as the situation in which the wild-type gene is present along with the mutant gene in partially diploid cells (F′):

(a) Mutation in the operator region

(b) Mutation in the promoter region

(c) Mutation in the regulator gene

In each case, will the operon be active or inactive in transcription, assuming that the mutation affects the regulation of the *theo* operon? Compare each response with the equivalent situation for the *lac* operon.

Solution: The *theo* operon is repressible and under negative control. When there is no amino acid present in the medium (or the environment), the product of the regulatory gene cannot bind to the operator region, and transcription proceeds under the direction of RNA polymerase. The enzymes necessary for the synthesis of the amino acid are produced, as is the regulator molecule. If the amino acid *is* present, either initially or after sufficient synthesis has occurred, the amino acid binds to the regulator, forming a complex that interacts with the operator region, causing repression of transcription of the genes within the operon.

The *theo* operon is similar to the tryptophan system, except that the regulator gene is within the operon rather than separate from it. Therefore, in the *theo* operon, the regulator gene is itself regulated by the presence or absence of the amino acid.

(a) As in the *lac* operon, a mutation in the *theo* operator gene inhibits binding with the repressor complex, and transcription occurs constitutively. The presence of an F′ plasmid bearing the wild-type allele would have no effect, since it is not adjacent to the structural genes.

(b) A mutation in the *theo* promoter region would no doubt inhibit binding to RNA polymerase and therefore inhibit transcription. This would also happen in the *lac* operon. A wild-type allele present in an F′ plasmid would have no effect.

(c) A mutation in the *theo* regulator gene, as in the *lac* system, may inhibit either its binding to the repressor or its binding to the operator gene. In both cases, transcription will be constitutive, because the *theo* system is repressible. Both cases result in the failure of the regulator to bind to the operator, allowing transcription to proceed. In the *lac* system, failure to bind the corepressor lactose would permanently repress the system. The addition of a wild-type allele would restore repressibility, provided that this gene was transcribed constitutively.

Problems and Discussion Questions

Mastering Genetics Visit for instructor-assigned tutorials and problems.

1. **HOW DO WE KNOW?** In this chapter, we focused on the regulation of gene expression in bacteria. Along the way, we found many opportunities to consider the methods and reasoning by which much of this information was acquired. From the explanations given in the chapter, what answers would you propose to the following fundamental questions?
   (a) How do we know that bacteria regulate the expression of certain genes in response to the environment?
   (b) What evidence established that lactose serves as the inducer of a gene whose product is related to lactose metabolism?
   (c) What led researchers to conclude that a repressor molecule regulates the *lac* operon?
   (d) How do we know that the *lac* repressor is a protein?
   (e) How do we know that the *trp* operon is a repressible control system, in contrast to the *lac* operon, which is an inducible control system?
2. **CONCEPT QUESTION** Review the Chapter Concepts list on p. 309. These all relate to the regulation of gene expression in bacteria. Write a brief essay that discusses why you think regulatory systems evolved in bacteria (i.e., what advantages do regulatory systems provide to these organisms?), and, in the context of regulation, discuss why genes related to common functions are found together in operons.
3. What are operon, *cis*-acting region, and *trans*-acting factors?
4. What would be the role of a repressor in positive versus negative control of gene expression?
5. For the *lac* genotypes shown in the following table, predict whether the structural genes (*Z*) are constitutive, permanently repressed, or inducible in the presence of lactose.

Genotype	Constitutive	Repressed	Inducible
$I^+O^+Z^+$			×
$I^-O^+Z^+$			
$I^-O^CZ^+$			
$I^-O^CZ^+/F'O^+$			
$I^+O^CZ^+/F'O^+$			
$I^SO^+Z^+$			
$I^SO^+Z^+/F'I^+$			

6. For the genotypes and conditions (lactose present or absent) shown in the following table, predict whether functional enzymes, nonfunctional enzymes, or no enzymes are made.

Genotype	Condition	Functional Enzyme Made	Nonfunctional Enzyme Made	No Enzyme Made
$I^+O^+Z^+$	No lactose			×
$I^+O^CZ^+$	Lactose			
$I^-O^+Z^-$	No lactose			
$I^-O^+Z^-$	Lactose			
$I^-O^+Z^+/F'I^+$	No lactose			
$I^+O^CZ^+/F'O^+$	Lactose			
$I^+O^+Z^-/F'I^+O^+Z^+$	Lactose			
$I^-O^+Z^-/F'I^-O^+Z^+$	No lactose			
$I^SO^+Z^-/F'O^+$	No lactose			
$I^+O^CZ^-/F'O^+Z^+$	Lactose			

7. The locations of numerous *lacI*$^-$ and *lacI*^{S} mutations have been determined within the DNA sequence of the *lacI* gene. Among these, *lacI*$^-$ mutations were found to occur in the 5′-upstream region of the gene, while *lacI*^{S} mutations were found to occur farther downstream in the gene. Are the locations of the two types of mutations within the gene consistent with what is known about the function of the repressor that is the product of the *lacI* gene?
8. Describe the experimental rationale that allowed the *lac* repressor to be isolated.
9. What properties demonstrate that the *lac* repressor is a protein? Describe the evidence that it indeed serves as a repressor within the operon system.
10. Predict the effect of the inducibility of the *lac* operon in bacteria that have loss-of-function mutations at (a) the *lacI binding* element in the operator, (b) the cAMP binding domain of the CAP protein.
11. Erythritol is a natural sugar abundant in fruits and fermenting foods. Pathogenic bacterial strains that catabolize erythritol contain four closely spaced genes, all involved in erythritol metabolism. One of the four genes (*eryD*) encodes a product that represses the expression of the other three genes. Erythritol catabolism is stimulated by erythritol. Present a regulatory model to account for the regulation of erythritol catabolism in such bacterial strains. Does this system appear to be under inducible or repressible control?
12. Describe the role of attenuation in the regulation of tryptophan biosynthesis.
13. Describe the structural and functional differences between attenuation regulating the tryptophan operon, riboswitches, and small noncoding RNAs.
14. A bacterial operon is responsible for the production of the biosynthetic enzymes needed to make the hypothetical amino acid tisophane (tis). The operon is regulated by a separate gene, *R*. The deletion of *R* causes the loss of enzyme synthesis. In the wild-type condition, when tis is present, no enzymes are made; in the absence of tis, the enzymes are made. Mutations in the operator gene (O^-) result in repression regardless of the presence of tis. Is the operon under positive or negative control? Propose a model for (a) repression of the genes in the presence of tis in wild-type cells and (b) the mutations.
15. Describe the structure of small noncoding RNAs and their origin in the bacterial genome.
16. Why is the CRISPR-Cas system of bacteria considered an adaptive immunity rather than an innate immunity?
17. Describe the CRISPR-Cas system of adaptive immunity and its mode of action.
18. In the publication that provided the first evidence of CRISPR-Cas as an adaptive immune system [Barrangou, R., et al. (2007). *Science*. 315:1709–1712], the authors state that CRISPR-Cas "provides a historical perspective of phage exposure, as well as a predictive tool for phage sensitivity." Explain how this is true using what you know about the CRISPR locus.

16

Regulation of Gene Expression in Eukaryotes

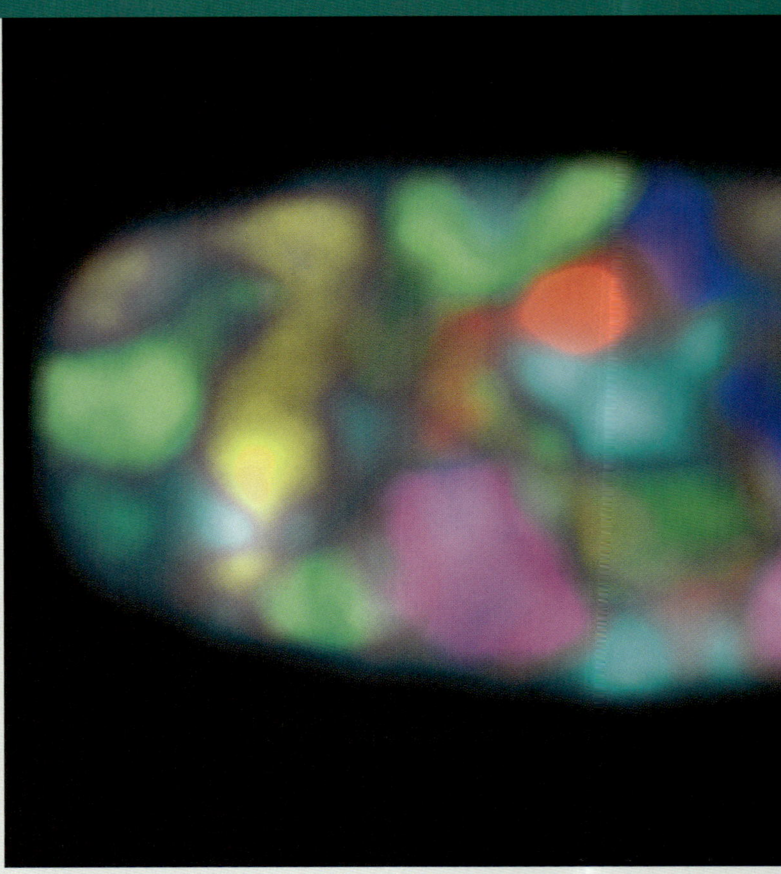

Chromosome territories in a human fibroblast cell nucleus. Each chromosome is stained with a different-colored probe.

CHAPTER CONCEPTS

- While transcription and translation are tightly coupled in bacteria, in eukaryotes, these processes are spatially and temporally separated, and thus independently regulated.
- Chromatin remodeling, as well as modifications to DNA and histones, play important roles in regulating gene expression in eukaryotes.
- Eukaryotic transcription initiation requires the assembly of transcription regulatory proteins on DNA sites known as promoters, enhancers, and silencers.
- Following transcription, there are several mechanisms that regulate gene expression, referred to as posttranscriptional regulation.
- Alternative splicing allows for a single gene to encode different protein isoforms with different functions.
- RNA-binding proteins regulate mRNA stability, degradation, localization, and translation.
- Noncoding RNAs may regulate gene expression by targeting mRNAs for destruction or translational inhibition.
- Posttranslational modification of proteins can alter their activity or promote their degradation.

Virtually all cells in a multicellular eukaryotic organism contain a complete genome; however, such organisms often possess different cell types with diverse morphologies and functions. This simple observation highlights the importance of the regulation of gene expression in eukaryotes. For example, skin cells and muscle cells differ in appearance and function because they express different genes. Skin cells express keratins, fibrous structural proteins that bestow the skin with protective properties. Muscle cells express high levels of myosin II, a protein that mediates muscle contraction. Skin cells do not express myosin II, and muscle cells do not express keratins.

In addition to gene expression that is cell-type specific, some genes are only expressed under certain conditions or at certain times. For example, when oxygen levels in the blood are low, such as at high altitude or after rigorous exercise, expression of the hormone erythropoietin is upregulated, which leads to an increase in red blood cell production and thus oxygen-carrying capacity.

Underscoring the importance of regulation, the misregulation of genes in eukaryotes is associated with developmental defects and disease. For instance, the overexpression of genes that regulate cellular growth can lead to uncontrolled cellular proliferation, a hallmark of cancer. Therefore, understanding the mechanisms that control gene expression in eukaryotes is of great interest and may lead to therapies for human diseases.

We will start this chapter by briefly comparing and contrasting eukaryotic gene expression with that of bacteria (see Chapter 15). Next, we will explore several molecular mechanisms that regulate gene expression in eukaryotes.

16.1 Organization of the Eukaryotic Cell Facilitates Gene Regulation at Several Levels

As you've previously seen (Chapter 15), in bacteria, the regulation of gene expression is often linked to the metabolic needs of the cell. For example, bacteria express genes to metabolize lactose when it is present in the environment. Gene expression in bacteria is largely controlled by mechanisms that exert positive or negative control over transcription and translation. While positive and negative regulation of transcription and translation are prominent regulatory mechanisms in eukaryotes as well, there are many important differences between these processes to consider, and several additional levels for gene regulation are found in eukaryotes. **Figure 16.1** compares differences in the main mechanisms of gene regulation in bacteria and eukaryotes, which include the following:

- Eukaryotic DNA, unlike that of bacteria, is associated with histones and other proteins to form chromatin. Eukaryotic cells decrease chromatin compaction to make genes accessible to transcription and increase chromatin compaction to inhibit transcription.
- In bacteria, the processes of transcription and translation both take place in the cytoplasm and are coupled. However, eukaryotic cells are organized with DNA in the nucleus and ribosomes in the cytoplasm. Thus, transcription and translation are separated spatially and temporally.
- Whereas bacterial mRNAs are translated directly, the mRNAs of many eukaryotic genes must be spliced, capped, and polyadenylated prior to export from the nucleus and translation in the cytoplasm. Each of these processes can be regulated in order to influence the numbers and types of mRNAs available for translation.
- Bacterial mRNAs are often translated immediately and degraded very rapidly. In contrast, eukaryotic mRNAs can have long half-lives and can be transported, localized, and translated in specific subcellular destinations.
- Proteins in bacteria and in eukaryotes can be posttranslationally modified by processes such as phosphorylation and methylation, which serve many functions,

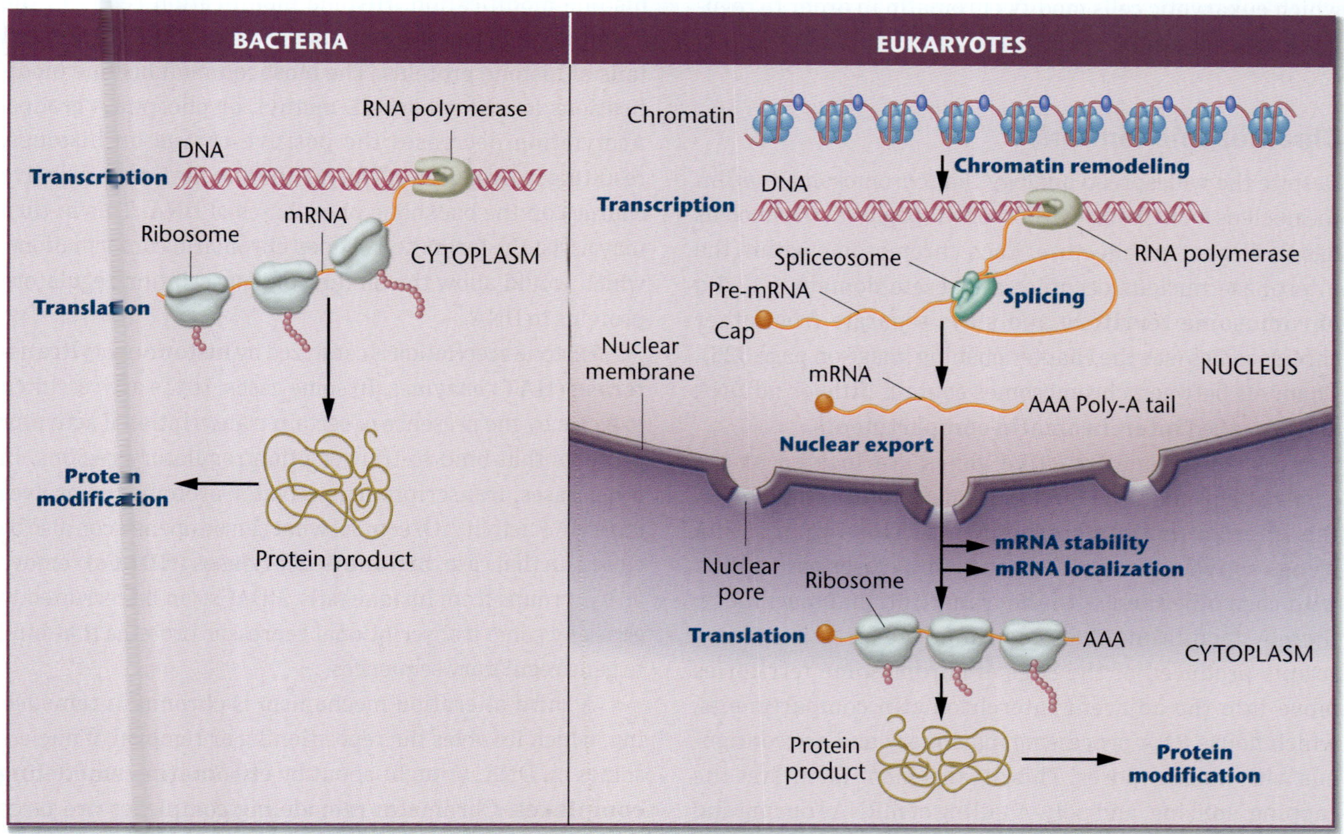

FIGURE 16.1 A comparison of gene regulation in bacteria (left) and eukaryotes (right).

including the regulation of protein activities. However, the repertoire of posttranslational modifications in eukaryotes is more extensive, leading to additional regulatory opportunities.

ESSENTIAL POINT

Compared to bacteria, the regulation of gene expression in eukaryotes is more complex and includes additional steps such as the regulation of chromatin packaging and the regulation of mRNA processing. ■

16.2 Eukaryotic Gene Expression Is Influenced by Chromatin Modifications

Recall from earlier in the text (see Chapter 11) that eukaryotic DNA is combined with histones and nonhistone proteins to form **chromatin.** The basic structure of chromatin is characterized by repeating DNA–histone complexes called **nucleosomes** that are wound into 30-nm fibers, which in turn form other, even more compact structures. The presence of these compact chromatin structures inhibits many processes, including DNA replication, repair, and transcription. In this section, we outline some of the ways in which eukaryotic cells modify chromatin in order to regulate gene expression.

Chromosome Territories

Despite the widespread analogy that chromosomes within the nucleus look like a bowl of cooked spaghetti, the nucleus has an elegant architecture. Each chromosome within the interphase nucleus occupies a discrete domain called a **chromosome territory** and stays separate from other chromosomes (see the chapter opening image on page 326). Channels between chromosomes contain little or no DNA and are called **interchromatin compartments.**

Transcriptionally active genes are located at the edges of chromosome territories next to the channels of the interchromatin compartments. This organization brings actively transcribed genes into closer association with each other and with the transcriptional machinery, thereby facilitating their coordinated expression. Transcripts produced at the edge of chromosome territories move into the adjacent interchromatin compartments, which house RNA processing machinery and are contiguous with nuclear pores. This arrangement facilitates the capping, splicing, and poly-A tailing of mRNAs during and after transcription, and the eventual export of mRNAs into the cytoplasm.

Nuclear architecture and transcriptional regulation are interdependent; changes in nuclear architecture affect transcription, and changes in transcriptional activity necessitate changes in chromosome organization.

Histone Modifications and Chromatin Remodeling

The tight association of DNA with histones and other chromatin-binding proteins inhibits access to the DNA by proteins involved in many functions including transcription. This inhibitory conformation is often referred to as "closed" chromatin. Before transcription can be initiated, the structure of chromatin must become "open" to RNA polymerase.

One way in which chromatin conformation can switch between "open" and "closed" is by changes to histone composition in nucleosomes (see Chapter 11). Most nucleosomes contain histone H2A, while some nucleosomes contain variant histones, such as H2A.Z. Whereas nucleosomes are generally a physical barrier to RNA polymerases and DNA-binding transcriptional regulators, nucleosomes containing the H2A.Z variant are not as stable and thus are less of a barrier. Interestingly, H2A.Z-containing nucleosomes are enriched at sequences that regulate gene expression, which suggests that the histone composition of nucleosomes can influence transcription.

A second mechanism of chromatin alteration involves histone modification. Histone modification refers to the covalent addition of functional groups to the N-terminal tails of histone proteins. The most common histone modifications are added acetyl, methyl, or phosphate groups. Acetylation decreases the positive charge on histones, resulting in a reduced affinity of the histone for the negative charges on the backbone phosphates of DNA. This in turn may assist the formation of open chromatin conformations, which would allow the binding of transcription regulatory proteins to DNA.

Histone acetylation is catalyzed by **histone acetyltransferase (HAT)** enzymes. In some cases, HATs are recruited to genes by the presence of certain transcriptional activator proteins that bind to transcription regulatory regions. In other cases, transcriptional activator proteins themselves have HAT activity. Of course, what can be opened can also be closed. In that case, **histone deacetylases (HDACs)** remove acetyl groups from histone tails. HDACs can be recruited to genes by some transcriptional repressor proteins that bind to gene regulatory sequences.

A third alteration mechanism is chromatin remodeling, which involves the repositioning or removal of nucleosomes on DNA, brought about by **chromatin remodeling complexes.** Chromatin remodeling complexes are large multi-subunit enzymes that use the energy of ATP hydrolysis to move and rearrange nucleosomes. Repositioning of nucleosomes makes regions of the chromosome accessible to transcription regulatory proteins and RNA polymerase.

See Special Topics Chapter ST1—Epigenetics for more information about chromatin modifications that regulate gene expression.

DNA Methylation

Another type of chromatin modification that plays a role in gene regulation in some eukaryotes is the enzyme-mediated addition or removal of methyl groups to or from bases in DNA. **DNA methylation** in eukaryotes most often occurs at position 5 of cytosine **(5-methylcytosine),** causing the methyl group to protrude into the major groove of the DNA helix. Methylation occurs most often on the cytosine of CG doublets in DNA, usually on both strands:

$$5' - {}^{m}CpG - 3'$$
$$3' - GpC_{m} - 5'$$

Methylatable CpG sequences are not randomly distributed throughout the genome, but tend to be concentrated in CpG-rich regions, called **CpG islands**, which are often located in or near promoter regions. Roughly 70 percent of human genes have a CpG island in their promoter sequence.

Evidence suggests that DNA methylation represses gene expression. For example, large transcriptionally inert regions of the genome, such as the inactivated X chromosome in female mammalian cells (see Chapter 5), are often heavily methylated. Conversely, blocking methylation of genes on the inactivated X chromosome leads to their expression.

By what mechanism might methylation affect gene regulation? Data from *in vitro* studies suggest that methylation can repress transcription by inhibiting the binding of transcription factors to DNA. Methylated DNA may also recruit repressive chromatin remodeling complexes and HDACs to gene-regulatory regions.

It is important to know that while cytosine methylation is clearly an important mechanism for gene regulation in some eukaryotes, it is not uniformly true for all eukaryotes. For example, while DNA methylation is an important gene regulatory mechanism in humans, mice, and many plants, DNA methylation is absent in yeast and the roundworm *Caenorhabditis elegans*.

See Special Topics Chapter ST1—Epigenetics for more information about DNA methylation and the regulation of gene expression.

ESSENTIAL POINT

Chromatin modifications, such as nucleosomal chromatin remodeling, histone modifications, and DNA methylation may either allow or inhibit the binding of transcriptional machinery to the DNA. ■

NOW SOLVE THIS

16.1 Cancer cells often have abnormal patterns of chromatin modifications. In some cancers, the DNA repair genes *MLH1* and *BRCA1* are hypermethylated on their promoter regions. Explain how this abnormal methylation pattern could contribute to cancer.

■ **HINT:** *This problem involves an understanding of the types of genes that are mutated in cancer cells. The key to its solution is to consider how DNA methylation affects gene expression of cancer-related genes.*

16.3 Eukaryotic Transcription Initiation Requires Specific *Cis*-Acting Sites

Earlier in the text (see Chapter 12), we noted that eukaryotes possess three different types of RNA polymerases (RNAPs). RNAP I and III transcribe ribosomal RNAs, transfer RNAs, and some small nuclear RNAs. RNAP II transcribes protein-coding genes and genes for some noncoding RNAs. For simplicity, in this chapter we will limit our discussion to regulation of genes transcribed by RNAP II, the best studied of the eukaryotic RNAPs.

Regulation of eukaryotic transcription requires the binding of many regulatory factors to specific DNA sequences located in and around genes, as well as to sequences located at great distances. In this section, we will discuss some of the DNA sequences—known as *cis*-acting elements—that are required for the accurate and regulated transcription of genes transcribed by RNAP II. As defined earlier in the text (see Chapter 12), ***cis*-acting DNA elements** are located on the same chromosome as the gene that they regulate. This is in contrast to ***trans*-acting factors** (such as DNA-binding proteins) that can influence the expression of a gene on any chromosome.

Promoters and Promoter Elements

A **promoter** is a region of DNA that is recognized and bound by the basic transcriptional machinery—including RNAP II and the general transcription factors (see Chapter 12). Promoters are required for transcription initiation and are located immediately adjacent to the genes they regulate. They specify the site or sites at which transcription begins (the **transcription start site**), as well as the direction of transcription along the DNA.

There are two subcategories within eukaryotic promoters. First, **core promoters** are the minimum part of the promoter needed for accurate initiation of transcription by

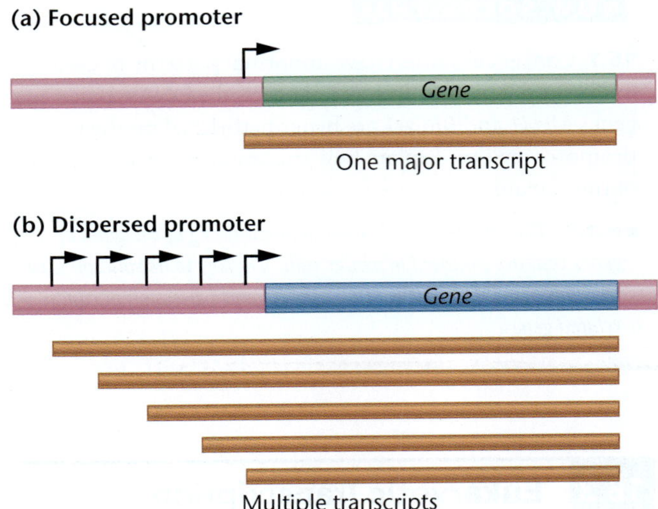

FIGURE 16.2 Focused and dispersed promoters. Focused promoters (a) specify one specific transcription initiation site. Dispersed promoters (b) specify weak transcription initiation at multiple start sites over an approximately 100-bp region. Transcription start sites and the directions of transcription are indicated with arrows.

RNAP II. Core promoters are sequences ~80 nucleotides long and include the transcription start site. Second, **proximal promoter elements** are generally located up to ~250 nucleotides upstream of the transcription start site and contain binding sites for sequence-specific DNA-binding proteins that modulate the efficiency of transcription.

Core promoters are classified in two ways with respect to transcription start sites. **Focused core promoters** specify transcription initiation at a single specific start site. In contrast, **dispersed core promoters** direct initiation from a number of weak transcription start sites located over a 50- to 100-nucleotide region (**Figure 16.2**). The major type of initiation for most genes of lower eukaryotes is focused transcription initiation, while about 70 percent of vertebrate genes employ dispersed promoters. Focused promoters are usually associated with genes whose transcription levels are highly regulated in terms of time or place. Dispersed promoters, in contrast, are associated with genes that are transcribed constitutively, so-called *housekeeping genes* whose expression is required in almost all cell types. Thus, a single transcription start site may facilitate precise regulation of some genes, whereas multiple start sites may allow for a steady level of transcription of genes that are required constitutively.

While it is not yet clear how dispersed promoters specify multiple transcriptional start sites, much more is known about the structure of focused promoters. Within focused promoters, there are numerous **core-promoter elements**—short nucleotide sequences that are bound by specific regulatory factors. Such core-promoter elements vary between species and between genes within a species. However, focused promoters contain several common core-promoter elements, which are shown in **Figure 16.3** and described in more detail below.

The Initiator element (**Inr**) encompasses the transcription start site, from approximately nucleotides −2 to −4, relative to the start site. In humans, the Inr consensus sequence is YYAN$^{A}/_{T}$YY (where Y indicates any pyrimidine nucleotide and N indicates any nucleotide). The transcription start site is the first A residue at +1. The **TATA box** is located at approximately −30 relative to the transcription start site and has the consensus sequence TATA$^{A}/_{T}$AAR (where R indicates any purine nucleotide). The TFIIB recognition element (**BRE**) is found at positions either immediately upstream or downstream from the TATA box. The motif ten element (**MTE**) and downstream promoter element (**DPE**) are located downstream of the transcription start site, at approximately +18 to +27 and +28 to +33, respectively.

Versions of the BRE, TATA box, and Inr elements appear to be universal components of all focused promoters. However, the MTEs and DPEs are found in only some of these promoters. This is not a comprehensive list of all known core-promoter elements; the important concept here is that the core-promoter elements serve as a platform for the assembly of RNAP II and the general transcription factors, which is a critical step in gene expression and will be discussed in Section 16.4.

In addition to core-promoter elements, many genes also contain proximal-promoter elements located upstream of the TATA box and BRE. Proximal-promoter elements act along with the core-promoter elements to increase the levels of basal transcription. For example, the CAAT box is a common proximal-promoter element. The **CAAT box** has the consensus sequence CAAT or CCAAT and is usually located about 70 to 80 base pairs upstream from the

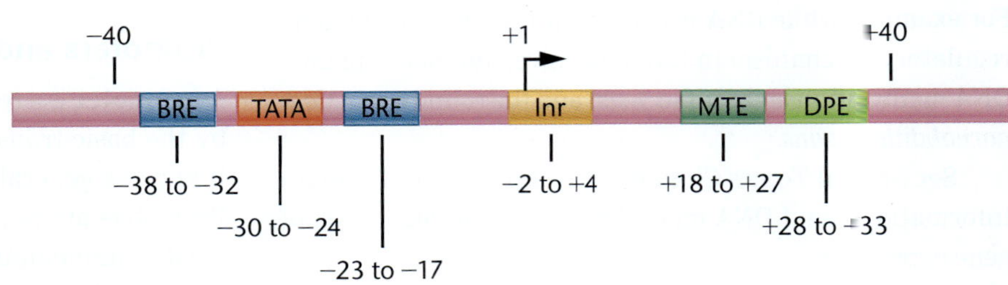

FIGURE 16.3 Core-promoter elements found in focused promoters. Core-promoter elements are usually located between −40 and +40 nucleotides, relative to the transcription start site, indicated as +1. BRE is the TFIIB recognition element, which can be found on one side or other of the TATA box. TATA is the TATA box, Inr is the Initiator element, MTE is the motif ten element, and DPE is the downstream promoter element.

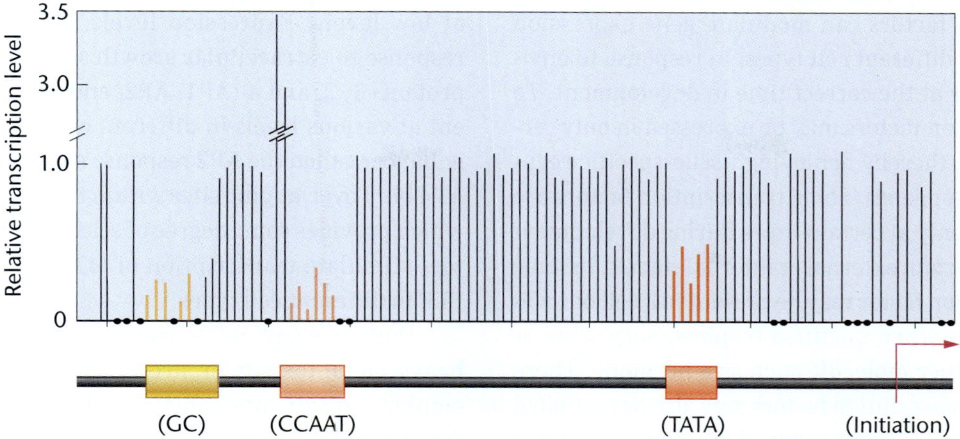

FIGURE 16.4 Summary of the effects on transcription levels of different point mutations in the promoter region of the β-globin gene. Each line represents the level of transcription produced in a separate experiment by a single-nucleotide mutation (relative to wild type) at a particular location. Dots represent nucleotides for which no mutation was tested. Note that mutations within specific elements of the promoter have the greatest effects on the level of transcription.

transcription start site. Mutations on either side of this element have no effect on transcription, whereas mutations within the CAAT sequence dramatically lower the rate of transcription. Thus, for genes with a CAAT box, it appears to be required for robust transcription. **Figure 16.4** summarizes the transcriptional effects of mutations in the CAAT box and other promoter elements. The **GC box** is another element often found in promoter regions and has the consensus sequence GGGCGG. It is located, in one or more copies, at about position −110 and is bound by transcription factors.

Enhancers and Silencers

In addition to promoters, transcription of eukaryotic genes is also influenced by DNA sequences called **enhancers.** While promoters are always found immediately upstream of a gene, enhancers can be located on either side of a gene, nearby or at some distance from the gene, or even within the gene. Some studies show that enhancers can be located as far away as a million base pairs from the genes they regulate. Like promoters, they are *cis* regulatory elements because they only serve to regulate genes on the same chromosome. However, there are several differences between promoters and enhancers.

1. While promoter sequences are essential for minimal or basal-level transcription, enhancers, as their name suggests, increase the rate of transcription. In addition, enhancers often confer time- and tissue-specific gene expression.
2. Whereas promoters must be immediately upstream of the genes they regulate, the position of an enhancer is not critical; it will function the same whether it is upstream, downstream, or within a gene.
3. Whereas promoters are orientation specific, an enhancer can be inverted, relative to the gene it regulates, without a significant effect on its action.

Another type of *cis*-acting regulatory element, the **silencer,** acts as a negative regulator of transcription. Silencers, like enhancers, are *cis*-acting short DNA sequence elements that may be located far upstream, downstream, or within the genes they regulate. Another similarity to enhancers is that they also often act in tissue- or temporal-specific ways to control gene expression.

ESSENTIAL POINT

Eukaryotic transcription is regulated by *cis*-acting regulatory sequences such as promoters, enhancers, and silencers. ■

16.4 Eukaryotic Transcription Initiation Is Regulated by Transcription Factors That Bind to *Cis*-Acting Sites

Promoters, enhancers, and silencers influence transcription initiation by acting as binding sites for transcription regulatory proteins, broadly termed **transcription factors.** In addition to the **general transcription factors (GTFs)** required for the basic process of transcription initiation (see Chapter 12), some transcription factors serve to *increase* the levels of transcription initiation and are known as **activators,** while others *reduce* transcription levels and are known as **repressors.**

Transcription factors can modulate gene expression as appropriate for different cell types, in response to environmental cues, or at the correct time in development. To do this, transcription factors may be expressed in only certain types of cells, thereby achieving tissue-specific regulation of their target genes. Some transcription factors are expressed in cells only at certain times during development or in response to certain external or internal signals. In some cases, a transcription factor may be present in a cell but will only become active when modified by phosphorylation or by binding to another molecule such as a hormone. These modifications to transcription factors may also be regulated in tissue- or temporal-specific ways. Finally, the inputs of multiple transcription factors binding to different enhancers and promoter elements are integrated to fine-tune the levels and timing of transcription initiation.

The Human Metallothionein 2A Gene: Multiple *Cis*-Acting Elements and Transcription Factors

The **human metallothionein 2A (*MT2A*) gene** provides an example of how the transcription of one gene can be regulated by the interplay of multiple promoter and enhancer elements and the transcription factors that bind them. The product of the *MT2A* gene is a protein that binds to heavy metals such as zinc and cadmium, thereby protecting cells from the toxic effects of high levels of these metals. The protein is also implicated in protecting cells from the effects of oxidative stress. The *MT2A* gene is expressed at a low or basal level in all cells but is transcribed at high levels when cells are exposed to heavy metals or stress hormones such as glucocorticoids.

The *cis*-acting regulatory elements controlling transcription of the *MT2A* gene include promoter, enhancer, and silencer elements (**Figure 16.5**). Core-promoter elements, such as the TATA box and Inr, are bound by several general transcription factors and RNAP II and are required for transcription initiation. The proximal-promoter element, GC, is bound by the SP1 factor, which is present at all times in most eukaryotic cells and stimulates transcription at low levels. Expression levels are also modulated in response to extracellular growth signals via the activator proteins 1, 2, and 4 (AP1, AP2, and AP4), which are present at various levels in different cell types. AP2 binds an enhancer called the AP2 response element (ARE). AP1 and AP4 bind overlapping sites within the basal element (BLE), which provides some degree of selectivity in how these factors stimulate transcription of *MT2A* when bound to the BLE in different cell types.

High levels of *MT2A* transcription are stimulated by heavy-metal toxicity or stress. The heavy-metal response employs an activator called metal-inducible transcription factor 1 (MTF-1). MTF-1 is normally found in the cytoplasm but translocates to the nucleus in the presence of heavy metals. Direct metal binding induces conformational changes that lead to MTF-1's nuclear translocation. In the nucleus, MTF-1 binds several metal response elements (MREs) that enhance *MT2A* transcription.

A different activator and enhancer mediate stress-induced transcription of *MT2A*: The glucocorticoid receptor is an activator that binds to the glucocorticoid response element (GRE). Under stressful conditions, vertebrates secrete a steroid hormone called glucocorticoid. Upon glucocorticoid binding, the glucocorticoid receptor, which is normally located in the cytoplasm, undergoes a conformational change that allows it to enter the nucleus, bind to the GRE, and enhance *MT2A* gene transcription. In addition to activation, transcription of the *MT2A* gene can be repressed by the actions of the repressor protein PZ120, which binds over the transcription start site.

The presence of multiple regulatory elements and transcription factors that bind to them allows the *MT2A* gene to be transcriptionally activated or repressed in response to subtle changes in both extracellular and intracellular conditions.

ESSENTIAL POINT

Transcription factors influence transcription by binding to promoters, enhancers, and silencers. ■

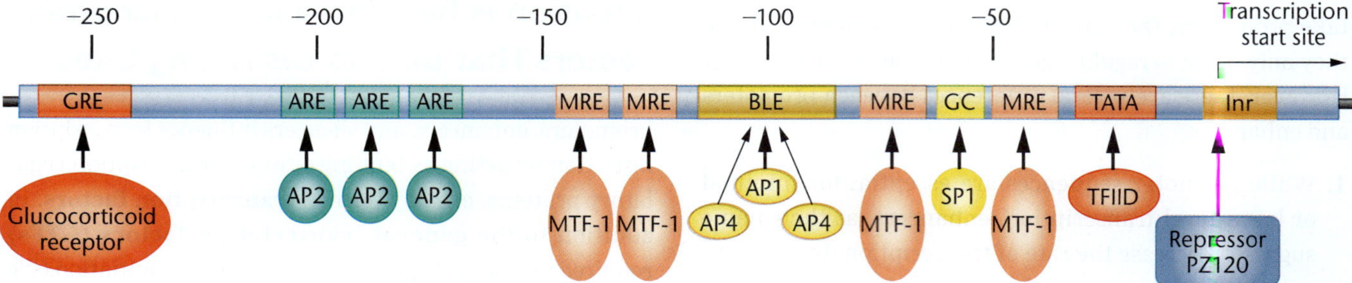

FIGURE 16.5 The human metallothionein 2A gene promoter and enhancer regions, containing multiple *cis*-acting regulatory sites. The transcription factors controlling both basal and induced levels of *MT2A* transcription are indicated below the gene, with arrows pointing to their binding sites.

16.5 Activators and Repressors Interact with General Transcription Factors and Affect Chromatin Structure

We have thus far discussed how chromatin remodeling and chromatin modifications are necessary to make *cis*-acting regulatory sequences accessible to binding by transcription factors. In this section we will see how these events come together to regulate the initiation of transcription by facilitating or inhibiting the binding of RNAP II to promoters, and also by regulating RNAP II activity.

Formation of the RNA Polymerase II Transcription Initiation Complex

A critical step in the initiation of transcription is the formation of a **pre-initiation complex (PIC).** The PIC consists of RNAP II and several general transcription factors (GTFs), which assemble onto the promoter in a specific order and provide a platform for RNAP II to recognize transcription start sites and to initiate transcription.

The GTFs that assist RNAP II at a core promoter are called **TFIIA** (*T*ranscription *F*actor for RNAP *IIA*), **TFIIB, TFIID, TFIIE, TFIIF, TFIIH,** and a large multi-subunit complex called **Mediator.** The GTFs and their interactions with the core promoter and RNAP II are outlined in **Figure 16.6** and described as follows.

The first step in the formation of a PIC (Step 1 in Figure 16.6) is the binding of TFIID to the TATA box. TFIID is a multi-subunit complex that contains **TBP** (*T*ATA *B*inding *P*rotein) and approximately 13 proteins called **TAFs** (*T*BP *A*ssociated *F*actors). In addition to binding the TATA box, TFIID binds other core-promoter elements such as Inr elements, DPEs, and MTEs. TFIIA interacts with TFIID and assists the binding of TFIID to the core promoter. Once TFIID has made contact with the core promoter, TFIIB binds to BREs near the TATA box (Step 2 in Figure 16.6). Finally, the other GTFs (Mediator, IIF, IIE, and IIH) help recruit RNAP II to the promoter (Steps 3 and 4). The fully formed PIC mediates the unwinding of promoter DNA at the start site and the transition of RNAP II from transcription initiation to elongation.

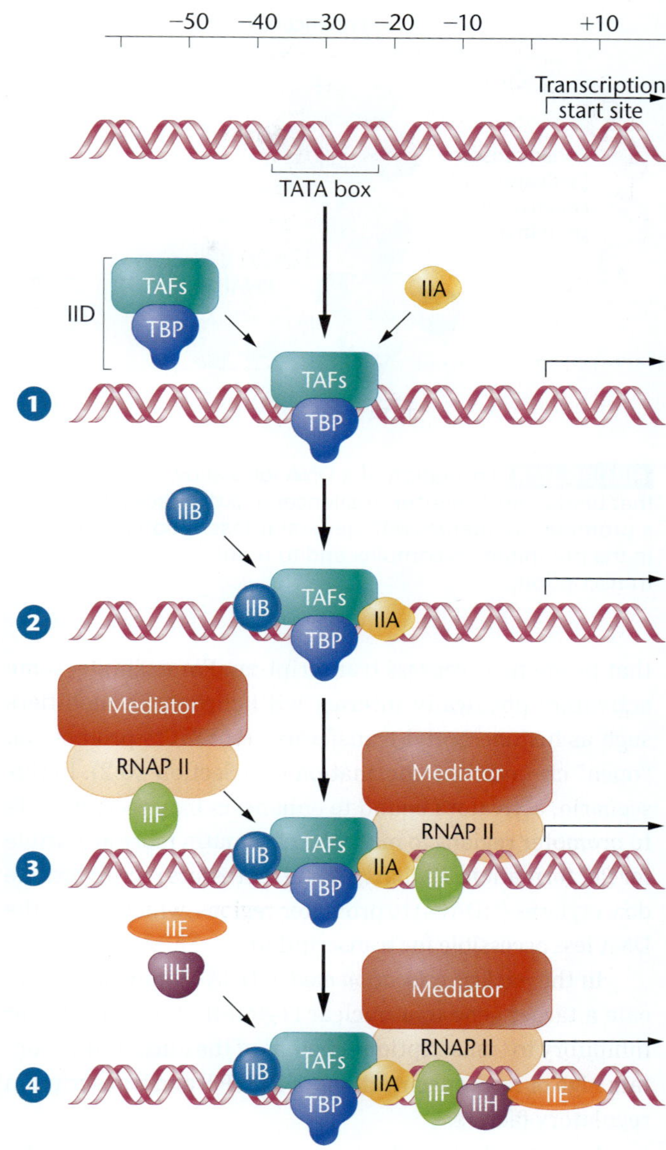

FIGURE 16.6 The assembly of general transcription factors (TFIIA, TFIIB, etc.; abbreviated as IIA, IIB, etc.) required for the initiation of transcription by RNAP II.

Mechanisms of Transcription Activation and Repression

Researchers have proposed several models by which activators and repressors modulate transcription. Common to these models is the formation of DNA loops that bring distant enhancers and silencers into close physical proximity with the promoter regions of the genes that they regulate.

In the *recruitment model*, DNA loops serve to recruit activators and GTFs to the promoter region, which increases the rate of PIC assembly and/or stability and stimulates release of RNAP II from the PIC. Direct interactions between activators and repressors with Mediator and TFIID may serve to close DNA loops between promoters and enhancers. In other cases, proteins called **coactivators** serve as a bridge between activators and promoter-bound GTFs. Large complexes of activators and coactivators that come together to direct transcriptional activation are called **enhanceosomes** (**Figure 16.7**). In a similar way, repressors bound at silencer elements may decrease the rate of PIC assembly and the release of RNAP II.

In the *chromatin alterations model*, interactions within DNA loops may facilitate changes in chromatin compaction

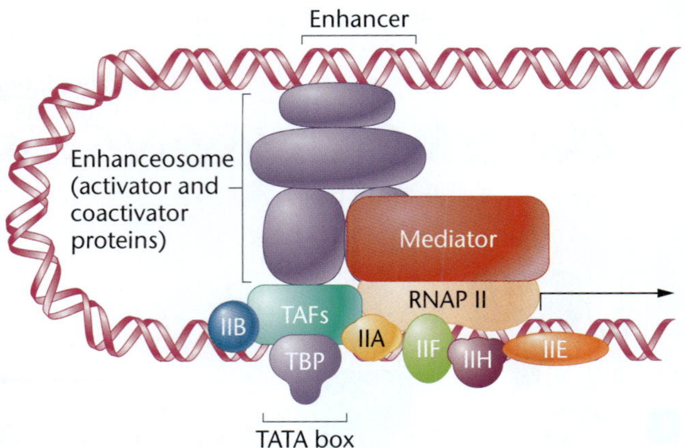

FIGURE 16.7 Formation of a DNA loop allows factors that bind to an enhancer or silencer at a distance from a promoter to interact with general transcription factors in the pre-initiation complex and to regulate the level of transcription.

that promote or repress transcription. For example, some activators physically interact with chromatin modifiers such as histone acetyl transferases (HATs) to promote an "open" chromatin conformation (see Section 16.2). In this scenario, activators bound to enhancers may recruit HATs to promoter regions to make the chromatin more accessible for PIC formation. Similarly, some repressors recruit histone deacetylases (HDACs) to promoter regions, which makes the DNA less accessible for transcription.

In the *nuclear relocation model*, DNA looping may relocate a target gene to a nuclear region that is favorable or inhibitory to transcription—regions of the nucleus that contain high or low concentrations of RNAP II and transcription regulatory factors.

Importantly, these three models are not mutually exclusive; some genes may be regulated by mechanisms that include all three models at the same time.

NOW SOLVE THIS

16.2 The hormone estrogen converts the estrogen receptor (ER) protein from an inactive molecule to an active transcription factor. The activated ER binds to *cis*-acting sites that act as enhancers, located near the promoters of a number of genes. In some tissues, the presence of estrogen appears to activate transcription of ER-target genes, whereas in other tissues, it appears to repress transcription of those same genes. Offer an explanation as to how this may occur.

■ **HINT:** *This problem involves an understanding of how transcription enhancers and silencers work. The key to its solution is to consider the many ways that trans-acting factors can interact at enhancers to bring about changes in transcription initiation.*

ESSENTIAL POINT

Activators and repressors act by enhancing or repressing the formation of a pre-initiation complex at the promoter, opening or closing chromatin, and/or relocating genes to specific nuclear sites. ■

16.6 Regulation of Alternative Splicing Determines Which RNA Spliceforms of a Gene Are Translated

As you learned earlier in the text (see Chapter 12), eukaryotic pre-mRNA transcripts are processed by the addition of a 5′-cap, the synthesis of a 3′ poly-A tail, and removal of noncoding introns through the process of splicing. However, the pre-mRNAs of many eukaryotic genes may be spliced in alternative ways to generate different **spliceforms** that include or omit different exons. This process, known as **alternative splicing,** enables a single gene to encode more than one variant of its protein product. These variants, known as **isoforms,** differ in the amino acids encoded by differentially included or excluded exons. Isoforms of the same gene may have different functions. Even small changes to the amino acid sequence of a protein may alter the active site of an enzyme, modify the DNA-binding specificity of a transcription factor, or change the localization of a protein within the cell. Thus, alternative splicing is important for the regulation of gene expression.

An elegant example of how protein activity can be modulated by alternative splicing is evidenced by the *Drosophila Mhc* gene, which encodes a motor protein responsible for muscle contraction. Different isoforms of this protein are expressed in different types of muscle with slightly different contractile properties. When an embryo-specific isoform is expressed in flight muscles, it slows the kinetic properties of the flight muscles and the flies beat their wings at a lower frequency! Thus, in this example we see evidence that alternative splicing regulates gene expression by specifying isoforms with functions that are specific to the cells they are expressed in.

Types of Alternative Splicing

There are many different ways in which a pre-mRNA may be alternatively spliced (**Figure 16.8**). One example involves **cassette exons**—such exons may be excluded from the mature mRNA by joining the 3′-end of the upstream exon to the 5′-end of the downstream exon. Skipping of cassette exons is the most prevalent type of alternative splicing in animals, accounting for nearly 40 percent of the alternative splicing events.

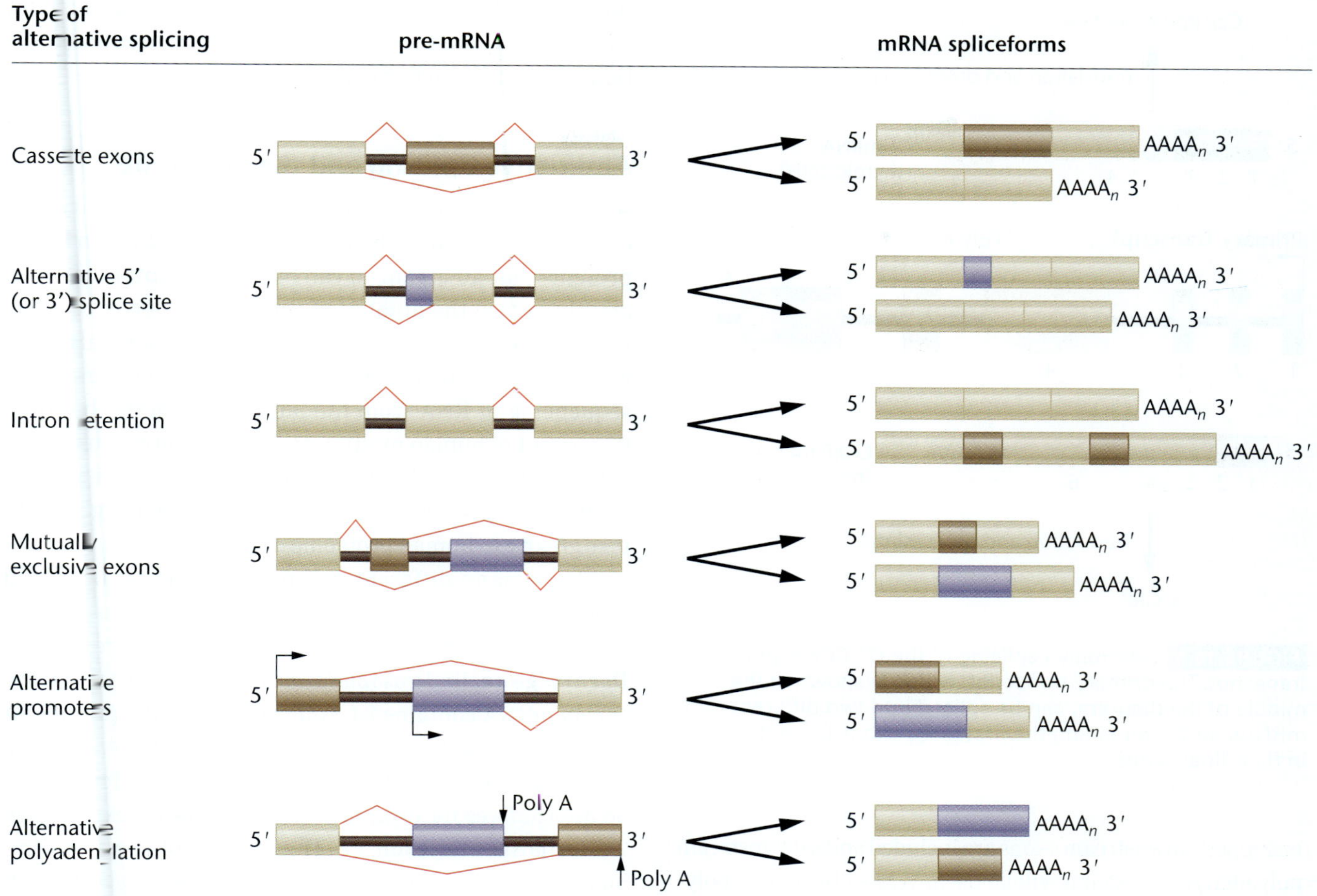

FIGURE 16.8 Different types of alternative splicing events. Exons are indicated by boxes with introns depicted by solid thick lines between them. Alternative splicing is indicated by thin red lines either above or below the pre-mRNA. Transcription start sites (bent arrows) and polyadenylation signals (Poly A) are indicated in the alternative promoters and polyadenylation examples.

In slightly over a quarter of the alternative splicing events in animals, splicing occurs at an **alternative splice site** within an exon that may be upstream or downstream of the normally used splice site. While some of these splice events are likely "noise," or errors in splice site selection by the spliceosome (Chapter 12), some instances of alternative splice site usage are important regulatory events.

Intron retention is the most common type of alternative splicing event in plants, fungi, and singed-celled eukaryotes known as protozoa but is rare in mammals. In some cases, introns, which are normally noncoding sequences, are included in the mature mRNAs and are translated, producing novel isoforms. In other cases, intron retention serves to negatively regulate gene expression at the posttranscriptional level; such mRNAs are degraded or are retained in the nucleus.

In rare cases, splicing is co-regulated for a cluster of two or more adjacent exons such that inclusion of one exon leads to the exclusion of the others in the same cluster. The use of these so-called **mutually exclusive exons** allows for swapping of protein domains encoded by different exons.

Pre-mRNAs with different 5′- and 3′-ends may be produced from the same gene due to different transcription initiation and termination sites. Some genes have **alternative promoters,** so they have more than one site where transcription may be initiated. Transcription from alternative promoters produces pre-mRNAs with different 5′-exons, which may be alternatively spliced to downstream exons. Tissue-specific expression of isoforms may result from different transcription factors recognizing different promoters of a gene in different tissues.

Spliceforms with different 3′-ends are produced by **alternative polyadenylation.** The polyadenylation signal (Chapter 12) is a sequence that directs transcriptional termination and addition of a poly-A tail. Thus, when a polyadenylation signal is transcribed, transcription is soon terminated and any downstream exon sequences are omitted. However, when an exon containing a polyadenylation signal

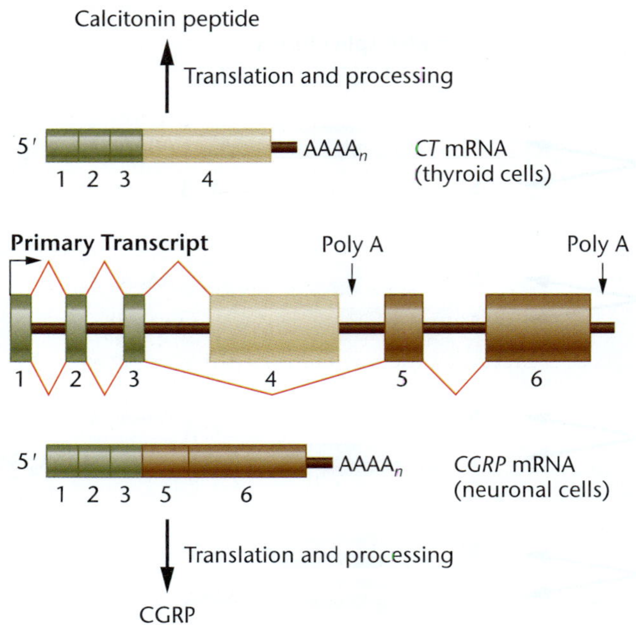

FIGURE 16.9 Alternative splicing of the *CT/CGRP* gene transcript. The primary transcript, which is shown in the middle of the diagram, can be spliced into two different mRNAs, both containing the first three exons but differing in their final exons.

is skipped, downstream exons are included and a downstream polyadenylation signal will be used. While alternative polyadenylation may produce spliceforms with different coding sequences, it also specifies different 3′ untranslated regions (UTRs) that are important for other posttranscriptional regulatory events discussed later in this chapter.

Figure 16.9 presents an example of alternative splicing of the pre-mRNA transcribed from the mammalian **calcitonin/calcitonin gene-related peptide (*CT/CGRP*)** gene. In thyroid cells, the *CT/CGRP* pre-mRNA is spliced to produce a mature mRNA containing the first four exons only. In these cells, the polyadenylation signal in exon 4 triggers transcription termination and addition of a poly-A tail. Thus, exons 5 and 6 are omitted. This mRNA is translated and processed into the calcitonin (CT) peptide, a 32-amino-acid peptide hormone that regulates calcium levels in the blood. In neurons, the *CT/CGRP* primary transcript is alternatively spliced to skip exon 4. Since the polyadenylation signal in exon 4 is quickly spliced out during transcription of the pre-mRNA, transcription continues, exons 5 and 6 are included, and the polyadenylation signal of exon 6 is recognized. The *CGRP* mRNA is translated and processed into a 37-amino-acid peptide hormone (CGRP) that stimulates the dilation of blood vessels. Through alternative splicing, the expression of the *CT/CGRP* gene is regulated such that two peptide hormones with different structures and functions are synthesized in different cell types.

Alternative Splicing and the Proteome

Alternative splicing increases the number of proteins that can be made from each gene. As a result, the number of proteins that an organism can make—its **proteome**—may greatly exceed the number of genes in the genome. Alternative splicing is found in plants, fungi, and animals but is especially common in vertebrates, including humans. Deep sequencing of RNA from human cells suggests that over 95 percent of human multi-exon genes undergo alternative splicing. While not all of these splicing events affect protein-coding sequences, it is clear that alternative splicing contributes greatly to human proteome diversity.

How many different polypeptides can be produced through alternative splicing of the same pre mRNA? One answer to that question comes from research on the ***Dscam* gene** in *Drosophila melanogaster*. The mature *Dscam* mRNA contains 24 exons; however, the pre-mRNA includes different alternative options for exons 4, 6, 9, and 17 (**Figure 16.10**). There are 12 alternatives for exon 4; 48 alternatives for exon 6; 33 alternatives for exon 9; and 2 alternatives for exon 17. The number of possible combinations that could be formed in this way suggests that, theoretically, the *Dscam* gene can produce 38,016 different proteins. Although this is an impressive number of isoforms, does the *Drosophila* nervous system require all these alternatives? Recent research suggests that it does.

During nervous system development, neurons must accurately connect with each other. Even in *Drosophila*, with only about 250,000 neurons, this is a formidable task. Neurons have cellular processes called axons that form connections with other nerve cells. The *Drosophila Dscam* gene encodes a protein that guides axon growth, ensuring

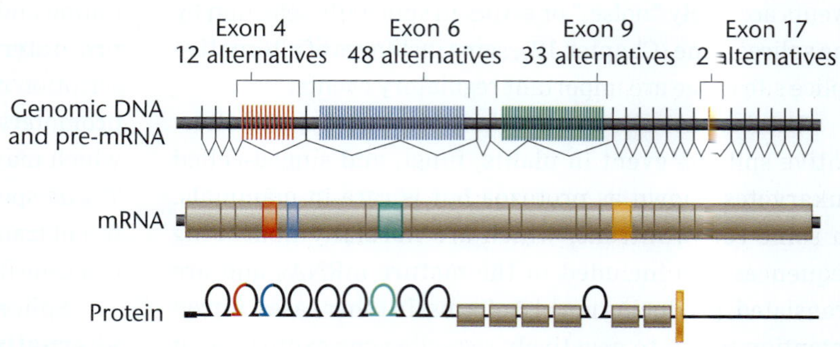

FIGURE 16.10 Alternative splicing of the *Drosophila Dscam* gene mRNA. Organization of the *Dscam* gene and the pre-mRNA. The *Dscam* gene encodes a protein that guides axon growth during development. Each mRNA will contain one of the 12 possible exons for exon 4 (red), one of the 48 possible exons for exon 6 (blue), one of the 33 possible exons for exon 9 (green), and one of the 2 possible exons for exon 17 (yellow).

that neurons are correctly wired together. Each neuron expresses a different subset of Dscam protein isoforms, and *in vitro* studies show that each Dscam isoform binds to the same isoform, but not to others. Therefore, it appears that the diversity of Dscam isoforms provides a molecular identity tag for each neuron, helping guide axons to the correct target and preventing miswiring of the nervous system.

The *Drosophila* genome contains about 14,000 protein-coding genes, but the *Dscam* gene alone encodes 2.5 times that many proteins. Because alternative splicing is far more common in vertebrates, the suite of proteins that can be produced from the human genome may be astronomically high. A large-scale mass spectrometry study of the human proteome found that the ~20,000 protein-coding genes in the human genome can produce at least 290,000 different proteins. See Chapter 18 for additional information on proteomes.

Regulation of Alternative Splicing

For pre-mRNAs that are alternatively spliced, how are specific splicing patterns selected? How does the spliceosome select one splice site instead of another? We know that this process is highly regulated, with some spliceforms only present in some cell types or under certain conditions.

Many **RNA-binding proteins (RBPs),** which are a class of proteins that bind to specific RNA sequences or RNA secondary structures, are involved in the regulation of alternative splicing. Since RBPs often exhibit tissue-specific expression, they are important regulators of tissue-specific alternative splicing. RBPs may act by binding and hiding splice sites to promote the use of alternative sites, by binding near alternative splice sites to recruit the spliceosome to such sites, and/or by directly interacting with the splicing machinery.

Alternative Splicing and Human Diseases

Since alternative splicing is an important mechanism for the regulation of gene expression, it is not surprising that defects in alternative splicing are associated with human diseases. Genetic disorders caused by mutations that disrupt RNA splicing are known as **spliceopathies.**

Myotonic dystrophy, abbreviated from the Greek term dystrophia myotonica as DM, is an autosomal dominant disorder that afflicts 1 in 8000 individuals. Patients with DM exhibit myotonia (inability to relax muscles), muscle wasting, insulin resistance, cataracts, intellectual disability, and cardiac muscle problems. Studies have shown that several of these symptoms are caused by widespread alternative splicing defects in muscle cells and neurons.

There are two forms of DM (DM1 and DM2), which are caused by mutations in different genes, but with similar outcomes that lead to splicing defects. DM1 is caused by expansion of a CTG repeat in the 3′-UTR of the *DMPK* gene. Unaffected individuals have 5 to 35 copies of the CTG repeat, whereas patients with DM1 have 150 to 2000 copies. The severity of the symptoms is directly related to the number of repeats. DM2 is caused by an expansion of a CCTG repeat sequence within the first intron of the *CNBP* gene (also known as *ZNF9*). Unaffected individuals have 11 to 26 repeats, while patients with DM2 have over 11,000 copies of this repeat; the severity of symptoms is not related to the number of repeats.

Interestingly, DM is not caused by defects in the proteins encoded by *DMPK* and *CNBP*. Rather, repeat-containing RNAs accumulate in the nucleus, instead of being exported to the cytoplasm, and are bound by proteins that regulate alternative splicing. In this way, these RNAs sequester splicing regulators and prevent them from regulating many RNAs that encode proteins important for muscle and neuron function. Strategies to degrade the repeat-containing RNAs, or to block the binding of the splicing regulators to the RNAs, are currently being researched for therapeutic purposes.

ESSENTIAL POINT

The pre-mRNAs of many eukaryotic genes undergo alternative splicing to produce different spliceforms encoding different protein isoforms, which may have different functions. Defects in alternative splicing are associated with several human diseases. ■

16.7 Gene Expression Is Regulated by mRNA Stability and Degradation

The **steady-state level** of an mRNA—meaning the total amount at any one point in time—is a function of the rate at which the gene is transcribed and the rate at which the mRNA is degraded. The steady-state level determines the amount of mRNA that is available for translation. mRNA stability can vary widely between different mRNAs, lasting a few minutes to several days, and can be regulated in response to the needs of the cell. Thus, the molecular mechanisms that control mRNA stability and degradation play important roles in the regulation of gene expression.

Mechanisms of mRNA Decay

RNA is susceptible to degradation, or decay, by **exoribonucleases**—enzymes that degrade RNA via the removal of terminal nucleotides. However, two features of mRNAs provide protection against exoribonucleases: a 7-methylguanosine (m^7G) cap at the 5′-end and a poly-A tail at the 3′-end. Maintenance or removal of the cap and poly-A tail are thus critical steps in determining the stability or decay of an mRNA.

Most eukaryotic mRNAs are degraded by **deadenylation-dependent decay** (**Figure 16.11**). This process is initiated by **deadenylases,** which are enzymes that shorten the poly-A tail. A newly synthesized mRNA has

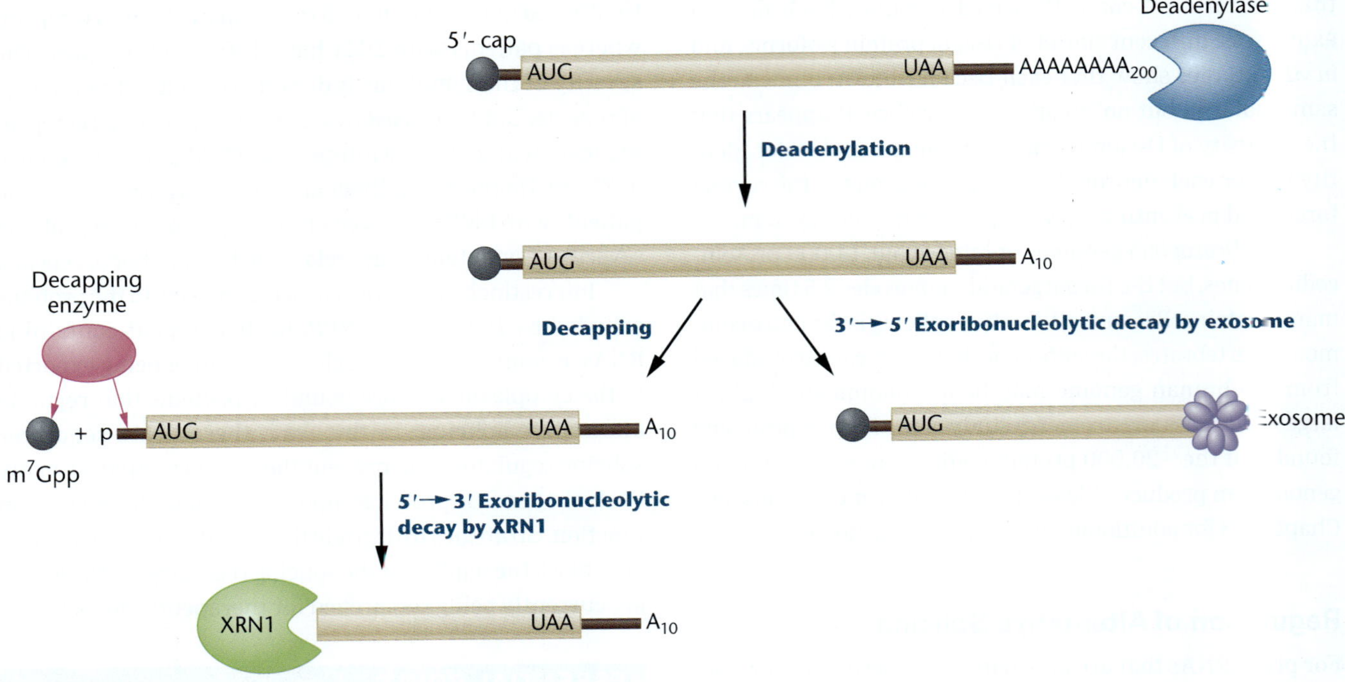

FIGURE 16.11 Deadenylation-dependent mRNA decay.

a poly-A tail that is about 200 nucleotides long. However, if deadenylation shortens it to less than ~30 nucleotides, the mRNA will be degraded. In some cases, an exoribonuclease-containing **exosome complex** destroys the mRNA in a 3′ to 5′ manner. In other cases, the shortened poly-A tail leads to the recruitment of **decapping enzymes,** which remove the 5′-cap and allow a specific exoribonuclease, **XRN1**, to destroy the mRNA in a 5′ to 3′ direction.

More rarely, mRNAs may meet their demise through **deadenylation-independent decay.** This pathway begins at the 5′-cap rather than the 3′ poly-A tail. Similar to deadenylation-dependent decay, decapping enzymes are recruited to remove the cap, and then the XRN1 exoribonuclease digests the mRNA in the 5′ to 3′ direction. In addition, in this pathway, mRNAs may also be cleaved internally by **endoribonucleases.** Following endoribonucleolytic attack, newly formed 5′- and 3′-ends are unprotected and subject to exoribonuclease digestion.

How are specific mRNAs targeted for decay by deadenylation-dependent or -independent decay? In many cases, RBPs have been identified that regulate stability and decay of specific mRNAs. However, we are still far from understanding the complex interactions of mRNAs and RBPs that determine mRNA fate.

mRNA Surveillance and Nonsense-Mediated Decay

Aberrant mRNAs can lead to nonfunctional proteins if translated. Eukaryotic cells have evolved several ways to eliminate these potentially harmful mRNAs. For example, mRNAs that lack poly-A tails or are improperly spliced may be retained in the nucleus to allow more time for processing or may be degraded by exoribonucleases.

mRNAs with a premature stop codon trigger an **mRNA surveillance** response. Premature stop codons may result from a nonsense mutation in the gene or be due to an RNA polymerase error during transcription. Translation of an mRNA with a premature stop codon would lead to a truncated and nonfunctional protein, which is a waste of cellular energy and resources and could possibly have toxic effects. However, **nonsense-mediated decay (NMD),** the most thoroughly studied mRNA surveillance pathway, efficiently eliminates mRNAs with premature stop codons.

How does NMD work? Research suggests that recognition of premature stop codons often occurs when translation terminates too far from the poly-A tail, upstream of an exon–exon junction, or upstream of other specific sequences. Once identified, mRNAs with premature stop codons are quickly degraded. In yeast and mammalian cells, decay is most often initiated by decapping enzymes or deadenylases, followed by rapid exoribonuclease digestion. In other species, such as *Drosophila*, NMD involves endoribonuclease attack near the premature stop codon and subsequent exoribonuclease digestion.

ESSENTIAL POINT

Modulation of mRNA stability and decay can eliminate aberrant mRNAs and regulate gene expression. ■

16.8 Noncoding RNAs Play Diverse Roles in Posttranscriptional Regulation

In addition to mRNAs that encode proteins, there are several types of **noncoding RNAs (ncRNAs)** that serve a variety of functions in the eukaryotic cell. You should already be familiar with rRNAs and tRNAs, which play important roles in translation (Chapter 13), and snRNAs, which mediate RNA splicing (Chapter 12). Here we will consider ncRNAs that serve as posttranscriptional regulators of gene expression.

RNA Interference and microRNAs

RNA interference (RNAi) is a mechanism by which ncRNA molecules guide the posttranscriptional silencing of mRNAs in a sequence-specific manner. The most important concept of RNAi is that since ncRNAs can associate with mRNAs through complementary base pairing, ncRNAs are able to target specific mRNAs and, via associated proteins, destroy the mRNAs or block their translation.

The ncRNAs involved in RNAi, broadly termed **small noncoding RNAs (sncRNAs),** are double-stranded RNAs that are 20 to 31 nucleotides long with 2-nucleotide overhangs at their 3′-ends. There are two main subtypes of sncRNAs: **small interfering RNAs (siRNAs)** and **microRNAs (miRNAs).** Although they arise from different sources, their mechanisms of action are similar.

siRNAs are derived from longer double-stranded RNA (dsRNA) molecules. These long dsRNAs may appear within cells as a result of virus infection or the expression of transposons, also called "jumping genes" (see Chapter 14), both of which may synthesize dsRNAs as part of their life cycles. RNAi may have evolved as a mechanism to recognize these dsRNAs and inactivate them, protecting the cell from external (viral) or internal (transposon) assaults. siRNAs can also be derived from lab-synthesized dsRNAs and introduced into cells for research or therapeutic purposes.

Whatever the source, when long dsRNAs are present in a eukaryotic cell, they are cleaved into approximately 22-nucleotide-long siRNAs by an enzyme called **Dicer** (**Figure 16.12**, left). These siRNAs then associate with the **RNA-induced silencing complex (RISC).** RISC contains an **Argonaute** family protein that binds RNA and has endoribonuclease or "slicer" activity. RISC cleaves and evicts one of the two strands of the double-stranded siRNA and retains the other strand as a single-stranded siRNA "guide" to recruit RISC to a complementary mRNA. RISC then cleaves the mRNA in the middle of the region of siRNA–mRNA complementarity (Figure 16.12, left, bottom). Cleaved mRNA

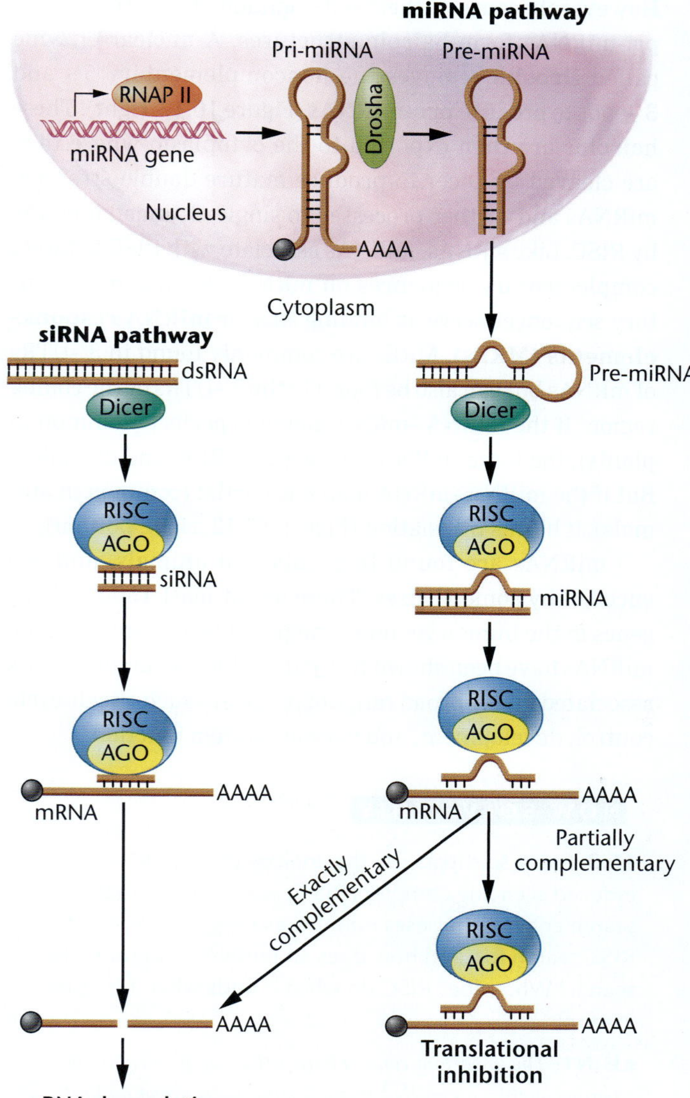

FIGURE 16.12 RNA interference pathways. Left: Double-stranded RNA is processed into siRNAs by Dicer. siRNAs then associate with RISC containing an Argonaute (AGO) family protein. RISC unwinds the siRNAs into single-stranded siRNAs and cleaves mRNAs complementary to the siRNA. Right: miRNA genes are transcribed as primary-miRNAs (pri-miRNAs), which are trimmed at the 5′- and 3′-ends by the nuclear enzyme Drosha to form pre-miRNAs, which are exported to the cytoplasm and processed by Dicer. These miRNAs then associate with RISC and mRNAs. If the miRNA and mRNA are perfectly complementary, the mRNA is destroyed; if there is a partial match, translation is inhibited.

fragments lacking a cap or a poly-A tail are then quickly degraded in the cell by exoribonucleases (see Section 16.7).

RNAi can also be mediated by miRNAs, which are produced from the transcription and processing of miRNA genes in eukaryote genomes. miRNA genes include self-complementary sequences and are transcribed into **primary miRNAs (pri-miRNAs),** which, like mRNAs, have a cap and a poly-A tail and may contain introns that are spliced out.

However, because of their self-complementary sequences, pri-miRNAs form hairpin structures. A nuclear enzyme called **Drosha** removes the noncomplementary 5′- and 3′-ends to produce **pre-miRNAs** (Figure 16.12, right). These hairpins are then exported to the cytoplasm where they are cleaved by Dicer to produce mature double-stranded miRNAs and further processed to single-stranded miRNAs by RISC. Like siRNAs, miRNAs associate with RISC to target complementary sequences on mRNAs. Such complementary sequences serve as binding sites or **miRNA response elements (MREs).** MREs are commonly found in 3′-UTRs of mRNAs but can also be found in the 5′-UTRs or the coding region. If the miRNA–mRNA match is perfect (common in plants), the target mRNA is cleaved by RISC and degraded. But if the miRNA–mRNA match is partial (common in animals), it blocks translation (Figure 16.12, right, bottom).

miRNAs are found in plants and animals, and are encoded by some viruses. There are at least 1500 miRNA genes in the human genome, and possibly as many as 5000. miRNAs have been shown to regulate diverse target mRNAs associated with a broad range of processes such as cell-cycle control, development, and nervous system function.

NOW SOLVE THIS

16.3 Some scientists use the analogy that the RNA-induced silencing complex (RISC) is a "programmable search engine" that uses miRNAs as programs. What does RISC search for, and how does an miRNA "program" the search? What does RISC do when it finds what it is searching for?

■ **HINT:** *The important concept here is that complementary base pairing enables an miRNA to guide RISC to its target for post-transcriptional regulation.*

RNA Interference in Medicine

The discovery of RNAi opened the door to RNA molecules being developed as potential pharmaceutical agents. In theory, any disease caused by overexpression of a specific gene, or even normal expression of an abnormal gene product, could be treated by RNAi. In early 2018, there were 30 ongoing clinical trials to treat viral infections like hepatitis and Ebola, as well as cancers, eye diseases, hemophilia, hypercholesterolemia, and even alcoholism. Some of these trials were in phase III, meaning that they were being administered to large groups to confirm effectiveness and monitor side effects.

In August of 2018, Alnylam Pharmaceuticals announced FDA approval of Patisiran, an siRNA and nanoparticle delivery system that treats transthyretin amyloidosis. This disorder is characterized by nervous system and cardiac problems due to a buildup of a mutant form of the transthyretin protein. Patisiran is the first ever RNAi-based drug to receive FDA approval.

Long Noncoding RNAs and Posttranscriptional Regulation

In addition to sncRNAs discussed previously, eukaryotic genomes also encode many **long noncoding RNAs (lncRNAs).** One obvious distinction is that lncRNAs are longer than sncRNAs and are often arbitrarily designated to be greater than 200 nucleotides in length. lncRNAs are produced in a similar fashion to mRNAs; they are modified with a cap and a poly-A tail, and they can be spliced. In contrast to mRNAs, they have no start and stop codons, indicating that they do not encode protein. The conservative estimate is that the human genome encodes ~ 17,000 lncRNAs.

lncRNAs have been linked to diverse regulatory functions. Some lncRNAs bind to chromatin-regulating complexes to influence chromatin modifications and alter patterns of gene expression (see Special Topics Chapter ST1—Epigenetics). Others regulate transcription by directly associating with transcription factors. However, in this section we will focus on the *posttranscriptional* roles of lncRNAs.

When lncRNAs are complementary to mRNAs or pre-mRNAs, the two can hybridize by base pairing. In some cases, this leads to regulation of alternative splicing. For example, an lncRNA that binds to splice sites for an exon can lead to its exclusion from the mature transcript. In other cases, lncRNA–mRNA hybridization produces a dsRNA that triggers an RNAi response. It is processed by Dicer into siRNAs that then target complementary mRNAs for destruction by RISC (see Figure 16.12). Other studies show that lncRNAs can bind to mRNAs in ways that regulate their stability, decay, and translation.

Some lncRNAs function as **competing endogenous RNAs (ceRNAs).** Conceptually, ceRNAs are "sponges" that "soak up" miRNAs due to the presence of complementary miRNA-binding sites in their sequence—the miRNA response elements (MREs) introduced earlier in this chapter. Thus, ceRNAs compete with mRNAs for miRNA binding. Whereas miRNAs downregulate their mRNA targets, ceRNAs are able to "derepress" the mRNA targets by sequestering miRNAs away from them. In other words, ceRNAs are decoys. The efficacy of a ceRNA depends on variables such as how many MREs it contains, how many copies of the ceRNA are expressed in the cell, and the affinity of its MREs for the miRNA.

ESSENTIAL POINT

Noncoding RNAs, such as microRNAs (miRNAs) and small interfering RNAs (siRNAs), can mediate sequence-specific degradation or translational inhibition of target mRNAs in a process called RNA interference (RNAi). Long noncoding RNAs (lncRNAs) have a variety of functions in the cell such as influencing alternative splicing and serving as decoys for miRNAs to allow translation of mRNAs that would otherwise be targeted by such miRNAs. ■

16.9 mRNA Localization and Translation Initiation Are Highly Regulated

We have already encountered several posttranscriptional mechanisms that impact translation. Alternative splicing determines which spliceforms may be translated, and mRNAs may be degraded or targeted by an miRNA to stop translation. However, translation may be regulated more directly as well.

Some mRNAs are localized to discrete regions of the cell where they are translated locally. This generates asymmetric protein distributions within the cell that enable different parts of the cell to have different functions. Similar to other posttranscriptional mechanisms, the regulation of mRNA localization and localized translation are governed by *cis*-regulatory sequences on the mRNA, and RNA-binding proteins.

One of the best-described RBP–mRNA interactions governs the localization and translational control of actin mRNAs in crawling cells. Following injury, fibroblasts migrate to the site of the wound and assist in wound healing. Fibroblasts control their direction of movement by controlling where within the cell they polymerize new cytoskeletal actin microfilaments. The "leading edge" of the cell where this actin polymerization occurs is called the lamellipodium.

Actin mRNA is localized to lamellipodia, and that localization is dependent on a 54-nucleotide element in the actin mRNA 3′-UTR termed a **zip code.** The zip code is a *cis*-regulatory element that serves as a binding site for an RBP called **zip code binding protein 1 (ZBP1).** ZBP1 initially binds the zip code of actin mRNAs in the nucleus. Once exported to the cytoplasm, ZBP1 blocks translation initiation by preventing the association of the large subunit (60*S*) of the ribosome. In addition, ZBP1 associates with cytoskeleton motor proteins to facilitate the transport of the mRNA to the lamellipodium. Once the mRNAs arrive at the final destination, a kinase called **Src** phosphorylates ZBP1, which disrupts RNA binding and allows translation initiation (**Figure 16.13**).

Since Src activity is limited to the cell periphery, this mechanism allows the transport of actin mRNAs in a translationally repressed state to the cell periphery, thus controlling where actin will be translated.

Actin mRNA localization is important in other cells as well, such as in neurons where localized translation of actin helps guide axon growth. In addition, many other mRNAs are also localized and translated in specific locations within the cell. mRNA localization and translational control are particularly important for nervous system function. In fact, defects in mRNA localization in neurons have been implicated in human disorders such as fragile-X syndrome, spinal muscular atrophy, and spinocerebellar ataxia.

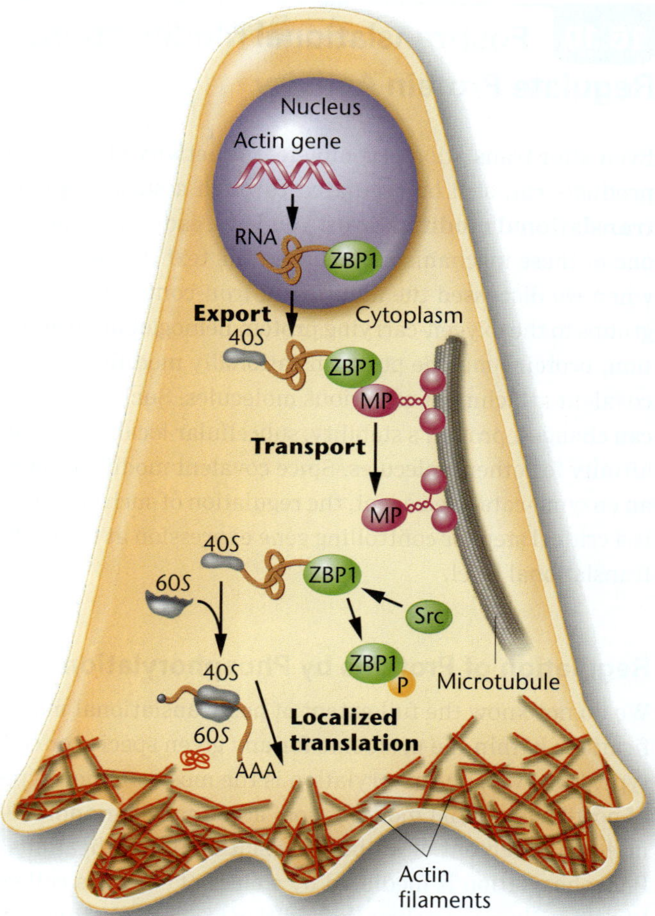

FIGURE 16.13 Localization and translational regulation of actin mRNA. The RNA-binding protein ZBP1 associates with actin mRNA in the nucleus and escorts it to the cytoplasm. ZBP1 blocks translation and binds cytoskeleton motor proteins (MP), which transport ZBP1 and actin mRNA to the cell periphery. At the cell periphery, ZBP1 is phosphorylated by Src and dissociates from actin mRNA, allowing it to be translated by a ribosome (40*S*/60*S*). Actin translation and polymerization at the leading edge direct cell movement.

NOW SOLVE THIS

16.4 Consider the example that actin mRNA localization is important for fibroblast migration. What would you predict to be the consequence of deleting the zip code sequence element of the actin mRNA?

■ **HINT:** *The key to answering this question is recalling that the zip code is a cis-acting element that is bound by an RNA-binding protein involved in its localization and translational control.*

ESSENTIAL POINT

mRNAs may not be translated immediately; some mRNAs are stored for later use and/or are localized to specific regions of the cell and then locally translated to create asymmetric protein distribution within the eukaryotic cell. ■

16.10 Posttranslational Modifications Regulate Protein Activity

Even after translation is complete, the activity of the gene products can still be regulated through a suite of **post-translational modifications.** You've already encountered one of these mechanisms earlier in the text (Chapter 13), when we discussed the addition of iron-containing heme groups to the oxygen-carrying protein hemoglobin. In addition, proteins may be posttranslationally modified by the covalent attachment of various molecules. Such additions can change a protein's stability, subcellular localization, or affinity for other molecules. Since covalent modification is an enzyme-catalyzed event, the regulation of such enzymes is a critical step for controlling gene expression at the post-translational level.

Regulation of Proteins by Phosphorylation

We do not know the full extent of posttranslational modifications within the proteome for any given species or cell type. However, phosphorylation is the most common type of posttranslational modification, accounting for approximately 65 percent of all posttranslational modifications. Phosphorylation is mediated by a class of enzymes called **kinases.** Kinases catalyze the addition of a phosphate group to serine, tyrosine, or threonine amino acid side chains. Such additions are reversible. **Phosphatases** are enzymes that remove phosphates. It is calculated that the human genome contains 518 kinase-encoding genes and 147 phosphatase-encoding genes. This suite of enzymes can be used in countless ways to regulate protein activity.

Phosphorylation usually induces conformational changes. These changes can have different effects depending on the type of target. For example, enzymes may be turned on or off by phosphorylation where conformational changes alter substrate binding. Transcription factors may be turned on or off by phosphorylation based on how the conformational changes impact its affinity for the target DNA sequence. In some cases, there is more than one phosphorylation site on a protein; phosphorylation in one site may activate, while phosphorylation at the other site may inactivate the protein.

Ubiquitin-Mediated Protein Degradation

One important way to regulate protein activity after translation is through the targeting of specific proteins for degradation. The principal mechanism by which the eukaryotic cell targets a protein for degradation is by covalently modifying it with **ubiquitin,** a small protein with 76 amino acids that is found in all eukaryotic cells. The fact that it is *ubiquitous* gives ubiquitin its name.

Ubiquitin is covalently attached to a target protein via a lysine side chain through a process called **ubiquitination.** Subsequently, lysine side chains in the attached ubiquitin molecule can be modified by the addition of other ubiquitin molecules. This process can be repeated to form long poly-ubiquitin chains, which serve as "tags" that mark the protein for destruction. Poly-ubiquitinated proteins are recognized by the **proteasome,** a multi-subunit protein complex with protease (protein cleaving) activity. The proteasome unwinds target proteins, removes their ubiquitin tags, and breaks the protein into small peptides about 7 to 8 amino acids long (**Figure 16.14**).

Since ubiquitinated proteins are quickly destroyed, the determination of which proteins get ubiquitinated is a major regulatory step. A class of enzymes, **ubiquitin ligases,** recognize and bind specific target proteins and catalyze the

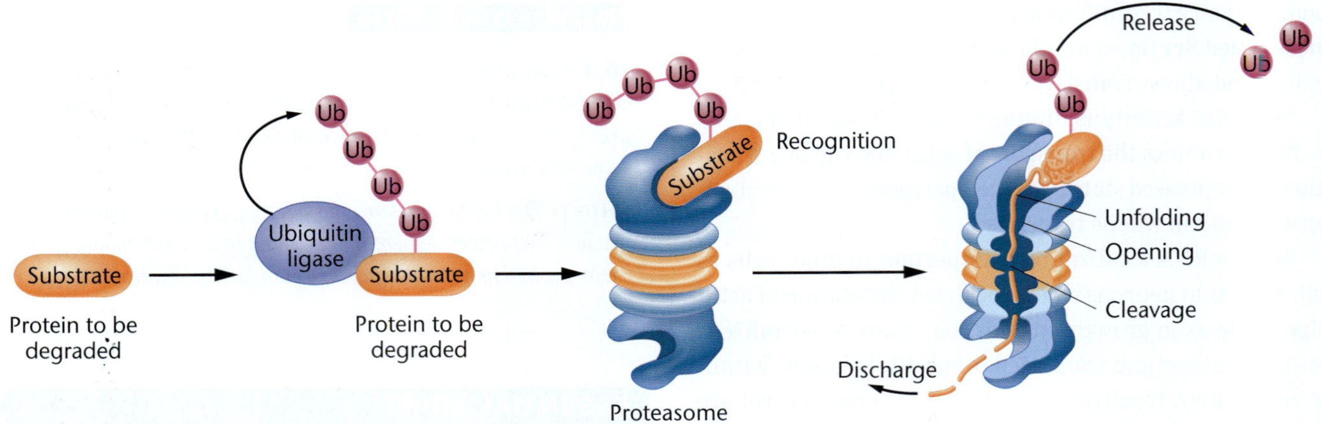

FIGURE 16.14 Ubiquitin-mediated protein degradation. Ubiquitin ligase enzymes recognize substrate proteins and catalyze the addition of ubiquitin (Ub) residues to create a long chain. Ubiquitinated proteins are then recognized by the proteasome, which removes ubiquitin tags, unfolds the protein, and proteolytically cleaves it into small polypeptides.

processive addition of ubiquitin residues (Figure 16.14). In turn, ubiquitin ligase activity can be regulated in many ways to serve the needs of the cell.

It is estimated that there are over 600 ubiquitin ligase–encoding genes in the human genome and that some human ubiquitin ligases interact with over 40 different substrate proteins. Overall, scientists estimate that human ubiquitin ligases target over 9000 different proteins, which accounts for approximately 40 percent of the protein-coding genes in the human genome. This suggests that ubiquitin-mediated protein degradation may be a broadly used mechanism to regulate biological function.

ESSENTIAL POINT

Following translation, protein activity can be modulated by post-translational modifications, such as phosphorylation or ubiquitin-mediated degradation. ■

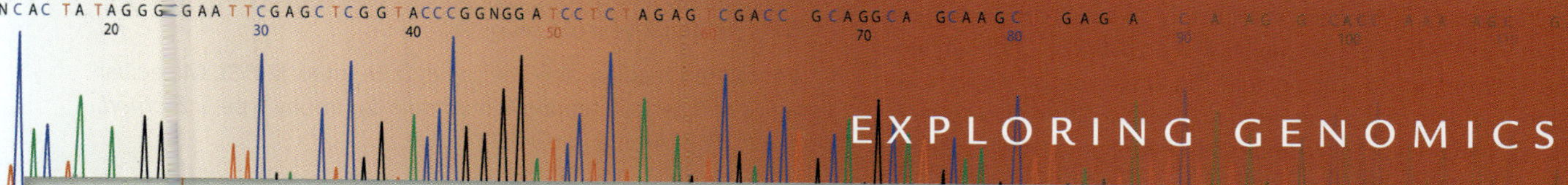

EXPLORING GENOMICS

Mastering Genetics Visit the Study Area: Exploring Genomics

Tissue-Specific Gene Expression

In this chapter, we discussed how gene expression can be regulated in complex ways. One aspect of regulation we considered is the way promoter, enhancer, and silencer sequences can govern transcriptional initiation of genes to allow for tissue-specific gene expression. All cells and tissues of an organism possess the same genome (with some genomic variation as you will learn in Chapter 18), and many genes are expressed in all cell and tissue types. However, muscle cells, blood cells, and all other tissue types express genes that are largely tissue specific (i.e., they have limited or no expression in other tissue types). In this exercise, we return to the National Center for Biotechnology Information (NCBI) site and use the search tool BLAST (Basic Local Alignment Search Tool) which you were introduced to in an earlier Exploring Genomics exercise (see Chapter 9). We will use BLAST to learn more about tissue-specific gene-expression patterns.

■ **Exercise – Tissue-Specific Gene Expression**

1. Access BLAST from the NCBI Web site at https://blast.ncbi.nlm.nih.gov/Blast.cgi.
2. The following are GenBank accession numbers for four different genes that show tissue-specific expression patterns. You will perform your searches on these genes.

   NM_021588.1
   NM_007391
   AY260853.1
   NM_004917

3. For each gene, carry out a nucleotide BLAST search using the accession numbers for your sequence query. (Refer to the Exploring Genomics feature in Chapter 9 if you need to refresh your memory on BLAST searches.) Because the accession numbers are for nucleotide sequences, be sure to use the "Nucleotide BLAST" program when running your searches. Once you enter "Nucleotide BLAST," under the "Choose Search Set" category, make sure the database is set to "Others (nr, etc.)" so that you are not searching an organism-specific database.
4. Once your BLAST search results appear, look at the top alignment for each gene. Clicking on the link for the top alignment will take you to the page showing the sequence alignment for this gene. To the far right of the page, if you scroll down, you will see a section called "Related Information." The "Gene" link provides a report on details related to this gene.

   Some alignments will display a link for "Map Viewer," which will take you to genome mapping information about the gene. The "UniGene" link will show you a UniGene report. For some genes, upon entering UniGene you may need to click a link above the gene name or the gene name itself in order to retrieve a UniGene report. Be sure to explore the "EST Profile" link under the "Gene Expression" category in each UniGene report. EST profiles will show a table of gene-expression patterns in different tissues.
5. Also explore the "GEO Profiles" link under the "Gene Expression" category of the UniGene reports, when available. These links will take you to a number of gene-expression studies related to each gene of interest. Explore these resources for each gene, and then answer the following questions:
   a. What is the identity of each sequence, based on sequence alignment? How do you know this?
   b. What species was each gene cloned from?
   c. Which tissue(s) are known to express each gene?
   d. Does this gene show regulated expression during different times of development?
   e. Which gene shows the most restricted pattern of expression by being expressed in the fewest tissues?

CASE STUDY A mysterious muscular dystrophy

A man in his early 30s suddenly developed weakness in his hands and neck, followed weeks later by burning muscle pain—all symptoms of late-onset muscular dystrophy. His internist ordered genetic tests to determine whether he had one of the most common adult-onset muscular dystrophies—myotonic dystrophy type 1 (DM1) or myotonic dystrophy type 2 (DM2). The tests detect mutations in the *DMPK* and *CNBP* genes, the only genes known to be associated with DM1 and DM2. While awaiting the results of the gene tests, the internist explained that the disease-causing mutations in these genes do not result in changes to the coding sequence. Rather, myotonic dystrophies result from increased, or expanded, numbers of tri- and tetranucleotide repeats in the 3′ untranslated region of the *DMPK* or *CNBP* genes. The doctor went on to explain that the presence of RNAs with expanded numbers of repeats leads to aberrant alternative splicing of other mRNAs, causing widespread disruption of cellular pathways. This discussion raises a number of interesting questions.

1. What is alternative splicing, where does it occur, and how could disrupting it affect the expression of the affected gene(s)?
2. What role might the expanded tri- and tetranucleotide repeats play in the altered splicing?
3. DM1 is characterized by a phenomenon known as genetic anticipation (see Chapter 4) where the age of onset tends to decrease and the severity of the symptoms tend to increase from one generation to the next due to expansion of the trinucleotide repeats in the *DMPK* gene. What are the implications of a diagnosis of DM1 in this patient with respect to his [illegible]-year-old son, and 2-year-old daughter?

For related reading, see Pavićević, D. S., et al. (2013). Molecular genetics and genetic testing in myotonic dystrophy type 1. *Biomed. Res. Int.* 2013. Article ID 391821.

INSIGHTS AND SOLUTIONS

1. As a research scientist, you have decided to study transcriptional regulation of a gene whose DNA has been cloned and sequenced. To begin your study, you obtain the DNA clone of the gene, which includes the coding sequence and at least 1 kb of upstream DNA. You then create a number of subclones of this DNA, containing various deletions in the gene's upstream region. These deletion templates are shown in the following figure.

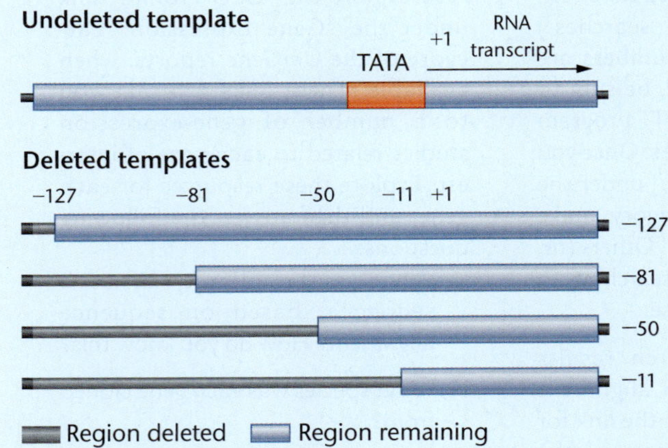

To test these DNA subclones for their ability to direct transcription of the gene, you prepare two different types of *in vitro* transcription systems. The first is a purified system containing RNAP II and the general transcription factors TFIIA, TFIIB, TFIID, TFIIE, TFIIF, and TFIIH. The second system consists of a crude nuclear extract, which is made by extracting most of the proteins from the nuclei of cultured cells. When you test your two transcription systems using each of your templates, you obtain the following results:

DNA Added	Purified System	Nuclear Extract
Undeleted	+	++++
−127 deletion	+	++++
−81 deletion	+	++++
−50 deletion	+	+
−11 deletion	*o*	*o*

\+ Low-efficiency transcription
++++ High-efficiency transcription
o No transcription

(a) Why is there no transcription from the −11 deletion template in both the crude extract and the purified system?

(b) How do the results for the nuclear extract and the purified system differ, for the *undeleted* template? How would you interpret this result?

(c) For each of the various deletion templates, compare the results obtained from both the nuclear extract and the purified systems.

(d) What do these data tell you about the transcription regulation of this gene?

Solution:

(a) The lack of transcription from the −11 template suggests that some essential DNA sequences are missing from this deletion template. As the −50 template does show some transcription in both the crude extract and purified system, it is likely that the essential missing sequences, at least for basal levels of transcription, lie between −50 and −11. As the TATA box is

(continued)

located in this region, its absence in the −11 template may be the reason for the lack of transcription.

(b) The undeleted template containing large amounts of upstream DNA is sufficient to promote high levels of transcription in a nuclear extract, but only low levels in a purified system. These data suggest that something is missing in the purified system, compared with the nuclear extract, and this component is important for high levels of transcription from this promoter. As crude nuclear extracts are not defined in content, it would not be clear from these data what factors in the extract are the essential ones.

(c) Both the −127 and −81 templates act the same way as the undeleted template in both the nuclear extract and the purified system—high levels of transcription in nuclear extracts but low levels in a purified system. In contrast, the −50 template shows only low levels of transcription in both systems. These results indicate that all of the sequences necessary for high levels of transcription in a crude system are located between −81 and −50.

(d) First, these data tell you that general transcription factors alone are not sufficient to specify high efficiencies of transcription from this promoter. The DNA sequence elements through which the general transcription factors work are located within 50 bp of the transcription start site. Second, the data tell you that the promoter for this gene is likely a member of the "focused" class of promoters, with one defined transcription start site and an essential TATA box. Third, high levels of transcription require sequences between −81 and −50 relative to the transcription start site. These sequences (enhancers) interact with some component(s) (transcriptional activators) of crude nuclear extracts.

2. Scientists estimate that more than 15 percent of disease-causing mutations involve errors in alternative splicing. However, there is an interesting case in which a mutation that deletes an exon results in increased protein production. Mutations that delete exon 45 of the 79-exon *dystrophin* gene are the most common cause of Duchenne muscular dystrophy (DMD), a disease associated with progressive muscle degeneration. However, some individuals with deletions of both exons 45 and 46 have Becker muscular dystrophy (BMD), a milder form of muscular dystrophy. Provide a possible explanation for why BMD patients, with a deletion of both exon 45 and 46, produce more dystrophin than DMD patients do.

Solution: Having a deletion of one exon has several possible effects on a gene product. One possibility is that the mRNA transcribed from the exon-deleted *dystrophin* gene is unstable, leading to a lack of dystrophin protein production. Even if the mRNA is stable, the resulting mutated dystrophin protein could be targeted for rapid degradation, leading to the absence of stable active protein. Another possibility is that the deletion of one exon creates a frameshift leading to a premature stop codon. As the *dystrophin* gene has 79 exons spanning over 2.6 million base pairs, a frameshift in exon 46 could create a stop codon near the middle of the gene, which would trigger nonsense-mediated mRNA decay. Any mRNA escaping degradation would encode a shorter than normal dystrophin protein, which would likely be nonfunctional.

It is possible that a deletion encompassing both exon 45 and 46 could restore the reading frame of the dystrophin protein in exon 47. The protein product of this gene would be missing amino acid sequences encoded by the two missing exons; however, the protein itself could still have some activity, partially preserving the wild-type phenotype.

Problems and Discussion Questions

Mastering Genetics Visit for instructor-assigned tutorials and problems.

1. **HOW DO WE KNOW?** In this chapter, we focused on the regulation of gene expression in eukaryotes. At the same time, we found many opportunities to consider the methods and reasoning by which much of this information was acquired. From the explanations given in the chapter:
   (a) How do we know that transcription and translation are spatially and temporally separated in eukaryotic cells?
   (b) How do we know that DNA methylation is associated with transcriptionally silent genes?
   (c) How do we know that core-promoter elements are important for transcription?
   (d) How do we know that the orientation of promoters relative to the transcription start site is important while enhancers are orientation independent?
   (e) How do we know that alternative splicing enables one gene to encode different isoforms with different functions?
   (f) How do we know that small noncoding RNA molecules can regulate gene expression?
2. **CONCEPT QUESTION** Review the Chapter Concepts list on p. 326. The third concept describes how transcription initiation requires the assembly of transcription regulatory proteins on DNA sites known as promoters, enhancers, and silencers. Write a short essay describing which types of *trans*-acting proteins bind to which type of *cis*-regulatory element, and how these interactions influence transcription initiation. ■
3. What features of eukaryotes provide additional opportunities for the regulation of gene expression compared to bacteria?
4. Describe the organization of the interphase nucleus. Include in your presentation a description of chromosome territories and interchromatin compartments.
5. How do histone variants contribute to the transcription of DNA?
6. Present an overview of the manner in which chromatin can be remodeled. Describe the manner in which these remodeling processes influence transcription.

7. What are the subcategories within eukaryotic promoters? How do enhancers and silencers differ from promoters?
8. Is it true that the number of proteins in a eukaryotic cell is higher than the number of genes? Explain.
9. Why would multiple transcriptional start sites help the regulation of housekeeping genes?
10. A point mutation has altered the codon coding for cysteine (UGU) in the middle of the coding sequence into a stop codon (UGA). What is this kind of a mutation called, and what is the fate of the resulting mRNA?
11. Focused promoters can contain several regulatory sequences that are conserved between genes and species. Some are upstream and some downstream from the transcription start site. Describe these core-promoter and proximal-promoter elements, their sequences, and their roles in transcription in eukaryotes.
12. Describe how the RNA polymerase II transcription initiation complex is assembled and how its activity can be regulated by enhancers and silencers.
13. List three types of alternative splicing patterns and how they lead to the production of different protein isoforms.
14. Explain how the use of alternative promoters and alternative polyadenylation signals produces mRNAs with different 5′- and 3′-ends.
15. The regulation of mRNA decay relies heavily upon deadenylases and decapping enzymes. Explain how these classes of enzymes are critical to initiating mRNA decay.
16. Nonsense-mediated decay is an mRNA surveillance pathway that eliminates mRNAs with premature stop codons. How does the cell distinguish between normal mRNAs and those with a premature stop?
17. In 1998, future Nobel laureates Andrew Fire and Craig Mello, and colleagues, published an article in *Nature* entitled, “Potent and specific genetic interference by double-stranded RNA in *Caenorhabditis elegans*.” Explain how RNAi is both “potent and specific.”
18. Present an overview of RNA interference (RNAi). How does the silencing process begin, and what major components participate?
19. Can you find any similarities between the CRISPR-Cas system of adaptive immunity in bacteria and the process of RNA interference in eukaryotes?
20. miRNAs target endogenous mRNAs in a sequence-specific manner. Explain, conceptually, how one might identify potential mRNA targets for a given miRNA if you only know the sequence of the miRNA and the sequence of all mRNAs in a cell or tissue of interest.
21. In principle, RNAi may be used to fight viral infection. How might this work?
22. Competing endogenous RNAs act as molecular “sponges.” What does this mean, and what do they compete with?
23. How and why are eukaryotic mRNAs transported and localized to discrete regions of the cell?
24. How is it possible that a given mRNA in a cell is found throughout the cytoplasm but the protein that it encodes is only found in a few specific regions?
25. How may the covalent modification of a protein with a phosphate group alter its function?
26. Ubiquitin ligases ubiquitinate proteins and target them for degradation. Why would the cell need to degrade a protein that it has just produced?
27. When challenged with a low oxygen environment, known as hypoxia, the body produces a hormone called erythropoietin (EPO), which then stimulates red blood cell production to carry more oxygen. Transcription of the gene encoding EPO is dependent upon the hypoxia-inducible factor (HIF), which is a transcriptional activator. However, HIF alone is not sufficient to activate EPO. For example, Wang et al. (2010. *PLOS ONE* 5: e10002) showed that HIF recruits another protein called p300 to an enhancer for the EPO gene. Furthermore, deletion of p300 significantly impaired transcription of the EPO gene in response to hypoxia. Given that p300 is a type of histone acetyl transferase, how might p300 influence transcription of the EPO gene?
28. The TBX20 transcription factor is important for the development of heart tissue. Deletion of the *Tbx20* gene in mice results in poor heart development and the death of mice well before birth. To better understand how TBX20 regulates heart development at a genetic level, Sakabe et al. (2012. *Hum. Mol. Genet.* 21:2194–2204) performed a transcriptome analysis in which they compared the levels of all mRNAs between heart cells from wild-type mice and mice with *Tbx20* deleted.
    (a) How might such a transcriptome analysis provide information about how TBX20 regulates heart development?
    (b) This study concluded that TBX20 acts as an activator of some genes but a repressor of other genes in cardiac tissue. How might a single transcription factor have opposite effects on the transcription of different genes?
29. Many viruses that infect eukaryotic cells express genes that alter the regulation of host gene expression to promote viral replication. For example, herpes simplex virus-1 (HSV-1) expresses a protein called ICP0, which is necessary for successful viral infection and replication within the host. Lutz et al. (2017. *Viruses* 9: 210) showed that ICP0 can act as a ubiquitin ligase and target the redundant transcriptional repressors ZEB1 and ZEB2, which leads to upregulation of the miR-183 cluster (a set of three miRNAs transcribed from the same locus).
    (a) What likely happens to ZEB1 and ZEB2 upon HSV-1 infection?
    (b) How may ICP0 expression in a host cell lead to upregulation of the miR-183 cluster?
    (c) Speculate on how miR-183 cluster upregulation may benefit the virus.

17

Recombinant DNA Technology

CHAPTER CONCEPTS

- Recombinant DNA technology creates combinations of DNA sequences from different sources.
- A common application of recombinant DNA technology is to clone a DNA segment of interest.
- Specific DNA segments are inserted into vectors to create recombinant DNA molecules that are transferred into eukaryotic or bacterial host cells, where the recombinant DNA replicates as the host cells divide.
- DNA libraries are collections of cloned DNA and were historically used to isolate specific genes.
- DNA segments can be quickly amplified and cloned millions of times using the polymerase chain reaction (PCR).
- Nucleic acids can be analyzed using a range of molecular techniques.
- Sequencing reveals the nucleotide composition of DNA, and major improvements in sequencing technologies have rapidly advanced many areas of modern genetics research, particularly genomics.
- Gene knockout methods and transgenic animals have become invaluable for studying gene function *in vivo*.
- Genome editing methods demonstrate value in basic research and exciting potential for clinical applications of genetic technology.
- Recombinant DNA technology has revolutionized our ability to investigate the genomes of diverse species and has led to the modern revolution in genomics.

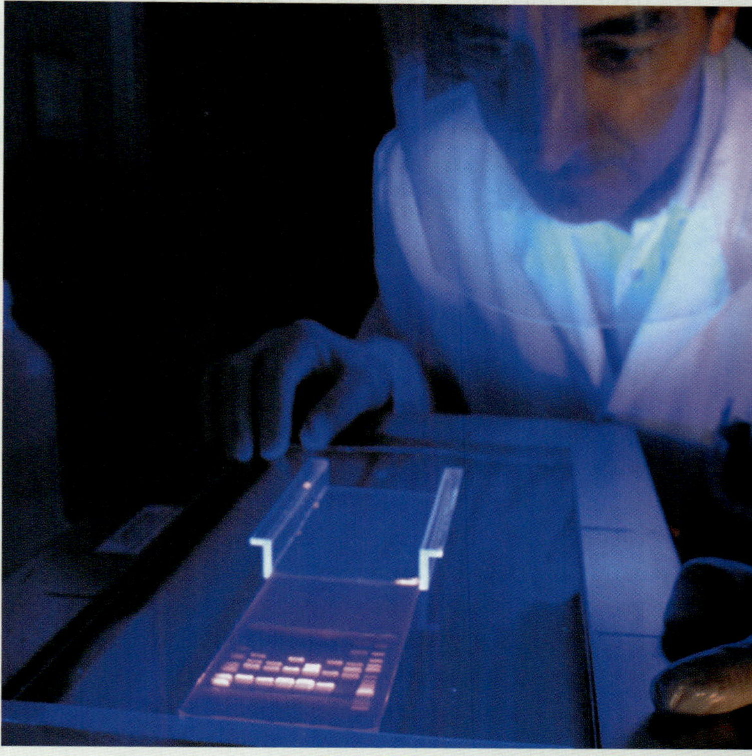

A researcher examines an agarose gel containing separated DNA fragments stained with the DNA-binding dye ethidium bromide and visualized under ultraviolet light.

Researchers of the mid- to late 1970s developed various techniques to create, replicate, and analyze **recombinant DNA** molecules—DNA created by joining together pieces of DNA from different sources. These techniques, called **recombinant DNA technology** and often known as "gene splicing" in the early days, marked a major advance in research in molecular biology and genetics, allowing scientists to isolate and study specific DNA sequences. For their contributions to the development of this technology, Daniel Nathans, Hamilton Smith, and Werner Arber were awarded the 1978 Nobel Prize in Physiology or Medicine.

The power of recombinant DNA technology is astonishing, enabling geneticists to identify and isolate a single gene or DNA segment of interest from a genome, and to produce large quantities of identical copies of this specific molecule. These identical copies, or **clones,** can then be manipulated for numerous purposes, including conducting research on the structure and organization of the DNA, studying gene expression, studying protein products to understand their structure and function, and producing important commercial products from the protein encoded by a gene.

The fundamental techniques involved in recombinant DNA technology subsequently led to the field of genomics, enabling scientists to sequence and analyze entire genomes. Note that some of the topics discussed in this chapter are explored in greater depth later in the text (see Special Topics Chapter 2—Genetic Testing, 3—Gene Therapy, 5—DNA Forensics, and

6—Genetically Modified Foods). In this chapter, we survey basic methods of recombinant DNA technology used to isolate, replicate, and analyze DNA.

17.1 Recombinant DNA Technology Began with Two Key Tools: Restriction Enzymes and Cloning Vectors

Although natural genetic processes such as crossing over produce recombined DNA molecules, the term *recombinant DNA* is reserved for molecules produced by artificially joining DNA obtained from different sources. We begin our discussion of recombinant DNA technology by considering two important tools used to construct and amplify recombinant DNA molecules: DNA-cutting enzymes called **restriction enzymes** and **cloning vectors.** The use of restriction enzymes and cloning vectors was largely responsible for advancing the field of molecular biology because a wide range of laboratory techniques are based on recombinant DNA technology.

Restriction Enzymes Cut DNA at Specific Recognition Sequences

Bacteria produce restriction enzymes as a defense mechanism against infection by bacteriophage. They *restrict* or prevent viral infection by degrading the DNA of invading viruses. More than 4300 restriction enzymes have been identified, and over 600 are commercially produced and available for use by researchers. A restriction enzyme recognizes and binds to DNA at a specific nucleotide sequence called a **recognition sequence** or **restriction site** (**Figure 17.1**). The enzyme then cuts both strands of the DNA within that sequence by cleaving the phosphodiester backbone. Scientists commonly refer to this as "digestion" of DNA. The usefulness of restriction enzymes is their ability to accurately and reproducibly cut DNA into fragments. Restriction enzymes represent sophisticated molecular scissors for cutting DNA into fragments of desired sizes. Restriction sites are distributed randomly in the genome and the size of the DNA fragments resulting from digestion with a given enzyme can be estimated based on the probable frequency of its recognition sequence. The actual fragment sizes produced will vary, however, because of variability in the number and locations of recognition sequences in relation to one another.

Recognition sequences exhibit a form of symmetry described as a **palindrome:** The nucleotide sequence reads the same on both strands of the DNA when read in the 5′ to 3′ direction. Each restriction enzyme recognizes its particular recognition sequence and cuts the DNA in a characteristic cleavage pattern (see Figure 17.1). The most common recognition sequences are four or six nucleotides long, but some contain eight or more nucleotides. Enzymes such as *Hin*dIII make offset cuts in the DNA strands, thus producing fragments with single-stranded overhanging ends called *cohesive ends* (or *"sticky" ends*), while others such as *Alu*I cut both strands at the same nucleotide pair, producing DNA fragments with double-stranded ends called *blunt-end* fragments.

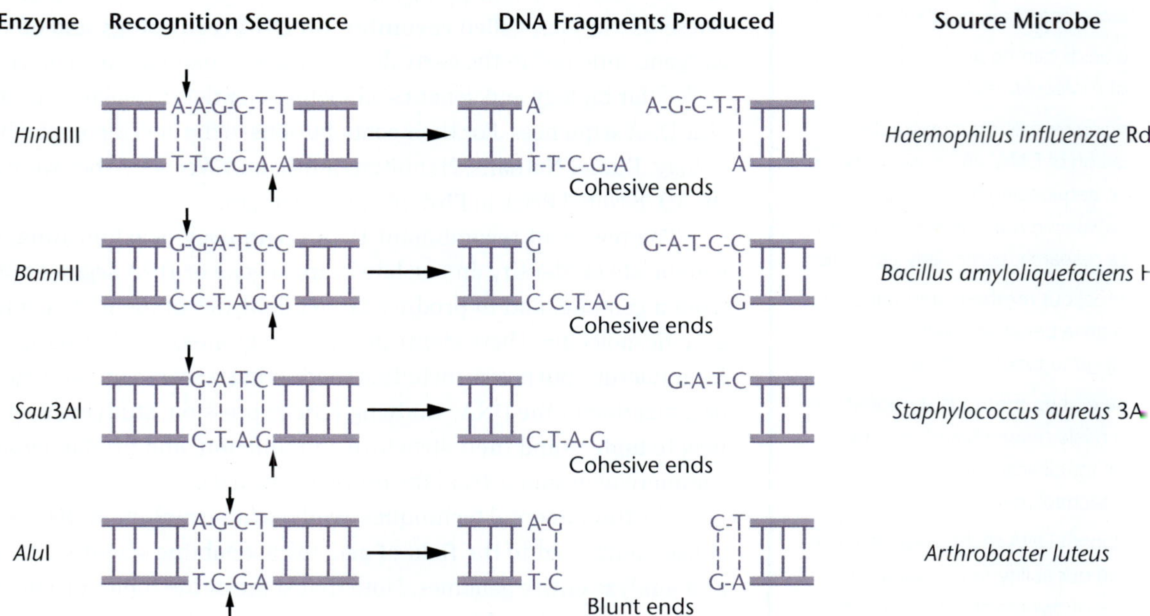

FIGURE 17.1 Common restriction enzymes, with their recognition sequence, DNA cutting patterns, and source microbes. Arrows indicate the location in the DNA cut by each enzyme.

One of the first restriction enzymes to be identified was isolated from *Escherichia coli* strain R and was designated *Eco*RI. DNA fragments produced by *Eco*RI digestion have cohesive ends because they can base-pair with complementary single-stranded ends on other DNA fragments cut using *Eco*RI. When mixed together, single-stranded ends of DNA fragments from different sources cut with the same restriction enzyme can **anneal,** or stick together, by hydrogen bonding of complementary base pairs in single-stranded ends (**Figure 17.2**). Addition of the enzyme **DNA ligase**—recall the role of DNA ligase in DNA replication as discussed earlier in the text (see Chapter 10)—to DNA fragments will seal the phosphodiester backbone of DNA to covalently join the fragments together to form recombinant DNA molecules.

Scientists often use restriction enzymes that create cohesive ends since the overhanging ends make it easier to combine fragments. Blunt-end ligation is more technically challenging because it is not facilitated by hydrogen bonding, but a scientist can ligate fragments digested at different sequences by different blunt-end generating enzymes.

ESSENTIAL POINT

Recombinant DNA technology was made possible by the discovery of proteins called restriction enzymes, which cut DNA at specific sequences, producing fragments that can be joined with other DNA fragments to form recombinant DNA molecules. ■

DNA Vectors Accept and Replicate DNA Molecules to Be Cloned

Scientists recognized that DNA fragments resulting from restriction-enzyme digestion could be copied or cloned if they also had a technique for replicating the fragments. Thus, a second key tool that allowed DNA cloning was the development of **cloning vectors,** DNA molecules that accept DNA fragments and replicate these fragments when vectors are introduced into host cells.

Many different vectors are available for cloning. Vectors differ in terms of the host cells they are able to enter and in the size of DNA fragment inserts they can carry, but most DNA vectors share several key properties.

- A vector contains several restriction sites that allow insertion of the DNA fragments to be cloned.
- Vectors must be capable of replicating in host cells independent of the host cell chromosome(s).
- To make it possible to distinguish between host cells that have taken up vectors and host cells that have not, the vector should carry a **selectable marker gene** (often an antibiotic resistance gene) or a **reporter gene** (a gene that encodes a protein which produces a visible effect, such as color or fluorescent light).
- Most vectors incorporate specific sequences that allow for sequencing inserted DNA.

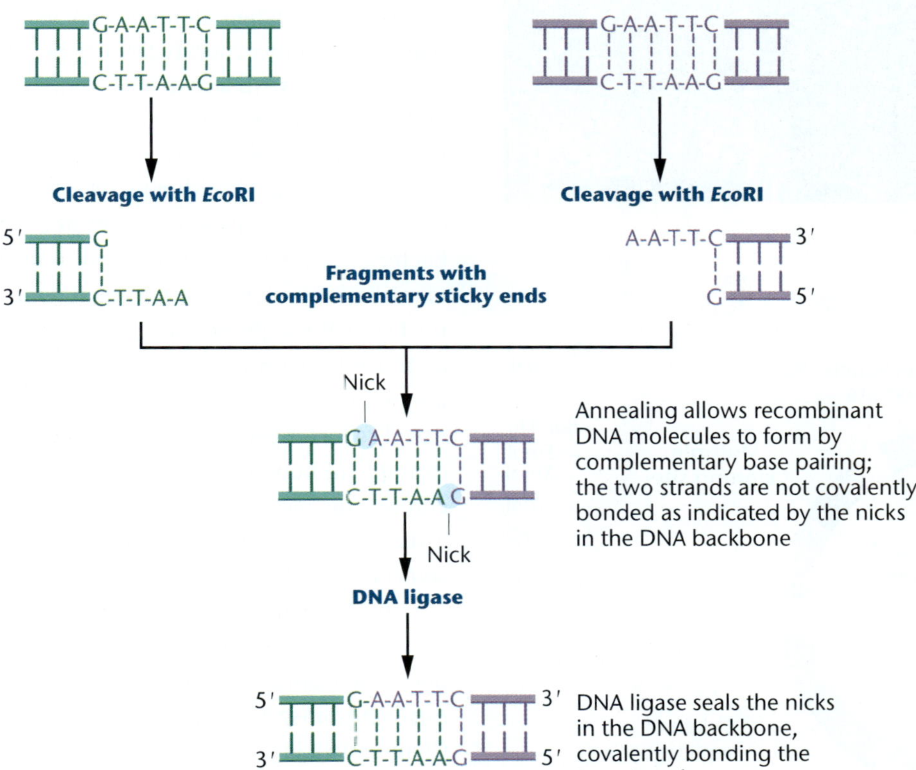
G-A-A-T-T-C
C-T-T-A-A-G
G-A-A-T-T-C
C-T-T-A-A-G
Cleavage with EcoRI
Cleavage with EcoRI
5′ G
3′ C-T-T-A-A
A-A-T-T-C 3′
G 5′
Fragments with complementary sticky ends
Nick
G A-A-T-T-C
C-T-T-A-A G
Nick
Annealing allows recombinant DNA molecules to form by complementary base pairing; the two strands are not covalently bonded as indicated by the nicks in the DNA backbone
DNA ligase
5′ G-A-A-T-T-C 3′
3′ C-T-T-A-A-G 5′
DNA ligase seals the nicks in the DNA backbone, covalently bonding the two strands

FIGURE 17.2 DNA from different sources is cleaved with *Eco*RI and mixed to allow annealing. The enzyme DNA ligase forms phosphodiester bonds between these fragments to create a recombinant DNA molecule.

Bacterial Plasmid Vectors

Genetically modified bacterial **plasmids,** derived from naturally occurring plasmids, were the first vectors developed, and they are still widely used for cloning. Recall from earlier discussions (Chapter 8) that plasmids are naturally occurring extrachromosomal, double-stranded, circular DNA molecules that replicate independently from the chromosomes within bacterial cells [**Figure 17.3(a)**]. These plasmids can be extensively modified by genetic engineering to serve as cloning vectors. Many commercially prepared plasmids are readily available with a range of useful features [**Figure 17.3(b)**]. One example is a region called the *multiple cloning site*, a short sequence that has been genetically engineered to contain a number of restriction sites for commonly used restriction enzymes. Multiple cloning sites allow scientists to clone a range of different fragments generated by many commonly used restriction enzymes.

(a)

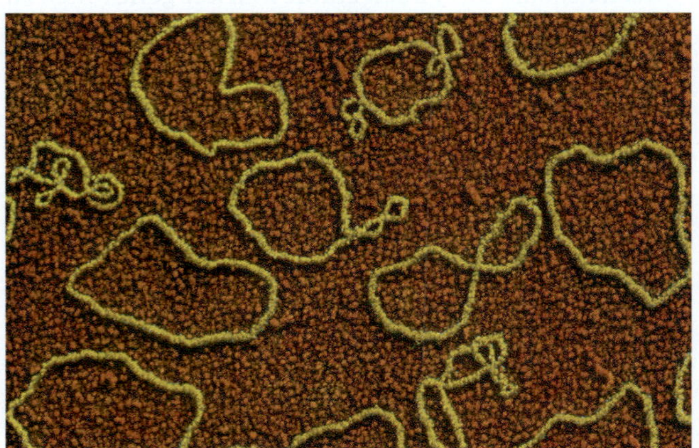

(b)

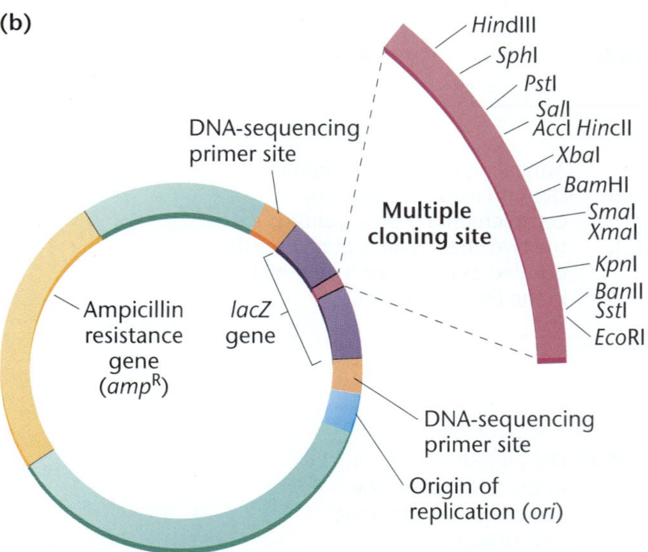

FIGURE 17.3 (a) Color-enhanced electron micrograph of plasmids isolated from *E. coli*. (b) Diagram of a typical DNA cloning plasmid.

Plasmids are introduced into bacteria by the process of **transformation** (see Chapter 8). Two main techniques are widely used for bacterial transformation. One approach involves treating cells with calcium ions and using a brief heat shock to introduce the plasmid DNA into cells. The other technique, called *electroporation*, uses a brief, but high-intensity, pulse of electricity to move plasmid DNA into bacterial cells.

Only one or a few plasmids generally enter a bacterial host cell by transformation. But because plasmids have an *origin of replication (ori)* site that allows for plasmid replication, it is possible to produce several hundred copies of a plasmid in a single host cell. This greatly enhances the number of DNA clones that can be produced.

Cloning DNA with a plasmid generally begins by cutting both the plasmid DNA and the DNA to be cloned with the same restriction enzyme (**Figure 17.4**). Typically, the plasmid is cut once within the multiple cloning site, converting the circular molecule into a linear vector. DNA restriction fragments from the DNA to be cloned are added to the linearized vector in the presence of DNA ligase. Sticky ends of DNA fragments anneal, joining the DNA to be cloned and the plasmid. DNA ligase is then used to create phosphodiester bonds to seal nicks in the DNA backbone, thus producing recombinant DNA, which is then introduced into bacterial host cells by transformation. Once inside the cell, plasmids replicate quickly to produce multiple copies.

However, when cloning DNA using plasmids, not all plasmids will incorporate DNA to be cloned. For example, a plasmid cut with a restriction enzyme generating sticky ends can close back on itself (self-ligate) if cut ends of the plasmid rejoin. Obviously, such nonrecombinant plasmids are not desired. Also, during transformation, not all host cells will take up plasmids. Therefore, it is important that bacterial cells containing recombinant DNA can be readily identified in a cloning experiment. One way this is accomplished is through the use of selectable marker genes, such as those that provide resistance to antibiotics (amp^R for ampicillin resistance, for example). Another strategy is the use of reporter genes such as *lacZ*. **Figure 17.5** provides an example of how the latter can be used to identify bacteria containing recombinant plasmids. This process is often referred to as **"blue-white" screening** for a reason that will soon become obvious.

In blue-white screening, a plasmid is used that contains the *lacZ* gene into which a multiple cloning site has been incorporated. The *lacZ* gene encodes the enzyme β-galactosidase, which, as you learned earlier in the text (see Chapter 15), is used to cleave the disaccharide lactose into its component monosaccharides glucose and galactose. Blue-white screening takes advantage of this enzymatic activity. Using this approach, one can easily identify transformed cells containing recombinant or nonrecombinant plasmids.

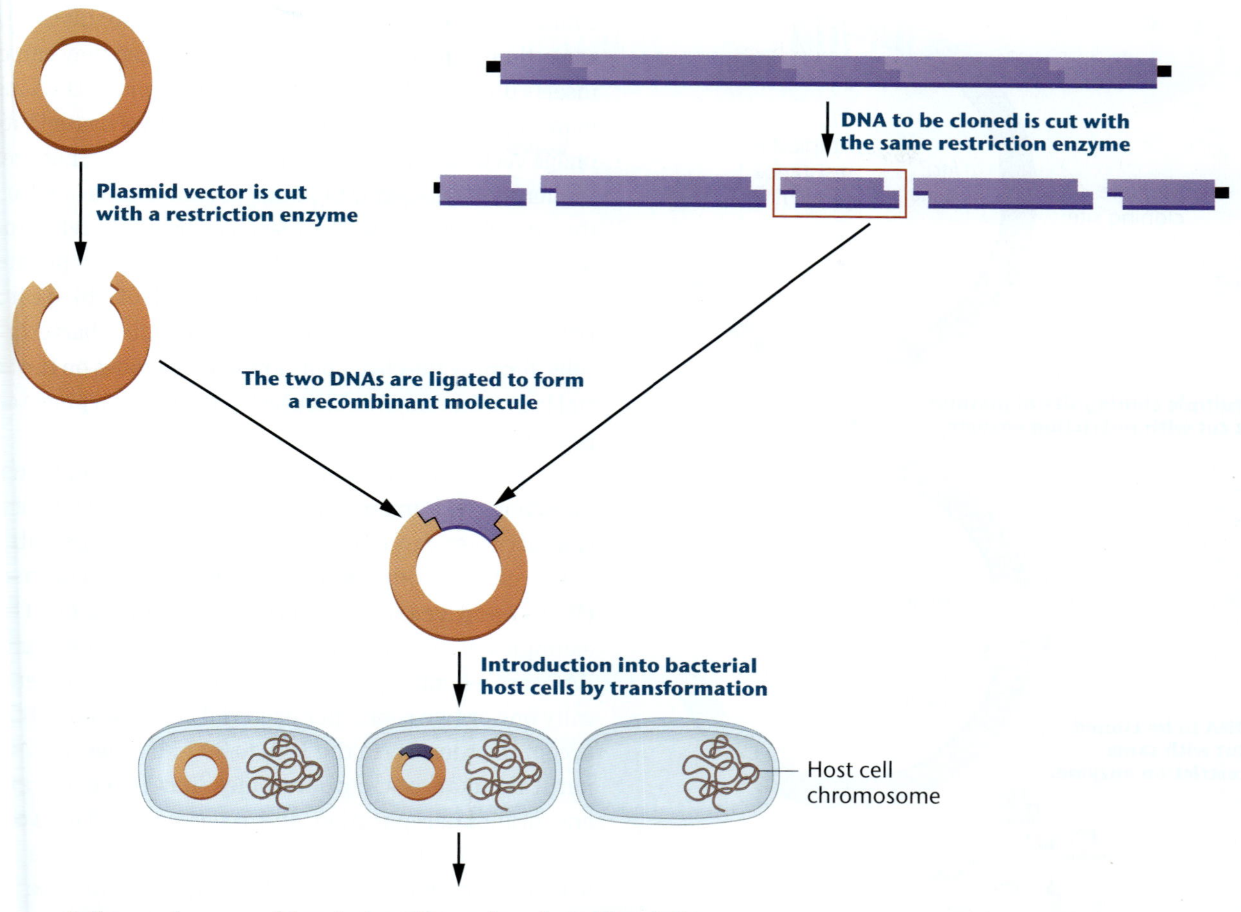

DNA to be cloned is cut with the same restriction enzyme
Plasmid vector is cut with a restriction enzyme
The two DNAs are ligated to form a recombinant molecule
Introduction into bacterial host cells by transformation
Host cell chromosome
Cells carrying recombinant plasmids can be selected by plating on agar containing antibiotics and color indicators such as X-gal (refer to Figure 17.5)

FIGURE 17.4 Cloning with a plasmid vector.

If a DNA fragment is inserted anywhere in the multiple cloning site, the *lacZ* gene is disrupted and will not produce functional copies of β-galactosidase. The agar plates used in the assay contain an antibiotic—ampicillin in this case. Nontransformed bacteria do not have the amp^R gene and are killed by the ampicillin.

These agar plates also contain a substance called X-gal (technically 5-bromo-4-chloro-3-indolyl-β-D-galactopyranoside), which is similar to lactose in structure. X-gal is, therefore, a substrate for β-galactosidase, and when it is cleaved, it turns blue. Bacterial cells that carry nonrecombinant plasmids (those that have self-ligated and thus do not contain inserted DNA) have a functional *lacZ* gene and produce β-galactosidase. As a result, these cells turn blue because the functional enzyme cleaves X-gal in the medium. By contrast, recombinant bacteria (with plasmids containing a DNA fragment inserted into the *lacZ* gene) will form white colonies when they grow on X-gal medium because functional β-galactosidase cannot be made (Figure 17.5, bottom). Bacteria in a colony are clones of each other—genetically identical cells with copies of recombinant plasmids. White colonies can be transferred to flasks of bacterial culture broth and grown in large quantities, after which it is relatively easy to isolate and purify recombinant plasmids from these cells.

Plasmids are still the workhorses for many applications of recombinant DNA technology, but they have a major limitation: Because they are small, they can only accept inserted pieces of DNA up to about 25 kilobases (kb) in size, and most plasmids can often only accept substantially smaller pieces. Therefore, as recombinant DNA technology developed and it became desirable to clone large pieces of DNA, other vectors were developed primarily for their ability to accept larger pieces of DNA and because they could be used with other types of host cells beside bacteria.

Other Types of Cloning Vectors

Phage vectors were among the earliest vectors used in addition to plasmids. These included genetically modified strains of bacteriophage λ—a double-stranded DNA virus. The genome of λ phage has been sequenced, and it has been modified to incorporate many of the important features of cloning vectors described earlier in this chapter, including a multiple cloning site. Phage vectors were popular for quite some time and are still in use today because they can carry inserts up to 45 kb—more than twice as long as DNA inserts in most plasmid vectors. DNA fragments are ligated into the phage vector to produce recombinant λ vectors that are subsequently packaged into phage protein heads *in vitro*, and then the phage are used to infect bacterial host cells growing on petri plates. Inside the bacteria, the vectors replicate and form many copies of infective phage, each of which carries a DNA insert. As they reproduce, they lyse their bacterial host cells, forming the clear spots known as plaques on the bacterial lawn (described in Chapter 8), from which phage can be isolated and the cloned DNA can be recovered.

The mapping and analysis of large eukaryotic genomes, including the human genome, require cloning vectors that can carry very large DNA fragments such as segments of an entire chromosome. **Bacterial artificial chromosomes (BACs)** and **yeast artificial chromosomes (YACs)** are examples of vectors that can be used for these purposes. BACs are essentially very large but low copy number (typically one or two copies per bacterial cell) plasmids that can accept DNA inserts in the 100- to 300-kb range. A YAC, like natural eukaryotic chromosomes, has telomeres at each end, origins of replication, and a centromere. These components are joined to selectable marker genes and to a cluster of restriction-enzyme recognition sequences for insertion of foreign DNA. Yeast chromosomes range in size from 230 kb to over 1900 kb, making it possible to clone DNA inserts from 100 to 1000 kb in YACs. The ability to clone large pieces of DNA in these vectors made them an important tool in the Human Genome Project (see Chapter 18).

Unlike the vectors described so far, **expression vectors** are designed to ensure mRNA expression of a cloned gene with the purpose of producing many copies of the gene's encoded protein in a host cell. This is an important distinction since most plasmids, phage vectors, and YACs only carry DNA and do not signal the cell to transcribe it into mRNA. Expression vectors are available for both bacterial and eukaryotic host cells and contain the appropriate sequences to initiate both transcription and translation of the cloned gene. For many research applications that involve studies

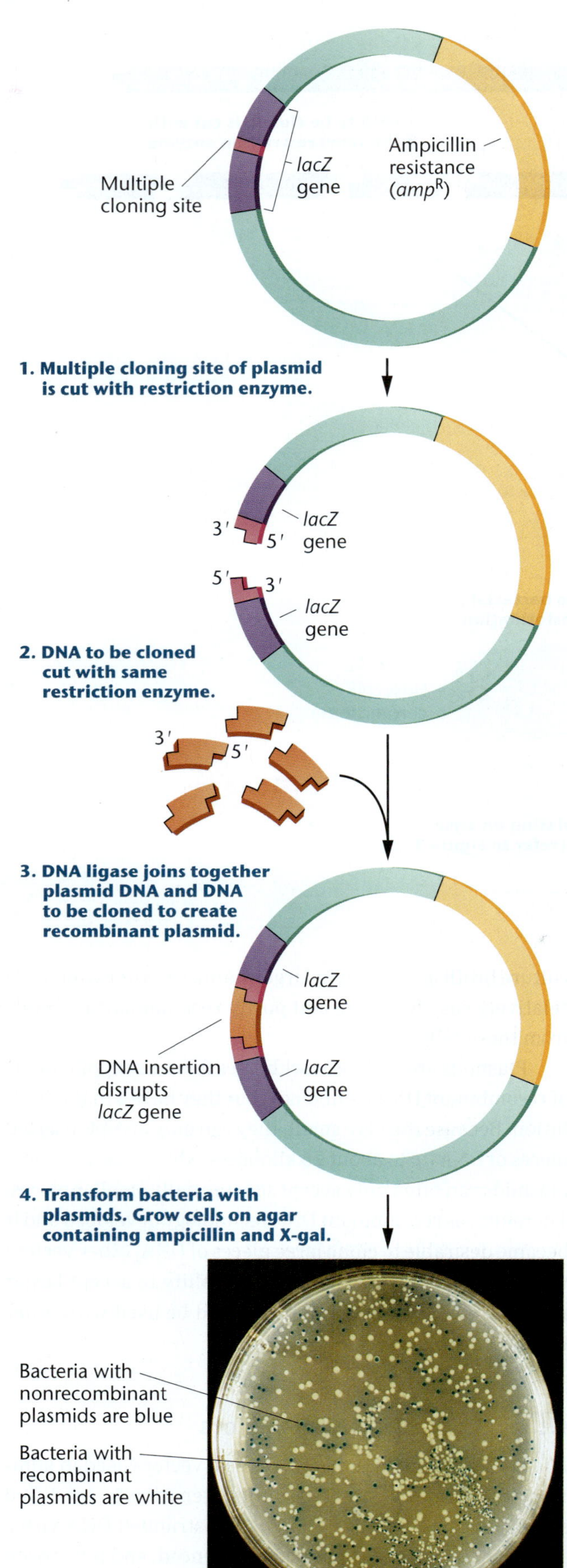
Multiple cloning site
lacZ gene
Ampicillin resistance (amp^R)
1. Multiple cloning site of plasmid is cut with restriction enzyme.
3′
5′
lacZ gene
5′
3′
lacZ gene
2. DNA to be cloned cut with same restriction enzyme.
3′
5′
3. DNA ligase joins together plasmid DNA and DNA to be cloned to create recombinant plasmid.
lacZ gene
DNA insertion disrupts *lacZ* gene
lacZ gene
4. Transform bacteria with plasmids. Grow cells on agar containing ampicillin and X-gal.
Bacteria with nonrecombinant plasmids are blue
Bacteria with recombinant plasmids are white

FIGURE 17.5 In blue-white screening, DNA inserted into the multiple cloning site of a plasmid disrupts the *lacZ* gene. Bacteria containing recombinant DNA are unable to metabolize X-gal, resulting in white colonies and allowing direct identification of colonies that carry DNA inserts to be cloned. (Bottom) Photo of a Petri dish showing the growth of bacterial cells after uptake of plasmids. Cells in blue colonies contain vectors without DNA inserts (nonrecombinant plasmids), whereas cells in white colonies contain vectors carrying DNA inserts (recombinant plasmids). Nontransformed cells did not grow into colonies due to the presence of ampicillin in the plating medium.

of protein structure and function, producing a recombinant protein in bacteria (or other host cells) and purifying the protein is a routine approach, although it is not always easy to properly express a protein that maintains its biological function. The biotechnology industry also relies heavily on expression vectors to produce commercially valuable protein products from cloned genes.

Introducing genes into plants is a common application that can be done in many ways. One widely used approach involves a species of soil bacterium and a type of plasmid called a **Ti plasmid**. We will discuss aspects of genetic engineering of food plants later in the text (see Special Topics Chapter 6—Genetically Modified Foods).

ESSENTIAL POINT

Vectors replicate autonomously in host cells and facilitate the cloning and manipulation of newly created recombinant DNA molecules. ■

17.2 DNA Libraries Are Collections of Cloned Sequences

Cloning DNA into smaller vectors, particularly plasmids, produces only relatively small DNA segments—representing just a single gene or even a portion of a gene. Even when several hundred genes are introduced into larger vectors such as BACs or YACs, one still needs a method for identifying the DNA pieces that were cloned. Consider this: In our cloning discussions so far, we have described how DNA can be inserted into vectors and cloned—a relatively straightforward process—but we have not discussed how a researcher knows what particular DNA sequence they have cloned. Simply cutting DNA and inserting it into vectors does not tell you what gene or sequences are being copied.

During the first several decades of DNA cloning, scientists created **DNA libraries,** which represent collections of cloned DNA. Depending on how a library is constructed, it may contain genes and noncoding regions of DNA. Generally speaking, there are two main types of libraries, genomic DNA libraries and complementary DNA (cDNA) libraries.

Genomic Libraries and cDNA Libraries

Ideally, a **genomic library** consists of many overlapping fragments of the genome, with at least one copy of every DNA sequence in an organism's chromosomes, which in summary span the entire genome. In making a genomic library, chromosomal DNA is extracted from cells or tissues and cut randomly with restriction enzymes, and the resulting fragments are inserted into vectors. Vectors in the genomic library may contain more than one gene or only a portion of a gene. Also, libraries built from eukaryotic cells will contain coding and noncoding segments of DNA such as introns.

Since some vectors (such as plasmids) can carry only a few thousand base pairs of inserted DNA, BACs and YACs were commonly used to accommodate the large sizes of DNA necessary to span the approximately 3 billion bp of human DNA (as in the Human Genome Project). As you will learn later in the text (see Chapter 18), **whole-genome sequencing** approaches (see Figure 18.1) and new sequencing methodologies are replacing traditional genomic DNA libraries because they effectively allow one to sequence an entire genomic DNA sample without the need for inserting DNA fragments into vectors and cloning them in host cells. But the concept of a DNA library is still important for a number of modern applications.

Complementary DNA (cDNA) libraries offer certain advantages over genomic libraries and continue to be a useful methodology for specific approaches to gene cloning and other applications. This is primarily because a cDNA library contains DNA copies—called cDNA—that are made from mRNA molecules isolated from cultured cells or a tissue sample. A cDNA library therefore represents only the genes being expressed in cells at the time the library was made—unlike a genomic library, which contains all of the DNA, coding and noncoding, in a genome. This is a key point: cDNA libraries provide a snapshot, or catalog, of just the genes that were transcriptionally active in a tissue at a particular time.

As a result, cDNA libraries have been particularly useful for identifying and studying genes expressed in certain cells or tissues under certain conditions: for example, during development, cell death, cancer, and other biological processes. One can also use these libraries to compare expressed genes from normal tissues and diseased tissues. For instance, this approach has been widely used to identify genes involved in cancer formation, such as those genes that contribute to progression from a normal cell to a cancer cell and genes involved in cancer cell metastasis (spreading).

The initial steps required to prepare a cDNA library are shown in **Figure 17.6**. Key to the technique is the process of **reverse transcription.** Because most eukaryotic mRNAs have a poly-A tail at the 3′-end, a short oligo(dT) molecule is annealed to this tail to serve as a primer for initiating DNA synthesis by the enzyme reverse transcriptase. Reverse transcriptase uses the mRNA as a template to synthesize a complementary DNA strand (cDNA) and forms a double-stranded mRNA/cDNA duplex. The mRNA is then partially digested with the enzyme RNAse H to produce gaps in the RNA strand. The 3′-ends of the remaining mRNA serve as primers for DNA polymerase I, which synthesizes a second DNA strand. The result is a double-stranded cDNA molecule that can be cloned into suitable vectors, usually plasmids.

Because one typically wouldn't know what restriction enzymes could be used to cut cDNA produced by the method just described, one usually needs to attach linker sequences

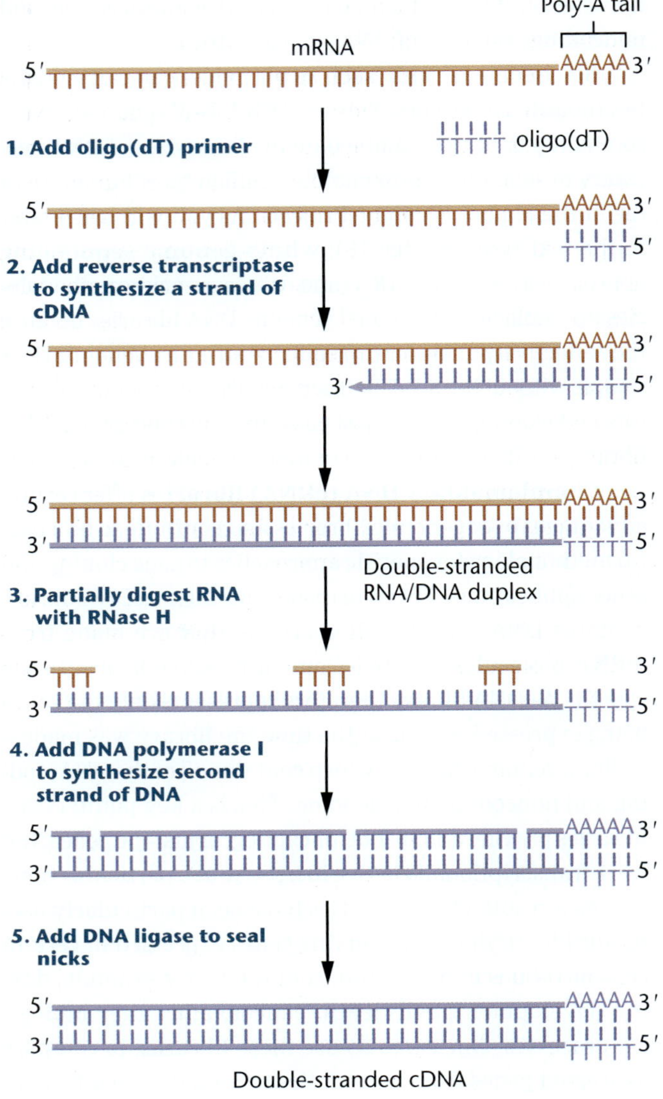

FIGURE 17.6 Producing cDNA from mRNA.

to the ends of the cDNA in order to insert it into a plasmid. Linkers are short double-stranded oligonucleotides containing a restriction-enzyme recognition sequence (e.g., for *Eco*RI). After attachment to the cDNAs, the linkers are cut with *Eco*RI and ligated to vectors treated with the same enzyme.

Specific Genes Can Be Recovered from a Library by Screening

Genomic and cDNA libraries often consist of several hundred thousand different DNA clones, much like a large book library may have many books but only a few of interest to your studies in genetics. So how can libraries be used to locate a specific gene of interest? To find a specific gene, we need to identify and isolate only the clone or clones containing that gene. We must also determine whether a given clone contains all or only part of the gene we are trying to study.

For several decades an approach called *library screening* was routinely used to sort through a library and isolate specific genes of interest. Many of the first genes to be cloned and sequenced were identified this way. Library screening usually involves use of a **probe,** any DNA or RNA sequence that is complementary to some part of the target gene or sequence to be identified in a library. The probe will bind (hybridize) to any complementary DNA sequences present in one or more clones.

Probes are derived from a variety of sources—often related genes isolated from another species can be used if enough of the DNA sequence is conserved. For example, genes from rats, mice, or even *Drosophila* that have conserved sequence similarity to human genes can be used as probes to identify human genes during library screening.

A probe must be labeled or tagged in some way so that it can be identified. Initially probes were labeled with radioactive isotopes, but modern applications use probes labeled with nonradioactive compounds that undergo chemical or color reactions to indicate the location of a specific clone in a library.

Although less central to research today, libraries still have their place in certain applications in the genetics lab. However, as you will learn later in the text (see Chapter 18), the basic methods of recombinant DNA technology, including DNA libraries, were the foundation for the development of more powerful whole-genome techniques that led to the **genomics** era of modern genetics and molecular biology. Genomic techniques, in which entire genomes are being sequenced without creating libraries, have largely replaced libraries at least for cloning and isolating one or a few genes at a time. We will also consider later (Chapter 18) how DNA sequence analysis using bioinformatics allows one to identify protein-coding and noncoding sequences in cloned DNA.

NOW SOLVE THIS

17.1 A plasmid that is both ampicillin and tetracycline resistant is cleaved with *Pst*I, which cleaves within the ampicillin resistance gene. The cut plasmid is ligated with *Pst*I-digested *Drosophila* DNA to prepare a genomic library, and the mixture is used to transform *E. coli* K12

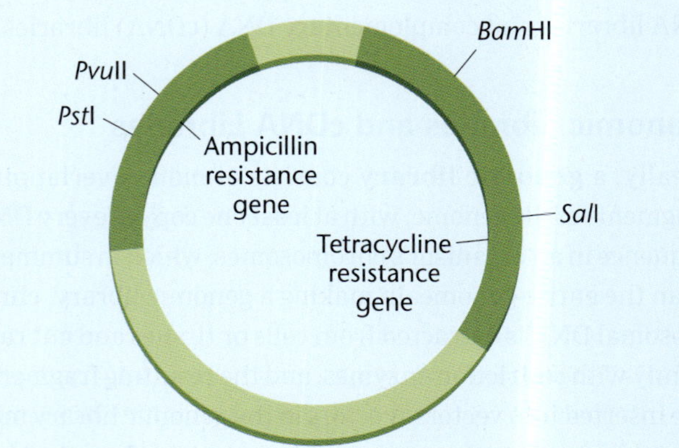

(a) Which antibiotic should be added to the medium to select cells that have incorporated a plasmid?

(b) If recombinant cells were plated on medium containing ampicillin or tetracycline and medium with both antibiotics, on which plates would you expect to see growth of bacteria containing plasmids with *Drosophila* DNA inserts?

(c) How can you explain the presence of colonies that are resistant to both antibiotics?

■ **HINT:** *This problem involves an understanding of antibiotic selectable marker genes in plasmids and antibiotic DNA selection for identifying bacteria transformed with recombinant plasmid DNA. The key to its solution is to recognize that inserting foreign DNA into the plasmid vector disrupts one of the antibiotic resistance genes in the plasmid.*

ESSENTIAL POINT

DNA libraries are collections of cloned DNA that can be screened or sequenced to isolate specific sequences of interest. ■

17.3 The Polymerase Chain Reaction is A Powerful Technique for Copying DNA

Cloning DNA using vectors and host cells is labor intensive and time consuming. In 1986, a technique called the **polymerase chain reaction (PCR)** was developed. PCR revolutionized recombinant DNA methodology and further accelerated the pace of biological research. The significance of this method was underscored by the awarding of the 1993 Nobel Prize in Chemistry to Kary Mullis, who developed the technique.

PCR is a rapid method of DNA cloning that extends the power of recombinant DNA and in many cases eliminates the need to use host cells for cloning. Genomic DNA or mitochondrial DNA can be used and the DNA to be cloned can come from many sources, including mummified remains, fossils, or forensic samples such as dried blood, semen, or hair. PCR is a method of choice for many applications, whether in molecular biology, human genetics, evolution, development, conservation, or forensics.

By copying a specific DNA sequence through a series of *in vitro* reactions, PCR can amplify target DNA sequences that are initially present in very small quantities in a population of other DNA molecules. When performing PCR, double-stranded target DNA to be amplified is placed in a tube with DNA polymerase, Mg^{2+} (an important cofactor for DNA polymerase), and the four deoxyribonucleoside triphosphates. In addition, some information about the nucleotide sequence of the target DNA is required. This sequence information is used to synthesize two oligonucleotide **primers:** short (typically about 20 nucleotides long) single-stranded DNA sequences, one complementary to the 5′-end of one strand of target DNA to be amplified and another primer complementary to the opposing strand of target DNA at its 3′-end.

When added to a sample of double-stranded DNA that has been denatured into single strands, the primers bind to complementary nucleotides at each end within the sequence to be cloned. DNA polymerase can then extend the 3′-end of each primer to synthesize second strands of the target DNA. One complete reaction process, called a *cycle*, doubles the number of DNA molecules in the reaction [**Figure 17.7(a)**]. Repetition of the process produces large numbers of copied target DNA very quickly [**Figure 17.7(b)**]. If desired, the PCR products can be cloned into plasmid vectors for further use.

The amount of amplified DNA produced is theoretically limited only by the number of times these cycles are repeated, although several factors prevent PCR reactions from amplifying very long stretches of DNA. Most routine PCR applications involve a series of three reaction steps in a cycle. These three steps are as follows:

1. **Denaturation:** The double-stranded DNA to be cloned is *denatured* into single strands by heating to 92 to 95°C for about 1 minute.

2. **Hybridization/Annealing:** The temperature of the reaction is lowered to between 45°C and 65°C, which allows primer binding, also called hybridization or annealing, to the denatured, single-stranded DNA. The primers serve as starting points for DNA polymerase to synthesize new DNA strands complementary to the target DNA. When selecting a hybridization temperature for an experiment, factors such as primer length, base composition of primers (GC-rich primers are more thermally stable than AT-rich primers), and whether or not all bases in a primer are complementary to bases in the target sequence are among primary considerations.

3. **Extension:** The reaction temperature is adjusted to between 65°C and 75°C, and DNA polymerase uses the primers as a starting point to synthesize new DNA strands by adding nucleotides to the ends of the primers in a 5′ to 3′ direction.

Each cycle results in amplification – a doubling of the number of DNA molecules present at the start of the cycle. PCR is, therefore, a "chain reaction" because the products of previous cycles serve as templates for each subsequent cycle. Each cycle takes 2 to 5 minutes and can be repeated immediately, so that in less than 3 hours, 25 to 30 cycles result in over a million-fold increase in the amount of DNA. Instruments called *thermocyclers*, or simply PCR machines, that can be programmed to carry out a predetermined number of cycles, automate the process. The large amounts of a

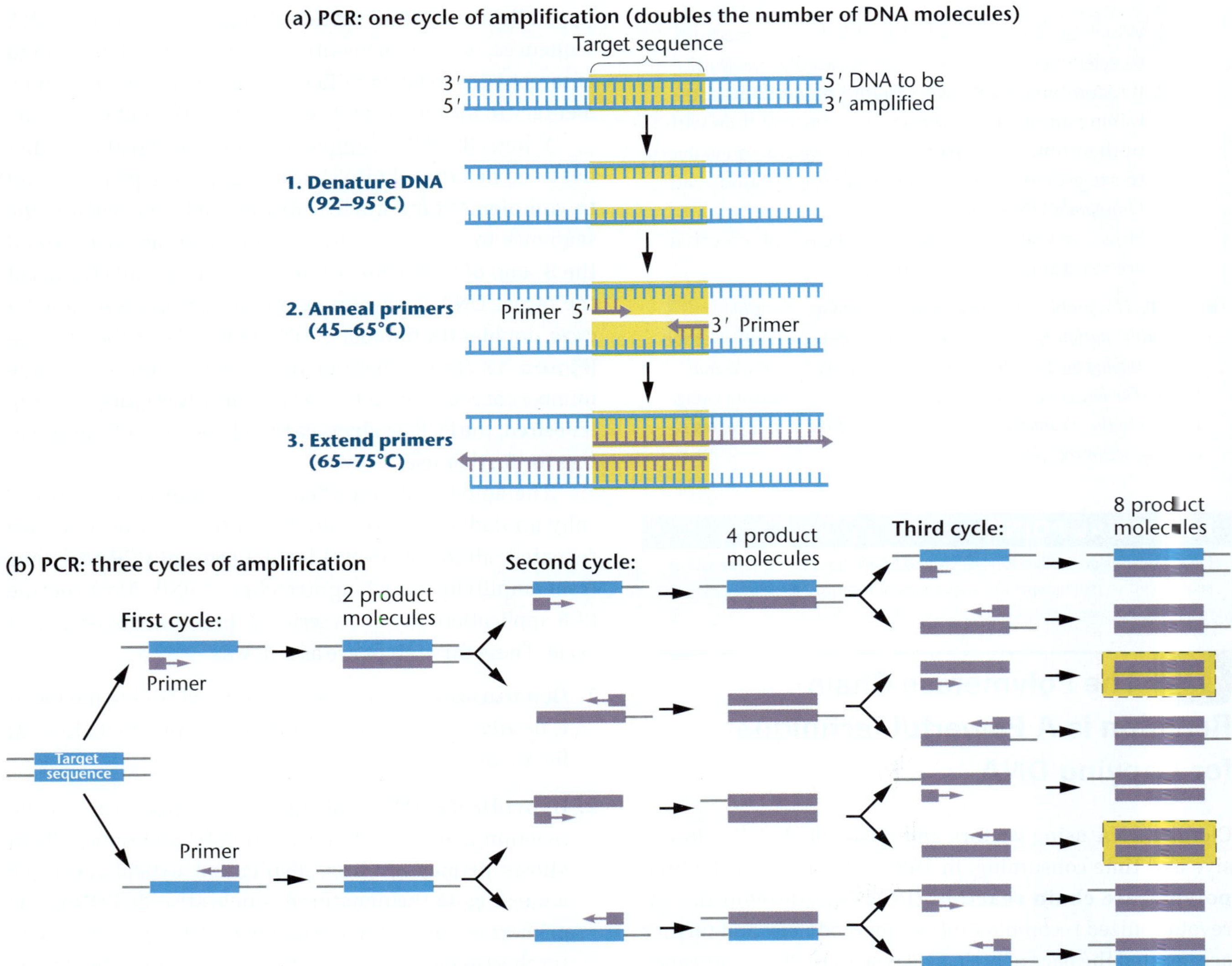

FIGURE 17.7 Steps in the polymerase chain reaction (PCR). (a) In this schematic representation, a relatively short sequence of DNA is shown being amplified. Notice that the first cycle produces amplified molecules with a strand that extends beyond the target sequence. (b) Repeated cycles of PCR can quickly amplify the target DNA sequence more than a millionfold. Products in part (b) that consist of only the target sequence are outlined and highlighted.

specific DNA sequence produced can be used for many purposes, including cloning into plasmid vectors, DNA sequencing, clinical diagnosis, and genetic screening.

PCR requires a special type of DNA polymerase. Multiple PCR cycles involve repeated heating and cooling of samples, which eventually lead to heat denaturation and loss of activity of most proteins. PCR reactions rely on thermostable forms of DNA polymerase that are capable of withstanding multiple heating and cooling cycles without significant loss of activity. PCR became a major tool when DNA polymerase was isolated from *Thermus aquaticus*, a bacterium living in habitats like the hot springs of Yellowstone National Park, where it was first discovered. Called *Taq* polymerase, this enzyme is capable of tolerating extreme temperature changes and was the first thermostable polymerase used for PCR.

Although PCR is a valuable research tool with many advantages over previously used techniques, it does have limitations. One is that some information about the nucleotide sequence of the target DNA must be known in order to synthesize primers. Another is its sensitivity: even minor contamination of the sample with DNA from other sources can cause problems. For example, cells shed from a researcher's skin can contaminate samples gathered from a crime scene, making it difficult to obtain accurate results. Also, PCR typically cannot be used to amplify particularly long segments of DNA. Normally, DNA polymerase in a PCR reaction extends primers only for relatively short distances. Because of this, scientists tend to amplify pieces of DNA that are only several thousand nucleotides in length, which is fine for most routine applications.

PCR Applications

PCR has become one of the most versatile and widely used techniques in modern genetics and molecular biology, with many different applications. DNA cloning using a PCR-based approach has several advantages over library cloning approaches. PCR is rapid and can be carried out in a few hours, rather than the days required for making and screening DNA libraries. PCR is also very sensitive and amplifies specific DNA sequences from vanishingly small DNA samples. PCR sensitivity is invaluable in several kinds of applications, including genetic testing, forensics, and molecular paleontology. With carefully designed primers, DNA samples that have been partially degraded, contaminated with other materials, or embedded in a matrix (such as the fossilized tree resin known as amber) can be amplified. This allows the study of samples from fossils or from single cells, such as those recovered at crime scenes, where a single hair or even a saliva-moistened postage stamp can serve as the sole source of the DNA. Later in the text (see Special Topics Chapter 5—DNA Forensics), we will discuss how PCR is used in human identification, including remains identification, and in forensic applications.

Another important application of PCR is as a diagnostic tool. As you will learn in Special Topics Chapter 2—Genetic Testing, gene-specific primers provide a way of using PCR for screening mutations involved in genetic disorders. PCR is also a key method for detecting bacteria and viruses (such as hepatitis or HIV) in humans, and pathogenic bacteria such as *E. coli* and *Staphylococcus aureus* in contaminated food.

Reverse transcription PCR (RT-PCR) is a powerful methodology for studying gene expression, that is, mRNA production by cells or tissues. In RT-PCR, RNA is isolated from cells or tissues to be studied, and reverse transcriptase is used to generate cDNA molecules, as described earlier when we discussed preparation of cDNA libraries. This reaction is followed by PCR to amplify cDNA with a set of primers specific for the gene of interest. Amplified cDNA fragments are then separated and visualized on an agarose gel. Because the amount of amplified cDNA in RT-PCR is based on the relative number of mRNA molecules in the starting reaction, RT-PCR can be used to evaluate relative levels of gene expression in different samples.

Finally, in discussing PCR approaches, one of the most valuable modern techniques is **quantitative real-time PCR (qPCR)** or simply **real-time PCR**. A key facet of this method is the ability to quantify the PCR product as it is made during an experiment (as the reactions occur in "real time") without having to run a gel. This, in turn, allows the calculation of the amount of template DNA originally present in a sample.

ESSENTIAL POINT

PCR allows DNA to be amplified, or copied, without cloning and is a rapid and sensitive method with wide-ranging applications. ■

NOW SOLVE THIS

17.2 You have just created the world's first genomic library from the African okapi, a relative of the giraffe. No genes from this genome have been previously isolated or described. You wish to isolate the gene encoding the oxygen-transporting protein β-globin from the okapi library. This gene has been isolated from humans, and its nucleotide sequence and amino acid sequence are available in databases. Using the information available about the human β-globin gene, what strategy can you use to isolate this gene from the okapi library?

■ **HINT:** *This problem asks you to design PCR primers to amplify the β-globin gene from a species whose genome you just cloned The key to its solution is to remember that you have at your disposal sequence data for the human β-globin gene and to consider that PCR experiments require the use of primers that bind to complementary bases in the DNA to be amplified.*

For more practice, see Problem 13.

17.4 Molecular Techniques for Analyzing DNA and RNA

A wide range of molecular techniques is available to almost anyone who does research involving DNA and RNA, particularly those who study the structure, expression, and regulation of genes. There are far too many techniques available than we can address in this chapter. In the following sections, we consider some of the techniques that are most commonly used to analyze DNA and RNA.

Agarose Gel Electrophoresis

One of the most routine techniques for analyzing DNA is **agarose gel electrophoresis,** a method that separates DNA fragments by size, with the smallest pieces moving farthest through the gel (see Chapter 9; refer to Figure 9.14). The fragments form a series of bands that can be visualized by treating the gel with DNA-binding stains such as *ethidium bromide* and illuminating it with ultraviolet light (**Figure 17.8**). This is usually the method of choice when smaller pieces of DNA need to be analyzed or isolated.

Before DNA sequencing and bioinformatics became routine, newly cloned DNA would be digested with enzymes and separated by gel electrophoresis. The digestion pattern of fragments generated could then be interpreted to determine the location of restriction sites for different enzymes, to create a *restriction map*. Restriction maps are now often created by simply using software to identify restriction-enzyme cutting sites in sequenced DNA. The Exploring Genomics

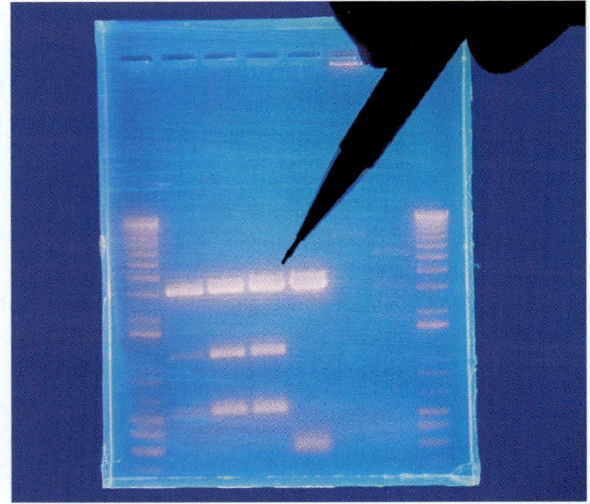

FIGURE 17.8 An agarose gel containing separated DNA fragments stained with a DNA-binding dye (ethidium bromide) and visualized under ultraviolet light. Smaller fragments migrate faster and farther than do larger fragments, resulting in the distribution shown. Molecular techniques involving agarose gel electrophoresis are routinely used in a wide range of applications.

exercise in this chapter involves a Web site, Webcutter, which is commonly used for generating restriction maps.

Nucleic Acid Blotting and Hybridization Techniques

Several of the techniques described in this chapter and elsewhere in the book rely on **hybridization** between complementary nucleic acid (DNA or RNA) molecules. One of the most widely used hybridization methods is called **Southern blot analysis** or simply **Southern blotting** (after Edwin Southern, who devised it). Southern blotting is another pioneering method that served essential roles in the early decades of DNA cloning such as identifying which clones in a library contained a given DNA sequence, identifying specific genes in genomic DNA digested with a restriction enzyme, and identifying the number of copies of a particular sequence or gene that are present in a genome. Modern DNA sequencing approaches have now replaced most of these application examples.

Gel electrophoresis can be used to characterize the number and molecular weights of fragments produced by restriction digestion of small genomes, when the number of fragments generated is relatively low. However, digestion of large genomes—such as the human genome, with more than 3 billion nucleotides—would produce so many different fragments that they would run together on a gel to produce a continuous smear. Southern blotting enables the identification of a particular DNA fragment of interest.

Southern blotting involves separation of DNA fragments by gel electrophoresis, transfer or "blotting" of fragments to a DNA binding membrane, and hybridization of these fragments to a labeled DNA or RNA probe. The membrane is washed to remove excess probe and overlaid with a piece of X-ray film for autoradiography or detected with a digital camera with chemiluminescence probes. Hybridization identifies specific DNA sequences present in the fragments, because only (**Figure 17.9**) fragments hybridizing to the probe are visualized.

Southern blotting led to the subsequent development of other widely used blotting approaches. RNA blotting was called **Northern blot analysis** or simply **Northern blotting.** Prior to the development of RT-PCR and real-time PCR, Northern blotting was commonly used to study gene expression (RNA production) by cells and tissues because it could both characterize and quantify the transcriptional activity of genes.

A related blotting technique for analyzing proteins was also developed. It is known as **Western blotting.** Thus part of the historical significance of Southern blotting is that it led to the development of other blotting methods that are key tools for studying nucleic acids and proteins.

Finally, as noted earlier in the text (see Chapter 9), **fluorescence *in situ* hybridization (FISH)** is a powerful tool that involves hybridizing a probe directly to a chromosome or RNA without blotting (see Figure 9.13 and Chapter 16 opening photograph). FISH can be carried out with isolated chromosomes on a slide or directly *in situ* in tissue sections or entire organisms, such as embryos. One type of application is in the field of developmental genetics, where FISH is used to identify which cell types in an embryo express different genes during specific stages of development (**Figure 17.10**).

Variations of the FISH technique are also being used to produce **spectral karyotypes** in which individual chromosomes can be detected using probes labeled with dyes

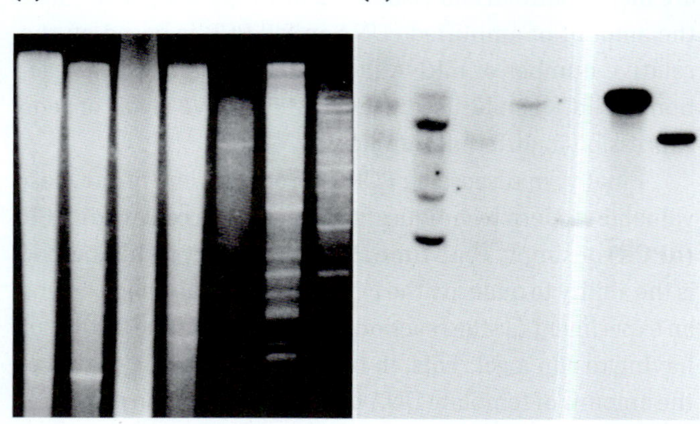

FIGURE 17.9 (a) Agarose gel stained with ethidium bromide to show DNA fragments. (b) Chemiluminescent image of a Southern blot prepared from the gel in part (a). Only those bands containing DNA sequences complementary to the probe show hybridization.

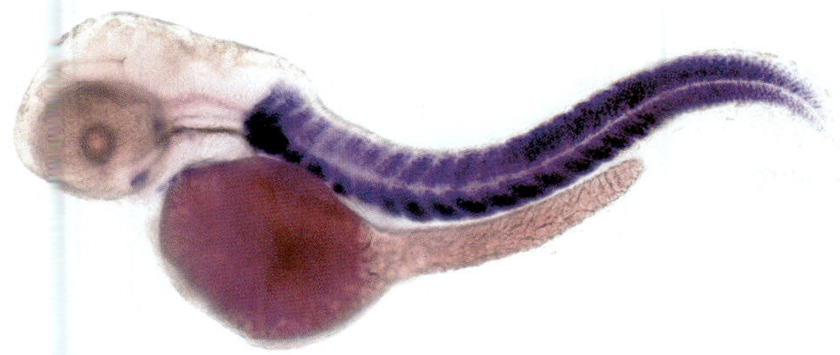

FIGURE 17.10 *In situ* hybridization of a zebrafish embryo 48 hours after fertilization. The probe used shows expression of *atp2a1* mRNA, which encodes a muscle-specific calcium pump, and is visualized as a dark blue stain. Notice that this staining is restricted to muscle cells surrounding the developing spinal cord of the embryo.

that will fluoresce at different wavelengths (see Chapter 6 opening photograph). Spectral karyotyping has proven to be extremely valuable for detecting deletions, translocations, duplications, and other anomalies in chromosome structure, such as chromosomal rearrangements, and for detecting chromosomal abnormalities in cancer cells.

ESSENTIAL POINT

DNA and RNA can be analyzed through a variety of methods that involve hybridization techniques. ■

17.5 DNA Sequencing Is the Ultimate Way to Characterize DNA at the Molecular Level

In a sense, cloned DNA, from a single gene to an entire genome, is completely characterized at the molecular level only when its nucleotide sequence is known. The ability to sequence DNA has greatly enhanced our understanding of genome organization and increased our knowledge of gene structure, function, and mechanisms of regulation.

Historically, the most commonly used method of DNA sequencing was developed by Fred Sanger and colleagues during the 1970s and is known as **dideoxy chain-termination sequencing** or simply **Sanger sequencing.** Because Sanger sequencing was an important foundational method for newer, more modern approaches to DNA sequencing, we will briefly discuss the technique here. A double-stranded DNA molecule whose sequence is to be determined is converted to single strands, one of which is used as a template for synthesizing a series of complementary strands. The template DNA is mixed with a primer that is complementary to either the target DNA or the vector, DNA polymerase, and the four deoxyribonucleotide triphosphates (dATP, dCTP, dGTP, and dTTP).

The key to the Sanger technique is the addition of a small amount of modified deoxyribonucleotides, called **dideoxynucleotides** (abbreviated ddNTPs) (**Figure 17.11**, inset box). Notice that a dideoxynucleotide has a 3′ hydrogen instead of a 3′ hydroxyl group. Dideoxynucleotides are called chain-termination nucleotides because they lack the 3′ oxygen required to form a phosphodiester bond with another nucleotide. If ddNTPs are included in a DNA synthesis reaction, the polymerase will occasionally (randomly) insert a ddNTP instead of a dNTP into a growing DNA strand. Once this occurs, synthesis terminates because DNA polymerase cannot add new nucleotides to a ddNTP due to its lack of a 3′ oxygen. The Sanger reaction takes advantage of this key modification.

This outcome is illustrated in Step 2 of Figure 17.11. Notice that the shortest fragment generated is a sequence that has added ddCTP to the 3′-end of the primer and the chain has terminated. Over time as the reaction proceeds, eventually a ddNTP will be inserted at every location in the sequence. The result is a population of newly synthesized DNA molecules each of which is terminated by a ddNTP and that differ in length by one nucleotide. The size difference allows for separation of reaction products by gel electrophoresis, which can then be used to determine the sequence.

When the Sanger technique was first developed, four separate reaction tubes, each with a different single ddNTP (e.g., ddATP, ddCTP, ddGTP, and ddTTP), were used. These reactions typically used either a radioactively labeled primer or a radioactively labeled ddNTP to permit analysis of the sequence following polyacrylamide gel electrophoresis and autoradiography. Historically, this approach involved large polyacrylamide gels in which each reaction was loaded on a separate lane of the gel and the ladder-like banding patterns revealed by autoradiography were read to determine the sequence. This original approach could typically read about 800 bases for each of 100 DNA molecules simultaneously. *Read length*—the amount of sequence that can be generated in a single individual reaction—and the total amount of DNA sequence generated in a sequence *run* (effectively, the read length times the number of reactions an instrument can process during a given period of time) together have become a hot area for innovation in sequencing technology.

Modifications of the Sanger technique in the mid-1980s led to technologies that allowed sequencing reactions to occur in a single tube. As shown in Figure 17.11, the four ddNTPs were each labeled with a different-colored fluorescent dye. These reactions were carried out in PCR-like fashion using cycling reactions that permit greater read and run capabilities. The reaction products were separated through a single, ultrathin-diameter polyacrylamide

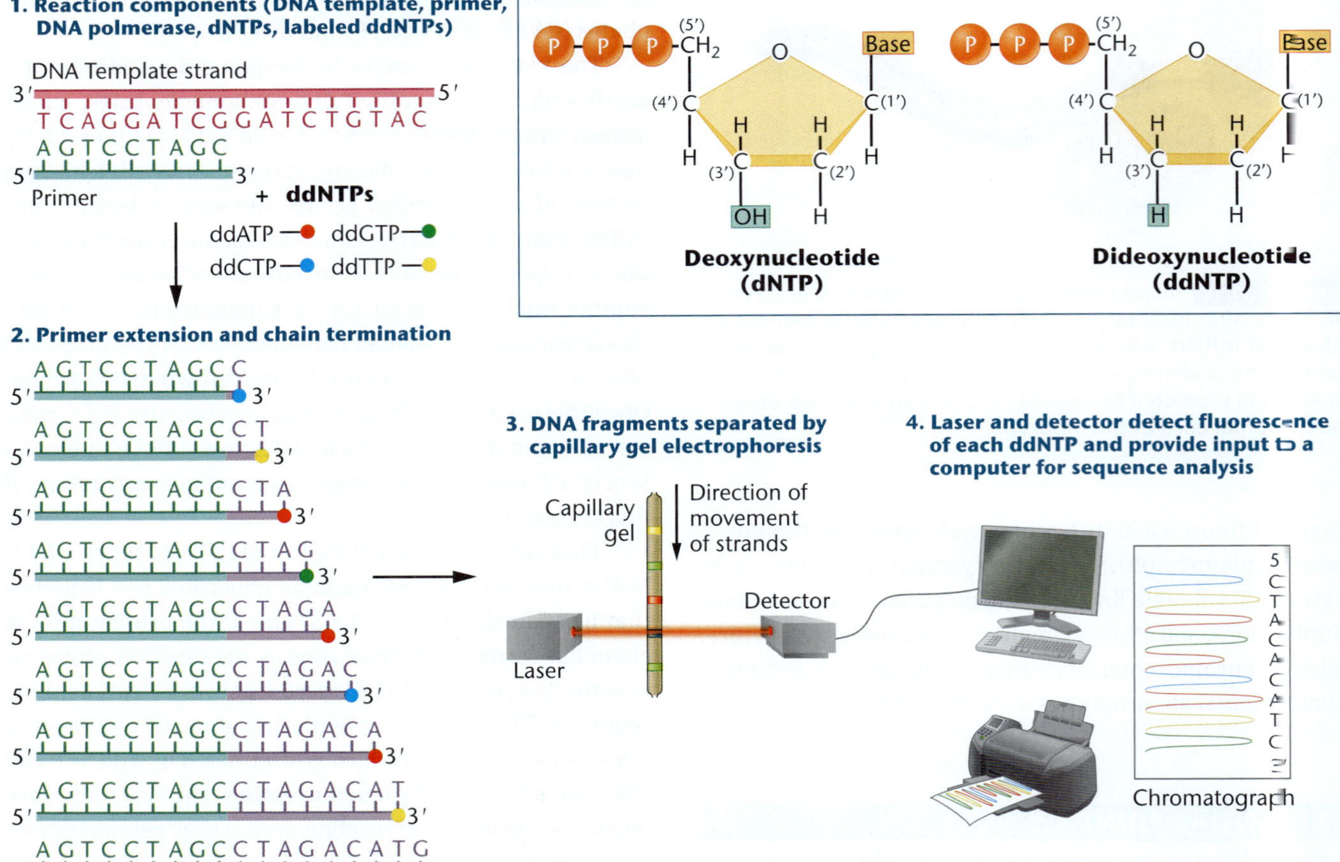

FIGURE 17.11 Computer-automated DNA sequencing using the chain-termination (modified Sanger) method. The inset box at upper right illustrates dideoxynucleotide (ddNTP) structure. (1) A primer is annealed to a sequence adjacent to the DNA being sequenced (usually near the multiple cloning site of a cloning vector). A reaction mixture is added to the primer–template combination. This includes DNA polymerase, the four dNTPs, and small molar amounts of ddNTPs labeled with fluorescent dyes. (2) All four ddNTPs are added to the same reaction tube. During primer extension, the polymerase occasionally (randomly) inserts a ddNTP instead of a dNTP, terminating the synthesis of the chain because the ddNTP does not have the OH group needed to attach the next nucleotide. Over the course of the reaction, all possible termination sites will have a ddNTP inserted, and thus all possible lengths of chains are produced. The products of the reaction are added to a single lane on a capillary gel (3), and the bands are read by a detector and imaging system (4) from the newly synthesized strand. In this case, the sequence obtained begins with 5′-CTAGACATG-3′ as seen in the chromatograph in step 4.

tube gel called a capillary gel (*capillary gel electrophoresis*). As DNA fragments move through the gel, they are scanned with a laser. The laser stimulates fluorescent dyes on each DNA fragment, which then emit different wavelengths of light for each ddNTP. Emitted light is captured by a detector that amplifies the signal and feeds this information into a computer to convert the light patterns into a DNA sequence that is technically called an electropherogram or chromatograph. The data are represented as a series of colored peaks, each corresponding to one nucleotide in the sequence.

For about two decades following the early 1990s, DNA sequencing was largely performed through computer-automated Sanger-reaction-based technology (shown in Figure 17.11) and referred to as **computer-automated high-throughput DNA sequencing**. These systems were a big improvement over manual Sanger systems because they could generate relatively large amounts of DNA sequence in relatively short periods of time. Computer-automated sequences could achieve read lengths of approximately 1000 bp with about 99.999 percent accuracy for about \$0.50 per kb. Automated DNA sequencers of this time period often contained multiple capillary gels (as many as 96) that were several feet long and could process several thousand bases of sequences, so many of these instruments made it possible to generate over 2 million bp of sequences in a day! Such systems became essential for the rapidly accelerating progress of the Human Genome Project. But by around 2005 and in the time since, sequencing technologies were to improve dramatically.

Sequencing Technologies Have Progressed Rapidly

Sanger sequencing approaches, including those involving computer-automated instruments, are rarely used today except for occasional sequencing of a relatively short piece of DNA or in labs that cannot afford more expensive sequencing instruments. When it comes to sequencing entire genomes, Sanger sequencing technologies and early-generation computer-automated approaches are outdated. Compared to newer technologies, the costs of those approaches were relatively high, and sequencing output, even with computer automation, was simply not high enough to support the growing demand for genomic data. This demand is being driven in large part by *personalized genomics* (see Chapter 18) and the desire to reveal the genetic basis of human diseases, which involves routine sequencing of complete individual human genomes.

The development of genomics has spurred a demand for sequencers that are capable of generating millions of bases of DNA sequences in a relatively short time. **Next-generation sequencing (NGS) technologies** (the second generation after Sanger methods) were the next big advance in DNA sequencing. NGS technologies dispensed with *first-generation* methods (the Sanger technique and capillary electrophoresis) in favor of sophisticated, parallel formats (simultaneous reaction formats) that synthesized DNA from tens of thousands of identical strands simultaneously and then used state-of-the-art fluorescence imaging techniques to detect newly synthesized strands and average sequence data across many molecules being sequenced. NGS technologies provided an unprecedented capacity for generating massive amounts of DNA sequence data rapidly and at dramatically reduced costs per base. NGS has become so routine that now when scientists talk about "sequencing" we are referring to NGS.

Ultimately, several companies emerged as winners in the race to commercialize NGS technology. The Illumina HiSeq, and variations of this instrument, is one of the most common NGS platforms currently used in cutting-edge sequencing laboratories today. This system uses a **sequencing-by-synthesis (SBS)** approach. DNA fragments serve as templates, and nucleotides labeled with different dyes are incorporated. Next, unincorporated nucleotides are washed away and incorporated nucleotides are imaged. This cycle is then repeated. In the SBS method developed by Illumina, the DNA fragments are attached to a solid support and then, using reactions similar to the Sanger method, the fluorescently tagged terminator nucleotides are added and detected. The fluorescent tags and terminator portions of the nucleotides are removed to allow another cycle of extension.

SBS methods can now generate about 600 Gb of data in 10 days—enough to sequence four complete human genomes, with each base sequenced an average of 30 times for accuracy! The instrumentation needed to run these platforms is expensive. For example, the Illumina HiSeq instrument costs about $650,000. But given the massive amounts of sequence data NGS methods can generate, the average cost per base is much lower than Sanger sequencing. Incidentally, NGS sequencing technologies also created major data management challenges for saving and storing large data files.

Shortly after NGS methods were commercialized, companies were announcing progress on **third-generation sequencing (TGS).** TGS methods are based on strategies that sequence a *single molecule* of single-stranded DNA, and at least four different approaches are being explored.

Pacific Biosciences' PacBio technology is one of the leaders in TGS and involves an approach known as *single-molecule sequencing in real time (SMRT)*. The PacBio instrument works by attaching single-stranded molecules of the DNA to be sequenced to a single molecule of DNA polymerase anchored to a substrate and then visualizing, in real time, the polymerase as it synthesizes a strand of DNA (see **Figure 17.12**). The DNA polymerase is confined within a

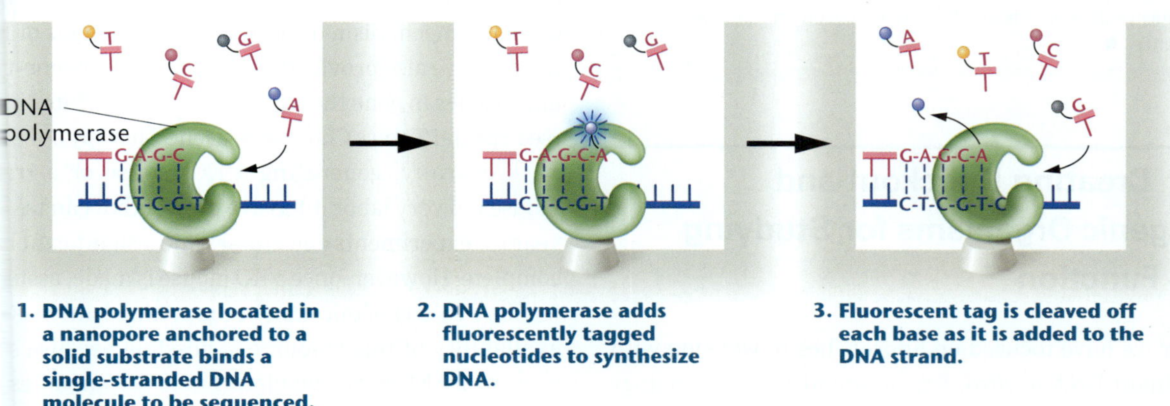

DNA polymerase
G-A-G-C
C-T-C-G-T
G-A-G-C-A
C-T-C-G-T
G-A-G-C-A
C-T-C-G-T-C
1. DNA polymerase located in a nanopore anchored to a solid substrate binds a single-stranded DNA molecule to be sequenced.
2. DNA polymerase adds fluorescently tagged nucleotides to synthesize DNA.
3. Fluorescent tag is cleaved off each base as it is added to the DNA strand.

FIGURE 17.12 Third-generation sequencing (TGS). A simplified version of one approach to TGS is shown here. In this example, a DNA polymerase molecule anchored within a nanopore binds to a single strand of DNA. As the polymerase incorporates fluorescently labeled nucleotides into a new DNA strand (shown in pink), each base emits a characteristic color that can be detected.

nanopore—a hole of about 10 nm in diameter located within a thin layer of metal on a glass substrate—a setup needed to detect the addition of individual nucleotides as they are added to the growing strand. The terminal phosphates of the nucleotides are tagged with a fluorescent dye, with each base assigned a characteristic color. Upon incorporation of a nucleotide, the tag is cleaved along with the phosphate—and flashes. Each colored flash is detected and recorded.

The PacBio was one of the first "long-read" instruments to reach the market. It generates read lengths of over 10,000 bp. Most TGS technologies still have somewhat high error rates for sequencing accuracy—about 15 percent errors per sequence generated. The list price for one PacBio sequencer is about $350,000, which is making this technology more affordable at least for biotechnology companies, pharmaceutical companies, and well-funded academic research laboratories.

A few years ago, Oxford Nanopore Technologies developed a portable, single-molecule sequencer called the MinION that is the size of a USB memory stick! Although accuracy of this sequencer limits its applications, there is no reason to think that the technology for highly accurate pocket-sized sequencers will not advance in the near future.

Rapid advances in sequencing technology were driven by the demands of genome scientists (particularly those working on the Human Genome Project) to rapidly generate more sequence with greater accuracy and at lower cost. Because of these technological advances, innovative approaches to genome sequencing are driving a range of new research and clinical applications. **RNA sequencing** has also emerged as a new technique that makes it possible to measure gene expression on a genome-wide scale. We will discuss this later in the book (in Chapter 18).

ESSENTIAL POINT

DNA sequencing technologies are changing rapidly. Next-generation and third-generation methods produce fairly large amounts of accurate sequence data in a short time at lower cost than traditional approaches. ■

17.6 Creating Knockout and Transgenic Organisms for Studying Gene Function

Thus far we have focused on approaches to working with recombinant DNA *in vitro*. Recombinant DNA technology has also made it possible to directly manipulate genes *in vivo* in ways that allow scientists to learn more about gene function *in living organisms*. These approaches also enable scientists to create genetically engineered plants and animals for research and for commercial applications. Here we discuss gene knockout technology and the creation of transgenic animals as examples of **gene targeting**—a collection of methods that have revolutionized research in genetics.

Gene Targeting and Knockout Animal Models

The concept behind gene targeting is to manipulate a specific allele, locus, or base sequence, through approaches that involve homologous recombination, to learn about the functions of a gene of interest.

In the 1980s, scientists devised a gene-targeting technique for creating **gene knockout** (often abbreviated as KO) organisms, specifically mice. The pioneers of knockout technology, Dr. Mario Capecchi of the University of Utah and colleagues Oliver Smithies of the University of North Carolina, Chapel Hill, and Sir Martin Evans of Cardiff University, United Kingdom, received the 2007 Nobel Prize in Physiology or Medicine for developing this technique.

A knockout is an example of a **loss-of-function mutation.** One can learn about gene function by creating a knockout that *disrupts or eliminates* copies of a specific gene or genes of interest and then asking, "What happens?" If physical, behavioral, biochemical, or other metabolic changes are observed in the KO animal, this would suggest that the gene of interest has some functional role or roles in the observed phenotypes. The KO techniques developed in mice led to similar technologies for making KOs in other animal species, including zebrafish, rats, pigs, fruit flies, as well as in plants.

Research in genetics, molecular biology, and biomedical fields have been revolutionized by the study of KO mice and other KO organisms. Our understanding of gene function has been advanced, and transgenic animals have been created, often resulting in animal models for many human diseases. KO animals are not limited to removal of a single gene: *double-knockout animals (DKOs)* and even *triple-knockout animals (TKOs)* are also possible. This approach is typically used when scientists want to study the functional effects of disrupting two or three genes thought to be involved in a related mechanism or pathway. Applications of KO technology have also provided the foundation for gene-targeting approaches in gene therapy that we discuss later in the text (see Special Topics Chapter 3—Gene Therapy).

Generally, generating a KO mouse or a transgenic mouse is a very labor intensive project that can take several years of experiments and crosses and a significant budget to complete. However, once a KO mouse is made, assuming it is fertile, a colony of mice can be maintained. Increased global availability of this resource is possible because scientists often share KO mice and biotechnology companies produce and sell them commercially.

A KO animal can be made in several ways but the same basic methods apply when making most KO animals (**Figure 17.13**). Because newer technologies such as CRISPR-Cas (see Section 17.7) are becoming the methods

1. Designing the targeting vector

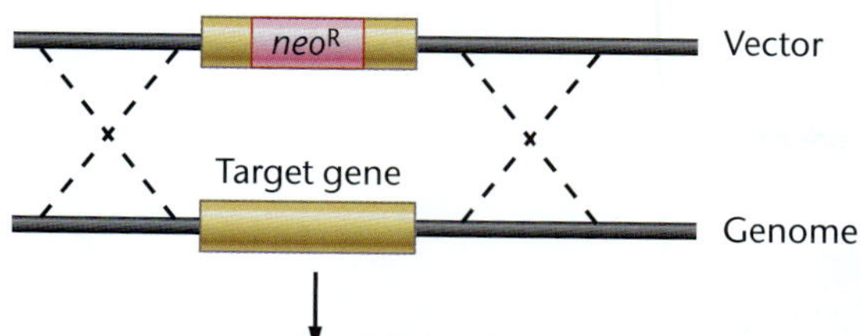

2. Transform ES cells with targeting vector and select cells for recombination

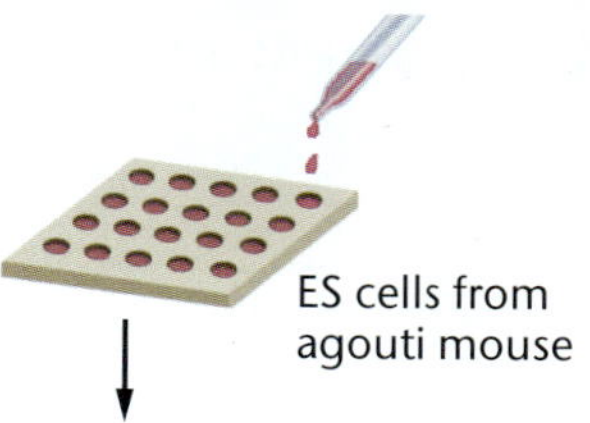

3. Microinject ES cells into blastocyst from black-colored mouse

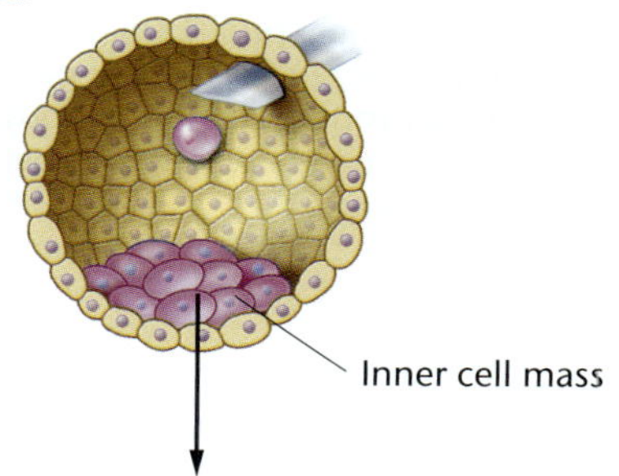

4. Transfer into pseudopregnant surrogate mother; birth of chimeras

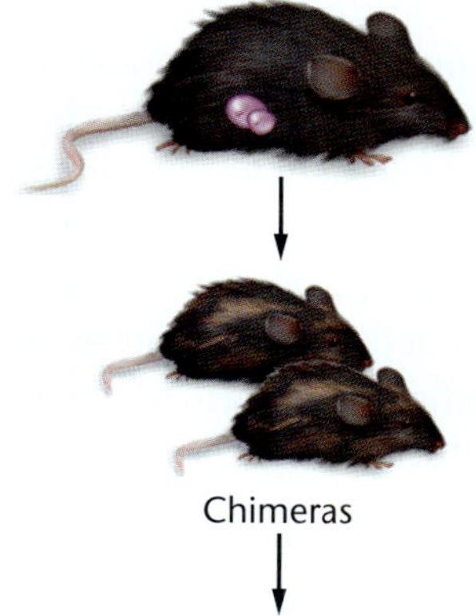

5. Chimeric mouse bred to black mouse to create mice heterozygous (+/−) for gene knockout

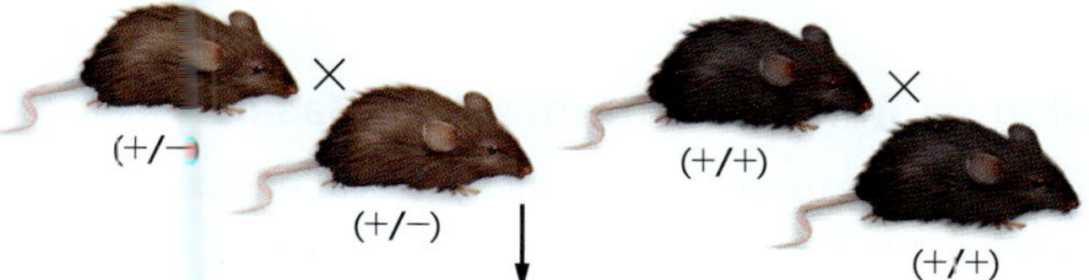

6 Breed heterozygous mice to produce mice homozygous (−/−) for gene knockout

FIGURE 17.13 A basic strategy for producing a knockout mouse.

of choice for making KO and transgenic animals, here we provide a very brief overview of traditional methods. To begin, the DNA sequence for the KO target gene and information about noncoding flanking genomic sequences must be known. A *targeting vector* is then constructed, which contains a copy of the gene of interest that has been mutated by insertion of a large segment of foreign DNA, typically a selectable marker gene. The inserted DNA disrupts the reading frame of the target gene so that a nonfunctional protein will be made if the mutated gene is expressed. The targeting vector is introduced into cells and undergoes homologous recombination with the genomic copy of the gene of interest (the target gene), disrupting or replacing it, thereby rendering it nonfunctional. The selectable marker gene helps scientists determine whether or not the targeting vector has been properly introduced into the genome.

There are several ways to introduce the targeting vector into cells. One popular approach involves using electroporation to deliver the vector into cultured **embryonic stem (ES) cells.** The ES cells are harvested from the inner cell mass of a mouse embryo at the blastocyst stage. Alternatively, the targeting vector is directly injected into the blastocyst with the hopes that it will enter ES cells in the inner cell mass. Sometimes it is possible to make KOs by isolating newly fertilized eggs from a female mouse (or female of another desired animal species) and microinjecting the targeting vector DNA directly into the diploid nucleus of the egg or into one of the haploid pronuclei prior to fusion [**Figure 17.14(a)**].

Only a small percentage of ES cells take up the targeting vector. In those that do, the actions of the endogenous enzyme recombinase catalyze homologous recombination between the targeting vector and the sequence for the gene of interest. Replacement of the original gene usually occurs on only one of the two chromosomes.

Scientists can select recombinant ES cells by treating them with a reagent that kill cells lacking the targeting vector. The selected cells are injected into mouse embryos at the blastocyst stage where they will be incorporated into the inner cell mass of the blastocyst. Several blastocysts are then placed into the uterus of a surrogate mother mouse, sometimes called a *pseudopregnant mouse*—a female that has been mated with an infertile male to stimulate production of pregnancy hormones. These, in turn, trigger physiological changes that make the uterus receptive to implantation of the blastocysts.

The surrogate will give birth to mice that are *chimeras*: Some cells in their body arise from injected KO stem cells, and others arise from endogenous inner cell mass cells of the recipient blastocyst. As long as germ cells develop from

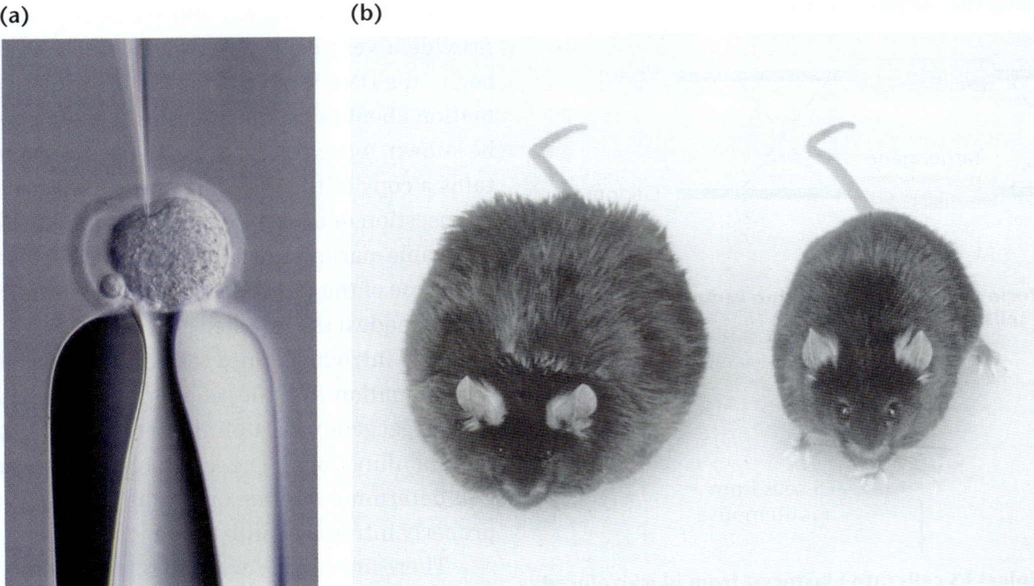

FIGURE 17.14 (a) Microinjecting DNA into a fertilized egg to create a knockout (or a transgenic) mouse. A fertilized egg is held by a suction or holding pipette (seen below the egg), and a microinjection needle delivers cloned DNA into the nucleus. (b) On the left is a knockout or null mouse (−/−) for both copies of the leptin (*Lep*) gene. The mouse on the right is wild type (+/+) for the *Lep* gene. Normal copies of the *Lep* gene produce a peptide hormone called leptin. The *Lep* knockout mouse weighs almost five times as much as its wild-type sibling.

recombinant ES cells, the mutated gene sequence will be inherited in all of the offspring generated by these mice. Typically, however, most F_1 generation KO mice produced this way are heterozygous (+/−) for the gene of interest and not homozygous for the KO. Sibling matings of F_1 animals can then be used to generate homozygous KO animals, referred to as *null mice* and given a −/− designation because they lack wild-type copies of the targeted gene of interest. As mentioned at the beginning of this section, KO animal models serve invaluable roles for learning about gene function, and they continue to be essential for biomedical research on disease genes [see **Figure 17.14(b)**].

Despite all of the work that goes into trying to produce a KO organism, sometimes viable offspring are never born. The KO is said to result in *embryonic lethality*. Knocking out a gene that is important during embryonic development will kill the mouse, often before researchers have a chance to study it. When this occurs, researchers typically examine embryos from the surrogate mouse to try to determine at what stage of embryonic development embryos are dying. This examination often reveals specific organ defects, which can be informative about the function of the KO gene.

If null mice for a particular gene of interest cannot be derived by traditional KO approaches, an alternative approach called *conditional knockout* can often provide a way to study such a gene. Conditional knockouts allow one to control the particular time in an animal's development that a target gene is disrupted. For example, if a target gene displays embryonic lethality, one can use a conditional KO to allow an animal to progress through development and be born before activating the disruption. Another advantage of conditional KOs is that target genes can also be turned off in a particular tissue or organ instead of the entire animal.

A worldwide collaboration to disable all (~20,000) protein-coding genes in the mouse genome was recently completed. The purpose of the initiative was to create a KO mouse resource that would help advance human disease research. Rather than using a gene-by-gene approach to make KOs, researchers used a high-throughput technique that involves knocking out thousands of genes in embryonic stem cells and then creating offspring in which specific genes of interest can then be turned off. Projects such as this provide a "library" of KO animals for the research community so that individual scientists can use these animals rather than having to go through the expense and technical challenges of making their own knockouts.

Making a Transgenic Animal: The Basics

Transgenic animals, also sometimes called **knock-in animals**, express, or overexpress, a particular gene of interest (the transgene)—in other words, turning genes on instead of off, the opposite of KOs. As with KOs, many of the prevailing techniques used to make transgenic animals were developed in mice; **Figure 17.15** illustrates two examples.

The method of creating a transgenic animal is conceptually simple, and many of the steps are similar to those involved in making a KO animal. But instead of trying to disrupt a

(a)

(b)

FIGURE 17.15 Examples of transgenic mice. (a) Transgenic mice incorporating the green fluorescent protein gene (*gfp*), a popular reporter gene, from jellyfish enable scientists to tag particular genes with green fluorescent protein. Thanks to the expression of *gfp*, which makes the transgenic mice glow green under ultraviolet light, scientists can track activity of the tagged genes, including activity in subsequent generations of mice generated from these transgenics. (b) The mouse on the left is transgenic for a rat growth hormone gene, cloned downstream from a mouse metallothionein promoter. When the transgenic mouse was fed zinc, the metallothionein promoter induced the transcription of the growth hormone gene, stimulating the growth of the transgenic mouse.

target gene, a vector with a functional transgene is created to undergo homologous recombination and enter into the host genome. In some applications, tissue-specific promoter sequences can be used so that transgene expression is limited to certain tissues. For example, in the biotechnology industry mammary-specific promoters are used so that recombinant products accumulate in milk for subsequent purification. It is often much easier to make a transgenic animal than a KO animal because vector incorporation into the host genome does not need to occur at a particular locus (just hopefully in a noncoding region) as is necessary when making a KO.

As with KOs, the vector with the transgene can be put into ES cells or injected directly into embryos or eggs. Likewise, a marker or reporter gene is included on the vector to help researchers identify successful transformation. In a relatively small percentage of embryos or eggs, the transgenic DNA becomes randomly inserted into the genome by recombination due to the action of naturally occurring DNA recombinases. The rest of the process is similar to making a KO: Embryos are implanted in surrogate mothers, and crosses from resulting progeny are used to derive mice that are homozygous for the transgene.

There are many variations in the types of transgenic organisms that can be created and in their roles in basic and applied research. In some experiments, a transgene is overexpressed in order to study its effects on the phenotype of the organism. Other experimental variations include creation of transgenic animals that express mutant genes or genes from a different species. For example, so-called *humanized mice*—transgenic mice that express human genes—have been created to study responses to different drugs for treating diseases, to understand the roles of genes in evolution, and to study embryonic development. Transgenic animals and plants are also created to produce commercially valuable biotechnology products.In Special Topics Chapter 6—Genetically Modified Foods, you will learn about examples of transgenic food crops.

As mentioned earlier in this section, newly developed technologies such as CRISPR-Cas are replacing these more traditional methods of gene manipulation. Therefore, we will conclude this chapter with a section on genome editing to precisely modify a particular gene or sequence in the genome.

ESSENTIAL POINT

Gene-targeting methods to create knockout animals and transgenic animals are widely used, valuable approaches for studying gene function *in vivo*. ■

17.7 Genome Editing with CRISPR-Cas

The process of **genome editing** involves removing, adding, or changing specific DNA sequences in the genome of living cells. The ability to edit genomes both specifically and efficiently has broad implications for research, biotechnology, and medicine. Essential to this process is a "programmable" nuclease that can be directed to cut the genome in a sequence-specific manner. Later in the text (see Special Topics Chapter 3—Gene Therapy), we discuss how genome-editing with *transcription activator-like effector nucleases (TALENs)* and *zinc finger nucleases (ZFNs)* can be used for gene therapy. However, the fastest and most efficient approach to genome editing is the **CRISPR (clustered regularly interspaced palindromic repeats)-Cas** system.

CRISPR-Cas was discovered by scientists trying to understand how bacteria fight viral infection and was later adopted as a genome-editing tool. See Chapter 15 for an explanation of how bacteria use CRISPR-Cas as an immune response system against foreign genetic material such as viral DNA.

The CRISPR-Cas9 Molecular Mechanism

The most widely used CRISPR-Cas system for genome editing uses a nuclease called **Cas9** from the bacterium *Streptococcus pyogenes*. Cas9 contains two nuclease domains and thus can create a double strand break in DNA. However, Cas9 will only cut DNA sequences if two specific parameters are satisfied [**Figure 17.16(a)**].

First, Cas9 will only cut DNA near a specific sequence called a **protospacer adjacent motif (PAM)**, which is 5′-NGG-3′, where N is any base. Cas9 recognizes the PAM sequence and cuts three base pairs upstream of it. Second, Cas9 requires a **single guide RNA (sgRNA)**—a short RNA molecule that is required for Cas9 activity and specificity. Cas9 is directed to cut sequences that are complementary to a 20-nucleotide region of the sgRNA. Thus, any site in the genome (near a PAM sequence) can be targeted by Cas9 with an sgRNA that is synthesized to be complementary to that site. The first demonstration of *in vivo* genome editing with CRISPR-Cas9 was performed on cultured mammalian cells in 2012 by introducing plasmid expression vectors carrying genes encoding Cas9 and an sgRNA with a specific targeting sequence. Since then, CRISPR-Cas has been used to edit the genomes of many different organisms using a wide range of Cas9 and sgRNA delivery methods, such as viruses, plasmids, and direct injection.

Using CRISPR-Cas9 to cut a genome at a precise location is only the first step of genome editing. Creating specific changes to the genome with intended consequences, such as disrupting a gene's function or correcting a mutation, requires a second step that takes advantage of the eukaryotic cell's double-strand DNA break repair mechanisms (see Chapter 14). Double-stranded breaks in the genome may be repaired by **nonhomologous end-joining (NHEJ)** or by **homology-directed repair (HDR).** NHEJ simply involves the ligation of broken DNA fragments. This process is error prone and often results in small insertions or deletions (*indels*) at the repair site. HDR is less error prone because it uses an undamaged homologous chromosome or sister chromatid as a template to correctly repair a broken chromosome.

When the goal of CRISPR-Cas9 genome editing is to disrupt a gene and create a nonfunctional allele, simply

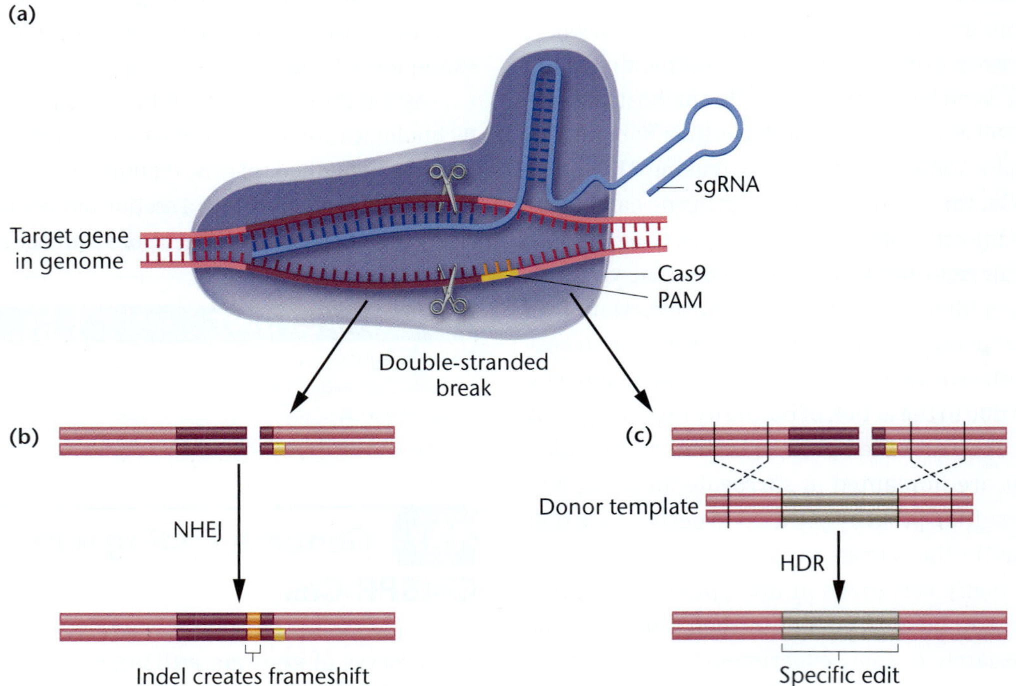

FIGURE 17.16 CRISPR-Cas9 genome editing. (a) An sgRNA guides Cas9 to cleave a target site adjacent to a PAM sequence. The double-stranded DNA break can be repaired by (b) NHEJ, which introduces insertions or deletions (indels), or by (c) HDR, which can make specific edits using an introduced donor template.

introducing Cas9 and an sgRNA into the cell is often sufficient. While the HDR mechanism may repair Cas9-induced double-stranded breaks correctly in some cells, in other cells the error-prone NHEJ pathway may introduce indels that result in a shift of the coding sequence reading frame, and thus lead to gene disruption [**Figure 17.16(b)**].

However, if the goal of CRISPR-Cas9 genome editing is to make a more precise edit, HDR can be "tricked" into using an artificial **donor template** (instead of the homologous chromosome) to make complex substitutions, deletions, or additions. The donor template is an experimentally introduced DNA molecule carrying a sequence with desired edits flanked by "homology arms" with sequences that match regions adjacent to the genomic target. Through the HDR mechanism, the target sequence in the genome is replaced by the sequence on the donor template [**Figure 17.16(c)**].

CRISPR-Cas Infidelity

CRISPR-Cas is clearly a powerful tool with immense potential, but it does have limitations. In some cases, Cas9 not only cuts at the intended target but also at off-target sites in the genome. Off-target edits may be due to an sgRNA having more than one perfect match in the genome or the sgRNA directing Cas9 to a sequence with one or few mismatches. To address this problem, several labs have tried modifying Cas9 to improve its specificity. Others have designed Web-based algorithms to improve sgRNA design. Others still have turned back to bacteria and archaea looking for CRISPR systems with alternative enzymes to Cas9 that have improved fidelity or other desirable traits. Some studies have shown that the Cpf1 nuclease, from the CRISPR system of bacteria in the *Francisella* and *Prevotella* genera, exhibits lower off-target editing than Cas9. Improving the specificity of CRISPR-Cas edits to the human genome will be important for the safety of medical applications of this technology.

CRISPR-Cas Technology Has Diverse Applications

CRISPR-Cas is an indispensable tool for basic genetics research. A fundamental objective that geneticists often pursue is to determine the function of an uncharacterized gene. A simple way to do this is to delete the gene and observe the phenotypic consequences. Although this is conceptually simple, the ability to efficiently and quickly delete a gene from the genome has only recently become possible with CRISPR-Cas technology. Beyond research, CRISPR-Cas is also beginning to have an impact on biotechnology and medicine.

Biotechnology is the use of living organisms to create a product or a process that helps improve the quality of life for humans or other organisms. CRISPR-Cas has greatly facilitated biotechnological innovation because it enables the rapid and cost-effective production of genetically modified organisms for various purposes. CRISPR-Cas biotechnological applications range from simple but useful innovations, such as the creation of tomatoes that ripen more quickly, to massive and controversial endeavors, such as "bringing back" the woolly mammoth by editing mammoth genes into the elephant genome. The woolly mammoth is not back yet, but there are already many examples of CRISPR-Cas edited or modified organisms that serve biotechnological purposes.

A major challenge in raising livestock is managing disease. For example, each year the pig farming industry in the United States loses over $600 million to a single disease—porcine respiratory and reproductive syndrome (PRRS). The porcine respiratory and reproductive syndrome virus (PRRSV) causes PRRS by infecting immune cells in the pig's lungs, which leads to respiratory complications and reproductive failure. Studies have determined that PRRSV gains entry into cells via the CD163 receptor. Therefore, researchers used CRISPR-Cas9 to remove the *CD163* gene from the pig genome. Pigs homozygous for a *CD163* deletion showed no clinical signs of the disease following exposure to the virus.

CRISPR-Cas technology is currently being used to modify food crops to introduce traits such as enhanced nutritional value, increased shelf life, and pest or drought resistance. For example, the biotech company DuPont Pioneer used CRISPR-Cas to modify the *ARGOS8* gene in corn to create a drought-resistant strain. Past studies had shown that *ARGOS8* expression improves drought resistance and that expression of this gene is low in corn. Therefore, scientists removed the native promoter of *ARGOS8* and introduced a promoter that directs stronger expression. Under drought conditions, *ARGOS8*-modified corn produced five bushels more per acre than unmodified corn. CRISPR-Cas has also been used to create mushrooms that resist browning after being sliced. (For additional information on the use of CRISPR-Cas in creating gene-edited food crops, see Special Topics Chapter 6—Genetically Modified Foods.) We are likely to see more CRISPR-Cas-derived foods in the grocery store soon.

Perhaps one of the most anticipated applications of CRISPR-Cas technology is for treating, or even curing, human genetic diseases—in other words, gene therapy. (For a broader discussion of gene therapy, see Special Topics Chapter 3—Gene Therapy.) Several clinical trials using CRISPR-Cas to treat cancers are currently under way, while numerous other CRISPR-Cas clinical trials are being planned for a wide variety of genetic disorders such as muscle, blood, and liver diseases, as well as heritable blindness, HIV infection, and various cancers.

ESSENTIAL POINT

The fastest and most efficient method of genome editing is the CRISPR-Cas system, which uses an sgRNA to direct sequence-specific cutting of the genome by the Cas9 nuclease. ■

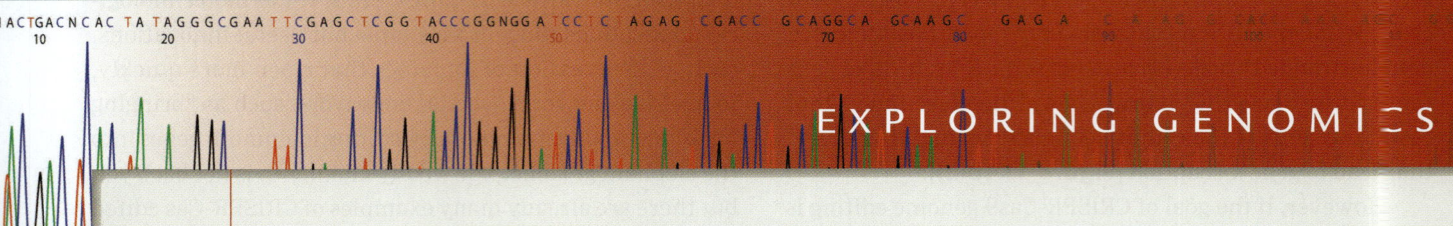

Manipulating Recombinant DNA: Restriction Mapping

Mastering Genetics Visit the Study Area: Exploring Genomics

As you learned in this chapter, restriction enzymes are sophisticated "scissors" that geneticists and molecular biologists routinely use to cut DNA for recombinant DNA experiments. A wide variety of online tools assist scientists working with restriction enzymes and manipulating recombinant DNA for different applications. Here we explore **Webcutter** and **Primer3**, two sites that make recombinant DNA experiments much easier.

■ Exercise I – Creating a Restriction Map in Webcutter

Suppose you had cloned and sequenced a gene and you wanted to design a probe approximately 600 bp long that could be used to analyze expression of this gene in different human tissues by Northern blot analysis. Internet sites such as Webcutter make it relatively easy to design experiments for manipulating recombinant DNA. In this exercise, you will use Webcutter to create a restriction map of human DNA with the enzymes *Eco*RI, *Bam*HI, and *Pst*I.

1. Access **Webcutter** at http://www.firstmarket.com/cutter/cut2.html. Go to the Study Area for *Essentials of Genetics*, and open the Exploring Genomics exercise for this chapter. Copy the sequence of cloned human DNA found there, and paste it into the text box in Webcutter.
2. Scroll down to "Please indicate which enzymes to include in the analysis." Click the button indicating, "Use only the following enzymes." Select the restriction enzymes *Eco*RI, *Bam*HI, and *Pst*I from the list provided, and then click "Analyze sequence." (*Note*: Use the command, control, or shift key to select multiple restriction enzymes.)
3. After examining the results provided by Webcutter, create a table showing the number of cutting sites for each enzyme and the fragment sizes that would be generated by digesting with each enzyme. Draw a restriction map indicating cutting sites for each enzyme with distances between each site and the total size of this piece of human DNA.

■ Exercise II – Designing a Recombinant DNA Experiment

Now that you have created a restriction map of your piece of human DNA, you need to ligate the DNA into a plasmid DNA vector that you can use to make your probe (molecular biologists often refer to this as subcloning). To do this you will need to determine which restriction enzymes would best be suited for cutting both the plasmid and the human DNA.

1. Referring back to the Study Area and the Exploring Genomics exercise for this chapter, copy the plasmid DNA sequence from Exercise I into the text box in Webcutter and identify cutting sites for the same enzymes you used in Exercise I. Then answer the following questions:
   a. What is the total size of the plasmid DNA analyzed in Webcutter?
   b. Which enzyme(s) could be used in a recombinant DNA experiment to ligate the plasmid to the *largest* DNA fragment from the human gene? Briefly explain your answer.
   c. What size recombinant DNA molecule will be created by ligating these fragments?
   d. Draw a simple diagram showing the cloned DNA inserted into the plasmid, and indicate the restriction-enzyme cutting site(s) used to create this recombinant plasmid.

CASE STUDY Ethical issues and genetic technology

In the 1970s, scientists realized that there might be unforeseen dangers and ethical issues with the use of recombinant DNA technology. A self-imposed moratorium on related research was implemented to develop safety protocols. As the Human Genome Project, designed to sequence and analyze the DNA of the human genome, came into existence in 1990, it was accompanied by the Ethical, Legal, and Social Implications (ELSI) program. ELSI was charged with identifying and addressing issues arising from genomic research.

This program focused mainly on privacy issues, the ethical use of genetic technology in medicine, and the design and conduct of genetic research, including gene therapy. The program led to the passage of federal legislation regulating the use of genetic information and the institution of guidelines limiting the scope of gene therapy. These guidelines prohibit germ-line therapy, which impacts future generations, and also prohibit gene therapy designed to enhance physical or mental aptitudes.

The recent development of CRISPR-Cas as a new genetic technology may allow for the removal of mutant alleles that cause devastating neurological disorders such as Huntington disease and prevent its transmission to future generations. Similar technology could be used to selectively eradicate the species of mosquito that transmits malaria, a painful and life-shortening disease that affects millions worldwide. With the development of these revolutionary methods, there are calls to redefine issues and to institute a new set of ethical guidelines for using these methods to eliminate genetic disorders and to revolutionize agriculture.

1. What undesirable or unforeseen consequences might occur in ecosystems if a species is eradicated using these new technologies?
2. Do we have the ethical right to alter the genomes of future generations of humans even if intervention eliminates lethal alleles?
3. Should these new technologies be regulated internationally to prevent their use by bioterrorists? How could violations be detected, and how could such regulations be enforced?

For related reading, see Rodriguez, E. (2016). Ethical issues in using Crispr/Cas9 system. *J. Clin. Res. Bioethics* 7:266 (doi:10.4172/2155-9627.1000266).

INSIGHTS AND SOLUTIONS

1. The recognition sequence for the restriction enzyme *Sau*3AI is GATC (see Figure 17.1); in the recognition sequence for the enzyme *Bam*HI—GGATCC—the four internal bases are identical to the *Sau*3AI sequence. The single-stranded ends produced by the two enzymes are identical. Suppose you have a cloning vector that contains a *Bam*HI recognition sequence and you also have foreign DNA that was cut with *Sau*3AI.

(a) Can this DNA be ligated into the *Bam*HI site of the vector, and if so, why?

(b) Can the DNA segment cloned into this sequence be cut from the vector with *Sau*3AI? With *Bam*HI? What potential problems do you see with the use of *Bam*HI?

Solution:

(a) DNA cut with *Sau*3AI can be ligated into the vector's *Bam*HI cutting site because the single-stranded ends generated by the two enzymes are identical.

(b) The DNA can be cut from the vector with *Sau*3AI because the recognition sequence for this enzyme (GATC) is maintained on each side of the insert. Cutting the cloned insert with *Bam*HI is more problematic. In the ligated vector, the conserved sequences are GGATC (left) and GATCC (right). The correct base for recognition by *Bam*HI will *follow* the conserved sequence (to produce GGATCC on the left) only about 25 percent of the time, and the correct base will *precede* the conserved sequence (and produce GGATCC on the right) about 25 percent of the time as well. Thus, *Bam*HI will be able to cut the insert from the vector ($0.25 \times 0.25 = 0.0625$), or only about 6 percent, of the time.

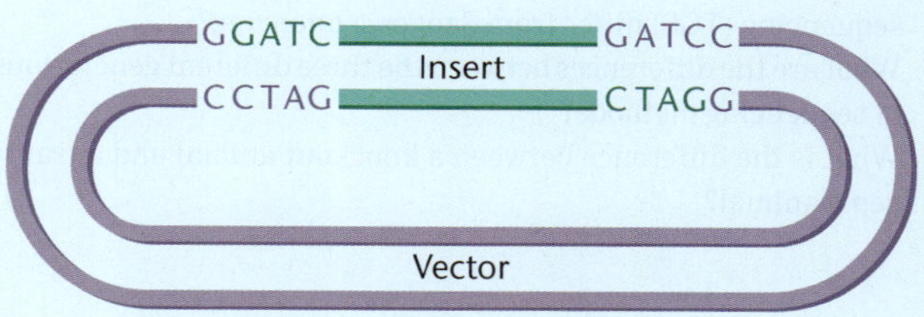

Problems and Discussion Questions

Mastering Genetics Visit for instructor-assigned tutorials and problems.

1. **HOW DO WE KNOW?** In this chapter we focused on how specific DNA sequences can be copied, identified, characterized, and sequenced. At the same time, we found many opportunities to consider the methods and reasoning underlying these techniques. From the explanations given in the chapter, what answers would you propose to the following fundamental questions?
   (a) In a recombinant DNA cloning experiment, how can we determine whether DNA fragments of interest have been incorporated into plasmids and, once host cells are transformed, which cells contain recombinant DNA?
   (b) What steps make PCR a chain reaction that can produce millions of copies of a specific DNA molecule in a matter of hours without using host cells?
   (c) How has DNA-sequencing technology evolved in response to the emerging needs of genome scientists?
   (d) How can gene knockouts, transgenic animals, and gene-editing techniques be used to explore gene function?
2. **CONCEPT QUESTION** Review the Chapter Concepts list on p. 347. All of these refer to recombinant DNA methods and applications. Write a short essay or sketch a diagram that provides an overview of how recombinant DNA techniques help geneticists study genes.
3. What roles do restriction enzymes, vectors, and host cells play in recombinant DNA studies? What role does DNA ligase perform in a DNA cloning experiment? How does the action of DNA ligase differ from the function of restriction enzymes?
4. The human insulin gene contains a number of sequences that are removed in the processing of the mRNA transcript. Bacterial cells cannot excise these sequences from mRNA transcripts, yet this gene can be cloned into a bacterial cell and produce insulin. Explain how this is possible.
5. Although many cloning applications involve introducing recombinant DNA into bacterial host cells, many other cell types are also used as hosts for recombinant DNA. Why?
6. Using DNA sequencing on a cloned DNA segment, you recover the nucleotide sequence shown below. Does this segment contain a palindromic recognition sequence for a restriction enzyme? If so, what is the double-stranded sequence of the palindrome, and what enzyme would cut at this sequence? (Consult Figure 17.1 for a list of restriction sites.)

   CAGTATGGATCCCAT

7. Cloning vectors were developed as DNA molecules that accept DNA fragments and replicate these fragments when vectors are introduced into host cells. What are their most important characteristics?
8. List the advantages and disadvantages of using plasmids as cloning vectors. What advantages do BACs and YACs provide over plasmids as cloning vectors?
9. Why would you rather use two different restriction enzymes with noncomplementary sticky ends to clone a DNA fragment into a vector?
10. What is a cDNA library, and for what purpose can it be used?
11. If you performed a PCR experiment starting with only one copy of double-stranded DNA, approximately how many DNA molecules would be present in the reaction tube after 15 cycles of amplification?
12. What advantages do cDNA libraries provide over genomic DNA libraries? Describe cloning applications where the use of a genomic library is necessary to provide information that a cDNA library cannot.
13. What is quantitative real-time PCR (qPCR)? Describe what happens during a qPCR reaction and how it is quantified.
14. We usually think of enzymes as being most active at around 37°C, yet in PCR the DNA polymerase is subjected to multiple exposures of relatively high temperatures and seems to function appropriately at 65–75°C. What is special about the DNA polymerase typically used in PCR?
15. How do next-generation sequencing (NGS) and third-generation sequencing (TGS) differ from Sanger sequencing?
16. What are the differences between the three different generations of sequencing methods?
17. What is the difference between a knockout animal and a transgenic animal?
18. One complication of making a transgenic animal is that the transgene might integrate at random into the coding region, or the regulatory region, of an endogenous gene. What might be the consequences of such random integrations? How might this complicate genetic analysis of the transgene?
19. Describe the Cre-Lox system for generating conditional knockout mice. What type of genes can be studied by using an inducible promoter to activate the *Cre* gene?
20. What techniques can scientists use to determine if a particular transgene has been integrated into the genome of an organism?
21. Gene targeting and genome editing are both techniques for removing or modifying a particular gene, each of which can produce the same ultimate goal. Describe some of the differences between the experimental methods used for these two techniques.
22. The CRISPR-Cas system has great potential but also raises many ethical issues about its potential applications because theoretically it can be used to edit any gene in the genome. What do you think are some of the concerns about the use of CRISPR-Cas on humans? Should CRISPR-Cas applications be limited for use on only certain human genes but not others? Explain your answers.
23. How is the CRISPR-Cas adaptive immunity system in bacteria modified for genome editing in eukaryotes?
24. What is the difference between nonhomologous end-joining (NHEJ) and homology-directed repair (HDR) in the context of genome editing?
25. What safety considerations must be taken before CRISPR-Cas is used to edit human embryos to cure disease?
26. Provide one example of a CRISPR-Cas application for biotechnology.
27. Why is genome editing by CRISPR-Cas advantageous over traditional methods for creating knockout or transgenic animals? Explain your answers.

```
AGGCCCAACAAGCACAGCCGGGAAGGAAAATGCGTTGTGGACCTCTGTGCCGATTCCTG
AGGCCCAAGAAGC-CATCCTGGGAAGGAAAATGCATTGGGGAACCCTGTGCGGATTCTTG

TGGCTTTGGCCCTATCTGTCCTGTGTTGAAGCTGTGCCAATCCGAAAAGTCCAGGATGAC
TGGCTTTGGCCCTATCTTTTCTATGTCCAAGCTGTGCCCATCCAAAAAGTCCAAGATGAC

ACCAAAACCCTCATCAAGACGATTGTCGCCAGGATCAATGACATTTCACACACGCAGTCT
ACCAAAACCCTCATCAAGACAATTGTCACCAGGATCAATGACATTTCACACACGCAGTCA

GTCTCCTCCAAACAGAGGGTCGCTGGTCTGGACTTCATTCCTGGGCTCCAACCAGTCCTG
GTCTCCTCCAAACAGAAAGTCACCGGTTTGGACTTCATTCCTGGGCTCCACCCCATCCTG

AGTTTGTCCAGGATGGACCAGACGTTGGCCATCTACCAACAGATCCTCAACAGTCTGCAT
ACCTTATCCAAGATGGACCAGACACTGGCAGTCTACCAACAGATCCTCACCAGTATGCCT

TCCAGAAATGTGGTCCAAATATCTAATGACCTGGAGAACCTCCGGGACCTTCTCCACCTG
TCCAGAAACGTGATCCAAATATCCAACGACCTGGAGAACCTCCGGGATCTTCTTCACGTG

CTGGCCTCCTCCAAGAGCTGCCCCTTGCCCCGGGCCAGGGGCCTGGAGACCTTTGAGAGC
CTGGCCTTCTCTAAGAGCTGCCACTTGCCCTGGGCCAGTGGCCTGGAGACCTTGGACAGC

CTGGGCGGCGTCCTGGAAGCCTCACTCTACTCCACAGAGGTGGTGGCTCTGAACAGACTG
CTGGGGGGTGTCCTGGAAGCTTCAGGCTACTCCACAGAGGTGGTGGCCCTGAGCAGGCTG
```

Alignment comparing the DNA sequence for the leptin gene from dogs (blue) and from humans (red). Vertical lines and shaded boxes indicate identical bases. *LEP* encodes a hormone that functions to suppress appetite. This type of analysis is a common application of bioinformatics and a good demonstration of comparative genomics.

18 Genomics, Bioinformatics, and Proteomics

CHAPTER CONCEPTS

- Genomics applies recombinant DNA, DNA sequencing methods, and bioinformatics to sequence, assemble, and analyze genomes.
- Disciplines in genomics encompass several areas of study, including structural and functional genomics, comparative genomics, and metagenomics, and have led to an "omics" revolution in modern biology.
- Bioinformatics merges information technology with biology and mathematics to store, share, compare, and analyze nucleic acid and protein sequence data.
- The Human Genome Project has greatly advanced our understanding of the organization, size, and function of the human genome.
- Fifteen years after completion of the Human Genome Project, a new era of genomics studies is providing deeper insights into the human genome.
- Comparative genomics analysis has revealed similarities and differences in genome size and organization.
- Metagenomics is the study of genomes from environmental samples and is valuable for identifying microbial genomes.
- Transcriptome analysis provides insight into patterns of gene expression and gene-regulatory activity of a genome.
- Proteomics focuses on the protein content of cells and on the structures, functions, and interactions of proteins.
- Synthetic genomes have been assembled, elevating interest in potential applications of engineering cells through synthetic biology.

In 1977, as recombinant DNA–based techniques, including DNA sequencing, were developing, Fred Sanger and colleagues launched the field of **genomic analysis** or simply **genomics,** the study of genomes, by sequencing the 5400-nucleotide DNA genome of the virus ϕX174. Over the past 40 years, genomic analysis has advanced so quickly that modern biological research is experiencing a genomics revolution. Genomics is one of the most rapidly advancing and exciting areas of modern genetics—providing scientists, and even the general public, with unprecedented information about genomes of different organisms, including humans, on a daily basis.

In this chapter, we will examine basic technologies used in genomic analysis, including **bioinformatics**—the use of computer hardware and software and mathematics applications to organize, share, and analyze data related to gene structure, gene sequence and expression, and protein structure and function. We discuss examples of genome data derived from different species including the human genome, and consider selected disciplines of genomics. We discuss *transcriptome analysis*, the study of genes expressed in a cell or tissue (the "transcriptome"), and *proteomics*, the study of proteins present in a cell or tissue. We conclude by presenting the concept of synthetic genomes and applications in synthetic biology.

18.1 Whole-Genome Sequencing Is Widely Used for Sequencing and Assembling Entire Genomes

A primary limitation of most recombinant DNA approaches is that they typically can identify only relatively small numbers of genes at a time. By contrast, genomics approaches allow the identification of many genes because entire genomes can be sequenced. The most widely used strategy for sequencing and assembling an entire genome involves variations of a method called **whole-genome sequencing (WGS),** also known as shotgun cloning or shotgun sequencing. In simple terms, this technique is analogous to you and a friend taking your respective copies of this genetics textbook and randomly ripping the pages into strips about 5 to 7 inches long. Each chapter represents a chromosome, and all the letters in the entire book are the "genome." Then you and your friend would go through the painstaking task of comparing the pieces of paper to find places that match, overlapping sentences—areas where there are similar sentences on different pieces of paper. Eventually, in theory, many of the strips containing matching sentences would overlap in ways that you could use to reconstruct the pages and assemble the order of the entire text.

Figure 18.1 shows a basic overview of WGS. First, multiple copies of an entire chromosome are cut into short, overlapping fragments, either by mechanical or enzymatic methods. For simplicity, here we present only an example using restriction enzymes. In separate digests, different restriction enzymes can be used so that chromosomes are cut at different sites. Alternatively, *partial digests* of DNA using the same restriction enzyme can be performed. To achieve a partial digest, the reaction mixture is incubated for only a short period of time so that not every target site in a particular DNA molecule is cut by the restriction enzyme. Restriction digests of whole chromosomes generate thousands to millions of overlapping DNA fragments. For example, a 6-bp cutter such as *Eco*RI creates about 700,000 fragments when used to digest the human genome!

One of the earliest bioinformatics applications to be developed for genomics was the use of algorithm-based

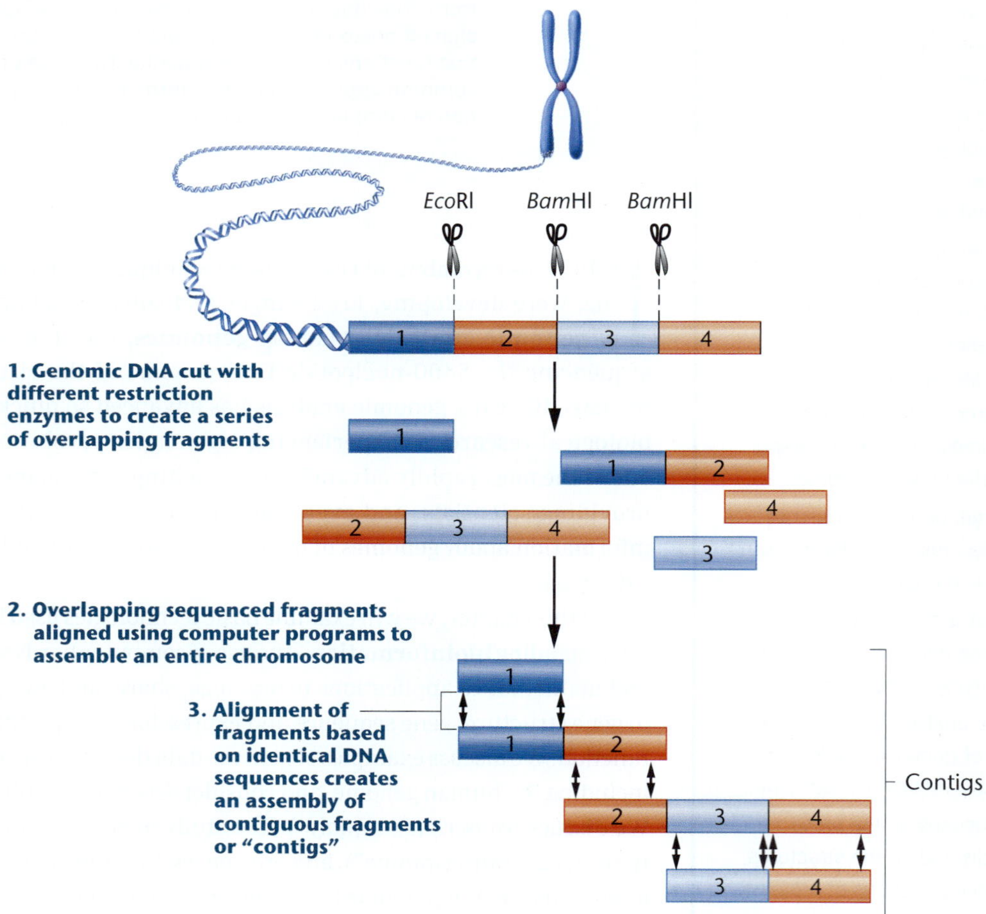
EcoRI
BamHI
BamHI
1
2
3
4
1. Genomic DNA cut with different restriction enzymes to create a series of overlapping fragments
2. Overlapping sequenced fragments aligned using computer programs to assemble an entire chromosome
3. Alignment of fragments based on identical DNA sequences creates an assembly of contiguous fragments or "contigs"
Contigs

FIGURE 18.1 An overview of whole-genome sequencing and assembly. One strategy (shown here) involves using restriction enzymes to digest genomic DNA into contigs, which are then sequenced and aligned using bioinformatics to identify overlapping fragments based on sequence identity. Notice that *Eco*RI digestion produces two fragments (contigs 1 and 2–4), whereas digestion with *Bam*HI produces three fragments (contigs 1–2, 3, and 4).

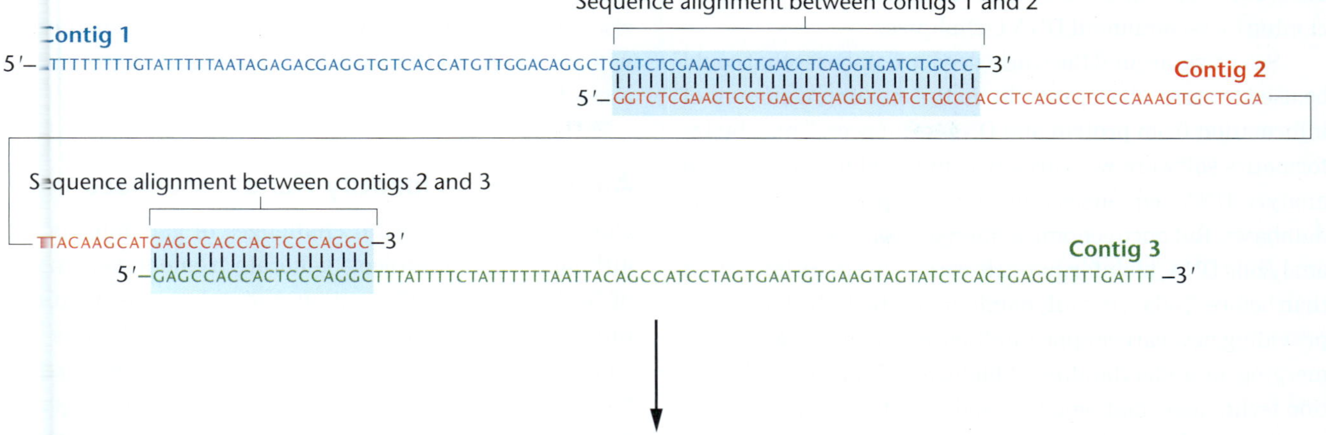

FIGURE 18.2 DNA-sequence alignment of contigs on human chromosome 2. Single-stranded DNA for three different contigs from human chromosome 2 is shown in blue, red, or green. The actual sequence from chromosome 2 is shown, but in reality, contig alignment involves fragments that are several thousand bases in length. Alignment of the three contigs allows a portion of chromosome 2 to be assembled. Alignment of all contigs for a particular chromosome would result in assembly of a completely sequenced chromosome.

software programs for creating a DNA-sequence **alignment,** in which similar sequences of bases are lined up for comparison. Alignment identifies overlapping sequences, allowing scientists to reconstruct their order in a chromosome.

Because these overlapping fragments are adjoining segments that collectively form one continuous DNA molecule within a chromosome, they are called **contiguous fragments,** or contigs. **Figure 18.2** shows an example of contig alignment and assembly for a portion of human chromosome 2. For simplicity, this figure shows relatively short sequences for each contig, which in actuality would be much longer. The figure is also simplified in that, in actual alignments, assembled sequences do not always overlap only at their ends.

The WGS method was developed by J. Craig Venter and colleagues at The Institute for Genome Research (TIGR, now named The J. Craig Venter Institute). In 1995, TIGR scientists used this approach to sequence the 1.83-million-bp genome of the bacterium *Haemophilus influenzae*. This was the first completed genome sequence from a free-living (i.e., nonviral) organism, and it demonstrated "proof of concept" that shotgun sequencing could be used to sequence an entire genome. Even after the genome for *H. influenzae* was sequenced, many scientists were skeptical that a shotgun approach would work on the larger genomes of eukaryotes. Now WGS approaches are the predominant method for sequencing genomes.

The major technological breakthrough that made genomics possible was the development of computer-automated sequencers. In the past 15 years, high-throughput sequencing has increased the productivity of DNA-sequencing technology over 500-fold. The total number of bases that could be sequenced in a single reaction was doubling about every 24 months. At the same time, this increase in efficiency brought about a dramatic decrease in cost, from about $1.00 to less than $0.001 per base pair. As we will discuss in Section 18.3, without question the development of high-throughput sequencing was essential for the Human Genome Project. And as you know from earlier in the text (see Chapter 17), next- and third-generation sequencers now enable genome scientists to produce a sequence more than 50,000 times faster than sequencers in 2000 with greater output, improved accuracy, and reduced cost.

ESSENTIAL POINT

Whole-genome sequencing enables scientists to assemble sequence maps of entire genomes. ■

18.2 DNA Sequence Analysis Relies on Bioinformatics Applications and Genome Databases

Genomics necessitated the rapid development of **bioinformatics,** the use of computer hardware and software and mathematics applications to organize, share, and analyze data related to gene structure, gene sequence and expression, and protein structure and function. However, even before WGS projects had been initiated, a large amount of sequence information from a range of different organisms was accumulating as a result of gene cloning by recombinant DNA techniques.

Scientists around the world needed databases that could be used to store, share, and obtain the maximum amount of information from protein and DNA sequences. Thus, bioinformatics software was already being used to compare and analyze DNA sequences and to create private and public databases. But once genomics emerged as a new approach for analyzing DNA, bioinformatics became even more important than before. Today, it is a dynamic area of biological research, providing new career opportunities for anyone interested in merging an understanding of biological data with information technology, mathematics, and statistical analysis.

Among the most common applications of bioinformatics are to:

- Compare DNA sequences, as in contig alignment, or compare sequences from individuals of different species
- Identify genes in a genomic DNA sequence
- Find gene-regulatory regions, such as promoters and enhancers
- Identify structural sequences, such as telomeric sequences, in chromosomes
- Predict the amino acid sequence of a putative polypeptide encoded by a cloned gene sequence
- Analyze protein structure, and predict protein functions on the basis of identified domains and motifs
- Deduce evolutionary relationships between genes and organisms on the basis of sequence information

High-throughput DNA-sequencing techniques were developed nearly simultaneously with the expansion of the Internet. As genome data accumulated, many DNA-sequence databases became freely available online. They are essential resources for archiving and sharing sequence information with other researchers and with the public. One of the largest genomic databases, called **GenBank,** is maintained by the National Center for Biotechnology Information (NCBI) in Washington, D.C. GenBank shares and acquires data from other databases in Japan and Europe. Containing more than 220 billion bases of sequence data from over 100,000 species, it is the largest publicly available database of DNA sequences—and it doubles in size roughly every 18 months! The Human Genome Nomenclature Committee, supported by the NIH, establishes rules for assigning names and symbols to newly cloned human genes. As sequences are identified and genes are named, each sequence deposited into GenBank is provided with an **accession number** that scientists can use to access and retrieve that sequence for analysis.

The NCBI is an invaluable source of public access databases and bioinformatics tools for analyzing genome data. You have already been introduced to NCBI and GenBank earlier in the text through several Exploring Genomics exercises. In the Exploring Genomics feature for this chapter, you will use NCBI and GenBank to compare and align contigs to assemble a chromosome segment.

Annotation to Identify Gene Sequences

One of the fundamental challenges of genomics is that, although genome projects generate tremendous amounts of DNA-sequence information, these data are of little use until they have been analyzed and interpreted. Thus, after a genome has been sequenced and compiled, scientists are faced with the task of identifying gene-regulatory sequences and other sequences of interest in the genome so that gene maps can be developed. This process, called **annotation,** relies heavily on bioinformatics, and a wealth of different software tools are available to carry it out.

One initial approach to annotating a sequence is to compare the newly sequenced genomic DNA to the known sequences already stored in various databases. The NCBI provides access to **BLAST (Basic Local Alignment Search Tool),** a very popular software application for searching through banks of DNA and protein sequence data. Using BLAST, we can compare a segment of genomic DNA to sequences throughout major databases such as GenBank to identify portions that align with or are the same as existing sequences. For WGS projects, simple BLAST alignments are insufficient and more complex algorithms are required to align the billions of reads of DNA sequence generated.

Figure 18.3 shows a representative example of a sequence alignment based on a BLAST search. Here a 280-bp contig from rat chromosome 12 (the query sequence) was used to search a mouse database to determine whether a sequence in the rat contig matched a known gene in mice. Notice that the rat contig is aligned with base pairs 174,612 to 174,891 of mouse chromosome 8 (the subject sequence). The accession number for the mouse sequence, NT_039455.6, is indicated at the top of the figure. BLAST searches calculate an *identity value*—determined by the sum of identical matches between aligned sequences divided by the total number of bases aligned. Gaps, indicating missing bases in one of the two sequences, are usually ignored in calculating similarity scores. The aligned rat and mouse sequences were 93 percent identical and showed no gaps in the alignment.

Notice that the BLAST report also provides an "Expect" value, or *E-value*, based on the number of matching sequences in the database that would be expected by chance. E-values take into account the length of the query sequence. Shorter sequences have a much greater likelihood of being present in the database by chance as compared to longer sequences. The lower the E-value (the closer it is to 0), the higher the significance of the match. DNA sequences that have E-values much less than 1.0 are considered to be significantly similar.

```
ref | NT_039455.6 | Mm8_39495_36
Mus musculus chromosome 8 genomic contig, strain C57BL/6J
Features in this part of subject sequence: insulin receptor
Score = 418 bits (226), Expect = 2e-114
Identities = 262/280 (93%), Gaps = 0/280 (0%)

Query  1       CAGGCCATCCCGAAAGCGAAGATCCCTTGAAGAGGTGGGCAATGTGACAGCCACTACACC  60
               ||||||||||||||||||||||||||||||||||||||| |||||||||||||| ||||
Sbjct  174891  CAGGCCATCCCGAAAGCGAAGATCCCTTGAAGAGGTGGGGAATGTGACAGCCACCACACT  174832

Query  61      CACACTTCCAGATTTTCCCAACATCTCCTCCACCATCGCGCCCACAAGCCACGAAGAGCA  120
               ||||||||||||||| |||||| ||||||| ||||| | ||||||||| || || |||||
Sbjct  174831  CACACTTCCAGATTTCCCCAACGTCTCCTCTACCATTGTGCCCACAAGTCAGGAGGAGCA  174772

Query  121     CAGACCATTTGAGAAAGTAGTAAACAAGGAGTCACTTGTCATCTCTGGCCTGAGACACTT  180
               ||| |||||||||||||| || ||||||||||||||||||||||||||||||||||||||
Sbjct  174771  CAGGCCATTTGAGAAAGTGGTGAACAAGGAGTCACTTGTCATCTCTGGCCTGAGACACTT  174712

Query  181     CACTGGGTACCGCATTGAGCTGCAGGCATGCAATCAGGACTCCCCAGAAGAGAGGTGCAG  240
               |||||||||||||||||||||||||||||||||||| || |||||||| |||||||||||
Sbjct  174711  CACTGGGTACCGCATTGAGCTGCAGGCATGCAATCAAGATTCCCCAGATGAGAGGTGCAG  174652

Query  241     CGTGGCTGCCTACGTCAGTGCCCGGACCATGCCTGAAGGT  280
                |||||||||||||||||||||||||||||||||||||||
Sbjct  174651  TGTGGCTGCCTACGTCAGTGCCCGGACCATGCCTGAAGGT  174612
```

FIGURE 18.3 BLAST results showing a 280-base sequence of a chromosome 12 contig from rats (*Rattus norvegicus*, the "query") aligned with a portion of chromosome 8 from mice (*Mus musculus*, the "subject") that contains a partial sequence for the insulin receptor gene. Vertical lines indicate exact matches. The rat contig sequence was used as a query sequence to search a mouse database in GenBank. Notice that the two sequences show 93 percent identity, strong evidence that this rat contig sequence contains a gene for the insulin receptor.

Because this mouse sequence on chromosome 8 is known to contain an insulin receptor gene (encoding a protein that binds the hormone insulin), it is highly likely that the rat contig sequence also contains an insulin receptor gene. We will return to the topic of similarity in Sections 18.4 and 18.7, where we consider how similarity between gene sequences can be used to infer function and to identify evolutionarily related genes through comparative genomics.

Hallmark Characteristics of a Gene Sequence Can Be Recognized during Annotation

Gene-prediction programs are used to annotate sequences. Yet even with bioinformatics, identifying a gene in a particular sequence of DNA is not always straightforward, particularly when one is studying genes that do not code for proteins. In fact, a reasonable question whenever one sequences a genome is, "Where are the genes?" In other words, how does one know what sequences of a genome are genes and which sequences are not genes or parts of a gene?

As we discussed earlier in the text (see Chapters 12 and 16), several hallmark features of genes exist, whether the genome under study is from a eukaryote or a bacterium (**Figure 18.4**). These characteristics can be identified in gene sequences by using bioinformatics software that incorporates search elements for them.

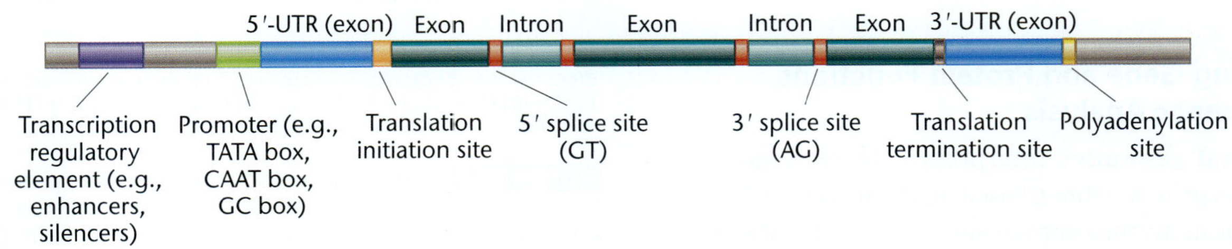

FIGURE 18.4 Characteristics of a protein-coding gene that can be used during annotation to identify a gene in an unknown sequence of genomic DNA. Most eukaryotic genes are organized into coding segments (exons) and noncoding segments (introns). When annotating a genome sequence to determine whether it contains a gene, it is necessary to distinguish between introns and exons; gene-regulatory sequences, such as promoters and enhancers; untranslated regions (UTRs); and gene termination sequences.

For instance, gene-regulatory regions found upstream of genes are marked by identifiable sequences such as promoters, enhancers, and silencers. Recall from earlier in the text (see Chapter 16) that TATA box, GC box, and CAAT box sequences are often present in the promoter region of eukaryotic genes. Protein-coding genes contain one or more **open reading frames (ORFs),** nucleotide sequences that, after transcription and mRNA splicing, are translated into the amino acid sequence of a protein. ORFs typically begin with an initiation sequence—usually ATG, which transcribes into the AUG start codon of an mRNA molecule—and end with a termination sequence—TAA, TAG, or TGA—which corresponds to the stop codons of UAA, UAG, and UGA in mRNA.

Within eukaryotic ORFs, recall also that splice sites between **exons** and **introns** can be identified, because they contain a predictable sequence (most introns begin with CT and end with AG). Annotation can sometimes be a little bit easier for bacterial genes than for eukaryotic genes because there are no introns in bacterial genes.

Finally, downstream elements, such as termination sequences and, in eukaryotes, a polyadenylation sequence that signals the addition of a poly-A tail to the 3′end of an mRNA transcript, are also important for annotation.

NOW SOLVE THIS

18.1 In a sequence encompassing 99.4 percent of the euchromatic regions of human chromosome 1, Gregory et al. [(2006) *Nature* 441:315–321] identified 3141 genes.

(a) How does one identify a gene within a raw sequence of bases in DNA?

(b) What features of a genome are used to verify likely gene assignments?

(c) Given that chromosome 1 contains approximately 8 percent of the human genome, and assuming that there are approximately 20,000 genes, would you consider chromosome 1 to be "gene rich"?

■ **HINT:** *This problem involves a basic understanding of bioinformatics and gene annotation approaches to determine how potential gene sequences can be identified in a stretch of sequenced DNA.*

Predicting Gene and Protein Functions by Sequence Analysis

Functional genomics interprets DNA sequences and establishes gene functions based on the projected RNAs or possible proteins they encode and, as well, identifies other components of the genome, such as gene-regulatory elements. Functional genomics also involves experimental approaches to confirm or refute computational predictions.

One computational approach to assigning functions to genes is to use *sequence similarity searches,* as described previously. Programs such as BLAST are used to search through databases to find alignments between the newly sequenced genome and genes that have already been identified, either in the same or in different species. (Figure 18.3). Inferring gene function from similarity searches is based on a relatively simple idea. If a genome sequence shows statistically significant similarity to the sequence of a gene whose function is known, then it is likely that the genome sequence encodes a protein with a similar or related function.

Another major benefit of similarity searches is that they are often able to identify **homologous genes,** genes that are evolutionarily related. Homologous genes in the same species are called **paralogs.** For example, in the globin gene family, the α- and β-globin subunits in humans are paralogs resulting from a gene-duplication event. Paralogs often have similar or identical functions.

If homologous genes in different species are thought to have descended from a gene in a common ancestor, the genes are known as **orthologs.** After the human genome was sequenced, many ORFs in it were identified as protein-coding genes based on their alignment with related genes of known function in other species. An example is the leptin gene, which was first discovered in mice and later identified in humans. **Figure 18.5** compares portions of the human (*LEP*) and the mouse (*Lep*) leptin genes, which are over 85 percent identical in sequence. This close match between the two sequences confirmed the leptin-coding function of the human DNA and, therefore, its identity as the human *LEP* gene.

A gene sequence can be used to predict a polypeptide sequence, which can then be analyzed for specific structural domains and motifs. Identification of **protein domains,** such as ion channels, membrane-spanning regions, DNA-binding regions, among others, can in turn be used to predict protein function.

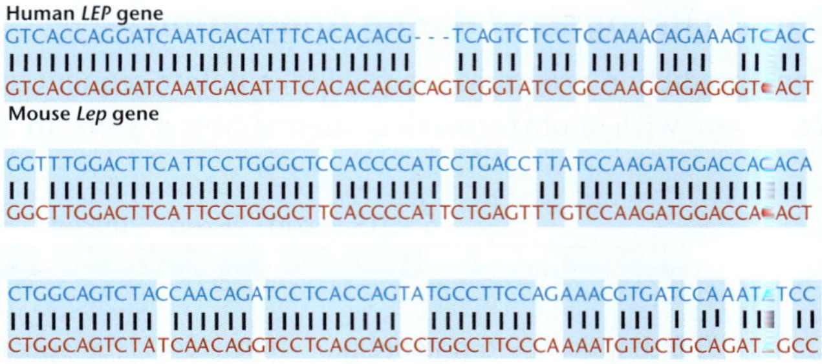

Human *LEP* gene

GTCACCAGGATCAATGACATTTCACACACG- - -TCAGTCTCCTCCAAACAGAAAGTCACC

GTCACCAGGATCAATGACATTTCACACACGCAGTCGGTATCCGCCAAGCAGAGGGT[illegible]ACT

Mouse *Lep* gene

GGTTTGGACTTCATTCCTGGGCTCCACCCCATCCTGACCTTATCCAAGATGGACCACACA

GGCTTGGACTTCATTCCTGGGCTTCACCCCATTCTGAGTTTGTCCAAGATGGACCA[illegible]ACT

CTGGCAGTCTACCAACAGATCCTCACCAGTATGCCTTCCAGAAACGTGATCCAAAT[illegible]TCC

CTGGCAGTCTATCAACAGGTCCTCACCAGCCTGCCTTCCCAAAATGTGCTGCAGAT[illegible]GCC

FIGURE 18.5 Comparison of the human *LEP* and mouse *Lep* genes. Partial sequences for these orthologs are shown with the human *LEP* gene on top (in blue) and the mouse *Lep* gene sequence below it (in red). Notice from the number of identical nucleotides, indicated by shaded boxes and vertical lines, that the nucleotide sequence for these two genes is very similar.

ESSENTIAL POINT

Bioinformatics can be used for sequence annotation to identify protein-coding and noncoding sequences of a gene, such as regulatory elements, and to predict gene function based on sequence analysis. ■

18.3 The Human Genome Project Revealed Many Important Aspects of Genome Organization in Humans

Now that you have a general idea of the basic strategies used for analyzing a genome, let's look at the largest genomics project completed to date. The **Human Genome Project (HGP)** was a coordinated international effort to determine the sequence of the human genome and to identify all the genes it contains.

Origins of the Project

The publicly funded Human Genome Project began in 1990 under the direction of James Watson, the co-discoverer of the double helix structure of DNA. Eventually the public project was headed by Dr. Francis Collins, who had previously led a research team involved in identifying the *CFTR* gene as the cause of cystic fibrosis. In the United States, the Collins-led HGP was coordinated by the Department of Energy and the National Center for Human Genome Research (now called the National Human Genome Research Institute), a division of the National Institutes of Health. It established a 15-year plan with a proposed budget of $3 billion to identify all human genes, originally thought to number between 80,000 and 100,000, to sequence and map them all, and to sequence the approximately 3 billion base pairs thought to be comprised by the 24 chromosomes (22 autosomes, plus X and Y) in humans. Other primary goals of the HGP included the following:

- Establish functional categories for all human genes.
- Analyze genetic variations among humans, including the identification of single-nucleotide polymorphisms (SNPs).
- Map and sequence the genomes of several model organisms used in experimental genetics, including *Escherichia coli, Saccharomyces cerevisiae, Caenorhabditis elegans, Drosophila melanogaster*, and *Mus musculus* (mouse).
- Develop new sequencing technologies, such as high-throughput computer-automated sequencers, to facilitate genome analysis.
- Disseminate genome information among both scientists and the general public.

Recognizing the impact that genetic information would have on society, the HGP also set up the **ELSI Program** (standing for Ethical, Legal, and Social Implications) to consider these types of issues arising from the HGP and to ensure that personal genetic information would be safeguarded and not used in discriminatory ways.

As the HGP grew into an international effort, scientists in 18 countries were involved in the project. Much of the work was carried out by the International Human Genome Sequencing Consortium, involving nearly 3000 scientists working at 20 centers in six countries (China, France, Germany, Great Britain, Japan, and the United States).

In 1999, a privately funded human genome project led by J. Craig Venter at *Celera Genomics* (aptly named from a word meaning "swiftness") was announced. Celera's goal was to use WGS and computer-automated high-throughput DNA sequencers to sequence the human genome more rapidly than the clone-by-clone approach used by the HGP. Recall that Venter and his colleagues had proven the potential of WGS in 1995 when they completed the genome for *H. influenzae*. Celera's announcement set off an intense competition between the two teams, each of which aspired to be first with the human genome sequence. This contest eventually led to the HGP finishing ahead of schedule and under budget after scientists from the public project began to use high-throughput sequencers and WGS strategies as well.

Major Features of the Human Genome

In June 2000, the leaders of the public and private genome projects met at the White House with President Bill Clinton and jointly announced the completion of a draft sequence of the human genome. In February 2001, they each published an analysis covering about 96 percent of the euchromatic region of the genome. The public project sequenced euchromatic portions of the genome 12 times and set a quality-control standard of a 0.01 percent error rate for their sequence. Although this error rate may seem very low, it still allows about 600,000 errors in the human genome sequence. Celera sequenced certain areas of the genome more than 35 times when compiling the genome.

The remaining work of completing the sequence involved filling in gaps clustered around centromeres, telomeres, and repetitive sequences (regions rich in CG base pairs can be particularly tough to sequence and interpret), correcting misaligned segments, and re-sequencing portions of the genome to ensure accuracy. In 2003 sequencing and error fixing were deemed sufficient to pass the international project's definition of completion—that the analysis contained fewer than 1 error per 10,000 nucleotides and that it covered 95 percent of the gene-containing portions of the genome. Yet even at the time of "completion" there were still some 350 gaps in the sequence that continued to be worked on.

And obviously the HGP did not sequence the genome of every person on Earth. The assembled sequence consists of haploid genomes pooled from different individuals so that they provide a reference genome representative of major, common elements widely shared among human populations. Examples of major features of the human genome are summarized in **Table 18.1**. As you can see, many unexpected observations have provided us with major new insights. The genome is not static! Genome variations, including the abundance of repetitive sequences scattered throughout the genome, verify that the genome is dynamic and reveal many evolutionary examples of sequences that have changed in structure and location. In many ways, the HGP has revealed just how little we know about our genome.

Two of the biggest surprises discovered by the HGP were that less than 2 percent of the genome codes for proteins and that there are only around 20,000 protein-coding genes. Scientists had originally estimated the number of genes to be about 100,000, based in part on a prediction that human cells produce about 100,000 proteins. This overestimate of gene number occurred, in part, because many genes code for multiple proteins through **alternative splicing.** Recall from earlier in the text (see Chapter 12), that alternative splicing patterns can generate multiple mRNA molecules, and thus multiple proteins, from a single gene. Initial estimates suggested that over 50 percent of human genes undergo alternative splicing to produce multiple transcripts and multiple proteins. Recent studies, however, suggest that ~94 to 95 percent of human pre-mRNAs containing multiple exons can potentially result in multiple different protein products.

There is still no consensus among scientists worldwide about the exact number of human genes, partly because it is unclear whether or not many of the presumed genes produce functional proteins. Currently, annotation predicts that the human genome encodes approximately 21,000 proteins. But (see Table 18.1), due to posttranslational modifications and other processes, the total number of proteins present in human cells is thought to be anywhere from approximately 200,000 to 1 million! Genome scientists continue to annotate the genome, and as mentioned earlier, functional genomics studies have important roles in determining whether or not computational predictions about the number of protein-coding and non–protein-coding genes are accurate.

TABLE 18.1 Major Features of the Human Genome

- The human genome contains ~3.1 billion nucleotides, but protein-coding sequences make up only about 2 percent of the genome.
- The genome sequence is ~99.9 percent similar in individuals of all nationalities. SNPs and copy number variations (CNVs) account for genome diversity from person to person.
- The genome is dynamic. At least 50 percent of the genome is derived from transposable elements, such as LINE and *Alu* sequences, and other repetitive DNA sequences.
- The human genome contains approximately 20,000 protein-coding genes, far fewer than the originally predicted number of 80,000–100,000 genes.
- The average size of a human gene is ~25 kb, including gene-regulatory regions, introns, and exons. On average, mRNAs produced by human genes are ~3000 nt long.
- Many human genes produce more than one protein through alternative splicing, thus enabling human cells to produce a much larger number of proteins (perhaps as many as 200,000) from only ~20,000 genes.
- More than 50 percent of human genes show a high degree of sequence similarity to genes in other organisms; however, more than 40 percent of the genes identified have no known molecular function.
- Genes are not uniformly distributed on the 24 human chromosomes. Gene-rich clusters are separated by gene-poor "deserts" that account for 20 percent of the genome. These deserts correlate with G bands seen in stained chromosomes. Chromosome 19 has the highest gene density, and chromosome 13 and the Y chromosome have the lowest gene densities.
- Chromosome 1 contains the largest number of genes, and the Y chromosome contains the smallest number.
- Human genes are larger and contain more and larger introns than genes of invertebrates, such as *Drosophila*. The largest known human gene encodes dystrophin, a muscle protein. This gene, associated in mutant form with muscular dystrophy (Chapter 14), is 2.5 Mb in length (Chapter 12), larger than many bacterial chromosomes. Most of this gene is composed of introns.
- The number of introns in human genes ranges from 0 (in histone genes) to 234 (in the gene for *titin*,which encodes a muscle protein).

It is clear that human genes encode an incredible diversity of proteins. During the HGP, functional categories were assigned for human genes, primarily on the basis of:

1. Functions determined previously (for example, from recombinant DNA cloning of human genes and known mutations involved in human diseases)
2. Comparisons to known genes and predicted protein sequences from other species
3. Predictions based on annotation and analysis of protein functional domains and motifs

Although functional categories and assignments continue to be revised, at the time the HGP was completed, the functions of over 40 percent of human genes were unknown. Determining human gene functions, deciphering complexities of gene-expression regulation and gene interaction, and uncovering the relationships between human genes and phenotypes are among the many ongoing challenges for genome scientists.

Individual Variations in the Human Genome

The HGP originally revealed that in all humans, regardless of racial and ethnic origins, the genomic sequence is approximately 99.9 percent the same. As we discuss in other chapters, most genetic differences among humans result from **single-nucleotide polymorphisms (SNPs)** and **copy number variations (CNVs).** Recall that SNPs are single-base changes in the genome and variations of many SNPs are associated with disease conditions, such as sickle-cell anemia and cystic fibrosis. Later in the text (see Special Topics Chapter 2—Genetic Testing), we will examine how SNPs can be detected and used for diagnosis and treatment of disease.

After the draft sequence of the human genome was completed, it initially appeared that most genetic variations between individuals (the 0.1 percent differences) were due to SNPs. While SNPs are important contributing factors to genome variation, structural differences that we discussed earlier in the text (see Chapter 11) such as deletions, duplications, inversions, and CNVs, which can span millions of base pairs of DNA, play much more important roles in genome variation than previously thought. Recall that CNVs are duplications or deletions of relatively large sections of DNA on the order of several hundred or several thousand base pairs. Many of the CNVs that vary the most among genomes appear to be at least 1 kilobase.

Although most human DNA is present in two copies per cell, one from each parent, CNVs are segments of DNA that are duplicated or deleted, resulting in variations in the number of copies of a DNA segment inherited by individuals. In some cases CNVs are major deletions at the exon level or involving entire genes; other deletions affect gene function by frameshifts in the reading code. CNVs that are duplicated can result in overexpression of a particular gene, yet many deleted and duplicated CNVs do not present clearly identifiable phenotypes.

Current estimates of the number of CNVs in an individual genome range from about 12 to perhaps 4 to 5 dozen per person. Some studies estimate that there may be as many as 1500 CNVs greater than 1 kb among the human genome. Other studies claim there are more than 1.5 million deletions of less than 100 bp that contribute to genome variation between individuals.

Accessing the Human Genome Project on the Internet

It is now possible to access databases and other sites on the Internet that display maps for all human chromosomes. You will visit a number of these databases in Exploring Genomics exercises. **Figure 18.6(a)** displays a partial gene map for chromosome 12 that was taken from an NCBI database called Genome Data Viewer. The first image shows an ideogram, or cytogenetic map, of chromosome 12. To the right of the ideogram is a column showing the contigs (arranged vertically) that were aligned to sequence this chromosome. The UniGene column displays a histogram representation of gene density on chromosome 12. Notice that relatively few genes are located near the centromere. Finally, gene symbols, loci, and gene names (by description) are provided for 20 selected genes. When accessing these maps on the Internet, one can magnify, or zoom in on, each region of the chromosome, revealing all genes mapped to a particular area.

You can see that most of the genes listed in Figure 18.6(a) have been assigned descriptions based on the functions of their products: Some are transmembrane proteins; some are enzymes such as kinases; some are receptors, including several involved in olfaction; and so on.

The HGP's most valuable contribution will perhaps be the identification of disease genes and the development of new treatment strategies as a result. Thus, extensive maps have been developed for genes implicated in human disease conditions. The disease gene map of chromosome 21 shown in **Figure 18.6(b)** indicates genes involved in amyotrophic lateral sclerosis (ALS), Alzheimer disease, cataracts, deafness, and several different cancers. In a later chapter (see Special Topics Chapter 2—Genetic Testing), we discuss implications of the HGP for the identification of genes involved in human genetic diseases and for disease diagnosis, detection, and gene therapy applications.

ESSENTIAL POINT

The HGP has revealed many surprises about human genetics, including gene number and the high degree of DNA sequence similarity both between individuals and between humans and other species, and showed that many genes encode multiple RNAs and proteins. ■

18.4 The "Omics" Revolution Has Created a New Era of Biological Research

The Human Genome Project and the development of genomics techniques have been largely responsible for launching a new era of biological research—the era of "omics." It seems

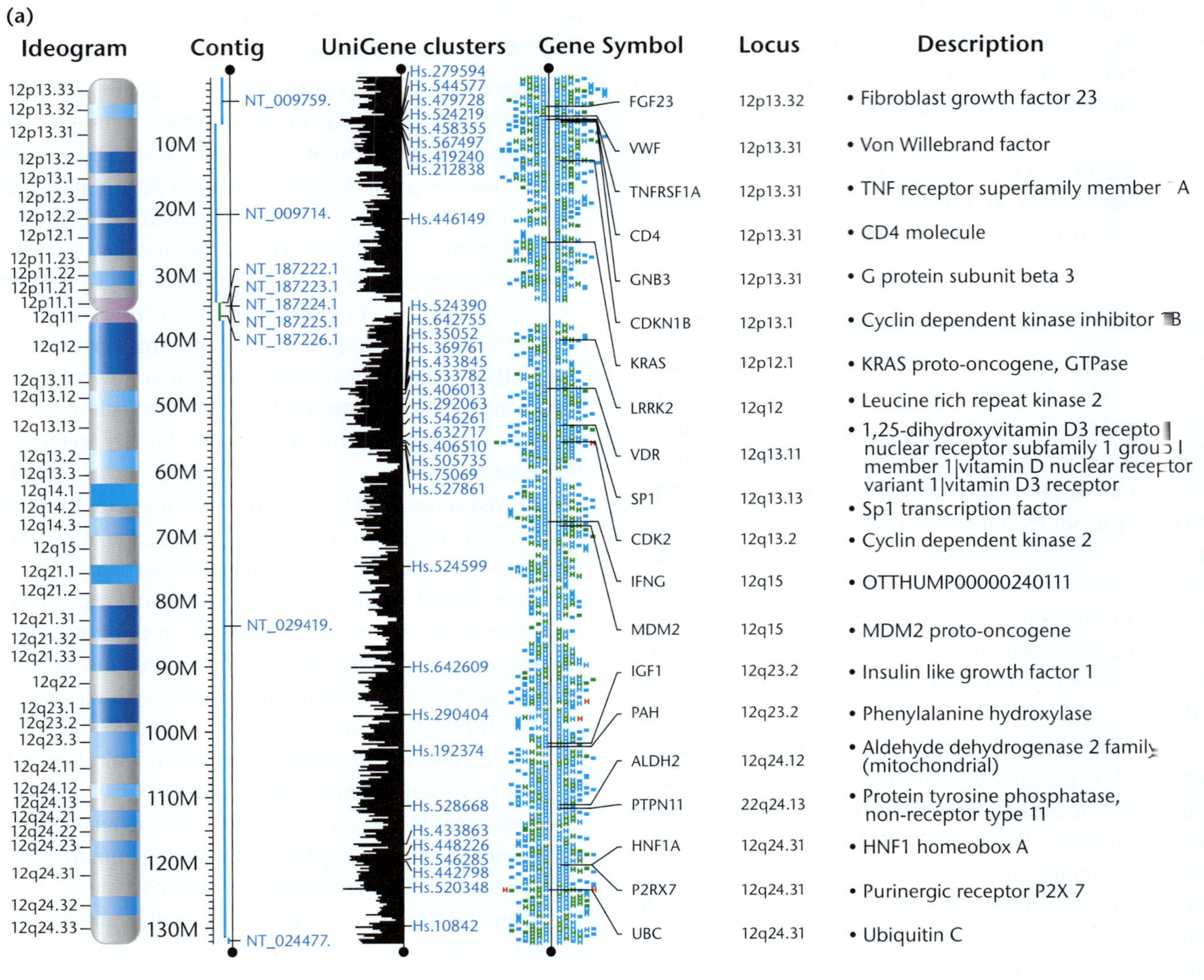

(b)

Chromosome 21
48 million bases

Coxsackie and adenovirus receptor
Amyloidosis cerebroarterial, Dutch type
Alzheimer disease, APP-related
Schizophrenia, chronic
Usher syndrome, autosomal recessive
Amyotrophic lateral sclerosis
Oligomycin sensitivity
Jervell and Lange-Nielsen syndrome
Long QT syndrome
Down syndrome cell-adhesion molecule
Homocystinuria
Cataract, congenital, autosomal dominant
Deafness, autosomal recessive
Myxovirus (influenza) resistance
Leukemia, acute myeloid

Myeloproliferative syndrome, transient
Leukemia transient of Down syndrome
Enterokinase deficiency
Multiple carboxylase deficiency
T-cell lymphoma invasion and metastasis
Mycobacterial infection, atypical
Down syndrome (critical region)
Autoimmune polyglandular disease, type 1
Bethlem myopathy
Epilepsy, progressive myoclonic
Holoprosencephaly, alobar
Knobloch syndrome
Hemolytic anemia
Breast cancer
Platelet disorder, with myeloid malignancy

FIGURE 18.6 (a) A gene map for chromosome 12 from the NCBI database Genome Data Viewer. (b) Partial map of disease genes on human chromosome 21. Maps such as this depict genes thought to be involved in human genetic disease conditions.

that every year, more areas of biological research are being described as having an omics connection. Some examples of "omics" are:

- **Proteomics:** the analysis of all the proteins in a cell or tissue
- **Metabolomics:** the analysis of proteins and enzymatic pathways involved in cell metabolism
- **Glycomics:** the analysis of the carbohydrates of a cell or tissue
- **Toxicogenomics:** the analysis of the effects of toxic chemicals on genes, including mutations created by toxins and changes in gene expression caused by toxins
- **Metagenomics:** the analysis of genomes of organisms collected from the environment
- **Pharmacogenomics:** the development of customized medicine based on a person's genetic profile for a particular condition
- **Transcriptomics:** the analysis of all expressed genes in a cell or tissue

We will consider several of these genomics disciplines in other parts of this chapter.

ESSENTIAL POINT

Genomics has led to a number of related "omics" disciplines that are rapidly changing how modern biologists study DNA, RNA, proteins, and many aspects of cell function. ■

After the HGP, What's Next?

Since completion of a reference sequence of the human genome, studies have continued at a very rapid pace and new areas for human genome research have emerged. One such area is the analysis of the epigenome, in which a Human Epigenome Project is creating hundreds of maps of epigenetic changes in different cell and tissue types and evaluating potential roles of epigenetics in complex diseases (see also Special Topics Chapter 1—Epigenetics). Another research area is the characterization of SNPs (the International HapMap Project) and CNVs for their role in genome variation, disease, and pharmacogenomics applications. Yet another theme includes human cancer genome projects. We will discuss aspects of one of these (Cancer Genome Atlas Project) later in the text (see Chapter 19). Here we consider several examples of genome research that are extensions of the HGP.

Personal Genome Projects

As we discussed earlier in this chapter (and in Chapter 17), next- and third-generation sequencing technologies are capable of generating sequence reads at higher speeds with greater accuracy and have greatly reduced the cost of DNA sequencing. Expectations for continued cost reductions along with continued technological advances are high (see **Figure 18.7**)and have led several companies to propose WGS for individual people—a **personal genomics** approach.

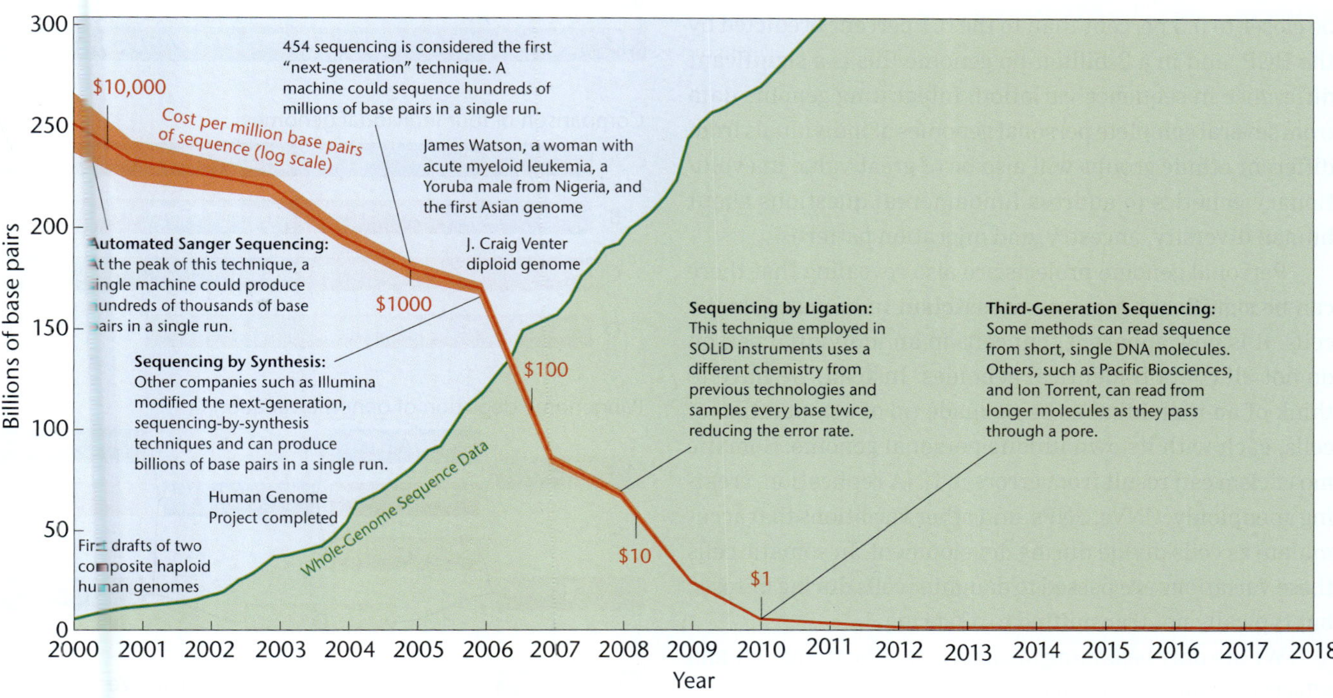

FIGURE 18.7 Human genome sequence explosion. Sequencing costs have steadily declined since 2000 due to innovations in sequencing technology. As a result, notice that the number of individual genomes sequenced has dramatically increased.

Several companies have now developed technology that can sequence a genome for less than $1000. Whether the $1000 mark represents the costs of reagents to sequence a genome or actual costs when sequence preparation, labor, and analysis of the genome are taken into account can be debated. Having somebody such as a geneticist analyze your personal genome data and consider how genome variations may affect your health may be expensive. So even if the cost of sequencing a genome is less than $1000, interpreting genome data to make sense for medical treatment may cost thousands of dollars more. Regardless of how the actual cost of sequencing a genome is calculated, the modern cost is substantially lower than the $3 billion cost proposed at the start of the HGP.

As of early 2018, an estimated 400,000 individual human genomes have been sequenced. In Special Topics Chapter 2—Genetic Testing, we will talk about the implication of personal genomics for genetic testing.

Somatic Genome Mosaicism and the Emerging Pangenome

The HGP pooled samples from multiple individuals to create a *haploid* reference genome. In contrast, personal genome projects (PGPs) sequence a *diploid* genome. Personal genome projects generate sequences of millions of short DNA fragments from *maternal and paternal chromosomes* that are mapped onto the reference genome. Such projects indicate that haploid reference genome comparisons often underestimate the extent of genome variation between individuals by five-fold or more.

Therefore, genome variation between individuals may be closer to 0.5 percent than to the 0.1 percent predicted by the HGP, and in a 3-billion-bp genome this is a significant difference in sequence variation. Integrating genome data from several complete personal genomes of individuals from different ethnic groups will also be of great value in evolutionary genetics to address fundamental questions about human diversity, ancestry, and migration patterns.

Personal genome projects are also revealing that there can be significant **genome mosaicism** in human somatic cells. It is now apparent that cells in an individual person do not all contain identical genomes. Instead, we have to think of an individual as being made up of a population of cells, each with its own unique personal genome. Somatic mosaicism can result from errors in DNA replication, creating aneuploidy, CNVs, SNPs, and other variations that accumulate as cells divide during development. In somatic cells these variations are passed to daughter cells during mitosis, but typically not transmitted to offspring.

We are only beginning to understand the frequency and effects of genetic mosaicism on health and disease.

Newer sequencing methods and expansion of personal genome projects are demonstrating significant **genomic variation** not just in humans but in most species. It has now become apparent that genomic variation among individuals is much more prevalent than can be determined by a reference sequence. In bacteria such as *Streptococcus*, for example, several dozen different genes can exist between isolates of the same strain. As a consequence, genome scientists in many fields have now replaced single reference genomes with the concept of the **pangenome** to describe all distinct genes and variations in a species (**Figure 18.8**).

Whole-Exome Sequencing

While we have thus far focused on sequencing the entire genome, it is worthwhile to recall that only about 2 percent of the genome sequence consists of protein-coding genes. Thus, in personal genomic analysis, the focus has shifted toward **whole-exome sequencing (WES),** that is, sequencing only the 180,000 exons in a person's genome. WES reveals mutations involved in disease by focusing only on protein-coding segments of the genome, thus there are more disease-related genetic variations in the exome than in other regions of the genome. WES can be done at a cost of much less than $1000 with greater than 100 × coverage (the percentage of bases in a DNA sequence that have been sequenced multiple times). Of course, a limitation of this approach is its failure to identify mutations in gene-regulatory regions that influence gene expression. As the cost of WGS continues to drop, scientists and clinicians trying to detect disease-causing mutations are debating whether it

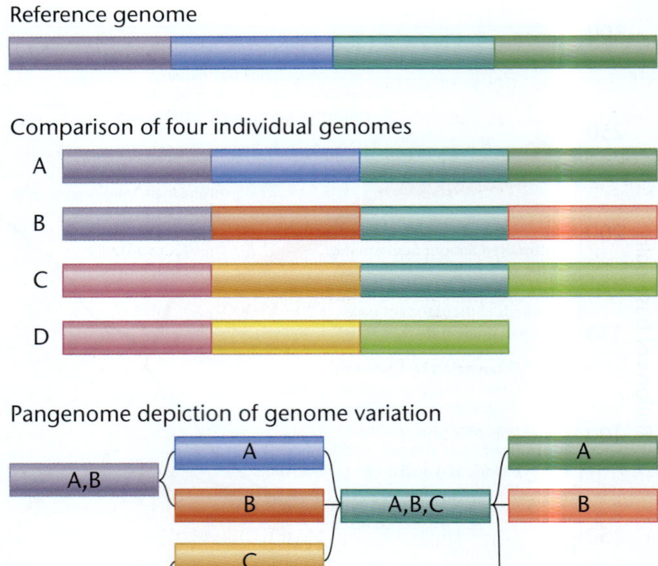

FIGURE 18.8 A pangenome attempts to visualize all genomic segments and gene variations found in a species. Notice that there are variations in individual genomes not represented in the reference genome, but these variations are included in the pangenome.

makes sense to simply sequence the entire genome or to just sequence exomes first to find mutations.

In 2015, after seven years of work, a group of scientists called the 1000 Genomes Project Consortium reported on the genomes of 2504 individuals from 26 populations representing Europe, East Asia, South Asia, Africa, and the Americas. WGS and WES data revealed nearly 88 million bi- and multi-allelic SNPs and many other structural variations such as CNVs. One interpretation of this work is that it reveals clear variations in individuals and associates particular diseases with geographic or ancestral background. Thus sequencing genomes of individuals from diverse populations can help us to better understand the spectrum of human genetic variation and to learn the causes of genetic diseases across diverse groups. We will come back to the topic of WES later in the text when we discuss genetic testing (see Special Topics Chapter 2—Genetic Testing).

Encyclopedia of DNA Elements (ENCODE) Project

In 2003, a few months after the announcement that the human genome had been sequenced, a group of about three dozen research teams around the world began the **Encyclopedia of DNA Elements (ENCODE) Project.** Using both experimental and computational approaches, a main goal of ENCODE was to identify and analyze all functional elements of the genome, including those that regulate the expression of human genes (such as transcriptional start sites, promoters, and enhancers). ENCODE projects have also been initiated for mouse, worm, and fly genomes.

Because only a relatively small percentage (less than 2 percent) of the human genome codes for proteins, ENCODE focused heavily not on genes but on all the rest of the sequences, commonly referred to as "junk" DNA. So what are all these other bases in the genome doing? The term *junk DNA* has always been a misnomer. We know that such sequences are important for chromosome structure, the regulation of gene expression, and other roles. Just because these sequences themselves do not code for protein does not mean that they are unimportant.

ENCODE studied gene expression in 147 different cell types because genome activity differs from cell to cell. After about a decade of research and a cost of $288 million, in 2012 a group of 30 research papers were published revealing the major initial findings of the ENCODE Project.

Selected highlights of what ENCODE revealed in initial papers and more recent papers include the following:

- The *majority*, ~80 percent, of the human genome is considered functional. In part, this is because large segments of the genome are transcribed into RNA, however, most do not encode proteins. These various RNAs include tRNA, rRNAs, and miRNAs, and long noncoding RNAs (lncRNAs)—defined as non–protein-coding transcripts longer than 200 nucleotides. A conservative estimate is that there may be over 17,000 genes for lncRNAs. It may turn out that the number of noncoding RNA sequences will outnumber protein-coding genes.
- The functional sequences also include gene-regulatory regions: ~70,000 promoter and nearly 400,000 enhancer regions.
- There are 20,687 protein-coding genes in the human genome.
- A total of 11,224 sequences are characterized as pseudogenes, previously thought to be inactive in all individuals. Some of these are inactive in most individuals but occasionally active in certain cell types of some individuals, which may eventually warrant their reclassification as active, transcribed genes.
- SNPs associated with disease are enriched within noncoding functional elements of the genome, often residing near protein-coding genes.

The ENCODE findings have broadly defined the functional roles of the genome to include encoding either proteins or noncoding RNAs and displaying biochemical properties such as the binding of regulatory proteins that influence transcription or chromatin structure. It is worth noting, however, that a relatively large body of geneticists and other scientists do not agree with ENCODE's definition of functional sequences. One reason cited is that ENCODE did not adequately address many of the repetitive sequences in the genome such as transposons, LINEs, SINEs, and other sequences such as telomeres and centromeres. There has also been significant debate about the value of ENCODE, given the cost of the project. Despite this, research teams are using information from ENCODE to identify risk factors for certain diseases, with the hopes of developing appropriate cures and treatments.

EVOLVING CONCEPT OF THE GENE

Based on the work of the ENCODE project, we now know that DNA sequences that have previously been thought of as "junk DNA" because they do not encode proteins, are nonetheless often transcribed into what we call noncoding RNA (ncRNA). Since the function of some of these RNAs is now being determined, we must consider whether the concept of the gene should be expanded to include DNA sequences that encode ncRNAs. At this writing, there is no consensus, but it is important for you to be aware of these current findings as you develop your final interpretation of a gene. ■

Nutrigenomics Considers Genetics and Diet

As evidence of the impact of genomics, a field of nutritional science called nutritional genomics, or **nutrigenomics**, has emerged. Nutrigenomics focuses on understanding the interactions between diet and genes. We have all had routine medical tests for blood pressure, blood sugar levels, and heart rate. Based on these tests, your physician may recommend that you change your diet and exercise more to lose weight or that you reduce your intake of sodium to help lower your blood pressure.

Now several companies claim to provide nutrigenomics tests that analyze your genome for genes thought to be associated with different medical conditions linked to nutrient metabolism. The companies then provide a customized nutrition report, recommending diet changes for improving your health and preventing illness, based on your genes! It is important to note that these tests have not yet been validated as accurate and they have not been approved by the U.S. Food and Drug Administration. It remains to be seen whether this approach as currently practiced is of valid scientific or nutritional value.

No Genome Left Behind and the Genome 10K Plan

Without question, new sequencing technologies that have been developed are an important part of the transformational effect the HGP has had on modern biology.

Recent headline-grabbing genomes that have been completed include:

- **Apple and tomato:** The apple, which has more than 57,000 genes, and the tomato, which has 31,760 genes, each have more genes than humans!
- **Potato:** This plant, which shares 92 percent of its DNA with tomatoes, turns out to be a fruit.
- **Chickpea:** It is one of the earliest cultivated legumes and the second most widely grown legume after the soybean.
- **Red-spotted newt:** This newt has a genome of almost 10 times the size of a human genome.

Modern sequencing technologies are asking some to consider the question, "What would you do if you could sequence everything?" In 2009, partners around the world, including genome scientists and zoo and museum curators, began work on sequencing 10,000 vertebrate genomes—the **Genome 10K Project.** Shortly after the HGP finished, the National Human Genome Research Institute (NHGRI) assembled a list of mammals and other vertebrates as priorities for genome sequencing in part because of their potential benefit for learning about the human genome through comparative genomics. Genome 10K Project will also provide insight into genome evolution and speciation.

Stone-Age Genomics

In yet another example of how genomics has taken over areas of DNA analysis, a number of labs around the world are involved in analyzing "ancient" DNA. These so-called **stone-age genomics** studies are generating fascinating data from miniscule amounts of ancient DNA obtained from bone and other tissues such as hair that are tens of thousands to about 700,000 years old and often involve samples from extinct species. Analysis of DNA from a 2400-year-old Egyptian mummy, bison, mosses, platypus, mammoths, Pleistocene-age cave bears and polar bears, coelacanths, and Neanderthals are some of the most prominent examples of stone-age genomics. In 2013, scientists reported the oldest intact genome sequence to be successfully analyzed to date. It came from a 700,000-year-old bone fragment from an ancient horse uncovered from the frozen ground in the Yukon Territory of Canada. This result is interesting in part because evolutionary biologists have used genomic data to estimate that ancient ancestors of modern horses branched off from other animal lineages around 4 million years ago—about twice as long ago as prior estimates.

A little over a decade ago, researchers published about 13 million bp of a sequence from a 27,000-year-old woolly mammoth found frozen and nearly intact in Siberia. This study revealed a ~98.5 percent sequence identity between mammoths and African elephants. Subsequent WGS studies by other scientists using mitochondrial and nuclear DNA from Siberian mammoths suggested that the mammoth genome differs from the African elephant by as little as 0.6 percent. These studies are also great demonstrations of how stable DNA can be under the right conditions, particularly when frozen.

In Section 18.5 we discuss recent work on the Neanderthal genome. Obtaining the genome of a human ancestor this old was previously unimaginable. This work is providing new insights into our understanding of human evolution.

ESSENTIAL POINT

Personal genome sequencing, including exome sequencing, and epigenome analysis will provide unparalleled insight into individual variations in genomes and both approaches have tremendous potential for the diagnosis of genetic diseases. ■

18.5 Comparative Genomics Provides Novel Information about the Human Genome and the Genomes of Model Organisms

Comparative genomics compares the genomes of different organisms to answer questions about genetics and other aspects of biology. It is a field with many research and practical applications, including gene discovery and the development of model organisms to study human diseases.

It also incorporates the study of gene and genome evolution and the relationship between organisms and their environment.

Comparative genomics uses a wide range of techniques and resources, such as the construction and use of nucleotide and protein databases, fluorescence *in situ* hybridization (FISH), and the creation of gene knockout animals. Comparative genomics can reveal genetic differences and similarities between organisms to provide insight into how those differences contribute to differences in phenotype, life cycle, or other attributes and to ascertain the evolutionary history of those genetic differences.

As of 2018, over 23,000 whole genomes have been sequenced—including many model organisms and a number of viruses. This is quite extraordinary progress in a relatively short time span! Among these organisms are yeast (*S. cerevisiae*, the first eukaryotic genome to be completely sequenced), bacteria such as *E. coli*, the nematode roundworm (*Caenorhabditis elegans*), the thale cress plant (*Arabidopsis thaliana*), mice (*M. musculus*), zebrafish (*Danio rerio*), and, of course, *Drosophila*. In recent years, genomes of chimpanzees, dogs, chickens, gorillas, sea urchins, honey bees, pigs, pufferfish, rice, and wheat have all been sequenced. These complete genome sequences have been invaluable for comparative genomics studies of gene function in these organisms and in humans. As shown in **Table 18.2**, the number of genes humans share with other species is very high, ranging from about 30 percent of the genes in yeast to ~80 percent in mice and ~98 percent in chimpanzees. The human genome even contains around 100 genes that are also present in many bacteria.

Comparative studies have demonstrated not only significant differences in genome organization among bacteria and eukaryotes but also many similarities between genomes of nearly all species. Analysis of the growing number of genome sequences confirms that all living organisms are related and descended from a common ancestor. A key piece of evidence for this is the observation that all organisms studied to date use similar gene sets for basic cellular functions, such as DNA replication, transcription, and translation.

Comparative genomics has further shown that many genes identified as being involved in human disease are also present in other species. For instance, approximately 60 percent of genes associated with nearly 300 human diseases are also present in *Drosophila*. These include genes involved in a variety cancers; cardiovascular disease; cystic fibrosis; and other conditions. These genetic relationships are the rationale for using model organisms to study inherited human disorders; the effects of the environment on genes; and the interactions of genes in complex diseases.

Next, we consider how comparative genomics studies of sea urchins, a model organism, and Neanderthals have revealed interesting elements of the human genome.

TABLE 18.2 Comparison of Selected Genomes

Organism (Scientific Name)	Approximate Size of Genome [in million (megabase, Mb) or billion (gigabase, Gb) base pairs] (Date Completed)	Diploid (2*n*) Chromosome Number	Number of Genes	Approximate Percentage (%) of Genes Shared with Humans
African clawed frog (*Xenopus laevis*)	3.1 Gb (2016)	36	~45,000	70
Bacterium (*Escherichia coli*)	4.6 Mb (1997)	1	4,403	Not determined
Chicken (*Gallus gallus*)	1 Gb (2004)	78	~20,000–23,000	60
Dog (*Canis familiaris*)	2.5 Gb (2003)	78	~18,400	75
Chimpanzee (*Pan troglodytes*)	3 Gb (2005)	48	~20,000–24,000	98
Fruit fly (*Drosophila melanogaster*)	165 Mb (2000)	8	~13,600	50
Human (*Homo sapiens*)	3.1 Gb (2004)	46	~20,000	100
Mouse (*Mus musculus*)	~2.5 Gb (2002)	40	~30,000	80
Pig (*Sus scrofa*)	~3 Gb (2012)	38	21,640	84
Rat (*Rattus norvegicus*)	~2.75 Gb (2004)	42	~22,000	80
Rhesus macaque (*Macaca mulatta*)	2.87 Gb (2007)	42	~20,000	93
Rice (*Oryza sativa*)	389 Mb (2005)	24	~41,000	Not determined
Roundworm (*Caenorhabditis elegans*)	97 Mb (1998)	12	19,099	40
Sea urchin (*Strongylocentrotus purpuratus*)	814 Mb (2006)	42	~23,500	60
Thale cress (plant) (*Arabidopsis thaliana*)	140 Mb (2000)	10	~27,500	Not determined
Zebrafish (*Danio rerio*)	1.4 Gb (2013)	50	~41,800	70
Yeast (*Saccharomyces cerevisiae*)	12 Mb (1996)	32	~5,700	30

Originally adapted from Palladino, M. A. (2006) *Understanding the Human Genome Project*, 2nd ed. Benjamin Cummings.

The Sea Urchin Genome

In 2006, researchers from the Sea Urchin Genome Sequencing Consortium completed the 814-million-bp genome of the sea urchin *Strongylocentrotus purpuratus*. Sea urchins are shallow-water marine invertebrates often studied by developmental biologists. Fossil records indicate that sea urchins appeared during the Early Cambrian period, around 520 million years ago (mya).

Sea urchins have an estimated 23,500 genes, including representatives of almost all major vertebrate gene families. Sequence alignment and homology searches demonstrate that the sea urchin contains many genes with important functions in humans, yet is missing genes that are important in flies and worms. For example, certain cytochrome P450 genes, which play a role in the breakdown of toxic compounds, are not found in sea urchins. The sea urchin genome also has an abundance (~25 to 30 percent) of **pseudogenes**—nonfunctional duplications of protein-coding genes. Sea urchins have a smaller average intron size than humans, supporting a general trend revealed by comparative genomics that intron size is correlated with overall genome size.

Urchins have nearly 1000 genes for sensing light and odor, indicative of great sensory abilities. In this respect, their genome is more typical of vertebrates than invertebrates. A number of orthologs of human genes involved in hearing and balance are present, as are many human-disease-associated orthologs, including protein kinases, transcription factors, innate immunity, and low-density lipoprotein receptors. Sea urchins and humans share approximately 7000 orthologs.

The Neanderthal Genome and Modern Humans

Svante Pääbo, of the Max Planck Institute for Evolutionary Anthropology in Germany, specializes in studying ancient genomes. In 1997, he and colleagues sequenced portions of Neanderthal (*Homo neanderthalensis*) mitochondrial DNA from an undated fossil. Nine years later, in late 2006, Pääbo's group along with a number of scientists in the United States reported the first sequence of nuclear DNA isolated from Neanderthal bone samples. Continuing this work, in 2010 he led an international team of scientists who reported completion of a rough draft of the Neanderthal genome that encompassed more than 4 billion bp of DNA. Bones from three females who lived in Vindija Cave in Croatia about 38,000 to 44,000 years ago were used to produce this draft sequence. In 2016, Pääbo and colleagues published a complete, high-quality (52-fold coverage) nuclear genomic sequence from a Neanderthal bone sample that is at least 50,000 years old. Interestingly, a sequence recently recovered from fossilized remains of a Neanderthal that lived over 400,000 years ago is thought to be the oldest Neanderthal DNA ever analyzed.

Because Neanderthals are members of the human family, and closer relatives to humans than chimpanzees, the Neanderthal genome provides an unprecedented opportunity to use comparative genomics to advance our understanding of evolutionary relationships between modern humans and our predecessors. In particular, scientists are interested in identifying areas in the genome where humans have undergone rapid evolution since splitting (diverging) from Neanderthals. Much of this analysis involves applying a comparative genomics approach to all three genomes.

The human and Neanderthal genomes are 99 percent identical. Comparative genomics has identified 78 protein-coding sequences in humans that seem to have arisen since the divergence from Neanderthals and that may have helped modern humans adapt. Some of these sequences are involved in cognitive development and sperm motility. Of the many genes shared by humans and Neanderthals is *FOXP2*, a gene that has been linked to speech and language ability. There are many genes that influence speech, so this finding does not mean that Neanderthals spoke as we do. But because Neanderthals had the same modern human *FOXP2* gene scientists have speculated that Neanderthals possessed linguistic abilities.

The realization that modern humans and Neanderthals lived in overlapping ranges as recently as 30,000 years ago has led to speculation about interactions between modern humans and Neanderthals. Genome studies suggest that interbreeding took place between Neanderthals and modern humans an estimated 45,000 to 80,000 years ago in the eastern Mediterranean. In fact, the genome of non-African *H. sapiens* contains approximately 1 to 4 percent of a sequence inherited from Neanderthals. These exciting studies, previously thought to be impossible, are having ramifications in many areas of the study of human evolution, and it will be interesting indeed to follow the progress of this work.

ESSENTIAL POINT

Studies in comparative genomics are revealing fascinating similarities and variations in genomes from different organisms. ■

18.6 Metagenomics Applies Genomics Techniques to Environmental Samples

Metagenomics, also called **environmental genomics,** is the discipline that uses WGS approaches to sequence genomes from entire communities of microbes in environmental samples of water, air, and soil. Oceans, glaciers, deserts, and virtually every other environment on Earth are being sampled for metagenomics projects. Human genome pioneer J. Craig Venter left Celera to form the J. Craig Venter Institute, and his group played a central role in developing metagenomics as an emerging area of genomics research.

One of the institute's major initiatives was a global expedition to sample marine and terrestrial microorganisms from around the world and to sequence their genomes. This project yielded over 1.2 million novel DNA sequences from 1800 microbial species, including 148 previously unknown bacterial species, and identified hundreds of photoreceptor genes.

A key benefit of metagenomics is its potential for teaching us more about millions of yet uncharacterized species of bacteria. Many new viruses, particularly bacteriophages, have been identified through studies of water and soil samples. Further, important new information about genetic diversity in microbes is emerging that is needed to be able to understand complex interactions between microbial communities and their environment and to allow phylogenetic classification of newly identified microbes. Metagenomics also has great potential for identifying genes with novel functions, some of which may have valuable applications in medicine and biotechnology.

The general method used in metagenomics often involves isolating DNA directly from an environmental sample without requiring cultures of the microbes or viruses. Such an approach is necessary because often it is difficult to replicate the complex array of growth conditions the microbes need to survive in culture.

An example of how metagenomics can provide novel insight into the microbial world around us is reflected by a recent study of the microbiota found in New York City subways. This project revealed that most microbes present are non-disease-causing bacteria normally prevalent on human skin and in the GI tract. Although, occasionally, pathogens such as *Bacillus anthracis* were identified. But almost half of the DNA sequenced did not match any organism in eukaryotic, bacterial, archaeal and viral genome databases!

The Human Microbiome Project

In 2007 the National Institutes of Health announced plans for the **Human Microbiome Project (HMP),** a $170 million effort to sequence the genomes of an estimated 600 to 1000 bacteria, viruses, yeasts, and other microorganisms that live on and inside humans. At the start of the project, microorganisms were thought to outnumber human cells by about 10 to 1, although this prediction is likely too high.

Many microbes, such as *E. coli* in the digestive tract, have important roles in human health, and, of course, other microbes make us ill. The HMP had several major goals, including:

- Determining if individuals share a core human microbiome.
- Understanding whether changes in the microbiome can be correlated with changes in human health.
- Developing new methods, including bioinformatics tools, to support analysis of the microbiome.
- Addressing ethical, legal, and social implications raised by human microbiome research.

The HMP involved about 200 scientists at 80 institutions who, in 2012, published a series of papers summarizing their findings. 242 healthy individuals in the United States participated in the studies and each person was sampled up to three times over nearly two years. Fifteen body sites of males and 18 sites of females were selected for analysis and WGS of genomes for microbes and viruses present at these sites was performed. In addition, sequences for 16*S* rRNA genes were used specifically to compare bacterial samples. From this work, more than 3000 reference sequences for microbes isolated from the human body were developed.

The HMP amassed more than 1000 times the sequencing data generated by the HGP. We have formulated the following concepts about the human microbiome:

- Sequence data identified an estimated 81 to 99 percent of the microbes and viruses distributed among body areas in human males and females.
- As many as 1000 bacterial strains may be present in each person.
- The microbiome starts at birth. Babies acquire bacteria from their mothers' microbiome.
- A surprise to HMP scientists, the microbiome can be substantially different from person to person. Based on variation between individuals, an estimated 10,000 bacterial species may be part of the total human microbiome.
- Although the microbiome of the human gut differs from person to person, it remains relatively stable over time in individuals.

Based on these findings, there is no single reference human microbiome to which people can be compared. Microbial diversity varies greatly from individual to individual, and a personalization of the microbiome occurs in individuals. For instance, comparing sequences of the microbiomes from two healthy people of equivalent age reveals microbiomes that can be quite different. There are, however, similarities in certain parts of the body; signature bacteria with characteristic genes are specific to certain locations.

Knowledge about the personalized nature of the microbiome is already proving valuable for improving human health and medicine, which may include microbiome-specific therapeutic drugs in the future. Criteria are being sought for a healthy microbiome, which is expected to help determine the role of bacteria in maintaining normal health. This may provide insight into how antibiotics can disturb a person's microbiome and why certain individuals are susceptible to certain diseases, especially chronic conditions such as psoriasis, irritable bowel syndrome, and potentially even obesity.

Related to this project, a team of researchers at the University of California, Los Angeles, analyzed DNA sequences from 101 college students, 49 of whom had acne and 52 of whom did not. Over 1000 strains of *Propionibacterium acnes* were isolated. Using WGS and bioinformatics, researchers clustered these strains into ten strain types (related strains). Six of these types were more common among acne-prone students, and one type appeared repeatedly in skin samples from students without acne. Sequence analysis of types associated with acne indicated groups of genes that may contribute to the skin disease. Further analysis of these strain types may help dermatologists develop new drugs targeted at killing acne-causing strains of *P. acnes*.

A *Venn diagram*, like the image shown in **Figure 18.9**, is a common way to represent overlapping data in metagenomics datasets. In this figure, overlapping circles indicate numbers of human gut microbiome genes from individuals with liver cirrhosis, Type 2 diabetes, and irritable bowel syndrome. Notice that each disease has a unique profile of microbial genes but that significant overlaps between microbial genes for each disease occur. Of the nearly 580,000 microbial genes, 403 were shared and thus considered common markers for all three diseases.

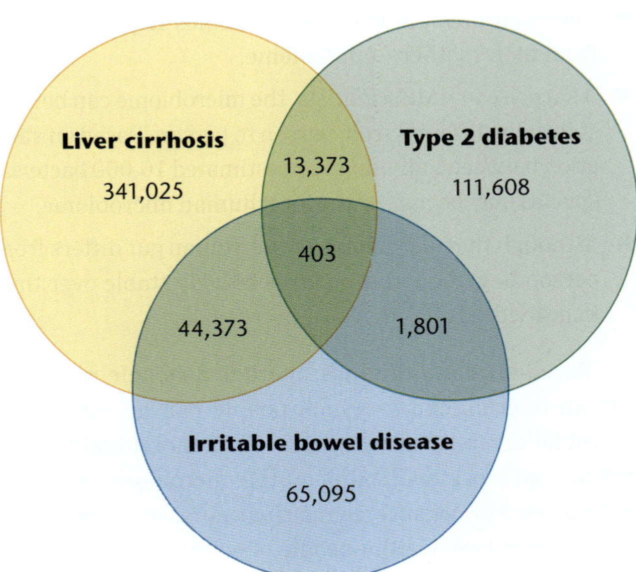

FIGURE 18.9 Venn diagram representation of gut microbial genes from patients with liver cirrhosis, Type 2 diabetes, and irritable bowel syndrome. Notice that different diseases show large numbers of unique genes with smaller numbers of shared genes.

The Case Study at the end of this chapter briefly discusses a clinical application focused on the importance of gut microbes for intestinal health.

ESSENTIAL POINT

Metagenomics sequences genomes from organisms in environmental samples, often identifying new sequences that encode proteins with novel functions. ■

18.7 Transcriptome Analysis Reveals Profiles of Expressed Genes in Cells and Tissues

Once any genome has been sequenced and annotated, a formidable challenge remains: that of understanding genome function by analyzing the genes it contains and the ways expressed genes are regulated. **Transcriptome analysis** (also called **transcriptomics** or global analysis of gene expression) studies the expression of genes in a genome both qualitatively—by identifying which genes are expressed or not expressed—and quantitatively—by measuring varying levels of expression for different genes. In other words, transcriptome analysis attempts to catalog and quantify the total RNA content of a cell, tissue, or organism.

Even though in theory all cells or tissue types of an organism possess the same genes, depending on each specific cell's function certain genes will be highly expressed, others expressed at low levels, and some not expressed at all. Transcriptome analysis reveals gene-expression profiles that, for the same genome, may vary from cell to cell or from tissue type to tissue type. Identifying where and when genes are expressed by a genome is essential for understanding how the genome functions.

Transcriptome analysis provides insights into (1) normal patterns of gene expression that are important for understanding how a cell or tissue type differentiates during development, (2) how gene expression dictates and controls the physiology of differentiated cells, and (3) mechanisms of disease development that result from or cause gene-expression changes in cells. Later in the text (see Special Topics Chapter 2—Genetic Testing), we will consider why transcriptome analysis is gradually becoming an important diagnostic tool in certain areas of medicine.

DNA Microarray Analysis

A number of different techniques can be used for transcriptome analysis. PCR-based methods—such as reverse transcription PCR (RT-PCR) and quantitative real-time PCR (qPCR) (described in Chapter 17) are useful because of their ability to detect genes expressed at low levels. For nearly two decades **DNA microarray analysis** has been widely used because it enables researchers to analyze all of a sample's expressed genes simultaneously (see **Figure 18.10**).

Most DNA microarrays, also known as **gene chips,** are prepared by "spotting" single-stranded DNA molecules onto glass slides using a computer-controlled high-speed robotic arm called an arrayer. Arrayers are fitted with a number of tiny pins. Each pin is immersed in a small amount of solution containing millions of copies of a different single-stranded DNA molecule. [For example, many microarrays are prepared with single-stranded complementary DNA (cDNAs) or expressed sequence tags (ESTs)—short fragments of DNA cloned from expressed genes.] The arrayer fixes the DNA onto the slide at specific locations (called spots, fields, or features) that will be scanned and recorded by a computer. A single microarray can have over 20,000 different spots of DNA (and over 1 million for exon-specific microarrays), each containing a unique sequence that serves as a probe for a different gene.

To use a microarray for transcriptome analysis, scientists typically begin by extracting mRNA from cells or tissues (Figure 18.10). The mRNA is usually then reverse transcribed to synthesize cDNA tagged with fluorescently

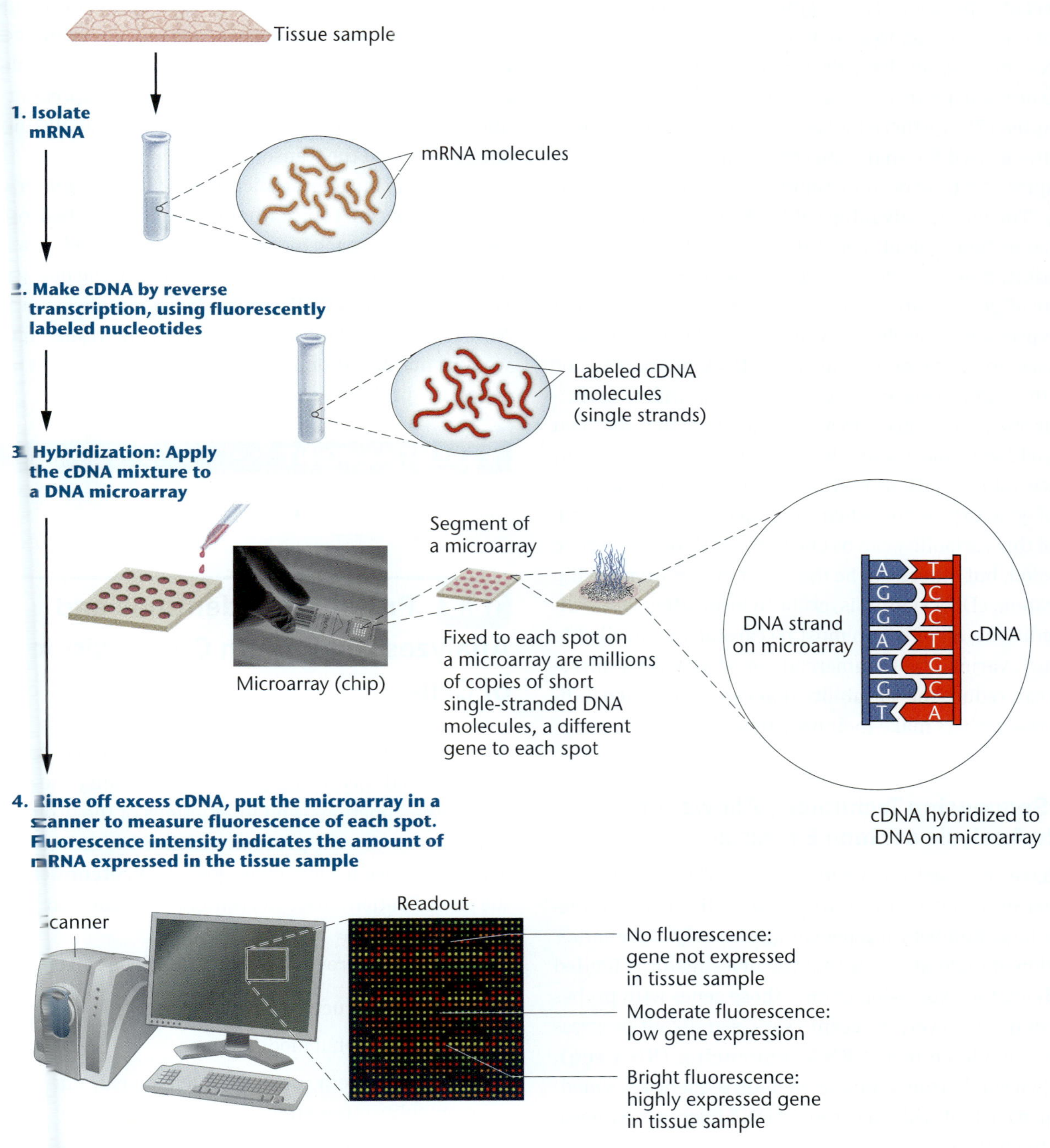

FIGURE 18.10 DNA microarray analysis for analyzing gene-expression patterns in a tissue.

labeled nucleotides. Microarray studies often involve comparing gene expression in different cell or tissue samples. In this case, cDNA prepared from one tissue is usually labeled with one color dye, red for example, and cDNA from another tissue is labeled with a different-colored dye, such as green. Labeled cDNAs are then denatured and incubated overnight with the microarray so that they will hybridize to spots on the microarray that contain complementary DNA sequences. Next, the microarray is washed to remove excess cDNA, and then scanned by a laser, which causes the cDNA hybridized to the microarray to fluoresce. The patterns of fluorescent spots reveal which genes are expressed in the tissue of interest, and the intensity of spot fluorescence indicates the relative level of expression. The brighter the spot, the more the particular mRNA is expressed in that tissue.

DNA microarrays have dramatically changed the way gene-expression patterns are analyzed. As discussed earlier (in Chapter 17), Northern blot analysis was one of the earliest methods used for analyzing gene expression. Then PCR techniques proved to be more rapid and have increased sensitivity. The biggest advantage of DNA microarrays is that they enable thousands of genes to be studied simultaneously. As a result, however, they can generate an overwhelming amount of gene-expression data. Over 1 million human gene-expression datasets are available in publicly accessible databases, and commercially available DNA microarrays for analyzing human gene expression have been widely used.

But one limitation of DNA microarrays is that they can often yield variable results. For example, one experiment under certain conditions may not always yield similar patterns of gene expression when the experiment is repeated. Some of this variability can be due to real differences in gene expression, but others can be the result of variation in chip preparation, cDNA synthesis, probe hybridization, or washing conditions, all of which must be carefully controlled to limit such variability. Commercially available DNA microarrays can reduce the variability that can result when individual researchers make their own arrays.

RNA Sequencing Technology Allows for *In Situ* Analysis of Gene Expression

As we have discussed, the significant value of gene expression microarrays has been the ability to quantify RNA expression for large numbers of genes simultaneously. But another limitation of microarrays is that the investigator is limited to studying the expression of only those genes with probes on the chip. However, in recent years, significant progress has been made on direct **RNA sequencing (RNA-seq),** also called whole-transcriptome shotgun sequencing, a modern approach that will likely render DNA microarrays obsolete in the future.

RNA-seq not only allows for quantitative analysis of all RNAs expressed in a particular tissue, but it also provides actual sequence data. And RNA-seq can also be carried out inside the cell (*in situ*). For this application, the cell itself can serve as a "gene chip."

Generally, most RNA-seq methods incorporate reverse-transcribing RNA *in situ*, sequencing cDNA, mapping sequences to a reference genome (to determine which sequences were transcribed from specific genes), and quantifying gene expression. Methods incorporating fluorescence *in situ* RNA-seq are enabling scientists also to visualize where specific RNAs are being transcribed in intact cells and tissues.

It is now also becoming possible to carry out DNA sequencing and RNA-seq on *individual* cells! Several groups are taking an integrated approach to sequence genomes and transcriptomes in the same cell. This strategy enables the correlation of genetic variability and mRNA expression variability simultaneously in single cells—thus analyzing both the genome and the transcriptome.

We will consider applications of RNA-seq later in the text (see Special Topics Chapter 2—Genetic Testing), but clearly this approach has already demonstrated great value for disease diagnosis including understanding how genome variations such as CNVs impact transcriptome expression. Now that we have discussed genomes and transcriptomes, we turn our attention to the ultimate end products of most genes, the proteins encoded by a genome.

ESSENTIAL POINT

Transcriptome analysis examines expression patterns for thousands of genes simultaneously. ■

18.8 Proteomics Identifies and Analyzes the Protein Composition of Cells

As genomes have been sequenced and studied, biologists have focused increasingly on understanding the complex structures, functions, and interactions of the proteins that genomes encode—the **proteome,** defined as the complete set of proteins encoded by a given genome. **Proteomics,** then, is the identification, characterization, and quantitative analysis of proteomes.

Proteomics provides information about many things:

- A protein's structure and function
- Posttranslational modifications
- Protein–protein, protein–nucleic acid, and protein–metabolite interactions
- Cellular localization of proteins

- Protein stability and aspects of translational and post-translational levels of gene-expression regulation
- Relationships (shared domains, evolutionary history) to other proteins

Proteomics projects have been used to characterize major families of proteins for some species. For example, about two-thirds of the *Drosophila* proteome has been well cataloged using proteomics.

Proteomics is also of clinical value because it allows comparison of proteins in normal and diseased tissues, which can lead to the identification of proteins as biomarkers for disease conditions. Proteomic analysis of mitochondrial proteins during aging, proteomic maps of atherosclerotic plaques from human coronary arteries, and protein profiles in saliva as a way to detect and diagnose diseases are examples of such work.

Reconciling the Number of Genes and the Number of Proteins Expressed by a Cell or Tissue

While Beadle and Tatum's one-gene:one-enzyme hypothesis was a worthy proposal in the 1940s (see Chapter 13), genomics has revealed that the link between gene and gene product is often much more complex. Genes can have multiple transcription start sites that produce several different types of RNA transcripts. Alternative splicing and editing of pre-mRNA molecules can generate dozens of different proteins from a single gene. As a result, proteomes are substantially larger than genomes. Sequencing of mRNAs from human tissues found that over 95 percent of protein-coding genes with more than one exon are alternatively spliced.

However, it is unclear how many different proteins are translated from this pool of transcripts. To address this, the *Human Proteome Map (HPM)*, published in 2014, aimed to catalog the human proteome in all its complexity. This project involved proteomic analysis of a wide range of human tissues and cell types using methods we will discuss in the next section. Based on the results of this study, we now know that the ~20,000 protein-coding genes in the human genome can produce at least 290,000 different proteins. The HPM accounted for ~85% of all annotated protein-coding genes in humans that currently exist in human proteomics databases. Refer to PDQ 20 to access the online database for the HPM.

The specific protein content (or profile) of a cell is determined in large part by its gene-expression patterns—its transcriptome. However, a number of other factors affect the proteome profile of a cell. To begin with, many proteins undergo co-translational or posttranslational modifications, such as cleavage of a signal sequence that targets a protein for an organelle pathway, a propeptide, or initiator methionine residues; linkage to carbohydrates and lipids; or the addition of chemical groups through methylation, acetylation, and phosphorylation; and other modifications. Over a hundred different mechanisms of posttranslational modification are known.

In addition, many proteins work via elaborate protein–protein interactions or as part of a large macromolecular complex. Furthermore, although every cell in the body contains an equivalent set of genes, not all cells express the same genes and, hence, the same proteins. Proteomics also considers proteins that a cell might acquire from another cell, not just the proteins encoded by the genome of the cell type being analyzed.

The early history of proteomics dates back to 1975 and the development of **two-dimensional gel electrophoresis (2DGE),** a technique for separating hundreds to thousands of proteins with high resolution. In this technique, proteins isolated from cells or tissues of interest are first loaded onto a polyacrylamide tube gel and separated by *isoelectric focusing*, which causes proteins to migrate based on their electrical charge in a pH gradient. In an electrical field, proteins will migrate through an established pH gradient until they reach the pH at which their net charge is zero (with equal numbers of positively-charged and negatively-charged side chains). Next, a second migration, perpendicular to the first, is performed, in which the proteins are further separated by their molecular weight using *sodium dodecyl sulfate polyacrylamide gel electrophoresis (SDS-PAGE)* (**Figure 18.11**).

It is not uncommon for a 2D gel loaded with a complex mixture of proteins to show several thousand spots, as in Figure 18.11, which displays the complex mixture of proteins in human platelets (thrombocytes). Particularly abundant protein spots in this gel have been labeled with the names of identified proteins. With thousands of different spots on the gel, how are the identities of the proteins ascertained?

In some cases, 2D gel patterns from experimental samples can be compared to gels run with reference standards containing known proteins with well-characterized migration patterns. Reference gels for different biological samples such as human plasma are available, and computer software programs can be used to align and compare the spots from different gels. In the early days of 2DGE, proteins were often identified by cutting spots out of a gel and sequencing the amino acids the spots contained. Only relatively short sequences of amino acids can typically be generated this way; rarely can an entire polypeptide be sequenced using this technique. BLAST and similar programs can be used to search protein databases containing amino acid sequences of known proteins. However, because of alternative splicing or posttranslational modifications, peptide sequences may not always match easily with the final product, and the identity of the protein may have to be confirmed by another approach.

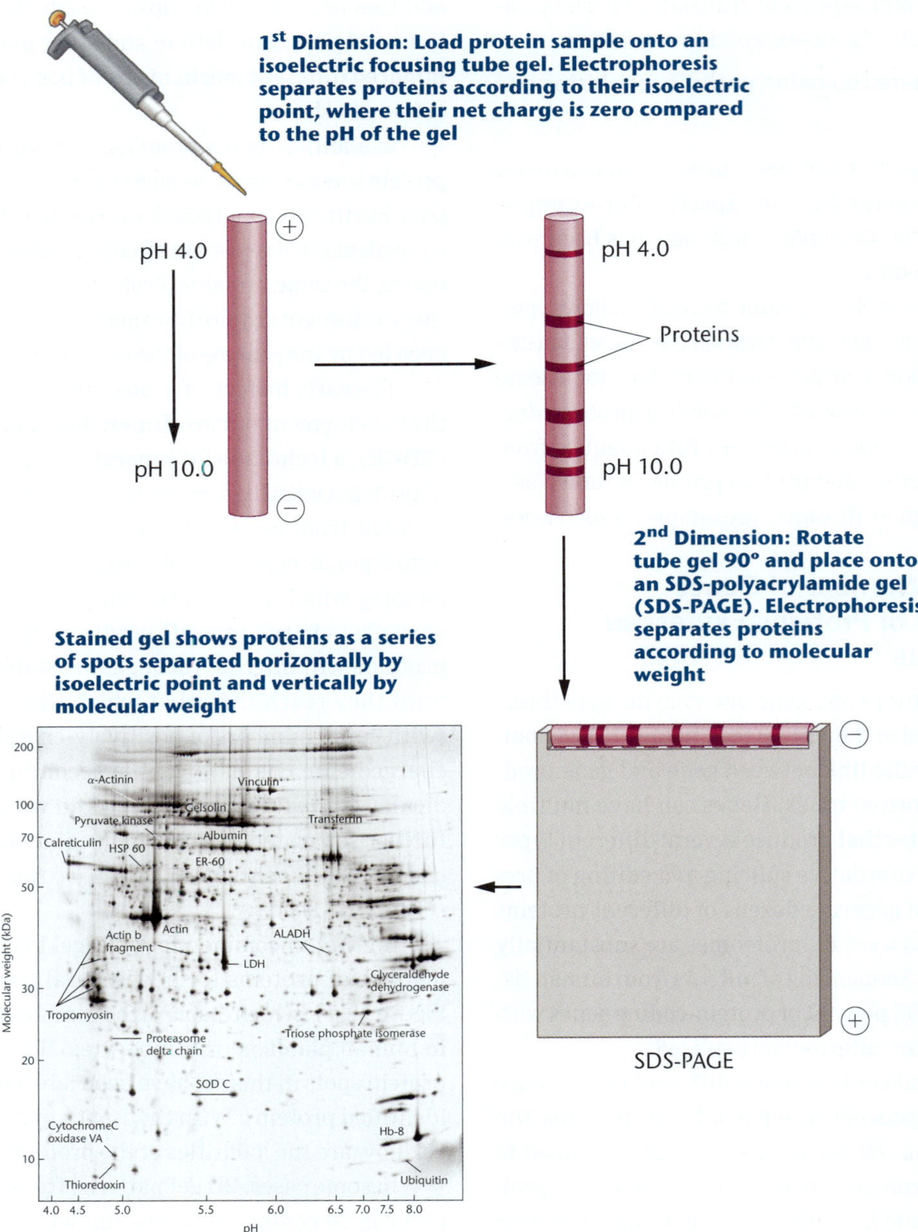

1st Dimension: Load protein sample onto an isoelectric focusing tube gel. Electrophoresis separates proteins according to their isoelectric point, where their net charge is zero compared to the pH of the gel
pH 4.0
pH 10.0
pH 4.0
Proteins
pH 10.0
2nd Dimension: Rotate tube gel 90° and place onto an SDS-polyacrylamide gel (SDS-PAGE). Electrophoresis separates proteins according to molecular weight
SDS-PAGE
Stained gel shows proteins as a series of spots separated horizontally by isoelectric point and vertically by molecular weight
α-Actinin
Vinculin
Gelsolin
Pyruvate kinase
Transferrin
Calreticulin
HSP 60
Albumin
ER-60
Actin b fragment
Actin
ALADH
LDH
Glyceraldehyde dehydrogenase
Tropomyosin
Proteasome delta chain
Triose phosphate isomerase
SOD C
CytochromeC oxidase VA
Hb-8
Thioredoxin
Ubiquitin
Molecular weight (kDa)
pH

FIGURE 18.11 Two-dimensional gel electrophoresis (2DGE) uses the different biochemical properties of proteins to separate proteins in a complex mixture. The 2D gel photo shows separations of human platelet proteins. Each spot represents a different polypeptide separated by isoelectric point, pH (*x*-axis), and molecular weight (*y*-axis). In this photo, some protein spots have been identified by name based on comparison to a reference gel or by determination of a protein sequence using mass spectrometry. Notice that many spots on the gel are unlabeled, indicating proteins of unknown identity.

Mass Spectrometry for Protein Identification

As important as 2DGE has been for protein analysis, **mass spectrometry (MS)** has been instrumental to the development of proteomics. Mass spectrometry techniques analyze ionized samples in gaseous form and measure the *mass-to-charge (m/z) ratio* of the different ions in a sample. Proteins analyzed by mass spectrometry generate characteristic *m/z* spectra. Their identities can be determined by correlating these spectra with an *m/z* database that contains known protein sequences. Certain forms of MS can provide peptide sequences directly from spectra. Some of the most valuable proteomics applications of this technology are to identify an unknown protein or proteins in a complex mix of proteins, to sequence peptides, to identify posttranslational modifications of proteins, and to characterize multiprotein complexes. Many other

NOW SOLVE THIS

18.2 Annotation of a proteome attempts to relate each protein to a function in time and space. Traditionally, protein annotation depended on an amino acid sequence comparison between a query protein and a protein with known function. If the two proteins shared a considerable portion of their sequence, the query would be assumed to share the function of the annotated protein. Following is a representation of this method of protein annotation involving a query sequence and three different human proteins. Note that the query sequence aligns to common domains within the three other proteins. What argument might you present to suggest that the function of the query is not related to the function of the other three proteins?

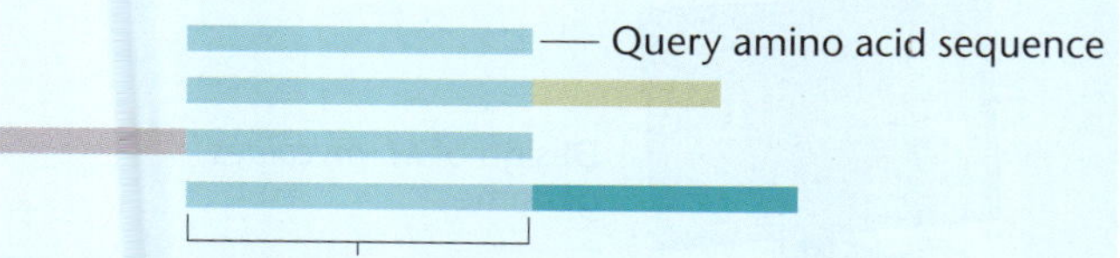

■ **HINT:** *This problem asks you to think about sequence similarities between four proteins and predict functional relationships. The key to its solution is to remember that although protein domains may have related functions, proteins can contain several different interacting domains that determine protein function.*

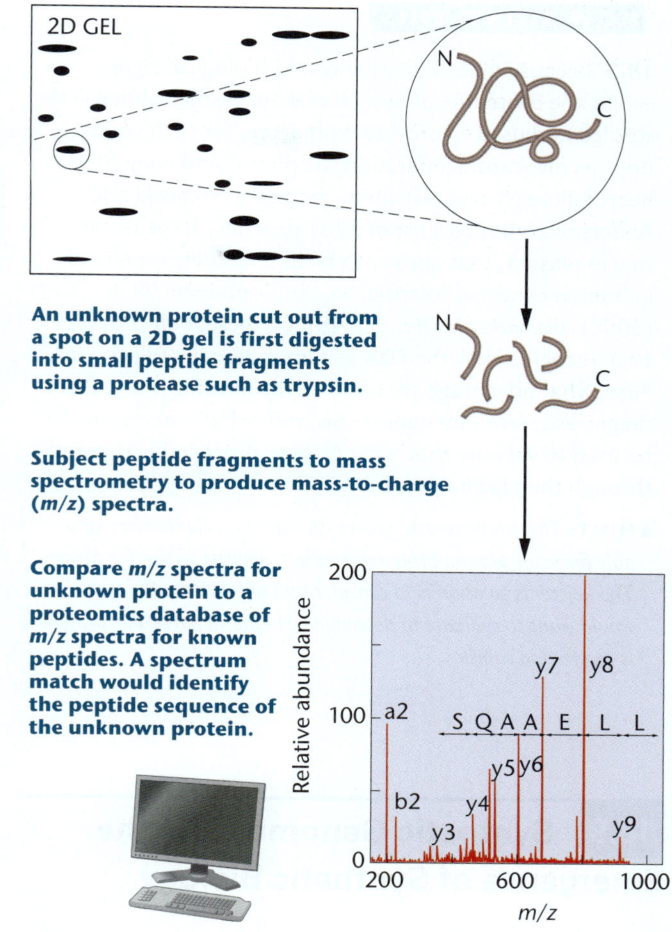

2D GEL
N
C
An unknown protein cut out from a spot on a 2D gel is first digested into small peptide fragments using a protease such as trypsin.
Subject peptide fragments to mass spectrometry to produce mass-to-charge (m/z) spectra.
Compare m/z spectra for unknown protein to a proteomics database of m/z spectra for known peptides. A spectrum match would identify the peptide sequence of the unknown protein.
Relative abundance
200
100
0
a2
b2
y3
y4
y5
y6
y7
y8
y9
S Q A A E L L
200
600
1000
m/z

FIGURE 18.12 Mass spectrometry for identifying an unknown protein isolated from a 2D gel. The peptide in this example was revealed to have the amino acid sequence serine (S)-glutamine (Q)-alanine (A)-alanine (A)-glutamic acid (E)-leucine (L)-leucine (L), shown in single-letter amino acid code.

biochemical methods can be used together with MS, and new MS techniques do not involve running gels. Here we simply provide an introduction to MS.

One common MS approach is *matrix-assisted laser desorption ionization (MALDI)*. This approach is ideally suited for identifying proteins and is widely used for proteomic analysis of tissue samples. In brief, MALDI employs an ultraviolet laser to heat, vaporize, and ionize peptide fragments. Released ions are then analyzed for mass; MALDI displays the *m/z* ratio of each ionized peptide as a series of peaks representative of the molecular masses of peptides in the mixture and their relative abundance (**Figure 18.12**). Because different proteins produce different sets of peptide fragments, MALDI produces a peptide "fingerprint" that is characteristic of the protein being analyzed.

For many MS approaches including MALDI, proteins are first extracted from cells or tissues of interest and usually separated by 2DGE. Protein spots are cut out of the gel, and proteins are purified out of each gel spot. Computer-automated high-throughput instruments are available that can pick all the spots out of a 2D gel, after which MALDI is used to identify the proteins in the different spots. Just about any source that provides a sufficient number of cells can be used: blood, whole tissues and organs, tumor samples, microbes, and many other substances. Many proteins involved in cancer have been identified by the use of MALDI to compare protein profiles in normal tissue and tumor samples.

The use of MALDI-generated *m/z* spectra databases to identify unknown samples is limited by database quality. An unknown protein from a 2D gel can be identified by MALDI only if proteomics databases have a MALDI spectrum for that protein. But as is occurring with genomics databases, proteomics databases with thousands of well-characterized proteins from different organisms are rapidly developing. The human proteome database described earlier in this section was developed from MS data.

ESSENTIAL POINT

Proteomics methods such as mass spectrometry are valuable for analyzing proteomes—the protein content of a cell. ■

NOW SOLVE THIS

18.3 Because of its accessibility and biological significance, the proteome of human plasma has been intensively studied and used to provide biomarkers for such conditions as myocardial infarction (troponin) and congestive heart failure (B-type natriuretic peptide). Polanski and Anderson compiled a list of 1261 proteins, some occurring in plasma, that appear to be differentially expressed in human cancers [Polanski, M., and Anderson, N. L. (2006). *Biomarker Insights* 2:1–48]. Of these, only 9 have been recognized by the FDA as tumor-associated proteins. First, what advantage should there be in using plasma as a diagnostic screen for cancer? Second, what criteria should be used to validate that a cancerous state can be assessed through the plasma proteome?

■ **HINT:** *This problem asks you to consider criteria that are valuable for using plasma proteomics as a diagnostic screen for cancer. The key to its solution is to consider proteomics data that you would want to evaluate to determine whether a particular protein is involved in cancer.*

18.9 Synthetic Genomes and the Emergence of Synthetic Biology

Many years ago, studying genomes led to a fundamental question: "What is the minimum number of genes necessary to support life?" Today, scientists are even more interested in this question because the answer is expected to set the groundwork to create artificial cells or designer organisms based on genes encoded by a **synthetic** or **artificial genome** constructed in the laboratory. To do so we need a better understanding of the minimum number of genes, often referred to as the "core genes," required to support life.

To help advance synthetic genome work, we can use the small genomes of obligate parasites. For example, the bacterium *Mycoplasma genitalium*, a human parasitic pathogen, is among the simplest self-replicating bacteria known and has served as a model for understanding the minimal elements of a genome necessary for a self-replicating cell. *M. genitalium* has a genome of 580 kb, has 525 genes, and is one of the smallest bacterial genomes known.

In contrast, the 1.8-Mb genome of *Haemophilus influenzae* (the first bacterial genome sequenced) has 1815 genes. By comparing the nucleotide sequences of the *M. genitalium* genes with the *H. influenzae* genes, scientists from the J. Craig Venter Institute (JCVI) identified 256 genes whose sequence was similar enough to consider that they arose from a common ancestral gene; that is, they are orthologous. Thus, comparative genomics estimated that at least 256 genes might represent the minimum gene set essential for life. But, could the number be more or less than 256? To answer this question, Venter and colleagues applied an experimental approach. They used transposon-based methods to selectively mutate each gene in *M. genitalium* using the following rationale. Mutations in genes that produce a lethal phenotype indicate that the genes are essential, but nonessential mutated genes would not be lethal. They found that of the 525 genes in *M. genitalium*, about 375 genes were essential, thus constituting the minimum gene set in this bacterium.

In 2010, the JCVI published the first report of a functional synthetic genome. In this approach they designed and chemically synthesized more than one thousand 1080-bp segments called cassettes covering the entire 1.08-Mb *Mycoplasma mycoides* genome (**Figure 18.13**).

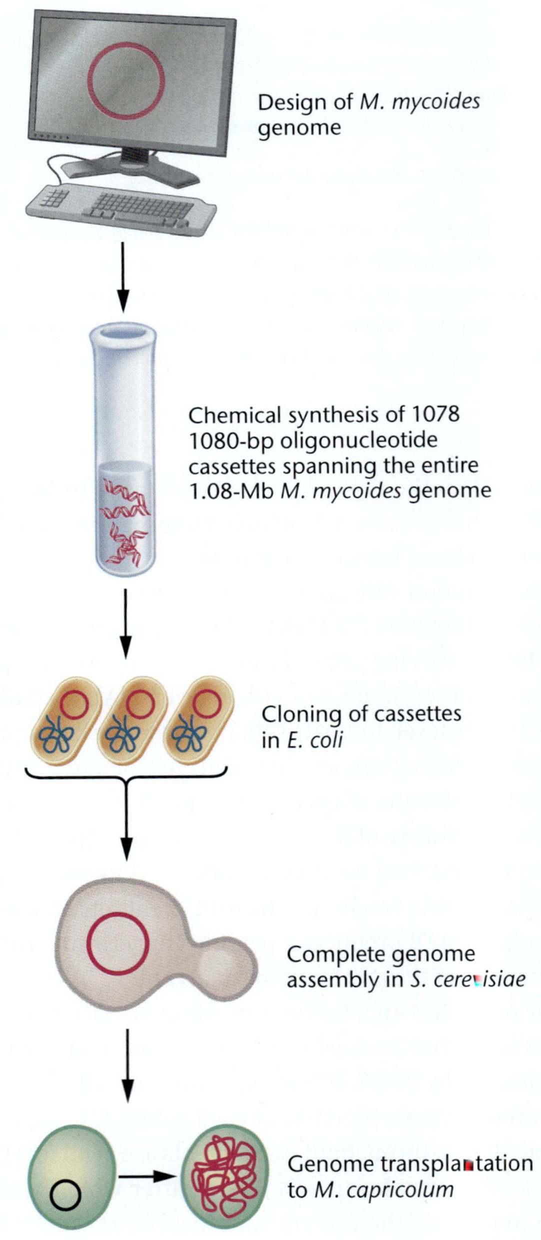
Design of M. mycoides genome
Chemical synthesis of 1078 1080-bp oligonucleotide cassettes spanning the entire 1.08-Mb M. mycoides genome
Cloning of cassettes in E. coli
Complete genome assembly in S. cerevisiae
Genome transplantation to M. capricolum

FIGURE 18.13 Building a synthetic version of the 1.08-Mb *Mycoplasma mycoides* genome.

A homologous recombination technique was used to assemble the cassettes into 11 separate 100-kb assemblies that were eventually combined to completely span the entire 1.08-Mb *M. mycoides* genome.

The entire assembled genome, called JCVI-syn1.0, was then transplanted into a close relative, *M. capricolum*, as recipient cells. *Genome transplantation* is effectively the true test of the functionality of a synthetic genome and an essential outcome is the complete phenotypic transformation of the recipient cells by the synthetic genome. Transplantation resulted in cells with the JCVI-syn1.0 genotype and the phenotype of a new strain of *M. mycoides*. Transformation of *M. capricolum* into JCVI-syn1.0 *M. mycoides* was verified, in part, because these cells were shown to express the *lacZ* gene, which was only present in the synthetic genome. Selection for tetracycline resistance and a determination that recipient cells also made proteins characteristic of *M. mycoides*, and not *M. capricolum*,were also used to verify strain conversion.

The synthetic genome effectively rebooted the *M. capricolum* recipient cells to change them from one form to another. When this work was announced, J. Craig Venter claimed, "This is equivalent to changing a Macintosh computer into a PC by inserting a new piece of software." This was tedious work, spanning over 15 years and ninety-nine percent of the experiments involved failed! Keep in mind also, that these experiments did not create life from an inanimate object since they were based on converting one living strain into another. But clearly these studies provided key "proof of concept" that synthetic genomes could be produced, assembled, and successfully transplanted to create a microbial strain encoded by a synthetic genome. Thus this research brought scientists closer to producing novel synthetic genomes incorporating genes for specific traits of interest. Yet this work still did not define the minimal genome and answer the question of how simple the genome can be.

In 2016, JCVI announced that 473 genes is the minimal bacterial genome. To make this determination, they created a synthetic version of the *M. mycoides* genome that was about half the size of JCVI-syn1.0 discussed earlier. This synthetic 531-kb genome, called JCVI-syn3.0, contained 473 genes, encoding 438 proteins and 35 RNAs. About one-third (149 genes) of the genes in JCVI-syn3.0 that are essential for life have no known function. And several of these 149 genes are present in other organisms including humans. Investigating their roles is of significant interest to the JCVI team.

In the future, gene-editing approaches such as CRISPR-Cas are expected to make it easier to alter genomes to address the minimal genome question. Several teams have already applied CRISPR-based functional analysis to bacteria such as *Bacillus subtilis*. Using CRISPR to knock out bacterial genes, geneticists can then screen for phenotypic changes as a way to identify essential genes in bacteria.

The Quest to Create a Synthetic Human Genome

The JCVI synthetic bacterial genome projects inspired researchers to address questions about the minimal genome and the identification of essential genes in more advanced eukaryotes, including yeast and humans. Using a single-celled yeast, *Saccharomyces cerevisiae*, scientists created synthetic versions of six chromosomes, which constituted a little more than one-third of the yeast genome. This work demonstrated that creating eukaryotic synthetic genomes is possible, but geneticists interested in the genes essential for a functional self-replicating human cell did not have the necessary tools to make significant progress experimentally. Thus, when the Human Genome Project was completed, creating a complete synthetic human genome was considered technically impossible.

In 2016 a group of scientists proposed an initiative, called the *Human Genome Project-Write (HGP-Write)*,to synthesize an entire human genome. Many have questioned the objectives for synthesizing a human genome and the ethical implications of doing so. With an estimated 10-year timeline and a cost of approximately $100 million, organizers hope HGP-Write will lead to technological advances in DNA synthesis while lowering costs, in the same way that the Human Genome Project resulted in advances in DNA sequencing and dramatic cost reductions. But by 2018, the project was scaled back. Rather than synthesizing a 3 billion bp human genome, the project will instead focus on recoding a genome to create cells that are resistant to infection by viruses.

Synthetic Biology for Bioengineering Applications

Venter's work with *M. mycoides* JCVI-syn1.0, a decade-long project that cost about $40 million, was hailed as a defining moment in the emerging field of **synthetic biology,** a discipline that applies engineering design principles to biological systems. What are other potential applications of synthetic genomes and synthetic biology? One of JCVI's goals is to create microorganisms that can be used to synthesize biofuels. Other possibilities include creating synthetic microbes with genomes engineered to (1) express gene products to degrade pollutants (bioremediation); (2) synthesize new biopharmaceutical products; (3) synthesize chemicals and fuels from sunlight and carbon dioxide; and (4) produce "semisynthetic" crops that contain synthetic chromosomes encoding genes for beneficial traits such as drought resistance or improved photosynthetic efficiency.

ESSENTIAL POINT

Synthetic genomes and synthetic biology offer the potential for geneticists to create engineered cells with novel characteristics that may have commercial value. ■

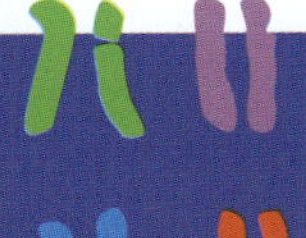

GENETICS, ETHICS, AND SOCIETY

Privacy and Anonymity in the Era of Genomic Big Data

Our lives are surrounded by Big Data. Enormous quantities of personal information are stored on private and public databases, revealing our purchasing preferences, search engine histories, social contacts, and even GPS locations.

Perhaps the most personal of all Big Data entries are genome sequences. Tens of thousands of individuals are now donating DNA for whole-genome sequencing—by both private gene-sequencing companies and public research projects. Most people who donate their DNA for sequence analysis do so with little concern. After all, what consequences could possibly come from access to gigabytes of As, Cs, Ts, and Gs? Surprisingly, the answer is—quite a lot.

One of the first inklings of genetic privacy problems arose in 2005, when a 15-year-old boy named Ryan Kramer tracked down his anonymous sperm-donor father using his own Y chromosome sequence data and the Internet. Ryan submitted a DNA sample to a genealogy company that generates Y chromosome profiles and puts people into contact with others who share similar genetic profiles, indicating relatedness. The search matched Ryan with two men with the same last name. Ryan combined the name information with the only information that he had about his sperm-donor father—date of birth, birth place, and college degree. Using an Internet search, he obtained the names of everyone born on that date in that place. One man with the same last name as his two Y chromosome matches also had the appropriate college degree. Ryan then contacted his sperm-donor father.

More recently, several published reports reveal the ease with which anyone's identity can be traced using DNA-sequence profiles and Internet searches. These searches can reveal people's identities and disease susceptibilities.

To many people, the implications of "genomic re-identification" are disturbing. Genomic information leaks could reveal personal medical information, physical appearance, and racial origins. They could also be used to synthesize DNA to plant at a crime scene or could be used in unforeseen ways in the future as we gain more information about what resides in our genome.

Your Turn

Take time to consider the following questions concerning the ethical challenges of ensuring genetic privacy.

1. What are some of the ethical arguments for and against maintaining genetic privacy and anonymity?

   For more information, see Hansson, M. G., et al. (2016). The risk of re-identification versus the need to identify individuals in rare disease research. *Eur. J. Hum. Genet.* 24:1553–1558.

2. Would you send a DNA sample to a private company for whole-genome sequencing? If not, why not? If so, what privacy assurances would you need before ordering your genome sequence?

   This topic is discussed in Niemiec, E., and Howard, H. C. (2016). Ethical issues in consumer genome sequencing: Use of consumers' samples and data. *Appl. Transl. Genom.* 8:23–30.

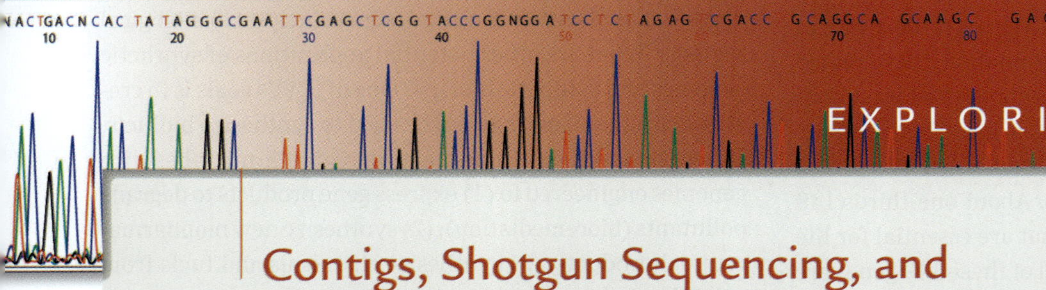

EXPLORING GENOMICS

Contigs, Shotgun Sequencing, and Comparative Genomics

Mastering Genetics Visit the Study Area: Exploring Genomics

In this chapter, we discussed how WGS can be used to assemble chromosome maps. Recall that in this technique chromosomal DNA is digested with different restriction enzymes (or mechanically sheared) to create a series of overlapping DNA fragments called contiguous sequences, or "contigs." The contigs are then subjected to DNA sequencing, after which bioinformatics-based programs are used to arrange the contigs in their correct order on the basis of short overlapping sequences of nucleotides.

In this Exploring Genomics exercise you will carry out a simulation of contig alignment to help you understand the underlying logic of this approach to creating sequence maps of a chromosome. For this purpose, you will use the **National Center for Biotechnology Information BLAST** site and apply a DNA alignment program.

- **Exercise I – Arranging Contigs to Create a Chromosome Map**

1. Access BLAST from the NCBI Web site at https://blast.ncbi.nlm.nih.gov/Blast.cgi. Locate and select the "Global Align" category under "Specialized searches." This feature allows you to compare two DNA sequences at a time to check for sequence similarity alignments.
2. Go to the Study Area, and open the Exploring Genomics exercise for this chapter. Listed are eight contig sequences, called Sequences A through H, taken from an actual human chromosome sequence deposited in GenBank. For this exercise we have used short fragments; however, in reality, contigs are usually several thousand base pairs long. To complete this exercise, copy one sequence into the "Enter Query Sequence" box and one sequence into the "Enter Subject Sequence" box and then run an alignment (by clicking on "Align"). Repeat these steps with other combinations of two sequences to determine which sequences overlap, and then use your findings to create a sequence map that places overlapping contigs in their proper order. Here are a few tips to consider:
   - Develop a strategy to be sure that you analyze alignments for all pairs of contigs.
   - Only consider alignment overlaps that show 100 percent sequence identity.
3. On the basis of your alignment results, answer the following questions, referring to the sequences by their letter codes (A through H):
   a. What is the correct order of overlapping contigs?
   b. What is the length, measured in number of nucleotides, of each sequence overlap between contigs?
   c. What is the total size of the chromosome segment that you assembled?
   d. Did you find any contigs that do not overlap with any of the others? Explain.
4. Run a nucleotide-nucleotide BLAST search with any one of the overlapping contigs to determine which chromosome these contigs were taken from, and report your answer.

CASE STUDY Your microbiome may be a risk factor for disease

A number of genes involved in susceptibility to inflammatory bowel disorders (IBDs), including Crohn disease and ulcerative colitis, have been identified. However, it is clear that other nongenetic risk factors often trigger the onset of these diseases. As noted in Section 18.6, the Human Microbiome Project has provided valuable insights about the roles of the gut microbiome and its impact on intestinal disorders, including IBD. It is known that the gut microbiome of those with IBD is different from those in remission, and it is also different from individuals who do not have IBD. These observations suggest that transfer of gut microbiota from healthy individuals via fecal microbiota transplantation (FMT) might be a successful treatment for IBD. This idea is supported by the use of FMT for a potentially life-threatening form of colitis caused by the bacterium *Clostridium difficile*. After successful clinical trials, the U.S. Federal Drug Administration (FDA) has classified FMT as an investigational new drug. However, until it is formally approved, FMT can only be used to treat *C. difficile* infections that are resistant to antibiotic therapy.

1. Suppose you had Crohn disease or ulcerative colitis and wanted to undertake FMT. Before undergoing the treatment, what genetic analyses might you consider to inform yourself about human genes, microbial genes, or the constitution of your gut microbiota and their correlation to or roles in Crohn disease or ulcerative colitis?
2. The use of FMT, whether in a physician's office or at home, raises a number of ethical issues. What might they be, and which of them would concern you the most?
3. Several Internet sources offer screened donor fecal samples for use in FMT. What risks would you assume in undertaking this therapy at home using these samples? If you are willing to use this therapy on yourself, would you use it on one of your children?

For related reading, see Daloiso, V., et al. (2015). Ethical aspects of fecal microbiota transplantation (FMT). *Eur. Rev. Med. Pharmacol. Sci.* 19(17):3173–3180.

INSIGHTS AND SOLUTIONS

1. One of the main problems in annotation is deciding how long a putative ORF must be before it is accepted as a gene. Shown on p. 543 are three different ORF scans of the same *E. coli* genome region—the region containing the *lacY* gene. Regions shaded in brown indicate ORFs. The top scan was set to accept ORFs of 50 nucleotides as genes. The middle and bottom scans accepted ORFs of 100 and 300 nucleotides as genes, respectively. How many putative genes are detected in each scan? The longest ORF covers 1254 bp; the next longest, 234 bp; and the shortest, 54 bp. How can we decide the actual number of genes in this region? In this type of ORF scan, is it more likely that the number of

(continued)

Insights and Solutions—continued

genes in the genome will be overestimated or underestimated? Why?

Solution: Generally, one can examine conserved sequences in other organisms to indicate that an ORF is likely a coding region. One can also match a sequence to previously described sequences that are known to code for proteins. The problem is not easily solved—that is, deciding which ORF is actually a gene. The shorter the ORFs scan, the more likely the overestimate of genes because ORFs longer than 200 are less likely to occur by chance. For these scans, notice that the 50-bp scans produce the highest number of possible genes, whereas the 300-bp scan produces the lowest number (1) of possible genes.

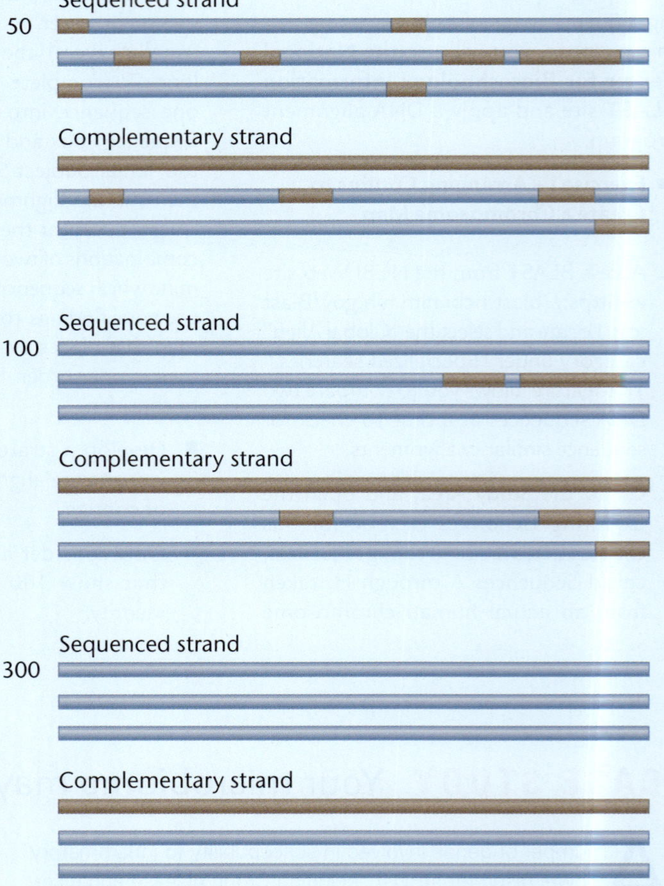

2. Sequencing of the heterochromatic regions (repeat-rich sequences concentrated in centromeres and telomeres) of the *Drosophila* genome indicates that within 20.7 Mb, there are 297 protein-coding genes (Misra et al. (2002). Annotation of the *Drosophila melanogaster* euchromatic genome: a systematic review *Genome Biol.* 3:research0083.1). Given that the euchromatic regions of the genome contain 13,379 protein-coding genes in 116.8 Mb, what general conclusion is apparent?

Solution: Gene density in euchromatic regions of the *Drosophila* genome is about one gene per 8730 base pairs, while gene density in heterochromatic regions is one gene per 70,000 bases (20.7 Mb/297). Clearly, a given region of heterochromatin is much less likely to contain a gene than the same-sized region in euchromatin.

Problems and Discussion Questions

Mastering Genetics Visit for instructor-assigned tutorials and problems.

1. **HOW DO WE KNOW?** In this chapter, we focused on the analysis of genomes, transcriptomes, and proteomes and considered important applications and findings from these endeavors. At the same time, we found many opportunities to consider the methods and reasoning by which much of this information was acquired. From the explanations given in the chapter, what answers would you propose to the following fundamental questions?
   (a) How do we know which contigs are part of the same chromosome?
   (b) How do we know if a genomic DNA sequence contains a protein-coding gene?
   (c) What evidence supports the concept that humans share substantial sequence similarities and gene functional similarities with model organisms?
   (d) How can proteomics identify differences between the number of protein-coding genes predicted for a genome and the number of proteins expressed by a genome?
   (e) How has the concept of a reference genome evolved to encompass a broader understanding of genomic variation in humans?
   (f) How have microarrays demonstrated that, although all cells of an organism have the same genome, some genes are expressed in almost all cells, whereas other genes show cell- and tissue-specific expression?
2. **CONCEPT QUESTION** Review the Chapter Concepts list on p. 371. All of these pertain to how genomics, bioinformatics, and proteomics approaches have changed how scientists study genes and proteins. Write a short essay that explains how recombinant DNA techniques were used to identify and study genes compared to how modern genomic techniques have revolutionized the cloning and analysis of genes. ■
3. What is functional genomics? How does it differ from comparative genomics?
4. Contrast WGS for gene identification to linkage map-based approaches that you learned about in Chapter 7.
5. What is bioinformatics, and why is this discipline essential for studying genomes? Provide two examples of bioinformatics applications.
6. Annotation involves identifying genes and gene-regulatory sequences in a genome. List and describe characteristics of a genome that are hallmarks for identifying genes in an unknown sequence. What characteristics would you look for in a bacterial genome? A eukaryotic genome?
7. Intron frequency varies considerably among eukaryotes. Provide a general comparison of intron frequencies in yeast and humans. What about intron size?
8. What are homologous genes, paralogous genes, and orthologous genes and how would you use BLAST to identify them? Could

you use BLAST to predict protein domains within those different types of genes?

9. Describe three major goals of the Human Genome Project.
10. Describe the human genome in terms of genome size, the percentage of the genome that codes for proteins, how much is composed of repetitive sequences, and how many genes it contains. Describe two other features of the human genome.
11. Recall that when the HGP was completed, more than 40 percent of the genes identified had unknown functions. The PANTHER database provides access to comprehensive and current functional assignments for human genes (and genes from other species). Go to http://www.pantherdb.org/data/. In the frame on the left side of the screen locate the "Quick links" and use the "Whole genome function views" link to a view of a pie chart of current functional classes for human genes. Mouse over the pie chart to answer these questions. What percentage of human genes encode transcription factors? Cytoskeletal proteins? Transmembrane receptor regulatory/adaptor proteins?
12. Sequence variation among individuals of the human species, regardless of their ethnic origin, is less than 1 percent. Despite such low variability, genetic differences can be very useful in forensics. Describe the different sources of genetic variation between individuals.
13. Through the Human Genome Project (HGP), a relatively accurate human genome sequence was published from combined samples from multiple individuals. It serves as a reference for a haploid genome. How do results from personal genome projects (PGP) differ from those of the HGP?
14. Explain differences between whole-genome sequencing (WGS) and whole-exome sequencing (WES), and describe advantages and disadvantages of each approach for identifying disease-causing mutations in a genome. Which approach was used for the Human Genome Project?
15. The term *paralog* is often used in conjunction with discussions of hemoglobin genes. What does this term mean, and how does it apply to hemoglobin genes?
16. Nutrigenomics tests are controversial by-products of the genomic revolution. Explain what they are and why they are polemic.
17. Metagenomics studies generate very large amounts of sequence data. Provide examples of genetic insight that can be learned from metagenomics.
18. What are DNA microarrays? How are they used?
19. Annotations of the human genome have shown that genes are not randomly distributed, but form clusters with gene "deserts" in between. These deserts correspond to the dark bands on G-banded chromosomes. Comparisons between the human transcriptome map and the genome sequence show that highly expressed genes are also clustered together. In terms of genome organization, how is this an advantage?
20. In Section 18.8 we briefly discussed The Human Proteome Map (HPM). An interactive Web site for the HPM is available at http://www.humanproteomemap.org. Visit this site, and then answer the questions in parts (a) and (b) and complete part (c).
    (a) How many proteins were identified in this project?
    (b) How many fetal tissues were analyzed?
    (c) Use the "Query" tab and select the "Gene family" dropdown menu to do a search on the distribution of proteins encoded by a pathway of interest to you. Search in fetal tissues, adult tissues, or both.
21. Systems biology models the complex networks of interacting genes, proteins, and other molecules that contribute to human genetic diseases, such as cancer, diabetes, and hypertension. These interactomes show the contribution of each piece towards the whole and where diseases overlap, and provide models for drug discovery and development. Describe some of the differences that might be seen in the interactomes of normal and cancerous cells taken from the same tissue, and explain how these differences could lead to drugs specifically targeted against cancer cells.
22. Whole-exome sequencing (WES) is helping physicians diagnose a genetic condition that has defied diagnosis by traditional means. The implication here is that exons in the nuclear genome are sequenced in the hopes that, by comparison with the genomes of nonaffected individuals, a diagnosis might be revealed.
    (a) What are the strengths and weaknesses of this approach?
    (b) If you were ordering WES for a patient, would you also include an analysis of the patient's mitochondrial genome?

19

The Genetics of Cancer

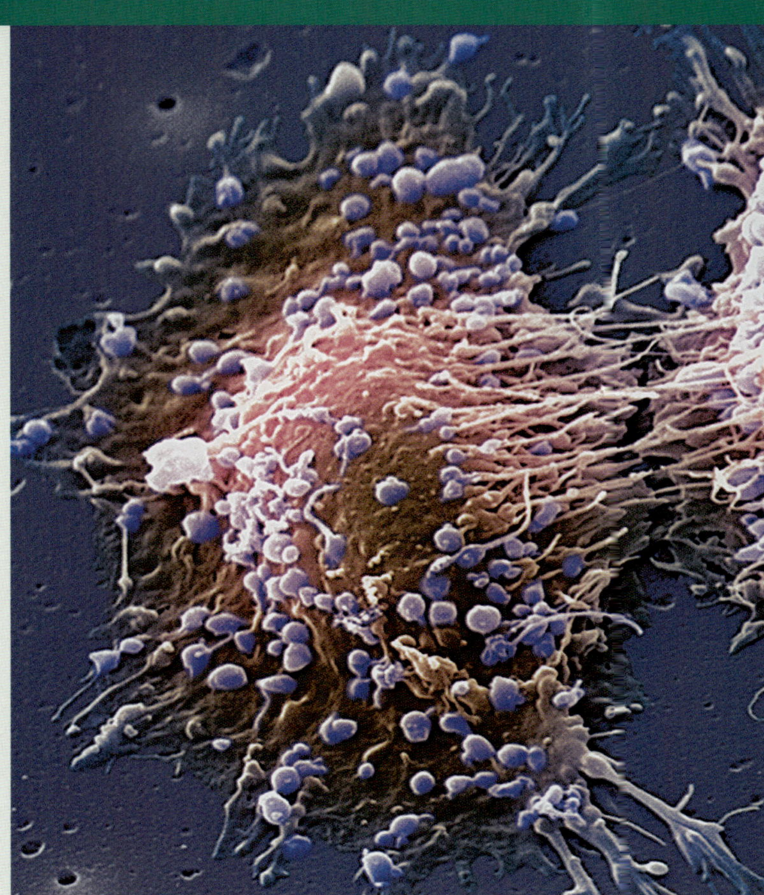

Colored scanning electron micrograph of two prostate cancer cells in the final stages of cell division (cytokinesis). The cells are still joined by strands of cytoplasm.

CHAPTER CONCEPTS

- Cancer is characterized by genetic defects in fundamental aspects of cellular function, including DNA repair, chromatin modification, cell-cycle regulation, apoptosis, and signal transduction.
- Most cancer-causing mutations occur in somatic cells; only about 5 to 10 percent of cancers have a hereditary component.
- Cancer cells in primary and secondary tumors are clonal, arising from one stem cell that accumulated several cancer-causing mutations.
- The development of cancer is a multistep process requiring multiple mutations and clonal expansions.
- Cancer cells show high levels of genomic instability, leading to the accumulation of multiple mutations, some in cancer-related genes.
- Alterations in epigenetic features such as DNA methylation and histone modifications play significant roles in the development of cancers.
- Mutations in proto-oncogenes and tumor-suppressor genes contribute to the development of cancers.
- Genetic changes that lead to cancer can be triggered by numerous natural and human-made carcinogens.

Cancer is the second most common cause of death in Western countries, exceeded only by heart disease. It strikes people of all ages, and one out of three people will experience a cancer diagnosis sometime in his or her lifetime. Each year, more than 14 million cases of cancer are diagnosed worldwide and more than 8 million people will die from the disease.

Over the last 40 years, scientists have discovered that cancer is a genetic disease at the somatic cell level, characterized by the presence of gene products derived from mutated or abnormally expressed genes. The combined effects of numerous abnormal gene products lead to the uncontrolled growth and spread of cancer cells. Although some mutated cancer genes may be inherited, most are created within somatic cells that then divide and form tumors. Completion of the Human Genome Project and numerous large-scale rapid DNA-sequencing studies have opened the door to a wealth of new information about the mutations that trigger a cell to become cancerous. This new understanding of cancer genetics is also leading to new gene-specific treatments, some of which are now in use or are entering clinical trials. Some scientists predict that gene-targeted therapies will replace chemotherapies within the next 25 years.

The goal of this chapter is to highlight our current understanding of the nature and causes of cancer. As we will see, cancer is a genetic disease that arises from the accumulation of mutations in genes controlling many basic aspects of cellular function. We will examine the relationship between genes and cancer and consider how mutations, chromosomal changes, epigenetics, and environmental agents play roles in the development of cancer. Please note that some of the topics discussed in this chapter are explored in greater depth elsewhere in the text (see Chapter 16 and Special Topics Chapter 1—Epigenetics.

19.1 Cancer Is a Genetic Disease at the Level of Somatic Cells

Perhaps the most significant development in our understanding of the causes of cancer is the realization that cancer is a genetic disease. Genomic alterations that are found in cancer cells range from single-nucleotide substitutions to large-scale chromosome rearrangements, amplifications, and deletions (**Figure 19.1**). However, unlike other genetic diseases, cancer is caused by mutations that arise predominantly in somatic cells. Only about 5 to 10 percent of cancers are associated with germ-line mutations that increase a person's susceptibility to certain types of cancer. Another important difference between cancers and other genetic diseases is that cancers rarely arise from a single mutation in a single gene. They arise instead from the accumulation of many mutations in many genes. The mutations that lead to cancer affect multiple cellular functions, including repair of DNA damage, cell division, apoptosis, cellular differentiation, migratory behavior, and cell–cell contact.

(a)

(b)

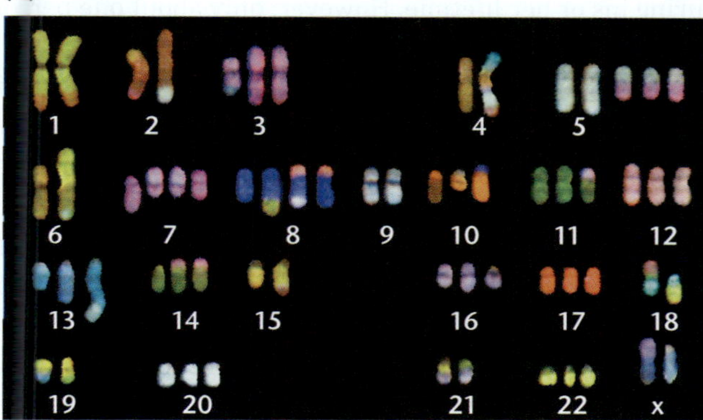

FIGURE 19.1 (a) Spectral karyotype of a normal cell. (b) Karyotype of a cancer cell showing translocations, deletions, and aneuploidy—characteristic features of cancer cells.

What Is Cancer?

Clinically, cancer defines a large number of complex diseases that behave differently depending on the cell types from which they originate and the types of genetic alterations that occur within each cancer type. Cancers vary in their ages of onset, growth rates, invasiveness, prognoses, and responsiveness to treatments. However, at the molecular level, all cancers exhibit common characteristics that unite them as a family.

All cancer cells share two fundamental properties: (1) abnormal cell growth and division (**proliferation**), and (2) defects in the normal restraints that keep cells from spreading and colonizing other parts of the body (**metastasis**). In normal cells, these functions are tightly controlled by genes that are expressed appropriately in time and place. In cancer cells, these genes are either mutated or are expressed inappropriately.

It is this combination of uncontrolled cell proliferation and metastatic spread that makes cancer cells dangerous. When a cell simply loses genetic control over cell growth, it may grow into a multicellular mass, a **benign tumor.** Such a tumor can often be removed by surgery and may cause no serious harm. However, if cells in the tumor also have the ability to break loose, enter the bloodstream, invade other tissues, and form secondary tumors (**metastases**), they become malignant. **Malignant tumors** are often difficult to treat and may become life threatening. As we will see later in the chapter, there are multiple steps and genetic mutations that convert a benign tumor into a dangerous malignant tumor.

ESSENTIAL POINT

Cancer cells show two fundamental properties: abnormal cell proliferation and a propensity to spread and invade other parts of the body (metastasis). ■

The Clonal Origin of Cancer Cells

Although malignant tumors may contain billions of cells, and may invade and grow in numerous parts of the body, all cancer cells in the primary and secondary tumors are clonal, meaning that they originated from a common ancestral cell that accumulated specific cancer-causing mutations. This is an important concept in understanding the molecular causes of cancer and has implications for its diagnosis.

Numerous data support the concept of cancer clonality. For example, reciprocal chromosomal translocations are characteristic of many cancers, including leukemias and lymphomas (two cancers involving white blood cells). Cancer cells from patients with Burkitt lymphoma show reciprocal translocations between chromosome 8 (with translocation breakpoints at or near the *c-myc* gene) and chromosomes 2, 14, or 22 (with translocation breakpoints at or near one of the immunoglobulin genes). Each patient with Burkitt lymphoma exhibits unique breakpoints in his or her *c-myc* and immunoglobulin gene DNA sequences; however, all lymphoma cells within that patient contain identical translocation breakpoints. This demonstrates that cells in a tumor arise from a single cell, and this cell passes on its genetic aberrations to its progeny.

Although all cancer cells within a tumor are clonal, containing the same core set of cancer-causing genes that arose in the ancestral tumor cell, not all cells in a tumor are genetically identical throughout their entire genomes. Next-generation sequencing studies reveal that tumors are composed of subpopulations, or subclones, each of which contains its own sets of distinctive mutations. We will discuss the origins and implications of cancer subclones later in the chapter.

Driver Mutations and Passenger Mutations

Scientists are now applying some of the recent advances in DNA sequencing to identify all the somatic mutations within tumors. These studies compare the DNA sequences of genomes from cancer cells and normal cells derived from the same patient. Data from these studies are revealing that tens of thousands of somatic mutations can be present in cancer cells. Researchers believe that only a handful of mutations in each tumor—called **driver mutations**—give a growth advantage to a tumor cell. The remainder of the mutations may be acquired over time, perhaps as a result of the high levels of DNA damage that occurs in cancer cells, but these mutations have no direct contribution to the cancer phenotype. These are known as **passenger mutations.** The total number of driver mutations that occur in any particular cancer is small—between 2 and 8.

It is now possible to sequence the genomes of individual tumor cells. These studies confirm that there is a great deal of genetic variation between individual cells and subclones within tumors. Most of these variations are due to the accumulation of different types of passenger mutations, with the few key driver mutations remaining constant between subclones. Although most of these passenger mutations may not initially confer a selective advantage on the cells that contain them, if environmental conditions change—such as during chemotherapy or radiotherapy—a passenger mutation that confers a new phenotype such as drug resistance will be selected for, leading to clonal expansion of that cell and the appearance of a new subclone within the tumor.

The Cancer Stem Cell Hypothesis

A concept that is related to the clonal origin of cancer cells is that of the cancer stem cell. Many scientists now believe that most of the cells within tumors do not proliferate. Those that do proliferate and give rise to all the cells within the tumor or within a tumor subclone are known as **cancer stem cells.** Stem cells are undifferentiated cells that have the capacity for self-renewal—a process in which the stem cell divides unevenly, creating one daughter cell that goes on to differentiate into a mature cell type and one that remains a stem cell. The cancer stem cell hypothesis is in contrast to the random or stochastic model that predicts that every cell within a tumor has the potential to form a new tumor.

Cancer stem cells have been identified in leukemias as well as in solid tumors of the brain, breast, colon, ovary, pancreas, and prostate.

It is possible that cancer stem cells may arise from normal adult stem cells within a tissue, or they may be created from more differentiated somatic cells that acquire properties similar to stem cells after accumulating numerous mutations and changes to chromatin structure.

Cancer as a Multistep Process, Requiring Multiple Mutations and Clonal Expansions

Although we know that cancer is a genetic disease initiated by driver mutations that lead to uncontrolled cell proliferation and metastasis, a single mutation is not sufficient to transform a normal cell into a tumor-forming (tumorigenic) malignant cell. If it were sufficient, then cancer would be far more prevalent than it is. In humans, mutations occur spontaneously at a rate of about 10^{-6} mutations per gene, per cell division, mainly due to the intrinsic error rates of DNA replication. Because there are approximately 10^{16} cell divisions in a human body during a lifetime, a person might suffer up to 10^{10} mutations per gene somewhere in the body, during his or her lifetime. However, only about one person in three will suffer from cancer.

The phenomenon of age-related cancer is another indication that cancer develops from the accumulation of several mutagenic events in a single cell. The incidence of most cancers rises exponentially with age. If a single mutation were sufficient to convert a normal cell to a malignant one, then cancer incidence would appear to be independent of age. Another indication that cancer is a multistep process is the delay that occurs between exposure to **carcinogens** (cancer-causing agents) and the appearance of the cancer. For example, there was an incubation period of 5 to 8 years between the time people were exposed to radiation from the atomic explosions at Hiroshima and Nagasaki and the onset of leukemias.

Each step in **tumorigenesis** (the development of a malignant tumor) appears to be the result of two or more genetic alterations that release a cancer stem cell from the controls that normally operate on proliferation and malignancy. Each step confers a selective advantage to the growth and survival of the cell and is propagated through successive **clonal expansions** leading to a fully malignant tumor.

The stepwise clonal evolution of tumors is illustrated by the development of colorectal cancer. Colorectal cancers are known to proceed through several clinical stages that are characterized by the stepwise accumulation of genetic defects in several genes (**Figure 19.2**). The first step is the conversion of a normal epithelial cell into a small cluster of cells known as an adenoma or polyp. This step requires inactivating mutations in the *adenomatous polyposis coli* (*APC*) gene, a gene that encodes a protein involved in the normal differentiation of intestinal cells. The *APC* gene is a tumor-suppressor gene, which will be discussed later in the chapter. The resulting adenoma grows slowly and is considered benign.

The second step in the development of colorectal cancer is the acquisition of a second genetic alteration in one of the cells within the small adenoma. This is usually a mutation in the *KRAS* gene, a gene whose product is normally involved with regulating cell growth. The mutations in *KRAS* that contribute to colorectal cancer cause the *KRAS* protein to become constitutively active, resulting in unregulated cell division. The cell containing the *APC* and *KRAS* mutations grows by clonal expansion to form a larger intermediate adenoma of approximately 1 cm in diameter. The cells of the original small adenoma (containing the *APC* mutation) are now vastly outnumbered by cells containing the two mutations.

The third step, which transforms a large adenoma into a malignant tumor (carcinoma), requires several more waves of clonal expansions triggered by the acquisition of defects in several genes, including *TP53*, *PI3K*, and *TGF-β*. The products of these genes control several important aspects of normal cell growth and division, such as apoptosis, growth signaling, and cell-cycle regulation—all of which we will discuss in more detail later in the chapter. The resulting carcinoma is able to further grow and invade the underlying tissues of the colon. A few cells within the carcinoma may break free of the tumor, migrate to other parts of the body, and form metastases.

ESSENTIAL POINT

The development of cancer is a multistep process, requiring mutations in several cancer-related genes. ■

19.2 Cancer Cells Contain Genetic Defects Affecting Genomic Stability, DNA Repair, and Chromatin Modifications

Cancer cells contain large numbers of mutations and chromosomal abnormalities. Many researchers believe that the fundamental defect in cancer cells is a derangement of the cells' normal ability to repair DNA damage. The resulting loss of genomic integrity leads to a general increase in the mutation rate for every gene in the genome, including cancer-causing driver mutations. The high level of genomic instability seen in cancer cells is known as the **mutator phenotype.** In addition, recent research has revealed that cancer cells contain aberrations in the types and locations of chromatin modifications, particularly DNA and histone methylation patterns.

Genomic Instability and Defective DNA Repair

Genomic instability in cancer cells is characterized by the presence of somatic point mutations and chromosomal

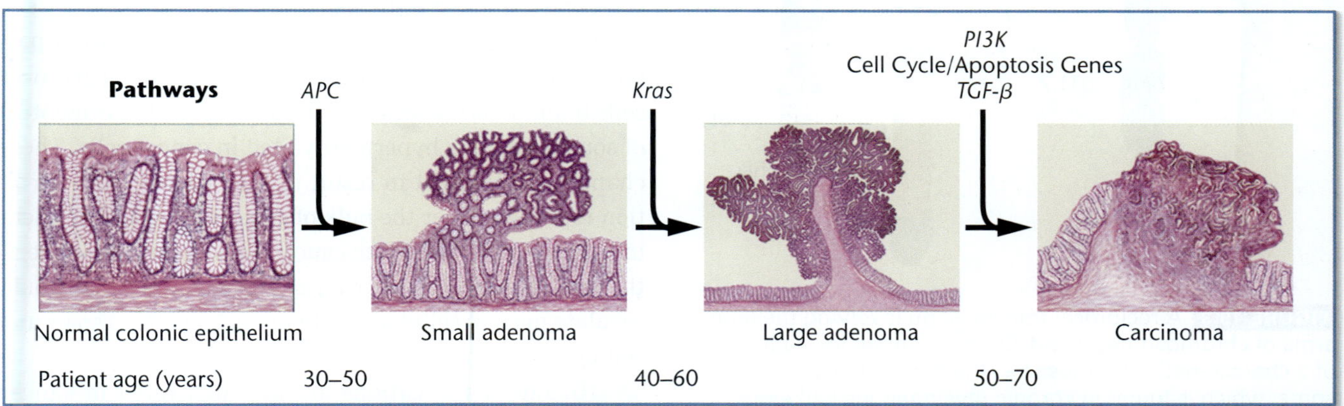

FIGURE 19.2 Steps in the development of colorectal cancers. Some of the genes that acquire driver mutations and cause the progressive development of colorectal cancer are shown above the photographs. These driver mutations accumulate over time and can take 40 years or more to result in the formation of a malignant tumor.

effects such as translocations, aneuploidy, chromosome loss, DNA amplification, and deletions (Figure 19.1). Cancer cells that are grown in cultures in the lab also show a great deal of genomic instability—duplicating, losing, and translocating chromosomes or parts of chromosomes. Often cancer cells show specific chromosomal defects that are used to diagnose the type and stage of the cancer. For example, leukemic white blood cells from patients with chronic myelogenous leukemia (CML) contain a specific translocation, in which the *C-ABL* gene on chromosome 9 is translocated into the *BCR* gene on chromosome 22. This translocation creates a structure known as the **Philadelphia chromosome** (**Figure 19.3**). The *BCR-ABL* fusion gene codes for a chimeric BCR-ABL protein. The normal ABL protein is a protein kinase that acts within signal transduction pathways, transferring growth factor signals from the external environment to the nucleus. The BCR-ABL protein is an abnormal signal transduction molecule in CML cells, which stimulates these cells to proliferate even in the absence of external growth signals.

A number of inherited cancers are caused by defects in genes that control DNA repair. For example, xeroderma pigmentosum (XP) is a rare hereditary disorder that is characterized by extreme sensitivity to ultraviolet (UV) light and other carcinogens. Patients with XP often develop skin cancer. Cells from patients with XP are defective in nucleotide excision repair, with mutations appearing in any one of seven genes whose products are necessary to carry out DNA repair. XP cells are impaired in their ability to repair DNA lesions such as thymine dimers induced by UV light.

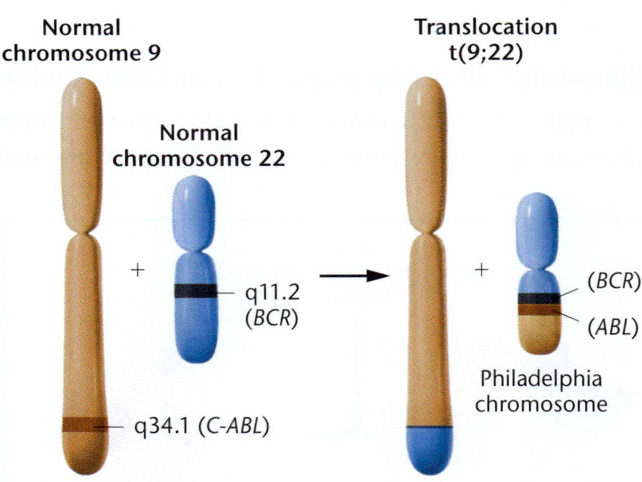

FIGURE 19.3 A reciprocal translocation involving the long arms of chromosomes 9 and 22 results in the formation of a characteristic chromosome, the Philadelphia chromosome, which is found in chronic myelogenous leukemia (CML) cells. The t(9;22) translocation results in the fusion of the *C-ABL* proto-oncogene on chromosome 9 with the *BCR* gene on chromosome 22. The fusion protein is a powerful hybrid molecule that allows cells to escape control of the cell cycle, contributing to the development of CML.

The relationship between XP and genes controlling nucleotide excision repair is also described earlier in the text (see Chapter 14).

Another example is hereditary nonpolyposis colorectal cancer (HNPCC), which is caused by mutations in genes controlling DNA repair. HNPCC is an autosomal dominant syndrome, affecting about 1 in every 200 to 1000 people. Patients affected by HNPCC have an increased risk of developing colon, ovary, uterine, and kidney cancers. Cells from patients with HNPCC show higher than normal mutation rates and genomic instability. At least eight genes are associated with HNPCC, and four of these genes control aspects of DNA mismatch repair. Inactivation of any of these four genes—*MSH2, MSH6, MLH1*, and *MLH3*—causes a rapid accumulation of genome-wide mutations and the subsequent development of cancers.

The observation that hereditary defects in genes controlling nucleotide excision repair and DNA mismatch repair lead to high rates of cancer lends support to the idea that the mutator phenotype is a significant contributor to the development of cancer.

Chromatin Modifications and Cancer Epigenetics

The field of cancer epigenetics is providing new perspectives on the genetics of cancer. **Epigenetics** is the study of chromosome-associated changes that affect gene expression but do not alter the nucleotide sequence of DNA. Epigenetic effects can be inherited from one cell to its progeny cells and may be present in either somatic or germ-line cells. DNA methylation and histone modifications such as acetylation and phosphorylation are examples of epigenetic modifications. The genomic patterns and locations of these modifications can affect gene expression. The effects of chromatin modifications and epigenetic factors on gene expression and cancer are discussed in more detail later in the text (see Special Topics Chapter 1—Epigenetics).

Cancer cells contain altered DNA methylation patterns. Overall, there is much less DNA methylation in cancer cells than in normal cells. At the same time, the promoters of some genes are hypermethylated in cancer cells. These changes are thought to result in the release of transcription repression over the bulk of genes that would be silent in normal cells—including cancer-causing genes—while at the same time repressing transcription of genes that would regulate normal cellular functions such as DNA repair and cell-cycle control.

Histone modifications are also disrupted in cancer cells. Genes that encode histone acetylases, deacetylases, methyltransferases, and demethylases are often mutated or aberrantly expressed in cancer cells. The large numbers of epigenetic abnormalities in tumors have prompted some scientists to speculate that there may be more epigenetic defects in cancer cells than there are gene mutations. In addition, because epigenetic modifications are reversible, it may be possible to treat cancers using epigenetic-based therapies.

ESSENTIAL POINT

Cancer cells show high rates of mutation, chromosomal abnormalities, genomic instability, and abnormal patterns of chromatin modifications. ■

NOW SOLVE THIS

19.1 In chronic myelogenous leukemia (CML), leukemic blood cells can be distinguished from other cells of the body by the presence of a functional BCR-ABL hybrid protein. Explain how this characteristic provides an opportunity to develop a therapeutic approach to a treatment for CML.

■ **HINT:** *This problem asks you to imagine a therapy that is based on the unique genetic characteristics of CML leukemic cells. The key to its solution is to remember that the BCR-ABL fusion protein is found only in CML white blood cells and that this unusual protein has a specific function thought to directly contribute to the development of CML. To help you answer this problem, you may wish to learn more about the cancer drug Gleevec (see https://www.cancer.gov/about-cancer/treatment/drugs/imatinibmesylate).*

19.3 Cancer Cells Contain Genetic Defects Affecting Cell-Cycle Regulation

One of the fundamental aberrations in all cancer cells is a loss of control over cell proliferation. Cell proliferation is the process of cell growth and division that is essential for all development and tissue repair in multicellular organisms. Although some cells, such as epidermal cells of the skin or blood-forming cells in the bone marrow, continue to grow and divide throughout an organism's lifetime, most cells in adult multicellular organisms remain in a nondividing, quiescent, and differentiated state. **Differentiated cells** are those that are specialized for specific functions, such as photoreceptor cells of the retina or muscle cells of the heart. In contrast, many differentiated cells, such as those in the liver and kidney, are able to grow and divide when stimulated by extracellular signals and growth factors. In this way, multicellular organisms are able to replace dead and damaged tissue. However, the growth and differentiation of cells must be strictly regulated; otherwise, the integrity of organs and tissues would be compromised by the presence of inappropriate types and quantities of cells. Normal regulation over cell proliferation involves a large number of gene products including those that control steps in the cell cycle.

In this section, we will review steps in the cell cycle, some of the genes that control the cell cycle, and how these genes, when mutated, contribute to the development of cancer.

The Cell Cycle and Signal Transduction

The cellular events that occur in sequence from one cell division to the next comprise the **cell cycle** (see Chapter 2).

In early to mid-**G1,** the cell makes a decision either to enter the next cell cycle or to withdraw from the cell cycle into quiescence. Continuously dividing cells do not exit the cell cycle but proceed through the **G1, S, G2,** and **M** phases; however, if the cell receives signals to stop growing, it enters the **G0** phase of the cell cycle. During G0, the cell remains metabolically active but does not grow or divide. Most differentiated cells in multicellular organisms can remain in this G0 phase indefinitely. Some, such as neurons, never reenter the cell cycle. In contrast, cancer cells are unable to enter G0, and instead, they continuously cycle. Their *rate* of proliferation is not necessarily any greater than that of normal proliferating cells; however, they are not able to become quiescent at the appropriate time or place.

Normal cells in G0 can often be stimulated to reenter the cell cycle by the presence of external growth signals such as growth factors and hormones that bind to cell-surface receptors. These receptors then relay the signal from the plasma membrane to the cytoplasm and the nucleus. The process of transmitting growth signals from the external environment to the cell nucleus is known as **signal transduction.** Ultimately, signal transduction stimulates the expression of genes whose products then allow cells to enter the cell cycle and divide. Cancer cells often have defects in signal transduction pathways. Sometimes, abnormal signal transduction molecules send continuous growth signals to the nucleus even in the absence of external growth factors. An example of abnormal signal transduction due to mutations in the *ras* gene is described in Section 19.4. In addition, malignant cells may not respond to external signals from surrounding cells—signals that would normally inhibit cell proliferation within a mature tissue.

Cell-Cycle Control and Checkpoints

In normal cells, progress through the cell cycle is tightly regulated, and each step must be completed before the next step can begin. There are at least three distinct points in the cell cycle at which the cell monitors external signals and internal equilibrium before proceeding to the next stage. These are the *G1/S, G2/M,* and *M checkpoints.* At the G1/S checkpoint, the cell monitors its size and determines whether its DNA has

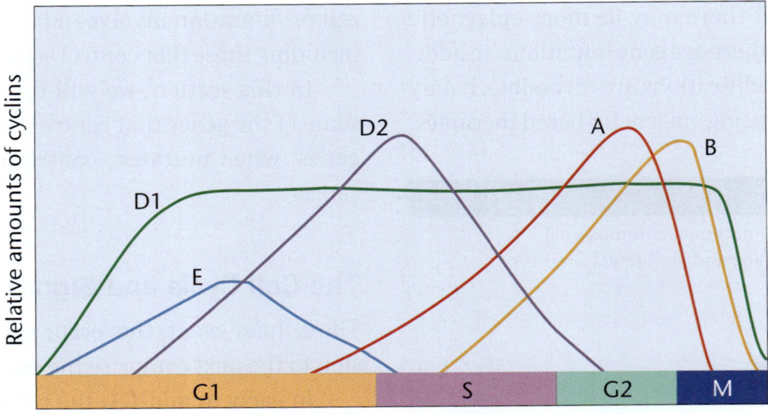

FIGURE 19.4 Relative expression times and amounts of cyclins during the cell cycle. Cyclin D1 accumulates early in G1 and is expressed at a constant level through most of the cycle. Cyclin E accumulates in G1, reaches a peak, and declines by mid–S phase. Cyclin D2 begins accumulating in the last half of G1, reaches a peak just after the beginning of S, and then declines by early G2. Cyclin A appears in late G1, accumulates through S phase, peaks at the G2/M transition, and is rapidly degraded. Cyclin B peaks at the G2/M transition and declines rapidly in M phase.

been damaged. If the cell has not achieved an adequate size, or if the DNA has been damaged, further progress through the cell cycle is halted until these conditions are corrected. If cell size and DNA integrity are normal, the G1/S checkpoint is traversed, and the cell proceeds to S phase. The second important checkpoint is the G2/M checkpoint, where physiological conditions in the cell are monitored prior to mitosis. If DNA replication or repair of any DNA damage has not been completed, the cell cycle arrests until these processes are complete. The third major checkpoint occurs during mitosis and is called the M checkpoint. At this checkpoint, both the successful formation of the spindle-fiber system and the attachment of spindle fibers to the kinetochores associated with the centromeres are monitored. If spindle fibers are not properly formed or attachment is inadequate, mitosis is arrested.

In addition to regulating the cell cycle at checkpoints, the cell controls progress through the cell cycle by means of two classes of proteins: *cyclins* and *cyclin-dependent kinases (CDKs)*. The cell accumulates and destroys cyclin proteins in a precise pattern during the cell cycle (**Figure 19.4**). When a cyclin is present, it binds to a specific CDK, triggering activity of the CDK/cyclin complex. The CDK/cyclin complex then selectively phosphorylates and activates other proteins that in turn bring about the changes necessary to advance the cell through the cell cycle. For example, in G1 phase, CDK4/cyclin D complexes activate proteins that stimulate transcription of genes whose products (such as DNA polymerase δ and DNA ligase) are required for DNA replication during S phase. Another CDK/cyclin complex, CDK1/cyclin B, phosphorylates a number of proteins that bring about the events of early mitosis, such as nuclear membrane breakdown, chromosome condensation, and cytoskeletal reorganization. Mitosis can only be completed, however, when cyclin B is degraded and the protein phosphorylations characteristic of M phase are reversed. Although a large number of different protein kinases exist in cells, only a few are involved in cell-cycle regulation.

Mutation or misexpression of any of the genes whose products control the cell cycle can contribute to the development of cancer. For example, if genes that control the G1/S or G2/M checkpoints are mutated, the cell may continue to grow and divide without repairing DNA damage. As these cells continue to divide, they accumulate mutations in genes whose products control cell proliferation or metastasis. Similarly, if genes that control progress through the cell cycle, such as those that encode the cyclins, are expressed at the wrong time or at incorrect levels, the cell may grow and divide continuously and may be unable to exit the cell cycle into G0. The result in both cases is that the cell loses control over proliferation and is on its way to becoming cancerous.

Control of Apoptosis

As already described, if DNA replication, repair, or chromosome assembly is defective, normal cells halt their progress through the cell cycle until the condition is corrected. This reduces the number of mutations and chromosomal abnormalities that accumulate in normal proliferating cells. However, if DNA or chromosomal damage is so severe that repair is impossible, the cell may initiate a second line of defense—a process called **apoptosis,** or *programmed cell death.* Apoptosis is a genetically controlled process whereby the cell commits suicide. Besides its role in preventing cancer, apoptosis is also initiated during normal multicellular development to eliminate certain cells that do not contribute to the adult organism. The steps in apoptosis are the same for damaged cells and for cells being eliminated during development: Nuclear DNA becomes fragmented, internal cellular structures are disrupted, and the cell dissolves into small spherical structures known as apoptotic bodies. In the final step, the immune system's phagocytic cells engulf the apoptotic bodies. A series of proteases called **caspases** are responsible for initiating apoptosis and for digesting intracellular components.

Apoptosis is genetically controlled by the regulation of the levels and activities of specific gene products that are necessary for apoptosis.

Some of the same genes whose products control cell-cycle checkpoints also initiate apoptotic pathways. These genes are mutated or inactivated in many cancers. As a result, the cell is unable to either repair its DNA or undergo apoptosis. This leads to the accumulation of even more mutations in genes that control growth, division, and metastasis.

ESSENTIAL POINT

Cancer cells have defects in cell-cycle progression, checkpoint controls, and programmed cell death. ■

19.4 Proto-oncogenes and Tumor-suppressor Genes Are Altered in Cancer Cells

Two categories of cancer-related genes are mutated or misexpressed in cancer cells—the proto-oncogenes and the tumor-suppressor genes (**Table 19.1**). **Proto-oncogenes** encode transcription factors that stimulate expression of other genes, signal transduction molecules that stimulate cell division, and cell-cycle regulators that move the cell through the cell cycle. Their products are important for normal cell functions, especially cell growth and division. When normal cells become quiescent and cease division, they repress the expression of most proto-oncogenes or modify the activities of their products. In cancer cells, one or more proto-oncogenes are altered in such a way that the activities of their products cannot be regulated in a normal fashion. This is sometimes due to mutations that result in an abnormal protein product. In other cases, proto-oncogenes may be overexpressed or expressed at an incorrect time due to mutations within gene-regulatory regions such as enhancer elements or due to alterations in chromatin structure that affect gene expression. If a proto-oncogene is continually in an "on" state, its product may constantly stimulate the cell to divide. When a proto-oncogene is mutated or abnormally expressed and contributes to the development of cancer, it is known as an **oncogene**—a cancer-causing gene. Oncogenes are proto-oncogenes that have experienced a gain-of-function alteration. As a result, only one allele of a proto-oncogene needs to be mutated or misexpressed to contribute to cancer. Hence, oncogenes confer a dominant cancer phenotype.

Tumor-suppressor genes are genes whose products normally regulate cell-cycle checkpoints or initiate the process of apoptosis. In normal cells, proteins encoded by tumor-suppressor genes halt progress through the cell cycle in response to DNA damage or growth-suppression signals from the extracellular environment. When tumor-suppressor genes are mutated or inactivated, cells are unable to respond normally at cell-cycle checkpoints or are unable to

TABLE 19.1 Some Proto-oncogenes and Tumor-suppressor Genes

	Normal Function	Alteration in Cancer	Associated Cancers
Proto-oncogene			
c-myc	Transcription factor, regulates cell cycle, differentiation, apoptosis	Translocation, amplification, point mutations	Lymphomas, leukemias, lung cancer, many types
c-kit	Tyrosine kinase, signal transduction	Mutation	Sarcomas
RARα	Hormone-dependent transcription factor, differentiation	Chromosomal translocations with *PML* gene, fusion product	Acute promyelocytic leukemia
Cyclins	Bind to CDKs, regulate cell cycle	Gene amplification, overexpression	Lung, esophagus, many types
Tumor-suppressor			
RB1	Cell-cycle checkpoints, binds E2F	Mutation, deletion, inactivation by viral oncogene products	Retinoblastoma, osteosarcoma, many types
TP53	Transcription regulation	Mutation, deletion, viruses	Many types
BRCA1, BRCA2	DNA repair	Point mutations	Breast, ovarian, prostate cancers

undergo programmed cell death if DNA damage is extensive. This leads to the accumulation of more mutations and the development of cancer. When both alleles of a tumor-suppressor gene are inactivated through mutation or epigenetic modifications, and other changes in the cell keep it growing and dividing, cells may become tumorigenic.

The following are examples of proto-oncogenes and tumor-suppressor genes that contribute to cancer when mutated or abnormally expressed. Genome-wide sequencing studies of cancer cells have identified approximately 200 oncogenes and tumor-suppressor genes, and more will likely be discovered as cancer research continues.

The *ras* Proto-oncogenes

Some of the most frequently mutated genes in human tumors are those in the ***ras* gene family.** These genes are mutated in more than 30 percent of human tumors. The *ras* gene family encodes signal transduction molecules that are associated with the cell membrane and regulate cell growth and division. Ras proteins normally transmit signals from the cell membrane to the nucleus, stimulating the cell to divide in response to external growth factors. Ras proteins alternate between an inactive (switched off) and an active (switched on) state by binding either guanosine diphosphate (GDP) or guanosine triphosphate (GTP). Mutations that convert the *ras* proto-oncogene to an oncogene prevent the Ras protein from hydrolyzing GTP to GDP and hence freeze the Ras protein into its "on" conformation, constantly stimulating the cell to divide.

The *TP53* Tumor-suppressor Gene

The most frequently mutated gene in human cancers—mutated in more than 50 percent of all cancers —is the *TP53* gene. This gene encodes a transcription factor (p53) that represses or stimulates transcription of more than 50 different genes.

Normally, the p53 protein is continuously synthesized but is rapidly degraded and therefore is present in cells at low levels. In addition, the p53 protein is normally bound to another protein called MDM2, which has several effects on p53. The presence of MDM2 on the p53 protein tags p53 for degradation and sequesters the transcriptional activation domain of p53. It also prevents the phosphorylations and acetylations that convert the p53 protein from an inactive to an active form.

Several types of cellular stress events bring about rapid increases in the nuclear levels of activated p53 protein. These include chemical damage to DNA, double-stranded breaks in DNA induced by ionizing radiation, and the presence of DNA-repair intermediates generated by exposure of cells to ultraviolet light. In response to these signals, MDM2 dissociates from p53, making p53 more stable and unmasking its transcription activation domain. Increases in the levels of activated p53 protein also result from increases in protein phosphorylation, acetylation, and other post-translational modifications. Activated p53 protein acts as a transcription factor that stimulates expression of the *MDM2* gene. As the levels of MDM2 increase, p53 protein is again bound by MDM2, returned to an inactive state, and targeted for degradation, in a negative feedback loop.

The activated p53 protein initiates several different responses to DNA damage including cell-cycle arrest followed by DNA repair and apoptosis if DNA cannot be repaired. These responses are accomplished by p53 acting as a transcription factor that stimulates or represses the expression of genes involved in each response.

In normal cells, activated p53 can arrest the cell cycle at the G1/S and G2/M checkpoints, as well as retard the progression of the cell through S phase. To arrest the cell cycle at the G1/S checkpoint, activated p53 protein stimulates transcription of a gene encoding the p21 protein. The p21 protein inhibits the CDK4/cyclin D1 complex, hence preventing the cell from moving from G1 phase into S phase. Activated p53 protein also regulates expression of genes that retard the progress of DNA replication, thus allowing time for DNA damage to be repaired during S phase. By regulating expression of other genes, activated p53 can block cells at the G2/M checkpoint, if DNA damage occurs during S phase.

Activated p53 can also instruct a damaged cell to commit suicide by apoptosis by activating or repressing the transcription of genes whose products control the process. In cancer cells that lack functional p53, apoptosis may not occur. This increases the number of cells that survive with damaged DNA, leading to more mutations in proto-oncogenes and tumor-suppressor genes.

Although the majority of *TP53* mutations inactivate the p53 protein, several mutations confer a gain of function. In these cases, mutant p53 increases the transcription of several genes whose products affect chromatin modifications, leading to genome-wide changes in histone methylation and acetylation and altered gene expression.

In summary, cells lacking functional p53 are unable to arrest at cell-cycle checkpoints or to enter apoptosis in response to DNA damage. As a result, they move unchecked through the cell cycle, regardless of the condition of the cell's DNA. This leads to high mutation rates and accumulation of mutations that lead to cancer. In addition, some mutated p53 proteins alter genome-wide patterns of chromatin modifications. Because of the importance of the *TP53* gene to genomic integrity, it is often referred to as the "guardian of the genome."

NOW SOLVE THIS

19.2 People with a genetic condition known as Li-Fraumeni syndrome inherit one mutant copy of the *TP53* gene. These people have a high risk of developing a number of different cancers, such as breast cancer, leukemia, bone cancer, adrenocortical tumors, and brain tumors. Explain how mutations in one cancer-related gene can give rise to such a diverse range of tumors.

■ **HINT:** *This problem involves an understanding of how tumor-suppressor genes regulate cell growth and behavior. The key to its solution is to consider which cellular functions are regulated by the p53 protein and how the absence of p53 could affect each of these functions. Also, read about loss of heterozygosity in Section 19.6.*

For more practice, see Problems 9–11.

19.5 Cancer Cells Metastasize and Invade Other Tissues

As discussed at the beginning of this chapter, uncontrolled growth alone is insufficient to create a life-threatening cancer. Cancer cells must also become malignant, acquiring the ability to disengage from the original tumor site, to enter the blood or lymphatic system, to invade surrounding tissues, and to develop into secondary tumors. To leave the site of the primary tumor and invade other tissues, tumor cells secrete proteases that digest components of the extracellular matrix and basal lamina which are composed of proteins and carbohydrates. The extracellular matrix surrounds and separates body tissues, forms the scaffold for tissue growth, and inhibits the migration of cells. The ability to invade the extracellular matrix is also a property of some normal cell types. For example, implantation of the embryo in the uterine wall during pregnancy requires cell migration across the extracellular matrix. In addition, white blood cells reach sites of infection by penetrating capillary walls. The mechanisms of invasion are probably similar in these normal cells and in cancer cells.

Metastatic tumors arise from one or several cancer stem cells within one or more subclones of the primary tumor. Once these cells have disengaged from the primary tumor and traversed tissue barriers, they enter the blood or lymphatic system. Only a small percentage of circulating cancer cells—about 0.01 percent—survive to establish metastatic tumors. Once a metastasis is established, its cells continue to mutate and undergo clonal selections and expansions, similarly to those that occurred in the primary tumor.

Metastasis is controlled by a large number of gene products, including cell-adhesion molecules, cytoskeleton regulators, and proteolytic enzymes. For example, epithelial tumors have a lower than normal level of the E-cadherin glycoprotein, which is responsible for cell–cell adhesion in normal tissues. Also, proteolytic enzymes such as metalloproteinases are present at higher than normal levels in many highly malignant tumors. For example, breast cancer cells that metastasize to bone abnormally express the metalloproteinase gene *MMP1*. Those that spread to the lungs overexpress the *MMP1* and *MMP2* genes. It has been shown that the level of aggressiveness of a tumor correlates positively with the levels of proteolytic enzymes expressed by the tumor. In addition, malignant cells are not susceptible to the normal controls conferred by regulatory molecules such as tissue inhibitors of metalloproteinases (TIMPs).

ESSENTIAL POINT

The ability of cancer cells to metastasize requires defects in gene products that control a number of functions such as cell adhesion and tissue invasion. ■

19.6 Predisposition to Some Cancers Can Be Inherited

Although the vast majority of human cancers are sporadic, a small fraction (approximately 5 to 10 percent) have a hereditary or familial component. At present, more than 50 forms of hereditary cancer are known (**Table 19.2**).

Most inherited cancer-susceptibility alleles occur in tumor-suppressor genes and, though transmitted in a Mendelian dominant fashion, are not sufficient in themselves to trigger development of a cancer. Usually, at least one other somatic mutation in the other copy of the gene must occur to contribute to tumorigenesis. In addition, mutations in other genes are usually necessary to fully express the cancer phenotype. For example, inherited mutations in the *RB1* tumor-suppressor gene predispose individuals to developing various cancers including retinoblastoma. Although the normal somatic cells of these patients are heterozygous for the *RB1* mutation, cells within their tumors contain mutations in both copies of the gene or loss of the wild-type gene allele. The phenomenon whereby the second, wild-type, allele is lost is known as **loss of heterozygosity.** This can occur through chromosome deletions or rearrangements. Although mutation or loss of heterozygosity is an essential first step in expression of these inherited cancers, further mutations in other proto-oncogenes, tumor-suppressor genes, or chromatin-modifying

TABLE 19.2 Some Inherited Predispositions to Cancer

Tumor Predisposition Syndrome	Chromosome	Gene Affected
Early-onset familial breast cancer	17q	*BRCA1*
Familial adenomatous polyposis	5q	*APC*
Familial melanoma	9p	*CDKN2*
Gorlin syndrome	9q	*PTCH1*
Hereditary nonpolyposis colon cancer	2p	*MSH2, 6*
Li-Fraumeni syndrome	17p	*TP53*
Multiple endocrine neoplasia, type 1	11q	*MEN1*
Multiple endocrine neoplasia, type 2	10q	*RET*
Neurofibromatosis, type 1	17q	*NF1*
Neurofibromatosis, type 2	22q	*NF2*
Retinoblastoma	13q	*pRb*
Von Hippel–Lindau syndrome	3p	*VHL*
Wilms tumor	11p	*WT1*

genes are necessary for the tumor cells to become fully malignant.

In the study of hereditary cancers, those genes that bear germ-line mutations that are associated with an increased risk of hereditary cancers are sometimes called cancer predisposition genes. In most cases, these genes overlap with known or suspected proto-oncogenes and tumor-suppressor genes that suffer somatic mutations in noninherited cancers.

The development of hereditary colon cancer illustrates how inherited mutations in only one allele of a tumor-suppressor gene can contribute to malignancy. In Section 19.1, we described how colorectal cancers develop through the accumulation of mutations in several genes, leading to a stepwise clonal expansion of cells and the development of carcinomas. Although the vast majority of colorectal cancers are sporadic, about 1 percent of cases result from a genetic predisposition to cancer known as familial adenomatous polyposis (FAP). In FAP, individuals inherit one mutant copy of the *APC* (adenomatous polyposis) gene located on the long arm of chromosome 5. Mutations include deletions, frameshift, and point mutations. The normal function of the *APC* gene product is to act as a tumor suppressor controlling growth and differentiation. The presence of a heterozygous *APC* mutation causes the epithelial cells of the colon to partially escape cell-cycle control, and the cells divide to form small clusters of cells called polyps or adenomas. People who are heterozygous for this condition develop hundreds to thousands of colon and rectal polyps early in life. Although it is not necessary for the second allele of the *APC* gene to be mutated in polyps at this stage, in the majority of cases, the second *APC* allele becomes mutant or lost in a later stage of cancer development. The remaining steps in development of colorectal carcinoma follow the same order as that shown in Figure 19.2.

ESSENTIAL POINT

Inherited mutations in cancer-susceptibility genes are not sufficient to trigger cancer. Other somatic mutations in proto-oncogenes or tumor-suppressor genes are necessary for the development of hereditary cancers. ■

NOW SOLVE THIS

19.3 Although tobacco smoking is responsible for a large number of human cancers, not all smokers develop cancer. Similarly, some people who inherit mutations in the tumor-suppressor genes *TP53* or *RB1* never develop cancer. Explain these observations.

■ **HINT:** *This problem asks you to consider the reasons why only some people develop cancer as a result of environmental factors or mutations in tumor-suppressor genes. The key to its solution is to consider the steps involved in the development of cancer and the number of abnormal functions in cancer cells. Also, consider how genetics may affect DNA repair functions.*

19.7 Environmental Agents Contribute to Human Cancers

Any substance or process that damages DNA has the potential to be carcinogenic. Unrepaired or inaccurately repaired DNA introduces mutations, which, if they occur in proto-oncogenes or tumor-suppressor genes, can lead to abnormal regulation of cell proliferation or disruption of controls over apoptosis or metastasis.

Our environment, both natural and human-made, contains abundant carcinogens. These include chemicals, radiation, some viruses, and chronic infections. In this section, we will examine some of these environmental agents that contribute to the development of cancer.

Natural Environmental Agents

Although most people perceive human-made, industrial chemicals to be the most significant contributors to cancer, they may contribute to less than 10 percent of cancers.

Some of the most mutagenic agents, and hence potentially the most carcinogenic, are natural substances and natural processes. For example, aflatoxin, a component of a mold that grows on peanuts and corn, is one of the most carcinogenic chemicals known. Most chemical carcinogens, such as nitrosamines, are components of synthetic substances and are found in some preserved meats; however, many are naturally occurring. For example, natural pesticides and antibiotics found in plants may be carcinogenic, and the human body itself creates alkylating agents in the acidic environment of the gut.

DNA lesions brought about by natural radiation, metabolism, and DNA replication contribute significantly to the development of cancer. Normal metabolism creates oxidative end products that can damage DNA, proteins, and lipids. It is estimated that the human body suffers about 10,000 damaging DNA lesions per day due to the actions of oxygen free radicals. DNA repair enzymes deal successfully with most of this damage; however, some damage may lead to permanent mutations. The process of DNA replication itself is mutagenic. Hence, substances such as growth factors or hormones that stimulate cell division are ultimately mutagenic and perhaps carcinogenic. Chronic inflammation due to infection also stimulates tissue repair and cell division, resulting in DNA lesions accumulating during replication. These mutations may persist, particularly if cell-cycle checkpoints are compromised due to mutations or inactivation of tumor-suppressor genes such as *TP53* or *RB1*.

As we learned in Chapter 14, both ultraviolet (UV) light and ionizing radiation (such as X rays and gamma rays) induce DNA damage. The UV radiation in sunlight is well accepted as an inducer of skin cancers. Ionizing radiation has clearly shown itself to be a carcinogen in studies of populations exposed to neutron and gamma radiation from atomic blasts such as those in Hiroshima and Nagasaki. Another significant environmental component, radon gas, may be responsible for about 50 percent of the ionizing radiation exposure of the U.S. population and could contribute to lung cancers in some populations.

Diet is often implicated in the development of cancer. It is estimated that approximately 20 percent of cancer deaths in the United States are linked to dietary influences in conjunction with obesity and physical inactivity. Consumption of red meat and animal fat is associated with some cancers, such as colon, prostate, and breast cancer. The mechanisms by which these substances may contribute to carcinogenesis may involve stimulation of cell division through hormones or creation of carcinogenic chemicals during cooking, processing, or digestion. Alcohol may cause inflammation of the liver and contribute to liver cancer. It is also linked to other cancers including breast, colon, and esophagus.

Human-Made Chemicals and Pollutants

Although lifestyle factors such as smoking make significant contributions to the development of cancers, exposure to human-made carcinogens in air, food, and water also contribute. For example, researchers estimate that approximately 15 percent of all lung cancer deaths are due to components of air pollution, such as particulate matter. The International Agency for Research on Cancer (IARC) has tested more than 900 natural and artificial chemicals and found that more than 400 of them show some degree of carcinogenic properties in laboratory or epidemiological studies. Despite these test results, only a small fraction of the approximately 80,000 industrial chemicals in use today have been tested for carcinogenicity.

At least two problems make it difficult to estimate the effects of human-made chemicals and pollutants on human cancer. First, even those chemicals that have been found to be carcinogenic in laboratory and animal tests may not have detectable effects in humans, as the dosages are much lower in the environment than in the lab. Second, some carcinogens may not show their effects unless found in low-level mixtures with other toxic materials or when exposures occur in certain susceptible subpopulations such as infants or pregnant women.

Epidemiologic studies in humans contribute some information, but, because humans do not live in controlled environments, it is difficult to isolate the effects of one chemical agent when humans are exposed to hundreds of environmental chemicals, in varying dosages, and have differing genetic backgrounds.

The IARC has been testing carcinogens for several decades and has classified approximately 100 chemical agents (natural and human-made) as "carcinogenic to humans." Other candidate carcinogens can only be classified as probable, possible, or unknown carcinogens.

Tobacco Smoke and Cancer

One of the most thoroughly studied environmental and lifestyle carcinogens is tobacco smoke. Tobacco smoking is associated with at least 17 different types of human cancer, including lung cancers and cancers of the oral cavity, bladder, liver, stomach and kidney. It is estimated that tobacco smoking kills more than six million people each year, worldwide. Seventy percent of lung cancer deaths can be linked to tobacco smoking.

TABLE 19.3 Human Viruses Associated with Cancers

Virus		Associated Cancers
DNA Viruses		
Epstein-Barr virus	EBV	Burkitt lymphoma, nasopharyngeal carcinoma, Hodgkin lymphoma
Hepatitis B virus	HBV	Hepatocellular carcinoma
Hepatitis C virus	HCV	Hepatocellular carcinoma, non-Hodgkin lymphoma
Human papilloma viruses 16, 18	HPV16, 18	Cervical cancer, anogenital cancers, oral cancers
Kaposi sarcoma–associated herpesvirus	KSHV	Kaposi sarcoma, primary effusion lymphoma
Retroviruses		
Human T-cell lymphotropic virus, type 1	HTLV-1	Adult T-cell leukemia and lymphoma
Human immunodeficiency virus, type 1	HIV-1	Immune suppression, leading to cancers caused by other viruses (KSHV, EBV, HPV)

Tobacco smoke contains a mixture of more than 4000 chemicals, and more than 60 of these are carcinogens. Well-known examples of these include benzene, arsenic, benzo[a]pyrene, cadmium, formaldehyde, and styrene. In addition, the particulate matter in smoke is a carcinogen.

Tobacco smoking triggers a large number of somatic mutations and epigenetic changes. Smoking one pack of cigarettes each day can create more than 150 mutations per year in the genomes of lung cells, as well as dozens of mutations in cells of the larynx, mouth, bladder, and liver. These mutations include base substitutions, insertions, deletions, copy-number aberrations in parts of chromosomes, and covalent bonding of reactive chemical adducts to DNA bases. These types of DNA damage can occur anywhere in the genome and may create driver mutations in proto-oncogenes or tumor-suppressor genes.

Lung cancer genomes from smokers also contain changes in DNA methylation patterns. Approximately 0.1 percent of CpG sequences that have been examined are either hypermethylated or hypomethylated. These changes, if present in regulatory regions of proto-oncogenes or tumor-suppressor genes, may contribute to altered gene expression in these cancers.

According to the World Health Organization and the American Cancer Society, smokers who quit tobacco smoking cut their risk of developing lung and other cancers by one-half within 5 years of quitting. They also see a reduced risk of developing coronary heart disease and diabetes.

Viruses and Human Cancers

It is thought that, worldwide, about 12 percent of human cancers are associated with viruses, making virus infection the second greatest risk factor for cancer, next to tobacco smoking. The most significant contributors to virus-induced cancers are listed in **Table 19.3**. Like other risk factors for cancer, including hereditary predisposition to certain cancers, virus infection alone is not sufficient to trigger human cancers. Other factors, including DNA damage or the accumulation of mutations in one or more of a cell's oncogenes and tumor-suppressor genes, are required to move a cell down the multistep pathway to cancer.

NOW SOLVE THIS

19.4 Cancer can arise spontaneously, but it can also be induced as a result of environmental factors such as sun exposure, infections, and tobacco smoking. If you were asked to help allocate resources to cancer research, what emphasis would you place on research to find cancer cures, compared to that placed on education about cancer prevention?

■ **HINT:** *This problem asks you to consider the outcomes of two different approaches to cancer research. The key to its solution is to think about the relative rates of environmentally induced and spontaneous cancers. [An interesting source of information on this topic is Ames, B. N. et al. (1995). The causes and prevention of cancer.* Proc. Natl. Acad. Sci. USA *92:5258–5265.]*

GENETICS, ETHICS, AND SOCIETY

Breast Cancer: The Ambiguities and Ethics of Genetic Testing

Breast cancer is the most common cancer among women. A woman's lifetime risk of developing breast cancer is about 12 percent. Approximately 5 to 10 percent of breast cancers are familial, defined by the early onset of the disease and the appearance of several cases of breast or ovarian cancer among near blood relatives. In 1994, scientists identified two genes that were linked to familial breast cancers: *BRCA1* and *BRCA2*. Women who inherit specific germ-line mutations in these genes have an approximately 70 percent chance of developing breast cancer and an approximately 26 to 40 percent chance of developing ovarian cancer. In men, mutations in these two genes lead to increased risks of both breast and prostate cancers.

BRCA1 and *BRCA2* genetic tests can detect more than 1000 different mutations that occur within the coding regions of these genes. Many patients at risk for familial breast cancer opt to undergo genetic testing. These patients feel that test results could help them take steps to prevent cancers, guide them in reproductive decisions, and provide information to relatives about their risks. But these potential benefits are fraught with uncertainties.

A woman whose *BRCA*-test results are negative may still have an approximately 12 percent risk of developing the disease (the population risk). Also, a negative *BRCA* genetic test does not eliminate the possibility that she carries an inherited mutation in another gene that increases breast cancer risk or that her *BRCA1* or *BRCA2* gene mutations exist in regions of the genes that are inaccessible to current genetic tests.

A woman whose test results are positive faces difficult choices. Her treatment options consist of close monitoring, prophylactic mastectomy or oophorectomy (removal of breasts and ovaries, respectively), or taking prophylactic drugs such as tamoxifen. Prophylactic surgery reduces her risk but does not eliminate it, as cancers can still occur in tissues that remain after surgery. Drugs such as tamoxifen can have serious side effects.

Genetic tests also affect the patient's entire family. People often experience fear, anxiety, and guilt on learning that they are carriers of a genetic disease. Confidentiality is also a major concern. Patients fear that their genetic test results may be leaked to insurance companies or employers, jeopardizing their prospects for jobs or affordable health and life insurance.

The unanswered scientific and ethical questions about *BRCA1* and *BRCA2* genetic testing are many and important. As we develop genetic tests for more and more diseases over the next few decades, our struggle with these questions will continue.

Your Turn

Take time, individually or in groups, to answer the following questions. Investigate the references and links, to help you understand some of the medical and ethical issues that surround genetic testing for breast cancer.

1. In 2018, the U.S. Food and Drug Administration authorized the first direct-to-consumer genetic tests for three mutations in the *BRCA1* and *BRCA2* genes. What are the potential positive and negative implications of this authorization? Would you opt to take these tests, even if you had no history of familial breast cancer?

   For background and links, see Janssens, C. (2018). Opinion: No, FDA didn't really approve 23 and Me's *BRCA* test. *The Scientist* (https://www.the-scientist.com/?articles.view/articleNo/52103/title/Opinion--No--FDA-Didn-t-Really-Approve-23andMe-s-BRCA-Test/).

2. If you tested positive for a mutation in one of the *BRCA* genes, what ethical and medical questions would you consider before informing family members, friends, or coworkers?

   For more information about these aspects of BRCA *gene testing, see* BRCA mutations: Cancer risks and genetic testing. National Cancer Institute (https://www.cancer.gov/about-cancer/causes-prevention/genetics/brca-fact-sheet).

CASE STUDY Cancer-killing bacteria

Ralph, a 57-year-old man, was diagnosed with colon cancer. His oncologist discussed the use of radiation and chemotherapy as treatments, both of which can cause debilitating side effects. Ralph decided to explore other options and went to a cancer clinic. He learned that researchers in a synthetic biology program were testing the use of genetically modified bacterial cells designed to selectively invade specific tumors and kill cancer cells, with no effects on normal cells. Ralph decided to participate and was informed that he would be part of a phase III trial, comparing the effects of the modified bacterial cell treatment against conventional chemotherapy. However, as part of the trial, he would be randomly assigned to receive one or the other treatment. He was disappointed to learn this, because he assumed that he would receive the bacterial therapy.

1. Informed consent is legally and ethically required before someone participates in a clinical trial. After potential participants receive information about the trial and what constitutes

(*continued*)

CASE STUDY Cancer-killing bacteria—continued

informed consent, research indicates that 25 percent of prospective participants do not understand that these trials are designed primarily to establish the efficacy of the treatment rather than directly benefit participants. What should investigators do to make sure that clinical trial participants understand that the trial is not primarily intended to help them?

2. If you were in Ralph's position, would you try radiation and chemotherapy instead, or enroll in the trial on the chance that you might receive the bacterial therapy, which may or may not be more effective than the conventional therapy?
3. If you agree to participate and then learn that you will not be receiving the bacterial treatment, would you be ethically bound to continue in the trial?

See Joffe,S., et al. (2001). Quality of informed consent in cancer clinical trials: A cross-sectional survey. *Lancet* 358(9295): 1772–1777.

INSIGHTS AND SOLUTIONS

1. In retinoblastoma, a mutation in one allele of the *RB1* tumor-suppressor gene can be inherited from the germ line, causing an autosomal dominant predisposition to the development of eye tumors. To develop tumors, a somatic mutation in the second copy of the *RB1* gene is necessary, indicating that the mutation itself acts as a recessive trait. Given that the first mutation can be inherited, in what ways can a second mutational event occur?

   Solution: In considering how this second mutation arises, we must look at several types of mutational events, including changes in nucleotide sequence and events that involve whole chromosomes or chromosome parts. Retinoblastoma results when both copies of the *RB1* locus are lost or inactivated. With this in mind, you must first list the phenomena that can result in a mutational loss or the inactivation of a gene.

   One way the second *RB1* mutation can occur is by a nucleotide alteration that converts the remaining normal *RB1* allele to a mutant form. This alteration can occur through a nucleotide substitution or through a frameshift mutation caused by the insertion or deletion of nucleotides during replication. A second mechanism involves the loss of the chromosome carrying the normal allele. This event would take place during mitosis, resulting in chromosome 13 monosomy and leaving the mutant copy of the gene as the only *RB1* allele. This mechanism does not necessarily involve loss of the entire chromosome; deletion of the long arm (*RB1* is on 13q) or an interstitial deletion involving the *RB1* locus and some surrounding material would have the same result. Alternatively, a chromosome aberration involving loss of the normal copy of the *RB1* gene might be followed by duplication of the chromosome carrying the mutant allele. Two copies of chromosome 13 would be restored to the cell, but the normal *RB1* allele would not be present. Finally, a recombination event followed by chromosome segregation could produce a homozygous combination of mutant *RB1* alleles.

2. Proto-oncogenes can be converted to oncogenes in a number of different ways. In some cases, the proto-oncogene itself becomes amplified up to hundreds of times in a cancer cell. An example is the *cyclin D1* gene, which is amplified in some cancers. In other cases, the proto-oncogene may be mutated in a limited number of specific ways, leading to alterations in the gene product's structure. The *ras* gene is an example of a proto-oncogene that becomes oncogenic after suffering point mutations in specific regions of the gene. Explain why these two proto-oncogenes (*cyclin D1* and *ras*) undergo such different alterations to convert them into oncogenes.

   Solution: The first step in solving this question is to understand the normal functions of these proto-oncogenes and to think about how either amplification or mutation would affect each of these functions.

   The cyclin D1 protein regulates progression of the cell cycle from G1 into S phase, by binding to CDK4 and activating this kinase. The cyclin D1/CDK4 complex phosphorylates a number of proteins including pRB, which in turn activate other proteins in a cascade that results in transcription of genes whose products are necessary for DNA replication in S phase. The simplest way to increase the activity of cyclin D1 would be to increase the number of cyclin D1 molecules available for binding to the cell's endogenous CDK4 molecules. This can be accomplished by several mechanisms, including amplification of the *cyclin D1* gene. In contrast, a point mutation in the *cyclin D1* gene would most likely interfere with the ability of the cyclin D1 protein to bind to CDK4; hence, mutations within the gene would probably repress cell-cycle progression rather than stimulate it.

   The *ras* gene product is a signal transduction protein that operates as an on/off switch in response to external stimulation by growth factors. It does so by binding either GTP (the "on" state) or GDP (the "off" state). Oncogenic mutations in the *ras* gene occur in specific regions that alter the ability of the Ras protein to exchange GDP for GTP. Oncogenic Ras proteins are locked in the "on" conformation, bound to GTP. In this way, they constantly stimulate the cell to divide. An amplification of the *ras* gene would simply provide more molecules of normal Ras protein, which would still be capable of on/off regulation. Hence, simple amplification of *ras* would less likely be oncogenic.

Problems and Discussion Questions

Mastering Genetics Visit for instructor-assigned tutorials and problems.

1. **HOW DO WE KNOW?** In this chapter, we focused on cancer as a genetic disease, with an emphasis on the relationship between cancer, the cell cycle, and DNA damage, as well as on the multiple steps that lead to cancer. At the same time, we found many opportunities to consider the methods and reasoning by which much of this information was acquired. From the explanations given in the chapter,
   (a) How do we know that malignant tumors arise from a single cell that contains mutations?
   (b) How do we know that cancer development requires more than one mutation?
   (c) How do we know that cancer cells often contain defects in DNA repair?
2. **CONCEPT QUESTION** Review the Chapter Concepts list on p. 400. These concepts relate to the multiple ways in which genetic alterations lead to the development of cancers. The sixth concept states that epigenetic effects including DNA methylation and histone modifications contribute to the genetic alterations leading to cancer. Write a short essay describing how epigenetic changes in cancer cells contribute to the development of cancers. ■
3. Define signal transduction and summarize its different stages.
4. Where are the major regulatory points in the cell cycle?
5. Describe kinases and cyclins. How do they interact to cause cells to move through the cell cycle?
6. What is different about the G0 phase between cancer cells and normal cells?
7. Can cancer be inherited or infectious?
8. How is apoptosis regulated, and what may be the connection between apoptotic pathways and checkpoint responses of the cells?
9. What is a proto-oncogene? Why does a mutation in a single copy of a proto-oncogene behave dominantly and how can it contribute to cancer?
10. Describe the ways in which the p53 tumor-suppressor protein can be activated in normal cells.
11. How does the p53 tumor-suppressor protein control cell-cycle checkpoints?
12. If a cell suffers damage to its DNA while in S phase, how can this damage be repaired before the cell enters mitosis?
13. Describe the *ras* proto-oncogene. How can it become an oncogene?
14. Do proto-oncogenes only have gain-of-function mutations but never loss-of-function ones to become oncogenes? Explain.
15. How do translocations such as the Philadelphia chromosome contribute to cancer?
16. What are the differences between cancer driver mutations and passenger mutations? How can a passenger mutation become a driver mutation?
17. What are the most significant environmental agents that contribute to human cancers?
18. How does the environment contribute to cancer?
19. What is loss of heterozygosity, and how does this process contribute to the development of cancers?
20. Explain how environmental agents such as chemicals and radiation cause cancer.
21. Radiotherapy (treatment with ionizing radiation) is one of the most effective current cancer treatments. It works by damaging DNA and other cellular components. In which ways could radiotherapy control or cure cancer, and why does radiotherapy often have significant side effects?
22. Genetic tests that detect mutations in the *BRCA1* and *BRCA2* oncogenes are widely available. These tests reveal a number of mutations in these genes—mutations that have been linked to familial breast cancer. Assume that a young woman in a suspected breast cancer family takes the *BRCA1* and *BRCA2* genetic tests and receives negative results. That is, she does not test positive for the mutant alleles of *BRCA1* or *BRCA2*. Can she consider herself free of risk for breast cancer?
23. What is the cancer stem cell hypothesis?
24. Explain the differences between a benign and malignant tumor.
25. As part of a cancer research project, you have discovered a gene that is mutated in many metastatic tumors. After determining the DNA sequence of this gene, you compare the sequence with those of other genes in the human genome sequence database. Your gene appears to code for an amino acid sequence that resembles sequences found in some serine proteases. Conjecture how your new gene might contribute to the development of highly invasive cancers.
26. A study by Bose and colleagues (1998. *Blood* 92: 3362–3367) and a previous study by Biernaux and others [1996. *Bone Marrow Transplant* 17: (Suppl. 3) S45–S47] showed that *BCR-ABL* fusion gene transcripts can be detected in 25 to 30 percent of healthy adults who do not develop chronic myelogenous leukemia (CML). Explain how these individuals can carry a fusion gene that is transcriptionally active and yet do not develop CML.
27. Those who inherit a mutant allele of the *RB1* tumor-suppressor gene are at risk for developing a bone cancer called osteosarcoma. You suspect that in these cases, osteosarcoma requires a mutation in the second *RB1* allele, and you have cultured some osteosarcoma cells and obtained a cDNA clone of a normal human *RB1* gene. A colleague sends you a research paper revealing that a strain of cancer-prone mice develops malignant tumors when injected with osteosarcoma cells, and you obtain these mice. Using these three resources, indicate which experiments you would perform to determine the following:
   (a) whether osteosarcoma cells carry two *RB1* mutations,
   (b) whether osteosarcoma cells produce any pRB protein,
   (c) if the addition of a normal *RB1* gene will change the cancer-causing potential of osteosarcoma cells.

20

Quantitative Genetics and Multifactorial Traits

A field of pumpkins, where size is under the influence of quantitative inheritance.

CHAPTER CONCEPTS

- Quantitative inheritance results in a range of measurable phenotypes for a polygenic trait.
- Polygenic traits most often demonstrate continuous variation.
- Quantitative inheritance can be explained in Mendelian terms whereby certain alleles have an additive effect on the traits under study.
- The study of polygenic traits relies on statistical analysis.
- Heritability values estimate the genetic contribution to phenotypic variability under specific environmental conditions.
- Twin studies allow an estimation of heritability in humans.
- Quantitative trait loci (QTLs) can be mapped and identified.

Up to this point in the text, most of our examples of phenotypic variation have been those that have been assigned to distinct and separate categories; for example, human blood type was A, B, AB, or O; squash fruit shape was spherical, disc-shaped, or elongated; and fruit fly eye color was red or white (see Chapter 4). Typically in these traits, a genotype will produce a single identifiable phenotype, although phenomena such as variable penetrance and expressivity, pleiotropy, and epistasis can obscure the relationship between genotype and phenotype.

However, many traits are not as distinct and clear-cut, including many that are of medical or agricultural importance. They show much more variation, often falling into a continuous range of multiple phenotypes. Most show what we call *continuous variation*, including, for example, height in humans, milk and meat production in cattle, and yield and seed protein content in various crops. Continuous variation across a range of phenotypes can be measured and described in quantitative terms, so this genetic phenomenon is known as **quantitative inheritance.** And because the varying phenotypes result from the input of genes at more than one, and often many, loci, the traits are said to be **polygenic** (literally "of many genes"). The genes involved are often referred to as **polygenes.**

To further complicate the link between the genotype and phenotype, the genotype generated at fertilization establishes a quantitative range within which a particular individual can fall. However, the final phenotype is often also influenced by environmental factors to which that individual is exposed. Human height, for example, is genetically influenced but is also affected by environmental factors such as nutrition. Quantitative (polygenic) traits whose phenotypes result from both gene action and environmental influences are termed **multifactorial, or complex traits.** Often these terms are used interchangeably. For consistency throughout the chapter, we will utilize the term *multifactorial* in our discussions.

In this chapter, we will examine examples of quantitative inheritance, multifactorial traits, and some of the statistical techniques used to study them. We will also consider how geneticists assess the relative importance of genetic versus environmental factors contributing to continuous phenotypic variation, and we will discuss approaches to identifying and mapping genes that influence quantitative traits.

20.1 Quantitative Traits Can Be Explained in Mendelian Terms

The question of whether continuous phenotypic variation could be explained in Mendelian terms caused considerable controversy in the early 1900s. Some scientists argued that, although Mendel's unit factors, or genes, explained patterns of discontinuous variation with discrete phenotypic classes, they could not account for the range of phenotypes seen in quantitative patterns of inheritance. However, geneticists William Bateson and G. Udny Yule, adhering to a Mendelian explanation, proposed the **multiple-factor** or **multiple-gene hypothesis,** in which many genes, each individually behaving in a Mendelian fashion, contribute to the phenotype in a *cumulative* or *quantitative* way.

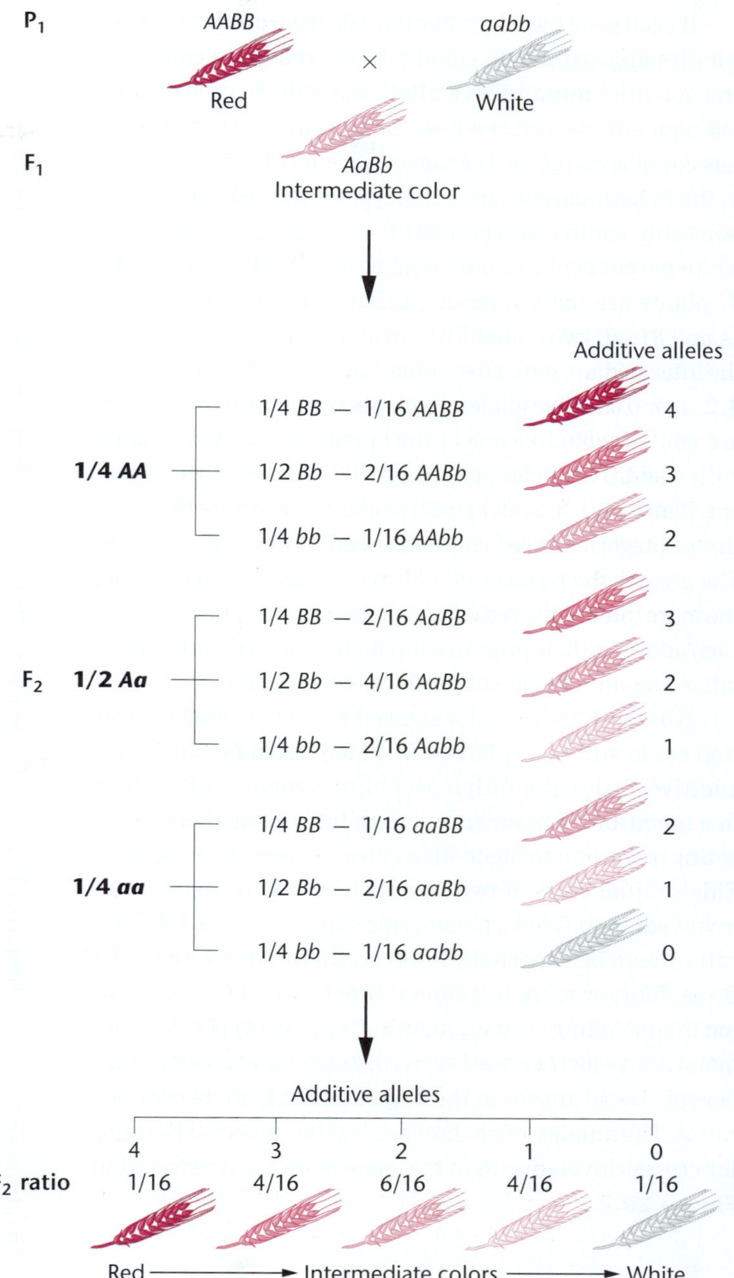

FIGURE 20.1 How the multiple-factor hypothesis accounts for the 1:4:6:4:1 phenotypic ratio of grain color when all alleles designated by an uppercase letter are additive and contribute an equal amount of pigment to the phenotype.

The Multiple-Gene Hypothesis for Quantitative Inheritance

The multiple-gene hypothesis was initially based on a key set of experimental results published by Hermann Nilsson-Ehle in 1909. Nilsson-Ehle used grain color in wheat to test the concept that the cumulative effects of alleles at multiple loci produce the range of phenotypes seen in quantitative traits. In one set of experiments, wheat with red grain was crossed to wheat with white grain (**Figure 20.1**). The F_1 generation demonstrated an intermediate pink color, which at first sight suggested incomplete dominance of two alleles at a single locus. However, in the F_2 generation, Nilsson-Ehle did not observe the typical segregation of a monohybrid cross. Instead, approximately 15/16 of the plants showed some degree of red grain color, while 1/16 of the plants showed white grain color. Careful examination of the F_2 revealed that grain with color could be classified into four different shades of red. Because the F_2 ratio occurred in sixteenths, it appears that two genes, each with two alleles, control the phenotype and that they segregate independently from one another in a Mendelian fashion.

If each gene has one potential **additive allele** that contributes approximately equally to the red grain color and one potential **nonadditive allele** that fails to produce any red pigment, we can see how the multiple-factor hypothesis could account for the various grain color phenotypes. In the P_1 both parents are homozygous; the red parent contains only additive alleles (*AABB* in Figure 20.1), while the white parent contains only nonadditive alleles (*aabb*). The F_1 plants are heterozygous (*AaBb*), contain two additive (*A* and *B*) and two nonadditive (*a* and *b*) alleles, and express the intermediate pink phenotype. Each of the F_2 plants has 4, 3, 2, 1, or 0 additive alleles. F_2 plants with no additive alleles are white (*aabb*) like one of the P_1 parents, while F_2 plants with 4 additive alleles are red (*AABB*) like the other P_1 parent. Plants with 3, 2, or 1 additive alleles constitute the other three categories of red color observed in the F_2 generation. The greater the number of additive alleles in the genotype, the more intense the red color expressed in the phenotype, as each additive allele present contributes equally to the cumulative amount of pigment produced in the grain.

Nilsson-Ehle's results showed how continuous variation could still be explained in a Mendelian fashion, with additive alleles at multiple loci influencing the phenotype in a quantitative manner, but each individual allele segregating according to Mendelian rules. As we saw in Nilsson-Ehle's initial cross, if two loci, each with two alleles, were involved, then five F_2 phenotypic categories in a 1:4:6:4:1 ratio would be expected. However, there is no reason why three, four, or more loci cannot function in a similar fashion in controlling various quantitative phenotypes. As more quantitative loci become involved, greater and greater numbers of classes appear in the F_2 generation in more complex ratios. The number of phenotypes and the expected F_2 ratios for crosses involving up to five gene pairs are illustrated in **Figure 20.2**.

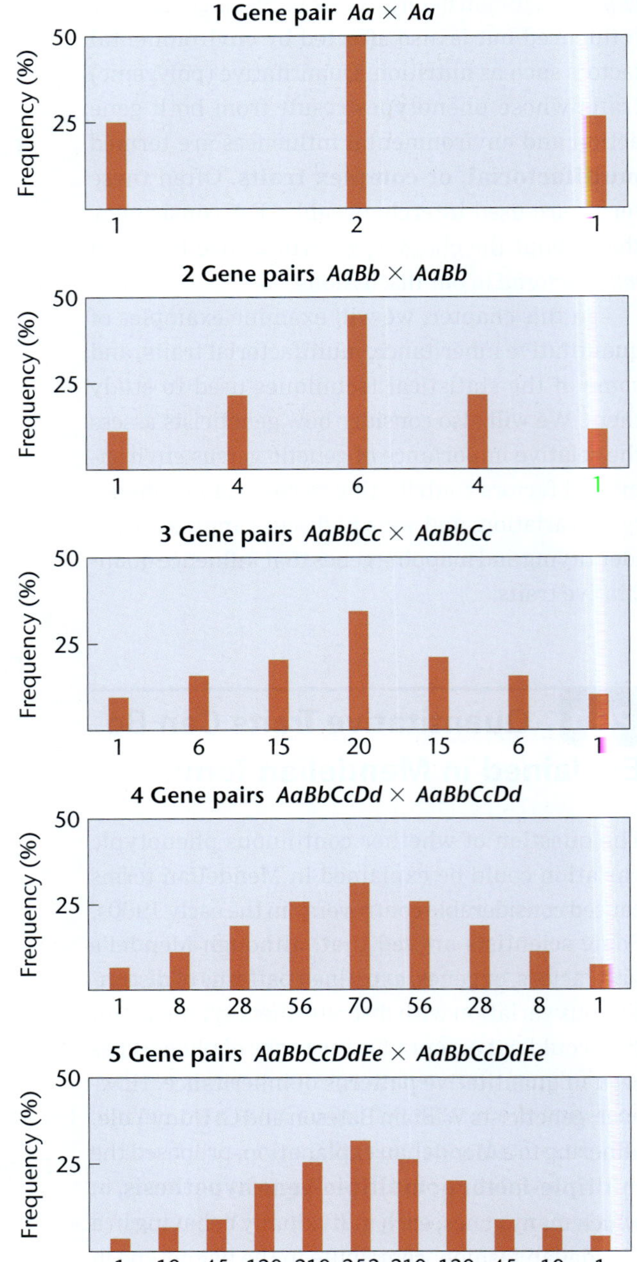

FIGURE 20.2 The genetic ratios (on the *x*-axis) resulting from crossing two heterozygotes when polygenic inheritance is in operation with 1 to 5 gene pairs. The histogram bars indicate the distinct F_2 phenotypic classes, ranging from one extreme (left end) to the other extreme (right end). Each phenotype results from a different number of additive alleles.

Additive Alleles: The Basis of Continuous Variation

The multiple-gene hypothesis consists of the following major points:

1. Phenotypic traits showing continuous variation can be quantified by measuring, weighing, counting, and so on.
2. Two or more gene loci, often scattered throughout the genome, account for the hereditary influence on the phenotype in an *additive way*. Because many genes may be involved, inheritance of this type is called *polygenic*.
3. Each gene locus may be occupied by either an *additive* allele, which contributes a constant amount to the phenotype, or a *nonadditive* allele, which does not contribute quantitatively to the phenotype.
4. The contribution to the phenotype of each additive allele, though often small, is approximately equal. While we now know this is not always true, we have made this assumption in the above discussion.
5. Together, the additive alleles contributing to a single quantitative character produce substantial phenotypic variation.

Calculating the Number of Polygenes

Various formulas have been developed for estimating the number of polygenes contributing to a quantitative trait. For example, if the ratio of F_2 individuals resembling *either* of the two extreme P_1 phenotypes can be determined, the number of polygenes (loci) involved (n) may be calculated as

$$1/4^n = \text{ratio of } F_2 \text{ individuals expressing either extreme phenotype}$$

In the example of the red and white wheat grain color summarized in Figure 20.1, 1/16 of the progeny are either red or white like the P_1 phenotypes. This ratio can be substituted on the right side of the equation to solve for n:

$$\frac{1}{4^n} = \frac{1}{16}$$

$$\frac{1}{4^2} = \frac{1}{16}$$

$$n = 2$$

Table 20.1 lists the ratio and the number of F_2 phenotypic classes produced in crosses involving up to five gene pairs.

For low numbers of polygenes (n), it is sometimes easier to use the equation

$$(2n + 1) = \text{the number of distinct phenotypic categories observed}$$

For example, when there are two polygenes involved ($n = 2$), then $(2n + 1) = 5$ and each phenotype is the result of 4, 3, 2, 1, or 0 additive alleles. If $n = 3$, $2n + 1 = 7$ and each phenotype is the result of 6, 5, 4, 3, 2, 1, or 0 additive alleles. Thus, working backwards with this rule and knowing the number of phenotypes, we can calculate the number of polygenes controlling them.

It should be noted, however, that both of these simple methods for estimating the number of polygenes involved in a quantitative trait assume not only that all the relevant alleles contribute equally and additively, but also that phenotypic expression in the F_2 is not affected significantly by environmental factors. As we will see later, for many quantitative traits, these assumptions may not be true.

TABLE 20.1 Determination of the Number of Polygenes (n) Involved in a Quantitative Trait

n	Individuals Expressing Either Extreme Phenotype	Distinct Phenotypic Classes
1	1/4	3
2	1/16	5
3	1/64	7
4	1/256	9
5	1/1024	11

ESSENTIAL POINT

Quantitative inheritance results in a range of phenotypes due to the action of additive alleles from two or more genes, as influenced by environmental factors. ■

NOW SOLVE THIS

20.1 A homozygous plant with 20-cm diameter flowers is crossed with a homozygous plant of the same species that has 40-cm diameter flowers. The F_1 plants all have flowers 30 cm in diameter. In the F_2 generation of 512 plants, 2 plants have flowers 20 cm in diameter, 2 plants have flowers 40 cm in diameter, and the remaining 508 plants have flowers of a range of sizes in between.

(a) Assuming that all alleles involved act additively, how many genes control flower size in this plant?

(b) What frequency distribution of flower diameter would you expect to see in the progeny of a backcross between an F_1 plant and the large-flowered parent?

■ **HINT:** *This problem provides F_1 and F_2 data for a cross involving a quantitative trait and asks you to calculate the number of genes controlling the trait. The key to its solution is to remember that unless you know the total number of distinct F_2 phenotypes involved, then the ratio (not the number) of parental phenotypes reappearing in the F_2 must be used in your determination of the number of genes involved.*

20.2 The Study of Polygenic Traits Relies on Statistical Analysis

Before considering the approaches that geneticists use to dissect how much of the phenotypic variation observed in a population is due to genotypic differences among individuals and how much is due to environmental factors, we need to consider the basic statistical tools they use for the task. It is not usually feasible to measure expression of a polygenic trait in every individual in a population, so a random subset of individuals is usually selected for measurement to provide a *sample*. It is important to remember that the accuracy of the final results of the measurements depends on whether the sample is truly random and representative of the population from which it was drawn. Suppose, for example, that a student wants to determine the average height of the 100 students in his genetics class, and for his sample he measures the two students sitting next to him, both of whom happen to be centers on the college basketball team. It is unlikely that this sample will provide a good estimate of the average height of the class, for two reasons: First, it is

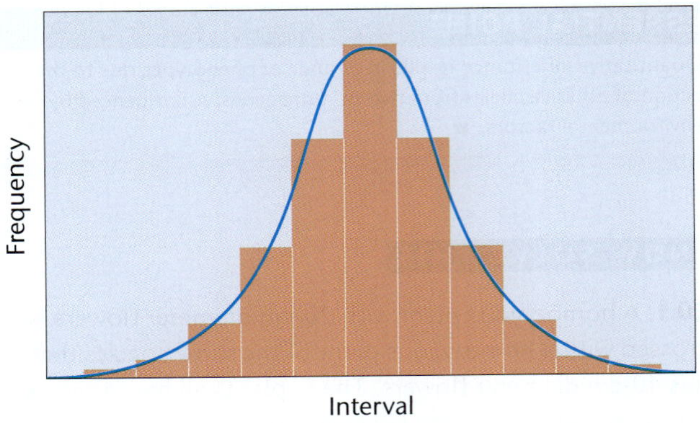

FIGURE 20.3 Normal frequency distribution, characterized by a bell-shaped curve.

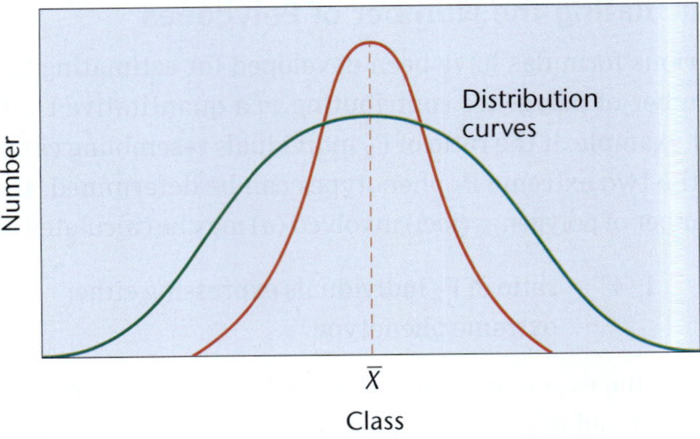

FIGURE 20.4 Two normal frequency distributions with the same mean but different amounts of variation.

too small; second, it is not a representative subset of the class (unless all 100 students are centers on the basketball team).

If the sample measured for expression of a quantitative trait is sufficiently large and also representative of the population from which it is drawn, we often find that the data form a *normal distribution*; that is, they produce a characteristic bell-shaped curve when plotted as a frequency histogram (**Figure 20.3**). Several statistical concepts are useful in the analysis of traits that exhibit a normal distribution, including the mean, variance, standard deviation, standard error of the mean, covariance, and correlation coefficient.

The Mean

The mean provides information about where the central point lies along a range of measurements for a quantitative trait. **Figure 20.4** shows the distribution curves for two different sets of phenotypic measurements. Each of these sets of measurements clusters around a central value (as it happens, they both cluster around the same value). This clustering is called a *central tendency*, and the central point is the *mean*.

Specifically, the **mean** ($\overline{X}$) is the arithmetic average of a set of measurements and is calculated as

$$\overline{X} = \frac{\Sigma X_i}{n}$$

where $\overline{X}$ is the mean, ΣX_i represents the sum of all individual values in the sample, and n is the number of individual values.

The mean provides a useful descriptive summary of the sample, but it tells us nothing about the range or spread of the data. As illustrated in Figure 20.4, a symmetrical distribution of values in the sample may, in one case, be clustered near the mean. Or a set of measurements may have the same mean but be distributed more widely around it. A second statistic, the variance, provides information about the spread of data around the mean.

Variance

The **variance (s^2)** for a sample is the average squared distance of all measurements from the mean. It is calculated as

$$s^2 = \frac{\Sigma (X_i - \overline{X})^2}{n - 1}$$

where the sum (Σ) of the squared differences between each measured value (X_i) and the mean ($\overline{X}$) is divided by one less than the total sample size ($n - 1$).

As Figure 20.4 shows, it is possible for two sets of sample measurements for a quantitative trait to have the same mean but a different distribution of values around it. This range will be reflected in different variances. Estimation of variance can be useful in determining the degree of genetic control of traits when the immediate environment also influences the phenotype.

Standard Deviation

Because the variance is a squared value, its unit of measurement is also squared (m^2, g^2, etc.). To express variation around the mean in the original units of measurement, we can use the square root of the variance, a term called the **standard deviation (s)**:

$$s = \sqrt{s^2}$$

Table 20.2 shows the percentage of individual values within a normal distribution that fall within different multiples of the standard deviation. The values that fall within

TABLE 20.2 Sample Inclusion for Various *s* Values

Multiples of *s*	Sample Included (%)
$\overline{X} \pm 1s$	68.3
$\overline{X} \pm 1.96s$	95.0
$\overline{X} \pm 2s$	95.5
$\overline{X} \pm 3s$	99.7

one standard deviation to either side of the mean represent 68 percent of all values in the sample. More than 95 percent of all values are found within two standard deviations to either side of the mean. This indicates that the standard deviation (s) can also be interpreted in the form of a probability. For example, a sample measurement picked at random has a 68 percent probability of falling within the range of one standard deviation.

Standard Error of the Mean

If multiple samples are taken from a population and measured for the same quantitative trait, we might find that their means vary. Theoretically, larger, truly random samples will represent the population more accurately, and their means will be closer to each other. To measure the accuracy of the sample mean, we use the **standard error of the mean ($S_{\overline{X}}$)**, calculated as

$$S_{\overline{X}} = \frac{s}{\sqrt{n}}$$

where s is the standard deviation and $\sqrt{n}$ is the square root of the sample size. Because the standard error of the mean is computed by dividing s by $\sqrt{n}$, it is always a smaller value than the standard deviation.

Covariance and Correlation Coefficient

Often geneticists working with quantitative traits find they have to consider two phenotypic characters simultaneously. For example, a poultry breeder might investigate the correlation between body weight and egg production in hens: Do heavier birds tend to lay more eggs? The covariance statistic measures how much variation is common to both quantitative traits. It is calculated by taking the deviations from the mean for each trait (just as we did for estimating variance) for each individual in the sample. This gives a pair of values for each individual. The two values are multiplied together, and the sum of all these individual products is then divided by one fewer than the number in the sample. Thus, the **covariance (cov_{XY})** of two sets of trait measurements, X and Y, is calculated as

$$\text{cov}_{XY} = \frac{\Sigma[(X_i - \overline{X})(Y_i - \overline{Y})]}{n - 1}$$

The covariance can then be standardized as yet another statistic, the **correlation coefficient (r).** The calculation is

$$r = \text{cov}_{XY}/S_X S_Y$$

where S_X is the standard deviation of the first set of quantitative measurements X, and S_Y is the standard deviation of the second set of quantitative measurements Y. Values for the correlation coefficient r can range from -1 to $+1$. Positive r values mean that an increase in measurement for one trait tends to be associated with an increase in measurement for the other, while negative r values mean that increases in one trait are associated with decreases in the other. Therefore, if heavier hens do tend to lay more eggs, a positive r value can be expected. A negative r value, on the other hand, suggests that greater egg production is more likely from less heavy birds. One important point to note about correlation coefficients is that even significant r values—close to $+1$ or -1—do not prove that a cause-and-effect relationship exists between two traits. Correlation analysis simply tells us the extent to which variation in one quantitative trait is associated with variation in another, not what causes that variation.

Analysis of a Quantitative Character

To apply these statistical concepts, let's consider a genetic experiment that crossed two different homozygous varieties of tomato. One of the tomato varieties produces fruit averaging 18 oz in weight, whereas fruit from the other averages 6 oz. The F_1 obtained by crossing these two varieties has fruit weights ranging from 10 to 14 oz. The F_2 population contains individuals that produce fruit ranging from 6 to 18 oz. The results characterizing both generations are shown in **Table 20.3**.

The mean value for the fruit weight in the F_1 generation can be calculated as

$$\overline{X} = \frac{\Sigma X_i}{n} = \frac{626}{52} = 12.04$$

The mean value for fruit weight in the F_2 generation is calculated as

$$\overline{X} = \frac{\Sigma X_i}{n} = \frac{872}{72} = 12.11$$

TABLE 20.3 Distribution of F_1 and F_2 Progeny Derived from a Theoretical Cross Involving Tomatoes

		Weight (oz)												
		6	7	8	9	10	11	12	13	14	15	16	17	18
Number of	F_1					4	14	16	12	6				
Individuals	F_2	1	1	2	0	9	13	17	14	7	4	3	0	1

Although these mean values are similar, the frequency distributions in Table 20.3 show more variation in the F_2 generation. The range of variation can be quantified as the *variance* s^2, calculated, as we saw on p. 420, as the sum of the squared differences between each value and the mean, divided by one less than the total number of observations:

$$s^2 = \frac{\Sigma (X_i - \overline{X})^2}{n - 1}$$

When the above calculation is made, the variance is found to be 1.29 for the F_1 generation and 4.27 for the F_2 generation. When converted to the standard deviation ($s = \sqrt{s^2}$), the values become 1.13 and 2.06, respectively. Therefore, the distribution of tomato weight in the F_1 generation can be described as 12.04 ± 1.13, and in the F_2 generation it can be described as 12.11 ± 2.06.

Assuming that both parental varieties are homozygous at the loci of interest and that the alleles controlling fruit weight act additively, we can estimate the number of loci involved in this trait. Since 1/72 of the F_2 offspring have a phenotype that overlaps one of the parental strains (72 total F_2 offspring; one weighs 6 oz, one weighs 18 oz; see Table 20.3), the use of the formula $1/4^n = 1/72$ indicates that n is between 3 and 4, providing evidence of the number of genes that control fruit weight in these tomato strains.

NOW SOLVE THIS

20.2 The following table shows measurements for fiber lengths and fleece weight in a small flock of eight sheep.

	Sheep Fiber Length (cm)	Fleece Weight (kg)
1	9.7	7.9
2	5.6	4.5
3	10.7	8.3
4	6.8	5.4
5	11.0	9.1
6	4.5	4.9
7	7.4	6.0
8	5.9	5.1

(a) What are the mean, variance, and standard deviation for each trait in this flock?
(b) What is the covariance of the two traits?
(c) What is the correlation coefficient for fiber length and fleece weight?
(d) Do you think greater fleece weight is correlated with an increase in fiber length? Why or why not?

■ **HINT:** *This problem provides data for two quantitative traits and asks you to make numerous statistical calculations, ultimately determining if the traits are correlated. The key to its solution is that once the calculation of the correlation coefficient (r) is completed, you must interpret that value—whether it is positive or negative, and how close to zero it is.*

ESSENTIAL POINT

Numerous statistical methods are essential during the analysis of quantitative traits, including the mean, variance, standard deviation, standard error, covariance, and the correlation coefficient. ■

20.3 Heritability Values Estimate the Genetic Contribution to Phenotypic Variability

The question most often asked by geneticists working with multifactorial traits and diseases is how much of the observed phenotypic variation in a population is due to genotypic differences among individuals and how much is due to environment. The term **heritability** is used to describe *the proportion of total phenotypic variation in a population that is due to genetic factors*. For a multifactorial trait in a given population, a high heritability estimate indicates that much of the variation can be attributed to genetic factors, with the environment having less impact on expression of the trait. With a low heritability estimate, environmental factors are likely to have a greater impact on phenotypic variation within the population.

The concept of heritability is frequently misunderstood and misused. It should be emphasized that *heritability indicates neither how much of a trait is genetically determined nor the extent to which an individual's phenotype is due to genotype*. In recent years, such misinterpretations of heritability for human quantitative traits have led to controversy, notably in relation to measurements such as intelligence quotients, or IQs. Variation in heritability estimates for IQ among different racial groups led to incorrect suggestions that unalterable genetic factors control differences in intelligence levels among humans of different ancestries. Such suggestions misrepresented the meaning of heritability and ignored the contribution of *genotype-by-environment interaction variance* (see p. 423) to phenotypic variation in a population. Moreover, heritability is not fixed for a trait. For example, a heritability estimate for egg production in a flock of chickens kept in individual cages might be high, indicating that differences in egg output among individual birds are largely due to genetic differences, as they all have very similar environments. For a different flock kept outdoors, heritability for egg production might be much lower, as variation among different birds may also reflect differences in their individual environments. Such differences could include how much food each bird manages to find and whether it competes successfully for a good roosting spot at night.

Thus, a heritability estimate tells us the proportion of *phenotypic variation* that can be attributed to *genetic variation within a certain population in a particular environment*. If we measure heritability for the same trait among different populations in a range of environments, we frequently find that the calculated heritability values have large standard errors. This is an important point to remember when considering heritability estimates. Parallel studies using different population bases are likely to yield different heritability estimates. For example, a mean heritability estimate of 0.65 for human height does not mean that your height is 65 percent due to your genes, but rather that in the populations sampled, on average, *65 percent of the overall variation in height could be explained by genotypic differences among individuals.*

With this subtle but important distinction in mind, we will now consider how geneticists divide the phenotypic variation observed in a population into genetic and environmental components. As we saw in the previous section, variation can be quantified as a sample variance: taking measurements of the trait in question from a representative sample of the population and determining the extent of the spread of those measurements around the sample mean. This gives us an estimate of the total **phenotypic variance** in the population **(V_P).** Heritability estimates are obtained by using different experimental and statistical techniques to partition V_P into **genotypic variance (V_G)** and **environmental variance (V_E)** components.

An important factor contributing to overall levels of phenotypic variation is the extent to which individual genotypes affect the phenotype differently depending on the environment. For example, wheat variety A may yield an average of 20 bushels an acre on poor soil, while variety B yields an average of 17 bushels. On good soil, variety A yields 22 bushels, while variety B averages 25 bushels an acre. There are differences in yield between the two genotypically distinct varieties, so variation in wheat yield has a genetic component. Both varieties yield more on good soil, so yield is also affected by environment. However, we also see that the two varieties do not respond to better soil conditions equally: The genotype of wheat variety B achieves a greater increase in yield on good soil than does variety A. Thus, we have differences in the interaction of genotype, with environment contributing to variation for yield in populations of wheat plants. This third component of phenotypic variation is **genotype-by-environment interaction variance ($V_{G\times E}$)** (**Figure 20.5**).

We can now summarize all the components of total phenotypic variance V_P using the following equation:

$$V_P = V_G + V_E + V_{G\times E}$$

In other words, total phenotypic variance can be subdivided into genotypic variance, environmental variance, and genotype-by-environment interaction variance. When obtaining heritability estimates for a multifactorial trait, researchers often assume that the genotype-by-environment interaction variance is small enough that it can be ignored or combined with the environmental variance. However, it is worth remembering that this kind of approximation is another reason heritability values are *estimates* for a given population in a particular context, not a *fixed attribute* for a trait.

Animal and plant breeders use a range of experimental techniques to estimate heritabilities by partitioning measurements of phenotypic variance into genotypic and environmental components. One approach uses inbred strains containing genetically homogeneous individuals with highly homozygous genotypes. Experiments are then designed to test the effects of a range of environmental conditions on phenotypic variability. Variation *between* different inbred strains reared in a constant environment is due predominantly to genetic factors. Variation *among* members of the same inbred strain reared under different conditions is more likely to be due to environmental factors. Other approaches involve analysis of variance for a quantitative trait among offspring from different crosses, or comparing expression of a trait among offspring and parents reared in the same environment.

(a)

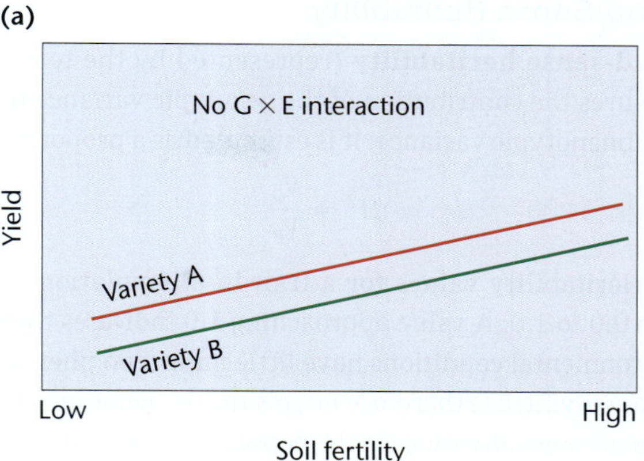

(b)

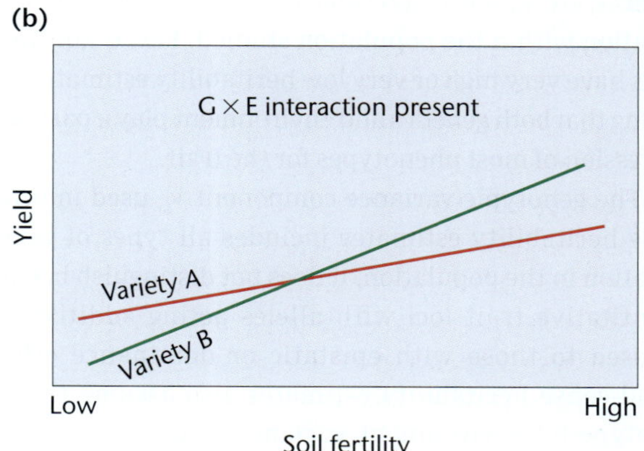

FIGURE 20.5 Differences in yield between two wheat varieties at different soil fertility levels. (a) No genotype-by-environment, or $G \times E$, interaction: The varieties show genetic differences in yield but respond equally to increasing soil fertility. (b) $G \times E$ interaction present: Variety A outyields B at low soil fertility, but B yields more than A at high-fertility levels.

Broad-Sense Heritability

Broad-sense heritability (represented by the term H^2) measures the contribution of the genotypic variance to the total phenotypic variance. It is estimated as a proportion:

$$H^2 = \frac{V_G}{V_P}$$

Heritability values for a trait in a population range from 0.0 to 1.0. A value approaching 1.0 indicates that the environmental conditions have little impact on phenotypic variance, which is therefore largely due to genotypic differences among individuals in the population. Low values close to 0.0 indicate that environmental factors, not genotypic differences, are largely responsible for the observed phenotypic variation within the population studied. Few quantitative traits have very high or very low heritability estimates, suggesting that both genetics and environment play a part in the expression of most phenotypes for the trait.

The genotypic variance component V_G used in broad-sense heritability estimates includes all types of genetic variation in the population. It does not distinguish between quantitative trait loci with alleles acting additively as opposed to those with epistatic or dominance effects. Broad-sense heritability estimates also assume that the genotype-by-environment variance component is negligible. While broad-sense heritability estimates for a trait are of general genetic interest, these limitations mean this kind of heritability is not very useful in breeding programs. Animal or plant breeders wishing to develop improved strains of livestock or higher-yielding crop varieties need more precise heritability estimates for the traits they wish to manipulate in a population. Therefore, another type of estimate, narrow-sense heritability, has been devised that is of more practical use.

Narrow-Sense Heritability

Narrow-sense heritability (h^2) is the proportion of phenotypic variance due to additive genotypic variance alone. Genotypic variance can be divided into subcomponents representing the different modes of action of alleles at quantitative trait loci. As not all the genes involved in a quantitative trait affect the phenotype in the same way, this partitioning distinguishes among three different kinds of gene action contributing to genotypic variance. **Additive variance, V_A,** is the genotypic variance due to the additive action of alleles at quantitative trait loci. **Dominance variance, V_D,** is the deviation from the additive components that results when phenotypic expression in heterozygotes is not precisely intermediate between the two homozygotes. **Interactive variance, V_I,** is the deviation from the additive components that occurs when two or more loci behave epistatically. The amount of interactive variance is often negligible, and so this component is often excluded from calculations of total genotypic variance.

The partitioning of the total genotypic variance V_G is summarized in the equation

$$V_G = V_A + V_D + V_I$$

and a narrow-sense heritability estimate based only on that portion of the genotypic variance due to additive gene action becomes

$$h^2 = \frac{V_A}{V_P}$$

Omitting V_I and separating V_P into genotypic and environmental variance components, we obtain

$$h^2 = \frac{V_A}{V_E + V_A + V_D}$$

Heritability estimates are used in animal and plant breeding to indicate the potential response of a population to artificial selection for a quantitative trait. Narrow-sense heritability, h^2, provides a more accurate prediction of selection response than broad-sense heritability, H^2, and therefore h^2 is more widely used by breeders.

ESSENTIAL POINT

Heritability is an estimate of the relative contribution of genetic versus environmental factors to the range of phenotypic variation seen in a quantitative trait in a particular population and environment. ■

Artificial Selection

Artificial selection is the process of choosing specific individuals with preferred phenotypes from an initially heterogeneous population for future breeding purposes. Theoretically, if artificial selection based on the same trait preferences is repeated over multiple generations, a population can be developed containing a high frequency of individuals with the desired characteristics. If selection is for a simple trait controlled by just one or two genes subject to little environmental influence, generating the desired population of plants or animals is relatively fast and easy. However, many traits of economic importance in crops and livestock, such as grain yield in plants, weight gain or milk yield in cattle, and speed or stamina in horses, are polygenic and frequently multifactorial. Artificial selection for such traits is slower and more complex. Narrow-sense heritability estimates are valuable to the plant or animal breeder because, as we have just seen, they estimate the proportion of total phenotypic variance for the trait that is due to additive genetic variance. Quantitative trait alleles with additive impact are those most easily manipulated by the breeder. Alleles at quantitative trait loci that generate dominance effects or interact epistatically (and therefore contribute to V_D or V_I) are less responsive to artificial selection. Thus, narrow-sense heritability, h^2, can be used to predict the impact of selection. The higher the estimated value for h^2 in a population, the more likely the breeder will observe a change in phenotypic range for the trait in the next generation after artificial selection.

Partitioning the genetic variance components to calculate h^2 and predict response to selection is a complex task requiring careful experimental design and analysis. The simplest approach is to select individuals with superior phenotypes for the desired quantitative trait from a heterogeneous population and breed offspring from those individuals. The mean score for the trait of those offspring ($M2$) can then be compared to that of: (1) the original population's mean score (M) and (2) the selected individuals used as parents ($M1$). The relationship between these means and h^2 is

$$h^2 = \frac{M2 - M}{M1 - M}$$

This equation can be further simplified by defining $M2 - M$ as the **selection response (*R*)**—the degree of response to mating the selected parents—and $M1 - M$ as the **selection differential (*S*)**—the difference between the mean for the whole population and the mean for the selected population—so h^2 reflects the ratio of the response observed to the total response possible. Thus,

$$h^2 = \frac{R}{S}$$

A narrow-sense heritability value obtained in this way by selective breeding and measuring the response in the offspring is referred to as an estimate of **realized heritability.**

As an example of a realized heritability estimate, suppose that we measure the diameter of corn kernels in a population where the mean diameter M is 20 mm. From this population, we select a group with the smallest diameters, for which the mean $M1$ equals 10 mm. The selected plants are interbred, and the mean diameter $M2$ of the progeny kernels is 13 mm. We can calculate the realized heritability h^2 to estimate the potential for artificial selection on kernel size:

$$h^2 = \frac{M2 - M}{M1 - M}$$

$$h^2 = \frac{13 - 20}{10 - 20}$$

$$= \frac{-7}{-10}$$

$$= 0.70$$

This value for narrow-sense heritability indicates that the selection potential for kernel size is relatively high.

The longest running artificial selection experiment known is still being conducted at the State Agricultural Laboratory in Illinois. Corn has been selected for both high and low oil content. After 76 generations, selection continues to result in increased oil content (**Figure 20.6**). With each cycle of successful selection, more of the corn plants accumulate a higher percentage of additive alleles involved in oil production.

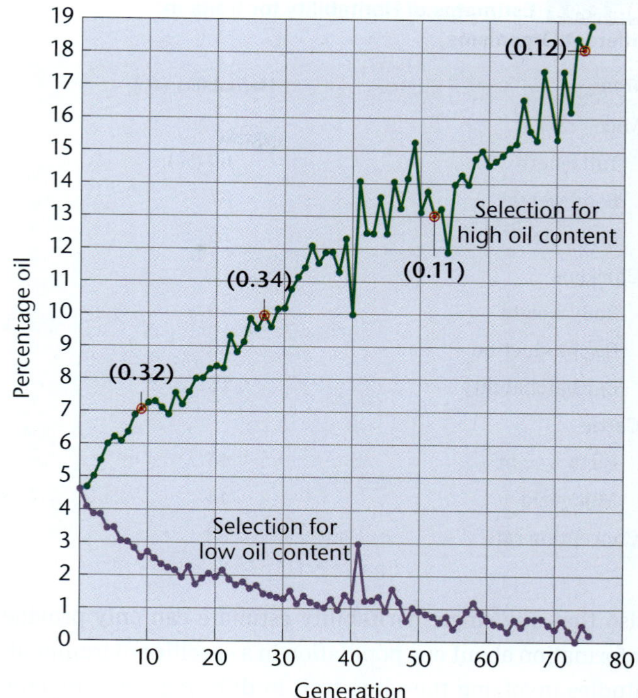

FIGURE 20.6 Response of corn selected for high and low oil content over 76 generations. The numbers in parentheses at generations 9, 25, 52, and 76 for the "high oil" line indicate the calculation of heritability at these points in the continuing experiment.

Consequently, the narrow-sense heritability h^2 of increased oil content in succeeding generations has declined (see parenthetical values at generations 9, 25, 52, and 76 in Figure 20.6) as artificial selection comes closer and closer to optimizing the genetic potential for oil production. Theoretically, the process will continue until all individuals in the population possess a uniform genotype that includes all the additive alleles responsible for high oil content. At that point, h^2 will be reduced to zero, and response to artificial selection will cease. The decrease in response to selection for low oil content shows that heritability for low oil content is approaching this point.

Table 20.4 lists narrow-sense heritability estimates expressed as percentage values for a variety of quantitative traits in different organisms. As you can see, these h^2 values vary, but heritability tends to be low for quantitative traits that are essential to an organism's survival. Remember, this does not indicate the absence of a genetic contribution to the observed phenotypes for such traits. Instead, the low h^2 values show that natural selection has already largely optimized the genetic component of these traits during evolution. Egg production, litter size, and conception rate are examples of how such physiological limitations on selection have already been reached. Traits that are less critical to survival, such as body weight, tail length, and wing length, have higher heritabilities because more genotypic variation for such traits is still present in the population. Remember

TABLE 20.4 Estimates of Heritability for Traits in Different Organisms

Trait	Heritability (h^2)
Mice	
Tail length	60 (%)
Body weight	37
Litter size	15
Chickens	
Body weight	50
Egg production	20
Egg hatchability	15
Cattle	
Birth weight	45
Milk yield	44
Conception rate	3

also that any single heritability estimate can only provide information about one population in a specific environment. Studies involving the same trait in differing environments most often yield different results. Therefore, narrow-sense heritability is a more valuable predictor of response to selection when estimates are calculated for many populations and environments and show the presence of a clear trend.

Limitations of Heritability Studies

While the above discussion makes clear that heritability studies are valuable in estimating the genetic contribution to phenotypic variance, the knowledge gained about heritability of traits must be balanced by awareness of some of the constraints inherent in such estimates:

- Heritability values provide no information about what genes are involved in traits.
- Heritability is measured in populations, and has only limited application to individuals.
- Measured heritability depends on the environmental variation present in the population being studied, and cannot be used to evaluate differences between populations.
- Future changes in environmental factors can affect heritability.

20.4 Twin Studies Allow an Estimation of Heritability in Humans

Human twins are useful subjects for examining how much phenotypic variance for a multifactorial trait is due to the genotype as opposed to the environment. In these studies, the underlying principle has been that **monozygotic (MZ),** or **identical, twins** are derived from a single zygote that divides mitotically and then spontaneously splits into two separate cells. Both cells give rise to a genotypically identical embryo. **Dizygotic (DZ),** or **fraternal, twins,** on the other hand, originate from two separate fertilization events and are only as genetically similar as any two siblings, with an average of 50 percent of their alleles in common. For a given trait, therefore, phenotypic differences between pairs of MZ twins will be equivalent to the environmental variance (V_E) (because the genotypic variance is zero). Phenotypic differences between DZ twins, however, display both environmental variance (V_E) and approximately half the genotypic variance (V_G). Comparing the extent of phenotypic variance for the same trait in MZ and DZ sets of twins provides an estimate of broad-sense heritability for the trait.

Twins are said to be **concordant** for a given trait if both express it or neither expresses it. If one expresses the trait and the other does not, the pair is said to be **discordant.** Comparison of concordance values of MZ versus DZ twins reared together illustrates the potential value for heritability assessment. (See the Now Solve This feature on p. 428, for example.)

Before any conclusions can be drawn from twin studies, the data must be examined carefully. For example, if concordance values approach 90 to 100 percent in MZ twins, we might be inclined to interpret that as a large genetic contribution to the phenotype of the trait. In some cases—for example, blood types and eye color—we know that this is indeed true. In the case of contracting measles, however, a high concordance value merely indicates that the trait is almost always induced by a factor in the environment—in this case, a virus.

It is more meaningful to compare the *difference* between the concordance values of MZ and DZ twins. If concordance values are significantly higher in MZ twins, we suspect a strong genetic component in the determination of the trait. In the case of measles, where concordance is high in both types of twins, the environment is assumed to be the major contributing factor. Such an analysis is useful because phenotypic characteristics that remain similar in different environments are likely to have a strong genetic component.

Large Scale Analysis of Twin Studies

For decades, researchers have used twin studies to examine the relative contributions of genotype and environment to the phenotypic variation observed in complex traits in humans. These traits involve the interplay of multiple genes with a network of environmental factors, and the genetic components of the resulting phenotypic variance can be difficult to study. The simplest way to assess the genetic contribution is to assume that the effect of each gene on a trait is independent of the effects of other genes. Because the effects of all genes are added together, this is called the *additive model*. However, in recent years, some geneticists have proposed that non-additive factors such as dominance and epistasis are more important than additive genetic effects. As a result, the relative roles of additive and non-additive factors are a subject of active debate.

In an attempt to resolve this issue, an international project recently examined the results of all twin studies performed in the last 50 years. This study, published in 2015, involved the compilation and analysis of the data for over 17,000 traits studied in more than 14 million twin pairs drawn from more than 2,700 published papers.

Several important general conclusions can be drawn from this landmark study. First, based on correlations between MZ and DZ twin pairs, which can be used to draw conclusions about how likely it is that genetic influences on a trait are mostly additive or non-additive, researchers concluded that the vast majority of traits follow a simple additive model, providing strong support for one of the foundations of heritability studies. This does not exclude the role of non-additive factors such as dominance and epistasis, but these factors most likely play a secondary role in heritability. Second, the results are consistent with the findings from **genome-wide association studies (GWAS)** that many complex traits are controlled by many genes, each with a small effect. Third, genetic variance is an important component of the individual variations observed in populations. In addition, the relative effects of genotypes and environmental factors are non-randomly distributed, making their contributions somewhat trait-specific.

The data from this study are available in a web-based application, Meta-analysis of Twin Correlations and Heritability (MaTCH), which can be used as a resource for the study of complex traits and the genetic and environmental components of heritability.

Twin Studies Have Several Limitations

Interesting as they are, human twin studies contain some unavoidable sources of error. For example, MZ twins are often treated more similarly by parents and teachers than are DZ twins, especially when the DZ siblings are of a different sex. This circumstance may inflate the environmental variance for DZ twins. Another possible error source is interactions between the genotype and the environment that produce variability in the phenotype. These interactions can increase the total phenotypic variance for DZ twins compared to MZ twins raised in the same environment, influencing heritability calculations. Overall, heritability estimates for human traits based on twin studies should therefore be considered approximations and examined very carefully before any conclusions are drawn.

Although they must often be viewed with caution, classical twin studies, based on the assumption that MZ twins share the same genome, have been valuable for estimating heritability over a wide range of traits including multifactorial disorders such as cardiovascular disease, diabetes, and mental illness, for example. These disorders clearly have genetic components, and twin studies provide a foundation for studying interactions between genes and environmental factors. However, results from genomics research have challenged the view that MZ twins are truly identical and have forced a reevaluation of both the methodology and the results of twin studies. Such research has also opened the way to new approaches to the study of interactions between the genotype and environmental factors.

The most relevant genomic discoveries about twins include the following:

- By the time they are born, MZ twins do not necessarily have identical genomes.
- Gene-expression patterns in MZ twins change with age, leading to phenotypic differences.

We will address these points in order. First, MZ twins develop from a single fertilized egg, where sometime early in development the resulting cell mass separates into two distinct populations creating two independent embryos. Until that time, MZ twins have identical genotypes. Subsequently, however, the genotypes can diverge slightly. For example, differences in *copy number variation (CNV)*—variation in the number of copies of numerous large DNA sequences (usually 1000 bp or more)—may arise, differentially producing genetically distinct populations of cells in each embryo (see Chapter 6 for a discussion of CNV). This creates a condition called *somatic mosaicism*, which may result in a milder disease phenotype in some disorders and may play a similar role in phenotypic discordance observed in some pairs of MZ twins.

At this point, it is difficult to know for certain how often CNV arises after MZ twinning, but one estimate suggests that such differences occur in 10 percent of all twin pairs. In those pairs where it does occur, one estimate is that such divergence takes place in 15 to 70 percent of the somatic cells. In one case, a CNV difference between MZ twins has been associated with chronic lymphocytic leukemia in one twin, but not the other.

The second genomic difference between MZ twins involves **epigenetics**—the chemical modification of their DNA and associated histones. An international study of epigenetic modifications in adult European MZ twins showed that MZ twin pairs are epigenetically identical at birth, but adult MZ twins show significant differences in the *methylation patterns* of both DNA and histones. Such epigenetic changes in turn affect patterns of gene expression. The accumulation of epigenetic changes and gene-expression profiles may explain some of the observed phenotypic discordance and susceptibility to diseases in adult MZ twins. For example, a clear difference in DNA methylation patterns is observed in MZ twins discordant for Beckwith–Wiedemann syndrome, a genetic disorder associated with variable developmental overgrowth of certain tissues and organs and an increased risk of developing cancerous and noncancerous tumors. Infants with Beckwith-Wiedemann syndrome are often larger than normal, and one in five dies early in life.

Other complex disorders displaying a genetic component are similarly being investigated using epigenetic analysis in twin studies. These include susceptibility to several neurobiological disorders, including schizophrenia and autism, as well as to the development of Type 1 diabetes, breast cancer, and autoimmune disease.

Progressive, age-related genomic modifications may be the result of MZ twins being exposed to different environmental factors, or from failure of epigenetic marking following DNA replication. These findings also indicate that concordance studies in DZ twins must take into account genetic as well as *epigenetic differences* that contribute to discordance in these twin pairs.

The realization that epigenetics may play an important role in the development of phenotypes promises to make twin studies an especially valuable tool in dissecting the interactions among genes and the role of environmental factors in the production of phenotypes. Once the degree of epigenetic differences between MZ and DZ twin pairs has been defined, molecular studies on DNA and histone modification can link changes in gene expression with differences in the concordance rates between MZ and DZ twins.

We will discuss the most recent findings involving epigenetics and summarize its many forms and functions later in the text (see Special Topics Chapter 1—Epigenetics).

ESSENTIAL POINT

Twin studies, while having some limitations, are useful in assessing heritabilities for polygenic traits in humans. ■

NOW SOLVE THIS

20.3 The following table gives the percentage of twin pairs studied in which both twins expressed the same phenotype for a trait (concordance). Percentages listed are for concordance for each trait in monozygotic (MZ) and dizygotic (DZ) twins. Assuming that both twins in each pair were raised together in the same environment, what do you conclude about the relative importance of genetic versus environmental factors for each trait?

Trait	MZ %	DZ %
Blood types	100	66
Eye color	99	28
Mental retardation	97	37
Measles	95	87
Hair color	89	22
Handedness	79	77
Idiopathic epilepsy	72	15
Schizophrenia	69	10
Diabetes	65	18
Identical allergy	59	5
Cleft lip	42	5
Club foot	32	3
Mammary cancer	6	3

■ **HINT:** *This problem asks you to evaluate the relative importance of genetic versus environmental contributions to specific traits by examining concordance values in MZ versus DZ twins. The key to its solution is to examine the difference in concordance values and to factor in what you have learned about the genetic differences between MZ and DZ twins.*

20.5 Quantitative Trait Loci Are Useful in Studying Multifactorial Phenotypes

Environmental effects, interaction among segregating alleles, and the large number of genes that may contribute to the phenotype of polygenically controlled complex traits make it difficult to: (1) identify all genes that are involved; and (2) determine the effect of each gene on the phenotype. However, because many quantitative traits are of economic or medical relevance, it is often desirable to obtain this information. In such studies, a chromosome region is identified as containing one or more genes contributing to

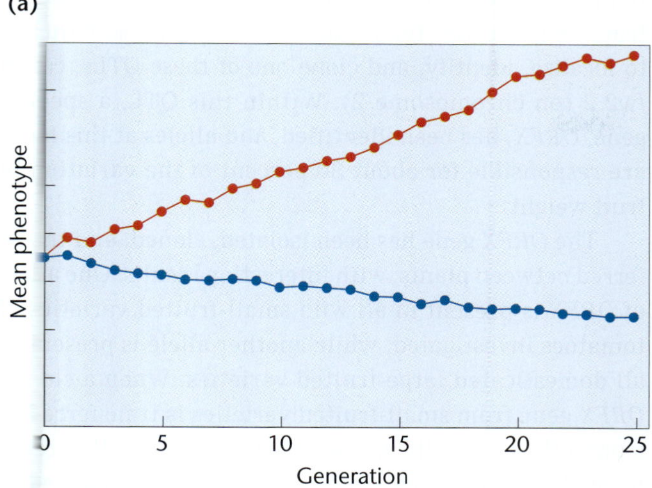

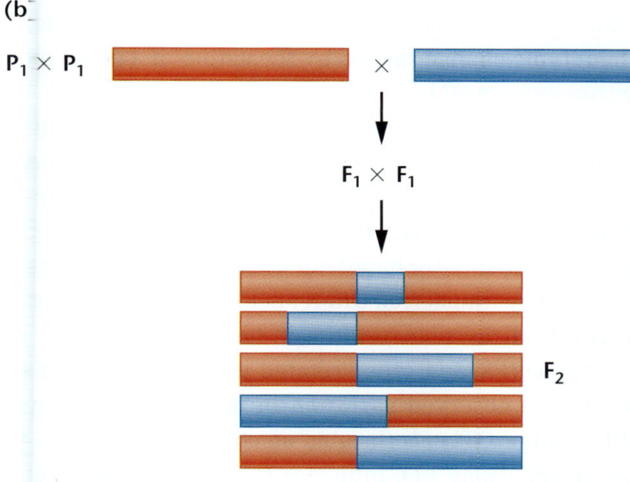

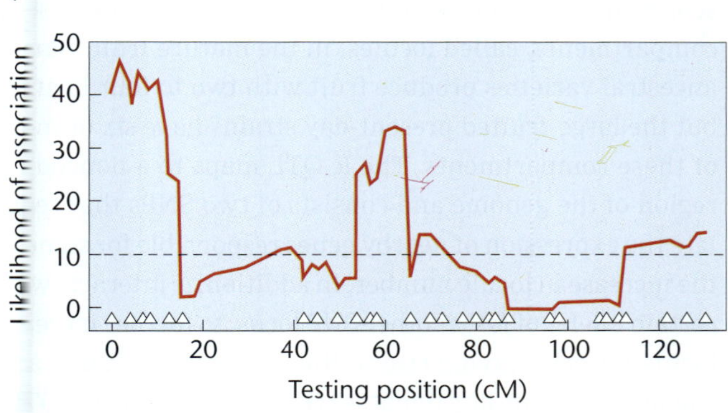

FIGURE 20.7 (a) Individuals from highly divergent lines created by artificial selection are chosen from generation 25 as parents. (b) The thick bars represent the genomes of individuals selected from the divergent lines as parents. These individuals are crossed to produce an F_1 generation (not shown). An F_2 generation is produced by crossing members of the F_1. As a result of crossing over, individual members of the F_2 generation carry different portions of the P_1 genome, as shown by the colored segments of the thick bars. DNA markers and phenotypes in individuals of the F_2 generation are analyzed. (c) Statistical methods are used to determine the probability that a DNA marker is associated with a QTL that affects the phenotype. The results are plotted as the likelihood of association against chromosomal location. Units on genetic maps are measured in centimorgans (cM), determined by crossover frequencies. Peaks above the horizontal line represent significant results. The data show five possible QTLs, with the most significant findings at about 10 cM and 60 cM.

a quantitative trait known as a **quantitative trait locus (QTL).*** When possible, the relevant gene or genes contained within a QTL are isolated and studied.

The modern approach used to find and map QTLs involves looking for associations between DNA markers and phenotypes. One way to do this is to begin with individuals from two lines created by artificial selection that are highly divergent for a phenotype (fruit weight, oil content, bristle number, etc.). For example, **Figure 20.7** illustrates a generic case of QTL mapping. Over many generations of artificial selection, two divergent lines become highly homozygous, which facilitates their use in QTL mapping. Individuals from each of the lines with divergent phenotypes [generation 25 in **Figure 20.7(a)**] are used as parents to create an F_1 generation whose members will be heterozygous at most of the loci contributing to the trait. Additional crosses, either among F_1 individuals or between the F_1 and the inbred parent lines, result in F_2 generations that carry different portions of the parental genomes [**Figure 20.7(b)**] with different QTL genotypes and associated phenotypes. This segregating F_2 is known as the **QTL mapping population.**

Researchers then measure phenotypic expression of the trait among individuals in the mapping population and identify genomic differences among individuals by using chromosome-specific DNA markers such as *restriction fragment length polymorphisms (RFLPs)*, *microsatellites*, and *single-nucleotide polymorphisms (SNPs)* (see Special Topics Chapter 2—Genetic Testing). Computer-based statistical analysis is used to search for linkage between the markers and a component of phenotypic variation associated with the trait. If a DNA marker (such as those markers described above) *is not* linked to a QTL, then the phenotypic mean score for the trait will not vary among individuals with different genotypes at that marker locus. However, if a DNA marker *is* linked to a QTL, then different genotypes at that marker locus will also differ in their phenotypic expression of the trait. When this occurs, the marker locus and the QTL are said to *cosegregate*. Consistent cosegregation establishes the presence of a QTL at or near the DNA marker along the chromosome—in other words, the marker and QTL are linked. When numerous QTLs for a given trait have been located, a genetic map is created, showing the probability that specific chromosomal regions are associated with the phenotype of interest [**Figure 20.7(c)**]. Further research using genomic techniques identifies genes in these regions that contribute to the phenotype.

* We utilize QTLs to designate the plural form, quantitative trait loci.

TABLE 20.5 QTLs for Quantitative Phenotypes

Organism	Quantitative Phenotype	QTLs Identified
Tomato	Soluble solids	7
	Fruit mass	13
	Fruit pH	9
	Growth	5
	Leaflet shape	9
	Height	9
Maize	Height	11
	Leaf length	7
	Grain yield	18
	Number of ears	10

Source: Used with permission of Annual Reviews of Genetics, from "Mapping Polygenes" by S.D. Tanksley, *Annual Review of Genetics*, Vol. 27:205–233, Table 1, December 1993. © Annual Reviews, Inc.

QTL mapping has been used extensively in agriculture, including plants such as corn, rice, wheat, and tomatoes (**Table 20.5**), and livestock such as cattle, pigs, sheep, and chickens.

Tomatoes are one of the world's major vegetable crops, and hundreds of varieties are grown and harvested each year. To aid in the creation of new varieties, hundreds of QTLs for traits including fruit size, shape, soluble solid content, and acidity have been identified and mapped to all 12 chromosomes in the tomato haploid genome. In addition, the genomes of several tomato varieties have been sequenced. We will describe studies focused on quantitative traits controlling fruit shape and weight as an example of QTL research.

While the cultivated tomato can weigh up to 1000 grams, fruit from the related wild species thought to be the ancestor of the modern tomato weighs only a few grams (**Figure 20.8**). In a study by Steven Tanksley, QTL mapping has identified more than 28 QTLs related to this thousand-fold variation in fruit weight. More than ten years of work was required to localize, identify, and clone one of these QTLs, called *fw2.2* (on chromosome 2). Within this QTL, a specific gene, *ORFX*, has been identified, and alleles at this locus are responsible for about 30 percent of the variation in fruit weight.

FIGURE 20.8 A theoretical wild species of tomato, similar in size to the tomato on the left, is regarded as the ancestor of all modern tomatoes, including the beefsteak tomato shown at the right.

The *ORFX* gene has been isolated, cloned, and transferred between plants, with interesting results. One allele of *ORFX* is present in all wild small-fruited varieties of tomatoes investigated, while another allele is present in all domesticated large-fruited varieties. When a cloned *ORFX* gene from small-fruited varieties is transferred to a plant that normally produces large tomatoes, the transformed plant produces fruits that are greatly reduced in weight. In the varieties studied by Tanksley's group, the reduction averaged 17 grams, a statistically significant phenotypic change caused by the action of a gene found within a single QTL.

Further analysis of *ORFX* revealed that this gene encodes a protein that negatively regulates cell division during fruit development. Differences in the time of gene expression and differences in the amount of transcript produced lead to small or large fruit. Higher levels of expression mediated by transferred *ORFX* alleles exert a negative control over cell division, resulting in smaller tomatoes.

Yet *ORFX* and other related genes cannot account for all the observed variation in tomato size. Analysis of two other QTLs, *lc* (located on chromosome 2) and *fas* (on chromosome 11), indicates that the development of extreme differences in fruit size resulting from artificial selection also involves an increase in the number of seed compartments, called locules, in the mature fruit. Small, ancestral varieties produce fruit with two to four locules, but the large-fruited present-day strains have six or more of these compartments. The *lc* QTL maps to a noncoding region of the genome and consists of two SNPs that regulate the expression of nearby genes responsible for some of the increase in locule number. In addition, *lc* interacts with certain alleles of *fas*, another SNP locus, to further increase locule number, giving rise to the wide range of sizes and shapes in present-day tomatoes (**Figure 20.9**). Thus, QTLs that affect fruit size in tomatoes work by controlling at least two developmental processes: cell division early in development (*ORFX*) and the determination of the number of seed compartments (*lc* and *fas*).

Expression QTLs Regulate Gene Expression

The discovery that QTLs can control gene expression led researchers to systematically hunt for genomic loci that regulate the expression of one or more genes involved in quantitative traits. These loci, called **expression QTLs (eQTLs)**,

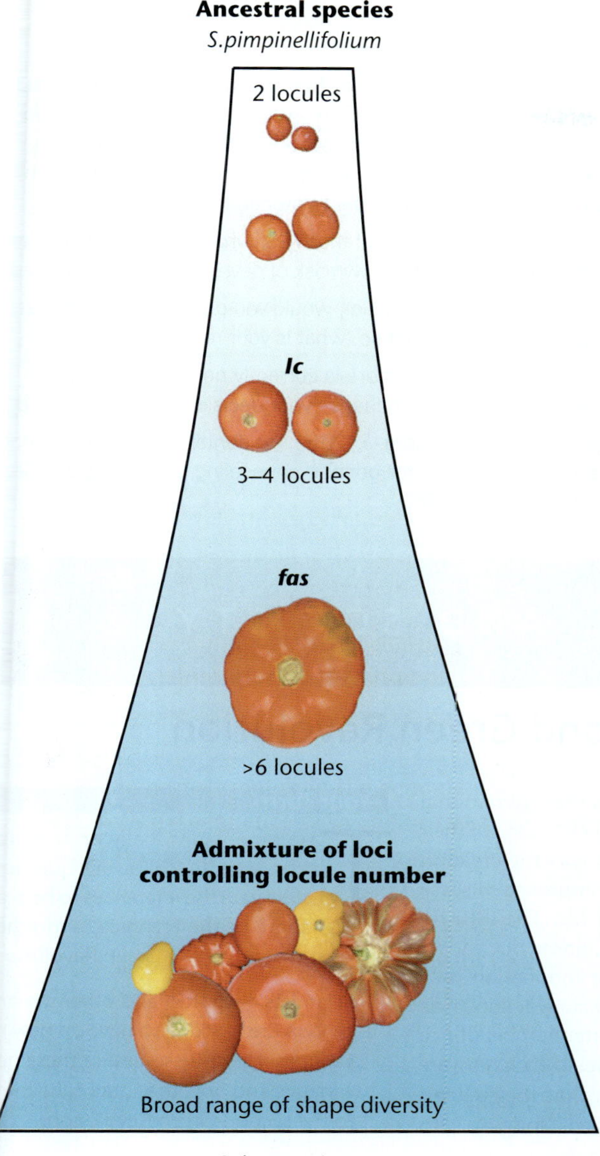

FIGURE 20.9 Changes in locule number during tomato domestication. The ancestral species *Solanum pimpinellifolium* contains two locules. At some point, a high-locule allele of *lc* was introduced and probably appeared before the introduction of the present-day *fas* allele, which further expanded locule number. These two QTLs are the major loci controlling locule number. As alleles of other loci controlling locule number were introduced into domesticated varieties, phenotypic diversity in the modern-day species *Solanum lycopersicum* expanded even further.

can be identified with the same methods used to find other QTLs. The first eQTLs were discovered in yeast but have now been identified in the genomes of many plants and animals. However, eQTLs differ from more classical QTLs in that the phenotype of variable gene expression can be regulated at any of the many steps along the path from DNA to protein (see Chapter 16 for a discussion of gene regulation in eukaryotes).

For example, most eQTLs identified to date are noncoding genomic variants, including SNPs or short indels (insertions/deletions) that affect transcription factor (TF) binding, the action of promoters and enhancers, and pre-mRNA processing. DNA sequence variations located in TF-binding sites are not only associated with differences in binding efficiency but are also linked to changes in DNA methylation, mRNA levels, and nucleosomal and chromatin changes including histone modifications.

Expression QTLs (eQTLs) and Genetic Disorders

We conclude this chapter by discussing the role that variation in the levels of gene expression plays in the phenotypic variation observed in complex disorders. In humans, variation resulting from the action of eQTLs encompasses a wide spectrum of phenotypes ranging from normal variations to disease states. The ability to study the expression of eQTLs and gene variability in the same individual helped identify gene/disease associations and the network of genes controlling those disorders. This approach has identified genes responsible for complex diseases such as asthma, cleft lip, Type 2 diabetes, and coronary artery disease. Asthma cases have risen dramatically over the last three decades and this disease is now a major public health concern. Genome-wide association studies (GWAS) have identified loci that confer susceptibility to asthma; however, the functions of many of these genes are unknown, and GWAS alone are unable to establish which alleles of these loci are responsible for susceptibility or the mechanism of their action.

To identify genes directly involved in asthma susceptibility, researchers collected lung specimens from over 1000 individuals and used lung-specific gene expression as a phenotype to study how genetic variants (DNA polymorphisms) are linked to both gene expression (eQTLs) and the asthma phenotype. Integration of the GWAS and the eQTL data identified 34 genes in six interconnected networks that constitute the gene set that causes asthma. Each network contains a single driver gene that controls the other genes in a network. These six driver genes are now candidates for drug discovery studies to develop therapies for this chronic and sometimes fatal disease. Similar approaches are likely to reveal the genetic networks that underlie other complex genetic disorders.

ESSENTIAL POINT

Quantitative trait loci, or QTLs, may be identified and mapped using DNA markers. ■

CASE STUDY A chance discovery

At an interview with a genetic counselor, the parents of a child with severe asthma learned that asthma is a complex disorder involving many genetic loci. The counselor explained that a method called whole genome sequencing (WGS) is now widely used in diagnosing and treating traits controlled by multiple loci and, in this case, could provide information to devise an effective therapy for their child. However, the parents were warned that because their child's entire genome was to be sequenced, information unrelated to asthma, but with potentially serious health consequences, might be discovered. After permission was granted, genome analysis created a panel of loci for therapy design. The analysis also revealed that the child carried two copies of an allele conferring an increased risk for Alzheimer disease. One copy of this allele increases the risk 4-fold; two copies raise the risk to 12-fold. Even though the child and both parents are at risk, current guidelines do not require that this finding be disclosed because it is unrelated to the primary reason for undertaking WGS. Knowing that disclosure was not legally required, but feeling she may have an ethical responsibility to divulge this information, the counselor was conflicted regarding how to proceed.

1. Based on the outcome of the WGS, what can the counselor tell the parents about their own risk of developing Alzheimer disease?
2. If you were the counselor, would you disclose this information to the parents, and if so, what is your reasoning?
3. Would the fact that there is currently no treatment for Alzheimer disease influence your decision about disclosure?

See Roche, M., and Berg, J. (2015). Incidental findings with genomic testing: Implications for genetic counseling practice. *Curr. Genet. Med. Rep.* 3:166–176.

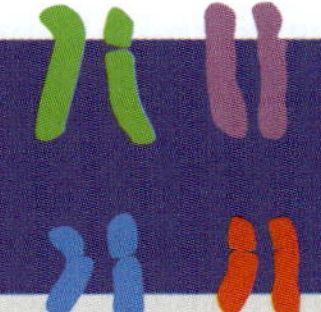

GENETICS, ETHICS, AND SOCIETY

Rice, Genes, and the Second Green Revolution

Of the 7 billion people now living on Earth, more than 800 million do not have enough to eat. This number is expected to grow by an additional 1 million people each year for the next several decades. How will we be able to feed the estimated 8 billion people on Earth by 2025?

The past gives us some basis for optimism. In the 1950s and 1960s, plant scientists set about to increase the production of crop plants, including the three most important grains—rice, wheat, and maize. These efforts became known as the *Green Revolution*. The approach was three-fold: (1) to increase the use of fertilizers, pesticides, and irrigation; (2) to bring more land under cultivation; and (3) to develop improved varieties of crop plants by intensive plant breeding.

The results were dramatic. Developing nations more than doubled their production of rice, wheat, and maize between 1961 and 1985. The Green Revolution saved millions of people from starvation and improved the quality of life for millions more; however, its impact is beginning to wane. If food production is to keep pace with the projected increase in the world's population, we will need a second Green Revolution.

Central to the success of the Green Revolution was research involving rice, upon which about half of the Earth's population depends. Advances were facilitated by the establishment in 1960 of the International Rice Research Institute (IRRI). One of their major accomplishments was the creation of a rice variety with improved disease resistance and higher yields. The IRRI research team crossed a Chinese rice variety (*Dee-geo-woo-gen*) and an Indonesian variety (*Peta*) to create a new cultivar known as IR8. IR8 produced a greater number of rice kernels per plant. However, IR8 plants were so top heavy with grain that they tended to fall over—a trait called "lodging." To reduce lodging, IRRI breeders crossed IR8 with a dwarf native variety to create semi-dwarf lines. Due in part to the adoption of the semi-dwarf IR8 lines, the world production of rice doubled in 25 years.

Despite the progress brought about by the introduction of IR8, scientists predict that we will need a further 40 percent increase in the annual rice harvest to keep pace with anticipated population growth during the next 30 years. Not only will we need higher yields, but also new rice varieties with greater disease resistance and tolerance to extreme climate changes, drought, salinity, and loss of soil fertility. Dozens of quantitative trait loci (QTLs) appear to contribute to these traits, making the task even more challenging. In the near future, scientists will need to introduce these traits into current dwarf varieties of domestic rice, using conventional breeding, genomics, and genetic engineering.

Your Turn

Take time, individually or in groups, to answer the following questions. Investigate the references to help you discuss some of the technical and ethical issues surrounding the Green Revolution.

1. Scientists from IRRI and other research centers are working to develop new rice varieties. Describe several of these new varieties and how they may contribute to the second Green Revolution.

Learn about some of the research projects supported by IRRI on the IRRI Web site (http://irri.org/our-work/our-research-networks). *Also, read about the C4 project at* Dayton, L. (2014). Blue-sky rice. *Nature* 514:S52–S54.

2. Despite its benefits, some critics question the long-term practical and ethical outcomes of the Green Revolution. What are some of these questions and criticisms? Which of these do you think has merit, and how can some of them be addressed?

A discussion of this topic can be found in Pingali, P. L. (2012). Green Revolution: Impacts, limits, and the path ahead. *Proc. Natl. Acad. Sci. USA* 09 (31):12302–12308. *Also see* Ellison, K., and Wellner, K. (2016). The Green Revolution: Research, ethics, and society at http://www.onlineethics.org/Resources/Cases/GreenRevolution.aspx.

INSIGHTS AND SOLUTIONS

1. In a certain plant, height varies from 6 to 36 cm. When 6-cm and 36-cm plants were crossed, all F_1 plants were 21 cm. In the F_2 generation, a continuous range of heights was observed. Most were around 21 cm, and 3 of 200 were as short as the 6-cm P_1 parent.

   (a) What mode of inheritance does this illustrate, and how many gene pairs are involved?

   (b) How much does each additive allele contribute to height?

   (c) List all genotypes that give rise to plants that are 31 cm.

Solution:

(a) Polygenic inheritance is illustrated when a trait is continuous and when alleles contribute additively to the phenotype. The 3/200 ratio of F_2 plants is the key to determining the number of gene pairs. This reduces to a ratio of 1/66.7, very close to 1/64. Using the formula $1/4^n = 1/64$ (where 1/64 is equal to the proportion of F_2 phenotypes as extreme as either P_1 parent), $n = 3$. Therefore, three gene pairs are involved.

(b) The variation between the two extreme phenotypes is

$$36 - 6 = 30 \text{ cm}$$

Because there are six potential additive alleles (*AABBCC*), each contributes

$$30/6 = 5 \text{ cm}$$

to the base height of 6 cm, which results when no additive alleles (*aabbcc*) are part of the genotype.

(c) All genotypes that include five additive alleles will be 31 cm (5 alleles × 5 cm/allele + 6 cm base height = 31 cm). Therefore, *AABBCc*, *AABbCC*, and *AaBBCC* are the genotypes that will result in plants that are 31 cm.

2. A plant of unknown phenotype and genotype from the population described in Problem 1 was testcrossed, with the following results

   1/4 11 cm
   2/4 16 cm
   1/4 21 cm

   An astute genetics student realized that the unknown plant could be only one phenotype but could be any of three genotypes. What were they?

Solution: When testcrossed (with *aabbcc*), the unknown plant must be able to contribute either one, two, or three additive alleles in its gametes in order to yield the three phenotypes in the offspring. Since no 6-cm offspring are observed, the unknown plant never contributes all nonadditive alleles (*abc*). Only plants that are homozygous at one locus and heterozygous at the other two loci will meet these criteria. Therefore, the unknown parent can be any of three genotypes, all of which have a phenotype of 26 cm:

AABbCc
AaBbCC
AaBBCc

For example, in the first genotype (*AABbCc*),

AABbCc × *aabbcc*
↓
1/4 *AaBbCc* 21 cm
1/4 *AaBbcc* 16 cm
1/4 *AabbCc* 16 cm
1/4 *Aabbcc* 11 cm

which is the ratio of phenotypes observed.

3. The mean and variance of corolla length in two highly inbred strains of *Nicotiana* and their progeny are shown in the following table. One parent (P_1) has a short corolla, and the other parent (P_2) has a long corolla. Calculate the broad-sense heritability (H^2) of corolla length in this plant.

Strain	Mean (mm)	Variance (mm)
P_1 short	40.47	3.12
P_2 long	93.75	3.87
$F_1(P_1 \times P_2)$	63.90	4.74
$F_2(F_1 \times F_1)$	68.72	47.70

Solution: The formula for estimating heritability is $H^2 = H_G/V_P$, where V_G and V_P are the genetic and phenotypic components of variation, respectively. The main issue in this problem is obtaining some estimate of two components of phenotypic variation: genetic and environmental factors. V_P is the combination of genetic and environmental variance. Because the two parental strains are true breeding, they are assumed to be homozygous, and the variance of 3.12 and 3.87 is considered to be the result of environmental influences. The average of these two values is 3.50. The F_1 is also genetically homogeneous and gives us an additional estimate of the impact of environmental factors. By averaging this value along with that of the parents,

$$\frac{4.74 + 3.50}{2} = 4.12$$

we obtain a relatively good idea of environmental impact on the phenotype. The phenotypic variance in the F_2 is the sum of the genetic (V_G) and environmental (V_E) components. We have estimated the environmental input as 4.12, so 47.70 minus 4.12 gives us an estimate of V_G of 43.58. Heritability then becomes 43.58/47.70, or 0.91. This value, when interpreted as a percentage, indicates that about 91 percent of the variation in corolla length is due to genetic influences.

Problems and Discussion Questions

Mastering Genetics Visit for instructor-assigned tutorials and problems.

1. **HOW DO WE KNOW?** In this chapter, we focused on a mode of inheritance referred to as quantitative genetics, as well as many of the statistical parameters utilized to study quantitative traits. Along the way, we found opportunities to consider the methods and reasoning by which geneticists acquired much of their understanding of quantitative genetics. From the explanations given in the chapter, what answers would you propose to the following fundamental questions:
   (a) How can we ascertain the number of polygenes involved in the inheritance of a quantitative trait?
   (b) What findings led geneticists to postulate the multiple-factor hypothesis that invoked the idea of additive alleles to explain inheritance patterns?
   (c) How do we assess environmental factors to determine if they impact the phenotype of a quantitatively inherited trait?
   (d) How do we know that monozygotic twins are not identical genotypically as adults?

2. **CONCEPT QUESTION** Review the Chapter Concepts list on p. 416. These all center on quantitative inheritance and the study and analysis of polygenic traits. Write a short essay that discusses the difference between the more traditional Mendelian and Neomendelian modes of inheritance (qualitative inheritance) and quantitative inheritance. ■

3. Define the following: (a) polygenic, (b) additive alleles, (c) monozygotic and dizygotic twins, (d) heritability, and (e) QTL.

4. A dark-red strain and a white strain of wheat are crossed and produce an intermediate, medium-red F_1. When the F_1 plants are interbred, an F_2 generation is produced in a ratio of 1 dark-red: 4 medium-dark-red: 6 medium-red: 4 light-red: 1 white. Further crosses reveal that the dark-red and white F_2 plants are true breeding.
   (a) Based on the ratios in the F_2 population, how many genes are involved in the production of color?
   (b) How many additive alleles are needed to produce each possible phenotype?
   (c) Assign symbols to these alleles and list possible genotypes that give rise to the medium-red and light-red phenotypes.
   (d) Predict the outcome of the F_1 and F_2 generations in a cross between a true-breeding medium-red plant and a white plant.

5. Height in humans depends on the additive action of genes. Assume that this trait is controlled by the four loci *R*, *S*, *T*, and *U* and that environmental effects are negligible. Instead of additive versus nonadditive alleles, assume that additive and partially additive alleles exist. Additive alleles contribute two units, and partially additive alleles contribute one unit to height.
   (a) Can two individuals of moderate height produce offspring that are much taller or shorter than either parent? If so, how?
   (b) If an individual with the minimum height specified by these genes marries an individual of intermediate or moderate height, will any of their children be taller than the tall parent? Why or why not?

6. A strain of plants has a mean height of 24 cm. A second strain of the same species from a different geographical region also has a mean height of 24 cm. When plants from the two strains are crossed together, the F_1 plants are the same height as the parent plants. However, the F_2 generation shows a wide range of heights; the majority are like the P_1 and F_1 plants, but approximately 4 of 1000 are only 12 cm high, and about 4 of 1000 are 36 cm high.
   (a) What mode of inheritance is occurring here?
   (b) How many gene pairs are involved?
   (c) How much does each gene contribute to plant height?
   (d) Indicate one possible set of genotypes for the original P_1 parents and the F_1 plants that could account for these results.
   (e) Indicate three possible genotypes that could account for F_2 plants that are 18 cm high and three that account for F_2 plants that are 33 cm high.

7. Erma and Harvey were a compatible barnyard pair, but a curious sight. Harvey's tail was only 6 cm long, while Erma's was 30 cm. Their F_1 piglet offspring all grew tails that were 18 cm. When inbred, an F_2 generation resulted in many piglets (Erma and Harvey's grandpigs), whose tails ranged in 4-cm intervals from 6 to 30 cm (6, 10, 14, 18, 22, 26, and 30). Most had 18-cm tails, while 1/64 had 6-cm tails and 1/64 had 30-cm tails.
   (a) Explain how these tail lengths were inherited by describing the mode of inheritance, indicating how many gene pairs were at work, and designating the genotypes of Harvey, Erma, and their 18-cm-tail offspring.
   (b) If one of the 18-cm F_1 pigs is mated with one of the 6-cm F_2 pigs, what phenotypic ratio would be predicted if many offspring resulted? Diagram the cross.

8. In the following table, average differences of height, weight, and fingerprint ridge count between monozygotic twins (reared together and apart), dizygotic twins, and nontwin siblings are compared:

Trait	MZ Reared Together	MZ Reared Apart	DZ Reared Together	Sibs Reared Together
Height (cm)	1.7	1.8	4.4	4.5
Weight (kg)	1.9	4.5	4.5	4.7
Ridge count	0.7	0.6	2.4	2.7

   Based on the data in this table, which of these quantitative traits has the highest heritability values?

9. Define the term *broad-sense heritability* (H^2). What is implied by a relatively high value of H^2? Express aspects of broad-sense heritability in equation form.

10. Describe the value of using twins in the study of questions relating to the relative impact of heredity versus environment.

11. Two different crosses were set up between carrots (*Daucus carota*) of different colors and carotenoid content (Santos, Carlos A. F. and Simon, Philipp W. 2002. *Horticultura Brasileira* 20). Analyses of the F_2 generations showed that four loci are associated with the α carotene content of carrots, with a broad-sense heritability of 90%. How many distinct phenotypic categories and genotypes would be seen in each F_2 generation, and what does a broad-sense heritability of 90% mean for carrot horticulture?

12. Corn plants from a test plot are measured, and the distribution of heights at 10-cm intervals is recorded in the following table:

Height (cm)	Plants (no.)
100	20
110	60
120	90
130	130
140	180
150	120
160	70
170	50
180	40

Calculate (a) the mean height, (b) the variance, (c) the standard deviation, and (d) the standard error of the mean. Plot a rough graph of plant height against frequency. Do the values represent a normal distribution? Based on your calculations, how would you assess the variation within this population?

13. The following variances were determined in a breed of cattle.

Trait	V_P	V_G	V_A
Body size	51.3	23.2	10.18
Milk production	42.5	16.8	12.32

(a) Calculate broad-sense (H^2) and narrow-sense (h^2) heritabilities for each trait in this cattle breed.

(b) What would you recommend to a breeder considering this breed for milk production?

14. The mean and the variance of body size in two highly inbred strains of mice (P_1 and P_2) and their progeny (F_1 and F_2) are shown here.

Strain	Mean (cm)	Variance
P_1	19	3.3
P_2	26	3.1
F_1	20	4.2
F_2	21	7.5

Calculate the broad-sense heritability (H^2) of body size in this species.

15. A hypothetical study investigated the vitamin A content and the cholesterol content of eggs from a large population of chickens. The variances (V) were calculated, as shown at the top of the next column:

Variance	Vitamin A	Cholesterol
V_P	123.5	862.0
V_E	96.2	484.6
V_A	12.0	192.1
V_D	15.3	185.3

(a) Calculate the narrow-sense heritability (h^2) for both traits.

(b) Which trait, if either, is likely to respond to selection?

16. In a herd of dairy cows the narrow-sense heritability for milk protein content is 0.76, and for milk butterfat it is 0.82. The correlation coefficient between milk protein content and butterfat is 0.91. If the farmer selects for cows producing more butterfat in their milk, what will be the most likely effect on milk protein content in the next generation?

17. A population of laboratory mice was weighed at the age of six weeks (full adult weight) and found to have a mean weight of 20 g. The narrow heritability of weight gain (h^2) is known to be 0.25 in this laboratory strain. If mice weighing 24 g are selected and mated at random, what is the expected mean weight of the next generation?

18. If the experiment was repeated by mating mice 4 g lighter than the mean (16 g), what would be the result? If repeated experiments were carried out always selecting mice that were 4 g lighter than the mean in the current generation, what would eventually happen?

19. In a herd of Texas Longhorn cattle, the mean horn length from tip to tip is 52" and h^2 is 0.2. Predict the mean horn length if cattle with horns 61" long are interbred.

20. In a population of 100 inbred, genotypically identical rice plants, variance for grain yield is 4.67. What is the heritability for yield? Would you advise a rice breeder to improve yield in this strain of rice plants by selection?

21. While most quantitative traits display continuous variation, there are others referred to as "threshold traits" that are distinguished by having a small number of discrete phenotypic classes. For example, Type 2 diabetes (adult-onset diabetes) is considered to be a polygenic trait, but demonstrates only two phenotypic classes: individuals who develop the disease and those who do not. Theorize how a threshold trait such as Type 2 diabetes may be under the control of many polygenes, but express a limited number of phenotypes.

21

Population and Evolutionary Genetics

These ladybird beetles, from the Chiricahua Mountains in Arizona, show considerable phenotypic variation.

CHAPTER CONCEPTS

- Most populations and species harbor considerable genetic variation.
- This variation is reflected in the alleles distributed among populations of a species.
- The relationship between allele frequencies and genotype frequencies in an ideal population is described by the Hardy–Weinberg law.
- Selection, migration, and genetic drift can cause changes in allele frequency.
- Mutation creates new alleles in a population gene pool.
- Nonrandom mating changes population genotype frequency but not allele frequency.
- A reduction in gene flow between populations, accompanied by selection or genetic drift, can lead to reproductive isolation and speciation.
- Genetic differences between populations or species are used to reconstruct evolutionary history.

In the mid-nineteenth century, Alfred Russel Wallace and Charles Darwin identified natural selection as the mechanism of evolution. In his book *On the Origin of Species*, published in 1859, Darwin provided evidence that populations and species are not fixed, but change, or evolve, over time due to natural selection. However, Wallace and Darwin could not explain either the origin of the variations that provide the raw material for evolution or the mechanisms by which such variations are passed from parents to offspring. Gregor Mendel published his work on the inheritance of traits in 1866, but it received little notice at the time. After the rediscovery of Mendel's work in 1900, twentieth-century biologists began a 30-year effort to reconcile Mendel's concept of genes and alleles with the theory of evolution by natural selection. As biologists applied the principles of Mendelian genetics to populations, both the source of variation (mutation and recombination) and the mechanism of inheritance (segregation of alleles) were explained. We now view evolution as a consequence of changes in genetic material through mutation and changes in allele frequencies in populations over time. This union of population genetics with the theory of natural selection generated a new view of the evolutionary process, called *neo-Darwinism*.

In addition to natural selection, other forces including mutation, migration, and drift, individually and collectively, alter allele frequencies and bring about evolutionary divergence that eventually may result in **speciation,** the formation of new species. Speciation is facilitated by environmental diversity. If a population is spread over a geographic range encompassing a number of ecologically distinct subenvironments with different selection pressures, the populations occupying these areas may adapt over time and become genetically distinct from one another. Genetically differentiated populations may remain in existence, become extinct, reunite with each other, or continue to diverge until they become reproductively isolated. Populations that are reproductively isolated are regarded as separate species. Genetic changes within populations can modify a species over time, transform it into another species, or cause it to split into two or more species.

Population geneticists investigate patterns of genetic variation within and among groups of interbreeding individuals. Changes in genetic structure form the basis for the evolution of a population. Thus, population genetics has become an important subdiscipline of evolutionary biology. In this chapter, we examine the population genetics processes of **microevolution,** which is defined as evolutionary change within populations of a species, and then consider how molecular aspects of these processes can be extended to **macroevolution,** which is defined as evolutionary events leading to the emergence of new species and other taxonomic groups.

21.1 Genetic Variation Is Present in Most Populations and Species

A **population** is a group of individuals belonging to the same species that live in a defined geographic area and actually or potentially interbreed. In thinking about the human population, we can define it as everyone who lives in the United States or in Sri Lanka, or we can specify a population as all the residents of a particular small town or village.

The genetic information carried by members of a population constitutes that population's **gene pool.** At first glance, it might seem that a population that is well adapted to its environment must have a gene pool that is highly homozygous because it would seem likely that the most favorable allele at each locus is present at a high frequency. In addition, a look at most populations of plants and animals reveals many phenotypic similarities among individuals. However, a large body of evidence indicates that, in reality, most populations contain a high degree of heterozygosity. This built-in genetic variation is not necessarily apparent in the phenotype; hence, detecting it is not a simple task. Nevertheless, the amount of variation within a population can be revealed by several methods.

Detecting Genetic Variation

The detection and use of genetic variation in individuals and populations began long before genetics emerged as a science. Millennia ago, plant and animal breeders began using artificial selection to domesticate plants and animals.

One of the more spectacular examples of how much variation exists in the gene pool of a species was the use of selective breeding to create hundreds of dog breeds in nineteenth-century England over a period of less than 75 years. Many people, seeing a Chihuahua (about 10 inches high) and a Great Dane (about 42 inches high) for the first time, might find it difficult to believe they are both members of the same species (**Figure 21.1**).

However, as genetic technology developed in the last century, the ability to detect and quantify genetic variation in genes, in individual genomes, and in the genomes of populations has grown exponentially.

Recombinant DNA Technology and Genetic Variation

After the development of recombinant DNA technology in the 1970s, efforts centered on detecting genetic variation in the sequence of single genes carried by individuals in a population.

In one such study, Martin Kreitman isolated, cloned, and sequenced copies of the alcohol dehydrogenase (*Adh*) gene from individuals representing five different populations of *Drosophila melanogaster*. The 11 cloned genes from these five populations contained a total of 43 nucleotide differences in the *Adh* sequence of 2721 base pairs (**Figure 21.2**). These variations are distributed throughout the gene: 14 in exon coding regions, 18 within introns, and 11 in untranslated flanking regions. Of the 14 variations in exons, only one leads to a phenotypic difference, an amino acid replacement in codon 192, resulting in the two known alleles of this gene. The other 13 nucleotide changes do not lead to amino acid replacements and are silent variations of this gene.

FIGURE 21.1 The size difference between a Chihuahua and a Great Dane illustrates the high degree of genetic variation present in the dog genome.

	Exon 3	Intron 3	Exon 4
Consensus *Adh* sequence:	C C C C	G G A A T	C T C C A* C T A G
Strain			
Wa-S	T T • A	C A • T A	A C • • • • • • • •
Fl1-S	T T • A	C A • T A	A C • • • • • • • •
Ja-S	• • • •	• • • • •	• • • T • T • C A
Fl-F	• • • •	• • • • •	• • G T C T C C •
Ja-F	• • A •	• • G • •	• • G T C T C C •

FIGURE 21.2 DNA sequence variation in parts of the *Drosophila Adh* gene in a sample of 11 laboratory strains derived from five natural populations. The dots represent nucleotides that are the same as the consensus sequence; letters represent nucleotide polymorphisms. An A/C polymorphism (A*) in codon 192 creates the two *Adh* alleles (F and S). All other polymorphisms are silent or noncoding.

Genetic Variation in Genomes

The development of **next generation sequencing (NGS)** technology has extended the detection of genomic variation from individuals to populations. The 1000 Genomes Project, which ran from 2008 through 2015, was a global effort to identify and catalog at least 95 percent of the common genetic variations carried by the 7 billion people now inhabiting the planet. The Project eventually sequenced the genomes of 2504 individuals from 26 populations using a combination of whole-genome sequencing, exome sequencing, and microarray genotyping.

Over 88 million genetic variants were identified in the human genome, including 84.7 million SNPs, 3.6 million indels (short insertion and deletions), and 60,000 structural variants [copy number variations (CNVs), Alu and LINE-1 insertions, etc.].

The Project's overall goal is to explore and understand the relationship between genotype and phenotype. In humans, this translates into using association studies to identify variants associated with disease. For example, in studies to date, no single variant has been associated with diabetes; this implies that a combination of heritable multiple rare variants is related to this common disorder. Eventually, researchers hope to associate specific genetic variants with cellular pathways and networks associated with complex disorders such as hypertension, cardiovascular disease, and neurological disorders associated with protein accumulation such as Alzheimer disease and Huntington disease.

Explaining the High Level of Genetic Variation in Populations

The finding that populations harbor considerable genetic variation at the amino acid and nucleotide levels came as a surprise to many evolutionary biologists. The early consensus had been that selection would favor a single optimal (wild-type) allele at each locus and that, as a result, populations would have high levels of homozygosity. This expectation was shown conclusively to be wrong, and considerable research and argument have ensued concerning the forces that maintain such high levels of genetic variation.

The **neutral theory** of molecular evolution, proposed by Motoo Kimura in 1968, proposes that mutations leading to amino acid substitutions are usually detrimental, with only a very small fraction being favorable. Some mutations are neutral; that is, they are functionally equivalent to the allele they replace. Mutations that are favorable or detrimental are preserved or removed from the population, respectively, by natural selection. However, the frequency of the neutral alleles in a population will be determined by mutation rates and random genetic drift, and not by selection. Some neutral mutations will drift to fixation in the population; other neutral mutations will be lost. At any given time, a population may contain several neutral alleles at any particular locus. The diversity of alleles at most loci does not, therefore, reflect the action of natural selection, but instead is a function of population size (larger populations have more variation) and the fraction of mutations that are neutral.

The alternative explanation for the surprisingly high genetic variation in populations is natural selection. There are several extensively documented examples in which enzyme or protein variations are maintained by adaptation to certain environmental conditions. The well-known resistance of sickle-cell anemia heterozygotes to infection by malarial parasites is such an example.

Fitness differences of a fraction of a percent would be sufficient to maintain such variations by selection, but at that level their presence would be difficult to measure. Current data are therefore insufficient to determine what fraction of molecular genetic variation is neutral and what fraction is subject to selection. The neutral theory nonetheless serves a crucial function: It points out that some genetic variation is expected simply as a result of mutation and drift. In addition, the neutral theory provides a working hypothesis for studies of molecular evolution. In other words, biologists must find positive evidence that selection is acting on allele frequencies at a particular locus before they can reject the simpler assumption that only mutation and drift are at work.

ESSENTIAL POINT

Genetic variation is widespread in most populations and provides a reservoir of alleles that serve as the basis for evolutionary changes within populations. ■

21.2 The Hardy–Weinberg Law Describes Allele Frequencies and Genotype Frequencies in Population Gene Pools

Often when we examine a single gene in a population, we find that different allele combinations of this gene result in individuals with different genotypes. For example, two alleles, *A* and *a*, of the *A* gene can be combined to produce three genotypes: *AA*, *Aa*, and *aa*. Key elements of population genetics depend on the calculation of allele frequencies and genotype frequencies in a gene pool, and the determination of how these frequencies change from one generation to the next. Population geneticists use these calculations to answer questions such as: How much genetic variation is present in a population? Are genotypes randomly distributed in time and space, or do discernible patterns exist? What processes affect the composition of a population's gene pool? Do these processes produce genetic divergence among populations that may lead to the formation of new species? Changes in allele frequencies in a population that do not directly result in species formation are examples of microevolution. In the following sections, we will discuss microevolutionary changes in population gene pools and then will consider macroevolution and the process of speciation.

The relationship between the relative proportions of alleles in the gene pool and the frequencies of different genotypes in the population was elegantly described in a mathematical model developed independently by the British mathematician Godfrey H. Hardy and the German physician Wilhelm Weinberg. This model, called the **Hardy–Weinberg law,** describes what happens to allele and genotype frequencies in an "ideal" population that is infinitely large, randomly mating, and not subject to evolutionary forces such as mutation, migration, or selection.

Calculating Genotype Frequencies

The Hardy–Weinberg model uses the principle of Mendelian segregation and simple probability to explain the relationship between allele and genotype frequencies in a population. We can demonstrate how this works by considering a single autosomal gene with two alleles, *A* and *a*, in a population where the frequency of *A* is 0.7 and the frequency of *a* is 0.3. Note that $0.7 + 0.3 = 1$, indicating that all the alleles of gene *A* present in the population are accounted for.

These allele frequencies mean that the probability that any female gamete will contain *A* is 0.7, and the probability that a male gamete will contain *A* is also 0.7. The probability that *both* gametes will contain *A* is $0.7 \times 0.7 = 0.49$. Thus we predict that in the offspring, the genotype *AA* will occur 49 percent of the time. The probability that a zygote will be formed from a female gamete carrying *A* and a male gamete carrying *a* is $0.7 \times 0.3 = 0.21$, and the probability of a female gamete carrying *a* being fertilized by a male gamete carrying *A* is $0.3 \times 0.7 = 0.21$, so the frequency of genotype *Aa* in the offspring is $0.21 + 0.21 = 0.42 = 42$ percent. Finally, the probability that a zygote will be formed from two gametes carrying *a* is $0.3 \times 0.3 = 0.09$, so the frequency of genotype *aa* is 9 percent. As a check on our calculations, note that $0.49 + 0.42 + 0.09 = 1.0$, confirming that we have accounted for all possible genotypic combinations in the zygotes. These calculations are summarized in **Figure 21.3.**

Calculating Allele Frequencies

Now that we know the frequencies of genotypes in the next generation, what will be the allele frequencies in this new generation? Under the Hardy–Weinberg law, we assume that all genotypes have equal rates of survival and reproduction. This means that in the next generation, all genotypes contribute equally to the new gene pool. The *AA* individuals constitute 49 percent of the population, and we can predict that the gametes they produce will constitute 49 percent of the gene pool. These gametes all carry allele *A*. Similarly, *Aa* individuals constitute 42 percent of the population, so we predict that their gametes will constitute 42 percent of the new gene pool. Half (0.5) of these gametes will carry allele *A*. Thus, the frequency of allele *A* in the gene pool is $0.49 + (0.5)\,0.42 = 0.7$. The other half of the gametes produced by *Aa* individuals will carry allele *a*. The *aa* individuals constitute 9 percent of the population, so their gametes will constitute 9 percent of the new gene pool. All these gametes carry allele *a*. Thus, we can predict that the allele *a* in

		Sperm	
		fr(*A*) = 0.7	fr(*a*) = 0.3
Eggs	fr(*A*) = 0.7	fr(*AA*) = 0.7 × 0.7 = 0.49	fr(*Aa*) = 0.7 × 0.3 = 0.21
	fr(*a*) = 0.3	fr(*aA*) = 0.3 × 0.7 = 0.21	fr(*aa*) = 0.3 × 0.3 = 0.09

FIGURE 21.3 Calculating genotype frequencies from allele frequencies. Gametes represent samples drawn from the gene pool to form the genotypes of the next generation. In this population, the frequency (fr) of the *A* allele is 0.7, and the frequency of the *a* allele is 0.3. The frequencies of the genotypes in the next generation are calculated as 0.49 for *AA*, 0.42 for *Aa*, and 0.09 for *aa*. Under the Hardy–Weinberg law, the frequencies of *A* and *a* remain constant from generation to generation.

the new gene pool is (0.5) 0.42 + 0.09 = 0.3. As a check on our calculation, note that 0.7 + 0.3 = 1.0, accounting for all the gametes in the gene pool of the new generation.

The Hardy–Weinberg Law and Its Assumptions

Because the Hardy–Weinberg law is a mathematical model, we use variables instead of numerical values for the allele frequencies in the general case. Imagine a gene pool in which the frequency of allele *A* is represented by p and the frequency of allele *a* is represented by q, such that $p + q = 1$. If we randomly draw male and female gametes from the gene pool and pair them to make a zygote, the probability that both will carry allele *A* is $p \times p$. Thus, the frequency of genotype *AA* among the zygotes is p^2. The probability that the female gamete carries *A* and the male gamete carries *a* is $p \times q$, and the probability that the female gamete carries *a* and the male gamete carries *A* is $q \times p$. Thus, the frequency of genotype *Aa* among the zygotes is $2pq$. Finally, the probability that both gametes carry *a* is $q \times q$, making the frequency of genotype *aa* among the zygotes q^2. Therefore, the distribution of genotypes among the zygotes is

$$p^2 + 2pq + q^2 = 1$$

These calculations are summarized in **Figure 21.4**. They demonstrate the two main predictions of the Hardy–Weinberg model:

1. Allele frequencies in our model population do not change from one generation to the next.
2. After one generation of random mating, genotype frequencies can be predicted from the allele frequencies.

In other words, there is no change in allele frequency, and for this locus, the population does not undergo any microevolutionary change. The theoretical population described by the Hardy–Weinberg model is based on the following assumptions:

1. Individuals of all genotypes have equal rates of survival and equal reproductive success—that is, there is no selection.
2. No new alleles are created or converted from one allele into another by mutation.
3. Individuals do not migrate into or out of the population.
4. The population is infinitely large, which in practical terms means that the population is large enough that sampling errors and other random effects are negligible.
5. Individuals in the population mate randomly.

These assumptions are what make the Hardy–Weinberg model so useful in population genetics research. By specifying the conditions under which the population does not evolve, the Hardy–Weinberg model can be used to identify the real-world forces that cause allele frequencies to change. Application of this model can also reveal "neutral genes" in a population gene pool—those not being operated on by the forces of evolution.

The Hardy–Weinberg model has three additional important consequences:

1. Dominant traits do not necessarily increase from one generation to the next.
2. **Genetic variability** can be maintained in a population, since, once established in an ideal population, allele frequencies remain unchanged.
3. Under Hardy–Weinberg assumptions, knowing the frequency of just one genotype enables us to calculate the frequencies of all other genotypes at that locus.

This is particularly useful in human genetics because we can calculate the frequency of heterozygous carriers for recessive genetic disorders even when all we know is the frequency of affected individuals.

	Sperm	
Eggs	fr(*A*) = *p*	fr(*a*) = *q*
fr(*A*) = *p*	fr(*AA*) = p^2	fr(*Aa*) = *pq*
fr(*a*) = *q*	fr(*aA*) = *qp*	fr(*aa*) = q^2

FIGURE 21.4 The general description of allele and genotype frequencies under Hardy–Weinberg assumptions. The frequency of allele *A* is *p*, and the frequency of allele *a* is *q*. After mating, the three genotypes *AA*, *Aa*, and *aa* have the frequencies p^2, $2pq$, and q^2, respectively.

NOW SOLVE THIS

21.1 The ability to taste the compound phenylthiocarbamide (PTC) is controlled by a dominant allele *T*. Individuals homozygous for the recessive allele *t* are unable to taste PTC. In a genetics class of 125 students, 88 can taste PTC and 37 cannot. Calculate the frequency of the *T* and *t* alleles in this population and the frequency of the genotypes.

■ **HINT:** *This problem involves an understanding of how to use the Hardy–Weinberg law. The key to its solution is to determine which allele frequency (p or q) you must estimate first when homozygous dominant and heterozygous genotypes have the same phenotype.*

21.3 The Hardy–Weinberg Law Can Be Applied to Human Populations

To show how allele frequencies are measured in a real population, let's consider a gene that influences an individual's susceptibility to infection by HIV-1, the virus responsible for acquired immunodeficiency syndrome (AIDS). A small number of individuals who make high-risk choices (such as having unprotected sex with HIV-positive partners) never become infected. Some of these individuals are homozygous for a mutant allele of a gene called *CCR5*.

The *CCR5* gene (**Figure 21.5**) encodes a protein called the C-C chemokine receptor-5 (CCR5). Chemokines are cell surface signaling molecules associated with the immune system. The CCR5 protein is also used by strains of HIV-1 to gain entry into cells. The mutant allele of the *CCR5* gene contains a 32-bp deletion, making the encoded protein shorter and nonfunctional, blocking the entry of HIV-1 into cells. The normal allele is called *CCR51* (also called *1*), and the mutant allele is called *CCR5-Δ32* (also called *Δ32*).

Individuals homozygous for the mutant allele (*Δ32/Δ32*) are resistant to HIV-1 infection. Heterozygous (*1/Δ32*) individuals are susceptible to HIV-1 infection but progress more slowly to AIDS. **Table 21.1** summarizes the genotypes possible at the *CCR5* locus and the phenotypes associated with each.

The discovery of the *CCR5-Δ32* allele generates two important questions: Which human populations carry this allele, and how common is it? To address these questions, teams of researchers surveyed members of several populations. Genotypes were determined by direct analysis of DNA (**Figure 21.6**). In one population, 79 individuals had genotype *1/1*, 20 had genotype *1/Δ32*, and 1 individual had genotype *Δ32/Δ32*. We can see that this population had 158 *1* alleles carried by the *1/1* individuals plus 20 *1* alleles carried by the *1/Δ32* individuals, for a total of 178. The frequency of the *CCR51* allele in the sample population is thus 178/200 = 0.89 = 89 percent. Twenty *1/Δ32* individuals and one *Δ32/Δ32* individual each carried a copy of the *CCR5-Δ32* allele. The frequency of the *CCR5-Δ32* allele is thus 22/200 = 0.11 = 11 percent. Notice that $p + q = 1$, confirming that we have accounted for all the alleles of the *CCR51* gene in the population. **Table 21.2** shows two methods for computing the frequencies of the alleles in the population surveyed.

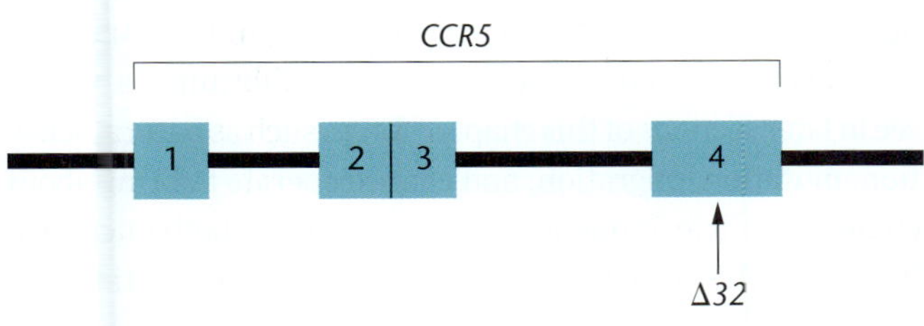

FIGURE 21.5 The organization of the *CCR5* gene in region 3p21.3 of human chromosome 3. The gene contains 4 exons and 2 introns (there is no intron between exons 2 and 3). The arrow shows the location of the 32-bp deletion in exon 4 that confers resistance to HIV-1 infection.

TABLE 21.1 ***CCR5* Genotypes and Phenotypes**

Genotype	Phenotype
1/1	Susceptible to sexually transmitted strains of HIV-1
1/Δ32	Susceptible but may progress to AIDS slowly
Δ32/Δ32	Resistant to most sexually transmitted strains of HIV-1

Can we expect the *CCR5-Δ32* allele to increase in human populations because it offers resistance to infection by HIV? This specific question is difficult to answer directly, but as we will see later in this chapter, when factors such as natural selection, mutation, migration, or genetic drift are present, the allele frequencies in a population may change from one generation to the next.

By determining allele frequencies over more than one generation, it is possible to determine whether the frequencies remain in equilibrium because the Hardy–Weinberg assumptions are operating. Populations that meet the Hardy–Weinberg assumptions are not evolving because allele frequencies (for the generations tested) are not changing. However, a population may be in Hardy–Weinberg equilibrium for the alleles being tested, but other genes may not be in equilibrium.

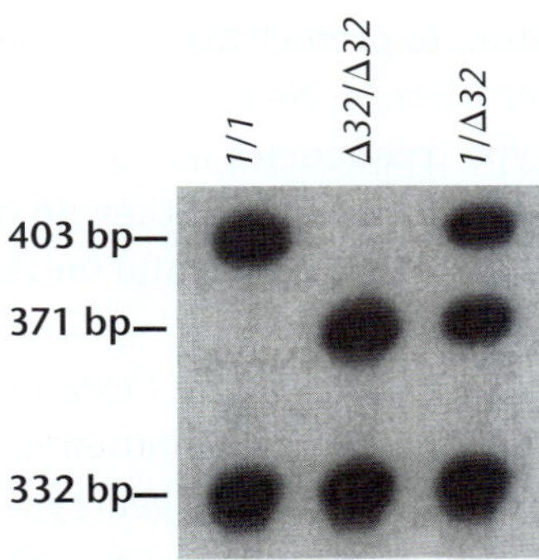

FIGURE 21.6 Allelic variation in the *CCR5* gene. Michel Samson and colleagues used polymerase chain reaction (PCR) to amplify a part of the *CCR5* gene containing the site of the 32-bp deletion, cut the resulting DNA fragments with a restriction enzyme, and ran the fragments on an electrophoresis gel. Each lane reveals the genotype of a single individual. The *1* allele produces a 332-bp fragment and a 403-bp fragment; the *Δ32* allele produces a 332-bp fragment and a 371-bp fragment. Heterozygotes produce three bands.

TABLE 21.2 Methods of Determining Allele Frequencies from Data on Genotypes

(a) Counting Alleles	Genotype			Total
	1/1	*1/Δ32*	*Δ32/Δ32*	
Number of individuals	79	20	1	100
Number of *1* alleles	158	20	0	178
Number of *Δ32* alleles	0	20	2	22
Total number of alleles	158	40	2	200

Frequency of *CCR5-1* in sample: 178/200 = 0.89 = 89%
Frequency of *CCR5-Δ32* in sample: 22/200 = 0.11 = 11%

(b) From Genotype Frequencies	Genotype			Total
	1/1	*1/Δ32*	*Δ32/Δ32*	
Number of individuals	79	20	1	100
Genotype frequency	79/100 = 0.79	20/100 = 0.20	1/100 = 0.01	1.00

Frequency of *CCR5-1* in sample: 0.79 + (0.5)0.20 = 0.89 = 89%
Frequency of *CCR5-Δ32* in sample: (0.5)0.20 + 0.01 = 0.11 = 11%

Testing for Hardy–Weinberg Equilibrium in a Population

One way to see if any of the Hardy–Weinberg assumptions do not hold in a given population is to determine whether the population's genotypes are in equilibrium. To do this, we first determine the genotype frequencies. This can be done directly from the phenotypes (if heterozygotes are recognizable), by analyzing proteins or DNA sequences, or indirectly, using the frequency of the HIV-1 resistant phenotype in the population to calculate genotype frequencies using the Hardy–Weinberg law. We can then calculate the allele frequencies from the genotype frequencies. Finally, the allele frequencies in the parental generation are used to predict the genotype frequencies in the next generation. According to the Hardy–Weinberg law, genotype frequencies are predicted to fit the $p^2 + 2pq + q^2 = 1$ relationship. If they do not, then one or more of the assumptions are invalid for the population in question.

To demonstrate, let's examine *CCR5* genotypes in a hypothetical population. Our population is composed of 283 individuals; of these, 223 have genotype *1/1*; 57 have genotype *1/Δ32*; and 3 have genotype *Δ32/Δ32*. These numbers represent the following genotype frequencies: *1/1* = 223/283 = 0.788, *1/Δ32* = 57/283 = 0.201, and *Δ32/Δ32* = 3/283 = 0.011, respectively. From the genotype frequencies, we can compute the *CCR5-1* allele frequency as 0.89 and the *CCR5-Δ32* allele frequency as 0.11. Once we know the allele frequencies, we can use the Hardy–Weinberg law to determine whether this population is in equilibrium. The allele frequencies predict the genotype frequencies in the next generation as follows:

- Expected frequency of genotype *1/1*:

$$p^2 = (0.89)^2 = 0.792$$

- Expected frequency of genotype *1/Δ32*:

$$2pq = 2(0.89)(0.11) = 0.196$$

- Expected frequency of genotype *Δ32/Δ32*:

$$q^2 = (0.11)^2 = 0.012$$

These expected frequencies are nearly identical to the frequencies observed in the parental generation. Our test of this population has failed to provide evidence that Hardy–Weinberg assumptions are being violated. The conclusion can be confirmed by using the whole numbers utilized in calculating the genotype frequencies to perform a χ^2 analysis (see Chapter 3). In this case, neither the genotype frequencies nor the allele frequencies are changing in this population, meaning that the population is in equilibrium. As we will see in later sections of this chapter, forces such as natural selection, mutation, migration, and chance operate to bring about changes in allele frequency. These forces drive both microevolution and the formation of new species (macroevolution).

ESSENTIAL POINT

Populations that are not in Hardy–Weinberg equilibrium may be undergoing changes in allele frequency owing to forces such as selection, drift, migration, or nonrandom mating. ■

NOW SOLVE THIS

21.2 Determine whether the following two sets of data represent populations that are in Hardy–Weinberg equilibrium.

(a) *CCR5* genotypes: *1/1*, 60 percent; *1/Δ32*, 35.1 percent; *Δ32/Δ32*, 4.9 percent

(b) Sickle-cell hemoglobin: *SS*, 75.6 percent; *Ss*, 24.2 percent; *ss*, 0.2 percent (*S* = normal hemoglobin allele; *s* = mutant hemoglobin allele)

HINT: *This problem involves an understanding of how to use the Hardy–Weinberg law to determine whether populations are in genetic equilibrium. The key to its solution is to first determine the allele frequencies based on the genotype frequencies provided.*

Calculating Frequencies for Multiple Alleles in Populations

Although we have used one-gene, two-allele systems as examples, many genes have several alleles, all of which can be found in a single population. The ABO blood group in humans (discussed in Chapter 4) is such an example. The locus *I* (isoagglutinin) has three alleles, I^A, I^B, and i, yielding six possible genotypic combinations (I^AI^A, I^BI^B, ii, I^AI^B, I^Ai, I^Bi). Remember that in this case I^A and I^B are codominant alleles and that both of these are dominant to i. The result is that homozygous I^AI^A and heterozygous I^Ai individuals are phenotypically identical, as are I^BI^B and I^Bi individuals, so we can distinguish only four phenotypic blood-type combinations: Type A, Type B, Type AB, and Type O.

By adding another variable to the Hardy–Weinberg equation, we can calculate both the genotype and allele frequencies for the situation involving three alleles. Let p, q, and r represent the frequencies of alleles I^A, I^B, and i, respectively. Note that because there are three alleles

$$p + q + r = 1$$

Under Hardy–Weinberg assumptions, the frequencies of the genotypes are given by

$$(p + q + r)^2 = p^2 + q^2 + r^2 + 2pq + 2pr + 2qr = 1$$

If we know the frequencies of blood types for a population, we can then estimate the frequencies for the three alleles of the ABO system. For example, in one population sampled, the following blood-type frequencies are observed: A = 0.53, B = 0.133, O = 0.26. Because the i allele is recessive, the population's frequency of Type O blood equals the proportion of the recessive genotype r^2. Thus,

$$r^2 = 0.26$$
$$r = \sqrt{0.26}$$
$$= 0.51$$

Using r, we can calculate the allele frequencies for the I^A and I^B alleles. The I^A allele is present in two genotypes, I^AI^A and I^Ai. The frequency of the I^AI^A genotype is represented by p^2 and the I^Ai genotype by $2pr$. Therefore, the combined frequency of Type A blood and Type O blood is given by

$$p^2 + 2pr + r^2 = 0.53 + 0.26$$

If we factor the left side of the equation and take the sum of the terms on the right,

$$(p + r)^2 = 0.79$$
$$p + r = \sqrt{0.79}$$
$$p = 0.89 - r$$
$$= 0.89 - 0.51 = 0.38$$

Having calculated p and r, the frequencies of allele I^A and allele i, we can now calculate the frequency for the I^B allele:

$$p + q + r = 1$$
$$q = 1 - p - r$$
$$= 1 - 0.38 - 0.51$$
$$= 0.11$$

The phenotypic and genotypic frequencies for this population are summarized in **Table 21.3**.

TABLE 21.3 Calculating Genotype Frequencies for Multiple Alleles in a Hardy–Weinberg Population Where the Frequency of Allele I^A = 0.38, Allele I^B = 0.11, and Allele i = 0.51

Genotype	Genotype Frequency	Phenotype	Phenotype Frequency
I^AI^A	$p^2 = (0.38)^2 = 0.14$	A	0.53
I^Ai	$2pr = 2(0.38)(0.51) = 0.39$		
I^BI^B	$q^2 = (0.11)^2 = 0.01$	B	0.12
I^Bi	$2qr = 2(0.11)(0.51) = 0.11$		
I^AI^B	$2pr = 2(0.38)(0.11) = 0.084$	AB	0.08
ii	$r^2 = (0.51)^2 = 0.26$	O	0.26

Calculating Heterozygote Frequency

A useful application of the Hardy–Weinberg law, especially in human genetics, allows us to estimate the frequency of heterozygotes in a population. To do this, we must first calculate the frequency of each allele in the population. Although homozygous unaffected and heterozygous individuals have the same phenotype, we can usually determine the frequency of a recessive trait by first identifying and counting individuals with the homozygous recessive phenotype in a sample of the population. Using this information and the Hardy–Weinberg law, we can then calculate both the allele and genotype frequencies for each genotype present in a population.

Cystic fibrosis, an autosomal recessive trait, has an incidence of about 1/2500 (0.0004) in people of northern European ancestry. Individuals with cystic fibrosis are easily distinguished from the population at large by such symptoms as extra-salty sweat, excess amounts of thick mucus in the lungs, and susceptibility to bacterial infections. Because this is a recessive trait, individuals with cystic fibrosis must be homozygous. Their frequency in a population is represented by q^2 (provided that mating has been random in the previous generation). The frequency of the recessive allele is therefore

$$q = \sqrt{q^2} = \sqrt{0.0004} = 0.02$$

Knowing that the frequency of the recessive allele is about 2 percent, we can calculate the frequency of the normal (dominant) allele because $p + q = 1$. Using this equation, the frequency of p is

$$p = 1 - q = 1 - 0.02 = 0.98$$

Now that the allele frequencies are known, we can calculate the frequency of the heterozygous genotype. In the Hardy–Weinberg equation, the frequency of heterozygotes is $2pq$. Thus,

$$\begin{aligned} 2pq &= 2(0.98)(0.02) \\ &= 0.04 = 4\% = 1/25 \end{aligned}$$

The results show that heterozygotes for cystic fibrosis are rather common (about 1/25 individuals, or 4 percent of the population), even though the frequency of homozygous recessives is only 1/2500, or 0.04 percent. However, keep in mind that this calculation of heterozygote frequency is an estimate because the population in question may not meet all Hardy–Weinberg assumptions.

In general, for a single locus with a dominant and recessive allele, the frequencies of all three genotypes (homozygous dominant, heterozygous, homozygous recessive) can be estimated once the frequency of either allele is known and Hardy–Weinberg assumptions are invoked. The relationship between genotype and allele frequency is shown in **Figure 21.7.** It is important to note that heterozygotes

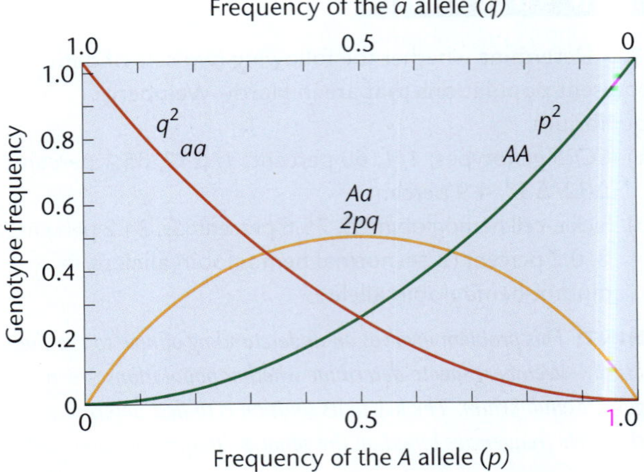

FIGURE 21.7 The relationship between genotype and allele frequencies derived from the Hardy–Weinberg equation.

increase rapidly in a population as the values of p and q move from 0 or 1.0 toward 0.5. This observation confirms our conclusion that when a recessive trait such as cystic fibrosis is rare, the majority of those carrying the allele are heterozygotes. In populations in which the frequencies of p and q are between 0.33 and 0.67, heterozygotes occur at a higher frequency than either homozygote.

NOW SOLVE THIS

21.3 If albinism occurs in 1/10,000 individuals in a population at equilibrium and is caused by an autosomal recessive allele *a*, calculate the frequency of:

(a) The recessive mutant allele

(b) The normal dominant allele

(c) Heterozygotes in the population

(d) Mating between heterozygotes

■ **HINT:** *This problem involves an understanding of the method of calculating allele and genotype frequencies. The key to its solution is to first determine the frequency of the albinism allele in this population.*

21.4 Natural Selection Is a Major Force Driving Allele Frequency Change

To understand evolution, we must understand the forces that transform the gene pools of populations and can lead to the formation of new species. Chief among the mechanisms transforming populations is **natural selection,** discovered independently by Alfred Russel Wallace and Charles Darwin. The Wallace–Darwin concept of natural selection can be summarized as follows:

1. Individuals of a species exhibit variations in phenotype, for example, differences in size, agility, coloration,

defenses against enemies, ability to obtain food, courtship behaviors, and flowering times.

2. Many of these variations, even small and seemingly insignificant ones, are heritable and are passed on to offspring.
3. Organisms tend to reproduce in an exponential fashion. More offspring are produced than can survive. This causes members of a species to engage in a struggle for survival, competing with other members of the community for scarce resources. Offspring also must avoid predators, and in sexually reproducing species, adults must compete for mates.
4. In the struggle for survival, individuals with particular phenotypes will be more successful than others, allowing the former to survive and reproduce at higher rates.

As a consequence of natural selection, populations and species change. Traits that promote differential survival and reproduction will become more common, and traits that confer a lowered ability for survival and reproduction will become less common. This means that over many generations, traits that confer a reproductive advantage will increase in frequency, which in turn will cause the population to become better adapted to its current environment. Over time, if selection continues, it may result in the appearance of new species.

Detecting Natural Selection in Populations

Recall that measuring allele frequencies and genotype frequencies using the Hardy–Weinberg law is based on several assumptions about an ideal population: large population size, lack of migration, presence of random mating, absence of selection and mutation, and equal survival rates of offspring.

However, if all genotypes do not have equal rates of survival or do not leave equal numbers of offspring, then allele frequencies may change from one generation to the next. To see why, let's imagine a population of 100 individuals in which the frequency of allele A is 0.5 and that of allele a is 0.5. Assuming the previous generation mated randomly, we find that the genotype frequencies in the present generation are $(0.5)^2 = 0.25$ for AA, $2(0.5)(0.5) = 0.5$ for Aa, and $(0.5)^2 = 0.25$ for aa. Because our population contains 100 individuals, we have 25 AA individuals, 50 Aa individuals, and 25 aa individuals.

Now let's suppose that individuals with different genotypes have different rates of survival: All 25 AA individuals survive to reproduce, 90 percent or 45/50 of the Aa individuals survive to reproduce, and 80 percent or 20/25 of the aa individuals survive to reproduce. When the survivors reproduce, each contributes two gametes to the new gene pool, giving us $2(25) + 2(45) + 2(20) = 180$ gametes. What are the frequencies of the two alleles in the surviving population? We have 50 A gametes from AA individuals, plus 45 A gametes from Aa individuals, so the frequency of allele A is $(50 + 45)/180 = 0.53$. We have 45 a gametes from Aa individuals, plus 40 a gametes from aa individuals, so the frequency of allele a is $(45 + 40)/180 = 0.47$.

These differ from the frequencies we started with. The frequency of allele A has increased, whereas the frequency of allele a has declined. A difference in survival or reproduction rate (or both) among individuals is an example of natural selection, which is the principal force that shifts allele frequencies within large populations. Natural selection is one of the most important factors in evolutionary change.

Fitness and Selection

Selection occurs whenever individuals with a particular genotype enjoy an advantage in survival or reproduction over other genotypes. However, selection may vary over a wide range, from much less than 1 percent to 100 percent. In the previous hypothetical example, selection was strong. Weak selection might involve just a fraction of a percent difference in the survival rates of different genotypes. Advantages in survival and reproduction ultimately translate into increased genetic contribution to future generations. An individual organism's genetic contribution to future generations is called its **fitness.** Genotypes associated with high rates of reproductive success are said to have high fitness, whereas genotypes associated with low rates of reproductive success are said to have low fitness.

Hardy–Weinberg analysis also allows us to examine fitness as a measure of the degree of natural selection. By convention, population geneticists use the letter w to represent fitness. Thus, w_{AA} represents the relative fitness of genotype AA, w_{Aa} the relative fitness of genotype Aa, and w_{aa} the relative fitness of genotype aa. Assigning the values $w_{AA} = 1$, $w_{Aa} = 0.9$, and $w_{aa} = 0.8$ would mean, for example, that all AA individuals survive, 90 percent of Aa individuals survive, and 80 percent of aa individuals survive, as in the previous hypothetical case.

Let's consider selection against deleterious alleles. Fitness values $w_{AA} = 1$, $w_{Aa} = 1$, and $w_{aa} = 0$ describe a situation in which a is a homozygous lethal allele. As homozygous recessive individuals die without leaving offspring, the frequency of allele a will decline. The decline in the frequency of allele a is described by the equation

$$q_g = \frac{q_0}{1 + gq_0}$$

where q_g is the frequency of allele a in generation g, q_0 is the starting frequency of a (i.e., the frequency of a in generation zero), and g is the number of generations that have passed.

Figure 21.8 shows what happens to a lethal recessive allele with an initial frequency of 0.5. At first, because of the high percentage of aa genotypes, the frequency of allele

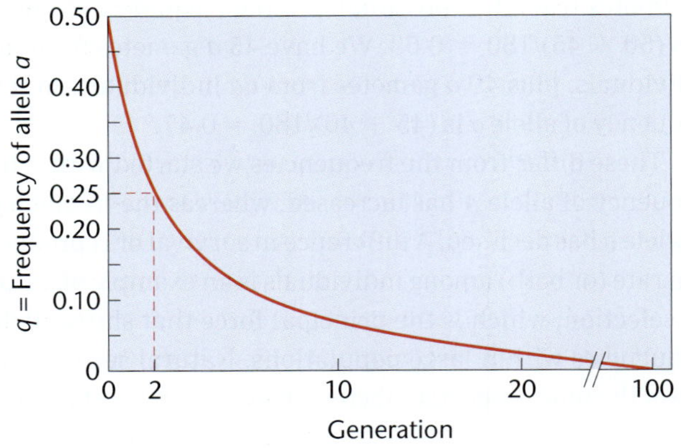

Generation	p	q	p^2	$2pq$	q^2
0	0.50	0.50	0.25	0.50	0.25
1	0.67	0.33	0.44	0.44	0.12
2	0.75	0.25	0.56	0.38	0.06
3	0.80	0.20	0.64	0.32	0.04
4	0.83	0.17	0.69	0.28	0.03
5	0.86	0.14	0.73	0.25	0.02
6	0.88	0.12	0.77	0.21	0.01
10	0.91	0.09	0.84	0.15	0.01
20	0.95	0.05	0.91	0.09	< 0.01
40	0.98	0.02	0.95	0.05	< 0.01
70	0.99	0.01	0.98	0.02	< 0.01
100	0.99	0.01	0.98	0.02	< 0.01

FIGURE 21.8 The change in the frequency of a lethal recessive allele, *a*. The frequency of *a* is halved in two generations and halved again by the sixth generation. Subsequent reductions occur slowly because the majority of *a* alleles are carried by heterozygotes.

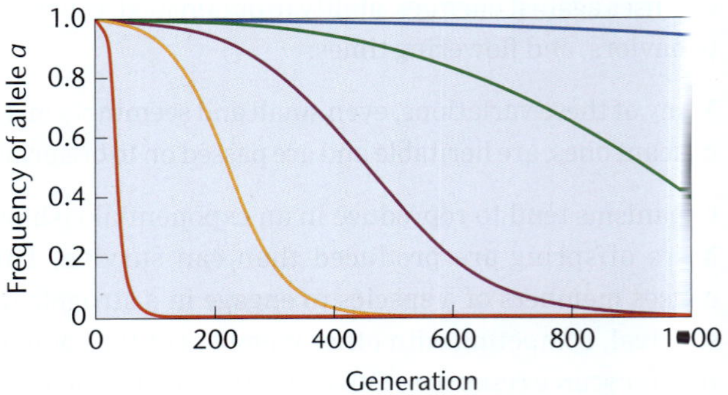

Selection Against Allele *a*					
	Strong ⟵				⟶ Weak
	— (red)	— (yellow)	— (purple)	— (green)	— (blue)
w_{AA}	1.0	1.0	1.0	1.0	1.0
w_{Aa}	0.90	0.98	0.99	0.995	0.998
w_{aa}	0.80	0.96	0.98	0.99	0.996

FIGURE 21.9 The effect of selection on allele frequency. The rate at which a deleterious allele is removed from a population depends heavily on the strength of selection.

a declines rapidly. The frequency of *a* is halved in only two generations. By the sixth generation, the frequency is halved again. By now, however, the majority of *a* alleles are carried by heterozygotes. Because *a* is recessive, these heterozygotes are not selected against. Consequently, as more time passes, the frequency of allele *a* declines ever more slowly. As long as heterozygotes continue to mate, it is difficult for selection to completely eliminate a recessive allele from a population.

Figure 21.9 shows the outcome of different degrees of selection against a nonlethal recessive allele, *a*. In this case, the intensity of selection varies from strong (red curve) to weak (blue curve), as well as intermediate values (yellow, purple, and green curves). In each example, the frequency of the deleterious allele, *a*, starts at 0.99 and declines over time. However, the rate of decline depends heavily on the strength of selection. When selection is strong and only 90 percent of the heterozygotes and 80 percent of the *aa* homozygotes survive (red curve), the frequency of allele *a* drops from 0.99 to less than 0.01 in about 85 generations. However, when selection is weak, and 99.8 percent of the heterozygotes and 99.6 percent of the *aa* homozygotes survive (blue curve), it takes 1000 generations for the frequency of allele *a* to drop from 0.99 to 0.93. Two important conclusions can be drawn from this example. First, over thousands of generations, even weak selection can cause substantial changes in allele frequencies; because evolution generally occurs over a large number of generations, selection is a powerful force in evolutionary change. Second, for selection to produce rapid changes in allele frequencies, the differences in fitness among genotypes must be large.

There Are Several Types of Selection

The phenotype is the result of the combined influence of the individual's genotype at many different loci and the effects of the environment. Selection can be classified as (1) directional, (2) stabilizing, or (3) disruptive.

In **directional selection** traits at one end of a spectrum of phenotypes present in the population become selected for or against, usually as a result of changes in the environment. A carefully documented example comes from research by Peter and Rosemary Grant and their colleagues, who study the medium ground finches (*Geospiza fortis*) of Daphne Major Island in the Galápagos Islands. These researchers discovered that the beak size of these birds varies over time in response to fluctuations in the environment (**Figure 21.10**). In 1977, a severe drought killed some 80 percent of the finches on the island.

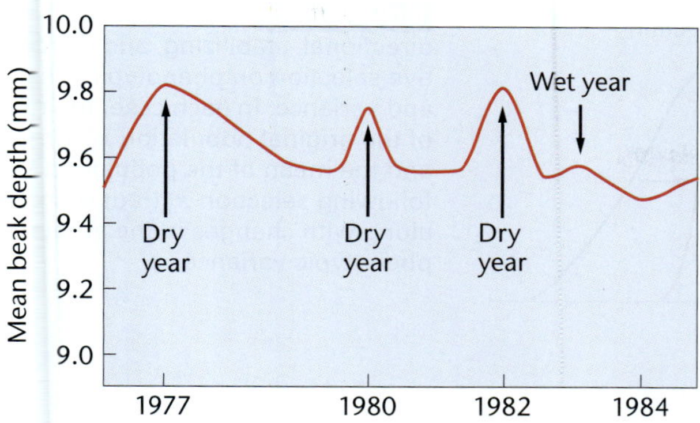

FIGURE 21.10 Beak size in finches during dry years increases because of strong selection. Between droughts, selection for large beak size is not as strong, and birds with smaller beak sizes survive and reproduce, increasing the number of birds with smaller beaks.

Big-beaked birds survived at higher rates than small-beaked birds because when food became scarce, the big-beaked birds were able to eat a greater variety of seeds, especially larger ones with hard shells. After the drought ended, more plants and seeds were available, selection was relaxed, and beak size declined. Droughts in 1980 and 1982 again saw differential survival and reproduction, shifting the average beak size toward the phenotypic extreme representing larger beak size.

Stabilizing selection, in contrast, selects for intermediate phenotypes, with those at both extremes being selected against. Over time, this will reduce the phenotypic variance in the population but without a significant shift in the mean. One of the clearest demonstrations of stabilizing selection is from a study of human birth weight and survival for 13,730 children born over an 11-year period. **Figure 21.11** shows the distribution of birth weight, the percentage of mortality at 5 weeks, and the percentage of births in the population (at right). Infant mortality increases on either side of the optimal birth weight of 7.5 pounds. Stabilizing selection acts to keep a population well adapted to its current environment.

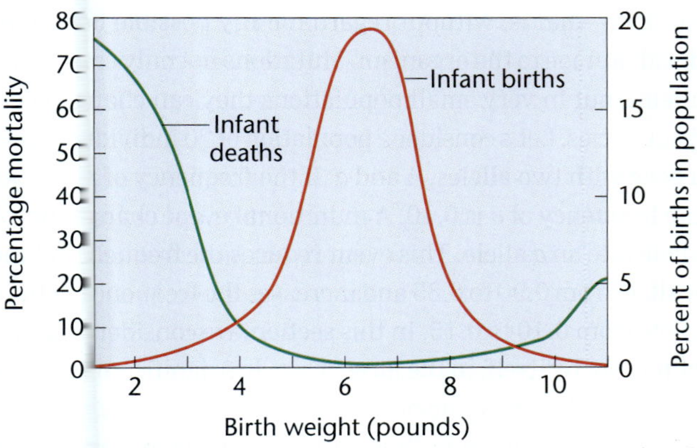

FIGURE 21.11 The relationship between birth weight and mortality in humans.

Disruptive selection is selection against intermediate phenotypes and selection for phenotypes at both extremes. It can be viewed as the opposite of stabilizing selection because the intermediate types are selected against. This will result in a population with an increasingly bimodal distribution for a trait, as we can see in **Figure 21.12.** In experiments using *Drosophila*, after several generations of artificial selection for bristle number, in which only flies with high- or low-bristle numbers were allowed to breed, most flies could be easily placed in a low- or high-bristle category. In natural populations, such a situation might exist for a population in a heterogeneous environment.

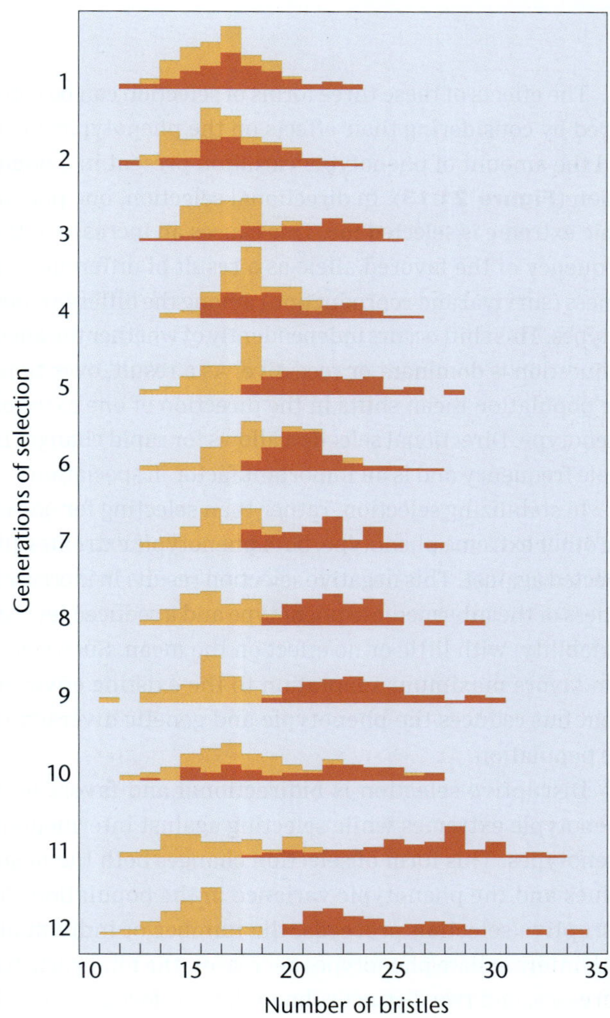

FIGURE 21.12 The effect of disruptive selection on bristle number in *Drosophila.* When individuals with the highest and lowest bristle numbers were selected, the population showed a nonoverlapping divergence in only 12 generations.

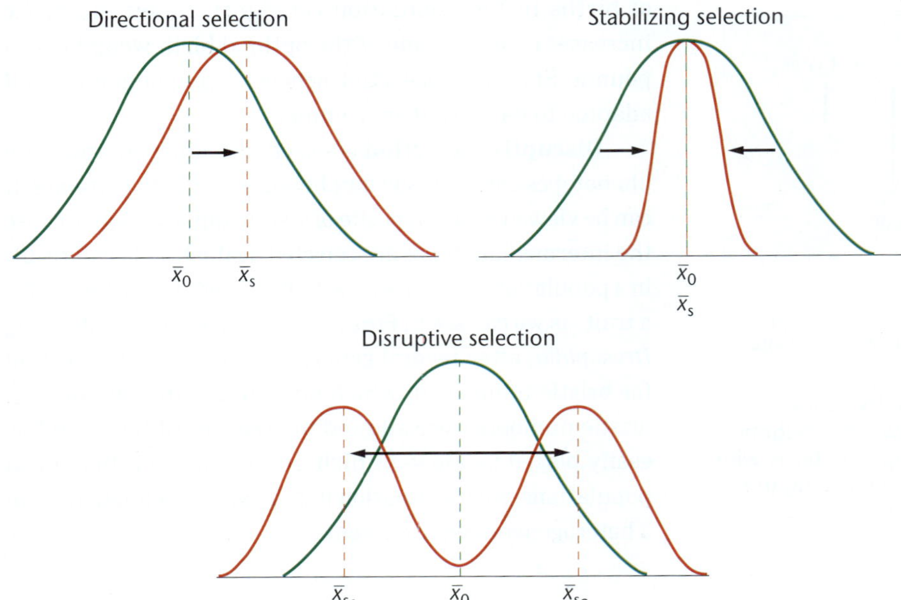

FIGURE 21.13 The impact of directional, stabilizing, and disruptive selection on phenotypic mean and variance. In each case, the mean of the original population $\overline{x_0}$ (green) and the mean of the population following selection $\overline{x_s}$ (red) is shown, along with changes in the amount of phenotypic variance.

The effects of these three forms of selection can be compared by considering their effects on the phenotypic mean and the amount of phenotypic variation present in a population (**Figure 21.13**). In directional selection, one phenotypic extreme is selected for. This causes an increase in the frequency of the favored allele as a result of differences in fitness (survival and reproduction) among the different phenotypes. This shift occurs independently of whether the allele in question is dominant or recessive. As a result, over time, the population mean shifts in the direction of one extreme phenotype. Directional selection allows for rapid changes in allele frequency and is an important factor in speciation.

In stabilizing selection, rather than selecting for one or the other extreme phenotype, both phenotypic extremes are selected against. This negative selection results in increased fitness of the intermediate phenotype and a reduced level of variability, with little or no effect on the mean. Such selection favors maximum adaptation to the existing environment but reduces the phenotypic and genetic diversity of the population.

Disruptive selection is bidirectional and favors both phenotypic extremes while selecting against intermediate phenotypes. This form of selection changes both the mean values and the phenotypic variance in the population. As disruptive selection proceeds, the number of individuals with intermediate phenotypes decreases, the total variance increases, and two distinct subpopulations form, each with its own mean value.

In conclusion, although each form of selection acts differentially on the phenotypic mean and variance of a population, they each play an important role in speciation by altering allele frequency.

ESSENTIAL POINT

The rate of change under natural selection depends on initial allele frequencies, selection intensity, and the relative fitness of different genotypes. ■

21.5 Mutation Creates New Alleles in a Gene Pool

Within a population, the gene pool is reshuffled each generation to produce new offspring. The enormous genetic variation present in the gene pool allows assortment and recombination to produce new combinations of genes already present in the gene pool. But assortment and recombination do not produce new alleles. **Mutation** alone acts to create new alleles. It is important to keep in mind that mutational events occur at random—that is, without regard for any possible benefit or disadvantage to the organism. Mutations not only create new alleles, but in very small populations they can change allele frequencies. Let's consider a population of 20 individuals, and a gene with two alleles, *A* and *a*. If the frequency of *A* is 0.90, the frequency of *a* is 0.10. A mutational event changes one *A* allele into an *a* allele. This event reduces the frequency of the *A* allele from 0.90 to 0.85 and increases the frequency of the *a* allele from 0.10 to 0.15. In this section, we consider whether mutation, by itself, in the larger case, is a significant factor in changing allele frequencies.

To determine whether mutation is a significant force in changing allele frequencies, we measure the rate at which mutations are produced. As in our example, most mutations are recessive, so it is difficult to observe mutation rates directly in diploid organisms. Indirect methods use probability and statistics or large-scale screening programs to estimate mutation rates. For certain dominant mutations, however, a direct method of measurement can be used. To ensure accuracy, several conditions must be met:

1. The allele must produce a distinctive phenotype that can be distinguished from similar phenotypes produced by recessive alleles.
2. The trait must be fully expressed or completely penetrant so that mutant individuals can be identified.
3. An identical phenotype must never be produced by nongenetic agents such as drugs or chemicals.

Suppose for a given gene that undergoes mutation to a dominant allele, 2 out of 100,000 births exhibit a mutant phenotype, but the parents are phenotypically normal. Because the zygotes that produced these births each carry two copies of the gene, we have actually surveyed 200,000 copies of the gene (or 200,000 gametes). If we assume that the affected births are each heterozygous, we have uncovered 2 mutant alleles out of 200,000. Thus, the mutation rate is 2/200,000 or 1/100,000, which in scientific notation is written as 1×10^{-5}. In humans, a dominant form of a skeletal disorder known as **achondroplasia** fulfills the requirements for measuring mutation rates. Individuals with this disorder have an enlarged skull and short arms and legs. They can be diagnosed by X-ray examination at birth. In a survey of almost 250,000 births, the mutation rate μ for achondroplasia has been calculated as

$$\mu = 1.4 \times 10^{-5} \pm 0.5 \times 10^{-5}$$

Knowing the rate of mutation, we can estimate the extent to which mutation can change allele frequencies from one generation to the next. We represent the normal allele as *d* and the allele for achondroplasia as *D*.

Instead of a population of 20 individuals, imagine a population of 500,000 individuals in which everyone has genotype *dd*. The initial frequency of *d* is 1.0, and the initial frequency of *D* is 0. If each individual contributes two gametes to the gene pool, the gene pool will contain 1,000,000 gametes, all carrying allele *d*. Although the gametes are in the gene pool, 1.4 of every 100,000 *d* alleles mutate into a *D* allele. The frequency of allele *d* is now $(1{,}000{,}000 - 14)/1{,}000{,}000 = 0.999986$, and the frequency of allele *D* is $14/1{,}000{,}000 = 0.000014$. From these numbers, it will clearly be a long time before mutation, by itself, causes any appreciable change in the allele frequencies in this population. In other words, mutation generates new alleles but, unless the population is very small or under very strong selection, mutation by itself does not alter allele frequencies at an appreciable rate.

21.6 Migration and Gene Flow Can Alter Allele Frequencies

The Hardy–Weinberg law assumes that migration does not take place. However, **migration,** or gene flow, occurs when individuals move between populations. Migration reduces the genetic differences between populations of a species and can increase the level of genetic variation in some populations.

Imagine a species in which a given locus has two alleles, *A* and *a*. There are two populations of this species, one on a mainland and one on an island. The frequency of *A* on the mainland is represented by p_m, and the frequency of *A* on the island is p_i. If there is migration from the mainland to the island, the frequency of *A* in the next generation on the island p_i' is given by

$$p_i' = (1 - m)p_i + mp_m$$

where *m* represents migrants from the mainland to the island.

As an example of how migration might affect the frequency of *A* in the next generation on the island p_i', assume that $p_i = 0.4$ and $p_m = 0.6$ and that 10 percent of the parents of the next generation are migrants from the mainland ($m = 0.1$). In the next generation, the frequency of allele *A* on the island will therefore be

$$\begin{aligned} p_i' &= [(1 - 0.1) \times 0.4] + (0.1 \times 0.6) \\ &= 0.36 + 0.06 \\ &= 0.42 \end{aligned}$$

In this case, the flow of genes from the mainland has changed the frequency of *A* on the island from 0.40 to 0.42 in a single generation.

These calculations reveal that the change in allele frequency attributable to migration is proportional to the differences in allele frequency between the donor and recipient populations *and* to the rate of migration. If either *m* is large or p_m is very different from p_i, then a rather large change in the frequency of *A* can occur in a single generation. If migration is the only force acting to change the allele frequency on the island, then equilibrium will be attained when $p_i = p_m$. These guidelines can often be used to estimate migration in cases where it is difficult to quantify. Even in large populations, over time, the effect of migration can substantially alter allele frequencies in populations, as shown for the I^B allele of the ABO blood group in **Figure 21.14.**

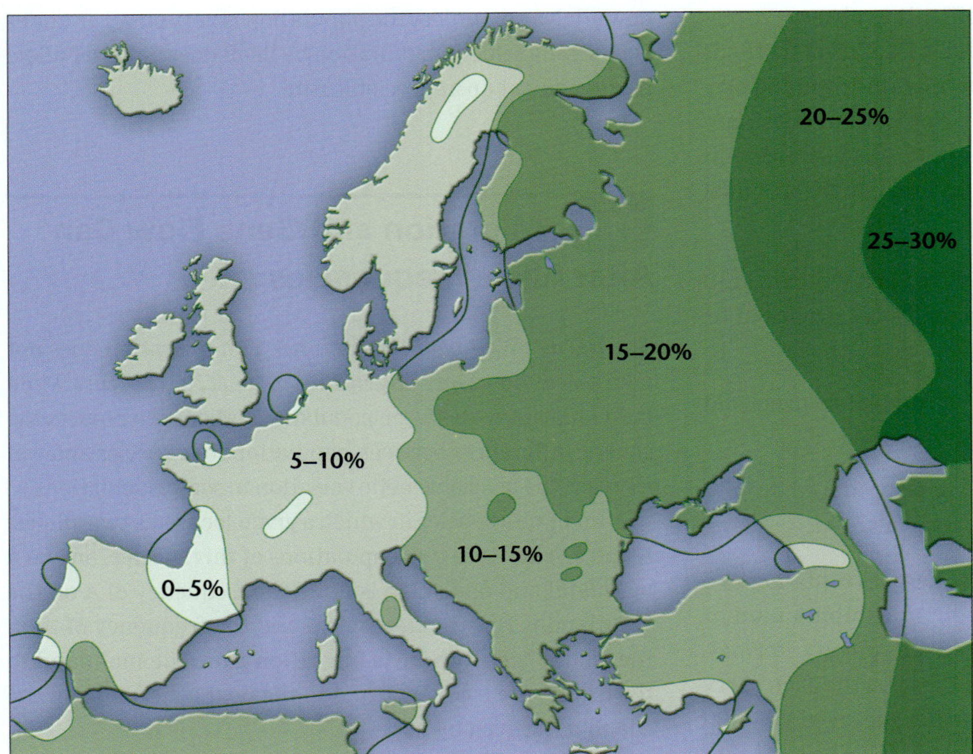

FIGURE 21.14 Migration as a force in evolution. The I^B allele of the *ABO* locus is present in a gradient from east to west. This allele shows the highest frequency in central Asia and the lowest in northeastern Spain. The gradient parallels the waves of Asian migration into Europe following the fall of the Roman Empire and is a genetic relic of human history.

21.7 Genetic Drift Causes Random Changes in Allele Frequency in Small Populations

In small populations, significant random fluctuations in allele frequencies are possible by chance alone, a situation known as **genetic drift.** In addition to small population size, drift can arise through the **founder effect,** which occurs when a population originates from a small number of individuals, usually by migration. Although the population may later increase to a large size, the genes carried by all members are derived from those of the founders (assuming no mutation, migration, or selection, and the presence of random mating). Drift can also arise via a **genetic bottleneck.** Bottlenecks develop when a large population remains in place, but undergoes a drastic but temporary reduction in numbers. Even though the population recovers, its genetic diversity has been greatly reduced. In summary, drift is a product of chance and can arise through small population size, founder effects, and bottlenecks. In the following section, we will examine how founder effects can affect allele frequencies.

Founder Effects in Human Populations

Allele frequencies in certain human populations demonstrate the role of genetic drift in natural populations. Native Americans living in the southwestern United States have a high frequency of oculocutaneous albinism (OCA). In the Navajo, who live primarily in northeastern Arizona, albinism occurs with a frequency of 1 in 1500 to 2000, compared with whites (1 in 36,000) and African-Americans (1 in 10,000). There are four different forms of OCA (OCA1–4), all with varying degrees of melanin deficiency in the skin, eyes, and hair. OCA2 is caused by mutations in the *P* gene, which encodes a plasma membrane protein. To investigate the genetic basis of albinism in the Navajo, researchers screened for mutations in the *P* gene. In their study, all Navajo with albinism were homozygous for a 122.5-kb deletion in the *P* gene, spanning exons 10 to 20. This deletion allele was not present in 34 individuals belonging to other Native American populations.

Using a set of PCR primers spanning the deletion, researchers identified homozygous affected individuals, heterozygous carriers, and homozygous normal individuals (**Figure 21.15**). They surveyed 134 normally pigmented Navajo and 42 members of the Apache, a tribe closely related to the Navajo. Based on this sample, the heterozygote frequency in the Navajo is estimated to be 4.5 percent. No carriers were found in the Apache population that was studied.

The 122.5-kb deletion allele causing OCA2 albinism was found only in the Navajo population and not in members of other Native American tribes in the southwestern United States, suggesting that the mutant allele is specific

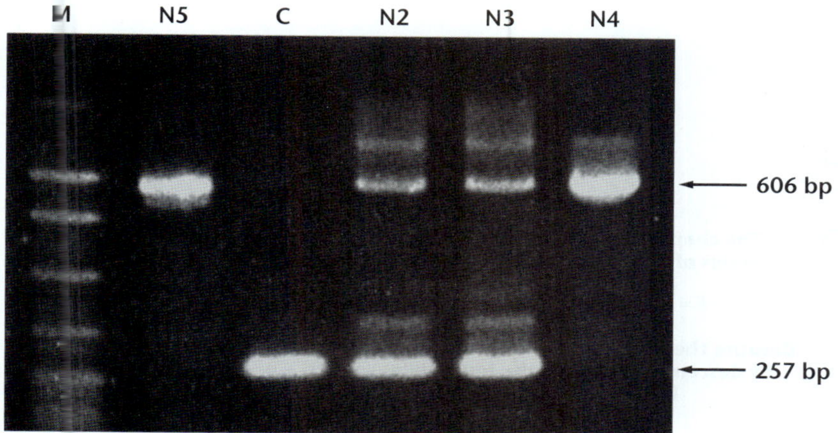

FIGURE 21.15 PCR screens of Navajo individuals affected with albinism (N4 and N5) and the parents of N4 (N2 and N3). Affected individuals (N4 and N5) have a single dense band at 606 bp; heterozygous carriers (N2 and N3) have two bands, one at 606 bp and one at 257 bp. The homozygous normal individual (C) has a single dense band at 257 bp. Each genotype produces a distinctive band pattern, allowing detection of heterozygous carriers in the population. Molecular size markers (M) are in the first lane.

to the Navajo and may have arisen in a single individual who was one of the small number of founders of the Navajo population. Workers originally estimated the age of the mutation to be between 400 and 11,000 years, but tribal history and Navajo oral tradition indicated that the Navajo and Apache became separate populations between 600 and 1000 years ago. Because the deletion is not found in the Apaches, it probably arose in the Navajo population after the tribes split. On this basis, the deletion is estimated to be 400 to 1000 years old and probably arose as a founder mutation.

21.8 Nonrandom Mating Changes Genotype Frequency but Not Allele Frequency

We have explored how populations that do not meet the first four assumptions of the Hardy–Weinberg law, in the form of selection, mutation, migration, and genetic drift, can have changes in allele frequencies. The fifth assumption is that members of a population mate at random; in other words, any one genotype has an equal probability of mating with any other genotype in the population. Nonrandom mating can change the frequencies of genotypes in a population. Subsequent selection for or against certain genotypes has the potential to affect the overall frequencies of the alleles they contain, but it is important to note that nonrandom mating *does not itself directly change allele frequencies.*

Nonrandom mating can take one of several forms. In **positive assortative mating**, similar genotypes are more likely to mate than dissimilar ones. This often occurs in humans: A number of studies have indicated that many people are more attracted to individuals who physically resemble them (and are therefore more likely to be genetically similar as well). **Negative assortative mating** occurs when dissimilar genotypes are more likely to mate; some plant species have inbuilt recognition systems that prevent fertilization between individuals with the same alleles at key loci. However, the form of nonrandom mating most commonly found to affect genotype frequencies in population genetics is **inbreeding.**

Inbreeding

Inbreeding occurs when mating individuals are more closely related than any two individuals drawn from the population at random; loosely defined, inbreeding is mating among relatives. For a given allele, inbreeding increases the proportion of homozygotes and decreases the proportion of heterozygotes in the population. A completely inbred population will theoretically consist only of homozygous genotypes. A high level of inbreeding can be harmful because it increases the probability that the number of individuals homozygous for deleterious and/or lethal alleles will increase in the population.

To describe the intensity of inbreeding in a population, Sewall Wright devised the **coefficient of inbreeding (*F*).** This coefficient quantifies the probability that the two alleles of a given gene present in an individual are identical *because they are descended from the same single copy of the allele in an ancestor*. If $F = 1$, all individuals in the population are homozygous, and both alleles in every individual are derived from the same ancestral copy. If $F = 0$, no individual has two alleles derived from a common ancestral copy.

One simple method of estimating F for a population is based on the inverse relationship between inbreeding and the frequency of heterozygotes: As the level of inbreeding increases, the frequency of heterozygotes declines. Therefore, F can be calculated as

$$F = \frac{H_e - H_o}{H_e}$$

where H_e is the expected heterozygosity based on the Hardy–Weinberg equation and H_o is the observed heterozygosity in the population. Note that if mating in the population is

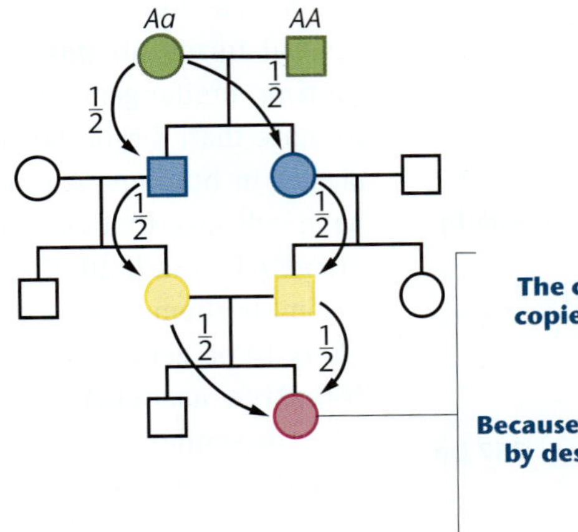

Aa
AA
1/2
1/2
1/2
1/2
1/2
1/2
The chance that this female will inherit two copies of her great-grandmother's *a* allele is
$F = \frac{1}{2} \times \frac{1}{2} \times \frac{1}{2} \times \frac{1}{2} \times \frac{1}{2} \times \frac{1}{2} = \frac{1}{64}$
Because the female's two alleles could be identical by descent from any of the four alleles carried by her great-grandparents,
$F = 4 \times \frac{1}{64} = \frac{1}{16}$

FIGURE 21.16 Calculating the coefficient of inbreeding *F* for the offspring of a first-cousin marriage.

completely at random, the expected and observed levels of heterozygosity will be equal and $F = 0$.

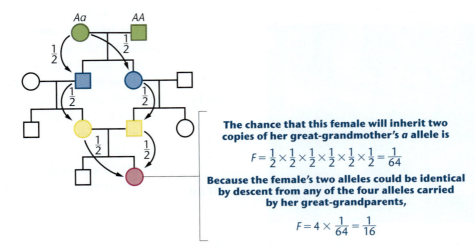

In humans, inbreeding (called *consanguineous marriage*) is related to population size, mobility, and social customs. A method that can be used for estimating *F* for an individual is shown in **Figure 21.16.** The fourth-generation female (shaded pink) is the daughter of first cousins (yellow). Suppose her great-grandmother (green) was a carrier of a recessive lethal allele, *a*. What is the probability that the fourth-generation female will inherit two copies of her great-grandmother's lethal allele? For this to happen, (1) the great-grandmother had to pass a copy of the allele to her son, (2) her son had to pass it to his daughter, and (3) his daughter had to pass it to her daughter (the pink female). Also, (4) the great-grandmother had to pass a copy of the allele to her daughter, (5) her daughter had to pass it to her son, and (6) her son had to pass it to his daughter (the pink female). Each of the six necessary events has an individual probability of 1/2, and they *all* have to happen, so the probability that the pink female will inherit two copies of her great-grandmother's lethal allele is $(1/2)^6 = 1/64$. However, to calculate an overall value of the inbreeding coefficient *F* for the pink female as a child of a first-cousin marriage, remember that she could also inherit two copies of any of the other three dominant alleles carried by her great-grandparents. Because any of four possibilities would give the pink female two alleles identical by descent from an ancestral copy, although not necessarily two copies of the lethal *a* allele,

$$F = 4 \times (1/64) = 1/16$$

ESSENTIAL POINT

Nonrandom mating in the form of inbreeding increases the frequency of homozygotes in the population and decreases the frequency of heterozygotes. ■

NOW SOLVE THIS

21.4 A prospective groom, who is unaffected, has a sister with cystic fibrosis (CF), an autosomal recessive disease. Their parents are unaffected. The brother plans to marry a woman who has no history of CF in her family. What is the probability that they will produce a CF child? They are both Caucasian, and the overall frequency of CF in the Caucasian population is 1/2500—that is, 1 affected child per 2500. (Assume the population meets the Hardy-Weinberg assumptions.)

■ **HINT:** *This problem involves an understanding of how recessive alleles are transmitted (see Chapter 3) and the probability of receiving a recessive allele from a heterozygous parent. The key to its solution is to first work out the probability that each parent carries the mutant allele.*

21.9 Speciation Can Occur through Reproductive Isolation

A **species** can be defined as a group of actually or potentially interbreeding organisms that is reproductively isolated in nature from all other such groups. In sexually reproducing organisms, speciation transforms the parental species into another species or divides a single species into two or more separate species (**Figure 21.17**). Changes in morphology or physiology and adaptations to ecological niches may also occur but are not necessary components of the speciation event.

Populations within a species may carry considerable genetic variation, present as differences in alleles or allele frequencies at a variety of loci. Genetic divergence of these populations that result in different allele frequencies and/or

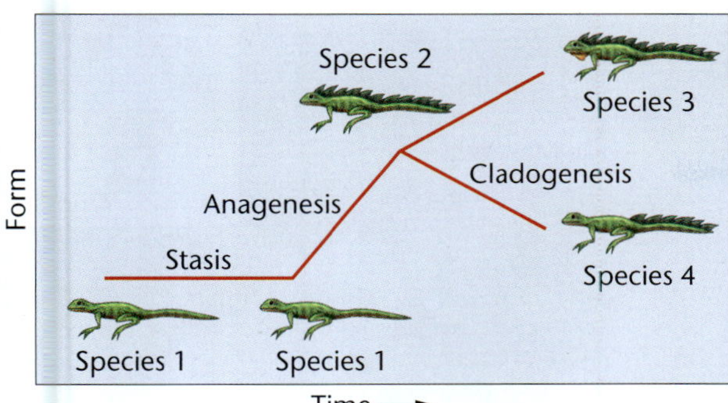

FIGURE 21.17 After a period with no change (stasis), species 1 is transformed into species 2, a process called *anagenesis*. Later, species 2 splits into two new species (species 3 and 4), a process called *cladogenesis*.

different alleles in their gene pools can reflect the action of forces such as natural selection, mutation, and genetic drift.

When gene flow between populations is reduced or absent, the populations may genetically diverge to the point that members of one population are no longer able to interbreed successfully with members of the other. When populations reach the point where they are reproductively isolated from one another, they have become different species. The genetic changes that result in reproductive isolation between or among populations and lead to the formation of new species or higher taxonomic groups define the process of macroevolution.

The biological barriers that prevent or reduce interbreeding between populations are called **reproductive isolating mechanisms** (**Table 21.4**). These mechanisms may be ecological, behavioral, seasonal, mechanical, or physiological.

Prezygotic isolating mechanisms prevent individuals from mating in the first place. Individuals from different populations may not find each other at the right time, may not recognize each other as suitable mates, or may try to mate but find that they are unable to do so because of differences in mating behavior.

Postzygotic isolating mechanisms create reproductive isolation even when the members of two populations are willing and able to mate with each other. For example, hybrid zygotes may be formed, but all or most of them may be inviable. Alternatively, the hybrids may be viable, but be sterile or suffer from reduced fertility. Yet again, the hybrids themselves may be fertile, but their progeny may have lowered viability or fertility. In all these situations, hybrids are genetic dead-ends.

Postzygotic isolating mechanisms lower the reproductive fitness of hybrid survivors. Selection favors the spread of alleles that lead to prezygotic isolating mechanisms, which in turn prevent interbreeding and the formation of hybrid zygotes and offspring. In animal evolution, one of the most effective prezygotic mechanisms is behavioral isolation involving courtship behavior.

TABLE 21.4 Reproductive Isolating Mechanisms

Prezygotic Mechanisms: Prevent fertilization and zygote formation.

1. **Geographic or ecological**. The populations live in the same regions but occupy different habitats.
2. **Seasonal or temporal**. The populations live in the same regions but are sexually mature at different times.
3. **Behavioral**. (Only in animals.) The populations are isolated by different and incompatible behavior before mating.
4. **Mechanical**. Cross-fertilization is prevented or restricted by differences in reproductive structures (genitalia in animals, flowers in plants).
5. **Physiological**. Gametes fail to survive in alien reproductive tracts.

Postzygotic Mechanisms: Fertilization takes place and hybrid zygotes are formed, but these are nonviable or give rise to weak or sterile hybrids.

1. **Hybrid nonviability or weakness**.
2. **Developmental hybrid sterility**. Hybrids are sterile because gonads develop abnormally or meiosis breaks down before completion.
3. **Segregational hybrid sterility**. Hybrids are sterile because of abnormal segregation into gametes of whole chromosomes, chromosome segments, or combinations of genes.

Changes Leading to Speciation

One form of speciation depends on the formation of geographic barriers between populations, which prevents gene flow between the isolated populations. Isolation allows the gene pools of these populations to diverge.

If the isolated populations later come into contact, several outcomes are possible. If reproductive isolating mechanisms are not in place, members of these populations will mate and will be regarded as one species. However, if reproductive isolating mechanisms have developed, the two populations will be regarded as separate species.

Formation of the Isthmus of Panama about 3 million years ago created a land bridge connecting North and South America and separated the Caribbean Sea from the Pacific Ocean. After identifying seven Caribbean species of snapping shrimp (**Figure 21.18**) and seven similar Pacific species, researchers matched them in pairs. Analysis of allele frequencies and mitochondrial DNA sequences confirmed that the ancestors of each pair were members of a single species. When the isthmus closed, each of the seven ancestral species was divided into two separate, isolated populations, one in the Caribbean and the other in the Pacific. But after 3 million years of separation, were members of these populations different species?

Males and females were paired together, and successful matings between Caribbean–Pacific couples versus those of Caribbean–Caribbean or Pacific–Pacific pairs were

FIGURE 21.18 A snapping shrimp (genus *Alpheus*).

determined. In three of the seven species pairs, transoceanic couples refused to mate altogether. Of the transoceanic pairs that mated, only 1 percent produced viable offspring, while 60 percent of same-ocean pairs produced viable offspring. We can conclude that 3 million years of separation has resulted in complete or nearly complete speciation, involving strong pre- and postzygotic isolating mechanisms for all seven species pairs.

The Rate of Macroevolution and Speciation

How much time is required for speciation? As we saw in the previous example, the time needed for genetic divergence and formation of new species can occur over a span of several million years. However, rapid speciation over much shorter time spans has been reported in a number of cases, including fishes in East African lakes, marine salmon, palm trees on isolated islands, polyploid plants, and fish in a volcanic lake in Nicaragua.

Lake Apoyo was formed within the last 23,000 years in the crater of a volcano (**Figure 21.19**). This small lake is home to two species of cichlid fish: the Midas cichlid, *Amphilophus citrinellus*, and the Arrow cichlid, *A. zaliosus*. The Midas is the most common cichlid in the region and is found in nearby lakes; the Arrow cichlid is found only in Lake Apoyo.

To establish the evolutionary origin of the Arrow cichlid, researchers used a variety of approaches, including phylogenetic, morphological, and ecological analyses. Sequence analysis of mitochondrial DNA established that the two species form a group derived from a common ancestor (a monophyletic group). Further genomic analysis of both species strengthened the conclusion that these two species are monophyletic and that *A. zaliosus* evolved from *A. citrinellus*. Members of the two species have distinctive phenotypes (**Figure 21.20**), including jaw specializations that reflect different food preferences. In addition, the two

FIGURE 21.19 Lake Apoyo in Nicaragua occupies the crater of an inactive volcano. The lake formed about 23,000 years ago. Two species of cichlid fish in the lake share a close evolutionary relationship.

(a)

(b)

FIGURE 21.20 The two species of cichlids in Lake Apoyo exhibit distinctive morphologies: (a) *Amphilophus citrinellus*. (b) *Amphilophus zaliosus*.

species are reproductively isolated, a conclusion substantiated by laboratory experiments. Using a molecular clock calibrated for cichlid mtDNA, researchers have estimated that *A. zaliosus* evolved from *A. citrinellus* sometime within the last 10,000 years. This estimate, and examples from other species, provides unambiguous evidence that, depending on the strength of selection and that of other parameters of the Hardy–Weinberg law, species formation can occur over a much shorter time scale than the usual range of 100,000–10,000,000 years.

21.10 Phylogeny Can Be Used to Analyze Evolutionary History

Speciation is associated with genetic divergence of populations. Therefore, we should be able to use genetic differences and similarities among present-day species to reconstruct their evolutionary histories. These relationships are most often presented in the form of phylogenetic trees (**Figure 21.21**), which show the ancestral relationships among a group of organisms. These groups can be species, or larger groups such as phyla. In a phylogenetic tree, branches represent the relationships among lineages over time. Depending on how the tree is constructed, the length of a branch can be derived from a time scale, showing the time between speciation events. Branch points, or nodes, show when a species split into two or more species. Each node represents a common ancestor of the species diverging at that node. The tips of the branches represent species (or a larger group) alive today (or those that ended in extinction).

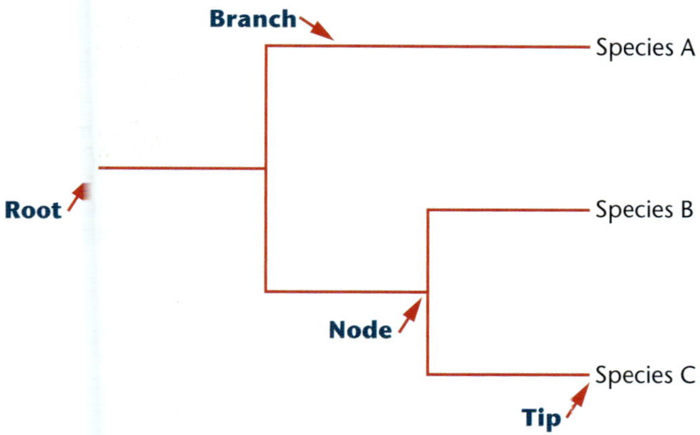

FIGURE 21.21 Elements of a phylogenetic tree showing the relationships among several species. The root represents a common ancestor to all species on the tree. Branches represent lineages through time. The points at which the branches separate are called nodes, and at the tips of the branches are the living or extinct species.

Groups that consist of an ancestral species and all its descendants are called monophyletic groups. The root of a phylogenetic tree represents the oldest common ancestor to all the groups shown in the tree. Trees can be constructed from differences in morphology of living organisms; fossils; and the molecular sequences of proteins, RNA, and DNA.

Constructing Phylogenetic Trees from DNA Sequences

Advances in DNA-sequencing technology have made genetic and genomic information from many species available, and today, most phylogenetic trees are constructed using DNA sequences.

Constructing a species-level phylogenetic tree using DNA sequences involves three steps:

1. DNA sequences representing a gene or genome of interest from a number of different species must be acquired. With the proliferation of DNA-sequencing projects, these are usually available from public databases.
2. The sequences must be aligned with each other so that the related parts of each sequence can be compared to see if they are the same or different. The sequences to be compared can be imported into software programs that maximize the number of aligned base pairs by inserting gaps as needed. As discussed earlier, more distantly related species have acquired more DNA differences because of the longer time that has elapsed since they last shared a common ancestor. More closely related species have fewer DNA differences because there has been less time for accumulation of DNA differences since they last shared a common ancestor.
3. These DNA differences are used to construct a phylogenetic tree, often beginning with the most closely related sequences and working backward through sequences that are less closely related.

Reconstructing Vertebrate Evolution by Phylogenetic Analysis

One of the most important steps in the evolutionary history of our species was the ancient transition of vertebrates from the ocean to the land. For more than a century, biologists have debated and argued about which group of lobe-finned fish crawled ashore as the ancestor of all terrestrial vertebrates (amphibians, reptiles, birds, and mammals). In past years, phylogenetic trees constructed from the fossil record, from living species, and from mitochondrial DNA sequences pointed to the lungfish [**Figure 21.22(a)**] as the closest living relative to terrestrial vertebrates, but could not rule out the possibility that vertebrates may have two common ancestors, the lungfish and another fish, the coelacanth [**Figure 21.22(b)**].

(a)

(b)

FIGURE 21.22 Phylogenetic evidence indicates that the lungfish (a) and not the coelacanth (b) is a common ancestor of amphibians, reptiles, birds, and mammals.

Recently, the coelacanth genome has been sequenced, and the data from this study have reopened the question of which group shares a common ancestor with our species and all other land vertebrates. Using sequence data from the coelacanth, the lungfish, and selected vertebrate species, researchers aligned and analyzed information from 251 protein-coding genes to construct a phylogenetic tree (**Figure 21.23**). The results strongly support earlier work indicating that terrestrial vertebrates are more closely related to the lungfish than to the coelacanth. Thus, the door has been closed on this important evolutionary question.

Molecular Clocks Measure the Rate of Evolutionary Change

In many cases, we would like to estimate not only which members of a set of species are most closely related, but also when their common ancestors lived. The ability to construct phylogenetic trees from protein and nucleic acid sequences led to the development of **molecular clocks,** which use the

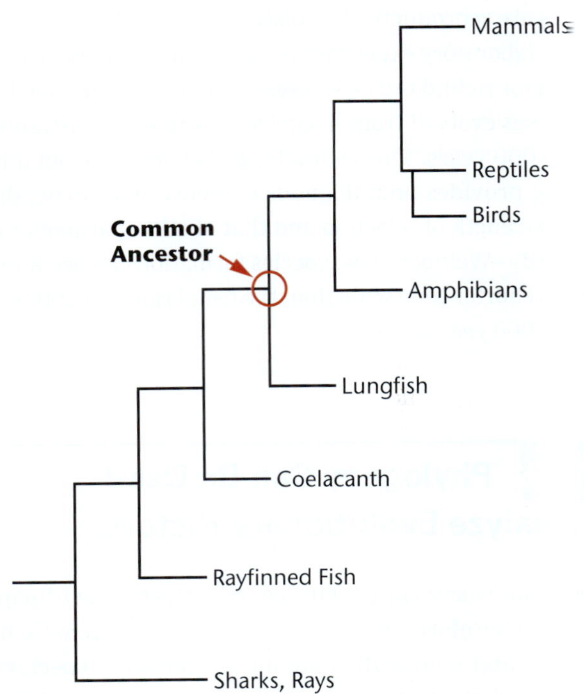

FIGURE 21.23 A phylogenetic tree of selected jawed vertebrates, including the lungfish and the coelacanth, shows that the lungfish shares the most recent common ancestor with these vertebrates.

rate of change in amino acid or nucleotide sequences as a way to estimate the time of divergence from a common ancestor.

To be useful, molecular clocks must be carefully calibrated. Molecular clocks can only measure changes in amino acids or nucleotides; they are linear over certain time scales, and times and dates must be added to the clock using independent evidence such as the fossil record. **Figure 21.24** shows a

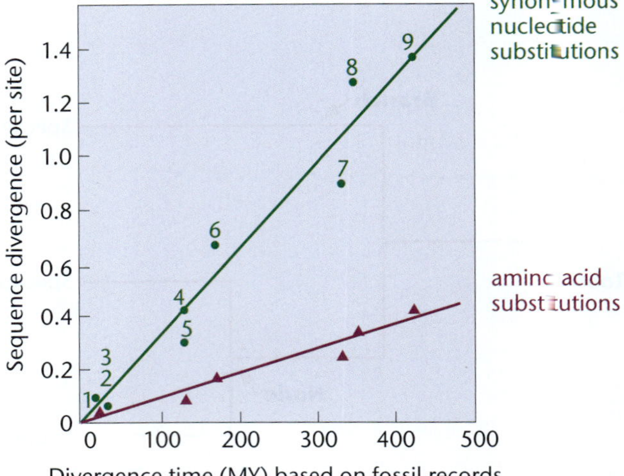

FIGURE 21.24 The relationship between the number of amino acid substitutions and the number of nucleotide substitutions for 4198 nuclear genes from 10 vertebrate species. Humans versus (1) chimpanzee, (2) orangutan, (3) macaque, (4) mouse, (5) cow, (6) opossum, (7) chicken, (8) western clawed frog, and (9) zebrafish. MY = millions of years.

molecular clock showing times of divergence from a common ancestor for humans and nine other vertebrate species based on the fossil record and molecular data. The results show that humans and zebrafish last shared a common ancestor about 450 million years ago, and humans last shared a common ancestor with chimpanzees about 7–10 million years ago.

The Complex Origins of the Human Genome

Current fossil, molecular, and genomic evidence indicates that our species, *Homo sapiens*, arose in Africa about 300,000 years ago from earlier species of *Homo*. When populations of *H. sapiens* first expanded out of Africa sometime between 50,000 and 70,000 years ago, parts of Europe and Asia were already occupied by members of other human (*Homo*) species. Advances in DNA-sequencing technology and new methods of DNA extraction that allow the recovery of genomic DNA from fossil remains have created a new field, called **paleogenomics,** which in turn, has revolutionized the study of human evolution. The genomes of two extinct groups who lived in the Middle East, Asia, and Europe, the Neanderthals and the Denisovans, have been sequenced and compared with the genomes of present-day humans. The results show that modern human populations outside Africa, including those of the Middle East, Europe, Asia, Australia/Oceania, and the Americas, carry sequences from these two groups.

Neanderthals coexisted with anatomically modern humans (*H. sapiens*) in regions of the Middle East and Europe, over a period of at least 30,000 years, providing an opportunity for interbreeding between these species. In fact, genes from extinct Neanderthals transferred to modern humans through interbreeding is estimated to represent 2–4 percent of the genome of non-African populations. However, because different individuals carry different portions of the Neanderthal genome; taken together, upward of 20 percent of the Neanderthal genome may be present in the genomes of modern non-African populations. Based on the size and distribution of Neanderthal DNA sequences in the genome of modern humans, it is estimated that mixing of these genomes occurred between 50,000 and 60,000 years ago.

In 2008, human fossils were discovered in a cave in the Altai Mountains near Denisova, Siberia (**Figure 21.25**). A complete mtDNA genome sequence showed that these fossils belonged to a group separate from both Neanderthals and our species and were named the Denisovans. A nuclear Denisovan genome sequence shows that they are more closely related to Neanderthals than to our species, and that Denisovans and Neanderthals separated from a common ancestor more than 430,000 years ago. In addition, the Denisovan genome contains sequences from another, as yet unknown, archaic group that made no contribution to the Neanderthal genome.

FIGURE 21.25 The cave in the Altai Mountains of Denisova, Siberia, where the Denisovan fossils were discovered.

Analysis of modern human populations shows that about 4 to 6 percent of the DNA in the genomes of Melanesian islanders in the South Pacific is derived from the Denisovans. Smaller amounts of Denisovan DNA are found in the genomes of Aboriginal Australians, as well as the genomes of Polynesians, Fujians, east Indonesians, and some populations of East Asia. As things stand now, we know that as a result of ancient interbreeding, some present-day individuals outside of Africa carry DNA from one or two other human groups.

Recent work has provided evidence for a transfer of genes into Han Chinese and Japanese East Asian populations from a second, separate Denisovan population. As things stand now, indications are that genetic information has been transferred from extinct human populations to modern non-African populations at least three times; once from Neanderthals, and twice from Denisovans (**Figure 21.26**).

Using the paleogenomic techniques currently available, we can expect exciting answers to questions about the similarities and differences between our genome and those of other human species, providing revolutionary insights into the evolution of our species and other human species that preceded us on this planet.

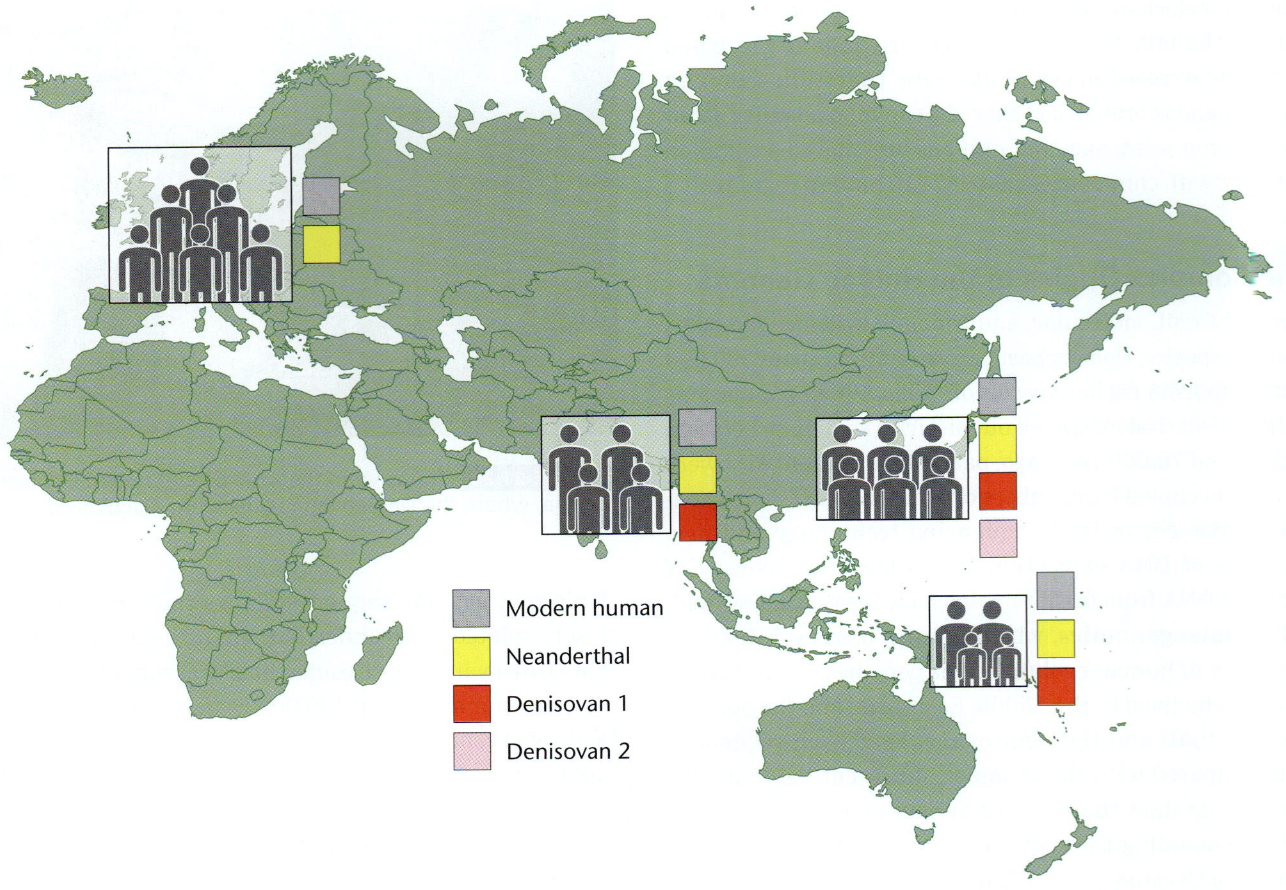

FIGURE 21.26 The genomes of non-African modern humans contain genes contributed by interbreeding with Neanderthals. In some present-day Asian populations, there is evidence of separate genetic contributions by two populations of Denisovans, distant relatives of the Neanderthals. Additional contributions to the genomes of modern-day humans may remain to be discovered.

GENETICS, ETHICS, AND SOCIETY

Tracking Our Genetic Footprints out of Africa

Approximately 2 million years ago, a large-brained, tool-using hominid called *Homo erectus* appeared in East Africa. By 1.7 million years ago, *H. erectus* had spread into Eurasia and South Asia. Most scientists agree that *H. erectus* likely developed into *H. heidelbergensis*—a species that became the ancestor to our species (in Africa), Neanderthals (in Europe), and Denisovans (in Asia). These groups disappeared 50,000 to 30,000 years ago—around the same time that anatomically modern humans (*H. sapiens*) appeared all over the world.

At present, the most widely accepted hypothesis explaining the presence of anatomically modern humans is the out-of-Africa hypothesis. This hypothesis is based on genetic data derived from mitochondrial, Y chromosome, and whole-genome sequencing of both archaic hominin fossils and modern human populations. The out-of-Africa hypothesis states that *H. sapiens* evolved from the descendants of *H. heidelbergensis* in Africa about 300,000 years ago. Around 50,000 years ago, a small band of *H. sapiens* (perhaps fewer than 1000) left Africa. By 40,000 years ago, they had reached Europe, Asia, and Australia. In the out-of-Africa model, *H. sapiens* interbred with Neanderthal and Denisovan populations, and then became the only species in the genus by about 30,000 years ago.

Most genetic evidence appears to support the out-of-Africa hypothesis. Humans all over the globe are extremely similar genetically. DNA sequences from any two humans chosen at random are 99.9 percent identical. More genetic identity exists between two persons chosen at random from a human population than between two chimpanzees chosen at random from a chimpanzee population. Interestingly, about 90 percent of the genetic differences that do exist occur between individuals rather than between populations. This unusually high degree of genetic relatedness in all humans supports the idea that our species arose recently from a small founding group of individuals. Other genetic data show that the highest levels of human genetic variation occur within African populations. This implies that the earliest branches of *H. sapiens* diverged in Africa and had a longer time to accumulate DNA mutations.

As with any explanation of human origins, the out-of-Africa hypothesis is actively debated. Some data suggest two or more out-of-Africa dispersals, as well as different timings of dispersals and migration routes. As methods for sequencing DNA from ancient fossils improve, it will soon be possible to fill the gaps in our understanding of the genetic pathways leading out of Africa and to resolve age-old questions about our origins.

Your Turn

Take time, individually or in groups, to consider the following questions. Investigate the references and links dealing with the ethical and technological aspects of how we understand the origins of modern humans.

1. Some genetic and archaeological evidence appears to support two separate dispersals of humans out of Africa. What are these data, and how might they be reconciled with the single-dispersal hypothesis?

Start your investigations by reading Nielsen, R. et al. (2017). Tracing the peopling of the world through genomics. *Nature* 541:302–310, *and* Tucci, S., and Akey, J. M. (2016). A map of human wanderlust. *Nature* 538:179–180.

2. Given that genetic studies show that all people on Earth are remarkably similar genetically, how did we come to develop the concept of racial differences? How has modern genomics contributed to the debate about the validity and definition of the term "race"?

For an interesting discussion of race, human variation, and genomics, see Lewontin, R. C. (2006). Confusions about human races, *on the Social Sciences Research Center Web site* (raceandgenomics.ssrc.org/Lewontin). *Also, see* Cooper, R. S. (2013). Race in biological and biomedical research. *Cold Spring Harb. Perspect. Med.* 3(11):a008573.

CASE STUDY A tale of two Olivias

Olivia S. was born with a rare recessive disorder called tyrosinemia. The next day, Olivia M. was born in a neighboring state with the same disorder. Tyrosinemia is caused by the lack of an enzyme in the degradation pathway of the amino acid tyrosine. Accumulation of metabolic intermediates causes progressive liver dysfunction and kidney problems. One-year-old Olivia S. is healthy and has no symptoms of the disorder. At the same age, Olivia M. developed total liver failure. Olivia S. was born in a state where newborns are tested for tyrosinemia, but Olivia M. was born in a state where newborns are not tested for this disorder. A week after diagnosis, Olivia S. was placed on a low-tyrosine diet and prescribed a drug to block the accumulation of metabolic intermediates. Olivia M. was not diagnosed until she was in liver failure; she then was placed on a low-tyrosine diet, was prescribed medication, and underwent a liver transplant. She faces a lifetime of anti-ejection drug therapy and may require a kidney transplant. In the United States, newborn screening programs are developed independently by each state and are often based on a cost–benefit analysis to decide which diseases are included in testing. In the United States, tyrosinemia occurs in only 1/100,000 births, and in this case, two states made different decisions about newborn testing for this disorder.

1. In a region of Quebec, Canada, 1 in 22 people are heterozygous for the mutant tyrosinemia allele. Using the frequency of heterozygotes, calculate the frequency of recessive homozygotes in this population. What might explain the difference between the frequency of tyrosinemia in the U.S. population and in this particular Canadian population?
2. Critics argue that a uniform panel of disorders should be used by all states in newborn testing. Aside from cost–benefit ratios, what would you regard as ethical guidelines for use in deciding which disorders to include or exclude in a newborn testing program?
3. Others argue that the current testing system should be replaced by whole genome sequencing for all newborns. What do you see as the ethical pros and cons of this position?

See Tarini, B., and Goldenberg, A. (2012). Ethical issues with newborn screening in the genomics era. *Ann. Rev. Genomics Hum Genet.* 13:381–393.

INSIGHTS AND SOLUTIONS

1. Tay–Sachs disease is caused by loss-of-function mutations in a gene on chromosome 15 that encodes a lysosomal enzyme. Tay–Sachs is inherited as an autosomal recessive condition. Among Ashkenazi Jews of Central European ancestry, about 1 in 3600 children is born with the disease. What fraction of the individuals in this population are carriers?

Solution: If we let p represent the frequency of the wild-type enzyme allele and q the total frequency of recessive loss-of-function alleles, and if we assume that the population is in Hardy–Weinberg equilibrium, then the frequencies of the genotypes are given by p^2 for homozygous normal, $2pq$ for carriers, and q^2 for individuals with Tay–Sachs. The frequency of Tay–Sachs alleles is thus

$$q = \sqrt{q^2} = \sqrt{\frac{1}{3600}} = 0.017$$

Since $p + q = 1$, we have

$$p = 1 - q = 1 - 0.017 = 0.983$$

Therefore, we can estimate that the frequency of carriers is

$$2pq = 2(0.983)(0.017) = 0.033$$

or about 1 in 30.

2. A single plant twice the size of others in the same population suddenly appears. Normally, plants of that species reproduce by self-fertilization and by cross-fertilization. Is this new giant plant simply a variant, or could it be a new species? How would you determine which it is?

Solution: One of the most widespread mechanisms of speciation in higher plants is polyploidy, the multiplication of entire sets of chromosomes. The result of polyploidy is usually a larger plant with larger flowers and seeds. There are two ways of testing the new variant to determine whether it is a new species. First, the giant plant should be crossed with a normal-sized plant to see whether the giant plant produces viable, fertile offspring. If it does not, then the two different types of plants would appear to be reproductively isolated. Second, the giant plant should be cytogenetically screened to examine its chromosome complement. If it has twice the number of its normal-sized neighbors, it is a tetraploid that may have arisen spontaneously. If the chromosome number differs by a factor of two and the new plant is reproductively isolated from its normal-sized neighbors, it is a new species.

Problems and Discussion Questions

Mastering Genetics Visit for instructor-assigned tutorials and problems.

1. **HOW DO WE KNOW?** Population geneticists study changes in the nature and amount of genetic variation in populations, the distribution of different genotypes, and how forces such as selection and drift act on genetic variation to bring about evolutionary change in populations and the formation of new species. From the explanation given in the chapter, what answers would you propose to the following fundamental questions?
   (a) How do we know how much genetic variation is in a population?
   (b) How do geneticists detect the presence of genetic variation as different alleles in a population?
   (c) How do we know whether the genetic structure of a population is static or dynamic?
   (d) How do we know when populations have diverged to the point that they form two different species?
   (e) How do we know the age of the last common ancestor shared by two species?
2. **CONCEPT QUESTION** Read the Chapter Concepts list on page 436. All these pertain to the principles of population genetics and the evolution of species. Write a short essay describing the roles of mutation, migration, and selection in bringing about speciation. ■
3. Price et al. [(1999). *J. Bacteriol.* 181:2358–2362] conducted a genetic study of the toxin transport protein (PA) of *Bacillus anthracis*, the bacterium that causes anthrax in humans. Within the 2294-nucleotide gene in 26 strains they identified five point mutations—two missense and three synonyms—among different isolates. Necropsy samples from an anthrax outbreak in 1979 revealed a novel missense mutation and five unique nucleotide changes among ten victims. The authors concluded that these data indicate little or no horizontal transfer between different *B. anthracis* strains.
   (a) Which types of nucleotide changes (missense or synonyms) cause amino acid changes?
   (b) What is meant by "horizontal transfer"?
   (c) On what basis did the authors conclude that evidence of horizontal transfer is absent from their data?
4. The genetic difference between two *Drosophila* species, *D. heteroneura* and *D. silvestris*, as measured by nucleotide diversity, is about 1.8 percent. The difference between chimpanzees (*Pan troglodytes*) and humans (*H. sapiens*) is about the same, yet the latter species is classified in a different genera. In your opinion, is this valid? Explain why.
5. The use of nucleotide sequence data to measure genetic variability is complicated by the fact that the genes of many eukaryotes are complex in organization and contain 5′ and 3′ flanking regions as well as introns. Researchers have compared the nucleotide sequence of two cloned alleles of the γ-*globin* gene from a single individual and found a variation of 1 percent. Those differences include 13 substitutions of one nucleotide for another and three short DNA segments that have been inserted in one allele or deleted in the other. None of the changes takes place in the gene's exons (coding regions). Why do you think this is so, and should it change our concept of genetic variation?

6. Consider rare disorders in a population caused by an autosomal recessive mutation. From the frequencies of the disorder in the population given, calculate the percentage of heterozygous carriers:
   (a) 0.0064
   (b) 0.000081
   (c) 0.09
   (d) 0.01
   (e) 0.10
7. What must be assumed in order to validate the answers in Problem 6?
8. In a population that meets the Hardy–Weinberg equilibrium assumptions, 81% of the individuals are homozygous for a recessive allele. What percentage of the individuals would be expected to be heterozygous for this locus in the next generation?
9. In a Drosophila laboratory population in equilibrium 9 percent of the individuals have rough eyes (recessive trait). Two randomly selected flies with normal eyes are mated. What is the probability their offspring will have rough eyes?
10. The frequency of two alleles in a population is $p = 0.6$ and $q = 0.4$, for *C* and *c* respectively. Determine the allele frequencies after one generation in the flowing scenarios:
    (a) $w_{CC} = 1, w_{Cc} = 0.9, w_{cc} = 0.8$
    (b) $w_{CC} = 1, w_{Cc} = 0.98, w_{cc} = 0.95$
    (c) $w_{CC} = 0.9, w_{Cc} = 1, w_{cc} = 0.54$
    (d) $w_{CC} = 0.99, w_{Cc} = 1, w_{cc} = 0$
11. If the initial allele frequencies are $p = 0.5$ and $q = 0.5$ and allele *a* is a lethal recessive, what will be the frequencies after 1, 5, 10, 25, 100, and 1000 generations?
12. Under what circumstances might a lethal dominant allele persist in a population?
13. Assume that a recessive autosomal disorder occurs in 1 of 10,000 individuals (0.0001) in the general population and that in this population about 2 percent (0.02) of the individuals are carriers for the disorder. Estimate the probability of this disorder occurring in the offspring of a marriage between first cousins. Compare this probability to the population at large.
14. One of the first Mendelian traits identified in humans was a dominant condition known as *brachydactyly*. This gene causes an abnormal shortening of the fingers or toes (or both). At the time, some researchers thought that the dominant trait would spread until 75 percent of the population would be affected (because the phenotypic ratio of dominant to recessive is 3:1). Show that the reasoning was incorrect.
15. Describe how populations with substantial genetic differences can form. What is the role of natural selection?
16. Achondroplasia is a dominant trait that causes a characteristic form of dwarfism. In a survey of 50,000 births, five infants with achondroplasia were identified. Three of these infants had parents with achondroplasia, while two did not. Calculate the mutation rate for achondroplasia, and express the rate as the number of mutant genes per given number of gametes.
17. A recent study examining the mutation rates of 5669 mammalian genes (17,208 sequences) indicates that, contrary to popular belief, mutation rates among lineages with vastly different generation lengths and physiological attributes are remarkably constant [Kumar, S., and Subramanian, S. (2002). *Proc. Natl. Acad. Sci. USA* 99:803–808]. The average rate is estimated at 12.2×10^{-9} per bp per year. What is the significance of this finding in terms of mammalian evolution?
18. What do genetic adaptations to certain environmental conditions and neutral alleles have in common?
19. A botanist studying water lilies in an isolated pond observed three leaf shapes in the population: round, arrowhead, and scalloped. Marker analysis of DNA from 125 individuals showed the round-leaf plants to be homozygous for allele *r1*, while the plants with arrowhead leaves were homozygous for a different allele at the same locus, *r2*. Plants with scalloped leaves showed DNA profiles with both the *r1* and *r2* alleles. Frequency of the *r1* allele was estimated at 0.81. If the botanist counted 20 plants with scalloped leaves in the pond, what is the inbreeding coefficient *F* for this population?
20. A farmer plants transgenic Bt corn that is genetically modified to produce its own insecticide. Of the corn borer larvae feeding on these Bt crop plants, only 10 percent survive unless they have at least one copy of the dominant resistance allele *B* that confers resistance to the Bt insecticide. When the farmer first plants Bt corn, the frequency of the *B* resistance allele in the corn borer population is 0.02. What will be the frequency of the resistance allele after one generation of corn borers have fed on Bt corn?
21. In an isolated population of 50 desert bighorn sheep, a mutant recessive allele *c* when homozygous causes curled coats in both males and females. The normal dominant allele *C* produces straight coats. A biologist studying these sheep counts four with curled coats. She also takes blood samples from the population for DNA analysis, which reveals that 17 of the sheep are heterozygous carriers of the *c* allele. What is the inbreeding coefficient *F* for this population?
22. To increase genetic diversity in the bighorn sheep population described in Problem 21, ten sheep are introduced from a population where the *c* allele is absent. Assuming that random mating occurs between the original and the introduced sheep, and that the *c* allele is selectively neutral, what will be the frequency of *c* in the next generation?
23. What is the relevance of gene flow in speciation?
24. Advances in medicine in the last century have made it possible to survive diseases that would otherwise have been lethal before. Recently gene therapy and future prospects of genome editing promise a further leap in the treatment of incurable diseases. Discuss their effect on natural selection of the human species.
25. Explain how each type of selection contributes to speciation.
26. What are the two groups of reproductive isolating mechanisms? Which of these is regarded as more efficient, and why?
27. A form of dwarfism known as Ellis–van Creveld syndrome was first discovered in the late 1930s, when Richard Ellis and Simon van Creveld shared a train compartment on the way to a pediatrics meeting. In the course of conversation, they discovered that they each had a patient with this syndrome. They published a description of the syndrome in 1940. Individuals with this syndrome have a short-limbed form of dwarfism and often have defects of the lips and teeth, and polydactyly (extra fingers). The largest pedigree for the condition was reported in an Old Order Amish population in eastern Pennsylvania by Victor McKusick and his colleagues (1964). In that population of 8000, the observed frequency is 2 per 1000. In all cases, parents of children with the syndrome were unaffected, and all cases can be traced to Samuel King and his wife, who arrived in the area in 1774. It is known that neither King nor his wife was affected with the disorder. There are no cases of the disorder in other Amish communities, such as those in Ohio or Indiana.
    (a) From the information provided, derive the most likely mode of inheritance of this disorder. Using the

Hardy–Weinberg law, calculate the frequency of the mutant allele in the population and the frequency of heterozygotes, assuming Hardy–Weinberg conditions.

(b) What is the most likely explanation for the high frequency of the disorder in the Pennsylvania Amish community and its absence in other Amish communities?

28. The original source of new alleles, upon which selection operates, is mutation, a random event that occurs without regard to selectional value in the organism. Although many model organisms have been used to study mutational events in populations, some investigators have developed abiotic molecular models. Soll et al. (2006. *Genetics* 175:267–275) examined one such model to study the relationship between both deleterious and advantageous mutations and population size in a ligase molecule composed of RNA (a ribozyme). Soll found that the smaller the population of molecules, the more likely it was that not only deleterious mutations but also advantageous mutations would disappear. Why would population size influence the survival of both types of mutations (deleterious and advantageous) in populations?

29. A number of comparisons of nucleotide sequences among hominids and rodents indicate that inbreeding may have occurred more often in hominid than in rodent ancestry. Bakewell et al. (2007. *Proc. Nat. Acad. Sci. [USA]* 104: 7489-7494) suggest that an ancient population bottleneck that left approximately 10,000 humans might have caused early humans to have a greater chance of genetic disease. Why would a population bottleneck influence the frequency of genetic disease?

30. Shown below are two homologous lengths of the alpha and beta chains of human hemoglobin. Consult a genetic code dictionary (Figure 12.7), and determine how many amino acid substitutions may have occurred as a result of a single nucleotide substitution. For any that cannot occur as a result of a single change, determine the minimal mutational distance.

Alpha: ala val ala his val asp asp met pro
Beta: gly leu ala his leu asp asn leu lys

31. Recent reconstructions of evolutionary history are often dependent on assigning divergence in terms of changes in amino acid or nucleotide sequences. For example, a comparison of cytochrome c shows 10 amino acid differences between humans and dogs, 24 differences between humans and moths, and 38 differences between humans and yeast. Such data provide no information as to the absolute times of divergence for humans, dogs, moths, and yeast. How might one calibrate the molecular clock to an absolute time clock? What problems might one encounter in such a calibration?

Epigenetics

Until recently, it was thought that most regulation of gene expression is coordinated by *cis*-regulatory elements along with DNA-binding proteins and transcription factors (Chapter 16). However, these classical regulatory mechanisms cannot fully explain how some phenotypes arise. For example, monozygotic twins have identical genotypes but often have different phenotypes. In other instances, although one allele of each gene is inherited maternally, and one is inherited paternally, only the maternal or paternal allele is expressed, while the other remains transcriptionally silent. Investigations of such phenomena have led to the emerging field of epigenetics, which is providing us with a molecular basis for understanding how heritable genomic alterations other than those encoded in the DNA sequence can alter patterns of gene expression and influence phenotypic variation (**Figure ST 1.1**).

Epigenetics can be defined as the study of phenomena and mechanisms that cause chromosome-associated heritable changes to gene expression that are not dependent on changes in DNA sequence.

Epigenetics can be defined as the study of phenomena and mechanisms that cause chromosome-associated heritable changes to gene expression that are not dependent on changes in DNA sequence.

The **epigenome** refers to the specific pattern of epigenetic modifications present in a cell at a given time. During its life span, an organism has one genome, which can be modified at different times to produce many different epigenomes.

Knowledge of the mechanisms of epigenetic modifications to the genome, how these modifications are maintained and transmitted, and their relationship to basic biological processes is important to enhance our understanding of reproduction and development, disease processes, and the evolution of adaptations to the environment, including behavior.

Current research efforts are focused on several aspects of epigenetics: (1) how an epigenomic state arises in developing and differentiated cells and (2) how these epigenetic states are transmitted via mitosis and meiosis, making them heritable traits. The fruits of these efforts will be a major focus of this chapter. In addition, because epigenetically controlled alterations to the genome are associated with common diseases such as cancer and diabetes, efforts are also directed toward developing drugs that can modify or reverse disease-associated epigenetic changes in cells.

Phenotype

Genome

Epigenome

FIGURE ST 1.1 The phenotype of an organism is the product of interactions between the genome and the epigenome (hatched areas). The genome is constant from fertilization throughout life, but cells, tissues, and the organism develop different epigenomes because of epigenetic reprogramming of gene activity in response to environmental stimuli. These reprogramming events lead to phenotypic changes throughout the life cycle.

ST 1.1 Molecular Alterations to the Genome Create an Epigenome

Unlike the genome, which is identical in all cell types of an organism, the epigenome is cell-type specific and changes throughout the life cycle in response to environmental cues. Like the genome, the epigenome can be transmitted to daughter cells by mitosis and to future generations by meiosis. In this section, we will examine mechanisms that shape the epigenome.

There are three major epigenetic mechanisms: (1) reversible modification of DNA by the addition or removal of methyl groups; (2) chromatin remodeling by the addition or removal of chemical groups to histone proteins; and (3) regulation of gene expression by noncoding RNA molecules. We now will look at each of these molecular activities in turn.

DNA Methylation and the Methylome

The set of methylated nucleotides present in an organism's genome at a given time is known as the **methylome.** The methylome is cell- and tissue-specific, but is not fixed, and changes as cells are called upon to respond to changing conditions. In mammals, **DNA methylation** takes place after DNA replication and during cell differentiation. This process involves the addition of a methyl group ($-CH_3$) to cytosine on the 5-carbon of the cytosine nitrogenous base (**Figure ST 1.2**), resulting in 5-methylcytosine (5mC), a reaction catalyzed by a family of enzymes called DNA methyltransferases (DNMTs). In humans, 5mC is present in about 1.5 percent of the genomic DNA.

Methylation takes place almost exclusively on cytosine bases located adjacent to a guanine base, a combination called a CpG dinucleotide:

$$\begin{array}{c} 5'-\overset{m}{C}pG-3' \\ 3'-Gp\underset{m}{C}-5' \end{array}$$

Many of these dinucleotide sites are clustered in regions called CpG islands, located in promoter and upstream sequences (**Figure ST 1.3**).

CpG islands and promoters adjacent to essential genes (housekeeping genes) and cell-specific genes are unmethylated, making them available for transcription. Genes with adjacent methylated CpG islands and methylated CpG sequences within promoters are transcriptionally silenced. The added methyl groups occupy the major groove of DNA and silence genes by blocking the binding of transcription factors and other proteins necessary to form transcription complexes.

However, the bulk of methylated CpG dinucleotides are not adjacent to genes; instead they are in the repetitive DNA sequences of heterochromatic regions of the genome, including the centromere. Methylation of these sequences contributes to silencing of transcription and replication of transposable elements such as LINE and SINE sequences which constitute a major portion of the human genome. (See Chapter 11 for a detailed discussion of repetitive sequences, heterochromatin, and chromosome organization.) Heterochromatic methylation is important in maintaining chromosome stability by preventing translocations and related abnormalities.

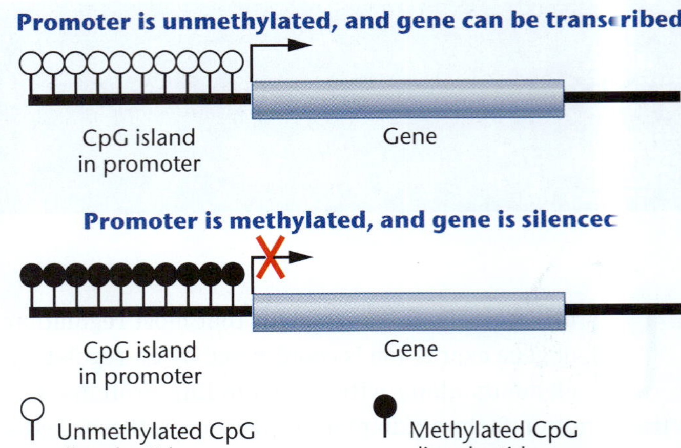

FIGURE ST 1.3 Methylation patterns of CpG dinucleotides in promoters control activity of adjacent genes. CpG islands outside and within genes also have characteristic methylation patterns, contributing to the overall level of genome methylation.

In mammals and other vertebrates, methylation of cytosine to form 5-methylcytosine (5mC) is the most common epigenetic modification of DNA. However, in the genomes of some eukaryotes, including the algae *Chlamydomonas reinhardtii* and the nematode *Caenorhabditis elegans,* 5mC is absent or, as in *Drosophila,* may be present at almost undetectable levels. Recent work has shown that although the methylomes of these and some other eukaryotes may not contain 5mC, they do contain adenine that has been methylated at its N6 position (6mA), a modification that may have epigenetic functions. At this early stage, further research is needed in these species to fully explore the details of how 6mA controls gene expression. In addition, the extent to which 6mA is present in the methylomes of other organisms including mammals, and has epigenetic functions, remains to be determined.

Cytosine → Methyltranferase → 5-Methylcytosine

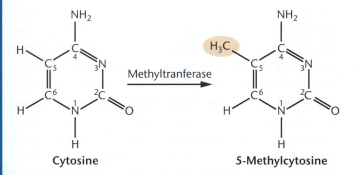

FIGURE ST 1.2 In DNA methylation, methyltransferase enzymes catalyze the transfer of a methyl group from a methyl donor to cytosine, producing 5-methylcytosine.

Histone Modification and Chromatin Remodeling

Interaction of DNA with proteins that facilitate transcription is controlled by two processes: (1) chromatin remodeling, which involves the action of ATP-powered protein complexes that move, remove, or alter nucleosomes, and (2) histone modifications, which are covalent posttranslational modifications of amino acids near the N-terminal ends of histone proteins. Together, these two processes activate or repress transcription and act as one of the primary methods of gene regulation.

Normally, DNA is wound tightly around nucleosomes to form chromatin (Chapter 11), which is further coiled and packaged to form chromosomes. In this state, regulatory regions of DNA and the genes themselves are unable to interact with proteins that facilitate transcription.

The N-terminal region of each histone extends beyond the nucleosome, forming a tail. Amino acids in these tails can be covalently modified in several ways, and a number of different proteins are involved in the process. These include proteins that add chemical groups to histones ("writers"), proteins that interpret those modifications ("readers"), and proteins that remove those chemical groups ("erasers").

Over 20 different chemical modifications can be made to histones, but the major changes include the addition of acetyl, methyl, and phosphate groups. Such additions alter the structure of chromatin, making genes on nucleosomes with modified histones accessible or inaccessible for transcription. Histone acetylation, for example, relaxes the grip of histones on DNA and makes genes available for transcription [**Figure ST 1.4(a)**]. Furthermore, acetylation is reversible. Removing (erasing) acetyl groups contributes to changing chromatin from an "open" configuration to a "closed" state, thereby silencing genes by making them unavailable for transcription [**Figure ST 1.4(b)**].

Histone modifications occur at specific amino acids in the N-terminal tail of histones 2A, 2B, H3, and H4. Many combinations of histone modifications are possible within and between histone molecules, and the sum of their complex patterns and interactions is called the **histone code.** The basic idea behind a histone code is that reversible enzymatic modification of histone amino acids (by writers and erasers) recruits nucleoplasmic proteins (readers) that either further modify chromatin structure or regulate transcription.

The code is represented in a shorthand as follows:

- Name of the histone (e.g., H3)
- Single-letter abbreviation for the amino acid (e.g., K for lysine)
- Position of the amino acid in the protein (e.g., 27)
- Type of modification (ac = acetyl, me = methyl, p = phosphate, etc.)
- Number of modifications (amino acids can be methylated one, two, or three times)

Thus, H3K27me3 represents a trimethylated lysine at position 27 from the N-terminus of histone H3.

Short and Long Noncoding RNAs

In addition to messenger RNA (mRNA), genome transcription produces several classes of **noncoding RNAs (ncRNAs),** which are transcribed from DNA but not translated into proteins. The ncRNAs related to epigenetic regulation include two groups: (1) short ncRNAs (less than 31 nucleotides) and (2) long ncRNAs (greater than 200 nucleotides). Both types

"Open" configuration.
DNA is unmethylated, and histones are acetylated.
Genes can be transcribed.

(a)

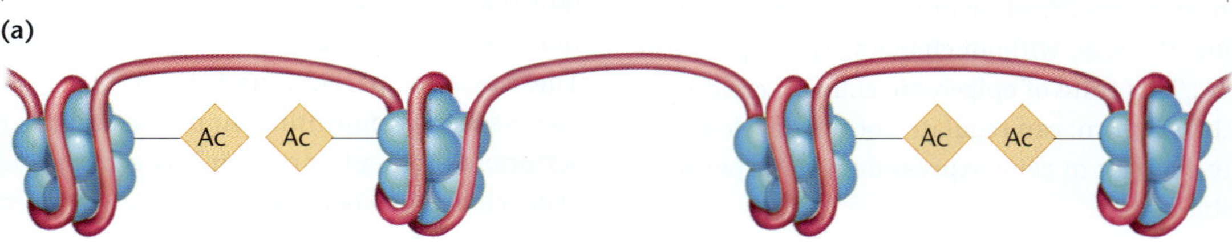

"Closed" configuration.
DNA is methylated at CpG islands (black circles), and histones are deacetylated.
Genes cannot be transcribed.

(b)

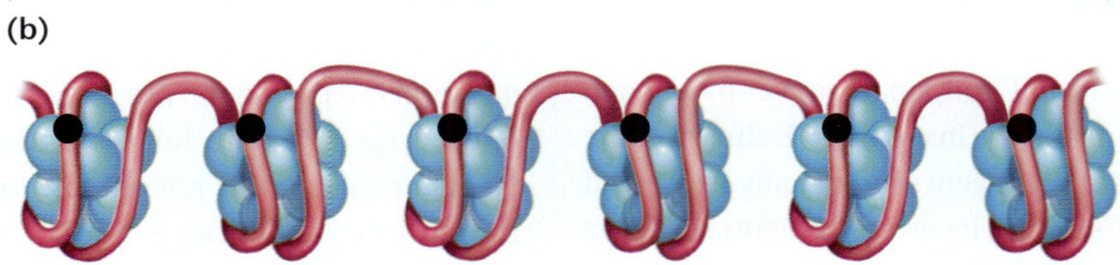

FIGURE ST 1.4 Epigenetic modifications to the genome alter the spacing of nucleosomes. (a) In the open configuration, chromatin remodeling shifts nucleosome positions, CpGs are unmethylated, and the genes on the DNA are available for transcription. (b) In the closed configuration, DNA is tightly wound onto the nucleosomes, CpGs are methylated, chemical groups have been removed from histones, and genes on the DNA are unavailable for transcription.

of ncRNAs have several roles, including the formation of heterochromatin, histone modification, site-specific DNA methylation, and gene silencing. They are also important in epigenetic regulatory networks.

There are three classes of short ncRNAs: miRNAs (microRNAs), siRNAs (short interfering RNAs), and piRNAs (piwi-interacting RNAs). miRNAs and siRNAs are transcribed as precursor molecules about 70 to 100 nucleotides long that contain a double-stranded stem-loop and single-stranded regions. After several processing steps that shorten the RNAs to lengths of 20 to 25 ribonucleotides, these RNAs act as repressors of gene expression. The origin of piRNAs is unclear, but they interact with proteins to form RNA-protein complexes that participate in epigenetic gene silencing in germ cells.

Long noncoding RNAs (lncRNAs) share many propertieswithmRNAs;theyoftenhave5′-capsand3′ poly-Atailsand are spliced. What distinguishes lncRNAs from coding (mRNA) transcripts is the lack of an extended open reading frame that codes for the insertion of amino acids into a polypeptide.

RNA genomic sequencing has identified more than 15,000 lncRNA genes in the human genome. lncRNAs are found in the nucleus and the cytoplasm, and through a variety of mechanisms, are involved in both transcriptional and post-transcriptional regulation of gene expression. As epigenetic initiators, lncRNAs bind to chromatin-modifying enzymes and direct their activity to specific regions of the genome. At these sites, the lncRNAs direct chromatin modification, altering the pattern of gene expression.

In summary, epigenetic modifications alter chromatin structure by several mechanisms: DNA methylation, reversible covalent modification of histones, and action of short and long RNAs, all without changing the sequence of genomic DNA. This suite of epigenetic changes creates an epigenome that, in turn, can regulate normal development and generate changes in gene expression as a response to environmental signals.

ST 1.2 Epigenetics and Monoallelic Gene Expression

Mammals inherit a maternal and a paternal copy of each gene, and aside from genes on the inactivated X chromosome in females, both copies of these genes are usually expressed at equal levels in the offspring. However, in some cases only one allele is transcribed, while the other allele is transcriptionally silent. This phenomenon is called **monoallelic expression (MAE).**

There are three major classes of MAE. In one class, genes are expressed in a *parent-of-origin pattern*; that is, certain genes show expression of only the maternal allele or the paternal allele, a phenomenon called **genomic imprinting.** The remaining two classes involve *random* monoallelic expression. First is the random inactivation of one X chromosome in the cells of mammalian females, which compensates for their increased dosage of X-linked genes (recall that mammalian males have only one X chromosome). Second, a randomly generated pattern of allele inactivation, independent of parental origin, is observed in a significant number of autosomal genes. We will look at each of these three classes in turn.

Parent-of-Origin Monoallelic Expression: Imprinting

Parentally imprinted genes are marked in male and female germ-line cells during gamete formation; the fertilized egg thus has different marks on the copies of certain genes that came from the mother or the father. How is this marking accomplished?

To begin with, the DNA carried by sperm and eggs are highly methylated. However, shortly after fertilization, most of the methylation marks are erased. This modification of the DNA resets embryonic cells to a pluripotent state, allowing them to undergo new epigenetic modifications to form the more than 200 cell types found in the adult body. About the same time the embryo is implanting in the wall of the uterus, cells take on tissue-specific epigenetic identities, and methylation patterns and histone modifications change rapidly to reflect those seen in differentiated cells.

Some genomic regions, however, escape these rounds of global demethylation and remethylation. The genes contained in these regions remain imprinted with the methylation marks of the maternal and/or paternal chromosomes. These original parental patterns of methylation produce allele-specific imprinting. Imprinted alleles remain transcriptionally silent during embryogenesis and later stages of development. For example, if the allele inherited from the father is imprinted, it is silenced, and only the allele from the mother is expressed.

In humans, imprinted genes are usually found in clusters on the same chromosome and can occupy more than 1000 kb of DNA. Because these genes are located near each other at a limited number of sites in the genome, mutation in one imprinted gene can often affect the function of adjacent or coordinately controlled imprinted genes, thereby amplifying the mutation's phenotypic impact. These mutations in imprinted genes can arise through changes in the DNA sequence or by resultant dysfunctional epigenetic changes, called **epimutations,** both of which can cause heritable changes in gene activity.

Occasionally, the imprinting process goes awry and is dysfunctional. In such cases, the imprinting defects can

TABLE ST 1.1 Some Imprinting Disorders in Humans

Disorder	Locus
Albright hereditary osteodystrophy	20q13
Angelman syndrome	15q11–q15
Beckwith–Wiedemann syndrome	11p15
Prader–Willi syndrome	15q11–q15
Silver–Russell syndrome	Chromosome 7
Uniparental disomy 14	Chromosome 14

cause human disorders such as Beckwith–Wiedemann syndrome, Prader–Willi syndrome, Angelman syndrome, and several other diseases (**Table ST 1.1**). However, given the number of imprinting-susceptible candidate genes and the possibility that additional imprinted genes remain to be discovered, the overall number of imprinting-related genetic disorders may be much higher.

In humans, most known imprinted genes encode growth factors or other growth-regulating genes. An autosomal dominant disorder associated with imprinting, Beckwith–Wiedemann syndrome (BWS) occurs in about 1 in 13,700 births and offers insight into how disruptions of epigenetic imprinting can lead to an abnormal phenotype. BWS is a prenatal overgrowth disorder typified by abdominal wall defects, enlarged organs, high birth weight, and a predisposition to cancer. The genes associated with BWS are in a cluster of epigenetically imprinted genes on the short arm of chromosome 11 (**Figure ST 1.5**). BWS is not caused by mutation in the DNA sequence of the gene, nor is it associated with any chromosomal aberrations. Instead, BWS is a disorder of imprinting and is caused by abnormal patterns of DNA methylation resulting in altered patterns of gene expression, and not by mutations that change the nucleotide sequence of the genes involved.

All the genes in this cluster are known to regulate growth during prenatal development. Two closely linked genes in this cluster are *insulin-like growth factor 2* (*IGF2*), whose encoded protein plays an important role in growth and development, and *H19*, which is transcribed into an ncRNA. These two genes are separated by an imprinting control region (ICR), which controls the expression of both genes. Normally, the ICR on the paternal copy of chromosome 11 is methylated, allowing expression of the paternal *IGF2* allele but maintaining the paternal *H19* allele in a silenced state [Figure ST 1.5(a)]. Reciprocally, on the maternal copy of chromosome 11, the ICR is unmethylated allowing for the expression of the maternal *H19* allele, while the maternal *IGF2* allele is maintained in a silenced state. In one form of BWS, both copies of the ICR are methylated [Figure ST 1.5(b)] and both the maternal and paternal *IGF2* alleles are transcribed, resulting in the overgrowth of tissues that are characteristic of this disease. The transcription of both *IGF2* alleles is accompanied by silencing of both copies

(a) Normal imprinting pattern

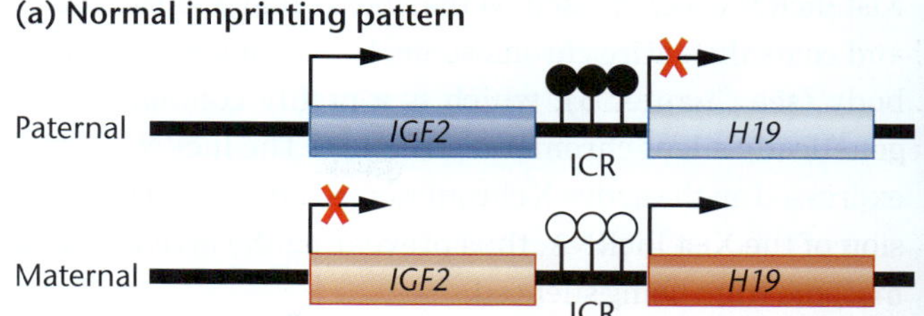

(b) BWS imprinting pattern

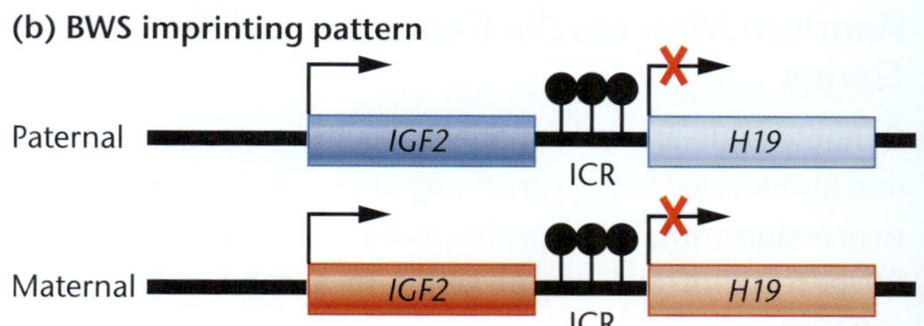

FIGURE ST 1.5 The imprinted region of human chromosome 11. (a) In normal imprinting, the ICR on the paternal chromosome is methylated (filled circles); the *IGF2* allele is active and the *H19* allele is silent. The ICR on the maternal chromosome is not methylated (open circles), and the *IGF2* allele is silent while the *H19* allele is active. (b) In one form of BWS, both the maternal and paternal ICRs are methylated (filled circles), both *IGF2* alleles are active, and both *H19* alleles are silent. The result is dysregulation of cell growth, resulting in the overgrowth of structures that are characteristic of BWS.

of the *H19* allele, further compounding the overgrowth of tissues.

The known number of imprinted genes represents less than 1 percent of the mammalian genome, but they play major roles in regulating growth and development during prenatal stages. Because they act so early in life, any external or internal factors that disturb the epigenetic patterns of imprinting or the expression of these imprinted genes can have serious phenotypic consequences.

Random Monoallelic Expression: Inactivation of the X Chromosome

The random inactivation of an X chromosome in cells of female mammals was the first example of epigenetic allele-specific regulation to be identified. At an early stage of development, about half of embryonic cells randomly inactivate the maternal X chromosome and the other half inactivate the paternal X chromosome, effectively silencing almost all the 900 or so genes on whichever homolog is inactivated. Once inactivated, the same X chromosome remains silenced in all cells descended from this progenitor cell.

How does X inactivation occur? Several lncRNAs play a key role in this process. Two of the major contributors are Xist (X inactive specific transcript), and Tsix (Xist spelled backward), which are sense and antisense transcripts of the same gene (transcribed in opposite directions). The

Xist lncRNA is expressed on the inactivated X chromosome and coats the entire chromosome, converting it into a Barr body (see Chapter 5), which is a highly condensed and genetically silent chromatin structure. The lncRNA Tsix is expressed on the active X chromosome and represses expression of the Xist lncRNA, thus preventing the active X chromosome from being silenced.

Random Monoallelic Expression of Autosomal Genes

Genome-wide analysis of allele-specific expression in mice and humans led to the surprising discovery that monoallelic expression (MAE) is a widespread event, involving 10 to 20 percent of autosomal genes in a range of different cell types.

Unlike imprinted genes, which are in clusters, autosomal MAE genes are scattered throughout the genome. Because autosomal MAE is a random process, four states of expression for a gene are possible in cells of a given tissue: (1) expression of both alleles (biallelic expression), (2) expression of only the maternal allele, (3) expression of only the paternal allele, or (4) expression of neither allele. These different patterns of expression, all present in the same tissue, can have an impact on the phenotype and may offer a molecular explanation for the incomplete penetrance of traits observed in some genetic disorders. (See Chapter 4 for a discussion of incomplete penetrance.)

By analyzing several epigenetic marks present in a wide range of cell types, researchers established that two modifications of histone 3, H3K27me3 and H3K36me3, explain most of the difference between cells with monoallelic expression of a given gene and cells with biallelic expression of the same gene. In MAE cells, the H3K27me3 marker, associated with gene silencing, is linked to the inactive allele, while the H3K36me3 marker, associated with transcription, is linked to the active allele. This chromatin signature is a powerful and reliable predictor of MAE activity in many cell types and offers a way of exploring the relationship between epigenetics and disease.

Assisted Reproductive Technologies (ART) and Imprinting Defects

In the United States, **assisted reproductive technologies (ART),** including *in vitro* fertilization (IVF), are now used in over 1 percent of all births. Over the past decade, several studies have suggested that children born using ART have an increased risk for imprinting errors (epimutations) caused by the manipulation of gametes or embryos.

For example, the use of ART results in a four- to ninefold increased risk of Beckwith–Wiedemann syndrome (BWS); in addition, there are increased risk for Prader–Willi syndrome (PWS) and Angelman syndrome (AS). Studies of children with BWS or AS conceived by IVF have shown that they have reduced levels or complete loss of maternal-specific methylation at known imprinting sites in the genome, confirming the role of epigenetics in these cases. ART procedures are done at times when the oocyte and the early embryo are undergoing epigenetic reprogramming. It appears that disturbances in epigenetic programming at sensitive times during development may be responsible for the increased risk of these disorders.

Although imprinting errors are uncommon in the general population (BWS occurs in only about 1 in 13,700 births), epimutations may be a significant risk factor for those conceived by ART.

ST 1.3 Epigenetics and Cancer

Originally it was thought that cancer is clonal in origin and begins in a single cell that has accumulated a suite of mutations that allow it to escape control of the cell cycle. Subsequent mutations allow cells of the tumor to become metastatic, spreading the cancer to other locations in the body where new malignant tumors appear. However, converging lines of evidence are now clarifying the importance of epigenetic changes in the initiation and maintenance of malignancy. These findings are helping researchers understand the properties of cancer cells that are difficult to explain by the action of mutant alleles alone. Evidence for the role of epigenetic changes in cancer has established epigenomic changes as a major pathway for the formation and spread of malignant cells.

DNA Methylation and Cancer

As far back as the 1980s, researchers observed that cancer cells had much lower levels of methylation than normal cells derived from the same tissue. Subsequent research by many investigators showed that complex changes in DNA methylation patterns are associated with cancer. These studies showed that genomic hypomethylation is a property of all cancers examined to date.

DNA hypomethylation reverses the silencing of genes, leading to unrestricted transcription of many gene sets—including those associated with the development of cancer. It also relaxes control over imprinted genes, causing cells to acquire new growth properties. Hypomethylation of repetitive DNA sequences in heterochromatic regions increases chromosome rearrangements and changes in chromosome number, both of which are characteristic of cancer cells.

TABLE ST 1.2 Some Human Cancer-Related Genes Inactivated by Hypermethylation

Gene	Locus	Function	Related Cancers
BRCA1	17q21	DNA repair	Breast, ovarian
APC	5q21	Nucleocytoplasmic signaling	Colorectal, duodenal
MLH1	3p21	DNA repair	Colon, stomach
RB1	13q14	Cell-cycle control point	Retinoblastoma, osteosarcoma
AR	Xq11–12	Nuclear receptor for androgen; transcriptional activator	Prostate
ESR1	6q25	Nuclear receptor for estrogen; transcriptional activator	Breast, colorectal

Even though cancer cells are characterized by global hypomethylation, selected regions of their genome are hypermethylated when compared to normal cells. Selective hypermethylation of promoter-associated CpG islands silences certain genes, including tumor-suppressor genes, often in a tumor-specific fashion (**Table ST 1.2**). Analysis of these patterns provides a way to identify tumor types and subtypes and predict the sites to which the tumor may metastasize.

For example, the promoter region of the breast cancer gene *BRCA1* is hypomethylated in normal cells but is hypermethylated and inactivated in many cases of breast and ovarian cancer. In another example, silencing of the DNA repair gene *MLH1* by hypermethylation is a key step in the development of some forms of colon cancer. *MLH1* illustrates how epimutations can be involved in tumor formation, either alone or in combination with genetic changes (**Figure ST 1.6**).

In fact, cancer is now viewed as a disease that usually results from the accumulation of both genetic *and* epigenetic changes (**Figure ST 1.7**). For example, in a bladder cancer cell line, one allele of a tumor-suppressor gene, *CDKN2A,* is mutated, and the other, normal, allele is silenced by hypermethylation. Because both alleles are inactivated (although by different mechanisms), cells can escape control of the cell cycle and divide continuously. Even more striking, in ovarian cancer, mutations in nine specific genes are predominant, but promoter hypermethylation is observed in 168 genes. These genes are

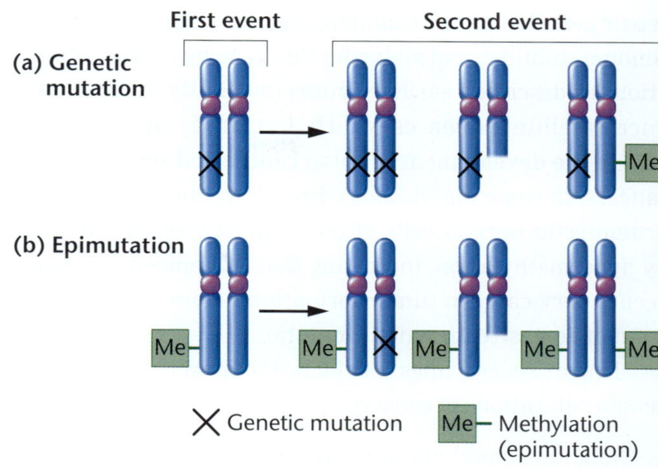

FIGURE ST 1.6 The role of epimutations versus genetic mutations in the initiation of cancer. (a) An inherited genetic mutation causes the loss of a tumor-suppressor allele. Several mechanisms can cause the loss or silencing of the second allele: mutation, chromosomal aberration, or an epimutation. (b) An epimutation silences one allele of a tumor-suppressor gene. The second allele can be lost through genetic mutation, chromosomal aberration, or silencing by an epigenetic event.

epigenetically silenced, and their reduced expression is linked to the development and maintenance of this cancer.

The broad pattern of hypermethylation seen in cancer cells and the many functions of the affected genes suggest that this phenomenon may result from a widespread deregulation of the methylation process rather than a targeted event.

At the present time, many of the mechanisms that cause epigenetic changes in cancer cells are not well understood, partly because the changes take place very early in the conversion of a normal cell to a cancerous one, and partly because by the time the cancer is detected, alterations in methylation patterns have already occurred. The DNA

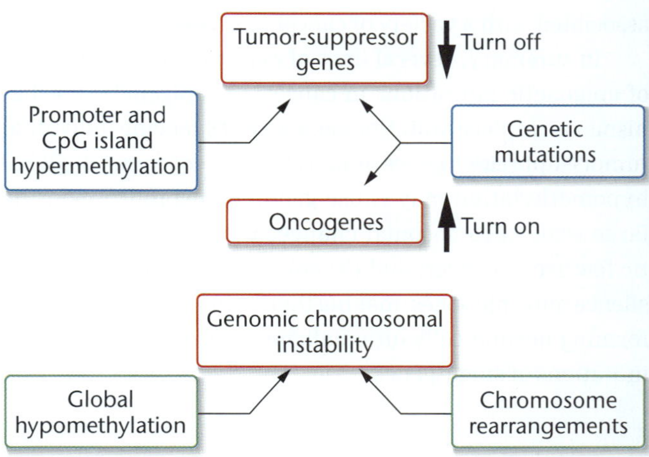

FIGURE ST 1.7 The development and maintenance of malignant growth in cancer involves the interaction of gene mutations, hypomethylation, hypermethylation, overexpression of oncogenes, and the silencing of tumor-suppressor genes.

repair gene *MLH1*, for example, plays an important role in genome stability, and silencing this gene by hypermethylation (as described earlier) causes instability in repetitive microsatellite sequences, which, in turn, is an important step in the development of colon cancer and several other cancers. In some individuals with colon cancer, the *MLH1* promoter in normal cells of the colon is already silenced by hypermethylation, indicating that this epigenetic event occurs very early in tumor formation before the development of downstream genetic mutations.

In sum, several lines of evidence support the role of epigenetic alterations in cancer:

1. Global hypomethylation may cause genomic instability and the large-scale chromosomal changes that are a characteristic feature of cancer.
2. Epigenetic mechanisms can replace mutations as a way of silencing individual tumor-suppressor genes or activating oncogenes.
3. Epigenetic modifications can silence multiple genes, making them more effective in transforming normal cells into malignant cells than sequential mutations of single genes.

Chromatin Remodeling and Histone Modification in Cancer

In addition to abnormal regulation of methylation, many cancers also have altered patterns of chromatin remodeling. One form of remodeling is controlled by the reversible covalent modification of histone proteins in nucleosome cores. Recall that this process involves three classes of enzymes: *writers* that add chemical groups to histones; *erasers* that remove these groups; and *readers* that recognize and read the epigenetic marks. Abnormal regulation of each of these enzyme classes results in disrupted histone profiles and is associated with a variety of cancer subtypes.

In summary, several lines of evidence support the role of epigenetic alterations in cancer: (1) epigenetic mechanisms can replace mutations as a way of silencing individual tumor-suppressor genes or activating oncogenes; (2) global hypomethylation may cause genomic instability and the large-scale chromosomal changes that are a characteristic feature of cancer; and (3) epigenetic modifications can silence multiple genes, making them more effective in transforming normal cells into malignant cells than sequential mutations of single genes.

Epigenetic Cancer Therapy

The fact that unlike genetic alterations, which are almost impossible to reverse, epigenetic changes are potentially reversible has inspired researchers to look for new classes of drugs to treat cancer. The focus of epigenetic therapy in the development of first-generation drugs has been the reactivation of genes silenced by methylation or histone modification, essentially reprogramming the pattern of gene expression in cancer cells.

The U.S. Food and Drug Administration has approved several epigenetic drugs, and another 18 or more drugs are in clinical trials. One approved drug, Vidaza (azacitidine), is used in the treatment of myelodysplastic syndrome, a precursor to leukemia, and acute myeloid leukemia. This drug is an analog of cytidine and is incorporated into DNA during replication during the S phase of the cell cycle. Methylation enzymes (methyltransferases) bind irreversibly to this analog, preventing methylation of DNA at many other sites, effectively reducing the amount of methylation in cancer cells.

Other drugs that inhibit histone deacetylases (HDACs) have been approved by the FDA for use in epigenetic therapy. Laboratory experiments with cancer cell lines indicate that inhibiting HDAC activity results in the re-expression of tumor-suppressor genes. HDAC inhibitors like Zolinza (vorinostat) are used to treat certain forms of lymphoma.

The development of epigenetic drugs for cancer therapy is still in its infancy. The approved epigenetic drugs are only moderately effective on their own and are best used in combination with other anticancer drugs. To develop more effective drugs, several important questions remain to be answered: What causes cancer cells to respond to certain epigenetic drugs? Which combinations of chromatin remodeling drugs, histone modification drugs, and conventional anticancer drugs are most effective on specific cancers? Which epigenetic markers will be effective in predicting sensitivity or resistance to newly developed drugs? Further research into the mechanisms and locations of epigenetic genome modification in cancer cells will allow the design of more potent drugs to target epigenetic events as a form of cancer therapy.

ST 1.4 Epigenetic Traits Are Heritable

Environmental agents including nutrition, exposure to chemicals, medical or recreational drugs, as well as social interactions and exercise can alter gene expression by affecting the epigenome. In humans it is difficult to determine the relative contributions of the environment as factors in altering the epigenome, but there is indirect evidence that changes in nutrition and exposure to agents that affect a developing fetus can have detrimental effects during adulthood.

During World War II, a famine in the western part of the Netherlands lasted from November 1944 to May 1945. During this time, daily food intake for adults was limited to 400 to 800 calories, well below the normal levels of 1800 to 2000 calories. Studies were conducted for decades afterward on the health of adult children of women who were pregnant or became pregnant during the famine. Overall, the findings show that the severity of health effects was correlated with prenatal time of exposure to famine conditions. Adults who were exposed early in prenatal development (an F_1 generation) had higher rates of several disorders—including obesity, heart disease, and breast cancer—and higher mortality rates than adults exposed later in development. In addition, as adults, there was increased risk for schizophrenia and other neuropsychiatric disorders for those with early exposure, perhaps related to nutritional deficiencies during development of the brain and nervous system. Some effects persisted in the F_2 generation, where adults had abnormal patterns of growth and increased rates of obesity. Other studies in China and Africa on the adult children of women who were pregnant or became pregnant during times of famine confirm the deleterious impact of poor maternal nutrition during pregnancy on offspring in subsequent generations.

More direct evidence for the role of environmental factors in modifying the epigenome comes from studies in experimental animals. One dramatic example of how epigenome modifications affect the phenotype comes from the study of coat color in mice, where color is controlled by the dominant allele *Agouti* (*A*). In homozygous *AA* mice, the allele is active only during a specific time during hair development, resulting in a yellow band on an otherwise black hair shaft, producing the agouti phenotype. A nonlethal mutant allele (A^{vy}) causes yellow pigment formation along the entire hair shaft, resulting in yellow fur color. This allele is the result of the insertion of a transposable element near the transcription start site of the *Agouti* gene. A promoter element within the transposon is responsible for this change in gene expression.

Researchers found that the degree of methylation in the transposon's promoter is related to the amount of yellow pigment deposited in the hair shaft and that the amount of methylation varies from individual to individual. The result is variation in coat color phenotypes even in genetically identical mice (**Figure ST 1.8**). In these mice, coat colors range from yellow (unmethylated promoter) to pseudoagouti (highly methylated promoter). In addition to a gradation in coat color, there is also a gradation in body weight. Yellow mice are more obese than the brown pseudoagouti mice and are more likely to be diabetic. Alleles such as A^{vy} that show variable expression from individual to individual in genetically identical strains caused by different patterns

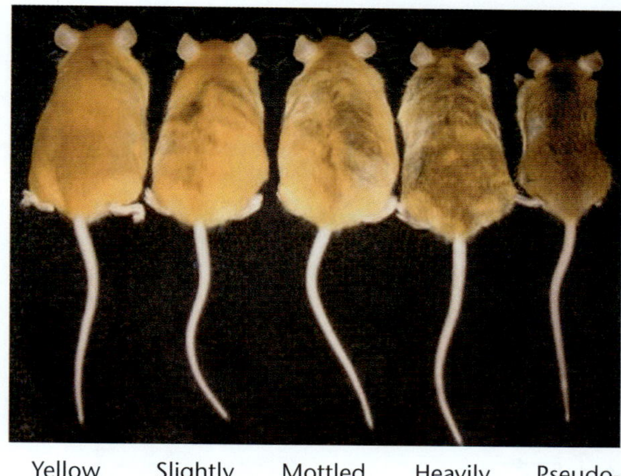

FIGURE ST 1.8 Variable expression of yellow phenotype in genetically identical mice caused by diet-related epigenetic changes in the A^{vy} allele.

of epigenetic modifications to the alleles are called *metastable epialleles*. "Metastable" refers to the variable nature of the epigenetic modifications, and "epiallele" refers to the heritability of the epigenetic status of the altered gene. In other words, the epigenetic modifications to the A^{vy} allele can be passed on to offspring; this is a clear example of transgenerational inheritance.

To evaluate the role of environmental factors in modifying the epigenome, the diet of pregnant A^{vy} mice was supplemented with methylation precursors, including folic acid, vitamin B_{12}, and choline. In the offspring, variation in coat color was reduced and shifted toward the pseudoagouti (highly methylated) phenotype. The shift in coat color was accompanied by increased methylation of the transposon's promoter. These findings have applications to epigenetic diseases in humans. For example, the risk of one form of colorectal cancer is linked directly to increased methylation of the DNA repair gene *MLH1*.

Stress-Induced Behavior Is Heritable

A growing body of evidence shows that epigenetic changes, including alterations in DNA methylation and histone modification, have important effects on behavioral phenotypes.

One of the most significant findings in the epigenetics of behavior is that stress-induced epigenetic changes that occur prenatally or early in life can influence behavior (and physical health) later in adult life and can potentially be transmitted to future generations A classic study showed that newborn rats raised with low levels of maternal nurturing (low-MN) did not adapt well to stress and anxiety-inducing situations in adulthood. In rats and humans, the hypothalamic region of the brain mediates stress reactions

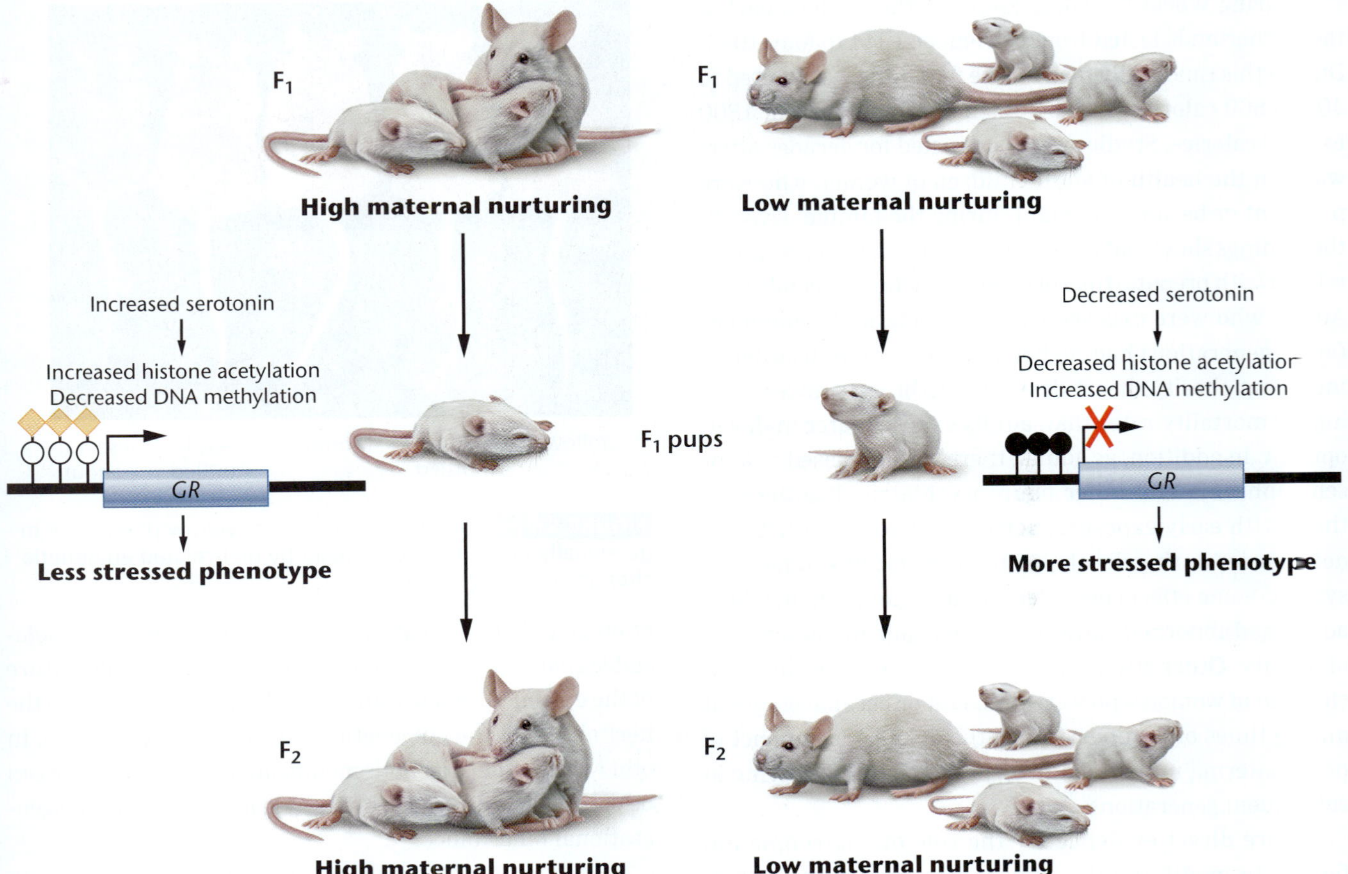

FIGURE ST 1.9 Style of maternal care is transmitted across rat generations through epigenetic events that take place early in postnatal life. High maternal nurturing induces high levels of serotonin in the brain, leading to DNA hypomethylation, histone acetylation, and increased expression of GR. In adulthood, high levels of GR expression increase adaptation to stress and, in females, passes on the high-MN phenotype. Rat pups experiencing low levels of maternal nurturing had higher levels of promoter methylation and reduced levels of GR expression. In adulthood, this led to poor stress adaptation and, in females, perpetuation of low levels of nurturing her pups.

by controlling levels of glucocorticoid hormones via the action of cell-surface glucocorticoid receptors (GRs).

In rats exposed to high levels of maternal nurturing care early in life (high-MN), GR expression is increased, and adults are stress adaptive. However, low-MN rats had reduced levels of GR transcription and were less able to adapt to stress. Differences in GR expression were associated with differences in histone acetylation and DNA methylation levels in the GR gene promoter. Low-MN rats had significantly higher levels of promoter methylation than high-MN rats (**Figure ST 1.9**).

Subsequent research showed that differences in DNA methylation are present in hundreds of genes across the genome, all of which show differential expression in low-MN and high-MN adults. Later studies showed that these behavioral phenotypes can be transmitted across generations. Female rats raised by more nurturing mothers are more attentive to their own newborns, whereas those raised by less nurturing mothers are much less attentive and less nurturing to their offspring.

ST 1.5 Epigenome Projects and Databases

As the role of the epigenome in disease has become increasingly clear, researchers across the globe have formed multidisciplinary projects to map all the epigenetic changes that occur in the normal genome and established databases to study the role of the epigenome in specific diseases. We will discuss some of these projects and their goals and will summarize the major findings of a few of the large-scale projects.

The NIH Roadmap Epigenomics Project was established to elucidate the role of epigenetic mechanisms in human biology and disease. The project has two main goals: (1) provide a set of at least 1000 reference epigenomes in a range of cell types from healthy and diseased individuals, and (2) delineate the epigenetic differences in conditions such as Alzheimer disease, autism, and schizophrenia.

In 2015, the project published an analysis of the first 111 reference genomes collected, representing the most comprehensive map of the human epigenome to date. One of the important results of this study establishes that genetic variants associated with several complex human disorders such as Alzheimer disease, cancer, and autoimmune disorders are enriched in tissue-specific epigenomic marks, identifying relevant cell types associated with these and other disorders.

The Human Epigenome Atlas, which is part of the Roadmap Project, collects and catalogs detailed information about epigenomic modifications at specific loci, in different cell types, different physiological states, and different genotypes. These data allow researchers to perform comparative analysis of epigenomic data across genomic regions or entire genomes.

The International Human Epigenome Consortium (IHEC) is a global program established to coordinate the collection of epigenome maps for 1000 human cell populations. Several projects are contributing to the program, each specializing in different cell types and/or approaches. The U.S. Reference Epigenome Mapping Centers are using stem cells and tissue samples from healthy donors, and the Germany-based DEEP Project is collecting 70 reference epigenomes of human and mouse tissues associated with metabolic and inflammatory diseases. The European BLUEPRINT Project is collecting epigenomic profiles from several different types of blood cells related to specific diseases.

To complement the efforts of IHEC in mapping the epigenomes of primary cell lines collected directly from tissues, the Encyclopedia of DNA Elements (ENCODE) project is focused on collecting epigenome maps for cell lines grown under laboratory conditions. To compare the epigenomes of normal cells with cancer cells, the International Cancer Genome Consortium (ICGC) is mapping the epigenomes and the transcriptome profiles of 50 different cancer types.

Although these projects are still in progress, the information already available strongly suggests that we are on the threshold of a new era in genetics, one in which we can study the development of disease at the genomic level and understand the impact of epigenetic factors on gene expression.

Visit the Study Area in MasteringGenetics for a list of further readings on this topic, including journal references and selected Web sites.

Review Questions

1. What are the major mechanisms of epigenetic genome modification?
2. What parts of the genome are reversibly methylated? How does this affect gene expression?
3. What are the roles of proteins in histone modification?
4. Describe how reversible chemical changes to histones are linked to chromatin modification.
5. What is the histone code?
6. What is the difference between silencing genes by imprinting and silencing by epigenetic modifications?
7. Why are changes in nucleosome spacing important in changing gene expression?
8. What is the role of imprinting in human genetic disorders?

Discussion Questions

1. Imprinting disorders do not involve changes in DNA sequence, but only the methylated state of the DNA, or the modification of histones. Does it seem likely that imprinting disorders could be treated prenatally or prevented by controlling the maternal environment in some way, perhaps by dietary changes?
2. Should fertility clinics be required by law to disclose that some assisted reproductive technologies (ART) can result in epigenetic diseases? How would you and your partner balance the risks of ART with the desire to have a child?
3. How can the role of epigenetics in cancer be reconciled with the idea that cancer is caused by the accumulation of mutations in tumor-suppressor genes and proto-oncogenes?
4. Several studies have shown that small RNAs from plants acquired through the diet circulate in the bloodstream and affect gene expression in animals, including humans. If these studies are confirmed, would this knowledge affect your food choices?

Genetic Testing

Earlier in the text (see Chapters 17 and 18), we reviewed essential concepts of recombinant DNA technology and genomic analysis. Because of the Human Genome Project and related advances in genomics, researchers have been making rapid progress in identifying genes involved in both single-gene diseases and complex genetic traits. As a result, **genetic testing**—the ability to analyze DNA, and increasingly RNA, for the purposes of identifying specific genes or sequences associated with different genetic conditions—has advanced very rapidly.

Genetic testing, including genomic analysis by DNA sequencing, is transforming medical diagnostics. Technologies for genetic testing have had major impacts on the diagnosis of disease and are revolutionizing medical treatments based on the development of specific and effective pharmaceuticals. In this Special Topics chapter we provide an overview of applications that are effective for the genetic testing of children and adults and examine historical and modern methods. We consider the impact of different genetic technologies on the diagnosis of human diseases and disease treatment. Finally, we consider some of the social, ethical, and legal implications of genetic testing.

"Genetic testing, including genomic analysis by DNA sequencing, is transforming medical diagnostics. Technologies for genetic testing have had major impacts on the diagnosis of disease and are revolutionizing medical treatments based on the development of specific and effective pharmaceuticals."

ST 2.1 Testing for Prognostic or Diagnostic Purposes

Genetic testing was one of the first successful applications of recombinant DNA technology, and currently more than 900 tests are in use that target a specific gene or sequence. Increasingly, scientists and physicians can directly examine an individual's DNA for mutations associated with disease, including through DNA sequencing, as we will discuss in Section ST 2.5. These tests usually detect gene alterations associated with single-gene disorders. But, only about 3900 genes have been linked to such disorders. Examples include sickle-cell anemia, cystic fibrosis, Huntington disease, hemophilia, and muscular dystrophy. Other tests have been developed for disorders that may involve multiple genes such as certain types of cancers.

Gene tests are used for prenatal, childhood, and adult prognosis and diagnosis of genetic diseases; to identify carriers; and to identify genetic diseases in embryos created by *in vitro* fertilization, among other applications. For genetic testing of adults, DNA from white blood cells is commonly used. Alternatively, many genetic tests can be carried out on cheek cells, collected by swabbing the inside of the mouth, or on hair cells. Some genetic testing can be carried out on gametes.

What does it mean when a genetic test is performed for *prognostic* purposes and how does this differ from a *diagnostic* test? A prognostic test predicts a person's likelihood of developing a particular genetic disorder. A diagnostic test for a genetic condition identifies a particular mutation or genetic change that causes the disease or condition. Sometimes a diagnostic test identifies a gene or mutation associated with a condition, but the test will not be able to determine whether the gene or mutation is the cause of the disorder or is a genetic variation that results from the condition.

ST 2.2 Prenatal Genetic Testing to Screen for Conditions

Although genetic testing of adults is increasing, over the past two decades more genetic testing has been used to detect genetic conditions in babies than in adults. In newborns, a simple prick of a baby's heel produces a few drops of blood that are used to check the newborn for many genetic disorders. In the United States, all states now require genetic testing, often called *newborn screening*, for certain medical conditions (the number of diseases screened for is set by the individual state, see Box 1). There are currently about 60 conditions that can be detected, although many of these tests detect proteins or other metabolites and are not DNA- or RNA-based genetic tests.

Prenatal genetic tests, performed before a baby is born, are used for certain disorders in which waiting until birth is not desirable. For prenatal testing, fetal cells are obtained by **amniocentesis** or **chorionic villus sampling (CVS). Figure ST 2.1** shows the procedure for amniocentesis, in which a small volume of the amniotic fluid surrounding the fetus is removed. Amniotic fluid contains fetal cells that can be used for karyotyping, genetic testing, and other procedures. For chorionic villus sampling, cells from the fetal portion of the placental wall (the chorionic villi) are sampled through a vacuum tube, and analyses can be carried out on this tissue. Captured fetal cells can then be subjected to genetic analysis by techniques that involve PCR (such as allele-specific oligonucleotide testing, described in Section ST 2.3) or DNA sequencing (described in Section ST 2.5). In Section ST 2.3 we discuss a fetal screening approach called preimplantation genetic diagnosis.

Noninvasive Procedures for Genetic Testing of Fetal DNA

Noninvasive procedures are also being developed for prenatal genetic testing of fetal DNA. These procedures reduce the risk to the fetus. Circulating in each person's bloodstream is DNA that is released from the person's dead and dying cells. This so-called *cell-free DNA (cfDNA)* is cut up into small fragments by enzymes in the blood. The blood of a pregnant woman also contains snippets of cfDNA from the fetus. It

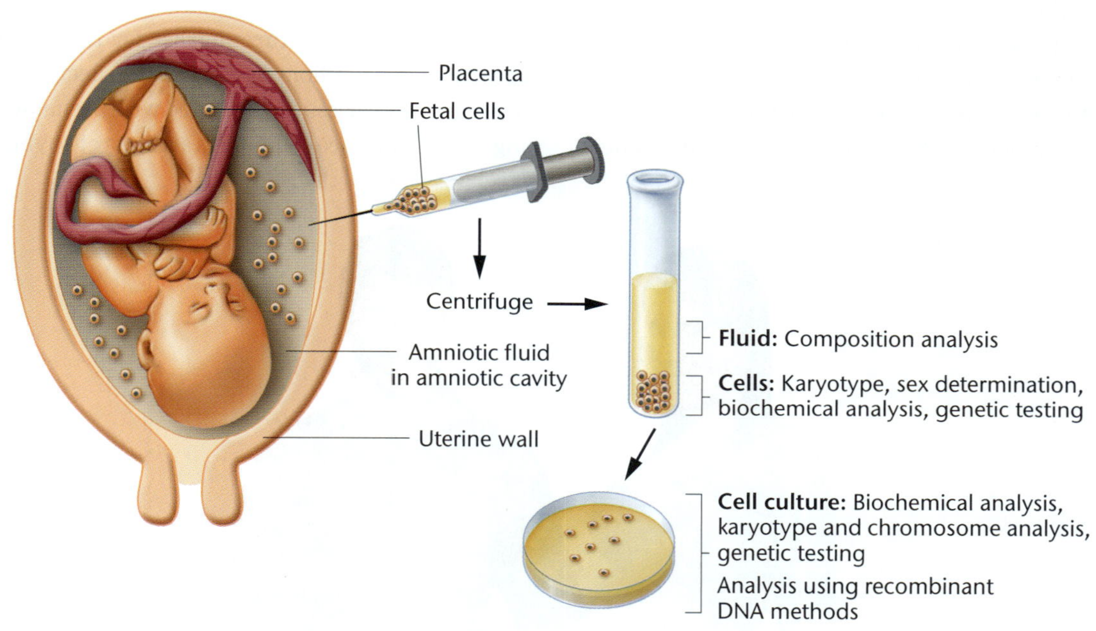

FIGURE ST 2.1 For amniocentesis, the position of the fetus is first determined by ultrasound, and then a needle is inserted through the abdominal and uterine walls to recover amniotic fluid containing fetal cells for genetic or biochemical analysis.

BOX 1

Recommended Uniform Screening Panel

The Recommended Uniform Screening Panel (RUSP) is a list of disorders recommended by the U.S. Department of Health and Human Services (HHS) for states to screen as part of newborn genetic testing programs. Currently, 35 conditions are on the RUSP. Most of these conditions involve metabolic disorders with a clear genetic component. Advocates for newborn screening have expressed frustration that it can take several years for a condition to be approved by HHS for inclusion on the RUSP. There are over 200 congenital disorders with U.S Food and Drug Administration (FDA) approvals, yet only 35 conditions are on the RUSP. Should RUSP approval happen faster? Should the RUSP be mandatory for all states? Newborn screening itself is fairly inexpensive, ~$1 per genetic test. The major cost associated with newborn screening is for the development of new drugs to treat genetic conditions. Thus there is interest from the biopharmaceutical industry to lobby the federal government for broader inclusion of conditions on the RUSP list, which will then lead to drug development. Do an Internet search for RUSP to visit the HHS Web site to see which conditions are included on the current screening list.

is estimated that ~3 to 6 percent of the DNA in a pregnant mother's blood belongs to her baby. It is now possible to analyze these traces of fetal DNA to determine if the baby has certain types of genetic conditions such as Down syndrome. Such tests require about a tablespoon of blood.

DNA in the blood is sequenced to analyze **haplotypes**—contiguous segments of DNA that do not undergo recombination during gamete formation—that distinguish which cfDNA segments are maternal and which are from the fetus (see **Figure ST 2.2**). If a fetal haplotype contained a specific mutation, this would also be revealed by sequence analysis. Nearly complete fetal genome sequences have been assembled from maternal blood. These are developed by sequencing cfDNA fragments from maternal blood and comparing those fragments to sequenced genomes from the mother and father. Bioinformatics software is then used to organize the genetic sequences from the fetus in an effort to assemble the fetal genome. Currently, this technology does not capture the entire fetal genome; it results in an assembled genome sequence with segments missing. It has been shown, however, that **whole genome sequencing (WGS)** (introduced in Chapter 18) of maternal plasma cfDNA can be used to accurately sequence the entire exome of a fetus.

Tests for fetal genetic analysis based on maternal blood samples started to arrive on the market in 2011. Sequenom of San Diego, California, was one of the first companies to launch such a test—*MaterniT® 21 PLUS*, a Down syndrome test that can also be used to test for trisomy 13 (Patau syndrome) and trisomy 18 (Edwards syndrome). MaterniT 21 PLUS analyzes 36-bp fragments of DNA to identify chromosome 21 from the fetus. Sequenom claims that this test is highly accurate with a false positive rate of just 0.2 percent. The test can be done as early as week 10 (about the same time CVS sampling can be performed, which is about 4 to 6 weeks earlier than amniocentesis can be performed). Several companies have followed the Sequenom approach. Nationwide, it has been estimated that the future market for these tests could be greater than $1 billion. As discussed in Section ST 2.7, there are many ethical issues associated with prenatal genetic testing. Most insurance companies are not yet paying for WGS of

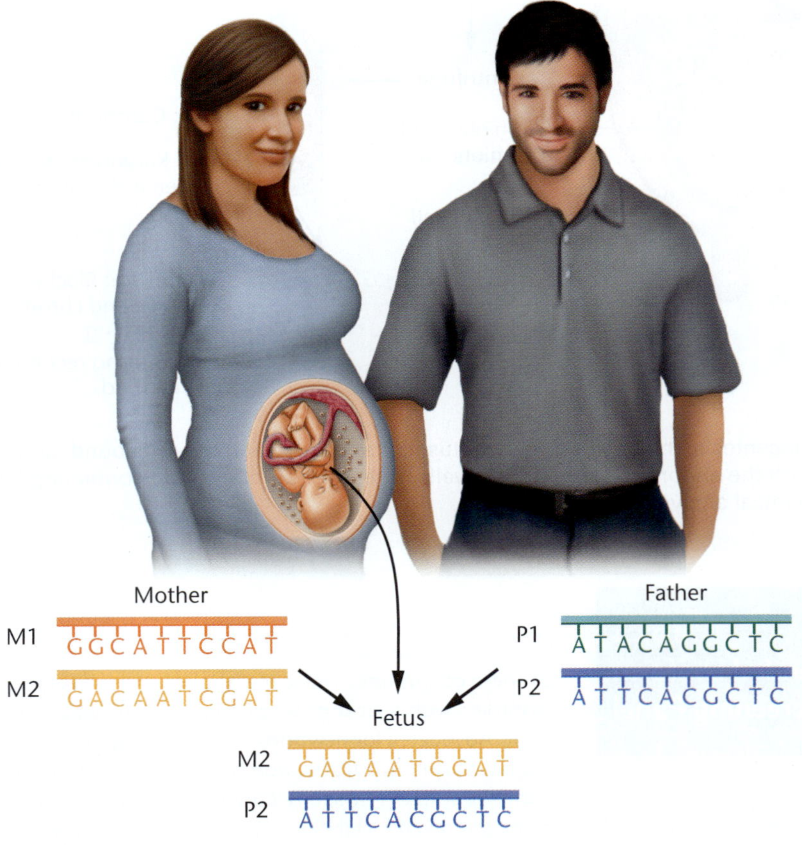

FIGURE ST 2.2 Deducing fetal genome sequences from maternal blood. For any given chromosome, a fetus inherits one copy of a haplotype from the mother (maternal copies, M1 or M2) and another from the father (paternal copies, P1 or P2). For simplicity, a single-stranded sequence of DNA from each haplotype is shown. These haplotype sequences can be detected by WGS. Here the fetus inherited haplotypes M2 and P2 from the mother and father, respectively. DNA from the blood of a pregnant woman would contain paternal haplotypes inherited by the fetus (P2, blue), maternal haplotypes that are not passed to the fetus (M1, orange), and maternal haplotypes that are inherited by the fetus (M2, yellow). The maternal haplotype inherited by the fetus (M2) would be present in excess amounts relative to the maternal haplotype that is not inherited (M1).

maternal blood, which can cost as much as $2000 for a single test. Recently, California agreed to subsidize noninvasive prenatal testing for women through the state's genetic diseases program, which screens ~400,000 women each year.

Originally these noninvasive tests were offered to women older than 35 years of age, or if they were identified as at risk based on family history of birth-related complications. Now these tests are being marketed to women with low-risk pregnancies as well, and their value, given the cost, has been questioned. Recent figures indicate that sales for these tests exceeded $600 million annually, and this number is estimated to increase four-fold by 2022.

In Section ST 2.7 we will discuss preconception testing and recent patents for computing technologies designed to predict the genetic potential of offspring (destiny tests).

ST 2.3 Genetic Testing Using Allele-Specific Oligonucleotides

When genetic testing of adults was initiated, one of the first methods was **restriction fragment length polymorphism (RFLP) analysis.** For example, historically, RFLP analysis was the primary method to detect **sickle-cell anemia.** As discussed in previous chapters, this disease is an autosomal recessive condition common in people with family origins in areas of West Africa, the Mediterranean basin, parts of the Middle East, and India. It is caused by a single amino acid substitution in the β-globin protein, as a consequence of a single-nucleotide substitution in the corresponding gene.

The single-nucleotide substitution also eliminates a restriction site in the β-globin gene for the restriction enzymes *Mst* II and *Cvn* I. As a result, the mutation alters the pattern of restriction fragments seen on Southern blots. These differences in restriction sites could be used to diagnose sickle-cell anemia prenatally and to establish the parental genotypes and the genotypes of other family members who may be heterozygous carriers of this condition.

But only about 5 to 10 percent of all point mutations can be detected by RFLP analysis because most mutations occur in regions of the genome that do not contain restriction enzyme sites. However, since the Human Genome Project (HGP) was completed and many more disease-associated mutations became known, geneticists employed PCR and synthetic oligonucleotides to detect these mutations, including the use of synthetic DNA probes known as **allele-specific oligonucleotides (ASOs).**

This rapid, inexpensive, and accurate technique is used to diagnose a wide range of genetic disorders caused by point mutations. In contrast to RFLP analysis, which is limited to cases for which a mutation changes a restriction site, ASOs detect single-nucleotide changes called **single-nucleotide polymorphisms (SNPs).**

An ASO is a short, single-stranded fragment of DNA designed to hybridize to a complementary specific allele in the genome. Under proper conditions, an ASO will hybridize only with its complementary DNA sequence and not with other sequences, even those that vary by as little as a single nucleotide.

Genetic testing using ASOs and PCR analysis is available to screen for many disorders in adults and newborns and for prenatal screening. In the case of sickle-cell anemia screening, DNA is extracted (either from a maternal blood sample or from fetal cells obtained by amniocentesis), and a region of the β-globin gene is amplified by PCR. A small amount of the amplified DNA is spotted onto strips of a DNA-binding membrane, and each strip is hybridized to an ASO synthesized to resemble the relevant sequence from either a normal or mutant β-globin gene. The ASO is tagged with a molecule that is either radioactive or fluorescent, to allow for visualization of the ASO hybridized to DNA on the membrane. **Figure ST 2.3** illustrates the principle behind this

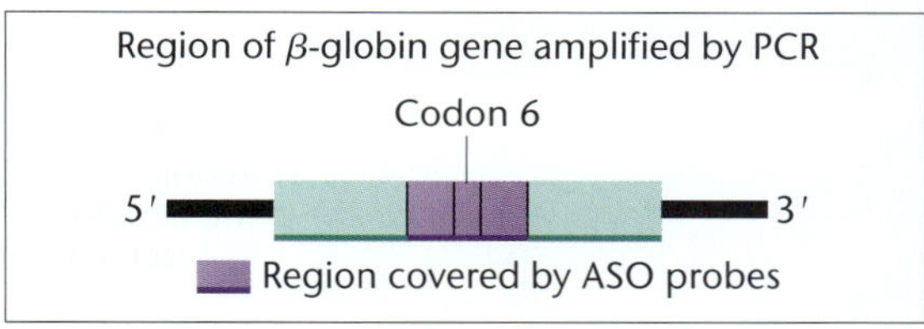

DNA is spotted onto binding membranes and hybridized with ASO probe

(a) Genotypes *AA* *AS* *SS*

Normal (β^A) ASO: 5′ – CTCCTG**A**GGAGAAGTCTGC – 3′

(b) Genotypes *AA* *AS* *SS*

Mutant (β^S) ASO: 5′ – CTCCTG**T**GGAGAAGTCTGC – 3′

FIGURE ST 2.3 Allele-specific oligonucleotide (ASO) testing for sickle-cell anemia. (a) Results observed if the three possible β-globin genotypes are hybridized to an ASO for the normal β-globin allele: *AA*-homozygous individuals have normal hemoglobin that has two copies of the normal β-globin gene and will show heavy hybridization; *AS*-heterozygous individuals carry one normal β-globin allele and one mutant allele and will show weaker hybridization; *SS*-homozygous sickle-cell individuals carry no normal copy of the β-globin gene and will show no hybridization to the ASO probe for the normal β-globin allele. (b) Results observed if DNA for the three genotypes are hybridized to the probe for the sickle-cell β-globin allele: no hybridization by the *AA* genotype, weak hybridization by the heterozygote (*AS*), and strong hybridization by the homozygous sickle-cell genotype (*SS*).

approach. This rapid, inexpensive, and accurate technique is used to diagnose a wide range of genetic disorders caused by point mutations.

Although ASO testing is highly effective, SNPs can affect ASO probe binding leading to false positive or false negative results that may not reflect a genetic disorder, particularly if precise hybridization conditions are not used. Sometimes DNA sequencing is carried out on amplified gene segments to confirm identification of a mutation.

Preimplantation Genetic Diagnosis

Because ASO testing makes use of PCR, only small amounts of DNA are required for analysis. As a result, ASO testing is ideal for **preimplantation genetic diagnosis (PGD),** also called preimplantation genetic testing or screening. PGD involves the genetic analysis of cells from embryos created by *in vitro* fertilization (IVF) (**Figure ST 2.4**). PGD has been used for over 25 years, typically when there is concern about a particular genetic defect. In the United States, ~25 percent of IVF attempts use PGD.

When sperm and eggs are mixed to create zygotes, the early-stage embryos are grown in culture. A single cell can be removed from an early-stage embryo using a vacuum pipette to gently aspirate one cell away from the embryo (Figure ST 2.4, top). This could possibly kill the embryo, but if it is done correctly, the embryo will often continue to divide normally. DNA from the single cell is then typically analyzed by fluorescence in situ hybridization (FISH; see Chapters 9 and 17) for chromosome analysis or by ASO

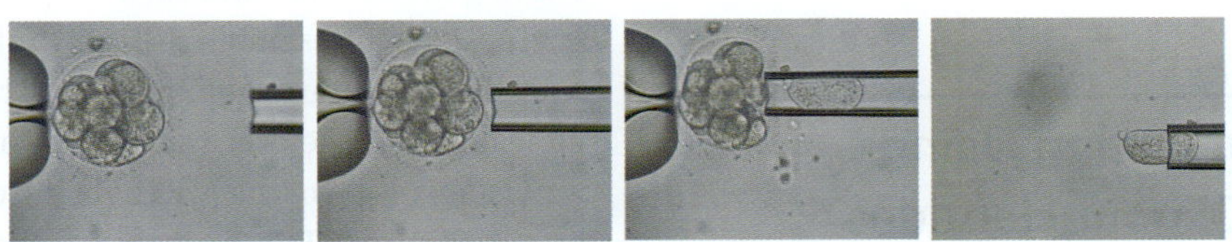

At the 8- to 16-cell stage, one cell from an embryo is gently removed with a suction pipette. The remaining cells continue to grow in culture.

DNA from an isolated cell is amplified by PCR with primers for the β-globin gene. Small volumes of denatured PCR products are spotted onto two separate DNA-binding membranes.

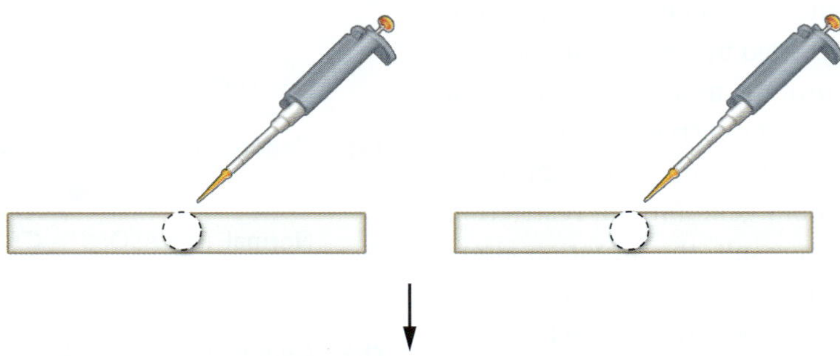

One membrane is hybridized to a probe for the normal β-globin allele (β^A), and the other membrane is hybridized to a probe for the mutant β-globin allele (β^S).

Membrane hybridized to a probe for the normal β-globin allele (β^A)
Membrane hybridized to a probe for the mutant β-globin allele (β^S)
In this example, hybridization of the PCR products to the probes for both the β^A and β^S alleles reveals that the cell analyzed by PGD has a carrier genotype ($\beta^A\beta^S$) for sickle-cell anemia.

FIGURE ST 2.4 A single cell from an early-stage human embryo created by *in vitro* fertilization can be removed and subjected to preimplantation genetic diagnosis (PGD) by ASO testing. DNA from the cell is isolated, amplified by PCR with primers specific for the gene of interest, and then subjected to ASO analysis. In this example, a region of the gene was amplified and analyzed by ASO testing to determine the sickle-cell genotype for this cell.

testing (Figure ST 2.4, bottom). The genotypes for each early-stage embryo can be tested to decide which embryos will be implanted into the uterus.

PGD can also involve removing 5 to 7 cells from the trophectoderm, the outermost layer of cells in a blastocyst-stage embryo which forms the placenta, as an attempt to determine if the embryo has a normal complement of chromosomes. But such analysis does not always reflect the genetics of the embryo, which develops from a cluster of cells within the blastocyst called the inner cell mass.

Any alleles that can be detected by ASO testing can be used during PGD. Sickle-cell anemia, cystic fibrosis, and dwarfism are often tested for by PGD, but alleles for many other conditions are also often analyzed. In theory, PGD should improve embryo implantation success rates and reduce miscarriages for couples—and success rates have improved, particularly for older women undergoing IVF. But this turns out not to be true for all couples because PGD cannot be used for identifying epigenetic changes that affect fertility. As we learn more about epigenetic influences on fertilization, it is expected that PGD will be expanded to incorporate epigenetic analysis in the future. Another limitation of PGD is that the genetics of a single cell may not provide a complete snapshot of the genetic health of an embryo.

Also, as you will learn in Section ST 2.5, it is now possible to carry out WGS on individual cells. This method is being applied for PGD of cells from an embryo created by IVF.

ST 2.4 Microarrays for Genetic Testing

ASO analysis is an effective method of screening for one, or a small number, of mutations within a gene. However, there is a significant demand for genetic tests that detect complex mutation patterns or previously unknown mutations in a single gene associated with genetic diseases and cancers. For example, the gene that is responsible for cystic fibrosis (the *CFTR* gene) contains 27 exons and encompasses 250 kb of genomic DNA. Of the 1000 known mutations of the *CFTR* gene, about half are point mutations, insertions, and deletions—and they are widely distributed throughout the gene. Similarly, over 500 different mutations are known to occur within the tumor suppressor *TP53* gene, and any of these mutations may be associated with, or predispose a patient to, a variety of cancers. In order to screen for mutations in these genes, comprehensive, high-throughput methods are required.

From Chapter 18, recall that one high-throughput screening technique is based on the use of **DNA microarrays** (also called DNA or gene chips). The numbers and types of DNA sequences on a microarray are dictated by the type of analysis that is required. For example, each spot or field (sometimes also called a feature) on a microarray might contain a DNA sequence derived from each member of a gene family, sequence variants from one or several genes of interest, or a sequence derived from each gene in an organism's genome.

In the recent past, DNA microarrays have been used for a wide range of applications, including the detection of mutations in genomic DNA and the detection of gene-expression patterns in diseased tissues. However, in the near future, whole genome sequencing, **exome sequencing,** and **RNA sequencing** are expected to replace most applications involving microarrays and render this technology obsolete.

But because of the impact microarrays have had on genetic testing, it is still valuable to discuss applications of this method. What makes DNA microarrays so useful is the immense amount of information that can be simultaneously generated from a single array. DNA microarrays the size of postage stamps (just over 1 cm square) can contain up to 500,000 different fields, each representing a different DNA sequence. Human genome microarrays containing probes for most human genes are available, including many disease-related genes, such as the *TP53* gene, which is mutated in a majority of human cancers, and the *BRCA1* gene, which, when mutated, predisposes women to breast cancer and men to breast and prostate cancer.

In addition to testing for mutations in single genes, DNA microarrays can include probes that detect SNPs. SNPs crop up in an estimated 15 million positions in the genome where these single-based changes reveal differences from one person to the next. SNP sequences as probes on a DNA microarray allow scientists to simultaneously screen thousands of mutations that might be involved in single-gene diseases as well as those involved in disorders exhibiting multifactorial inheritance. This technique, known as **genome scanning,** makes it possible to analyze a person's DNA for dozens or even hundreds of disease alleles, including those that might predispose the person to heart attacks, asthma, diabetes, Alzheimer disease, and other genetically defined disease subtypes. Genome scans are now occasionally used when physicians encounter patients with chronic illnesses where the underlying cause cannot be diagnosed.

In contrast to genome scanning microarrays that detect mutations in DNA, **gene-expression microarrays** detect gene-expression patterns for specific genes. This can be an effective approach for diagnosing genetic diseases because the progression of a tissue from a healthy to a diseased state is almost always accompanied by changes in mRNA expression of hundreds to thousands of genes. Gene-expression microarrays may contain probes for only a few specific genes thought to be expressed differently in different cell types or may contain probes representing each gene in the genome. Although microarray techniques provide novel information about gene expression, it should be emphasized that DNA microarrays do not directly provide us with information about protein levels in a cell or tissue. We often infer what predicted protein levels may be based on mRNA expression patterns, but this may not always be accurate.

In one type of gene-expression microarray analysis, mRNA is isolated from two different cell or tissue types—for example, normal cells and cancer cells arising from the same cell type (**Figure ST 2.5**). The mRNA samples contain transcripts from each gene that is expressed in that cell type. Some genes are expressed at higher levels than others. The expression level of each mRNA can be used to develop a *gene-expression profile* that is characteristic of the cell type. To do this, isolated mRNA molecules are converted into cDNA molecules, using reverse transcriptase. The cDNAs from the normal cells are tagged with fluorescent dye-labeled nucleotides (for example, green), and the cDNAs from the cancer cells are tagged with a different fluorescent dye-labeled nucleotide (for example, red).

The labeled cDNAs are mixed together and applied to a DNA microarray. The cDNA molecules bind to complementary single-stranded probes on the microarray but not to other probes. Keep in mind that each field or feature does not consist of just one probe molecule but rather contains thousands of copies of the probe. After washing off the nonbinding cDNAs, scientists scan the microarray with a laser, and a computer captures the fluorescent image pattern for analysis. The pattern of hybridization appears as a series of colored dots, with each dot corresponding to one field of the microarray.

This color pattern representation of results is often referred to as a *heat map*, because the color (or intensity of brightness) of a particular spot provides a sensitive measure of the relative levels of each cDNA in the mixture. In the example shown in Figure ST 2.5, if an mRNA is present only in normal cells, the probe representing the gene encoding that mRNA will appear as a green dot because only "green" cDNAs have hybridized to it. Similarly, if an mRNA is present only in the cancer cells, the microarray probe for that gene will appear as a red dot. If both samples contain the same cDNA, in the same relative amounts, both cDNAs will hybridize to the same field, which will appear yellow. Intermediate colors indicate that the cDNAs are present at different levels in the two samples.

Gene-expression microarray analysis has revealed that certain cancers have distinct patterns of gene expression and that these patterns correlate with factors such as the cancer's stage, clinical course, or response to treatment. For example, scientists examined gene expression in both normal white blood cells and in cells from a white blood cell cancer known as *diffuse large B-cell lymphoma (DLBCL)*. About 40 percent of patients with DLBCL respond well to chemotherapy and have long survival times. The other 60 percent respond poorly to therapy and have short survival.

The investigators assayed the expression profiles of 18,000 genes and discovered that there were two types of DLBCL, with almost inverse patterns of gene expression

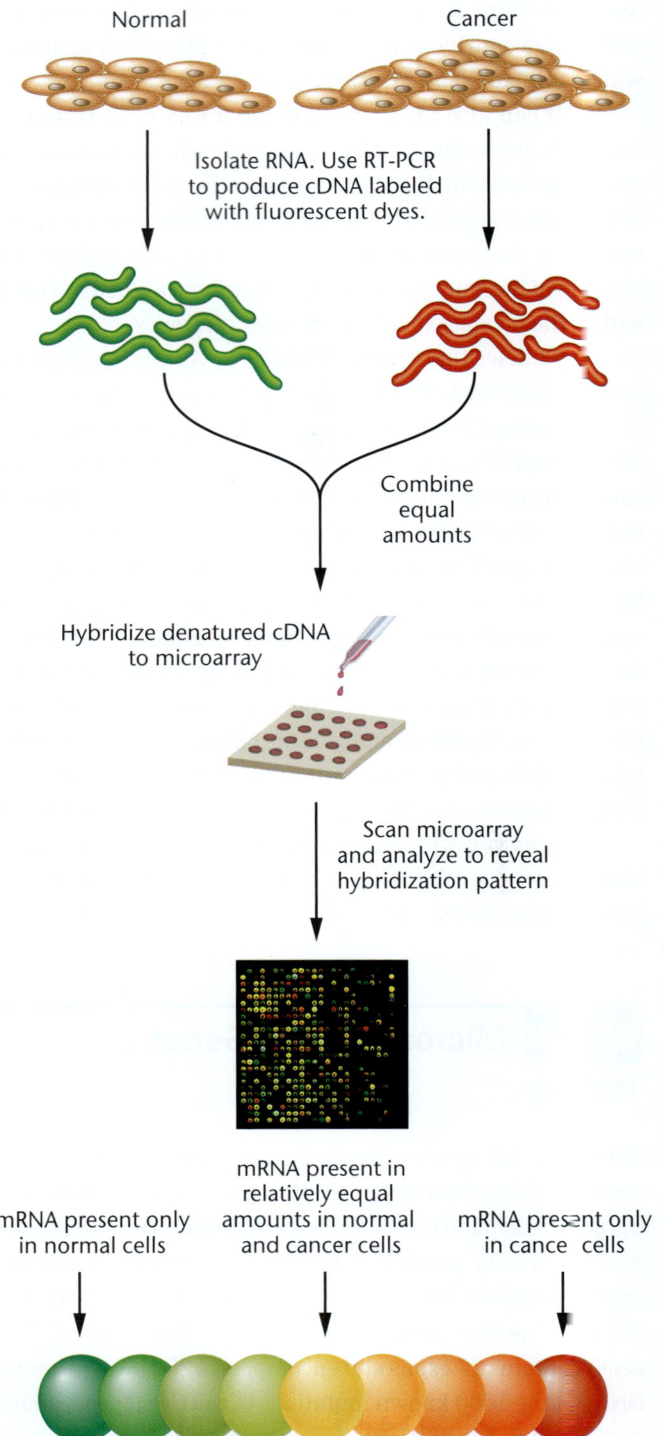

FIGURE ST 2.5 Microarray procedure for analyzing gene expression in normal and cancer cells. The method shown here is based on a two-channel microarray in which cDNA samples from the two different tissues are competing for binding to the same probe sets. The colors of the dots on an expression microarray represent levels of gene expression in a "heat map." In the heat map shown here, green dots represent genes expressed only in one cell type (i.e., the normal cells), and red dots represent genes expressed only in another cell type (i.e., the cancer cells). Intermediate colors represent different levels of expression of the same gene in the two cell types. (Only a small portion of the complete DNA microarray is shown.)

(**Figure ST 2.6**). One type of DLBCL, called GC B-like, had an expression pattern dramatically different from that of a second type, called activated B-like. Patients with the activated B-like pattern of gene expression had much lower survival rates than patients with the GC B-like pattern. The researchers concluded that DLBCL is actually two different diseases with different outcomes.

Once this type of profiling analysis is introduced into routine clinical use, it may be possible to adjust therapies for each group of cancer patients and to identify new specific treatments based on gene-expression profiles. Similar gene-expression profiles have been generated for many other cancers, including breast, prostate, ovarian, and colon cancer, providing tremendous insight into both substantial and subtle variations in genetic diseases.

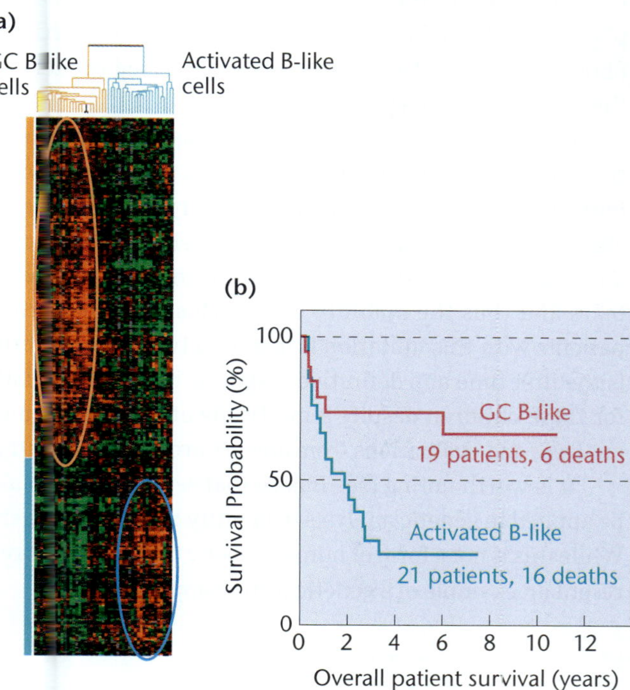

FIGURE ST 2.6 (a) Gene-expression microarray results analyzing 8,000 genes expressed in DLBCL lymphocytes. Each row represents a summary of the gene expression from one particular gene; each column represents data from one cancer patient's sample. In this heat map, the colors represent ratios of relative gene expression compared to normal control cells. Red represents expression greater than the mean level in controls, green represents expression lower than in the controls, and the intensity of the color represents the magnitude of difference from the mean. In this summary analysis, the cancer patients' samples are grouped by how closely their gene-expression profiles resemble each other. The cluster of cancer patients' samples marked with an orange oval at the top of the figure are GC B-like DLBCL cells. The blue oval cluster contains samples from cancer patients within the activated B-like DLBCL group. (b) Gene-expression profiling and survival probability. Patients with activated B-like profiles have a poorer chance of survival (16 in 21) than those with GC B-like profiles (6 in 19). Data such as these demonstrate the value of microarray analysis for diagnosing disease conditions.

Several companies are now promoting **nutrigenomics** services in which they claim to use gene-expression analysis to identify allele polymorphisms and gene-expression patterns for genes involved in nutrient metabolism. For example, polymorphisms in genes such as that for apolipoprotein A (*APOA1*), involved in lipid metabolism, and for methylenetetrahydrofolate reductase (*MTHFR*), involved in metabolism of folic acid, have been implicated in cardiovascular disease. Nutrigenomics companies claim that analysis of a patient's DNA for genes such as these enables them to judge whether allele variations or gene-expression profiles warrant dietary changes to potentially improve that person's health and reduce the risk of diet-related diseases.

ST 2.5 Genetic Analysis of Individual Genomes by DNA Sequencing

Because of the relatively low cost of quickly and accurately sequencing individual genomes—what we call **personal genomics**—the ways that scientists and physicians evaluate a person's genetic information is rapidly changing. WGS is being utilized in medical clinics at an accelerating rate. Many major hospitals around the world are setting up clinical sequencing facilities for use in screening for the causes of rare diseases (see Box 2).

Recently, WGS has provided new insights into the genetics of anorexia, Alzheimer disease, and autism, among other disorders. Already there have been some very exciting success stories whereby WGS of individual genomes has led to improved treatment of diseases in children and adults. For example, native Newfoundlanders have one of the highest incidences in the world of *arrhythmogenic right ventricular dysplasia/cardiomyopathy (ARVD/C)*, a rare condition in which affected individuals often have no symptoms but then die suddenly from irregular electrical impulses within the heart.

Through individual genome sequencing, a mutation in the *AVRD5* gene has been identified as the cause of such cases of premature death. Of those with this mutation, approximately 50 percent of males and 5 percent of females die by age 40, and 80 percent of males and 20 percent of females die by age 50. Individuals carrying this mutation are now being implanted with internal cardiac defibrillators that can restart their hearts if electrical impulses stop or become irregular.

Diseases that are caused by multiple genes are much harder to diagnose and treat based on sequencing data. For example, WGS of individuals affected by **autism spectrum disorder (ASD)** has revealed the involvement of more than 100 different genes. The genetics of ASD is particularly complex because of the broad range of phenotypes associated

BOX 2

Undiagnosed Diseases Network

The National Institutes of Health has an initiative called the *Undiagnosed Diseases Network* (https://www.genome.gov/27550959/undiagnosed-diseases-network-udn/). Its goal is to use whole exome and whole genome sequencing to help diagnose rare and mysterious disease conditions of an unknown genetic basis. In this program, exome sequences from an individual with a disorder can be compared to exome sequences from healthy family members and reference sequences to identify mutations that may be involved in the disease. To date, the program has diagnosed over 40 cases.

Unfortunately, even though a mutation responsible for a rare condition may have been identified, a cure or drug treatment has often not been developed for the disorder.

with this disorder. While WGS of individuals affected by ASD has revealed inherited mutations, it has also identified sporadic *de novo* mutations. In the future, sequence-based knowledge of these mutations may help physicians develop patient-specific treatment strategies.

Recall our introduction of the concept of **whole exome sequencing (WES)** (see Chapter 18). This alternative to WGS has also produced promising results in clinical settings. For example, from the time he was born, Nicholas Volmer had to live with unimaginable discomfort from an undiagnosed condition that was causing intestinal fistulas (holes from his gut to outside of his body) that were leaking body fluids and feces and requiring constant surgery. By 3 years of age, Nicholas had been to the operating room more than 100 times. A team at the Medical College of Wisconsin decided to have Nicholas's exome sequenced. Applying bioinformatics to compare his sequence to that of the general population, they identified a mutation in a gene on the X chromosome called *X-linked inhibitor of apoptosis (XIAP)*. *XIAP* is known to be linked to another condition that can often be corrected by a bone marrow transplant. In 2010 a bone marrow transplant saved Nicholas's life and largely restored his health. Shortly thereafter the popular press described Nicholas as the first child saved by DNA sequencing.

Recently WGS and WES have been used to identify mutations of the *NGLY1* gene (which encodes a protein processing enzyme) that are associated with a very rare condition in children sharing certain development delays, liver disease, and a phenotype notable because of the inability to produce tears and thus the inability to cry. This list of confirmed patients with this mutation is less than two dozen, but this is the first time any definitive diagnosis has been available for these children despite a multitude of different clinical analyses and evaluations from doctors around the country.

It is worth noting (see Box 3) that sequencing can also be applied in genetic analysis to identify human pathogens. While this is not a form of human genetic testing, it does represent an example of a genetic testing application.

BOX 3

Genetic Analysis for Pathogen Identification During Infectious Disease Outbreaks

Increasingly, next-generation sequencing (NGS) and third-generation sequencing (TGS) methods are being used for pathogen identification. Whole genome sequencing (WGS) is a quick way to identify pathogenic bacteria, viruses, or other microorganisms, and it is also very valuable for tracking genetic variations in microorganisms. This allows public health officials to track the evolution of changes in microbes to help develop proactive approaches to combat outbreaks of new, deadly strains. Routinely, seasonal strains of influenza are now analyzed by WGS, and vaccines are developed based on a genetic analysis of prevailing strains. NGS provided a large body of data on evolving strains of Ebola virus during the 2014–15 outbreak in West Africa, information that was essential for advancing vaccine development.

From the first confirmed hospitalized case of a woman with Ebola in Sierra Leone in late May 2014, blood samples from patients testing positive for Ebola were shipped to a lab in Cambridge, Massachusetts, for sequencing. By mid-June 149 blood samples from 78 patients had been processed for deep sequencing. Alignment of these sequences yielded 78 confirmed cases of Ebola and provided 99 full-length genomes. New mutations were mapped to a reference genome. From genomic analysis, it was estimated that strains from the current outbreak diverged from a Central African version of the Ebola virus less than 10 years ago.

Avian influenza, Middle East respiratory syndrome (MERS), and Zika virus are among other pathogens that have been identified and characterized by WGS.

Genetic Analysis of Single Cells by DNA and RNA Sequencing

We now have the ability to sequence the genome from a single cell! **Single-cell sequencing (SCS)** typically involves isolating genomic DNA from a single cell that is then subjected to *whole genome amplification (WGA)* by PCR to produce sufficient DNA to be sequenced. Amplification of the genome to produce enough DNA for sequencing without introducing errors remains a major challenge that researchers are working on so that SCS can become a more reliable and accurate technique for genetic testing.

Genomic sequencing from single cells is valuable for analyzing both *somatic cell mutations* (for example, mutations that arise in somatic cells such as in a skin cancer, which are not heritable) as well as *germ-line mutations* (heritable mutations that are transmitted to offspring via gametes). Sequencing genomes from individual egg or sperm cells, especially for couples undergoing *in vitro* fertilization, can identify carrier conditions or specific germ-line mutations that could result in a genetic disorder in the offspring.

SCS allows scientists to explore genetic variations from cell to cell. These studies are revealing that different mutant genes can vary greatly between individual cells. In particular, cancer cells from a tumor often show genetic diversity, a fact that is increasingly being appreciated by researchers and clinicians. Understanding variations in genetic diversity and gene expression by individual cells within a tumor could lead to better and more specific treatment options.

The contribution of individual cells to the phenotype of a tissue or organ affected by a genetic disorder is increasingly of interest. **RNA sequencing (RNA-seq)** is becoming a powerful tool for transcriptome-wide analysis of genes expressed by cells within a population, thus allowing researchers to differentiate genetic variations between cells.

Until recently, researchers or clinicians had to analyze the genome and transcriptome from cells independently. But now, it is possible to isolate DNA and RNA from the same cells, sequence the DNA, and, thanks to **single-cell RNA sequencing (scRNA-seq),** sequence the RNA. This enables a comparison of the genes present in a cell and the relative levels of expression for each transcript encoded by the genome.

Many disease treatments are designed to target cells, such as those in a tumor, as if all cells are homogeneous in genotype and phenotype. In fact, often such cells are quite heterogeneous genetically. scRNA-seq is now being applied to reveal the heterogeneity of cell types in tumors and other conditions, to then help plan better treatment approaches based on genetics of the cell types and their relative abundance.

Sequencing DNA and RNA from the same cell type typically requires the use of PCR to amplify genomic DNA (to sequence DNA) or mRNA, which is reverse transcribed into cDNA and then subsequently incorporated into a library and sequenced. scRNA-seq also then provides a quantitative transcriptome analysis in which the relative levels of RNA expressed in a cell can be determined. scRNA-seq provides quantitative data about RNA expression, similar to gene-expression microarray analysis. But scRNA-seq reveals all transcripts expressed in a cell, whereas the transcripts analyzed by microarrays are limited by the probes present on the array. These are among the reasons why scRNA-seq is likely to eventually replace microarrays for transcriptome analysis.

As an example of scRNA-seq, **Figure ST 2.7** presents an analysis of innate lymphoid cells (ILCs). ILCs are a relatively recently identified group of immune cells that reside in bone marrow and other tissues of the body. ILCs can differentiate into a variety of different immune cell types. They resemble T lymphocytes (T cells), although they lack antigen-specific immune response capability, and play important roles in immunity and the regulation of inflammation. Abnormalities in ILCs are involved in conditions such as autoimmune diseases, allergic responses, and asthma. As a result, ILCs have emerged as important cellular targets for medical interventions designed to manipulate the immune system—approaches often called *immunotherapy*.

In the example shown in Figure ST 2.7, scRNA-seq of mouse bone marrow progenitor cells enabled identification of different subsets of ILCs by their RNA-expression patterns. In this experiment, levels of RNA expression are color-coded in "heat map" fashion similar to the way gene-expression results are displayed for microarray analysis.

Computational analysis applies algorithms that result in *clustering*—the grouping of cells based on similar patterns of gene-expression data. Such analysis reveals similarities in gene expression but significant differences in the transcriptomes of ILCs that, by phenotype, might appear to be the same. For example, in Figure ST 2.7(b) notice how the RNA-seq profiles of RNAs expressed by cells in cluster C10 are very different than the profiles of RNAs expressed by cells in cluster C3. This delineation of ILCs based on transcriptome analysis reveals important genetic differences and offers the potential to develop new therapeutic approaches to manipulate these cells and maximize their immune responses.

Screening the Genome for Genes or Mutations You Want

While we have thus far focused on genetic testing and identifying genes involved in disease, genomic analysis is also revealing genome diversity that confers beneficial attributes or phenotypes to humans. This can have a role

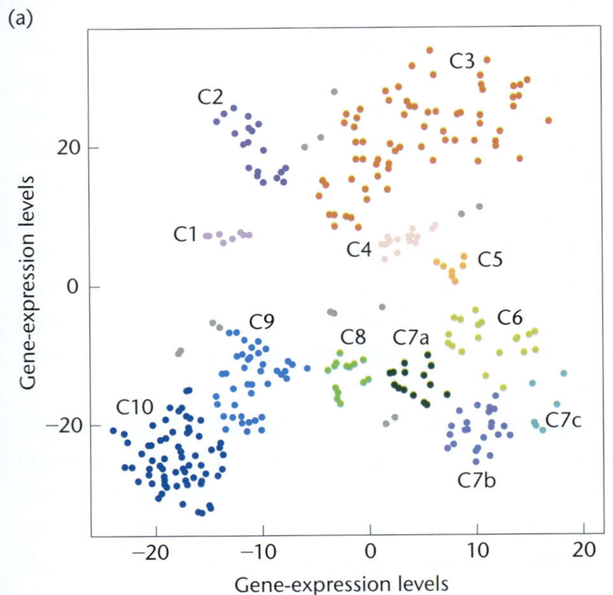

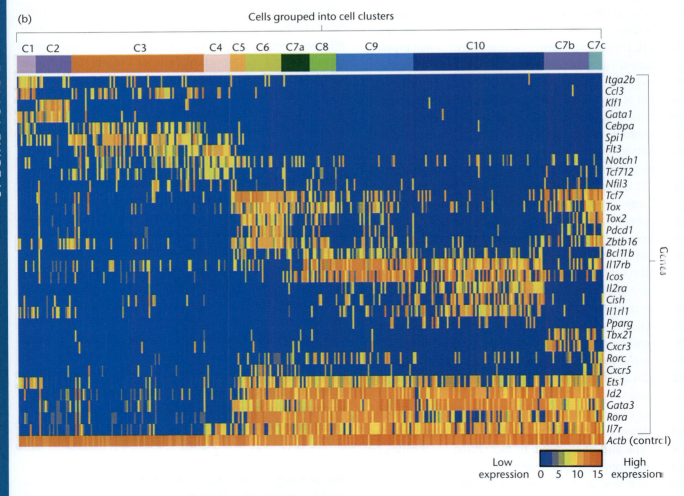

FIGURE ST 2.7 scRNA-seq analysis of innate lymphoid progenitor cells (ILCs). (a) A data plot of 325 ILCs from mouse bone marrow that were subjected to scRNA-seq. This data plot groups the ILCs into similar clusters based on shared gene-expression patterns. Each dot represents an individual cell, plotted against its expression levels for all genes analyzed by RNA-seq, and is color-coded based on 10 genetically distinct clustering assignments (C1 to C10). (b) A heat map displays RNA expression levels for selected genes in different cell clusters. Columns represent scRNA-seq data for each of the 325 different individual cells seen in part (a), grouped into the same 10 color-coded clusters. Each row displays a heat map for the expression levels of a specific gene (gene names shown to the far right) in all 325 cells. As a control, the bottom row of the heat map displays data for expression of β-actin mRNA; notice its relatively equal and high expression in all ILCs.

in medical diagnosis because it allows scientists and physicians to understand why genetic differences account for resistance to certain diseases in some individuals compared to others.

For instance, researchers at the Broad Institute in Boston were studying elderly, overweight individuals who by all conventional medical diagnostic approaches should have shown symptoms of diabetes, yet they were not diabetic. Instead of seeking diabetes-causing mutations, the Broad group employed genomic analysis to search for mutations associated with protection from diabetes. Their efforts were rewarded when they determined that individuals with loss-of-function mutations in the *SLC30A8* gene (*solute carrier family 30, member 8* gene, which encodes a zinc transport protein involved in insulin secretion) are 65 percent less likely to develop diabetes even when they have highly associated risk factors such as obesity.

In this spirit, increasingly geneticists are analyzing "natural or healthy knockouts"—the fortunate few individuals who may lack a specific gene or have a mutation in a disease-causing gene that provides a health benefit, such as protecting against development of a particular disease. Identifying such genetic variations may make it easier to help combat infections and disease.

For example, through mutation many viruses that infect humans can evade drugs used to combat them. But these same viruses are defenseless against a rare mutation in the human gene *ISG15*. Individuals with mutations in *ISG15* fight off many if not most viruses. (Estimates suggest that less than 1 person in 10 million has this mutation.) *ISG15* mutations knock out a function that helps to dampen inflammation, so individuals with this mutation have a heighted inflammatory system, which helps fight off viruses. It is thought that this elevated response prevents viruses from replicating to levels that typically cause illness.

Based on this knowledge, researchers are trying to develop drugs that might mimic the effects of the *ISG15* mutation as a future treatment strategy. As you will learn in Special Topics Chapter 3—Gene Therapy, this strategy has been used to mimic naturally occurring mutations in the *CCR5* receptor gene that provides a rare subset of individuals (a percentage of northern Europeans) with complete immunity to HIV infection.

A project called the Exome Aggregation Consortium (ExAC) is cataloging genetic variation from exome sequences of more than 60,000 individuals from diverse ancestries with the purpose of identifying naturally occurring gene knockouts. People with such knockouts or those who carry disease-causing genes but don't develop a particular illness are of significant interest to geneticists. Clearly this is an area of research that will continue to advance rapidly.

ST 2.6 Genome-Wide Association Studies Identify Genome Variations That Contribute to Disease

Many of the genetic testing approaches we have discussed so far have focused on analyzing genes in individuals or relatively small numbers of people. Microarray-based genomic analysis and WGS have led geneticists to employ a powerful strategy called **genome-wide association studies (GWAS)** in their quest to analyze populations of people for disease genes.

GWAS of relatively large populations of people for diagnostic or prognostic purposes often enables scientists to identify multiple genes that may influence disease risk. During the past decade there has been a dramatic expansion in the number of GWAS being reported. For example, GWAS for autism, obesity, diabetes, macular degeneration, myocardial infarction, arthritis, hypertension, several cancers, bipolar disease, autoimmune diseases, Crohn disease, schizophrenia (a recent publication noted more than 100 genetic loci contributing to disease risk), amyotrophic lateral sclerosis, and multiple sclerosis are among the many GWAS that have been widely publicized in the scientific literature and popular press. Behavioral traits such as intelligence have also been analyzed by GWAS. For example, recently, a high-profile, controversial genome-wide association study reported genetic markers influencing cognitive ability and attempted to relate these markers to differences in educational attainment between people. Other studies have identified more than 50 genes that may influence intelligence.

In a genome-wide association study, the genomes from several hundred, or several thousand (if available), unrelated individuals with a particular disease are analyzed, and the results are compared with genomes of individuals without the disease. The goal is to identify genetic variations that may confer risk of developing the disease. Many GWAS involve large-scale use of SNP microarrays that can probe on the order of 500,000 SNPs to evaluate results from different individuals. Other approaches of GWAS use WGS to look for specific gene differences, evaluate CNVs, or search for changes in the epigenome, such as methylation patterns. By determining which SNPs, CNVs, or epigenome changes are present in individuals with the disease, scientists can calculate the disease risk associated with each variation. Analysis of GWAS results requires statistical analysis to predict the relative potential impact (association or risk) of a particular genetic variation on the development of a disease phenotype.

Figure ST 2.8 shows a typical representation of one way that results from GWAS are commonly reported. The scatterplot representation, called a Manhattan plot, is used

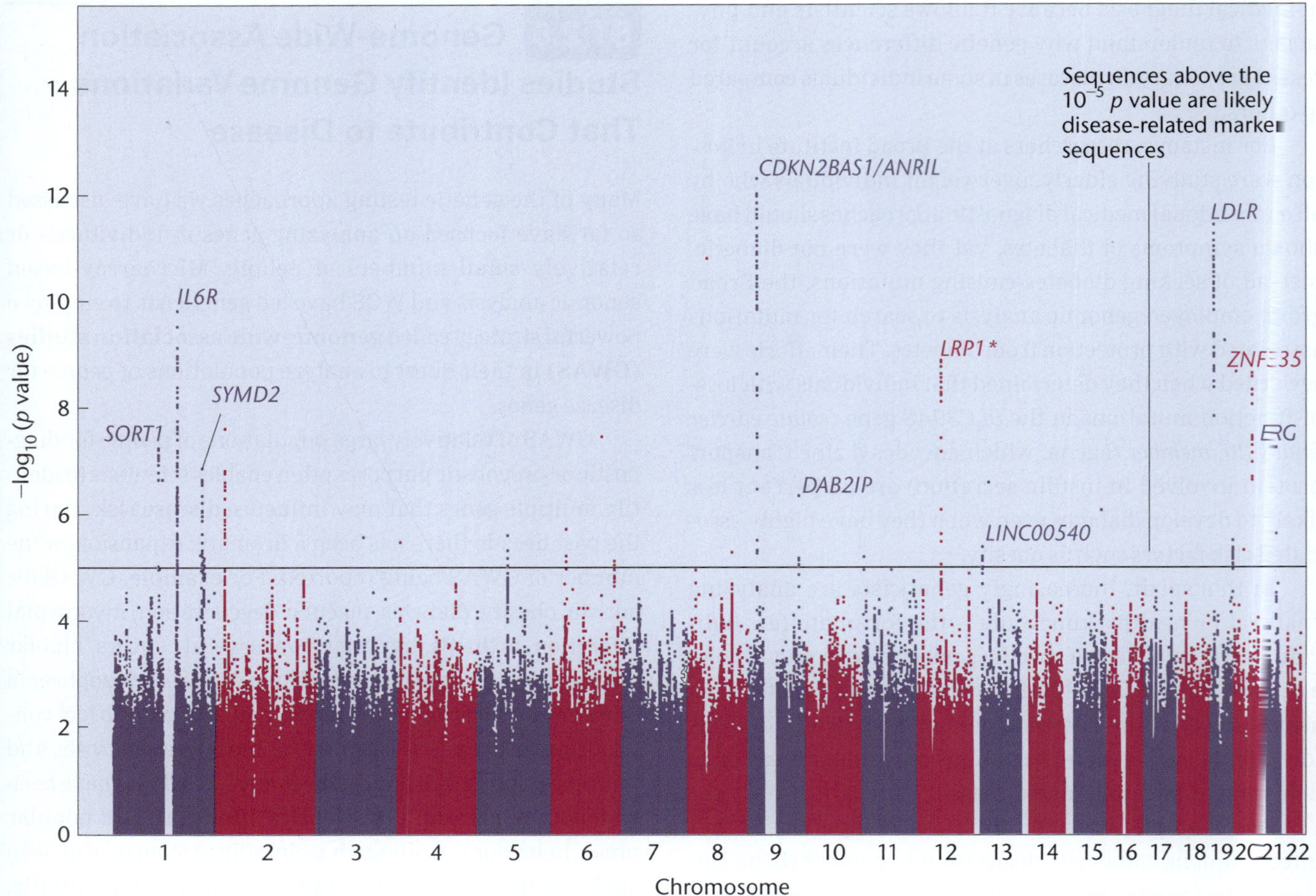

FIGURE ST 2.8 A genome-wide association study for Type 2 diabetes revealed 386,371 genetic markers, clustered here by chromosome number. Markers above the black line appeared to be significantly associated with the disease.

to display data with a large number of data points. Particular positions in the genome are plotted on the *x*-axis; in this case loci on each chromosome are plotted in a different color. The results of a genotypic association test are plotted on the *y*-axis. There are several ways that associations can be calculated. Shown here is a negative log of *p* values for loci determined to be significantly associated with a particular condition. The top line of this plot establishes a threshold value for significance. Marker sequences with significance levels exceeding the threshold *p* value of 10^{-5}, corresponding to 5.0 on the *y*-axis, are likely disease-related sequences.

One prominent study that brought the potential of GWAS to light involved research on 4587 patients in Iceland and the United States with a history of myocardial infarction (MI), and 12,767 control patients. This work was done with microarrays containing 305,953 SNPs. Among the most notable results, the study revealed variations in two tumor-suppressor genes (*CDKN2A* and *CDKN2B*) on chromosome 9. Twenty-one percent of individuals with a history of MI were homozygous for deleterious mutations in both genes, and these individuals showed a 1.64 odds ratio of MI compared with noncarriers, including individuals homozygous for wild-type alleles. These variations were correlated with those in people of European descent, but interestingly, these same mutant alleles are not prominent in African-Americans. Does this mean that these genetic markers are not MI risk factors among the latter ethnicity?

Examples such as this raise questions and ethical concerns about patients' emotional responses to knowing about genetic risk data. For example, GWAS often reveal dozens of DNA variations, but many variations have only a modest effect on risk. How does one explain to a person that he or she has a gene variation that changes a risk difference for a particular disease from 12 to 16 percent over an individual's lifetime? What does this information mean? Similarly, if the sum total of GWAS for a particular condition reveals about 50 percent of the risk alleles, what are the other missing elements of heritability that may contribute to developing a complex disease? In some cases, risk data revealed by GWAS may help patients and physicians develop diet and exercise plans designed to minimize the potential for developing a particular disease.

GWAS are showing us that, unlike single-gene disorders, complex genetic disease conditions involve a multitude of genetic factors contributing to the total risk for developing a condition. We need such information to make meaningful progress in disease diagnosis and treatment, which is ultimately a major purpose of what GWAS are all about.

ST 2.7 Genetic Testing and Ethical, Social, and Legal Questions

Applications of genetic testing raise important ethical, social, and legal issues that must be identified, debated, and resolved. Here we present a brief overview of some current ethical debates concerning genetic testing.

Genetic Testing and Ethical Dilemmas

When the Human Genome Project was first discussed, scientists and the general public voiced concerns about how genome information would be used and how the interests of both individuals and society can be protected. To address these concerns, the **Ethical, Legal, and Social Implications (ELSI) Research Program** was established by the National Human Genome Research Institute [a division of the National Institutes of Health (NIH)]. The ELSI Program focuses on four areas: (1) privacy and fairness in the use and interpretation of genetic information, (2) the transfer of genetic knowledge from the research laboratory to clinical practice, (3) ways to ensure that participants in genetic research know and understand the potential risks and benefits of their participation and give informed consent, and (4) enhancement of public and professional education.

The majority of the most widely applied genetic tests that have been used to date have provided patients and physicians with information that improves quality of life. One example involves prenatal testing for phenylketonuria (PKU) and implementing dietary restrictions to diminish the effects of the disease. But many of the potential benefits and consequences of genetic testing are not always clear. For example,

- We have the technologies to test for genetic diseases for which there are no effective treatments. *Should we test people for these disorders?*
- With current genetic tests, a negative result does not necessarily rule out future development of a disease, nor does a positive result always mean that an individual will get the disease. *How can we effectively communicate the results of testing and the actual risks to those being tested?*
- *What information should people have before deciding to have a genome scan or a genetic test for a single disorder or to have their whole genome sequenced?*
- Sequencing fetal genomes from the maternal bloodstream has revealed examples of mutations in the fetal genome (for example, a gene involved in Parkinson disease). *How might parents and physicians use this information?*
- Because sharing patient data through electronic medical records is a significant concern, *what issues of consent need to be considered?*
- *How can we protect the information revealed by genetic tests?*
- *How can we define and prevent genetic discrimination?*

Let's consider a specific example. In 2011, a case in Boston revealed the dangers of misleading results based on genetic testing. A prenatal ultrasound of a pregnant woman revealed a potentially debilitating problem (Noonan syndrome) involving the spinal cord of the woman's developing fetus. Physicians ordered a DNA test, which came back positive for a gene variant in a database that listed the gene as implicated in Noonan syndrome. The parents chose to terminate the pregnancy. Months later it was learned that the locus linked to Noonan was not involved in the disease, yet there was no effective way to inform the research and commercial genetic testing community.

To minimize these kinds of problems in the future, the NIH National Center for Biotechnology Information (NCBI) has developed a database called ClinVar (see www.ncbi.nlm.nih.gov/clinvar/), which integrates data from clinical genetic testing labs and research literature to provide an updated resource for researchers and physicians.

Disclosure of *incidental results* is another ethically challenging issue. When someone has his or her genome sequenced or has a test done involving a particular locus thought to be involved in a disease condition, the analysis sometimes reveals other mutations that could be of significance to the patient. Researchers and clinicians are divided on whether such information should be disclosed to the patient or whether patients should be asked for consent to receive all results from such tests. For example, a recent study considered 26 pregnant women who underwent prenatal genetic testing and learned they had genes associated with certain cancers and cognitive disorders as well as sex-chromosome abnormalities associated with reduced fertility. Again, this raises ethical issues about what type of consent women should consider when having these tests. Should these results be disclosed to these women? What do you think?

Earlier in this chapter we discussed preimplantation genetic diagnosis (PGD), which provides couples with the ability to screen embryos created by *in vitro* fertilization for genetic diseases. As we learn more about genes involved in human traits, will other, non-disease-related genes be screened for by PGD? Will couples be able to select embryos with certain genes encoding desirable traits for height, weight, intellect, or other physical or mental characteristics? What do you think of using genetic testing to purposely select for an embryo with a genetic disorder? There have been several well-publicized cases of couples seeking to use prenatal diagnosis or PGD to select for embryos with dwarfism and deafness.

As identification of genetic traits becomes more routine in clinical settings, physicians will need to ensure genetic privacy for their patients. There are significant concerns about how genetic information could be used in negative ways by employers, insurance companies, governmental agencies, or the general public. Genetic privacy and prevention of genetic discrimination will be increasingly important in the coming years. In 2008, the **Genetic Information Nondiscrimination Act (GINA)** was signed into law in the United States. This legislation is designed to prohibit the improper use of genetic information in health insurance and employment, but not life insurance.

Direct-to-Consumer Genetic Testing and Regulating the Genetic Test Providers

The past decade has seen dramatic developments in **direct-to-consumer (DTC) genetic tests.** A simple Web search will reveal many companies offering DTC genetic tests. As of 2018, there were over 2000 diseases for which such tests are now available (in 1993 there were about 100 such tests). Most DTC tests require that a person mail a saliva sample, hair sample, or cheek cell swab to the company. For a range of pricing options, DTC testing companies largely use SNP-based tests such as ASO tests to screen for different mutations. For example, in 2007, Myriad Genetics, Inc., began a major DTC marketing campaign of its tests for *BRCA1* and *BRCA2*. Mutations in these genes increase the risk of developing breast and ovarian cancer. DTC testing companies report absolute risk, the probability that an individual will develop a disease, but how such risks are calculated is highly variable and subject to certain assumptions.

Such tests are controversial for many reasons. For example, the test is purchased online by individual consumers and requires no involvement of a physician or other health-care professionals such as a nurse to administer the test or a genetic counselor to interpret the results. There are significant questions about the quality, effectiveness, and accuracy of such products because currently the DTC testing industry is largely self-regulated. The FDA does not regulate DTC genetic tests. There is at present no comprehensive way for patients to make comparisons and evaluations about the range of tests available and their relative quality.

Most companies make it clear that they are not trying to diagnose or prevent disease, nor are they offering health advice, so what is the purpose of the information that test results provide to the consumer? Web sites and online programs from DTC testing companies provide information on what advice a person should pursue if positive results are obtained. But is this enough? If results are not understood, might negative tests not provide a false sense of security? Just because a woman is negative for *BRCA1* and *BRCA2* mutations *does not* mean that one cannot develop breast or ovarian cancer. Refer to Discussion Question 6 for an example of a personal decision that actress Angelina Jolie made based on the results of a genetic test.

Whether the FDA will oversee DTC genetic tests in the future is unclear. However, at the time of publication of this edition, the FDA has not revealed any definitive plans to regulate or oversee DTC genetic tests. But because some DTC genetic testing companies, such as 23andMe, offer health-related analyses or health reports, they do fall under FDA regulations. The FDA continues to issue warnings to DTC testing companies to provide what the FDA considers to be appropriate health-related interpretations of genetic tests. For example, in 2017 for the first time the FDA approved a DTC saliva test from 23andMe that can test for genetic mutations associated with ten conditions including Parkinson disease and Alzheimer disease. There are varying opinions on the regulatory issue. Some believe that the FDA has no business regulating DTC tests and that consumers should be free to purchase products based on their personal needs or interests. Others insist that the FDA must regulate DTCs in the interest of protecting consumers.

Genetic Testing and Patents

Intellectual property (IP) rights are being debated as an aspect of the ethical implications of genetic engineering, genomics, and biotechnology in many ways. For example, patents on IP for isolated genes, recombinant cell types, and GMOs can be potentially lucrative for the patent-holders but may also pose ethical and scientific problems. Why is protecting IP important for companies? Consider this issue. If a company is willing to spend millions or billions of dollars and several years doing research and development (R&D) to produce a valuable product, then shouldn't it be afforded a period of time to protect its discovery so that it can recover R&D costs and made a profit on its product?

Genes in their natural state as products of nature cannot be patented. Consider the possibilities for a human gene that has been cloned and then patented by the scientists who did the cloning. The person or company holding the patent could require that anyone attempting to do research with the patented gene pay a licensing fee for its use. Should a diagnostic test or therapy result from the research, more fees and royalties may be demanded, and as a result the costs of a genetic test may be too high for many patients to afford. But limiting or preventing the holding of patents for genes or genetic tools could reduce the incentive for pursuing the research that produces such genes and tools, especially for companies that need to profit from their research.

Patenting of genetic tests is under increased scrutiny in part because of concerns that a patented test can create monopolies in which patients cannot get a second opinion if only one company holds the rights to conduct a particular genetic test. Recent analysis has estimated that as many as 64 percent of patented tests for disease genes make it very difficult or impossible for other groups to propose a different way to test for the same disease.

In 2010 a landmark case brought by the American Civil Liberties Union against Myriad Genetics contended that Myriad could not patent the *BRCA1* and *BRCA2* gene sequences used to diagnose breast cancer. Myriad's BRACAnalysis® product has been used to screen over a million women for *BRCA1* and *BRCA2* during its period of patent exclusivity. A U.S. District Court judge ruled Myriad's patents invalid on the basis that DNA in an isolated form is not fundamentally different from how it exists in the body. Myriad was essentially accused of having a monopoly on its tests, which have existed for a little over a decade based on its exclusive licenses in the United States.

This case went to the Supreme Court in 2013, which rendered a 9–0 ruling against Myriad, stripping it of five of its patent claims for the *BRCA1* and *BRCA2* genes, largely based on the view that natural genes are a product of nature and just because they are isolated does not mean they can be patented. The Court ruled that cDNA sequences produced in a lab can continue to be patentable. Myriad still holds about 500 valid claims related to *BRCA* gene testing.

Whole Genome Sequence Analysis Presents Many Ethical Questions

In the next decade and beyond, it is expected that WGS analysis of adults and babies will become increasingly common in clinical settings. The Newborn Sequencing in Genomic Medicine and Public Health (NSIGHT) research program initiated by the NIH is under way to sequence the genomes of more than 1500 babies. Both infants with illnesses and babies who are healthy will be part of this study. This initiative will allow scientists to carry out comparative genomic analyses of specific sequences to help identify genes involved in disease conditions.

Screening of newborns is important to help prevent or minimize the impacts of certain disorders. Each year routine blood tests from a heel prick of newborn babies reveal rare genetic conditions in several thousand infants in the United States alone. A small number of states allow parents to opt out of newborn testing. In the future, should DNA sequencing at the time of birth be universally required? Do we really know enough about which human genes are involved in disease to help prevent disease in children? Estimates suggest that sequencing can identify approximately 15 to 50 percent of children with diseases that currently cannot be diagnosed by other methods. What is the value of having sequencing data for healthy children?

Personal genomics adds another layer of complexity to this discussion. When people donate their DNA for WGS projects, should they have access to the raw data from their sequence analysis? Currently, such volunteers are refused access to their genetic data.

As exciting as this period of human genetics and medicine is becoming, most of the WGS studies of individuals are happening in a largely unregulated environment. This raises significant ethical concerns especially with respect to DNA collection, the variability and quality control of DNA handling protocols, sequence analysis, storage, and confidentiality of genetic information.

Preconception Testing, Destiny Predictions, and Baby-Predicting Patents

Companies are now promoting the ability to do *preconception* testing and thus make "destiny predictions" about the potential phenotypes of hypothetical offspring based on computational methods for analyzing sequence data of parental DNA samples. The company 23andMe has been awarded a U.S. patent for a computational method called the *Family Traits Inheritance Calculator* to use parental DNA samples to predict a baby's traits, including eye color and the risk of certain diseases. This patent includes applications of technologies to screen sperm and ova for IVF.

More than 5 million babies have been born by IVF. Woman are having children at a later age, couples involved in IVF may be same sex or transgender, and the use of sperm and egg donors, as well as surrogates, is also increasing. Currently, gender selection of embryos generated by IVF is very common. But could preconception testing lead to the selection of "designer babies" as people look to customize their offspring? Fear of *eugenics* surrounds these conversations, particularly as genetic analysis starts moving away from disease conditions to nonmedical traits such as hair color, eye color, other physical traits, and potentially behavioral traits. The patent has been awarded for a process that will compare the genotypic data of an egg provider and a sperm provider to suggest gamete donors that might result in a baby or hypothetical offspring with particular phenotypes of interest to a prospective parent. What do you think about this?

A company called GenePeeks claims to have a patent-pending technology for reducing the risk of inherited disorders by "digitally weaving" together the DNA of prospective parents. GenePeeks plans to sequence the DNA of sperm donors and women who want to get pregnant to inform women about donors who are most genetically compatible for the traits they seek in offspring. Their proprietary computing technology is intended to use sequence data to examine "virtual" eggs and sperm from donor–client pairings to estimate the likelihood of about 10,000 specific diseases in hypothetical offspring from prospective parents. Will technologies such as this become widespread and attract consumer demand in the future? What do you think? Would you want this analysis done before deciding whether to have a child with a particular person?

Review Questions

1. What are some of the key differences and applications of genetic testing for prognostic versus diagnostic purposes?
2. Which of the examples of genetic testing below are prognostic tests? Which are diagnostic?
   (a) Individual sequencing (personal genomics) identifies a mutation associated with Alzheimer disease.
   (b) ASO testing determines that an individual is a carrier for the mutant allele (β^S) found in sickle-cell anemia.
   (c) DNA sequencing of a breast tumor reveals mutations in the *BRCA1* gene.
   (d) Genetic testing in a healthy teenager identifies an SNP correlated with autism.
   (e) An adult diagnosed with Asperger syndrome (AS) has a genetic test that reveals a SNP in the *GABRB3* gene that is significantly more common in people with AS than the general population.
3. In the United States, what are the federal requirements for genetic screening of newborns?
4. How does a positive ASO test for sickle-cell anemia determine that an individual is homozygous recessive for the mutation that causes sickle-cell anemia?
5. Does genetic analysis by ASO testing allow for detection of epigenetic changes that may contribute to a genetic disorder? Explain your answer.
6. What is the primary purpose of preimplantation genetic diagnosis (PGD)?
7. How can we correlate the genome with RNA expression data in a tissue or a single cell, and why is this important for the purposes of disease diagnosis and treatment?
8. From microarray analysis, how do we know what genes are being expressed in a specific tissue?
9. Sequencing the human genome and personal genomics is improving our understanding of normal and abnormal cell behavior. How are these approaches dramatically changing our understanding and treatment of complex diseases such as cancer?
10. What is the main purpose of genome-wide association studies (GWAS)? How can information from GWAS be used to inform scientists and physicians about genetic diseases?
11. What are incidental results, and how does this relate to genetic testing?

SPECIAL TOPIC 2

Discussion Questions

1. Maternal blood tests for three pregnant women revealed they would be having boys, yet subsequent ultrasound images showed all three were pregnant with girls. In each case Y chromosome sequences in each mother's blood originated from transplanted organs they had received from men! This demonstrates one dramatic example of a limitation of genetic analysis of maternal blood samples. What kind of information could have been collected from each mother in advance of these tests to better inform physicians prior to performing each test?
2. A couple with European ancestry seeks genetic counseling before having children because of a history of cystic fibrosis (CF) in the husband's family. ASO testing for CF reveals that the husband is heterozygous for the Δ508 mutation and that the wife is heterozygous for the *R117* mutation. You are the couple's genetic counselor. When consulting with you, they express their conviction that they are not at risk for having an affected child because they each carry different mutations and cannot have a child who is homozygous for either mutation. What would you say to them?
3. As genetic testing becomes widespread, medical records will contain the results of such testing. Who should have access to this information? Should employers, potential employers, or insurance companies be allowed to have this information? Would you favor or oppose having the government establish and maintain a central database containing the results of individuals' genome scans?
4. Might it make sense someday to sequence every newborn's genome at the time of birth? What are the potential advantages and concerns of this approach?
5. The family of a sixth-grade boy in Palo Alto, California, was informed by school administrators that he would have to transfer out of his middle school because they believed his mutation of the *CFTR* gene, which does not produce any symptoms associated with cystic fibrosis, posed a risk to other students at the school who have cystic fibrosis. After missing 11 days of school, a settlement was reached to have the boy return to school. What ethical problems might you associate with this example?
6. In 2013 the actress Angelina Jolie elected to have prophylactic double-mastectomy surgery to prevent breast cancer based on a positive test for mutation of the *BRCA1* gene. What are some potential positive and negative consequences of this high-profile example of acting on the results of a genetic test?

7. The National Institutes of Health created the *Genetic Testing Registry (GTR)* to increase transparency by publicly sharing information about the utility of their tests, research for the general public, patients, health-care workers, genetic counselors, insurance companies, and others. The Registry is intended to provide better information to patients, but companies involved in genetic testing are not required to participate. Should company participation be mandatory? Why or why not? Explain your answers.
8. Should the FDA regulate direct-to-consumer genetic tests, or should these tests be available as a "buyer beware" product?
9. Would you have your genome sequenced, if the price was affordable? Why or why not? If you answered yes, would you make your genome sequence publicly available? How might such information be misused?
10. Following the tragic shooting of 20 children at a school in Newtown, Connecticut, in 2012, Connecticut's state medical examiner requested a full genetic analysis of the killer's genome. What do you think investigators might be looking for? What might they expect to find? Might this analysis lead to oversimplified analysis of the cause of the tragedy?
11. Private companies are offering personal DNA sequencing along with interpretation. What services do they offer? Do you think that these services should be regulated, and if so, in what way? Investigate one such company, 23andMe, at http://www.23andMe.com, before answering these questions.
12. In 2010, a U.S. District Judge ruled to invalidate Myriad Genetics' patents on the *BRCA1* and *BRCA2* genes. Judge Sweet noted that since the genes are part of the natural world, they are not patentable. Myriad Genetics also holds patents on the development of a direct-to-consumer test for the *BRCA1* and *BRCA2* genes. Do you agree with the ruling to invalidate the patenting of the *BRCA1* and *BRCA2* genes? If you were asked to judge the patenting of the direct-to-consumer test for the *BRCA1* and *BRCA2* genes, how would you rule?

Gene Therapy

Although drug treatments can be effective in controlling symptoms of genetic disorders, the ideal outcome of medical treatment is to cure a disease. This is the goal of **gene therapy**—the delivery of therapeutic genes into a patient's cells to correct genetic disease conditions caused by a faulty gene or genes. The earliest attempts at gene therapy focused on the delivery of normal, *therapeutic copies* of a gene to be expressed in such a way as to override or negate the effects of the disease gene and thus minimize or eliminate symptoms of the genetic disease. But in recent years newer methods for removing, correcting, inhibiting, or silencing defective genes have shown promise. However, no approach to gene therapy has generated more excitement for its potential than genome-editing applications involving CRISPR-Cas (see Chapter 17). We will consider several recent examples of genome editing to target specific genes.

Gene therapy is one of the goals of **translational medicine**—taking a scientific discovery, such as the identification of a disease-causing gene, and translating the finding into an effective therapy, thus moving from the laboratory bench to a patient's bedside to treat a disease. In theory, the delivery of a therapeutic gene is rather simple, but in practice, gene therapy has been very difficult to execute. In spite of over 25 years of trials, this field has not lived up to its expectations. However, gene therapy is currently experiencing a fast-paced resurgence, with several high-profile new successes and potentially exciting new technologies sitting on the horizon. Gene therapy is an example of **precision medicine** (see Special Topic Chapter 7—Genomics and Precision Medicine). It is hoped that gene therapy will soon become part of mainstream medicine. The treatment of a human genetic disease by gene therapy is the ultimate application of genetic technology. In this Special Topics chapter we will explore how gene therapy is executed, and we will highlight selected examples of successes and failures as well as discuss new approaches to gene therapy. Finally, we will consider ethical issues regarding gene therapy.

"The treatment of a human genetic disease by gene therapy is the ultimate application of genetic technology."

ST 3.1 What Genetic Conditions Are Candidates for Treatment by Gene Therapy?

Two essential criteria for gene therapy are that the gene or genes involved in causing a particular disease have been identified and that the gene can be cloned or synthesized in a laboratory. As a result of the Human Genome Project, the identification of human disease genes and their specific DNA sequences has greatly increased the number of candidate genes for gene therapy trials. Almost all of the early gene therapy trials and most gene therapy approaches have focused on treating conditions caused by a single gene.

The cells affected by the genetic condition must be readily accessible for treatment by gene therapy. For example, blood disorders such as leukemia, hemophilia, and other conditions have been major targets of gene therapy because it is relatively routine to manipulate blood cells outside of the body and return them to the body in comparison to treating cells in the brain and spinal cord, skeletal or cardiac muscle, and organs with heterogeneous populations of cells such as the pancreas.

In the past decade, every major category of genetic disease has been targeted by gene therapy (**Figure ST 3.1**). A majority of recently approved clinical trials are for cancer treatment. Gene therapy approaches are also currently being investigated for the treatment of hereditary blindness; hearing loss; neurodegenerative diseases including Alzheimer disease, Parkinson disease, and amyotrophic lateral sclerosis (ALS); cardiovascular disease; muscular dystrophy; hemophilia; and infectious diseases, such as HIV; among many other conditions, including depression and drug and alcohol addiction. Worldwide, over 2300 approved gene therapy clinical trials have occurred or recently been initiated (see Box 1).

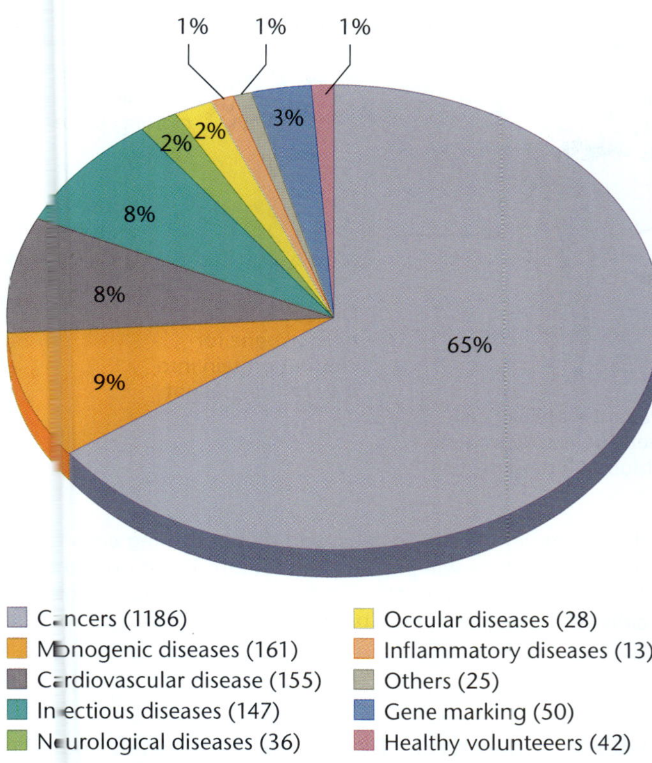

FIGURE ST 3.1 Graphic representation of different genetic conditions being treated by gene therapy clinical trials worldwide. Notice that cancers are the major target for treatment.

In the United States, proposed gene therapy clinical trials must first be approved by review boards at the institution where they will be carried out, and then the protocols must be approved by the Food and Drug Administration (FDA).

ST 3.2 How Are Therapeutic Genes Delivered?

In general, there are two broad approaches for delivering therapeutic genes to a patient being treated by gene therapy, *ex vivo gene therapy* and *in vivo gene therapy* (**Figure ST 3.2**). In *ex vivo* gene therapy, cells from a person with a particular genetic condition are removed, treated in a laboratory by adding either normal copies of a therapeutic gene or a DNA or RNA sequence that will inhibit expression of a defective gene, and then these cells are transplanted back into the person. Genetically altered cells treated in this manner can be transplanted back into the patient without fear of immune system rejection because these cells were derived from the patient initially.

In vivo gene therapy does not involve removal of a person's cells. Instead, therapeutic DNA is introduced directly into affected cells of the body. One of the major challenges of *in vivo* gene therapy is restricting the delivery of therapeutic genes to only the intended tissues and not to all tissues throughout the body.

Viral Vectors for Gene Therapy

For both *in vitro* and *ex vivo* approaches, the key to successful gene therapy is having a delivery system to transfer genes into a patient's cells. Because of the relatively large molecular size and electrically charged properties of DNA, most human cells do not take up DNA easily. Therefore, delivering therapeutic DNA molecules into human cells is challenging. Since the early days of gene therapy, genetically engineered viruses as vectors have been the main tools for delivering therapeutic genes into human cells. Viral vectors for gene therapy are engineered to carry therapeutic DNA as their payload so that the virus infects target cells and delivers the therapeutic DNA without causing damage to cells.

In a majority of gene therapy trials around the world, scientists have used genetically modified *retroviruses* as vectors. Recall from Chapter 9 that retroviruses such as HIV contain an RNA genome that scientists use as a template for the synthesis of a complementary DNA molecule. **Retroviral vectors** are created by removing replication and disease-causing genes from the virus and replacing them with a cloned human gene. After the altered RNA has been packaged into the virus, the recombinant viral vector

BOX 1
ClinicalTrials.gov

One of the best resources on the Web for learning about ongoing clinical trials, including current gene therapy trials, is ClinicalTrials.gov. The site can easily be searched to find a wealth of resources about ongoing gene therapy trials throughout the United States that are of interest to you. To find a gene therapy clinical trial, use the "Search for Studies" box and type in the name of a disease and "gene therapy." This search string will take you to a page listing active gene therapy clinical trials, with links to detailed information about the trial.

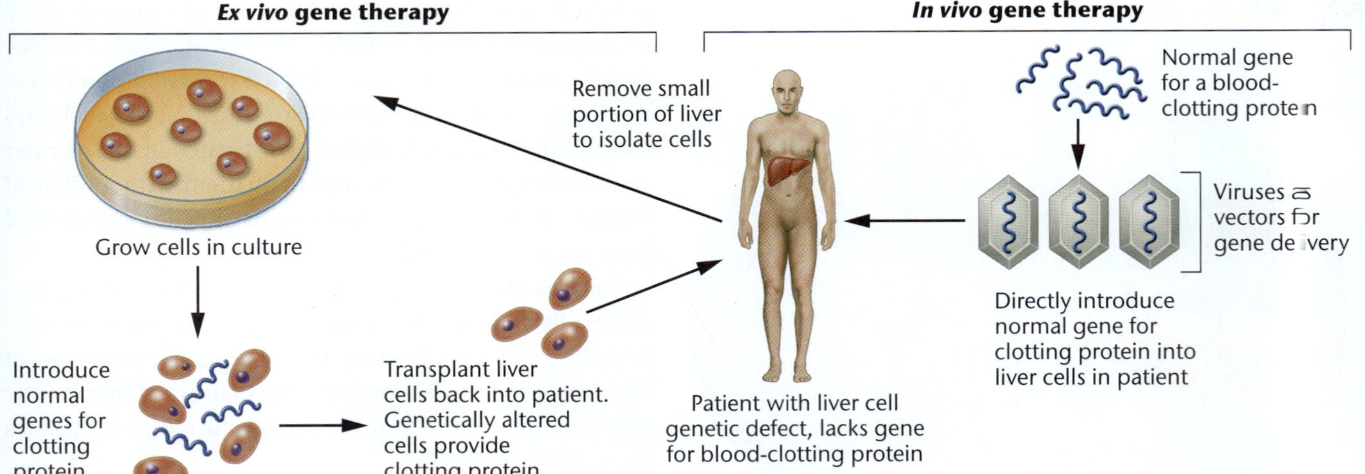

Ex vivo gene therapy
In vivo gene therapy
Remove small portion of liver to isolate cells
Grow cells in culture
Introduce normal genes for clotting protein
Transplant liver cells back into patient. Genetically altered cells provide clotting protein
Patient with liver cell genetic defect, lacks gene for blood-clotting protein
Normal gene for a blood-clotting prote n
Viruses as vectors for gene de very
Directly introduce normal gene for clotting protein into liver cells in patient

FIGURE ST 3.2 *Ex vivo* and *in vivo* gene therapy for a patient with a liver disorder. *Ex vivo* gene therapy involves isolating cells from the patient, introducing normal copies of a therapeutic gene (encoding a blood-clotting protein in this example) into these cells, and then returning cells to the body where they will produce the required clotting protein. *In vivo* approaches involve introducing DNA directly into cells while they are in the body.

containing the therapeutic human gene is used to infect a patient's cells. Technically, virus particles are carrying RNA copies of the therapeutic gene. Once inside a cell, the virus cannot replicate itself, but the therapeutic RNA is reverse transcribed into DNA, which enters the nucleus of cells and *integrates* into the genome of the host cells' chromosome. If the inserted therapeutic gene is properly expressed, it produces a normal gene product that may be able to ameliorate the effects of the mutation carried by the affected individual.

One advantage of retroviral vectors is that they provide long-term expression of delivered genes because they integrate the therapeutic gene into the genome of the patient's cells. But a major problem with retroviral vectors is that they have produced random, unintended alterations in the genome, in some cases due to *insertional mutations*. Retroviral vectors generally integrate their genome into the host-cell genome at random sites. Thus, there is the potential for retroviral integration that randomly inactivates genes in the genome or gene-regulatory regions such as a promoter sequence.

In many early gene therapy trials, **adenovirus vectors** were the retrovirus vector of choice. An advantage of these vectors is that they are capable of carrying large therapeutic genes. But because many humans produce antibodies to adenovirus vectors, they can mount immune reactions that render the virus and its therapeutic gene ineffective or cause significant side effects to the patient. A related virus called **adeno-associated virus (AAV)** is now widely used as a gene therapy vector [**Figure ST 3.3(a)**]. In its native form, AAV infects about 80 to 90 percent of humans during childhood, causing symptoms associated with the common cold. Disabled forms of AAV are popular for gene therapy because the virus is nonpathogenic, so it usually does not elicit a major response from the immune system of treated patients. AAV also does not typically integrate into the host-cell genome, so there is little risk of the insertional mutations that have plagued retroviruses, although modified forms of AAV have been used to deliver genes to specific sites on individual chromosomes. Most forms of AAV deliver genes into the host-cell nucleus where it forms small hoops of DNA called *episomes* that are expressed under the control of promoter sequences contained within the viral genome. But because therapeutic DNA delivered by AAV does not usually become incorporated into the genome, it is not replicated when host cells divide. This is fine for certain cells in the brain or the retina that do not divide, but treating rapidly dividing cells typically requires repeated, ongoing applications to be successful [Figure ST 3.3(a)].

Work with **lentivirus vectors** is an active area of gene therapy research [**Figure ST 3.3(b)**]. Lentivirus is a retrovirus that can accept relatively large pieces of genetic material. Another positive feature of lentivirus is that it is capable of infecting nondividing cells, whereas other viral vectors often infect cells only when they are dividing. It is still not possible to control where lentivirus integration occurs in the host-cell genome, but the virus does not appear to gravitate toward gene-regulatory regions the way that other retroviruses do. Thus the likelihood of causing insertional mutations appears to be much lower than for other vectors.

The human immunodeficiency virus (HIV) responsible for acquired immunodeficiency syndrome (AIDS) is

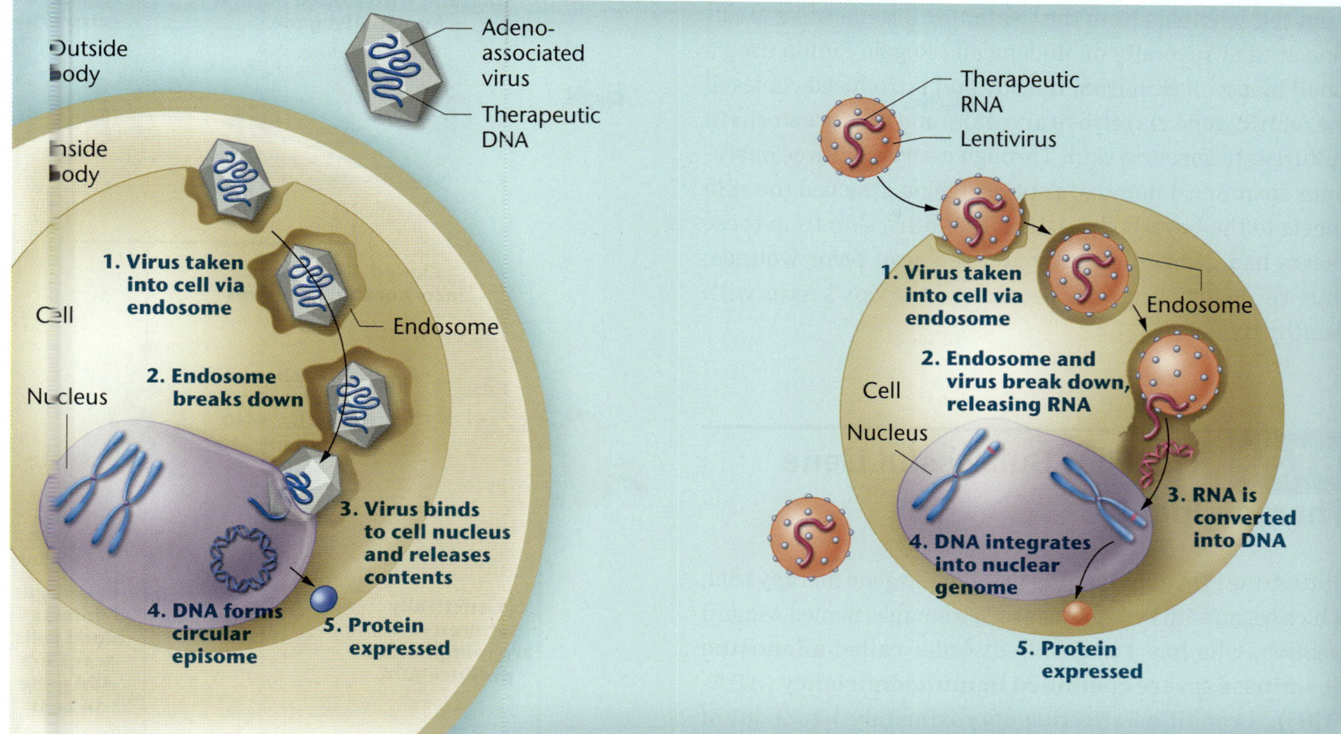

FIGURE ST 3.3 Delivering therapeutic genes.(a) Nonintegrating viruses such as modified adeno-associated virus (AAV) deliver therapeutic genes without integrating them into the genome of target cells. Delivered DNA resides as minichromosomes (episomes), but over time as cells divide, these nonintegrating hoops of DNA are gradually lost. (b) Integrating viruses include lentivirus, an RNA retrovirus that delivers therapeutic genes into the cytoplasm where reverse transcriptase converts RNA into DNA. DNA then integrates into the genome, ensuring that therapeutic DNA will be passed into daughter cells during cell division.

a type of lentivirus. It may surprise you that HIV could be used as a vector for gene therapy. For any viral vector, scientists must be sure that the vector has been genetically engineered to render it inactive. Modified forms of HIV, strains lacking the genes necessary for reconstitution of fully functional viral particles, are being used for gene therapy trials. HIV has evolved to infect certain types of T lymphocytes (T cells) and macrophages, making it a good vector for delivering therapeutic genes into the bloodstream.

Nonviral Delivery Methods

Scientists continue to experiment with various *in vivo* and *ex vivo* strategies for trying to deliver so-called naked DNA into cells without the use of viral vectors. Nonviral methods include chemically assisted transfer of genes across cell membranes, nanoparticle delivery of therapeutic genes, and fusion of cells with artificial lipid vesicles called *liposomes*. Short-term expression of genes through "gene pills" is being explored, whereby a pill delivers therapeutic DNA to the intestines where the DNA is absorbed by cells that express the therapeutic protein and secrete it into the bloodstream.

Stem Cells for Delivering Therapeutic Genes

Increasingly, viral and nonviral vectors are being used to deliver therapeutic genes into **stem cells,** usually *in vitro*, and then the stem cells are either introduced into the patient or differentiated *in vitro* into mature cell types before being transplanted into a patient. In particular, *hematopoietic stem cells (HSCs),* which are found in bone marrow and give rise to blood cells, are widely used for gene therapy in adults and children. There are many advantages to using HSCs: they are easily accessible and often taken from the marrow of the patient to be treated to avoid complications with tissue rejection by the immune system when they are reintroduced; they replicate quickly *in vitro*; they are fairly long lived; and they differentiate into both red blood cells and white blood cells (leukocytes). And as you will also learn, stem cells are being used in CRISPR-Cas and gene-silencing approaches, thus demonstrating their value for different gene therapy applications.

In late 2017 one of the most extraordinarily successful applications of stem cell therapy was reported. This approach combined gene therapy with stem cells to regenerate the epidermis of a 7-year-old boy with a genetic condition called junctional epidermolysis bullosa, who had lost nearly two-thirds of his skin due to bacterial infections. A team of stem cell scientists from the Center for Regenerative Medicine at the University of Modena and Reggio Emilia used a small biopsy of skin from the boy and introduced copies of the *LAMB3* gene via retroviral vectors and then grew cells in culture into sheets of skin. Through a series of three operations spanning 4 months, a surgical team attached the skin sheets to the boy's body. After 21 months, skin from these sheets had grown normally and replaced prior wounds, thus reconstructing ~80 percent of the boy's skin with healthy tissue.

ST 3.3 The First Successful Gene Therapy Trial

In 1990 the FDA approved the first human gene therapy trial, which began with the treatment of a young girl named Ashanti DeSilva, who has a heritable disorder called **adenosine deaminase severe combined immunodeficiency (ADA-SCID),** a condition affecting approximately 1 to 9 out of every 1 million live births. Individuals with SCID have no functional immune system and usually die from what would normally be minor infections. Ashanti has an autosomal form of SCID caused by a mutation in the gene encoding the enzyme *adenosine deaminase*. Her gene therapy began when clinicians isolated some of her white blood cells, called T cells [**Figure ST 3.4**]. These cells, which are key components of the immune system, were mixed with a retroviral vector carrying an inserted copy of the normal *ADA* gene. The virus infected many of the T cells, and a normal copy of the *ADA* gene was inserted into the genome of some T cells.

After being mixed with the vector, the T cells were grown in the laboratory and analyzed to make sure that the transferred *ADA* gene was expressed (**Figure ST 3.4**). Then a billion or so genetically altered T cells were injected into Ashanti's bloodstream. Repeated treatments were required to produce a sufficient number of functioning T cells. In addition, Ashanti also periodically received injections of purified ADA protein throughout this process, so the exact effects of gene therapy were difficult to discern. Ashanti continues to receive supplements of the ADA enzyme to allow her to lead a normal life.

Subsequent gene therapy treatments for SCID have focused on using bone marrow stem cells called hematopoietic stem cells (HSCs), and *in vitro* approaches to repopulate the number of ADA-producing T cells. In the past decade alone, over 100 people (mostly children) have received gene therapy for ADA-SCID, and most have been treated successfully and are disease-free. ADA-SCID treatment is still considered the most successful example of gene therapy.

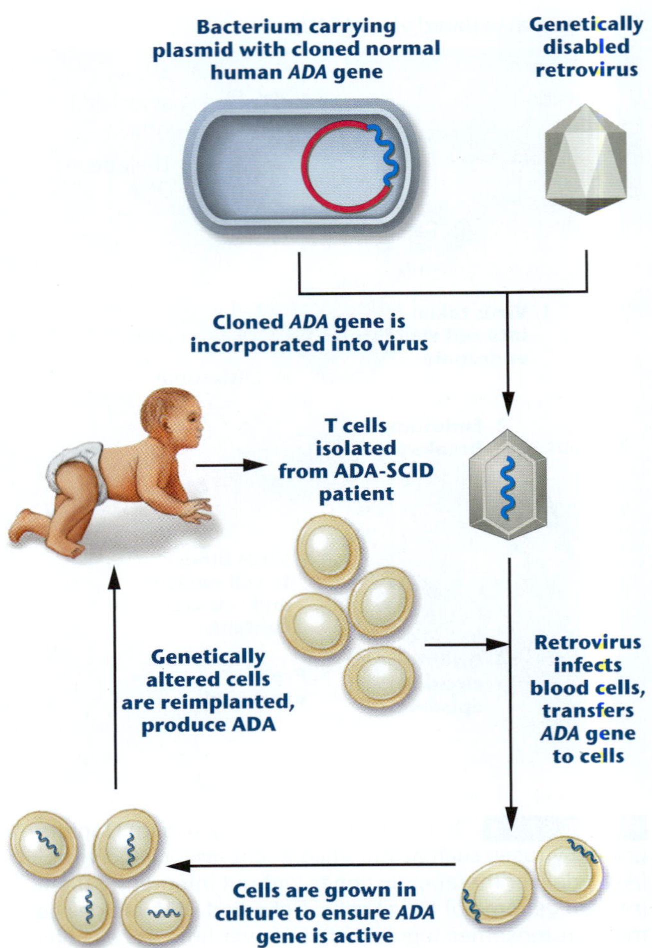

FIGURE ST 3.4 The first successful gene therapy trial. To treat ADA-SCID using gene therapy, a cloned human *ADA* gene was transferred into a viral vector, which was then used to infect white blood cells removed from the patient. The transferred *ADA* gene was incorporated into a chromosome, after which the cells were cultured to increase their numbers. Finally, the cells were inserted back into the patient, where they produce ADA, allowing the development of an immune response.

ST 3.4 Gene Therapy Setbacks

From 1990 to 1999, more than 4000 people underwent gene therapy for a variety of genetic disorders. These trials often failed and thus led to a loss of confidence in gene therapy. In the United States, optimism for gene therapy plummeted even further in 1999 when teenager Jesse Gelsinger died while undergoing a test for the safety of gene therapy to treat a liver disease called *ornithine transcarbamylase* (OTC) deficiency. Large numbers of adenovirus vectors bearing the *OTC* gene were injected into his hepatic artery. The vectors were expected to target his liver, enter liver cells, and trigger the production of OTC protein. In turn, it was hoped that the OTC protein might correct his genetic defect and cure him of his liver disease.

Researchers had previously treated 17 people with the therapeutic virus, and early results from these patients were promising. But the 18th patient, Jesse Gelsinger, within hours of his first treatment, developed a massive immune reaction. He developed a high fever, his lungs filled with fluid, multiple organs shut down, and he died four days later of acute respiratory failure. Jesse's severe response to the adenovirus may have resulted from how his body reacted to a previous exposure to the virus used as the vector for this protocol.

In the aftermath of the tragedy, several government and scientific inquiries were conducted. Investigators learned that in the clinical trial scientists had not reported other adverse reactions to gene therapy, and that some of the scientists were affiliated with private companies that could benefit financially from the trials. It was also determined that serious side effects seen in animal studies were not explained to patients during informed-consent discussions. The FDA subsequently scrutinized gene therapy trials across the country, halted a number of them, and shut down several gene therapy programs. Other groups voluntarily suspended their gene therapy studies. Tighter restrictions on clinical trial protocols were imposed to correct some of the procedural problems that emerged from the Gelsinger case. Jesse's death had dealt a severe blow to the struggling field of gene therapy—a blow from which it was still reeling when a second tragedy hit.

The outlook for gene therapy brightened momentarily in 2000, when a group of French researchers reported what was hailed as the first large-scale success in gene therapy. Children with a fatal X-linked form of ADA-SCID (X-SCID, also known as "bubble boy" disease) developed functional immune systems after being treated with a retroviral vector carrying a normal gene. But elation over this study soon turned to despair, when it became clear that 5 of the 20 patients in two different trials developed leukemia as a direct result of their therapy. One of these patients died as a result of the treatment, while the other four went into remission from the leukemia. In two of the children examined, their cancer cells contained the retroviral vector, inserted near a gene called *LMO2*. This *insertional mutation* activated the *LMO2* gene, causing uncontrolled white blood cell proliferation and development of leukemia. The FDA immediately halted 27 similar gene therapy clinical trials, and once again gene therapy underwent a profound reassessment.

On a positive note, long-term survival data from trials in the UK to treat X-SCID and ADA-SCID using HSCs from the patients' bone marrow for gene therapy have shown that 14 of 16 children have had their immune system restored at least nine years after the treatment. These children formerly had life expectancies of less than 20 years. In the past five years alone, more than 40 patients have been treated for ADA-SCID in three well-developed programs in Italy, the United Kingdom, and the United States. All individuals treated have survived, and 75 percent of those treated are disease-free.

Problems with Gene Therapy Vectors

Most of the problems associated with gene therapy, including the Jesse Gelsinger case and the French X-SCID trial, have been traced to the viral vectors used to transfer therapeutic genes into cells. These vectors have been shown to have several serious drawbacks.

- First, integration of retroviral genomes, including the human therapeutic gene into the host cell's genome, occurs only if the host cells are replicating their DNA. But only a small number of cells in any tissue are dividing and replicating their DNA.
- Second, the injection of massive quantities of most viral vectors, but particularly adenovirus vectors, is capable of causing an adverse immune response in the patient, as happened in Jesse Gelsinger's case.
- Third, insertion of viral genomes into host chromosomes can activate or mutate an essential gene, as in the case of the French patients. Viral integrase, the enzyme that allows for viral genome integration into the host genome, interacts with chromatin-associated proteins, often steering integration toward transcriptionally active genes.
- Fourth, AAV vectors cannot carry DNA sequences larger than about 5 kb, and retroviruses cannot carry DNA sequences much larger than 10 kb. Many human genes exceed the 5- to 10-kb size range.
- Finally, there is a possibility that a fully infectious virus could be created if the inactivated vector were to recombine with another unaltered viral genome already present in the host cell.

To overcome these problems, new viral vectors and strategies for transferring genes into cells are being developed in an attempt to improve the action and safety of vectors. No new technology has had a greater impact on gene therapy than gene targeting, especially by CRISPR-Cas. Fortunately, gene therapy has experienced a resurgence in part because of several promising new trials and successful treatments.

ST 3.5 Recent Successful Trials by Conventional Gene Therapy Approaches

Treating Retinal Blindness

In recent years, patients being treated for blindness have greatly benefited from gene therapy approaches. Congenital retinal blinding conditions affect about 1 in 2000 people worldwide, many of which are the result of a wide range of genetic defects. Over 165 different genes have been implicated in various forms of retinal blindness. Currently there are over two dozen active gene therapy trials for at least 10 different retinal diseases.

Successful gene therapy has been achieved in subsets of patients with **Leber congenital amaurosis (LCA),** a degenerative disease of the retina that affects 1 in 50,000 to 1 in 100,000 infants each year and causes severe blindness. Gene therapy treatments for LCA were originally pioneered in dogs. Based on the success of these treatments, the protocols were adapted and applied to human gene therapy trials.

LCA is caused by alterations to photoreceptor cells (rods and cones), light-sensitive cells in the retina, due to 18 or more genes. One gene in particular, *RPE65*, has been the gene therapy target of choice. The protein product of the *RPE65* gene metabolizes retinol, which is a form of vitamin A that allows the rod and cone cells of the retina to detect light and transmit electrical signals to the brain. In one of the earliest trials, young adult patients with defects in the *RPE65* gene were given injections of the normal gene incorporated into an AAV vector. Several months after a single treatment, many adult patients, while still legally blind, could detect light, and some of them could read lines of an eye chart. This treatment approach for LCA was based on injecting AAV-carrying *RPE65* at the back of the eye directly under the retina. The therapeutic gene enters about 15 to 20 percent of cells in the retinal pigment epithelium, the layer of cells just beneath the visual cells of the retina.

Adults treated by this approach have shown substantial improvements in a variety of visual functions tests, but the greatest improvement has been demonstrated in children, all of whom have gained sufficient vision to allow them to be ambulatory. Researchers think the success in children has occurred because younger patients have not lost as many photoreceptor cells as older patients. In January 2018, the FDA approved a *RPE65* gene therapy approach by Spark Therapeutics as the first treatment to target a genetic disease caused by mutations in a single gene.

Because of the small size of the eye and the relatively small number of cells that need to be treated, the prospects for gene therapy to become routine treatment for eye disorders appears to be very good. Retinal cells are also very long lived; thus, AAV delivery approaches can be successful for long periods of time even if the gene does not integrate.

Successful Treatment of Hemophilia B

A very encouraging gene therapy trial in England successfully treated a small group of adults with hemophilia B, a blood disorder caused by a deficiency in the coagulation protein human factor IX. This, and other similar trials, are based largely on approaches derived from successful gene therapy to treat hemophilia B in dogs. Currently, most hemophilia B patients are treated several times each week with infusions of concentrated doses of the factor IX protein. In the gene therapy trial, six adult patients received, *in vivo*, a single dose of an adenovirus vector (AAV8) carrying normal copies of the human factor IX gene introduced into liver cells. Of six patients treated, four were able to stop factor IX infusion treatments after the gene therapy trial. Several other trials of this AAV treatment approach are under way, and expectations are high that a gene therapy cure for hemophilia B is close to becoming a routine reality.

HIV as a Vector Shows Promise in Recent Trials

Researchers at the University of Paris and Harvard Medical School reported that two years after gene therapy treatment for *β*-**thalassemia,** a blood disorder involving the *β*-globin gene that reduces the production of hemoglobin, a young man no longer needed transfusions and appeared to be healthy. A modified, disabled HIV was used to carry a copy of the normal *β*-globin gene. Although this trial resulted in activation of the growth factor gene called *HMGA2*, reminiscent of what occurred in the French X-SCID trials, activation of the transcription factor did not result in an overproduction of hematopoietic cells or create a condition of preleukemia.

In 2013, researchers at the San Raffaele Telethon Institute for Gene Therapy in Milan, Italy, first reported two studies using lentivirus vectors derived from HIV in combination with HSCs to successfully treat children with either **metachromatic leukodystrophy (MLD)** or **Wiskott–Aldrich syndrome (WAS).** MLD is a neurodegenerative disorder affecting storage of enzymes in lysosomes and is caused by mutation in the arylsulfatase A (*ARSA*) gene that results in an accumulation of fats called sulfatides. These are toxic to neurons, causing progressive loss of the myelin sheath (demyelination) surrounding neurons in the brain, leading to a loss of cognitive functions and motor skills. There is no cure for MLD. Children with MLD appear healthy at birth but eventually develop MLD symptoms.

Researchers used an *ex vivo* approach with a lentivirus vector to introduce a functional *ARSA* gene into bone marrow–derived HSCs from each patient and then infused treated HSCs back into patients. Four years after the start of a trial involving 10 patients with MLD, data from 6 patients analyzed 18 to 24 months after gene therapy indicated that the trials are safe and effective. Treatment halted disease progression as determined by magnetic resonance images of the brain and through tests of cognitive and motor skills. Patients with MLD in the first group treated have already lived past the expected lifetime normally associated with this disease. Additional patients are now being treated. This approach took over 15 years of research and a team of over 70 people, including researchers and clinicians, which is indicative of the teamwork approach typical of gene therapy trials.

The trial was technically complicated because it required that HSCs travel through the bloodstream and release the ARSA protein that is taken up into neurons. A major challenge was to create enough engineered cells to produce a sufficient quantity of therapeutic ARSA protein to counteract the neurodegenerative process.

Similar results were reported for treating patients with WAS, an X-linked condition resulting in defective platelets that make patients more vulnerable to infections, frequent bleeding, autoimmune diseases, and cancer. Genome sequencing of MLD and WAS patients treated in these trials showed no evidence of genome integration near oncogenes. Similarly, patients showed no evidence of HSC overproduction, suggesting that this lentivirus delivery protocol produced a safe and stable delivery of the therapeutic genes.

ST 3.6 Genome-Editing Approaches to Gene Therapy

The gene therapy approaches and examples we have highlighted thus far have focused on the addition of a therapeutic gene that functions along with the defective gene. However, rapid progress is being made with **genome editing**—the removal, correction, and/or replacement of a mutated gene. Genome editing by CRISPR-Cas, in particular, has shown great potential and provided renewed optimism for scientists and physicians involved in gene therapy as well as patients.

DNA-Editing Nucleases

For nearly 20 years, scientists have been working on modifications of restriction enzymes and other nucleases to engineer proteins capable of editing the genome with precision, including the ability to edit one or a few bases or to replace specific genes. The concept is to combine a nuclease with a sequence-specific DNA binding domain that can be precisely targeted for digestion. In 1996 researchers fused DNA-binding proteins with a zinc-finger motif and DNA cutting domain from the restriction enzyme *Fok*I to create enzymes called **zinc-finger nucleases (ZFNs).** The zinc-finger motif is found in many transcription factors and consists of a cluster of two cysteine and two histidine residues that bind zinc atoms and interact with specific DNA sequences. By coupling zinc-finger motifs to DNA cutting portions of a polypeptide, ZFNs provide a mechanism for modifying sequences in the genome in a sequence-specific targeted way.

The DNA-binding domain of the ZFN can be engineered to attach to any sequence in the genome. The zinc fingers bind with a spacing of 5 to 7 nucleotides, and the nuclease domain of the ZFN cleaves between the binding sites.

Another category of DNA-editing nucleases called **transcription activator-like effector nucleases (TALENs)** was created by adding a DNA-binding motif identified in transcription factors from plant pathogenic bacteria known as transcription activator-like effectors (TALEs) to nucleases to create TALENs. TALENs also cleave as dimers. The DNA-binding domain is a tandem array of amino acid repeats, with each TALEN repeat binding to a specific single base pair. The nuclease domain then cuts the sequence between the dimers, a stretch that spans about 13 bp.

ZFNs and TALENs have shown promise in animal models and cultured cells for gene replacement approaches that involve removing a defective gene from the genome. ZFNs, TALENs, and CRISPR-Cas, as we will discuss shortly, all create double-stranded breaks in the DNA and then are mended by either nonhomologous end joining or homologous recombination. These enzymes can create site-specific double-stranded cleavage in the genome. When coupled with certain integrases, ZFNs and TALENs may lead to genome editing by cutting out defective sequences and using recombination to introduce homologous sequences into the genome that replace defective sequences. Although this technology has not yet advanced sufficiently for reliable use in humans, there have been several promising trials.

For example, ZFNs are actively being used in clinical trials for treating patients with HIV. Scientists are exploring ways to deliver immune system–stimulating genes that could make individuals resistant to HIV infection or cripple the virus in HIV-positive persons. In 2007, Timothy Brown, a 40-year-old HIV-positive American, had a relapse of acute myeloid leukemia and received a stem cell transplant. Because he was HIV-positive, Brown's physician selected a donor with a mutation in both copies of the *CCR5* gene, which encodes an HIV co-receptor carried on the surface of T cells to which HIV must bind to enter T cells (specifically CD4+ cells). People with naturally occurring mutations in both copies of the *CCR5* gene are resistant to most forms of HIV. Brown relapsed again and received another stem cell transplant from the *CCR5*-mutant donor. Eventually, the cancer was contained, and by 2010, levels of HIV in his body were still undetectable even though he was no longer receiving immune-suppressive treatment. Brown is generally considered to be the first person to have been cured of an HIV infection.

This example encouraged researchers to press forward with a gene therapy approach to modify the *CCR5* gene of HIV patients. In the first genome-editing trial to treat people with HIV, T cells were removed from HIV-positive men, and ZFNs were used to disrupt the *CCR5* gene *ex vivo*. The modified cells were then reintroduced into patients. In five of six patients treated, immune-cell counts rose substantially and viral loads also decreased following the therapy. To date, more than 90 people have been treated by this approach.

What percentage of immune cells would have to be treated this way to significantly inhibit spread of the virus is still not known.

Recently, researchers working with human cells used TALENs to remove defective copies of the *COL7A1* gene, which causes recessive dystrophic epidermolysis bullosa (RDEB), an incurable and often fatal disease that presents as excessive blistering of the skin, pain, and severely debilitating skin damage. Researchers at the University of Minnesota used a TALEN to cut DNA near a mutation in the *COL7A1* gene in fibroblast cells taken from a patient with RDEB and supplied these cells with a functional copy of the *COL7A1* gene. These cells were then converted into a type of stem cell called *induced pluripotent stem cells (iPSCs)*, which were then differentiated into skin cells that expressed the correct protein. This is a promising result, and researchers now plan to transplant these skin cells into patients in an attempt to cure them of RDEB. Another group has recently taken a similar approach using TALENs to repair cultured cells in order to correct the mutation in Duchenne muscular dystrophy (DMD). Researchers are optimistic that this approach can soon be adapted to treat patients.

CRISPR-Cas Method Revolutionizes Genome-Editing Applications and Renews Optimism in Gene Therapy

No method has created more excitement than the genome-editing technique known as **CRISPR-Cas** (clustered regularly interspaced short palindromic repeats—CRISPR-associated proteins). Identified in bacterial cells, the CRISPR-Cas system functions to provide bacteria and archaea immunity against invading bacteriophages and foreign plasmids (see Chapter 15). However, CRISPR-Cas has been adapted as a tool for genome editing in eukaryotic cells (see Chapter 17), which has revolutionized gene therapy strategies.

CRISPR-Cas is based on delivering a single-stranded guide RNA sequence (sgRNA) and a Cas endonuclease. The sgRNA is complementary to the target gene sequence in the genome and directs the endonucleas to cut that sequence. One commonly used nuclease is called Cas9 (**Figure ST 3.5**). At the same time as Cas9 and the sgRNA are delivered, a DNA donor template coding for a replacement sequence is also delivered. The sgRNA-Cas9 complex binds to the target DNA sequence through complementary base pairing, and Cas9 generates a double-stranded break in the DNA. In order for Cas9 to cut a target sequence, there must be a specific three-nucleotide sequence adjacent to the complementary sequence called a protospacer adjacent motif (PAM). As cells repair the DNA damage caused by Cas9, repair enzymes incorporate donor template DNA into the genome, thus replacing the target DNA sequence. See Chapter 17 for a detailed description of the CRISPR-Cas9 editing mechanism.

While ZFNs and TALENs are sufficient for genome editing, CRISPR-Cas is more efficient and easier to design. Although ZFN and TALEN target specificity is provided by protein–DNA interactions, which can be difficult to engineer, CRISPR-Cas specificity is simply determined by complementary base pairing of the sgRNA and the target sequence. Within months of the technique being widely

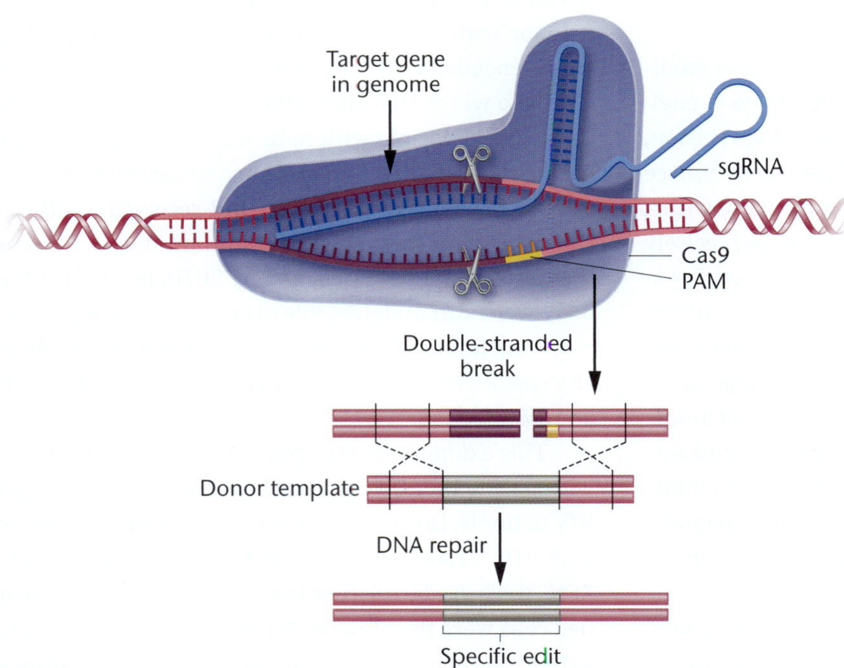

FIGURE ST 3.5 The CRISPR-Cas system allows for genome editing by targeting specific sequences in the genome.

available, researchers around the world used CRISPR-Cas to target specific genes in human cells, mice, rats, bacteria, fruit flies, yeast, zebrafish, and dozens of other organisms. In one of the first reported applications of CRISPR-Cas for gene therapy, a team from the Massachusetts Institute of Technology (MIT) cured mice of a rare liver disorder, type I tyrosinemia, through genome editing. In tyrosinemia, a condition affecting about 1 in 100,000 people, mutation of the *FUH* gene encoding the enzyme fumarylacetoacetase prevents breakdown of the amino acid tyrosine. After an *in vivo* approach with a one-time treatment, roughly 1 in 250 liver cells accepted the CRISPR-Cas–delivered replacement of the mutant gene with a normal copy of the gene. But about 1 month later these cells proliferated and replaced diseased cells, taking over about one-third of the liver, which was sufficient to allow mice to metabolize tyrosine and show no effects of disease. Mice were subsequently taken off a low-protein diet and a drug normally used to disrupt tyrosine production.

In Special Topic Chapter 7—Genomics and Precision Medicine—we will discuss a promising way to treat cancer called **immunotherapy.** This approach involves the genetic engineering of **T cells** (cells of the immune system), which can recognize, bind to, and destroy tumor cells. However, cancer cells have ways of evading T cells or of preventing their activation. The principle behind engineered T-cell therapies is to create T cells that are better equipped to find and target tumor cells for destruction. Two strategies for immunotherapy are to create recombinant **T-cell receptors (TCRs)** that specifically recognize antigens on or within cancer cells or **chimeric antigen receptor (CAR)–T cells** engineered to express receptors that can directly recognize antigens on the surface of the tumor cell without requiring T-cell activation by antigen-presenting cells. Immunotherapy has shown great promise for the treatment of certain forms of leukemia, and its applications are one area where the ease of genome editing by CRISPR-Cas may improve its efficacy.

In 2017, the FDA approved the first ever immunotherapy—Novartis' CTL019 CAR-T treatment for children and young adults with B-cell acute lymphoblastic leukemia (ALL). ALL is the most common childhood cancer in the United States, and patients who relapse or fail to respond to chemotherapy have a low survival rate. But over 80% of 63 patients treated with CTL109 went into remission almost immediately after treatment began, and most remained cancer-free 6 months after treatment. CTL019 (brand name, Kymriah™) is approved for ALL patients up to 25 years of age, and as of May 2018, is also approved for another cancer called diffuse large B-cell lymphoma. Kymriah also made headlines for its sticker price of approximately $475,000 for a single treatment and estimates that the total cost of care with this drug could exceed $1.5 million.

More recently, a team from the University of Pennsylvania; the University of California, San Francisco; and the MD Anderson Cancer Center at the University of Texas plans to treat cancer patient T cells with a different immunotherapy strategy involving CRISPR-Cas. The T cells will have their *TCR* gene inactivated, and a recombinant *TCR* gene will be introduced to help these cells target destruction of tumor cells and to avoid nontumor cells. The trial will also use CRISPR-Cas to edit a gene called *program cell death 1 (PD-1)*, which expresses a protein on the surface of T cells that can often be neutralized by cancer cells to minimize immune responses and ward off T-cell destruction of a tumor. The edited T cells with the mutant *PD-1* gene are expected to be able to recognize and attack lung tumor cells. This trial was recruiting patients as of June 2018 and was expected to start in late 2018 in the United States.

In 2016, a team from China reported the first CRISPR-Cas trial of human patients, designed to treat an aggressive form of cancer called metastatic non–small-cell lung cancer. An *ex vivo* approach was used to edit *PD-1* in T cells, which were then reintroduced into patients with the hope that these cells would target tumors in the lung for destruction without being disabled by the cancer cells. The main purpose of this initial trial is to determine whether this approach is safe. The success of this trial, which has enrolled at least 86 patients, had not been reported at the time this edition of *Essentials of Genetics* was published. Several more genome-editing trials are scheduled to begin at different centers around the world targeting kidney, bladder, and prostate cancer, among others.

Additional headline-grabbing examples of successful CRISPR-Cas applications in mice and humans that highlight the potential of this approach for gene therapy include:

- AAV delivery of CRISPR-Cas9 to remove a defective exon from the *Dmd* gene in a mouse model (*mdx* mice) of DMD significantly restored muscle function in treated mice. It has been estimated that this approach has the potential to cure approximately 80 percent of human cases of DMD.
- After successful trials in mice, *in vitro* repair of the human β-globin (*HBB*) gene in HSCs was used to treat sickle-cell disease and β-thalassemia in humans. It is expected that these preclinical studies will soon lead to the delivery of edited HSCs in humans.
- CRISPR-Cas9 targeting and replacement of the defective clotting factor IX gene in liver cells was used to cure mice of hemophilia B.
- HIV-infected patients are being recruited for a clinical study in China where CRISPR-Cas9 editing will be used to disable the *CCR5* gene (which we discussed as a ZFN treatment) to block HIV infection in modified cells.
- A CRISPR-Cas approach to edit the *CEP290* gene to treat LCA type 10 (recall that this a form of blindness) is being developed in a joint effort by the pharmaceutical

company Allergan and the genome-editing company Editas Medicine.

- CRISPR-Cas editing of mutations involved in genetic forms of hearing loss in mice have shown early potential.

Finally, the use of CRISPR-Cas to edit the human germ line (sperm and egg cells) and human embryos has been one of the most controversial potential applications of this technology, although similar concerns existed when ZFNs and TALENs were first being used. In 2015, a team of Chinese scientists reported using CRISPR-Cas9 to edit the *HBB* gene in 86 human embryos. The embryos were generated by *in vitro* fertilization but were donated for research because they were triploid and thus would only survive a few days. Two days after CRISPR-Cas treatment, 71 embryos had survived, but only 4 of them carried the intended change to the *HBB* gene. Unexpectedly, many other embryos had acquired mutations in genes other than *HBB* as a result of the treatment. From this the Chinese research team concluded that genome-editing technology is not sufficiently developed for use in embryos.

In 2016, the second published report of genome editing of human (triploid) embryos, also from a team in China, described the use of CRISPR-Cas9 to introduce a mutation in the *CCR5* gene to confer resistance to HIV infection. In this study, 4 of 26 embryos were successfully edited, but others contained undesirable mutations as a result of the treatment. While this demonstrated some proof of concept for creating HIV-resistant embryos, like the 2015 work, it also clearly showed that genome editing of embryos is neither precise nor safe at this point. However, a 2017 study on viable diploid embryos, created by *in vitro* fertilization for the purpose of embryo-editing research, reported a high degree of success in correcting a disease-causing mutation. Other researchers have disputed the claims of this study, and it has yet to be replicated.

These and other studies have stimulated significant ethical discussions about genetic engineering of embryos (which we discuss further in the final section of this chapter).

Also, as is the case with other gene therapy approaches, there is concern about CRISPR-Cas creating mutations at nontarget locations in the genome and also unwanted deletions and rearrangements in a target gene. But so far, CRISPR-Cas is clearly the most promising tool for genome editing and gene therapy that has ever been developed, and in a short time, the pace of progress with this technique in animals and humans has been remarkable. Stay tuned!

RNA-Based Therapeutics

Over the past decade, RNA-based therapeutics such as antisense RNAs, RNA interference, and microRNAs for gene therapy have received a great deal of attention. This is partly because these methods can be designed to be highly specific for a target RNA of interest to block or upregulate gene expression and are versatile because they can also be used to alter mRNA splicing, to target noncoding RNAs, and to express an exogenous RNA among other examples.

Attempts have been made to use **antisense oligonucleotides** to inhibit translation of mRNAs from defective genes (see **Figure ST 3.6**), thus blocking or "silencing" gene expression, but this approach to gene therapy has generally not yet proven to be reliable. The emergence of **RNA interference (RNAi)** as a powerful gene-silencing tool *in vitro* for research has reinvigorated interest in gene therapy approaches by gene silencing. RNAi is a form of gene-expression regulation. In animals, short, double-stranded RNA molecules are delivered into cells where the enzyme Dicer chops them into 21- to 25-nt-long pieces called **small interfering RNAs (siRNAs).** siRNAs then join with an enyzme complex called the **RNA-indued silencing complex (RISC),** which shuttles the siRNAs to their target mRNA, where they bind by complementary base pairing. The RISC complex can block siRNA-bound mRNAs from being translated into protein or can lead to degradation of siRNA-bound mRNAs so that they cannot be translated into protein (**Figure ST 3.6**).

A main challenge to RNAi-based therapeutics so far has been *in vivo* delivery of double-stranded RNA or siRNA. RNAs degrade quickly in the body. It is also hard to get RNA to cross lipid bilayers to penetrate cells in the target tissue. And how does one deliver RNA-based therapies to cancer cells but

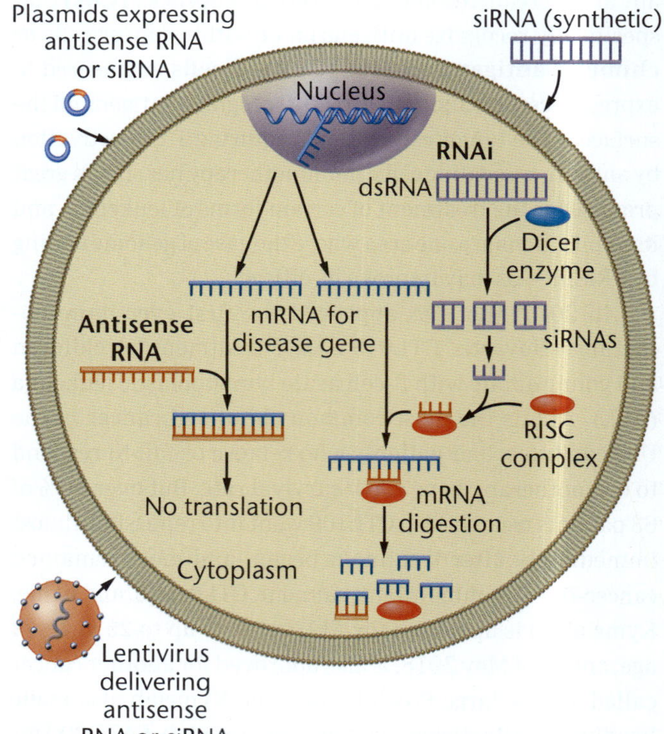

FIGURE ST 3.6 Antisense RNA and RNA interference (RNAi) approaches to silence genes for gene therapy. Antisense RNA technology and RNAi are two ways to silence gene expression and turn off disease genes.

not to noncancerous, healthy cells? Two common delivery approaches are to inject the siRNA directly or to deliver them via a DNA plasmid vector that is taken in by cells and transcribed to make double-stranded RNA which Dicer can cleave into siRNAs. Lentivirus, liposome, and attachment of siRNAs to cholesterol and fatty acids are other approaches being used to deliver siRNAs (Figure ST 3.6). When delivered in liposomes or attached to lipids, siRNAs are taken into the cell by endocytosis, but because of their charge, another challenge is getting therapeutic RNA out of the endosome and into the cytoplasm.

The same approaches used to deliver antisense RNAs and siRNAs can also be used to deliver vectors encoding **microRNAs (miRNAs)** or miRNAs themselves. (Recall that we discussed miRNAs in Chapter 16.) In recent years a tremendous body of research literature has developed on the roles of miRNAs in silencing gene expression naturally in cells. The application of miRNAs for gene therapy is only in initial stages of development.

More than a dozen clinical trials involving RNAi are under way in the United States. Several RNAi clinical trials to treat blindness are showing promising results. One RNAi strategy to treat a form of blindness called macular degeneration targets a gene called *VEGF*. The VEGF protein promotes blood vessel growth. Overexpression of this gene, causing excessive production of blood vessels in the retina, leads to impaired vision and eventually blindness. In 2018, the FDA approved the drug Onpattro, by Alnylam Pharmaceuticals, Inc., for the treatment of a peripheral nerve disease in adults called polyneuropathy. Onpattro is the first FDA approval of a siRNA gene therapy treatment. It targets mRNA for the *TTR* gene, interfering with production of an abnormal form of the protein transthyretin (TTR) that contributes to polyneuropathy. Other disease candidates for treatment by RNAi include several different cancers, diabetes, liver diseases, multiple sclerosis, and arthritis.

But many are predicting that RNA-based therapies will become much more successful and more widely adopted within the next decade. Antisense RNA is the oldest RNA-based therapeutic approach, but it initially did not live up to expectations. In the mid-2000s, many companies dropped this approach to gene therapy because of problems associated with delivering antisense RNA oligonucleotides across cell membranes and keeping them from being degraded while in the circulatory system. However, recent advances in RNA oligonucleotide chemistry have helped overcome these hurdles, and hundreds of clinical trials involving antisense RNAs are in the planning stages.

By late 2016, within a six-month span, antisense oligonucleotide trials were approved by the FDA for familial cholesterolemia, Duchenne muscular dystrophy, and spinal muscular atrophy (SMA). For example, the antisense oligonucleotide called Spinraza® (nusinersen), produced by Ionis Pharmaceuticals of Carlsbad, California, was approved as the first treatment for SMA. Affecting 1 in 10,000 to 12,000 children born, this disease is characterized by the loss of motor neurons in the spinal cord resulting in progressive muscle weakness and is a leading genetic cause of death for infants. In 2018, SMA was added to the Recommended Uniform Screening Panel for genetic testing which we discussed in Special Topics Chapter 2—Genetic Testing (see ST 2 Box 1).

Patients with SMA have a mutation in the *SMN1* gene that prevents production of the functional SMN protein required for normal motor neuron development. Spinraza is delivered into the cerebrospinal fluid. It is an 18-nt antisense oligonucleotide that targets *SMN2*, a homolog of *SMN1*. *SMN2* is normally mis-spliced at exon 7 to produce a truncated, largely nonfunctional SMN protein.

Spinraza binds to pre-mRNA of the *SMN2* gene, altering splicing to include exon 7 in the mature transcript leading to translation of a functional copy of the SMN protein. The drug showed such promise in two trials that the trials were halted early and considered successful because it was deemed unethical to continue to deny the drug to SMN-affected children in placebo groups. This antisense approach to alter mRNA splicing for gene therapy has also generated excitement because it has potential for treating Huntington disease, ALS, and other neurological conditions.

ST 3.7 Future Challenges and Ethical Issues

Despite the progress that we have noted thus far, many questions remain to be answered before widespread application of gene therapy for the treatment of genetic disorders becomes routine:

- What is the proper route for gene delivery in different kinds of disorders? For example, what is the best way to treat brain or muscle tissues?
- What percentage of cells in an organ or a tissue need to express a therapeutic gene to alleviate the effects of a genetic disorder?
- What amount of a therapeutic gene product must be produced to provide lasting improvement of the condition, and how can sufficient production be ensured? Currently, many approaches provide only short-lived delivery of the therapeutic gene and its protein.
- Will it be possible to use gene therapy to treat diseases that involve multiple genes?
- Can expression or the timing of expression of therapeutic genes be controlled in a patient so that genes can be turned on or off at a particular time or as necessary?
- Will genome-editing approaches become more widely used for gene therapy trials?

- Will genome editing emerge as the safest and most reliable method of gene therapy, rendering other approaches obsolete, or will a combination of approaches (vector and nonvector delivery, RNA-based therapeutics, and genome editing) be necessary depending on the genetic condition being treated?

For many people, the question remains whether gene therapy can ever recover from past setbacks and fulfill its promise as a cure for genetic diseases. Clinical trials for any new therapy are potentially dangerous, and often, animal studies will not accurately reflect the reaction of individual humans to the methodology leading to the delivery of new genes. However, as the history of similar struggles encountered with life-saving developments such as the use of antibiotics and organ transplants show, there will be setbacks and even tragedies, but step by small step, we will move toward a technology that could—someday—provide reliable and safe treatment for severe genetic diseases.

Ethical Concerns Surrounding Gene Therapy

Gene therapy raises several ethical concerns, and many forms of gene therapy are sources of intense debate. At present, in the United States, all gene therapy trials are restricted to using somatic cells as targets for gene transfer. This form of gene therapy is called **somatic gene therapy;** only one individual is affected, and the therapy is done with the permission and informed consent of the patient or family.

Two other forms of gene therapy have not been approved, primarily because of the unresolved ethical issues surrounding them. The first is called **germ-line therapy,** whereby germ cells (the cells that give rise to the gametes—i.e., sperm and eggs) or mature gametes are used as targets for gene transfer or genome editing. In this approach, the transferred or edited gene is incorporated into all the future cells of the body, including the germ cells. As a result, individuals in future generations will also be affected, without their consent. Recently, ethical discussions about germ-line therapy have been accelerated by several reports of CRISPR-Cas being used for genome editing of human embryos *in vitro* (none were transferred to a uterus to develop). A report from the U.S. National Academy of Sciences and the National Academy of Medicine is recommending that germ-line therapy trials only be considered for serious conditions for which there is no reasonable alternative treatment option, and where both the risk–benefit options and broad oversight are available. Is this kind of procedure ethical? Do we have the right to make this decision for future generations? Thus far, the concerns have outweighed the potential benefit in the United States where Congress has barred the FDA from approving clinical trials of germ-line gene therapy. Across the globe, the legality of editing human embryos varies widely. Some countries have banned the editing of human embryos, some have set some restrictions, and others have no restrictions at all. Box 3 mentions gene doping, which is an example of **enhancement gene therapy,** whereby people may be "enhanced" for some desired trait.

Gene therapy is currently a fairly expensive treatment (see Box 2). For rare conditions, the fewer the people treated, the more expensive the treatment will be. But what is the right price for a cure? It remains to be seen how health-care insurance providers will view gene therapy. But if gene therapy treatments provide a health-care option that drastically improves the quality of life for patients for whom there are few other options, it is likely that insurance companies will reimburse patients for treatment costs.

Finally, *whom* to treat by gene therapy is yet another ethically provocative consideration. In the Jesse Gelsinger case mentioned earlier, the symptoms of his OTC deficiency

BOX 2

Glybera: The First Commercial Gene Therapy to be Approved in the West Lasted Only Five Years

In 2012, a gene therapy product called Glybera® (alipogene tiparvovec), developed by Amsterdam-based company uniQure, made history when the European Medicines Agency of the European Union approved it as the first gene therapy trial to win commercial approval in the Western world. Glybera is an AAV vector system for delivering therapeutic copies of the *LPL* gene to treat patients with a rare disease called lipoprotein lipase deficiency (LPLD, also called familial hyperchylomicronemia). LPLD patients have high levels of triglycerides in their blood. Elevated serum triglycerides are toxic to the pancreas and cause a severe form of pancreatic inflammation called pancreatitis. Because, the U.S. FDA requested additional clinical trials before it would consider approval of Glybera, uniQure discontinued plans to seek approval by the FDA. The success of Glybera trials in Europe signaled what many researchers hoped would be the beginning of many new gene therapy approvals in Europe and the United States. But despite its promise, by 2017, Glybera had failed to be widely used by any European country and since inception had only been used in one patient. Glybera failed, in part, because at a cost of over $1 million per treatment, it was one of the most expensive drugs in history. Thus uniQure announced it would not renew European marketing authorization after October 2017.

BOX 3

Gene Doping for Athletic Performance?

Gene therapy is intended to provide treatments or cures for genetic disease. But should it also apply for those seeking enhancement gene therapy to improve athletic performance? As athletes seek a competitive edge, will gene therapy as a form of "gene doping" to improve performance be far behind?

We already know that in animal models enhanced muscle function can be achieved by gene addition. For example, adding copies of the insulin-like growth factor (*IGF-1*) gene to mice improves aspects of muscle function. The kidney hormone erythropoietin (EPO) increases red blood cell production, which leads to a higher oxygen content of the blood and thus improved endurance. Synthetic forms of EPO are banned in Olympic athletes. Several groups have proposed using gene therapy to deliver the *EPO* gene into athletes "naturally."

Since 2004 the World Anti-Doping Agency (WADA) has included gene doping through gene therapy as a prohibited method in sanctioned competitions. However, methods to detect this are not well established. If techniques for gene therapy become more routine, many feel it is simply a matter of time before gene doping will be the next generation of performance-enhancement treatments. Obviously, many legal and ethical questions will arise if this becomes a reality.

were minimized by a low-protein diet and drug treatments. Whether it was necessary to treat Jesse by gene therapy is a question that has been widely debated.

Jesse Gelsinger volunteered for the study to test the safety of the treatment for those with more severe disease. If a benefit was shown, it would have relieved him of an intense treatment regimen. Whether he should have been selected for the safety study is, of course, a matter to be debated. His tragic death due to unforeseen complications could not have been predicted at the time.

Review Questions

1. What is gene therapy?
2. Compare and contrast *ex vivo* and *in vivo* gene therapy as approaches for delivering therapeutic genes.
3. When treating a person by gene therapy, is it necessary that the therapeutic gene becomes part of a chromosome (integration) when inserted into cells? Explain your answer.
4. Describe two ways that therapeutic genes can be delivered into cells.
5. Explain how viral vectors can be used for gene therapy, and provide two examples of commonly used viral vectors. What are some of the major challenges that must be overcome to develop safer and more effective viral vectors for gene therapy?
6. During the first successful gene therapy trial in which Ashanti DeSilva was treated for SCID, did the therapeutic gene delivered to Ashanti replace the defective copy of the ADA gene? Why were white blood cells chosen as the targets for the therapeutic gene?
7. Explain an example of a successful gene therapy trial. In your answer be sure to consider: a description of the disease condition that was treated, the mutation or disease gene affected, the therapeutic gene delivered, and the method of delivery use for the therapy.
8. What is genome editing, and how does this approach differ from traditional gene therapy approaches?
9. What are some of the reasons why the development of CRISPR-Cas may be the technical breakthrough that will make gene therapy a safe and more common treatment for many genetic conditions?
10. How do ZFNs work?
11. Describe two gene-silencing techniques, and explain how they may be used for gene therapy.

Discussion Questions

1. Discuss the challenges scientists face in making gene therapy a safe, reliable, and effective technique for treating human disease conditions.
2. Who should be treated by gene therapy? What criteria are used to determine if a person is a candidate for gene therapy? Should gene therapy be used for cosmetic purposes or to improve athletic performance?
3. The lifetime costs for treatment of conditions such as hemophilia A and sickle-cell disease can be several million dollars. Many immunotherapies and Glybera (see Box 2) treatment cost over $1 million per patient. What is the appropriate way to determine the price for a gene therapy treatment? Who should pay? The patient? Insurance?
4. Should CRISPR-Cas or other techniques be used for editing human germ cells and/or embryos? What safety concerns might need to be addressed to support genome editing in humans?
5. Describe future challenges and ethical issues associated with gene therapy.

Advances in Neurogenetics: The Study of Huntington Disease

As the result of groundbreaking advances in molecular genetics and genomics made since the 1970s, new fields in genetics and related disciplines have emerged. One new field is **neurogenetics**—the study of the genetic basis of normal and abnormal functioning of the nervous system, with emphasis on brain functions. Research in this field includes the genes associated with neurodegenerative disorders, with the ultimate goal of developing effective therapies to combat these devastating conditions. Of the many such diseases, including Alzheimer disease, Parkinson disease, and amyotrophic lateral sclerosis (ALS), **Huntington disease (HD)** stands out as a model for the genetic investigation of neurodegenerative disorders. Not only is it monogenic and 100 percent penetrant, but nearly all analytical approaches in molecular genetics have been successfully applied to the study of HD, validating its significance as a model for these diseases.

HD is an autosomal dominant disorder characterized by adult onset of defined and progressive behavioral changes, including uncontrolled movements (chorea), cognitive decline, and psychiatric disturbances, with death occurring within 10 to 15 years after symptoms appear. HD was one of the first examples of complete dominance in human inheritance, with no differences in phenotypes between homozygotes and heterozygotes. In the vast majority of cases, symptoms do not develop until about age 45. Overall, HD currently affects about 25,000 to 30,000 people in North America.

"Driving with my father through a wooded road leading from Easthampton to Amagansett, we suddenly came upon two women, mother and daughter, both bowing, twisting, grimacing. I stared in wonderment, almost in fear. What could it mean?"

The disease is named after George Huntington, a nineteenth-century physician. He was not the first to describe the disorder, but his account was so comprehensive and detailed (see Box 1) that the disease eventually took on his name. Further, his observation of transgenerational cases in several families precisely matched an autosomal dominant pattern of inheritance. Shortly after the rediscovery of Mendel's work in the early twentieth century, pedigree analysis confirmed that HD is inherited as an autosomal dominant disorder.

We will begin our consideration of Huntington disease by discussing the successful efforts to map, isolate, and clone the HD gene. We will then turn our attention to what we know about the molecular and cellular mechanisms associated with the disorder, particularly those discovered during the study of transgenic model systems. Finally, we will consider how this information is being used to develop a range of therapies.

ST 4.1 The Search for the Huntington Gene

Mapping the gene for Huntington disease was one of the first attempts to employ a method from a landmark 1980 paper by Botstein, White, and Davis in which the authors proposed that DNA sequence variations in humans could be detected as differences in the length of DNA fragments produced by cutting DNA with restriction enzymes. These differences, known as restriction fragment length polymorphisms (RFLPs), could be visualized using Southern blots (see Special Topics Chapter 2—Genetic Testing for a discussion of RFLPs, and Chapter 17 for a discussion of Southern blots). The authors estimated that a collection of about 150 RFLPs distributed across the genome could be used with pedigrees to detect linkage anywhere in the genome between an RFLP marker and a disease gene of interest. In practical terms, this meant that it would be possible to map a disease gene with no information about the gene, its gene product, or its function—an approach referred to as reverse genetics.

Finding Linkage between Huntington Disease and an RFLP Marker

In the early 1980s, Huntington disease research was largely driven by the Hereditary Disease Foundation, established by the family of Leonore Wexler, who, along with her three brothers, died of Huntington disease. One daughter, Nancy, after learning about the proposal to map disease genes using DNA markers, used her awareness of a large population affected with Huntington disease in Venezuela to organize

BOX 1

George Huntington and His Namesake Disease

George Huntington first encountered the disease that would later bear his name as an 8-year-old boy riding in a horse-drawn carriage with his father, a local physician from Long Island, New York:

Driving with my father through a wooded road leading from Easthampton to Amagansett, we suddenly came upon two women, mother and daughter, both bowing, twisting, grimacing. I stared in wonderment, almost in fear. What could it mean? My father paused to speak with them, and we passed on. . . . From this point, my interest in the disease has never wholly ceased.

Later in 1872 as a physician at age 22, he published a paper providing a definitive description of this disorder, which at the time was known to affect the nervous system, causing uncontrollable twitches and limb movements called *chorea* (a word that means "dance" in ancient Greek). After his description, this disorder came to be known as Huntington's chorea and is now called Huntington disease. From thorough observations of patients and their families, Huntington arrived at several important conclusions:

1. The disorder is hereditary. Huntington accurately described a pattern of autosomal dominant inheritance, several decades before Mendel's paper was brought to wider attention.
2. Progressive cognitive deficits and dementia are an important part of the disease.
3. The disease has an adult onset and is incurable.

Unfortunately, what Huntington concluded about this disease in the nineteenth century remains true today. In spite of the many advances made since the gene for Huntington disease was identified in 1993, there is no treatment yet available to slow or reverse the inevitable and relentless progression of this terminal disease.

trips to collect pedigree information and to obtain blood samples for DNA linkage studies.

About the same time Nancy Wexler began working on the Venezuelan pedigree, James Gusella began collecting RFLP markers to map the gene for Huntington disease. One of the RFLP markers developed in Gusella's lab, called G8, identified four possible patterns of DNA fragments when human DNA was cut with the restriction enzyme *Hin*dIII. These patterns, called haplotypes, were named A, B, C, and D. Using this marker and DNA from a large Venezuelan HD pedigree provided by Nancy Wexler, Gusella's team concluded there was linkage between the gene for Huntington disease and haplotype C (**Figure ST 4.1**). For confirmation, they sent their results to an expert in linkage analysis,

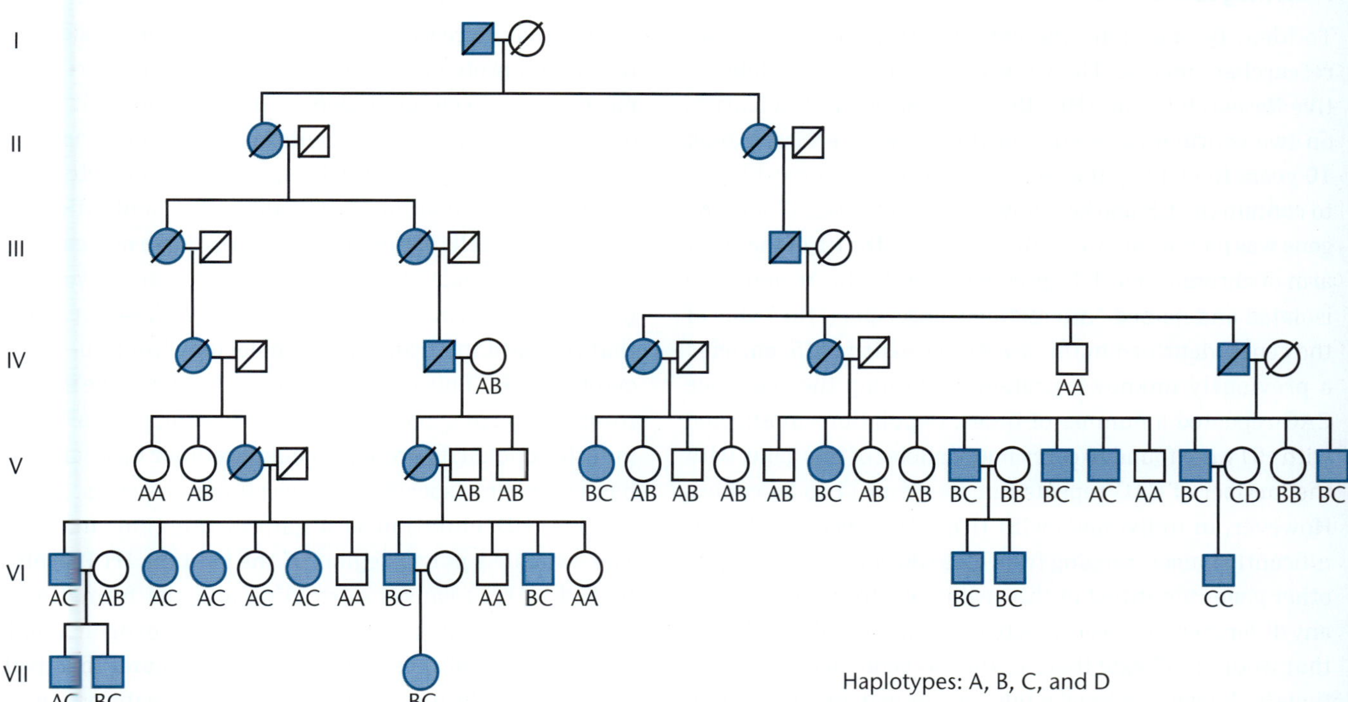

FIGURE ST 4.1 A part of the Venezuelan pedigree used in the search for linkage between RFLP markers and Huntington disease. Filled symbols indicate affected individuals. Deceased individuals are marked by diagonal slashes. In this pedigree, haplotype C of the G8 marker is coinherited with HD in all cases, indicating that the RFLP marker and the mutant HD allele are on the same chromosome.

who verified that the evidence for linkage between HD and haplotype C was overwhelming and that Gusella and his colleagues had discovered linkage to HD.

Once the G8 probe was linked to the HD gene, the next task was to determine which human chromosome carried the G8 marker and the gene for Huntington disease.

Assigning the HD Gene to Chromosome 4

A collection of somatic cell hybrids (see Chapter 14 for a description of this method) can be used to map DNA markers or genes to specific chromosomes. In this case, a panel of 18 mouse–human somatic cell hybrid cell lines, each of which contained a unique combination of human chromosomes, was used for mapping the G8 probe. On Southern blots from these hybrid cells digested with *Hin*dIII, G8 fragments were seen in all cells carrying human chromosome 4 and never seen when chromosome 4 was absent.

These results established that the G8 marker and the gene for Huntington disease were both on chromosome 4. This was the first time that an RFLP marker was used to map an autosomal disease gene to a specific chromosome. This discovery launched a whole new branch of genetics, called *positional cloning* (sometimes called reverse genetics).

The Identification and Cloning of the Huntington Gene

To identify and clone the gene for Huntington disease, researchers formed The Huntington's Disease Collaborative Research Group (HDCRG), consisting of 58 scientists on two continents. In spite of this massive effort, it took 10 years to identify the gene. First, the region most likely to contain the G8 marker (now renamed as *D4S10*) and the gene was narrowed to a small region near the tip of the short arm of chromosome 4. Expressed genes in this region were isolated and named "interesting transcripts (ITs)." One of the genes identified in this screening, called *IT15*, encoded a previously unknown protein containing the sequence CAG repeated a number of times. Populations unaffected with HD were found to carry many alleles of this gene, with the number of CAG repeats ranging from 11 to 34 copies. However, in individuals with HD, CAG repeats were significantly longer, ranging from 42 to 66 copies. None of the other genes identified in this region of chromosome 4 had any differences between affected and unaffected individuals that would implicate them as the HD gene. Because variations in the size of trinucleotide repeats had previously been identified as the causes of myotonic dystrophy and fragile-X syndrome, researchers proposed that variation in the number of CAG repeats was the cause of HD and that *IT15* encoded the HD gene.

A paper authored by all 58 members of the HDCRG ended the decade-long search for the gene, now called *HTT*, and its encoded protein, which they named huntingtin. Subsequent analysis of CAG repeat lengths in populations of unaffected and affected individuals clarified the relationship between repeat length and the onset of HD.

ST 4.2 The *HTT* Gene and Its Protein Product

Huntington disease is caused by the expansion of a CAG repeat and is one of 14 known trinucleotide repeat disorders. Nine of these, including HD, are caused by the expansion of CAG repeats, each of which codes for the insertion of the amino acid glutamine in the protein product. Thus, these genetic conditions are known as polyglutamine or polyQ disorders (Q is the one-letter abbreviation for glutamine). In addition to carrying mutant alleles with expansion of CAG repeats, other polyQ disorders have many symptoms in common with HD, including adult onset, behavioral changes, neurodegeneration, and premature death.

The *HTT* gene encodes a large protein that is 348 to 350 kDa in size. In normal alleles, a region near the 5′-end of the gene contains 6 to 35 CAG repeats encoding a stretch of glutamines in the protein product. Disease-causing mutant alleles contain an expanded number of CAG repeats (>36) that increase the number of glutamine residues in the mutant protein. The normal HTT protein is expressed in most, if not all, cells of the body and is associated with many different cellular compartments and organelles, including the plasma membrane, nucleus, cytoskeleton, cytoplasm, endoplasmic reticulum, Golgi complexes, and mitochondria. In brain cells of the striatum and caudate nucleus, HTT is present at synapses. The HTT protein domains involved in protein–protein interactions are similar to those in proteins that regulate the transport of molecules from nucleus to cytoplasm. HTT also has a role in facilitating nerve impulse transmission at synaptic junctions. In sum, normal HTT is multifunctional, is present in many cellular locations, and has a number of specific roles in cellular processes.

The HD mutation is a gain-of-function mutation. The extended polyQ region of the mutant HTT protein (called mHTT) causes misfolding and the formation of protein aggregates held together by hydrogen bonds. PolyQ regions of these aggregates bind to and inactivate regulatory molecules (**Figure ST 4.2**), disrupting a number of cellular functions, causing neurodegeneration in the striatum and caudate nucleus. In addition, toxic peptide fragments generated by breakdown of aggregates are transported into the nucleus where they accumulate

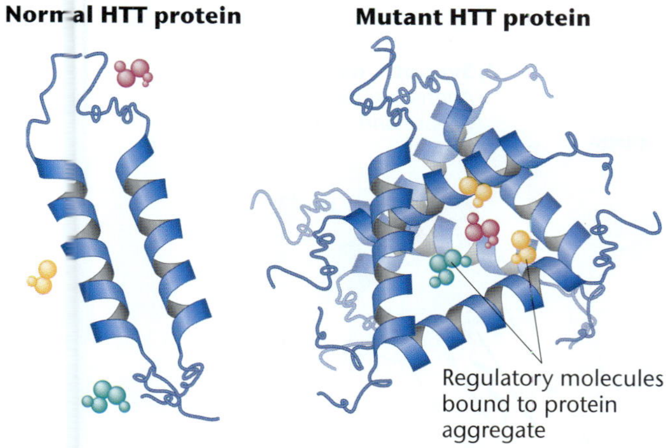

FIGURE ST 4.2 In HD, misfolded mHTT proteins clump together to form aggregates that disrupt cellular functions, partly by binding and sequestering regulatory molecules essential for normal cellular tasks.

and disrupt transcription and nucleocytoplasmic transport. The net result of these cellular changes is a gradually increasing degradation of cellular function that culminates in degeneration and death of nerve cells in specific brain regions.

ST 4.3 Molecular and Cellular Alterations in Huntington Disease

Although HD is caused by the mutation of a single gene, which was isolated and characterized over 25 years ago, the mechanisms by which mutant forms of the HTT protein (mHTT) cause HD are still largely unknown. In spite of decades of research employing a wide number of techniques and model systems, the full range of functions carried out by the normal HTT protein and those of the mHTT protein have not been completely elucidated. The straightforward view is that the mutant allele encodes a toxic protein that causes cell death initially in the striatal region of the brain. Increasing loss of cells in the striatum and other regions results in progressive and degenerative changes in muscle coordination and behavior. Death usually occurs 10 to 15 years after symptoms appear.

Unraveling the functions of the normal and mutant versions of HTT is proving to be extremely difficult because HTT interacts with more than 180 different proteins. Protein–protein network maps indicate that network proteins are involved in many cellular processes including transcription, protein folding and degradation, synaptic transmission, and mitochondrial function. In the following four sections, we will review the malfunction of each of the processes in the presence of mHTT. **Figure ST 4.3** supports each brief discussion.

Transcriptional Disruption

The effects of mHTT on the transcriptome are one of the key molecular events in HD. Expression of mHTT causes the formation of heterochromatin, effectively closing off transcription of genes located in affected chromosome regions. In addition, smaller soluble mHTT fragments interact with molecular components of transcription and obstruct the functions of the promoter-binding transcription factors and the proteins necessary for transcription initiation. The net result is reduced promoter accessibility and transcription initiation across the genome.

Impaired Protein Folding and Degradation

The correct folding of proteins depends on the action of proteins called chaperones that mediate folding. In HD, transcriptional disruption inactivates several families of chaperones, which leads to the accumulation of incorrectly folded proteins in the cytoplasm. The result is disruption of normal folding and a slowdown in degradation of misfolded proteins.

In normal cells, misfolded proteins are tagged by a small protein (ubiquitin) and directed to a cellular structure called the proteasome (see Chapter 13 for a discussion of protein misfolding and proteasomes) where they are degraded. In brain cells of individuals with HD, ubiquitin binds to aggregated mHTT in the cytoplasm, which is then targeted for degradation, but the proteasome system is inhibited by an unknown mechanism. As a result, mHTT aggregates accumulate and impair cellular functions, triggering apoptosis and cell death. Together, the inhibition of chaperone function and proteasome function causes a collapse of normal protein function and turnover in brain cells of individuals with HD.

Synaptic Dysfunction

In individuals with HD, subtle changes in motor function resulting from synaptic dysfunction can appear decades before the onset of neuronal death. To investigate synaptic defects, researchers constructed transgenic *Drosophila* strains carrying a mutant human HD allele. In transgenic flies, the synaptic vesicles carrying neurotransmitters were much smaller than normal. As a result, synaptic transmission was disrupted, causing behavioral changes in locomotion.

Impaired Mitochondrial Function

mHTT binds to the outer mitochondrial membrane and impairs electron transport, reducing the amount of ATP available to the cell. Disruption of the electron transport

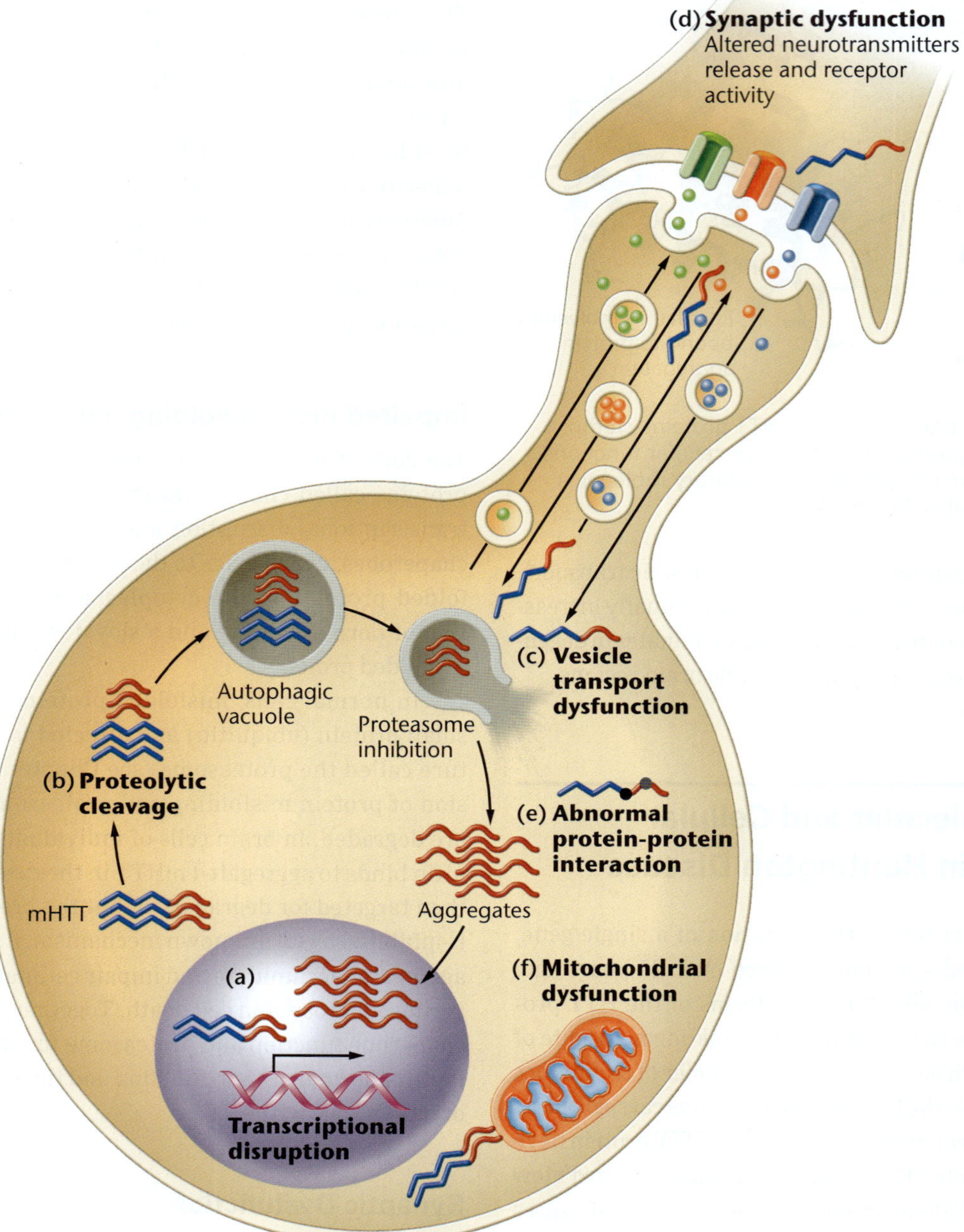

FIGURE ST 4.3 The major pathways of cellular functions disrupted in HD. The letters (a-f) within the neuron correspond to the disrupted processes within the nucleus and cytoplasm (a) Transcriptional disruption occurs by blocking access of transcription factors and interfering with histone acetylation. (b) Impaired protein degradation by cleavage is caused by blocking the loading of ubiquitin-tagged protein aggregates into vesicles called autophagosomes and by hindering transport to proteasomes. (c and d) Altered synaptic function results from blockage of synaptic vesicle transport by mHTT. (e) Abnormal protein-protein interactions prevent refolding and impair degradation of improperly folded proteins (f) Disruption of mitochondrial function by binding of mHTT to the outer membrane triggers a cascade of problems with function, transport, biogenesis, and maintenance.

chain also increases the levels of reactive oxygen species including free radicals, which cause widespread oxidative damage to cellular structures.

Within neurons, mitochondria migrate to synapses when rates of nerve impulse transmissions increase. Mitochondrial movement is inhibited by aggregation of N-terminal mHTT fragments that physically block migration along microtubules. This reduces the energy available for transmission of nerve impulses at synapses.

In sum, the damage to mitochondria caused by expression of mHTT includes the disruption of ATP production, promotion of oxidative damage within mitochondria and the cytoplasm, lowered synaptic transmission, and reduction of mitochondrial numbers to a level that can no longer support the core activities of cells, which eventually triggers apoptosis and cell death.

Although progress has been made in defining the defects that follow the expression and accumulation of mHTT, several important questions about the underlying mechanisms of HD remain unanswered. For example, it is not known whether any single disruption of cell function is sufficient to cause neurodegeneration or cell death, or whether one or more of these pathways must interact to bring about these results. To answer these and other questions, researchers turned to the use of transgenic animal models of HD.

ST 4.4 Transgenic Animal Models of Huntington Disease

Shortly after the *HTT* gene was cloned, researchers constructed transgenic model organisms to analyze the disease process at the molecular level.

Animal models of human behavioral disorders present an opportunity to separate behavioral phenotypes into their components. This makes it possible to study the developmental, structural, and functional neuronal mechanisms related to these behaviors that are difficult or impossible to do in humans. In addition, behavior in animal models can be studied in controlled conditions that limit the impact of environmental factors. Although it is possible to construct transgenic models of human HD in a wide range of organisms, including yeast, *Caenorhabditis elegans*, and *Drosophila melanogaster*, the mouse is the most widely used model organism for these studies. Researchers favor mice because humans and mice share about 90 percent of their genes and because a wide range of strains with specific behavioral phenotypes are available.

Using Transgenic Mice to Study Huntington Disease

The first mouse model of HD was constructed using the promoter sequence and first exon of the human mutant *HTT* allele, which contains an expanded CAG repeat. Examination of transgenic mouse brains a year after gene transfer showed abnormalities in the levels of neurotransmitter receptors and the presence of protein aggregates, a significant finding that was later confirmed to exist in the brains of humans with HD.

Soon after, researchers began to examine the relationship between CAG repeat length and disease progression. Transgenic mice carrying human HD genes with 16, 48, or 89 CAG repeats were monitored from birth to death to determine the age of onset and stages of abnormal behavior. Mice carrying 48- or 89-repeat human *HD* genes showed behavioral abnormalities as early as 8 weeks, and by 20 weeks they showed problems with motor coordination.

At various ages, brains of wild-type and transgenic mice carrying these mutant alleles were examined for changes in structure. Degenerating neurons and cell loss were evident in mice carrying 48 and 89 repeats, but no changes were seen in brains of wild-type mice or of those carrying a 16-repeat transgene (**Figure ST 4.4**).

There are now more than 20 different mouse models of HD, which are used to examine changes in molecular and cellular processes or in brain structure that occur before or just after the onset of symptoms and to develop experimental treatments to slow or reverse cell loss. Transgenic models for HD allow researchers to administer treatment at specific times in disease progression and to evaluate the outcome of treatments in the presymptomatic stages of HD, something that is not possible in humans with HD.

In one important study, A. Yamamoto and colleagues constructed a transgenic mouse with inducible expression of the *HTT* gene. In this case, researchers constructed a human *HTT* exon 1 fragment containing 94 CAG repeats with an adjacent promoter that could be switched off when the antibiotic doxycycline was added to the drinking water. When the gene was switched off shortly after motor symptoms of HD developed, protein aggregates in the brain were rapidly degraded and disappeared, along with the abnormal motor symptoms. This provided the first clue that treatment in the early stages of the disease might be effective in controlling or reversing the symptoms in humans.

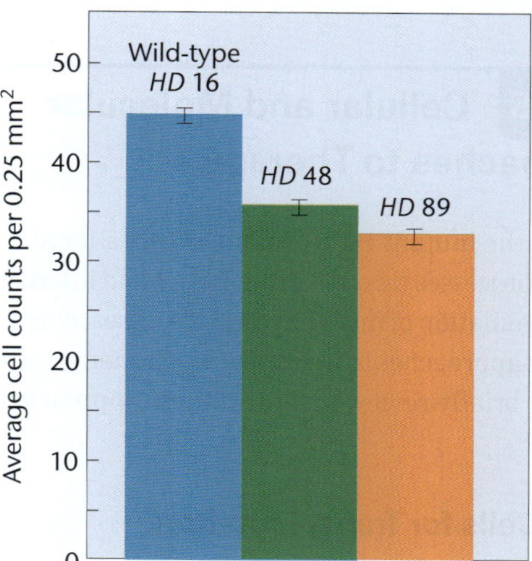

FIGURE ST 4.4 Relative levels of neuronal loss in HD transgenic mice. Cell counts show a significant reduction of neurons in the striatum in the brains of *HD* 48 mutants (middle column) and *HD* 89 mutants (right column). Cell loss in this brain region is also found in humans with HD, making these transgenic mice valuable models to study the course of this disease.

Transgenic Large-Animal Models of Huntington Disease

Despite the important discoveries made in mice, the main problem with transgenic mouse models of HD is that mice do not show the same pattern of neurodegeneration seen in humans. In addition, mice have a smaller brain, a shorter life span (two to three years), and different physiology than humans. To overcome some of these problems, large-animal models, including sheep, mini-pigs, and a number of nonhuman primates are being developed to study the mechanisms of disease and the testing of drugs for human therapies.

Transgenic sheep were one of the first large-animal models developed to study HD. One transgenic sheep model carries a full-length human *HTT* gene with 73 CAG repeats. Given the relatively long life span of sheep (>10 years), this model is being used to study the long-term development of HD. In addition, because the size and structure of sheep brains are similar to those of humans, it is possible to do MRI and PET scans that can be directly compared to the results of scans in humans affected with HD.

A recently developed pig model carrying a human HD gene with an expanded set of CAG repeats shows a pattern of neurodegeneration that mirrors the brain changes seen in affected humans. These pigs also develop the characteristic abnormal movements seen in humans with HD. Researchers who developed this model anticipate that it will be valuable in testing treatments for HD. Continued development of large-animal models will hopefully lead to a more detailed understanding of the mechanism of HD and the development of effective therapies.

ST 4.5 Cellular and Molecular Approaches to Therapy

Because the mutant HTT protein affects a large number of cellular processes through protein–protein interactions and the accumulation of mHTT aggregates, researchers are using multiple approaches to investigate treatment strategies. We will now briefly review several of these approaches.

Stem Cells for Transplantation

Stem cells are undifferentiated somatic cells with two properties: (a) the ability to renew their numbers by mitosis, and (b) the ability to differentiate and form tissue-specific specialized cell types. Research on HD uses human stem cells for studies of HD mechanisms, drug screening, drug testing, genetic correction of mHTT accumulation, and as donor cells for transplantation.

HD is associated with transcriptional repression of many gene sets, including the gene that encodes brain-derived neurotrophic factor (BDNF), which is essential for the survival and function of cells in the striatum. Loss of this factor contributes to the death of striatal brain cells in HD. Human mesenchymal stem cells (MSCs) can be genetically programmed to overexpress BDNF. Injection of human MSCs modified to overexpress BDNF into the brains of a transgenic mouse strain carrying a full-length human *HTT* gene with 128 CAG repeats significantly increased the production of new nerve cells, increased life span, and reduced HD-associated behaviors. Further development of MSCs as a delivery system in other mouse models and large-animal models will be required before the system is ready for human clinical trials, which are expected to begin as soon as additional animal studies are completed.

Gene Silencing to Reduce mHTT Levels

Because mHTT initiates a chain of events that lead to cell death and the onset of HD symptoms, therapies that reduce or eliminate the expression or accumulation of mHTT, either alone or in combination with other therapies, might be especially effective in treating HD.

A therapy using synthetic zinc-finger nucleases (ZFNs) to repress transcription of m*HTT* alleles (see Special Topic Chapter 3—Gene Therapy for a discussion of ZFNs) in mouse neuronal cell lines and a cell line from an individual with HD carrying a mutant allele with 45 CAG repeats showed that ZFN treatment significantly repressed m*HTT* expression with no effects on the expression of other genes containing CAG repeats.

The ZFN construct was then tested on a transgenic mouse strain carrying 115 to 160 CAG repeats by injection into the striatal region of one side of the brain. A control vector was injected into the other side of the brain. Two weeks later, levels of mRNA from the normal and mutant alleles were assayed in tissue from each side of the brain. Overall, on the ZFN-injected side, there was a 40 to 60 percent reduction of m*HTT* mRNA compared with the control side, with no effect on expression of the normal allele on either side (**Figure ST 4.5**). Mice injected on both sides of the brain with the ZFN construct showed no differences in behavioral tests compared with normal mice. This was an important proof of principle that ZFN repression of m*HTT* transcription reduces m*HTT* mRNA and protein levels and may be an effective therapy for HD.

Instead of inhibiting transcription, gene-silencing techniques can be used to intervene in gene expression after transcription but before translation takes place. Two widely used methods of gene silencing use antisense oligonucleotides (ASOs) and RNA interference (RNAi). ASOs are short single-stranded DNAs (8 to 50 nucleotides long) that bind to target mRNAs by complementary base pairing

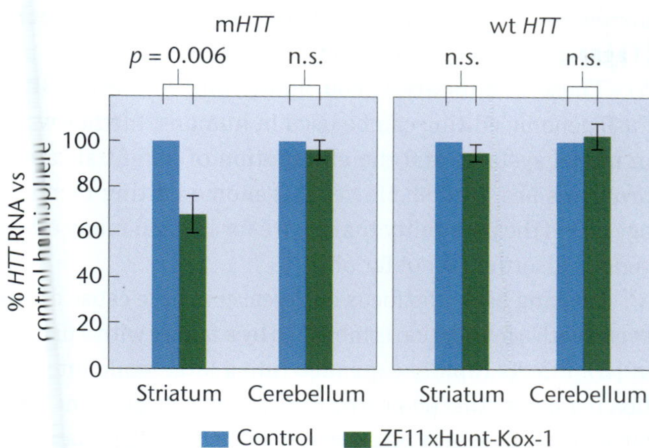

FIGURE ST 4.5 Zinc-finger nuclease repression of m*HTT* expression in transgenic mice. Left panel: Expression of m*HTT* is repressed by ~40 percent by injection of the ZFN construct into the striatum but has no effect on expression in the noninjected cerebellum. The *p* value of 0.006 indicates there is a significant difference between values in the striatum. The abbreviation n.s. means there is no significant difference in the values shown. Right panel: The expression of the normal *HTT* allele is unaffected in both brain regions.

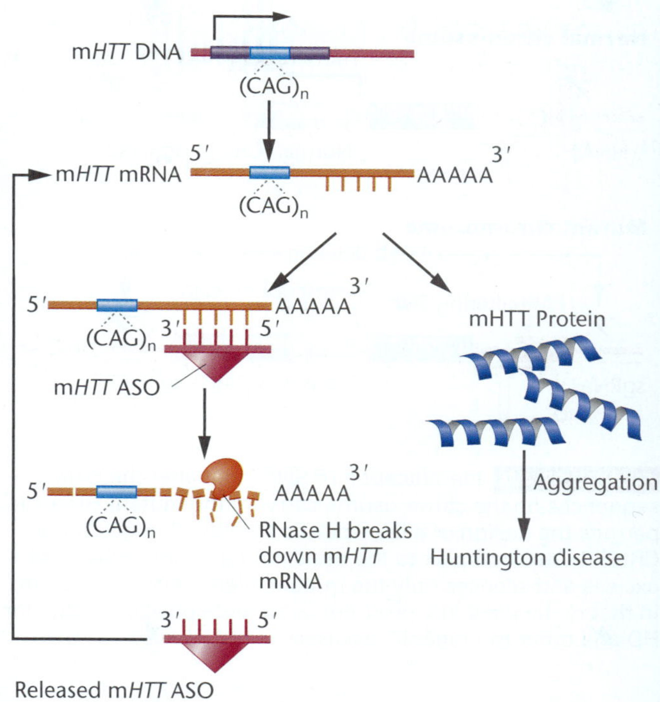

FIGURE ST 4.6 An ASO constructed to bind to m*HTT* mRNA forms a DNA–RNA hybrid, which then attracts RNase to degrade the m*HTT* mRNA, inhibiting translation of the HTT protein and halting disease progression. The released ASO is free to bind other m*HTT* mRNAs and continue the cycle of degradation.

(**Figure ST 4.6**). The mRNA strand of the resulting DNA–RNA hybrid is degraded by RNase H, a cytoplasmic enzyme. The intact ASO is released and can bind other copies of the target mRNA, marking them for degradation by RNase H.

A ground-breaking study by Lu and Yang transfused an ASO complementary to a human m*HTT* allele into the cerebrospinal fluid of transgenic mice carrying a full-length human m*HTT* allele with 97 CAG repeats. m*HTT* RNA and mutant HTT protein levels were selectively reduced for up to 12 weeks, with a rebound to near normal levels by 16 weeks. With two weeks of continuous infusion at the age of six months, treated mice still showed improved motor coordination and behavior nine months after infusion and five months after m*HTT* mRNA and protein levels rebounded to pretreatment levels. Similar ASO studies in nonhuman primates showed a 25 to 68 percent reduction in levels of *HTT* mRNA in brain regions involved in HD, with no adverse effects.

Human Phase I clinical trials on ASO therapy for HD began in 2015 in Canada and Europe. The drug is being infused into the cerebrospinal fluid, which surrounds and bathes the brain. This trial is designed to evaluate the safety of ASO infusion. If this and subsequent clinical trials are successful, the first treatment to directly target the cause of HD will soon be available for the thousands of people affected with this devastating disease.

Genome Editing in Huntington Disease

Over the last decade, there has been rapid development of methods for genome editing. While details of genome editing techniques differ, conceptually they all work in a similar way. A nuclease is guided to cut a specific DNA sequence, which then allows for the replacement or deletion of all or part of a given gene. Because of its specificity and ease of use, the CRISPR-Cas9 system (see Special Topic Chapter 3—Gene Therapy for a discussion of genome editing) is the most widely used method. The enzyme Cas9 can be directed to cut DNA at specific sequences at nearly any location in the genome by means of a single-stranded guide RNA (sgRNA). The guide RNA is complementary to the sequence to be cut and directs cutting by the Cas9 nuclease. Since HD is a dominant disorder, in theory, this disorder can be treated by using this technology to edit and silence the mutant allele.

Using CRISPR/Cas9 technology and human cells carrying a mutant and a normal *HTT* allele, Jong-Min Lee and colleagues edited the disease-associated mutant *HTT* allele, (**Figure ST 4.7**) removing a 44-kb DNA fragment, completely inactivating the mutant allele, while leaving the normal allele intact. This experiment shows that inactivation of a disease gene can be individually tailored to edit a mutant allele carried by an affected individual and, in theory, can be used to inactivate disease alleles of any gene by editing.

A research team led by Xiao-Jiang Li used CRISPR genome editing to reverse symptoms of HD in transgenic mice carrying exon 1 of a mutant human *HTT* allele containing 140 CAG repeats. They injected one vector carrying Cas9 into the striatum along with another vector that carried an

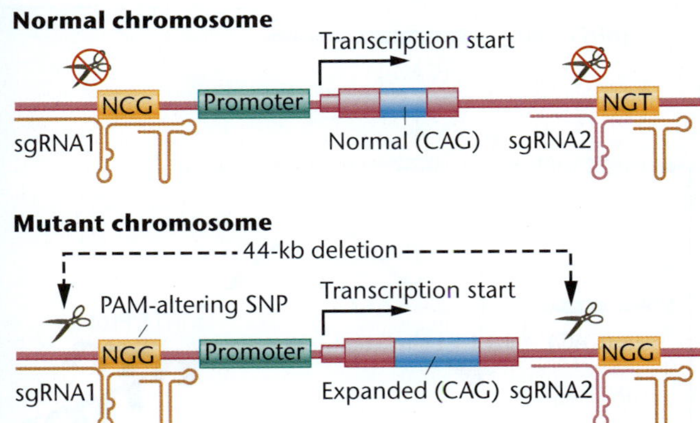

FIGURE ST 4.7 Identification of SNPs that alter the PAM sequences on the chromosome carrying the mutant HD allele permits the design of allele-specific sgRNAs that guide the CRISPR-Cas9 complex to the mutant locus. Cutting by Cas9 excises and silences only the mutant allele. This method can, in theory, be used in a program of patient-specific therapy for HD and other monogenic diseases.

sgRNA to direct Cas9 to target the mutant *HTT* allele to edit out the expanded CAG repeat region. Three weeks after injection, analysis of striatal cells showed that production of the mutant human HTT protein was suppressed and the number of aggregated protein clumps was reduced.

In subsequent experiments, sgRNAs and Cas9 were injected into 9-month-old mice. Over the next three months, these mice showed improvements in motor skills including balance, mobility, and muscle coordination. In addition, increases in motor skills were correlated with the amount of aggregated protein cleared from striatal cells.

These results are encouraging, but before CRISPR-Cas9 genome editing can be used in humans, further work in model systems and the elimination of potential safety problems are needed. However, genome-editing technology offers the possibility that a cure for HD and many other genetic disorders is not far off.

Looking back, the focus on researching the cause of HD began with a foundation established by a family with a history of HD. Efforts rapidly expanded into a large-scale international program that pioneered the use of genetics and molecular genetics to identify an expanded stretch of polyglutamines in the huntingtin protein as the cause of this disorder. Along the way, researchers developed an integrative strategy that combined old and new methods including pedigree analysis, RFLP markers, somatic cell genetics, Southern blots, new cloning vectors, predictive genetic testing, population genetics, and other methods that are now universally used in genetic research. Nevertheless, in spite of the progress made in more than 30 years of work, there are still no therapies that can halt, reverse, or prevent the onset and progression of this devastating neurogenetic disorder. Looking forward, recent results from the combined use of human cells, animal models, genome editing, and clinical trials suggest that this last barrier may fall in the near future. If this is the case, the approach used in HD research will stand as a paradigm for understanding the structure and function of the brain and the development of therapeutic methods for a range of genetic disorders.

Review Questions

1. What are RFLP markers and how were they used to identify which chromosome carries the gene for Huntington disease?
2. Why was information from Nancy Wexler's large pedigrees necessary in locating the *HTT* gene?
3. What holds mHTT aggregates together, and how do these aggregates disrupt cellular function?
4. Why are the results from the inducible mouse model of HD so important?
5. Based on the results from mouse models, is it necessary to have the whole mutant protein (mHTT) present to generate protein aggregates and death of brain cells?
6. What do the results from creating transgenic mice carrying only exon 1 of the mutant allele tell us about the disease?
7. How does binding of mHTT protein to mitochondria disrupt mitochondrial function?
8. Summarize the approaches to therapy designed to reduce the level of the mHTT protein in cells. How can this therapy be delivered to target cells?

Discussion Questions

1. There are nine known progressive neurodegenerative disorders that all share expanded numbers of the CAG codon, which inserts extra glutamine residues into the coding regions of specific genes. Genes carrying such mutations are typically gain-of-function mutations and often share a common mechanism of pathogenesis. Why would such genes be gain-of-function? Speculate on why such diseases may be caused by a common mechanism.
2. Most of the research efforts on Huntington disease have focused on the mechanisms by which the mutant form of the HTT protein causes cell death or the clinical symptoms of this disorder. For the development of effective therapies, how important is it to understand the functions of the normal HTT protein?
3. If protein aggregates can spread from cell to cell, would the use of stem cell transplants be an effective therapy?
4. Why is there an inverse correlation between the number of CAG repeats in a mutant allele and the onset of symptoms?
5. Discuss the ethical issues raised by the use of a diagnostic test for those at risk for HD.

DNA Forensics

Forensic science (or *forensics*) uses technological and scientific approaches to answer questions about the facts of criminal or civil cases. Prior to 1986, forensic scientists had a limited array of tools with which to link evidence to specific individuals or suspects. These included some reliable methods such as blood typing and fingerprint analysis, but also many unreliable methods such as bite mark comparisons and hair microscopy.

Since the first forensic use of **DNA profiling** in 1986 (Box 1), **DNA forensics** (also called **forensic DNA fingerprinting** or **DNA typing**) has become an important method for police to identify sources of biological materials. DNA profiles can now be obtained from saliva left on cigarette butts or postage stamps, pet hairs found at crime scenes, or bloodspots the size of pinheads. Even biological samples that are degraded by fire or time are yielding DNA profiles that help the legal system determine identity, innocence, or guilt. Investigators now scan large databases of stored DNA profiles in order to match profiles generated from crime scene evidence. DNA profiling has proven the innocence of people who were convicted of serious crimes and even sentenced to death. Forensic scientists have used DNA profiling to identify victims of mass disasters such as the Asian Tsunami of 2004 and the September 11, 2001, terrorist attacks in New York. They have also used forensic DNA analysis to identify endangered species and animals trafficked in the illegal wildlife trade.

The applications of DNA profiling extend beyond forensic investigations. These include paternity and family relationship testing, identification of plant materials, verification of military casualties, and evolutionary studies.

In this Special Topics chapter, we will explore how DNA profiling works and how the results of profiles are interpreted. We will learn about DNA databases, the potential problems associated with DNA profiling, and the future of this powerful technology.

> "Even biological samples degraded by fire or time are yielding DNA profiles that help determine identity, innocence, or guilt."

ST 5.1 DNA Profiling Methods

VNTR-Based DNA Fingerprinting

The era of DNA-based human identification began in 1985, with Dr. Alec Jeffreys's publication on DNA loci known as **minisatellites,** or **variable number tandem repeats (VNTRs).** As described earlier in the text (see Chapter 11), VNTRs are located in noncoding regions of the genome and are made up of DNA sequences of between 15 and 100 base pairs (bp) long, with each unit repeated a number of times. The number of repeats found at each VNTR locus varies from person to person, and hence VNTRs can be from 1 to 20 kilobases (kb) in length, depending on the person. For example, the VNTR

5′-GACTGCCTGCTAAGAT**GACTGCCTGCTAAGAT**GACTGCCTGCTAAGAT-3′

is composed of three tandem repeats of a 16-nucleotide sequence (highlighted in bold).

VNTRs are useful for DNA profiling because there are as many as 30 different possible alleles (repeat lengths) at any VNTR in a population. This creates a large number of possible genotypes. For example, if one examined four different VNTR loci within a population, and each locus had 20 possible alleles, there would be approximately 2 billion possible genotypes in this four-locus profile.

To create a VNTR profile (also known as a DNA fingerprint), scientists extract DNA from a tissue sample and digest it with a restriction enzyme that cleaves on either side of the VNTR repeat region (**Figure ST 5.1**). The digested DNA is separated by gel electrophoresis and subjected to Southern blot analysis (which is described in detail in Chapter 17). Briefly, separated DNA is transferred from the gel to a membrane and hybridized with a radioactive probe that recognizes DNA sequences within the VNTR region. After exposing the membrane to X-ray film, the pattern of bands is measured, with larger VNTR repeat alleles remaining near the top of the gel and smaller VNTRs, which migrate more rapidly through the gel, being closer to the bottom. The pattern of bands is the same for a given individual, no matter what tissue is used as the source of the DNA. If enough VNTRs are analyzed, each person's DNA profile will be unique (except, of course, for identical twins) because of the huge number of possible VNTRs and alleles. In practice, scientists analyze about five or six loci to create a DNA profile.

A significant limitation of VNTR profiling is that it requires a relatively large sample of DNA (10,000 cells or about 50 μg of DNA)—more than is usually found at a typical

BOX 1

The Pitchfork Case: The First Criminal Conviction Using DNA Profiling

In the mid-1980s, the bodies of two schoolgirls, Lynda Mann and Dawn Ashworth, were found in Leicestershire, England. Both girls had been raped, strangled, and their bodies left in the bushes. In the absence of useful clues, the police questioned a local intellectually disabled porter named Richard Buckland. During interrogation, Buckland confessed to the murder of Dawn Ashworth; however, police did not know whether he was also responsible for Lynda Mann's death. In 1986, in order to identify the second killer, the police asked Dr. Alec Jeffreys of the University of Leicester to analyze the crime scene evidence using a new method of DNA analysis called VNTR profiling. Dr. Jeffreys's VNTR analysis revealed a match between the DNA profiles from semen samples obtained from both crime scenes, suggesting that the same person was responsible for both rapes. However, neither of the DNA profiles matched those from a blood sample taken from Richard Buckland. Having eliminated their only suspect, the police embarked on the first mass DNA dragnet in history by requesting blood samples from every adult male in the region. Although 4000 men offered samples, one did not. Colin Pitchfork, a bakery worker, paid a friend to give a blood sample in his place, using forged identity documents. Their plan was detected when their conversation was overheard at a local pub. The conversation was reported to police, who then arrested Pitchfork, obtained his blood sample, and sent it for analysis. His DNA profile matched the profiles from the semen samples left at both crime scenes. Pitchfork confessed to the murders, pleaded guilty, and was sentenced to life in prison. The Pitchfork Case was not only the first criminal case resolved by forensic DNA profiling, but also the first case in which DNA profiling led to the exoneration of an innocent person.

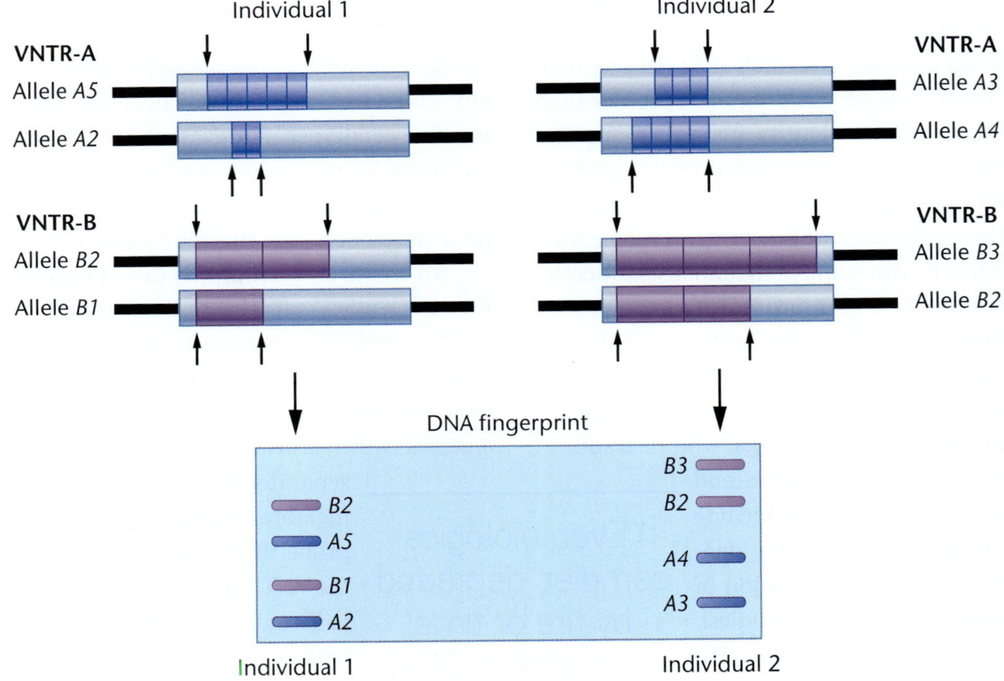

FIGURE ST 5.1 DNA fingerprint at two VNTR loci for two individuals. VNTR alleles at two loci (*A* and *B*) are shown for two different individuals. Arrows mark restriction-enzyme cutting sites that flank the VNTRs. Restriction-enzyme digestion produces a series of fragments that can be separated by gel electrophoresis and detected as bands on a Southern blot (bottom). The number of repeats at each locus is variable, so the overall pattern of bands is distinct for each individual. The DNA fingerprint profile shows that these individuals share one allele (*B2*).

crime scene. In addition, the DNA must be relatively intact (nondegraded). As a result, VNTR profiling has been used most frequently when large tissue samples are available—such as in paternity testing. Although VNTR profiling is still used in some cases, it has mostly been replaced by more sensitive methods, as described next.

Autosomal STR DNA Profiling

The development of the **polymerase chain reaction (PCR)** revolutionized DNA profiling. As described in Chapter 17, PCR is an *in vitro* method that uses specific primers and a heat-tolerant DNA polymerase to amplify specific regions of DNA. Within a few hours, this method can generate a

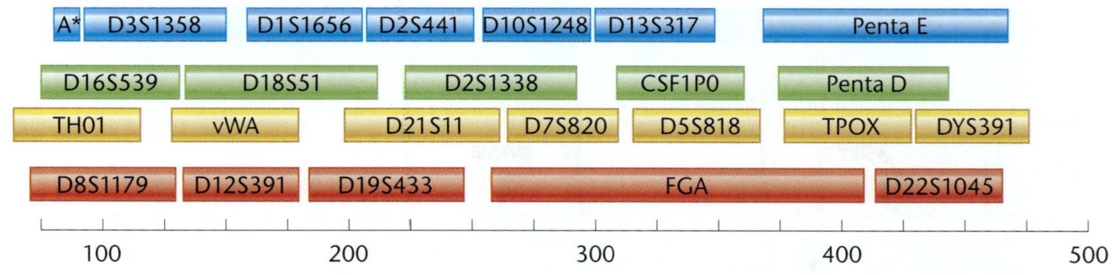

FIGURE ST 5.2 Relative size ranges and fluorescent dye labeling colors of 24 STR products generated by a commercially available DNA profiling kit. The scale at the bottom of the diagram indicates DNA fragment sizes in base pairs.

millionfold increase in the quantity of DNA within a specific sequence region. Using PCR-amplified DNA samples, scientists are able to generate DNA profiles from trace samples (e.g., the bulb of single hairs or a few cells from a bloodstain) and from samples that are old or degraded (such as a bone found in a field or an ancient Egyptian mummy).

The majority of human forensic DNA profiling is now done by amplifying and analyzing regions of the genome known as **microsatellites,** or **short tandem repeats (STRs).** STRs are similar to VNTRs, but the repeated motif is shorter—between two and nine base pairs, repeated from 7 to 40 times. For example, one locus known as D8S1179 is made up of the four base-pair sequence TCTA, repeated 7 to 20 times, depending on the allele. There are 19 possible alleles of the locus that are found within a population. Although hundreds of STR loci are present in the human genome, only a subset is used for DNA profiling. At the present time, the FBI and other U.S. law enforcement agencies use 20 STR loci as a core set for forensic analysis. Most European countries now use 12 STR loci as a core set.

Several commercially available kits are used for forensic DNA analysis of STR loci. The methods vary slightly, but generally involve the following steps. As shown in **Figure ST 5.2**, each primer set is tagged by one of four fluorescent dyes—represented here as blue, green, yellow, or red. Each primer set is designed to amplify DNA fragments, the sizes of which vary depending on the number of repeats within the region amplified. For example, the primer sets that amplify the TH01, vWA, D21S11, D7S820, D5S818, TPOX, and DYS391 STR loci are all labeled with a fluorescent tag indicated as yellow. The sizes of the amplified DNA fragments allow scientists to differentiate between the yellow-labeled products. For example, the amplified products from the D21S11 locus range from about 200 to 260 bp in length, whereas those from the TPOX locus range from about 375 to 425 bp, and so on.

After amplification, the DNA sample will contain a small amount of the original template DNA sample and a large amount of fluorescently labeled amplification products (**Figure ST 5.3**). The sizes of the amplified fragments are measured by **capillary electrophoresis.** This method uses thin glass tubes that are filled with a polyacrylamide gel material similar to that used in slab gel electrophoresis. The amplified DNA sample is loaded onto the top of the capillary tube, and an electric current is passed through the tube. The negatively charged DNA fragments migrate through the gel toward the positive electrode, according to their sizes. Short fragments move more quickly through the gel, and larger ones more slowly. At the bottom of the tube, a laser detects each fluorescent fragment as it migrates through the tube. The data are analyzed by software that calculates both the

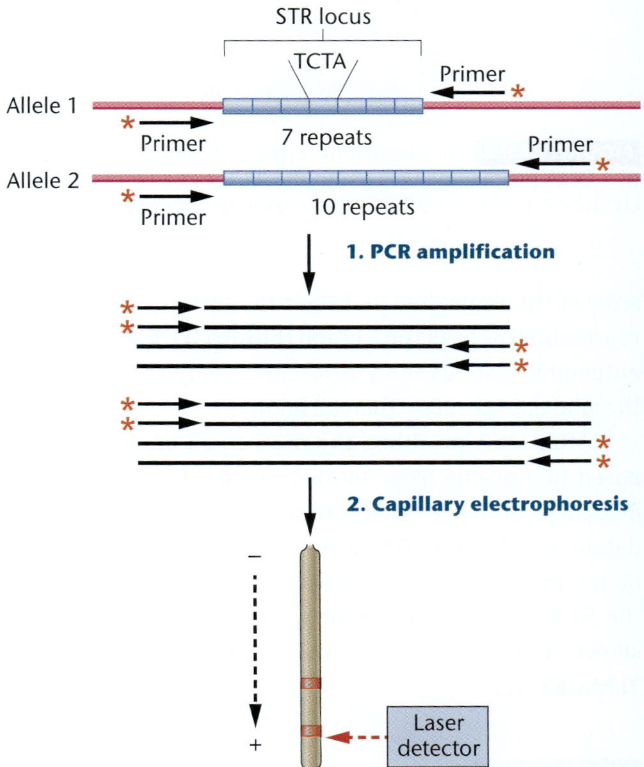

FIGURE ST 5.3 Steps in the PCR amplification and analysis of one STR locus (D8S1179). In this example, the person is heterozygous at the D8S1179 locus: One allele has 7 repeats and one has 10 repeats. Primers are specific for sequences flanking the STR locus and are labeled with an orange fluorescent dye. The double-stranded DNA is denatured, the primers are annealed, and each allele is amplified by PCR in the presence of all four dNTPs and Taq DNA polymerase. After amplification, the labeled products are separated according to size by capillary electrophoresis, followed by fluorescence detection.

SPECIAL TOPICS 5

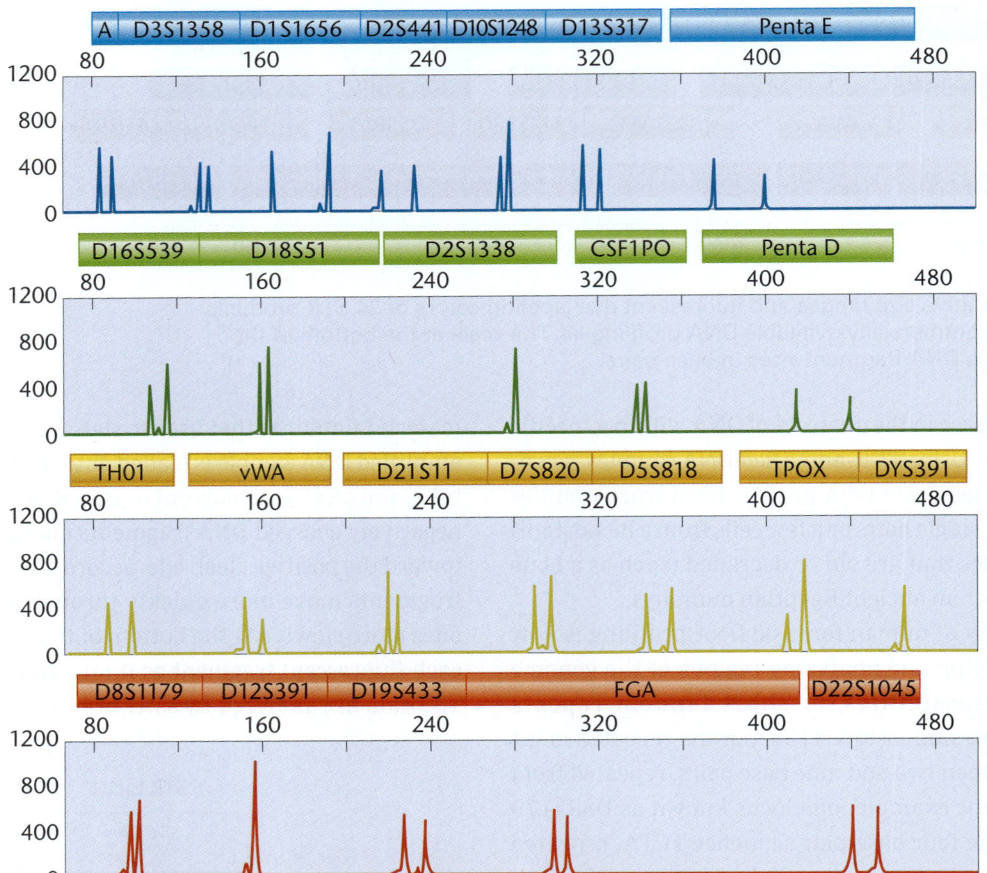

FIGURE ST 5.4 An electropherogram showing the results of a DNA profile analysis using the 24-locus STR profile kit shown in Figure ST 5.2. Heterozygous loci show up as double peaks, and homozygous loci as single, higher peaks. The sizes of each allele can be calculated from the peak locations relative to the size axis shown at the top of each panel.

sizes of the fragments and their quantities, and these are represented as peaks on a graph (**Figure ST 5.4**). Typically, automated systems analyze dozens of samples at a time, and the analysis takes less than an hour.

After DNA profiling, the profile can be directly compared to a profile from another person, from crime scene evidence, or from other profiles stored in DNA profile databases (**Figure ST 5.5**). The STR profile genotype of an individual is expressed as the number of times the STR sequence is repeated. For example, the profiles shown in Figure ST 5.5 would be expressed as shown in **Table ST 5.1.**

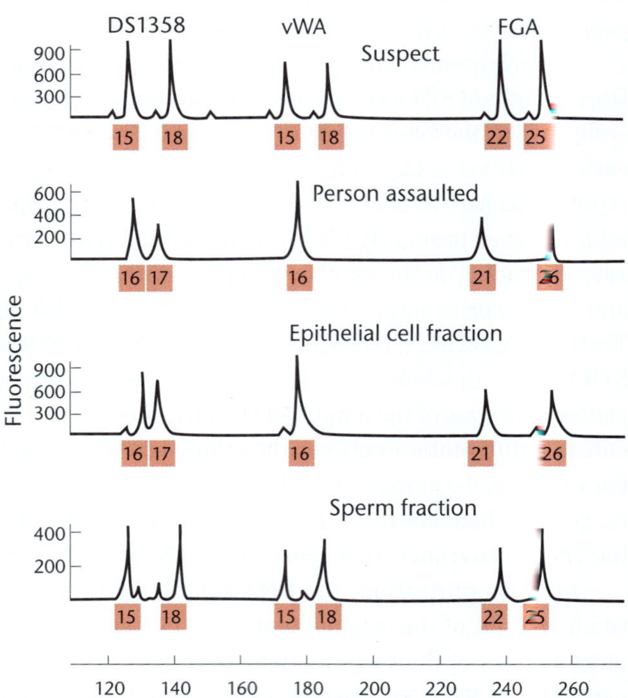

FIGURE ST 5.5 An electropherogram showing the STR profiles of four samples from a rape case. Three STR loci were examined from samples taken from a suspect (male), the person who was sexually assaulted (female), and two fractions from a vaginal swab taken from the female. The *x*-axis shows the DNA size ladder, and the *y*-axis indicates relative fluorescence intensity. The number below each allele indicates the number of repeats in each allele, as measured against the DNA size ladder. Notice that the STR profile of the sperm sample taken from the female matches that of the suspect.

TABLE ST 5.1 STR Profile Genotypes from the Four Profiles Shown in Figure ST 5.5

	Profile Genotype from			
STR Locus	Suspect	Person Assaulted	Epithelial Cells	Sperm Fraction
D3S1358	15, 18	16, 17	16, 17	15, 18
vWA	15, 18	16, 16	16, 16	15, 18
FGA	22, 25	21, 26	21, 26	22, 25

Scientists interpret STR profiles using statistics, probability, and population genetics, and these methods will be discussed in the section Interpreting DNA Profiles.

Y-Chromosome STR Profiling

In many forensic applications, it is important to differentiate the DNA profiles of two or more people in a mixed sample. For example, vaginal swabs from rape cases usually contain a mixture of female somatic cells and male sperm cells. In addition, some crime samples may contain evidence material from a number of male suspects. In these types of cases, STR profiling of Y-chromosome DNA is useful. There are more than 200 STR loci on the Y chromosome that are useful for DNA profiling; however, fewer than 20 of these are used routinely for forensic analysis. PCR amplification of Y-chromosome STRs uses specific primers that do not amplify DNA on the X chromosome.

One limitation of Y-chromosome DNA profiling is that it cannot differentiate between the DNA from fathers and sons or from male siblings. This is because the Y chromosome is directly inherited from the father to his sons, as a single unit. The Y chromosome does not undergo recombination, meaning that less genetic variability exists on the Y chromosome than on autosomal chromosomes. Therefore, all patrilineal relatives share the same Y-chromosome profile. Even two apparently unrelated males may share the same Y profile, if they also share a distant male ancestor.

Although these features of Y-chromosome profiles present limitations for some forensic applications, they are useful for identifying missing persons when a male relative's DNA is available for comparison. They also allow researchers to trace paternal lineages in genetic genealogy studies.

Mitochondrial DNA Profiling

Another important addition to DNA profiling methods is **mitochondrial DNA (mtDNA)** analysis. Between 200 and 1700 mitochondria are present in each human somatic cell. Each mitochondrion contains one or more 16-kb circular DNA chromosomes. Mitochrondria are passed from the human egg cell to the zygote during fertilization; however, as sperm cells contribute few if any mitochondria to the zygote, they do not contribute these organelles to the next generation. Therefore, all cells in an individual contain multiple copies of specific mitochondrial variants derived from the mother. Like Y-chromosome DNA, mtDNA undergoes little if any recombination and is inherited as a single unit.

Scientists create mtDNA profiles by amplifying regions of mtDNA that show variability between unrelated individuals and populations. After PCR amplification, the DNA sequence within these regions is determined by automated DNA sequencing. Scientists then compare the sequence with sequences from other individuals or crime samples, to determine whether or not they match.

The fact that mtDNA is present in high copy numbers in cells makes its analysis useful in cases where samples are small, old, or degraded. mtDNA profiling is particularly useful for identifying victims of mass murders or disasters, such as the Srebrenica massacre of 1995 and the World Trade Center attacks of 2001, where reference samples from relatives are available. The main disadvantage of mtDNA profiling is that it is not possible to differentiate between the mtDNA from maternal relatives or from siblings. Like Y-chromosome profiles, mtDNA profiles may be shared by two apparently unrelated individuals who also share a distant ancestor—in this case a maternal ancestor. Researchers use mtDNA profiles in scientific studies of genealogy, evolution, and human population migrations.

Single-Nucleotide Polymorphism Profiling

Single-nucleotide polymorphisms (SNPs) are single-nucleotide differences between two DNA molecules. They may be base-pair changes or small insertions or deletions. SNPs occur randomly throughout the genome, approximately every 500 to 1000 nucleotides. This means that there are potentially millions of loci in the human genome that can be used for profiling. However, as SNPs usually have only two alleles, many SNPs (50 or more) must be used to create a DNA profile that can distinguish between two individuals as efficiently as STRs.

Scientists analyze SNPs by using specific primers to amplify the regions of interest. The amplified DNA regions are then analyzed by a number of different methods such as automated DNA sequencing or hybridization to immobilized probes on DNA microarrays that distinguish between DNA molecules with single-nucleotide differences.

Forensic SNP profiling has one major advantage over STR profiling. Because a SNP involves only one nucleotide of a DNA molecule, the theoretical size of DNA required for a PCR reaction is the size of the two primers and one more nucleotide (i.e., about 50 nucleotides). This feature makes SNP analysis suitable for analyzing DNA samples that are severely degraded. Despite this advantage, SNP profiling has not yet become routine in forensic applications. More frequently, researchers use SNP profiling of Y-chromosome and mtDNA loci for lineage and evolution studies.

DNA Phenotyping

An emerging and controversial method, known as *DNA phenotyping,* is gaining popularity as a new DNA forensics tool. Unlike DNA profiling, which is used to confirm or exclude sample identities, DNA phenotyping uses DNA sequence information to reveal a person's physical features and ancestral origins.

Currently, DNA phenotyping methods can predict a person's eye, hair, and skin colors based on their DNA SNP patterns. For example, scientists have found six SNPs in six genes that are related to blue and brown eye color. Using statistical models based on these six SNPs, it is possible to predict with 95 percent accuracy whether a person has brown or blue eyes. Using 22 SNPs associated with 11 genes, it is possible to predict with 90 percent accuracy whether a person has black hair and 80 percent accuracy whether a person has red or brown hair. Skin color predictions involve 36 SNPs associated with 15 genes, with prediction accuracies similar to those for hair colors. Both biological sex and geographic ancestry can also be accurately determined from a person's DNA sequence.

Some researchers and private companies have taken DNA phenotyping well beyond prediction of these features. Their algorithms claim to predict three-dimensional facial structures which allow them to compile full-color photographic representations of a person's face, based only on their DNA sample.

At the present time, DNA phenotyping has not been validated sufficiently to be presented in court. However, police are using the method to help identify unknown missing persons and to provide leads in cold cases.

ST 5.2 Interpreting DNA Profiles

After a DNA profile is generated, its significance must be determined. In a typical forensic investigation, a profile derived from a suspect is compared to a profile from an evidence sample or to profiles already present in DNA databases. If the suspect's profile does not match that of the evidence profile or database entries, investigators can conclude that the suspect is not the source of the sample(s) that generated the other profile(s). However, if the suspect's profile matches the evidence profile or a database entry, the interpretation becomes more complicated. In this case, one could conclude that the two profiles either came from the same person—or they came from two different people who share the same DNA profile by chance. To determine the significance of any DNA profile match, it is necessary to estimate the probability that the two profiles are a random match.

The *profile probability* or *random match probability* method gives a numerical probability that a person chosen at random from a population would share the same DNA profile as the evidence or suspect profiles. The following example demonstrates how to arrive at a profile probability (**Table ST 5.2**).

The first locus examined in this DNA profile (D5S818) has two alleles: 11 and 13. Population studies show that the 11 allele of this locus appears at a frequency of 0.361 in this population and the 13 allele appears at a frequency of 0.141. In population genetics, the frequencies of two different alleles at a locus are given the designation p and q, following the Hardy—Weinberg law described earlier in the text (see Chapter 21). We assume that the person having this DNA profile received the 11 and 13 alleles at random from each parent. Therefore, the probability that this person received allele 11 from the mother and allele 13 from the father is

TABLE ST 5.2 A Profile Probability Calculation Based on Analysis of Five STR Loci

STR Locus	Alleles from Profile	Allele Frequency from Population Database*	Genotype Frequency Calculation
D5S818	11	0.361	$2pq = 2 \times 0.361 \times 0.141 = 0.102$
	13	0.141	
TPOX	11	0.243	$p^2 = 0.243 \times 0.243 = 0.059$
	11	0.243	
D8S1179	13	0.305	$2pq = 2 \times 0.305 \times 0.031 = 0.019$
	16	0.031	
CSF1PO	10	0.217	$p^2 = 0.217 \times 0.217 \times 0.047$
	10	0.217	
D19S433	13	0.253	$2pq = 2 \times 0.253 \times 0.369 = 0.187$
	14	0.369	
Genotype frequency from this 5-locus profile = $0.102 \times 0.059 \times 0.019 \times 0.047 \times 0.187 = 0.0000009 = 9 \times 10^{-7}$			

*A U.S. Caucasian population database [Butler, J. M., et al. (2003). *J. Forensic Sci.* 48:908–911].

expressed as $p \times q = pq$. In addition, the probability that the person received allele 11 from the father and allele 13 from the mother is also pq. Hence, the total probability that this person would have the 11, 13 genotype at this locus, by chance, is $2pq$. As we see from Table ST 5.2, $2pq$ is 0.102 or approximately 10 percent. It is obvious from this sample that using a DNA profile of only one locus would not be very informative, as about 10 percent of the population would also have the D5S818 11, 13 genotype.

The discrimination power of the DNA profile increases when we add more loci to the analysis. The next locus of this person's DNA profile (TPOX) has two identical alleles—the 11 allele. Allele 11 appears at a frequency of 0.243 in this population. The probability of inheriting the 11 allele from each parent is $p \times p = p^2$. As we see in the table, the genotype frequency at this locus would be 0.059, which is about 6 percent of the population. If this DNA profile contained only the first two loci, we could calculate how frequently a person chosen at random from this population would have the genotype shown in the table, by multiplying the two genotype probabilities together. This would be $0.102 \times 0.059 = 0.006$. This analysis would mean that about 6 persons in 1000 (or 1 person in 166) would have this genotype. The method of multiplying all frequencies of genotypes at each locus is known as the *product rule*. It is the most frequently used method of DNA profile interpretation and is widely accepted in U.S. courts.

By multiplying all the genotype probabilities at the five loci, we arrive at the genotype frequency for this DNA profile: 9×10^{-7}. This means that approximately 9 people in every 10 million (or about 1 person in a million), chosen at random from this population, would share this 5-locus DNA profile.

The Uniqueness of DNA Profiles

As we increase the number of loci analyzed in a DNA profile, we obtain smaller probabilities of a random match. Theoretically, if a sufficient number of loci were analyzed, we could be *almost* certain that the DNA profile was unique. At the present time, law enforcement agencies in North America use a core set of 20 STR loci to generate DNA profiles. Using this 20-loci set, the probability that two people selected at random would have identical genotypes at these loci would be approximately 1×10^{-28}.

Although this would suggest that most DNA profiles generated by analysis of the 20 core STR loci would be unique on the planet, several situations can alter this interpretation. For example, identical twins share the same DNA, and their DNA profiles will be identical. Identical twins occur at a frequency of about 1 in 250 births. In addition, siblings can share one allele at any DNA locus in about 50 percent of cases and can share both alleles at a locus in about 25 percent of cases. Parents and children also share alleles, but are less likely than siblings to share both alleles at a locus. When DNA profiles come from two people who are closely related, the profile probabilities must be adjusted to take this into account. The allele frequencies and calculations that we describe here are based on assumptions that the population is large and has little relatedness or inbreeding. If a DNA profile is analyzed from a person in a small interrelated group, allele frequency tables and calculations may not apply.

DNA Profile Databases

Many countries throughout the world maintain national DNA profile databases. The first of these databases was established in the United Kingdom in 1995 and now contains more than 6 million profiles. In the United States, both state and federal governments have DNA profile databases. The entire system of databases along with tools to analyze the data is known as the **Combined DNA Index System (CODIS)** and is maintained by the FBI. As of August 2018, there were more than 17 million DNA profiles stored within the CODIS system. These include the *convicted offender database,* which contains DNA profiles from individuals convicted of certain crimes, and the *forensic database,* which contains profiles generated from crime scene evidence. In addition, some states have DNA profile databases containing profiles from suspects and from unidentified human remains and missing persons.

DNA profile databases have proven their value in many different situations. As of August 2018, use of CODIS databases had resulted in more than 400,000 profile matches that assisted criminal investigations and missing persons searches (Box 2). Despite the value of DNA profile databases, they remain a concern for many people who question the privacy and civil liberties of individuals versus the needs of the state.

ST 5.3 Technical and Ethical Issues Surrounding DNA Profiling

Although DNA profiling is sensitive, accurate, and powerful, it is important to be aware of its limitations. One limitation is that most criminal cases have either no DNA evidence for analysis or DNA evidence that would not be informative to the case. In some cases, potentially valuable DNA evidence exists but remains unprocessed and backlogged. Another serious problem is that of human error. There are cases in which innocent people have been convicted of violent crimes based on DNA samples that had been inadvertently switched during processing. DNA evidence samples from crime scenes are often mixtures derived from any number of people present at the crime scene or even from people who were not present, but whose biological material (such as hair or saliva) was indirectly introduced to the site (Box 3). Crime scene evidence is often degraded, yielding partial DNA profiles that are difficult to interpret.

BOX 2

The Kennedy Brewer Case: Two Bite-Mark Errors and One Hit

In 1992 in Mississippi, Kennedy Brewer was arrested and charged with the rape and murder of his girlfriend's 3-year-old daughter, Christine Jackson. Although a semen sample had been obtained from Christine's body, there was not sufficient DNA for profiling. Forensic scientists were also unable to identify the ABO blood group from the bloodstains left at the crime scene. The prosecution's only evidence came from a forensic bite-mark specialist who testified that the 19 "bite marks" found on Christine's body matched imprints made by Brewer's two top teeth. Even though the specialist had recently been discredited by the American Board of Forensic Odontology, and the defense's expert dentistry witness testified that the marks on Christine's body were actually postmortem insect bites, the court convicted Brewer of capital murder and sexual battery and sentenced him to death.

In 2001, more sensitive DNA profiling was conducted on the 1992 semen sample. The profile excluded Brewer as the donor of the semen sample. It also excluded two of Brewer's friends, and Y-chromosome profiles excluded Brewer's male relatives. Despite these test results, Brewer remained in prison for another five years, awaiting a new trial. In 2007, the Innocence Project took on Brewer's case and retested the DNA samples. The profiles matched those of another man, Justin Albert Johnson, a man with a history of sexual assaults who had been one of the original suspects in the case. Johnson subsequently confessed to Christine Jackson's murder, as well as to another rape and murder—that of a 3-year-old girl named Courtney Smith. Levon Brooks, the ex-boyfriend of Courtney's mother, had been convicted of murder in the Smith case, also based on bite-mark testimony by the same discredited expert witness.

On February 15, 2008, all charges against Kennedy Brewer were dropped, and he was exonerated of the crimes. Levon Brooks was subsequently exonerated of the Smith murder in March of 2008.

Since 1989, more than 350 people in the United States have been exonerated of serious crimes, based on DNA profile evidence. Seventeen of these people had served time on death row. In more than 150 of these exoneration cases, the true perpetrator has been identified, often through searches of DNA databases.

BOX 3

A Case of Transference: The Lukis Anderson Story

On November 30, 2012, police discovered the body of Raveesh Kumra at his home in Monte Sereno, California. Kumra's house had been ransacked, and he had suffocated from the tape used to gag him. Police collected DNA samples from the crime scene and performed DNA profiling. Several suspects were identified through matches to DNA database entries. One match, to a sample taken from Kumra's fingernails, was that of Lukis Anderson, a homeless man who was known to police. Based on the DNA profile match, Anderson was arrested, charged with murder, and jailed. He remained in jail, with a death sentence over his head, for the next five months.

The authorities believed that they had a solid case. The crime scene DNA profile was a perfect match to Anderson's DNA profile, and the lab results were accurate. Prosecutors planned to pursue the death penalty. The only problem for the prosecution was that Anderson could not have been involved in the murder, or even present at the crime scene.

On the night of the murder, Anderson had been intoxicated and barely conscious on the streets of San Jose and had been taken to the hospital, where he remained for the next 12 hours. Given his iron-clad alibi, authorities were forced to release Anderson. But they remained baffled about how an innocent person's DNA could have been found on a murder victim—one whom Anderson had never even met.

Several months after Anderson's release, prosecutors announced that they had solved the puzzle. The paramedics who had treated Anderson and taken him to the hospital had then responded to the call at Kumra's house, where they had inadvertently transferred Anderson's DNA onto Kumra's fingernails. It is not clear how the transfer had occurred, but likely Anderson's DNA had been present on the paramedics' equipment or clothing.

If Lukis Anderson had not been in the hospital with an irrefutable alibi, he may have faced the death sentence based on DNA evidence. His story illustrates how too much confidence in the power of DNA evidence can lead to false accusations. It also points to the robustness of DNA, which can remain intact, survive disinfection, and be transferred from one location to another, under unlikely circumstances.

One of the most disturbing problems with DNA profiling is its potential for deliberate tampering. DNA profile technologies are so sensitive that profiles can be generated from only a few cells—or even from fragments of synthetic DNA. There have been cases in which criminals have introduced biological material to crime scenes, in an attempt to affect forensic DNA profiles. It is also possible to manufacture artificial DNA fragments that match STR loci of a person's DNA profile. In 2010, a research paper[1] reported methods for synthesizing DNA of a known STR profile, mixing the DNA with body fluids, and depositing the sample on crime scene items. When subjected to routine forensic analysis, these artificial samples generated perfect STR profiles. In the future, it may be necessary to develop methods to detect the presence of synthetic or cloned DNA in crime scene samples. It has been suggested that such detections could be done, based on the fact that natural DNA contains epigenetic markers such as methylation.

Many of the ethical questions related to DNA profiling involve the collection and storage of biological samples and DNA profiles. Such questions deal with who should have their DNA profiles stored on a database and whether police should be able to collect DNA samples without a suspect's knowledge or consent.

Another ethical question involves the use of DNA profiles that partially match those of a suspect. There are cases in which investigators search for partial matches between the suspect's DNA profile and other profiles in a DNA database. On the assumption that the two profiles arise from two genetically related individuals, law enforcement agencies pursue relatives of the person whose profile is stored in the DNA database. Testing in these cases is known as *familial DNA testing*. Should such searches be considered scientifically valid or even ethical?

As described previously, it is now possible to predict some facial features and geographic ancestries of persons based on information in their DNA sample—a method known as *DNA phenotyping*. Should this type of information be used to identify or convict a suspect?

As DNA profiling becomes more sophisticated and prevalent, we should carefully consider both the technical and ethical questions that surround this powerful new technology.

[1] Frumkin, D., et al. (2010). Authentication of forensic DNA samples. *Forensic Sci. Int. Genet.* 4:95–103.

Review Questions

1. What is VNTR profiling, and what are the applications of this technique?
2. Why are short tandem repeats (STRs) the most commonly used loci for forensic DNA profiling?
3. Describe capillary electrophoresis. How does this technique distinguish between input DNA and amplified DNA?
4. What are the advantages and limitations of Y-chromosome STR profiling?
5. How does SNP profiling differ from STR profiling, and what are the advantages of SNP profiling?
6. Explain why mitochondrial DNA profiling is often the method of choice for identifying victims of massacres and mass disasters.
7. What is a "profile probability," and what information is required in order to calculate it?
8. Describe the database system known as CODIS. What determines whether a person's DNA profile will be entered into the CODIS system?
9. What is DNA phenotyping, and how do law enforcement agencies use this profiling method?
10. What are three major limitations of forensic DNA profiling?

Discussion Questions

1. Given the possibility that synthetic DNA could be purposely introduced into a crime scene in order to implicate an innocent person, what methods could be developed to distinguish between synthetic and natural DNA?
2. Different countries and jurisdictions have different regulations regarding the collection and storage of DNA samples and profiles. What are the regulations within your region? Do you think that these regulations sufficiently protect individual rights?
3. If you were acting as a defense lawyer in a murder case that used DNA profiling as evidence against the defendant, how would you explain to the jury the factors that might alter their interpretation of the crime scene DNA profile?
4. The phenomena of somatic mosaicism and chimerism are more prevalent than most people realize. For example, pregnancy and bone marrow transplantation may lead to a person's genome becoming a mixture of two different genomes. Describe how DNA forensic analysis may be affected by chimerism and what measures could be used to mitigate any confusion during DNA profiling. Find out more about genetic chimerism in an article by Zimmer, C., DNA double take, *New York Times*, September 16, 2013.

Genetically Modified Foods

Throughout the ages, humans have used selective breeding techniques to create plants and animals with desirable genetic traits. By selecting organisms with naturally occurring or mutagen-induced variations and breeding them to establish the phenotype, we have evolved varieties that now feed our growing populations and support our complex civilizations.

Although we have had tremendous success shuffling genes through selective breeding, the process is a slow one. When recombinant DNA technologies emerged in the 1970s and 1980s, scientists realized that they could modify agriculturally significant organisms in a more precise and rapid way by identifying and cloning genes that confer desirable traits, then introducing these genes into organisms. Genetic engineering of animals and plants promised an exciting new phase in scientific agriculture, with increased productivity, reduced pesticide use, and enhanced flavor and nutrition.

Beginning in the 1990s, scientists created a large number of **genetically modified (GM) food** varieties. The first one, approved for sale in 1994, was the Flavr Savr tomato—a tomato that stayed firm and ripe longer than non-GM tomatoes. Soon afterward, other GM foods were developed: papaya and zucchini with resistance to virus infection, canola containing the tropical oil laurate, corn and cotton plants with resistance to insects, and soybeans and sugar beets with tolerance to agricultural herbicides.

Although many people see great potential for GM foods—to help address malnutrition in a world with a growing human population and climate change—others question the technology, oppose GM food development, and sometimes resort to violence to stop the introduction of GM varieties.

Some countries have outright bans on all GM foods, whereas others embrace the technologies. Opponents cite safety and environmental concerns, whereas some scientists and commercial interests extol the almost limitless virtues of GM foods. The topic of GM food attracts hyperbole and exaggerated rhetoric, information, and misinformation—on both sides of the debate.

> "Genetic engineering of animals and plants promised an exciting new phase in scientific agriculture, with increased productivity, reduced pesticide use, and enhanced flavor and nutrition."

So, what are the truths about GM foods? In this Special Topics chapter, we will introduce the science behind GM foods and examine the promises and problems of the new technologies. We will look at some of the controversies and present information to help us evaluate the complex questions that surround this topic.

ST 6.1 What Are GM Foods?

GM foods are derived from **genetically modified organisms (GMOs),** specifically plants and animals of agricultural importance. GMOs are defined as organisms whose genomes have been altered in ways that do not occur naturally. Although the definition of GMOs sometimes includes organisms that have been genetically modified by selective breeding, the most commonly used definition refers to organisms modified through genetic engineering or recombinant DNA technologies. Genetic engineering allows one or more genes to be cloned and transferred from one organism to another—either between individuals of the same species or between those of unrelated species. It also allows an organism's endogenous genes to be altered in ways that lead to enhanced or reduced expression levels. When genes are transferred between unrelated species, the resulting organism is called **transgenic.** The term **cisgenic** is sometimes used to describe gene transfers within a species. In contrast, the term **biotechnology** is a general term, encompassing a wide range of methods that manipulate organisms or their components—such as isolating enzymes or producing wine cheese, or yogurt. Genetic modification of plants or animals is one aspect of biotechnology.

It is estimated that GM crops are grown in approximately 30 countries on 11 percent of the arable land on Earth. The majority of these GM crops (almost 90 percent) are grown in five countries—the United States, Brazil, Argentina, Canada, and India. Of these five, the United States accounts for approximately half of the acreage devoted to GM crops. According to the U.S. Department of Agriculture, 93 percent of soybeans and 90 percent of maize grown in the United States are from GM crops. In the United States, more than 70 percent of processed foods contain ingredients derived from GM crops.

Approximately 200 different GM crop varieties are approved for use as food or livestock feed in the United States. However, only about two dozen are widely planted. **Table ST 6.1** lists some of the common GM food crops available for planting in the United States. Only one GM food animal, the AquAdvantage salmon, has been approved for consumption (Box 1).

Herbicide-Resistant GM Crops

Herbicide-tolerant varieties are the most widely planted of GM crops, making up approximately 70 percent of all GM crops. The majority of these varieties contain a bacterial gene that confers tolerance to the broad-spectrum herbicide **glyphosate**—the active ingredient in commercial herbicides such as Roundup®.

Farmers who plant glyphosate-tolerant crops can treat their fields with glyphosate, even while the GM crop is growing. This approach is more efficient and economical than mechanical weeding and reduces soil damage caused by repeated tillage. It is suggested that there is less environmental impact when using glyphosate, compared with having to apply higher levels of other, more toxic, herbicides.

TABLE ST 6.1 Some GM Crops Approved for Food, Feed, or Cultivation in the United States*

Crop	Number of Varieties	GM Characteristics
Soybeans	19	Tolerance to glyphosate herbicide Tolerance to glufosinate herbicide Reduced saturated fats Enhanced oleic acid Enhanced omega-3 fatty acid
Maize	68	Tolerance to glyphosate herbicide Tolerance to glufosinate herbicide Bt insect resistance Enhanced ethanol production
Cotton	30	Tolerance to glyphosate herbicide Bt insect resistance
Potatoes	28	Bt insect resistance
Canola	23	Tolerance to glyphosate herbicide Tolerance to glufosinate herbicide Enhanced lauric acid
Papaya	4	Resistance to papaya ringspot virus
Sugar beets	3	Tolerance to glyphosate herbicide
Rice	3	Tolerance to glufosinate herbicide
Zucchini squash	2	Resistance to zucchini, watermelon, and cucumber mosaic viruses
Alfalfa	2	Tolerance to glyphosate herbicide
Plum	1	Resistance to plum pox virus

* Information from the International Service for the Acquisition of Agri-Biotech Applications, www.isaaa.org.

Insect-Resistant GM Crops

The second most prevalent GM modifications are those that make plants resistant to agricultural pests.

The most widely used GM insect-resistant crops are the **Bt crops.** ***Bacillus thuringiensis*** (Bt) are soil-dwelling bacterial strains that produce crystal (Cry) proteins that are toxic to certain species of insects. These Cry proteins are encoded by the bacterial *cry* genes and form crystal structures during sporulation. The Cry proteins are toxic to Lepidoptera (moths and butterflies), Diptera (mosquitoes and flies), Coleoptera (beetles), and Hymenoptera (wasps and ants). Insects must ingest the bacterial spores or Cry proteins in order for the toxins to act. Within the high pH of the insect gut, the crystals dissolve and are cleaved by insect protease enzymes. The Cry proteins bind to receptors on the gut wall, leading to breakdown of the gut membranes and death of the insect.

Each insect species has specific types of gut receptors that will match only a few types of Bt Cry toxins. As there are more than 200 different Cry proteins, it is possible to select a Bt strain that will be specific to one pest type.

Bt spores have been used for decades as insecticides in both conventional and organic gardening, usually applied in liquid sprays. Sunlight and soil rapidly break down the Bt insecticides, which have not shown any adverse effects on groundwater, mammals, fish, or birds. Toxicity tests on humans and animals have shown that Bt causes few negative effects.

To create Bt crops, scientists introduce one or more cloned *cry* genes into plant cells using methods described in the next section. The GM crop plants will then manufacture their own Bt Cry proteins, which will kill the target pest species when it eats the plant tissues.

GM Crops for Direct Consumption

To date, most GM crops have been designed to help farmers increase yields. Also, most GM food crops are not consumed directly by humans, but are used as animal feed or as sources of processed food ingredients such as oils, starches, syrups,

BOX 1

The Tale of GM Salmon—Downstream Effects?

In 2015, the AquAdvantage salmon became the first GM animal to be approved for human consumption.

The AquAdvantage salmon is an Atlantic salmon that is genetically modified to grow twice as fast as its non-GM cousins, reaching marketable size in one and a half years rather than the usual three years. Scientists at AquaBounty Technologies in Massachusetts created the variety by transforming an Atlantic salmon with a single gene encoding the Chinook salmon growth hormone. The gene was cloned downstream of the antifreeze protein gene promoter from an eel. This promoter stimulates growth hormone synthesis in the winter, a time when the fish's own growth hormone gene is not expressed. The rapid growth of the GM salmon allows fish farmers to double their productivity.

AquaBounty will produce the salmon eggs at a facility in Canada and grow the salmon in a containment facility in Panama. To ensure that the fish will not escape the facilities, the company promises to sell only fertilized eggs that are female, triploid, and sterile. The facilities have tanks that are located inland and have sufficient filters to ensure that eggs and small fish cannot escape.

Despite these assurances, environmental groups are planning to fight the sale of GM salmon. Some grocery chains in the United States have banned GM fish.

Critics of the new GM salmon point out that the technique used to create sterile triploids (pressure-shocking the fertilized eggs) still allows a small percentage of fertile diploids to remain in the stock. If GM salmon could escape, breed, and introduce transgenes into wild populations, there could be unknown negative downstream effects on fish ecosystems.

The AquAdvantage salmon grows twice as fast as a non-GM Atlantic salmon, reaching market size in half the time.

and sugars. For example, 98 percent of the U.S. soybean crop is used as livestock feed. The remainder is processed into a variety of food ingredients, such as lecithin, textured soy proteins, soybean oil, and soy flours. However, a few GM foods have been developed for direct consumption. Examples are rice, squash, and papaya.

One of the most famous and controversial examples of GM foods is **Golden Rice**—a rice variety designed to synthesize beta-carotene (the precursor to **vitamin A**) in the rice grain endosperm.

Vitamin A deficiency is a serious health problem in more than 60 countries, particularly countries in Asia and Africa. The World Health Organization estimates that 190 million children and 19 million pregnant women are vitamin A deficient. Between 250,000 and 500,000 children with vitamin A deficiencies become blind each year, and half of these will die within a year of losing their sight. As vitamin A is also necessary for immune system function, deficiencies lead to increases in many other conditions, including diarrhea and virus infections.

Several approaches are being taken to alleviate the vitamin A deficiency status of people in developing countries. These include supplying high-dose vitamin A supplements and growing fresh fruits and vegetables in home gardens. These initiatives have had partial success, but the expense of delivering education and supplementation has impeded the effectiveness of these programs.

In the 1990s, scientists began to apply recombinant DNA technology to help solve vitamin A deficiencies in people with rice-based diets. Although the rice plant naturally produces beta-carotene in its leaves, it does not produce it in the rice grain endosperm, which is the edible part of the rice. The beta-carotene precursor, geranylgeranyl-diphosphate, is present in the endosperm, but the enzymes that convert it to beta-carotene are not synthesized (**Figure ST 6.1**).

In the first version of Golden Rice, scientists introduced the genes *phytoene synthase* (*psy*) cloned from the daffodil plant and *carotene desaturase* (*crtI*) cloned from the bacterium *Erwinia uredovora* into rice plants. The bacterial *crtI* gene was chosen because the enzyme encoded by this gene can perform the functions of two of the missing rice enzymes, thereby simplifying the transformation process. The resulting plant produced rice grains that were a yellow color due to the presence of beta-carotene (**Figure ST 6.2**).

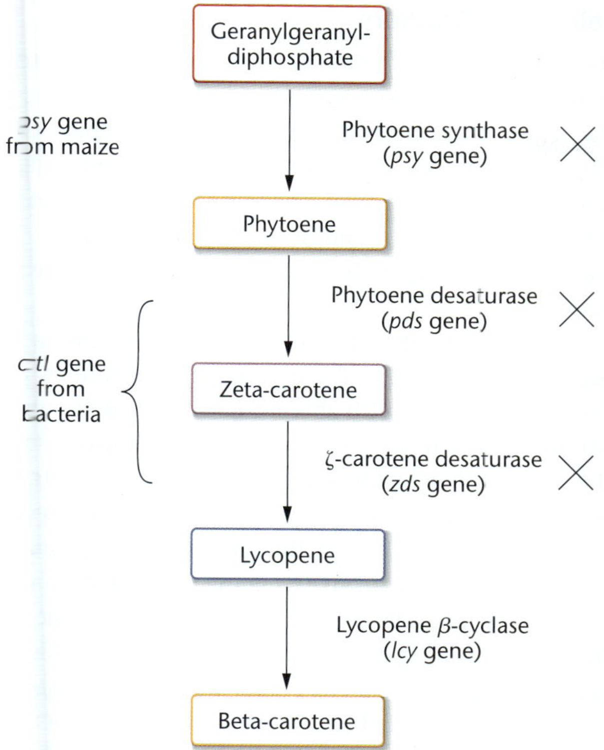

FIGURE ST 6.1 Beta-carotene pathway in Golden Rice 2. Rice plant enzymes and genes involved in beta-carotene synthesis are shown on the right. The enzymes that are not expressed in rice endosperm are indicated with an "X." The genes inserted into Golden Rice 2 are shown on the left.

FIGURE ST 6.2 Non-GM and Golden Rice 2. Golden Rice 2 contains high levels of beta-carotene, giving the rice endosperm a yellow color. The intensity of the color reflects the amount of beta-carotene in the endosperm.

This strain synthesized modest levels of beta-carotene—but only enough to potentially supply 15 to 20 percent of the recommended daily allowance of vitamin A. In the second version of the GM plant, called Golden Rice 2, the daffodil *psy* gene was replaced with the *psy* gene from maize. Golden Rice 2 produced beta-carotene levels that were more than 20-fold greater than those in Golden Rice. In the next section we describe the methods used to create Golden Rice 2.

Clinical trials have shown that the beta-carotene in Golden Rice 2 is efficiently converted into vitamin A in humans and that about 150 grams of uncooked Golden Rice 2 (which is close to the normal daily rice consumption of children aged 4 to 8 years) would supply all of the childhood daily requirement for vitamin A.

At the present time, Golden Rice 2 is undergoing field, biosafety, and efficacy testing in preparation for approval by government regulators in Bangladesh and the Philippines. If Golden Rice 2 proves useful in alleviating vitamin A deficiencies and is approved for use, seed will be made available at the same price as non-GM seed and farmers will be allowed to keep and replant seed from their own crops.

Despite the promise of Golden Rice 2, controversies remain. Critics of GM foods suggest that Golden Rice could make farmers too dependent on one type of food or might have long-term health or environmental effects. These and other controversies surrounding GM foods are discussed in subsequent sections of this chapter.

ST 6.2 Methods Used to Create GM Plants

Most GM plants have been created using one of two approaches: the **biolistic method** or ***Agrobacterium tumefaciens*–mediated transformation** technology. Both methods target plant cells that are growing *in vitro*. Scientists can generate plant tissue cultures from various types of plant tissues, and these cultured cells will grow either in liquid cultures or on the surface of solid growth media. When grown in the presence of specific nutrients and hormones, these cultured cells will form clumps of cells called calluses, which, when transferred to other types of media, will form roots. When the rooted plantlets are mature, they are transferred to soil medium in greenhouses where they develop into normal plants.

The *biolistic method* is a physical method of introducing DNA into cells. Particles of heavy metals such as gold are coated with the DNA that will transform the cells; these particles are then fired at high speed into plant cells *in vitro*, using a device called a **gene gun.** Cells that survive the bombardment may take up the DNA-coated particles, and the

DNA may migrate into the cell nucleus and integrate into a plant chromosome. Plants that grow from the bombarded cells are then selected for the desired phenotype.

Although biolistic methods are successful for a wide range of plant types, a much improved transformation rate is achieved using *Agrobacterium-mediated technology*. *Agrobacterium tumefaciens* (also called *Rhizobium radiobacter*) is a soil microbe that can infect plant cells and cause tumors. These characteristics are conferred by a 200-kb tumor-inducing plasmid called a **Ti plasmid.** After infection with *Agrobacterium*, the Ti plasmid integrates a segment of its DNA known as transfer DNA (T-DNA) into random locations within the plant genome (**Figure ST 6.3**). To use the Ti plasmid as a transformation vector, scientists remove the T-DNA segment and replace it with cloned DNA of the genes to be introduced into the plant cells.

In order to have the newly introduced gene expressed in the plant, the gene must be cloned next to an appropriate promoter sequence that will direct transcription in the required plant tissue. For example, the beta-carotene pathway genes introduced into Golden Rice were cloned next to a promoter that directs transcription of the genes in the rice endosperm. In addition, the transformed gene requires appropriate transcription termination signals and signal sequences that allow insertion of the encoded protein into the correct cell compartment.

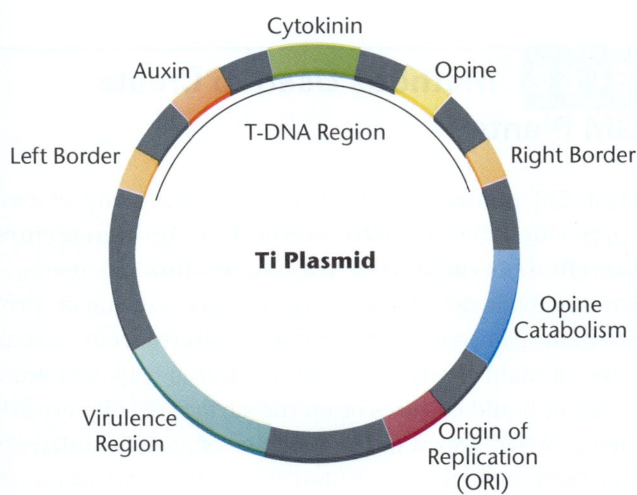

FIGURE ST 6.3 Structure of the Ti plasmid. The 250-kb Ti plasmid from *Agrobacterium tumefaciens* inserts the T-DNA portion of the plasmid into the host cell's nuclear genome and induces tumors. Genes within the virulence region code for enzymes responsible for transfer of T-DNA into the plant genome. The T-DNA region contains auxin and cytokinin genes that encode hormones responsible for cell growth and tumor formation. The opine genes encode compounds used as energy sources for the bacterium. The T-DNA region of the Ti plasmid is replaced with the gene of interest when the plasmid is used as a transformation vector.

Selectable Markers

The rates at which T-DNA successfully integrates into the plant genome and becomes appropriately expressed are low. Often, only one cell in 1000 or more will be successfully transformed. Before growing cultured plant cells into mature plants to test their phenotypes, it is important to eliminate the background of nontransformed cells. This can be done using either positive or negative selection techniques.

An example of negative selection involves use of a **marker gene** such as the hygromycin-resistance gene. This gene, together with an appropriate promoter, can be introduced into plant cells along with the gene of interest. The cells are then incubated in culture medium containing hygromycin—an antibiotic that also inhibits the growth of eukaryotic cells. Only cells that express the hygromycin-resistance gene will survive. It is then necessary to verify that the resistant cells also express the cotransformed gene. This is often done by techniques such as PCR amplification using gene-specific primers. Plants that express the gene of interest are then tested for other characteristics, including the phenotype conferred by the introduced gene of interest.

An example of positive selection involves the use of a selectable marker gene such as that encoding **phosphomannose isomerase (PMI).** This enzyme is common in animals but is not found in most plants. It catalyzes the interconversion of mannose 6-phosphate and fructose 6-phosphate. Plant cells that express the *pmi* gene can survive on synthetic culture medium that contains only mannose as a carbon source. Cells that are cotransformed with the *pmi* gene under control of an appropriate promoter and the gene of interest can be positively selected by growing the plant cells on a mannose-containing medium. This type of positive selection was used to create Golden Rice 2. Studies have shown that purified PMI protein is easily digested, nonallergenic, and nontoxic in mouse oral toxicity tests. A variation in positive selection involves use of a marker gene whose expression results in a visible phenotype, such as deposition of a colored pigment.

The following description illustrates the method used to engineer Golden Rice 2.

Golden Rice 2

To create Golden Rice 2, scientists cloned three genes into the T-DNA region of a Ti plasmid. The resulting Ti plasmid, called pSYN12424, is shown in **Figure ST 6.4**. The first gene was the *carotene desaturase* (*crtI*) gene from *Erwinia uredovora,* fused between the rice *glutelin* gene promoter (*Glu*) and the *nos* gene terminator region (*nos*). The *Glu* promoter directs transcription of the fusion gene specifically in the rice endosperm. The *nos* terminator was cloned from

FIGURE ST 6.4 T-DNA region of Ti plasmid pSYN12424. The plasmid's genes, promoters, and termination signals are described in the text.

the *Agrobacterium tumefaciens nopaline synthase* gene and supplies the transcription termination and polyadenylation sequences required at the 3′-end of plant genes. The second gene was the *phytoene synthase* (*psy*) gene cloned from maize. The maize *psy* gene has approximately 90 percent sequence similarity to the rice *psy* gene and is involved in carotenoid synthesis in maize endosperm. This gene was also fused to the *Glu* promoter and the *nos* terminator sequences in order to obtain proper transcription initiation and termination in rice endosperm. The third gene was the selectable marker gene *phosphomannose isomerase* (*pmi*), cloned from *E. coli*. In the Golden Rice 2 Ti plasmid, the *pmi* gene was fused to the maize *polyubiquitin* gene promoter (*Ubi1*) and the *nos* terminator sequences. The *Ubi1* promoter is a constitutive promoter, directing transcription of the *pmi* gene in all plant tissues.

To introduce the pSYN12424 plasmid into rice cells, researchers established embryonic rice cell cultures and infected them with *Agrobacterium tumefaciens* that contained pSYN12424 (**Figure ST 6.5**). The cells were then placed under selection, using culture medium containing only mannose as a carbon source. Surviving cells expressing the *pmi* gene were then stimulated to form calluses that were grown into plants. To confirm that all three genes were present in the transformed rice plants, samples were taken and analyzed by the polymerase chain reaction (PCR) using gene-specific primers. Plants that contained one integrated copy of the transgenic construct and synthesized beta-carotene in their seeds were selected for further testing.

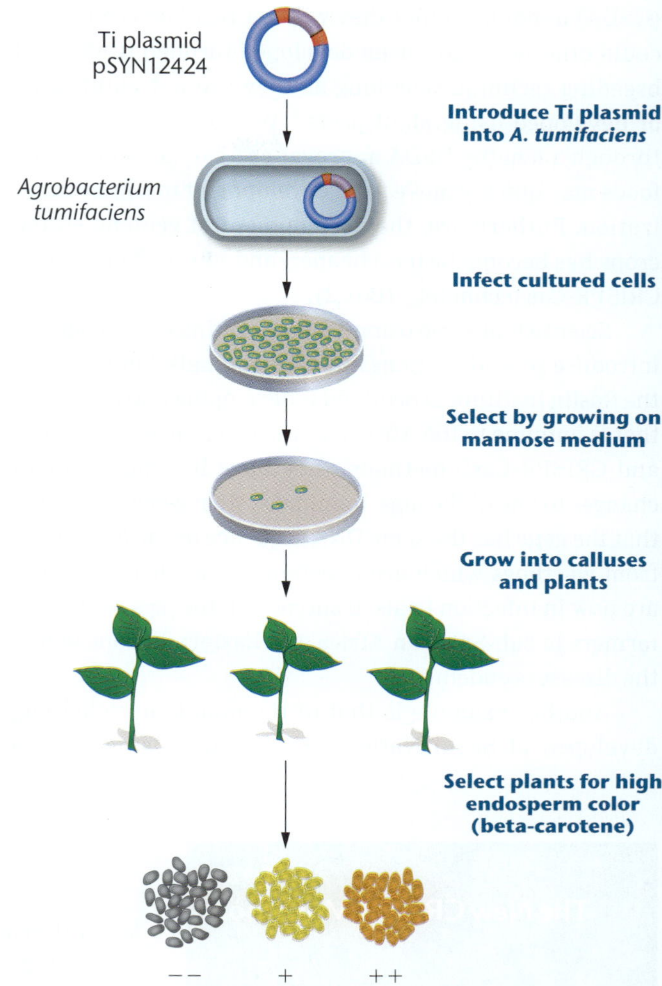

FIGURE ST 6.5 Method for creating Golden Rice 2. Rice plant cells were transformed by pSYN12424 and selected on mannose-containing medium, as described in the text. Plants that produced high levels of beta-carotene in rice grain endosperm (+ +), based on the intensity of the grain's yellow color, were selected for further analysis.

Genome Editing and GM Foods

The previously described methods are those that have been used to create the majority of GM plants. In the last few years, several new and revolutionary methods of genome modification have entered the field of GM foods. Collectively, these are known as **genome-editing methods.** They include zinc-finger nuclease (ZFN), transcription activator-like effector nuclease (TALEN), and CRISPR-Cas techniques. These methods are described in detail in Chapter 17 and Special Topics Chapter 3—Gene Therapy.

Genome-editing methods have had significant effects on the speed at which scientists can induce genetic changes in plants and animals as well as on the types of changes that are possible. Genome editing allows researchers to create precise nucleotide or single-gene mutations or deletions without introducing foreign DNA into the organism. To create genome-edited plants or animals, scientists typically mutate or inactivate only one or two of the organism's endogenous genes.

One example of a genome-edited food is a potato developed by the biotechnology company Calyxt, Inc. Using TALEN methods, they inactivated the *vacuolar invertase* gene that encodes an enzyme responsible for degrading sugars in cold-stored potatoes. This gene inactivation resulted in a potato with an increased storage life as well as one that does not produce harmful acrylamides when the potato is fried. The genome-edited potato is currently in field trials.

Because genome-edited organisms contain no transgene material, they have not been considered genetically modified by most regulatory agencies and therefore do not have the same oversight as other GM foods. For example, in March of 2018 the United States Department of Agriculture

(USDA) announced that they will not regulate crops "that could otherwise have been developed through traditional breeding techniques as long as they are not plant pests or developed using plant pests." Without the need to go through a lengthy USDA approval process, genome-edited foods may quickly move from development to commercialization. Futhermore, the development of genome-edited crops has become faster, cheaper, and more efficient with CRISPR-Cas technology (Box 2).

Scientists are also using genome-editing technologies to introduce gene alterations into farm animals. For example, the Roslin Institute in Scotland is developing a strain of pigs that is immune to the African swine fever virus. Using ZFN and CRISPR-Cas9 methods, they have introduced small changes to one of the pigs' immune system genes (*RELA*) so that the gene has the same DNA sequence as the *RELA* gene from warthogs which are resistant to the virus. The pigs are now in infection trials. If successful, the pigs could help farmers in sub-Saharan Africa and Eastern Europe where the disease is endemic.

Another example is that of the double-muscled pig, developed at Seoul National University. Using TALEN methods, researchers introduced mutations into both copies of the myostatin (*MSTN*) gene, inactivating it. Myostatin is responsible for inhibiting the growth of muscle cells. When the gene is inactivated, muscle tissue grows to produce muscle-enhanced animals. Such animals produce higher yields of lean meat.

Although genome-edited crops and animals are currently not regulated in the same way as GM foods, some countries are reviewing their guidelines and new regulations may be introduced in the near future.

ST 6.3 GM Foods Controversies

GM foods may be the most contentious of all products of modern biotechnology. Advocates of GM foods state that the technologies have increased farm productivity, reduced pesticide use, preserved soils, and have the potential to feed growing human populations. Critics claim that GM foods are unsafe for both humans and the environment; accordingly, they are applying pressure on regulatory agencies to ban or

BOX 2
The New CRISPR Mushroom

The common white-button mushroom (*Agaricus bisporus*) has become the first genome-edited crop, created using the *CRISPR-Cas9* method, to be cleared for commercialization and human consumption. In April 2016, the U.S. Department of Agriculture (USDA) declared that the new mushroom, genome-edited to reduce browning, would not be subject to its regulation.

The nonbrowning mushroom was developed by Dr. Yinong Yang of Pennsylvania State University. Dr. Yang's team used the CRISPR-Cas9 genome-editing method (described in Chapter 17) to inactivate one member of the mushroom's *polyphenol oxidase (PPO)* gene family, resulting in a 30 percent reduction in the levels of PPO enzyme activity. The PPO enzyme is responsible for the browning effect when mushrooms are bruised or cut. The new mushroom has a longer shelf life and resists browning caused by processing and mechanical harvesting.

The CRISPR mushroom is not regulated by the USDA because it does not contain foreign DNA from plasmids, viruses, or genes from other species. In addition, the genetic changes were small (a few nucleotides) and precise, at known locations in the genome.

Will the genome-edited mushroom be available in grocery stores soon? This will depend on whether mushroom farmers decide to grow the genome-edited variety, which will likely depend on consumer attitudes toward genome-edited foods

The CRISPR-edited mushroom resists the browning that occurs when mushrooms are cut or bruised during processing or storage, as seen here.

severely limit the extent of GM food use. These campaigns have affected regulators and politicians, resulting in a patchwork of regulations throughout the world. Often the debates surrounding GM foods are highly polarized and emotional, with both sides in the debate exaggerating their points of view and selectively presenting the data. So, what are the truths behind these controversies?

One point that is important to make as we try to answer this question is that *it is not possible to make general statements about all "GM foods."* Each GM crop or organism contains different genes from different sources, attached to different expression sequences, accompanied by different marker or selection genes, and inserted into the genome in different ways and in different locations. In addition, the recent proliferation of genome-edited food organisms further complicates the situation. Many advocates and regulatory agencies state that genome-edited foods are not GM foods, as they do not fit the previous definitions of GMOs. GM foods are created for different purposes and are used in ways that are both planned and unplanned. Each construction is unique and therefore needs to be assessed separately.

We will now examine two of the main GM foods controversies: those involving human health and safety, and environmental effects.

Health and Safety

GM food advocates often state that there is no evidence that GM foods currently on the market have any adverse health effects, either from the presence of toxins or from potential allergens. These conclusions are based on two observations. First, humans have consumed several types of GM foods for more than 20 years, and no reliable reports of adverse effects have emerged. Second, the vast majority of toxicity tests in animals, which are required by government regulators prior to approval, have shown no negative effects. A few negative studies have been published, but these have been criticized as poorly executed or nonreproducible.

Critics of GM foods counter the first observation in several ways. First, as described previously, few GM foods are eaten directly by consumers. Instead, most are used as livestock feed, and the remainder form the basis of purified food ingredients. To date, no adverse effects of GM foods in livestock have been detected. In addition, the processing of many food ingredients removes most, if not all, plant proteins and DNA. Hence, ingestion of GM food-derived ingredients may not be a sufficient test for health and safety. Second, GM food critics argue that there have been few human clinical trials to directly examine the health effects of most GM foods. One notable exception is Golden Rice 2, which has undergone two small clinical trials. They also say that the toxicity studies that have been completed are performed in animals—primarily rats and mice—and most of these are short-term toxicity studies.

Supporters of GM foods answer these criticisms with several other arguments. The first argument is that short-term toxicity studies in animals are well-established methods for detecting toxins and allergens. The regulatory processes required prior to approval of any GM food demand data from animal toxicity studies. If any negative effects are detected, approval would not be given. Supporters also note that several dozen long-term toxicity studies have been published that deal with GM crops such as glyphosate-resistant soybeans and Bt corn, and none of these has shown long-term negative effects on test animals. Those few studies reporting negative effects have been shown to have serious design flaws and their conclusions are considered unreliable. GM food advocates note that human clinical trials are not required for any other food derived from other genetic modification methods such as selective breeding. During standard breeding of plants and animals, genomes may be mutagenized with radiation or chemicals to enhance the possibilities of obtaining a desired phenotype. This type of manipulation has the potential to introduce mutations into genes other than the ones that are directly selected. Also, plants and animals naturally exchange and shuffle DNA in ways that cannot be anticipated. These include interspecies DNA transfers, transposon integrations, and chromosome modifications. These events may result in unintended changes to the physiology of organisms—changes that could potentially be as great as those arising in GM foods.

Environmental Effects

Critics of GM foods point out that GMOs that are released into the environment have both documented and potential consequences for the environment—and hence may indirectly affect human health and safety. GM food advocates argue that these potential environmental consequences can be identified and managed. Here, we will describe two different aspects of GM foods as they may affect the natural environment and agriculture.

1. Emerging herbicide and insecticide resistance. Many published studies report that the planting of herbicide-tolerant and insect-resistant GM crops has reduced the quantities of herbicides and insecticides that are broadly applied to agricultural crops. As a result, the effects of GM crops on the environment have been assumed to be positive. However, these positive effects may be transient, as herbicide and insecticide resistance is beginning to emerge.

   Since glyphosate-tolerant crops were introduced in the mid-1990s, more than 24 glyphosate-resistant weed species have appeared in the United States. Resistant weeds have been found in 18 other countries, and in some cases, the presence of these weeds is affecting crop yields. One reason for the rapid rise of resistant weeds is that farmers have abandoned other weed-management practices in favor of using a single broad-spectrum herbicide. This strong selection pressure has brought the rapid evolution of weed species bearing gene variants that confer herbicide resistance. Scientists point out that herbicide resistance is not limited to the use of GM crops. Weed populations will evolve resistance to any herbicide used to control them, and the speed of evolution will be affected by the extent to which the herbicide is used.

Since 1996, more than eight different species of insect pests have evolved some level of resistance to Bt insecticidal proteins. In order to slow down the development of Bt resistance, several strategies are being followed. The first is to develop varieties of GM crops that express two Bt toxins simultaneously. Several of these varieties are already on the market and are replacing varieties that express only one Bt *cry* gene. The second strategy involves the use of "refuges" surrounding fields that grow Bt crops. These refuges contain non-GM crops. Insect pests grow easily within the refuges, which place no evolutionary pressure on the insects for resistance to Bt toxins. The idea is for these nonselected insects to mate with any resistant insects that appear in the Bt crop region of the field. The resulting hybrid offspring will be heterozygous for any resistance gene variant. As long as the resistance gene variant is recessive, the hybrids will be killed by eating the Bt crop. In fields that use refuges and plant GM crops containing two Bt genes, resistance to Bt toxins has been delayed or is absent. As with emerging herbicide resistance, farmers are also encouraged to combine the use of Bt crops with conventional pest control methods.

2. The spread of GM crops into non-GM crops. There have been several documented cases of GM crop plants appearing in uncultivated areas in the United States, Canada, Australia, Japan, and Europe. For example, GM sugar beet plants have been found growing in commercial top soils. GM canola plants have been found growing in ditches and along roadways, railway tracks, and in fill soils, far from the fields in which they were grown.

   One of the major concerns about the escape of GM crop plants from cultivation is the possibility of **outcrossing** or **gene flow**—the transfer of transgenes from GM crops into sexually compatible non-GM crops or wild plants, conferring undesired phenotypes to the other plants. Gene flow between GM crops and adjacent non-GM crops is of particular concern for farmers who want to market their crops as "GM-free" or "organic" and for farmers who grow seed for planting.

   Gene flow of GM transgenes has been documented in GM and non-GM canola as well as sugar beets, and in experiments using rice, wheat, and maize.

   It is thought that the presence of glyphosate-resistant transgenes in wild plant populations is not likely to be an environmental risk and would confer no positive fitness benefits to the hybrids. The presence of glyphosate-resistant genes in wild populations would, however, make it more difficult to eradicate the plants. This is illustrated in a case of escaped GM bentgrass in Oregon, where it has been difficult to get rid of the plants because it is no longer possible to use the relatively safe herbicide glyphosate. The potential for environmental damage may be greater if the GM transgenes did confer an advantage—such as insect resistance or tolerance to drought or flooding.

   In an attempt to limit the spread of transgenes from GM crops to non-GM crops, regulators are considering a requirement to separate the crops so that pollen would be less likely to travel between them. Each crop plant would require different isolation distances to take into account the dynamics of pollen spreading. Several other methods are being considered. For example, one proposal is to make all GM plants sterile using RNAi technology. Another is to introduce the transgenes into chloroplasts. As chloroplasts are inherited maternally, their genomes would not be transferred via pollen. All of these containment methods are in development stages and may take years to reach the market.

ST 6.4 The Future of GM Foods

Over the last 20 years, GM foods have revealed both promise and problems. GM advocates are confident that the next generation of GM food, especially those created using genome-editing technologies, will show even more promising prospects—and may also address many of the problems.

Research is continuing on ways to fortify staple crops with nutrients to address diet problems in poor countries. For example, Australian scientists are adding genes to bananas that will not only provide resistance to Panama disease—a serious fungal disease that can destroy crops—but also increase the levels of beta-carotene and other nutrients, including iron. Other GM crops in the pipeline include plants engineered to resist drought, high salinity, nitrogen starvation, and low temperatures.

Researchers are also devising more creative ways to protect plants from insects and diseases. One project introduces into wheat a gene that encodes a pheromone that acts as a chemical alarm signal to aphids. If successful, this approach could protect the wheat plants from aphids without using toxins. Another project involves cassava, which is a staple crop for many Africans and is afflicted by two viral diseases—cassava mosaic virus and brown streak virus—that stunt growth and cause root rot. Although some varieties of cassava are resistant to these viruses, the life cycle of cassava is so long that it would be difficult to introduce resistance into other varieties using conventional breeding techniques. Scientists plan to transform plants with genes from resistant cassava. This type of cisgenic gene transfer is more comparable to traditional breeding than transgenic techniques.

In the future, GM foods will likely include additional GM animals. In one project, scientists have introduced a DNA sequence into chickens that protects the birds from spreading avian influenza. The sequence encodes a hairpin RNA molecule with similarity to a normal viral RNA that binds to the viral polymerase. The presence of the hairpin RNA inhibits the activity of the viral polymerase and interferes with viral propagation. If this strategy proves useful *in vivo*, the use of these GM chickens would not only reduce the incidence of avian influenza in poultry production, but also reduce the transmissibility of avian influenza viruses to humans.

Although these and other GM foods show promise for increasing agricultural productivity and decreasing disease, the political pressure from anti-GM critics remains a powerful force. An understanding of the science behind these technologies will help us all to evaluate the future of GM foods.

Review Questions

1. How do genetically modified organisms compare with organisms created through selective breeding?
2. Can current GM crops be considered as transgenic or cisgenic? Why?
3. Of the approximately 200 GM crop varieties that have been developed, only a few are widely used. What are these varieties, and how prevalent are they?
4. Describe the mechanisms by which the Cry proteins from *Bacillus thuringiensis* act as insecticides.
5. What measures have been taken to alleviate vitamin A deficiencies in developing countries? To date, how successful have these strategies been?
6. What is Golden Rice 2, and how was it created?
7. Describe how plants can be transformed using biolistic methods. How does this method compare with *Agrobacterium tumefaciens*–mediated transformation?
8. How do positive and negative selection techniques contribute to the development of GM crops?
9. Describe two major differences between GM foods such as Golden Rice 2 and genome-edited foods such as the CRISPR mushroom.

Discussion Questions

1. What are the laws regulating the development, approval, and use of GM foods in your region and nationally?
2. Do you think that foods containing GM ingredients should be labeled as such? What would be the advantages and disadvantages to such a strategy?
3. One of the major objections to GM foods is that they may be harmful to human health. Do you agree or disagree, and why?

Genomics and Precision Medicine

Over the last decade, the terms *precision medicine* and *personalized medicine* have entered public consciousness as emerging, and likely revolutionary, approaches to disease prevention, diagnosis, and treatment. In 2015, the United States announced the Precision Medicine Initiative—a $215 million investment into the molecular tools required to bring precision medicine into routine clinical use. In the same year, the United Kingdom announced its Precision Medicine Catapult—aimed to accelerate the development and application of precision medicine technologies.

So, what are precision medicine and personalized medicine? **Precision medicine** can be defined as an individualized, molecular approach to disease diagnosis and treatment—one that examines a patient's individual genomic, proteomic, gene expression, and other molecular profiles and applies that information to select precise disease treatments and to develop new treatments and drugs. Precision medicine classifies patients into subpopulations based on their molecular profiles, and then directs each group into a treatment regimen that will bring about maximum benefit. Although often used interchangeably with precision medicine, **personalized medicine** is defined as a way to design specific, even unique, treatments for each individual, also based on their unique molecular profiles. Personalized medicine can be considered a part of precision medicine.

"Precision medicine classifies patients into subpopulations based on their molecular profiles, and then directs each group into a treatment regimen that will bring about maximum benefit."

In this Special Topic chapter, we will examine some of the new developments in precision medicine, with an emphasis on pharmacogenomics and precision oncology. The types of genetic tests currently used in precision medicine are described in Special Topics Chapter 2—Genetic Testing.

ST 7.1 Pharmacogenomics

Perhaps the most developed area in precision medicine is pharmacogenomics. **Pharmacogenomics** is the study of how an individual's genetic makeup determines the body's response to drugs. It also involves the development and use of drugs that are specifically targeted to a patient's genetic profile. The term *pharmacogenetics* is often used interchangeably with pharmacogenomics but refers to the study of how sequence variation within specific genes affects an individual's drug responses.

In this section, we examine two ways in which precision medicine is changing the development and use of drugs: by optimizing drug responses and by developing molecularly targeted drugs.

Optimizing Drug Responses

Every year, approximately 2 million people in the United States have serious side effects from pharmaceutical drugs and of these, approximately 100,000 will die. In addition, many patients do not respond to drug treatment as well as expected, due in part to their genetic makeup and the genomic variants that are associated with their diseases.

Sequence variations in dozens of genes affect a person's reactions to drugs. The proteins encoded by these gene variants control many aspects of drug metabolism, such as the interactions of drugs with carriers, cell-surface receptors, and transporters; with enzymes that degrade or modify drugs; and with proteins that affect a drug's storage or excretion.

Examples of genes that are involved in drug metabolism are members of the cytochrome P450 gene family. People with some cytochrome P450 gene variants metabolize and eliminate drugs slowly, which can lead to accumulations of the drug and overdose side effects. In contrast, other people have variants that cause drugs to be eliminated quickly, leading to reduced effectiveness. An example is the *CYP2D6* gene, which encodes debrisoquine hydroxylase. This enzyme is involved in the metabolism of approximately 25 percent of all pharmaceutical drugs, including acetaminophen, clozapine, beta blockers, tamoxifen, and codeine. There are more than 70 variant alleles of this gene. Some variants reduce the activity of the encoded enzyme, and others can increase it. Approximately 80 percent of people are homozygous or heterozygous for

TABLE ST 7.1 Examples of Drug Responses Affected by Gene Variants*

Gene	Drug Affected	Description
TPMT	Mercaptopurine, thioguanine, azathioprine	People with low levels of TPMT enzyme develop toxic side effects after taking thiopurine drugs for the treatment of leukemia or inflammatory conditions.
HLA-B	Allopurinol, carbamazepine, abacavir	Alleles of *HLA-B* are associated with allergic reactions to these drugs used to treat gout, epilepsy, and HIV, respectively.
CYP2D6	Codeine, tramadol, tricyclic antidepressants	Numerous alleles in the population affect metabolism of many drugs leading to underdoses and overdoses.
VKORC1	Warfarin	Warfarin anticoagulant inactivates VKORC1 protein. Variants in the *VKORC1* gene produce less protein, resulting in overdose effects at normal warfarin dosages.
CYP2C9	Warfarin	This gene encodes a liver enzyme that oxidizes warfarin. Variants metabolize warfarin less efficiently, leading to overdoses.
CYP2C19	Tricyclic antidepressants, clopidogrel, voriconazole	CYP2C19 protein is a liver enzyme that metabolizes 10–15% of drugs. Alleles result in poor metabolizers to ultra-metabolizers.
SLCO1B1	Simvastatin	This gene encodes a liver transporter. Variants are less efficient at removal of statins, which are used to control cholesterol levels.

*For more information, visit the PharmGKB Web site (https://www.pharmgkb.org/index.jsp).

the wild-type *CYP2D6* gene and are known as extensive metabolizers. Approximately 10 to 15 percent of people are homozygous for alleles that decrease activity (poor metabolizers), and the remainder of the population have duplicated genes (ultra-rapid metabolizers). Examples of other gene variants that influence the effectiveness of drugs are presented in **Table ST 7.1**.

One of the primary goals of precision medicine is to provide screening to patients prior to treatment so that the choice of drug and its dosage can be tailored to the patient's genomic profile. Normally, physicians order a single-gene test only when a specific drug needs to be prescribed or when a prescribed drug is not performing as expected. Currently, tests for genetic variants in about 20 genes are available. These tests predict reactions to approximately 100 drugs representing about 18 percent of all prescriptions in the United States. Several research hospitals have initiated programs to bring extensive genomic screening to all patients prior to treatment, and prior to development of future diseases—an approach called preemptive screening (Box 1).

Developing Targeted Drugs

Another goal of pharmacogenomics is to develop drugs that are targeted to the genetic profiles of specific subpopulations of patients. The most advanced applications are in the treatment of cancers. Large-scale sequencing studies show that each tumor is genetically unique. This genomic variability has been exploited to develop new drugs that specifically target cancer cells that may express mutant proteins or overexpress others.

One of the first success stories in precision targeted therapeutics was that of the ***HER-2*** gene and the drug **Herceptin®** in breast cancer. The *HER-2* gene codes for a transmembrane tyrosine kinase receptor protein. These receptors are located within the cell membranes of normal breast epithelial cells and, when bound to other growth factor receptors and ligands on the cell surface, they send signals to the cell nucleus that result in the transcription of genes whose products stimulate cell growth and division.

In about 25 percent of invasive breast cancers, the *HER-2* gene is amplified and the protein is overexpressed on the cell surface. The presence of *HER-2* overexpression is associated with increased tumor invasiveness, metastasis, and cell proliferation, as well as a poorer patient prognosis.

Based on this knowledge, Genentech Corporation in California developed a monoclonal antibody known as trastuzumab (or Herceptin) that binds to the extracellular region of the HER-2 receptor, inhibiting HER-2 signaling, triggering cell-cycle arrest, and leading to destruction of the cancer cell.

Because Herceptin acts only on cancer cells that have amplified *HER-2* genes, it is important to know the HER-2 status of each tumor. A number of molecular assays have been developed to determine the gene and protein status of breast cancer cells. These include immunohistochemistry (IHC) and fluorescence *in situ* hybridization (FISH) assays. In IHC assays, an antibody that binds to the HER-2 protein is added to fixed tissue on a slide. The presence of bound antibody is then detected with a stain and observed under the microscope [**Figure ST 7.1(a)**]. The FISH assay (which is described in Chapters 9 and 17) assesses the number of *HER-2* genes by comparing the fluorescence signal from a HER-2 probe with a control signal from another gene that is not amplified in the cancer cells [**Figure ST 7.1(b)**].

Herceptin has had a major effect on the treatment of HER-2 positive breast cancers. When Herceptin is used in combination with chemotherapy, there is a 25 to 50 percent

BOX 1

Preemptive Pharmacogenomic Screening: The PGEN4Kids Program

Beginning in 2011, St. Jude Children's Research Hospital in Memphis, Tennessee, has offered a clinical research program entitled **PGEN4Kids.** The goal of this program is to provide patients and clinicians with pharmacogenomic screening tests for thousands of variants in hundreds of genes that may be involved in drug responses. The program is available to all incoming hospital patients, and 97 percent of patients have signed up for the program.

Using DNA derived during a single blood sample, patients are tested for approximately 2000 variants in 230 genes whose products are expected to be linked to drug responses. These tests are a combination of multigene genotyping arrays and quantitative PCR tests (described in Chapter 17). Because the effects of most of these gene variants are still unclear, the data from most of these gene tests will remain in the program's research files. However, at the present time, test results on variants in seven of the genes known to affect reactions to 23 drugs have been presented to patients and made accessible to clinicians through the patient's electronic health records. As future research reveals more associations between gene variants and drug effects, the patients' electronic health records will be automatically updated with information and alerts, to inform clinicians about how to prescribe these drugs for each individual patient. In one study, the PGEN4Kids program found that 78 percent of enrolled patients had at least one gene variant that could affect their reaction to a drug. The program also includes ongoing clinician education to introduce new information about gene–drug interactions as new research data become available.

This type of genetic screening, known as "preemptive screening" is expected to be cheaper, faster, and more efficient than ordering separate tests every time a patient is prescribed a potentially high risk drug or has had an adverse drug reaction. In 2012, a research study of patients at Vanderbilt University Medical Center revealed that almost 400 adverse drug reactions could have been avoided if clinicians had had access to a preemptive pharmacogenomic screening program.

increase in survival, compared with the use of chemotherapy alone. Herceptin has also been found effective in the treatment of other HER-2 overexpressing cancer cells, including those of the stomach and gastroesophageal junction.

There are now dozens of drugs that are targeted to the genetic status of the cancer cells (**Table ST 7.2**). For example, about 40 percent of colon cancer patients respond to the drugs **Erbitux®** (cetuximab) and **Vectibix®** (panitumumab). These two drugs are monoclonal antibodies that bind to **epidermal growth factor receptors (EGFRs)** on the surface of cells and inhibit the EGFR signal transduction pathway. To work, cancer cells must express EGFR on their surfaces and must also have a wild-type *K-ras* gene. The presence of EGFR protein can be assayed using

(a)

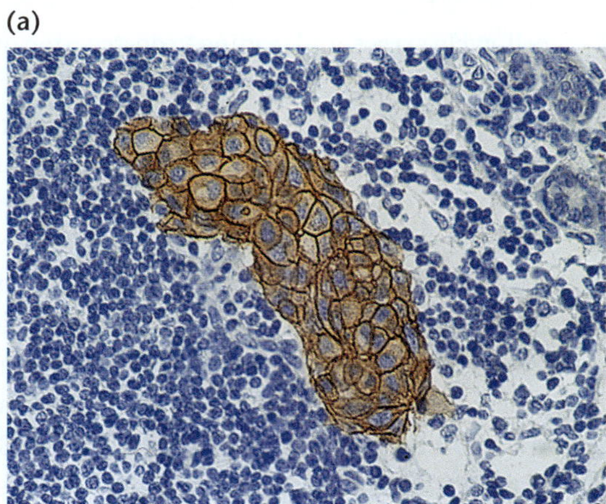

(b)

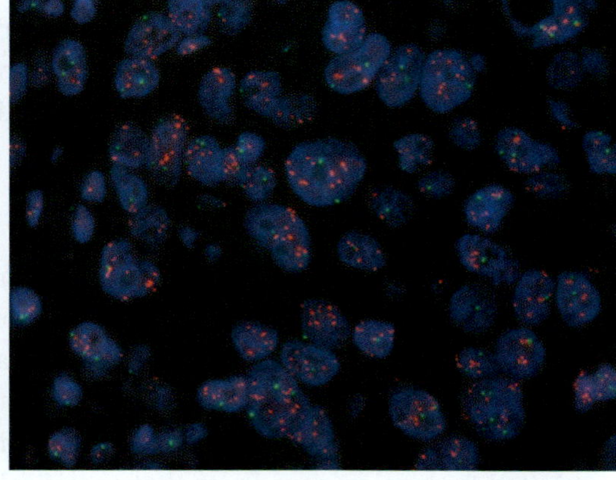

FIGURE ST 7.1 Protein and gene-amplification assays to determine HER-2 levels in cancer cells. (a) Normal and breast cancer cells within a biopsy sample, stained by HER-2 immunohistochemistry. Cell nuclei are stained blue. Cancer cells that overexpress HER-2 protein stain brown at the cell membrane. (b) Cancer cells assayed for *HER-2* gene copy number by fluorescence *in situ* hybridization. Cell nuclei are stained blue. *HER-2* gene DNA appears bright red. Chromosome 17 centromeres stain green. The degree of *HER-2* gene amplification is expressed as the ratio of red-staining foci to green-staining foci.

TABLE ST 7.2 Examples of Drugs That Specifically Target Proteins Mutated or Abnormally Expressed in Cancer Cells

Drug	Cancer Types	Target	Description
Imatinib (Gleevec)	Ph+ CML and ALL	BCR-ABL kinase	Imatinib binds to and inhibits BCR-ABL, which is encoded by the *bcl-abl* fusion gene located on the Philadelphia chromosome.
Olaparib (Lynparza)	BRCA1/2 mutated ovarian cancer	Poly ADP ribose polymerase (PARP)	BRCA1/2-defective cancers rely on PARP for DNA repair. Olaparib inhibits PARP repair, blocking cell division.
Trametinib (Mekinist)	Melanoma	Mitogen-activated protein kinase (MEK)	Trametinib inhibits mutated constitutively active MEK pathways, resulting in cell-cycle arrest and increased apoptosis.
Crizotinib (Xalkori)	NSCLC	EML4-ALK fusion kinase	Crizotinib inhibits fusion kinase activity, reducing cancer cell growth and invasion.
Vismodegib (Erivedge)	Basal-cell carcinoma	Smoothen receptor	Inhibits transcription factors that are necessary for expression of tumor genes.

Ph+ CML = Philadelphia chromosome-positive chronic myelogenous leukemia
Ph+ ALL = Philadelphia chromosome-positive acute lymphoblastic leukemia
NSCLC = non-small-cell lung carcinoma

a staining test and observation of cancer cells under a microscope. Mutations in the *K-ras* gene can be detected using assays based on the polymerase chain reaction (PCR) method, which is described earlier in the text (see Chapter 17).

ST 7.2 Precision Oncology

One of the promises of precision medicine is to treat cancer patients with therapies that target specific gene mutations and gene expression defects in their tumors, leading to effective remissions and even cures. To support these promises, advances in exomic and whole-genome sequencing methods are making these technologies more cost effective for the diagnosis of many diseases including cancers. Large research programs, such as The Cancer Genome Atlas project, are mapping the genomes of thousands of tumor types to identify mutations and expression profiles for which targeted drugs can be developed. Targeted therapies and diagnostics also benefit from high throughput proteomic and metabolomic assays.

As described in the previous section, many cancer drugs targeted to specific genetic and gene expression profiles are already being used, sometimes with dramatic effects (Box 2). So far, the percentage of patients that can be successfully treated with precision cancer drugs is small. One clinical trial showed that only 6.4 percent of enrolled patients could be matched to a targeted drug based on their tumor's genomic profile. Another challenge is to deal with tumor resistance. To circumvent resistance, it will be necessary to use multiple treatment approaches simultaneously—both targeted and generalized.

Beyond the use of targeted drugs, researchers are also making progress in the use of other targeted modalities, including targeted cancer immunotherapies, which are described next.

Targeted Cancer Immunotherapies

Some of the most promising new developments in precision medicine are in the field of cancer **immunotherapy.** These therapies harness the patient's own immune system to kill tumors, and some have brought remarkable therapeutic effects in clinical trials and triggered billions of dollars of investment into their development. In this section, we will describe two of the most promising precision cancer immunotherapies—*adoptive cell transfer* and *engineered T-cell* methods.

To understand how these therapies work, we need to briefly review how the immune system, particularly **T cells,** defends against the development of cancer. As summarized in Box 3 and **Figure ST 7.2**, the immune system consists of cell types and chemical signals constituting the innate and the adaptive systems.

Both adoptive cell transfer and engineered T-cell methods exploit **cytotoxic T lymphocytes** (CTLs) to recognize specific antigens on the surface of cancer cells, bind to the cells, and destroy them. Box 4 and **Figure ST 7.3** summarize the steps involved in normal T-cell recognition and destruction of cancer cells.

BOX 2

Precision Cancer Diagnostics and Treatments: The Lukas Wartman Story

During his final year of medical school in 2002, Dr. Lukas Wartman began to experience symptoms of fatigue, fever, and bone pain. After months of tests, he was diagnosed with adult acute lymphoblastic leukemia (ALL). Following two years of chemotherapies, his cancer went into remission for three years. When the ALL recurred, his doctors treated him with intensive chemotherapy and a bone marrow transplant, which put him back into remission for another three years. After his second relapse, all attempts at treatment failed and he was rapidly deteriorating.

At the time of his second relapse, Dr. Wartman was working as a physician-scientist at Washington University, researching the genetics of leukemia. His colleagues, including Dr. Timothy Ley, associate director of the Washington University Genome Institute, decided to rush into a last-minute effort to save him. Using the university's sequencing facilities and supercomputers, the research team sequenced the entire genome of both his normal and his cancer cells. They also analyzed his RNA types and expression levels using RNAseq technologies.

As they had expected, Dr. Wartman's cancer cells contained many gene mutations. Unfortunately, there were no known drugs that would attack the products of these mutated genes. The RNA sequence analysis, however, revealed unexpected results.

It showed that the fms-related tyrosine kinase 3 (*FLT3*) gene, although having a normal DNA sequence, was overexpressed in his cancer cells—perhaps due to mutations in the gene's regulatory regions. The *FLT3* gene encodes a protein kinase that is involved in normal hematopoietic cell growth and differentiation, and its overexpression would be a potentially important contributor to Dr. Wartman's cancer. Equally informative, and fortunate, was that the drug sunitinib (Sutent®) was known to inhibit the FLT3 kinase and had been approved for use in the treatment of some kidney and gastrointestinal cancers.

Dr. Wartman decided to try sunitinib. Unfortunately, the drug cost $330 per day, and Dr. Wartman's insurance company refused to pay for it. In addition, the drug company Pfizer refused to supply the drug to him under its compassionate use program. Despite these setbacks, he collected enough money to buy a week's worth of sunitinib. Within days of starting treatment, his blood counts were approaching normal. Within two weeks his bone marrow was free of cancer cells. At this juncture, Pfizer reversed its decision and supplied Dr. Wartman with the drug. In addition, he underwent a second bone marrow transplant to help ensure that the cancer would not return. Although Dr. Wartman's long-term prognosis is still uncertain, his successful experience with precision cancer treatment has given him hope and has spurred research into the regulation of the *FLT3* gene in other cancers.

Cancer cells express many proteins that are specific to the tumor and have the capacity to be recognized by the patient's immune system as nonself antigens. These nonself antigens result from abnormal gene expression and mutations in the coding regions of both cancer driver and passenger genes. For example, 30 percent of human cancers contain mutated *ras*-family genes (such as *K-ras* and *H-ras*), which act as cancer driver genes. Many different point mutations can occur in these genes, each encoding an altered protein that is not found in normal cells. Cancer cells also contain up to hundreds of mutations in passenger genes whose products are not involved in the cancer phenotype, but also encode mutated, and hence nonself, proteins. Collectively, the novel, nonself antigens that are contained within their proteins are known as **neoantigens.**

Although T cells are known to associate with tumors and are able to recognize tumor neoantigens, they are often not able to destroy tumor cells. These tumor-associated T cells are also known as **tumor-infiltrating lymphocytes (TILs).**

Cancers use many different strategies to suppress T-cell responses. These strategies include the synthesis of molecules that bind to T cells and repress their activity. Interestingly, some effective new drugs called *checkpoint inhibitors* help T cells avoid these checkpoint molecules, thereby enhancing the tumor-killing ability of TILs. Another way that tumors avoid immune system activity is that they are often abnormal in their expression of cell-surface major histocompatibility complex (MHC) molecules, which are essential to stimulate antigen-presenting cells, which in turn are necessary to stimulate T cells to recognize and kill cells that bear nonself antigens (see Box 4). A third way that tumors avoid immune responses is through the presence of tumor-associated **regulatory T cells** called T-regs (including **suppressor T cells**), whose role is to repress the activities of activated T cells. The presence of other tumor-infiltrating cells such as **macrophages** and **monocytes** also repress the activities of T cells.

To circumvent and overwhelm the mechanisms that cancers use to repress anticancer immune responses, scientists have developed the following personalized T-cell–based therapies.

Adoptive cell transfer **Adoptive cell transfer (ACT)** involves removing TILs from a patient's tumor, selecting those that specifically recognize tumor antigens,

BOX 3

Cell Types in the Innate and Adaptive Immune Systems

The immune system is made up of a large number of cell types and chemicals that protect the body from external and internal "nonself" entities such as bacteria, viruses, toxins, and tumor cells (**Figure ST 7.2**).

The **innate system** acts rapidly and nonspecifically to these agents, engulfing or degrading them. Some components also assist cells of the adaptive system.

The **adaptive (or acquired) system** destroys pathogens, tumor cells, and molecules such as toxins by recognizing and acting specifically against each entity. It does this by recognizing specific "nonself" molecules called **antigens.** Cells of the adaptive system develop a memory of previous contact with nonself antigens, allowing them to quickly replicate and respond to a subsequent appearance of the antigen.

The adaptive system has two branches. The humoral branch involves B lymphocytes (B cells) that synthesize antibodies directed at specific antigens. The cell-mediated branch consists of T lymphocytes (T cells) including cytotoxic T cells and helper T cells. These cells recognize specific antigens on the surface of or inside cells that are infected or cancerous. Cytotoxic T cells then contact the cell, release cytotoxic molecules, and trigger apoptosis of the target cell.

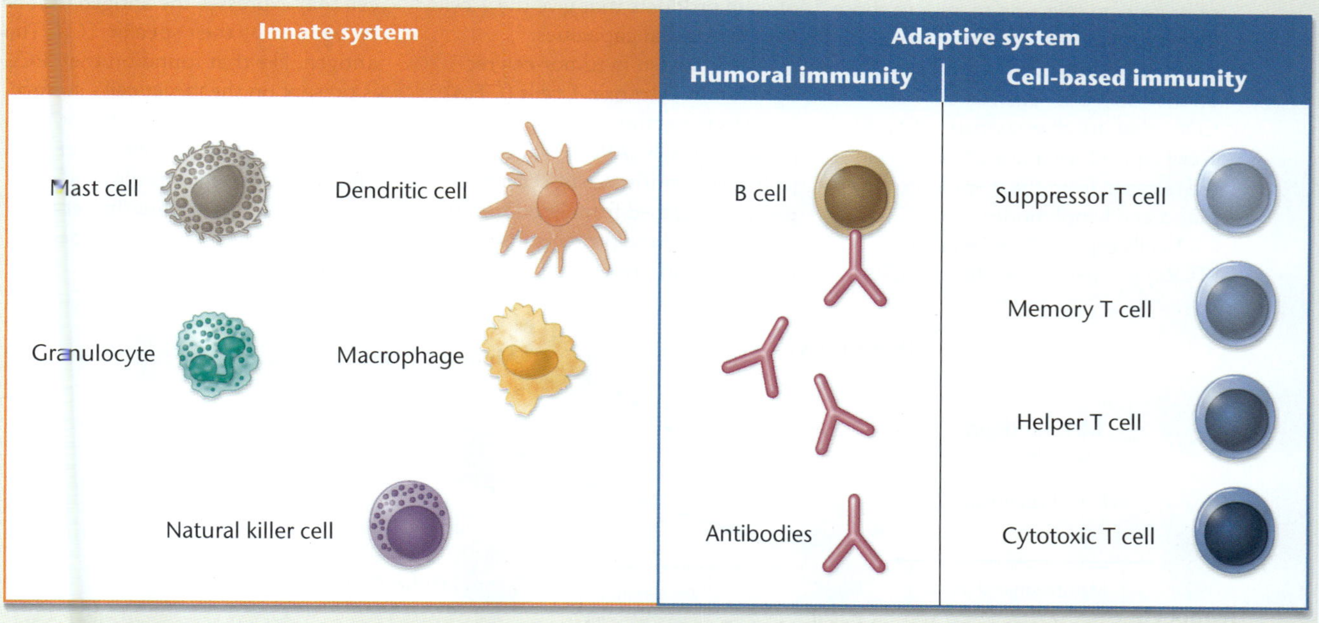

FIGURE ST 7.2 Cell types of the innate and adaptive immune systems.

amplifying these specific TILs *in vitro*, and reintroducing them back into the patient.

The steps in ACT are summarized in **Figure ST 7.4**. In the first step, tumor specimens that contain TILs are removed from the patient and digested into small samples containing one or several cells. Each sample is grown in a culture dish in the presence of tumor material and IL-2, a growth factor for T cells. As the T cells grow in the dish, those with reactivity to the tumor cells destroy the tumor cells within two to three weeks. These T cells are selected and retested for their tumor-destroying activity in coculture assays. Positive T cells are then grown to high numbers (10^{11} cells) in the lab, in the presence of several growth-stimulatory factors. The process requires about six weeks from obtaining the tumor specimen to harvesting the amplified reactive T-cell preparation. At this point, the patient is treated with chemotherapy to rid the body of immune system cells such as T-regs and macrophages that repress the activity of activated T cells. Then, the patient is reinfused with the amplified T cells and IL-2. The adoptive T cells can continue to expand up to 1000-fold after reinfusion. In some patients, the tumor-reactive T cells can be found in the circulation months after the initial infusion, where they make up as much as 80 percent of the T-cell population. The persistence of the adoptive T cells correlates with a positive antitumor effect.

The results of some ACT clinical trials have produced positive results, particularly in patients with metastatic melanoma—a cancer that normally has a poor outcome. For example, in trials conducted by the National Cancer Institute, after only one treatment, the outcome was

BOX 4

Steps in Cytotoxic T-cell Recognition, Activation, and Destruction of Cancer Cells

T cells (T lymphocytes) play a major role in cell-mediated immunity, including the recognition and destruction of cancer cells. Immature thymocytes, the precursors of T cells, originate in the bone marrow and move to the thymus where they progress through developmental stages to become naïve, inactive T cells. During their residence in the thymus, T cells undergo selection to remove those that recognize self antigens. They also begin to express specific glycoproteins on their surfaces—such as CD4 and CD8—that are characteristic of each T-cell type. Once matured, the T cells are released from the thymus into the blood and lymph nodes.

T cells express T-cell receptors (TCRs) on their cell surfaces. TCRs are transmembrane protein complexes composed of two polypeptide chains. Both of these chains have an amino-terminal variable region and a constant region, similar to antibodies. Variable regions make contact with antigens that are present on both antigen-presenting cells and target cells (infected or tumor cells). Intracellular domains of TCRs contain regions that send signals to the T cell after the variable regions make contact with a specific antigen. These signals activate the T cell, leading to T-cell proliferation and expression of gene products that give the T cells their functional capacities.

The first step in tumor-cell recognition by cytotoxic T cells (CTLs) involves the actions of antigen-presenting cells (APCs). These cells, primarily dendritic cells, scavenge proteins released from lysed tumor cells, digest these proteins into short peptides, and present the peptides on their cell surfaces, in association with cell-surface molecules known as major histocompatibility complex (MHC) molecules. MHC molecules are present on the surfaces of most cells in the body. APCs bearing MHC-antigen complexes travel to the lymph nodes where they contact naïve T cells bearing TCRs that have binding affinity for each specific antigen. After contact with the MHC antigen on an APC cell, the activated T cell proliferates and expresses "effector" gene products.

Once activated, CTLs travel through the body, until they contact a cell that expresses the antigen that is recognized by their specific TCR. This antigen, like that found on the APCs, is present on the target cell's surface in association with an MHC molecule. CTLs bind tightly to their target cells and then release molecules such as perforin and granzymes that enter the target cells, inducing cell death through apoptosis.

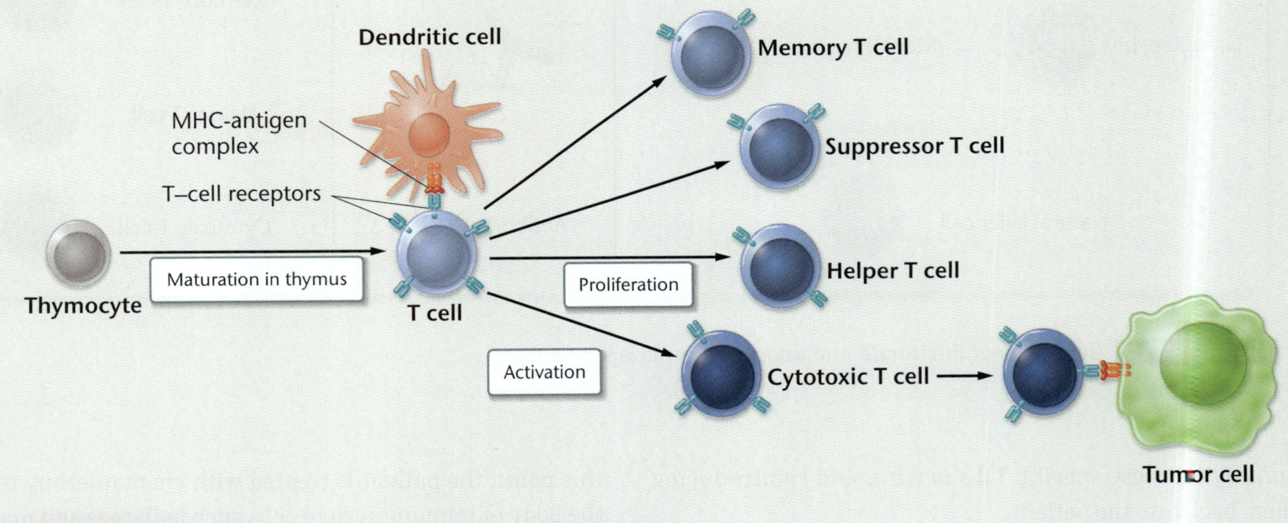

FIGURE ST 7.3 Steps in the maturation and activation of T cells.

remarkable. Between 53 and 72 percent of patients showed positive responses to the treatment, 22 percent showing complete regressions and 20 percent having no recurrence of their cancer up to 10 years later.

The promising results of ACT clinical trials on metastatic melanoma have encouraged attempts to use this method to treat other malignancies, such as cervical cancer and some blood cancers. To extend ACT therapies to patients who may not have activated TILs within their tumors that recognize unique neoantigens on those tumors, scientists are developing other ways to target the immune system to cancer. One of these methods is described next.

Immunotherapy with genetically engineered T Cells The principle behind genetically engineered T-cell therapies is to create recombinant **T-cell receptors (TCRs)** that specifically recognize antigens on cancer cells. The DNA sequences that encode these engineered TCRs are then introduced *in vitro* into a patient's normal,

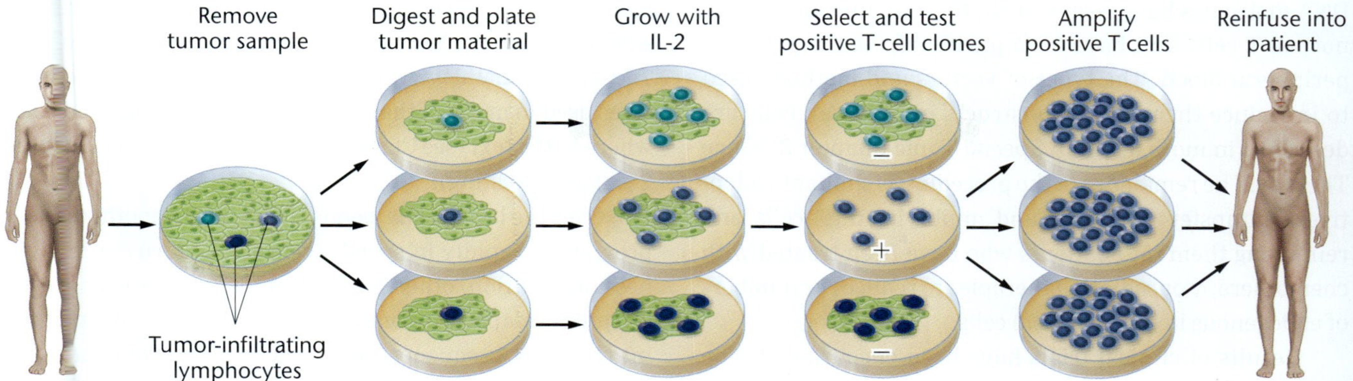

FIGURE ST 7.4 Summary of adoptive cell transfer method.

naïve T cells which then express these TCRs on their surfaces. The TCR-transduced T cells are then selected, amplified, and reinfused into the patient. The synthetic TCR genes encode either TCRs that are structurally similar to natural TCRs or **chimeric antigen receptors (CARs)** that can directly recognize antigens on the tumor cell without requiring T-cell activation by antigen-presenting cells. In this subsection we will describe CAR T-cell therapies.

CAR proteins are fusions of several proteins derived from a variety of sources. (The structures of normal TCRs and CARs are shown in **Figure ST 7.5**.) The extracellular portion of a CAR consists of the variable regions of immunoglobulin heavy and light chains, separated by a linker sequence. These variable regions fold in such a way that they mimic the specificity of an antibody that recognizes a specific antigen—such as a tumor neoantigen. The variable antibody regions are preceded by a signal peptide to direct the CAR to the surface of the T cell in which it is expressed. A spacer region allows the variable regions to orient themselves to bind to antigens on the cancer cell. A transmembrane region anchors the CAR in the T-cell membrane, and the intracellular region is responsible for sending various activation signals to the T cell, after the variable antibody regions have made contact with an antigen. The activation signals include instructions to proliferate, differentiate, produce cytokines, and kill the target cell.

To create CARs, scientists clone DNA fragments encoding each of the regions described above into a single linear recombinant DNA molecule, which encodes the entire CAR fusion protein. The DNA fragment encoding immunoglobulin variable regions is usually cloned from cells that secrete monoclonal antibodies. The cells that produce these monoclonal antibodies have previously been screened and selected for their reactivity against the desired neoantigens found on the surface of cancer cells. Once the chimeric

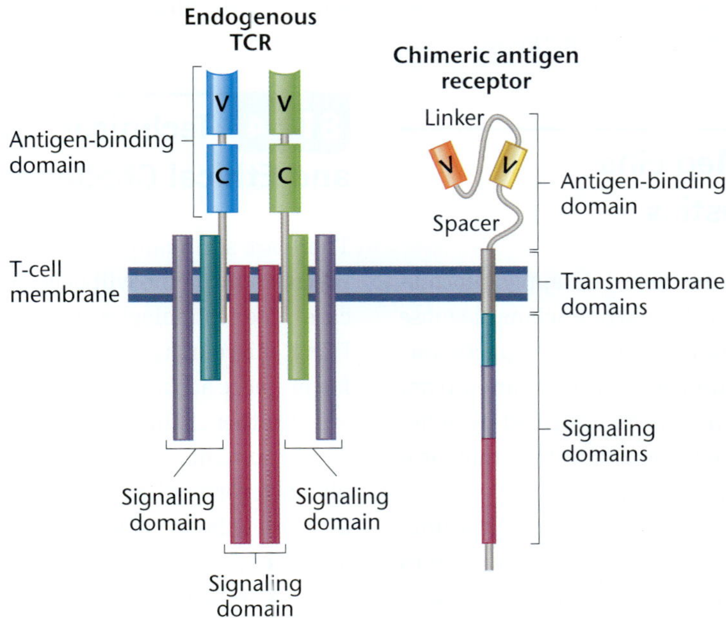

FIGURE ST 7.5 Structures of an endogenous T-cell receptor (TCR) and a recombinant chimeric antigen receptor (CAR). Variable regions of antigen-binding domains are labelled V and constant regions as C.

DNA molecules have been cloned, they are introduced into normal T cells that have been purified from the patient's peripheral blood. The types of vectors and methods used to introduce therapeutic constructs into human cells are described in more detail in Special Topic Chapter 3—Gene Therapy. The remainder of the procedure is similar to adoptive cell transfer—screening and amplifying the T cells, and reinfusing them into patients who have been treated with chemotherapy or radiation therapies to reduce the numbers of endogenous immune system cells.

Results of clinical trials have been encouraging. The majority of trials have tested CAR T cells for effectiveness in treating B-cell cancers such as leukemias and lymphomas. The CAR T cells recognize the CD19 surface proteins that are expressed on B cells but not on other cells. The response rates have varied between 70 and 100 percent with reports of long-term remissions of up to several years. Several clinical trials have tested CAR T-cell therapies against solid tumors, with less encouraging results.

As of October 2017, two CAR T-cell therapies have been approved by the US Food and Drug Administration (FDA). In August 2017, a CAR T-cell therapy called Kymriah™ was approved for treating children with acute lymphoblastic leukemia (ALL) who had relapsed twice or did not respond to earlier treatments. In clinical trials of these patients, Kymriah treatment produced complete remissions in 83 percent of patients. In October 2017, the FDA approved a CAR T-cell therapy called Yescarta™ for treating adults with some types of large B-cell lymphomas.

Despite promising results with CAR T-cell therapies, these treatments have several serious side effects in many patients. These include systemic inflammatory responses, neurotoxicity, and eventual tumor resistance. Researchers expect that these side effects will become manageable as more experience is gained with these new therapies.

ST 7.3 Precision Medicine and Disease Diagnostics

The ultimate goal of precision medicine is to apply information from a patient's full genome to help physicians diagnose disease and select treatments tailored to that particular patient. Not only will this information be gleaned from genome sequencing, but it will also be informed by gene-expression information derived from transcriptomic, proteomic, metabolomic, and epigenetic tests.

Presently, the most prevalent use of genomic information for disease diagnostics is genetic testing that examines specific disease-related genes and gene variants. Most existing genetic tests detect the presence of mutations in single genes that are known to be linked to a disease. Currently, more than 45,000 genetic tests are available. A comprehensive list of genetic tests can be viewed on the NIH Genetic Testing Registry at www.ncbi.nlm.nih.gov/gtr/. The technologies used in many of these genetic tests are presented elsewhere in the text (see Chapter 18 and Special Topic Chapter 2—Genetic Testing).

Over the last decade, genome sequencing methods have progressed rapidly in speed, accuracy, and cost effectiveness. In addition, other "omics" technologies such as transcriptomics and proteomics are providing major insights into how DNA sequences lead to gene expression and, ultimately, to phenotype. (Refer to Chapter 18 for descriptions of techniques and data emerging from human omics technologies.) As these technologies become more rapid and cost effective, they will begin to make important contributions to precision medicine.

Although the application of omics technologies to precision diagnostics has not yet entered routine medical care, several proof-of-principle cases have illustrated the ways in which whole-genome analysis may be applied in the future. They also reveal some of the limitations that must be overcome before genome-based medicine becomes commonplace and practical.

One such case study is described in Box 5. This study combined data from whole-genome sequencing, transcriptomics, proteomics, and metabolomics profiles from a single patient at multiple time points over a 14-month period. This in-depth, multilevel personal profiling followed the patient through both healthy and diseased states, as he contracted two virus infections and a period of Type 2 diabetes. This research points out how complex changes in gene expression may affect phenotype and shows the importance of looking beyond the raw sequence of DNA. It also indicates that gene-expression profiles can be monitored by current technologies and may be applied in the future as part of personalized medical testing.

ST 7.4 Technical, Social, and Ethical Challenges

There are still many technical hurdles to overcome before precision medicine will become a standard part of medical care. The technologies of genome sequencing, omics profiling, microarray analysis, and SNP detection need to be faster, more accurate, and cheaper.

Another challenge will be the storage and interpretation of vast amounts of genomic and other omics data. For example, each personal genome generates the letter-equivalent of 200 large phone books, which must be stored in databases and mined for relevant sequence variants. Then, meaning must be assigned to each sequence variant. To undertake these kinds of analyses, scientists need to gather data from large-scale population genotyping studies that will link

BOX 5

Beyond Genomics: Personal Omics Profiling

A study published by a research team led by Dr. Michael Snyder of Stanford University provides an example of how multiple omics technologies can be used to examine one person's healthy and diseased states.[1]

Blood samples were taken from a healthy individual (Dr. Snyder) at 20 time points over a 14-month period. Dr. Snyder's whole-genome sequence was generated at each time point using two different methods and backed up by exome sequencing using three different methods. In addition, his genome sequence was compared to that of his mother. Concurrently, whole-transcriptome sequencing, proteomic profiling, and metabolomics assays were performed.

Using RNAseq technologies, the researchers monitored the numbers and types of more than 19,000 mRNAs and miRNAs transcribed from more than 12,000 genes over 20 time points. The data showed that sets of genes were coordinately regulated in response to conditions such as RSV infection and glucose levels. The researchers also found that RNA species underwent differential splicing and editing during changes in physiological states. Editing events included changes of adenosine to inosine and cytidine to uridine, and many of these RNA edits altered the amino acid sequences of translated proteins.

The researchers also profiled the levels of more than 6000 proteins and metabolites over the time course of the study. Like the RNA data, the protein and metabolite data showed coordinated changes that occurred throughout virus infections as well as glucose level changes. Some of these changes were shared between RNA, protein, and metabolites, and others were unique to each category. The medical significance of these patterns is not clear but will be addressed in future studies.

Dr. Snyder's genome sequence revealed a number of SNPs that are known to be associated with elevated risks for coronary artery disease, basal cell carcinoma, hypertriglyceridemia, and Type 2 diabetes. A mutation in the *TERT* gene, which is involved in telomere replication, is associated with an increased risk for aplastic anemia.

These omics assays were followed by a series of medical tests. Dr. Snyder had no signs of aplastic anemia, and his telomere lengths were close to normal. Similarly, his mother, who shared his mutation in the *TERT* gene, had no symptoms of aplastic anemia. Medical tests revealed he did have elevated triglyceride levels, which he subsequently controlled using medication. Blood glucose levels were initially normal but became abnormally high after he became infected with respiratory syncytial virus (RSV). In response to these data, Dr. Snyder modified his diet and exercise regime and later brought his blood glucose down to normal levels. An analysis of drug response gene variants revealed that he should have good responses to diabetes drugs, should he need them in the future.

The detailed analysis of Dr. Snyder's genome and gene expression profiles generated approximately 500,000 gigabytes of data.

Since the study was published, Dr. Snyder has expanded his research to examine another 100 people, for their genomic, proteomic, transcriptomic, and other profiles. The goal is to analyze these big-data collections to predict diseases and match profiles with targeted therapies.

[1] Chen, R. et al. (2012). Personal omics profiling reveals dynamic molecular and medical phenotypes. *Cell* 148:1293–1307.

sequence variants to phenotype, disease susceptibility, or drug responses. Experts suggest that such studies will take the coordinated efforts of public and private research teams more than a decade to complete. Scientists will also need to develop efficient automated systems and algorithms to deal with this massive amount of information. Precision medicine will also need to integrate information about environmental, personal lifestyle, and epigenetic factors.

Another technical challenge for precision medicine is the development of automated health information technologies. Health-care providers will need to use electronic health records to store, retrieve, and analyze each patient's genomic profile, as well as to compare this information with constantly advancing knowledge about genes and disease. Currently, approximately 10 percent of hospitals and physicians in the United States have access to these types of information technologies.

Precision medicine raises a number of societal concerns. To make precision medicine available to everyone, the costs of genetic tests, as well as the genetic counselling that accompanies them, must be reimbursed by insurance companies, even in cases where there are no prior diseases or symptoms. Regulatory changes are required to ensure that genetic tests and genomic sequencing are accurate and that the data generated are reliably stored in databases that guarantee the patient's privacy.

Precision medicine also requires changes to medical education. In the future, physicians will be expected to use genomics information as part of their patient management. For this to be possible, medical schools will need to train future physicians to interpret and explain genetic data. In addition, more genetic counsellors and genomics specialists will be required. These specialists will need to understand genomics and disease, as well as to manipulate bioinformatic data. As of 2017, there were only about 4000 licensed genetic counsellors in the United States.

The ethical aspects of precision medicine are also diverse and challenging. For example, it is sometimes argued that the costs involved in the development of genomics and precision medicine are a misallocation of limited resources. Some argue that science should solve larger problems facing humanity, such as the distribution of food and clean water, before allocating resources on precision medicine. Similarly, some critics argue that such highly specialized and expensive medical care will not be available to everyone and represents a worsening of economic inequality. There are also concerns about how we will protect the privacy of genome information that is contained in databases and private health-care records.

Most experts agree that we are at the beginning of a precision medicine revolution. Information from genetics and genomics research is already increasing the effectiveness of drugs and enabling health-care providers to predict diseases prior to their occurrence. In the future, precision and personalized medicine will touch almost every aspect of medical care. By addressing the upcoming challenges of precision medicine, we can guide its use for the maximum benefit to the greatest number of people.

Review Questions

1. What is pharmacogenomics, and how does it differ from pharmacogenetics?
2. Describe how the drug Herceptin works. What types of gene tests are ordered prior to treatment with Herceptin?
3. How do the cytochrome P450 proteins affect drug responses? Give two examples.
4. What are some of the ways that cancer cells avoid the killing effects of T cells?
5. What is adoptive cell transfer, and how is it being used to treat cancer?
6. What are chimeric antigen receptors, and how are they constructed?
7. Why is it necessary to examine gene-expression profiles, in addition to genome sequences, for effective precision medicine?
8. Using the information available on the NIH Genetic Testing Registry Web site, describe three single-gene tests that are currently used in disease diagnostics.

Discussion Questions

1. In this chapter, we present two case studies (Boxes 2 and 5) that use precision genomics analysis to predict and treat diseases. These cases have shown how precision medicine may evolve in the future and have inspired enthusiasm; however, they have also triggered concerns. What are some of these concerns, and how can they be addressed?
2. What are the biggest technical challenges that must be overcome before precision medicine becomes a routine component of medical care? What do you think is the most difficult of these challenges and why?
3. How can we ensure that a patient's privacy is maintained as genome information accumulates within medical records? How would you feel about allowing your genome sequence to be available for use in research?
4. As gene tests and genomic sequences become more commonplace, how can we prevent the emergence of "genetic discrimination" in employment and medical insurance?

APPENDIX

Solutions to Selected Problems and Discussion Questions

Chapter 1

2. Your essay should include a description of the impact of recombinant DNA technology on the following: plant and animal husbandry and production, drug development, medical advances, and understanding gene function.

4. The genotype of an organism is defined as the specific allelic or genetic constitution of an organism, or, often, the allelic composition of one or a limited number of genes under investigation. The observable feature of those genes is called the phenotype.

6. A gene is a portion of DNA that encodes the information required to make a specific protein. The DNA is transcribed to make a messenger RNA, which is then translated by ribosomes into a chain of amino acids. The chain then folds up into a specific structure and is processed into the finished protein.

8. The central dogma of molecular genetics refers to the relationships among DNA, RNA, and proteins. The processes of transcription and translation are integral to understanding these relationships. Because DNA and RNA are discrete chemical entities, they can be isolated, studied, and manipulated in a variety of experiments that define modern genetics.

10. Restriction enzymes (endonucleases) cut double-stranded DNA at particular base sequences. Vectors are carrier DNA molecules that can enter and replicate in cells. When vector and target DNAs are cleaved with the same enzyme, complementary ends are created, which can be combined and ligated to form intact double-stranded structures. Once recombinant molecules enter cells, they can replicate, producing many copies or clones.

12. Unique transgenic plants and animals can be patented, as the U.S. Supreme Court ruled in 1980. Supporters of organismic patenting argue that it is needed to encourage innovation and allow the costs of discovery to be recovered. Capital investors assume that there is a likely chance that their investments will yield positive returns. Others argue that natural substances should not be privately owned and that once they are owned by a small number of companies, free enterprise will be stifled.

14. Several cancers originate in conserved cellular mechanisms. Some of these processes, such as DNA repair (where *mutL* in *E. coli* and *MLH1* in humans play the same role), are often carried out by the same genes in *E. coli* and humans. *E. coli* has a very short generation time, and it is possible to generate and study the effect of different mutations in a short duration of time in such simple model organisms.

16. For approximately 60 years discoveries in genetics have guided our understanding of living systems, aided rational drug design, and dominated many social discussions. Genetics provides the framework for universal biological processes and helps explain species' stability and diversity. Given the central focus of genetics in so many of life's processes, it is understandable why so many genetic scientists have been awarded the Nobel Prize.

Chapter 2. Answers to Now Solve This

2.1. **(a)** 32 chromatids **(b)** 16 daughter chromosomes moving to each pole

2.2. **(a)** eight tetrads **(b)** eight dyads **(c)** eight monads

2.3. Not necessarily. If crossing over occurred in meiosis I, then the chromatids in the secondary oocyte are not identical.

Solutions to Problems and Discussion Questions

2. Mitosis maintains chromosomal constancy, so there is no change in chromosome number or kind in the two daughter cells. By contrast, meiosis provides for a reduction in chromosome number and an opportunity for exchange of genetic material between homologous chromosomes. This leads to the production of numerous potentially different haploid (n) cells. During oogenesis, only one of the four meiotic products is functional; however, all four meiotic products of spermatogenesis are potentially functional. Errors during either mitosis or meiosis (such as nondisjunction events) can lead to cells with too many or too few chromosomes.

4. Chromosomes that are homologous share many properties including *overall length, position of the centromere (metacentric, submetacentric, acrocentric, telocentric), banding patterns,* and *type and location of genes. Diploidy* is a term often used in conjunction with the symbol $2n$. It means that both members of a homologous pair of chromosomes are present. *Haploidy* refers to the presence of a single copy of each homologous chromosome and is symbolized as n.

6. Because a major section of Chapter 2 deals with mitosis, it would be best to deal with this question by reviewing the appropriate section in the text and examining the corresponding figures.

8. Plant cells have a chromatin-driven spindle assembly pathway unlike animal cells that have centrosomes. Many of the components necessary for microtubule polymerization are present in plant cells even though centrosomes are absent.

10. **(a)** *Synapsis* is the point-by-point pairing of homologous chromosomes during prophase of meiosis I.

(b) *Bivalents* are those structures formed by the synapsis of homologous chromosomes: There are two chromosomes (and four chromatids) that make up a bivalent.

(c) *Chiasmata* is the plural form of chiasma and refers to the structure, when viewed microscopically, of crossed chromatids.

(d) *Crossing over* is the exchange of genetic material between chromatids. It is a method of providing genetic variation through the breaking and rejoining of chromatids.

(e) *Sister chromatids* are "post-S phase" structures of replicated chromosomes. Sister chromatids are genetically identical (except where mutations have occurred) and are originally attached to the same centromere.

(f) *Tetrads* are synapsed homologous chromosomes and are thereby composed of four chromatids.

(g) *Dyads* are replicated chromosomes, composed of two sister chromatids joined by a centromere.

(h) *Monads*, or daughter chromosomes, result when the centromeres divide at anaphase II of meiosis.

12. During meiosis I, chromosome number is reduced to haploid complements. This is achieved by synapsis of homologous chromosomes and their subsequent separation. During mitosis, chromosome number and content are maintained. If homologous chromosomes paired, it would be mechanically difficult to produce two identical daughter nuclei. However, by having chromosomes unpaired at metaphase of mitosis, only centromere division is required for daughter cells to eventually receive identical chromosomal complements.

14. Cell cycle stages follow a specific order of events. In addition, a new stage is never initiated without the completion of the prior stage. This is due to sequential activation of cyclin and cyclin-dependent kinases (CDK). For example, mitotic cyclin/CDKs can only be active in mitosis and never in G1, S, or G2. This regulation is established by cell cycle specific transcription, post-translational modifications, regulating cellular localization of cyclin, CDK partners, and cyclin/CDK inhibitors.

16. There would be 2^8(256) possible combinations.

18. One half of each tetrad will have a maternal homolog: $(1/2)^{23}$.

20. In multicellular plants, the diploid sporophyte stage predominates. Meiosis results in the formation of the haploid male and female gametophyte stage, producing pollen and ovules. Following fertilization, the sporophyte is formed.

22. The checkpoints monitor the DNA and the quality of certain events of the cell cycle, such as the attachment of spindles to kinetochores. DNA breaks or modified bases can trigger DNA repair mechanisms, whereas unattached spindles can activate the spindle checkpoint. To repair the mistakes, activated checkpoints halt the progress of the cell cycle by activating CDK complex inhibitors.

24. 50, 50, 50, 100, 200

26. At the end of prophase I, maternal and paternal copies of each homologous chromosome (A^m and A^p, B^m and B^p, and C^m and C^p) will have completed synapsis and begun to pull apart. Eight possible combinations of products will occur at the end of anaphase I: A^m or A^p, B^m or B^p, C^m or C^p.

28. Eight (2 × 2 × 2) combinations are possible.

< > A^m or A^p
< > B^m or B^p
< > C^m or C^p

30. B chromosomes move as dyads into one of the secondary oocytes after meiosis II. This results in a cell with two monads for chromosome B (and single monads for A and C). The zygote complement after fusion with a normal gamete (single monads for each chromosome) will be: two copies of chromosome A, three copies of chromosome B, and two copies of chromosome C.

Chapter 3. Answers to Now Solve This

3.1. *P* = checkered; *p* = plain. Genotypes of all individuals:

		Progeny	
	P_1 Cross	**Checkered**	**Plain**
(a)	*PP* × *PP*	*PP*	
(b)	*PP* × *pp*	*Pp*	
(c)	*pp* × *pp*		*pp*
	$F_1 \times F_1$ Cross	**Checkered**	**Plain**
(d)	*PP* × *pp*	*Pp*	
(e)	*Pp* × *pp*	*Pp*	*pp*
(f)	*Pp* × *Pp*	*PP*, *Pp*	*pp*
(g)	*PP* × *Pp*	*PP*, *Pp*	

3.2. Symbolism as before:

w = wrinkled seeds *g* = green cotyledons
W = round seeds *G* = yellow cotyledons

Examine each characteristic (seed shape vs. cotyledon color) separately.

(a) Notice a 3:1 ratio for seed shape; therefore, *Ww* × *Ww*, and no green cotyledons; therefore, *GG* × *GG* or *GG* × *Gg*. Putting the two characteristics together gives *WwGG* × *WwGG* or *WwGG* × *WwGg*.

(b) *WwGg* × *wwGg*.

(c) *WwGg* × *WwGg*.

(d) *WwGg* × *wwgg*.

3.3. (a)

First gene	Second gene	Third gene	Genotype ratios	Phenotype ratios
1/4 *AA*	1/2 *BB*	1/2 *CC*	= 1/16 *AABBCC*	12/16 *A–B–C–*
		1/2 *Cc*	= 1/16 *AABBCc*	
	1/2 *Bb*	1/2 *CC*	= 1/16 *AABbCC*	
		1/2 *Cc*	= 1/16 *AABbCc*	
2/4 *Aa*	1/2 *BB*	1/2 *CC*	= 2/16 *AaBBCC*	
		1/2 *Cc*	= 2/16 *AaBBCc*	
	1/2 *Bb*	1/2 *CC*	= 2/16 *AaBbCC*	
		1/2 *Cc*	= 2/16 *AaBbCc*	
1/4 *aa*	1/2 *BB*	1/2 *CC*	= 1/16 *aaBBCC*	4/16 *aaB–C–*
		1/2 *Cc*	= 1/16 *aaBBCc*	
	1/2 *Bb*	1/2 *CC*	= 1/16 *aaBbCC*	
		1/2 *Cc*	= 1/16 *aaBbCc*	

(b)

First gene	Second gene	Third gene	Genotype ratios	Phenotype ratios
1/2 *Aa*	1 *BB*	1/4 *CC*	= 1/8 *AaBBCC*	3/8 *AaBBC–*
		2/4 *Cc*	= 2/8 *AaBBCc*	
		1/2 *Cc*	= 1/8 *AaBBcc*	1/8 *AaBBcc*
1/2 *aa*	1 *BB*	1/4 *CC*	= 1/8 *aaBBCC*	3/8 *aaBBC–*
		2/4 *Cc*	= 2/8 *aaBBCc*	
		1/2 *Cc*	= 1/8 *aaBBcc*	1/8 *aaBBCC*

(c) There will be eight (2^n) different kinds of gametes from each of the parents and therefore a 64-box Punnett square. Doing this problem by the forked-line method helps considerably.

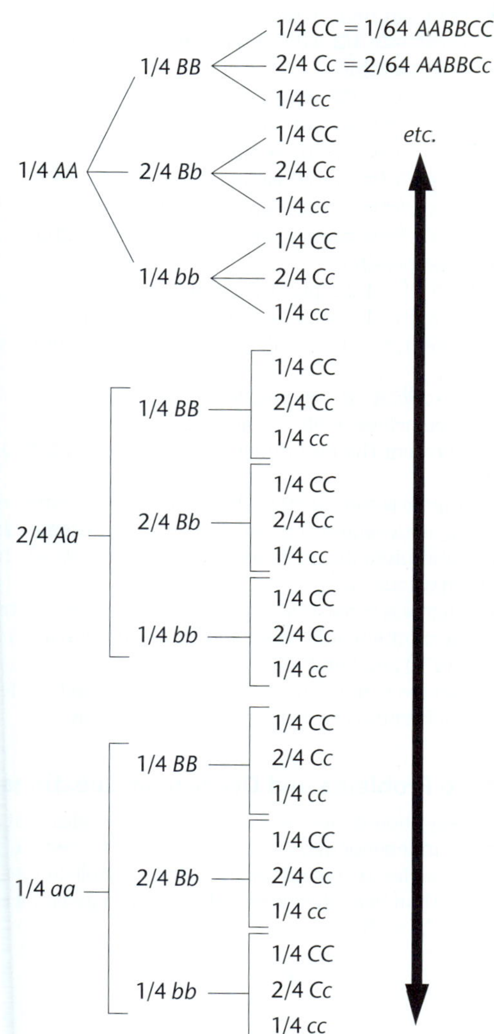

For the phenotypic frequencies, set up the problem in the following manner:

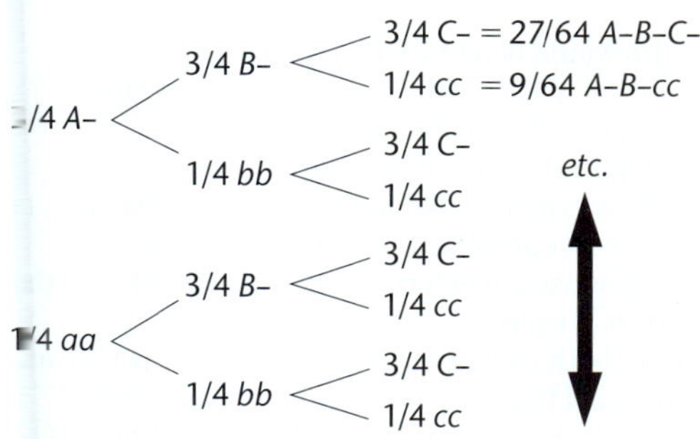

3.4. **(a)** $\chi^2 = 0.47$
For 3 degrees of freedom (because there are four classes in the χ^2 test) the probability is greater than 0.90. We fail to reject the null hypothesis and conclude that the observed values do not differ significantly from the expected values.
(b) The χ^2 value is 0.35. The p value for 1 degree of freedom is greater than 0.50 and less than 0.90. We fail to reject the null hypothesis.
(c) The χ^2 value is 0.01. The p value for 1 degree of freedom is greater than 0.90. We fail to reject the null hypothesis.

3.5. The gene is inherited as an autosomal recessive. Notice that two normal individuals II-3 and II-4 have produced a daughter (III-2) with myopia. Symbolism: m = myopic and M = normal vision.

I-1 (*mm*), I-2 (*Mm* or *MM*), I-3 (*Mm*), I-4 (*Mm*)
II-1 (*Mm*), II-2 (*Mm*), II-3 (*Mm*), II-4 (*Mm*), II-5 (*mm*), II-6 (*MM* or *Mm*), II-7 (*MM* or *Mm*)
III-1 (*MM* or *Mm*), III-2 (*mm*), III-3 (*MM* or *Mm*)

Solutions to Problems and Discussion Questions

2. Your essay should include the following points: (1) Factors (genes) occur in pairs. (2) Some genes have dominant and recessive alleles. (3) Alleles segregate from each other during gamete formation. When homologous chromosomes separate from each other at anaphase I, alleles will go to opposite poles of the meiotic apparatus. (4) One gene pair separates independently from other gene pairs. Different gene pairs on the same homologous pair of chromosomes (if far apart) or on nonhomologous chromosomes will separate independently from each other during meiosis.

4. A = without albinism, a = with albinism.
(a) The parents are both unaffected; therefore, they could be either *AA* or *Aa*. Because they produced a child with albinism, the parents must both be heterozygous (*Aa*).
(b) The unaffected male could have either the *AA* or *Aa* genotype. The female must be *aa*. Because all of the children are unaffected, one would consider it more likely that the male is *AA* instead of *Aa*. However, it *is* possible for the male to be *Aa*.

6. Mendel procured plants with individual true-breeding traits from local seed merchants. He also restricted his examination to one or very few pairs of contrasting traits in each experiment, and kept accurate quantitative records of these. All these highlight Mendel's success in laying down the principles of transmission genetics.

10. (1) Mendel's paired unit factors correlate with pairs of genes, which are located on homologous chromosomes. (2) Variants of genes (alleles) exhibit a dominant/recessive relationship (for example, round and wrinkled pea shape). (3) The movement of homologous chromosomes to opposite poles of the spindle apparatus during anaphase I of meiosis correlates with the segregation of Mendel's paired unit factors during gamete formation. (4) In meiosis, alignment and separation of one homologous chromosome pair does not affect that of another pair, just as Mendel's factors assorted independently of one another.

12. First, assign meaningful gene symbols.

E = gray body color V = long wings
e = ebony body color v = vestigial wings

(a) P_1: *EEVV* × *eevv*
F_1: *EeVv* (gray, long)
F_2: This will be the result of a Punnett square with 16 boxes with the following outcome:

Phenotype	Ratio	Genotype	Ratio
gray, long	9/16	*EEVV*	1/16
		EEVv	2/16
		EeVV	2/16
		EeVv	4/16
gray, vestigial	3/16	*EEvv*	1/16
		Eevv	2/16
ebony, long	3/16	*eeVV*	1/16
		eeVv	2/16
ebony, vestigial	1/16	*eevv*	1/16

(b) P_1: *EEvv* × *eeVV*

F_1: *EeVv* (gray, long)

The F_2 ratios will be the same as in part (a).

(c) P_1: *EEVV* × *EEvv*

F_1: *EEVv* (gray, long)

F_2:

Phenotype	Ratio	Genotype	Ratio
gray, long	3/4	*EEVV*	1/4
		EEVv	2/4
gray, vestigial	1/4	*EEvv*	1/4

14. Symbols:

Pod shape	Pod color
P = inflated	Y = green
p = pinched	y = yellow

P_1: *PPyy* × *ppYY*
F_1: *PpYy* cross to *ppyy*
(which is a typical testcross)
The offspring will occur in a typical 1:1:1:1 as

1/4 *PpYy* (inflated, green)
1/4 *PpYy* (inflated, yellow)
1/4 *ppYy* (pinched, green)
1/4 *ppyy* (pinched, yellow)

Fifty percent of the pods will be true breeding yellow (*ppyy* and *Ppyy* are 25 percent each).

16. (a, b) For the test of a 3:1 ratio, the χ^2 value is 33.3, with an associated p value of less than 0.01 for 1 degree of freedom. For the test of a 1:1 ratio, the χ^2 value is 25.0, again with an associated p value of less than 0.01 for 1 degree of freedom. Based on these probability values, both null hypotheses should be rejected.

18. 1/8

20. The parents are *AaBbCC* and *AABbCc*. The proportions for each trait are to be determined independently at first.

For *A*: *AA* and *Aa* offspring would show dominant traits. Therefore, all the offspring would be dominant for *A* (Probability = 1).

For *B*: the *bb* offspring would show recessive trait, that is, 1/4 of the offspring.

The probability of both occurring simultaneously = 1 × 1/4 = 1/4.

22. (a) There are two possibilities. The trait could be dominant, in which case I-1 is heterozygous, as are II-2 and II-3, and the remaining individuals are homozygous recessive. Alternatively, the trait is recessive and I-1 is homozygous and I-2 is heterozygous. Under the condition of recessiveness, both II-1 and II-4 would be heterozygous; II-2 and II-3 would be homozygous.

(b) Recessive: Parents *Aa*, *Aa*

(c) Recessive: Parents *Aa*, *Aa*

24. (a) For set I, $\chi^2 = 2.15$, with p being between 0.2 and 0.05. We fail to reject the null hypothesis of no significant difference between the expected and observed values. For set II, $\chi^2 = 21.43$ and $p < 0.001$. A significant difference between the observed and expected values is found, so the null hypothesis is rejected.

(b) Clearly, with an increase in sample size, a different conclusion is reached. In fact, most statisticians recommend that the expected values in each class should not be less than 10. In most cases, more confidence is gained as the sample size increases; however, depending on the organism or experiment, there may be practical limits on sample size.

Chapter 4. Answers to Now Solve This

4.1. (a) $c^kc^a \times c^dc^a \Longrightarrow$ 2/4 sepia; 1/4 cream; 1/4 albino

(b) If the cream parent is homozygous, the cross is:
$c^kc^a \times c^dc^d \Longrightarrow$ 1/2 sepia; 1/2 cream
If the cream parent is heterozygous, the cross is:
$c^kc^a \times c^dc^a \Longrightarrow$ 1/2 sepia; 1/4 cream; 1/4 albino

(c) Crosses possible:
$c^kc^k \times c^dc^d \Longrightarrow$ all sepia
$c^kc^k \times c^dc^a \Longrightarrow$ all sepia
$c^kc^d \times c^dc^d \Longrightarrow$ 1/2 sepia; 1/2 cream
$c^kc^d \times c^dc^a \Longrightarrow$ 1/2 sepia; 1/2 cream
$c^kc^a \times c^dc^d \Longrightarrow$ 1/2 sepia; 1/2 cream
$c^kc^a \times c^dc^a \Longrightarrow$ 1/2 sepia; 1/4 cream; 1/4 albino

(d) Crosses possible:
$c^kc^a \times c^dc^d$ 1/2 sepia; 1/2 cream
$c^kc^a \times c^dc^a$ 1/2 sepia; 1/4 cream; 1/4 albino

4.2. A = pigment; a = pigmentless (colorless), B = purple; b = red
AaBb × *AaBb*: $A{-}B{-}$ = purple; $A{-}bb$ = red;
$aaB{-}$ = colorless; *aabb* = colorless

4.3. Let a represent the mutant gene and A represent its normal allele.

(a) This pedigree is consistent with an X-linked recessive trait because the male would contribute an X chromosome carrying the a mutation to the *aa* daughter. The mother would have to be heterozygous *Aa*.

(b) This pedigree is consistent with an X-linked recessive trait because the mother could be *Aa* and transmit her a allele to her one son (a/Y) and her A allele to her other son.

(c) This pedigree is not consistent with an X-linked mode of inheritance because the *aa* mother has an A/Y son.

Solutions to Problems and Discussion Questions

2. Your essay should include a description of alleles that do not function independently of each other or that reduce the viability of a class(es) of offspring. With multiple alleles, there are more than two alternatives of a gene at a given locus.

4. Parents: $I^Ai \times I^Bi$

	I^B	i
I^A	I^AI^B (AB)	I^Ai (A)
i	I^Bi (B)	ii (O)

The ratio would be 1(A):1(B):1(AB):1(O).

6. Flower color: *RR* = red; *Rr* = pink; *rr* = white
Flower shape: P = personate; p = peloric
Plant height: D = tall; d = dwarf
RRPPDD × *rrppdd* $\Longrightarrow$ *RrPpDd* (pink, personate, tall)
2/4 pink × 3/4 personate × 3/4 tall = 18/64

8. (a) All progeny are heterozygous for each gene, so the phenotype is purple.

(b) Since both parents are homozygous for p, CRp are always inherited together. Hence, all progeny is red.

(c) Using forked line method, we get: 9/32 purple, 9/32 red, 14/32 colorless.

10. In a progeny of 1000, 500 females with red eyes, 250 males with red eyes, and 250 males with white eyes.

12. This situation is similar to sex-influenced pattern baldness in humans. Consider two alleles that are autosomal and let
BB = beardless in both sexes
Bb = beardless in females
Bb = bearded in males
bb = bearded in both sexes
P_1 : female: *bb* (bearded) × male: *BB* (beardless)
F_1 : *Bb* = female beardless; male bearded

Because half of the offspring are males and half are females, one could, for clarity, rewrite the F_2 as:

	1/2 females	1/2 males
1/4 *BB*	1/8 Beardless	1/8 Beardless
2/4 *Bb*	2/8 Beardless	2/8 Bearded
1/4 *bb*	1/8 Bearded	1/8 Bearded

One could test the above model by crossing F_1 (heterozygous) beardless females with bearded (homozygous) males. Comparing these results with the reciprocal cross would support the model if the distributions of sexes with phenotypes were the same in both crosses.

14. Symbolism: Normal wing margins = X^+; scalloped = X^{sd}

(a) P_1: $X^{sd}/X^{sd} \times X^+/Y$
F_1: 1/2 X^+/X^{sd} (female, normal)
1/2 X^{sd}/Y (male, scalloped)
F_2: 1/4 X^+/X^{sd} (female, normal)
1/4 X^{sd}/X^{sd} (female, scalloped)
1/4 X^+/Y (male, normal)
1/4 X^{sd}/Y (male, scalloped)

(b) P_1: $X^+/X^+ \times X^{sd}/Y$
F_1: 1/2 X^+/X^{sd} (female, normal)
1/2 X^+/Y (male, normal)
F_2: 1/4 X^+/X^+ (female, normal)
1/4 X^+/X^{sd} (female, normal)
1/4 X^+/Y (male, normal)
1/4 X^{sd}/Y (male, scalloped)

If the *scalloped* gene were not X-linked, then all the F_1 offspring would be wild type (phenotypically) and a 3:1 ratio of normal to scalloped would occur in the F_2.

16. (a) P_1: X^v/X^v; $+/+ \times X^+/Y$; bw/bw
F_1: 1/2 X^+/X^v; $+/bw$ (female, normal)
1/2 X^v/Y; $+/bw$ (male, vermilion)
F_2: 3/16 = female, normal
1/16 = female, brown eyes
3/16 = female, vermilion eyes
1/16 = female, white eyes
3/16 = male, normal
1/16 = male, brown eyes
3/16 = male, vermilion eyes
1/16 = male, white eyes

(b) P_1: X^+/X^+; $bw/bw \times X^v/Y$; $+/+$
F_1: 1/2 X^+/X^v; $+/bw$ (female, normal)
1/2 X^+/Y; $+/bw$ (male, normal)
F_2: 6/16 = female, normal
2/16 = female, brown eyes
3/16 = male, normal
1/16 = male, brown eyes
3/16 = male, vermilion eyes
1/16 = male, white eyes

(c) P_1: X^v/X^v; $bw/bw \times X^+/Y$; $+/+$
F_1: 1/2 X^+/X^v; $+/bw$ (female, normal)
1/2 X^v/Y; $+/bw$ (male, vermilion)
F_2: 3/16 = female, normal
1/16 = female, brown eyes
3/16 = female, vermilion eyes
1/16 = female, white eyes
3/16 = male, normal
1/16 = male, brown eyes
3/16 = male, vermilion eyes
1/16 = male, white eyes

18. (a) There are three independently assorting gene pairs operating in this problem: two gene pairs are involved in the inheritance of one trait, while one gene pair is involved in the other.

(b) Based on phenotypic ratios, croaking is due to one (dominant/recessive) gene pair and eye color is due to two gene pairs that interact (recessive epistasis).

(c) Symbolism: Croaking: $R-$ = utterer; rr = mutterer. Eye color: Since the most frequent phenotypic class is blue eye, let $A-B-$ represent those genotypes. For the purple class, a 3/16 group uses the $A-bb$ genotypes. The 4/16 class (green) would be the $aaB-$ and the $aabb$ groups.

(d) The cross is:

$$AABBrr \times AAbbRR$$

All F_1 are $AABbRr$; blue-eyed and utterers. The F_2 will follow a pattern of a 9:3:3:1 ratio:
9/16 $AAB-R-$ = blue-eyed, utterer
3/16 $AAB-rr$ = blue-eyed, mutterer
3/16 $AAbbR-$ = purple-eyed, utterer
1/16 $AAbbrr$ = purple-eyed, mutterer

20. Two independently assorting gene pairs interact to produce coat color. The 12:3:1 F_2 ratio indicates that dominant epistasis has modified a typical 9:3:3:1 ratio. The following genotypic classifications fit the results:

$A-B-$ = solid white (9/16)
$aaB-$ = solid white (3/16)
$A-bb$ = black and white spotted (3/16)
$aabb$ = solid black (1/16)
One could obtain *AAbb* true-breeding black and white spotted cattle.

22. Symbolism:

$A-B-$ = black	$A-bb$ = golden
$aabb$ = golden	$aaB-$ = brown

(a) $AAB- \times aaBB$ (other configurations are possible, but each must give all offspring with *A* and *B* dominant alleles)
(b) $AaB- \times aaBB$ (other configurations are possible, but both parents cannot be *Bb*)
(c) $AABb \times aaBb$
(d) $AABB \times aabb$ or $Aabb$ or $AAbb$
(e) $AaBb \times Aabb$
(f) $AaBb \times aabb$
(g) $aaBb \times aaBb$
(h) $AaBb \times AaBb$
Those genotypes that will breed true are:
black = *AABB*
golden = all (all are genotypically *bb*)
brown = *aaBB*

24. (a) CC = chestnut $C^{CR}C^{CR}$ = cremello CC^{CR} = palomino
(b) The F_1 resulting from matings between cremello and chestnut horses would be expected to be all palomino. The F_2 would be expected to fall in a 1:2:1 ratio as in the third cross in part (a).

26. C1: 1/2 $\text{i}\text{I}^\text{B}$, $\text{L}^\text{N}\text{L}^\text{N}$ (phenotype B, N) or 1/2 $\text{i}\text{I}^\text{A}$, $\text{L}^\text{N}\text{L}^\text{M}$ (phenotype A, N)
C2: 1/2 $\text{i}\text{I}^\text{A}$, $\text{L}^\text{N}\text{L}^\text{M}$ (phenotypes A, MN) or 1/2 $\text{i}\text{I}^\text{B}$, $\text{L}^\text{N}\text{L}^\text{M}$ (phenotypes B, MN)
C3: 1/4 $\text{I}^\text{A}\text{I}^\text{B}$, $\text{L}^\text{M}\text{L}^\text{N}$; 1/8 $\text{I}^\text{A}\text{I}^\text{B}$, $\text{L}^\text{M}\text{L}^\text{M}$; 1/8 $\text{I}^\text{A}\text{I}^\text{B}$, $\text{L}^\text{N}\text{L}^\text{N}$; 1/8 $\text{I}^\text{A}\text{I}^\text{A}$, $\text{L}^\text{M}\text{L}^\text{N}$; 1/16 $\text{I}^\text{A}\text{I}^\text{A}$, $\text{L}^\text{M}\text{L}^\text{M}$; 1/16 $\text{I}^\text{A}\text{I}^\text{A}$, $\text{L}^\text{N}\text{L}^\text{N}$; 1/8 $\text{I}^\text{B}\text{I}^\text{B}$, $\text{L}^\text{M}\text{L}^\text{N}$; 1/16 $\text{I}^\text{B}\text{I}^\text{B}$, $\text{L}^\text{M}\text{L}^\text{M}$; or 1/16 $\text{I}^\text{B}\text{I}^\text{B}$, $\text{L}^\text{N}\text{L}^\text{N}$ (phenotypes AB, MN; AB, MM; AB, NN; AA, MN; AA, MM; AA, NN; BB, MN; BB, MM; or BB, NN, respectively)

28. Coat color and tail length are two independently assorting genes. For coat color, the silver coat is recessive. The platinum coat is dominant and is always present as heterozygous (homozygous is lethal). For tail length, the long tail is dominant over the short tail.

30. Maternal effect genes encode products that are not carried over for more than one generation, unlike organelle heredity. Crosses that illustrate the transient nature of a maternal effect could include the following:
Female $Aa \times$ male $aa \rightarrow$ all offspring of the A phenotype (even though half of the offspring are genotypically *aa*).

Take a female that has the A phenotype (but an *aa* genotype) from the above cross and mate her to a heterozygous (*Aa*) male. All offspring will be of the recessive (a) phenotype because all of the offspring will reflect the *genotype* of the mother, not her *phenotype*. This cross illustrates that maternal effects last only one generation. Had the phenotype been inherited through organelle heredity, the maternal phenotype would have persisted in each generation.

32. **(a)** The reduced ratio is 12 white, 3 orange, and 1 brown, which corresponds to a dihybrid cross with dominant epistasis. The genotypes for this dihybrid cross are as follows: F_1 = *AaBb*; F_2 = 12 white *A–B–* or *aaB–*:3 orange *A–bb*:1 brown *aabb*.

(b) To solve the problem for three gene pairs, direct your attention to text Figure 3.9, which shows the phenotype ratios for a trihybrid cross and see what you can come up with.

34. Beatrice, Alice of Hesse, and Alice of Athlone are carriers. There is a 1/2 chance that Princess Irene is a carrier.

Chapter 5. Answers to Now Solve This

5.1. **(a)** Something is missing from the male-determining system at the level of the genes, gene products, or receptors, and so on, and the loss is correlated with CMD1.

(b) The *SOX9* gene, or its product, is probably involved in male development. Perhaps it is activated by *SRY*, which would explain the lack of expression in development of female gonads.

(c) There is probably some evolutionary relationship between the *SOX9* gene and *SRY*. There is considerable evidence that many other genes and pseudogenes are also homologous to *SRY*.

(d) Normal female sexual development does not require the *SOX9* gene or gene product(s).

5.2. The gene that produces orange or black coat colors is X-linked. Because of X chromosome inactivation in mammals, scientists would be interested in determining whether the nucleus taken from Rainbow (donor) would continue to show such inactivation. If the inactivated X chromosome *remained inactivated*, CC would only express the active X chromosome allele and would express patches of only a single color (orange or black). However, if the inactive X *reverted to an active state* in the zygote and then went through inactivation, CC would develop as a calico with both orange and black patches of different size and distribution than seen in Rainbow. In either case, CC would have a different patch pattern from her genetic mother based on random X inactivation.

Solutions to Problems and Discussion Questions

2. Your essay should include various aspects of sex chromosomes that contain genes responsible for sex determination. Mention should also be made of those organisms in which autosomes play a role in concert with the sex chromosomes.

4. In sexual determination, genes signal developmental pathways whereby the sexes are generated. Sexual differentiation is the complex set of responses by cells, tissues, and organs to those genetic signals.

6. In *Drosophila* sex is determined by the balance of female determinants on the X chromosome(s) and male determinants on the autosomes, mediated by the Sxl gene. The Y chromosome does not determine sex, but is required for sperm production, therefore, XO individuals are sterile males.

8. The first evidence was found during the study of individuals with Klinefelter and Turner syndromes in the late 1950s. Individuals with Klinefelter syndrome have more than one X chromosome (their karyotype is generally XXY) while people with Turner syndrome result in a karyotype with only 45 chromosomes, including just one X chromosome. Even if individuals with Klinefelter syndrome have two (or more) X chromosomes, the presence of the Y chromosome is enough to lead to maleness. The absence of the Y chromosome prevents masculinization in Turner syndrome.

10. **(a)** female $X^{rw}Y$ × male X^+X^+

F_1:	females:	X^+Y (normal)
	males:	$X^{rw}X^+$ (normal)
F_2:	females:	X^+Y (normal)
		$X^{rw}Y$ (reduced wing)
	males:	$X^{rw}X^+$ (normal)
		X^+X^+ (normal)

(b) female $X^{rw}X^{rw}$ × male X^+Y

F_1:	females:	$X^{rw}X^+$ (normal)
	males:	$X^{rw}Y$ (reduced wing)
F_2:	females:	$X^{rw}X^+$ (normal)
		$X^{rw}X^{rw}$ (reduced wing)
	males:	X^+Y (normal)
		$X^{rw}Y$ (reduced wing)

12. The zygote develops masculinized characteristics because of the hormones and transcription factors that are produced by the activity of the Y chromosome. Even though genetically the zygote is female, it transforms into a sterile female with masculinized reproductive organs.

14. If some male offspring had white eyes and normal wings and some female offspring were wild type, one might suspect that the attached-X had become unattached.

16. A *Barr body* is a differentially staining chromosome seen in some interphase nuclei of mammals with two X chromosomes. There will be one fewer Barr body than the number of X chromosomes. The Barr body is an X chromosome that is considered to be genetically inactive.

18. The *Lyon hypothesis* states that the inactivation of the X chromosome occurs at random early in embryonic development. Such X chromosomes are "marked" in some way, such that all clonally related cells have the same X chromosome inactivated.

20. Like XX females, triplo-X genotypes have only one active X chromosome. Their retinas would be mosaic and they may suffer from red–green color blindness.

22. Autosomal genes in males and females are expressed from two copies (identical or similar) of each gene. X-linked genes may also be required for normal cellular and organismic processes that function in *both* sexes. Since males carry only a single copy of the X chromosome, the output of these genes must be balanced in some manner, even if the gene products have nothing to do with sex determination or sex differentiation.

24. Nondisjunction could have occurred either at meiosis I or meiosis II in the mother, thus giving the X_wX_wY complement in the offspring.

26. One can conclude that the general architecture of sex determination in fowl is comparable to that in humans (the presence of heteromorphic chromosomes and a sex-determining gene, plus the need for dosage compensation); however, the specific mechanism is somewhat reversed. In chickens, the sex-determining gene, *DMRT1*, is located on the Z chromosome, which is found in both males and females, whereas, in humans, the *SRY* gene is located on the male-specific chromosome. The gene dosage to induce testis development also differs; a single copy of *SRY* is sufficient, whereas two copies of *DMRT1* are necessary. Furthermore, artificial dosage compensation in ZZ embryos results in the feminization of developing gonads.

Chapter 6. Answers to Now Solve This

6.1. Because both she and her father express hemophilia, it is likely that the woman inherited her sole X chromosome from her father and no sex chromosome from her mother. This means that nondisjunction occurred in the mother, either during meiosis I or meiosis II, producing an egg with no X chromosome.

6.2. The sterility of interspecific hybrids is often caused from a high proportion of univalents in meiosis I, producing mostly inviable gametes. Viable gametes are rare, and the likelihood of two of them "meeting" is remote. The horticulturist might attempt to reverse the sterility by treating the hybrid with colchicine, which, if successful, would double the chromosome number, so each chromosome would have a homolog with which to pair during meiosis.

6.3. Rare double crossovers *within the boundaries* of a paracentric inversion heterozygote produce only minor departures from the standard chromosomal arrangement as long as the crossovers involve the same two chromatids. With two-strand double crossovers, the second crossover negates the first. However, three-strand and four-strand double crossovers have consequences that lead to anaphase bridges as well as a high degree of genetically unbalanced gametes.

Solutions to Problems and Discussion Questions

2. Your essay can draw from the information presented in the text that provides examples of deletions, duplications, inversions, translocations, and copy number variations.

4. **(a)** monosomy, Turner syndrome, female; **(b)** trisomy, Klinefelter syndrome, male; **(c)** trisomy, Down syndrome, female; **(d)** trisomy, Patau syndrome, male

6. Chromosomes can break spontaneously or due to chemicals and may abnormally fuse with other nonhomologous chromosomes leading to loss or rearrangement of the genetic material. Telomere shortening can be another reason for chromosome ends to fuse with each other.

8. In addition to increased genetic variation, polyploid plants can be physically larger than their diploid relatives due to gene redundancy. This may provide faster growth and larger flowers and fruits with improved economic value. One important disadvantage could be infertility. Fertility of polyploid plants depends on the ability to generate balanced gametes. Two homologs of each chromosome are required for successful meiosis and therefore, fertilization.

10. Organisms with one inverted chromosome and one noninverted homolog are called inversion heterozygotes. Pairing of two such chromosomes is possible only through inversion loops. In case of pairing within the inversion loop, abnormal chromatids consisting of deletions and duplications will be produced.

12. Both genome and gene duplications provide new evolutionary opportunities and contribute to species diversity and phenotypic variation. Multiple copies of genes provide room for accumulating mutations which may turn out to be beneficial for the organism. Also see the text for Ohno's hypothesis and supporting evidence.

14. Given the basic haploid complement of nine unique chromosomes (see figure below), other forms with the "n multiples" are said to be euploid, diploid (for $2n$), or polyploid (for multiples above 2)—and all are autoploid. Karyotypes would share the same basic chromosome set, with the appropriate number of copies of each chromosome (for example, four copies of each in an autotetraploid). Individuals with 27 chromosomes are triploids ($3n$) and are likely to be sterile because there are trivalents at meiosis I, which cause a relatively high number of unbalanced gametes to be formed.

16. The F_1 individuals would all have a genotype of *WWww* and, therefore, green seeds. Of the 36 possible progeny in the F_2 generation, 35 will bear green seeds, since only a single dominant allele is needed to produce green seeds, and only one, the fully recessive genotype (*wwww*), will have white seeds.

18. Parental genotypes: Blue: B_1B_1 and B_2B_2. Possible gametes: B_1B_1, B_1B_2 (x4), and B_2B_2. White: all gametes will carry the same allele combination: B_2B_2 (x6). The resulting ratios and allele combinations in the progeny will be: 1/6 $B_1B_1B_2B_2$ (blue phenotype), 2/3 $B_1B_2B_2B_2$ (light blue phenotype), and 1/6 $B_2B_2B_2B_2$ (white phenotype).

20. **(a)** reciprocal translocation

(b)

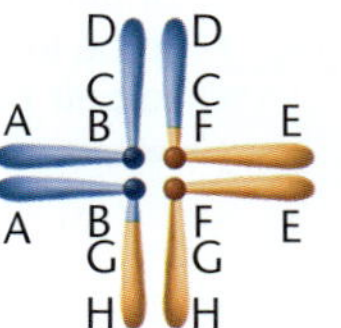

(c) No, this is not surprising because all chromosomal segments are present and there is no apparent loss of chromosomal material. If, however, the breakpoints for the translocation occurred within genes, then an abnormal phenotype might have resulted. In addition, a gene's function is sometimes influenced by its position—its neighboring genes, in other words. If such "position effects" occurred, then a different phenotype might have resulted.

22. Alternate segregation pattern may generate gametes with normal and non-translocated chromosomes 1 and 4. However, adjacent segregation-1 pattern would never generate any such normal gamete.

24. Following is a description of breakage/reunion events that illustrate a translocation in the relatively small, similarly sized chromosomes 19 (metacentric) and 20 (metacentric/submetacentric). The case described here is shown occurring before S phase duplication. The parents were each heterozygous for the translocation; that is, they each have one normal chromosome 19, one normal chromosome 20, and one Robertsonian t(19;20). This is the only way they could produce both "homozygous" and "heterozygous" children. Stillbirths might have resulted from genetically unbalanced gametes caused by various segregation products of meiosis. Since the likelihood of such a translocation is fairly small in a general

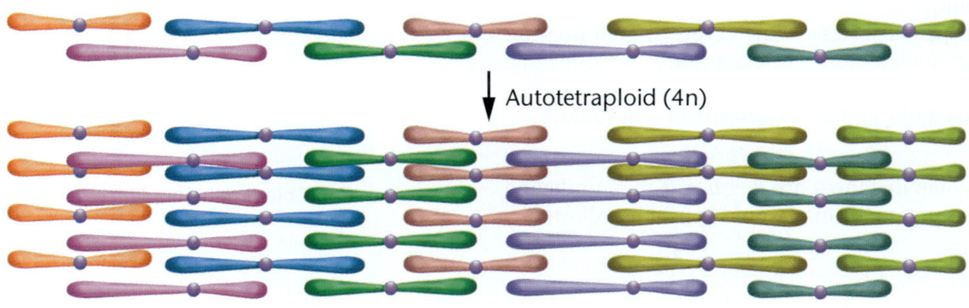

population, inbreeding played a significant role in allowing the translocation to "meet itself."

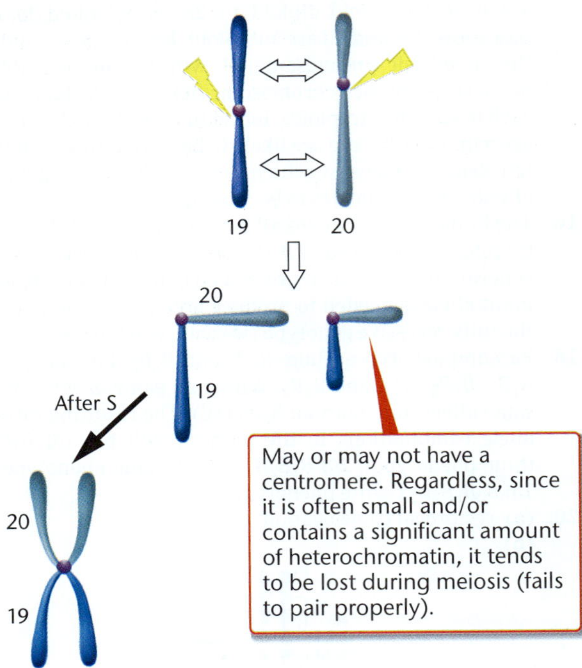
19
20
20
19
After S
20
19
May or may not have a centromere. Regardless, since it is often small and/or contains a significant amount of heterochromatin, it tends to be lost during meiosis (fails to pair properly).

26. This female will produce meiotic products of the following four types:

normal: 18 + 21   *translocated plus* 21: 18/21 + 21

translocated: 18/21   *deficient:* 18 only

Fertilization with a normal 18 + 21 sperm cell will produce the following offspring:
normal: 46 chromosomes
translocation carrier: 45 chromosomes 18/21 + 18 + 21
trisomy 21: 46 chromosomes 18/21 + 21 + 21
monosomic: 45 chromosomes 18 + 18 + 21, lethal

Chapter 7. Answers to Now Solve This

7.1. **(a)** 1 *AaBb* : 1 *Aabb* : 1 *aaBb* : 1 *aabb*
(b) 1 *AaBb* : 1 *Aabb* : 1 *aaBb* : 1 *aabb*
(c) If the arrangement is *AB*/*ab* × *ab*/*ab* then the two types of offspring will be as follows:

1 *Ab*/*ab* : 1 *aB*/*ab*

If, however, the arrangement is *Ab*/*aB* × *aabb*, the offspring would occur as follows:

1 *Ab*/*ab* : 1 *aB*/*ab*

7.2. The most frequent classes are *PZ* and *pz*, which represent the parental (noncrossover) groups. This indicates that the original parental arrangement in the testcross was *PZ*/*pz* × *pz*/*pz*. Adding the crossover percentages together (6.9 + 7.1) gives 14 percent, so the map distance between the two genes is 14 map units (14 mu).

7.3. Examine the progeny list to see which types are not present. In this case, the double crossover classes are the following: + + *c* and *a b* +

(a) Gene *b* is in the middle, and the arrangement is as follows.

+ *b c*/*a* + +

(b) The calculated map distances are *a* − *b* = 7 map units and *b* − *c* = 2 map units, so the map would appear as follows:

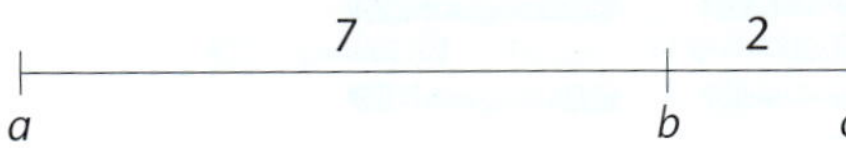
7
2
a
b
c

(c) The progeny phenotypes that are missing are − + *c* and *a b* +. In a population of 1000 offspring, (0.07 × 0.02 × 1000) = 1.4 (or 0.7 each) would be expected. It is likely that the sample size was too small to reliably detect such infrequently occurring genotypes.

Solutions to Problems and Discussion Questions

2. Your essay should include methods of detection through crosses with appropriate, distinguishable markers, and in most cases, the frequency of crossing over is directly related to the distance between genes.

4. With some qualification, especially around the centromeres and telomeres, one can say that crossing over is somewhat randomly distributed over the length of the chromosome. Two loci that are far apart are more likely to experience a crossover event between them than are two loci that are close together.

6. The crossover frequency between any two loci becomes more variable the farther apart they are, because it is not possible to detect all crossover events. For example, double crossovers would result in the original arrangement of alleles.

8. First, one of the parents must be heterozygous at all loci being mapped. If crossover in an organism is limited to one sex, the heterozygous individual must be of the sex in which crossing over occurs. Second, the cross must be set up so that the phenotypes of the offspring readily reveal their genotypes. The best arrangement is one in which a heterozygous individual is crossed with another that is fully recessive for the genes being mapped. Finally, sufficient progeny must be generated to ensure observation of rare crossover products.

10. The heterozygous parent in the test cross is *RY*/*ry* with the two dominant alleles on one chromosome and the two recessives on the homolog. The map distance between the *R* and *Y* loci is 10 map units.

12. The map for parts (a) and (b) is the following:

d..........*b*..........*pr*...........*vg*..........*c*...........*adp*
31   48   54   67   75   83

Map units

The expected map units between *d* and *c* would be 44, between *d* and *vg* 36, and between *d* and *adp* 52. However, because there is a theoretical maximum of 50 map units possible between two loci in any one cross, the latter distance would be below the 52 determined by simple subtraction.

14. **(a)** P_1 females: *corr b pr*/*corr b pr*
F_1 females: *corr b pr*/+ + +
F_1 tester male: *corr b pr*/*corr b pr*

(b) Use method I or II for determining the sequence of genes. From the offspring, it is clear that double crossovers are *pr corr* + and + + *b*:

$$\frac{corr\ b\ \ pr}{+\ \ +\ \ +}$$

$$corr - b = \frac{100 + 96 + 1}{1652} \times 100$$
$$= 11.9 \text{ (map units)}$$
$$b - pr = \frac{49 + 54 + 1}{1652} \times 100$$
$$= 6.3 \text{ (map units)}$$

corr-------------------------------*b*----------------*pr*

11.9 *mu*   6.3 *mu*

(c) Coefficient of coincidence (C) $= \dfrac{\text{Observed DCO}}{\text{Expected DCO}}$
$= \dfrac{0.0006}{0.0075} = 0.08.$

(d) Interference (I) = 1 − C = 1.00 − 0.08 = 0.92. There is positive interference as there are fewer crossovers than expected.

16 Because two of the genes are linked and 20 map units apart on chromosome III, and one is on chromosome II, the problem is a combination of linkage and independent assortment. In the first situation, the gametes from the heterozygous F_1 females can be determined by using a modification of the forked-line method. The *dumpy* locus will give 0.5 + and 0.5 *dp* to the gametes because of independent assortment (on a different chromosome), and the other two loci will segregate with 20 percent being the recombinants and 80 percent being the parentals. The progeny of this first testcross are as follows:

0.20 wild type
0.05 ebony
0.05 pink
0.20 pink, ebony
0.20 dumpy
0.05 dumpy, ebony
0.05 dumpy, pink
0.20 dumpy, pink, ebony

In the reciprocal parental cross, heterozygous males are used in the testcross; therefore, there will be no crossover classes. The progeny of this second testcross are as follows:

0.25 wild type
0.25 pink, ebony
0.25 dumpy
0.25 dumpy, pink, ebony

The results would change because of the absence of crossing over in males.

18. Realize that there are four chromatids in each tetrad and that a single crossover involves only two of the four chromatids. Noninvolved chromatids must be added to the noncrossover classes. Do all the crossover classes first; then add up the noncrossover chromatids. For example, in the first crossover class (20 between a and b), notice that there will be 40 chromatids that were not involved in the crossover. These 40 must be added to the *abc* and + + + classes.

a b c = 168
+ + + = 168
a + + = 20
+ b c = 20
+ + c = 10
a b + = 10
+ b + = 2
a b c = 2

20. (a)

P_1:	$D + +/+ + +$	×	$+ e\ p/+ e\ p$
F_1:	$D + +/+e\ p$	×	$+ e\ p/+ e\ p$
F_2:	$D + +/+e\ p$		Dichaete
	$+ e\ p/ + e\ p$		ebony, pink
	$De + / + e\ p$		Dichaete, ebony
	$+ + p/+ e\ p$		pink
	$D + p/+e\ p$		Dichaete, pink
	$+ e+/+ e\ p$		ebony
	$De\ p/+ e\ p$		Dichaete, ebony, pink
	$+ + +/+ e\ p$		wild type

(b) By comparing the parental classes with the double crossover classes, determine that *pink* is the middle gene. The gene distances are as follows:

$$D - p = \frac{12 + 13 + 2 + 3}{1000} \times 100$$
$$= 3.0 \text{ map units}$$

$$p - e = \frac{84 + 96 + 2 + 3}{1000} \times 100$$
$$= 18.5 \text{ map units}$$

22. The relative distance between the A and B genes might not be accurate. The map could be improved if recombination data for genes located between the A and B loci were included and if a larger sample size were used.

24. RFLPs and microsatellites are DNA markers, i.e. short segments of DNA whose sequence and location are known, making them useful landmarks for mapping purposes. RFLPS are polymorphic sites that are generated when specific DNA sequences are recognized and cut by restriction enzymes. Microsatellites are short repetitive sequences that are found throughout the genome, and they vary in the number of repeats at any given site.

Chapter 8. Answers to Now Solve This

8.1.

Hfr Strain	Order											
1	T	C	H	R	O							
2			H	R	O	M	B					
3		C	H	R	O	M	⟹					
4						M	B	A	K	T		
5							B	A	K	T	C	⟹
Overall	T	C	H	R	O	M	B	A	K			

8.2. In the first dataset, the transformation of each locus, a^+ and b^+, occurs at a frequency of 0.031 and 0.012, respectively. To decide if there is linkage between a and b, determine whether the frequency of double transformants a^+b^+ is greater than that expected for two independent events. Multiplying 0.031 × 0.012 gives 0.00037, or approximately 0.04 percent, which is the same as the frequency reported. From this information, one would consider no linkage between these two loci. Notice that this frequency is approximately the same as the frequency in the second experiment, where the loci are transformed independently.

Solutions to Problems and Discussion Questions

2. Your essay should include a description of conjugation, transformation, transduction, and the potential recombination that occurs as a result of these processes, and a summary of the consequences.

4. (a) The requirement for physical contact between bacterial cells during conjugation was established by placing a filter in a U-tube such that the medium can be exchanged but the bacteria cannot come in contact. Under this condition, conjugation does not occur.

(b) Early experiments suggested directionality, with one strain being the donor and the other being the recipient. Strains identified as donors could convert recipients into donors, but the converse was never true. Additional experimentation revealed that each donor strain contained a mobile fertility factor (F factor). Further confirmation came from studies using Hfr bacteria.

(c) An F^+ bacterium contains a circular, double-stranded, structurally independent DNA molecule that can direct transfer of DNA from one cell to another.

6. For each Hfr type, the point of insertion and the direction of transfer are fixed. Breaking the conjugation tube at different times produces exconjugants with correspondingly different lengths of the donor chromosome transferred. Thus, mapping of genes is based on time.

8. The F factor can integrate into the host bacterial chromosome, forming an Hfr bacterium. In the process of returning to its independent state, the F factor occasionally excises imprecisely, picking up a piece of the adjacent bacterial chromosome and becoming an F′. When this F′ is transferred to a bacterium with a complete chromosome, the recipient cell becomes diploid for the bacterial gene(s) carried on that F′. The cell is said to be a partial diploid, or merozygote.

10. The translation machinery of the infected bacterium provides the necessary materials for protein synthesis.

12. A single plaque originates from the replicative activity of a single bacteriophage.

14. The initial density of bacteriophage suspension is 1.7×10^8 phage/mL.

16. A prophage is the latent, noninfective state of the bacteriophage chromosome when it is incorporated into the host bacterial chromosome.

18. The cotransduction of two bacterial genes is more likely if they appear on the same DNA fragment than if they are located on two separate DNA fragments. The genes are more likely to be found on a single DNA fragment the closer they are to each other on the intact bacterial chromosome. Cotransduction studies of two or three linked genes can be used to determine the gene order.

20. (a) Since all bacteria lysed, the initial concentration of the phage stock is greater than 10^5 phage/mL.
(b) This result indicates that the initial concentration of the phage stock is around $14/0.1 \times 10^6$ or 1.4×10^8 phage/mL.
(c) Since no plaques were detected, the initial concentration of phages is less than 10^9 phage/mL. This is consistent with the calculations performed in part (b).

Chapter 9. Answers to Now Solve This

9.1. In an *in vitro* transformation experiment, the general differential labeling design of the Hershey–Chase experiment would *theoretically* be appropriate: Some substance, if labeled, could show up in the recipient bacterium or its progeny. However, in practice, there would be complications, chief among them that only one strand is taken up and fragment length is relatively small. As a result, the majority of the ^{32}P label would be recovered outside the cell and any DNA that did enter the cell would be difficult to detect. Overall, this could potentially lead to the conclusion that the ^{32}P-labeled material (DNA) was unable to enter recipient cells.

9.2. Guanine = 17.5%, adenine and thymine each = 32.5%.

9.3. Rubella is a single-stranded RNA virus.

Solutions to Problems and Discussion Questions

2. Your essay should include a description of structural aspects, including sugar and base content comparisons. In addition, you should mention complementation aspects, strandedness, flexibility, and conformation.

4. Griffith observed that a bacterial mixture containing heat-killed cells of a virulent strain of *Diplococcus pneumoniae* and live cells of an avirulent strain killed the injected mice and led to recovery of live cells of the virulent strain. He concluded that the heat-killed virulent bacteria transformed the avirulent strain into a virulent strain. By contrast, Avery and coworkers, using an *in vitro* system, systematically searched for the transforming principle originating from the heat-killed pathogenic strain and determined it to be DNA.

6. Transformation resulted when heat-killed virulent cells of *Diplococcus pneumoniae* (IIIS) were injected with living avirulent ones (IIR) in mice and living virulent bacteria (*S*) appeared. Furthermore, just soluble extract of the killed *R* bacteria could lead to transformation *in vitro*. Transfection was achieved when only DNA from a T2 bacteriophage was enough to drive phage multiplication in *E. coli*, stripped of its cell wall. Transformation indicated that a chemical substance was responsible for a physiological modification across generations. Transfection demonstrated that DNA serves as the genetic material.

8. Early, indirect evidence included the observation that DNA is localized only in subcellular structures where genetic functions occur (the nucleus, chloroplasts, and mitochondria) whereas proteins are found throughout the cell. Further, DNA content and ploidy in various cell types (sperm and somatic cells) are related. Moreover, the *action* and *absorption* spectra of ultraviolet light are correlated, supporting the interpretation that DNA was the genetic material. Direct evidence comes from a variety of observations, including gene transfer, which has been facilitated by recombinant DNA techniques.

10. The structure of deoxyadenylic acid is given below and in the text. Linkages among the three components require the removal of water (H_2O).

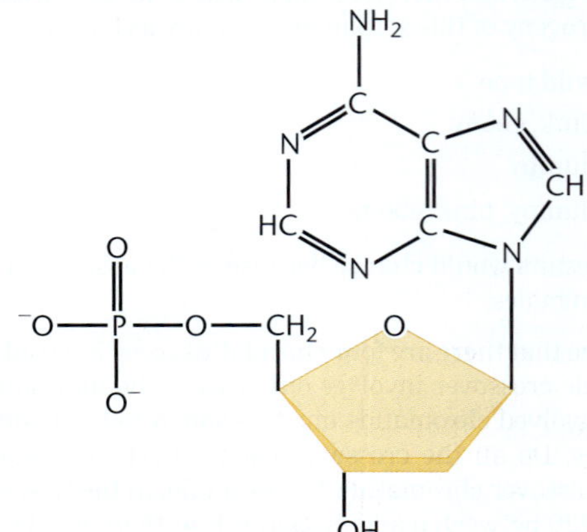

12.

Uracil:	2,4-dioxypyrimidine,
Thymine:	2,4-dioxy-5-methylpyrimidine,
Adenine:	6-aminopurine,
Guanine:	2-amino-6-oxypurine

14. There are two antiparallel polynucleotide chains, which coil around each other to form a double helix. Each chain is formed by phosphodiester linkages between the five-carbon sugars and the phosphates. Bases are stacked, 0.34 nm apart with 10 bases per turn. Hydrogen bonds hold the two polynucleotide chains together. There are two hydrogen bonds between the A-T pair and three between the G-C pair. The double helix is approximately 2 nm in diameter, with a topography of major and minor grooves. The hydrophobic bases are located in the center of the molecule, whereas the hydrophilic phosphodiester backbone is on the outside.

16. The data given show that A = C and T = G, which do not normally pair, making it unlikely that a tight helical structure would form. This observation would likely lead to the interpretation of DNA as a single-stranded or other non-hydrogen bonded structure. Such structures would not produce the regular X-ray diffraction pattern observed by Wilkins and Franklin, which indicated a double-helical structure.

18. No. The following are the structural differences between DNA and RNA.
(a) DNA has deoxyribose, RNA has ribose,
(b) DNA has thymine, RNA has uracil instead of thymine,
(c) DNA is double-stranded, RNA is generally single-stranded (except in ds RNA viruses).

20. Z-DNA was identified as a novel DNA conformation in X-ray diffraction studies on poly G-C synthetic DNA fragments.
Defining characteristics: anti-parallel strands forming a left-handed double helix; a diameter of 18 Å; 12 bp per turn in zig-zag conformation; the major groove present in B-DNA is almost absent.
Although it is unknown if this DNA form is present in living cells, it has been suggested that its structure could provide specific interaction with other molecules.

22. *A hyperchromic effect* is the increased absorption of UV light as double-stranded DNA (or RNA) is converted to a single-stranded form. As illustrated in the text, the change in absorption is quite significant, with structures of higher G-C content *melting* at a higher temperature than A-T rich nucleic acids. If one monitors the UV absorption with a spectrophotometer during the melting process, the hyperchromic shift can be observed. The T_m is the point (temperature) on the profile at which half (50 percent) of the sample is denatured.

24. Electrophoretic separation of nucleic acids is based on their differential migration toward the anode due to their negative charge. Separation is achieved using a porous semisolid media that allows faster migration of smaller molecules.

26. (1) As shown, the extra phosphate is not normally expected.
(2) In the adenine ring, a nitrogen is at position 8 rather than position 9, and a carbon is at position 9 rather than position 8.
(3) The bond from the C-1′ of the sugar should form with the N at position 9 (N-9) of the adenine.
(4) There should be a double bond between C-4 and C-5 of adenine.
(5) The dinucleotide is a "deoxy" form; therefore, each C-2′ should not have a hydroxyl group. Notice the hydroxyl group at C-2′ on the sugar of the adenylic acid.
(6) At the C-5 position on the thymine residue, there should be a methyl group.
(7) There are too many bonds between the N-3 and C-2 of thymine.
(8) There are too few bonds between the C-5 and C-6 of thymine (there should be a double bond there).
(9) The extra hydroxyl group on C-5′ of the sugar of the thymidylic acid should not be there (the C-5′ hydroxyl group is involved in the bond with the phosphate group).

28. If thymine gets converted to uracil, the A = T base pair will get converted to A = U base pair. Since uracil is unmethylated, it will make the DNA molecule more susceptible to damage. If cytosine gets converted to uracil, the DNA double helix will become unstable as uracil cannot form base pairs with guanine.

30. (a, b) Given the chemical similarities to terrestrial DNA, it is probable that the alien DNA follows a similar structural plan. The X-ray diffraction pattern suggests a wider helix than terrestrial DNA. Because there are equal amounts of A, T, and H, one could suggest that they are hydrogen bonded to one another; the same may be said for C, G, and X. Given the molar equivalence of erythrose and phosphate, an alternating sugar-phosphate-sugar backbone, would seem likely. A model of a triple helix would be consistent with these data.
(c) Hypoxanthine and xanthine each interact with a purine–pyrimidine pair. To achieve the constant diameter indicated, both would need to be of the same class, either purine or pyrimidine (as also suggested by the similarity of their names). In fact, hypoxanthine and xanthine are both purines.

32. Even though RNA molecules, like DNA molecules, have the same charge-to-mass ratios, they are single stranded and can exist in a variety of shapes. Complementary intrastrand base pairing can make more compact structures compared to the more relaxed, open conformation of molecules that do not base pair. As with DNA, during electrophoresis, smaller and more compact molecules migrate faster than do relaxed, open structures. Therefore, RNA molecules must be denatured to eliminate secondary structural variables to be able to perform electrophoretic size comparisons.

Chapter 10. Answers to Now Solve This

10.1. After one round of replication in the ^{14}N medium, the conservative scheme can be ruled out. Two rounds of replication are required to rule out the dispersive mode.

10.2. If the DNA double helix contained parallel (rather than anti-parallel) strands and the polymerase were able to accommodate such parallel strands, there would be continuous synthesis and no Okazaki fragments. Alternative hypotheses are the following: the DNA existed only as a single strand; synthesis was initiated from two origins located at opposite ends of a linear molecule, producing a continuous complementary strand for each template; the DNA was a circular molecule.

Solutions to Problems and Discussion Questions

2. Your essay should describe replication as the process of making daughter nucleic acids from existing ones. Synthesis refers to the precise series of steps, components, and reactions that allow such replication to occur.

4. The technique is a centrifugation through a density gradient of a heavy metal salt. Any molecule will reach equilibrium and position itself when it reaches the position in the gradient matching its own density. It can be used to separate molecules of different density.

6. The model proposed a triple helical DNA structure. Because the semiconservative scheme predicts that *half* of the DNA in each daughter double helix is labeled, it would be difficult to envision a scheme in which three strands are replicated in such a semiconservative manner. It is possible that either the conservative or dispersive scheme would fit more appropriately.

8. The indirect approach described in the text was the analysis of *base composition*. Comparison of the bases comprising the template with those of the product showed that, within experimental error, expectations were met and the DNA replicated faithfully.

10. The first strain may show an inhibition to replication since the RNase may have destroyed the RNA primer that is necessary for the polymerase to continue with the replication. The second strain may have a mutation in the DNA polymerase that negates the requirement of a free 3′-OH group. An RNA primer would not be necessary in that case.

12. All three enzymes can *elongate* an existing DNA strand, assuming there is a template strand, but none can *initiate* DNA synthesis. Polymerization occurs in the 5′ to 3′ direction for all three. In addition, each enzyme has 3′ to 5′ exonuclease activity, which provides proofreading capabilities. DNA Pol I is unique in its possession of 5′ to 3′ exonuclease activity, which is suitable for its role in the removal of RNA primers. Pol I is highly abundant and stable, whereas Pol III is not. Pol I is a single polypeptide, whereas Pol III consists of multiple subunits and is required for chromosomal replication. DNA pol II is believed to be involved in repair synthesis.

14. After initiation at a given point, *unidirectional* synthesis replicates strands in one direction only, whereas *bidirectional* synthesis replicates strands in both directions. Synthesis using the leading strand template proceeds in the direction of the replication fork, producing a long, unbroken complementary strand. It is said to be *continuous*. Synthesis using the lagging strand template proceeds in the direction opposite that of the replication fork and produces multiple, short fragments (Okazaki fragments) that must be ligated together to form a long complementary strand. This type of synthesis is said to be *discontinuous*.

16. **(a)** Okazaki fragments are relatively short DNA fragments (1000 to 2000 bases in bacteria) that are synthesized in a discontinuous fashion on the lagging strand during DNA replication. Such fragments appear to be necessary because template DNA is not available for 5′ to 3′ synthesis until some degree of continuous DNA synthesis occurs on the leading strand in the direction of the replication fork. The isolation of such fragments provides support for the scheme of replication shown in the text.

(b) DNA ligase is required to form phosphodiester linkages in nicks that are generated when DNA polymerase I removes an RNA primer and meets newly synthesized DNA ahead of it. This action seals the two strands together to form a continuous single strand.

(c) Primer RNA is formed by primase to serve as an initiation point for the production of DNA strands on a DNA template. None of the DNA polymerases is capable of initiating synthesis without a free 3′-hydroxyl group. The primer RNA provides that group and thus can be used by DNA polymerase III.

18. In bacteria, it is the ori sequence. They have a single origin of replication. In yeast, it is autonomously replicating sequence (ARS), which is an AT-rich region where the origin recognition complex (ORC) can bind and initiate replication at several locations. Origins in mammalian cells are unrelated to any specific DNA sequence. There are several origins on each chromosome where replication can start.

20. **(a)** No repair from DNA polymerase I and/or DNA polymerase III.

(b) No DNA ligase activity and/or no DNA polymerase I activity.

(c) No primase activity; other possibilities include no DnaA protein or faulty helicase.

(d) Only DNA polymerase I activity (recall that DNA Pol III has high processivity, due to the sliding clamp subunit).

(e) No DNA gyrase activity.

22. If replication is conservative, the first autoradiograms (see metaphase I in the text) would have the label distributed on only one side (in one chromatid) of the metaphase chromosome.

24. Telomeres cannot shorten indefinitely without eventually eroding genetic information. Telomerase activity is present in germ-line tissue to maintain telomere length from one generation to the next. It is also necessary in stem cells and other proliferating tissues.

Chapter 11. Answers to Now Solve This

11.1. A circular chromosome, with no free ends present, avoids the problem faced by linear chromosomes, namely, complete replication of terminal sequences.

11.2. ^{3}H-thymidylic acid is a precursor of DNA. Since polytene chromosomes are formed by multiple rounds of DNA replication without strand separation, you would expect grains along the entire length of each polytene chromosome.

11.3. Volume of the nucleus $= 4/3\,\pi r^3 = 5.23 \times 10^{11}$ nm^3
Volume of the chromosome $= \pi r^2 \times$ length $= 1.9 \times 10^{11}$ nm^3
Therefore, the percentage of the volume of the nucleus occupied by the chromatin is $\text{Vol}_{\text{nucleus}}/\text{Vol}_{\text{chromosome}} \times 100 =$ about 36.3 percent

Solutions to Problems and Discussion Questions

2. Your essay should include a description of overall chromosomal configuration (linear, circular, strandedness, etc.) as well as association with chromosomal proteins. In addition, it should describe higher-level structures, such as condensation in the case of eukaryotic chromosomes.

4. Polytene chromosomes are formed from numerous DNA replications and pairings of homologs, in the absence of either strand separation or cytoplasmic division.

6. Lampbrush chromosomes are homologous pairs of chromosomes held together by chiasmata, with numerous loops of DNA protruding from a central axis of chromomeres. They are located in oocytes in the diplotene stage of the first prophase of meiosis.

8. X-ray diffraction studies of chromatin with and without histones suggested regular, repeating structural units. Digestion of chromatin with micrococcal nuclease gave DNA fragments of approximately 200 base pairs or multiples of such segments. Observations of chromatin by electron microscopy showed regularly spaced beadlike structures (nucleosomes). Kornberg's analysis of histone–DNA interactions led to his proposal that nucleosomes consisted of two types of histone tetramers associated with 200 bp of DNA. Further nuclease studies using increased digestion time revealed shortened DNA fragments (147 bp) and suggested the presence of linker DNA. Finally, neutron-scattering analysis of crystallized nucleosome core particles led to a detailed model of the nucleosome.

10. As chromosome condensation occurs, five or six nucleosomes coil together to form a 30-nm fiber called a solenoid. These fibers form a series of loops that further condense into the chromatin fiber, which is then coiled into chromosome arms making up each chromatid.

12. There are three main categories of repetitive sequences. (1) Satellite DNA is highly repetitive short sequences that are found in heterochromatic regions of centromeres and telomeres. (2) Middle repetitive tandem repeats can be found at a single locus (multiple-copy genes) or dispersed throughout the genome (minisatellites and microsatellites). Sizes of the repeated sequence range from very small (STRs) to about 100 bp (VNTRs) to very large (multiple-copy genes). (3) Transposable sequences are varied in structure and size and are generally interspersed throughout the entire genome.

14. SINE = short interspersed elements, a moderately repetitive sequence class; LINE = long interspersed elements. They are called "repetitive" because multiple copies exist.

16. Use the formulas $\pi r^2 h$ for the volume of a cylinder (DNA) and $(4/3)\pi r^3$ for the volume of a sphere (viral head).

Volume of DNA $= 1.57 \times 10^5$ nm^3

Volume of capsid $= 2.67 \times 10^5$ nm^3
Because the capsid head has a greater volume than does the DNA, the DNA will fit into the capsid.

18. The approximate size of the *Alu* sequence is 300 bp, and there are 10^6 copies. Assuming a random distribution, dividing the size of the genome (3×10^9 base pairs) by the (size $\times$ number of copies) of the *Alu* element (3×10^6) gives an average of 1 kb *between* the *Alu* sequences.

Chapter 12. Answers to Now Solve This

12.1. **(a)** GGG $= 3/4 \times 3/4 \times 3/4 = 27/64$
GGC $= 3/4 \times 3/4 \times 1/4 = 9/64$
GCG $= 3/4 \times 1/4 \times 3/4 = 9/64$
CGG $= 1/4 \times 3/4 \times 3/4 = 9/64$
CCG $= 1/4 \times 1/4 \times 3/4 = 3/64$
CGC $= 1/4 \times 3/4 \times 1/4 = 3/64$
GCC $= 3/4 \times 1/4 \times 1/4 = 3/64$
CCC $= 1/4 \times 1/4 \times 1/4 = 1/64$

(b) Glycine: GGG and one G_2C (adds up to 36/64)
Alanine: one G_2C and one C_2G (adds up to 12/64)
Arginine: one G_2C and one C_2G (adds up to 12/64)
Proline: one C_2G and CCC (adds up to 4/64)

12.2. Because of a triplet code, a trinucleotide sequence will, once initiated, remain in the same reading frame and produce the same code all along the sequence regardless of the initiation site. If a tetranucleotide is used, there will be three different reading frames, each of which produces a polypeptide that contains a repeating sequence of four amino acids. The sequences in all four polypeptides will be the same (repeats of aa_1-aa_2-aa_3-aa_4) except that the starting amino acid will have changed.

12.3. **(a)** Apply complementary bases, substituting U for T:
Transcripts below are written in the 5′ to 3′ direction
Sequence 1: 5′-AUGGCAAAAAAG-3′
Sequence 2: 5′-AGUUAUUGAUGU-3′
Sequence 3: 5′-AGAACCCUUGUA-3′

(b) Sequence 1: Met-Ala-Lys-Lys
Sequence 2: Ser-Tyr-(termination)
Sequence 3: Arg-Thr-Leu-Val

(c) The coding strand has the same sequence as the mRNA, except T is substituted for U: 5′-ATGGCAAAAAAG-3′

Solutions to Problems and Discussion Questions

2. Your essay should include a description of the nature and structure of the genetic code, the enzymes and logistics of transcription, and the chemical nature of RNA polymerization.

4. This sequence can be read as three possible repeating triplets—UUC, UCU, and CUU—depending on the initiation point. Hence, three different polypeptide homopolymers are produced, containing either phenylalanine (phe), serine (ser), or leucine (leu).

6. Given that AGG = Arg, information from the AG copolymer and understanding of the wobble hypothesis indicates that AGA also codes for Arg and that GAG must therefore code for Glu. Coupling this information with that of the AAG copolymer, one can see that GAA must also code for Glu and AAG must code for Lys.

8. List the amino acid substitutions; then from the code table apply the codons to the original amino acids. Select codons for the substitutions that provide single base changes. These are shown, with the base change underlined, in the following table.

Original		Substitutions
leucine	⟹	isoleucine
CU (U, C or A) and UUA		AU (U, C or A)
serine	⟹	arginine
AG (U or C)		AG (A or G)
lysine	⟹	arginine
AA (A or G)		AG (A or G)
glutamic acid	⟹	glutamine
GA (A or G)		CA (A or G)

10. The enzyme generally functions in the degradation of RNA; however, in an *in vitro* environment, with high concentrations of the ribonucleoside diphosphates, the direction of the reaction can be forced toward polymerization. *In vivo*, the concentration of ribonucleoside diphosphates is low and the degradative process is favored.

12. Applying the coding dictionary, the following sequences are "decoded":
Sequence 1: Met-Val-Val-Arg-Arg-Arg-Arg-Arg-Stop
Sequence 2: Met-Val-Val-Glu-Glu-Glu-Glu-Asp-Glu
There is a frameshift mutation in sequence 2 (nucleotide 10, C, is missing). The shift in the reading frame caused by this deletion results in a different sequence of amino acids from that point. This gives rise to a longer protein as the original stop codon is no longer recognized.

14. For UCA, changing the wobble base has no effect, but for AGU, it may result in a change to arginine. For the GGU codon, four of the six remaining single nucleotide substitutions result in an amino acid substitution, but two of them result in a premature stop codon. For the AGU codon, the remaining six single nucleotide substitutions result in an amino acid substitution at position 33.

16. Leu: UUA, UUG, CUU, CUC, CUA, CUG. Hence, there is a preponderance of these codons.
Ala: GCU, GCC, GCA, GCG,
Tyr: UAU, UAC.

18. In bacteria, transcription of all types is performed by a single type of RNA polymerase. RNA polymerase in bacteria is cytoplasmic, as bacteria have no nucleus. In eukaryotes, there are three different RNA polymerases: RNAP I (located in the nucleolus, transcribes rRNAs), RNAP II (located in the nucleoplasm, transcribes snRNAs and mRNAs), and RNAP III (located in the nucleoplasm, transcribes 5SrRNAs and tRNAs). Bacterial RNA polymerases are composed of 5–6 subunits, whereas eukaryotic RNA PII consists of 12 subunits. Bacterial RNA polymerases interact through their α subunit with the promoter, whereas eukaryotic RNA PII cannot bind directly to the core promoter and requires the general TFII transcription factors.

20. two; one

22.

Proline:	C_3 and one of the C_2A triplets
Histidine:	one of the C_2A triplets
Threonine:	one C_2A triplet and one A_2C triplet
Glutamine:	one of the A_2C triplets
Asparagine:	one of the A_2C triplets
Lysine:	A_3

24. **(a)** Alternative splicing occurs when copies of a pre-mRNA are spliced in more than one way, yielding various combinations of exons in the final mRNA products. Upon translation of a group of alternatively spliced mRNAs, a series of different, but related, proteins can be produced. The evolutionary advantage provided is the ability to encode more proteins without an increase in gene number.

(b) Some tissues might be more prone to develop alternative splicing if they depend on a number of related protein functions.

Chapter 13. Answers to Now Solve This

13.1. One can conclude that the tRNA and not the amino acid is involved in recognition of the codon.

13.2. There are two codons for glutamic acid: GAA and GAG. With two of the codons for valine being GUA and GUG, a single base change in the second codon position (from A in glutamic acid to U in valine) could cause the Glu to Val switch. Likewise, a single base change in the first codon position (from G in glutamic acid to A in lysine) could convert GAA to AAA or GAG to AAG, causing a Glu to Lys switch. Glutamic acid (the amino acid normally present at position 6 of the β chain) is negatively charged, whereas valine is nonpolar and carries no net charge, and lysine is positively charged. Given these significant charge changes, one would predict some, if not considerable, influence on protein structure and function.

Solutions to Problems and Discussion Questions

2. Your essay should include descriptions of the ribosome and of translation, in which a functional ribosome, in conjunction with mRNA, orders amino acids and forms peptide bonds between them.

4. A mutation in a germ-line cell will be transmitted to the following generation in half of the progeny. However, if the mutation is recessive and on an autosome (or sex chromosome linked, but in the homogametic sex), no phenotype will be manifested if the other copy of the gene has a wild-type function. If the mutation is sex chromosome linked, it may manifest a phenotype in the heterogametic sex.

6. The sequence of base triplets in mRNA constitutes the sequence of codons. A three-base portion of the tRNA constitutes the anticodon.

8. The important regions present in a t-RNA molecule are the acceptor arm, D arm, anticodon arm and loop, variable loop, TΨC arm, and amino acid binding site from 5′ to 3′ direction.

10. The concept of the gene has been constantly evolving. The one-gene:one-protein hypothesis evolved into the one-gene:one-polypeptide hypothesis. The ribosome is composed of rRNAs belonging to non-coding RNAs that never get translated into proteins. Thus, "one-gene:one-RNA" is a more accurate statement for ribosomes.

12. Sickle-cell anemia is considered to be a *molecular* disease because it is well understood at the molecular level. It is considered to be a *genetic* disease because it is inherited from one generation to the next. Sickle-cell anemia is not contagious, unlike many diseases caused by microorganisms. Furthermore, infectious diseases would not necessarily follow family pedigrees, whereas genetic diseases do. Finally, infectious diseases can often be cured, whereas inherited diseases cannot.

14. Given that each nucleotide is 0.34 nm long, we can calculate that each triplet is about 1 nm in length. If the diameter of a ribosome is 20 nm, approximately 20 codons could be found in that ribosome.

16. The four levels of protein structure are: primary, the linear sequence of amino acids; secondary, regular structures formed by interactions between amino acids located close to each other in the primary sequence; tertiary, three-dimensional folding due to interactions among amino acid side chains; and quaternary, association of two or more polypeptide chains. Because all higher levels are dependent on the amino acid sequence (primary structure), it is the primary structure that is most influential in determining protein structure and function.

18. Protein domains are discrete regions within a protein, usually between 50 and 300 amino acids. These regions are the structural units, or modules of protein. They may be composed of a mixture of secondary structures like alpha-helices and beta-pleated sheets and adopt specific ternary structures. Some proteins may contain just one domain while others may possess multiple. In many cases, a single domain is associated with an individual function of a protein like DNA binding activity.

20. **(a)** F_1: $AABbCC$ = speckled

F_2: 3 $AAB{-}CC$ = speckled

1 $AAbbCC$ = yellow

(b) F_1: $AABbCc$ = speckled

F_2: 9 $AAB{-}C{-}$ = speckled

3 $AAB{-}cc$ = green

3 $AAbbC{-}$ = yellow } 4

1 $AAbbcc$ = yellow

(c) F_1: $AaBBCc$ = speckled

F_2: 9 $A{-}BBC{-}$ = speckled

3 $A{-}BBcc$ = green

3 $aaBBC{-}$ = colorless } 4

1 $aaBBcc$ = colorless

22. The pathway would be as follows:

AA----->IGP----->I----->TRP

And the metabolic blocks are created at the following locations:

trp 8 trp 2 trp 3 trp 1

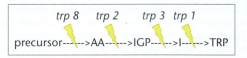

precursor------>AA------>IGP------>I------>TRP

Chapter 14. Answers to Now Solve This

14.1. The phenotypic influence of any base change is dependent on a number of factors including its location in coding or noncoding regions, its potential in dominance or recessiveness, and its interaction with other base sequences in the genome.

14.2. If a gene is incompletely penetrant, it may be present in a population and only express itself under certain conditions. It is unlikely that the gene for hemophilia behaved in this manner. If a gene's expression is suppressed by another mutation in an individual, it is possible that offspring may inherit a given gene and not inherit its suppressor. Such offspring would have hemophilia. Since all genetic variations must arise at some point, it is possible that the mutation in Queen Victoria's family was a new germ-line mutation, arising in her father. Lastly it is possible (but not very probable) that her mother was heterozygous and, by chance, no other individuals in her family received the mutant gene.

14.3. Any agent that inhibits DNA replication, through mutation and/or DNA crosslinking, will suppress the cell cycle and might be useful in cancer therapy. Guanine alkylation often leads to mismatched bases, which can be repaired by a variety of repair mechanisms. DNA crosslinking can be repaired as well, using recombinational mechanisms. Thus, for such agents to be successful in cancer therapy, suppressors of DNA repair systems are often used in conjunction with certain cancer drugs.

14.4. Ethylmethane sulfonate (EMS) alkylates the keto groups at the sixth position of guanine and at the fourth position of thymine. In each case, base-pairing affinities are altered and transition mutations result. Altered bases are not readily repaired if mismatch repair systems are nonfunctional, and once the transition to normal bases occurs through replication, such mutations avoid repair altogether.

Solutions to Problems and Discussion Questions

2. Your essay should include a brief description of the genomic differences between eukaryotes and bacteria and the ways that ploidy influences the phenotypic effects of mutations in one copy of a gene. You should also include a summary of repair pathways that operate predominantly in bacteria or predominantly in eukaryotes, as well as a description of the differences in repair pathways that are shared by both types of organisms.

4. A mutation in a germ-line cell will be transmitted to the following generation in half of the progeny. However, if the mutation is recessive and on an autosome (or sex chromosome linked, but in the homogametic sex), no phenotype will be manifested if the other copy of the gene has a wild-type function. If the mutation is sex chromosome linked, it may manifest a phenotype in the heterogametic sex.

6. A gene is likely to be the product of perhaps a billion years of evolution. Each gene and its product function in an environment that has coevolved. A coordinated output of each gene product is required for life. Deviations, caused by mutation, are likely to be disruptive because of the complex and interactive environment in which each gene product must function. However, on occasion a beneficial variation occurs.

8. A silent mutation is a point mutation in an open reading frame that does not alter the amino acid encoded, due to degeneracy of the genetic code. A neutral mutation is one that occurs in noncoding DNA and does not affect gene products or gene expression.
10. All three agents are mutagenic because they cause base substitutions, specifically transitions, but by different mechanisms. Deaminating agents oxidatively convert an amino group to a keto group. Alkylating agents donate an alkyl group to the amino or keto groups of nucleotides. Base analogs such as 5-bromouracil and 2-amino purine are incorporated as thymine and adenine, respectively, yet they base-pair with guanine and cytosine, respectively.
12. Double strand break repair requires sister chromatids and is thus likely to be functional in the S/G2 phase.

    Non-homologous end joining is activated prior to DNA replication in the G1 phase since it does not require a homologous chromosome.
14. Because mammography involves the use of X rays, which are known to be mutagenic, it has been suggested that frequent mammograms may do harm. Although data show that screening decreases overall mortality from breast cancer, the side effects of exposure to diagnostic radiation are often difficult to identify and quantify.
16. The Ames test uses a number of different strains of the bacterium *Salmonella typhimurium* that have been selected for their ability to reveal the presence of specific types of mutations. The assay measures the frequency of reverse mutations that occur within the mutant gene, yielding wild-type bacteria (his^+ revertants). The revertants are screened by their ability to grow in medium devoid of histidine.
18. Types: LTR retrotransposons and non-LTR retrotransposons, depending on the presence of long terminal repeats.

    Structure: Sequence flanked by short direct repeats (DR), which represent target site duplications that are a consequence of transposition. The DNA sequence of the retrotransposon codes for a reverse transcriptase and an integrase.

    Mechanisms: 1. Transcription of retrotransposon DNA by the cell molecular machinery. 2. Translation of reverse transcriptase and integrase transcripts. 3. RNA is reverse transcribed into double-stranded DNA. 4. Integration into the host genome.
20. The faulty repair process would progress as expected, with the wild-type sequence laid into the appropriate place. However, in the absence of an active DNA ligase, the nick will not be sealed and, therefore, the repair would not be completed.
22. Complementation groups:

XP1	*XP4*	*XP5*
XP2		*XP6*
XP3		*XP7*

    The groupings indicate that there are at least three genes that form products necessary for unscheduled DNA synthesis. All of the cell lines that are in the same complementation group are defective in the same product.
24. First, although less likely, one might suggest that transposons are more likely to insert in noncoding regions of the genome. One might also suggest that they are more stable in such regions. Second, and more likely, it is possible that transposons insert randomly and that selection eliminates those that have interrupted coding regions of the genome.

Chapter 15. Answers to Now Solve This

15.1. **(a)** The *lac Z* gene product will likely be extremely altered or truncated. A cell with this mutation will be unable to grow on a lactose medium because it cannot produce functional β-galactosidase. The reading frames for the *lac Y* gene and the *lac A* gene will be unaffected, because each has its own initiator codon.

(b) The *A* gene product will likely be impaired, but this will not influence the cell's use of lactose as a carbon source.

15.2. **(a)** With no lactose and no glucose, the operon is off. The *lac* repressor is bound to the operator and, although CAP is bound to its binding site, it will not override the action of the repressor.

(b) With lactose added to the medium, the *lac* repressor is inactivated and the operon is transcribing the structural genes. With no glucose, the CAP is bound to its binding site, thus enhancing transcription.

(c) With no lactose present in the medium, the *lac* repressor is bound to the operator region, and since glucose inhibits adenyl cyclase, the CAP protein will not interact with its binding site. The operon is therefore "off."

(d) With lactose present, the *lac* repressor is inactivated; however, since glucose is also present, CAP will not interact with its binding site. Under this condition transcription is severely diminished and the operon can be considered to be "off."

Solutions to Problems and Discussion Questions

2. Your essay should include a description of the evolutionary advantages of the efficient response to environmental resources and challenges (antibiotics, for example) when such resources are present. Also include the advantages of having related functions in operons.
4. Under negative control, gene expression occurs unless it is downregulated by a repressor, for example, binding to the regulatory region. Under positive control, a regulatory molecule induces expression. In such a case, a repressor could be permanently present inhibiting the inducer, preventing it from binding the regulatory region.
6. $I^+ O^+ Z^+$ = **no enzyme made.**

   $I^+ O^C Z^+$ = **functional enzyme made.**

   $I^- O^+ Z^-$ = **nonfunctional enzyme made.**

   $I^- O^+ Z^-$ = **nonfunctional enzyme made.**

   $I^- O^+ Z^+/F'\ I^+$ = **no enzyme made.**

   $I^+ O^C Z^+/F'\ O^+$ = **functional enzyme made.**

   $I^+ O^+ Z^-/F'\ I^+ O^+ Z^+$ = both a **functional and a nonfunctional enzyme made.**

   $I^- O^+ Z^-/F'\ I^+ O^+ Z^+$ = **no enzyme made.**

   $I^S O^+ Z^+/F'\ O^+$ = no **enzyme made.**

   $I^+ O^C Z^+/F'\ O^+ Z^+$ = **functional enzyme to be made.** Notice that the presence of lactose in the medium will allow for synthesis of functional enzyme from the plasmid as well.
8. *E. coli* cells with the *lac* I^q mutation make ten times more repressor than do wild-type cells, which facilitated repressor protein isolation. Repressor in cell extracts from *lac* I^q cells was allowed to bind to a radioactively labeled gratuitous inducer (IPTG) during equilibrium dialysis. The material that bound the labeled inducer (the repressor) was purified and shown to be heat labile and to have other characteristics of protein. Extracts of *lac* I^- cells did not bind the labeled IPTG.
10. **(a)** Activated CAP facilitates the binding of RNA polymerase to the *lac* promoter. Absence of a functional *crp* gene would compromise the positive control exhibited by CAP.

    (b) Without a CAP binding site, the outcome would be similar to that described in part (a): there would be a reduction in the inducibility of the *lac* operon.
12. Attenuation functions to reduce the synthesis of tryptophan when it is in full supply, by reducing transcription of the *tryptophan* operon.
14. Because the deletion of the regulatory gene, *R*, causes a loss of synthesis of the enzymes, the regulatory gene product can be viewed as one exerting *positive control*.

    (a) In wild-type cells, in the absence of tisophane (tis), the regulatory protein (R) is free to bind to the operator and promote transcription. When tis is present, it binds R and induces a conformational change that prevents R from binding to the operator region.

(b) Mutations in the operator would prevent binding of R, even in the absence of tis, thereby preventing transcription.

16. Innate immunity describes cellular defense mechanisms that are not targeted to a specific pathogen. By contrast, adaptive immunity describes an evolving mechanism whereby exposure to a pathogen results in improved defense upon subsequent exposure to that same pathogen. The CRISPR-Cas system, which allows bacteria to become resistant to a given phage, exemplifies adaptive immunity.

18. With each infection by a different type of phage, the bacterial cell adds viral-derived spacers to its CRISPR locus, with newer sequences closer to the leader sequence and older sequences more distant. Such stored "trophies," therefore, provide scientists with both the identities of previously encountered phage and the order in which they were encountered. Because these spacers are mobilized to direct a protective response by the bacterial cell, scientists can use them to predict the cell's immunities and sensitivities.

Chapter 16. Answers to Now Solve This

16.1 Tumor suppressors, such as *BRCA1*, and DNA repair proteins, such as *MLH1*, normally act to prevent accumulation of DNA damage in cells. If either of these genes were suppressed, cells with damaged DNA would continue to proliferate and the frequency of mutation would continue to increase. These subsequent mutations might occur in tumor-suppressor genes or proto-oncogenes, thereby promoting cancer.

16.2 General transcription factors associate with a promoter to stimulate transcription of a specific gene. Some *trans*-acting factors, when bound to enhancers (for example the ER protein), interact with coactivators (estrogen, in this example) to enhance transcription by forming an enhanceosome, which stimulates transcription initiation. Transcription can be repressed by a similar mechanism when proteins bind to silencer DNA elements and generate repressive chromatin structures. The same molecule may bind to a different chromosomal regulatory site (enhancer or silencer), depending on the molecular environment of a given tissue type.

16.3 RISC searches for target mRNAs, guided by single-stranded miRNA molecules. These guided RNAs recruit RISC to the proper targets by binding to complementary sequences in the mRNA. If the miRNA and message are perfectly matched, RISC cleaves the target mRNA, leading to message degradation. If the miRNA is a partial match, translation is blocked.

16.4 Without the zip code sequence, the actin mRNA would not be bound by ZBP1 and, consequently, neither silenced nor transported. As a result, actin would be produced in the cytoplasm near the nucleus and not in the lamellipodium.

Solutions to Problems and Discussion Questions

2. Your essay should include a description of the identity and nature of each type of *cis* element, the action of the *trans*-acting protein(s), and the consequence of this action.

4. Chromosomes occupy discrete domains called chromosome territories, which are separated by channels, called interchromatin compartments. Active genes are found at territory edges and transcripts are moved into the interchromatin compartments for processing. These compartments adjoin to the nuclear pores, which facilitates transport of mRNA from the nucleus.

6. One form of remodeling is the use of histone variants, such as H2A.Z, which change the strength of DNA–nucleosome interactions. In addition, large remodeling complexes can be recruited to promoters to reposition nucleosomes along the DNA. Alterations that create a more open chromatin conformation will allow transcription, whereas downregulation of transcription will result if a more closed conformation is created.

8. The statement is correct. As a result of alternative splicing, one primary mRNA can be spliced into several mature mRNAs of varying lengths, giving rise to different proteins after translation.

10. The mutation will result in a premature STOP codon. These kinds of mutations are called nonsense mutations. The mRNA may be translated into a truncated protein that may be nonfunctional. Alternately, the mRNA may undergo nonsense mediated decay.

12. The assembly of the RNA polymerase II transcription initiation complex is a multistep process:

1. Binding of TFIID to the TATA box: TFIID is composed of the TATA binding protein (TBP) and approximately 13 proteins (TBP associated factors, TAFs). This process is aided by TFIIA.
2. Binding of TFIIB to BRE
3. Recruitment of RNAP and the mediator to the promoter by TFIIF, TFIIH, and TFIIE. Once this process is completed, the pre-initiation complex (PIC) is formed and can mediate unwinding of the promoter, and the polymerase can catalyze RNA elongation.

Regulation: Enhancers and silencers can interact with the PIC even across large distances. This is achieved through the formation of DNA loops that bring together the promoter and the regulatory elements. This process, aided by TFIID and the mediator complex, may also involve chromatin alterations through histone acetylation as well as DNA looping.

14. Promoters mark transcriptional start sites, located at the 5′-end of the transcript. A gene with multiple alternative promoters could, therefore, produce transcripts with different 5′-ends. Depending on which promoter-containing exon is used and its location relative to other exons in the gene, other sequence differences could also exist.

Similarly, polyadenylation signals mark transcriptional stop sites at the 3′-ends of genes. When alternative polyadenylation signals are used, the exon containing the first signal is skipped, allowing previously excluded exons to be transcribed and resulting in a message with a different 3′-end.

16. Normally, stop codons are located near the poly-A tail of the message or downstream of exon–exon junctions. Termination signals that do not meet these criteria are considered by the cell as premature.

18. In general terms, a cytoplasmic protein, Dicer, processes double-stranded small noncoding RNA (sncRNA) molecules to produce shorter dsRNAs. These associate with RISC (RNA-induced silencing complex), where an Argonaut-family protein cleaves and discards one of the two strands. The retained strand guides RISC to the complementary target message, where the complex acts to prevent expression. The specific mechanism of silencing depends on the type of sncRNA used.

20. Since miRNA must base-pair with its target to direct RISC to the correct message, you can determine the sequence that is complementary to the miRNA and search for that sequence among the cellular mRNAs, whose sequences are also known.

22. ceRNAs contain miRNA response elements (MREs)—sequences that are complementary to miRNA and that serve as miRNA binding sites. When present, ceRNAs will compete with mRNA targets for binding of miRNA molecules, rendering the miRNA less effective.

24. It is likely that the mRNA is bound by an RBP that prevents translation. A modification of the RBP that changes its conformation (phosphorylation, for example) would be needed to remove the RBP and allow translation. The modifying enzyme would need to be localized to specific subcellular locations to result in a dispersed mRNA distribution but localized protein production.

26. In many cases, degradation may be required to stop the activity of a protein. For example, while mitosis is dependent on B-type cyclins the cell needs to degrade them to exit mitosis.

28. **(a)** As a transcription factor, TBX20 helps to either upregulate or downregulate expression of genes involved in heart development. When the *Tbx20* gene is deleted, expression of genes normally regulated by TBX20 will change. Comparison of the transcriptomes of the wild-type and mutant cells will reflect these changes. A transcript normally activated by TBX20 would be more prevalent in wild-type cells than in *Tbx20* cells. Likewise, a transcript that is less prevalent in wild-type cells is normally suppressed by TBX20.

(b) Opposite responses by different genes to the same transcription factor could be driven by the presence of different cofactors, by differences in the timing of activation vs. repression, or even by the presence of different isoforms of TBX20.

Chapter 17. Answers to Now Solve This

17.1. **(a)** Either antibiotic will select cells that have taken up a plasmid, but the populations of cells selected by each will differ.

(b) Cells containing plasmids with the *Drosophila* insert will grow only on plates containing tetracycline.

(c) Resistance to both antibiotics by a transformed bacterium could be due to incomplete cleavage of the original plasmid by the *Pst*I enzyme or to re-ligation of the cut ends of plasmids with no insert.

17.2. Using the human nucleotide sequence, identify regions of the β-globin gene that are relatively conserved among mammals. Select sequences from these regions to create PCR primers and amplify the sequences from human DNA, thereby creating a probe to screen the okapi library. The probe will hybridize to complementary sequences in the library, thus identifying the library clone (or clones) that contain the sequence of interest. Alternatively, isolate DNA from the okapi library and use the primers to amplify the sequence directly. Next, sequence the amplified DNA and compare the nucleotide and deduced amino acid sequences against their human counterparts.

Solutions to Problems and Discussion Questions

2. Your essay should include an appreciation for the relative ease with which sections of DNA can be inserted into various vectors as well as for the amplification and isolation of such DNA. You should also include the possibilities of modifying recombinant molecules.

4. Even though the human gene coding for insulin contains introns, the processed mRNA does not. Plasmids containing a cDNA copy of the insulin gene are free of introns, so no processing issue surfaces.

6. This segment contains the palindromic sequence of GGATCC, which is recognized by the restriction enzyme *Bam*HI. It also contains the sequence of GATC, which is recognized by the restriction enzyme *Sau*3AI. The double-stranded sequences are as follows:

GGATCC	GATC
CCTAGG	CTAG
*Bam*HI	*Sau*3AI

8. Plasmids are small, so they are relatively easy to separate from the host bacterial chromosome. They have relatively few restriction sites and can be engineered fairly easily (i.e., polylinkers and reporter genes added). However, plasmids suffer from the limitation that they can only accept small DNA fragments and that they can use only bacteria as hosts. BACs are artificial bacterial chromosomes that can be engineered for certain qualities such as carrying large DNA fragments. YACs (yeast artificial chromosomes) are extensively used to clone DNA in yeast. They accept extremely large DNA inserts, ranging in size from 100 to 1000 kb. As a eukaryotic organism, yeast processes RNA and proteins, allowing the study of eukaryotic genes.

10. A cDNA library is a collection of cDNA clones made by reverse transcribing mRNA from the cells/tissues of interest. It is useful as a 'snapshot' for identifying which genes are expressed in a given tissue at a given stage of development.

12. A cDNA library provides DNAs from RNA transcripts and is, therefore, useful in identifying what are likely to be functional DNAs. If one desires an examination of noncoding as well as coding regions, a genomic library would be more useful.

14. *Taq* polymerase is a heat-stable DNA polymerase that can tolerate extreme temperature changes.

16. The original method, Sanger sequencing, is based on the addition of modified and differently labeled deoxyribonucleotides to a polymerization reaction over a single template to be sequenced. The next generation of sequencing is based on parallel synthesis and fluorescent imaging of incorporated dNTPs. The third generation of sequencing is based on the direct imaging of a single molecule being polymerized.

18. If a transgene integrates into a coding or regulatory region of the genome, it is likely to alter the function of that sequence and, as a result, change the phenotype of the organism in an unexpected fashion. Since two different genetic events would have occurred (disruption of one gene by addition of another), it could be difficult to tease apart the effects of the one event from those of the other.

20. A change in the phenotype of the organism could indicate successful incorporation of a transgenic construct. Also, in many cases, samples of DNA can be isolated and tested either by PCR amplification or by Southern blot analysis.

22. One concern that has already been raised is the potential of creating so-called designer babies, whose genomes are edited for nonmedical reasons, such as to produce physical traits that are seen as desirable to parents. Limiting use of CRISPR-Cas to certain human genes (those that result in medical conditions) could also raise a number of ethical questions, among them: Which diseases "merit" treatment? Who will be charged with making such decisions? Will discriminatory consequences result for untreated individuals? The ethics of genome editing will certainly need to be discussed, and no doubt, guidelines will need to be established.

24. Repair of CRISPR-Cas9 editing by NHEJ, an error-prone pathway, would likely disrupt the function of the target gene, whereas repair by HDR, when a suitable donor template is provided, would allow precise substitutions as well as additions or deletions.

26. One example discussed in the text is the precise modification of the corn *ARGOS8* gene. Scientists removed the weak endogenous promoter and replaced it with a stronger promoter, thereby increasing expression and making the modified corn more drought tolerant.

Chapter 18. Answers to Now Solve This

18.1. **(a)** To annotate a gene, identify gene-regulatory sequences found upstream of genes (promoters, enhancers, and silencers), downstream elements (termination sequences), and in-frame triplet nucleotides that are part of the coding region of the gene. In addition, 5′ and 3′ splice sites that are used to distinguish exons from introns as well as polyadenylation sites are also used in annotation.

(b) Similarity to other annotated sequences often provides insight as to a sequence's function and might serve to substantiate a particular genetic assignment. Direct sequencing of cDNAs from various tissues and developmental stages aids in verification.

(c) The 3141 genes identified on chromosome 1 constitute 15.7 percent of the total number of genes in the human genome (estimated to be 20,000). Since chromosome 1 contains 8 percent of the human genome and almost 16 percent of the genes, it would appear that chromosome 1 is gene rich.

18.2. Structural and chemical factors determine the function of a protein, so it is possible to have several proteins that share a considerable amino acid sequence identity, but not be functionally identical. The *in vivo* function of a protein is determined by secondary and tertiary structures, as well as local surface chemistries in active or functional sites. The nonidentical sequences of the four proteins might have considerable influence on protein folding, overall structure, and, therefore, function. Note that the query matches to different site positions within the target proteins. A number of other factors suggesting different functions include associations with other molecules (cytoplasmic, membrane, or extracellular), chemical nature and position of binding domains, posttranslational modification, and signal sequences.

18.3. **Advantages:** First, blood is relatively easy to obtain in a pure state, so its components can be analyzed without fear of tissue-site contamination. Second, blood is intimately exposed to virtually all cells of the body and might therefore carry chemical markers to certain abnormal cells it represents. It represents, theoretically, an ideal probe into the human body. **Validation criteria:** When blood is removed from the body, its proteome changes, and those changes are dependent on a number of environmental factors. Thus, what might be a valid diagnostic under one set of conditions might not be so under others. In addition, the serum proteome is subject to change depending on the genetic, physiologic, and environmental state of the patient. Age and sex are additional variables that must be considered. *Validation* of a plasma proteome for a particular cancer would be strengthened by demonstrating that the stage of development of the cancer correlates with a commensurate change in the proteome in a relatively large, statistically significant pool of patients. As well, the types of changes in the proteome should be reproducible and, at least until complexities are clarified, involve tumorigenic proteins. It would be helpful to have comparisons with archived samples of each individual at a disease-free time.

Solutions to Problems and Discussion Questions

2. Your essay should include a description of traditional recombinant DNA technology, which involved cutting and splicing genes, as well as modern methods of synthesizing genes of interest, PCR amplification, microarray analysis, etc.

4. Whole-genome sequencing involves randomly cutting the genome into numerous smaller segments and determining their sequences. Overlapping sequences are used to identify segments that were once contiguous, eventually producing the entire sequence. Ultimately, entire chromosomes can be sequenced and the positions of genes, both known and presumptive, precisely mapped. Linkage mapping, by contrast, relies on recombination frequencies observed in the progeny of specific crosses. As a result, only relative map distances between known genes can be determined. While this method can also produce a map of an entire chromosome, all distances are relative and only known genes with detectable phenotypes are mapped.

6. Hallmarks to annotation in both bacteria and eukaryotes are the identification of upstream gene regulatory sequences (such as promoters) and downstream elements (termination sequences), as well as triplet nucleotides that are part of the coding region of the gene. In eukaryotes, upstream elements would also include enhancers and silencers, and downstream elements would also include a polyadenylation signal sequence (enhancers and silencers are also possible). In addition, 5′ and 3′ splice sites that distinguish exons from introns are also used in annotation.

8. Homologous genes are evolutionarily related. Paralogous genes are evolutionarily related (homologous) within the same species. Orthologous genes can be found in different species that derive from a common ancestor.

BLAST can definitely be used to identify different types of genes if there are enough sequences available for alignment. If a protein domain has known evolutionarily conserved amino acid sequences, a BLAST alignment of sequences with that domain will predict its structure.

10. The human genome is composed of more than 3 billion nucleotides of which about 2 percent code for genes. There appears to be approximately 20,000 protein-coding genes; however, there is still uncertainty as to the total number. Genes are unevenly distributed over chromosomes, with clusters of gene-rich regions separated by gene-poor ones (deserts). Human genes tend to be larger and contain more and larger introns than in invertebrates. It is estimated that at least half of the genes generate products by alternative splicing. The human genome is dynamic, containing an abundance of repetitive sequence scattered throughout.

12. Sources of genetic variation: SNPs: variations in single nucleotides (0.1 percent of the difference). Structural differences: CNVs: deletions and duplications of large regions of repetitive DNA. Estimated to be up to 5 dozen per individual (accounting for the majority of the variation). Other structural differences include deletions, duplications, inversions, and aneuploidy.

14. Whole-genome sequencing provides sequences of entire genomes, including noncoding regions, whereas whole-exome sequencing provides sequence information for exons only. Since there are more disease-related variations in the exome than in other regions of the genome, WES is more likely to identify these mutations than is WGS. However, only WGS is able to identify mutations in regulatory regions that lead to disease.

16. Nutrigenomics tests are provided by companies on request. They are based on the analysis of the genomes of individuals who request them. They evaluate genes thought to be involved in conditions related to metabolism and provide these results to the customer with a set of nutritional recommendations that should help them improve their health and prevent illness. These tests are controversial as they have not been validated and approved by any governmental health organizations.

18. Most microarrays consist of a glass slide that is spotted with thousands of different single-stranded DNAs (probes). A single microarray can have as many as 20,000 different spots of DNA, each containing a unique sequence. Researchers use microarrays to compare patterns of gene expression in tissues under different conditions or to compare gene-expression patterns in normal and diseased tissues. In addition, microarrays can be used to identify pathogens.

20. (a) Over 290,000 nonredundant peptides were identified from multiple organs, tissues, and cell types from clinically healthy individuals.

(b) Seven fetal tissues were used.

(c) A wide variety of searches can be performed.

22. (a) The strength of WES is that one might get lucky and identify a coding issue in a gene that has relevance to the patient's condition. However, there are several weaknesses to this approach. Given the multitude of genetic variations known to exist, using this approach might be akin to looking for a needle in a haystack. In addition, many important regulatory and structural components in the genome are outside the exon pool and would not be detected by this method.

(b) Depending on the symptoms and the nature of the disorder, examination of the mitochondrial genome (which is highly variable between individuals) might be advisable because the list of human conditions known to involve mitochondrial defects is growing. However, the high degree of genetic variability of mtDNA might make identification of meaningful differences difficult.

Chapter 19. Answers to Now Solve This

19.1. Being able to distinguish leukemic cells from healthy cells allows not only the design of a targeted therapy, but also the quantification of responses to that therapy. CML cells produce a hybrid protein, so it may be possible to develop an immunotherapy, based on the uniqueness of the BCR-ABL protein. Alternatively, one could create a drug that would inactivate the BCR-ABL protein without affecting the function of any of the other protein kinases. This is exactly what scientists at Ciba-Geigy (now Novartis) did in the production of Gleevec.

19.2. *TP53* is a tumor suppressor gene that protects cells from multiplying with damaged DNA. It is present in its mutant state in more than 50 percent of all tumors. *TP53* mediates the immediate control of a critical and universal cell cycle-checkpoint and its action is not limited to specific cell types. As a result, mutation of this gene will influence a wide range of cell types and give rise to a diversity of tumors.

19.3. Even if a major "cancer-causing" gene is transmitted, other genes, often new mutations, are usually necessary to drive a cell toward tumor formation. Full expression of the cancer phenotype is likely to be the result of interplay among a variety of genes and therefore likely to show variable penetrance and expressivity.

19.4. Unfortunately, it is common to spend enormous amounts of money dealing with diseases after they occur rather than concentrating on disease prevention. Too often, pressure from special-interest groups or lack of political will retards advances in education and prevention. Obviously, it is less expensive, in terms of both human suffering and money, to seek preventive measures for as many diseases as possible. However, having gained some understanding of the mechanisms of disease, in this case, cancer, it must also be stated that no matter what preventive measures are taken, it will be impossible to completely eliminate disease from the human population. It is extremely important, however, that we increase efforts to educate and protect the human population from as many hazardous environmental agents as possible. A balanced, multipronged approach seems appropriate.

Solutions to Problems and Discussion Questions

2. Your essay should describe the general influence of genetics in cancer. Epigenetic factors alter gene output, making it likely that such factors could cause cancer.

4. The major regulatory points of the cell cycle include the following:

late G1 (G1/S)

the border between G2 and mitosis (G2/M)

mitosis (M)

6. During the cell cycle, a normal cell can decide to enter S phase and divide again or remain quiescent. When a cell has received signals to stop dividing, it will enter G0. Normal cells can enter G0 while cancer cells cannot, therefore they never stop dividing. To exit G0, a normal cell may receive growth factor signals not required by cancer cells.

8. Apoptosis is regulated by a cascade of caspases. In case of extensive or prolonged stress to the cell (such as extensive DNA damage), checkpoint responses may shift from cell-cycle arrest to apoptosis (such as p53-induced p21 and G1/S arrest or p53-induced BAX and apoptosis).

10. Under normal conditions, the p53 protein is bound by another protein, MDM2, which keeps p53 in an inactive state. MDM2 not only marks p53 for degradation, it also blocks the p53 transcription activation domain and prevents posttranslational modifications that activate p53. Upon DNA damage or cellular stress, MDM2 dissociates from p53, which both stabilizes the protein and exposes the transcription activation domain. Increases in acetylation and phosphorylation of p53 also contribute to increases in activity.

12. If DNA replication or repair of any DNA damage has not been completed, the cell cycle arrests prior to mitosis until these processes are complete. Activated p53 protein regulates expression of genes that retard the progress of DNA replication and that prevent cells from passing the G2/M checkpoint, thereby allowing time for DNA damage to be repaired.

14. They can have loss-of-function mutations. There is actually no mechanism for the selective mutagenesis of oncogenes. However, a loss-of-function mutation will simply inactivate the oncogenic activity; therefore, the cell is not likely to survive.

16. Driver mutations are those mutations in a tumor cell that confer a growth advantage upon that cell. Passenger mutations do not directly induce a cancerous state. However, any passenger mutation that provides a growth advantage after a change in environmental conditions, will become a driver mutation.

18. The environment, both natural and man-made, contains a large number of carcinogens. These include chemicals, radiation, tobacco smoke, some viruses, and chronic infections. They contribute to carcinogenesis by causing point mutations, double-stranded breaks in DNA, etc.

20. Certain environmental agents such as chemicals and X rays cause DNA damage that can lead to mutations. If these mutations affect cell-cycle checkpoint controls, they can lead to cancer.

22. No, she will still have the same risk as the general female population (about 12 percent). In addition, it is possible that genetic tests will not detect all breast cancer mutations.

24. A benign tumor is a mass of proliferating cells that is usually localized to a given anatomical site. Malignant tumors are those in which some cells are able to break away, invade other tissues, and form tumors at one or more secondary sites.

26. As with many forms of cancer, a single genetic alteration is not the only requirement. It is possible that the cells that produce these transcripts are extremely low in number and are unable to establish themselves as key stem cells for clonal expansion, a necessary precursor to cancer formation. It is also possible that the transcripts produced do not result in a functional fusion protein.

Chapter 20. Answers to Now Solve This

20.1. (a) Since ~ 1/256 of the F_2 plants are 20 cm and ~ 1/256 are 40 cm, there must be four gene pairs involved in determining flower size.

(b) The backcross is: $AaBbCcDd \times AABBCCDD$. The frequency distribution in the backcross would be

1/16 = 40 cm	4/16 = 32.5 cm
4/16 = 37.5 cm	1/16 = 30 cm
6/16 = 35 cm	

20.2. (a) The mean for each is:

$$\textit{mean sheep fiber length} = 7.7 \text{ cm}$$
$$\textit{mean fleece weight} = 6.4 \text{ kg}$$

The variance for each is

$$\textit{variance sheep fiber length} = 6.097$$
$$\textit{variance fleece weight} = 3.12$$

The standard deviations are:

$$\textit{sheep fiber length} = 2.469$$
$$\textit{fleece weight} = 1.766$$

(b) The covariance for the two traits is 30.36/7, or 4.34.

(c) The correlation coefficient is +0.998.

(d) The correlation coefficient indicates that there is a very high correlation between fleece weight and fiber length, and it is unlikely that this correlation is by chance. Even though correlation does not mean cause and effect, it would seem logical that as you increased fiber length, you would also increase fleece weight. It is probably safe to say that the increase in fleece weight is directly related to an increase in fiber length.

20.3. Compare the expression of traits in monozygotic and dizygotic twins. A higher concordance value for monozygotic twins indicates a significant genetic component for a given trait. Notice that for traits including blood type, eye color, and mental retardation, there is a fairly substantial difference between MZ and DZ groups. However, for measles, the difference is not as large, indicating a greater role of the environment. Hair color has a substantial genetic component as do idiopathic epilepsy, schizophrenia, diabetes, allergies, cleft lip, and club foot. The genetic component to mammary cancer is present but minimal according to these data.

Solutions to Problems and Discussion Questions

2. Your essay should include a description of various ratios typical of Mendelian genetics as compared with the more blending, continuously varying expressions of Neomendelian modes of inheritance. It should contrast discontinuous inheritance and continuous patterns.

4. **(a)** There are two alleles at each locus for a total of four alleles.

(b, c) We can say that each additive allele provides an equal unit amount to the phenotype and the colors differ from each other by multiples of that unit amount. The number of additive alleles needed to produce each phenotype is as follows:

1/16	= dark red	= 4 alleles	*AABB*
4/16	= medium-dark red	= 3 alleles	2 *AABb* 2 *AaBB*
6/16	= medium red	= 2 alleles	*AAbb* 4 *AaBb* *aaBB*
4/16	= light red	= 1 allele	2 *aaBb* 2 *Aabb*
1/16	= white	= 0 alleles	*aabb*

(d) F_1 = all light red
F_2 = 1/4 medium red   2/4 light red   1/4 white

6. **(a)** Height is determined by quantitative trait loci.

(b) There are four gene pairs involved.

(c) There are nine categories and eight increments between them. Since there is a difference of 24 cm between the extremes, each additive allele contributes 24 cm/8 = 3 cm.

(d) The shortest extreme, representing the completely recessive genotype (*aabbccdd*), is 12 cm tall, and each parental plant is 24 cm tall; therefore each parental plant needs to have four additive alleles to have a height of 24 cm. Since the plants are inbred, this means that each must be homozygous at two loci. The F_2 distribution suggests that all gene pairs in the F_1 individuals are heterozygous (*AaBbCcDd*) and independently assorting. Therefore, the parental plants must be homozygous dominant at different gene pairs. There are many possible sets of parents that would fit this description. An example is

AABBccdd × *aabbCCDD*

(e) Two additive alleles are needed for a height of 18 cm. Three possibilities are:
AAbbccdd, *AaBbccdd*, *aaBbCcdd*

Likewise, any plant with seven uppercase letters will be 33 cm tall. Three examples are:
AABBCCDd, *AABBCcDD*, *AABbCCDD*

8. Average differences between MZ twins reared either apart or together that are substantially smaller than differences for DZ twins or siblings reared together indicate that genetics plays a major role in determining the trait. For the data presented, ridge count and height appear to have the highest heritability values.

10. Monozygotic twins are derived from the splitting of a single fertilized egg and are, therefore, of identical genetic makeup. When such twins are raised in the same versus different settings, an estimate of relative hereditary and environmental influences can often be made

12. **(a)** The mean is 140 cm.

(b). Use the formula for variance as given in the text: $s^2 = \frac{\Sigma(X_i - \overline{X})^2}{n-1}$ for each data point. To avoid adding the same numbers repeatedly for each height category, you can multiply each squared difference by the number of plants in that category: For example, for the first group (100 cm), we would have $(100 - 140)^2 \times 20 = 32{,}000$. Do this for the remaining groups, and then perform the summation step, which gives a total of 284,000. Dividing this sum by $(n - 1)$ gives 284,000/759. So $s^2 = V = 374.18$.

(c) The *standard deviation* is the square root of the variance, or 19.34.

(d) The *standard error* of the mean is the standard deviation divided by the square root of *n*, or about 0.70. The plot approximates a normal distribution. Variation is continuous.

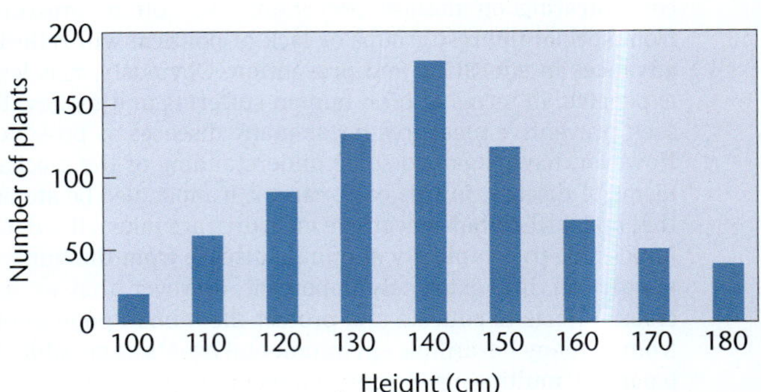

14. The formula for estimating broad sense heritability is $H^2 = V_G/V_P = 3.8/7.5$ or 0.50. This value, when viewed in percentage form, indicates that about 50 percent of the variation in body size is due to genetic influences.
V_G is the genotypic and V_P the phenotypic component of the variation. V_P is a combination of genetic and environmental variance. Both parental strains are inbred. We assume they are homozygous, therefore, variance is the result of environment and we can average both ($V_E = 3.2$). The F_1 is also genetically homogeneous and gives us an additional estimation of the environmental factors [$V_E = (3.2 + 4.2)/2 = 3.7$]. The phenotypic variance in the F_2 is the sum of both V_G (genetic) and V_E (environmental) components. Therefore, $V_G = 7.5 - 3.7 = 3.8$.

16. Given that both narrow-sense heritability values are relatively high, it is likely that a farmer would be able to alter both milk protein content and butterfat by selection. The value of 0.91 for the correlation coefficient between protein content and butterfat suggests that if one selects for butterfat, protein content will increase.

18. The mean would be

$$0.25 = (M2 - 20)/(19 - 20) = 19\text{ g}.$$

Repeated experiments would result in progressively less weight loss as h^2 decreased to 0 as the mice reached the minimum weight possible to breed.

20. Because the rice plants are genetically identical, V_G is zero and $H^2 = V_G/V_P$ = zero. Broad-sense heritability is a measure in which the phenotypic variance is due to genetic factors. In this case, with genetically identical plants, H^2 is zero, and the variance observed in grain yield is due to the environment. Selection would not be effective in this strain of rice.

22. Even though the disease is considered to be polygenic, with multiple genes contributing in an additive fashion, a particular number of additive alleles might be necessary for the trait to be expressed and, once expressed, additional alleles do not alter expression appreciably.

Chapter 21. Answers to Now Solve This

21.1. $q^2 = 37/125 = 0.296$, therefore $q = 0.544$
$p = 1 - q$, therefore $p = 0.456$
Frequency of $AA = p^2 = (0.456)^2 = 0.208$ or 20.8%
Frequency of $Aa = 2pq = 2(0.456)(0.544) = 0.496$ or 49.6%
Frequency of $aa = q^2 = (0.544)^2 = 0.296$ or 29.6%

21.2. **(a)** For the *CCR5* analysis, first determine p and q, and then determine the equilibrium genotype values: $p = 0.7755$ and $q = 0.2245$
The equilibrium values are
Frequency of $I/I = p^2 = 0.6014$ or 60.14%
Frequency of $I/\Delta32 = 2pq = 0.3482$ or 34.82%
Frequency of $\Delta32/\Delta32 = q^2 = 0.0504$ or 5.04%
Comparing these equilibrium values with the observed distribution strongly suggests that the observed values are drawn from a population in Hardy–Weinberg equilibrium.

(b) For the sickle-cell analysis, $p = 0.877$ and $q = 0.123$.
The equilibrium values are:
Frequency of $SS = p^2 = 0.7691$ or 76.91%
Frequency of $Ss = 2pq = 0.2157$ or 21.57%
Frequency of $ss = q^2 = 0.0151$ or 1.51%
Comparing these equilibrium values with the observed values suggests that the observed values may be drawn from a population that is not in equilibrium. Notice that there are more heterozygotes and fewer homozygotes (especially of the *ss* types) than predicted.

21.3. **(a)** $q = 0.01$
(b) $p = 1 - q$ or 0.99
(c) $2pq = 2(0.01)(0.99) = 0.0198$ (or about 1/50)
(d) $2pq \times 2pq = 0.0198 \times 0.0198 = 0.000392$ (or about 1/2550)

21.4. The probability that the woman (with no family history of CF) is heterozygous is 98/2500. The probability that the man is heterozygous is 2/3. The probability that a child with CF will be produced by two heterozygotes is 1/4. Therefore, the overall probability of the couple producing a CF child is $98/2500 \times 2/3 \times 1/4 = 0.00653$, or about 1/153.

Solutions to Problems and Discussion Questions

2. Your essay should explain how mutation and migration (especially when the immigrants have different gene frequencies compared to the host population) can cause gene frequencies to change in a population. You should also describe selection as resulting from the biased passage of gametes to the next generation. Be sure to include a discussion of the mechanisms of reproductive isolation.

4. There must be evidence that gene flow does not occur among the groups being called different species. Classifications above the species level (genus, family, etc.) are not based on such empirical data and are somewhat arbitrary, based on traditions that extend far beyond DNA sequence information. In addition, recall that DNA sequence divergence is not always directly proportional to morphological, behavioral, or ecological divergence.

6. For each of these values, Take the square root to determine q, and then compute p. Using these values, calculate $2pq$ to determine the number of heterozygotes.
(a) $q = 0.08$; $2pq = 0.1472$ or 14.72%
(b) $q = 0.009$; $2pq = 0.01784$ or 1.78%
(c) $q = 0.3$; $2pq = 2(0.7)(0.3) = 0.42$ or 42%
(d) $q = 0.1$; $2pq = 0.18$ or 18%
(e) $q = 0.316$; $2pq = 0.4323$ or 43.23%

8. Eighteen percent of the individuals would be expected to be heterozygous for this locus in the next generation.

10. **(a)** $q_{g+1} = [0.8(0.4)^2 + 0.9(0.6)(0.4)]/[1(0.6)^2 + 0.9(2)(0.6)(0.4) + 0.8(0.4)^2]$
$q_{g+1} = 0.37$; $p_{g+1} = 0.63$
(b) $q_{g+1} = 0.39$ $p_{g+1} = 0.61$
(c) $q_{g+1} = 0.37$ $p_{g+1} = 0.63$
(d) $q_{g+1} = 0.29$ $p_{g+1} = 0.71$

12. Selection against a dominant lethal allele is high, so it is unlikely that it will be present at a high frequency, if at all. However, if the allele shows incomplete penetrance or late age of onset (after reproductive age), it may remain in a population.

14. The frequency of an allele is determined by a number of factors, including the fitness it confers, mutation rate, and input from migration. The distribution of a gene among individuals is determined by mating (population size, inbreeding, etc.) and environmental factors (selection, etc.). If a population is in Hardy–Weinberg equilibrium, the distribution of genotypes occurs at or around the $p^2 + 2pq + q^2 = 1$ expression. Equilibrium does not mean 25 percent *AA*, 50 percent *Aa*, and 25 percent *aa*. This confusion often stems from the 1:2:1 (or 3:1) ratio seen in Mendelian crosses.

16. 2/100,000 or 2×10^{-5}

18. Neutral alleles are a source of genetic variation since at any given time, a population may contain several neutral alleles at any particular locus. Genetic adaptations to certain environmental conditions can also favor variation if heterozygosis is advantageous in that situation.

20. $q_{g+1} = 0.852$ and $p_{g+1} = 0.148$

22. $p_i' = 0.792$ and $q_i' = 0.208$

24. Any therapy already developed or future therapy to treat late onset diseases beyond the reproductive age would have no effect on natural selection of the human species as the disease has not manifested itself before the individual has been able to reproduce and pass on their alleles. If the disease has early onset or is age independent (for example as a viral infection) clearly the therapy is increasing the fitness of the individual and as a consequence is affecting natural selection. At the same time, it may help increase genetic variation through preservation of previously detrimental alleles and possibly lead to an increased level of less fit alleles in the gene pool. Gene therapy and genome editing would have the same effect as long as it is performed in somatic cells.

26. Reproductive isolating mechanisms are grouped into prezygotic and postzygotic. Prezygotic mechanisms are more efficient because they occur before resources are expended in the processes of mating.

28. In small populations, large fluctuations in gene frequency occur because random gametic sampling may not include all the genetic variation in the parents. Two factors can cause the extinction of a particular mutation in small populations.

First, sampling error may allow fixation of one form and the elimination of others. If a mutation (either advantageous or deleterious) occurs, it is possible that it might not be represented if the founding population is small. Second, if a deleterious mutation becomes fixed, it can lead to extinction of that population.

30. With one exception (the final amino acid in the sequences), all the amino acid substitutions require only one nucleotide change. The last change, Pro (CCN) to Lys (AAA, G), requires two changes (the minimal mutational distance).

Special Topic 1

Review Question Answers

2. In mammals, methylation of cytosine residues, when it occurs, takes place at CpG-rich regions in promoters and upstream sequences. The effect is to block transcription, thereby silencing the genes. The majority of methylated sequences in mammals are found in heterochromatin, where they promote chromatin stability by preventing replication and transcription of transposable elements.

4. Reversible modifications of histone tails alter the ability of histones (and therefore nucleosomes) to bind tightly to DNA. Modifications that weaken this interaction make the chromatin more accessible to interactions with other proteins, such as the transcriptional machinery. Conversely, modifications that strengthen this interaction make the chromatin less accessible.

6. Imprinting usually involves specific genes that are restricted in number and that are altered by passage through meiosis. Whereas most genomic methylation patterns are reversed at fertilization, imprinting is not. A gene that is marked by a maternally derived imprint will not have a paternally derived imprint and vice versa. Imprinted sequences are the same in all cells of an organism and, in somatic cells, are permanent. Imprinted alleles are transcriptionally silent in all cells. Epigenetic modifications (methylation, histone modification, interaction with ncRNAs), by contrast, are generally reversible and can be responsive to environmental signals.

8. Most imprinted genes are found clustered on chromosomes and generally encode growth factors or genes that regulate such factors. Mutations in the imprinting of genes (epimutations) are caused either by changes in DNA sequence or by dysfunctional epigenetic changes. Both can result in heritable changes in gene activity. Dysfunctional imprinting is especially dangerous, since changes in methylation for one imprinted gene often affect the expression of adjacent or coordinately controlled imprinted genes within a cluster. Imprinted genes generally act early in development, so defects in their expression have serious consequences. Some diseases caused by faulty imprinting include Prader–Willi syndrome, Angelman syndrome, and Beckwith–Wiedemann syndrome.

Discussion Question Answers

2. Studies show there is a four- to nine-fold increased risk of a child conceived via ART having Beckwith–Wiedemann syndrome. Increased incidences of Prader–Willi syndrome and Angelman syndrome have also been reported. Children conceived by IVF that have one of these syndromes have reduced levels or loss of maternal-specific imprinting. Furthermore, the time frame of ART manipulations coincides with the timing of epigenetic reprogramming, giving weight to the hypothesis that the manipulation of embryos interferes with imprinting. Given these data, it would seem reasonable that such information should be provided to prospective parents of an ART child. Each couple would need to reach a decision based on available science and their own value and belief sets.

4. Some RNAs (such as miRNAs and lncRNAs) are known to downregulate gene expression. According to the study, some foods could be the source of regulatory RNAs that circulate in body fluids of humans. Further studies would need to show from which plants the regulatory RNAs are derived, which genes are affected, and at what levels these RNAs are required to significantly alter gene expression before informed decisions can be made regarding safety. At this point it might be premature to design a dietary regimen based on such a frail understanding of the role of plant miRNAs in humans.

Special Topic 2

Review Question Answers

2. Scenarios (a) and (d) are prognostic, while scenarios (b), (c), and (e) are diagnostic.

4. DNA from an individual that is homozygous recessive for the sickle-cell allele will hybridize strongly to the probe for the mutant allele (the β^S probe) but not to the wild-type probe (the β^A probe).

6. Preimplantation genetic diagnosis is used to screen the genomes of embryos created by *in vitro* fertilization for chromosomal abnormalities or for a specific genetic defect, to determine which of the IVF embryos will be introduced into the uterus for implantation.

8. RNA populations from different tissues will give different patterns of hybridization, a so-called transcriptome analysis, identifying which genes are expressed and at what level.

10. GWAS attempt to identify genes and mutations that influence disease risk and to make substantial contributions to the diagnosis and treatment of genetic diseases.

Discussion Question Answers

2. Since both mutations occur in the CF gene, children who possess both mutant alleles will develop CF—neither allele will be able to produce a wild-type CF protein. Since both parents are heterozygotes, there is a 25 percent probability that any child will inherit both mutant alleles and develop CF.

4. Widespread screening of newborns would allow the identification of a virtually infinite number of variables associated with the human genome that might be of scientific and personal interest. Some new bases for certain disease states might be identified. However, disadvantages would be the likely stigmatizing of certain individuals and numerous issues of privacy invasion and discrimination.

6. Positive consequences include increasing awareness of the disease, the genetic test, and possible treatment options. Negative consequences include rash decisions made by individuals based solely on the decision of a highly public persona.

8. FDA regulation is one way to decrease the distribution of misinformation that could be associated with such a test. It would also ensure a standard of practice among all companies that could improve the reliability of the results and increase consumer protection.

10. Investigators were likely looking for some reason to explain the killer's behavior—possibly SNPs correlated with a particular mental illness or a tendency to violence, or perhaps even a specific genetic mutation. Given the present state of understanding of behavioral genetics, however, it is unlikely that any definitive conclusions could be reached by such an analysis.

12. Since a gene is a product of the natural world, it does not conform to U.S. patent laws, which govern patentable matter, and is, therefore, not patentable. However, the direct-to-consumer test for the *BRCA1* and *BRCA2* genes is original in its process or development, so it should be patentable.

Special Topic 3

Review Question Answers

2. In *ex vivo* gene therapy, a potential genetic correction takes place in cells that have been removed from the patient and are subsequently reintroduced into the patient. *In vivo* gene therapy treats cells of the body through the introduction of DNA into affected cells of the patient.

4. In many cases, therapeutic DNA hitches a ride with genetically engineered viruses, such as retrovirus or adenovirus vectors. Nonviral delivery methods may use chemical assistance to cross cell membranes, nanoparticles, or cell fusion with artificial vesicles.

6. White blood cells, T cells in this case, were used because they are key players in the mounting of an immune response, which Ashanti was incapable of developing. A normal copy of the *ADA* gene was engineered into a retroviral vector, which then infected many of her T cells. Those cells that expressed the *ADA* gene were then injected into Ashanti's bloodstream, and some of them populated her bone marrow. At the time of Ashanti's treatment, targeted gene therapy was not possible, so integration of the *ADA* gene into Ashanti's genome probably did not replace her defective gene.

8. Genome editing involves the removal, correction, or replacement of a defective gene, whereas traditional gene therapy involves the addition of a therapeutic gene that coexists with the defective copy. Genome editing can alter one or several bases of a gene or replace the gene entirely. To some extent, genome editing is designed to alleviate one of the major pitfalls of gene therapy, random DNA integration.

10. ZFNs, or zinc-finger nucleases, consist of a DNA cutting domain from a restriction endonuclease and a DNA-binding domain containing a zinc-finger motif, which can be engineered to recognize any sequence. The zinc-finger motifs bind to the DNA recognition sequence five to seven nucleotides apart from each other, and the nuclease domain cuts the DNA between them. The resulting double-strand break can be repaired by homologous recombination or nonhomologous end-joining.

Discussion Question Answers

2. Currently, gene therapy is considered to be acceptable for the relief of genetic disease states. Given its expense, however, its use is the subject of considerable debate. It remains to be seen whether insurance companies will embrace what might be considered experimental treatments. Use of gene therapy to enhance the competitive status of individuals (genetic enhancement or gene doping) is presently viewed as cheating by most organizations and the public. Finally, it is unlikely that germ-line therapy will be viewed favorably by the public or scientific communities; however, this and other issues mentioned here will be the subject of considerable future debate.

4. Germ-line and embryo therapy is controversial for a number of reasons, but primarily because any genomic changes made will persist through future generations. In addition, there are concerns about human experimentation in general, and experimentation involving embryos, which are unable to give consent, in particular. There are also safety issues to consider, including that the treatment would need to be shown to be effective and efficient. Side effects, specifically effects of mistargeting, would need to be documented, quantified, and addressed. Further, the potential for altered viability of treated germ cells and embryos would also need to be addressed.

Special Topic 4

Review Question Answers

2. Not only did Nancy Wexler and her team gather extensive pedigree information, but she also collected blood samples from some individuals in each pedigree, which allowed James Gusella to establish that the HD locus and the C haplotype of the G8 RFLP marker are coinherited.

4. The results of experiments using an inducible system showed that protein aggregation in the brain and the resulting motor symptoms could be reversed when the mutant *HTT* gene was silenced shortly after the appearance of motor symptoms. This suggested that the disease might be controlled or reversed if treated during the early stages.

6. Early mouse models showed that the presence of the promoter and a copy of the first exon seems to be necessary and sufficient for disease manifestation.

8. Several approaches have been tested in mice and in human culture cells and include the following:

(1) Stem cell therapy has been tested in transgenic mice carrying a full-length copy of the human *HTT* gene (*HTT* tg mice). These were injected with human stem cells engineered to overexpress brain-derived neurotrophic factor, a required protein for survival of striatal brain cells.

(2) Two strategies for gene silencing were tested in mice. (a) Repression of m*HTT* transcription by gene editing using ZFNs: The ZFN constructs were delivered into *HTT* tg mice by injection into the striatal region of one side of the brain. (b) Degradation of target mRNA using ASOs complementary to the mutant allele: ASOs were delivered to *HTT* tg mice by transfusion into the cerebrospinal fluid.

(3) Genome editing using CRISPR-Cas9: Separate constructs for Cas9 and for sgRNAs specific for the m*HTT* gene were injected into the striatum of tg mice carrying exon 1 of the human m*HTT* gene.

Discussion Question Answers

2. Prioritizing understanding of the functions and mechanisms of the mutant HTT protein is necessary for the development and assessment of treatment strategies. However, it is critical for any potential treatment to specifically target the mutant protein and not also inhibit the normal protein. Therefore, understanding of the function and mechanisms of the *normal* HTT protein is equally important.

4. Trinucleotide repeat diseases, such as HD, show genetic anticipation, in which the disease appears earlier and with greater severity in each successive generation. Anticipation has been correlated with increases in the number of repeats in each such disease.

Special Topic 5

Review Question Answers

2. STRs are like VNTRs, but the repeat portion is shorter, between two and nine base pairs, repeated from 7 to 40 times. With the development of the polymerase chain reaction, even trace amounts of DNA samples can be analyzed relatively quickly. A core set of STR loci, 20 in the United States and 12 in Europe, is most often used in forensic applications.

4. Typically males contain a Y chromosome and females do not (exceptions include transgender and mosaic individuals); therefore, gender separation of a mixed tissue sample is easily achieved by Y chromosome profiling. In addition, STR profiling is possible for over 200 loci; however, because of the relative stability of DNA in the Y chromosome, it is difficult to differentiate between DNA from fathers and sons or male siblings or any male from the same patrilineage.

6. Like the Y chromosome, mtDNA is relatively stable because it undergoes very little, if any, recombination. There are many mitochondria present in each cell, resulting in a high copy number of mtDNA per cell, which makes mitochondrial profiling especially useful and the technique of choice in situations where samples are small, old, or degraded, which is often the case in catastrophes.

8. The FBI maintains the Combined DNA Index System (CODIS), which is a collection of state and federal DNA databases and the analytical tools needed to mine the data. DNA profiles are collected from convicted offenders, forensic investigations, and in some states, those suspected of crimes as well as from unidentified human remains and missing persons.
10. In many criminal cases, there is no DNA evidence or the DNA evidence is not useful. For those cases for which DNA *is* useful, samples can remain unprocessed. Human mistakes—or deliberate human interference—can introduce errors into DNA evidence. Frequently samples from crime scenes are not pure, but instead contain mixtures of DNA from multiple people. In addition, evidence can be degraded and will produce only a partial profile, which can be difficult to interpret.

Discussion Question Answers

2. Using your search engine of choice, search for "regulations regarding the collection and storage of DNA samples and profiles." Even without specifying your state of residence, there are many sites that provide this type of information. Another strategy is to access the National Conference of State Legislatures Web site and navigate to "Welcome to the DNA Laws Database."
4. Somatic mosaicism and chimerism involve a mixture of cell types with different origins, within one individual. A chimeric individual contains a mixed population of cells; therefore, a DNA sample taken from one tissue site may not match a DNA sample taken from another site. This can lead to a conflicting set of results when it comes to matching an individual's DNA profile to that from a crime scene sample. Taking DNA samples from various sites on an individual might be useful in mitigating such confusion. In addition, in STR DNA profiling, mosaicism might be detectable at the electrophoresis/analysis stage by the presence of additional peaks or peak height imbalances.

Special Topic 6

Review Question Answers

2. Genetic engineering allows genetic material to be transferred within and between species and to alter expression levels of genes. A transgenic organism is one that involves the transfer of genetic material between different *species*, whereas the term cisgenic is used in cases where gene transfers occur within a species. Currently, most genetically modified (GM) organisms are transgenic; however, this may change as gene-editing techniques are increasingly used.
4. *Bacillus thuringiensis* (Bt) produces crystal (Cry) proteins that are toxic to certain orders of insects. When ingested by susceptible insects, the Cry proteins bind to receptors in the gut wall, which leads to a breakdown of the gut membranes and the death of the insect.
6. Golden Rice 2 is the second-generation transgenic rice crop, developed to synthesize beta-carotene. Its production involved the introduction of three genes into the T-DNA region of a Ti plasmid: a bacterial *carotene desaturase* (*crtI*) gene, which is under the control of an endosperm promoter; the maize *phytoene synthase* (*psy*) gene, also under the control of an endosperm promoter; and the bacterial *phosphomannose isomerase* (*pmi*) gene, which is under the control of a constitutive promoter. Researchers transformed established embryonic rice cell cultures and employed a positive selection method by growing plant cells on a mannose-containing medium. Surviving cells (those that expressed the *pmi* gene) were stimulated to form calluses, which were grown into plants. The desired transgenic constructs were verified using the polymerase chain reaction.
8. In positive selection, conditions are arranged such that only the organism of interest can grow (mannose selection for example). In negative selection, a condition is arranged that inhibits the growth of organisms that are of no interest (antibiotic resistance, for example).

Discussion Question Answers

2. Generally, many feel a "right to know" would allow consumers to make educated choices about the food they consume. They would consider it an advantage to be able to judge the safety of a given food if they had information about the possibility that it contains GM components. Others wonder about the usefulness of a GM label if there is little information provided as to how the food has been modified. Of what value would it be to know that food was genetically modified if the science and specifics about the modifications were not included? How much background knowledge would be needed by the consumer to be able to interpret such information?

Special Topic 7

Review Question Answers

2. Herceptin is used to treat breast cancers that have amplified the *HER-2* gene and overexpress its product, a tyrosine kinase receptor protein. Herceptin acts by binding the extracellular domain of the receptor and prevents signaling that would normally trigger cell proliferation. This inhibition leads to cell-cycle arrest and death of the cancer cell.
4. Repression of T cells by cancer cells can be indirect. Abnormal expression of MHC molecules allows cancer cells to avoid recognition by antigen-presenting cells, thereby preventing activation of T cells. Cancer cells can also repress T-cell activity directly, for example, by synthesizing molecules that bind to and repress T cells. Another direct strategy is to take advantage of the presence of tumor-associated regulatory T cells (T-regs) or tumor-infiltrating cells (macrophages and monocytes), all of which suppress T-cell activity.
6. Chimeric antigen receptors are genetically engineered proteins that are expressed on T cells and allow them to recognize tumor cells without activation by antigen-presenting cells. The chimeric proteins contain five key domains: a signal peptide, the variable regions, a linker region, a transmembrane domain, and the intracellular signaling domain. The protein is made by cloning DNA fragments encoding each domain into a single molecule and introducing this construct into naïve T cells of the patient.
8. As of late October 2018, there are 55,308 tests described on this Web site, most of which are used diagnostically. One example of single-gene testing occurs for patients with Campomelic dysplasia, a disease characterized by abnormal skeletal growth, cervical spine instability, progressive scoliosis, and hearing loss. The associated gene is *SOX9*, located on the long arm of chromosome 17, at region 24.3. The two molecular tests available are (1) analysis for deletions or duplications, using a method called Multiplex Ligation-Dependent Probe Amplification (MLPA) and (2) sequencing of the coding region by Sanger sequencing of both strands, looking for gene variants. By exploring the entries on the Web site, you should be able to identify many more examples of single-gene testing.

Discussion Question Answers

2. There are a number of technical challenges to be overcome. Current methodologies need to be faster, more accurate, and less expensive. In addition, coordinating the storage of and access to the vast amounts of data that will be generated will be daunting. Once the data are generated, scientists

and health professionals will need to ensure that all relevant information is linked. In addition, nonmedical or nongenetic information, such as the effects of environment, life style, and epigenetic contributions, will need to be included. The gap between data collection and interpretation of complex interactions will need to be closed, as well. This will likely require changing education parameters for medical students and the training of more genetics counselors and genomics specialists.

4. At present, genetic discrimination does exist; however, recent developments in health-care laws seek to minimize such discrimination by medical insurance companies and employers through GINA. It remains to be seen whether genetic discrimination in the workplace will continue.

Glossary

7-methylguanosine (m^7G) cap A guanosine residue that has been modified with a methyl group (CH_3) at the 7 position of the base, and has been added to the 5′ end of a eukaryotic mRNA through a 5′ to 5′ triphosphate bridge. The cap is important for mRNA stability, m export from the nucleus, and translation initiation.

A-DNA An alternative form of right-handed, double-helical DNA. Its helix is more tightly coiled than the more common B-DNA, with 11 base pairs per full turn. In the A form, the bases in the helix are displaced laterally and tilted in relation to the longitudinal axis.

accession number An identifying number or code assigned to a nucleotide or amino acid sequence for entry and cataloging in a database.

acrocentric chromosome A chromosome or chromosome fragment with no centromere.

activators A class of transcription factors that bind to enhancers or proximal-promoter elements to increase the transcription of a target gene.

adeno-associated virus (AAV) Virus that infects humans and some primate species. Usually does not cause disease but triggers a mild immune response. Modified versions of AAV are used as a gene therapy vector.

adenovirus vectors Most commonly used gene therapy vectors for delivering therapeutic genes; derived by rendering harmless adenovirus—a DNA virus that often infects the respiratory and intestinal tracts in children.

additive variance (V_A) Genetic variance attributed to the substitution of one allele for another at a given locus. This variance can be used to predict the rate of response to phenotypic selection in quantitative traits.

agarose gel electrophoresis Method for using electrical current to separate DNA or RNA molecules through agarose as a matrix.

alignment Sequences of DNA, RNA, or amino acids are compared and aligned based on sequence similarities.

allele-specific oligonucleotides (ASO) Short nucleotide chains, synthesized in the laboratory, usually 15–20 bp in length, that under carefully controlled conditions will hybridize only to a perfectly matching complementary sequence.

allele One of the possible alternative forms of a gene, often distinguished from other alleles by phenotypic effects.

allopolyploidy Polyploid condition formed by the union of two or more distinct chromosome sets with a subsequent doubling of chromosome number.

allotetraploid An allopolyploid containing two genomes derived from different species.

alternative splicing Generation of different protein molecules from the same pre-mRNA by incorporation of a different set and order of exons into the mRNA product.

aminoacyl (A) site One of the three sites, or pockets, in the large and small subunits of the ribosome that may be occupied by tRNA during translation. This is the first to be occupied by each charged tRNA corresponding to the triplet codon of an mRNA.

aminoacyl tRNA synthetases Enzymes that catalyze the attachment of an amino acid to the appropriate tRNA.

amniocentesis A procedure in which fluid and fetal cells are withdrawn from the amnion, a membrane surrounding the fetus; used for genetic testing of the fetus.

amphidiploid An autopolyploid condition composed of four copies of the same genome.

anaphase The stage of cell division (mitosis and meiosis) in which chromosomes begin moving to opposite poles of the cell.

aneuploidy A condition in which the chromosome number is not an exact multiple of the haploid set.

anneal Joining together of DNA fragments by hydrogen-bonding.

annotation Analysis of genomic nucleotide sequence data to identify the protein-coding genes, the non-protein-coding genes, and the regulatory sequences and function(s) of each gene.

anticodon In a tRNA molecule, the nucleotide triplet that binds to its complementary codon triplet in an mRNA molecule.

antisense oligonucleotide (ASO) A short, single-stranded DNA or RNA molecule complementary to a specific sequence.

apoptosis A genetically controlled program of cell death, activated as part of normal development or as a result of cell damage.

Argonaute A family of proteins found within the RNA-induced silencing complex (RISC) with endonuclease activity associated with the destruction of target mRNAs.

artificial genome Genome produced in the laboratory usually by chemical processes for DNA synthesis as opposed to a naturally occurring genome; also called a synthetic genome.

artificial selection Selection practiced artificially during research efforts. See *selection*.

assisted reproductive technologies (ART) The set of technologies used to treat infertility and assist couples in achieving pregnancy.

attenuation A regulatory process in some bacterial operons that terminates transcription prematurely, thus reducing the production of the mRNA encoding the structural genes in that operon.

autism spectrum disorder (ASD) A range of symptoms and neurodevelopmental disorders characterized by specific social, cognitive, and communication behaviors or deficits.

autonomously replicating sequences (ARSs) Origins of DNA replication, about 100 nucleotides in length, found in yeast chromosomes. ARS elements are also present in organelle DNA.

autopolyploidy Polyploid condition resulting from the duplication of one diploid set of chromosomes.

autoradiography Production of a photographic image by radioactive decay. Used to localize radioactively labeled compounds within cells and tissues or to identify radioactive probes in various blotting techniques. See also *Southern blotting*.

auxotroph A mutant microorganism or cell line that through mutation has lost the ability to synthesize one or more substances required for growth.

B-DNA The conformation of DNA most often found in cells and which serves as the basis of the Watson–Crick double-helical model. There are 10 base pairs per full turn of its right-handed helix, with the nucleotides stacked 0.34 nm apart. The helix has a diameter of 2.0 nm.

bacterial artificial chromosomes (BACs) Cloning vectors derived from bacterial chromosomes; designed to replicate larger fragments of cloned DNA than plasmids.

bacteriophage A virus that infects bacteria, using it as the host for reproduction. Often referred to as a phage.

balancer chromosome Chromosome containing one or more inversions that suppress crossing over with its homolog and which carries a dominant marker that is usually lethal when homozygous.

Barr body Densely staining DNA-positive mass seen in the somatic nuclei of mammalian females. Discovered by Murray Barr, this body represents an inactivated X chromosome. Also called a *sex chromatin body*.

base analogs Purine or pyrimidine bases that differ structurally from one normally used in biological systems but whose chemical behavior is the same. For example, 5-bromouracil, which "looks like" thymidine, substitutes for it, and after incorporation into a DNA molecule, can lead to mutations.

base substitution A single base change in a DNA molecule that produces a mutation. There are two types of substitutions: *transitions*, in which a purine is substituted for a purine, or a pyrimidine for a pyrimidine; and *transversions*, in which a purine is substituted for a pyrimidine or vice versa.

bioinformatics A field that focuses on the design and use of software and computational methods for the storage, analysis, and management of biological information such as nucleotide or amino acid sequences.

biotechnology Commercial and/or industrial processes that utilize biological organisms or products.

bipotential gonads In mammalian embryos, a precursor to gonads are the the gonadal or genital ridges, described as bipotential gonads because they can develop into male or female gonads based on genetic and hormonal influences.

bivalents Synapsed homologous chromosomes in the first prophase of meiosis.

BLAST (Basic Local Alignment Search Tool) Any of a family of search engines designed to compare or query nucleotide or amino acid sequences against sequences in databases. BLAST also calculates the statistical significance of the matches.

"blue-white" screening DNA cloning technique used to distinguish host bacterial cells containing recombinant plasmids (white colonies) from host cells containing nonrecombinant plasmids (blue colonies).

Bombay phenotype A rare variant of the ABO antigen system in which affected individuals do not have A or B antigens and thus appear to have blood type O, even though their genotype may carry unexpressed alleles for the A and/or B antigens.

broad-sense heritability (H^2) The contribution of the genotypic variance responsible for the phenotypic variation of a trait observed in a population.

bromodeoxyuridine (BrdU) A mutagenically active base analog of thymidine in which the methyl group at the 5′ position in thymine is replaced by bromine.

CAAT box A highly conserved DNA sequence found in the promoter region of eukaryotic genes upstream of the transcription start site. This sequence is recognized and bound by transcription factors.

cancer stem cells Tumor-forming cells in a cancer that can give rise to all the cell types in a particular form of cancer. These cells have the properties of normal stem cells: self-renewal and the ability to differentiate into multiple cell types.

capillary electrophoresis A group of analytical methods that separates large and small charged molecules in a capillary tube by their size-to-charge ratio. Detection and analysis of separated components takes place in the capillary usually by use of ultraviolet (UV) light.

carcinogens Physical or chemical agents that cause cancer.

Cas9 A CRISPR-associated nuclease from a type II CRISPR-Cas system that directs target DNA cleavage. Cas9 from the bacterium *Streptococcus pyogenes* is a well-studied protein that has been adapted as a tool for genome editing in eukaryotes.

catabolite repression The selective inactivation of an operon by a metabolic product of the enzymes encoded by the operon.

catabolite-activating protein (CAP) A catabolite-activating protein; a protein that binds cAMP and regulates the activation of inducible operons.

cell cycle The sequence of growth phases in a cell; divided into G0, G1 (gap I), S (DNA synthesis), G2 (gap II), and M (mitosis). A cell may temporarily or permanently withdraw from the cell cycle, in which case it is said to enter the G0 stage.

cell theory The theory that all organisms are made of cells and that all cells come from pre-existing cells.

cell-cycle checkpoints Regulated transitions from one stage to another during the cell cycle.

CEN region The DNA region of centromeres critical to their function. In yeasts, fragments of chromosomal DNA, about 120 bp in length, that when inserted into plasmids confer the ability to segregate during mitosis.

central dogma of molecular genetics The classical concept that genetic information flow progresses from DNA to RNA to proteins. Although exceptions are now known, this idea is central to an understanding of gene function.

centriole A cytoplasmic organelle composed of nine groups of microtubules, generally arranged in triplets. Centrioles function in the generation of cilia and flagella and serve as foci for the spindles in cell division.

centromere The specialized heterochromatic chromosomal region at which sister chromatids remain attached after replication, and the site to which spindle fibers attach to the chromosome during cell division. The location of the centromere determines the shape of the chromosome during the anaphase portion of cell division. Also known as the primary constriction.

centrosome The region of the cytoplasm containing a pair of centrioles.

chance deviation The inherent error in a predictive statistical model, occurring strictly due to chance.

chaperone A protein that facilitates the folding of a polypeptide into the three-dimensional shape of a functional protein.

checkpoints See *cell-cycle checkpoints*.

chi-square (χ^2) analysis A statistical test to determine whether or not an observed set of data is equivalent to a theoretical expectation.

chiasmata (sing., chiasma) The crossed strands of nonsister chromatids seen in the first meiotic division. Regarded as the cytological evidence for exchange of chromosomal material, or crossing over.

chimeric antigen receptor (CAR)T cells Engineered T-cell receptors that can target the immune system to attack cancer cells in immunotherapy approaches.

chloroplast A self-replicating cytoplasmic organelle containing chlorophyll. The site of photosynthesis.

chorionic villus sampling (CVS) A technique of prenatal diagnosis in which chorionic fetal cells are retrieved and used to detect cytogenetic and biochemical defects in the embryo.

chromatin The complex of DNA, RNA histones, and nonhistone proteins that make up uncoiled chromosomes, characteristic of the eukaryotic interphase nucleus.

chromatin remodeling complexes Large multi-subunit enzyme complexes that use the energy of ATP hydrolysis to move and rearrange nucleosomes within chromatin.

chromatin remodeling A process in which the structure of chromatin is altered by a protein complex, resulting in changes in the transcriptional state of genes in the altered region.

chromomere A coiled, bead-like region of a chromosome, most easily visualized during cell division. The aligned chromomeres of polytene chromosomes are responsible for their distinctive banding pattern.

chromosome In bacteria, a DNA molecule containing the organism's genome; in eukaryotes, a DNA molecule complexed with RNA and proteins to form a threadlike structure containing genetic information arranged in a linear sequence; a structure that is visible during mitosis and meiosis.

chromosome theory of inheritance The idea put forward independently by Walter Sutton and Theodor Boveri that chromosomes are the carriers of genes and the basis for the Mendelian mechanisms of segregation and independent assortment.

chromosome-banding technique Technique for the differential staining of mitotic or meiotic chromosomes to produce a characteristic banding pattern; or selective staining of certain chromosomal regions such as centromeres, the nucleolus organizer regions, and GC- or AT-rich regions.

chromosome territory The discrete region that a chromosome occupies in the eukaryotic nucleus.

***cis*-acting DNA element** A DNA sequence that regulates the expression of a gene located on the same chromosome. This contrasts with a *trans*-acting element where regulation is under the control of a sequence on the homologous chromosome.

cisgenic A genetically modified organism that contains a genetic material that was transferred from a member of the same species.

cistron That portion of a DNA molecule (a gene) that encodes a single-polypeptide chain; defined by a genetic test as a region within which two mutations cannot complement each other.

classical genetics See *forward genetics*.

clone Identical molecules, cells, or organisms derived from a single ancestor by asexual or parasexual methods; for example, a DNA segment that has been inserted into a plasmid or chromosome of a phage or a bacterium and replicated to produce many copies, or an organism with a genetic composition identical to that used in its production.

cloning vectors See *vector*.

coactivators Proteins that do not directly bind to DNA but that work together with transcriptional activators to activate the transcription of a target gene.

coding strand In a double-stranded DNA molecule, the strand opposite and complementary to the template strand, which is not transcribed during transcription. So named because it is very similar in sequence to the RNA transcript.

codominance A condition in which the phenotypic effects of a gene's alleles are fully and simultaneously expressed in the heterozygote.

codon A triplet of messenger RNA (mRNA) nucleotides that specifies a particular amino acid or a start or stop signal in the genetic code. Sixty-one codons specify the amino acids used in proteins, and three codons, called stop codons (UAG, UAA, UGA), signal termination of growth of the polypeptide chain. One codon (AUG) acts as a start codon in addition to specifying an amino acid.

coefficient of coincidence (*C*) A ratio of the observed number of double crossovers divided by the expected number of such crossovers.

coefficient of inbreeding (*F*) The probability that two alleles present in a zygote are descended from a common ancestor.

cohesin A protein complex that holds sister chromatids together during mitosis and meiosis and facilitates attachments of spindle fibers to kinetochores.

colicins Proteins encoded by bacterial plasmids that are highly toxic to other bacterial strains that do not harbor the responsible plasmid.

colinearity The linear relationship between the nucleotide sequence in a gene (or the RNA transcribed from it) and the order of amino acids in the polypeptide chain specified by the gene.

Combined DNA Index System (CODIS) A standardized set of 13 short tandem repeat (STR) DNA sequences used by law enforcement and government agencies in preparing DNA profiles.

comparative genomics Comparing genomes from different organisms to evaluate genetic and evolutionary similarity and other elements

competence In bacteria, the transient state or condition during which the cell can bind and internalize exogenous DNA molecules, making transformation possible.

competing endogenous RNAs (ceRNAs) RNAs expressed in the cell, usually lncRNAs, that serve as decoys for the binding of miRNAs thus enabling the miRNA target genes to be expressed.

complementarity Chemical affinity between nitrogenous bases of nucleic acid strands as a result of hydrogen bonding. Responsible for the base pairing between the strands of the DNA double helix and between DNA and RNA strands during gene expression in cells and during the use of molecular hybridization techniques.

complementary DNA (cDNA) libraries A collection of cloned cDNA sequences.

complementation group An array of mutations that all test negatively when assayed in a complementation test. Thus, they are alleles in the same gene.

complex trait A quantitative trait whose phenotypic variation is the result of the interaction of additive alleles from multiple genes and environmental factors. Also called a *multifactorial trait*.

computer-automated high-throughput DNA sequencing Computer automated DNA sequencing technique that can produce large amounts (high-throughput) of DNA sequence in relatively short periods of time compared to manual sequencing.

concordance Condition when identical twins both express a trait or neither of them express that trait.

conditional mutation A mutation expressed only under a certain condition; that is, a wild-type phenotype is expressed under certain (permissive) conditions and a mutant phenotype under other (restrictive) conditions.

conjugation Temporary fusion of two single-celled organisms for the sexual transfer of genetic material.

consanguineous Related by a common ancestor within the previous few generations.

consensus sequence The sequence of nucleotides in DNA or amino acids in proteins most often present in a particular gene or protein.

constitutive mutations Mutations in bacterial operons that results in the continuous transcription of the structural genes in the operon.

contigs A continuous DNA sequence reconstructed from overlapping DNA sequences derived by cloning or sequence analysis.

contiguous fragments See *contigs*.

copy number variation (CNV) Any DNA segments larger than 1 kb that are repeated a variable number of times in the genome.

core enzyme The subunits of an enzyme necessary for catalytic activity. For DNA polymerase III, the core enzyme consists of three subunits: alpha, beta, and theta that confer catalytic activity to the holoenzyme.

core promoter The minimum part of a promoter needed for accurate initiation of transcription. Contains several core-promoter elements spanning a region of approximately 80 nucleotides including the transcription start site.

correlation coefficient (*r*) In quantitative genetic studies, a statistical value describing the degree of association between two interrelated traits.

covariance (cov_{XY}) In quantitative genetic studies, a statistical value describing how much observed variation is common to two interrelated traits.

CpG island A short region of regulatory DNA found upstream of genes that contain unmethylated stretches of sequence with a high frequency of C and G nucleotides.

CRISPR (clustered regularly interspaced palindromic repeats)-Cas system The adaptive immunity mechanism present in many bacteria, which utilizes CRISPR RNAs to guide Cas nucleases to invading complementary DNAs and destroy them. The CRISPR-Cas mechanism has also been exploited to introduce specific mutations in eukaryotic genomes.

CRISPR-associated (*cas*) genes Genes located physically near the CRISPR locus in bacteria and archaea that are necessary for at least one of the three steps of the CRISPR-Cas mechanism: spacer acquisition, crRNA biogenesis, and target interference.

CRISPR-derived RNAs (crRNAs) RNA transcripts from a CRISPR locus in a bacteria or archaea genome are processed into a mature crRNA, which then guides a Cas nuclease (such as Cas9) to cleave complementary DNA sequences.

crossing over The exchange of chromosomal material (parts of chromosomal arms) between homologous chromosomes by breakage and reunion. The exchange of material between nonsister chromatids during meiosis is the basis of genetic recombination.

crRNA biogenesis One of the steps in the CRISPR-Cas mechanism in which RNAs are transcribed and processed by Cas proteins.

cyclic adenosine monophosphate (cAMP) An important regulatory molecule in both bacteria and eukaryotes.

cyclins In eukaryotic cells, a class of proteins that are synthesized and degraded in synchrony with the cell cycle and regulate passage through stages of the cycle.

cytokinesis The division or separation of the cytoplasm at the end of cell division (mitosis and meiosis).

deadenylases Eukaryotic enzymes that degrade the poly-A tail of an mRNA in a 3′ to 5′ direction.

deadenylation-dependent decay A eukaryotic mRNA degradation pathway that is initiated by shortening the poly-A tail of an mRNA and subsequent degradation by either the exosome in the 3′ to 5′ direction or decapping of the mRNA and degradation in the 5′ to 3′ direction by the XRN1 exoribonuclease.

deadenylation-independent decay A eukaryotic mRNA degradation pathway that is initiated by decapping of an mRNA followed by degradation of the mRNA in the 5′ to 3′ direction by the XRN1 exoribonuclease.

decapping enzymes Eukaryotic enzymes that remove the m^7G cap at the 5′ end of an mRNA.

deletion A chromosomal mutation, also referred to as a deficiency, involving the loss of chromosomal material.

Dicer A ribonuclease that participates in the RNA interference (RNAi) pathway by cleaving long double-stranded RNAs and pre-microRNAs into small interfering RNAs (siRNAs) and microRNAs (miRNAs), respectively. The siRNAs and miRNAs are approximately 20–25 base pairs long with two-base 3′ overhangs and associate with the RNA-induced silencing complex (RISC) to facilitate silencing of complementary mRNAs.

dideoxy chain-termination sequencing One of the first DNA sequencing techniques; also referred to as Sanger sequencing. Technology relies on modified nucleotides called dideoxynu cleotides (ddNTPs), which terminate a newly synthesized strand of DNA when incorporated during a sequencing reaction.

dideoxynucleotide A nucleotide containing a deoxyribose sugar lacking a hydroxyl group. It stops further chain elongation when incorporated into a growing polynucleotide and is used in the Sanger method of DNA sequencing.

dihybrid cross A genetic cross involving two characters in which the parents possess different forms of each character (e.g., yellow, round × green, wrinkled peas).

diploid The condition when cells contain homologous pairs of each chromosome, one derived from the paternal parent and one from the maternal parent.

direct-to-consumer (DTC) genetic tests Genetic tests that can be ordered or purchased (through the Internet, for example) by an individual without the involvement of a physician, hospital or other health care provider.

directional selection A selective force that changes the frequency of an allele in a given direction, either toward fixation or toward elimination.

discordance Condition when one member of a set of identical twins expresses a trait, while the other member does not.

disjunction The separation of chromosomes during the anaphase stage of cell division.

disruptive selection Simultaneous selection for phenotypic extremes in a population, usually resulting in the production of two phenotypically discontinuous strains.

dizygotic twins Twins produced from separate fertilization events; two ova fertilized independently. Also known as *fraternal twins.*

DNA fingerprinting A molecular method for identifying an individual member of a population or species. A unique pattern of DNA fragments is obtained by restriction enzyme digestion followed by Southern blot hybridization using minisatellite probes. See also *DNA profiling; STR sequences.*

DNA gyrase One of a class of enzymes known as topoisomerases that converts closed circular DNA to a negatively supercoiled form prior to replication, transcription, or recombination. The enzyme acts during DNA replication to reduce molecular tension caused by supercoiling.

DNA helicase An enzyme that participates in DNA replication by unwinding the double helix near the replication fork.

DNA libraries Collections of cloned DNA in host cells; cDNA or genomic libraries are common types of libraries. Can be screened to identify particular genes of interest.

DNA ligase An enzyme that forms a covalent bond between the 5′ end of one polynucleotide chain and the end of another polynucleotide chain. It is also called polynucleotide-joining enzyme.

DNA methylation The enzymatically controlled process of transferring a methyl group from a donor molecule to a base in DNA.

DNA microarray analysis An ordered arrangement of DNA sequences or oligonucleotides on a substrate (often glass). Microarrays are used in quantitative assays of DNA–DNA or DNA–RNA binding to measure profiles of gene expression (for example, during development or to compare the differences in gene expression between normal and cancer cells).

DNA polymerase An enzyme that catalyzes the synthesis of DNA from deoxyribonucleotides utilizing a template DNA molecule.

DNA profiling A method for identification of individuals that uses variations in the length of short tandem repeating (STR). DNA sequences that are widely distributed in the genome.

DNA transposons Mobile genetic elements that are major components of many eukaryotic and prokaryotic genomes. They are excised from one site in the genome and inserted into another, often causing mutations.

DNA typing In law enforcement, the use of a standard set of short tandem repeats (STRs) to identify an individual. Also known as *forensic DNA fingerprinting*. See also *DNA profiling.*

dominant mutation A mutation in one allele that confers a mutant phenotype in a diploid organism, even in the presence of a wild-type allele.

dominant negative mutation A mutation whose gene product acts in opposition to the normal gene product, usually by binding to it to form dimers.

donor template Used for genome editing by CRISPR-Cas; an experimentally introduced DNA molecule carrying a sequence with desired edits flanked by "homology arms" with sequences that match regions of the genome adjacent to the target sequence to be edited.

dosage compensation A genetic mechanism that equalizes the levels of expression of genes at loci on the X chromosome. In mammals, this is accomplished by random inactivation of one X chromosome, leading to Barr body formation.

Down syndrome critical region (DSCR) A hypothetical region of human chromosome 21 housing dosage-sensitive genes that in the trisomic condition causes Down syndrome.

driver mutation A mutation in a cancer cell that contributes to tumor progression.

Drosha A nuclear RNA-specific nuclease involved in the maturation of microRNAs. It removes 5′ and 3′ non-self-complementary regions of a primary miRNA to produce a pre-miRNA.

duplication A chromosomal aberration in which a segment of the chromosome is repeated.

electrophoresis A technique that separates a mixture of molecules by their differential migration through a stationary medium (such as a gel) under the influence of an electrical field.

ELSI Program A program established by the National Human Genome Research Institute in 1990 as part of the Human Genome Project to sponsor research on the ethical, legal, and social implications of genomic research and its impact on individuals and social institutions.

embryonic stem (ES) cells Cells derived from the inner cell mass of early blastocyst mammalian embryos. These cells are pluripotent, meaning they can differentiate into any of the embryonic or adult cell types characteristic of the organism.

Encyclopedia of DNA Elements (ENCODE) Project A worldwide consortium of researchers using experimental approaches and bioinformatics to identify and analyze functional elements of the genome (such as transcriptional start sites, promoters, and enhancers) that regulate expression of human genes.

endonuclease An enzyme that hydrolyzes internal phosphodiester bonds in a single- or double-stranded polynucleotide chain.

endoplasmic reticulum (ER) A membranous organelle system in the cytoplasm of eukaryotic cells. In rough ER, the outer surface of the membranes is ribosome-studded; in smooth ER, it is not.

endoribonucleases Eukaryotic enzymes that cleave an RNA at an internal site creating new 5′ and 3′ ends.

enhancement gene therapy Gene therapy for the purpose of enhancing a desired trait and not for disease treatment or prevention.

enhanceosome A complex of transcriptional activators and coactivators that directs the transcriptional activation of a target gene.

enhancer A DNA sequence that enhances transcription and the expression of structural genes. Enhancers can act over a distance of thousands of base pairs and can be located upstream, downstream, or internal to the gene they affect, differentiating them from promoters.

environmental genomics See *metagenomics.*

enzyme A protein or complex of proteins that catalyzes a specific biochemical reaction by lowering the energy of activation that would otherwise be required to initiate the reaction.

epigenesis The idea that an organism or organ arises through the sequential appearance and development of new structures, in contrast to *preformationism,* which holds that development is the result of the assembly of structures already present in the egg.

epigenetics The study of the effects of reversible chemical modifications to DNA and/or histones on the pattern of gene expression. Epigenetic modifications do not alter the nucleotide sequence of DNA.

epigenome The set of chemical modifications made to DNA and histones that are present in each cell at a specific time period.

epimutations Heritable changes in gene expression that are associated with changes in the pattern of DNA methylation and not with any changes in the DNA sequence.

epistasis The nonreciprocal interaction between nonallelic genes such that one gene influences or interferes with the expression of another gene, leading to a specific phenotype.

Ethical, Legal, Social Implications (ELSI) Program A program established by the National Human Genome Research Institute in 1990 as part of the Human Genome Project to sponsor research on the ethical, legal, and social implications of genomic research and its impact on individuals and social institutions.

euchromatin Chromatin or chromosomal regions that are lightly staining and relatively uncoiled during the interphase portion of the cell cycle. Euchromatic regions contain most of the structural genes.

eukaryotes Organisms having true nuclei and membranous organelles and whose cells divide by mitosis and meiosis.

euploidy A condition in which a cell has a chromosome number that is an exact multiple of the haploid number.

excision repair Removal of damaged DNA segments followed by repair. Excision can include the removal of individual bases (base excision repair) or of a stretch of damaged nucleotides (nucleotide excision repair).

exit (E) site One of the three sites, or pockets, in the large and small subunits of the ribosome that may be occupied by tRNA during translation. This is the site where an uncharged tRNA, which has already contributed its amino acid to the growing polypeptide chain, leaves the ribosome.

exome sequencing A DNA-sequencing method in which only the protein-coding regions (exons) of the genome are sequenced.

exons The DNA segments of a gene that contain the sequences that, through transcription and translation, are eventually represented in the final polypeptide product.

exonuclease An enzyme that breaks down nucleic acid molecules by breaking the phosphodiester bonds at the 3′- or 5′-terminal nucleotides.

exoribonucleases Eukaryotic enzymes that degrade RNA via the removal of terminal nucleotides.

exosome complex A eukaryotic enzyme with exoribonuclease activity that degrades mRNAs in the 3′ to 5′ direction.

expression QTLs (eQTLs) Genomic loci that affect expression of one or more genes involved in a quantitative trait.

expression vectors Plasmids or phages carrying promoter regions designed to cause expression of inserted DNA sequences.

expressivity The degree to which a phenotype for a given trait is expressed.

extranuclear inheritance Transmission of traits by genetic information contained in the DNA of cytoplasmic organelles such as mitochondria and chloroplasts.

F^- cell A bacterial cell that does not contain a fertility factor and that acts as a recipient in bacterial conjugation.

F^+ cell A bacterial cell that contains a fertility factor and that acts as a donor in bacterial conjugation.

F_1 generation The first filial generation; the progeny resulting from the first cross in a series.

F_2 generation The second filial generation; the progeny resulting from a cross of the F_1 generation.

F factor An episomal plasmid in bacterial cells that confers the ability to act as a donor in conjugation.

F pilus On bacterial cells possessing an F factor, a filament-like projection that plays a role in conjugation.

FISH See *fluorescence* in situ *hybridization.*

fitness A measure of the relative survival and reproductive success of a given individual or genotype.

fluorescence *in situ* hybridization (FISH) A method of *in situ* hybridization that utilizes probes labeled with a fluorescent tag, causing the site of hybridization to fluoresce when viewed using ultraviolet light.

folded-fiber model A model of eukaryotic chromosome organization in which each sister chromatid consists of a single chromatin fiber composed of double-stranded DNA and proteins wound together like a tightly coiled skein of yarn.

forensic DNA fingerprinting In law enforcement, the use of a standard set of short tandem repeats (STRs) to identify an individual. Also known as *DNA typing*. See also *DNA profiling.*

forensic science The use of laboratory scientific methods to obtain data used in criminal and civil law cases.

forward genetics The classical approach used to identify a gene controlling a phenotypic trait in the absence of knowledge of the gene's location in the genome or its DNA sequence. Accomplished by isolating mutant alleles and mapping the gene's location, most traditionally using recombination analysis. Once mapped, the gene may be cloned and further studied at the molecular level. An approach contrasted with *reverse genetics.*

founder effect A form of genetic drift. The establishment of a population by a small number of individuals whose genotypes carry only a fraction of the different alleles present in the parental population.

fragile site A heritable gap, or nonstaining region, of a chromosome that can be induced to generate chromosome breaks.

frameshift mutation A mutational event leading to the insertion or deletion (indels) of a number of base pairs in a gene that is not a multiple of three. This shifts the codon reading frame in all codons that follow the mutational site.

fraternal twins See *dizygotic (DZ) twins.*

functional genomics Analysis of DNA sequence data to propose functions for sequenced DNA such as protein-coding and non-coding genes, regulatory elements, etc.

G0 stage A nondividing but metabolically active state (G-zero) that cells may enter from the G1 phase of the cell cycle.

G1 (gap I) stage The phase during the cell cycle between G0 and the S phase, during which the cell develops and grows.

G2 (gap II) stage The phase during the cell cycle following the S phase, during which the cell, having replicated its DNA, prepares for mitosis.

gain-of-function mutation A type of mutation in which the gene product takes on a new function and produces a phenotype different from that of the normal allele and from any loss-of-function alleles.

gamete A specialized reproductive cell with a haploid number of chromosomes.

GC box In eukaryotes, a region in a promoter containing a 5′-GGGCGG-3′ sequence, which is a binding site for transcriptional regulatory proteins.

gene The fundamental physical unit of heredity, whose existence can be confirmed by allelic variants and which occupy a specific chromosomal locus. A DNA sequence coding for a single polypeptide or an RNA molecule.

gene chips See *DNA microarray analysis.*

gene-expression microarrays Microarrays used to detect gene-expression patterns for specific genes.

gene family A number of closely related genes derived from a common ancestral gene by duplication and sequence divergence over evolutionary time.

gene flow The exchange of genes between two populations; brought about by the dispersal of gametes resulting from the migration of individuals between populations.

gene interaction The production of novel phenotypes by the interaction of alleles of different genes.

gene knockout Generation of a *null mutation* in a gene that is subsequently introduced into an organism using transgenic techniques, causing a loss of function in the targeted gene. Often used in mice. See also *gene targeting*.

gene pool The total of all alleles possessed by the reproductive members of a population.

gene redundancy The presence of several genes in an organism's genome that all have variations of the same function.

gene targeting A transgenic technique used to create and introduce a specifically altered gene into an organism. In mice, gene targeting often involves the induction of a specific mutation in a cloned gene that is subsequently introduced into the genome of a gamete involved in fertilization. The organism produced is bred to produce adults homozygous for the mutation, for example, the creation of a *gene knockout*.

gene therapy A therapeutic approach for providing a normal copy of a gene, replacement of a defective gene, or supplementing a gene for treating or curing a genetic disorder.

general transcription factors (GTFs) A class of DNA-binding proteins that bind to specific sites within a gene's promoter and are required to load RNA polymerase onto the DNA, and for the initiation of transcription.

genetic anticipation The phenomenon in which the severity of symptoms in genetic disorders increases from generation to generation and the age of onset decreases from generation to generation. It is caused by the expansion of trinucleotide repeats within or near a gene and was first observed in myotonic dystrophy.

genetic background The impact of the collective genome of an organism on the expression of a gene under investigation.

genetic bottleneck A drastic reduction in the size and genetic variation present in a population. These events can be triggered by environmental disasters (volcanic eruptions, earthquakes, floods, etc.) or human activity (habitat destruction, hunting to near-extinction, etc.). Also known as a population bottleneck.

genetic code The deoxynucleotide triplets that encode the 20 amino acids or specify initiation or termination of translation.

genetic drift Random variation in allele frequency from generation to generation, most often observed in small populations.

Genetic Information Nondiscrimination Act (GINA) Legislation signed into law in the United States in 2008 and designed to prohibit the improper use of genetic information in health insurance and employment.

genetic testing The analysis of DNA, and increasingly RNA, for the purposes of identifying specific genes or sequences associated with different genetic conditions.

genetically modified organism (GMO) A plant or animal whose genome has been altered in ways that do not occur naturally, most often using recombinant DNA and the techniques of genetic engineering.

genetics The branch of biology concerned with the study of inherited variation. More specifically, the study of the origin, transmission, and expression of genetic information.

genic balance theory In *Drosophila*, sex is determined by the ratio of X-chromosomes and autosomes.

genome The set of hereditary information encoded in the DNA of an organism, including both the protein-coding and non–protein-coding sequences.

genome editing Method for removing, replacing, or modifying a specific sequence in the genome; CRISPR-Cas is one of the most effective approaches to genome editing.

genome mosaicism DNA sequences differences in the genomes of somatic cells within the same individual; thus, the genome in cells of a tissue such as skin may be a mosaic comprised of sequence variations that vary among skin cells.

genome scanning Approach using SNP sequences as probes on a DNA microarray to simultaneously screen thousands of mutations that might be involved in a genetic disease.

Genome 10K Project Genomics project to sequence 10,000 vertebrate genomes.

genome-wide association studies (GWAS) Analysis of genetic variation across an entire genome, searching for linkage (associations) between variations in DNA sequences and a genome region encoding a specific phenotype.

genomic analysis General term describing a range of approaches for studying the genome (all DNA in an organisms' cells).

genomic imprinting The process by which the expression of an allele depends on whether it has been inherited from a male or a female parent. Also referred to as parental imprinting.

genomic library A collection of clones that contains all the DNA sequences of an organism's genome.

genomic variation Individual changes or variations in the human (reference) genome.

genomics A subdiscipline of genetics created by the union of classical and molecular biology with the goal of sequencing and understanding genes, gene interaction, genetic elements, as well as the structure and evolution of genomes.

genotype The allelic or genetic constitution of an organism; often, the allelic composition of one or a limited number of genes under investigation.

germ-line therapy Gene therapy involving germ cells or mature gametes as targets for gene transfer.

Goldberg–Hogness box A short nucleotide sequence 20–30 bp upstream from the initiation site of eukaryotic genes to which RNA polymerase II binds. The consensus sequence is TATAAAA. Also known as a *TATA box*.

gonadal (genital) ridges In mammalian embryos, a precursor to gonads are the gonadal or genital ridges, described as bipotential gonads because they can develop into male or female gonads based on genetic and hormonal influences.

gratuitous inducer A molecule such as IPGT that is a chemical analogue of lactose, which in the lactose operon in bacteria induces the transcription of the structural genes, but itself is not metabolized by the gene products of the operon.

H substance The carbohydrate group present on the surface of red blood cells to which the A and/or B antigen may be added. When unmodified, it results in blood type O.

haploid number (*n*) The number of homologous chromosome pairs characteristic of an organism or species.

haploinsufficiency In a diploid organism, a condition in which an individual possesses only one functional copy of a gene with the other inactivated by mutation. The amount of protein produced by the single copy is insufficient to produce a normal phenotype, leading to an abnormal phenotype. In humans, this condition is present in many autosomal dominant disorders.

haplotypes A set of alleles from closely linked loci carried by an individual that are inherited as a unit.

Hardy–Weinberg law The principle that genotype frequencies will remain in equilibrium in an infinitely large, randomly mating population in the absence of mutation, migration, and selection.

harlequin chromosomes Paired human sister chromatids stained to reveal sister chromatid exchanges.

hemizygous Having a gene present in a single dose in an otherwise diploid cell. Usually applied to genes on the X chromosome in heterogametic males.

hemoglobin (Hb) An iron-containing, oxygen-carrying multimeric protein occurring chiefly in the red blood cells of vertebrates.

heritability For a given trait, a measure of the proportion of total phenotypic variation in a population that is due to genetic factors.

heterochromatin The heavily staining, late-replicating regions of chromosomes that are prematurely condensed in interphase.

heteroduplex A double-stranded nucleic acid molecule in which each polynucleotide chain has a different origin. It may be produced as an intermediate in a recombinational event or by the *in vitro* reannealing of single-stranded complementary molecules.

heterogametic sex The sex that produces gametes containing unlike sex chromosomes. In mammals, the male is the heterogametic sex.

heterogeneous trait A mutant phenotype that may occur as the result of a mutation in any one of many genes required for normal expression of the trait during development, e.g., hereditary deafness.

heterokaryon A somatic cell containing nuclei from two different sources.

heteromorphic chromosomes Dissimilar chromosome pairs that characterize one sex or the other in a wide range of species, e.g., XY in mammals and ZW in chickens.

heteroplasmy Variation in the DNA within organelles such as mitochondria and chloroplasts within the same cell.

heterozygote An individual with different alleles at one or more loci. Such individuals will produce unlike gametes and therefore will not breed true.

Hfr Strains of bacteria exhibiting a high frequency of recombination. These strains have a chromosomally integrated F factor that is able to mobilize and transfer part of the chromosome to a recipient F^- cell.

high-frequency recombination See *Hfr*.

high-throughput DNA sequencing A collection of DNA-equencing methods that outperform the standard (Sanger) method of DNA sequencing by a factor of 100–1000 and reduce sequencing costs by more than 99 percent. Also called *next-generation sequencing*.

histone acetyltransferase (HAT) A class of enzymes that catalyze the addition of an acetyl group to histone proteins, which leads to an open chromatin configuration that is permissive for gene expression.

histone code Chemical modifications of amino acids in histone tails (the N-terminal ends of histone molecules, projecting from nucleosomes). These modifications influence DNA–histone interactions and promote or repress transcription.

histone deacetylase (HDAC) A class of enzymes that catalyzes the removal of acetyl groups from histone proteins, which leads to a closed chromatin configuration that is refractory to gene expression.

histones Positively charged proteins complexed with DNA in the nucleus. They are rich in the basic amino acids arginine and lysine and function in coiling DNA to form nucleosomes.

holoenzyme For proteins with multiple subunits, the complex formed by the union of all subunits necessary for all functions of the enzyme.

homogametic sex The sex that produces gametes with identical sex-chromosome content; in mammals, the female is homogametic.

homologous chromosomes Chromosomes that synapse or pair during meiosis and that are identical with respect to their genetic loci and centromere placement.

homologous genes Genes related through evolution.

homology-directed repair (HDR) A genomic DNA repair mechanism that is induced by double-strand DNA breaks. In this type of repair, an intact molecule of DNA, such as a homologous chromosome in diploid eukaryotes, is used as a template for repair of a damaged homolog.

homozygote An individual with identical alleles for a gene or genes of interest. These individuals will produce identical gametes (with respect to the gene or genes in question) and will therefore breed true.

homunculus In the incorrect theory of preformationism, the idea that the egg or sperm contained a fully formed human called the homunculus, and that development consisted only of enlarging the preformed individual.

horizontal gene transfer The nonreproductive transfer of genetic information from one organism to another, across species and higher taxa (even domains). This mode is contrasted with vertical gene transfer, which is the transfer of genetic information from parent to offspring. In some species of bacteria and archaea, up to 5 percent of the genome may have originally been acquired through horizontal gene transfer.

mutation hot spots Genome regions where mutations are observed with a high frequency. These include a predisposition toward single-nucleotide substitutions or unequal crossing over.

Human Genome Project (HGP) International effort to identify all human genes and to sequence an estimated 3 billion based pairs of the entire human genome; also included goals to sequence genomes for model organisms; to evaluate genetic variation in humans; and to address ethical, legal and social issues among other goals.

Human Microbiome Project (HMP) National Institutes of Health project to sequence the genomes of microorganisms that reside inside and on humans.

hydrogen bond A weak electrostatic attraction between a hydrogen atom covalently bonded to an oxygen or a nitrogen atom and an atom that contains an unshared electron pair.

identical twins See *monozygotic twins*.

immunoglobulin (Ig) The class of serum proteins having the properties of antibodies.

immunotherapy Therapies that stimulate, suppress, or modify the actions of components of the patient's immune system, in order to treat diseases such as cancer.

imprinting See *genomic imprinting*.

inbreeding Mating between closely related organisms.

incomplete dominance Expressing a heterozygous phenotype that is distinct from the phenotype of either homozygous parent. Also called *partial dominance*.

induced mutations Mutations that result from the actions of exogenous agents, either natural or artificial.

inducible enzyme system An enzyme system under the control of an inducer, a regulatory molecule that acts to block a repressor and allow transcription.

***in situ* hybridization** A cytological technique for pinpointing the chromosomal location of DNA sequences complementary to a given nucleic acid or polynucleotide.

intellectual property (IP) General term referring to the ownership held by a person or company to ideas, products, or plans developed through creativity and innovation.

interchromatin compartments The regions between chromosome territories in the nucleus that are largely void of DNA.

interference (*I*) A measure of the degree to which one crossover affects the incidence of another crossover in an adjacent region of the same chromatid. Negative interference increases the chance of another crossover; positive interference reduces the probability of a second crossover event.

interphase In the cell cycle, the interval between divisions.

intrinsic termination In bacteria, one mechanism by which transcription is terminated. Following the transcription of a stretch of U residues, the weak A-U base-pairing between the DNA template and the RNA leads to dissociation and transcription termination.

intron Any segment of DNA that lies between coding regions in a gene. Introns are transcribed but are spliced out of the RNA product and are not represented in the polypeptide encoded by the gene. Short for intervening sequence.

inversion A chromosomal aberration in which a chromosomal segment has been reversed.

inversion loop The chromosomal configuration resulting from the synapsis of homologous chromosomes, one of which carries an inversion.

isoforms A set of related proteins that are encoded by different alternative splice forms of the same gene.

karyokinesis The process of nuclear division.

karyotype The chromosome complement of a cell or an individual. An arrangement of metaphase chromosomes in a sequence according to length and centromere position.

kinases A broad class of enzymes that phosphorylate a substrate molecule such as a protein, nucleic acid, carbohydrate, or lipid.

kinetochore A protein structure that assembles on the centromere during mitosis and meiosis. It is the site of microtubule attachment during cell division.

knock-in animals See *transgenic animals*.

Kozak sequence A short nucleotide sequence adjacent to the initiation codon that is recognized as the translational start site in eukaryotic mRNA.

lagging strand During DNA replication, the strand synthesized in a discontinuous fashion, in the direction opposite of the replication fork. See also *Okazaki fragment*.

lampbrush chromosomes Meiotic chromosomes characterized by extended lateral loops. Although most intensively studied in amphibians, these structures occur in meiotic cells of organisms ranging from insects to humans.

leader sequence That portion of an mRNA molecule from the 5′ end to the initiating codon, often containing regulatory or ribosome-binding sites.

leading strand During DNA replication, the strand synthesized continuously in the direction of the replication fork.

lentivirus vectors RNA-based (retroviruses) viruses that can cause illness in humans. HIV is one example. Disabled and engineered lentivirus is used as a gene therapy vector.

linkage The condition in which genes are present on the same chromosome, causing them to be inherited as a unit, provided that they are not separated by crossing over during meiosis.

locus (pl., loci) The place on a chromosome where a particular gene is located.

long interspersed elements (LINEs) Long, repetitive sequences found interspersed in the genomes of higher organisms.

long noncoding RNAs (lncRNAs) RNAs longer than 200 nucleotides that are noncoding transcripts and have various functions including epigenetic modification of DNA and regulation of gene-specific transcription.

long terminal repeat (LTR) Identical sequences found at both ends of a retroviral DNA.

loss of heterozygosity The loss of a wild-type allele at a heterozygous locus that also contains a mutant allele. Most commonly, loss of heterozygosity occurs through deletion of a chromosomal region or a recombination event that converts the wild-type allele to the mutant allele sequence.

loss-of-function mutation A mutation that produces alleles encoding proteins with reduced or no function.

Lyon hypothesis The proposal that there is random inactivation of the maternal or paternal X chromosome in somatic cells of mammalian females early in development. All daughter cells will have the same X chromosome inactivated as in the cell they descended from, producing a mosaic pattern of expression of X chromosome genes.

lysogeny The process by which the DNA of an infecting phage becomes repressed and integrated into the chromosome of the bacterial cell it infects.

male-specific region of the Y chromosome (MSY) Region of the Y-chromosome that does not recombine with the X-chromosome; contains a majority of genes specific to the Y-chromosome.

malignant tumors Tumors that have acquired the ability to spread to distant sites in the body and form new tumors.

mass spectrometry (MS) Technique that can be applied to proteomics and involves analyzing protein sample in gaseous forms to measure mass to charge ratios ions in a protein sample as as way to identify the amino acid sequence in a sample.

maternal effect Phenotypic effects in offspring attributable to genetic information in the oocyte derived from the maternal genome.

mean ($\overline{X}$) In statistics, the arithmetic average.

Mediator During transcription initiation, this multiprotein complex serves as a co-activator and interacts with general transcription factors and RNA polymerase II.

meiosis The process of cell division in gametogenesis or sporogenesis during which the diploid number of chromosomes is reduced to the haploid number.

merozygote A partially diploid bacterial cell containing, in addition to its own chromosome, a chromosome fragment introduced into the cell by transformation, transduction, or conjugation.

messenger RNA (mRNA) An RNA molecule transcribed from DNA and translated into the amino acid sequence of a polypeptide.

metacentric chromosome A chromosome with a centrally located centromere and therefore chromosome arms of equal lengths.

metafemale In *Drosophila*, a poorly developed female of low viability with a ratio of X chromosomes to sets of autosomes that exceeds 1.0. Previously called a *superfemale*.

metagenomics The study of DNA recovered from organisms collected from the environment as opposed to those grown as laboratory cultures. Often used for estimating the diversity of organisms in an environmental sample.

metamale In *Drosophila*, a poorly developed male of low viability with a ratio of X chromosomes to sets of autosomes that is below 0.5. Previously called a *supermale*.

metaphase The stage of cell division (mitosis and meiosis) in which condensed chromosomes lie in a central plane between the two poles of the cell and during which the chromosomes become attached to the spindle fibers.

metastases Secondary tumors that develop from cancer cells that migrate from the primary tumor and spread to distant sites in the body.

metastasis The process by which cancer cells spread from the primary tumor and establish malignant tumors in other parts of the body.

methylation Enzymatic transfer of methyl groups from S-adenosylmethionine to biological molecules, including phospholipids, proteins, RNA, and DNA. Methylation of DNA is associated with the regulation of gene expression and with epigenetic phenomena such as imprinting.

methylome The pattern of nucleic acid methylation present at a particular time in a genome or a specific cell type.

microfilaments Actin-containing microfibers that are a part of the structural framework of the cytoplasm.

microRNA (miRNA) Single-stranded RNA molecules approximately 20–23 nucleotides in length that regulate gene expression by participating in the degradation of mRNA.

microsatellite A short, highly polymorphic DNA sequence of 1–4 base pairs, widely distributed in the genome, that are used as molecular markers in a variety of methods. Also called *simple sequence repeats (SSRs)*.

microtubules Fibers composed of tubulin, bunches of which are part of the structural framework of the cytoplasm (the cytoskeleton) and also which compose spindle fibers that facilitate chromosome migration during mitosis and meiosis.

minimal medium A medium containing only the essential nutrients needed to support the growth and reproduction of wild-type strains of an organism. Usually comprised of inorganic components that include a carbon and nitrogen source.

minisatellite Series of short tandem repeat sequences (STRs) 10–100 nucleotides in length that occur frequently throughout the genome of eukaryotes. Because the number of repeats at each locus is variable, the loci are known as variable number tandem repeats (VNTRs). Used in the preparation of DNA fingerprints and DNA profiles. See also *variable number tandem repeats (VNTRs); STR sequences*.

miRNA response elements (MREs) Sequences in RNAs that are complementary or partially complementary to miRNAs and thus serve as binding sites for miRNAs.

mismatch repair (MMR) A form of excision repair of DNA in which the repair mechanism is able to distinguish between the strand with the error and the strand that is correct.

missense mutation A mutation that changes a codon to that of another amino acid and thus results in an amino acid substitution in the translated protein. Such changes can make the protein nonfunctional.

mitochondria Self-reproducing, DNA-containing, cytoplasmic organelles in eukaryotes involved in generating the high-energy compound ATP. They are the so-called powerhouse of the cell.

mitochondrial DNA (mtDNA) Double-stranded, self-replicating circular DNA found in mitochondria that encodes mitochondrial ribosomal RNAs, transfer RNAs, and proteins used in oxidative respiratory functions of the organelle.

mitosis A form of cell division producing two progeny cells identical genetically to the parental cell—that is, the production of two cells from one, each having the same chromosome complement as the parent cell.

model organisms In genetics, model organisms are non-human species with well characterized genetics that are studied to understand basic biological processes. The expectation is that the process in the model organism can be extrapolated to other species, including humans.

molecular clock In evolutionary studies, a method that counts the number of differences in DNA or protein sequences as a way of measuring the time elapsed since two species diverged from a common ancestor.

monoallelic expression (MAE) Process in which transcription occurs only from one of two homologous alleles in a diploid cell.

monohybrid cross A genetic cross involving only one character (e.g., $AA \times aa$).

monosomy An aneuploid condition in which one member of a chromosome pair is missing; having a chromosome number of $2n - 1$.

monozygotic twins Twins produced from a single fertilization event; the first division of the zygote produces two cells, each of which develops into an embryo. Also known as *identical twins*.

mosaics Refers to individuals with cell types that contain different karyotypes (for example, in Turner syndrome, cells with 45,X and cells with 46,XY) and thus often show mixed phenotypes for a particular genetic condition compared with non-mosaics (for example, 45,X individuals with Turner syndrome).

mRNA See *messenger RNA*.

multifactorial trait See *complex trait*.

multigene families A set of genes descended from a common ancestral gene usually by duplication and subsequent sequence divergence. The globin genes are an example of a multigene family.

multiple alleles In a population of organisms, the presence of three or more alleles of the same gene.

multiple-factor hypothesis Proposal describing the inheritance of a phenotypic character controlled by many genes, each behaving in a Mendelian fashion and contributing to the phenotype in a cumulative or quantitative way.

mutagens Any agent that causes an increase in the rate of mutation.

mutation The process that produces an alteration in DNA or chromosome structure; in genes, the source of new alleles.

mutation rate The frequency with which mutations take place at a given locus or in a population.

mutator phenotype The high levels of genomic instability that occur in cancer cells.

N-formylmethionine (fMet) A molecule derived from the amino acid methionine by attachment of a formyl group to its terminal amino group. This is the first monomer used in the synthesis of all bacterial polypeptides. Also known as *N-formyl methionine*.

narrow-sense heritability The proportion of phenotypic variance in a population due to additive genotypic variance.

natural selection Differential reproduction among members of a species owing to variable fitness conferred by genotypic differences.

neoantigens Novel, non-self antigens present on cancer cells but not found on the patient's normal cells.

neutral mutation A mutation with no perceived immediate adaptive significance or phenotypic effect.

neutral theory The idea that most of the genetic variation present within and between species is caused by random fixation of alleles that are not acted on by natural selection and are therefore selectively neutral or nearly neutral mutations.

next-generation sequencing (NGS) technologies See *high-throughput DNA sequencing*.

noncoding RNAs (ncRNAs) RNAs that do not encode polypeptides.

nondisjunction A cell division error in which homologous chromosomes (in meiosis) or the sister chromatids (in mitosis) fail to separate and migrate to opposite poles; responsible for defects such as monosomy and trisomy.

nonhomologous end joining (NHEJ) A genomic DNA repair mechanism that is induced by double-strand DNA breaks. This type of repair is error prone because broken ends of DNA molecules are randomly ligated together, which may lead to insertions, deletions, translocations, or inversions.

noninvasive prenatal genetic diagnosis (NIPGD) A noninvasive method of fetal genotyping that uses a maternal blood sample to analyze thousands of fetal loci using fetal cells or fetal DNA fragments present in the maternal blood.

nonsense codons The nucleotide triplets (UGA, UAG, and UAA) in an mRNA molecule that signal the termination of translation.

nonsense mutation A mutation that changes a codon specifying an amino acid into a termination codon, leading to premature termination during translation of mRNA.

nonsense-mediated decay (NMD) A eukaryotic mRNA degradation pathway that is triggered by the presence of a stop codon too far upstream of the poly-A tail of the mRNA.

nonsister chromatids Non-identical chromatids visible during mitosis and meiosis where each chromatid represents one or the other of the two members of a homologous pair of chromosomes. See *sister chromatids*.

Northern blot analysis Electrophoresis, blotting, and hybridization technique for detecting and quantifying RNA expression.

Northern blotting An analytic technique in which RNA molecules are separated by electrophoresis and transferred by capillary action to a nylon or nitrocellulose membrane. Specific RNA molecules can then be identified by hybridization to a labeled nucleic acid probe.

nucleoid The DNA-containing region within the cytoplasm in bacterial cells.

nucleolar organizer region (NOR) A chromosomal region containing the genes for ribosomal RNA (rRNA); most often found in physical association with the *nucleolus*.

nucleolus The nuclear site of ribosome biosynthesis and assembly; associated with or formed in association with the DNA comprising the *nucleolar organizer region*.

nucleoside In nucleic acid chemical nomenclature, a purine or pyrimidine base covalently linked to a ribose or deoxyribose sugar molecule.

nucleosome In eukaryotes, a complex consisting of four pairs of histone molecules wrapped by two turns of a DNA molecule. The major structure associated with the organization of chromatin in the nucleus.

nucleotide In nucleic acid chemical nomenclature, a nucleoside covalently linked to one or more phosphate groups. Nucleotides containing a single phosphate linked to the 5′ carbon of the ribose or deoxyribose are the building blocks of nucleic acids.

nucleus The membrane-bound cytoplasmic organelle of eukaryotic cells that contains the chromosomes and nucleolus.

neutral theory The idea that most of the genetic variation present within and between species is caused by random fixation of alleles that are not acted on by natural selection and are therefore selectively neutral or nearly neutral mutations.

null allele A mutant allele that produces no functional gene product. Usually inherited as a recessive trait.

null hypothesis (H_0) Used in statistical tests, the hypothesis that there is no real difference between the observed and expected datasets. Statistical methods such as chi-square analysis are used to test the probability associated with this hypothesis.

nutrigenomics The study of how food and components of food affect gene expression.

Okazaki fragment The short, discontinuous strands of DNA produced on the lagging strand during DNA synthesis.

oligonucleotide A linear sequence of about 10–20 nucleotides connected by 5′-3′ phosphodiester bonds.

oncogene A gene whose activity promotes uncontrolled proliferation in eukaryotic cells. Usually a mutant gene derived from a *proto-oncogene*.

open reading frame (ORF) A nucleotide sequence organized as triplets that encodes the amino acid sequence of a polypeptide, including an initiation codon and a termination codon.

operator region In bacterial DNA, a region that interacts with a specific repressor protein to regulate the expression of an adjacent gene or gene set.

operon A genetic unit consisting of one or more structural genes encoding polypeptides, and an adjacent operator gene that regulates the transcriptional activity of the structural gene or genes.

ordered genetic code The pattern of triplet code sequences whereby chemically similar amino acids often share one or two middle bases.

orthologs Genes with sequence similarity found in two or more related species that arose from a single gene in a common ancestor.

p arm Shorter of the two arms extending from the centromere of the chromosome.

paleogenomics The extraction and analysis of DNA from the fossils of extinct organisms.

palindrome In genetics, a sequence of DNA base pairs that read the same on complementary strands. Because the strands run antiparallel to one another in DNA, the base sequences on the two strands read the same backwards and forward when read from the 5′ end. For example:

5′-GAATTC-3′
3′-CTTAAG-5′

Palindromic sequences are noteworthy as recognition and cleavage sites for restriction endonucleases.

pangenome An attempt to show or display a genome inclusive of known major genome variations (as opposed to a reference genome that displays the most commonly observed sequence without variations).

paracentric inversion A chromosomal inversion that does not include the region containing the centromere.

paralogs Two or more genes in the same species derived by duplication and subsequent divergence from a single ancestral gene.

parental gamete A gamete whose chromosomes have undergone no genetic recombination.

partial dominance Expressing a heterozygous phenotype that is distinct from the phenotype of either homozygous parent. Also called *incomplete dominance*.

passenger mutations Mutations that accumulate in cancer cells that have no direct contribution to the development or progression of the cancer.

pedigree In human genetics, a diagram showing the ancestral relationships and transmission of genetic traits over several generations in a family.

penetrance The frequency, expressed as a percentage, with which individuals of a given genotype manifest at least some degree of a specific mutant phenotype associated with a trait.

peptide bond The covalent bond between the amino group of one amino acid and the carboxyl group of another amino acid.

peptidyl (P) site The middle of three sites, or pockets, in the large and small subunits of the ribosome that may be occupied by tRNA during translation. The growing polypeptide chain is attached to tRNA in this site and during each step of translation, peptide bond formation occurs, extending the polypeptide by one amino acid.

pericentric inversion A chromosomal inversion that involves both arms of the chromosome and thus the centromere.

Personal Genome Project A project to enroll 100,000 individuals to share their genome sequence, personal information, and medical history with researchers and the general public to increase understanding of the contribution of genetic and environmental factors to genetic traits.

personal genomics Whole-genome sequencing of individual genomes.

personalized medicine Devising and applying unique and specific therapies for each individual, based on the individual's unique molecular profiles. A part of precision medicine.

pharmacogenomics The study of how genetic variation influences the action of pharmaceutical drugs in individuals.

phenotype The overt appearance of a genetically controlled trait.

Philadelphia chromosome The product of a reciprocal translocation in humans that contains the short arm of chromosome 9, carrying the *C-ABL* oncogene, and the long arm of chromosome 22, carrying the *BCR* gene.

phosphatases A broad class of enzymes that remove a phosphate group from a substrate molecule such as a protein, nucleic acid, carbohydrate, or lipid.

phosphodiester bond In nucleic acids, covalent bonds by which a phosphate group links adjacent nucleotides, extending from the 5′ carbon of one pentose sugar (ribose or deoxyribose) to the 3′ carbon of the pentose sugar in the neighboring nucleotide. Phosphodiester bonds create the backbone of nucleic acid molecules.

photolyase See *photoreactivation enzyme (PRE)*.

photoreactivation enzyme (PRE) An enzyme that cleaves the cross-linking bonds in thymine dimers. Also called *photolyase*.

plaque On an otherwise opaque bacterial lawn, a clear area caused by the growth and reproduction of a single bacteriophage.

plaque assay A quantitative assay using a serial dilution of a solution containing a large unknown number of bacteriophages that determines the original density (phage per ml).

plasmid An extrachromosomal, circular DNA molecule that replicates independently of the host chromosome.

pleiotropy A condition in which a single mutation causes multiple phenotypic effects.

pluripotent See *totipotent*.

point mutation A mutation that can be mapped to a single locus. At the molecular level, a mutation that results in the substitution of one nucleotide for another.

poly-A binding protein A protein that binds to the poly-A tail on the 3′ end of eukaryotic mRNAs to stabilize them, help them be exported from the nucleus, and initiate translation.

poly-A polymerase An enzyme that catalyzes the addition of adenosine residues to the 3′ end of a eukaryotic mRNA to synthesize the poly-A tail.

polyadenylation signal sequence A conserved AAUAAA sequence in eukaryotic mRNAs. Once transcribed, the nascent transcript is cleaved roughly 10-35 base pairs downstream.

polycistronic mRNA A messenger RNA molecule that encodes the amino acid sequence of two or more polypeptide chains in adjacent structural genes. Characteristic of bacteria.

polymerase chain reaction (PCR) A method for amplifying DNA segments that depends on repeated cycles of denaturation, primer annealing, and DNA polymerase–directed DNA synthesis.

polynucleotide A linear sequence of 20 or more nucleotides, joined by 5′-3′ phosphodiester bonds. See also *oligonucleotide*.

polypeptide A molecule composed of amino acids linked together by covalent peptide bonds. This term is used to denote the amino acid chain before it assumes its functional three-dimensional configuration and is called a protein.

polyploidy A condition in which a cell or individual has more than two haploid sets of chromosomes.

polyribosome A structure composed of two or more ribosomes associated with an mRNA engaged in translation. Also called a *polysome*.

polytene chromosome Literally, a many-stranded chromosome; one that has undergone numerous rounds of DNA replication without separation of the replicated strands, which remain in exact parallel register. The result is a giant chromosome with aligned chromomeres displaying a characteristic banding pattern, often studied in *Drosophila* larval salivary gland cells.

population A local group of actually or potentially interbreeding individuals belonging to the same species.

position effect A change in expression of a gene associated with a change in the gene's location within the genome.

posttranscriptional modification Changes made to pre-mRNA molecules during conversion to mature mRNA. These include the addition of a methylated cap at the 5′ end and a poly-A tail at the 3′ end, excision of introns, and exon splicing.

posttranslational modification The processing or modification of the translated polypeptide chain by enzymatic cleavage, addition of phosphate groups, carbohydrate chains, or lipids.

postzygotic isolation mechanism A factor that prevents or reduces inbreeding by acting after fertilization to produce nonviable, sterile hybrids or hybrids of lowered fitness.

pre-initiation complex (PIC) Prior to transcription initiation in eukaryotes, this complex of RNA polymerase II and general transcription factors loads onto the promoter.

pre-miRNAs A stage of miRNA processing that occurs after a miRNA gene is transcribed into a primary miRNA and then is cleaved by Drosha to remove 5′ and 3′ ends.

precision medicine An individualized approach to disease diagnosis and treatment, which uses various molecular profiles to select precise therapies and to develop new treatments.

preformationism The idea that development is the result of the assembly of structures already present in the egg, in contrast to *epigenesis*, which holds that an organism or organ arises through the sequential appearance and development of new structures.

preimplantation genetic diagnosis (PGD) The removal and genetic analysis of unfertilized oocytes, polar bodies, or single cells from an early embryo (3–5 days old).

prenatal genetic tests Genetic tests carried out on a baby before birth (during the prenatal period of development).

prezygotic isolation mechanism A factor that reduces inbreeding by preventing courtship, mating, or fertilization.

Pribnow box In bacterial genes, a 6-bp sequence to which the sigma (σ) subunit of RNA polymerase binds upstream from the beginning of transcription. The consensus sequence for this box is TATAAT. Also referred to as the -10 site.

primary miRNAs (pri-miRNAs) Transcriptional product of a microRNA gene containing a 5′ methylated cap, a 3′ polyadenylated tail, and a hairpin structure derived from self-complementary sequences.

primary structure The sequence of amino acids in a polypeptide chain.

primary sex ratio (PSR) Ratio of males to females at fertilization, often expressed in decimal form (e.g., 1.06).

primer In nucleic acids, a short length of RNA or single-stranded DNA required for initiating synthesis directed by polymerases.

prion An infectious pathogenic agent devoid of nucleic acid and composed of a protein, PrP, with a molecular weight of 27,000–30,000 Da. Prions are known to cause scrapie, a degenerative neurological disease in sheep; bovine spongiform encephalopathy (BSE, or mad cow disease) in cattle; and similar diseases in humans, including kuru and Creutzfeldt–Jakob disease.

proband An individual who is the focus of a genetic study leading to the construction of a pedigree tracking the inheritance of a genetically determined trait of interest. Formerly known as a *propositus*.

probe A macromolecule such as DNA or RNA that has been labeled and can be detected by an assay such as autoradiography or fluorescence microscopy. Probes are used to identify target molecules, genes, or gene products.

processivity The ability of an enzyme to carry out consecutive reactions before dissociating from its substrate. In the case of DNA polymerase III, processivity is the number of nucleotides added to a template strand before the enzyme falls off the template.

product law In statistics, the law holding that the probability of two independent events occurring simultaneously is equal to the product of their independent probabilities.

prokaryotes Organisms lacking a nuclear membrane, true chromosomes, and membranous organelles, e.g., bacteria and blue-green algae.

prometaphase Stage of cell division (mitosis and meiosis) during which the spindle fibers are assembled and attach to the centromeres of chromosomes, which begin their migration to the equatorial plate.

promoters A region of a gene where RNA polymerase binds and initiates transcription, located in the 5′ direction from the coding sequence.

proofreading A molecular mechanism for scanning and correcting errors in replication, transcription, or translation.

prophage A bacteriophage genome integrated into a bacterial chromosome that is replicated along with the bacterial chromosome. Bacterial cells carrying prophages are said to be *lysogenic* and to be capable of entering the *lytic cycle*, whereby phage particles are produced.

prophase The initial stage of cell division (mitosis and meiosis) during which the nuclear envelope breaks down, the nucleolus disintegrates, centrioles are formed and migrate to opposite ends of the cell, cytoplasmic microtubules are organized into spindle fibers, and diffuse chromatin fibers begin to condense into chromosomes.

proteasome A multi-subunit protein complex with protease (protein cleaving) activity that is used to degrade unneeded, damaged, or misfolded proteins.

protein A molecule composed of one or more polypeptides, each composed of amino acids covalently linked together. Proteins demonstrate *primary*, *secondary*, *tertiary*, and often, *quaternary structure*.

protein domain Amino acid sequences with specific conformations and functions that are structurally and functionally distinct from other regions of the same protein.

proteome The entire set of proteins expressed by a cell, tissue, or organism at a given time.

proteomics The study of the expressed proteins present in a cell at a given time.

proto-oncogene A gene that functions to initiate, facilitate, or maintain cell growth and division. Proto-oncogenes can be converted to *oncogenes* by mutation.

protoplast A bacterial or plant cell with the cell wall removed. Sometimes called a *spheroplast*.

protospacer adjacent motif (PAM) A short (2-6 basepair) DNA segment immediately adjacent to the DNA sequence that is targeted by a Cas nuclease of the CRISPR-Cas system. In the well-described CRISPR-Cas system of *Streptococcus pyogenes*, the PAM sequence is 5′-NGG-3′ immediately downstream of the target sequence on the non-complementary strand of the DNA.

prototroph A strain (usually of a microorganism) that is capable of growth on a defined, minimal medium. Wild-type strains are usually regarded as prototrophs and contrasted with *auxotrophs*.

pseudoautosomal region (PARs) A region on the human Y chromosome that is also represented on the X chromosome. Genes found in this region of the Y chromosome have a pattern of inheritance that is indistinguishable from genes on autosomes.

pseudogene A nonfunctional gene with sequence homology to a known structural gene present elsewhere in the genome. It differs from the functional version by insertions or deletions and by the presence of flanking direct-repeat sequences of 10–20 nucleotides.

puff A localized uncoiling and swelling in a polytene chromosome, usually regarded as a sign of active transcription.

q arm Longer of the two arms extending from the centromere of the chromosome.

QTL mapping population F_2 populations from crosses of inbred strains that carry different parts of the P_1 parental genomes and associated QTL genotypes and phenotypes.

quantitative inheritance See *polygenic inheritance*.

quantitative real-time PCR (qPCR) A variation of PCR (polymerase chain reaction) that uses fluorescent probes to quantitate the amount of DNA or RNA product present after each round of amplification.

quantitative trait loci (QTLs) Two or more genes that act on a single polygenic trait in a quantitative way.

quaternary structure Types and modes of interaction between two or more polypeptide chains within a protein molecule.

R plasmid A bacterial plasmid that carries antibiotic resistance genes. Most R plasmids have two components: an r-determinant that carries the antibiotic resistance genes and the resistance transfer factor (RTF).

real-time PCR Method for amplifying and quantifying PCR products without the need for gel electrophoresis; quantification of PCR products occurs at the end of each PCR cycle and is captured by software; thus, analysis occurs in "real-time."

recessive mutation A mutation that results in a wild-type phenotype when present in a diploid organism and the other allele is also wild-type.

reciprocal cross A pair of crosses in which the genotype of the female in one is present as the genotype of the male in the other, and vice versa.

recognition sequence See *restriction site*.

recombinant DNA DNA molecules created by joining together pieces of DNA from different sources.

recombinant DNA technology A collection of methods used to create DNA molecules by *in vitro* ligation of DNA from two different organisms, and the replication and recovery of such recombinant DNA molecules.

recombinant gamete A gamete containing a new combination of alleles produced by crossing over during meiosis.

recombination The process that leads to the formation of new allele combinations on chromosomes.

replication fork The Y-shaped region of a chromosome at the site of DNA replication.

replication The process whereby DNA is duplicated.

replicon The unit of DNA replication, beginning with DNA sequences necessary for the initiation of DNA replication. In bacteria, the entire chromosome is a replicon.

reporter gene A recombinant DNA tool that detects gene expression. The regulatory sequence of a gene of interest is fused to a coding sequence that confers an easily observable phenotype, such as fluorescence, and is inserted into an organism to learn when, where, and under what conditions the gene of interest is expressed.

repressors Proteins that bind to DNA sequences known as silencers to reduce or block gene transcription.

reproductive isolating mechanisms The set of environmental, physiological, behavioral, or mechanical impediments that prevent or reduce interbreeding between populations.

resistance transfer factor (RTF) A component of R plasmids that confers the ability to transfer the R plasmid between bacterial cells by conjugation.

restriction endonuclease DNA cutting enzymes that cleave or "digest" DNA at specific sequences.

restriction enzymes See *restriction endonuclease.*

restriction fragment length polymorphism (RFLP) Variation in the length of DNA fragments generated by restriction endonucleases. These variations are caused by mutations that create or abolish cutting sites for restriction enzymes. RFLPs are inherited in a codominant fashion and are extremely useful as genetic markers.

restriction map Physical map displaying the number, order, and distances between restriction enzyme digestion sites in a particular segment of DNA.

restriction site A DNA sequence, often palindromic, recognized by a restriction endonuclease. The enzyme binds to and cleaves DNA at the restriction site.

retrotransposons Mobile genetic elements that are major components of many eukaryotic genomes. They are copied by means of an RNA intermediate and inserted at other chromosomal sites, often causing mutations.

retroviral vectors Gene delivery vectors derived from retroviruses. Often used in gene therapy.

retrovirus A type of virus that uses RNA as its genetic material and employs the enzyme reverse transcriptase during its life cycle.

reverse genetics An experimental approach used to discover gene function after the gene has been identified, isolated, cloned, and sequenced. The cloned gene may be knocked out (e.g., by *gene targeting*) or have its expression altered (e.g., by *RNA interference* or *transgenic overexpression*) and the resulting phenotype studied. An approach contrasted with *forward genetics.*

reverse transcriptase A polymerase that facilitates reverse transcription using RNA as a template to transcribe a single-stranded DNA molecule.

reverse transcription PCR (RT_PCR) Method for synthesizing DNA from RNA through the actions of the enzyme reverse transcriptase.

rho-dependent termination In bacteria, one mechanism by which transcription is terminated. The rho protein breaks DNA-RNA base pairing with its helicase activity, which leads to RNA dissociation and transcription termination.

ribosomal RNA (rRNA) The RNA molecules that are the structural components of the ribosomal subunits. In bacteria, these are the 16*S*, 23*S*, and 5*S* molecules; in eukaryotes, they are the 18*S*, 28*S*, and 5*S* molecules. See also *Svedberg coefficient unit (S).*

ribosome A ribonucleoprotein organelle consisting of two subunits, each containing RNA and protein molecules. Ribosomes are the site of translation of mRNA codons into the amino acid sequence of a polypeptide chain.

riboswitch An RNA-based intracellular sensor that binds to a small ligand, such as a metabolite, modulating control of gene expression.

ribozymes RNAs that catalyze specific biochemical reactions.

RNA editing Alteration of the nucleotide sequence of an mRNA molecule after transcription and before translation. There are two main types of editing: substitution editing, which changes individual nucleotides, and insertion/deletion editing, in which individual nucleotides are added or deleted.

RNA interference (RNAi) Inhibition of gene expression in which a protein complex (RNA-induced silencing complex, or RISC) containing a complementary (or partially complementary) RNA strand binds to an mRNA, leading to degradation or reduced translation of the mRNA.

RNA polymerase An enzyme that catalyzes the formation of an RNA polynucleotide strand using the base sequence of a DNA molecule as a template.

RNA sequencing (RNA-seq) Technique for determining the nucleotide sequence of RNA molecules.

RNA splicing The processing of a nascent transcript of RNA by the removal of introns and the joining together of exons.

RNA-binding proteins (RBPs) A class of proteins that bind to specific RNA sequences or RNA secondary structures and influence many posttranscriptional regulatory mechanisms such as alternative splicing, RNA decay, RNA stability, RNA transport and localization, and translation.

RNA-induced silencing complex (RISC) A protein complex containing an Argonaute family protein with endonuclease activity. siRNAs and miRNAs guide RISC to complementary mRNAs to cleave them.

Robertsonian translocation A chromosomal aberration created by breaks in the short arms of two acrocentric chromosomes followed by fusion of the long arms of these chromosomes at the centromere. Also called *centric fusion.*

S phase The "synthesis" portion of the cell cycle following the G1 phase during which DNA is replicated.

Sanger sequencing DNA sequencing by synthesis of DNA chains that are randomly terminated by incorporation of a nucleotide analog (dideoxynucleotides) followed by sequence determination by analysis of resulting fragment lengths in each reaction.

satellite DNA DNA that forms a minor band when genomic DNA is centrifuged in a cesium salt gradient. This DNA usually consists of short sequences repeated many times in the genome.

secondary structure The α-helical or β-pleated-sheet formations in a polypeptide, dependent on hydrogen bonding between certain amino acids.

secondary sex ratio The ratio of males to females at birth, usually expressed in decimal form (e.g., 1.05).

selectable marker gene Marker gene such as one that encodes a gene for antibiotic resistance allowing specific cells to be chosen or selected by phenotype or characteristics (for example, antibiotic resistance) provided by the marker gene.

selfing In plant genetics, the fertilization of a plant's ovules by pollen produced by the same plant. Reproduction by self-fertilization.

semiconservative replication A mode of DNA replication in which a double-stranded molecule replicates in such a way that the daughter molecules are each composed of one parental (old) and one newly synthesized strand.

sequencing-by-synthesis (SBS) Next-generation sequencing platform that uses DNA fragments to synthesize new strands which are sequenced.

severe combined immunodeficiency A collection of genetic disorders characterized by a lack of immune system response; both cell-mediated and antibody-mediated responses are missing.

sex chromatin body See *Barr body.*

sex ratio See *primary sex ratio* and *secondary sex ratio.*

sex-determining region Y (SRY) Essential gene on the human Y chromosome that controls male sexual development; encodes TDF protein.

sex-influenced inheritance A phenotypic expression conditioned by the sex of the individual. A heterozygote may express one phenotype in one sex and an alternate phenotype in the other sex (e.g., pattern baldness in humans).

sex-limited inheritance A trait that is expressed in only one sex even though the trait may not be X-linked or Y-linked.

Shine–Dalgarno sequence The nucleotides AGGAGG that serve as a ribosome-binding site in the leader sequence of bacterial genes. The 16*S* RNA of the small ribosomal subunit contains a complementary sequence to which the mRNA binds.

short interspersed elements (SINEs) Repetitive sequences found in the genomes of higher organisms. The 300-bp *Alu* sequence is a SINE element.

short tandem repeats (STRs) Short tandem repeats 2–9 base pairs long found within minisatellites. These sequences are used to prepare DNA profiles in forensics, paternity identification, and other applications.

shotgun cloning The cloning of random fragments of genomic DNA into a vector (a plasmid or phage), usually to produce a library from which clones of specific interest can be selected for use, as in sequencing.

shugoshins A class of proteins involved in maintaining cohesion of the centromeres of sister chromatids during mitosis and meiosis.

sigma (σ) factor In RNA polymerase, a polypeptide subunit that recognizes the DNA binding site for the initiation of transcription.

signal transduction An intercellular or intracellular molecular pathway by which an external signal is converted into a functional biological response.

silencers A DNA sequence that reduces or blocks the transcription and the expression of genes. Silencers can act over a distance of thousands of base pairs and can be located upstream, downstream, or internal to the gene they affect.

silent mutation A mutation that alters the sequence of a codon but does not result in a change in the amino acid at that position in the protein.

single-cell RNA sequencing (scRNA-seq) RNA sequencing approach to analyze gene expression in individual cells.

single-cell sequencing (SCS) DNA sequencing approach to analyze the genome in individual cells.

single guide RNA (sgRNA) An engineered hybrid RNA molecule that combines sequences of the crRNA and tracrRNA of type II CRISPR-Cas systems into a single RNA. This makes CRISPR-Cas-mediated genome editing more convenient because two separate components are integrated into one.

single-nucleotide polymorphism (SNP) A variation in one nucleotide pair in DNA, as detected during genomic analysis. Present in at least 1 percent of a population, a SNP is useful as a genetic marker.

single-stranded binding proteins (SSBs) In DNA replication, proteins that bind to and stabilize the single-stranded regions of DNA that result from the action of unwinding proteins.

sister chromatid exchange (SCE) A crossing-over event in meiotic or mitotic cells involving the reciprocal exchange of chromosomal material between sister chromatids joined by a common centromere. Such exchanges can be detected cytologically after BrdU incorporation into the replicating chromosomes.

sister chromatids A pair of identical chromatids visible during mitosis and meiosis that are formed following replication of DNA of one member of a homologous chromosome pair.

sliding clamp loader A component of the DNA polymerase III holoenzyme consisting of five subunits that attach a circular protein complex to the polymerase in an ATP-dependent reaction.

sliding DNA clamp A component of the DNA polymerase III holoenzyme composed of multiple copies of the beta subunit that forms a circular structure attached to the polymerase, which promotes processivity.

small interfering RNAs (siRNAs) Small (or short) interfering RNAs. Short 20–25 nucleotide double-stranded RNA sequences with two 3′ overhanging nucleotides; they are processed by Dicer and participate in transcriptional and/or posttranscriptional mechanisms of gene regulation.

small noncoding RNAs (sRNAs) Any of a number of short RNAs that are noncoding transcripts that associate with the RNA-induced silencing complex (RISC) to regulate transcription or to regulate mRNAs posttranscriptionally.

small nuclear RNA (snRNA) Abundant species of small RNA molecules ranging in size from 90 to 400 nucleotides that in association with proteins form RNP particles known as snRNPs or *snurps*. Located in the nucleoplasm, snRNAs have been implicated in the processing of pre-mRNA and may have a range of cleavage and ligation functions.

somatic cell hybridization A technique involving the fusion of somatic cells first utilized in the 1960s to assign human genes to their respective chromosomes.

somatic gene therapy Gene therapy involving somatic cells as targets for gene transfer.

somatic mutation A nonheritable mutation occurring in a somatic cell.

Southern blot analysis Electrophoresis, blotting, and hybridization technique for detecting specific DNA fragments; pioneered by Edwin Southern.

Southern blotting Developed by Edwin Southern, a technique in which DNA fragments produced by restriction enzyme digestion are separated by electrophoresis and transferred by capillary action to a nylon or nitrocellulose membrane. Specific DNA fragments can be identified by hybridization to a complementary radioactively labeled nucleic acid probe using the technique of *autoradiography*.

spacer acquisition One of the steps in the CRISPR-Cas mechanism in which spacer DNA sequences, often derived from invading bacteriophage genomes, are inserted into the CRISPR locus of a host bacterial or archaea genome.

spacer DNA DNA sequences found between genes. Usually, these are repetitive DNA segments.

speciation The process by which new species arise.

species A group of actually or potentially interbreeding individuals that is reproductively isolated from other such groups.

spectral karyotype A display of all the chromosomes in an organism as a karyotype with each chromosome stained in a different color.

spheroplast See *protoplast*.

spindle fibers Cytoplasmic fibrils formed during cell division that attach to and are involved with separation of chromatids at the anaphase stage of mitosis and meiosis as well as their movement toward opposite poles in the cell.

spliceforms Different mRNAs produced from the same gene through the process of alternative splicing.

spliceopathies A class of human diseases caused by defects in RNA splicing.

spliceosome The nuclear macromolecule complex within which splicing reactions occur to remove introns from pre-mRNAs.

spontaneous generation The incorrect idea that living organisms can be created directly from nonliving material rather than by descent from other living organisms.

spontaneous mutation A random mutation that is not induced by a mutagenic agent.

spore A unicellular body or cell encased in a protective coat. Produced by some bacteria, plants, and invertebrates, spores are capable of surviving in unfavorable environmental conditions and give rise to a new individual upon germination. In plants, spores are the haploid products of meiosis.

Src A protein kinase that phosphorylates many target proteins to regulate their activity, for example, regulating the activity of the RNA-binding protein ZBP1.

stabilizing selection Preferential reproduction of individuals with genotypes close to the mean for the population. A selective elimination of genotypes at both extremes.

standard deviation (*s*) A quantitative measure of the amount of variation present in a sample of measurements from a population calculated as the square root of the variance.

standard error of the mean (*S*) Statistical calculation that compares variation between the means of multiple samples in a population.

stem cells Undifferentiated cells that can self-replicate and differentiate to develop into a variety of different cell types.

Stone-Age genomics Sequencing and analysis of ancient DNA samples.

structural gene A gene that encodes the amino acid sequence of a polypeptide chain.

submetacentric chromosome A chromosome with the centromere placed so that one arm of the chromosome is slightly longer than the other.

sum law The law that holds that the probability of one of two mutually exclusive outcomes occurring, where that outcome can be achieved by two or more events, is equal to the sum of their individual probabilities.

supercoiling In reference to the tertiary structure of DNA, the underwinding (creating negative supercoils) or overwinding (creating positive supercoils) that occurs when the helix is strained.

suppressor mutation A second mutation that reverts or relieves the effects of a previous mutation. Suppressor mutations can occur either within the same gene that contained the first mutation, or elsewhere in the genome.

Svedberg coefficient *(S)* A unit of measure for the rate at which particles (molecules) sediment in a centrifugal field. This rate is a function of several physicochemical properties, including size and shape. A rate of 1×10^{-13} seconds is defined as 1 Svedberg coefficient unit.

synapsis The pairing of homologous chromosomes at meiosis.

synthetic biology A scientific discipline that combines science and engineering to research the complexity of living systems and to construct biological-based systems that do not exist in nature.

synthetic genome A genome assembled from chemically synthesized DNA fragments that is transferred to a host cell without a genome.

T cell White blood cell type with varied and essential roles in the immune system; also called T-lymphocytes.

T-cell receptors (TCRs) Receptor protein on the surface of T-cells (T-lymphocytes) that can recognize and bind to foreign particles (antigens) during an immune response. Engineered versions of TCRs can be used for immunotherapy.

TALENs See *transcription activator-like effector nucleases.*

target interference One of the steps in the CRISPR-Cas mechanism in which crRNAs guide a Cas nuclease to cleave target DNA sequences.

TATA box See *Goldberg–Hogness box.*

tautomeric shift A reversible isomerization in a molecule, brought about by a shift in the location of a hydrogen atom. In nucleic acids, tautomeric shifts in the bases of nucleotides can cause changes in other bases at replication and are a source of mutations.

telocentric chromosome A chromosome in which the centromere is located at its very end.

telomerase The enzyme that adds short, tandemly repeated DNA sequences to the ends of eukaryotic chromosomes.

telomeres The heterochromatic terminal regions of a chromosome.

telophase The stage of cell division (mitosis and meiosis) in which the daughter chromosomes have reached the opposite poles of the cell and reverse the stages characteristic of prophase, re-forming the nuclear envelopes and uncoiling the chromosomes. Telophase ends during cytokinesis, which divides the cytoplasm and splits the parental cell into two daughter cells.

temperate phage A bacteriophage that can become a prophage, integrating its DNA into the chromosome of the host bacterial cell and making the latter lysogenic.

temperature-sensitive mutation A conditional mutation that produces a mutant phenotype at one temperature and a wild-type phenotype at another.

template strand In a double stranded DNA molecule, the strand that is transcribed by RNA polymerase during transcription.

termination factor, rho (ρ) In bacterial rho-dependent transcriptional termination, the rho protein breaks DNA-RNA base pairing with its helicase activity, which leads to RNA dissociation and transcription termination.

tertiary structure The three-dimensional conformation of a polypeptide chain in space, specified by the polypeptide's primary structure. The tertiary structure achieves a state of maximum thermodynamic stability.

testcross A cross between an individual whose genotype at one or more loci may be unknown and an individual who is homozygous recessive for the gene or genes in question.

testis-determining factor (TDF) Protein encoded by SRY gene; causes bipotential gonads of an embryo to form testes.

tetrad The four chromatids that make up paired homologs in the prophase of the first meiotic division. In eukaryotes with a predominant haploid stage (some algae and fungi), a tetrad denotes the four haploid cells produced by a single meiotic division.

third-generation sequencing (TGS) DNA-sequencing technologies based largely on methods for sequencing individual molecules of single-stranded DNA.

Ti plasmid A bacterial plasmid used as a vector to transfer foreign DNA to plant cells.

DNA topoisomerase A class of enzymes that converts DNA from one topological form to another, leading to *topoisomers.* During replication, a topoisomerase, *DNA gyrase,* facilitates DNA replication by reducing molecular tension caused by supercoiling upstream from the *replication fork.*

***trans*-acting factor** A gene product (usually a diffusible protein or an RNA molecule) that acts to regulate the expression of a target gene.

transcription Transfer of genetic information from DNA by the synthesis of a complementary RNA molecule using one strand of the DNA as a template.

transcription activator-like effector nucleases (TALENs) Artificial DNA-cleaving enzymes created by combining DNA-binding motifs (transcription activator-like effectors or TALES) from plant pathogenic bacteria to a DNA-cutting domain from a nuclease. This method can produce restriction enzymes for any DNA sequence.

transcription factor A DNA-binding protein that binds to specific sequences adjacent to or within the promoter region of a gene; regulates gene transcription.

transcriptional activators Proteins that bind to DNA sequences known as enhancers to increase the rate of gene transcription.

transcriptional repressors Proteins that bind to DNA sequences known as silencers to reduce or block gene transcription.

transcriptomics The set of mRNA molecules present in a cell at any given time.

transduction Virally mediated bacterial recombination. Also used to describe the transfer of eukaryotic genes mediated by a retrovirus.

transfection In bacteria, the infection process occurring when cells treated with lysozyme, removing their outer walls, are exposed to viral DNA or RNA, leading to the multiplication of viral particles.

transfer RNA (tRNA) A small ribonucleic acid molecule with an essential role in *translation.* tRNAs contain: (1) a three-base segment (anticodon) that recognizes a codon in mRNA; (2) a binding site for the specific amino acid corresponding to the anticodon; and (3) recognition sites for interaction with ribosomes and with the enzyme that links the tRNA to its specific amino acid.

transformation Heritable change in a cell or an organism brought about by exogenous DNA. Known to occur naturally and also used in *recombinant DNA* studies.

transgenic animal See *transgenic organism.*

transgenic organism An organism whose genome has been modified by the introduction of external DNA sequences into the germ line.

translation The derivation of the amino acid sequence of a polypeptide from the base sequence of an mRNA molecule in association with a ribosome and tRNAs.

translational medicine Also called translational genetics. Refers to moving research discoveries in genetics and other disciplines from the laboratory bench to the bedside to improve human health by disease prevention and the treatment of diseases.

translocation A chromosomal mutation associated with the reciprocal or nonreciprocal transfer of a chromosomal segment from one chromosome to another. Also denotes the movement of mRNA through the ribosome during translation.

transmission genetics The field of genetics concerned with heredity and the mechanisms by which genes are transferred from parent to offspring.

transposable element (TE) A DNA segment that moves to other sites in the genome, essentially independent of sequence homology. Usually, such elements are flanked at each end by short inverted repeats of 20–40 base pairs. Insertion into a structural gene can produce a mutant phenotype. Insertion and excision of transposable elements depend on two enzymes, transposase and resolvase. Such elements have been identified in both prokaryotes and eukaryotes.

transversion A base change in a DNA molecule that substitutes a purine for a pyrimidine or vice versa.

trinucleotide repeat A tandemly repeated cluster of three nucleotides (such as CTG) within or near a gene. Certain diseases (myotonic dystrophy, Huntington disease) are caused by expansion in copy number of such repeats.

trisomy The condition in which a cell or an organism possesses two copies of each chromosome except for one, which is present in three copies (designated $2n + 1$).

tubulin Protein making up microtubules characteristic of mitosis and meiosis.

tumor-suppressor gene A gene that encodes a product that normally functions to suppress cell division. Mutations in tumor-suppressor genes result in the activation of cell division and tumor formation.

tumorigenesis The development of a tumor.

two-dimensional gel electrophoresis (2DGE) Biochemistry technique for separating peptides or proteins by mass and charge.

ubiquitin ligase A eukaryotic enzyme that catalyzes the addition of a ubiquitin molecule to specific target proteins to tag them for degradation by the proteasome.

ubiquitin A small eukaryotic protein that when covalently attached to other proteins, tags them for degradation by the proteasome.

ubiquitination The process of adding a unit of ubiquitin to a target protein, which serves as a tag for degradation of that target protein by the proteasome.

unit factors The term used by Mendel to describe hereditary factors controlling specific traits. In modern terms, alleles of a specific gene.

variable number tandem repeats (VNTRs) Short, repeated DNA sequences (of 2–20 nucleotides) present as tandem repeats between two restriction enzyme sites. Variation in the number of repeats creates DNA fragments of differing lengths following restriction enzyme digestion. Used in early versions of *DNA fingerprinting*.

variance (s^2) A statistical measure of the variation of values from a central value, calculated as the square of the standard deviation.

vector In recombinant DNA, an agent such as a phage or plasmid into which a foreign DNA segment will be inserted and used to transform host cells.

vertical gene transfer The transfer of genetic information from parents to offspring generation after generation.

virulent phage A bacteriophage that infects, replicates within, and lyses bacterial cells, releasing new phage particles.

Western blotting An analytical technique in which proteins are separated by gel electrophoresis and transferred by capillary action to a nylon membrane or nitrocellulose sheet. A specific protein can be identified through hybridization to a labeled antibody.

whole-exome sequencing (WES) Method for sequencing only the RNA-coding (exons) of a genome.

whole-genome sequencing (WGS) High-throughput techniques for sequencing all of the DNA in a genome and organizing sequencing data to produce a complete genome sequence.

wobble hypothesis An idea proposed by Francis Crick, stating that the third base in an anticodon in tRNA that can align in several ways to allow it to recognize more than one base in the codons of mRNA.

X-inactivation center (Xic) Region of the p-arm of the X-chromosomes in humans which produces a non-coding RNA (Xist) that is essential for X-inactivation mechanisms.

X-inactive specific transcript (*XIST*) A locus in the X chromosome inactivation center that controls inactivation of the X chromosome in mammalian females.

X-linkage The pattern of inheritance resulting from genes located on the X chromosome.

XRN1 A eukaryotic exoribonuclease involved in mRNA degradation.

Y chromosome The sex chromosome in species where the male is heterogametic (XY).

yeast artificial chromosomes (YACs) A cloning vector in the form of a yeast artificial chromosome, constructed using chromosomal components including telomeres (from a ciliate) and centromeres, origin of replication, and marker genes from yeast. YACs are used to clone long stretches of eukaryotic DNA.

zinc-finger nucleases (ZFNs) Sequence-specific DNA cleaving enzymes consisting of a zinc-finger motif (a cluster of two cysteine and two histidine residues that bind zinc atoms). ZFNs provide a mechanism for modifying sequences in the genome in a sequence-specific targeted way.

zip code Sequences found in many mRNAs that serve as binding sites for RNA-binding proteins that influence transcript localization within the cell and translational control.

zip code binding protein 1 (ZBP1) One of a family of highly conserved RNA-binding proteins that play important roles in the localization, stability, and translational control of mRNAs.

zygote The diploid cell produced by the fusion of haploid gametic nuclei.

ZZ/ZW Sex-determining chromosomes in animals such as chickens, frogs, and certain fish, where the maleness is determined by homomorphic chromosomes (ZZ) and females by heteromorphic chromosomes (ZW).

Credits

PHOTOS

Chapter 1 01-COa, Sinclair Stammers/Science Source; 01-COb, Mark Smith/Science Source; 01-COc, Darwin Dale/Science Source; F01-01, Wellcome Library, London; F01-02, Biophoto Associates/Science Source; F01-03a, Photo Researchers/Science Source; F01-03b, Photo Researchers/Science Source; F01-07, Eye of Science/Science Source; F01-08, Photo courtesy of The Roslin Institute, The University of Edinburgh, Roslin, Scotland, UK; F01-09a, Redmond O. Durrell/Alamy Stock Photo; F01-09b, Hermann Eisenbeiss/Science Source; F01-10a, Dr Jeremy Burgess/Science Source; F01-10b, David McCarthy/Science Source

Chapter 2 02-CO, Waheeb Heneen; F02-02, CNRI/Science Source; F02-04, Leonard Lessin/Science Source; 02-07a, Waheeb Heneen; F02-07b, Waheeb Heneen; F02-07c, Waheeb Heneen; F02-07d, Waheeb Heneen; F02-07e, Waheeb Heneen; F02-07f_1, Waheeb Heneen; F02-07f_2, Waheeb Heneen F02-12a, Biophoto Associates/Science Source; F02-12b, Andrew Syred/Science Source; F02-12c Biophoto Association/Science Source

Chapter 3 03-CO, National Library of Medicine; F03-01, Elenamiv/Shutterstock, Martin Shields/Alamy Stock Photo

Chapter 4 04-CO, Juniors Bildarchiv GmbH/Alamy Stock Photo; F04-01, John Kaprielian/Science Source; F04-03_1, Jackson Laboratory; F04-03_2, Jackson Laboratory; F04-08, RoJo Images/Shutterstock; F04-12, Prisma by Dukas Presseagentur GmbH/Alamy Stock Photo; F04-13, Volodymyr Martyniuk/Shutterstock; F04-14, Stanislav Fridkin/Shutterstock; F04-15_1, Joel Eissenberg; F04-15_2, Joel Eissenberg; F04-15_3, Joel Eissenberg; F04-16a, P. Wegner/Arco Images/AGE Fotostock; F04-16b, Dr. William S. Klug; F04-17, Tikta Alik/Shutterstock; F04-18a, W. C. Copeland et al. (2003). POS5 Gene of Saccharomyces cerevisiae Encodes a Mitochondrial NADH Kinase Required for Stability of Mitochondrial DNA. Eukaryotic Cell, 2(4):809-820; F04-18b, W. C. Copeland et. al. (2003). POS5 Gene of Saccharomyces cerevisiae Encodes a Mitochondrial NADH Kinase Required for Stability of Mitochondrial DNA. Eukaryotic Cell, 2(4):809-820; UNF04-02_1, Ralph Somes; UNF04-02_2, Ralph Somes; UNF04-02_3, Ralph Somes; UNF04-02_4, Ralph Somes; UNF04-05_1, Shout It Out Design/Shutterstock; UNF04-05_2, Zuzule/Shutterstock; UNF04-05_3, Julia Remezova/Shutterstock

Chapter 5 05-CO, Wessex Reg. Genetics Centre, Wellcome Images; 05-02a, Catherine G. Palmer; F05-02b, Catherine G. Palmer; F05-04a, Michael Abbey/Science Source; F05-04b, Michael Abbey/Science Source; F05-06a, Sari Oneal/Shutterstock; F05-06b, Dr. William S. Klug; UNF05-01, Texas A&M University/Newscom; F05-08a, Dr. Maria Gallegos, Associate Professor at California State University, East Bay; F05-08b, Dr. Maria Gallegos, Associate Professor at California State University, East Bay

Chapter 6 06-CO, NHI/National Human Genome Research Institute; F06-02a, Illustration from Genetic Counseling Aids, 3rd Edition, Copyright 1995, Permission for use granted by Greenwood Genetic Center; F06-02b, Design Pics Inc/Alamy Stock Photo; F06-04, Pr Philippe Vago, ISM/Science Photo Library; F06-08, National Cotton Council of America; F06-11a, L. Willatt, East Anglian Regional Genetics Service/Science Source; F06-11b, 5p- Society; F06-13, Mary Lilly/The Observatories for the Carnegie Institution for Science; F06-18, Christine Harrison

Chapter 7 07-CO, Stanley Sessions; F07-14, Arturo Londono, ISM/Science Photo Library

Chapter 8 08-CO, Dr. L. Caro/Science Source; F08-10a, Dr. Gopal Murti/Science Source; F08-11_1, M. Wurtz/Biozentrum, University of Basel/Science Source; F08-13, Christine Case

Chapter 9 09-CO, Kenneth Eward/Science Source; F09-03, Meckes/Ottawa/Science Source; F09-10, Arizona Board of Regents/ASU Ask A Biologist; F09-13, Ventana Medical Systems, Inc. F09-14, Permission Granted by John Wiley and Sons, from Gel Electrophoresis by R. Westermeier, Serva Electrophoresis GmbH, Heidelberg, Germany, ELS (February 2013)

Chapter 10 10-CO, Dr Gopal Murti/Science Source; F10-105a, J.H. Taylor, (1963). The Replication and Organization of DNA in Chromosomes, Molecular Genetics, Part 1, pp. 74 and 75, Academic Press, New York, 1963; F10-05b, J.H. Taylor, (1963). The Replication and Organization of DNA in Chromosomes, Molecular Genetics, Part 1, pp. 74 and 75, Academic Press, New York, 1963; F10-14, Victoria Foe

Chapter 11 11-CO, Don. W. Fawcett/Science Source; F11-01a, Dr. Michel Wurtz/Biozentrum, University of Basel/Science Source; F11-01b, Dr. William S. Klug; F11-02, Biology Pics/Science Source; F11-03, G. Murti/Science Source; F11-04, Dr. Richard Kolodner; F11-05, Don W. Fawcett/Science Source; F11-06a, Republished with permission of The Company of Biologists, Ltd., Identification of the Drosophila interband-specific protein Z4 as a DNA-binding zinc-finger protein determining chromosomal structure by Harald Eggert, et. al., Journal of Cell Science 17: 4253-4264 2004; permission conveyed through Copyright Clearance Center, Inc; F11-06b, Republished with permission of The Company of Biologists, Ltd., Identification of the Drosophila interband-specific protein Z4 as a DNA-binding zinc-finger protein determining chromosomal structure by Harald Eggert, et. al., Journal of Cell Science 17: 4253-4264 2004; permission conveyed through Copyright Clearance Center, Inc; F11-07b, John Ellison; F11-08a, Omikron/Science Source; F11-08b, Dr. William S. Klug; F11-09, Victoria Foe; F11-13, Dr. William S. Klug

Chapter 12 12-CO, Professor Oscar L. Miller/Science Source

Chapter 13 13-CO, From the front cover of Science, Vol. 292, Issue 5518, pp. 870–872, May 4, 2001; F13-09a, Copyright holder is Cold Spring Harbor Laboratory Press; F13-09b, Elena Kiseleva, ICiG, Russia; F13-11, AKU Society; F13-12, SciePro/Shutterstock; F13-16, Ken Eward/Science Source

Chapter 14 14-CO, Gerry Neuffer; F14-12, HADJ/SIPA/Newscom

Chapter 16 16-CO, Bolzer A., et.al. (2005). Three-Dimensional Maps of All Chromosomes in Human Male Fibroblast Nuclei and Prometaphase Rosettes. PLoS Biology, 3(5): e157; F16-08, Reprinted by permission from Macmillan Publishers Ltd: Alternative splicing and evolution: diversification, exon definition and function. Nature Reviews Genetics by Keren, H. et. al., 11: 345–355 (May 2010); F16-14, Hochstrasser M. (2009). Origin and function of ubiquitin-like proteins. Nature, 458(7237):422-9.

Chapter 17 17-CO, Pascal Goetgheluck/Science Source; F17-03a, Dr. Gopal Murti/Science Source; F17-05_2, Reproduced with permission from Sigma-Aldrich Co. LLC; F17-08, Scott Sinklier/Alamy Stock Photo; F17-09a, Proceedings of the National Academy of Sciences; F17-09b, Proceedings of the National Academy of Sciences; F17-10, Gerald B. Downes; F17-14a, RAGUET/PHANIE/AGE Fotostock; F17-14b, Michael A. Palladino; F17-15a, Eye of Science/Science Source; F17-15b, Photograph courtesy of Ralph Brinster. University of Pennsylvania School of Veterinary Medicine

Chapter 18 F18-10_1, Volker Steger/Science Source; F18-10_2, dra_schwartz/E+/Getty Images; F18-11, Republished with permission of ELECTROPHORESIS. SWISS-2DPAGE: A database of two-dimensional gel electrophoresis images. Appel, Ron D. et.al. vol. 14(1) pp. 1232–1238 (1993)]; permission conveyed through Copyright Clearance Center, Inc.

Chapter 19 19-CO, SPL/Science Source; F19-01a, Hesed M. Padilla-Nash/Antonio Fargiano/National Institute of Health Genetics; F19-01b, Hesed M. Padilla-Nash/Antonio Fargiano/National Institute of Health Genetics; F19-02, Elizabeth Cook

Chapter 20 20-CO, 2009fotofriends/Shutterstock; F20-08, Rekemp/iStock/Getty Images

Chapter 21 21-CO, Simon Booth/Science Source; F21-01, Pixshots/Shutterstock; F 21-06, Reprinted by permission from Macmillan Publishers Ltd: Resistance to HIV-1 infection in Caucasian individuals bearing mutant alleles of the CCR- 5 chemokine receptor gene by Samson M., et al., Nature 382(6593):725 (August 1996); F21-15, Republished with permission of Elsevier; from A 122.5-Kilobase Deletion of the P Gene Underlies the High Prevalence of Oculocutaneous Albinism Type 2 in the Navajo Population. Yi Z, et. al., American Journal of Human Genetics. 72(1):62-72, 2003; permission conveyed through Copyright Clearance Center, Inc; F21-18, Smithsonian Institution Libraries; F21-20a, Niels Poulsen DK/Alamy Stock Photo; F21-20b, Niklas Lindqvist/National Ciklid Society; F21-22a, Mchugh Tom/Science Source/Getty Images; F21-22b, Peter Scoones/The Image Bank/Getty Images; F21-25, RIA Novosti/Science Source

Special Topic 1 ST F01-08, Dolinoy et al. (2006). Maternal genistein alters coat color and protects Avy mouse offspring from obesity by modifying the fetal epigenome. Environmental

Health Perspectives, 114(4):567-72; ST F01-09, Reproduced with permission from the Annual Review of Genomics and Human Genetics, Volume 9 © 2008 by Annual Reviews, http://www.annualreviews.org

Special Topic 2 ST F02-05, Affymetrix/Thermo Fisher Scientific; ST F02-06, Reprinted by permission from Macmillan Publishers Ltd: Distinct types of diffuse large B-cell lymphoma identified by gene expression profiling. Ash A. Alizadeh et al. Nature, 403, 503-511 (February 2000)

Special Topic 6 ST UNF06-01, AquaBounty Technologies; ST F06-02, Courtesy of International Rice Research Institute; ST UNF06-02, Kheng Guan Toh/Shutterstock

Special Topic 7 ST F07-01a, Reproduced with permission of Dako Denmark A/S, a subsidiary of Agilent Technologies, Inc., Santa Clara, California, USA. All rights reserved; ST F07-01b, Reproduced with permission of Dako Denmark A/S, a subsidiary of Agilent Technologies, Inc., Santa Clara, California, USA. All rights reserved

TEXT

Chapter 12 F12-11, Nobel Media Ab 2017; F12-12, Nobel Media Ab 2017; T12-01, After M. Nirenberg and J. H. Matthaei (1961)

Chapter 18 F18-07, Reprinted by permission from Macmillan Publishers Ltd: Nature News Feature, Human Genome at ten: The sequence explosion. Nature, 464:670- 671, (2010 March); F18-08, Republished with permission of Oxford University Press, from SplitMEM: a graphical algorithm for pan-genome analysis with suffix skips Marcus, S. Lee, H., Schatz, M.C., vol.30(24):3476-3483 (2014); permission conveyed through Copyright Clearance Center, Inc.; F18-09, Qin, N., Yang, F., et al. Alterations of the human gut microbiome in liver cirrhosis. Nature, 513(7516):59-64

Chapter 19 F19-02a-d, Elizabeth Cook

Chapter 20 F20-05, Data from Tanksley, S.D., (1993). Mapping Polygenes. Annual Review of Genetics, 27:205–253. Table 1; F20-07, Reprinted by permission from Macmillan Publishers Ltd: Quantitative trait loci in Drosophila by Mackay, TF. Nature Reviews Genetics, 2:11-20 (January 2001); F20-09, Republished with permission of American Society for Microbiology, from Increase in tomato locule number is controlled by two single nucleotide polymorphisms located near WUSCHEL. Plant Physiology, by Munos, S. et al.,156(4):2244-2254, August 2011; permission conveyed through Copyright Clearance Center, Inc.

Chapter 21 F21-02, Kreitman, M. (1983). Nucleotide polymorphism at the alcohol dehydrogenase locus of Drosophila melanogaster. Nature, 304(5925):412-417; F21-24, Reproduced with permission from the Annual Review of Genomics and Human Genetics, Volume 11 © 2010 by Annual Reviews, http://www.annualreviews.org; F21-26, Browning, Sharon R. et al. "Analysis of Human Sequence Data Reveals Two Pulses of Archaic Denisovan Admixture." Cell, Volume 173 , Issue 1, 53 - 61.e9 (March 22, 2018)

Special Topic 1 ST F01-05, Adapted by permission from Macmillan Publishers Ltd: Rosanna Weksberg, Cheryl Shuman, and J Bruce Beckwith. Beckwith–Wiedemann syndrome, European Journal of Human Genetics (2010) 18, 8–14, p. 10, fig.2; ST F01-06, Banno, Kouji, et al. (2013). Epimutation in DNA Mismatch Repair (MMR) Genes. Biochemistry, Genetics and Molecular Biology "New Research Directions in DNA Repair", book edited by Clark Chen, May 22, 2013 under CC BY 3.0 license.

Special Topic 2 ST F02-07a, Reprinted by permission from Macmillan Publishers Ltd: Single-cell RNA-seq identifies a PD-1hi ILC progenitor and defines its development pathway. Yu et al., Nature, 539, 102-106, (2016); ST02-07b, Reprinted by permission from Macmillan Publishers Ltd: Single-cell RNA-seq identifies a PD-1hi ILC progenitor and defines its development pathway]. Yu et al., Nature, 539, 102-106, (2016); ST F02-08, Jones, G. T., et.al. (2017). Meta-Analysis of Genome-Wide Association Studies for Abdominal Aortic Aneurysm Identifies Four New Disease-Specific Risk Loci, Circulation Research. 120(2):341-353.

Special Topic 3 ST F03-03, Targeting DNA. After 20 years of high-profile failure, gene therapy is finally well on its way to clinical approval. Jef Akst. Reprinted with permission of The Scientist from "Targeting DNA" July, 2002

Special Topic 4 ST F04-02, Prion-like Behavior in the Huntingtin Protein by Ocords, Protein Aggregation (May 2015); ST F04-03, C. Zuccato, M. Valenza, and E. Cattaneo, Molecular Mechanisms and Potential Therapeutical Targets in Huntington's Disease. Physiological Reviews, Volume 90 Issue 3 July 2010, Pages 905-981; ST F04-05, Permission granted by PNAS from Synthetic zinc finger repressors reduce mutant huntingtin expression in the brain of R6/2 mice by Garriga-Canut M, et.al. 109(45):E3136-45. (Oct. 2012); ST0 F4-07, Jong-Min Lee

Special Topic 5 ST F05-02, Reproduced with permission from Promega Corporation; ST F05-04, Reproduced with permission from Promega Corporation; ST Tbl05-02, Reprinted with permission from the Journal of Forensic Sciences, Vol 48, Issue 4, Copyright ASTM International, 100 Barr Harbor Drive, West Conshohocken, PA 19428.

Index

Note: Pages numbers followed by f, t, and b indicate figures, tables, and boxes, respectively.

A

B

C

D

E

F

G

H

I

J

K

L

M

N

O

P

Q

R

S

T

U

V

W

X

Y

Z